# Remote Sensing Handbook, Volume I (Six Volume Set)

Volume I of the Six Volume *Remote Sensing Handbook*, Second Edition, is focused on satellites and sensors including radar, light detection and ranging (LiDAR), microwave, hyperspectral, unmanned aerial vehicles (UAVs), and their applications. It discusses data normalization and harmonization, accuracies, and uncertainties of remote sensing products, global navigation satellite system (GNSS) theory and practice, crowdsourcing, cloud computing environments, Google Earth Engine, and remote sensing and space law. This thoroughly revised and updated volume draws on the expertise of a diverse array of leading international authorities in remote sensing and provides an essential resource for researchers at all levels interested in using remote sensing. It integrates discussions of remote sensing principles, data, methods, development, applications, and scientific and social context.

## FEATURES

- Provides the most up-to-date comprehensive coverage of remote sensing science.
- Discusses and analyzes data from old and new generations of satellites and sensors.
- Provides comprehensive methods and approaches for remote sensing data normalization, standardization, and harmonization.
- Includes numerous case studies on advances and applications at local, regional, and global scales.
- Presents remote sensing data and product accuracies, errors, and uncertainties.
- Outlines and enumerates on remote sensing and space law.
- Introduces advanced methods in remote sensing such as machine learning, cloud computing, and AI.
- Highlights scientific achievements over the last decade and provides guidance for future developments.

This volume is an excellent resource for the entire remote sensing and GIS community. Academics, researchers, undergraduate and graduate students, as well as practitioners, decision makers, and policymakers, will benefit from the expertise of the professionals featured in this book, and their extensive knowledge of new and emerging trends.

# Remote Sensing Handbook, Volume I (Six Volume Set)

## Sensors, Data Normalization, Harmonization, Cloud Computing, and Accuracies

Second Edition

Edited by Prasad S. Thenkabail, PhD

CRC Press is an imprint of the
Taylor & Francis Group, an **informa** business

Designed cover image: © Prasad S. Thenkabail

Second edition published 2025
by CRC Press
2385 NW Executive Center Drive, Suite 320, Boca Raton FL 33431

and by CRC Press
4 Park Square, Milton Park, Abingdon, Oxon, OX14 4RN

*CRC Press is an imprint of Taylor & Francis Group, LLC*

First edition published by CRC Press 2016



*Library of Congress Cataloging-in-Publication Data*
Names: Thenkabail, Prasad Srinivasa, 1958– editor.
Title: Remote sensing handbook / edited by Prasad S. Thenkabail ; foreword by Compton J. Tucker.
Description: Second edition. | Boca Raton, FL : CRC Press, 2025. | Includes bibliographical references and index. | Contents: v. 1. Remotely sensed data characterization, classification, and accuracies — v. 2. Image processing, change detection, GIS and spatial data analysis — v. 3. Agriculture, food security, rangelands, vegetation, phenology, and soils — v. 4. Forests, biodiversity, ecology, LULC, and carbon — v. 5. Water, hydrology, floods, snow and ice, wetlands, and water productivity — v. 6. Droughts, disasters, pollution, and urban mapping.
Identifiers: LCCN 2024029377 (print) | LCCN 2024029378 (ebook) | ISBN 9781032890951 (hbk ; v. 1) | ISBN 9781032890968 (pbk ; v. 1) | ISBN 9781032890975 (hbk ; v. 2) | ISBN 9781032890982 (pbk ; v. 2) | ISBN 9781032891019 (hbk ; v. 3) | ISBN 9781032891026 (pbk ; v. 3) | ISBN 9781032891033 (hbk ; v. 4) | ISBN 9781032891040 (pbk ; v. 4) | ISBN 9781032891453 (hbk ; v. 5) | ISBN 9781032891477 (pbk ; v. 5) | ISBN 9781032891484 (hbk ; v. 6) | ISBN 9781032891507 (pbk ; v. 6)
Subjects: LCSH: Remote sensing—Handbooks, manuals, etc.
Classification: LCC G70.4 .R4573 2025 (print) | LCC G70.4 (ebook) | DDC 621.36/780285—dc23/eng/20240722
LC record available at https://lccn.loc.gov/2024029377
LC ebook record available at https://lccn.loc.gov/2024029378

ISBN: 978-1-032-89095-1 (hbk)
ISBN: 978-1-032-89096-8 (pbk)
ISBN: 978-1-003-54114-1 (ebk)

DOI: 10.1201/9781003541141

Typeset in Times
by Apex CoVantage, LLC

# Contents



## PART I Earth-Observing Satellites and Sensors from Different Eras and Their Characteristics

# PART II Global Navigation Satellite Systems (GNSS) and Their Characteristics

# PART V Vegetation Index Standardization and Cross-Calibration of Data from Multiple Sensors

## PART VI Crowdsourcing and Remote Sensing Data



## PART VII Cloud Computing and Remote Sensing

## *PART VIII Google Earth for Remote Sensing*

## PART IX Accuracies, Errors, and Uncertainties of Remote Sensing Derived Products



## PART X Remote Sensing Law



## PART XI Summary and Synthesis of Volume I

# Foreword

Satellite remote sensing has progressed tremendously since the first Landsat was launched on June 23, 1972. Since the 1970s, satellite remote sensing and associated airborne and in situ measurements have resulted in geophysical observations for understanding our planet through time. These observations have also led to improvements in numerical simulation models of the coupled atmosphere-land-ocean systems at increasing accuracies and predictive capabilities. This was made possible by data assimilation of satellite geophysical variables into simulation models, to update model variables with more current information. The same observations document the Earth's climate and have driven consensus that *Homo sapiens* are changing our climate through greenhouse gas emissions.

These accomplishments are the work of many scientists from a host of countries and a dedicated cadre of engineers who build and operate the instruments and satellites that collect geophysical observation data from satellites, all working toward the goal of improving our understanding of the Earth. This edition of *Remote Sensing Handbook* (Second Edition, Volumes I–VI) is a compendium of information for many research areas of the Earth system that have contributed to our substantial progress since the 1970s. The remote sensing community is now using multiple sources of satellite and in situ data to advance our studies of planet Earth. In the following paragraphs, I will illustrate how valuable and pivotal satellite remote sensing has been in climate system study since the 1970s. The chapters in *Remote Sensing Handbook* provide other specific studies on land, water, and other applications using Earth observation data of the past 60+ years.

The Landsat system of Earth-observing satellites led the way in pioneering sustained observations of our planet. From 1972 to the present, at least one and frequently two Landsat satellites have been in operation (Wulder et al. 2022; Irons et al. 2012). Starting with the launch of the first NOAA-NASA Polar Orbiting Environmental Satellites NOAA-6 in 1978, improved imaging of land, clouds, and oceans and atmospheric soundings of temperature were accomplished. The NOAA system of polar-orbiting meteorological satellites has continued uninterrupted since that time, providing vital observations for numerical weather prediction. These same satellites are also responsible for the remarkable records of sea surface temperature and land vegetation index from the Advanced Very High-Resolution Radiometers (AVHRR) that now span more than 46 years as of 2024, although no one anticipated valuable climate records from these instrument before the launch of NOAA-6 in 1978 (Cracknell 2001). AVHRR instruments are expected to remain in operation on the European MetOps satellites into 2026 and possibly beyond.

The successes of data from the AVHRR led to the MODerate resolution Imaging Spectrometer (MODIS) instruments on NASA's Earth Observing System of satellite platforms that improved substantially upon the AVHRR. The first of the EOS platforms, Terra, was launched in 2000, and the second of these platforms, Aqua, was launched in 2002. Both of these platforms are nearing their operational end-of-life and many of the climate data records from MODIS will be continued with the Visible Infrared Imaging Suite (VIIRS) instrument on the Joint Polar Satellite System (JPSS) meteorological satellites of NOAA. The first of these missions, the NPOES Preparation Project, was launched in 2012 with the first VIIRS instrument that is operating currently along with similar instruments on JPSS-1 (launched in 2017) and JPSS-2 (launched in 2022). However, unlike the morning/afternoon overpasses of MODIS, the VIIRS instruments are all in an afternoon overpass orbit. One of the strengths of the MODIS observations was morning and afternoon data from identical instruments.

Continuity of observations is crucial for advancing our understanding of the Earth's climate system. Many scientists feel the crucial climate observations provided by remote sensing satellites are among the most important satellite measurements because they contribute to documenting the current state of our climate and how it is evolving. These key satellite observations of our climate are second in importance only to the polar orbiting and geostationary satellites needed for numerical weather prediction that provide natural disaster alerts.

The current state of the art for remote sensing is to combine different satellite observations in a complementary fashion for what is being studied. Climate study is an example of using disparate observations from multiple satellites coupled with in situ data to determine if climate change is occurring and where it is occurring, and to identify the various component processes responsible.

1. **Planet warming quantified by satellite radar altimetry**. Remotely sensed climate observations provide the data to understand our planet and to identify what forces shape our climate. The primary sea level climate observations come from radar altimetry that started in late 1992 with TOPEX-Poseidon and has been continued by Jason-1, Jason-2, Jason-3, and Sentinel-6 to provide an uninterrupted record of global sea level. Changes in global sea level provide unequivocal evidence that our planet is warming, cooling, or staying at the same temperature. Radar altimetry from 1992 to date has shown global sea level increases of ~3.5 mm/y; hence our planet is warming (Figure 0.1). Sea level rise has two components, ocean thermal expansion and ice melt from the ice sheets of Greenland and Antarctica, and to a lesser extent for glacier concentrations in places like the Gulf of Alaska and Patagonia. The combination of GRACE and GRACE Follow-On gravity measurements quantifies the ice mass losses of Greenland and Antarctica to a high degree of accuracy. Combining the gravity data with the flotilla of almost 4,000 Argo floats provides the temperature data with the depth necessary to quantify ocean temperatures and isolate the thermal component of sea level rise.
2. **Our Sun is remarkedly stable in total solar irradiance**. Observations of total solar irradiance have been made from satellites since 1979 and show total solar irradiance has

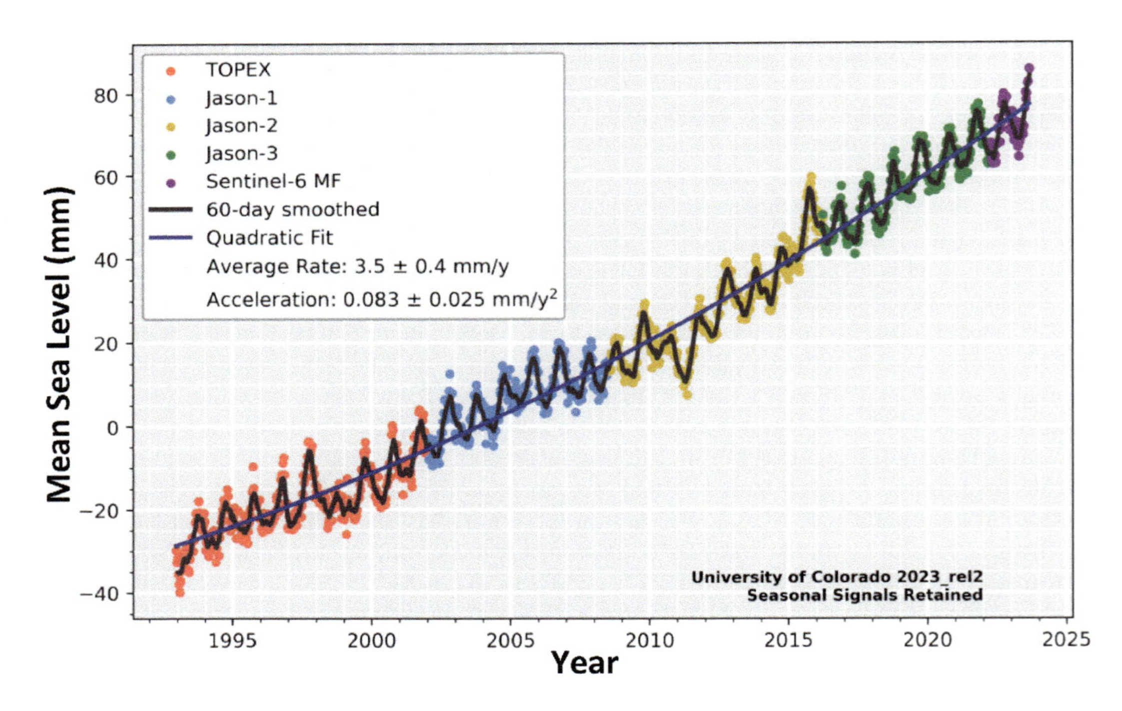


FIGURE 0.1 Seasonal sea level from five satellite radar altimeters from later 1992 to the present. Sea level is the unequivocal indicator of the Earth's climate—when sea level rises, the planet is warming; when sea level falls, the planet is cooling. (Nerem et al. 2018 updated to 2023; https://sealevel.colorado.edu/data/total-sea-level-change.)

varied only ±1 part in 500 over the past 35 years, establishing that our Sun is not to blame for global warming (Figure 0.2).

3. **Determining ice sheet contributions to sea level rise**. Since 2002 gravity observations from the Gravity Recovery and Climate Experiment Satellite, or GRACE, mission and the GRACE Follow-On mission have been measured. GRACE data quantify ice mass changes from the Antarctic and Greenland ice sheets that constitute 98% of the ice mass on land (Luthcke et al. 2013). GRACE data are truly remarkable—their retrieval of variations in the Earth's gravity field is quantitatively and directly linked to mass variations. With GRACE data we are able for the first time to determine the mass balance with time of the Antarctic and Greenland ice sheets and concentrations of glaciers on land. GRACE data show sea level rise is 60% explained by ice sheet mass loss (Figure 0.3). GRACE data have many other uses, such as changes in groundwater storage. See www.csr.utexas.edu/grace/.

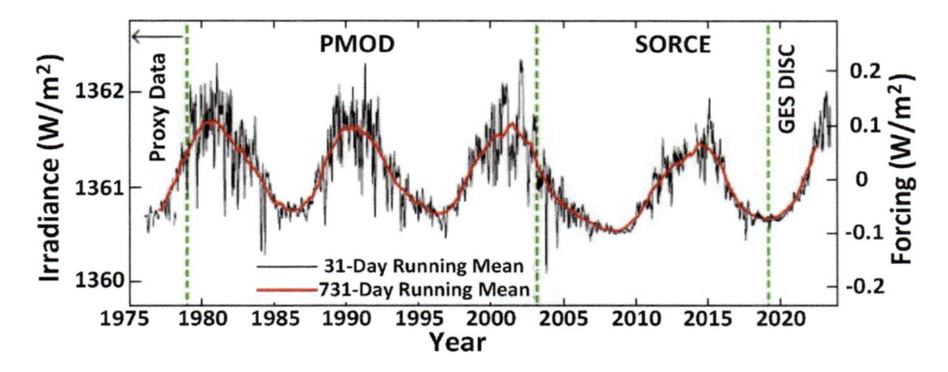


FIGURE 0.2 The Sun is not to blame for global warming, based on total solar irradiance observations from satellites. The few Watts/m² solar irradiance variations covary with the sunspot cycle. The luminosity of the Sun varies 0.2% over the course of the 11-year solar and sunspot cycle. The SORCE TSI dataset continues these important observations with improved accuracy on the order of ±0.035 (Kopp et al. 2024, and from https://lasp.colorado.edu/sorce/data/tsi-data/.)

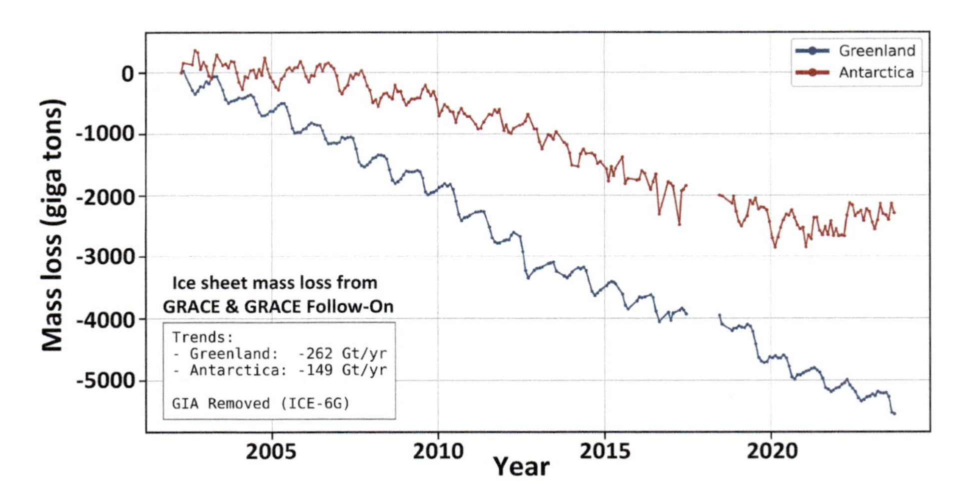


FIGURE 0.3 Sixty percent sea level rise is explained by mass balance of melting of ice measured by GRACE and GRACE Follow-On satellites. Ice mass variations are from 2003 to 2023 for the Antarctica and the Greenland ice sheets using gravity data (Croteau et al. 2021 updated to 2023). The Antarctic and Greenland Ice Sheets constitute 98% of the Earth's land ice.

4. **Forty percent sea level rise explained by thermal expansion in the planet's oceans measured by in situ ~ 3,700 Argo drifting floats**. The other contributor to sea level rise is the thermal expansion or "steric" component of our planet's oceans. To document this necessitates using diving and drifting floats or buoys in the Argo network to record temperature with depth (Romerich et al. 2009 and Figure 0.4). Argo floats are deployed from ships; they then submerge and descend slowly to 1,000 m depth, recording temperature, pressure, and salinity as they descend. At 1,000 m depth, they drift for ten days continuing their measurements of temperature and salinity. After ten days, they slowly descend to 3,000 m and then ascend to the surface, all the time recording their measurements. At the surface, each float transmits all the data collected on the most recent excursion to a geostationary satellite and then descends again to repeat this process.

   Argo temperature data show that 40% of sea level rise results from warming and thermal expansion of our oceans. Combining radar altimeter data, GRACE and GRACE Follow-On data, and Argo data provide confirmation of sea level rise and show what is responsible for it and in what proportions. With total solar irradiance being near constant, what is driving global warming can be determined. Analysis of surface in situ air temperature coupled with lower tropospheric air temperature and stratospheric temperature data from remote sensing infrared and microwave sounders show the surface and near-surface is warming while the stratosphere is cooling. This is an unequivocal confirmation that greenhouse gases are warming the planet.

   Combining sea level radar altimetry, GRACE and GRACE Follow On gravity data to quantify ice sheet mass loses, and Argo floats to measure ocean temperatures with depth enables reconciliation of sea level increases with mass loss of ice sheets and ocean thermal expansion. The ice and steric expansion explains 95% of sea level rise (Figure 0.5).
5. **The global carbon cycle**. Many scientists are actively working to study the Earth's carbon cycle and there are several chapters in this *Remote Sensing Handbook* (Volumes I–VI) on various components under study.

Carbon cycles through reservoirs on the Earth's surface in plants and soils, exists in the atmosphere as gases such as carbon dioxide ($CO_2$), and exists in ocean water in phytoplankton and in

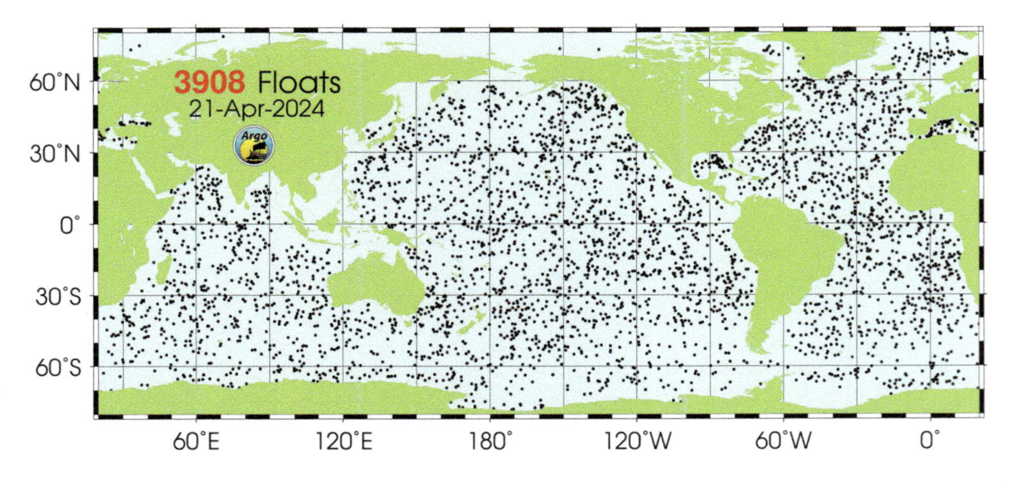


FIGURE 0.4 Forty percent sea level rise explained by thermal expansion in the planet's oceans measured in situ by ~3,908 drifting floats that were in operation on April 21, 2024. These floats provide the data needed to document thermal expansion of the oceans. (Roemmich & the Argo Float Team 2009, updated to 2024, and www.argo.ucsd.edu/.)

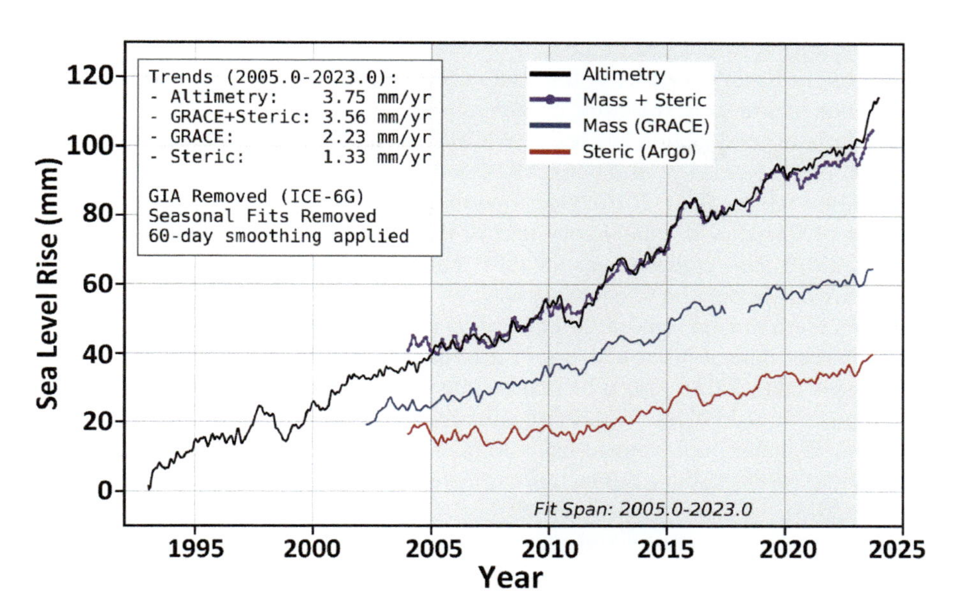


**FIGURE 0.5** Sea level rise with the gravity tide mass loss and the Argo thermal expansion quantities added to the plot of global mean sea level. The GRACE and GRACE Follow On ice sheet gravity term and Argo thermal expansion term together explain 95% of sea level rise. (Croteau et al. 2021 updated to 2023.)

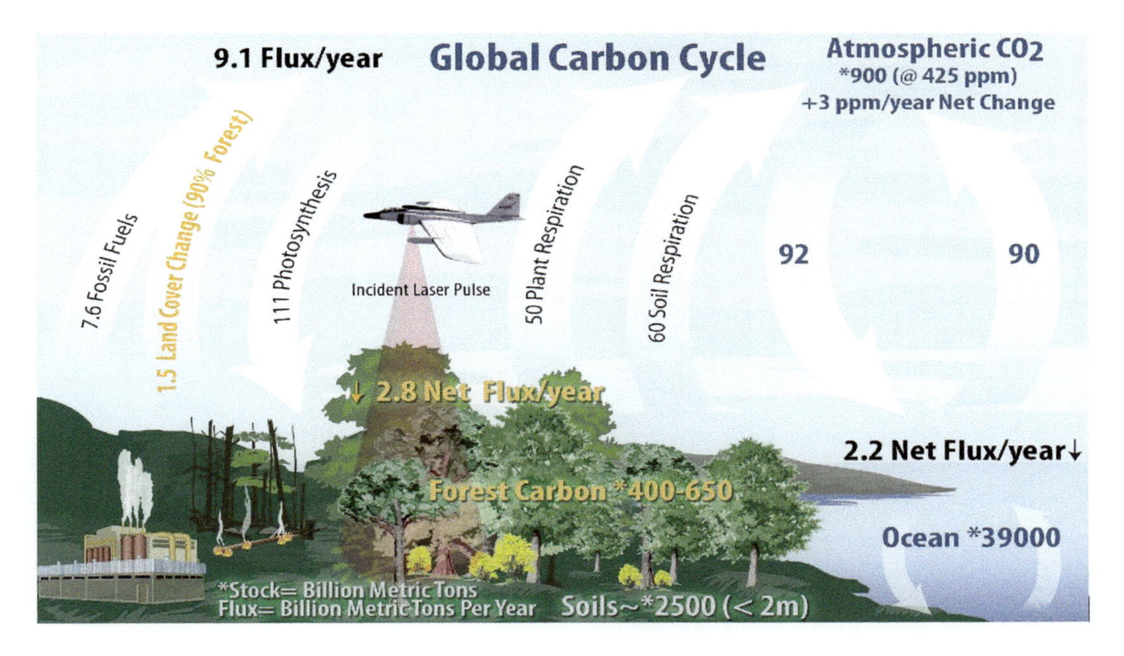


**FIGURE 0.6** Global carbon cycle measurements from a multitude of satellite sensors. A representation of the global carbon cycle showing our best estimates of carbon fluxes and carbon reservoirs as of 2024. A series of satellite observations are needed simultaneously to understand the carbon cycle and its role in the Earth's climate system (Ciais et al. 2014 updated to 2023). The major unknowns in the global carbon cycle are fluxes between different reservoirs, oceanic gross primary production, carbon in soils, and the carbon in woody vegetation.

marine sediments. $CO_2$ is released to the atmosphere from the combustion of fossil fuels, by land cover changes on the Earth's surface, by respiration of green plants, and by decomposition of carbon in dead vegetation and in soils, including carbon in permafrost.

Land gross primary production has been a MODIS product that is extended into the VIIRS era (Running et al. 2004; Román et al. 2024). MODIS data also provide burned area and $CO_2$ emissions from wildfire (Giglio et al. 2016). Oceanic gross primary production will be provided by the Plankton, Aerosol, Cloud, and ocean Ecosystem, or PACE, satellite that was launched in early 2024 (Gorman et al. 2019). This complements the GPP land portion of the carbon cycle and will enable global gross primary production to be determined by MODIS-VIIRS and PACE.

Furthermore, Harmonized Landsat-8, Landsat-9, and Sentinel-2 30 m data (HLS) provide multispectral time series data at 30 m with a revisit frequency of three days at the equator (Crawford et al. 2023; Masek et al. 2018). This will enable time series improvements in spatial detail to 30 m from the 250 m scale of MODIS. The revisit time of Sentinel-2 with 10 m data is five days at the equator, which is a major improvement from 30 m. Multispectral time series observations are the basis for providing gross primary production estimates on land that are also used for food security (Claverie et al. 2018).

Refinements in satellite multispectral spatial resolution to the 50 cm to 3–4 m scale provided by commercial satellite data have enabled tree carbon to be determined from large areas of trees outside of forests. NASA has started using commercial satellite data to complement MODIS, Landsat, and other observations. One of the uses for Planet 3–4 m and Maxar <1 m data has been for mapping trees outside of forests (Brandt et al. 2020; Reiner et al. 2023; Tucker et al. 2023). Tucker et al. (2023) mapped ten billion trees at the 50 cm scale over 10 million $km^2$ and converted them into carbon at the tree level with allometry. The value of Planet and Maxar (formerly Digital Globe) data allows carbon studies to be extended into areas with discrete trees and Huang et al. (2024) has successfully mapped one tree species across the entire Sahelian and Sudanian Zones of Africa.

The height of trees is an important measurement to determine their carbon content. For areas of contiguous tree crowns, GEDI and ICESat laser altimetry (Magruder et al. 2024) coupled with Landsat and Sentinel-2 observations enable improved estimates of carbon in these forests (Claverie et al. 2018).

The key to closing several uncertainties in the carbon cycle is to quantify fluxes among the various components. Passive $CO_2$ retrieval methods from the Greenhouse gases Observing SATellite (GOSAT) (Noël et al. 2021) and the Orbiting Carbon Observatory-2 (OCO-2) (Jacobs et al. 2024) are inadequate to provide this. Passive methods are not possible at night, in all seasons, and require specific Sun-target-sensor viewing perspectives and conditions. A recent development of the Aerosol and Carbon dioxide Detection Lidar (ACDL) instrument (Dai et al. 2023) by our Chinese colleagues offers a tenfold coverage improvement in $CO_2$ retrievals over those provided by OCO-2, and 20-fold coverage improvement over GOSAT. The reported uncertainty of ACDL is on the order of $\pm 0.6$ ppm.

Understanding the carbon cycle requires a "full court press" of satellite and in situ observations, because all of these observations must be made at the same time. Many of these measurements have been made over the past 30 to 40 years but new measurements are needed to quantify carbon storage in vegetation, to quantify $CO_2$ fluxes, to quantify land respiration, and to improve numerical carbon models. Similar work needs to be performed for the role of clouds and aerosols in climate and to improve our understanding of the global hydrological cycle.

The remote sensing community has made tremendous progress over the last six decades as captured in various chapters of *Remote Sensing Handbook* (Second Edition, Volumes I–VI). Handbook chapters provide comprehensive understanding of land and water studies through detailed methods, approaches, algorithms, synthesis, and key references. Every type of remote sensing data obtained from systems such as optical, radar, LiDAR, hyperspectral, and hyperspatial are presented and discussed in different chapters. Chapters in this volume address remote sensing data characteristics, within and between sensor calibrations, classification methods, and accuracies by utilizing wide array of remote sensing data from numerous platforms over last five decades. Volume I also brings in new remote sensing technologies such as radio occultation and reflectometry from the global navigation satellite system, or GPS, satellites, crowdsourcing, drones, cloud computing, artificial intelligence, machine learning, hyperspectral, radar, and remote sensing law. The chapters in *Remote Sensing Handbook* (Second Edition, Six Volumes) are written by leading remote sensing scientists of the world and ably edited by Dr. Prasad S. Thenkabail, senior scientist (ST) at U.S. Geological Survey (USGS) in Flagstaff, Arizona. The importance and the value of *Remote Sensing Handbook* is clearly demonstrated by the need for a second edition. The *Remote Sensing Handbook* (First Edition, Volumes I–III) was published in 2014, and now after ten years *Remote Sensing Handbook* (Second Edition, Volumes I–VI) with 91 chapters and nearly 3,500 pages is published. It is certainly monumental work in remote sensing science, and for this I want to complement Dr. Prasad S. Thenkabail. Remote sensing is now important to a large number of scientific disciplines beyond our community, and I recommend *Remote Sensing Handbook* (Second Edition, six volumes) to not only remote sensers but also to the entire scientific community.

We can look forward in the coming decades to improving our quantitative understanding of the global carbon cycle, understanding the interaction of clouds and aerosols in our radiation budget, and understanding the global hydrological cycle.

**by Compton J. Tucker**
*Satellite Remote Sensing Beyond 2025*
*NASA/Goddard Space Flight Center*
*Earth Science Division*
*Greenbelt, Maryland 20771 USA*

## REFERENCES

Brandt, M., Tucker, C.J., Kariryaa, A., et al. 2020. An unexpectedly large count of trees in the West African Sahara and Sahel. *Nature* 587:78–82. https://doi.org/10.1038/s41586-020-2824-5.

Ciais, P., et al. 2014. Current systematic carbon-cycle observations and the need for implementing a policy-relevant carbon observing system. *Biogeosciences* 11(13):3547–3602.

Claverie, M., Ju, J., Masek, J.G., Dungan, J.L., Vermote, E.F., Roger, J.-C., Skakun, et al. 2018. The Harmonized Landsat and Sentinel-2 surface reflectance data set. *Remote Sensing of Environment* 219:145–161. https://doi.org/10.1016/j.rse.2018.09.002.

Cracknell, A. 2001. The exciting and totally unanticipated success of the AVHRR in applications for which it was never intended. *Advances in Space Research* 28:233–240. https://doi.org/10.1016/S0273-1177(01)00349-0.

Crawford, C.J., Roy, D.P., Arab, S., Barnes, C., Vermote, E., Hulley, et al. 2023. The 50-year Landsat collection 2 archive, *Science of Remote Sensing* 8:100103, ISSN 2666–0172, https://doi.org/10.1016/j.srs.2023.100103. (www.sciencedirect.com/science/article/pii/S2666017223000287).

Croteau, M.J., Sabaka, T.J., Loomis, B.D. 2021. GRACE fast mascons from spherical harmonics and a regularization design trade study. *Journal Geophysical Res earch: Solid Earth* 126:e2021JB022113. https://doi.org/10.1029/2021JB022113.10.1029/2021JB022113.

Dai, G., Wu, S., Sun, K., Long, W., Liu, J., Chen, W. 2023. Aerosol and Carbon dioxide Detection Lidar (ACDL) overview. Presentation at ESA-JAXA EarthCare Workshop, November.

Giglio, L., Schroeder, W., Justice, C.O. 2016. The collection 6 MODIS active fire detection algorithm and fire products. *Remote Sensing of Environment* 178:31–41. https://doi.org/10.1016/j.rse.2016.02.054.

Gorman, E.T., Kubalak, D.A., Patel, D., Dress, A., Mott, D.B., Meister, G., Werdell, P.J. 2019. The NASA Plankton, Aerosol, Cloud, ocean Ecosystem (PACE) mission: An emerging era of global, hyperspectral Earth system remote sensing. *Sensors, Systems, and Next Generation Satellites* XXIII:11151. https://doi.org/10.1117/12.2537146.

Huang, K., et al. 2024. Mapping every adult baobab (Adansonia digitata L.) across the Sahel to uncover the co-existence with rural livelihoods. *Nature Ecology and Evolution* https://doi.org/10.21203/rs.3.rs-3243009/v1.

Irons, J.R., Dwyer, J.L., Barsi, J.A. 2012. The next Landsat satellite: The Landsat data continuity mission. *Remote Sensing of Environment* 122:11–21. https://doi.org/10.1016/j.rse.2011.08.026.

Jacobs, N., et al. 2024. The importance of digital elevation model accuracy in $X_{CO2}$ retrievals: Improving the Orbiting Carbon Observatory-2 Atmospheric Carbon Observations from Space version 11 retrieval product. *Atmospheric Measurement Techniques* 17(5):1375–1401. https://doi.org/10.5194/amt-17–1375–2024.

Kopp, G., Nèmec, N.E., Shapiro, A. 2024. Correlations between total and spectral solar irradiance variations. *Astrophysical Journal* 964(1). https://doi.org/10.3847/1538–4357/ad24e5.

Luthcke, S.B., Sabaka, T.J., Loomis, B.D., Arendt, A.A., McCarthy, J.J., Camp, J. 2013. Antarctica, Greenland and Gulf of Alaska land-ice evolution from an iterated GRACE global mascon solution. *Journal of Glaciology* 59(216):613–631. https://doi.org/10.3189/2013JoG12J147.

Magruder, L.A., Farrell, S.L., Neuenschwander, A., Duncanson, L., Csatho, B., Kacimi, S., et al. 2024. Monitoring Earth's climate variables with satellite laser altimetry. *Nature Reviews of Earth and Environment* 5(2):120–136. https://doi.org/10.1038/s43017-023-00508-8.

Masek, J., Ju, J., Roger, J.-C., Skakun, S., Claverie, M., Dungan, J. 2018. Harmonized Landsat/Sentinel-2 products for land monitoring, IGARSS 2018–2018. *IEEE International Geoscience and Remote Sensing Symposium*, Valencia, Spain, pp. 8163–8165, https://doi.org/10.1109/IGARSS.2018.8517760.

Nerem, R.S., Beckley, B.D., Fasullo, J.T., Mitchum, G.T. 2018. Climate-change—driven accelerated sea-level rise detected in the altimeter era. *Proceeding of the National Academy of Sciences* 115(9):2022–2025. https://doi.org/10.1073/pnas.1717312115.

Noël, S., et al. 2021. $XCO_2$ retrieval for GOSAT and GOSAT-2 based on the FOCAL algorithm. *Atmospheric Measurement Techniques* 14(5):3837–3869. https://doi.org/10.5194/amt-14–3837-2021.

Reiner, F., et al. 2023. More than one quarter of Africa's tree cover is found outside areas previously classified as forest. *Nature Commun ications*. https://doi.org/10.1038/s41467-023-37880-4.

Roemmich, D., the Argo Steering Team. 2009. Argo: The challenge of continuing 10 years of progress. *Oceanography* 22(3):46–55.

Román, M., et al. 2024. Continuity between NASA MODIS collection 6.1 and VIIRS collection 2 land products. *Remote Sensing of Environment* 302. https://doi.org/10.1016/j.rse.2023.113963.

Running, S.W., Nemani, R.R., Heinsch, F.A., Zhao, M.S., Reeves, M., Hashimoto, H. 2004. A continuous satellite-derived measure of global terrestrial primary production. *Bioscience* 54(6): 547–560. https://doi.org/10.1641/0006-3568.

Tucker, C., Brandt, M., Hiernaux, P., Kariryaa, et al. 2023. Sub-continental-scale carbon stocks of individual trees in African drylands. *Nature* 615:80–86. https://doi.org/10.1038/s41586-022-05653-6.

Wulder, M.A., Roy, D.P., Radeloff, V.C., Loveland, T.R., Anderson, M.C., Johnson, et al. 2022. Fifty years of Landsat science and impacts. *Remote Sensing of Environment* 280:113195, ISSN 0034–4257, https://doi.org/10.1016/j.rse.2022.113195. (www.sciencedirect.com/science/article/pii/S0034425722003054).

# Preface

The overarching goal of this six-volume, 91-chapter, about 3,500-page *Remote Sensing Handbook* (Second Edition, Volumes I–VI) was to capture and provide the most comprehensive state of the art of remote sensing science and technology development and advancement in the last 60+ years, by clearly demonstrating the (1) scientific advances, (2) methodological advances, and (3) societal benefits achieved during this period, as well as to provide a vision of what is to come in years ahead. The book volumes are, to date and to my best knowledge, the most comprehensive documentation of the scientific and methodological advances that have taken place in understanding remote sensing data, methods, and a wide array of applications. Written by 300+ leading global experts in the area, each chapter (1) focuses on specific topic (e.g., data, methods, and specific set of applications), (2) reviews existing state-of-the-art knowledge, (3) highlights the advances made, and (4) provides guidance for areas requiring future development. Chapters in the book cover a wide array of subject matter concerning remote sensing applications. *Remote Sensing Handbook* (Second Edition, Volumes I–VI) is planned as reference material for a broad spectrum of remote sensing scientists to understand the fundamentals as well as the latest advances, and a wide array of applications for land and water resource practitioners, natural and environmental practitioners, professors, students, and decision makers.

Special features of the six-volume *Remote Sensing Handbook* (Second Edition) include the following:

1. Participation of an outstanding group of remote sensing experts, an unparalleled team of writers for such a book project
2. Exhaustive coverage of a wide array of remote sensing science: data, methods, applications
3. Each chapter led by a luminary and most chapters written by writing teams that further enriched the chapters
4. Broadening the scope of the book to make it ideal for expert practitioners as well as students
5. Global team of writers, global geographic coverage of study areas, and wide array of satellites and sensors
6. Plenty of color illustrations

Chapters in the book cover the following aspects of remote sensing:

State of the art on satellites, sensors, science, technology, and applications
Methods and techniques
Wide array of applications such as land and water applications, natural resources management, and environmental issues
Scientific achievements and advancements over the last 60+ years
Societal benefits
Knowledge gaps
Future possibilities in the 21st century

Great advances have taken place over the last 60+ years in the study of planet Earth from remote sensing, especially using data gathered from the multitude of Earth Observation (EO) satellites launched by various governments as well as private entities. A large part of the initial remote sensing technology was developed and tested during the two world wars. In the 1950s, remote sensing slowly began its foray into civilian applications. But, during the years of cold war, remote sensing applications both in civilian and military increased swiftly. But, it was also an age when remote sensing was the domain of very few top experts, often having multiple skills in engineering, science, and computer technology. From the 1960s onward, there have been many governmental agencies that have initiated civilian remote sensing. The National Aeronautics and Space Administration (NASA) of the United States has been at the forefront of many of these efforts. Others who have provided leadership in civilian remote sensing include, but are not limited to, European Space Agency (ESA) of the European Union; Indian Space Research Organization (ISRO) of India; the Centre national d'études spatiales (CNES) of France; the Canadian Space Agency (CSA), Canada; the Japan Aerospace Exploration Agency (JAXA), Japan; the German Aerospace Center (DLR), Germany; the China National Space Administration (CNSA), China; the United Kingdom Space Agency (UKSA), UK; and Instituto Nacional de Pesquisas Espaciais (INPE), Brazil. Many private entities such as the Planet Labs PBC have launched and operate satellites. These government and private agencies and enterprises have launched, and continue to launch and operate, a wide array of satellites and sensors that capture data of planet Earth in various regions of electromagnetic spectrum and in various spatial, radiometric, and temporal resolutions, routinely and repeatedly. However, the real thrust for remote sensing advancement came during the last decade of the 20th century and the beginning of the 21st century. These initiatives included the launch of series of new generation EO satellites to gather data more frequently and routinely, release of pathfinder datasets, web-enabling the data for free by many agencies (e.g., USGS release of the entire Landsat archives as well as real-time acquisitions of the world for free by making them web accessible), and providing processed data ready to users (e.g., the Harmonized Landsat and Sentinel-2 or HLS data, surface reflectance products of MODIS). Other efforts like Google Earth made remote sensing more popular and brought in a new platform for easy visualization and navigation of remote sensing data. Advances in computer hardware and software made it possible to handle big data. Crowdsourcing, web access, cloud computing such as in Google Earth Engine (GEE) platform, machine learning, deep learning, coding, artificial intelligence, mobile apps, and mobile platforms (e.g., drones) added new dimension to how remote sensing data is used. Integration with global positioning systems (GPS) and global navigation satellite systems (GNSS), and inclusion of digital secondary data (e.g., digital elevation, precipitation, temperature) in analysis has made remote sensing much more powerful. Collectively, these initiatives provided new vision in making remote sensing data more popular, more widely understood, and increasingly used for diverse applications, hitherto considered difficult. Freely available archival data when combined with more recent acquisitions has also enabled quantitative studies of change over space and time. ***Remote Sensing Handbook* (Volumes I–VI) is targeted to capture these vast advances in data, methods, and applications, so a remote sensing student, scientist, or a professional practitioner will have the most comprehensive, all-encompassing reference material in one place.**

Modern-day remote sensing technology, science, and applications are growing exponentially. This growth is as a result of combination of factors that include (1) advances and innovations in data capture, access, processing, computing, and delivery (e.g, big data analytics, harmonized and normalized data, inter-sensor relationships, web-enabling of data, cloud computing such as in Google Earth Engine (GEE), crowdsourcing, mobile apps, machine learning, deep learning, coding in Python and Java Script, and artificial intelligence); (2) an increasing number of satellites and sensors gathering data of the planet, repeatedly and routinely, in various portions of the electromagnetic spectrum as well as in array of spatial, radiometric, and temporal resolutions; (3) efforts at integrating data from multiple satellites and sensors (e.g., Sentinels with Landsat); (4) advances in data normalization, standardization, and harmonization (e.g., delivery of data in surface reflectance, inter-sensor calibration); (5) methods and techniques for handling very large data volumes (e.g., global mosaics); (6) quantum leap in computer hardware and software capabilities (e.g., ability to process several terabytes of data); (7) innovation in methods, approaches, and techniques leading to sophisticated algorithms (e.g., spectral matching techniques, neural network perceptron); and (8) development of new spectral indices to quantify and study specific land and water parameters (e.g., hyperspectral vegetation indices or HVIs). As a result of these all-round developments, remote sensing science is today very mature and is widely used in virtually every discipline of Earth sciences for quantifying, mapping, modeling, and monitoring our planet Earth. Such rapid advances are captured in a number of remote sensing and Earth science journals. However, students, scientists, and practitioners of remote sensing science and applications have significant difficulty in gathering a complete understanding of the various developments and advances that have taken place because of their vastness spread across the last 60+ years. Thereby, the chapters in *Remote Sensing Handbook* are designed to give a whole picture of scientific and technological advances of the last 60+ years.

Today, the science, art, and technology of remote sensing is truly ubiquitous and increasingly part of everyone's everyday life, often without the user even knowing it. Whether looking at your own home or farm (e.g., Figure 0.7), helping you navigate when you drive, visualizing a phenomenon occurring in a distant part of the world (e.g., Figure 0.7), monitoring events such as droughts and floods, reporting weather, detecting and monitoring troop movements or nuclear sites, studying deforestation, assessing biomass carbon, addressing disasters like earthquakes or tsunamis, and a host of other applications (e.g, precision farming, crop productivity, water productivity, deforestation, desertification, water resources management), remote sensing plays a key role. Already, many new innovations are taking place. Companies such as the Planet Labs PBC and Skybox are capturing very high spatial resolution imagery and even videos from space using many microsatellite (CubeSat) constellations. Planet Labs also will soon launch hyperspectral satellites called Tanager. There are others (e.g., Pixxel, India) who have launched and continue to launch constellations of hyperspectral or other sensors. China is constantly putting a wide array of satellites into orbit. Just as the smartphone and social media connected the world, remote sensing is making the world our

**FIGURE 0.7** Google Earth can be used to seamlessly navigate and precisely locate any place on Earth, often with very high spatial resolution data (VHRI; sub-meter to 5 m) from satellites such as IKONOS, Quickbird, and Geoeye (Note: this image is from one of the VHRI). Here, the editor in chief (EiC) of this *Remote Sensing Handbook* (Volumes I–VI) (Thenkabail) located his village home and surroundings, which has land cover such as secondary rainforests, lowland paddy farms, areca nut plantations, coconut plantations, minor roads, walking routes, open grazing lands, and minor streams (typically, first and second order) (Note: land cover based on ground knowledge of the EiC). The first primary school attended by the EiC is located precisely. Precise coordinates (13 45 39.22 Northern latitude, 75 06 56.03 Eastern longitude) of Thenkabail's village house on the planet are located, and the date of image acquisition (March 1, 2014) is noted. Google Earth Images are used for visualization as well as for numerous science applications such as accuracy assessment, reconnaissance, determining land cover, establishing land use, and for various ground surveys.

backyard (e.g., Figure 0.7). No place goes unobserved and no event gets reported without an image. True liberation for any technology and science comes when it is widely used by common people who often have no idea on how it all comes together, but understand the information provided intuitively. That is already happening (e.g., how we use smartphones is significantly driven by satellite data-driven maps and GPS-driven locations). These developments make it clear that not only do we need to understand the state of the art but also have a vision on where the future of remote sensing is headed. Thereby, in a nutshell, the goal of *Remote Sensing Handbook* (Volumes I–VI) is to cover the developments and advancement of six distinct eras (listed here) in terms of data characterization and processing as well as myriad land and water applications:

**Pre-civilian remote sensing era of pre-1950s:** World War I and II when remote sensing was a military tool

**Technology demonstration era of 1950s and 1960s:** Sputnik-I and NOAA AVHRR era of the 1950s and 1960s

**Landsat era of 1970s:** when first truly operational land remote sensing satellite (Earth Resources Technology Satellite or ERTS, later renamed Landsat) was launched and operated

**Earth observation era of 1980s and 1990s:** when a number of space agencies began launching and operating satellites (e.g., Landsat 4,5 by USA, SPOT-1,2 by France; IRS-1a, 1b by India)

**Earth observation and new millennium era of 2000s:** when data dissemination to user's became as important as launching, operating, and capturing data (e.g., MODIS terra/acqua, Landsat-8, Resourcesat)

**Twenty-first century era starting 2010s:** when new generation micro/Nano satellites or CubeSats (e.g., Planet Labs PBC, Skybox), hyperspectral satellite sensors (e.g., Tanager-1, DESIS, PRISMA, EnMAP, upcoming NASA SBG) add to increasing constellation of multi-agency sensors (e.g., Sentinels, Landsat-8, 9, upcoming Landsat-Next)

Motivation to take up editing the six Volume *Remote Sensing Handbook* (second edition) wasn't easy. It is a daunting work and requires an extraordinary commitment over two to three years. After repeated requests from Ms. Irma Shagla-Britton, manager and leader for Remote Sensing and GIS books of Taylor and Francis/CRC Press, and considerable thought, I finally agreed to take the challenge in 2022. Having earlier edited the three-volume *Remote Sensing Handbook*, published in 2014, I was pleased that the books were of considerable demand for a second edition. This was enough motivation. Further, I wanted to do something significant at this stage of my career that will make a considerable contribution to the global remote sensing community. When I edited the first edition during the 2012–2014 period, I was still recovering post colon cancer surgery and chemotherapy. But this second edition is a celebration of my complete recovery from the dreaded disease. I have not only fully recovered but never felt so completely full of health and vigor. This, naturally, gave me the sufficient energy and enthusiasm required to back my motivation to edit this monumental six-volume *Remote Sensing Handbook*. At least for me, this is the *Magnus opus* that I feel proud to have accomplished and feel confident of the immense value for students, scientists, and professional practitioners of remote sensing who are interested in a standard reference on the subject. They will find these six volumes of *Remote Sensing Handbook*: "Complete and comprehensive coverage of the state-of-the-art remote sensing, capturing the advances that have taken place over last 60+ years, which will set the stage for a vision for the future."

Above all, I am indebted to some 300+ authors and co-authors of the chapters who have spent so much of their creative energy to work on the chapters, deliver them on time, and patiently address all edits and comments. These are amongst the very best remote sensing scientists from around the world. Extremely busy people, making time for the book project and making outstanding contributions. I went back to everyone who contributed to *Remote Sensing Handbook* (First Edition, three volumes) published in 2014 and requested them to revise their chapters. Most of the lead authors of the chapters agreed to revise, which was reassuring. However, some were not available, due to retirement or for other reasons. In such cases, I adopted two strategies: (1) invite a few new chapter authors to make up for this gap, and (2) update the chapters myself in other cases. I am convinced this strategy worked very well to ensure capturing the latest information and to maintain the integrity of every chapter. What was also important was to ensure that the latest advances in remote sensing science were adequately covered. The authors of the chapters amazed me by their commitment and attention to detail. First, the quality of each of the chapters was of the highest standards. Second, with very few exceptions, chapters were delivered on time. Third, edited chapters were revised thoroughly and returned on time. Fourth, all my requests on various formatting and quality enhancements were addressed. My heartfelt gratitude to these great authors for their dedication to quality science. It has been my great honor and privilege to work with these dedicated legends. Indeed, I call them my "heroes" in the true sense. These are highly accomplished, renowned, pioneering scientists of the highest merit in remote sensing science, and I am ever grateful to have their time, effort, enthusiasm, and outstanding intellectual contributions. I am indebted to their kindness and generosity. In the end, we had 300+ authors writing 91 chapters.

Overall, ***Remote Sensing Handbook* (Volumes I–VI)** took about two years, from the time book chapters and authors were identified to final publication of the book. The six volumes of *Remote Sensing Handbook* were designed in such a way that a reader can have all six volumes as standard reference or have individual volumes to study specific subject areas. The six volumes are:

***Remote Sensing Handbook, Second Edition, Vol. I***
***Volume I**: Sensors, Data Normalization, Harmonization, Cloud Computing, and Accuracies—9781032890951*

***Remote Sensing Handbook, Second Edition, Vol. II***
***Volume II**: Image Processing, Change Detection, GIS and Spatial Data Analysis—9781032890975*

***Remote Sensing Handbook, Second Edition, Vol. III***
***Volume III**: Agriculture, Food Security, Rangelands, Vegetation, Phenology, and Soils—9781032891019*

***Remote Sensing Handbook, Second Edition, Vol. IV***
***Volume IV**: Forests, Biodiversity, Ecology, LULC, and Carbon—9781032891033*

***Remote Sensing Handbook, Second Edition, Vol. V***
***Volume V**: Water Resources: Hydrology, Floods, Snow and Ice, Wetlands, and Water Productivity—9781032891453*

***Remote Sensing Handbook, Second Edition, Vol. VI***
***Volume VI**: Droughts, Disasters, Pollution, and Urban Mapping—9781032891484*

**There are 18, 17, 17, 12, 13, and 14 chapters, respectively, in the six volumes**.

A wide array of topics covered in the six volumes.

The topics covered in the **chapters of Volume I** include: (1) satellites and sensors, (2) global navigation satellite systems (GNSS), (3) remote sensing fundamentals, (4) data normalization, harmonization, and standardization, (5) vegetation indices and their within and across sensor calibration, (6) crowdsourcing, (7) cloud computing, (8) Google Earth Engine supported remote sensing, (9) accuracy assessments, and (10) remote sensing law.

The topics covered in the **chapters of Volume II** include: (1) digital image processing fundamentals and advances; (2) digital image classifications for applications such as urban, land use, and land cover; (3) hyperspectral image processing methods and approaches; (4) thermal infrared image processing principles and practices; (5) image segmentation; (6) object-oriented image analysis (OBIA), including geospatial data integration techniques in OBIA; (7) image segmentation in specific applications like land use and land cover; (8) LiDAR digital image processing; (9) change detection; and (10) integrating geographic information systems (GIS) with remote sensing in geoprocessing workflows, democratization of GIS data and tools, fronters of GIScience, and GIS and remote sensing policies.

The topics covered in the **chapters of Volume III** include: (a) vegetation and biomass, (b) agricultural croplands, (c) rangelands, (d) phenology and food security, and (e) soils.

The topics covered in the **chapters of Volume IV** include: (1) forests, (2) biodiversity, (3) ecology, (4) land use and land cover, and (5) carbon. Under each of the preceding broad topics, there are one or more chapters.

The **chapters of Volume V** focus on hydrology, water resources, ice, wetlands, and crop water productivity. The chapters are broadly classified into (1) geomorphology, (2) hydrology and water resources, (3) floods, (4) wetlands, (5) crop water use and crop water productivity, and (6) snow and ice.

The **chapters of Volume VI** focus on water resources, disasters, and urban remote sensing. The chapters are broadly classified into (1) droughts and drylands, (2) disasters, (3) volcanoes, (4) fires, and (5) nightlights.

There are many ways to use the *Remote Sensing Handbook* (Second Edition, six volumes). A lot of thought went in organizing the volumes and chapters. So, you will see a "flow" from chapter to chapter and volume to volume. As you read through the chapters, you will see how they are interconnected and how reading all of them provides you with greater in-depth understanding. You will also realize, as someone deeply interested in one of the topics, that you will have greater interest in one volume. Having all six volumes as reference material is ideal for any remote sensing expert, practitioner, or student. However, you can also refer to individual volumes based on your interest. We have also made great attempts to ensure chapters are self-contained. That way, you can focus on a chapter and read it through, without having to be overly dependent on other chapters. Taking this perspective, a small amount of material (~5 to 10%) may be repeated across chapters. This is done deliberately. For example, when you are reading a chapter on LiDAR or Radar, you don't want to go all the way back to another chapter to understand characteristics of these data. Similarly, certain indices (e.g., vegetation condition index (VCI) or temperature condition index (TCI)) that are defined in one chapter (e.g., on drought) may be repeated in another chapter (also on drought). Such minor overlaps help the reader avoid going back to another chapter to understand a phenomenon or an index or a characteristic of a sensor. However, if you want a lot of details of these sensors or indices or phenomenon, then you will have to read the appropriate chapter where there is in-depth coverage of the topic.

Each volume has a summary chapter (the last chapter of each volume). The summary chapter can be read two ways: (1) either as last chapter to recapture the main points of each of the chapters, or (2) or as an initial overview to get the first feeling for what is in the volume, before diving into to read each chapter in detail. I suggest the readers do it both ways: read it first before reading chapters in detail to gather an idea on what to expect in each chapter and then read it at the end to recapture what is being read in each of the chapters.

It has been a great honor as well as humbling experience to edit the *Remote Sensing Handbook* (Volumes I–VI). I truly enjoyed the effort, albeit felt overwhelmed at times with never-ending work. What an honor to work with luminaries in your field of expertise. I learned a lot from them and am very grateful for their support, encouragement, and deep insights. Also, it has been a pleasure working with the outstanding professionals of Taylor and Francis/CRC Press. There is no joy greater than being immersed in pursuit of excellence, knowledge gain, and knowledge capture. At the same time, I am happy it is over. If there will be a third edition in a decade or so from now, it will be taken up by someone else (individually or as a team) and certainly not me!

I expect the book to be a standard reference of immense value to any student, scientist, professional, and practical practitioners of remote sensing. Any book that has the privilege of 300+ of the best brains of truly outstanding and dedicated remote sensing scientists ought to be a *Magnum Opus* deserving to be standard reference on the subject.

**Dr. Prasad S. Thenkabail, PhD**
*Editor in Chief (EiC)*
*Remote Sensing Handbook (Second Edition, Volumes I–VI)*

**Volume I: Sensors, Data Normalization, Harmonization, Cloud Computing, and Accuracies**
**Volume II: Image Processing, Change Detection, GIS, and Spatial Data Analysis**
**Volume III: Agriculture, Food Security, Rangelands, Vegetation, Phenology, and**
**Volume IV: Forests, Biodiversity, Ecology, LULC and Carbon**
**Volume V: Water Resources: Hydrology, Floods, Snow and Ice, Wetlands, and Water Productivity**
**Volume VI: Droughts, Disasters, Pollution, and Urban Mapping**

# About the Editor

**Dr. Prasad S. Thenkabail**, PhD, is a senior scientist with the U.S. Geological Survey (USGS), specializing in remote sensing science for agriculture, water, and food security. He is a world-recognized expert in remote sensing science with multiple major contributions in the field sustained for 40+ years. Dr. Thenkabail has conducted pioneering research in hyperspectral remote sensing of vegetation, global croplands mapping for water and food security, and crop water productivity. His work on hyperspectral remote sensing of agriculture and vegetation are widely cited. His papers on hyperspectral remote sensing are first of its kind and, collectively, they have (1) determined optimal hyperspectral narrowbands (OHNBs) in study of agricultural crops; (2) established hyperspectral vegetation indices (HVIs) to model and map crop biophysical and biochemical quantities; (3) created framework and sample data for the global hyperspectral imaging spectral libraries of crops (GHISA); (4) developed methods and techniques of overcoming Hughes' phenomenon; (5) demonstrated the strengths of hyperspectral narrowband (HNB) data in advancing classification accuracies relative to multispectral broadband (MBB) data; (6) showed advances one can make in modeling crop biophysical and biochemical quantities using HNB and HVI data relative to MBB data; and (7) created a body of work in understanding, processing, and utilizing HNB and HVI data in agricultural cropland studies. This body of work has become a widely referred reference worldwide. In studies of global croplands for food and water security, he has led the release of the world's first 30-m Landsat Satellite-derived global cropland extent product at 30 m (GCEP30; https://www.usgs.gov/apps/croplands/app/map) (Thenkabail et al. 2021; https://lpdaac.usgs.gov/news/release-of-gfsad-30-meter-cropland-extent-products/) and Landsat-derived global rainfed and irrigated area product at 30 m (LGRIP30; https://lpdaac.usgs.gov/products/lgrip30v001/) (Teluguntla et al., 2023). Earlier, he led producing the world's first global irrigated area map (https://lpdaac.usgs.gov/products/gfsad1kcdv001/ and https://lpdaac.usgs.gov/products/gfsad1kcmv001/) using multi-sensor satellite data at nominal 1 km spatial resolution. The global cropland datasets using satellite remote sensing demonstrates a "paradigm shift" in global cropland mapping using remote sensing through big data analytics, machine learning, and petabyte-scale cloud computing on the Google Earth Engine (GEE). The LGRIP30 and GCEP30 products are released through NASA's LP DAAC and published in USGS professional paper 1868 (Thenkabail et al., 2021). He has been principal investigator of many projects over the years, including the NASA-funded global food security support analysis data in the 30-m (GFSAD) project (www.usgs.gov/wgsc/gfsad30).

His career scientific achievements can be gauged by successfully making the list of the world's top 1% of scientists as per the Stanford study ranking world's scientists from across 22 scientific fields and 176 subfields based on deep analysis evaluating about ten million scientists based on SCOPUS data from Elsevier from 1996 to 2023 (Ioannidis and John, 2023; Ioannidis et al., 2020). Dr. Thenkabail was recognized as Fellow of the American Society of Photogrammetry and Remote Sensing (ASPRS) in 2023. Dr. Thenkabail has published more than 150 peer-reviewed scientific papers and edited 15 books. His scientific papers have won several awards over the years, demonstrating world-class highest quality research. These include: 2023 Talbert Abrams Grand Award, the highest scientific paper award of the ASPRS (with Itiya Aneece); 2015 ASPRS ERDAS award for best scientific paper in remote sensing (with Michael Marshall); 2008 John I. Davidson ASPRS President's Award for Practical Papers (with Pardha Teluguntla); and 1994 Autometric Award for the outstanding paper in remote sensing (with Dr. Andy Ward).

Dr. Thenkabail' s contributions to series of leading edited books places him as a world leader in remote sensing science. There are three seminal book-sets with a total of 13 volumes that he edited that have demonstrated his major contributions as an internationally acclaimed remote sensing scientist. These are (1) *Remote Sensing Handbook* (Second Edition, six-volume book-set, 2024) with 91 chapters and nearly 3,000 pages and for which he is the sole editor; (2) *Remote Sensing Handbook* (First Edition, three-volume book-set, 2015) with 82 chapters and 2,304 pages and for which he is the sole editor; and

(3) *Hyperspectral Remote Sensing of Vegetation* (four-volume Book-set, 2018) with 50 chapters and 1,632 pages that he edited as the chief editor (co-editors: Prof. John Lyon and Prof. Alfredo Huete).

Dr. Thenkabail is at the center of rendering scientific service to the world's remote sensing community over long periods of service. This includes serving as editor in chief (2011–present) of *Remote Sensing Open Access Journal*; associate editor (2017–present) of *Photogrammetric Engineering and Remote Sensing* (PE&RS); Editorial Advisory Board (2016–present) of the International Society of Photogrammetry and Remote Sensing (ISPRS); and Editorial Board Member of Remote Sensing of Environment (2007–2016).

The USGS and NASA selected him as one of the three international members on the Landsat Science Team (2006–2011). He is an Advisory Board member of the online library collection to support the United Nations' Sustainable Development Goals (UN SDGs), and currently scientist for the NASA and ISRO (Indian Space Research Organization) Professional Engineer and Scientist Exchange Program (PESEP) program for 2022–2024. He was the chair, International Society of Photogrammetry and Remote Sensing (ISPRS) Working Group WG VIII/7 (land cover and its dynamics) from 2013–2016; played a vital role for USGS as global coordinator, Agricultural Societal Beneficial Area (SBA), Committee for Earth Observation (CEOS) (2010–2013) during which he co-wrote the global food security case study for the CEOS for the *Earth Observation Handbook* (EOS), Special Edition for the UN Conference on Sustainable Development, presented in Rio de Janeiro, Brazil; was the co-lead (2007–2011) of IEEE "Water for the World" initiative, a nonprofit effort funded by IEEE which worked in coordination with the Group on Earth Observations (GEO) in its GEO Water and GEO Agriculture initiatives.

Dr. Thenkabail worked as a postdoctoral researcher and research faculty at the Center for Earth Observation (YCEO), Yale University (1997–2003), and led remote sensing programs in three international organizations including the following:

- International Water Management Institute (IWMI), 2003–2008
- International Center for Integrated Mountain Development (ICIMOD), 1995–1997
- International Institute of Tropical Agriculture (IITA), 1992–1995

He began his scientific career as a scientist (1986–1988) working for the National Remote Sensing Agency (NRSA) (now renamed National Remote Sensing Center, or NRSC), Indian Space Research Organization (ISRO), Department of State, Government of India.

Dr. Thenkabail's work experience spans over 25 countries including East Asia (China), South-East Asia (Cambodia, Indonesia, Myanmar, Thailand, Vietnam), Middle East (Israel, Syria), North America (United States, Canada), South America (Brazil), Central Asia (Uzbekistan), South Asia (Bangladesh, India, Nepal, and Sri Lanka), West Africa (Republic of Benin, Burkina Faso, Cameroon, Central African Republic, Cote d'Ivoire, Gambia, Ghana, Mali, Nigeria, Senegal, and Togo), and Southern Africa (Mozambique, South Africa). Dr. Thenkabail is regularly invited as keynote speaker or invited speaker at major international conferences and at other important national and international forums every year.

Dr. Thenkabail obtained his PhD in agricultural engineering from The Ohio State University, USA, in 1992. He has a master's degree in hydraulics and water resources engineering, and a bachelor's degree in civil engineering (both from India). He has 168 publications including 15 books; 175+ peer-reviewed journal articles, book chapters, and professional papers/monographs; and 15+ significant major global and regional data releases.

## REFERENCES

### Scientific Papers

https://scholar.google.com/citations?user=9IO5Y7YAAAAJ&hl=en

## USGS Professional Paper, Data and Product Gateways, Interactive Viewers

Thenkabail, P.S., Teluguntla, P.G., Xiong, J., Oliphant, A., Congalton, R.G., Ozdogan, M., Gumma, M.K., Tilton, J.C., Giri, C., Milesi, C., Phalke, A., Massey, R., Yadav, K., Sankey, T., Zhong, Y., Aneece, I., and Foley, D. (2021). Global cropland-extent product at 30-m resolution (GCEP30) derived from Landsat satellite time-series data for the year 2015 using multiple machine-learning algorithms on Google Earth Engine cloud. U.S. Geological Survey Professional Paper 1868, 63 pages, https://doi.org/10.3133/pp1868 (research paper). https://lpdaac.usgs.gov/news/release-of-gfsad-30-meter-cropland-extent-products/ (download data, documents). www.usgs.gov/apps/croplands/app/map (view data interactively).

Teluguntla, P., Thenkabail, P., Oliphant, A., Gumma, M., Aneece, I., Foley, D., and McCormick, R. (2023a). Landsat-Derived Global Rainfed and Irrigated-Cropland Product @ 30-m (LGRIP30) of the World (GFSADLGRIP30WORLD). The Land Processes Distributed Active Archive Center (LP DAAC) of NASA and USGS. p. 103. https://lpdaac.usgs.gov/news/release-of-lgrip30-data-product/ (download data, documents)

## Books

### Remote Sensing Handbook (Second Edition, Six Volumes, 2024)

Thenkabail, Prasad. 2024. Remote Sensing Handbook (Second Edition, Six Volume Book-set), *Volume I: Sensors, Data Normalization, Harmonization, Cloud Computing, and Accuracies*. Taylor and Francis Inc./CRC Press, Boca Raton, London, New York. 978-1-032-89095-1—CAT# T132478. Print ISBN: 9781032890951. eBook ISBN: 9781003541141. Pp. 581.

Thenkabail, Prasad. 2024. Remote Sensing Handbook (Second Edition, Six Volume Book-set), *Volume II: Image Processing, Change Detection, GIS, and Spatial Data Analysis*. Taylor and Francis Inc./CRC Press, Boca Raton, London, New York. 978-1-032-89097-5—CAT# T133208. Print ISBN: 9781032890975. eBook ISBN: 9781003541158. Pp. 464.

Thenkabail, Prasad. 2024. Remote Sensing Handbook (Second Edition, Six Volume Book-set), *Volume III: Agriculture, Food Security, Rangelands, Vegetation, Phenology, and Soils*. Taylor and Francis Inc./CRC Press, Boca Raton, London, New York. 978-1-032-89101-9—CAT# T133213. Print ISBN: 9781032891019; eBook ISBN: 9781003541165. Pp. 788.

Thenkabail, Prasad. 2024. Remote Sensing Handbook (Second Edition, Six Volume Book-set), *Volume IV: Forests, Biodiversity, Ecology, LULC, and Carbon*. Taylor and Francis Inc./CRC Press, Boca Raton, London, New York. 978-1-032-89103-3—CAT# T133215. Print ISBN: 9781032891033. eBook ISBN: 9781003541172. Pp. 501.

Thenkabail, Prasad. 2024. Remote Sensing Handbook (Second Edition, Six Volume Book-set), *Volume V: Water, Hydrology, Floods, Snow and Ice, Wetlands, and Water Productivity*. Taylor and Francis Inc./CRC Press, Boca Raton, London, New York. 978-1-032-89145-3—CAT# T133261. Print ISBN: 9781032891453. eBook ISBN: 9781003541400. Pp. 516.

Thenkabail, Prasad. Remote Sensing Handbook (Second Edition, Six Volume Book-set), *Volume VI: Droughts, Disasters, Pollution, and Urban Mapping*. Taylor and Francis Inc./CRC Press, Boca Raton, London, New York. 978-1-032-89148-4—CAT# T133267. Print ISBN: 9781032891484; eBook ISBN: 9781003541417. Pp. 467.

### Hyperspectral Remote Sensing of Vegetation (First Edition, Four Volumes, 2018)

Thenkabail, P.S., Lyon, G.J., and Huete, A. (Editors) 2018. *Hyperspectral Remote Sensing of Vegetation* (Second Edition, Four-Volume set).

Volume I: *Fundamentals, Sensor Systems, Spectral Libraries, and Data Mining for Vegetation*. CRC Press-Taylor and Francis Group, Boca Raton, London, New York. p. 449, Hardback ID: 9781138058545; eBook ID: 9781315164151.

Volume II: *Hyperspectral Indices and Image Classifications for Agriculture and Vegetation*. CRC Press-Taylor and Francis Group, Boca Raton, London, New York. p. 296. Hardback ID: 9781138066038; eBook ID: 9781315159331.

Volume III: *Biophysical and Biochemical Characterization and Plant Species Studies*. CRC Press-Taylor and Francis Group, Boca Raton, London, New York. p. 348. Hardback: 9781138364714; eBook ID: 9780429431180.

Volume IV: Advanced Applications in Remote Sensing of Agricultural Crops and Natural Vegetation. CRC Press-Taylor and Francis Group, Boca Raton, London, New York. p. 386. Hardback: 9781138364769; eBook ID: 9780429431166.

## Remote Sensing Handbook (First Edition, Three Volumes, 2015)

Thenkabail, P.S., (Editor in Chief) (2015). *Remote Sensing Handbook.*

Volume I: *Remotely Sensed Data Characterization, Classification, and Accuracies*. Taylor and Francis Inc./CRC Press, Boca Raton, London, New York. ISBN 9781482217865—CAT# K22125. Print ISBN: 978-1-4822-1786-5; eBook ISBN: 978-1-4822-1787-2. p. 678.

Volume II: *Land Resources Monitoring, Modeling, and Mapping with Remote Sensing*. Taylor and Francis Inc./CRC Press, Boca Raton, London, New York. ISBN 9781482217957—CAT# K22130. p. 849.

Volume III: *Remote Sensing of Water Resources, Disasters, and Urban Studies*. Taylor and Francis Inc./CRC Press, Boca Raton, London, New York. ISBN 9781482217919—CAT# K22128. p. 673.

## Hyperspectral Remote Sensing of Vegetation (First Edition, Single Volume, 2013)

Thenkabail, P.S., Lyon, G.J., and Huete, A. (Editors) (2012). *Hyperspectral Remote Sensing of Vegetation*. CRC Press-Taylor and Francis group, Boca Raton, London, New York. p. 781 (80+ pages in color). www.crcpress.com/product/isbn/9781439845370.

## Remote Sensing of Global Croplands for Food Security (First Edition, Single Volume, 2009)

Thenkabail. P., Lyon, G.J., Turral, H., and Biradar, C.M. (Editors) (2009). *Remote Sensing of Global Croplands for Food Security* (CRC Press-Taylor and Francis Group, Boca Raton, London, New York. p. 556 (48 pages in color). Published in June 2009.

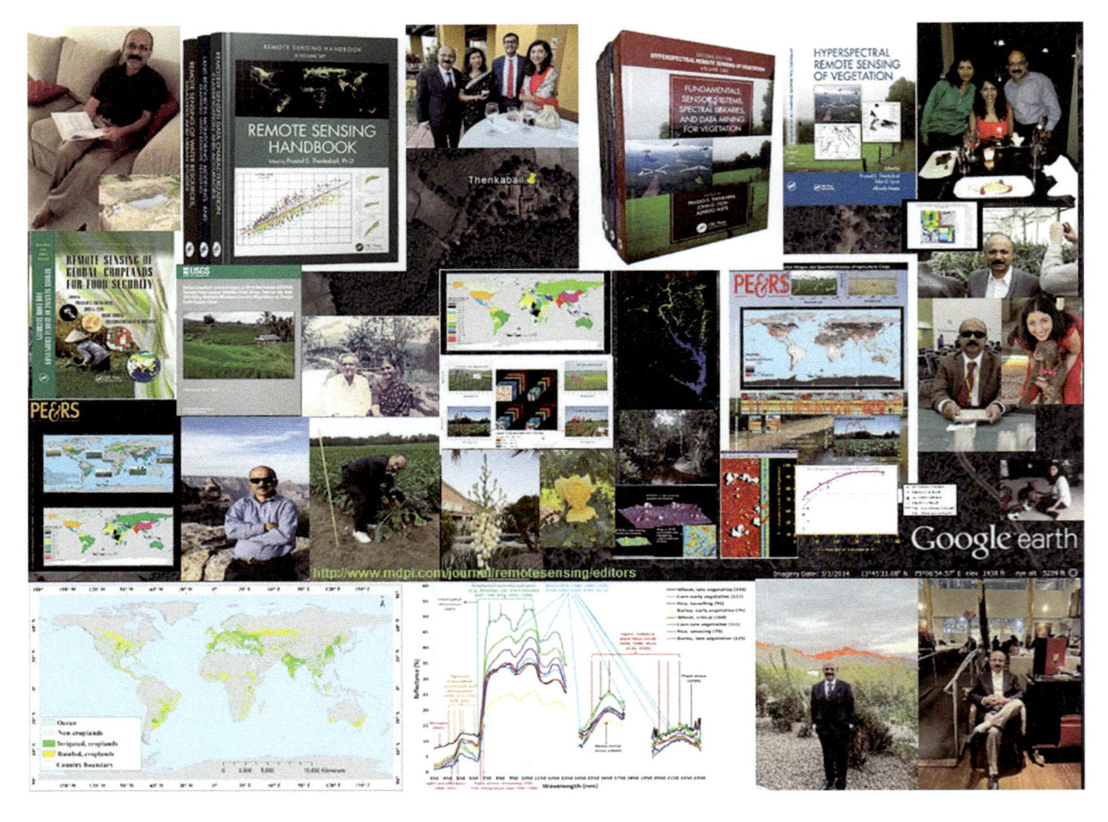


**FIGURE ABOVE:** Few snap-shots in life and work of Editor-in-Chief Dr. Prasad S. Thenkabail.

# Contributors

**O. Antemijczuk**
Institute of Computer Science
Faculty of Automatic Control, Electronics and Computer Science
Silesian University of Technology
Gliwice, Poland

**Marc P. Armstrong**
Department of Geographical and Sustainability Sciences
The University of Iowa
Iowa, USA

**John E. Bailey**
Professor
Arizona State University
Phoenix, Arizona, USA

**Frédéric Baret**
Research Director
INRA-EMMAH/CAPTE
Avignon, France

**P.J. Blount**
Lecturer in Law
Cardiff University
Cardiff, UK

**Gyanesh Chander**
Science Data Systems Branch
NASA Goddard Space Flight Center (GSFC)
Greenbelt, Maryland

**Russell G. Congalton**
Department of Natural Resources and the Environment
University of New Hampshire
Durham, New Hamshire

**K.A. Cyran**
Institute of Computer Science
Faculty of Automatic Control, Electronics and Computer Science
Silesian University of Technology
Gliwice, Poland

**Fabio Dell'Acqua**
Dipartimento di Ingegneria Industriale e dell'Informazione
Università di Pavia
Pavia, Italy

**Silvio Dell'Acqua**
Editor
Laputa blog on cultural and historical facts
Pavia, Italy

**Andrew Dempster**
Professor
School of Electrical Engineering and Telecommunications
UNSW
Sydney, Australia

**James P. Fitzgerald**
Department of Land Management
Rudiger Gens
Geophysical Institute
University of Alaska Fairbanks
Fairbanks, Alaska

**Mohinder S. Grewal**
Distinguished Professor Emeritus
Electrical Engineering
California State University
Fullerton, California

**M. Grygierek**
LG Nexera Business Solutions AG
Vienna, Austria

**Haixu He**
China University of Geosciences
Wuhan, China

**James W. Hegeman**
Department of Computer Science
The University of Iowa
Iowa City, Iowa

**Xiaohui Huang**
Lecturer
School of Computer Science
China University of Geosciences
Wuhan, China

**Alfredo Huete**
School of Life Sciences
University of Technology
Sydney, Australia

**Aolin Jia**
Department of Environment Research and Innovation
Luxembourg Institute of Science and Technology (LIST)
Belvaux, Luxembourg

**Jiabao Li**
China University of Geosciences
Wuhan, China

**Ao Long**
China University of Geosciences
Wuhan, China

**Yan Ma**
Associate Professor
Aerospace Information Research Institute
Chinese Academy of Sciences
Beijing, China

**Timothy J. Malthus**
Research Team Leader—Aquatic Remote Sensing
CSIRO Oceans and Atmosphere Flagship
Ecosciences Precinct
Dutton Park, Australia

**Debasmita Misra**
Professor, Geological Engineering
College of Engineering & Mines
University of Alaska Fairbanks
Fairbanks, Alaska

**Tomoaki Miura**
Department of Natural Resources and Environmental Management
University of Hawaii
Manoa, Hawaii

**Arnab Muhuri**
Earth Observation and Modeling (EOM)
Department of Geography
Christian-Albrechts-Universität zu Kiel
Kiel, Germany

**D. Myszor**
Institute of Computer Science
Faculty of Automatic Control, Electronics and Computer Science
Silesian University of Technology
Gliwice, Poland
and
LG Nexera Business Solutions AG
Vienna, Austria

**Kenta Obata**
Department of Information Science and Technology
Aichi Prefectural University
Japan

**Natascha Oppelt**
Earth Observation and Modeling (EOM)
Department of Geography
Christian-Albrechts-Universität zu Kiel
Kiel, Germany

**Sudhanshu S. Panda**
Professor, GIS/Environmental Science
Institute for Environmental Spatial Analysis
University of North Georgia
Oakwood, Georgia

**Mahesh N. Rao**
Data Management Lead
United States Fish and Wildlife Services Ecological Services
Falls Church, Virginia

**Chris Rizos**
Professor
School of Civil and Environmental Engineering
UNSW
Sydney, Australia

**Jordi Cristóbal Rosselló**
Efficient Use of Water in Agriculture Program
Institute of Agrifood Research and Technology Fruitcentre

Parc Científic i Tecnològic Agroalimentari de Lleida
and
Department of Geography
Autonomous University of Barcelona
Bellaterra, Spain

**Vivek B. Sardeshmukh**
Department of Computer Science
The University of Iowa
Iowa City, Iowa

**Michael D. Steven**
School of Geography
University of Nottingham
Nottingham, United Kingdom

**Ramanathan Sugumaran**
Department of Geographical and Sustainability Sciences
The University of Iowa
Iowa, USA
and
John Deere
Moline, Illinois

**Philippe M. Teillet**
Department of Physics and Astronomy
University of Lethbridge
Lethbridge, Alberta, Canada

**Prasad S. Thenkabail**
Senior Scientist (ST)
United States Geological Survey (USGS)
Western Geographic Science Center
Flagstaff, Arizona

**Dongdong Wang**
Department of Geographical Sciences
University of Maryland
College Park, Maryland

**Lizhe Wang**
Professor and Vice Chancellor
China University of Geosciences
Wuhan, China

**Sheng Wang**
China University of Geosciences
Wuhan, China

**M. Wierzchanowski**
Institute of Computer Science
Faculty of Automatic Control, Electronics and Computer Science
Silesian University of Technology
Gliwice, Poland
and
LG Nexera Business Solutions AG
Vienna, Austria

**Josh Williams**
Westwood High School
Round Rock, Texas

**Jining Yan**
Associate Professor
School of Computer Science
China University of Geosciences
Wuhan, China

**Hiroki Yoshioka**
Department of Information Science and Technology
Aichi Prefectural University
Japan

**Kegen Yu**
Professor
School of Environment Science and Spatial Informatics
China University of Mining and Technology
Xuzhou, China

**Xiaohan Zhang**
China University of Geosciences
Wuhan, China

# Acknowledgments

*Remote Sensing Handbook* (Second Edition, Volumes I–VI) brought together a galaxy of highly accomplished, renowned remote sensing scientists, professionals, and legends from around the world. The lead authors were chosen by me after careful review of their accomplishments and sustained publication record over the years. The chapters in the second edition were written/revised over a period of two years. All chapters were edited and revised.

Gathering such a galaxy of authors was the biggest challenge. These are all extremely busy people, and committing to a book project that requires substantial work is never easy. However, almost all those whom I requested agreed to write a chapter specific to their area of specialization, and only a few I had to convince to make time. The quality of the chapters should convince readers why these authors are such highly rated professionals and why they are so successful and accomplished in their fields of expertise. They not only wrote very high-quality chapters but also delivered them on time, addressed any editorial comments in a timely manner without complaints, and were extremely humble and helpful. Their commitment for quality science is what makes them special. I am truly honored to have worked with such great professionals.

I would like to mention the names of everyone who contributed and made ***Remote Sensing Handbook* (Second Edition, Volumes I–VI) possible**. In the end, we had 91 chapters, a little over 3,000 pages, and a little more than 400 authors. My gratitude goes to each one of them. These are well-known **"who is who"** in remote sensing science in the world. List of all authors are provided here. The names of the authors are organized chronologically for each volume and the chapters. Each lead author of the chapter is in bold type. **The names of the 400+ authors who contributed to six volumes are as follows:**

**Volume I:** Sensors, Data Normalization, Harmonization, Cloud Computing, and Accuracies—18 chapters written by 53 authors (editor in chief: Prasad S. Thenkabail):

**Drs. Sudhanshu S. Panda,** Mahesh N. Rao, Prasad S. Thenkabail, Debasmita Misra, and James P. Fitzgerald; **Mohinder S. Grewal**; **Kegen Yu**, Chris Rizos, and Andrew Dempster; **D. Myszor**, O. Antemijczuk, M. Grygierek, M. Wierzchanowski, K.A. Cyran; **Natascha Oppelt** and Arnab Muhuri; **Philippe M. Teillet**; **Philippe M. Teillet** and Gyanesh Chander; **Rudiger Gens** and Jordi Cristóbal Rosselló; **Aolin Jia** and Dongdong Wang; **Tomoaki Miura**, Kenta Obata, Hiroki Yoshioka, and Alfredo Huete; **Michael D. Steven**, Timothy J. Malthus, and Frédéric Baret; **Fabio Dell'Acqua** and Silvio Dell'Acqua; **Ramanathan Sugumaran**, James W. Hegeman, Vivek B. Sardeshmukh and Marc P. Armstrong; **Lizhe Wang**, Jining Yan, Yan Ma, Xiaohui Huang, Jiabao Li, Sheng Wang, Haixu He, Ao Long, and Xiaohan Zhang; **John E. Bailey** and Josh Williams; **Russell G. Congalton**; **P.J. Blount**; **Prasad S. Thenkabail**.

**Volume II:** Image Processing, Change Detection, GIS and Spatial Data Analysis: 17 chapters written by 64 authors (editor in chief: Prasad S. Thenkabail):

**Sunil Narumalani** and Paul Merani; **Mutlu Ozdogan**; **Soe W. Myint**, Victor Mesev, Dale Quattrochi, and Elizabeth A. Wentz; **Jun Li**, Paolo Gamba, and Antonio Plaza; **Qian Du**, Chiranjibi Shah, Hongjun Su, and Wei Li; **Claudia Kuenzer**, Philipp Reiners, Jianzhong Zhang, Stefan Dech; **Mohammad D. Hossain** and Dongmei Chen; **Thomas Blaschke**, Maggi Kelly, Helena Merschdorf; **Stefan Lang** and Dirk Tiede; **James C. Tilton**, Selim Aksoy, and Yuliya Tarabalka; **Shih-Hong Chio**, Tzu-Yi Chuang, Pai-Hui Hsu, Jen-Jer Jaw, Shih-Yuan Lin, Yu-Ching Lin, Tee-Ann Teo, Fuan Tsai, Yi-Hsing Tseng, Cheng-Kai Wang, Chi-Kuei Wang, Miao Wang, and Ming-Der Yang; **Guiying Li**, Mingxing Zhou, Ming Zhang, Dengsheng Lu; **Jason A. Tullis**, David P. Lanter, Aryabrata Basu, Jackson D. Cothren, Xuan Shi, W. Fredrick Limp, Rachel F. Linck, Sean G. Young, Jason Davis, and Tareefa S. Alsumaiti; **Gaurav Sinha**, Barry J. Kronenfeld, and Jeffrey C. Brunskill; **May Yuan**; **Stefan Lang**, Stefan Kienberger, Michael Hagenlocher, and Lena Pernkopf; **Prasad S. Thenkabail**.

**Volume III:** Agriculture, Food Security, Rangelands, Vegetation, Phenology, and Soils—17 chapters written by 110 authors (editor in chief: Prasad S. Thenkabail):

**Alfredo Huete**, Guillermo Ponce-Campos, Yongguang Zhang, Natalia Restrepo-Coupe, Xuanlong Ma; **Juan Quiros-Vargas,** Bastian Siegmann, Juliane Bendig, Laura Verena Junker-Frohn, Christoph Jedmowski, David Herrera, Uwe Rascher; **Frédéric Baret; Lea Hallik**, Egidijus Šarauskis, Ruchita Ingle, Indrė Bručienė, Vilma Naujokienė, Kristina Lekavičienė; **Clement Atzberger** and Markus Immitzer; **Agnès Bégué**, Damien Arvor, Camille Lelong, Elodie Vintrou, and Margareth Simoes; **Pardhasaradhi Teluguntla,** Prasad S. Thenkabail, Jun Xiong, Murali Krishna Gumma, Chandra Giri, Cristina Milesi, Mutlu Ozdogan, Russell G. Congalton, James Tilton, Temuulen Tsagaan Sankey, Richard Massey, Aparna Phalke, and Kamini Yadav; **Yuxin Miao**, David J. Mulla, Yanbo Huang; **Baojuan Zheng**, James B. Campbell, Guy Serbin, Craig S.T. Daughtry, Heather McNairn, and Anna Pacheco; **Prasad S. Thenkabail**, Itiya Aneece, Pardhasaradhi Teluguntla, Richa Upadhyay, Asfa Siddiqui, Justin George Kalambukattu, Suresh Kumar, Murali Krishna Gumma, Venkateswarlu Dheeravath; **Matthew C. Reeves,** Robert Washington-Allen, Jay Angerer, Raymond Hunt, Wasantha Kulawardhana, Lalit Kumar, Tatiana Loboda, Thomas Loveland, Graciela Metternicht, Douglas Ramsey, Joanne V. Hall, Trenton Benedict, Pedro Millikan, Angus Retallack, Arjan J.H. Meddens, William K. Smith, Wen Zhang; **E. Raymond Hunt Jr,** Cuizhen Wang, D. Terrance Booth, Samuel E. Cox, **Lalit Kumar**, and Matthew C. Reeves; Lalit Kumar, Priyakant Sinha, Jesslyn F Brown, R. Douglas Ramsey, Matthew Rigge, Carson A Stam, Alexander J. Hernandez, E. Raymond Hunt, Jr. and Matt Reeves; **Molly E. Brown**, Kirsten de Beurs, Kathryn Grace; **José A. M. Demattê**, Cristine L. S. Morgan, Sabine Chabrillat, Rodnei Rizzo, Marston H. D. Franceschini, Fabrício da S. Terra, Gustavo M. Vasques, Johanna Wetterlind, Henrique Bellinaso, Letícia G. Vogel; **E. Ben-Dor**, J. A. M. Demattê; **Prasad S. Thenkabail**.

**Volume IV:** Forests, Biodiversity, Ecology, LULC and Carbon: **12 chapters written by 71 authors listed here (editor in chief: Prasad S. Thenkabail)**

**E. H. Helmer**, Nicholas R. Goodwin, Valéry Gond, Carlos M. Souza Jr., and Gregory P. Asner; **Juha Hyyppä**, Xiaowei Yu, Mika Karjalainen, Xinlian Liang, Anttoni Jaakkola, Mike Wulder, Markus Hollaus, Joanne C. White, Mikko Vastaranta, Jiri Pyörälä, Tuomas Yrttimaa, Ninni Saarinen, Josef Taher, Juho-Pekka Virtanen, Leena Matikainen, Yunsheng Wang, Eetu Puttonen, Mariana Campos, Matti Hyyppä, Kirsi Karila, Harri Kaartinen, Matti Vaaja, Ville Kankare, Antero Kukko, Markus Holopainen, Hannu Hyyppä, Masato Katoh, Eric Hyyppä; **Gregory P. Asner**, Susan L. Ustin, Philip A. Townsend, and Roberta E. Martin; **Sylvie Durrieu**, Cédric Véga, Marc Bouvier, Frédéric Gosselin, Jean-Pierre Renaud, Laurent Saint-André; **Thomas W. Gillespie**, Morgan Rogers, Chelsea Robinson, Duccio Rocchini; **Stefan LANG**, Christina CORBANE, Palma BLONDA, Kyle PIPKINS, Michael FÖRSTER; **Conghe Song**, Jing Ming Chen, Taehee Hwang, Alemu Gonsamo, Holly Croft, Quanfa Zhang, Matthew Dannenberg, Yulong Zhang, Christopher Hakkenberg, Juxiang Li; **John Rogan** and Nathan Mietkiewicz; **Zhixin Qi**, Anthony Gar-On Yeh, Xia Li, Qianwen Lv; **R.A. Houghton; Wenge Ni-Meister; Prasad S. Thenkabail**.

**Volume V:** Water Resources: Hydrology, Floods, Snow and Ice, Wetlands, and Water Productivity—13 chapters written by 60 authors (editor in chief: Prasad S. Thenkabail):

**James B. Campbell** and Lynn M. Resler; **Sadiq I. Khan**, Ni-Bin Chang, Yang Hong, Xianwu Xue, Yu Zhang; **Santhosh Kumar Seelan; Allan S. Arnesen**, Frederico T. Genofre, Marcelo P. Curtarelli, and Matheus Z. Francisco; **Allan S. Arnesen**, Frederico T. Genofre, Marcelo P. Curtarelli, and Matheus Z. Francisco; **Sandro Martinis**, Claudia Kuenzer, and André Twele; **Le Wang**, Jing Miao, Ying Lu; **Chandra Giri; D. R. Mishra**, X. Yan, S. Ghosh, C. Hladik, J. L. O'Connell, H. J. Cho; **Murali Krishna Gumma**, Prasad S. Thenkabail, Pranay Panjala, Pardhasaradhi Teluguntla,

Birhanu Zemadim Birhanu, Mangi Lal Jat; **Trent W. Biggs**, Pamela Nagler, Anderson Ruhoff, Triantafyllia Petsini, Michael Marshall, George P. Petropoulos, Camila Abe, Edward P. Glenn; **Antônio Teixeira**, Janice Leivas; Celina Takemura, Edson Patto, Edlene Garçon, Inajá Sousa, André Quintão, Prasad Thenkabail, and Ana Azevedo; **Hongjie Xie,** Tiangang Liang, Xianwei Wang, Guoqing Zhang, Xiaodong Huang, and Xiongxin Xiao; **Prasad. S. Thenkabail**.

**Volume VI:** Droughts, Disasters, Pollution, and Urban Mapping—14 chapters written by 53 authors (editor in chief: Prasad S. Thenkabail):

**Felix Kogan** and Wei Guo; **F. Rembold**, M. Meroni, O. Rojas, C. Atzberger, F. Ham and E. Fillol; **Brian D.Wardlow**, Martha A. Anderson, Tsegaye Tadesse, Mark S. Svoboda, Brian Fuchs, Chris R. Hain, Wade T. Crow, and Matt Rodell; **Jinyoung Rhee**, Jungho Im, and Seonyoung Park; **Marion Stellmes**, Ruth Sonnenschein, Achim Röder, Thomas Udelhoven, Gabriel del Barrio, and Joachim Hill; **Norman Kerle; Stefan LANG**, Petra FÜREDER, Olaf KRANZ, Brittany CARD, Shadrock ROBERTS, Andreas PAPP; **Robert Wright; Krishna Prasad Vadrevu** and Kristofer Lasko; **Anupma Prakash**, Claudia Kuenzer, Santosh K. Panda, Anushree Badola, Christine F. Waigl; **Hasi Bagana**, Chaomin Chena, and Yoshiki Yamagata; **Yoshiki Yamagata**, Daisuke Murakami, Hajime Seya, and Takahiro Yoshida; **Qingling Zhang**, Noam Levin, Christos Chalkias, Husi Letu and Di Liu; **Prasad S. Thenkabail**.

The authors not only delivered excellent chapters, but they also provided valuable insights and inputs for me in many ways throughout the book project.

I was delighted when **Dr. Compton J. Tucker**, senior Earth scientist, Earth Sciences Division, Science and Exploration Directorate, NASA Goddard Space Flight Center (GSFC) agreed to write the foreword for the book. For anyone practicing remote sensing, Dr. Tucker needs no introduction. He has been a "godfather" of remote sensing and has inspired a generation of remote sensing scientists. I have been a student of his without ever really being one. I mean, I have not been his student in a classroom, but have followed his legendary work throughout my career. I remember reading his highly cited paper (now with citations nearing 7,700!):

- Tucker, C.J. (1979) "Red and Photographic Infrared Linear Combinations for Monitoring Vegetation," *Remote Sensing of Environment*, **8(2)**, 127–150.

I first read this paper in 1986 when I had just joined the National Remote Sensing Agency (NRSA; now NRSC), Indian Space Research Organization (ISRO). Ever since, Dr. Tucker's pioneering works have been a guiding light for me. After getting his PhD from the Colorado State University in 1975, Dr. Tucker joined NASA GSFC as a postdoctoral fellow in 1975 and became a full-time NASA employee in 1977. Ever since, he has conducted several path-finding research studies. He has used NOAA AVHRR, MODIS, SPOT Vegetation, and Landsat satellite data for studying deforestation, habitat fragmentation, desert boundary determination, ecologically coupled diseases, terrestrial primary production, glacier extent, and how climate affects global vegetation. He has authored or co-authored more than 280 journal articles that have been **cited more than 93,000 times**, is an adjunct professor at the University of Maryland, is a consulting scholar at the University of Pennsylvania's Museum of Archaeology and Anthropology and has appeared in more than 20 radio and TV programs. He is a Fellow of the American Geophysical Union and has been awarded several medals and honors, including NASA's Exceptional Scientific Achievement Medal, the Pecora Award from the U.S. Geological Survey, the National Air and Space Museum Trophy, the Henry Shaw Medal from the Missouri Botanical Garden, the Galathea Medal from the Royal Danish Geographical Society, and the Vega Medal from the Swedish Society of Anthropology and Geography. He was the NASA representative to the U.S. Global Change Research Program from 2006 to 2009. He is instrumental in releasing the AVHRR 33-year (1982–2014) **Global Inventory Monitoring and Modeling Studies** (GIMMS) data. **I strongly recommend that everyone read his excellent foreword before reading the book.** In the foreword, Dr. Tucker demonstrates the importance of data from Earth Observation (EO) sensors from orbiting satellites to maintaining a reliable and consistent climate record. Dr. Tucker further highlights the importance of continued measurements of these variables of our planet in the new millennium through new, improved, and innovative EO sensors from Sun-synchronous and/or geostationary satellites.

I want to acknowledge with thanks for the encouragement and support received by my U.S. Geological Survey (USGS) colleagues. I would like to mention the late Mr. Edwin Pfeifer, Dr. Susan Benjamin (my director at the Western Geographic Science Center), Dr. Dennis Dye, Mr. Larry Gaffney, Mr. David F. Penisten, Ms. Emily A. Yamamoto, Mr. Dario D. Garcia, Mr. Miguel Velasco, Dr. Chandra Giri, Dr. Terrance Slonecker, Dr. Jonathan Smith, Timothy Newman, and Zhouting Wu. Of couse, my dear colleagues at USGS, Dr. Pardhasaradhi Teluguntla, Dr. Itiya Aneece, Mr. Adam Oliphant, and Mr. Daniel Foley, have helped me in numerous ways. I am ever grateful for their support and significant contributions to my growth and this body of work. Throughout my career, there have been many postdoctoral level scientists who have worked with me closely and contributed in my scientific growth in different ways. They include Dr. Murali Krishna Gumma, head of Remote Sensing at the International Crops Research Institute for the Semi-Arid Tropics; Dr. Jun Xiong, Geo ML ≠ ML with GeoData, Climate Corp., Dr. Michael Marshall, associate professor, University of Twente, Netherlands; Dr. Isabella Mariotto, former USGS postdoctoral researcher; Dr. Chandrashekar Biradar, country director, India for World Agroforestry; and numerous others. I am thankful for their contributions. I know I am missing many names: too numerous to mention them all, but my gratitude for them is the same as the names I have mentioned here.

There is a very special person I am very thankful for: the late Dr. Thomas Loveland. I first met Dr. Loveland at USGS, Sioux Falls, for an interview to work for him as a scientist in the late 1990s when I was still at Yale University. But even though I was selected, I was not able to join him as I was not a U.S. citizen at that time and working for USGS required that. He has been my mentor and pillar of strength over two decades, particularly during my Landsat Science Team days (2006–2011) and later once I joined USGS in 2008. I have watched him conduct Landsat Science Team meetings with great professionalism, insights, and creativity. I remember him telling my PhD advisor on me being hired at USGS: "We don't make mistakes!" During my USGS days, he was someone I could ask for guidance and seek advice and he would always be there to respond with kindness and understanding. And, above all, share his helpful insights. It is too sad that we lost him too early. I pray for his soul. Thank you, Tom, for your kindness and generosity.

Over the years, there are numerous people who have come into my professional life who have helped me grow. It is a tribute to their guidance, insights, and blessings that I am here today. In this regard, I need to mention a few names as gratitude: (1) Prof. G. Ranganna, my master's thesis advisor in India at the National Institute of Technology (NIT), Surathkal, Karnataka, India. Prof. Ranganna is 92 years old (2024) and I met him few months back and to this day, he remains my guiding light on how to conduct oneself with fairness and dignity in professional and personal conduct. Prof. Ranganna's trait of selflessly caring for his students throughout his life is something that influenced me to follow. (2) Prof. E.J. James, former director of the Center for Water Resources Development and Management (CWRDM), Calicut, Kerala, India. Prof. James was my master's thesis advisor in India, whose dynamic personality in professional and personal matters had an influence on me. Dr. James's always went out of his way to help his students in spite of his busy schedule. 3. The late Dr. Andrew Ward, my PhD advisor at The Ohio State University, Columbus, Ohio. He funded my PhD studies in the U.S. through grants. Through him I learned how to write scientific papers and how to become a thorough professional. He was a tough task master, your worst critic (to help you grow), but also a perfectionist who helped you grow as a peerless professional, and above all a very kind human being at the core. He would write you long on the flaws in your research, but then help you out of it by making you work double the time! To make you work harder, he would tell you, "You won't get my sympathy." Then when you accomplished the task, he would tell you, "You have paid back for your scholarship many times over!!" (4) Dr. John G. Lyon, also my PhD advisor at The Ohio State University, Columbus, Ohio. He was a peerless motivator, encouraged you to believe in yourself. (5) Dr. Thiruvengadachari, scientist at the National Remote Sensing Agency (NRSA), which is now the National Remote Sensing Center (NRSC), India. He was my first boss at the Indian Space Research Organization (ISRO) and through him I learned the initial steps in remote sensing science. I was just 25 years of age then and had joined NRSA after my Master of Engineering (hydraulics and water resources) and Bachelor of Engineering (civil engineering) degrees. The first day in office, Dr. Thiruvengadachari asked me how much remote sensing I knew. I told him "Zero" and instantly thought he will ask me to leave the room. But his response was "very good!" and he gave me a manual on remote sensing from Laboratory for Applications of Remote Sensing (LARS), Purdue University, to study. Those were the days where there was no formal training in remote sensing in universities. So, my remote sensing lessons began working practically on projects and one of our first projects was "drought monitoring for India using NOAA AVHRR data." This was an intense period of learning the fundamentals of remote sensing science for me by practicing on a daily basis. Data came in 9 mm tapes, data was read on massive computing systems, image processing was done mostly working on night shifts by booking time on centralized computing, fieldwork was conducted using false color composite (FCC) outputs and topographic maps (there was no global positioning systems or GPS), geographic information system (GIS) was in its infancy, and a lot of calculations were done using calculators, as we had just started working in IBM 286 computers with floppy disks. So, when I decided to resign my NRSA job and go to the United States to do my PhD, Dr. Thiruvengadachari told me, "Prasad, I am losing my right hand, but you can't miss this opportunity." Those initial wonderful days of learning from Dr. Thiruvengadachari will remain etched in my memory. I am also thankful to my very good old friend Shri C.J. Jagadeesha, who was my colleague at NRSA/NRSC, ISRO. He was a friend who encouraged me to grow as a remote sensing scientist through our endless rambling discussions over tea in Iranian restaurants outside NRSA those days and elsewhere.

I am ever grateful to my former professors at The Ohio State University, Columbus, Ohio, USA: the late Prof. Carolyn Merry, Dr. Duane Marble, and Dr. Michael Demers. They taught, and/or encouraged, and/or inspired, and/or gave me opportunities at the right time. The opportunity to work for six years at the Center for Earth Observation of the Yale University (YCEO) was incredibly important. I am thankful to Prof. Ronald G. Smith, director of YCEO for the opportunity, guidance, and kindness. At YCEO I learned and advanced myself as a remote sensing scientist. The opportunities I got working for the International Institute of Tropical Agriculture (IITA) based in Nigeria and the International Water Management Institute (IWMI) based in Sri Lanka, where I worked on remote sensing science pertaining to a number of applications such as agriculture, water, wetlands, food security, sustainability, climate, natural resources management, environmental issues, droughts, and biodiversity water, were extremely important in my growth as a remote sensing scientist—especially from the point of view of understanding the real issues on the ground in real-life situations. Finding solutions and applying one's theoretical understanding to practical problems and seeing them work has its own nirvana.

As it is clear from the preceding, it is of great importantance to have guiding pillars of light at crucial stages of your education. That is where you become what you become in the end, grow, and make your own contributions. I am so blessed to have had these wonderful guiding lights come into my professional life at right time of my career (which also influenced me positively in my personal life). From that firm foundation, I could build on from what I learned, and through the confidence of knowledge and accomplishments pursue my passion for science and do several significant pioneering research projects throughout my career.

**I mention all the preceding as a gratitude for my ability today to edit such a monumental *Remote Sensing Handbook* (Second Edition, Volumes I–VI).**

I am very thankful to Ms. Irma Shagla-Britton, manager and leader for Remote Sensing and GIS books at Taylor and Francis/CRC Press. Without her consistent encouragement to take on this responsibility of editing *Remote Sensing Handbook*, especially in trusting me to accomplish this momentous work over so many other renowned experts, I would never have gotten to work on this in the first place. Thank you, Irma. Sometimes you need to ask several times, before one can say yes to something!

I am very grateful to my wife (Sharmila Prasad), daughter (Spandana Thenkabail), and son-in-law (Tejas Mayekar) for their usual unconditional understanding, love, and support. My wife and daughter have always been pillars of my life, now joined by my equally loving son-in-law. I learned the values of hard work and dedication from my revered parents. This work wouldn't come through without their life of sacrifices to educate their children and their silent blessings. My father's vision in putting emphasis on education and sending me to the best of places to study in spite of our family's very modest income, and my mother's endless hard work are my guiding light and inspiration. Of couse, there are many, many others to be thankful for, but too many to mention here. Finally, it must be noted that work of this magnitude, editing monumental *Remote Sensing Handbook* (Second Edition, Volumes I–VI) continuing from the three-volume first edition, requires blessings of almighty. I firmly believe nothing happens without the powers of the universe blessing you and providing needed energy, strength, health, and intelligence. To that infinite power my humble submission of everlasting gratefulness.

It has been my deep honor and great privilege to have edited the *Remote Sensing Handbook* (Second Edition, Volumes I–VI) after having edited the three-volume first edition that was published in 2014. Now, after ten years, we will have six-volume second edition in the year 2024. A huge thanks to all the authors, publisher, family, friends, and everyone who made this huge task possible.

**Dr. Prasad S. Thenkabail, PhD**
*Editor in Chief*
*Remote Sensing Handbook (Second Edition, Volumes I–VI)*

**Volume I:** Sensors, Data Normalization, Harmonization, Cloud Computing, and Accuracies
**Volume II**: Image Processing, Change Detection, GIS and Spatial Data Analysis
**Volume III**: *Agriculture*, Food Security, Rangelands, Vegetation, Phenology, and
**Volume IV:** Forests, Biodiversity, Ecology, LULC and Carbon
**Volume V:** Water Resources: Hydrology, Floods, Snow and Ice, Wetlands, and Water Productivity
**Volume VI:** Droughts, Disasters, Pollution, and Urban Mapping

# Part I

## Earth-Observing Satellites and Sensors from Different Eras and Their Characteristics

# 1 Remote Sensing Systems—Platforms and Sensors

## *Aerial, Satellite, UAV, Optical, RADAR, and LiDAR*

*Sudhanshu S. Panda, Mahesh N. Rao, Prasad S. Thenkabail, Debasmita Misra, and James P. Fitzgerald*

## LIST OF ACRONYMS

| | |
|---|---|
| ACRIM | Active Cavity Radiometer Irradiance Monitor |
| ADEOS | Advanced Earth Observing Satellite |
| AEP | Architecture Evolution Plan |
| AIRS | Atmospheric Infrared Sounder |
| AIRSAR | Airborne Synthetic Aperture Radar |
| Aladin | Atmospheric Laser Doppler Instrument |
| ALI | Advanced Land Imager |
| ALOS | Advanced Land Observing Satellite (from JAXA) |
| AMS | Airborne Multispectral Scanner |
| AOI | Area of Interest |
| ASAR | Advanced Synthetic Aperture Radar (from ESA) |
| ASFA | Aquatic Sciences and Fisheries Abstracts (FAO) |
| ASTER | Advanced Spaceborne Thermal Emission and Reflection Radiometer |
| ATM | Airborne Thematic Mapper |
| ATSR | Along Track Scanning Radiometer |
| AUV | Autonomous Underwater Vehicles |
| AVHRR | Advanced Very High Resolution Radiometry |
| AVIRIS | Airborne Visible/Infra-Red |
| AVNIR | Advanced Visible and Near-Infrared Radiometer |
| AWiFS | Advanced Wide Field Sensor (IRS) |
| BDRF | Bi-Directional Reflectance Function |
| BV | Brightness Value |
| CASI | Compact Airborne Spectrographic Imager |
| CBERS-2 | China-Brazil Earth Resources Satellite |
| C-CAP | coastal change analysis program |
| CCD | Charge Couple Device |
| CCRS | Canadian Center for Remote Sensing |
| CCT | Computer Compatible Tape |
| CEO | Centre for Earth Observation |
| CEOS | Committee on Earth Observation Satellites |
| CERES | Clouds and the Earth's Radiant Energy System |
| CIR | color infrared |

DOI: 10.1201/9781003541141-2

| | |
|---|---|
| CMOS | Complementary Metal-Oxide Semiconductor |
| CNES | Centre National d'Etudes Spatiales (French space agency) |
| DEM | Digital Elevation Model |
| DGPS | Differential Global Positioning |
| DIDSON | Dual-Frequency Identification Sonar |
| DLR | German Aerospace Research Establishment |
| DN | Digital Number (pixel value) |
| DOQQ | digital orthophoto (quarter) quad |
| DTM | Digital Terrain Model |
| EDC | EROS Data Center |
| EME | Electro-Magnetic Energy |
| EMR | Electromagnetic Radiation |
| EMS | Electromagnetic Spectrum |
| ENVISAT | Polar orbiting Environmental Satellite by ESA |
| EOS | Earth Observing System |
| EOSAT | Earth Observation Satellite |
| EOSDIS | Earth Observing System Data and Information System |
| EOSP | Earth Observing Scanning Polarimeter |
| EROS | Earth Resources Observation Systems |
| ERS | European Remote Sensing Satellite |
| ERS-1 | Earth Remote Sensing Satellite |
| ERTS | Earth Resources Technology Satellite |
| ESA | European Space Agency |
| ESRI | Environmental Systems Research Institute |
| ETM+ | Enhanced Thematic Mapper Plus |
| EUMETSAT | European Organization for the Exploitation of Meteorological Satellites |
| FAS | Foreign Agricultural service |
| FGDC | Federal Geographic Data Committee |
| FIR | Far Infrared |
| FLIR | Forward Looking Infra-Red |
| FORMOSAT | Taiwanese Satellite Operated by Taiwanese National Space Organization NSPO. |
| FOV | Field Of View |
| FSA | USDA-Farm Service Agency |
| G3OS | Global Observing Systems of GCOS, GOOS, and GTOS |
| GAP | National Gap Analysis Program |
| GCP | Ground Control Point |
| GCP | Ground Control Point |
| GEMS | Global Environment Monitoring System |
| GEO | Group on Earth Observations |
| GEOSS | Global Earth Observation System of Systems |
| GIS | Geographic Information System |
| GLAS | Geoscience Laser Altimetry System |
| GNSS | Global Navigation Satellite System |
| GOES | Geostationary Orbiting Earth Satellites |
| GOES | Geostationary Operational Environmental Satellite |
| GOMOS | Global Ozone Monitoring by Occultation of Stars |
| GOOS | Global Ocean Observing System |
| GOS | Global Observing System (of the WMO) |
| GOMS | Geostationary Operational Meteorological Satellite |
| GPM | Global Precipitation Measurement |
| GPS | global positioning system |

| | |
|---|---|
| GSD | Ground sample Distance |
| GSFC | Goddard Space Flight Center |
| GSTC | Geospatial Service and Technology Center |
| HIRS | High-Resolution Infrared Sounder |
| Hyperion | First Spaceborne Hyperspectral Sensor Onboard Earth Observing-1(EO-1) |
| IFOV | instantaneous field-of-view |
| IKONOS | High-Resolution Satellite Operated by GeoEye |
| IPCC | Intergovernmental Panel on Climate Change |
| IRS | Indian Remote Sensing Satellite |
| IVOS | Infrared and Visible Optical Sensors |
| JASON | The satellite series planned as follow on to TOPEX/Poseidon; JASON-1 is the first of these |
| JAXA | Japan Aerospace Exploration Agency |
| JERS | Japanese Earth Resources Satellite |
| JPSS | Joint Polar Satellite System |
| KOMFOSAT | Korean Multipurpose Satellite. Data Marketed by SPOT Image |
| LANDSAT | Land Remote-Sensing Satellite |
| LASER | Light Amplification by Stimulated Emission of Radiation |
| LCCP | land cover characterization program |
| LCT | Laser Communication Terminal |
| LiDAR | Laser Image Detection and Ranging |
| LISS | Linear Imaging Self Scanner (IRS) |
| LRA | Laser Reflectometry Array |
| LWIR | Long-Wave Infrared |
| MERIS | Medium Resolution Imaging Spectrometer |
| METEOSAT | Meteorological Satellite by ESA |
| MIR | Middle Infrared |
| MIRAS | Microwave Imaging Radiometer with Aperture Synthesis |
| MISR | Multi-Angle Imaging Spectroradiometer |
| MLS | Microwave Limb Scanner |
| MODIS | Moderate Resolution Imaging Spectroradiometer |
| MOS | Marine Observation Satellite |
| MRLC | multi-resolution land characteristics consortium |
| MSS | Multi-Spectral Scanner |
| MTF | Modular Transfer Function |
| MTSAT | Multi-purpose Transport SATellite by JAXA |
| NAIP | National Agricultural Imagery Program |
| NALC | North American Landscape Characterization Project |
| NAOS | North American Atmospheric Observing System |
| NCAR | National Center for Atmospheric Research (USA) |
| NASA | National Aeronautics and Space Administration |
| NAPP | National Aerial Photography Program |
| NAWQA | National Water Quality Assessment Program (a USGS program) |
| NCDC | National Climatic Data Center |
| nDSM | Normalized Digital Surface Model |
| NDVI | Normalized Difference Vegetation Index |
| NED | National Elevation Dataset |
| NEXRAD | Next-Generation Radar |
| NHAP | National High Altitude Program |
| NIMA | National Image and Mapping Agency |
| NIR | Near Infrared |

| | |
|---|---|
| NLCD | National Land Cover Data |
| NOAA | National Oceanic and Atmospheric Administration |
| NPOESS | National Polar Orbiting Environmental Satellite System |
| NRCS | national resources conservation service (a USDI agency) |
| NRI | national resource inventory |
| NRSC | National Remote Sensing Centre (UK) |
| NSSDC | National Space Science Data Center |
| NWS | National Weather Service |
| OCTS | Ocean Color and Temperature Scanner |
| OCX | Operational Control System |
| OSCAR | Ocean Surface Current Analysis Real-time |
| OSTM | Ocean Surface Topography Mission |
| OSTM | Ocean Surface Topography Mission |
| PALSAR | Phased Array L-band Synthetic Aperture Radar |
| PBS | Public Broadcasting System |
| QUICKBIRD | Satellite from DigitalGlobe, a private company in USA |
| RADAR | Radio Detection and Ranging |
| RADARSAT | Canadian radar satellite |
| RAPID EYE | Satellite constellation from Rapideye, a German company |
| RESOURSESAT | Satellite launched by the India |
| RGB | Red, Green, Blue |
| RKA | Russian Federal Space Agency |
| S/N | Signal to Noise ratio |
| SAR | Synthetic Aperture Radar |
| SeaWiFS | Sea-viewing Wide Field-of-view Sensor |
| SEVIRI | Spinning Enhanced Visible and Infrared Imager |
| SIRS | Satellite Infrared Spectrometer |
| SLAR | Side-Looking Airborne Radar |
| SODAR | Sound Detection And Ranging |
| SONAR | Sound Navigation And Ranging |
| SPOT | Systeme Probatoire D'Observation De La Terre (France) |
| SSM | Special Sensor Microwave |
| SSM/I | Special Sensor Microwave/Imager |
| SSM/T | Special Sensor Microwave/Temperature |
| SSM/T-2 | Special Sensor Microwave/Water Vapor |
| SSS | Sea Surface Salinity |
| SWIR | Short Wave Infrared Sensor |
| SST | Sea Surface Temperature |
| SWAT | Soil and Water Assessment Tool |
| TES | Tropospheric Emission Spectrometer |
| TIMS | Thermal Infrared Multispectral Scanner |
| TIR | Thermal Infrared |
| TM | Thematic Mapper |
| TOMS | Total Ozone Mapping Spectrometer |
| TOPEX | Ocean Topography Experiment |
| TOPSAR | Topographic Synthetic Aperture Radar |
| TRMM | Tropical Rainfall Measurement Mission |
| UAS | Unmanned Aircraft Systems |
| UAV | Unmanned Aerial Vehicle |
| UHF | Ultra High frequency |
| USA | United States of America |

| | |
|---|---|
| USGS | United States Geological Survey |
| UV | Ultraviolet |
| VIS | Visible Spectrum |
| VMS | Vessel Monitoring Systems |
| VNIR | Visible Near-Infrared Sensor |
| WFOV | Wide Field of View |
| WRS | Worldwide Reference System |
| X-SAR | X-Band Synthetic Aperture Radar |

## 1.1 INTRODUCTION

### 1.1.1 Remote Sensing Definition and History

The American Society of Photogrammetry and Remote Sensing defined remote sensing as the measurement or acquisition of information of some property of an object or phenomenon, by a recording device that is not in physical or intimate contact with the object or phenomenon under study (Colwell et al., 1983). Environmental Systems Research Institute (ESRI) in its geographic information system (GIS) dictionary defines remote sensing as

> collecting and interpreting information about the environment and the surface of the earth from a distance, primarily by sensing radiation that is naturally emitted or reflected by the earth's surface or from the atmosphere, or by sending signals transmitted from a device and reflected back to it.
>
> *(ESRI, 2014)*

The usual source of passive remote sensing data is the measurement of reflected or transmitted electromagnetic radiation (EMR) from the Sun across the electromagnetic spectrum (EMS); this can also include acoustic or sound energy, gravity, or the magnetic field from or of the objects under consideration. In this context, the simple act of reading this text is considered remote sensing. In this case, the eye acts as a sensor and senses the light reflected from the object to obtain information about the object. It is the same technology used by a handheld camera to take a photograph of a person or a distant scenic view. Active remote sensing, however, involves sending a pulse of energy and then measuring the returned energy through a sensor (e.g., Radio Detection and Ranging [RADAR], Light Detection and Ranging [LiDAR]). Thermal sensors measure emitted energy by different objects. Thus, in general, passive remote sensing involves the measurement of solar energy reflected from the Earth's surface, while active remote sensing involves synthetic (man-made) energy pulsed at the environment and the return signals are measured and recorded.

Remote sensing functions in harmony with other *spatial* data collection techniques or tools of the *mapping sciences*, including cartography and GIS (Jensen, 2009). Remote sensing is a tool or technique similar to mathematics where each sensor is used to measure from a distance the amount of EMR exiting an object, or geographic area, then extracting valuable information from the data using mathematical and statistical algorithms (Jensen, 2009). In fact, the science and art of obtaining reliable measurements using photographs is called photogrammetry (American Society of Photogrammetry, 1952, 1966). According to Aronoff (2004), remote sensing is the technology, science, and art of obtaining information about objects from a distance over regions too costly, dangerous, or remote for human observers to directly access. This includes target areas beyond the limits of human ability. Avery and Berlin (1992) simply defined remote sensing as the "technique of reconnaissance from a distance" in which information about objects are obtained through the analysis of data collected by special instruments that are not in physical contact with the objects of investigation. Therefore, remote sensing is different from in situ observation that involves use of sensing instruments in direct contact with the objects under consideration but not proximal in situ sensing, where data are collected from the in situ objects with controlled spectra. In situ observations of the

Earth's land cover and other physical phenomena are often time-consuming, costly, and impossible to have a full spatial coverage. In contrast, remote sensing acquires information about an object or phenomenon on the Earth's surface and even beneath the surface without making physical contact with the object by use of propagated signals, for example, EMR (Lillesand et al., 2004).

The very first successful photographic image of nature, considered as a permanent photograph or remotely sensed image, was captured by a French native Joseph Nicephore Niepce (Gernsheim and Gernsheim, 1952). Modern remote sensing started in 1858 at Paris, France, with Gaspard-Felix Tournachon first taking aerial photographs of the city of Paris from a hot air balloon (Briney, 2014). Remote sensing continued to grow from there but at a slow rate. During the U.S. Civil War (1861–1865), messenger pigeons, kites, and unmanned balloons were flown over enemy territory with cameras attached to them, which is considered as one of the very first planned usages of remote sensing technology (Trenear-Harvey, 2009; Briney, 2014). During World Wars I and II, the first government-organized air photography or photogrammetry missions were developed for military surveillance and later adopted by many other countries for other applications, including land surveying (Briney, 2014). After World War II, during the Cold War era, aerial photo reconnaissance became much more prominent and prevalent between the United States and the Soviet Union to collect information about each other. In December 1954, U.S. President Dwight Eisenhower approved the U-2 reconnaissance program (Brugioni and Doyle, 1997) for World War II reconnaissance from space. In recent years, tremendous advancements in technology have resulted in a rapid growth in the remote sensing industry (Jensen, 2009). During the last few decades, growth in the civilian sector has far surpassed the defense and military applications. However, the recent years have seen new applications of miniature remote sensors or camera systems that are mounted on both manned and unmanned aerial platforms, which are used by law enforcement and military sectors for surveillance purposes (Briney, 2014). Unmanned aerial vehicles (UAVs) are one of the most important advancements in aerial remote sensing in recent times and used even for fighting stealth wars (Russo et al., 2006). UAVs are now controlled from home base through onboard Global Positioning System (GPS) and forward-looking infrared (FLIR) and/or videography (FLIR) technology (Jensen, 2009). Other advances in remote sensing technology include RADAR, LiDAR, sound navigation and ranging (SONAR), sonic detection and ranging (SODAR), microwave synthetic aperture radar (SAR), infrared sensors, hyperspectral imaging, spectrometry, Doppler radar, and space probe sensors, in addition to improvements in conventional aerial photography, hyperspectral imaging, and imaging spectroradiometer.

### 1.1.2 Data Collection by Remote Sensing and Usage

Remote sensing in comparison to other methods of data collection is much more advantageous, as it provides an overview of the Earth's phenomena that allows users to discern patterns and relationships not apparent from the ground (Aronoff, 2004). EMR (usually solar energy or any form of energy) is reflected or emitted from the object and is detected by the devices called remote sensors or simply *sensors* (similar to cameras or scanners) fitted on platforms such as aircrafts, satellites, or ships. Traditional remote sensing involves two basic processes: data acquisition and data analysis. Figure 1.1 illustrates the data collection and processing involved in a typical remote sensing operation.

According to Lillesand et al. (2004), the elements of remote sensing data acquisition involve (1) energy sources such as solar energy, self-produced heat/light energy, or sound energy; (2) propagation of energy through the atmosphere; (3) energy interaction with the Earth's surface features; (4) retransmission of energy through the atmosphere; (5) sensing platforms like airborne, spaceborne, and onboard ship sensors; and (6) resultant data in the form of pictorial or digital format. There are numerous factors that affect the remote sensing data collection process, such as (1) solar position, (2) atmospheric condition, (3) weather and meteorology, (4) season of data collection, (5) ground condition, (6) sensor characteristics, and (7) sensor position. As one can

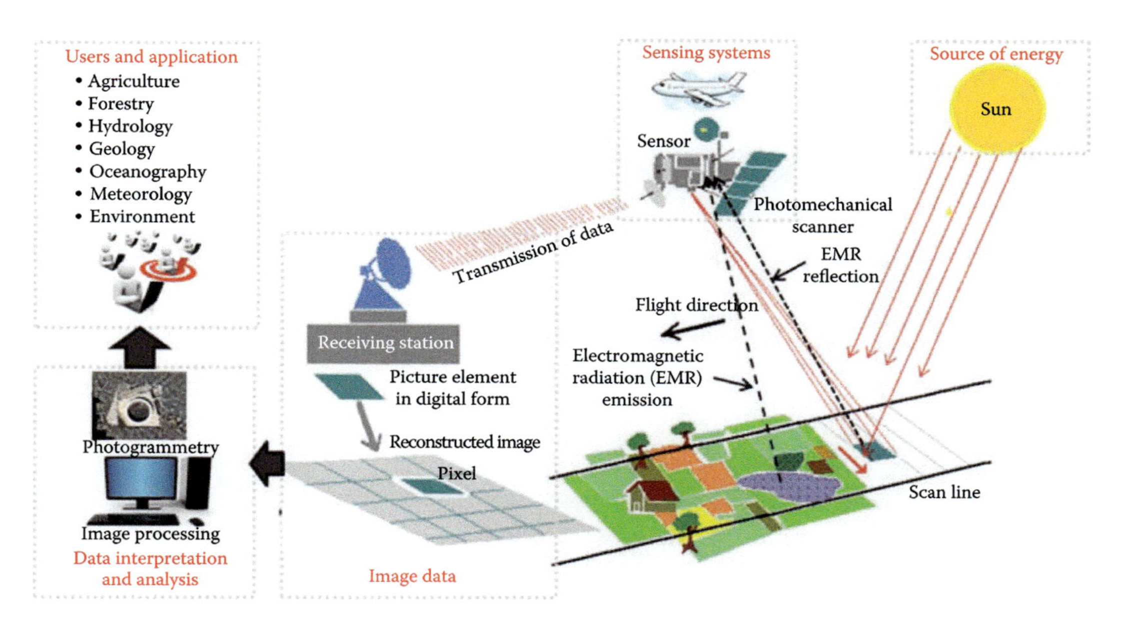


**FIGURE 1.1** Illustration of remote sensing data collection for user application. Sensors onboard aircraft or satellite platforms collect and record reflected energy from target features. The reflected data are collected across the electromagnetic spectrum and are used for various studies such as agriculture and forestry.

realize, most of the aforementioned factors directly relate to the intensity of the solar energy in terms of enhancing or attenuating it. On the other hand, remote sensing data analysis involves (1) data interpretation using various interpreting/viewing devices like computers and software, (2) analyzing the data in collaboration with other geospatial data, and (3) applying the interpreted and analyzed data for Earth management decision support.

Basically, in terms of the energy source and data collection mode, there are two types of remote sensing: (1) passive remote sensing, when reflected or emitted energy from an object or phenomenon is recorded by sensors mounted on airborne or spaceborne platforms, and (2) active remote sensing, when reflectance of synthetic light (nonsolar) that is actively pulsed or emitted from an aircraft, satellite, or any other energy-producing/recording platform is recorded (Schott, 2007; Schowengerdt, 2007). Passive remote sensing systems such as film photography have become obsolete due to several disadvantages, including low exposure time, low spectral range (i.e., usually below 700 nm only), low dynamic range (saturation/minimum limit) of only around 100, and other image characteristic issues, including linearity and photometric accuracy. Currently, most systems use charge-coupled device (CCD) and complementary metal-oxide semiconductor (CMOS)—very common in digital cameras and camcorders that convert light energy into electrons, which is then measured and converted into radiometric intensity values (McGlone, 2004). However, the active remote sensing systems such as RADAR, LiDAR, SONAR, and GPS detect not only the backscattered or energy reflecting off of Earth objects but also record time lag and intensity (Schott, 2007; Schowengerdt, 2007). Thus, unlike passive remote sensing, which only detects the location of an object, active remote sensing uses the time delay between transmission and reception of energy pulse to establish the location, height, depth, speed, and direction of an object. Figure 1.2 illustrates the difference between active and passive remote sensing with examples of different platforms.

National Aeronautics and Space Administration (NASA) reference publication #1139 by Bowker et al. (1985) lists different factors affecting apparent reflectance, which subsequently determines the object or phenomena on the Earth or in the atmosphere. These factors are (1) viewing geometry

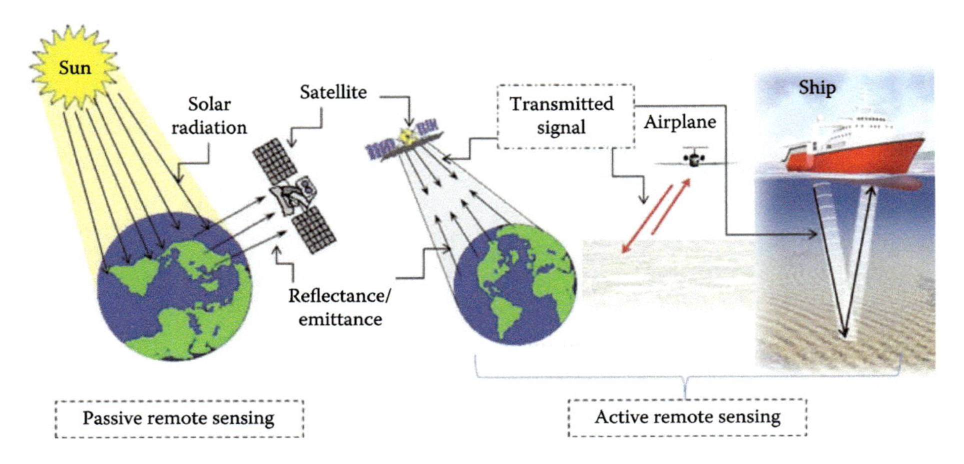


FIGURE 1.2 Illustration of passive versus active remote sensing with different platforms. In passive remote sensing, the Sun acts as the source of energy. Sensors then measure the reflected solar energy off the targets. Active remote sensing sends a pulse of synthetic energy and measures the energy reflected off the targets.

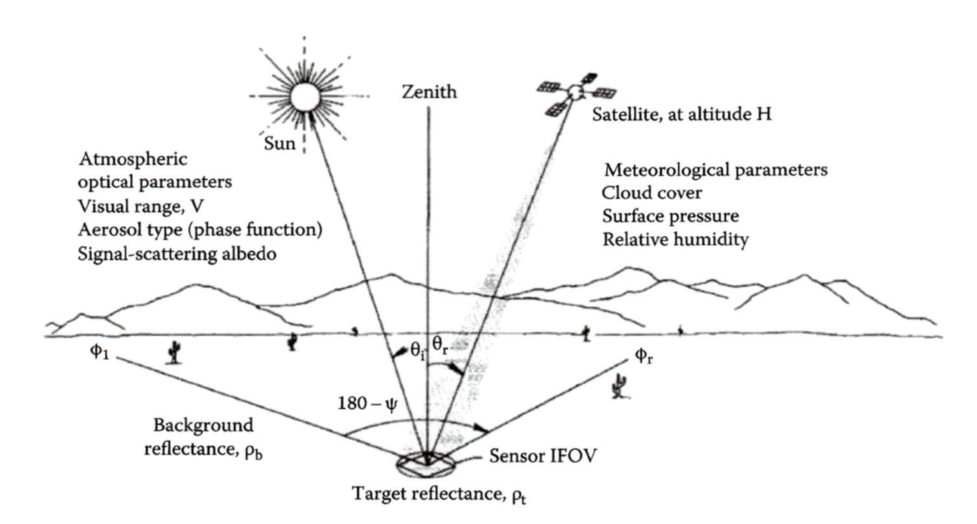


FIGURE 1.3 Factors affecting apparent reflectance in passive remote sensing. (Adapted and modified from Bowker, D.E. et al., *Spectral Reflectances of Natural Targets for Use in Remote Sensing Studies*, NASA reference publication #1139, Hampton, VA, 1985.) Note that the solar energy passes through the atmosphere and hits the target (any object on the planet). Then, depending on the type of the object, a certain portion of the energy reflects off the target, again passes back through the atmosphere, and is captured by the sensor. Such reflected energy can be captured in various portions of the electromagnetic spectrum (EMS) (e.g., visible, near infrared, shortwave infrared). Also, there can be several wave bands within each portion of the EMS.

(solar zenith angle, viewing angle, azimuthal angle, relative azimuthal angle, and altitude of the sensor), (2) meteorological parameters (relative humidity, cloud cover, and surface pressure), (3) atmospheric optical parameters (aerosol optical thickness, atmospheric visual range, aerosol types, and single-/multiple-scattering albedo), and (4) target and background parameters (target size, target reflectance, background reflectance, instantaneous field of view [IFOV], and bidirectional reflectance function [BDRF]). Figure 1.3, extracted from Bowker et al. (1985), pictorially depicts these factors affecting target reflectance measured by a passive remote sensing system.

### 1.1.3 Principles of Electromagnetic Spectrum in Remote Sensing

Light or radiant flux in the form of electromagnetic energy (EME), which includes visible light, infrared, radio waves, heat, ultraviolet (UV) rays, and X-rays, is the primary form of energy used in remote sensing (Lillesand et al., 2004). EMR is a carrier of EME by transmitting the oscillation of the electromagnetic field through space or matter based on Maxwell's equation (Murai, 1993). EMR has characteristics of both wave and particle motion. EMR essentially follows the basic wave theory ($c = \nu\lambda$), which describes that EME travels in a harmonic or sinusoidal fashion (Figure 1.4) at the velocity of light ($c = 2.998 \times 108$ m/s); the distance from one wave peak to another is wavelength, $\lambda$, and the number of peaks passing a fixed point in space per unit of time is wave frequency, $\nu$ (Lillesand et al., 2004).

According to the Murai (1993), EMR consists of four elements: (1) frequency/wavelength, (2) transmission direction, (3) amplitude, and (4) polarization. These four features are important for remote sensing as each corresponds to different features. Wavelength/frequency corresponds to the color of an object in the visible region, which is represented by a unique characteristic curve, aka reflectance curve correlating wavelength and the radiant energy from the object (Figure 1.5).

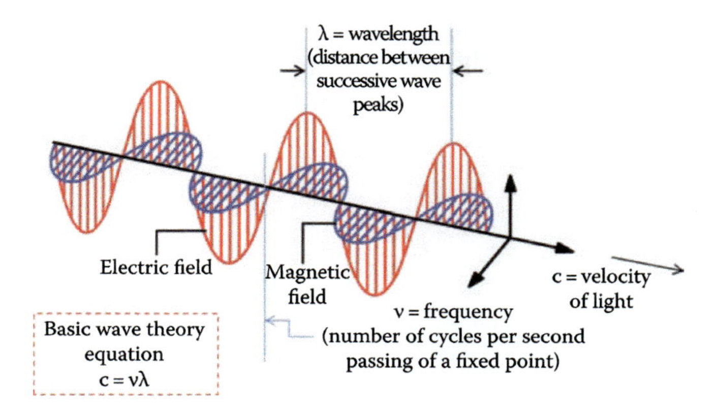


**FIGURE 1.4** Illustration of an electromagnetic wave.

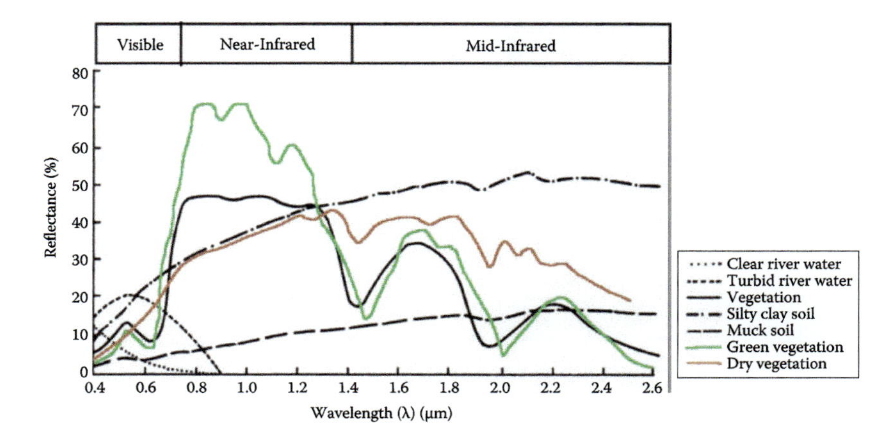


**FIGURE 1.5** Typical reflectance curve of different objects generated with electromagnetic radiation frequency characteristics that are useful to distinguish/interpret objects. (Adapted and modified from Murai, S., *Remote Sensing Note*, JARS, Tokyo, Japan, 1993.) Note how energy is reflected or absorbed in various portions of the spectrum that forms the basis of remote sensing. For example, healthy vegetation absorbs heavily in the red band (especially around 0.68 μm) but reflects heavily in near-infrared 0.76–0.90 μm.

Transmission direction and amplitude, corresponding to the direction of propagation and magnitude of the waves, are influenced by the spatial location and shape of the objects. Finally, the plane of polarization, that is, orientation of the electric field of the radiant energy, is influenced by the geometric shape of the objects under investigation (Murai, 1993). Therefore, EMR provides detailed information about the object(s) under investigation based on the spectral characteristics of the objects.

NASA reference publication #1139 by Bowker et al. (1985) is one of the most comprehensive articles on spectral reflectance curves and curve development processes of natural targets for use in remote sensing studies. Most investigations based on remote sensing data are models that are developed between the amount of electromagnetic or light energy reflected, emitted, transmitted, or backscattered from the object at different frequencies of the EMS and the biophysical/chemical properties of the object or phenomena under investigation (Jensen, 2009). The EMS is divided into several wavelength (frequency) regions, such as gamma rays ($10^{-6}$–$10^{-5}$ μm), X-ray ($10^{-5}$–$10^{-2}$ μm), UV (0.1–0.4 μm), visible (0.4–0.7 μm), infrared (0.7–103 μm), microwaves (103–106 μm), and radio waves (>106 μm). Panchromatic (i.e., grayscale) and color (i.e., red, green, blue [RGB]) imaging systems have dominated electro-optical sensing in the visible region of the EMS (Figure 1.6), which describe the efficacy of a remotely sensed imagery (Shaw and Burke, 2003). Figure 1.7 depicts the bandwidth (wavelength) of different bands associated with the spectral regions, along with the different remote sensing systems that acquire data in these bands.

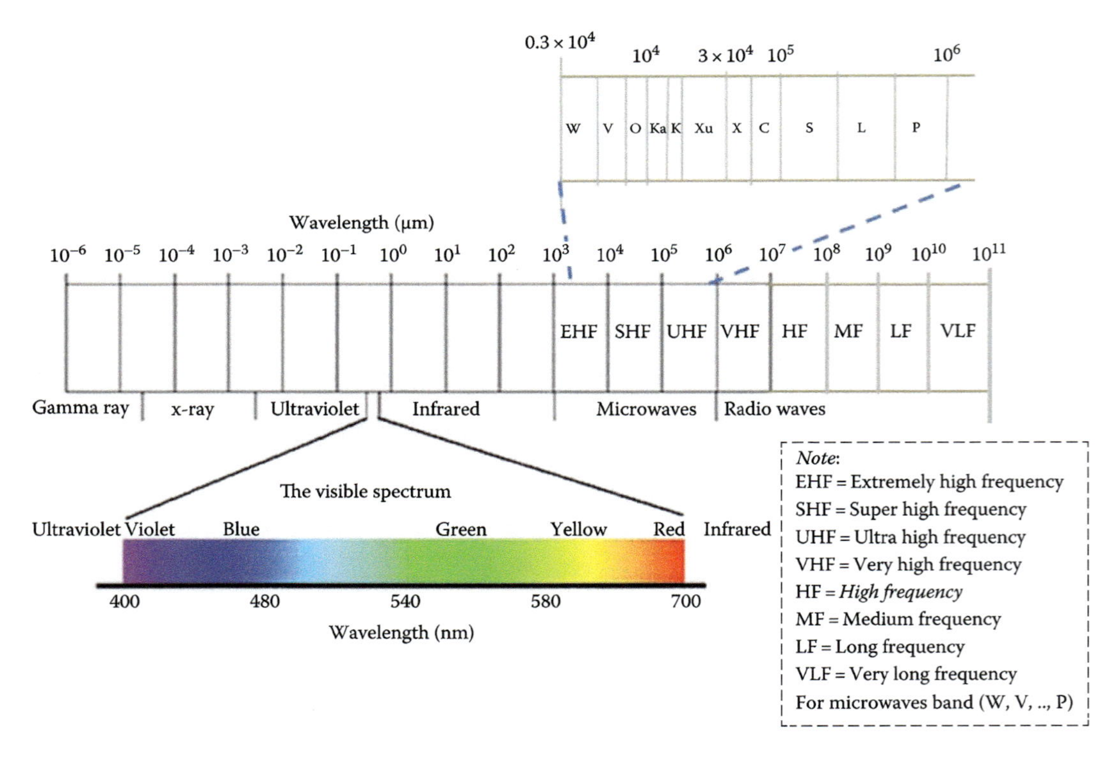


**FIGURE 1.6** Electromagnetic spectrum on which remote sensing systems are based. For example, visible portion of electromagnetic radiation is in 400–700 nm. The data can be captured in narrowband (e.g., 1 nm) or over broad wave bands (e.g., a single band over 630–690 nm).

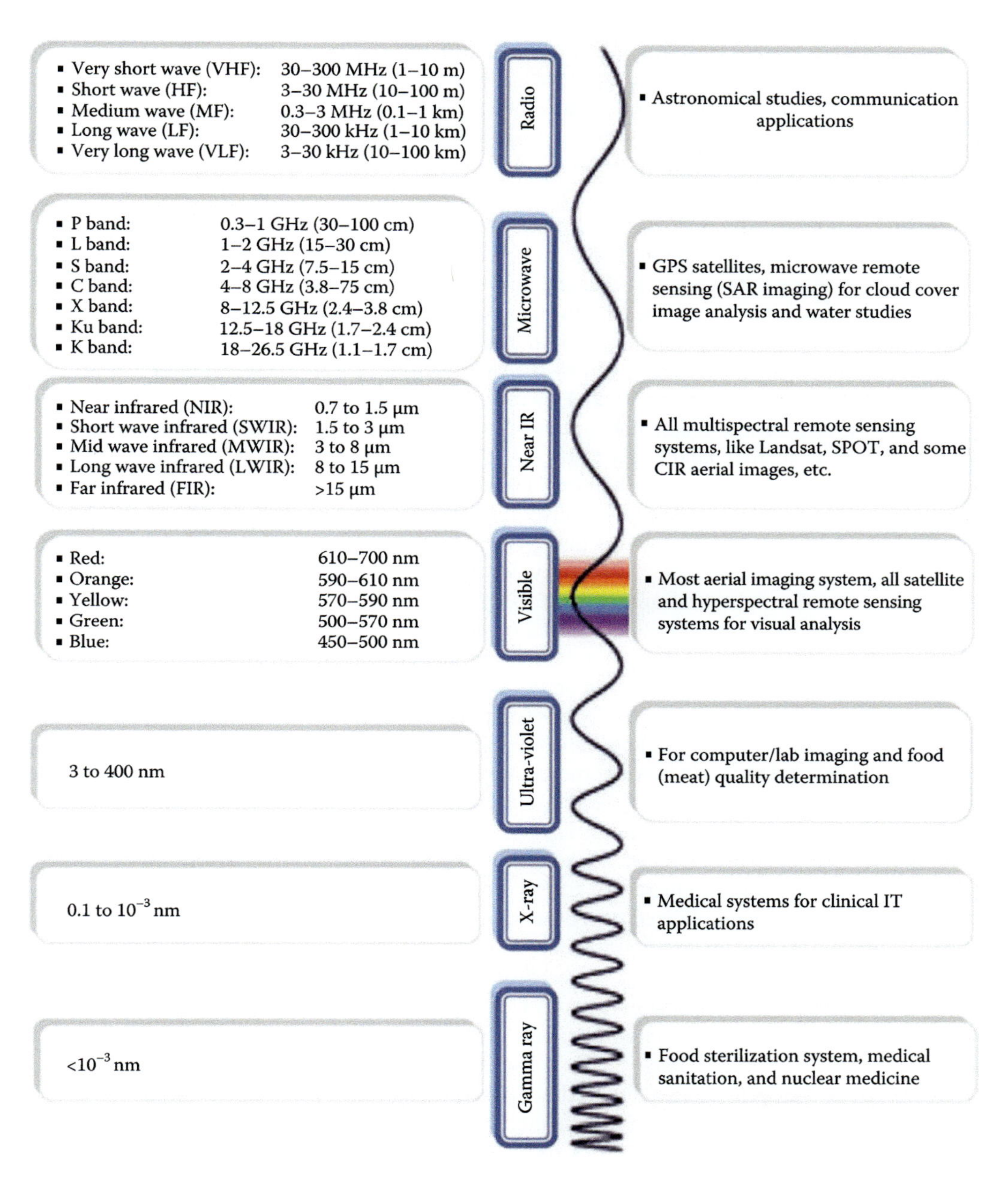


**FIGURE 1.7** Different bands (bandwidth) associated with the electromagnetic radiation regions and their remote sensing systems. Gathering data in various spectral ranges is extremely important for characterizing Earth systems. For example, radar penetrates the clouds and it is important to gather data during cloudy days when optical remote sensing is infeasible.

## 1.2 REMOTE SENSING PLATFORMS AND SENSOR CHARACTERISTICS

Traditional remote sensing has its roots in aircraft-based platforms where photographic instruments capture Earth features as either single images or dual overlapping stereo images. The landscape images provided the basis for some of the early photogrammetric and stereo analysis of images, which are still considered as foundations for aerial surveys. The aircrafts used for airborne remote sensing are usually single- or twin-engine propeller or turboprop platforms. However, small jets (Learjet) are used for high-altitude reconnaissance or mapping surveys. These aerial platforms usually fly at about 1,200–12,000 ft above the terrain with flying speeds of about 100 knots. Aerial platforms provide the main advantage of responding to the need of the user application. For example, in an emergency forest fire and flooding event, an aerial survey can be quickly planned and implemented. Thus, an aerial platform is a quick response system that one can easily adapt to changes in weather conditions. Moreover, airborne platforms enable near-real-time review of acquired data and provide great control of data quality. In contrast to aerial platforms, satellite platforms are enormously expensive and complex, and the development and deployment of sensors as payload on satellites can take five to ten years. Unlike aerial platforms, satellite platforms due to orbital characteristics have limitations of revisiting the same land area at user-needed time interval. This revisit cycle that relates to the time taken by the satellite to take a subsequent image acquisition of the same land area might not be practical. However, once a satellite-based imaging sensor becomes operational, the spaceborne image data are usually very consistent and allow end users to develop robust applications such as land cover change analysis, particularly for large tracts of the Earth's surface. Hence, depending on the resolutions provided by the onboard sensors, satellite platforms offer a wide array of scale-based mapping. Over the past few decades, numerous satellites have been launched by various private enterprises and government organizations to acquire remote sensing data. Remote sensors are energy-sensitive devices mounted on the particular satellite to view and take images of the Earth in different bands of the EMS (Avery and Berlin, 1992). In addition to the two types of sensors (discussed in Section 1.1.2) used in remote sensing, passive and active sensors are classified into various types based on their scanning and imaging mechanism.

One of the main advantages of using active systems relate to the characteristics of the active energy (RADAR, LiDAR, etc.) used, which is solar energy independent and least affected by atmospheric constituents. This provides huge advantages for remote sensing over regions such as tropical environments where clouds and rain are frequent weather events that interfere with traditional optical/passive systems. Furthermore, active sensors do not require solar energy and hence can operate during night. Additionally, active systems have better capability of sensing vegetation and soil attributes that are dependent on moisture content, thus having immense value in various applications, including hydrology, geology, glaciology, forestry, and agriculture. The disadvantages include lower spectral characteristics, complicated data processing, massive data volume, and higher cost.

Passive and active sensors are divided into nonscanning and scanning types, which in turn are divided into imaging and nonimaging sensors (Murai, 1993). Microwave radiometer, magnetic sensor, gravimeter, Fourier spectrometer, microwave altimeter, laser water depth meter, laser distance meter, etc., are examples of nonimaging scanners (Murai, 1993). Figure 1.8 graphically depicts the sensor types.

According to NASA's Earth Observation System Data and Information System (EOSDIS), accelerometer, radiometer, imaging radiometer, spectrometer, spectroradiometer, hyperspectral radiometer, and sounders are examples of passive remote sensors. Linear Imaging Self-Scanning Sensor (LISS), advanced very-high-resolution radiometer (AVHRR), Coastal Zone Color Scanner (CZCS), Sea-Viewing Wide Field-of-View Sensor (SeaWiFS), Moderate-Resolution Imaging Spectroradiometer (MODIS), Active Cavity Radiometer Irradiance Monitor (ACRIM II and III), Advanced Spaceborne Thermal Emission and Reflection Radiometer (ASTER), Imaging Infrared Radiometer (IIR), Clouds and the Earth's Radiant Energy System (CERES), special sensor microwave radiometer (SMMR), airborne visible/infrared imaging spectrometer (AVIRIS), Polarization and Directionality of the Earth's Reflectance (POLDER), and Atmospheric Infrared Sounders (AIRSs) are some of the prominent passive sensors used in remote sensing data collection in the present day (EOSDIS, 2014). As discussed in the previous section, RADAR, ranging instruments, scatterometer, LiDAR, laser altimeter, and sounders are examples of active sensors (EOSDIS, 2014). Table 1.1 provides a detailed list of remote sensors with the instrument name, type of sensor, platform, data center, and other descriptions about the sensors.

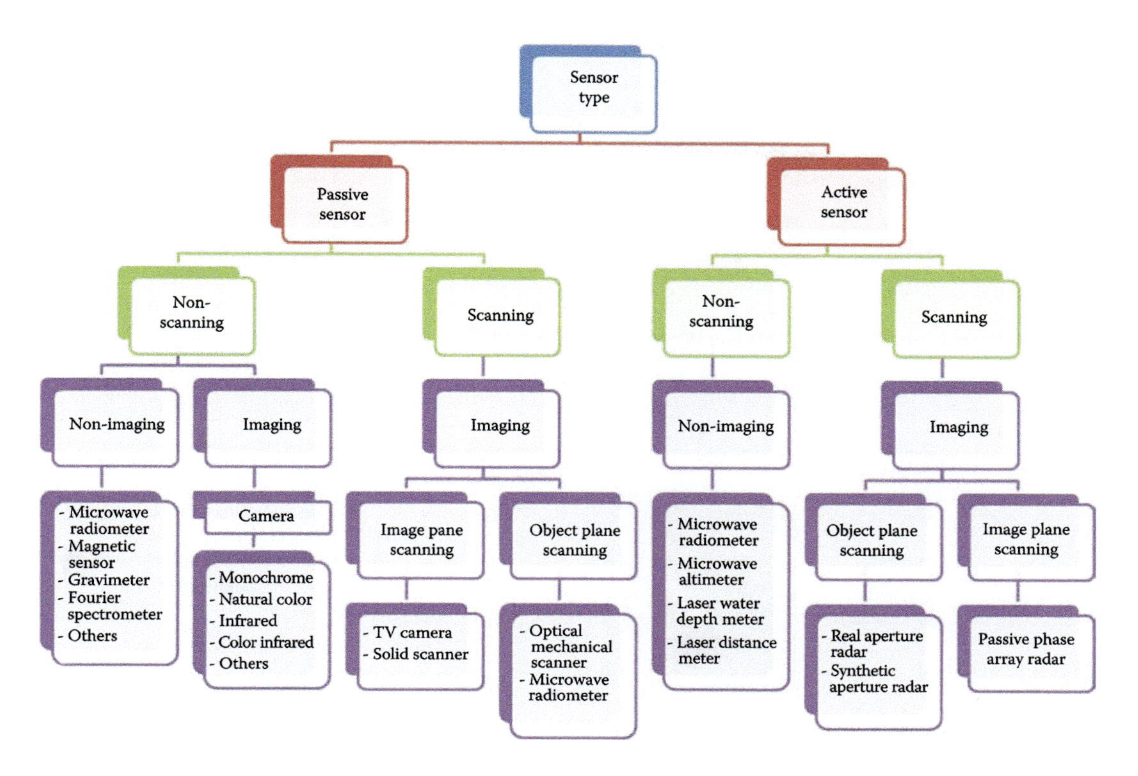


**FIGURE 1.8** Remote sensing sensor–type classification. (Adapted and modified from Murai, S., *Remote Sensing Note*, JARS, Tokyo, Japan, 1993.)

**TABLE 1.1**
**List of Selected Active and Passive Sensors and Their Types**

| Instrument | Type | Platform | Data Center | Description |
|---|---|---|---|---|
| **Passive sensors** | | | | |
| ***Power radiometers and imagers*** | | | | |
| ACRIM II | Total power radiometer | UARS | LaRC ASDC | Measures total solar irradiance. |
| ACRIM III | Total power radiometer | ACRIMSAT | LaRC ASDC | Measures total solar irradiance. |
| TIM | Total power radiometer | SORCE | GES DISC | Measures total solar irradiance. |
| LIS | Imager | TRMM | GHRC DAAC | Detects intracloud and cloud-to-ground lightning, day and night. |
| WFC | Wide Field Camera | CALIPSO | LaRC ASDC | Fixed, nadir-viewing imager with a single spectral channel covering the 620-670 nm region. |
| ***Multispectral instruments*** | | | | |
| AMPR | Microwave radiometer | ER-2 and DC-8 | GHRC DAAC | Cross track scanning total power microwave radiometer with four channels centered at 10.7, 19.35, 37.1, and 85.5 GHz (FIRE ACE, Teflon-B, TRMM-LBA, CAMEX-4. TCSP, TC4 Projects). |

(*Continued*)

**TABLE 1.1** *(Continued)*

**List of Selected Active and Passive Sensors and Their Types**

| Instrument | Type | Platform | Data Center | Description |
|---|---|---|---|---|
| AMSR-E | Multichannel microwave radiometer | Aqua | NSIDC DAAC GHRC DAAC | Measures precipitation, oceanic water vapor, cloud water, near-surface wind speed, sea and land surface temperature, soil moisture, snow cover, and sea ice. Provides spatial resolutions of 5.4, 12, 21, 25, 38, and 56 km and a 0.25° resolution. |
| ASTER | Multispectral radiometer | Terra | LP DAAC ORNLDAAC | Measures surface radiance, reflectance, emissivity, and temperature. Provides spatial resolutions of 15, 30, and 90 m. |
| AVHRR | Multispectral radiometer | NOAA/POES | GES DISC NSIDC DAAC ORNL DAAC PO.DAAC | Four or six bands, depending on platform. Telemetry resolutions are 1.1 km (HRPT data), 4 km (Pathfinder V5 and GAC data), 5, and 25 km spatial resolution. |
| CERES | Broadband scanning radiometer | Aqua Terra TRMM NPP | LaRC ASDC | Four to six channels (shortwave, longwave, total). Measures atmospheric and surface energy fluxes. Provides 20 km resolution at nadir. |
| ECOSTRESS | TIR, SWIR Radiometer | NASA | LPDACC | Five spectral bands in TIR (8.3-12.1 pm) and one band SWIR (1.6 pm) for geolocation. Measures evapotranspiration, evaporation stress index and water use efficiency. It is of 69 m x 38 m spatial resolution. |
| IIR | IIR | CALIPSO | LaRC ASDC | Nadir-viewing, nonscanning imager having a 64 km swath with a pixel size of 1 km. Provides measurements at three channels in the TIR window region at 8.7, 10.5, and 12.0 mm. |
| MAS | Imaging spectrometer | NASA ER-2 aircraft | GES DISC GHRC DAAC LaRC ASDC ORNL DAAC | Fifty spectral bands that provide spatial resolution of 50 m at typical flight altitudes. |
| MISR | Imaging spectrometer | Terra | LaRC ASDC ORNL DAAC | Obtains precisely calibrated images in four spectral bands, at nine different angles, to provide aerosol, cloud, and land surface data. Provides spatial resolution of 250 m to 1.1 km. |
| MODIS | Imaging spectroradiometer | Aqua Terra | GES DISC GHRC DAAC LP DAAC MODAPS NSIDC DAAC OBPG ORNL DAAC PO.DAAC | Measures many environmental parameters (ocean and land surface temperatures, fire products, snow and ice cover, vegetation properties and dynamics, surface reflectance and emissivity, cloud and aerosol properties, atmospheric temperature and water vapor, ocean color and pigments, and ocean biological properties). Provides moderate spatial resolutions of 250 m (bands 1 and 2), 500 m (bands 3-7), and 1 km (bands 8-36). |
| SSM/I | Multispectral microwave radiometer | DMSP | GHRCDAAC LaRC ASDC NSIDC DAAC PO.DAAC ORNL DAAC | Has seven channels and four frequencies. Measures atmospheric, ocean, and terrain microwave brightness temperatures, which are used to derive ocean near-surface wind speed, atmospheric integrated water vapor, and doud/rain liquid water content and sea ice extent and concentration. |

**TABLE 1.1** *(Continued)*
**List of Selected Active and Passive Sensors and Their Types**

| Instrument | Type | Platform | Data Center | Description |
|---|---|---|---|---|
| SMMR | Multispectral microwave radiometer | NIMBUS-7 | GES DISC LaRC ASDC NSIDC DAAC PO.DAAC | Ten channels. Measures SSTs, ocean near-surface winds, water vapor and cloud liquid water content, sea ice extent, sea ice concentration, snow cover, snow moisture, rainfall rates, and differential of ice types. |
| TMI | Multispectral microwave radiometer | TRMM | GES DISC GHRC DAAC | TMI measures the intensity of radiation at five separate frequencies: 10.7, 19.4, 21.3, 37, and 85.5 GHz. TMI measures microwave brightness temperatures, water vapor, cloud water, and rainfall intensity. |
| ***Hyperspectral instruments*** | | | | |
| AVIRIS | Imaging spectrometer | Aircraft | ORNL DAAC | 224 contiguous channels, approximately 10 nm wide. Measurements are used to derive water vapor, ocean color, vegetation classification, mineral mapping, and snow and ice cover (BOREAS Project). |
| SOLSTICE | Spectrometer | SORCE | GES DISC | Measures the solar spectral irradiance of the total solar disk in the UV wavelengths from 115 to 430 nm. |
| Gaofen-5 | AHSI, VIMS, GMI, AIUS, EMI, DPC | GF-5 Satellite | CNSA | AHSI is a330-channel (150 Visible and NIR; 180 SWIR) imaging spectrometer with an approximately 30m spatial resolution covering the 0.4-2.5pm spectral range. Supports high spatial and temporal environmental monitoring and resource discovery with a 5 nanometer and higher spectral resolution in visible (5nm), NIR (5nm), and SWIR(10nm) ranges (0.4-2.5 pm) |
| ***Polarimetric instruments*** | | | | |
| POLDER | Polarimeter | Aircraft | ORNL DAAC | Measures the polarization and the directional and spectral characteristics of the solar light reflected by aerosols, clouds, and the Earth's surface (BOREAS Project). |
| PSR | Microwave polarimeter | Aircraft | GHRC DAAC | Measures wind speed and direction (CAMEX-3 Project). |
| ***Ranging and sounding instruments*** | | | | |
| ACC | Accelerometer | GRACE | PO.DAAC | The Onera SuperSTAR Accelerometer measures the nongravitational forces acting on the GRACE satellites. |
| AIRS | Sounder | Aqua | GES DISC | Measures air temperature, humidity, clouds, and surface temperature. Provides spatial resolution of ~13.5 km in the IR channels and ~2.3 km in the visible. Swath retrieval products are at 50 km resolution. |
| AMSU | Sounder | Aqua | GES DISC GHRC DAAC | Has 15 channels. Measures temperature profiles in the upper atmosphere. Has a cloud filtering capability for tropospheric temperature observations. Provides spatial resolution of 40 km at nadir. |

*(Continued)*

**TABLE 1.1** *(Continued)*
**List of Selected Active and Passive Sensors and Their Types**

| Instrument | Type | Platform | Data Center | Description |
|---|---|---|---|---|
| HAMSR | Sounder | DC-8 | GHRC DAAC | Measures vertical profiles of temperature and water vapor, from the surface to 100 mb in 2-4 km layers (CAMEX-4, NAMMA Projects). |
| HIRDLS | Sounder | Aura | GES DISC | Measures infrared emissions at the Earth's limb in 21 channels to obtain profiles of temperature, ozone, CFCs, various other gases affecting ozone chemistry, and aerosols at 1 km vertical resolution. In addition, HIRDLS measures the location of polar stratospheric clouds. |
| MLS | Sounder | Aura | GES DISC | Five broadband radiometers and 28 spectrometers measure microwave thermal emission from the Earth's atmosphere to derive profiles of ozone, SO2, N2O, OH group, and other atmospheric gases, temperature, pressure, and cloud ice. |
| MOPITT | Sounder | Terra | LaRC ASDC ORNLDAAC | Measures carbon monoxide and methane in the troposphere. Is able to collect data under cloud-free conditions. Provides horizontal resolution of ~22 km and vertical resolution of ~4 km. |
| OMI | Multispectral radiometer | Aura | GES DISC | 740 wavelength bands in the visible and UV. Measures total ozone and profiles of ozone, N2O, SO2, and several other chemical species. |
| TES | Imaging Spectrometer | Aura | LaRC ASDC | High-resolution imaging infrared Fourier transform spectrometer that operates in both nadir and limb-sounding modes. Provides profile measurements of ozone, water vapor, carbon monoxide, methane, nitric oxide, nitrogen dioxide, nitric acid, $CO^2$ and ammonia. |
| **Active sensors** | | | | |
| ***Radar and laser (LiDAR)*** | | | | |
| ALT-A, -B | Radar altimeter | TOPEX/ Poseidon | PO.DAAC | Dual-frequency altimeter that measures height of the satellite above the sea (satellite range), wind speed, wave height, and ionospheric correction. |
| CALIOP | Cloud and aerosol LiDAR | CALIPSO | LaRC ASDC | Two-wavelength polarization-sensitive LiDAR that provides high- resolution vertical profiles of aerosols and clouds. |
| Altimeters: radar and laser (LiDAR) GLAS | Laser altimeter | ICESat | NSIDC DAAC | The main objective is to measure ice sheet elevations and changes in elevation through time. Secondary objectives include measurement of cloud and aerosol height profiles, land elevation and vegetation cover, and sea ice thickness. |
| Poseidon-1 | Radar altimeter | TOPEX/ Poseidon | PO.DAAC | Single-frequency altimeter that measures height of the satellite above the sea (satellite range), wind speed, and wave height. |

**TABLE 1.1** *(Continued)*

**List of Selected Active and Passive Sensors and Their Types**

| Instrument | Type | Platform | Data Center | Description |
|---|---|---|---|---|
| Poseidon-2 | Radar altimeter | Jason-1 | PO.DAAC | Measures sea level, wave height, wind speed, and ionospheric correction. |
| SAR | SAR | ERS-1<br>ERS-2<br>JERS-1<br>RADARSAT-1<br>PALSAR<br>UAVSAR | ASF SDC<br>NSIDC DAAC<br>ORNL DAAC | Provides high-resolution surface imagery at 7-240 m. Multiple polarizations are utilized by some SAR instruments. |
| KBR | Ranging instrument | GRACE | PO.DACC | The dual-frequency KBR instrument measures the range between the GRACE satellites to extremely high precision. |
| *Active and Passive* | *Combined* | | | |
| SMAP | Active Radar Passive Radiometer | SMAP | ASF<br>NSIDC | The Soil Moisture Active Passive (SMAP) measures global soil moisture and its L-Band radiometer measures sea surface salinity and sea surface wind speed. SMAP spatial resolution varies from 3-36 km. |
| ***Scatterometers*** | | | | |
| NSCAT | Radar scatterometer | ADEOS-I | PO.DAAC | Dual Fan-Beam Ku Band that measures ocean vector winds at a nominal grid resolution of 25 km. |
| SASS | Radar scatterometer | Seasat | PO.DAAC | Dual Fan-Beam Ku Band that measures ocean vector winds at a nominal grid resolution of 25 km. |
| Seawinds | Radar scatterometer | QuikSCAT<br>ADEOS-II | PO.DAAC | Dual Pencil-Beam Ku Band that measures ocean vector winds at a nominal grid resolution of 25 km. |
| ***Sounding instruments*** | | | | |
| CLS | LiDAR | ER-2 | LaRC ASDC | Determines vertical cloud structure (FIRE Project). |
| LASE | LiDAR | DC-8 | GHRC DAAC | Measures water vapor, aerosols, and clouds throughout the troposphere (CAMEX-4, TCSP, NAMMA Projects). |
| PR | Phased array radar | TRMM | GES DISC<br>ORNL DAAC | Measures 3D distribution of rain and ice. Provides horizontal resolution of 250 m and vertical resolution of 5 km. |
| VIL | LiDAR | Ground | LaRC ASDC<br>ORNL DAAC | Determines vertical cloud structure (FIFE, FIRE, and BOREAS Projects). |

*Source:* Adapted and modified from Earth Observing System Data and Information System (EOSDIS), Earth Data, 2023, https://earthdata.nasa.gov/, accessed on October 15, 2023.

Remote sensing optical sensors are characterized by spectral, radiometric, and geometric performance (Murai, 1993). Observation range of the electromagnetic wave, center wavelength of a band, changes at both ends of a band, sensitivity of a band, polarization sensitivity, and ratio of sensitivity difference between different bands are some of the spectral characteristics that define the types of optical sensors (Murai, 1993). Detection accuracy, signal-to-noise ratio (S/N), dynamic range, quantization level, sensitivity difference between pixels, linearity of sensitivity, and noise equivalent power are a few radiometric characteristics of optical sensors (Murai, 1993). Field of view (FOV), IFOV, registration between different spectral bands, modular transfer function (MTF), and optical distortions are examples of geometric characteristics of optical sensors that classify the sensors (Murai, 1993). These sensor characteristics determine the spatial, spectral, radiometric, and temporal resolutions of remote sensing data.

### 1.2.1 Image Characteristics

Different sensors continuously scan the Earth's surface to produce imagery. Each image is a matrix of pixels or picture elements defined by columns and rows as shown in Figure 1.9. Each pixel records a numeric value representative of the brightness or intensity level of the reflected energy as discussed in Section 1.2. The image data produced by different sensors on satellite systems have unique characteristics that relate to the sensor's resolutions—spatial, spectral, radiometric, and temporal. Radiometric resolution refers to the data depth indicative of the sensitivity of the sensor to incoming energy. About 8-bit data have higher contrast (0–255 digital number [DN] range) and higher radiometric resolution than a 6-bit sensor that provides data in a lower range (0–63 DN range) and thus lower contrast. For example, Landsat 7 ETM+ sensor provides 8-bit data for individual bands, while the latest version of Landsat satellite series, Landsat 8, provides 12-bit data with DN values ranging from 0 to 4,095, and Landsat 9 provides even higher radiometric resolution of 14-bits (DN values of 0 to 16,383) for the operational land imager 2 (OLI-2). Temporal resolution refers to the revisit frequency/time of the sensor to a specific location on the Earth's surface. For example, Landsat revisit time is 16 days over a specific geographic location on Earth. Table 1.2 provides a list of selected remote sensing systems (sensors) with detailed data characteristics including the spatial, spectral (band ranges), and frequency of revisit over a location on Earth.

**TABLE 1.2**
**Data Characteristics of Selective Satellite Sensors**

| | Image Characteristics | | | | |
|---|---|---|---|---|---|
| **Sensor** | **Spatial (m)** | **Spectral (#)** | **Radiometric (bit)** | **Temporal (Days)** | **Band Range (pm)** |
| ***c oarse-resohition sensors*** | | | | | |
| AVHRR | 1000 | 4 | 11 | Daily | 0.58–0.68 |
| | | | | | 0.725–1.1 |
| | | | | | 5.55–3.95 |
| | | | | | 10.30–10.95 |
| | | | | | 10.95–11.65 |
| MODIS | 250; 500; 1000 | 36 | 12 | Daily | 0.62–0.67 |
| | | | | | 0.84–0.876 |
| | | | | | 0.459–0.479 |
| | | | | | 0.545–0.565 |
| | | | | | 1.23–1.25 |
| | | | | | 1.63–1.65 |
| | | | | | 2.11–2.16 |
| ***Mnltispectml sensors*** | | | | | |
| Landsat 1,2,3 MSS | 56 × 79 | 4 | 6 | 16 | 0.5–0.6 |
| | | | | | 0.6–0.7 |
| | | | | | 0.7–0.8 |

**TABLE 1.2** *(Continued)*
**Data Characteristics of Selective Satellite Sensors**

| | Image Characteristics | | | | |
|---|---|---|---|---|---|
| Sensor | Spatial (m) | Spectral (#) | Radiometric (bit) | Temporal (Days) | Band Range (pm) |
| | | | | | 0.8–1.1 |
| Landsat 4,5 TM 30 | | 7 | 8 | 16 | 0.45–0.52 |
| | | | | | 0.52–0.60 |
| | | | | | 0.63–0.69 |
| | | | | | 0.76–0.90 |
| | | | | | 1.55–1.74 |
| | | | | | 10.4–12.5 |
| | | | | | 2.08–2.35 |
| Landsat 7 ELM+ 30 | | 8 | 8 | 16 | 0.45–0.52 |
| | | | | | 0.52–0.60 |
| | | | | | 0.63–0.69 |
| | | | | | 0.50–0.75 |
| | | | | | 0.75–0.90 |
| | | | | | 10.4–12.5 |
| | | | | | 1.55–1.75 |
| | | | | | 0.52–0.90 (p) |
| Landsat 8 15; 30; 100 | | 11 | 12 | 16 | 0.43–0.45 |
| | | | | | 0.45–0.51 |
| | | | | | 0.53–0.59 |
| | | | | | 0.64–0.67 |
| | | | | | 0.85–0.88 |
| | | | | | 1.57–2.29 |
| | | | | | 0.50–0.68 (p) |
| | | | | | 1.36–1.38 |
| | | | | | 10.60–11.19 |
| | | | | | 11.50–12.51 |
| Landsat 9 15; 30; 100 | | 11 | 14 | 16 | 0.43–0.45 |
| | | | | | 0.45–0.51 |
| | | | | | 0.53–0.59 |
| | | | | | 0.64–0.67 |
| | | | | | 0.85–0.88 |
| | | | | | 1.57–1.65 |
| | | | | | 2.11–2.29 |
| | | | | | 0.50–0.68 (p) |
| | | | | | 1.36–1.38 |
| | | | | | 10.60–11.19 |
| | | | | | 11.50–12.51 |
| ASTER 15; 30; 90 | | 15 | 8 | 16 | 0.52–0.63 |
| | | | | | 0.63–0.69 |
| | | | | | 0.76–0.86 |
| | | | | | 1.60–1.70 |
| | | | | | 2.145–2.185 |
| | | | | | 2.185–2.225 |
| | | | | | 2.235–2.285 |

(*Continued*)

**TABLE 1.2** *(Continued)*
**Data Characteristics of Selective Satellite Sensors**

| | Image Characteristics | | | | |
|---|---|---|---|---|---|
| **Sensor** | **Spatial (m)** | **Spectral (#)** | **Radiometric (bit)** | **Temporal (Days)** | **Band Range (pm)** |
| | | | | | 2.295–2.365 |
| | | | | | 2.360–2.430 |
| | | | | | 8.125–8.475 |
| | | | | | 8.475–8.825 |
| | | | | | 8.925–9.275 |
| | | | | | 10.25–10.95 |
| | | | | | 10.95–11.65 |
| ALI | 30 | 10 | 12 | 16 | 0.48–0.69 (p) |
| | | | | | 0.433–0.453 |
| | | | | | 0.450–0.515 |
| | | | | | 0.425–0.605 |
| | | | | | 0.633–0.690 |
| | | | | | 0.775–0.805 |
| | | | | | 0.845–0.890 |
| | | | | | 1.200–1.300 |
| | | | | | 1.550–1.750 |
| | | | | | 2.080–2.350 |
| SOPT-5 | 2.5-20 | 5 | 8 | 2-3 | 0.48–0.71 (p) |
| | | | | | 0.5–0.59 |
| | | | | | 0.61–0.68 |
| | | | | | 0.78–0.89 |
| | | | | | 1.58–1.75 |
| IRS-1A, IRS-1B | 36; 73 | 4 | 8 | 22 | 0.45–0.52 |
| | | | | | 0.52–0.68 |
| | | | | | 0.62–0.68 |
| | | | | | 0.77–0.86 |
| IRS-1C | 5.8 (p); 23.5; 70.5 (B5) | 5 | 8 | 24 | 0.50–0.75 (p) |
| | | | | | 0.52–0.59 |
| | | | | | 0.62–0.68 |
| | | | | | 0.77–0.86 |
| | | | | | 1.55–1.70 |
| IRS-P6 AWiFS | 56 | 4 | 10 | 16 | 0.52–0.59 |
| | | | | | 0.62–0.68 |
| | | | | | 0.77–0.86 |
| | | | | | 1.55–1.70 |
| JERS-1 | 18 | 8 | 8 | 44 | 0.52–0.60 |
| | | | | | 0.63–0.69 |
| | | | | | 0.76–0.86 (2 bands) |
| | | | | | 1.60–1.71 |
| | | | | | 2.01–2.12 |
| | | | | | 2.13–2.25 |
| | | | | | 2.27–2.40 |
| CBERS-2 | 20 (p) | 5 | 8 | 26 | 0.51–0.73 (p) |

**TABLE 1.2 *(Continued)***
**Data Characteristics of Selective Satellite Sensors**

| | Image Characteristics | | | | |
|---|---|---|---|---|---|
| Sensor | Spatial (m) | Spectral (#) | Radiometric (bit) | Temporal (Days) | Band Range (pm) |
| CBERS-3B | 20 (MS) | | | | 0.45–0.52 |
| CBERS-3 | 5 (p) | | | | 0.52–0.59 |
| CBERS-4 | 20 (MS) | | | | 0.63–0.69 |
| | | | | | 0.77–0.89 |
| ***Hyperspectral sensor*** | | | | | |
| Hyperion | 30 | 196 | 16 | 16 | 196 effective calibrated bands |
| | | | | | VNIR (bands 8–57) 0.427–0.925 |
| | | | | | SWIR (band 79–224) 0.93–2.4 |
| ***Hyperspatial sensor*** | | | | | |
| IKONOS | 1-4 | 4 | 11 | 5 | 0.445–0.516 |
| | | | | | 0.506–0.595 |
| | | | | | 0.632–0.698 |
| | | | | | 0.757–0.853 |
| QuickBird | 0.61-2.44 | 4 | 11 | 5 | 0.45–0.52 |
| | | | | | 0.52–0.60 |
| | | | | | 0.63–0.69 |
| | | | | | 0.76–0.89 |
| RESOURCESAT | 5.8 | 3 | 10 | 24 | 0.52–0.59 |
| | | | | | 0.62–0.68 |
| | | | | | 0.77–0.86 |
| Rapid Eye A-E | 6.5 | 5 | 12 | 1-2 | 0.44–0.51 |
| | | | | | 0.52–0.59 |
| | | | | | 0.63–0.68 |
| | | | | | 0.69–0.73 |
| | | | | | 0.77–0.89 |
| World View | 0.55 | 1 | 11 | 1.7-5.9 | 0.45–0.51 |
| FORMOSAT-2 | 2-8 | 5 | 11 | Daily | 0.45–0.52 |
| | | | | | 0.52–0.60 |
| | | | | | 0.63–0.69 |
| | | | | | 0.76–0.9 |
| | | | | | 0.45–0.9 (p) |
| KOMPSAT-2 | 1-4 | 5 | 10 | 3-28 | 0.45–0.52 |
| | | | | | 0.52–0.60 |
| | | | | | 0.63–0.69 |
| | | | | | 0.76–0.9 |
| | | | | | 0.5–0.9 (p) |

*Source:* Adopted from Melesse et al. (2007).
*Note:* (p) is the anchromatic band. (MS) is multispectral bands.

#### 1.2.1.1 Spatial Resolution

Spatial resolution implies the unit ground area for which the sensor records the reflected energy providing a unique DN or brightness value (BV) (refer to Figure 1.9). For example, a Landsat sensor senses a ground space that is 30 m × 30 m. Usually, the image spatial resolution is equal to

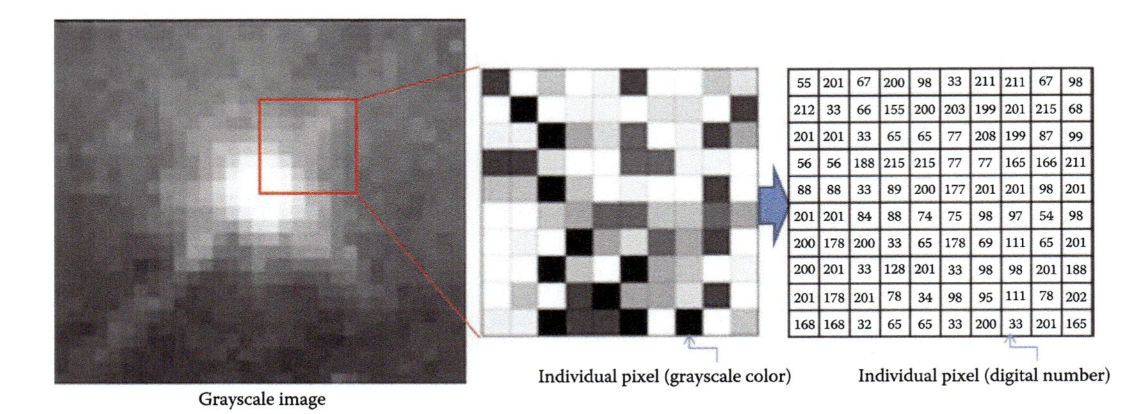


**FIGURE 1.9** Basic matrix configuration of an image showing a pixel (picture element). Each pixel denotes an intensity value (e.g., reflected energy, radar backscatter). Each pixel can have data values between 0 and 256 for 8-bit (quantization) data, 0 and 4096 for 12-bit data, and so on.

the ground sample distance (GSD), which is the smallest discernible detail in an image (Gonzalez and Woods, 2002). The GSD could be smaller than the spatial resolution of an image acquired by a remote sensing system, and they vary from a fraction of a meter to tens of meters (Shaw and Burke, 2003). The image spatial resolution is recognized primarily by the sensor aperture and platform altitude. Platform altitude is loosely constrained by the class of sensor platform, that is, either spaceborne or airborne (Shaw and Burke, 2003). Sensor aperture size, particularly for spaceborne systems, determines the cost of the remote sensing system; many times, to cut cost, the aperture size is made bigger, thus providing low-spatial-resolution image. According to Shaw and Burke (2003), the best detection performance is expected when the angular resolution of the sensor, specified in terms of the GSD, is commensurate with the footprint of the targets of interest. For example, if the application requires observation of global land cover, then low-resolution imagery is suitable. On the other hand, if the application requires local land cover information such as water stress in a corn field, then high-resolution imagery is ideal. According to Belward and Skoien (2014), land cover datasets at 1 km resolution (National Oceanic and Atmospheric Administration [NOAA] AVHRR) are *low*-resolution data when compared with a 30 m Landsat data product but become acceptable when land cover information is gathered for a global assessment perspective or when it is used in a climate model running with a 100 km cell size. In addition to sensor optical characteristics, the flight height of the satellite or the airplane determines the spatial resolution. The higher the location of the satellite in space, the lower will be the image resolution.

Sensor specifications are usually provided by satellite operator and/or sponsoring space agency and/or manufacturer, reproduced by three agencies (National Space Science Data Center [NSSDC], Committee on Earth Observation Satellites [CEOS], and Ocean Surface Current Analysis Real Time [OSCAR]) (Belward and Skoien, 2014). Some remote sensors provide data at two or more spatial resolutions with panchromatic spatial resolution being higher than the multispectral scanner (MSS) sensors. MODIS, Earth Observing System (EOS) Terra, Landsat, and Systeme Probatoire D'Observation De La Terre (SPOT) are the examples of dual spatial resolution–based image acquisition. Some SAR missions, such as Canada's RADARSAT, also provide multiple-resolution images at different band frequencies (Belward and Skoien, 2014). In general, spatial resolution can be categorized into five broad classes: 0.5–4.9 m (very high resolution), 5.0–9.9 m (high resolution), 10.0–39.9 m (medium resolution), 40.0–249.9 m (moderate resolution), and 250 m–1.5 km (low resolution). Any such grouping is somewhat arbitrary; the low-resolution class acknowledges the threshold of 250 m established for global monitoring of land transformations (Townshend and Justice, 1988), and the moderate-/medium-resolution classes include imagery available through *free-and-open* data policies. The high- and very-high-resolution classes reference commercial distinctions. The upper limit of 50 cm is set because the U.S. government licensing limits unrestricted distribution of spatial data at this resolution. Table 1.2 provides the spatial resolution information of a selected list of remote sensing systems (sensors).

Generally, panchromatic bands of satellites are of very high spatial resolution. For example, SPOT 1 acquired the first 10 m resolution images in 1986; then in 1995, Indian Remote Sensing Satellite (IRS-1C) image exceeded the 6 m spatial resolution. In 1999, IKONOS satellite provided imagery with 1 m spatial resolution, the 0.5 m spatial resolution imagery was obtained by WorldView-1 after that, and finally, the highest resolution imagery of 41 cm was obtained with GeoEye in 2007. Lately, 30 cm satellite-based best spatial resolution imageries are available through SkySat (https://eos.com/blog/spatial-resolution/). KOMPSAT-7 mission's advanced Earth imaging sensor system with high-resolution (AEISS-HR) satellite has an ultra-high resolution of 0.3 m for its panchromatic bandwidth of 0.5–0.9 μm. Multi-angle imaging spectroradiometer (MISR) of NASA's Terra satellite mission provides a spatial resolution of 0.275 m of its all four bands of 0.446, 0.557, 0.671, and 0.866 μm. Refer to the University of Twente's ITC website (https://webapps.itc.utwente.nl/sensor/default.aspx?view=allsensors) for a quarriable list of satellites of different specifications. All these high-spatial-resolution satellite images are acquired with panchromatic bands. However, there is a common misconception that ultra-high-spatial resolution images are better for environmental monitoring. One must make an informed decision with relation to the intensity (details) of the research and cost constraints to find out if a medium-spatial resolution imagery suffice the need of exploration.

#### 1.2.1.2 Spectral Resolution

Satellite sensors measure EMR in different portions of the EMS. Spectral resolution is the ability of the sensor to resolve spectral features and *bands* into their separate components in the EMS. It also describes the ability of a sensor to distinguish between wavelength intervals in EMS. Thus, in general, spectral resolution determines the number of bands a satellite sensor can sense. And so, higher-spectral-resolution imagery will provide more spectral information when compared to lower-spectral-resolution imagery. The spectral information is particularly useful in applications dealing with mapping and modeling biophysical properties of objects such as water quality, plant vigor, and soil nutrients. As shown in Figure 1.10, U.S. Landsat has higher spectral resolution than the French SPOT sensor. MODIS sensor has a high spectral resolution because it senses in 36 wavelength regions of the EMS, in comparison to Landsat TM, which senses in only seven EMR regions. Similarly, Hyperion sensor provides hyperspectral capabilities of resolving 220 spectral bands in the 0.4–2.6 μm range. The position, number, and width of spectral bands in an image determine the degree to which individual targets can be discriminated. Multispectral imagery has a higher degree of individual target discrimination power than a panchromatic single-band image. Table 1.2 provides a list of spectral bands available for selective remote sensing sensors.

Gaofen-5 (GF-5) is a fifth-generation satellite of a series sent to space by the China High-Resolution Earth Observation System (CHEOS) satellites of the China National Space Administration (CNSA). This satellite carries six payloads: two hyperspectral/multispectral payloads for terrestrial Earth observation and four atmospheric observation payloads (https://nssdc.gsfc.nasa.gov/nmc/spacecraft/display.action?id=2018-043A). Amazingly, the advanced hyperspectral imager (AHSI) is a 330-channel (150 Visible and NIR; 180 SWIR) imaging spectrometer with an approximately 30 m spatial resolution covering the 0.4–2.5 μm spectral range. It supports high spatial and temporal environmental monitoring and resource discovery with a 5 nm and higher spectral resolution in visible (5 nm), NIR (5 nm), and SWIR(10 nm) ranges (0.4–2.5 μm). Newer satellites like GF-5 are trying to emulate the laboratory spectroscopic imaging, which can

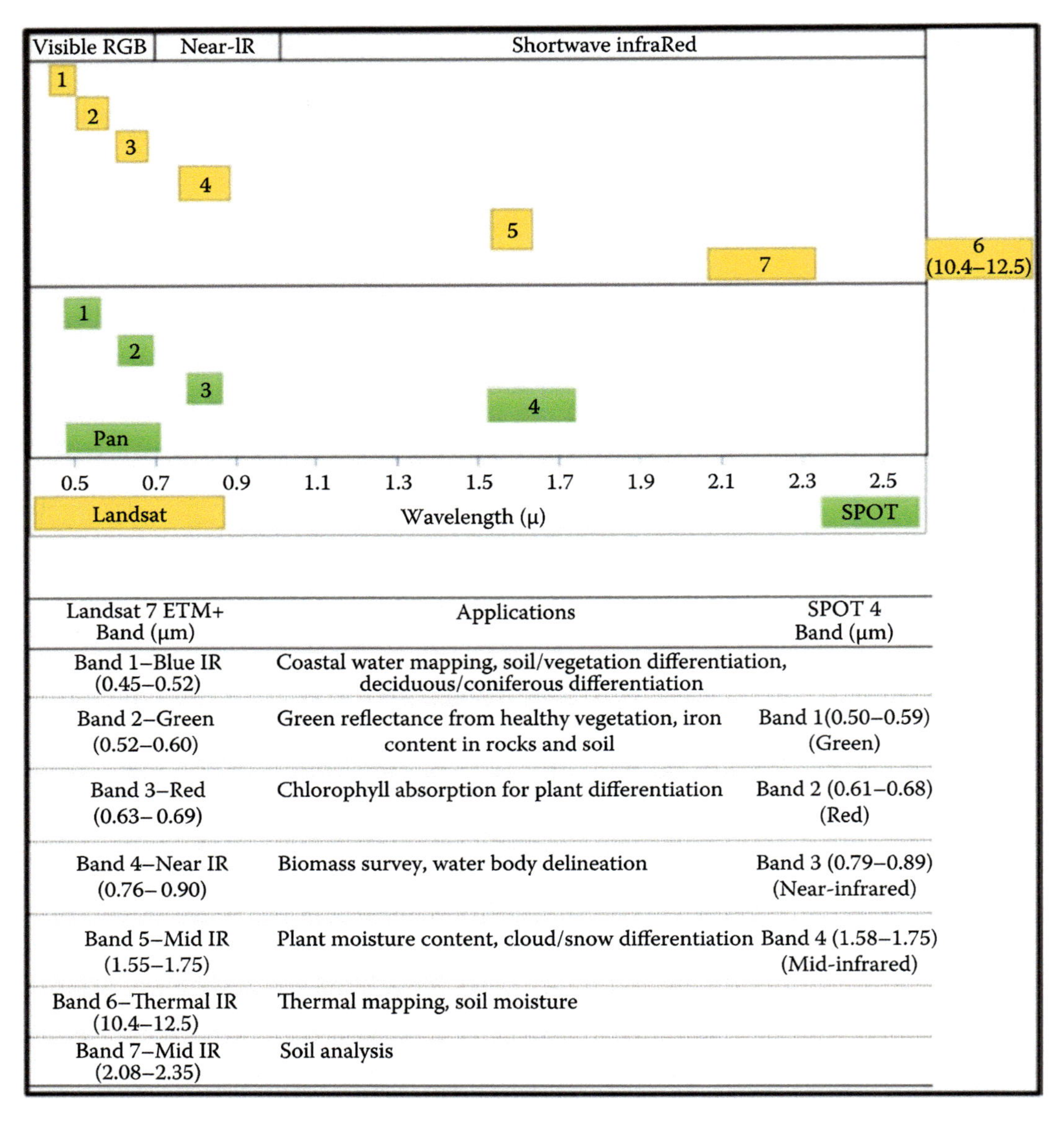


| Landsat 7 ETM+ Band (μm) | Applications | SPOT 4 Band (μm) |
|---|---|---|
| Band 1–Blue IR (0.45–0.52) | Coastal water mapping, soil/vegetation differentiation, deciduous/coniferous differentiation | |
| Band 2–Green (0.52–0.60) | Green reflectance from healthy vegetation, iron content in rocks and soil | Band 1(0.50–0.59) (Green) |
| Band 3–Red (0.63–0.69) | Chlorophyll absorption for plant differentiation | Band 2 (0.61–0.68) (Red) |
| Band 4–Near IR (0.76–0.90) | Biomass survey, water body delineation | Band 3 (0.79–0.89) (Near-infrared) |
| Band 5–Mid IR (1.55–1.75) | Plant moisture content, cloud/snow differentiation | Band 4 (1.58–1.75) (Mid-infrared) |
| Band 6–Thermal IR (10.4–12.5) | Thermal mapping, soil moisture | |
| Band 7–Mid IR (2.08–2.35) | Soil analysis | |

**FIGURE 1.10** General comparison of Landsat and Systeme Probatoire D'Observation De La Terre (SPOT) spectral bands for Earth observation application.

have spectral signature each single nanometer. The meteorologic satellite, Infrared Atmospheric Sounding Interferometer—Next Generation hyperspectral satellite even has a whoofing 16,921 bands with a spatial resolution of 25 km. The huge numbers of spectral bands help climate scientists to predict weather with accuracy.

Landsat Next (USGS, 2024), the new Landsat mission in the offing to be launched in late 2030, even plans to include spectral resolution of 26 bands at 10–20 m (VSWIR) improved spatial resolution with an aim to enhance Landsat series' existing applications to unlock new applications like water quality and aquatic health assessments, crop production, soil conservation, forest management and monitoring, climate and snow dynamics research, mineral mapping, etc. (USGS, 2024; Landsat Science, 2023).

Spectral reflectance curves (Figure 1.5) and discussion about EMS in Section 1.2 explain the advantages of discerning Earth objects or phenomena from the images using the spectral band information. One of the important advantages of separately sensing different spectral bands relates to the ability of combining the bands in various ways to enhance the visual information from the image. These combinations are easily implemented using software routines that stack three image bands and assign the reflected data values to three fundamental colors (red, green, and blue) to create image composites. Thus, when we have higher spectral resolution in our satellite data, more Earth's surface features can be discerned using different band combinations resulting in different color composites. In addition to visual enhancements, higher spectral resolution provides higher capabilities to combine band information using *band algebra* to extract additional biophysical information about the Earth's surface such as plant vigor and soil moisture status. These are implemented as spectral indices that mathematically combine band data into derivate, for example, normalized difference vegetation index (NDVI). Figure 1.10 shows some of the important applications of individual bands in satellite images, for example, Landsat 7 ETM+ and SPOT 4. Thus, higher-spectral-resolution images are better for feature extraction and decision support.

#### 1.2.1.3 Radiometric Resolution

Satellite sensors capture reflected energy (range of spectral region) from the Earth's surface (ground area unit) and the intensity of the energy sensed is quantified into a DN value as a percentage of the bit-based range (e.g., for 8-bit data range, the data range is 28,256, represented as 0–255 and 100% intensity of the range sensed is represented as 255, DN value). ESRI GIS Dictionary (2014) defines radiometric resolution as "the sensitivity of a sensor to incoming reflectance. Radiometric resolution refers to the number of divisions of bit depth (for example, 255 for 8-bit, 65,536 for 16-bit, and so on) in data collected by a sensor." While the pixel arrangement describes the spatial structure of an image, the radiometric characteristics describe the actual information content in a remotely sensed image. When an image is acquired on film or by a sensor, its sensitivity to the magnitude of the EMR determines the radiometric resolution (CCRS, 2014). The radiometric resolution of an imaging system describes its ability to differentiate slight differences in energy. The finer the radiometric resolution of a sensor, the more sensitive it is in detecting small differences in reflected or emitted energy (CCRS, 2014). As can be seen in Figure 1.11, a sensor with low radiometric resolution (1 bit) can resolve the incoming reflectance into two levels of contrast, while sensors with higher radiometric resolution provide higher levels of energy quantization. These results not only provide better contrast to the display but also facilitate the numerical analysis of the subtle differences in the radiometry. The downside to this is the high data volumes and storage issues associated with high-radiometric-resolution imagery. Even the Landsat series of satellites by NASA is enhancing its radiometric resolution over the years. Until Landsat 7, the radiometric resolution was 8-bit, Landsat 8 was of 12-bit, and the latest Landsat 9 is of 14-bit. Table 1.2 provides the radiometric resolution of many selective remote sensing sensors.

#### 1.2.1.4 Temporal Resolution

Satellite revisit period refers to the length of time it takes for the satellite to complete one entire orbit cycle around the Earth, for example, Landsat satellite takes 16 days. The revisit period of a satellite sensor is usually several days and also known as the temporal resolution of the sensor. Table 1.2 provides the temporal resolution of many selective remote sensing sensors. With an ever-changing global environment due to events such as urban sprawl, deforestation, and natural disasters, including drought, flood, landslides, wildfire, avalanches, and earthquakes, it is extremely essential for end users to conduct temporal analysis. There is an increase in studies pertaining to land use forecasting more so now than earlier due to increased awareness and implications of rapid land cover–land use changes occurring in the landscape. Temporal image analysis is highly essential for these studies. Better temporal resolution of images such as RapidEye, WorldView, MODIS, and even SPOT 5 can provide rapid

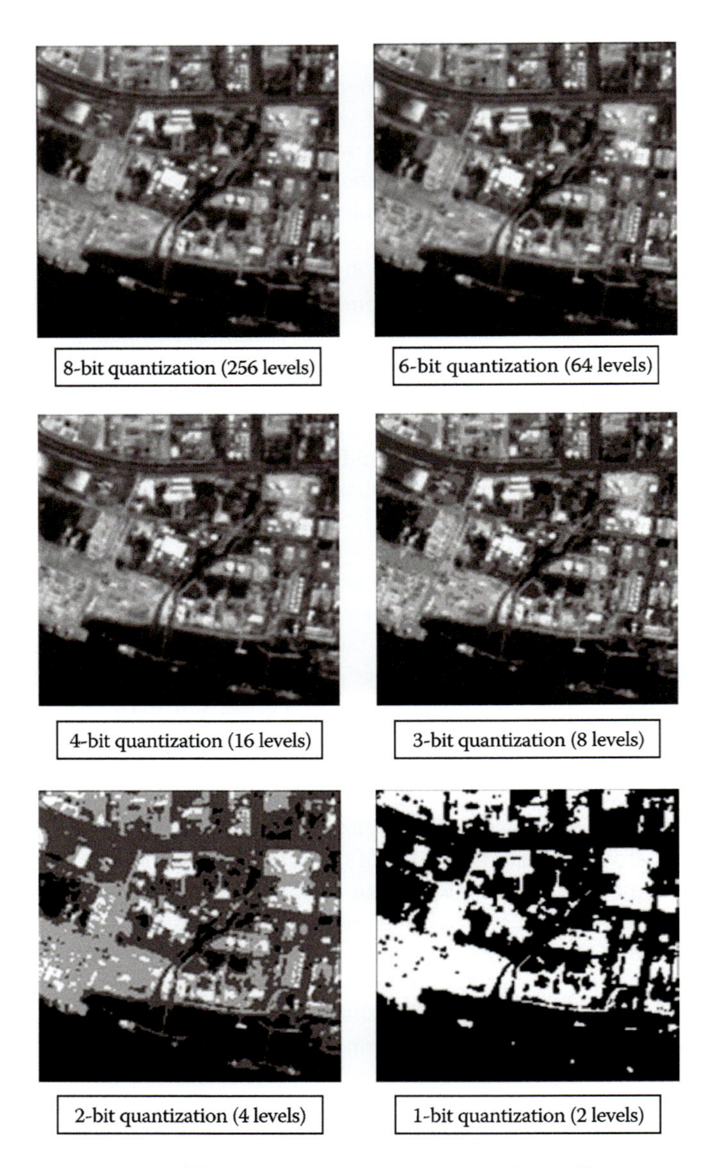


**FIGURE 1.11** Image captured at different levels of energy quantization to illustrate concept of radiometric resolution. Center for Remote Imaging, Sensing, and Processing (CRISP) (2014). (From www. crisp.nus.edu. sg/~research/tutorial/image.htm.) The more the quantization, the greater is the detail of the information captured, thus enabling greater separability between features.

land cover change information in scenarios such as tracking the spread of wildfire or damages caused by any natural disaster. Aerial imaging can be conducted daily over a study area and even more than once in a day. Therefore, aerial imaging does not have a specific temporal resolution. The advantages of temporal image resolution are shown in Figure 1.12a and b, in which IRS-1 LISS-A satellite imageries of two different years, 1989 and 1996, were used to classify the land use of a remote watershed in Orissa (India) and perform land use change analysis to determine a positive environmental effect (increase in forest cover and increase in changes from low-density forest to high-density forest cover) over seven years using social awareness in forest preservation (forming village-level forest preservation committee and implementing with earnest) through the state department of forestry and village committees of the watershed (Panda et al., 2004). It is to be noted that space organizations like NASA

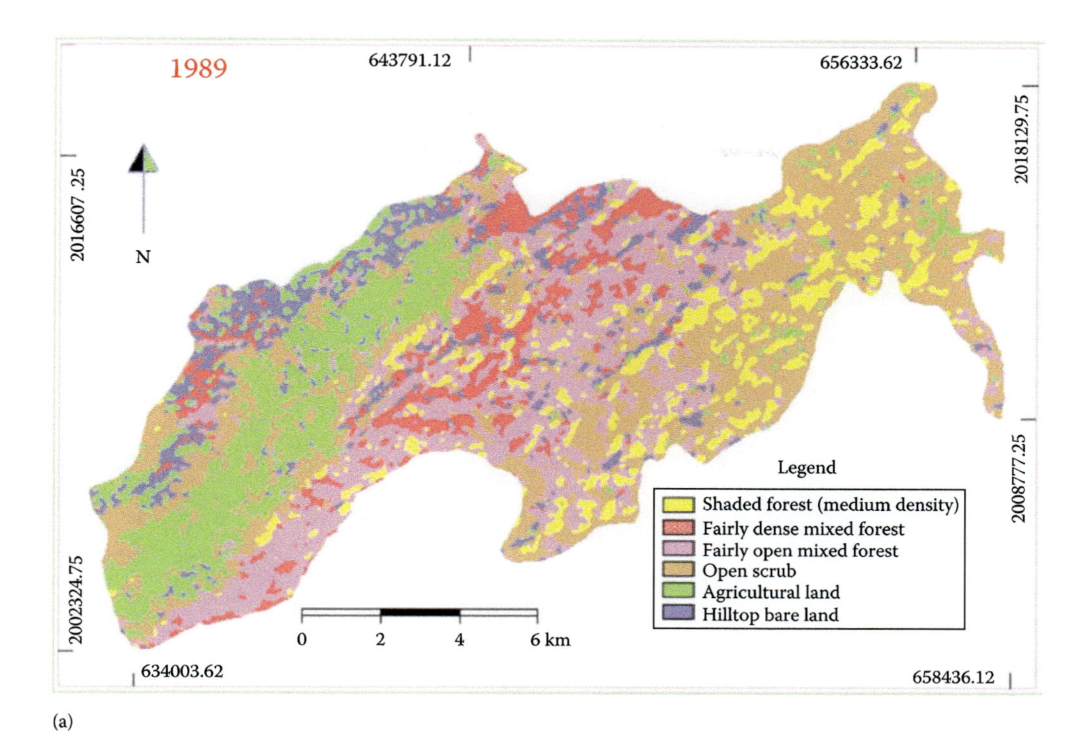


(a)

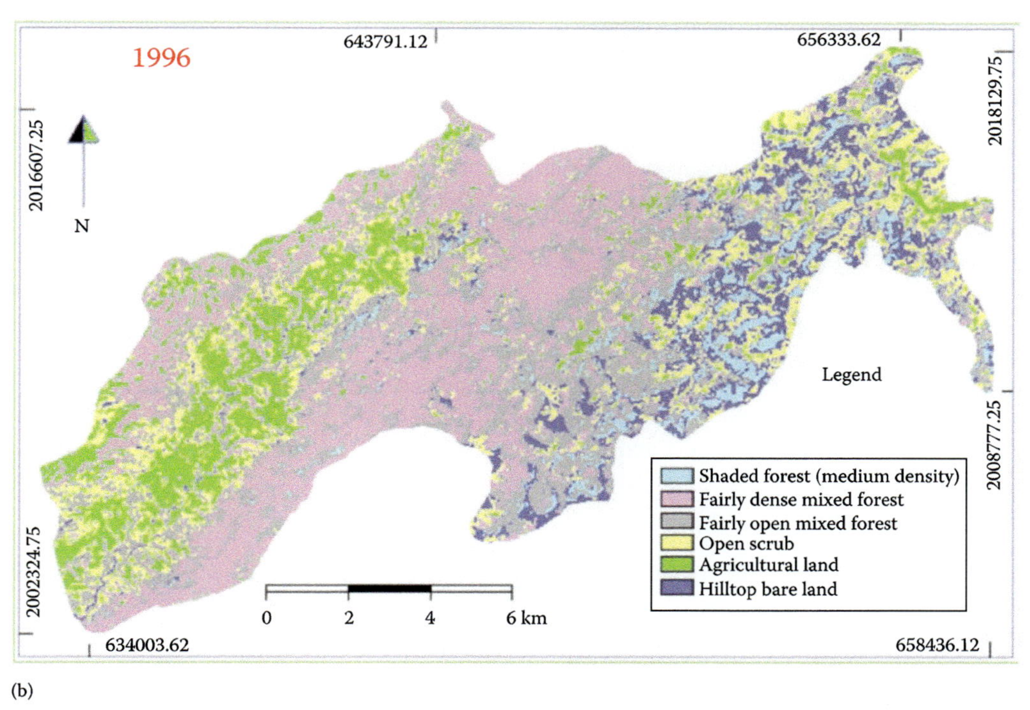


(b)

**FIGURE 1.12** Land use change analysis within a watershed in Orissa (India) using Indian Remote Sensing Satellite 1 Linear Imaging Self-Scanning Sensor A satellite imagery: (a) 1989 classified land use map with six classes (From IRS-1A-LISS2 satellite image); (b) 1996 classified land use map, also with six classes (from IRS-1B-LISS2 satellite image); (c) land use change matrix analysis showing 36 classes (six classes of 1989 × 6 classes of 1996) in the map—Class 11 means Class 1 (medium density shaded forest) in year 1989 remains the same Class 1 in the year 1996 and so on (from GIS analysis of two date satellite images).

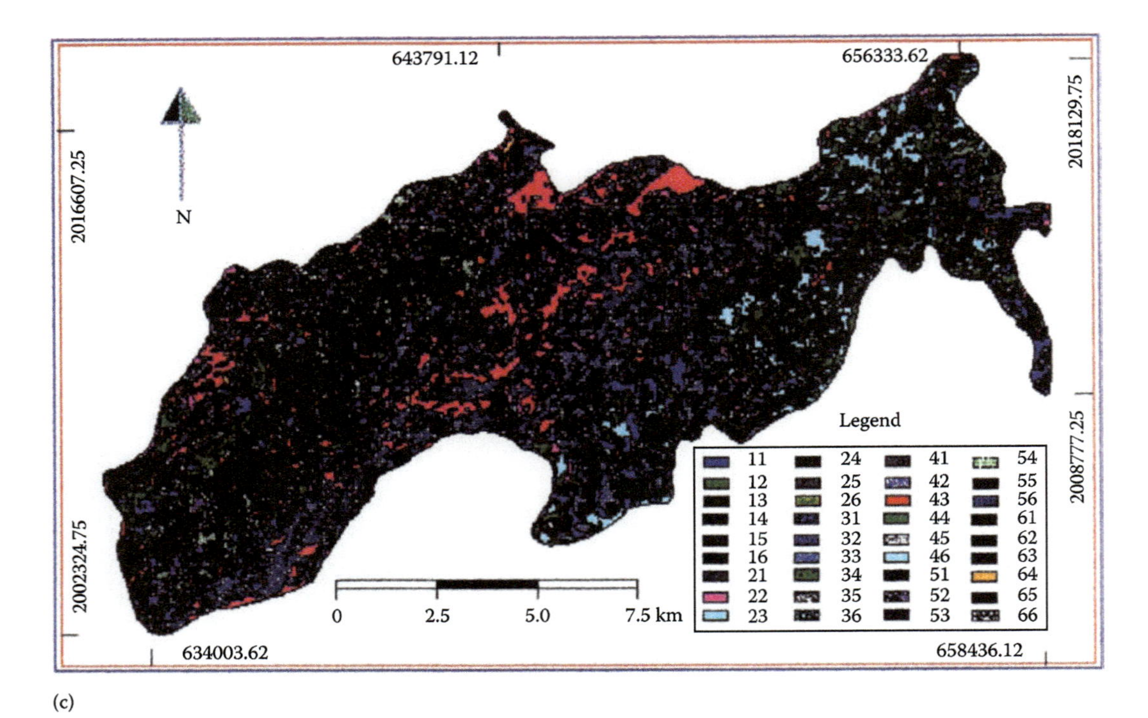


(c)

FIGURE 1.12 *(Continued)*

are constantly working on improving the temporal resolution of newer upcoming satellites and even improving the temporal resolution of current satellites. Vergopolan et al. (2021) improved the spatial and temporal resolution of SMAP satellite data to 30 m and 6 hours, respectively, from 3–36 km spatial resolution and 2–3 image acquisition repeating periods of two to three days. They used the image fusion approach of STAR-FM to enhance the spatial and temporal resolution so that forest and crop production managers can obtain soil moisture information of forest floor and agricultural fields on a daily basis, which is a need for better management decisions.

Landsat Next, new generation satellite, plans to have an improved temporal resolution of six days, a significant upgrade from the 16-day repeat interval of earlier Landsat satellites (Landsat Science, 2023). With the modification, the Landsat program will support global, synoptic, and repetitive coverage of Earth's land surfaces at a scale where natural and human-induced changes can be detected, differentiated, characterized, and monitored over time for a well-manageable decision support system development (Landsat Science, 2023).

### 1.2.2 Categorization of Satellites by Country of Ownership

It is a matter of national pride to have an Earth-observing program, and the acquired satellite data are very important for applications pertaining to resource inventory and management. Based on the study of Baker et al. (2001), the civilian remote sensing systems are government owned, the commercial systems are government licensed but privately owned, and the military/intelligence gathering platforms are government owned but with highly restricted access to the systems and products. According to Martin (2012), South Africa's scientific and engineering capacities were enhanced by the experience of designing, building, and launching SumbandilaSat; at the same time, the country obtained a great deal of social benefit due to the Earth-observing program. Belward (2012) opined that national space programs and their success not only enhance the nation's self-esteem but also reinforce national identity and sense of purpose for the country. This is why many countries in the world are in the forefront of the Earth observation missions. Table 1.3 presents a comprehensive list

**TABLE 1.3**
**Selected List of Earth-Observing Programs of Different Countries in the World**

| Country | Agency Name | Prominent Satellite Programs |
|---|---|---|
| France | Centre National d'Etudes Spatiales (CNES) | • SPOT |
| | | • Pleiades |
| | | • TOPEX/Poseidon |
| United States | National Aeronautics and Space Administration (NASA) | • Thermosphere, Ionosphere Mesosphere, Energetics and Dynamics (TIMED) |
| | | • TOPEX/Poseidon |
| | | • Upper Atmosphere Research Satellite |
| | | • Landsat |
| | | • NOAA-AVHRR |
| | | • SRTM |
| | | • Vanguard |
| Russia | Russian Federal Space Agency (RKA; Roscosmos) | • Elektro-L |
| | | • Monitor-E |
| | | • Resurs-DKl |
| | | • Resurs-P No. 1 |
| Japan | Japan Aerospace Exploration Agency (JAXA) | • MOS-1 (Momo-1) |
| | (former NASDA) | • MOS-lb (Momo-lb) |
| | | • JERS-1 (Fuyo-1) |
| | | • ADEOS (Midori) |
| | | • ADEOS-II (Midori II) |
| | | • GOSAT (Ibuki) |
| | | • ALOS (Daichi) PALSAR, AVNIR-2, and PRISM |
| India | Indian Space Research Organisation (ISRO) | • Oceansat-2, September 23, 2009 |
| | | • IMS-1, April 28, 2008 |
| | | • Cartosat-2A, April 28, 2008 |
| | | • Cartosat-2, January 10, 2007 |
| | | • IRS-P5 (Cartosat-1), May 5, 2005 |
| | | • IRS-P6 (RESOURCESAT-1), October 17, 2003 |
| | | • IRS-P4, May 27, 1999 |
| | | • IRS-P3, March 21, 1996 |
| | | • IRS-P2, October 15, 1994 |
| | | • IRS-1D, September 29, 1997 |
| | | • IRS-1C, December 28, 1995 |
| | | • IRS-1B, August 29, 1991 |
| | | • IRS-1A, March 17, 1988 |
| Canada | Canadian Space Agency | • RADARSAT |
| China | Chinese Space Agency | • Yogan-21 |
| | | • Gaofen |
| Europe | European Space Agency (ESA) | • ENVISAT |
| | | • ERS (1 and 2) |
| | | • CryoSat-2 |
| Brazil | Brazilian Space Agency (AEB) | • CBERS-1 |

(*Continued*)

**TABLE 1.3** *(Continued)*
**Selected List of Earth-Observing Programs of Different Countries in the World**

| Country | Agency Name | Prominent Satellite Programs |
|---|---|---|
| | | • CBERS-2 |
| | | • CBERS-2B |
| Argentina | Argentina Space Agency (CONAE) | • SAC-A |
| | | • SAC-B |
| | | • SAC-C |
| | | • SAC-D |
| | | • SAOCOM |
| Indonesia | Lembaga Penerbangan dan Antariksa Nasional (LAPAN, Indonesia) | • Lapan (TUBsat) |
| Belarus | National Academy of Sciences of the Republic of Belarus | • BelKA |
| Pakistan | SUPARCO (Space and Upper Atmosphere Research | • Badr-1 |
| | Commission) | • Badr-B |
| | | • Paksat-1R |
| | | • Pakistan Remote Sensing Satellite |
| Sweden | Swedish National Space Board | • Munin |
| Bolivia | Bolivarian Agency for Space Activities | • VRSS-1 |
| South Korea | | • KOMPSAT-2 |
| Thailand | | • Thaichote |
| Turkey | | • Gokturk-2 (2012) (IMINT), Turkish Armed Forces, Intelligence |
| | | • RASAT (2011), mapping |
| | | • BILSAT-1 (2003-2006), part of the Disaster Monitoring Constellation Project |
| | | • Gokturk-1 (2013) (IMINT), Turkish Ministry of National Defense, Intelligence |

of countries that have Earth-observing programs and the associated operational satellite series in the program.

Some developing countries, with fewer resources to have a space program of their own, collaborate with other countries with excellent space technology knowhows. In 2018, China helped Pakistan with its first optical remote sensing satellite PRSS-1, and a smaller observation craft PakTES-1A to help the country's populace for resource management in a smart and efficient manner (Ahsan and Khan, 2019). In 2019, China completed the in-orbit delivery of Venezuelan Remote-Sensing Satellite (VRSS-2), Sudan Remote-Sensing Satellite (SRSS-1), and the Algerian Communications Satellite (Alcomsat-1), enhancing the country's pride in joining the space satellite club for better resource management decision support. Countries like Saudi Arabia, Argentina, Brazil, Canada, Nigeria, Belarus, Ukraine, and even Luxembourg are exploring the possibility of sending remote sensing satellites to space with the help of other space-technology-smart countries like China, the United States, India, and France to name a few.

### 1.2.3 Types of Sensing Technologies Based on Applications

The NASA NSSDC's master catalog (https://nssdc.gsfc.nasa.gov/nmc/spacecraft/query) lists 7,262 spacecraft that have been launched between October 4, 1957, and March 31, 2015 (NSSDC, 2014), and the latest search on the website (December 25, 2023) yielded 18,417 spacecrafts launched so far, an astounding 250% increase during the last eight-and-a-half-year period. They were listed in the following categories based on their application: astronomy (344), Earth science (1,738), planetary sciences (381), solar physics (208), space physics (675), microgravity (76), human crew (371), life science (100), communication (8,892), engineering (429), navigation and global positioning (562), resupply/refurbishment/repair (293), surveillance and other military (2,445), and technology applications (1,428). The number in parentheses depicts the number of spacecraft launched with application mission objective. These numbers are updated on the site frequently. Among these, 1,738 spacecraft were launched for Earth science application so far, almost double the number in the last eight and a half years. The most increase in spacecrafts during this last eight and a half years was in the area of communication (400%) and technology application (700%), which may be attributed to the fact that commercial companies are thronging into satellite-related business because of its efficacy in technological advancement and communication management decision support. As discussed earlier, Earth science as a broad discipline encompasses land use–land cover issues, including surface (biophysical or chemical features) or underground (minerals) feature identification, but, clearly, not all 890 missions can be used for global land cover mapping. All of these near-polar-orbiting Earth science application–oriented satellites are searchable in the NSSDC website. For example, upon clicking on the satellite Aryabhata in the search result, the following description shows:

> This spacecraft, named after the famous Indian astronomer, was India's first satellite and was completely designed and fabricated in India. It was launched by a Soviet rocket from a Soviet cosmodrome. The spacecraft was quasispherical in shape containing 26 sides and contained three experiments for the measurement of cosmic X-rays, solar neutrons, and Gamma rays, and an ionospheric electron trap along with a UV sensor. The spacecraft weighed 360 kg, used solar panels on 24 sides to provide 46 watts of power, used a passive thermal control system, contained batteries, and a spin-up gas jet system to provide a spin rate of not more than 90 rpm. There was a set of altitude sensors comprised of a tri-axial magnetometer, a digital elevation solar sensor, and four azimuth solar sensors. The data system included a tape recorder at 256 b/s with playback at 10 times that rate. The PCM-FM-PM telemetry system operated at 137.44 MHz. The necessary ground telemetry and telecommand stations were established at Shar Centre in Sriharikota, Andhra Pradesh.

The link page to the satellite also includes an alternate name of the spacecraft, other brief facts, funding agency names, and discipline names for which the spacecraft was launched. The reader is directed to the NASA website (http://nssdc.gsfc.nasa.gov/nmc) for additional metadata information about the Earth-observing satellites.

On another note, among several types, the three most important types of satellite orbits are (1) near-polar orbits, (2) Sun-synchronous orbits, and (3) geosynchronous orbits. In near-polar orbits, the satellites' inclination to the Earth is nearly 90°. Thus, the satellite can virtually see every part of the Earth as the Earth rotates beneath the satellite. It takes approximately 90 minutes for the near-polar-orbiting satellites to complete one orbit and is very useful for atmospheric (stratosphere) measurements like greenhouse gas concentration, ozone concentration, temperature, and water vapor (Montenbruck and Gill, 2000). Sun-synchronous satellites like Landsat can pass over a section of the Earth at the same time of day due to these orbits (Montenbruck and Gill, 2000). These satellites, in general, orbit the Earth at an altitude of approximately 700–800 km. As some of these satellites take pictures of the Earth, they work best with bright sunlight, that is, passing over a section of Earth during late morning to until 2–3 p.m. or early afternoon. When Sun-synchronous satellites measure longwave radiation, they work best in complete darkness and hence pass over the Earth section at night. In geosynchronous orbits, also known as geostationary orbits, satellites circle the Earth at the same rate as the Earth spins (Montenbruck and Gill, 2000). These orbits make the satellites stay over a location of the Earth constantly throughout and observe almost the full hemisphere of the Earth.

According to Jensen (2009), remote sensing systems can directly measure the fundamental biological and/or physical characteristics of Earth features without using other ancillary data. The following are a few example of how remotely sensed data (MSS, hyperspectral, LiDAR, etc.) can help find water and nutrient stress in agricultural crop fields and support farmers to schedule irrigation and fertilizer applications for increased crop yields (Panda et al., 2011a, 2011b): finding the growth stages of blueberry crops and conducting site-specific crop management to enhance the crop yield (Panda et al., 2009, 2010a, 2010b, 2011a, 2011b; Panda and Hoogenboom, 2013), finding and assessing the drainage characteristics of low-gradient coastal watershed for watershed management decision support (Amatya et al., 2013), and estimating the deciduous forest structure for forest management decision support (Defibaugh et al., 2013). Engman and Chauhan (1995) show the advantage of microwave remote sensing to accurately measure soil moisture amounts. Panda et al. (2012a, 2012b) and Amatya et al. (2011) used Landsat data for estimating soil moisture, stomatal conductance, leaf area index (LAI), and canopy temperature of pine forest. Hale et al. (2012), Phillips et al. (2012), Rylee et al. (2012), and Ertberger et al. (2012) used remote sensing and other watershed geospatial data for urban and rural watershed development decision support, while Cash and Panda (2012) used high-resolution National Agriculture Imagery Program (NAIP) imagery, SSURGO soil data, and LiDAR data to develop a hydrologic model using Soil and Water Assessment Tool (SWAT). Numerous studies have already been conducted with the use of remote sensing to detect, interpret, and analyze the Earth's biophysical characteristics for management decision support. More studies for environmental and Earth resource management using remote sensing are provided later in individual sections. Jensen (2009) has developed a database showing the potential remote sensing systems for detection, interpretation, and analysis of various biophysical variables (Table 1.4).

## 1.3 REMOTE SENSING PLATFORMS

There are hundreds of different sensors available for specific usage depending on the airplane or the Earth-observing satellite platform. These sensors can be sorted by the sensor type and are categorized into the following: (1) acoustic, sound, and vibration; (2) automotive and transportation; (3) chemical; (4) electric current, electric potential, magnetic, and radio; (5) environment, weather,

**TABLE 1.4**
**Potential Remote Sensing Systems for Detection, Interpretation, and Analysis of Biophysical Variables**

| Biophysical Variables | Potential Remote Sensing Systems |
|---|---|
| • Geodetic control | • GPS |
| • Location from orthocorrected imagery | • Analog and digital stereoscopic aerial photography, space imaging |
| | • IKONOS, DigitalGlobe, QuickBird, OrbView-3, SPOT, Landsat (Thematic Mapper, Enhanced TM+), Indian IRS-ICD, European ERS-1 and ERS-2 microwave and ENVISAT MERIS, MODIS, LiDAR, Canadian RADARSAT-1 and -2, etc. |
| ***Topography/bathymetry*** | |
| • DEM | • GPS, stereoscopic aerial photography, and other location identification satellite systems |

**TABLE 1.4** *(Continued)*

**Potential Remote Sensing Systems for Detection, Interpretation, and Analysis of Biophysical Variables**

| Biophysical Variables | Potential Remote Sensing Systems |
|---|---|
| • Digital Bathymetric Model | • SONAR, bathymetric LiDAR, stereoscopic aerial photography |
| ***Vegetation*** | |
| • Chlorophyll a and b | • Color aerial photography, Landsat ETM+, IKONOS, QuickBird, and most other MSS sensor-based satellites |
| • Canopy structure | |
| • Biomass | • Stereoscopic aerial photography, LiDAR, RADARSAT, IFSAR |
| • LAI | • CIR aerial photography, most other MSS (Landsat, QuickBird, etc.) and hyperspectral systems |
| • Absorbed photosynthetically active radiation | (AVIRIS, HyMap, CASI), AVHRR, multiple imaging spectrometer (MISR) |
| • Evapotranspiration | |
| ***Surface temperature*** | |
| • Land, water, and atmosphere | • ASTER, AVHRR, GOES, Hyperian, MISR, MODIS, SeaWiFS, Airborne TIR |
| *Soil and rocks* | |
| • Moisture | • ASTER, passive microwave (SSM/1), RADARSAT, MISR, ALMAZ, Landsat, ERS-1 and ERS-2, |
| • Mineral composition | Intermap STAR-3i |
| • Taxonomy | • ASTER, MODIS, hyperspectral systems (AVIRIS, HyMap) |
| • Hydrothermal alteration | • High-resolution color and CIR aerial photography, airborne hyperspectral systems |
| | • Landsat, ASTER, MODIS, airborne hyperspectral systems |
| *Surface roughness* | • Aerial photography, ALMAZ, ERS-1 and ERS-2, RADARSAT, Intermap STAR-3i, IKONOS, QuickBird, ASTER, etc. |
| ***Atmosphere*** | |
| • Aerosols (optical depth) | • MISR, GOES, AVHRR, MODIS, CERES, MOPITT, MERIS |
| • Clouds | • GOES, AVHRR, MODIS, CERES, MOPITT, MERIS, UARS |
| • Precipitation | • TRMM, GOES, AVHRR, SSM/1, MERIS |
| • Water vapor (precipitable) | • GOES, MODIS, MERIS |
| • Ozone | • MODIS |
| ***Water*** | |
| • Color | • CIR aerial photography, most other MSS (Landsat, QuickBird, etc.) and hyperspectral systems (AVIRIS, HyMap, CASI), bathymetric LiDAR, MISR, Hyperion, TOPEX/Poseidon, MERIS, AVHRR, CERES, etc. |
| • Surface hydrology, suspended minerals | |
| • Chlorophyll | |
| • Dissolved organic matter | |
| ***Snow and sea ice*** | |
| • Extent and characteristics | • CIR aerial photography, most other MSS (Landsat, QuickBird, etc.) and hyperspectral systems (AVIRIS, HyMap, CASI) |
| • Bidirectional reflectance distribution function | • MISR, MODIS, CERES |

*Source:* Adapted and modified from Jensen, J.R., *Remote Sensing of the Environment: An Earth Resource Perspective 2/e*, Pearson Education India, Delhi, India, 2009, pp. 11-12.

moisture, and humidity; (6) fluid flow; (7) ionizing radiation and subatomic particles; (8) navigation instruments; (9) position, angle, displacement, distance, speed, and acceleration; (10) optical, light, imaging, and photon; (11) pressure; (12) force, density, and level; (13) thermal, heat, and temperature; (14) proximity and presence; (15) sensor technology; and (16) other sensors. However, this chapter is focused on Earth observation remote sensing, covering both active and passive remote sensing technologies as discussed in Section 1.1.2. These Earth-observing satellites and aerial imaging systems are grouped into several types based on their operational principles within the EMS and means of obtaining images. They are optical, radar, microwave, hyperspectral, sonar, etc.

Typical remote sensing systems operate in the visible (0.4–0.7 μm) and IR (0.7–1000 μm) portion of the EMS, known as optical remote sensing (Aronoff, 2004). Other sensors such as microwave sensors operate in the microwave region of the EMS (0.3 mm–1 m). RADAR uses radio waves (104–1011 μm) to determine the range, altitude, direction, and speed of an object or phenomena on Earth. On the other hand, hyperspectral imaging sensors, although operate in the visible/NIR ranges of the EMS, are marked by very high spectral resolutions resulting in many numbers of spectral bands (refer to Figure 1.6).

Another active sensor LiDAR is a remote sensing technology in which the signal (return) distances are measured based on the lag time of the pulsed signal. Based on the different returns recorded, it is possible to accurately measure Earth objects such as canopy height and ground surface elevation. SONAR is another active remote sensing technology platform. It uses sound propagation usually in water for depth mapping, object detection, and navigation underwater. These different platform-based remote sensing technologies have their individual applications and advantages in real-world problem solving and decision support.

### 1.3.1 Aerial Imaging

Aerial photography is a remote sensing system in which photographs of the Earth's surface are taken from an elevated position, mostly by airplanes flying within a 10-mile height from the ground. Platforms for aerial photography include fixed-wing aircraft, helicopters, multirotor unmanned aircraft systems (UASs), balloons, blimps and dirigibles, rockets, kites, parachutes, stand-alone telescoping, and vehicle-mounted poles and in recent times military operated drones. Recently, radio-controlled model aircrafts—drones, for example, are very popular in taking images of the Earth's surface from a much lower height. These aerial images are of high resolution and have a wide application in environmental and natural resource management fields.

The U.S. Geological Survey (USGS) initiated novel programs called the National High-Altitude Program (NHAP) and the National Aerial Photography Program (NAPP) in the 1980s to acquire aerial photograph for the conterminous 48 states to support operational and mapping needs of the local, state, and federal agencies (USGS, 1992a, 1992b). The black-and-white and color-infrared (CIR) products from these programs provide an invaluable resource for historical assessment of land cover/use at map scales ranging from 1:58,000 to 1:40,000.

Building on the efforts of the USGS, the USDA Farm Service Agency (FSA) launched and administered the NAIP aimed at acquiring high-resolution color and near-infrared (NIR) imagery during the agricultural growing seasons in the conterminous United States. These annual image data products are orthorectified and GIS-ready and are primarily used by USDA to maintain the Common Land Unit database. NAIP imagery is one of the major resources for the geospatial community working on vegetation mapping and precision agriculture due to its high resolution (1 and 2 m). More information on the NAIP imagery can be obtained from www.fsa.usda.gov/FSA. Use of aerial images in urban management (Hodgson et al., 2003; Cleve et al., 2008), site-specific crop management (Jhang et al., 2002; Panda et al., 2009, 2010a, 2010b, 2011a, 2011b), forest and ecological resource management (Suarez et al., 2005; Morgan et al., 2010; Rao, 2013), water resource management (Ritchie et al., 2003; Jha et al., 2007), and many other natural resource management is abundant. As discussed previously, photogrammetry or aerial stereo photography was the earliest form of remote sensing. Even today, image and terrain analysts use stereographic pairs of aerial photographs to make topographic maps by image and terrain analysts for digital elevation—based land management decision support such as road construction, watershed delineation, stream flow direction and flow accumulation mapping, and traffic ability applications. According to Dornaika and Hammoudi (2009) and Bulstrode et al. (1986), stereoscopy is a photogrammetric technique for creating or enhancing the illusion of depth in an image by means of stereopsis for binocularly vision called Stereogram. Adjacent but overlapping aerial photos are called stereopairs and are needed to determine parallax and stereo/3D viewing. Advances in the hardware and software sector simulate such procedures in the digital and visualization domains, called soft-copy photogrammetry.

### 1.3.2 Optical Remote Sensing

As discussed earlier, optical remote sensing uses visible, near infrared and short-wave infrared sensors to acquire images of the Earth's surface (Figure 1.13). Optical remote sensing systems are classified into several types, depending on the number of spectral bands used in the imaging process, such as (1) panchromatic (PAN) imaging system, (2) multispectral (MSS) imaging system, (3) superspectral imaging system, and (4) hyperspectral imaging system. In panchromatic imaging system, the sensor is a single-channel detector sensitive to radiation within a broad wavelength. For example, IKONOS PAN image covering a broad bandwidth of 0.45–0.9 μm, which is the range of multispectral B, G, R, and IR bands. Similarly, the SPOT PAN image has a band range of 0.51–0.73 μm, which covers most of the multispectral region of B and G bands. PAN image spatial resolution is mostly better than the multispectral images of the same sensor, for example, IKONOS PAN and SPOT PAN spatial resolutions are 1 and 10 m, respectively, whereas the spatial resolutions of their MSS bands are 4 and 20 m. In multispectral imaging system, the sensor uses a multichannel detector with a few spectral bands like B, G, R, and IR within a narrow wavelength band (refer to Table 1.2). Table 1.2 provides a detailed list of MSS and PAN bands of some select optical sensors. In contrast to MSS imaging system, in a superspectral imaging system, there are many more spectral channels

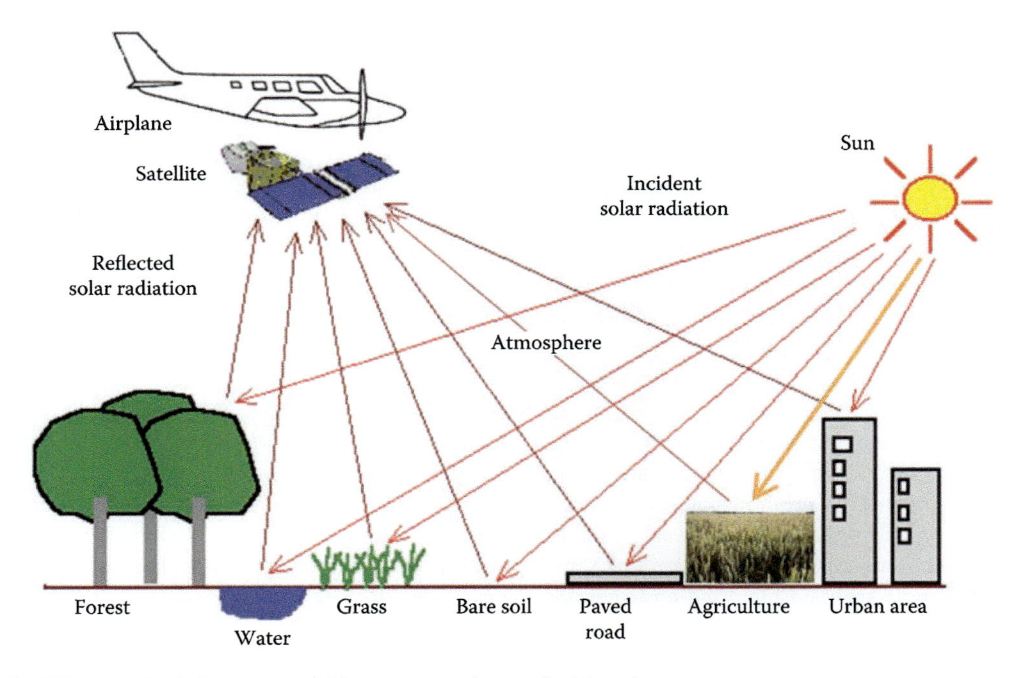


**FIGURE 1.13** Basic image acquisition process of an optical imaging system.

(typically >10). The bands have narrower bandwidths than MSS imaging system, enabling the finer spectral characteristics of the targets to be captured by the sensor. MODIS and medium-resolution imaging spectrometer (MERIS) are a few examples of the superspectral systems. MODIS superspectral imaging system has 36 bands ranging from 0.405 to 14.385 µm with each band having narrower bandwidths; MODIS band 11 and band 12 have bandwidths of 0.526–0.536 and 0.546–0.556 µm, respectively, whereas the bandwidth of Landsat MSS imaging system has little wider bandwidths. For instance, bands 4 and 5 of Landsat MSS have bandwidths of 0.76–0.90 and 1.55–1.75 µm, respectively. However, hyperspectral imaging systems acquire images in about 100 or more contiguous spectral bands. Hyperspectral imaging system is discussed in a later section.

According to Avery and Berlin (1992), unlike photographic cameras that record radiation reflected from ground scene directly onto films, electro-optical sensors use nonfilm detectors, which convert the reflected and/or emitted radiation from the object or phenomena on ground/Earth to proportional electrical signals that are used to construct 2D images for conventional viewing. Avery and Berlin (1992) categorized electro-optical sensors into video camera, vidicon camera, across-track scanner, and along-track scanner. According to Murai (1993), optical sensors are characterized by spectral (spectral band, bandwidth, central wavelength, response sensitivity at the edge of band, spectral sensitivity at outer wavelength, and sensitivity of polarization), radiometric (radiometry of the sensor, sensitivity in noise equivalent power, dynamic range, S/N, and other quantification noise), and geometric (FOV, IFOV, band-to-band registration, MTF, geometric distortions, and alignment of optical elements) performance. Table 1.5 lists optical sensor–based remote sensing satellites with their spectral band information. The table also contains satellites based on other different sensing platforms as discussed in this section. The description column of the table explains about the sensor platform of the satellite. In addition to the mechanics of image acquisition, Figures 1.14a and b illustrate a comprehensive working principle of optical remote sensing, which includes across-track/whisk broom, push broom, and multispectral scanning processes. As the names suggest, in a whisk broom system, the sensor detectors scan and record the reflected energy using oscillation mirrors that move across the satellite track, while the push broom system employ an array of detectors that are perpendicular to the satellite track. Landsat sensor systems until Landsat 8 OLI use the whisk broom system. Sensors that use the push broom design include SPOT, IRS, QuickBird, OrbView, and IKONOS.

### 1.3.3 Hyperspectral Remote Sensing

Hyperspectral imaging system is an optical imaging system but is known as an *imaging spectrometer* as the system acquires an image from a satellite or airplane in about 100 or more contiguous spectral bands. Imaging spectrometry is discussed in a later section. The targets on Earth are better identified and characterized by a hyperspectral imaging system than other optical imaging systems because of the system's precise spectral information. Hyperspectral imaging or field spectrometry has potential applications in precision agriculture or site-specific crop management (e.g., monitoring the health and growth of crops and measuring water stress, nutrient stress, or stress from crop diseases in crop fields), intense forest management (e.g., forest species differentiation, timber readiness for harvesting determination and analysis, wildfire susceptibility determination and assessment), and coastal management (e.g., monitoring of water quality parameters such as phytoplankton, organic matter, sediment pollution, and bathymetry changes).

Hyperspectral imaging produces an image where each pixel has full spectral information with imaging narrow spectral bands over a contiguous spectral range of the EMS. Hyperspectral imaging is a very-fast-growing area in remote sensing that expands and improves the efficiency/capability of multispectral satellite imaging (Lees and Ritman, 1991; Goetz, 2009). Hyperspectral imagers are used in various applications since the late 1980s including mineralogy, biology (food safety), defense, and environmental measurements (Clark et al., 1990; Kim et al., 2001; Chang, 2003; Lu, 2003; Van Wagtendonk et al., 2004; Asner et al., 2007; Gowen et al., 2007). Hyperion, AVIRIS,

**TABLE 1.5**
**List of Selected Optical Remote Sensing Satellites along with Their Spectral Band Information**

| Name of Sensor | Description | Mission | No. of Bands |
|---|---|---|---|
| **ACe -FTS** | Atmospheric Chemistry Experiment Fourier Transform Spectrometer | SCISAT | |
| **ACS** | Atmospheric Correction Sensor | RESOURCESAT-3 | |
| **ADCS/SAr SAT** | Advanced Data Collection System/ Search and Rescue Satellite Aided Tracking | JPSS-1 (NPOESS) | 4 |
| **Ae ISS** | Advanced Earth Imaging Sensor System | KOMPSAT-3 (Arirang 3) | 9 |
| **ALI** | Advanced Land Imager | EO-1 | 3 |
| **AlSat** | Standard DMC sensor | AlSat-1 | |
| **APS** | Aerosol Polarimetry Sensor | GLORY | |
| **Arg OS** | | Metop-A, Metop-B, SARAL/AltiKa | 4 |
| **AWFI** | Wide Field Imaging Camera | CBERS-3, CBERS-4 | 4 |
| **AWiFS** | Advanced Wide Field Sensor | RESOURCESAT-1, RESOURCESAT-2 | 3 |
| **Be IJINg -1-MS** | Standard DMC sensor | BEIJING-1 | 4 |
| **BILSAT-MS** | | BILSAT | 1 |
| **BILSAT-PAN** | | BILSAT | |
| **Camera** | Multispectral camera with black-and-white and color imaging | Svea | 4 |
| **CCD** | Multispectral camera | CBERS, CBERS-2, CBERS-3, CBERS-4 | |
| **CCD/TDI** | Selectable-multispectral possibility | EROS C | 1 |
| **CCD/TDI** | Selectable | EROS B | |
| **CCSP** | Cloud Camera Sensor Package | GLORY | 3 |
| **Cere S** | Clouds and Earth's Radiant Energy System | Aqua, JPSS-1 (NPOESS), Suomi NPP (SNPP), Terra, TRMM | 1 |
| **Chinese Mapping Telescope** | | BEIJING-1 | 1 |
| **CIr C** | Compact infrared camera | ALOS-2 | 9 |
| **COBAN** | | BILSAT | 6 |
| **CZCS** | Coastal Zone Color Scanner | Nimbus-7 | 3 |
| **DMC2** | Standard DMC sensor | UK-DMC 2 | 2 |
| **DOr IS** | Doppler Orbitography and Radiopositioning Integrated by Satellite | CryoSat-2, ENVISAT, Jason-1, Jason-2 (OSTM), SPOT 2, SPOT 3, SPOT 4, SPOT 5, TOPEX/Poseidon | 1 |
| **e OC** | Electro-optical camera | KOMPSAT-1 | |
| **e -OP1** | Panchromatic and 5 band color imager | NigeriaSat-2 | |
| **e -OP2** | Wide area coverage imager | NigeriaSat-2 | 13 |
| **er B** | Earth Radiation Budget | Nimbus-7 | 4 |
| **er Be** | Earth Radiation Budget Experiment | NOAA-10 | 1 |
| **er OS A** | | EROS A | 8 |
| **e TM** | Enhanced Thematic Mapper | Landsat 7 | |
| **ECOSTRESS** | Prototype HyspIRI Thermal Infrared Radiometer | NASA-JPL | 6 |
| **Fr S** | Strip map mode | RISAT-1 | 2 |

(*Continued*)

## TABLE 1.5 *(Continued)*
## List of Selected Optical Remote Sensing Satellites along with Their Spectral Band Information

| Name of Sensor | Description | Mission | No. of Bands |
|---|---|---|---|
| **ger B** | Geostationary Earth Radiation Budget | MSG-1, MSG-2, MSG-3 | 4 |
| **g IS-MS** | Visible and NIR | GeoEye-1 (OrbView-5) | 1 |
| **g IS-PAN** | Panchromatic | GeoEye-1 (OrbView-5) | 36 |
| **g LI** | Global Imager | ADEOS-II | 5 |
| **g Oe S Imager** | Multichannel instrument to sense radiant and solar-reflected energy | GOES-14 | 6 |
| **g OMe** | Global Ozone Monitoring Experiment | ERS-2 | 3 |
| **g OMOS** | Global Ozone Monitoring by Occultation of Stars | ENVISAT | |
| **gu VI** | Global Ultraviolet Imager | TIMED | 1 |
| **HCS** | High-sensitivity camera | SAC-C | 7 |
| **He PD** | High-energy particle detector | KOMPSAT-1 | |
| **Hir I** | High-resolution optical imager | Pleiades HR 1A, Pleiades HR 1B | |
| **HPO** | High-performance optical sensor | ASNARO-1 | 1 |
| **Hr C** | High-resolution camera | PROBA-1 | 4 |
| **Hrg** | High-resolution geometric | SPOT 5 | 1 |
| **Hr S** | High-resolution stereoscopic | SPOT 5 | 1 |
| **Hr TC** | High-resolution technological camera | SAC-C | 4 |
| **Hr V** | High-resolution visible | SPOT 1, SPOT 2, SPOT 3 | 4 |
| **Hr VIr** | High-resolution visible and infrared | SPOT 4 | 2 |
| **HSr S** | Hot Spot Recognition Sensor System | BIRD | 64 |
| **HySI** | Hyper Spectral Camera with 64 fixed bands | IMS 1 | |
| **IDee** | Instrument for Detection of High-Energy Electrons | TARANIS | 3 |
| **IKAr -De LTA** | | PRIRODA-MIR | 5 |
| **IKAr -N** | | PRIRODA-MIR | 2 |
| **IKAr -P** | | PRIRODA-MIR | |
| **IMA** | Multispectral Imager operating in 5 bands | Argo (RapidEye 6) | 5 |
| **Imager** | | GOES-10, GOES-11, GOES-13 = GOES-East, GOES-8, GOES-9 | 5 |
| **Imager-M** | | GOES-12 = GOES-South America | 1 |
| **IMg** | Interferometer Monitor for Greenhouse Gases | ADEOS | |
| **IMM** | Instrument for magnetic measurements | TARANIS | 1 |
| **IMS** | Ionosphere Measurement Sensor | KOMPSAT-1 | 4 |
| **IPe I** | Ionosphere Plasma and Electrodynamics Instrument | FORMOSAT-1 | 4 |
| **Ir MSS** | Medium-resolution scanner | CBERS-3, CBERS-4 | 4 |
| **Ir -MSS** | Infrared multispectral scanner | CBERS, CBERS-2 | |
| **ISu AL** | Imager of Sprites and Upper Atmospheric Lightning | FORMOSAT-2 | 5 |

**TABLE 1.5** *(Continued)*
**List of Selected Optical Remote Sensing Satellites along with Their Spectral Band Information**

| Name of Sensor | Description | Mission | No. of Bands |
|---|---|---|---|
| **JAMI** | Japanese Advanced Meteorological Imager | MTSAR-1R | 2 |
| **KBr** | K-Band Ranging System | GRACE | 4 |
| **KOMPSAT-MSC** | B&W panchromatic, MSS and merged 1 m resolution images | KOMPSAT-2 | 1 |
| **KVr -1000** | | Cosmos | 256 |
| **LAC** | LEISA Atmospheric Corrector, corrects high- spatial-resolution multispectral data for atmospheric effects on surface reflectance | EO-1 | |
| **Laser reflector** | | CryoSat-2 | 6 |
| **LIMS** | Limb infrared monitor of the atmosphere | Nimbus-7 | 1 |
| **LIS** | Lightning Imaging Sensor | TRMM | 4 |
| **LISS-1** | Linear Imagine Self-Scanning System | IRS-1A, IRS-1B | 4 |
| **LISS-2** | Linear Imagine Self-Scanning System | IRS-1A, IRS-1B, IRS-P2 | 4 |
| **LISS-3** | Linear Imagine Self-Scanning System | IRS-1C, IRS-1D | 4 |
| **LISS-3*** | Linear Imagine Self-Scanning System | RESOURCESAT-1 | 3 |
| **LISS-4** | Linear Imagine Self-Scanning System | RESOURCESAT-1 | 4 |
| **LISS-3** | Linear Imagine Self-Scanning System | RESOURCESAT-2 | 3 |
| **LISS-4** | Linear Imagine Self-Scanning System | RESOURCESAT-2 | 4 |
| **LISS-III-WS** | Wide-swath sensor | RESOURCESAT-3 | 4 |
| **MAC** | MS | RazakSAT (MACSAT), RazakSAT (MACSAT) | |
| **MAe STr O** | Measurements of Aerosol Extinction in the Stratosphere and Troposphere Retrieved by Occultation | SCISAT | |
| **MCP** | Micro Camera and Photometer | TARANIS | 15 |
| **Mer IS** | Medium-resolution imaging spectrometer | ENVISAT | |
| **Me XIC** | Multiexperimental Interface Controller | TARANIS | 5 |
| **MMr S** | Multispectral Medium-Resolution Scanner | SAC-C | |
| **MODI** | Moderate-Radiation Visible and NIR Imager | FengYun-3A | 4 |
| **MOMS-2P** | Modular Optoelectronic Multispectral Scanner | PRIRODA-MIR | 64 |
| **MOPITT** | Measurement of Pollution in the Troposphere | Terra | 4 |
| **MS camera** | Derived from SPOT camera, has thermoelastic stability | Deimos 1, Kanopus-Vulkan, THEOS | 12 |
| **MSC** | Multispectral | VENUS | 13 |
| **MSI** | Multispectral instrument | EarthCARE, Sentinel-2 | 4 |
| **MSS** | Multispectral scanner | AlSat-2A, Landsat 4, Landsat 5 | 5 |

*(Continued)*

**TABLE 1.5** *(Continued)*

**List of Selected Optical Remote Sensing Satellites along with Their Spectral Band Information**

| Name of Sensor | Description | Mission | No. of Bands |
|---|---|---|---|
| **MSS (LS 1-3)** | Multispectral scanner Landsat 1, 2, 3 | Landsat 1, Landsat 2, Landsat 3 | 3 |
| **MSu -e 1** | | Resurs-O1 | 3 |
| **MSu -e 2** | | PRIRODA-MIR | 5 |
| **MSu -g S** | Multispectral Scanner Geostationary | Electro-L/GOMS 2 | 5 |
| **MSu -SK** | | PRIRODA-MIR | 5 |
| **MSu -SK (r esurs)** | | Resurs-O1-3 | 5 |
| **MSu -SK1** | | Resurs-O1 | 16 |
| **MTI** | Multispectral Thermal Imager | MTI | 3 |
| **Mu X** | Multispectral CCD Camera | CBERS-3, CBERS-4 | 4 |
| **Mx** | Four band multispectral CCD camera | IMS 1 | 5 |
| **NAOMI** | New AstroSat Optical Modular Instrument | SPOT 6, SPOT 7, VNREDSat-1A | 3 |
| **NigeriaSat-1** | Standard DMC sensor | NigeriaSat-1 | 1 |
| **NSCAT** | NASA Scatterometer | ADEOS | 7 |
| **OCI** | Ocean Color Imager | FORMOSAT-1 | 8 |
| **OCM** | Ocean Color Monitor | IRS-P4 (Oceansat), Oceansat-2 | 12 |
| **OCTS** | Ocean Color and Temperature Scanner | ADEOS | 3 |
| **OIS** | Optical Imaging System | RASAT | |
| **OLCI** | Ocean and Land Color Instrument | Sentinel-3 | 9 |
| **OLI** | Operational Land Imager | Landsat 8 (LDCM) | 3 |
| **OLS** | Operational Linescan System | DMSP-16 | 3 |
| **OMI** | Ozone Monitoring Instrument | AURA | |
| **OMPS** | Ozone Mapping and Profiler Suite | JPSS-1 (NPOESS), Suomi NPP (SNPP) | 8 |
| **OPS** | Optical Sensor | JERS-1 | 6 |
| **Optical Imaging (g OKTur K-2)** | | GOKTURK-2 | 5 |
| **OrbView-3** | | OrbView-3 | 5 |
| **OSA** | Optical Sensor Assembly | IKONOS | |
| **OSIr IS** | Optical Spectrograph and Infrared Imaging System | ODIN | 6 |
| **OSMI** | Ocean Scanning Multispectral Imager | KOMPSAT-1 | 1 |
| **OTD** | Optical Transient Detector | OrbView-1 | 4 |
| **OZON-M** | | PRIRODA-MIR | 4 |
| **PAMeLA** | | Resurs-DK1 | 1 |
| **PAN** | | Cartosat-2B, CBERS-3, CBERS-4, IRS-1C, IRS-1D | 1 |
| **PAN Camera** | Panchromatic camera | Cartosat-2 (IRS-P7), Kanopus-Vulkan | 1 |
| **PAN Telescope** | | THEOS | 1 |
| **PAN-A** | Panchromatic aft pointing | Cartosat-1 | 1 |
| **PAN C** | Panchromatic camera | Cartosat-2A | 1 |
| **PAN-F** | Panchromatic forward pointing | Cartosat-1 | 9 |
| **POAM-II** | Polar Ozone and Aerosol Measurement | SPOT 3 | 9 |
| **POAM-III** | Polar Ozone and Aerosol Measurement | SPOT 4 | 15 |

**TABLE 1.5 *(Continued)***
**List of Selected Optical Remote Sensing Satellites along with Their Spectral Band Information**

| Name of Sensor | Description | Mission | No. of Bands |
|---|---|---|---|
| **POLDer** | Polarization and Directionality of the Earth's Reflectance | ADEOS, ADEOS-II, Parasol | 2 |
| **Poseidon 2 altimeter** | | Jason-1 | |
| **POSe IDON-3** | Altimeter | Jason-2 (OSTM) | 1 |
| **Pr ISM** | Three panchromatic sensors for stereomapping | ALOS | 4 |
| **QuickBird** | High-resolution PAN, 61 cm (nadir) to 72 cm (25° off nadir); MS, 2.44-2.88 m | QuickBird | 1 |
| **r -400** | | PRIRODA-MIR | 1 |
| **r adar altimeter** | | GFO | |
| **r adar sensor** | | SMAP | |
| **r adiometer** | L-band | SMAP | 4 |
| **r BV** | Return Beam Vidicon Camera | Landsat 1 | 5 |
| **re IS** | Records in 5 spectral bands on VIR and NIR | RapidEye | 1 |
| **r IS** | Reflector in space | ADEOS | 5 |
| **r OCSAT-2** | | FORMOSAT-2 | 4 |
| **r SI** | Remote sensing instrument | FORMOSAT-5 | 1 |
| **SAM II** | Stratospheric Aerosol Measurement | Nimbus-7 | 1 |
| **SASS** | Seasat-A Satellite Scatterometer | Seasat | 12 |
| **SBu V/2** | Solar Backscatter Ultraviolet Instrument | NOAA-11, NOAA-12, NOAA-14, NOAA-15, NOAA-16, NOAA-17, NOAA-18 (NOAA-N), NOAA-19 (NOAA-N Prime) | 6 |
| **SBu V/TOMS** | Solar Backscatter Ultraviolet/Total Ozone Mapping—failed in 1993 | Nimbus-7 | |
| **SCAr AB** | Scanning Radiative Budget Instrument | Megha-Tropiques | |
| **Scatterometer** | | Oceansat-2 | 8 |
| **SeaWiFS** | Sea-Viewing Wide Field-of-View Sensor | OrbView-2 | 1 |
| **Seawinds** | | ADEOS-II, QuikSCAT | |
| **See** | Solar Extreme Ultraviolet Experiment | TIMED | |
| **Se M** | Space Environment Monitor | GOES-14 | 7 |
| **Se M** | To measure solar radiation in the x-ray and extreme ultraviolet (EUV) region | GOES-13 = GOES-East | 12 |
| **Se VIr I** | Spinning Enhanced Visible and Infrared Imager | MSG-1, MSG-2, MSG-3 | 35 |
| **Sg LI** | Second-Generation Global Imager | GCOM-C1 | 1 |
| **SIM** | Spectral Irradiance Monitor | SORCE | 1 |
| **SIr -C** | | SRTM | 3 |
| **SLIM6** | Surrey Linear Imager Multispectral 6 channels — but 3 spectral bands | NigeriaSat-X | 1 |
| **SOLSTICe A and B** | Solar-Stellar Irradiance Comparison Experiment | SORCE | 19 |

*(Continued)*

**TABLE 1.5** *(Continued)*
**List of Selected Optical Remote Sensing Satellites along with Their Spectral Band Information**

| Name of Sensor | Description | Mission | No. of Bands |
|---|---|---|---|
| **Sounder** | | GOES-10, GOES-11, GOES-12 = GOES-South America, GOES-13 = GOES-East, GOES-8, GOES-9 | |
| **SSu LI** | Ultraviolet Limb Imager | DMSP-16 | |
| **SXI** | Solar X-ray Imager | GOES-14 | |
| **SXI Solar X-ray Imager** | To monitor the sun's x-rays | GOES-13 = GOES-East | 4 |
| **TANSO-CAI** | Thermal and NIR Sensor for Carbon Observation, Cloud and Aerosol Imager | GOSAT (Ibuki) | 1 |
| **TDI** | Panchromatic | AlSat-2A | |
| **TIDI** | TIMED Doppler Interferometer | TIMED | 1 |
| **TIM** | Total Irradiance Monitor | SORCE | |
| **TIP** | Tiny Ionosphere Photometer | FORMOSAT-3 (COSMIC) | 2 |
| **TIr S** | Thermal Infrared Sensor | Landsat 8 (LDCM) | 1 |
| **TK-350** | | Cosmos | 7 |
| **TM** | Thematic Mapper | Landsat 4, Landsat 5 | 1 |
| **TopSat** | | TopSat | |
| **TOr Package** | Tracking, Occupation, and Ranging | TanDEM-X | 2 |
| **Travers SAr** | | PRIRODA-MIR | 1 |
| **TV Camera** | | PRIRODA-MIR | 3 |
| **u K-DMC** | Standard DMC sensor | UK-DMC | 4 |
| **Vege TATION** | | SPOT 4, SPOT 5 | 4 |
| **Vg T-P** | Vegetation instrument on PROBA-V | PROBA-V | 5 |
| **VIr S** | Visible and Infrared scanner | TRMM | 2 |
| **WAOSS-B** | Wide-Angle Optoelectronic Stereo Scanner | BIRD | 2 |
| **Water Vapor r adiometer** | | GFO | 2 |
| **WFI** | Wide Field Imager | CBERS, CBERS-2 | 3 |
| **WiFS** | Wide Field Sensor | IRS-1C, IRS-1D, IRS-P3 | 5 |
| **WindSat** | | Coriolis | 1 |
| **WorldView-1** | Provides highly detailed imagery for precise map creation and in-depth image analysis | WorldView-1 | 8 |
| **WV 3 MSS** | WorldView-3 multispectral sensor | WorldView-3 | 1 |
| **WV 3 PAN** | WorldView-3 panchromatic sensor | WorldView-3 | 8 |
| **WV 3 SW** | WorldView-3 shortwave infrared sensor | WorldView-3 | 8 |
| **WV110** | Standard 4 colors + new 4 colors | WorldView-2 | 1 |
| **WV60** | PAN band for WorldView-2 | WorldView-2 | |
| **Xgre** | X-ray and Gamma Relativistic Electron Detector | TARANIS | 1 |
| **XPS** | XUV photometer | SORCE | |

*Source:* Adapted and modified from ITC, ITC's database of satellites and sensors, 2023, http://www.itc.nl/research/products/sensordb/allsensors.aspx. accessed on Oct 15, 2023.

HyMap, GF-5, and compact airborne spectrographic imager (CASI) are examples of a high-spatial-resolution hyperspectral system. Hyperion high-resolution (30 m) hyperspectral imaging system acquires images in 220 contiguous spectral bands with very high radiometric accuracy within a bandwidth range of 0.4–2.5 μm. With these large numbers (220) of bands, complex land characteristics can be identified. Kruse et al. (2003) used Hyperion imagery to map the minerals in and around Cuprite, NV, with high accuracy.

Figure 1.15 is an example of AVIRIS hyperspectral imaging system data taken over Yellowstone National Park acquired in 224 continuous bands having a 10 nm band pass over the spectral wavelength range of 350–2,500 nm (from visible light to NIR). AVIRIS collects 20 m wide pixels at approximately 14 m spacing. The sensor swath width is approximately 10.5 km. Kokaly et al. (2003) used AVIRIS data from Yellowstone National Park to map the vegetation types in the park with very high accuracy. Kokaly et al. (1998) used AVIRIS imaging spectrometry data to characterize and map the biology and mineralogy of Yellowstone National Park. Above all, commercial companies also acquire hyperspectral imageries on demand. Table 1.6 contains the selective list of hyperspectral imaging system satellites with pertinent specifications on number of bands, spatial resolution, and applications.

Over the past years, the demands for better mapping and characterization of Earth objects and related phenomena have increased, resulting in a higher need to better understand the biophysical interactions of radiation, atmospheric effects, and albedo properties of surface materials. Ground-based remote sensing using spectroradiometers provide an innovative approach to measuring surface reflectance of materials while removing the interfering effects of atmospheric path radiance, absorption, and scattering effects (Clark et al., 2002). Thus, spectroscopy serves as a tool to map specific material and mineral for environmental assessments including vegetation health studies under field or laboratory conditions (Yang et al., 2005; Swayze et al., 2014). Hyperspectral imaging and imaging spectroscopy are the technologies used to acquire a spectrally resolved image of an object or scene (Butler and Laqua, 1996). Also, spectrometry is used as lab/computer imaging technology for food quality—meat deterioration (Panigrahi et al., 2003; Savenije et al., 2006), fruit sweetness analysis (Guthrie and Walsh, 1997; Nicolai et al., 2007), leaf chlorophyll measurement (Shibata, 1957), etc.—and thus images acquired through space imaging spectrometry platforms would analyze the Earth objects and phenomena in the most qualitative means.

Ge et al. (2022) used the 300 channel GF-5 hyperspectral imagery with a strategy that includes bootstrap methods, fractional order derivative (FOD) techniques, and decision-level fusion models to remotely determine the soil salinity in the Ebinur Lake oasis in northwestern China. They obtained a strong correlation between soil salinity and more diagnostic bands from the ultra-hyperspectral (5 nm) imagery with optimal estimated model $R^2 = 0.95$, root mean square error (RMSE) = 3.20 dS/m, and a ratio of performance to interquartile distance (RPIQ) = 5.96. Ye et al. (2020) assessed GF-5 AHSI imagery for lithological mapping in comparison with Shortwave Infrared Airborne Spectrographic Imager (SASI) data through the use of a multi-scale 3D deep convolutional neural network (M3D-DCNN), a hybrid spectral CNN (HybridSN), and a spectral–spatial unified network (SSUN) and obtained classification accuracy greater than 90% on all datasets. That explains that ultra-hyperspectral satellite imageries can act like laboratory spectroscopic process and detect even finer information from soil as well as of geology. As explained earlier in Section 1.2.1.2, GF-5 satellite's AHSI is the first hyperspectral sensor that concurrently offers broad coverage and a broad spectrum of bands at a small bandwidth interval of only 5 nm to support large spatial environmental monitoring with efficiency.

### 1.3.4 RADAR and SONAR Remote Sensing

The earliest development of radar technology was in 1886 by a German physicist named Heinrich Hertz. A Russian physicist, Alexander Popov in 1985 showed an application of radar technology in

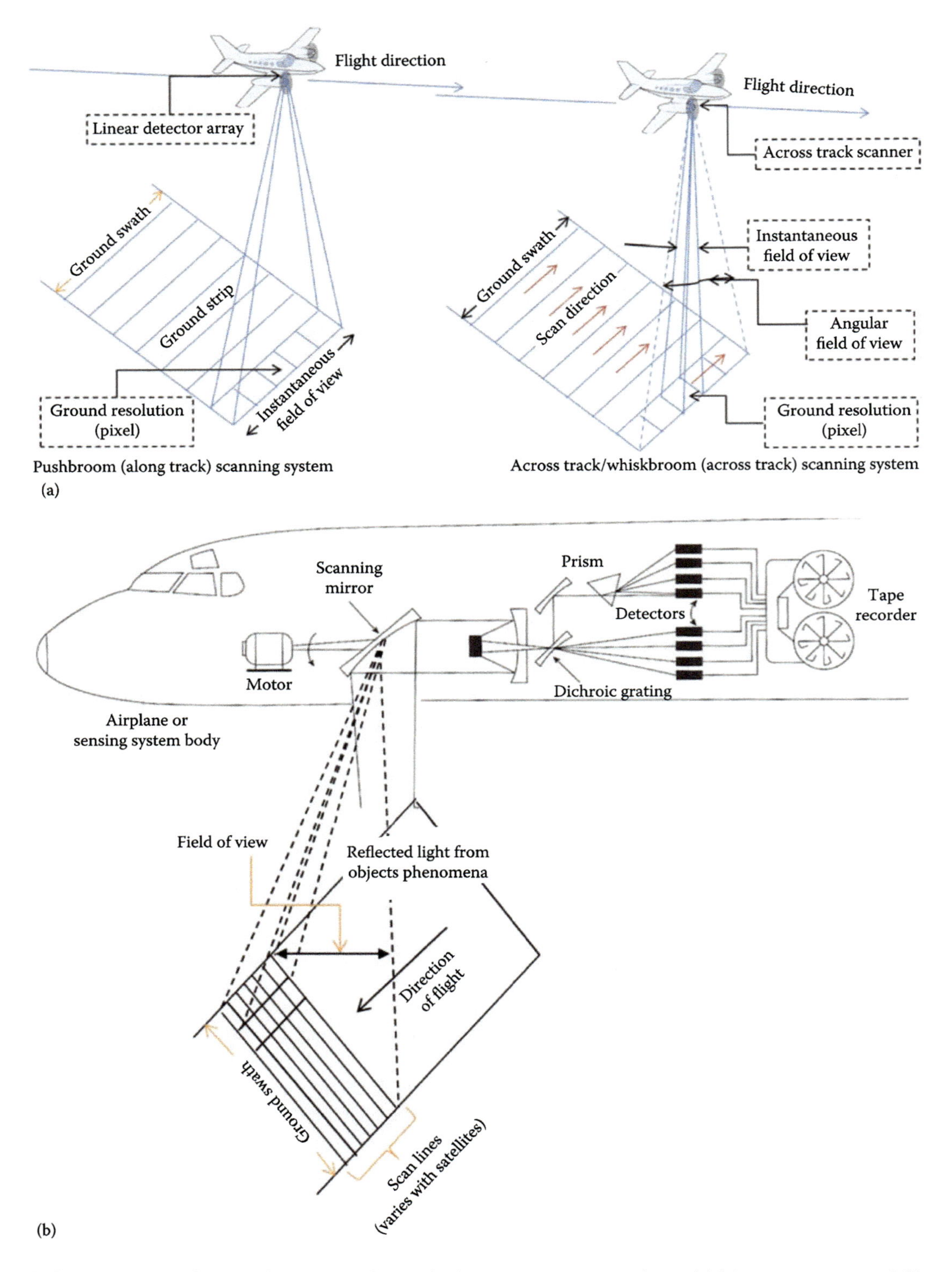


**FIGURE 1.14** Basic operating system of an optical sensor: (a) across-track or whisk broom scanner, and (b) multispectral scanner, onboard an airplane or satellite acquiring images. (Adapted and modified from Avery, T.E. and Berlin, G.L., *Fundamentals of Remote Sensing and Airphoto Interpretation*, Prentice Hall, Upper Saddle River, NJ, 1992.)

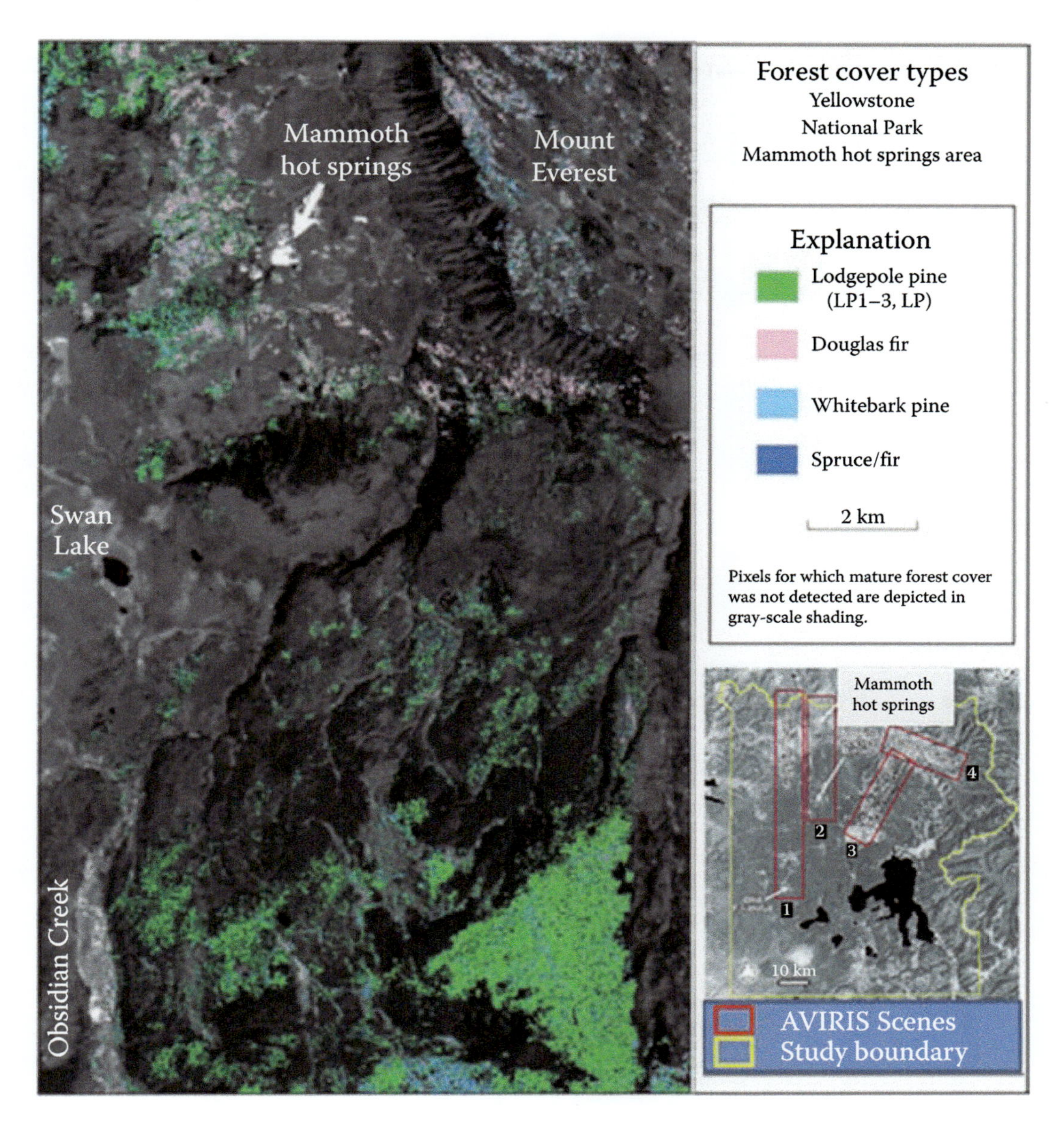


**FIGURE 1.15** AVIRIS image coverage for Yellowstone National Park collected on August 7, 1996, overlaid on Landsat TM imagery. (From http://speclab.cr.usgs.gov/national.parks/Yellowstone/ynppaper.html.)

detecting far lightning strikes. While the early applications of RADAR remote sensing were focused on defense and military applications primarily for reconnaissance surveys of tropical environments, recent applications in the past few decades have seen more diverse applications. Conventional radar such as Doppler radar is mostly associated with weather forecasting, aerial traffic control, and subsequent early warning. Doppler radar is used by local law enforcements' monitoring of speed limits and in enhanced meteorological data collection such as wind speed, direction, precipitation location, and intensity. Interferometric SAR is used to produce precise digital elevation models (DEMs) of large-scale terrain using RADARSAT, TerraSAR-X, Magellan, etc. Refer to Table 1.3 for specifications about RADAR data. In the United States, the National Weather Service (NWS) uses Next-Generation Radar (NEXRAD) satellites to detect atmospheric events such as precipitation and wind

**TABLE 1.6**
**List of Selected Hyperspectral Sensor–Based Satellites along with Detailed Specification**

| Name of Sensor | Description | Mission | Platform (Airborne/ Space borne | | No. of Bands | Spatial Resolution | Revisit Period (Days) | Operation Since | Country/ Agency |
|---|---|---|---|---|---|---|---|---|---|
| | | | Air | Space | | | | | |
| **AVIr IS** | Airborne Visible/Infrared Imaging Spectrometer | JPL Earth Remote Sensing | X | | 224 | 20 m and less | At wish | 1987 | NASA JPL, United States |
| **HSI** | Hyperspectral imager | EnMAP | | X | 244 | 30 m | 4 | 2008 | Germany |
| **ACe -FTS** | Atmospheric Chemistry (Ozone) Experiment Fourier Transform Spectrometer | SCISAT | | X | | n/a | 15 times/ day | 2003 | Canada |
| **CHr IS** | Compact High-Resolution Imaging Spectrometer | PROBA-1 | | X | 153 (5 modes) | 17 m/34 m | | 2002 | ESA, Europe |
| **e PTOMS** | Earth Probe Total Ozone Mapping Spectrometer | Earth Probe | | X | 6 | n/a | n/a | 1996 | NASA, United States |
| **Hyperion** | High-resolution hyperspectral imager with 220 spectral bands (from 0.4 to 2.5 μm) | EO-1 | | X | 220 | 30 | | 2001 | USGS, United States |
| **ILAS** | Improved Limb Atmospheric Spectrometer | ADEOS | | X | 2 | 2,000-13,000 m (IFOV) | | 2002 | JAXA, Japan |
| **ILAS-II** | Improved Limb Atmospheric Spectrometer-II | ADEOS-II | | X | 4 | 1,000 m (IFOV) | | 2002 | JAXA, Japan |
| **ISTOK-1** | Infrared Spectrometer | PRIRODA-MIR | | X | 64 | 750 m | | 1996 | RKA, Russia |
| **MISr** | Multi-angle Imaging Spectro Radiometer | Terra | | X | 4 | 275 m | 9 | 1999 | JPL, United States |
| **MODIS** | Moderate-Resolution Imaging Spectroradiometer (PFM on Terra, FM1 on Aqua) | Aqua, Terra | | X | 36 | 250, 500, 1,000 m | 2 | 1999 | NASA, United States |

| | | | | | | | | |
|---|---|---|---|---|---|---|---|---|
| **MOS** | Modular Optoelectronic Scanning Spectrometer | IRS-P3 | X | 18 | 500 m | 3 (approx.) | 1996 | DLR, Germany |
| **MOS-A** | Modular Optoelectronic Scanning Spectrometer | PRIRODA-MIR | X | 4 | 2.87 km | 3 (approx.) | 1996 | DLR, Germany |
| **MOS-B** | Modular Optoelectronic Scanning Spectrometer | PRIRODA-MIR | X | 13 | 0.7 x 0.65 km | 3 (approx.) | 1996 | DLR, Germany |
| **SCIAMACHY** | Scanning Imaging Absorption Spectrometer for Atmospheric Chartography | ENVISAT | X | 8 | 30 x 27-30 x 240 km (IFOV) | 3 | 2002 | ESA, Europe |
| **SSJ/5** | Precipitating Particle Spectrometer | DMSP-16 | X | | | | 2003 | U.S. Air Force, United States |
| **SSu SI** | Special Sensor Ultraviolet Spectrographic Imager | DMSP-16 | X | | | | 2003 | U.S. Air Force, United States |
| **TANSO-FTS** | Thermal and NIR Sensor for Carbon Observation, Fourier transform spectrometer | Greenhouse Gas Observing Satellite (GOSAT) (Ibuki) | X | 4 | 0.5 km | 3 | 2009 | JAXA, Japan |
| **Te S** | Tropospheric Emission Spectrometer | AURA | X | 12 | 0.5 km | 16 | 2004 | NASA, United States |
| **TOMS** | Total Ozone Mapping Spectrometer (see also SBUV/TOMS) | ADEOS | X | 6 | 47 km x 3.1 km (IFOV) | 41 | 1996-1997 (out of service) | ILRS, NASA |

*Source:* Adapted and modified from ITC, ITC's database of satellites and sensors, 2023, http://www.itc.nl/research/products/sensordb/allsensors.aspx. accessed on Oct 15, 2023.

movement and thus becomes useful in tracking tornados, thunderstorms, and other weather-related hazards. Figure 1.16 is an example of a freely accessible NEXRAD data analysis by NWS to detect probable tornado sighting. NASA's next-generation dual-frequency precipitation radar whose mission was Global Precipitation Measurement (GPM) can accurately measure rain and snow worldwide every three hours. The European Space Agency (ESA) SAR imagery is the most popular microwave radar imagery used for land cover mapping in cloudy, humid, and wet areas similar to coastal regions. Microwaves can penetrate the water vapor in the atmosphere and collect discernable images for analysis. SAR imageries are very useful when conventional multispectral satellite imaging cannot acquire cloud-free images. Figure 1.17 is an example of a SAR image acquired to distinguish the land inundation in China due to the Three Gorges Dam. Table 1.7 shows a list of radar-based satellite information.

In addition to applications in weather and climate analysis, RADAR data are used in various applications such as precision agriculture (Brisco et al., 1989), mapping soil water distribution with ground-penetrating RADAR (Dobson and Ulaby, 1998; Huisman et al., 2003; D'urso and Minacapilli, 2006), forestry application (Leckie and Ranson, 1998), ecosystem studies (Waring et al., 1995; Kasischke et al., 1997), geomorphic and hydrologic applications (Lewis and Henderson, 1998; Glenn and Carr, 2004), snow and ice mapping and analysis (Hall et al., 2012), urban remote sensing (Dong et al., 1997; Chen et al., 2003), and land use and land cover mapping (Haack and Bechdol, 2000).

The multiple look direction (ascending and descending) characteristic of RADARSAT is especially helpful in distinguishing differently oriented linear features such as the traces of fracture and faults (Riopel, 2000). Some of the fracture traces that might be hidden in one look direction due to a shadowing effect become visible from the opposite look direction. Natural corner reflection is another special property of radar. If any lineament and fracture along a small fault is present, the backscatter energy from the corner will be highlighted, revealing faults or other lineaments that would be otherwise hidden (Riopel, 2000).

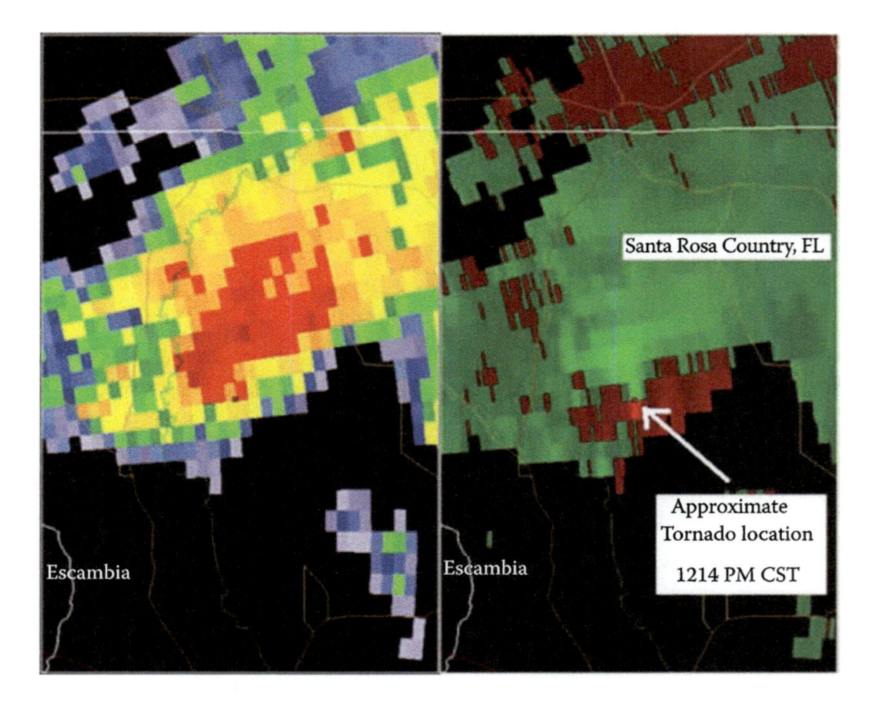


FIGURE 1.16 Sample of a Next-Generation Radar imagery from a tornado event.

FIGURE 1.17 On October 21, 2009, a scene of the Three Gorges Dam, China, was captured using the TerraSAR-X sensor. It is an X-band radar sensor, operating in different modes and recording images with different swath widths, resolutions, and polarizations. (From www.geo-airbusds.com.)

SMAP (Soil Moisture Active Passive), a new-generation Earth science satellite sent recently to space by NASA-JPL, is a RADAR and radiometer instrumentation system that has a very ambitious mission of improving weather forecasts; monitoring droughts; predicting floods; assisting in crop productivity; and providing details in water, energy, and carbon cycles. It is part of NASA's Earth Campaign. Its two instruments map the soil moisture (radiometer and RADAR at 9 km spatial resolution) and determine the freeze and thaw state (radiometer at 3 and 36 km spatial resolution) of the same area (https://smap.jpl.nasa.gov/observatory/overview/). SMAP satellite data is found to be very useful in many climate change-induced geohazard mitigation proactive management decision support systems development such as flood potential mapping (Perry et al., 2022), wildfire susceptibility determination (Perry & Panda, 2022), permafrost melting and impact analysis on increased evapotranspiration (Panda et al., 2024b), forest evapotranspiration estimation for forest growth analyses (Panda et al., 2024a), drought monitoring (Shrum et al., 2018), and many more. Like the LiDAR data, 2D active remote sensing data shows promise in Earth environmental management.

### 1.3.5 LASER and RADAR Altimetry Imaging

LASER and RADAR altimeters on satellites have provided a wide range of data on detecting the bulges of water caused by gravity in the ocean. It helps map features on the seafloor to a resolution of approximately 1 mile. The Light Amplification by Stimulated Emission of Radiation (LASER) and RADAR altimeters measure the height and wavelength of ocean waves, wind speeds, wind direction, surface ocean currents and their directions, etc. (Brenner et al., 2007; Giles et al., 2007; Connor et al., 2009). Connor et al. (2009) used satellite microwave altimeters boarded on ENVISAT/

**TABLE 1.7**
**List of Selected RADAR Sensor–Based Satellites along with Detailed Specification**

| Name of Sensor | Description | Mission | Platform (Airborne/Spaceborne) | | No. of Bands | Spatial Resolution/IFOV | Revisit Period (Days) | Operation Since | Agency/Country |
|---|---|---|---|---|---|---|---|---|---|
| | | | Air | Space | | | | | |
| **ALMAZ-1** | SAR | ALMAZ-1 | | X | 1 | 15 m | | 1991–1992 a Out of Service | NPO Mashinostroyenia, Russia |
| **Active Phased Array** | SAR-X band | TanDEM-X | | X | 1 (SAR-X) | | 11 | 2010 | DLR, Germany |
| **ALT** | Dual-Frequency Radar Altimeter | TOPEX/Poseidon | | X | 2 | 6 km | 10 | 1992–2005 b Proposed | NASA, United States |
| **ASAr** | Advanced SAR | ENVISAT | | X | 1 (SAR-C) | 30 (150) m | 35 | 2002–2012 (out of service) | ESA, Europe |
| **COSI** | Corea SAR Instrument | KOMPSAT-5 (Arirang 5) | | X | 1 | 1, 3, 20 m | 28 | 2013 | KARI, South Korea |
| **CPr** | Cloud Profiling RADAR | EarthCARE | | X | 1 | | 25 | 2015 (proposed) | ESA, Europe |
| **DPr** | Dual-Frequency Precipitation Radar | GPM Core | | X | 2 | 5 km | | 2014 | NASA, United States |
| **g MI** | Microwave Radar Instrument | GPM Core | | X | 13 | 6–26 km (IFOV) | | 2014 | NASA, United States |
| **HQSAr** | High-Quality SAR | TerraSAR-X | | X | 1 SAR-X) | 1 m | 11 | 2007 | DLR, Germany |
| **LBI** | L-band SAR | SAOCOM-1A | | X | 1 (SAR-L) | 10 m | 16 | 2015 (proposed) | CONAE, Argentina |
| **MMSAr** | Multimode SAR | TecSAR | | X | 1 SAR-X) | 0.1 m | | 2008 | IAI, Israel |
| **Mr S, Cr S** | Scan SAR mode | RISAT-1 | | X | 1 (SAR-C) | 25 m | 25 | 2012 | ISRO, India |
| **PALSAr** | L-band SAR | ALOS | | X | 1 (SAR-L) | 10 m | 46 | 2006–2011 (out of service) | JAXA, Japan |
| **PALSAr -2** | L-band SAR-2 | ALOS-2 | | X | 1 (SAR-L) | 3 m | 14 | 2014 | JAXA, Japan |
| **Pr** | Precipitation radar | TRMM | | X | 2 | 25 km | 0.5 | 1997 | NASA, United States |

| | | | | | | | | |
|---|---|---|---|---|---|---|---|---|
| **r ADAr SAT-2** | Radar | RADARSAT-2 | X | 1 (SAR-C) | 3–100 m | 24 | 2007 | MDA Geospatial Services Inc. |
| **SAr (Jer S-1)** | SAR | Japanese Earth Resource satellite (JERS-1) | X | 1 (SAR-L) | 18 m | 44 | 1992–1998 (out of service) | JAXA, Japan |
| **SAr (r ADAr SAT-1)** | SAR | RADARSAT-1 | X | 1 | 8.4–100 m | 24 | 1995–2007 (out of service) | MDA Geospatial Services Inc. |
| **SAr (Seasat)** | SAR | Seasat | X | 1 | 25 m | | 1978–1978 (out of service) | NASA, United States |
| **SAr (Sentinel-1A)** | C-band SAR on Sentinel-1A | Sentinel-1A | X | 1 (SAR-C) | 5 m | 12 | 2014 | ESA, Europe |
| **SAr 2000 SAr -L** | RADAR L-band SAR | COSMO-SkyMed<br>MapSAR | X<br>X | 1 | 1 m<br>3 m | 16 | 2007<br>2011 | ASI, Italy INPE, Brazil |
| **SIr AL** | Interferometric Radar Altimeter | CryoSat-2 | X | | 250 m | 369 with 30 days subcycle | 2010 | ESA, Europe |
| **Sr AL** | SAR Radar Altimeter | Sentinel-2 | X | 13 | 10, 20, 60 m | 5 | 2007 (proposed) | ESA, Europe |
| **SMAP** | SMAP Radar SMAP Radiometer | SMAP | X | 3 | 3, 9, 36 km | 2–3 days | 2015 | NASA, United States |
| **Sr AL** | SAR Radar Altimeter | Sentinel-3 | X | | 0.5–1 km | 27 | 2017 (proposed) | ESA, Europe |
| **X-SAr** | SAR | SRTM | X | 1 | 25-90 m | | 2000–2000 (out of service) | NASA, United States |

*Source:* Adapted and modified from ITC, ITC's database of satellites and sensors, 2023, http://www.itc.nl/research/products/sensordb/allsensors.aspx. accessed on Oct 15, 2023.

RA-2 to measure the Arctic Sea ice depth and height accurately. ESA-operated CryoSat-2 interferometric radar altimeter acquires accurate measurements of the thickness of floating sea ice for annual variations and surveys the surface of ice sheets accurately enough to detect small changes. This supports the study of global warming and climate change effect on present-day global land cover changes and especially diminishing snow and ice cover. Figure 1.18 shows a graphical depiction of how satellite altimetry works with Jason-2 Ocean Surface Topography Mission (OSTM). Jason-2 OSTM measures sea surface height through imaging that ultimately determines ocean circulation, climate change, and sea-level rise. Table 1.8 contains a selected list of LASER and RADAR altimetry imaging sensor platform with their ancillary information.

### 1.3.6 LiDAR Remote Sensing

LiDAR technology, a type of active remote sensing, was developed in the early 1960s following the invention of LASER and was initially used to measure distance by illuminating a target with LASER. The first LiDAR application was in the field of meteorology; the National Center for Atmospheric Research (NCAR) used LiDAR to measure the cloud distance from ground surface (Goyer and Watson, 1963). The 1971 use of LiDAR technology by the Apollo mission to map the surface of the Moon made the technology well known among the general public. Soon LiDAR became a very common tool for geospatial technology researchers and users in the present century when LiDAR was first used by airplanes to map the land topography.

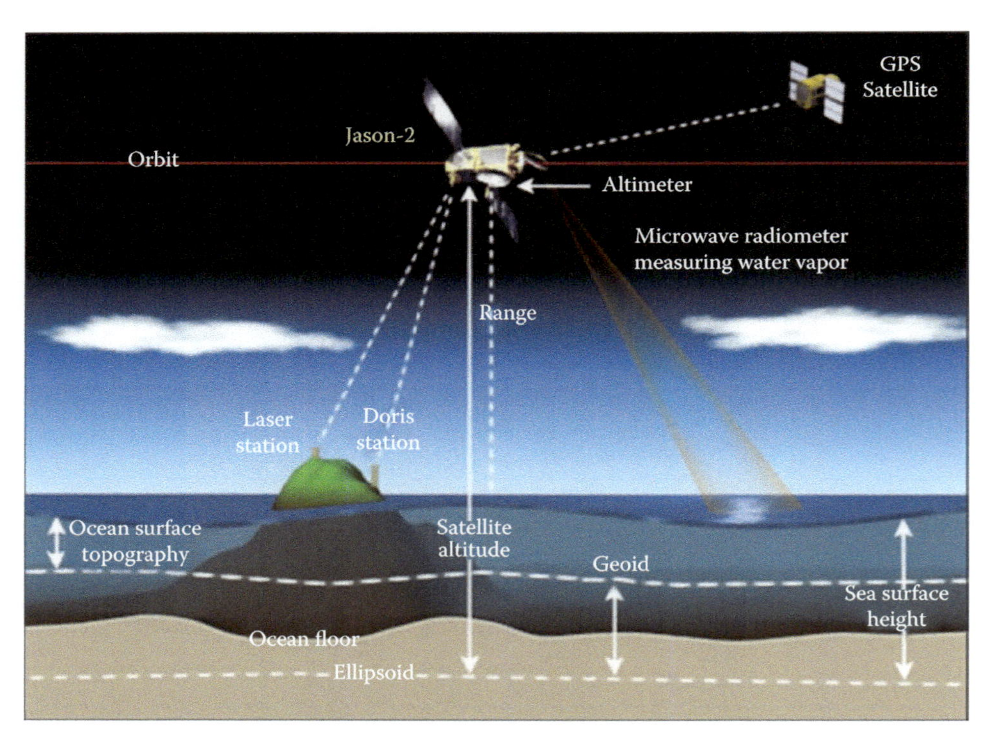


FIGURE 1.18 Graphical depiction of how Jason-2 satellite altimetry works to monitor the ocean. (From www.ppi.noaa.gov/bom_chapter3_ satellite_management/National Oceanic and Atmospheric Administration—Office of Program Planning and Integration. 2015. Chapter 3—NOAA Operations—Satellite Management. Retrieved from www.ppi.noaa.gov/bom_chapter3_satellite_management/ on 3/30/15.) Please read the foreword to gather the importance of radar altimetry in recording climate impacts such as sea-level rise.

**TABLE 1.8**
**List of Selected LASER and RADAR Altimetry Sensor–Based Satellites along with Detailed Specification (Airborne/Spaceborne)**

| Name of Sensor | Description | Mission | Air Space | No. of Bands | Spatial Resolution | Revisit Period (Days) | Operation Since | Agency/ Country |
|---|---|---|---|---|---|---|---|---|
| **ALT** | Dual-Frequency Radar Altimeter | TOPEX/ Poseidon | X | 2 | 6 km | 10 | 1992–2005 (out of service) | NASA, United States |
| **ALT (Seasat)** | Altimeter | Seasat | X | 1 (SAR-K) | 1.6 km | | 1978–1978 (out of service) | NASA, United States |
| **AltiKa** | High-resolution altimeter including bifrequency radiometric function | SARAL/ AltiKa | X | 2 | 2 km | 35 | 2013 | CNES, France |
| **g LAS** | GLAS | ICESat | X | 2 | 100 m–1 km | Monthly/ seasonal | 2003–2010 (out of service) | NASA, United States |
| **r A** | Radio altimeter | ERS-1 | X | 2 | 16 km | 35 | 1991–2000 (out of service) | ESA, Europe |
| **r A** | Radio altimeter | ERS-2 | X | 2 | 16 km | 35 | 1995–2011 (out of service) | ESA, Europe |
| **SIr AL** | SAR Interferometric Radar Altimeter | CryoSat-2 | X | 1 (Radar altimeter) | | 369 with 30 day subcycle | 2010 | ESA, Europe |
| **Sr AL** | SAR Radar Altimeter | Sentinel-3 | X | 1 (Radar altimeter) | 0.5–1 km | 27 | 2017 (proposed) | ESA, Europe |

*Source:* Adapted and modified from ITC, ITC's database of satellites and sensors, 2023, www.itc.nl/research/products/sensordb/allsensors.aspx, accessed on October 15, 2023.

Satellite-based LiDAR data acquisition is spearheaded by NASA, ESA, Russian Federal Space Agency (RKA), and German Aerospace Research Establishment (DLR) with sensors such as laser reflectometry array (LRA) on Jason-2 (OSTM) and Geoscience Laser Altimeter System (GLAS) for Ice, Cloud, and land Elevation Satellite (ICESat) mission, atmospheric laser Doppler instrument (ALADIN) and ATLID EarthCARE, Priroda-Mir mission, and laser communication terminal (LCT) for TanDEM-X mission, respectively (Table 1.9). Table 1.9 also provides the satellite specifications on selected LiDAR remote sensing satellites.

Present-day aerial LiDAR data are collected from airplanes flying around 2–5 miles above the ground surface and provide a minimum of 30 cm resolution data. The data are collected by focusing and scanning a LASER beam from an airplane to the objects on the ground surface, which then acquires the data as different returns (Panda et al., 2012a, 2012b; Amatya et al., 2013; Panda and Hoogenboom, 2013). Table 1.9 includes a list of LiDAR-based satellite information. LiDAR technology is becoming popular since the start of the millennium due to its advantage in mapping the Earth topography along with object heights on the Earth's surface, thus supporting image classification process tremendously (Dubayah et al., 2000). The NOAA mission of collecting LiDAR data

**TABLE 1.9**

**Selected LiDAR Sensor–Based Satellites along with Detailed Specification**

| Name of Sensor | Description | Mission | Platform (Spaceborne) | No. of Bands | Spatial Resolution | Revisit Period (Days) | Operation Since | Agency/Country |
|---|---|---|---|---|---|---|---|---|
| **ALADIN** | ALADIN | ADM-Aeolus | X | | | 7 | 2015 | ESA, Europe |
| **ALISSA** | L'atmosphere par Lidar Sur Saliout | PRIRODA-MIR | X | 1 | 150 m | | 1996 | RKA, Russia |
| **ATLID** | High-spectral-resolution LiDAR | EarthCARE | X | 2 (S and X) | | 25 | 2016 | ESA, Europe, and JAXA, Japan |
| **CryoSat-2** | Band Laser Reflector, RADAR altimeter | ESA Polar sea ice monitoring | X | | | | | ESA, Europe |
| **GEDI** | Global Ecosystem Dynamics Investigation | Earth forest biomas, carbon cycle, and biodiversity monitoring | X On Int. Space Station | 1 | 25m | 1 | 2018 | NASA, United States |
| **g LAS** | GLAS | ICESat | X | 2 | 70 m and 3 cm vertical | 91 | 2003–2010 (out of service) | NASA, United States |
| **IPDA LiDAR** | Integrated Path Differential Absorption LiDAR | | X | 2 | | 28 | 2028 (planned) | DLR-Germany |
| **LCT** | LCT | TanDEM-X | X | 11 | | | 2010 | DLR, GFZ, Infoterra, Germany |
| **Lr A (Jason)** | Laser Retroreflector Array | Jason-2 (OSTM) | X | 10 | | | 2008 | NASA, United States |

*Source:* Adapted and modified from ITC, ITC's database of satellites and sensors, 2023, www.itc.nl/research/products/sensordb/allsensors.aspx, accessed on October 15, 2023.

of the entire U.S. coast and some ecologically sensitive areas in the United States made LiDAR a well-known technology. NOAA Coastal Services Center's Digital Coast Data Access Viewer (www.csc.noaa.gov/dataviewer/#) provides free access to LiDAR data, including some recent acquisitions of 2009 and later, for a major portion of the United States.

LiDAR remote sensing instrument provides point cloud data. The crude point cloud data are processed and each laser shot is converted to a position in a 3D frame of reference with spatially coherent cloud of points. In this processing stage, some LiDAR data provide texture or color information for each point Renslow, 2012; Rao, 2013). The processed 3D spatial and spectral information contained in the dataset allows great flexibility to perform manipulations to extract the required information from the point cloud data Renslow, 2012; Rao, 2013). Thereafter, visualization, segmentation, classification, filtering, transformations, gridding, and mathematical operations are conducted on the data to obtain required information of Earth objects or phenomena as discussed. The first return of the LiDAR data is generally from the tallest features, that is, tallest tree canopy or top of high-rise buildings, the intermediate returns are from the canopy of the small trees and shrubs, and the final return is from the ground surface. These individual return data are processed to get height information of the features discussed earlier. Figure 1.19 displays the point cloud LiDAR data

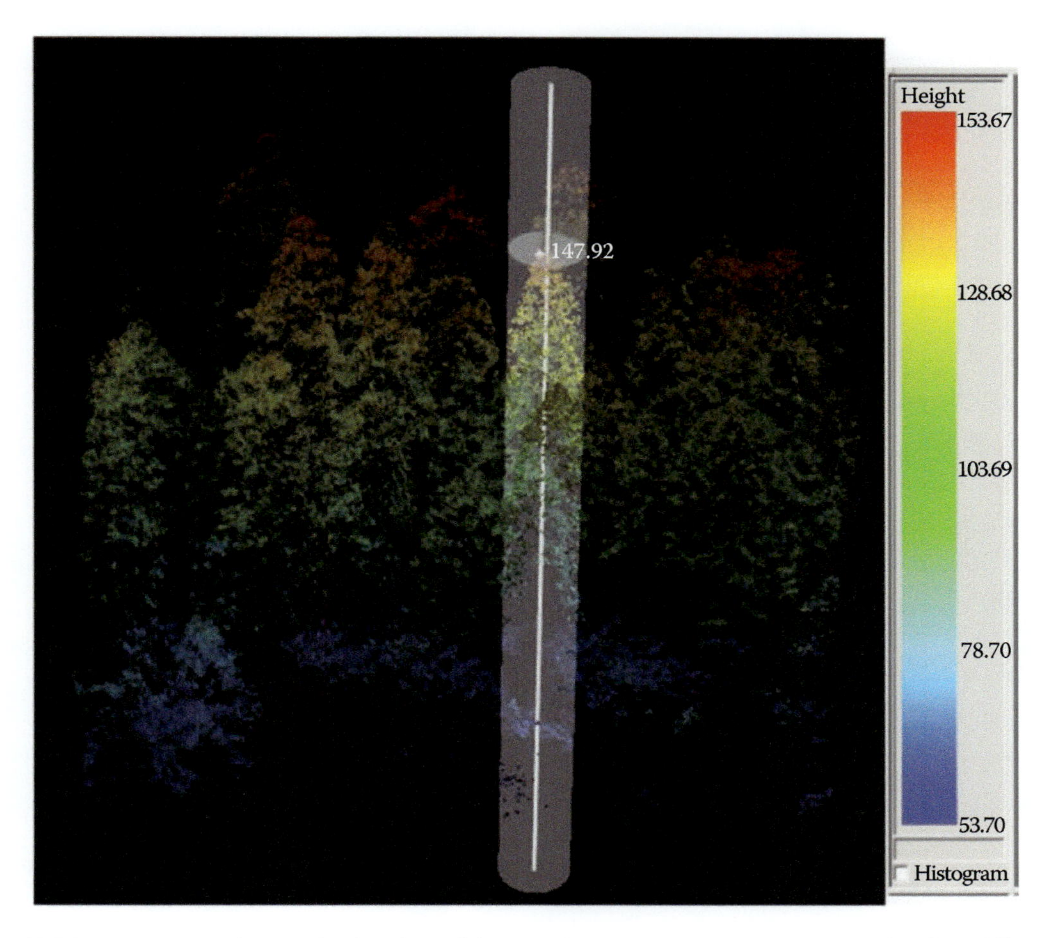


**FIGURE 1.19** A 3D display of point cloud light detection and ranging (LiDAR) data showing different returns (colorized based on height as visualized using the LiDAR Data Viewer of the USFS Fusion software). Also note the interactive measuring cylinder with marker can be moved about the virtual space for measuring height.

showing different returns colorized based on height as visualized using LiDAR Data Viewer of the USGS Fusion software (http://forsys.cfr.washington.edu/fusion/fusionlatest.html). Using interpolation and smoothing algorithms, the point cloud is rendered as a grid surface, which can be easily manipulated in a GIS using map algebra operations to produce canopy height, ground elevations, etc. The normalized digital surface model (nDSM) is developed in the LiDAR processing software Quick Terrain Modeler (Applied Imagery, Chevy Chase, MD) from the ground return data to produce the accurate DEM (topography) data of the ground surface (Panda et al., 2012a, 2012b). Panda et al. (2012a, 2012b) with their wetland change and cause recognition study of Georgia, United States, coast found a major discrepancy with the 10-m DEM data available through USGS's National Elevation Dataset (NED). The problem was found while comparing with the DEM created with 30-cm resolution LiDAR for Camden County, GA. They found an average change of ground elevation of 0.9 m with a range of −0.17 to 1.5 m when comparing the elevation at permanent locations. Figure 1.20 shows the DEM developed for Camden County, GA, using the 2010 LiDAR data,

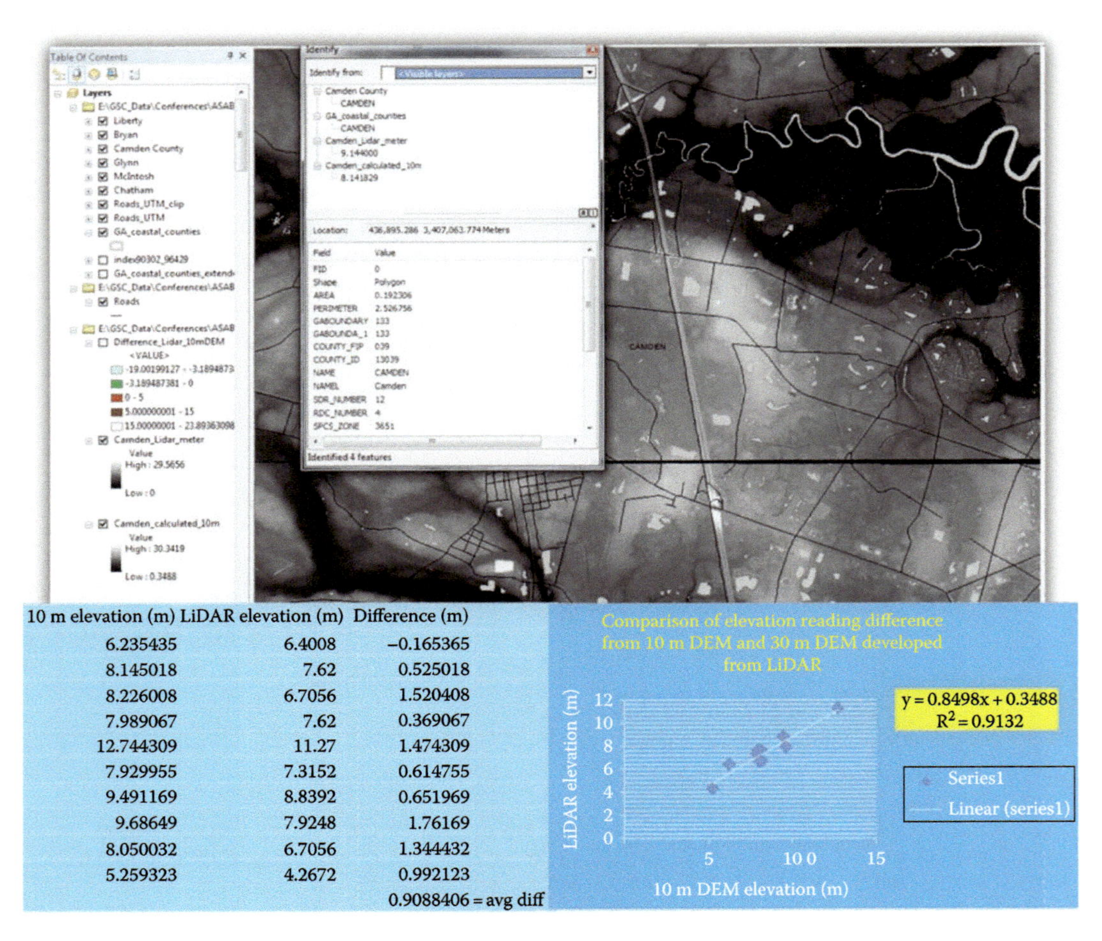

| 10 m elevation (m) | LiDAR elevation (m) | Difference (m) |
|---|---|---|
| 6.235435 | 6.4008 | −0.165365 |
| 8.145018 | 7.62 | 0.525018 |
| 8.226008 | 6.7056 | 1.520408 |
| 7.989067 | 7.62 | 0.369067 |
| 12.744309 | 11.27 | 1.474309 |
| 7.929955 | 7.3152 | 0.614755 |
| 9.491169 | 8.8392 | 0.651969 |
| 9.68649 | 7.9248 | 1.76169 |
| 8.050032 | 6.7056 | 1.344432 |
| 5.259323 | 4.2672 | 0.992123 |
| | | 0.9088406 = avg diff |



**FIGURE 1.20** Comparison of 10 m digital elevation model (DEM) of National Elevation Dataset and DEM developed with airborne LiDAR of Camden County, GA. (From Panda, S.S. et al., Wetland change and cause recognition in Georgia coastal plain, in: Presented in the *International American Society of Agricultural and Biological Engineers (ASABE) Conference 2012*, July 29–August 2, 2012, Dallas, TX, Paper #1338205, 2012a; Panda, S.S. et al., Stomatal conductance and leaf area index estimation using remotely sensed information and forest speciation, in: Presented in the *Third International Conference on Forests and Water in Changing Environment 2012*, September 18–20, 2012, Fukuoka, Japan, 2012b).

the 10 m DEM of the county created in 2000 (based on the raster metadata), the comparison table of elevation differences, and the algorithm developed to modify the 10 m DEM of the study area.

Panda and Hoogenboom (2013) used 30 cm LiDAR from a blueberry orchard in Pierce County, GA, to obtain the height of pine trees, existing structures, and blueberry plants with ground elevation to be used in the object-based image analysis (OBIA)—based image segmentation with eCognition software for blueberry orchard SSCM. Yu et al. (2010) in their study of urban building density determination using airborne LiDAR data explained the development of nDSM and subsequent DEM from the LiDAR data for decision support. Kelly et al. (2014) completed a study of coastal marsh management decision support by using 30 cm LiDAR tiles. The DEM of South Carolina coast was developed to delineate the areas of saltwater intrusion (with the highest possible tide heights) calculated with the latest climate change analysis. Additionally, the researchers used nDSM from LiDAR imagery to classify the NAIP imagery using eCongnition's OBIA image segmentation procedure to produce an output of the coastal forest plant species. These applications show the potential of LiDAR in environmental (forestry, ecology, agriculture, urban, and others) management decision support systems.

Similarly, Rao (2012) demonstrated applications of LiDAR-based mapping and height estimates for snags and coarse woody debris using threshold filtering of the point cloud data. Furthermore, LiDAR data, owing to its properties to penetrate canopy, have high application in improving wildland fire mapping through better estimates of canopy bulk density and base height (Riano et al., 2004; Lee and Lucas, 2007; Erdody and Moskal, 2010), canopy cover (Lefsky et al., 2002; Hall et al., 2005), shrub height (Riano et al., 2007), and aboveground biomass (Edson and Wing, 2011; Vuong and Rao, 2013) and carbon stocks (Asner et al., 2010; Goetz and Dubayah, 2011). Other studies have documented better mapping and classification outputs using an integrated approach where multispectral data are fused with LiDAR data (Garcia et al., 2011; Rao and Miller, 2012). Innovative approaches to map natural resources at global scales have been initiated in the recent past such as using the GLAS (see Figure 1.21) aboard the ICESat in conjunction with MODIS percent tree cover

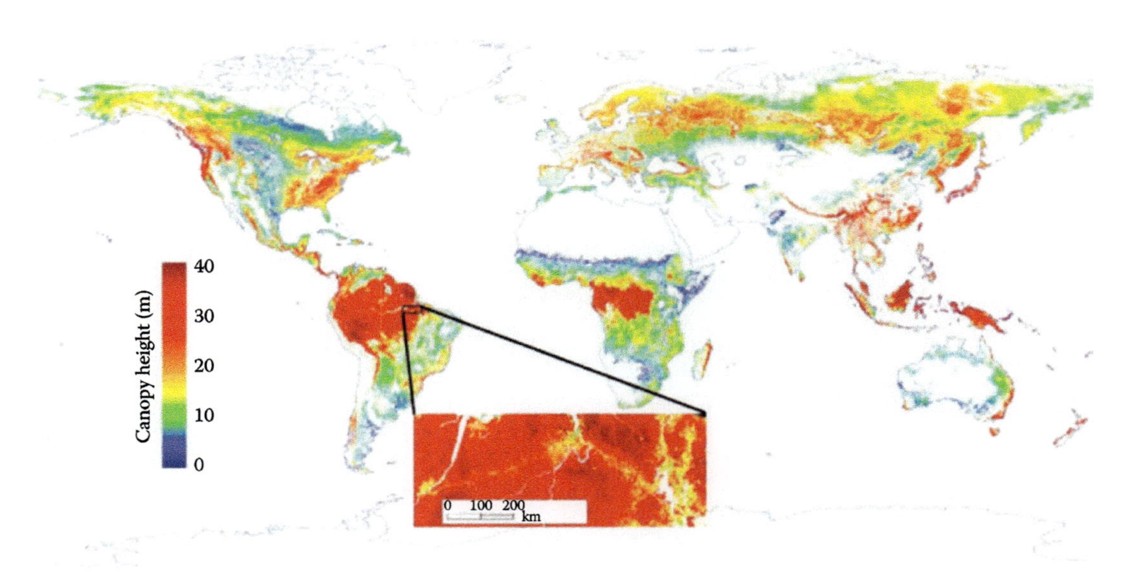


**FIGURE 1.21** Wall-to-wall mapping of canopy height implemented using the Geoscience Laser Altimeter System (GLAS) instrument on the Ice, Cloud, and land Elevation Satellite. GLAS data model. The inset shows the sensitivity and capability of the model to disturbance gradient in the Amazon. The GLAS laser transmits short pulses (4 ns) of infrared light at 1064 nm and visible green light at 532 nm 40 times/s. (From Simard, M. et al., *J. Geophys. Res.*, 116, G04021, 2011, doi: 10.1029/2011JG001708.)

product (MOD44B) to map forest canopy height globally with spaceborne LiDAR (Simard et al., 2011). LiDAR is also used to detect and measure the concentration of various chemicals in the atmosphere including the use of camera LiDAR instrument to accurately measure the particulates (Barnes et al., 2003). These and many other studies show the potential of LiDAR data in environmental (forestry, ecology, agriculture, urban, and others) management and decision support systems, particularly in context of climate change issues at regional and global levels.

Currently, LiDAR is also extensively used by law enforcement for tracking of vehicle speed military for weapon ranging, mine detection for countermine warfare, and laser-illuminated homing of projectiles; in mining industry to calculate the ore volume (3Dlasermapping, 2014); in physics and astronomy by measuring the distance to the reflectors placed on the Moon and detecting snow on the Mars atmosphere (NASA-Phoenix Mars Lander, 2014); in the field of robotics for the perception of the environment as well as object classification and safe landing of robots and manned vehicles (Amzajerdian et al., 2011); in spaceflight, surveying using mobile LiDAR instruments (Figure 1.22a); in adaptive cruise control; and in many other high-end energy-producing sources like wind farms and solar farms for wind velocity and turbulence measurement and optimizing solar photovoltaic systems by determining shading losses.

Terrestrial LiDAR application in surveying land topography, stream channel profile development, and even city infrastructure mapping has gained a lot of ground. The instrument is either mounted on a vehicle (truck) or used on a tripod to monitor the 360° 3D view by mapping it continuously. Bateman et al. (2023) used the tripod-mounted terrestrial LiDAR equipment (Figure 1.22b) to map the stream channel profile of Tumbling Creek in the University of Georgia, Gainesville campus, on a temporal basis. Between three surveys using the 360° 3D view-based mapping they determine the potential area of aggradation, as shown in Figure 1.23. Figure 1.23a and b show the changes observed between survey 1 and 2 and between survey 2 and 3, respectively. It is easy to carry and set up to perform topographic mapping at ease.

Global Ecosystem Dynamics Investigation (GEDI) LiDAR is the latest mission by NASA in collaboration with the University of Maryland to construct three-dimensional maps of forest canopy with its height and branch and leaf distribution patterns. The three lasers of the instrument are installed aboard the International Space Station. This LiDAR mission is an excellent step toward continuously monitoring Earth's forest biomass, the carbon cycle, plant and animal's habitat and biodiversity, and their changes. GEDI collects data between 51.6° N and 51.6° N latitudes of the Earth missing the polar regions. GEDI products are distributed through NASA's Land Processes

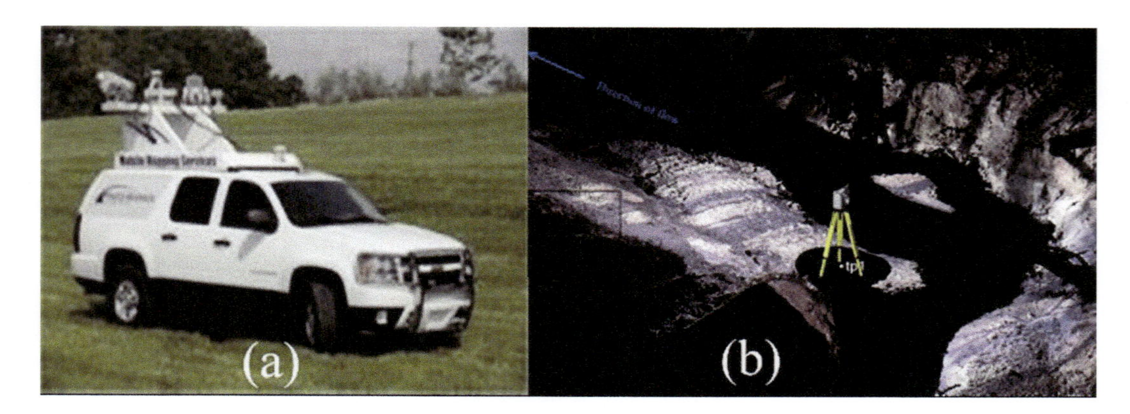


FIGURE 1.22 Terrestrial mobile light detection and ranging for land surveying: (a) vehicle mounted (from Photo Science Geospatial Solutions, Lexington, KY, www.photoscience.com/services/mobile-mapping) and (b) tripod mounted and being used in stream profile development for change monitoring. (Source: Bateman et al., 2023.)

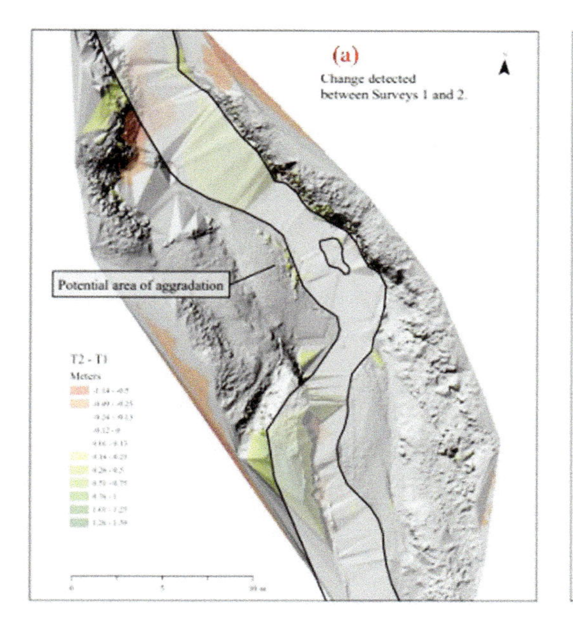


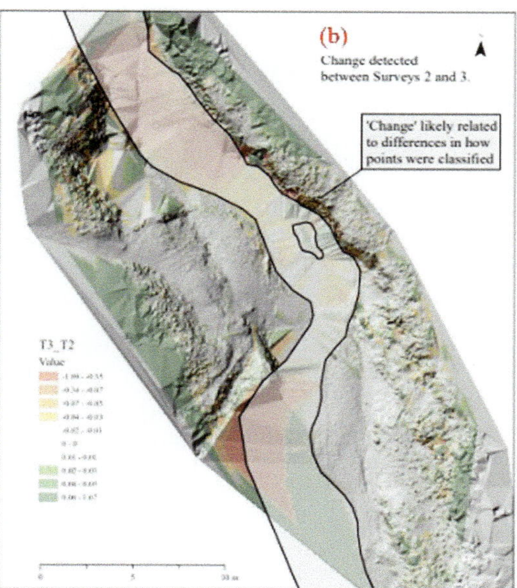


**FIGURE 1.23** Stream channel profile and bank changes analyses ((a) changes detected between surveys 1 and 2 and (b) changes detected between surveys 2 and 3) using tripod-mounted terrestrial LiDAR equipment (Figure 1.22b). (Source: adapted and modified from Bateman et al., 2023.)

Distributed Archive Center (LP DAAC). GEDI products are of ~30 m spatial resolution and are produced nominally daily temporal resolution (https://lpdaac.usgs.gov/data/get-started-data/collection-overview/missions/gedi-overview/). This GEDI mission will revolutionize how we study our Earth's ecosystem dynamics with the lower level (L1 & L2) data available to download at LP DAAC (https://lpdaac.usgs.gov/product_search/) and higher level (L3 & L4) data available at ORNL DAAC (https://daac.ornl.gov/cgi-bin/dsviewer.pl?ds_id=2299) locations, respectively. Figure 1.24 provides a glimpse of the efficacy of the GEDI LiDAR data acquired from space to monitor our ecosystem, explaining the average biomass (Mg/ha) yield amount averaged for 1 km grid. It is understandable how much efficient a biodiversity or simply a forest manager would be once they had easy access to the data on a daily interval. Global Ecosystems management is getting better with GEDI LiDAR.

### 1.3.7 Microwave Remote Sensing

Unlike LiDAR remote sensing, microwave sensing encompasses both active and passive forms of remote sensing. Atmospheric scattering affects shorter optical wavelengths but longer wavelengths of EMS are not affected by atmospheric scattering (Aggarwal, 2003). Therefore, longer-wavelength microwave radiation can penetrate through cloud cover, haze, dust, and all but the heaviest rainfall, and thus microwave remote sensing can be useful for geospatial community as data can be collected with microwave remote sensing systems in almost all weather and environmental conditions. RADAR remote sensing is the example of active microwave remote sensing. The radar units transmit short pulses or bursts of microwave radiation when sunlight is unavailable and then obtain reflectance signal from the Earth-based objects (Avery and Berlin, 1992). Therefore, microwave remote sensing system can be operable in the night. When the microwave remote sensors collect naturally emitted microwave energy, though small, from the objects without sending pulse from the system, it is passive microwave sensing, which is similar in concept to thermal remote sensing (Avery and Berlin, 1992). The naturally emitted microwave energy is related to the temperature

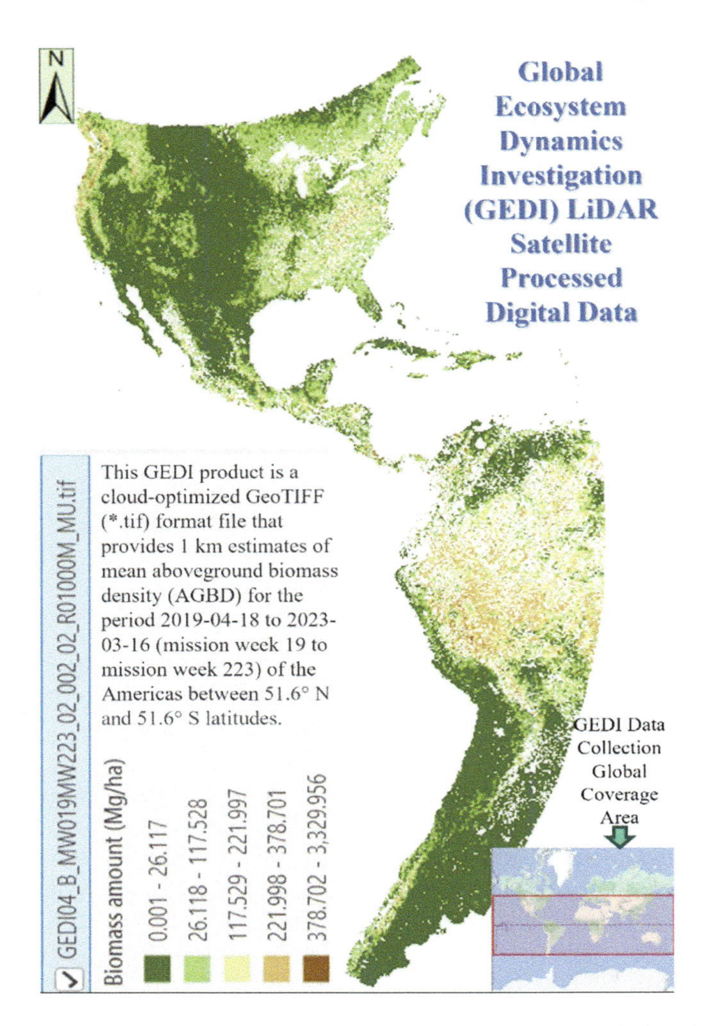


**FIGURE 1.24** GEDI LiDAR data processed to show the estimated biomass variations in the United States and parts of Central and South America. Level 4 data downloaded from LP DAAC (https://lpdaac.usgs.gov/product_search/) and processed in ArcGIS Pro 3.2.2 (ESRI™).

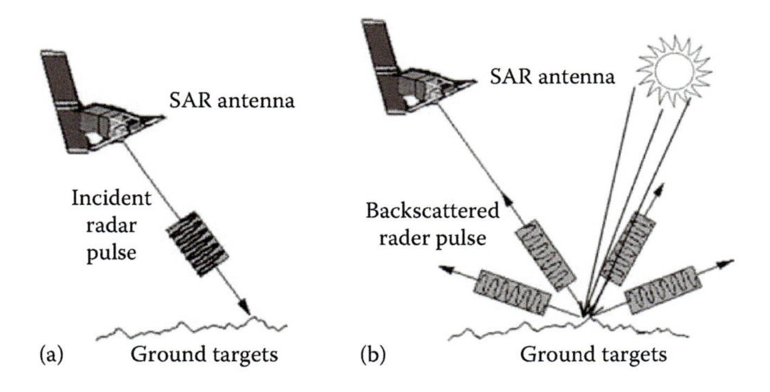


**FIGURE 1.25** (a) Active microwave remote sensing: microwave pulses are incident upon objects from the synthetic aperture radar (SAR) satellite in no-sunlight condition. (b) Passive microwave remote sensing: microwave radiations are reflected from the objects on groundwater receiving light energy from the Sun.

and moisture properties of the emitting object or surface. Figure 1.25 shows the image acquisition principle of both active and passive remote sensing systems.

Microwave remote sensors efficiently monitor targets such as soil moisture content (Engman and Chauhan, 1995; Wagner et al., 2007), salinity (Wilson et al., 2001; Metternicht and Zinck, 2003), temperature of the estuary and sea surface temperature (SST) (Blume et al., 1978; Klemas, 2011), and rain, snow, ice, and sea surface condition (Fung and Chen, 2010), water vapor, cloud, oil slick, and different gaseous presence in the atmosphere like $CO_2$, ozone, and $NO_x$. Microwave scatterometer, microwave altimeter, and imaging radar are the examples of active microwave sensing, while microwave, radiometers, and scanners are all examples of passive microwave sensing technology. Table 1.10 shows the typical microwave sensor–based satellites and their specifications. Table 1.11 summarizes passive and active sensing–based microwave sensors and their target areas of applications for Earth resources measurement.

### 1.3.8 SONAR and SODAR Remote Sensing

SONAR is a technique that uses sound propagation to navigate and communicate with or detect objects on or under the surface of the water (Figure 1.26). Two types of technology share the name *sonar: passive* sonar is essentially listening for the sound made by vessels; *active* sonar is emitting pulses of sound energy and listening for the echoes. Blackinton et al. (1983) used SONAR technology to map the seafloor with the development of SeaMARC II. Kidd et al. (1985) used long-range side-scan SONAR to map sediment distributions over wide expansions of the ocean floor. Fairfield and Wettergreen (2008) recently used multibeam SONAR to map the ocean floor. As sound can propagate through water, the sensing system uses it to know the objects underneath water. Sonar may also be used in air for robot navigation. Skarda et al. (2011) used SONAR technology—based fish-finder instrument to map the bottom of two retention ponds to study the advantages of retention pond in soil conservation. Similarly, Dual-Frequency Identification Sonar (DIDSON) uses sound to produce images using high- and low-frequency modes in the range of about 15–40 m and have demonstrated benefit in fish inventorying (Belcher et al., 2001; Moursund et al., 2003; Tiffan et al., 2004).

SODAR (sonic detection and ranging) is upward-looking, in-air sonar used for atmospheric investigations. SODAR remote sensing systems are like a LiDAR/RADAR system in which sound waves are propagated instead of light or radio waves in order to detect the object of interest (Bailey, 2000). Doppler SODAR (Figure 1.27) is a widely used remote sensing system for weather forecasting (Goel and Srivastava, 1990; Beyrich, 1997). Wind farms are using fulcrum compact-beam multiple-axis 3D SODAR to increase efficiency of wind power production. Satellite platform–based SONAR systems are operated by the NOAA, NASA, U.S. Department of Defense, ESA, and Space Agency of France (Centre National d'Etudes Spatiales [CNES]). Table 1.12 provides selected list of these SONAR satellite systems with their specifications. These SONAR systems are mostly used for ocean phenomena study, atmospheric measurement, and ice and snow studies.

### 1.3.9 Global Positioning System

The GPS, now known as Global Navigation Satellite System (GNSS), is a space-based satellite navigation system that provides location (coordinates and elevation) and time information in all weather conditions, anywhere on or near the Earth, where there is an unobstructed line of sight to four or more GPS satellites (Van Diggelen, 2009). GPS satellites use the trilateration technique to know the location of a receiver by coordinating signals received from four satellites. It is an application of active remote sensing system as unique signals (pseudorandom code) in longwaves, L1 = 1575.42 MHz (19 cm wavelength) and L2 = 1227.6 MHz (24.4 cm wavelength), are transmitted to the receivers along with accurate time stamps and other satellite information. These signals are intercepted by the GPS receivers where the timing code is translated to interpret the precise time it took the signal to reach the receiver. This precise time is used with the signal speed to calculate distance or

**TABLE 1.10**
**List of Selected Microwave Sensor–Based Satellites along with Detailed Specification**

| Name of Sensor | Description | Mission | Platform (Airborne/Spaceborne) Air | Space | No. of Bands | Spatial Resolution | Revisit Period (Days) | Operation Since | Agency/Country |
|---|---|---|---|---|---|---|---|---|---|
| **AMI** | Active Microwave | ERS-1 | | X | 1 | 30 m | 35 | 1991–2000 (out of service) | ESA, Europe |
| **AMI** | Active Microwave | ERS-2 | | X | 1 | 30 m | 35 | 1995–2011 (out of service) | ESA, Europe |
| **AMr** | Advanced Microwave Radiometer | Jason-2 (OSTM) | | X | | 3.3 cm sea-level measurement accuracy | 10 | 2001 | JPL, NASA, United States |
| **AMSr** | Advanced Microwave Scanning Radiometer | ADEOS-II | | X | 8 | 5, 10 × 1.6 km (IFOV) | 4 | 2002–2003 (out of service) | JAXA, Japan |
| **AMSr 2** | Advanced Microwave Scanning Radiometer | GCOM-W1 (SHIZUKU) | | X | 7 | 0.15, 0.35, 0.75, 0.65, 1.2, 1.8 km | 2 | 2002 | NASA, United States |
| **AMSr -e** | Advanced Microwave Scanning Radiometer for EOS | Aqua | | X | 6 | 5, 10 × 1.4 km (IFOV) | 4 | 2002 | NASA, United States |
| **AMSu -A** | Advanced Microwave Sounding (temperature) | Aqua, Metop-A, Metop-B, NOAA-15, NOAA-16, NOAA-17, NOAA-18 (NOAA-N), NOAA-19 (NOAA-N Prime) | | X | 15 | 5 × 2.343 km (IFOV) | 29 (Metop) and 11 (NOAA) | 2002 | NOAA, United States |
| **AMSu -B** | Advanced Microwave Sounding | Aqua, NOAA-15, NOAA-16, NOAA-17 | | X | 5 | 15 × 2.343 km (IFOV) | 11 | 1998 | NOAA, United States |
| **CrIMSS** | Cross track Infrared and Advanced Technology Microwave Sounder | JPSS-1 (NPOESS) | | X | 3 | 0.025, 0.0125, 0.0625 m | | 2017 (proposed) | NOAA, United States |
| **CrIMSS** | Cross track Infrared and Advanced Technology Microwave Sounder | Suomi NPP (SNPP) | | X | 3 | 0.025, 0.0125, 0.0625 m | | 2011 | NASA, United States |
| **g MI** | Microwave Radar Instrument | GPM Core | | X | 13 | | | 2014 (proposed) | NOAA, United States |
| **JMr** | Jason-1 Microwave Radiometer | Jason-1 | | X | 3 | | 10 | 2001–2013 (out of service) | NASA, United States |
| **MADr AS** | Microwave Imager | Megha-Tropiques | | X | 9 | 6 × 40 km | 1–6 times a day | 2011 | CNES, France |

| | | | | | | | | |
|---|---|---|---|---|---|---|---|---|
| **MHS** | Microwave Humidity Sounder | Metop-A, Metop-B, NOAA-18 (NOAA-N), NOAA-19 (NOAA-N Prime) | X | 5 | 16.3 × 1.078 km (IFOV) | 29 (Metop) and 11 (NOAA) | 2006, 2012, 2005, and 2009, respectively | ESA, Europe |
| **MIr AS** | Microwave Imaging Radiometer Aperture Synthesis | SMOS | X | 1 (SAR-L) | 35 km | 3 | 2009 | ESA, Europe |
| **MIS** | Microwave Imager/Sounder | JPSS-1 (NPOESS) | X | 1 (SAR-L) | | | 2017 (proposed) | NOAA, United States |
| **MIS** | Microwave Imager/Sounder | Suomi NPP (SNPP) | X | 1 (SAR-L) | | 16 | 2011 | NOAA, |
| **MLS** | Microwave Limb | AURA | X | 5 | 3 km | | 2004 | United States NASA, |
| **MSMr** | Sounder Multifrequency | IRS-P4 | X | 8 | 360 × 236 m | 2 | 1999–2010 (out of service) | United States ISRO, India |
| **MSr** | Scanning Microwave Radiometer | (Oceansat) | | | 32 km | 17 | 1987–1995 (out of service) | |
| | Microwave Scanning | MOS-1 | X | 2 | | | | JAXA, Japan |
| **MSr** | Radiometer Microwave Scanning | MOS-1b | X | 2 | 32 km | 17 | 1990–1996 (out of service) | JAXA, Japan |
| **MSu** | Radiometer Microwave Sounding | NOAA-10, | X | 4 | | 11 | 1986–2001 (all out of service) | NOAA, |
| | Unit | NOAA-11, NOAA-12, NOAA-14 | | | | | | United States |
| **MSu -e** | Microwave Sounding | RESURS-O1–3 | X | 3 | 34 m | 18 | 1994SRC Planeta, | |
| **MWr** | Microwave Radiometer | ENVISAT | X | 2 | 1040 × 1200 m for marine and 2630 × 300 m for land and coastal application | 35 | 2002–2012 (outESA, Europe of service) | |
| **MWr** | Microwave Radiometer | Sentinel-3 | X | 2 | 0.3 km | 27 | 2015 (proposed) | ESA, Europe |
| **MWr I** | Microwave Radiation Imager | FengYun-3A | X | | 1 km, 250 m | | 2008 | NSMC, China |
| **SAPHIr** | Microwave Sounding Instrument | Megha-Tropiques | X | | 10 km at NADIR | 1–6 times a day | 2011 | CNES, France |
| **SMMr** | Scanning Multichannel Microwave Radiometer | Nimbus-7 | X | 5 | 27–149 km | 6 | 1978–1994 (out of service) | NASA, |
| **SMMr** | Scanning Multichannel Microwave Radiometer | Seasat | X | 5 | 27–149 km | 6 | 1978–1978 (out of service) | United States NASA, |
| **TMI** | TRMM Microwave Imager | TRMM | X | 5 | 4.4 km | 0.5 | 1997 | United States NASA, |
| **TMr** | TOPEX Microwave Radiometer | TOPEX/Poseidon | X | 3 | 23.5–44.6 km | 10 | 1992–2005 (out of service) | United States NASA, United States |

**TABLE 1.11**
**Microwave Remote Sensors and Their Application Areas**

| Sensor | | Target |
|---|---|---|
| *Passive sensor* | | Near-sea-surface wind |
| | | SST |
| | | Sea condition |
| | | Salinity and sea ice condition |
| | Microwave radiometer | Water vapor, cloud water content |
| | | Precipitation intensity |
| | | Air temperature |
| | | Ozone, aerosol, NO., other atmospheric constituents |
| | | Soil moisture content |
| | | Surface roughness |
| | | Lake and ocean ice distribution |
| | | Snow distribution |
| | | Biomass |
| | | SST |
| | Microwave | Sea condition |
| | | Salinity and sea ice condition |
| | | Water vapor |
| | | Precipitation intensity |
| | | Wind direction and velocity |
| | | Sea surface topography, geoid |
| | | Ocean wave height |
| | Microwave altimeter | Change of ocean current |
| | | Mesoscale eddy, tide, etc. |
| | | Wind velocity |
| | | Image of ground surface |
| | | Ocean wave |
| | Imaging radar | Near-sea-surface wind |
| | | Topography and geology |
| | | Submarine topography |
| | | Ice monitoring |

*Source:* Adapted and modified from ITC, ITC's database of satellites and sensors, 2023, www.itc.nl/research/products/sensordb/allsensors.aspx, accessed on October 15, 2023.

range. The intersection of the three ranges from three satellites is used to accurately determine the location of the GPS receiver.

There are three segments involved in GPS data collection: (1) space segment, (2) control segment, and (3) user segment. Figure 1.28 shows the GPS geosynchronous satellite constellations and their orbit. The control segment consists of five ground monitoring stations that continuously monitor the satellite trajectory and atmospheric conditions and help develop an error model for the atmospheric interference. The error correction based on the model is uploaded back to the GPS satellites via the master control facility at Colorado Springs, Colorado, United States. The receiver segment is the most important part because it involves the end user application needs. With the advancement of technology, we have centimeter accuracy GPS receivers that can process signal correction information (kinematic) on the fly, thereby facilitating accurate location data collection. Follow this

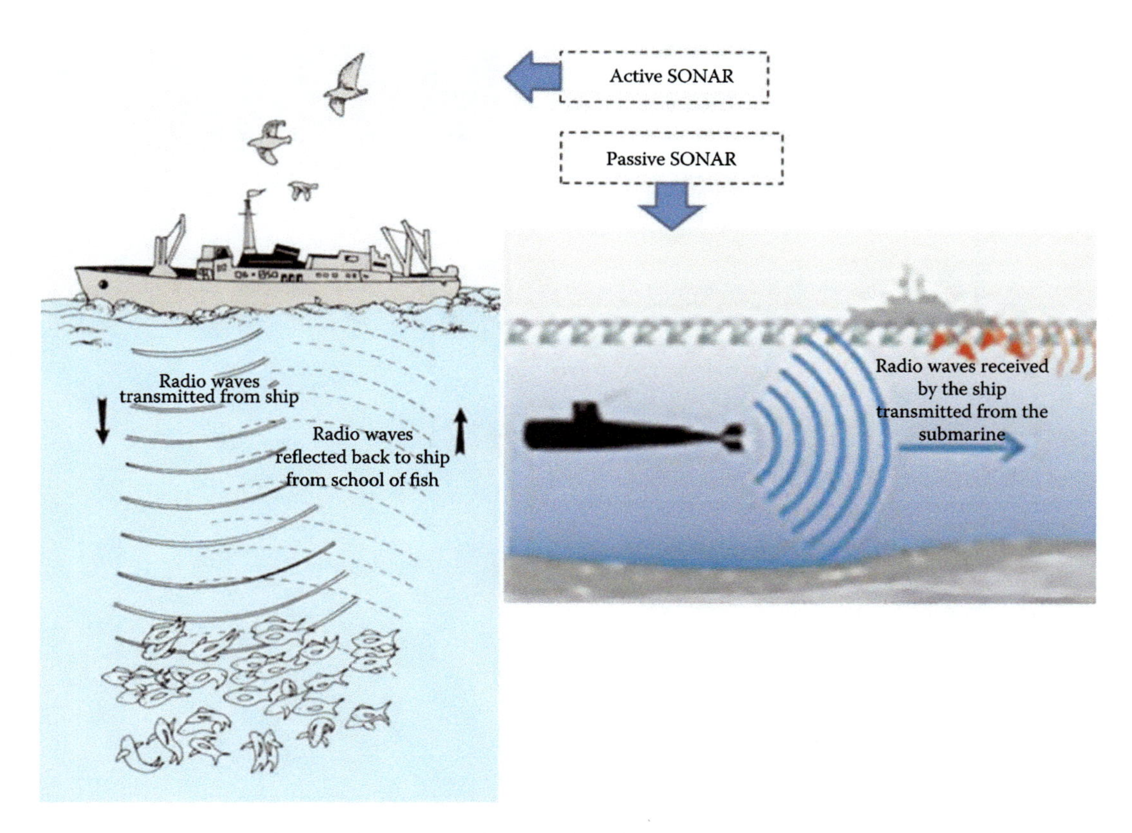


FIGURE 1.26 Example of active and passive sound navigation and ranging (SONAR) detection system emitting ultrasound and radio waves to find the location of fish in the ocean and receiving radio waves from undersea submarine and detecting its location. SONAR is used in navigation, communication, and detection of objects on surface and underwater.

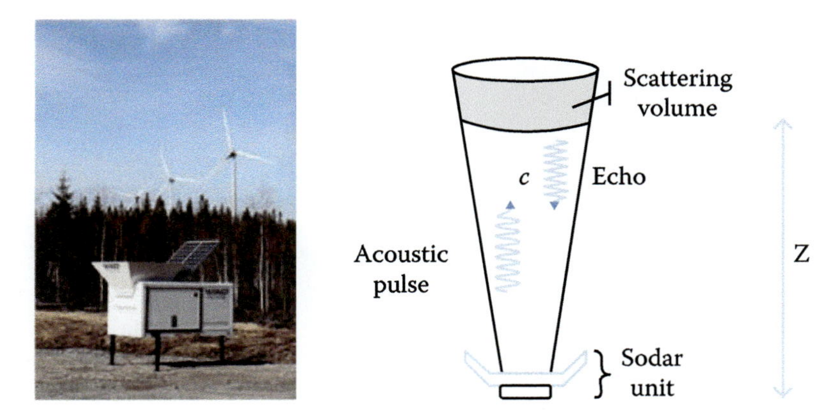


FIGURE 1.27 Wind measurement with a phased array SODAR SFAS from Scintec Corporation (Louisville, CO) and the working principle of SODAR sensing system.

**TABLE 1.12**
**List of Selected SONAR Technology–Based Satellites along with Detailed Specification**

| Name of Sensor | Description | Mission | Platform (Spaceborne) | No. of Bands | Spatial Resolution | Revisit Period (Days) | Operation Since | Agency/ Country |
|---|---|---|---|---|---|---|---|---|
| **AIr S** | AIS | Aqua | X | 7 | 13.5 km | 0.5 | 2002 | NASA, United States |
| **AMSu -A** | AIS | Aqua | X | 15 | 50 km | | 2002 | NASA, United States |
| **AMSu -A** | AIS | Metop-A | X | 15 | 50 km | 29 | 2006 | ESA, Europe |
| **AMSu -A** | AIS | Metop-B | X | 15 | 50 km | 29 | 2012 | ESA, Europe |
| **AMSu -A** | AIS | NOAA-15 | X | 15 | 50 km | 11 | 1998 | NOAA, United States |
| **AMSu -A** | AIS | NOAA-16 | X | 15 | 50 km | 11 | 2000 | NOAA, United States |
| **AMSu -A** | AIS | NOAA-17 | X | 15 | 50 km | 11 | 2002 | NOAA, United States |
| **AMSu -A** | AIS | NOAA-18 (NOAA-N) | X | 15 | 50 km | 11 | 2005 | NOAA, United States |
| **AMSu -A** | Advanced Microwave Sounding | NOAA-19 (NOAA-N Prime) | X | 15 | 50 km | 11 | 2009 | NOAA, United States |
| **HSB** | Humidity Sounder for Brazil | Aqua | X | 4 | 13.5 km | 0.5 | 2002 | NASA, United States |
| **MHS** | Microwave Humidity Sounder | Metop-A | X | 5 | 50 km | 29 | 2006 | ESA, Europe |
| **MHS** | Microwave Humidity Sounder | Metop-B | X | 5 | 50 km | 29 | 2012 | ESA, Europe |
| **MHS** | Microwave Humidity Sounder | NOAA-18 (NOAA-N) | X | 5 | 50 km | 11 | 2005 | NOAA, United States |
| **MHS** | Microwave Humidity Sounder | NOAA-19 (NOAA-N Prime) | X | 5 | 50 km | 11 | 2009 | NOAA, United States |
| **MLS** | Microwave Limb Sounder | AURA | X | 5 | 3 km | 16 | 2004 | NASA, United States |

| | | | | | | | | |
|---|---|---|---|---|---|---|---|---|
| **MSu** | Microwave Sounding Unit | NOAA-10 | X | 4 | 105 km | 11 | 1986–2001 (out of service) | NOAA, United States |
| **MSu** | Microwave Sounding Unit | NOAA-11 | X | 4 | 105 km | 11 | 1988–2004 (out of service) | NOAA, United States |
| **MSu** | Microwave Sounding Unit | NOAA-12 | X | 4 | 105 km | 11 | 1991–2007 (out of service) | NOAA, United States |
| **MSu** | Microwave Sounding Unit | NOAA-14 | X | 4 | 105 km | 11 | 1994–2007 (out of service) | NOAA, United States |
| **SAPHIr** | Microwave Sounding Instrument | Megha-Tropiques | X | | 1° | 1–6 times/day | 2011 | CNES, France |
| **SSMIS** | Special sensor microwave imager/ sounder | DMSP-16 | X | 4 | 13–43 m | | 2003 | USDD, United States |
| **SSu** | Stratospheric Sounder Unit | NOAA-11 | X | 3 | 147.3 km | 11 | 1988–2004 (out of service) | NOAA, United States |
| **SSu** | Stratospheric Sounder Unit | NOAA-14 | X | 3 | 147.3 km | 11 | 1997–2007 (out of service) | NOAA, United States |

*Source:* Adapted and modified from ITC, ITC's database of satellites and sensors, 2023, www.itc.nl/research/products/sensordb/allsensors.aspx, accessed on October 15, 2023.

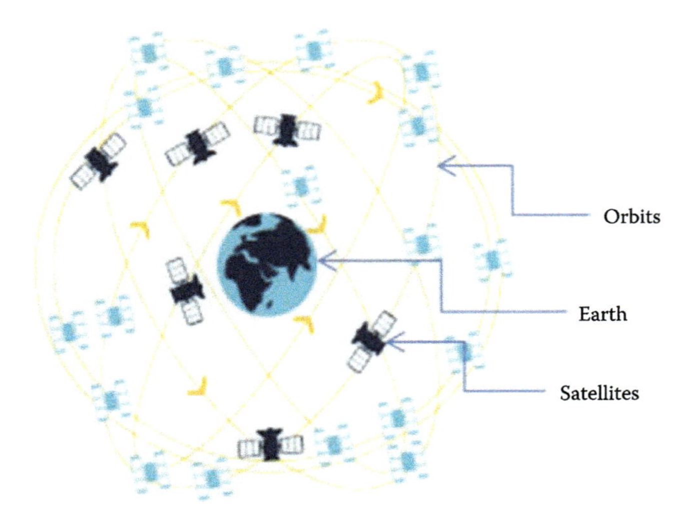


FIGURE 1.28 Global Positioning System space segment showing 24 satellites orbiting the Earth at an altitude of 20,200 km in six orbits with four satellites in each. (From www.gps.gov/multimedia/images/constellation.jpg.)

(www.gps.gov/) site for more information about GPS. GPS is described in detail in another chapter of this book.

While the history of the GPS is rooted in military applications, over the past several decades, the civilian applications of GPS have far exceeded the military applications. The literature is replete with numerous applications of GPS data that include telecommunications, commerce and retailing industry, and recreation, in addition to the traditional applications in natural resource management and planning. The GPS.gov—the official U.S. government information portal about GPS and related topics—provides a good treatise about the diverse applications of GPS in everyday life limited only by the human imagination. More importantly, the website provides an important repository of information pertaining to the GPS modernization program that includes new series of satellite acquisitions (GPS IIR, GPS IIF, and GPS III). The improvements bring about new features such as the Architecture Evolution Plan (AEP) and the Next-Generation Operational Control System (OCX) that are geared to deliver enormous improvements in the robustness and accuracy of the entire GPS system in the coming years (Figure 1.29). These infrastructure and signal developments are designed not only to increase operational capacity but also to provide better geometry in *shadow* environments of urban canyons and mountainous terrain.

The obvious benefits of location-based information collected through GPS are immense and beyond the scope of this chapter to discuss. While these benefits are invaluable for applications involving field surveys, navigation, and basic GIS mapping of what features are located where, there are additional benefits that directly relate to remote sensing applications. Some of these include geometric rectification and registration of image data by way of ground control points (GCPs) collected through high-accuracy GPS units. Similarly, location information collected through ground-truthing surveys using high-accuracy GPS units plays an important role in image classification and accuracy assessment. Using the location information collected using GPS, better *training* sites could be used for algorithm *learning* phases during image classification procedures. Furthermore,

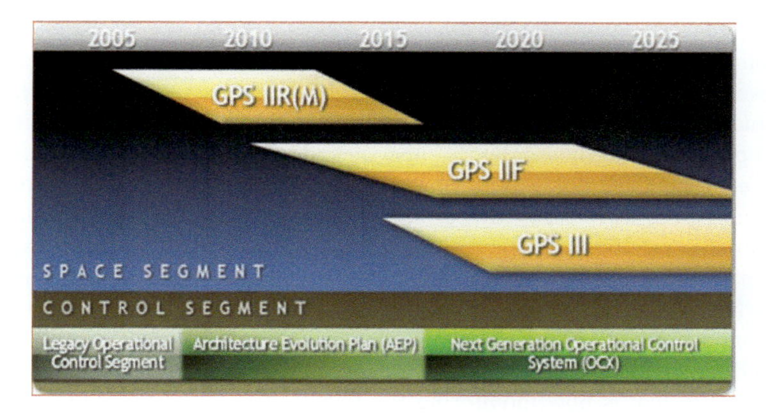


**FIGURE 1.29** Global Positioning System modernization program schedule. The graphic provides links to the important information about developments in the space segment (www.gps.gov/systems/gps/space/#IIRM) and control segment (www.gps.gov/systems/gps/control/), including Architecture Evolution Plan (www.gps.gov/systems/gps/control/#AEP) and Operational Control System (www.gps.gov/systems/gps/control/#OCX). (From GPS.gov.)

as a postclassification routine, the field-collected GPS data are useful as a *reference* information about the landscape feature or biophysical phenomenon that could be used in the accuracy assessment exercise of comparing and validating outputs generated from image classification algorithms (Powell and Matzke, 2004; Congalton and Green, 2008).

With the inclusion of other global navigation satellites as clusters from Russia (Globalnaya Navigazionnaya Sputnikovaya Sistema, GLOSNASS, 24+ operational satellites), Europe (Galileo, 24+ satellites), India (Indian Regional Navigation Satellite System/Navigation Indian Constellation, IRNSS/NavIC, seven satellites), China (BeiDou Navigation Satellite System, BDS, 35 satellites), Japan (Quasi-Zenith Satellite System, QZSS, seven satellites), the United States' GPS is renamed commercially as GNSS (www.gps.gov/systems/gnss/). The accuracy of GNSS measurement has become much better, close to centimeter accuracy. Many reference stations are available all over the world to provide better positional accuracy. They are being used to boost management efficiency and productions in various sectors, including farming, construction, maritime/terrestrial transportation, construction, mining, surveying, package delivery, logistical supply chain management, communication networking, financial markets, power grids, search and rescue efforts, emergency service delivery, and above all recreation, to name a few. Global environmental/biodiversity management has gotten much better through tracking faunas, locating geohazards, managing mitigation efforts, etc. With the advent of RTK GIS and DGPS, GNSS will get better over time.

## 1.4 NEWER SATELLITE SENSOR PLATFORMS AND UPCOMING MISSION

Some of the prominent newer satellites sent to space in the last ten years are (1) ALOS 2, a RADAR imaging satellite system operated by the Japan Aerospace Exploration Agency (JAXA). The primary functions are for land resource studies, disaster monitoring, and environmental research. (2) GPM Core Observatory satellite is a core mission of NASA and JAXA to provide real-time accurate information of rain and snow every three hours, along with soil moisture, carbon cycle, winds, and aerosol estimation. (3) Hodoyoshi 3 and 4 is an experimental Earth-observing microsatellite built by the University of Tokyo. (4) IRNSS 1B is a satellite-based navigation system developed in India to be compatible with GPS and Galileo (European navigation system). (5) Sentinel-1A is an ESA two-satellite constellation with the prime objective of land and ocean monitoring with C-band SAR continuity. (6) SkySat 2 is a commercial Earth-observing satellite by

Skybox Imaging to acquire very-high-resolution images. (7) TSAT is a 2U CubeSat microsatellite communication network from Taylor University, Upland, IN, used to research about a nanosat network to observe space weather with a plasma probe, three-axis magnetometer, and three UV photodiodes and is part of the NASA ELaNa 5 CubeSat initiative. (8) UKube is the U.K. Space Agency's pilot program with the miniature cube-shaped satellite that allows the United Kingdom to test cutting-edge new technologies in space. (9) Velox 1 is a small satellite from Singapore's Nanyang Technological University with a high-resolution camera for Earth imaging and later split into two smaller satellites—N-sat and P-sat—to conduct communications experiments. (10) WorldView-3 is a 0.5 m panchromatic and 1.8 m multispectral resolution Earth observation imaging satellite by Digital Globe. (11) NASA's SMAP (Soil Moisture Active Passive) mission is slated to provide global soil moisture measurement and soil freeze/thaw state using a combination of active radar and passive radiometer. (12) NASA-JPL's ECOSTRESS is a unique SWIR and TIR bands-based satellite that is now monitoring the Earth's ecosystem through evapotranspiration measurement of Earth features. (13) NASA's ISS mounted GEDI LiDAR based satellite system is a revolution that provides 30 m spatial resolution with one day repeatability to monitor forest biomass and other ecosystem (biodiversity and climatic impact) changes of major portions of the Earth. (14) NASA's ever popular Landsat 9 with the modified 14-Digit radiometric resolution version from the earlier version of Landsat 8 is one of the sought after digital remote sensing products globally (15) Gaofen-5, the latest satellite of the GF series by CNSA is one of the best improvements in hyperspectral imaging. The satellite has 330 bands at a 5-nanometer spectral interval and higher spectral resolution in visible (5 nm), NIR (5 nm), and SWIR (10 nm) ranges (0.4–2.5 μm). From this present trend of latest satellite launches, it is presumed that the upcoming satellite launches will be with the research-related nanosatellite platforms, satellites doing dual duties after expiration of successful first job assigned to satellites, and satellites launched with very-high-resolution imaging capabilities.

Newer satellites from older programs like NASA, CNES, JAXA, and ESA are being programmed for future space launches with minimum two new improvements, that is, collecting data with larger swath so that revisit time of a satellite will be faster and including more bands into the system to help collect data for better and improved Earth observation. For example, according to Astrium Inc. (www.astrium-geo.com/en/147-spot-6-7-satellite-imagery), France's CNES is expanding SPOT imaging satellite series to 2024, and the latest series of SPOT 6 and SPOT 7 emphasize on collecting data by covering larger areas, that is, up to 6 million km$^2$ area/day, an area larger than the entire European Union. SPOT 6 and 7 satellites also include four-day weather forecasting to the satellite's tasking. NASA's Landsat Next is on the horizon and is expected to launch in late 2030 or early 2031. This freely available digital resource will ensure the continuity of the longest space-based record of Earth's land surface, but with better spatial, spectral, and temporal resolutions from earlier versions. Landsat Next will have 26-nad constellation with more than double bands from the previous Landsat 9 that was launched on September 27, 2021 from Space Launch Complex-3E at Vandenberg Space Force Base on an Atlas V 401 launch vehicle. The triplet constellation would improve the temporal resolution to 6 days with an improved spatial resolution of 10 m, Joint Polar Satellite System (JPSS) is the next-generation satellite system of the United States and a collaborative program between NOAA and its acquisition agent NASA. It will help weather and natural hazard forecasting at a newer level.

Oblique aerial photograph is the latest development in aerial imaging area, which has high potential growth in future years. Figure 1.30 shows an example of oblique aerial photograph. Oblique aerial photographs or a combination of oblique and vertical photographs, which are widely used for high-density urban land use mapping, are reviewed by Petrie (2009), and according to him, oblique aerial photography have been used since the 1960s, but its advantages in environmental management and other spatial analyses are observed more recently. More images will be acquired with oblique imaging platform in the next years.

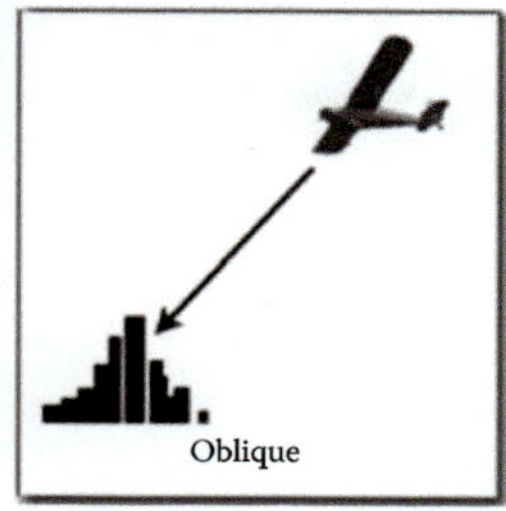


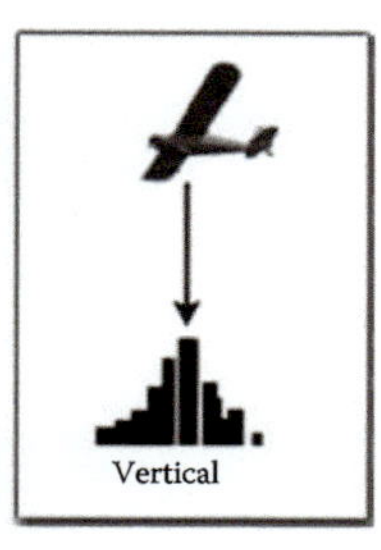


FIGURE 1.30 Examples of an oblique aerial image and graphics of the image acquisition processes.

## 1.5 FUTURE OF REMOTE SENSING AND EVOLVING MICROSATELLITES

According to a study by U.S.-based Forecast International, to address the need for better imaging data driven by aerospace and defense requirements, more satellites will be launched from high-resolution imagery acquiring platforms. For this purpose, UAS and UAV sensor platforms would serve the purpose. As discussed in Section 1.4, the present trends in satellite launches are from very high-resolution, research-oriented nanosatellite platforms; the UAS and UAV platforms would suffice the need. According to Everaerts (2008), the ISPRS Congress in Istanbul passed a resolution that UAVs provide a new controllable platform for remote sensing and permit data acquisition in inaccessible and dangerous environments. UAS or UAVs are the prominent part of the entire system of flying an aircraft and acquiring ultraspatial-resolution imagery. The aircraft is controlled from the ground control station with a reliable communication network through air traffic control. Over the last four years, the number of UAV systems increased by leaps and bounds in remote sensing and mapping (Everaerts, 2008). Everaerts (2008) also state that much of the work in the use of UAVs in remote sensing is in research stage presently and the future of remote sensing in essence depends on the growth of UAS- and UAV-based remote sensing. As in most countries aviation regulations are adapted to include UAS and UAV systems into the general airspace, these microsatellite systems will become the preferred platforms of future remote sensing. The following is an example of the utility of UAVs for Earth observation and analysis in most advanced manner.

One of the popular fully autonomous UAV eBee (Figure 1.31a) manufactured by a Swiss sensor manufacturer senseFly (SenseFly SA, Cheseaux-Lausanne, Switzerland) can acquire 1.5 cm spatial resolution aerial images (photos) that can be transformed into 2D orthomosaics and 3D models. The eBee can cover up to 12 $km^2$ (4.6 $miles^2$) in a single flight. When it flies over smaller areas at lower altitudes, it can acquire high resolution imagery at 1.5 cm/pixel with overlap (senseFly, 2014). It uses a built-in 12-megapixel cameras, one for visible spectrum (VIS) (R-, G-, and B-bands) and another for NIR-band image acquisition to create aerial maps for supporting like wildlife, crops, and traffic management to name a few. Thermal infrared (TIR) cameras can be mounted in the UAV for TIR image acquisition. The intuitive eMotion software of the senseFly makes it easy to plan and simulate your mapping mission, which also generates a full flight path by calculating the UAV's required altitude and projected trajectory (senseFly, 2014). Sharma and Husley (2014) of the University of North Georgia used eBee as a research and teaching tool. Figure 1.31b shows the simulated flight paths of the UAV created by Sharma and Husley (2014) to acquire aerial images at 1-inch spatial resolution for their study. On the site, it is launched by hand, after which it autonomously follows a flight path along with the simulated image acquisition schedule mapped out in advance by the user, via the included eMotion 2 software. Users can take

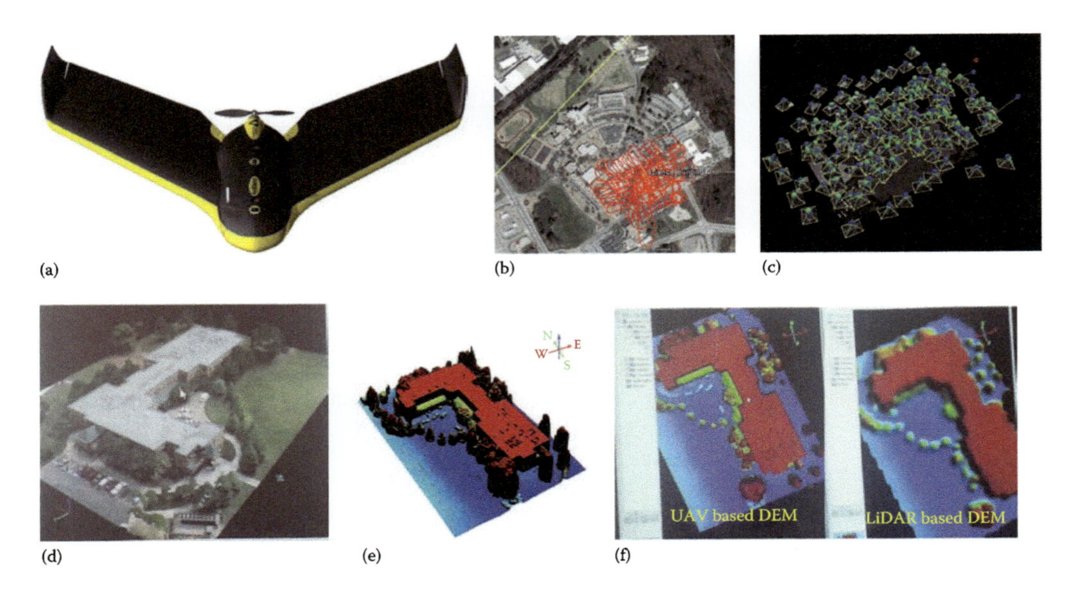


**FIGURE 1.31** Example of an unmanned aerial vehicle (UAV) (senseFly's eBee) used for ultrahigh-resolution orthophoto collection and ancillary derivative data development: (a) eBee UAV, (b) flight path simulated by the Institute of Environmental Spatial Analysis (IESA), (c) all acquired aerial orthophotos by the UAV over UNG Gainesville campus, (d) mosaicked 2D orthophoto of the campus (partial), (e) stereoscopic analysis–based 3D point cloud, and (f) comparison of digital elevation models developed by imaging through UAV (eBee) and LiDAR technology. (From www.sensefly.com/drones/ebee.html and Sharma and Husley, 2014.)

control of the UAV at any point and pilot it remotely in real time (senseFly, 2014). Upon completion of its flight, it can land itself. The eBee website (www.sensefly.com/drones/ebee.html) contains other important specifications (weight, wingspan, material, propulsion battery, camera, maximum flight time, nominal cruise speed, radio link range, maximum coverage, wind resistance, GSD, relative orthomosaic/3D model accuracy, absolute vertical/horizontal accuracy [with or without GCPs], multidrone operation, automatic 3D flight planning, linear landing accuracy, etc.) about the UAV that can be useful for readers to make decisions to own one and use for Earth observation and analysis in a very advanced manner.

The eBee's acquired data are processed by senseFly's postflight Terra 3D software to produce 2D orthomosaics, 3D point clouds, triangle models, and DEMs. Figure 1.31c is the example of all the aerial photos acquired by Sharma and Husley (2014) over the university campus, and they mosaicked the acquired images using in-house software of the UAV to produce the 2D (Figure 1.31d), stereoscopic analysis–based 3D point cloud (Figure 1.31e) and subsequent highly accurate DEM. Sharma and Husley (2014) compared the DEM with the DEM created by LiDAR data acquired over the campus (Figure 1.31f). Triangulation between the points together with the onboard GPS camera helped in the image development with high accuracy (1.5–4.5 cm). UAV aka drones are getting intelligent through the machine learning application. Panda et al. (2023) used a rotating wing DJI Mavic 3 Pro (DJI Inc., Shenzhen, China) (Figure 1.32a) to remotely map the soil characteristics (Figure 1.32b), distinguish the weed from a forage (Sericea Lespedeza) crop using the Leaf Chlorophyll Index (LCI) from image spectral band analysis (Figure 1.32c), and, and confirm the diffentiation of weed from crop by 3D mapping (Figure 1.32d) with the aid of a RTK GNSS system in a research field. The accuracy of weed detection and soil spatial electrical conductivity variation was more than 90%. They used the DJI Argas T16 drone to spay pesticide on site specific basis. This is just a small example how the UAV/UAS technology is making us more and more efficient in remote environmental management of all kinds. UAVs are equipped with hyperspectral cameras to

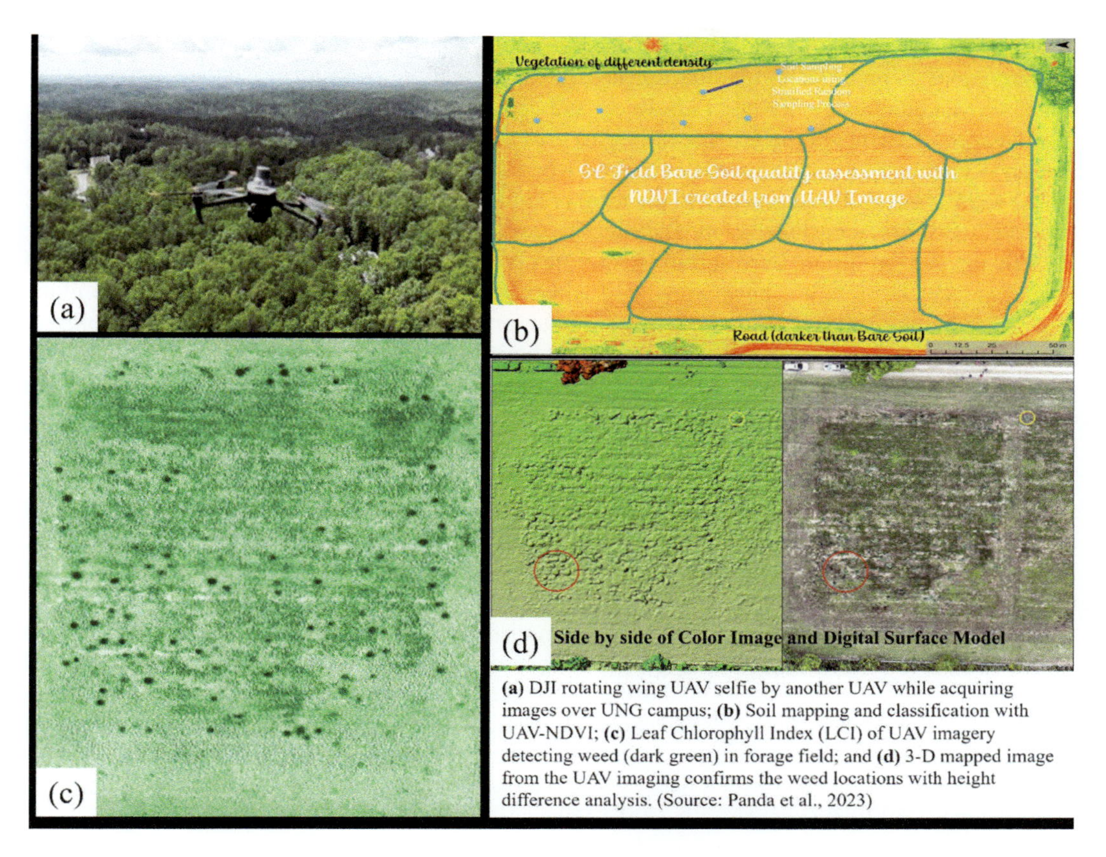


**FIGURE 1.32** Example of the advance application of UAV in crop production management with the development of image indices, 3D mapping, and even performing crop management farm mechanization tasks with ease and efficiency: (a) DJI rotating wing UAV selfie by another UAV while acquiring images over UNG campus; (b) soil mapping and classification with UAV-NDVI; (c) Leaf Chlorophyll Index (LCI) of UAV imagery detecting weed (dark green) in forage field; and (d) 3D mapped image from the UAV imaging confirms the weed locations with height difference analysis. (Source: Panda et al., 2023.)

add in environmental management as developed at Headwall Inc. (https://headwallphotonics.com/product-category/remote-sensing/).

The future trend in remote sensing is to have full free-and-open data access, even for higher-resolution data (Belward and Skoien, 2014). NASA Landsat mission paved the way in providing free imagery to the public via the internet (http://earthexplorer.usgs. gov/), and others with commercially driven programs such as Planet Labs and Skybox are following to make their data available free to academics and nongovernment organizations, and more will follow in the future (Butler, 2014). This approach of free data sharing is a core element of the Group on Earth Observation's Global Earth Observing System of System's goal of data sharing and data management (Withee et al., 2004; GEO, 2012).

## 1.6 DISCUSSIONS

In addition to the rich visual information in remote sensing data that can be directly interpreted to obtain the geographic information about landscape features (e.g., the boundary extent of forest cover), remote sensing data also provide additional biophysical information about the object (e.g., forest health, biomass). The EMR reflected, emitted, or backscattered from an object or geographic area is used as a *surrogate* for the actual property under investigation (Jensen, 2009). The EME measurements must be calibrated and turned into information using visual and/or digital image processing techniques. According to Belward and Skoien (2014), to meet the food, fuel, freshwater, and fiber requirements of the Earth's eight-plus billion humans, obtaining correct land cover information is vital and its importance is enhanced when the world is facing severe consequences from global warming and climate change in the form of land degradation such as desertification, wildfire, flooding, landslides, coastal erosion, and other hazards.

A major portion of our Earth-orbiting satellites provide a unique vantage point from which to map, measure, and monitor how, when, and where land resources are changing across the globe (Townshend et al., 2008). The importance of remote sensing has increased abundantly in the present century due to growing and shifting human population, and therefore, the resulting land use pattern has changed with urbanization and resource use for food, fiber, and fuel production (De Castro et al., 2013). Additionally, with growing impacts of climate change, newer approaches to resource conservation aimed at sustainability such as biodiversity offsetting (McKenney and Kiesecker, 2010) and carbon markets (Mollicone et al., 2007) are increasingly using remote sensing data to study and analyze the spatial distribution of land use. Land cover mapping using remote sensing can be implemented at different scales ranging from global (Loveland et al., 1999; Friedl et al., 2002; Bartholome and Belward, 2005; Arino et al., 2007; Gong et al., 2013) to more regional and localized levels (Cihlar, 2000; Panda et al., 2010a, 2010b).

To better plan and balance resource management issues, global land cover mapping with remote sensing supports decision makers and policy analysts at national and international scales (Belward and Skoien, 2014). The success of land use management plans depends primarily on reliable information concerning how, when, and where about resource use and its changing pattern. A great deal of the reliable information leading to land use characterization for ecological habitat assessment and planning can be accomplished using geospatial technologies such as remote sensing and GIS (Osborne et al., 2001; Rao et al., 2005, 2007; Valavanis et al., 2008; Vierling et al., 2008).

It is to be noted that remote sensing data are available in raster format on a square pixel basis. Though a crude representation of the nonhomogeneous Earth environment, the pixel-based representation of individual environmental spatial parameters of the Earth helps in algorithmically connecting through geospatial models to obtain proactive management decision support. In such geospatial models, even the environmental spatial factors available in vector format are rasterized to similar pixel size as other parameters. Image or raster calculators can be used to algorithmically determine the contribution of each parameter for one pixel coverage to determine probability, susceptibility, vulnerability, suitability, and other ability toward an event like geohazards occurring, flora and fauna habitat and production, and environmental changes to name a few.

Figure 1.33 is a simple potent example of how the efficiency of remote sensing data can be exploited to obtain useful proactive management decisions to mitigate one of the geohazards, floods, which is a big bother for humanity now in the present climate change scenario (Perry et al., 2022). Overbeck et al. (2024) used WorldView-2 satellite imageries in two different dates (2012/2013 and 2022/2023) to determine bathymetric changes due to coastal erosion in the shallow water near Golovin and Wales, Alaska. Two multispectral satellite images were taken in succession in 2013 over Golovin and one image over Wales in 2012 GMT and later acquired over the same study areas in 2022 and 2023. They found the green band (510–580 nm) reflected the bathymetry of the region with less noise than other bands. They calibrated the sensor radiances to depth measurements taken with a single-beam sonar system and obtained the resultant digital elevation models (DEM) with standard errors of 0.013 and 0.006 m for Golovin, and 0.007 m for Wales. Zhang et al. (2023) confirms through their study in Ganquan Island and Qilianyu Island that Sentinel-2, Gaofen-2, Gaofen-6, and Worldview-2 shown promise in estimating nearshore bathymetry with high accuracy ($R^2 > 0.9$). These are few examples of how remote sensing can do wonders in environmental management.

During June 10–13, 2022, an "atmospheric river" event—a system of warm and extremely wet air that usually originates in the tropics—struck the Yellowstone National Park region, Wyoming

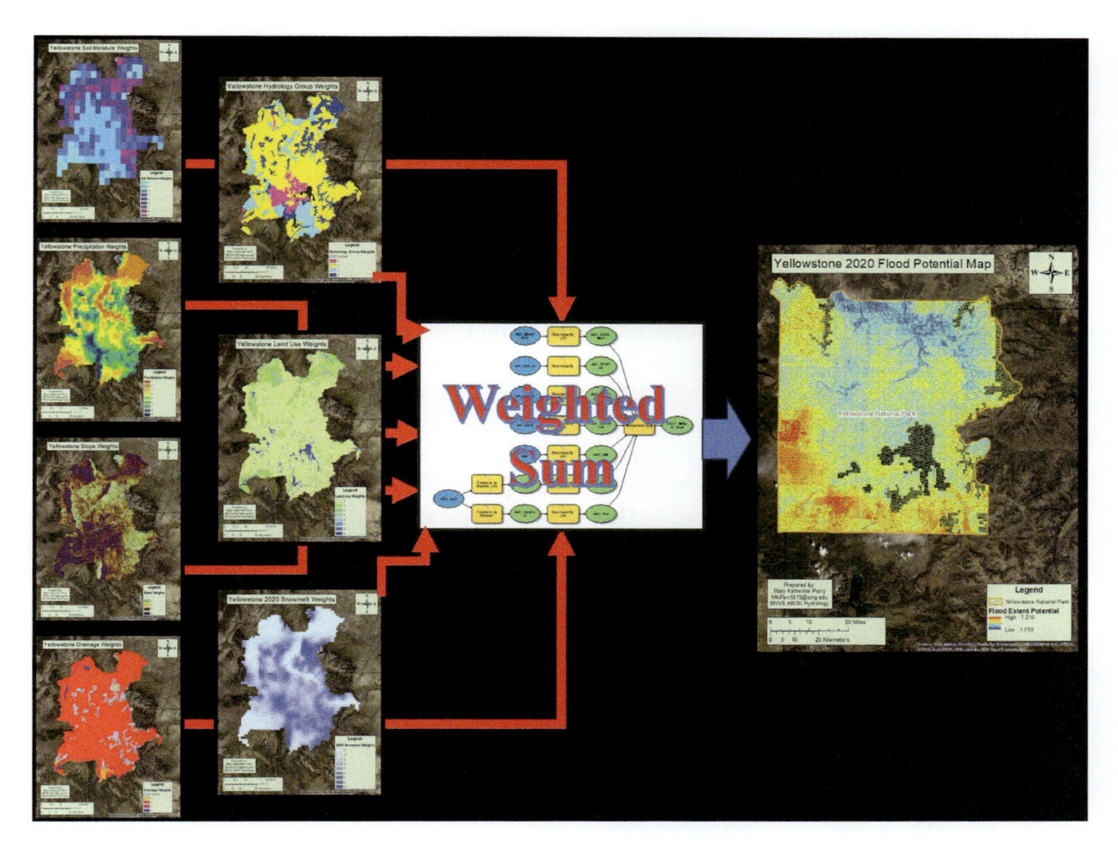


FIGURE 1.33 Yellowstone National Park spatial flood potential map developed with seven environmental contributing factors in raster format created with satellite imageries and other geospatial data processing on a resampled 3 m pixel spatial unit. (Note: the raster names are on the top of each (left) prepared scaled potential parameter raster (with halo) and the final automated geospatial model architecture (middle) is inscribed with the main tool (Weighted Sum) to algorithmically combined to obtain spatial flood potential map of the study area (right).

(Poland et al., 2022). As snow melts in dry heat and melts faster in the rain, the extreme rain, 2 to 3 inches worth, combined with the warm temperatures causing a rapid snowmelt, and eventually gave rise to unprecedented flooding, feet higher than the historic record. The communities surrounding the park were cut off, and large sections of the park were washed out, including many roads (Poland et al., 2022). The study (Figure 1.32) tried to piece together seven flooding potential environmental parameters, such as (1) soil moisture (obtained from SMAP satellite data), (2) snow melt (obtained from the NOAA National Operational Hydrologic Remote Sensing Center [NOHRSC]), (3) land cover (NLCD, from processed Landsat data), (4) slope (from DEM, created with LiDAR and other Topographic Mission satellite data), (5) precipitation (developed with RADAR data modeled with in situ instrumentation data), (6) Drainage (gSSURGO soil database), and (7) Hydrologic group (gSSURGO soil database). It is to be noted that the snow melt record obtained through NOAA NOHRSC is created with the SNOw Data Assimilation System (SNODAS) model, which integrates modeled snow estimates with satellite (passive microwave measurements), airborne, and ground observations to create 1 km resolution gridded datasets of snow-covered area, SWE, and other snow variables for the continental United States and are available daily from October 2003 through the present (www.nohrsc.noaa.gov/). Even gSSURGO soil database is developed as a combination of soil survey, land cover, and topographic data where remotely sensed images are a major contributing factor. It should explain the advantages of remote sensing in everyday management decision support and it is getting better when ultra-high resolution (spatial, spectral, radiometric, and temporal) remote sensing data are made available to stakeholders, researchers, managers, and even lay end users for free and online in most cases.

## 1.7 SUMMARY AND CONCLUSION

This chapter provides an exhaustive overview of the remote sensing satellites, sensors, and their characteristics. *First*, the chapter provides definition and an understanding of remote sensing from best known sources. This common understanding is required for any student and practitioner of remote sensing. *Second*, the principles of EMS are enumerated. *Third*, the spatial, spectral, radiometric, and temporal resolution are explained. *Fourth*, over 18,000 satellites have been launched and operated by various governments and private enterprises from the mid-1950s to the present. The characteristics of many of the important satellite and sensor systems have been described. *Fifth*, remote sensing is both active and passive and is gathered in various EMSs such as the visible, NIR, shortwave infrared, mid-infrared (MIR), TIR, microwave, and radio waves (high-frequency waves). Radar, SAR, LiDAR, SONAR, and SODAR technologies operate with microwave and radio wave regions of the EMS, while optical and hyperspectral imaging operates with visible to TIR region of EMS. The importance of gathering these data across such a wide range of EMS has been highlighted. *Sixth*, remote sensing data gathering is done in various modes such as nadir, off nadir, hyperspatial, hyperspectral, and multispectral. All of these have various implications on what we study and the level of these accuracies as a result of such acquisitions. This aspect has been implied throughout the chapter. *Seventh*, remote sensing data are gathered in various platforms: ground based, platform mounted, airborne, spaceborne, undersea, and UAV. The chapter focuses on spaceborne but does discuss several other platforms. Reading through these sections, it becomes clear that remote sensing has truly evolved in both sensor design and acquisition platforms. *Eighth*, the chapter provides a window into what to expect in the near future through upcoming newer satellites and sensors. *Ninth*, the evolution of microsatellites has been highlighted. Like miniaturization of computing technology, satellites and sensors are undergoing revolutionary technological innovations, which make them provide smaller, cheaper, and yet better-quality data. *Tenth*, the chapter captures in a nutshell key developments in remote sensing from 1858 to the present and provides a glimpse on where the future of remote sensing satellites, sensors, and data is headed. Finally, the authors recommend readers to watch the Public Broadcasting Station (PBS) NOVA documentary, "Earth from Space" (www.youtube.com/watch?v=aU0GhTmZhrs) to understand the amazing ability of remote sensing satellites to monitor, measure, and analyze the Earth's resources (oceans, land, and atmosphere) for prudent decision support to ensure a sustainable life on it.

## REFERENCES

3Dlasermapping. 2014. Volume measuring. http://3dlasermapping.com/index.php/mining-monitoring-applications/volume-measuring. Accessed on July 2, 2014.

Aggarwal, S. 2003. Principles of remote sensing. In: *Satellite Remote Sensing and GIS Applications in Agricultural Meteorology*, p. 23.

Ahsan, A., and Khan, A. 2019. Pakistan's journey into space. *Astropolitics* 17(1), 38–50.

Amatya, D., Panda, S. S., Cheschair, G., Nettles, J., Appleboom, T., and Skaggs, W. 2011. Evaluating evapotranspiration and stomatal conductance of matured pine using geospatial technology. In: *American Geophysical Union Conference 2011*, San Fransisco, CA, December 5–9.

Amatya, D., Trettin, C., Panda, S. S., and Ssegane, H. 2013. Application of LiDAR data for hydrologic assessments of low-gradient coastal watershed drainage characteristics. *Journal of Geographic Information System* 5(2), 175–191. doi: 10.4236/jgis.2013.52017.

American Society of Photogrammetry and Remote Sensing. 1952, 1966. *Manuals of Photogrammetry*. Bethesda, MD: ASPRS.

Amzajerdian, F., Pierrottet, D., Petway, L., Hines, G., and Roback, V. 2011. LiDAR systems for precision navigation and safe landing on planetary bodies. In: *International Symposium on Photoelectronic Detection and Imaging 2011*, Beijing, China. International Society for Optics and Photonics, p. 819202.

Arino, O., Gross, D., Ranera, F., Bourg, L., Leroy, M., Bicheron, P., Latham, J., et al. 2007. GlobCover: ESA service for global land cover from MERIS. In: *Proceedings of the IEEE International Geoscience and Remote Sensing Symposium, 2007 (IGARSS'07)*, Barcelona, Spain, pp. 2412–2415. doi: 10.1109/IGARSS.2007.4423328.

Aronoff, S. 2004. *Remote Sensing for GIS Managers*. Redlands, CA: ESRI Press.

Asner, G. P., Boardman, J., Field, C. B., Knapp, D. E., Kennedy-Bowdoin, T., Jones, M. O., and Martin, R. E. 2007. Carnegie airborne observatory: In-flight fusion of hyperspectral imaging and waveform light detection and ranging for three-dimensional studies of ecosystems. *Journal of Applied Remote Sensing* 1(1), 013536–013536.

Asner, G. P., Powell, G. V., Mascaro, J., Knapp, D. E., Clark, J. K., Jacobson, J., Kennedy-Bowdoin, T., et al. 2010. High-resolution forest carbon stocks and emissions in the Amazon. *Proceedings of the National Academy of Sciences* 107(38), 16738–16742.

Avery, T. E. and Berlin, G. L. 1992. *Fundamentals of Remote Sensing and Airphoto Interpretation*. Upper Saddle River, NJ: Prentice Hall.

Bailey, D. T., 2000. *Meteorological Monitoring Guidance for Regulatory Modeling Applications*. Environmental Protection Agency, Office of Air Quality Planning. EPA document EPA-454/R-99-005.

Baker, J. C., O'Connell, K. M., and Williamson, R., eds. 2001. *Commercial Observation Satellites*. Santa Monica, CA: Rand Corporation, p. 668.

Barnes, J. E., Bronner, S., Beck, R., and Parikh, N. C. 2003. Boundary layer scattering measurements with a charge-coupled device camera lidar. *Applied Optics* 42(15), 2647–2652.

Bartholome, E. M. and Belward, A. S. 2005. GLC.2000: A new approach to global land cover mapping from earth observation data. *International Journal of Remote Sensing* 26(9), 1959–1977.

Bateman McDonald, J., Nye, D., Hand, S., Hartman, M., McGee, M., Odum, E., Strand, W., Teague, J., and Wild, M. 2023. *Repeatability of a Terrestrial LiDAR Survey of a Stream Channel: Balus Creek*. Oakwood, GA: Project Developed at Institute for Environmental Spatial Analysis, University of North Georgia.

Belcher, E. O., Matsuyama, B., and Trimble, G. M. 2001. Object identification with acoustic lenses. In: *Proceedings of Oceans 2001 Conference*, Honolulu, HI: Marine Technical Society/IEEE, November 5–8, pp. 6–11.

Belward, A. S. 2012. Europe's relations with the wider world: A unique view from space. In: Venet, C. and Baranes, B., eds. *European Identity through Space: Space Activities and Programmes as a Tool to Reinvigorate the European Identity*. Vienna, Austria: Springer, p. 318.

Belward, A. S. and Skoien, J. O. 2014. Who launched what, when and why; trends in global land-cover observation capacity from civilian earth observation satellites. *ISPRS Journal of Photogrammetry and Remote Sensing*.

Beyrich, F. 1997. Mixing height estimation from SODAR data: A critical discussion. *Atmospheric Environment* 31(23), 3941–3953.

Blackinton, J. G., Hussong, D. M., and Kosalos, J. G. 1983. First results from a combination side-scan sonar and seafloor mapping system (SeaMARC II). In: *Proceedings of the 15th Annual Offshore Technology Conference*, Dallas, TX, pp. 307–314.

Blume, H. J. C., Kendall, B. M., and Fedors, J. C. 1978. Measurement of ocean temperature and salinity via microwave radiometry. *Boundary-Layer Meteorology* 13(1–4), 295–308.

Bowker, D. E., Davis, R. E., Myrick, D. L., Stacy, K., and Jones, W. T. 1985. Spectral reflectances of natural targets for use in remote sensing studies. NASA reference publication # 1139, Hampton, VA.

Brenner, A. C., DiMarzio, J. P., and Zwally, H. J. 2007. Precision and accuracy of satellite radar and laser altimeter data over the continental ice sheets. *IEEE Transactions on Geoscience and Remote Sensing* 45(2), 321–331.

Briney, A. 2014. An overview of remote sensing. http://geography.about.com/od/geographictechnology/a/remotesensing.htm. Accessed on May 2, 2014.

Brisco, B., Brown, R. J., and Manore, M. J. 1989. Early season crop discrimination with combined SAR and TM data. *Canadian Journal of Remote Sensing* 15(1), 44–54.

Brugioni, D. A. and Doyle, F. J. 1997. Arthur C. Lundahl: Founder of the image exploitation discipline. In: *Corona between the Sun and the Earth: The First NRO Reconnaissance Eye in Space*. Bethesda, MD: American Society for Photogrammetry and Remote Sensing. pp. 159–166.

Bulstrode, C. J. K., Goode, A. W., and Scott, P. J. 1986. Stereophotogrammetry for measuring rates of cutaneous healing: A comparison with conventional techniques. *Clinical Science* 71(4), 437–443.

Butler, D. 2014. Many eyes on earth. *Nature* 505, 143–144.

Butler, L. R. P. and Laqua, K. 1996. Nomenclature, symbols, units and their usage in spectrochemical analysis: IX. Instrumentation for the spectral dispersion and isolation of optical radiation. *Spectrochimica Acta Part B: Atomic Spectroscopy* 51(7), 645–664.

Canadian Center for Remote Sensing (CCRS). 2014. Fundamentals of remote sensing. www.nrcan.gc.ca/sites/www.nrcan.gc.ca/files/earthsciences/pdf/resource/tutor/fundam/pdf/fundamentals_e.pdf. Accessed on July 5, 2014.

Cash, M. and Panda, S. S. 2012. Urban stream water quality control with green island provision. In: Presented in *Southeast Lake and Watershed Management Conference*, Columbus, GA, May 13–15.

Center for Remote Imaging, Sensing, and Processing (CRISP). 2014. Electromagnetic waves. www.crisp.nus.edu. sg/~research/tutorial/em.htm. Accessed on May 2, 2014.

Chang, C. I. (Ed.). 2003. *Hyperspectral Imaging: Techniques for Spectral Detection and Classification* (Vol. 1). New York, NY: Springer Science & Business Media.

Chen, C. M., Hepner, G. F., and Forster, R. R. 2003. Fusion of hyperspectral and radar data using the IHS transformation to enhance urban surface features. *ISPRS Journal of Photogrammetry and Remote Sensing* 58(1), 19–30.

Cihlar, J. 2000. Land cover mapping of large areas from satellites: Status and research priorities. *International Journal of Remote Sensing* 21(6–7), 1093–1114.

Clark, R. N., King, T. V. V., Klejwa, M., and Swayze, G. A. 1990. High spectral resolution spectroscopy of minerals. *Journal of Geophysical Research* 95(B8), 12653–12680.

Clark, R. N., Swayze, G. A., Livo, K. E., Kokaly, R. F., King, T. V. V., Dalton, J. B., Vance, J. S., Rockwell, B. W., Hoefen, T., and McDougal, R. R. 2002. Surface reflectance calibration of terrestrial imaging spectroscopy data: A tutorial using AVIRIS. In: *Proceedings of the 10thAirborne Earth Science Workshop*, Pasadena, CA: JPL Publication 02–1.

Cleve, C., Kelly, M., Kearns, F. R., and Moritz, M. 2008. Classification of the wildland—urban interface: A comparison of pixel-and object-based classifications using high-resolution aerial photography. *Computers, Environment and Urban Systems* 32(4), 317–326.

Colwell, R. N., Ulaby, F. T., Simonett, D. S., Estes, J. E., and Thorley, G. A. 1983. *Manual of Remote Sensing. Interpretation and Applications* (Vol. 2). Falls Church, VA: American Society of Photogrammetry. 2143pp.

Congalton, R. G. and Green, K. 2008. *Assessing the Accuracy of Remotely Sensed Data: Principles and Practices*. Boca Raton, FL: CRC Press.

Connor, L. N., Laxon, S. W., Ridout, A. L., Krabill, W. B., and McAdoo, D. C. 2009. Comparison of ENVISAT radar and airborne laser altimeter measurements over Arctic sea ice. *Remote Sensing of Environment*, 113(3), 563–570.

De Castro, P., Adinolfi, F., Capitanio, F., Di Falco, S., Di Mambro, A. (Eds.) 2013. *The Politics of Land and Food Scarcity*. Abingdon, UK: Routledge, p. 154.

Defibaugh, Y., Chávez, J., and Tullis, J. A. 2013. Deciduous forest structure estimated with LiDAR-optimized spectral remote sensing. *Remote Sensing* 5(1), 155–182.

Dobson, M. C. and Ulaby, F. T. 1998. Mapping soil moisture distribution with imaging radar. In: *Principles and Applications of Imaging Radar.Manual of Remote Sensing* (Vol. 2). New York, NY: John Wiley, pp. 407–433.

Dong, Y., Forster, B., and Ticehurst, C. 1997. Radar backscatter analysis for urban environments. *International Journal of Remote Sensing* 18(6), 1351–1364.

Dornaika, F. and Hammoudi, K. 2009. Extracting 3D polyhedral building models from aerial images using a featureless and direct approach. In: *MVA*, IAPR Conference on Machine Vision Applications, Yokohama, JAPAN, May 20–22, pp. 378–381.

Dubayah, R. O., Knox, R. G., Hofton, M. A., Blair, J. B., and Drake, J. B. 2000. Land surface characterization using lidar remote sensing. In: M. Hill, and R. Aspinall, eds., *Spatial Information for Land Use Management*. Singapore: International Publishers Direct, pp. 25–38.

D'urso, G. and Minacapilli, M. 2006. A semi-empirical approach for surface soil water content estimation from radar data without a-priori information on surface roughness. *Journal of Hydrology* 321(1), 297–310.

Earth Observing System Data and Information System (EOSDIS). 2014. Earth data. https://earthdata.nasa.gov/. Accessed on July 5, 2014.

Edson, C. and Wing, M. G. 2011. Airborne light detection and ranging (LiDAR) for individual tree stem location, height, and biomass measurements. *Remote Sensing*, 3, 2494–2528. doi: 10.3390/rs3112494.

Engman, E. T. and Chauhan, N. 1995. Status of microwave soil moisture measurements with remote sensing. *Remote Sensing of Environment* 51(1), 189–198.

Erdody, T. L. and Moskal, L. M. 2010. Fusion of LiDAR and imagery for estimating forest canopy fuels. *Remote Sensing of Environment* 114, 725–737.

Ertberger, C., Jaume, A., and Panda, S. S. 2012. Estimation and evaluation of bacterial loadings of the Upper Chattahoochee watershed. In: Presented in *Southeast Lake and Watershed Management Conference*, Columbus, GA, May 13–15.

ESRI. 2014. GIS dictionary. http://support.esri.com/en/knowledgebase/GISDictionary/term/remote%20sensing. Accessed on July 5, 2014.

Everaerts, J. 2008. The use of unmanned aerial vehicles (UAVs) for remote sensing and mapping. *The International Archives of the Photogrammetry, Remote Sensing and Spatial Information Sciences* 37, 1187–1192.

Fairfield, N. and Wettergreen, D. 2008. Active localization on the ocean floor with multibeam sonar. In: *Proceedings of the IEEE/MTS OCEANS Conference and Exhibition*, pp. 1–10.

Friedl, M. A., McIver, D. K., Hodges, J. C. F., Zhang, X. Y., Muchoney, D., Strahler, A. H., Woodcock, C. E., et al. 2002. Global land cover mapping from MODIS: Algorithms and early results. *Remote Sensing of Environment* 83(1), 287–302.

Fung, A. K. and Chen, K. S. 2010. *Microwave Scattering and Emission Models for Users*. Norwood, MA: Artech House.

García, M., Riaño, D., Chuvieco, E., Salas, J., and Danson, F. M. 2011. Multispectral and LiDAR data fusion for fuel type mapping using Support Vector Machine and decision rules. *Remote Sensing of Environment* 115(6), 1369–1379.

Ge, X., Ding, J., Teng, D., Xie, B., Zhang, X., Wang, J., . . . and Wang, J. 2022. Exploring the capability of Gaofen-5 hyperspectral data for assessing soil salinity risks. *International Journal of Applied Earth Observation and Geoinformation* 112, 102969.

GEO. 2012. *The Group on Earth Observation 2012–2015 Work Plan*, GEO Secretariat. Geneva, Switzerland: WMO, p. 79. www.earthobservations.org/. Accessed on July 2, 2014.

Gernsheim, H. and Gernsheim, A. 1952. Re-discovery of the world's first photograph. *Photographic Journal*, 118.

Giles, K. A., Laxon, S. W., Wingham, D. J., Wallis, D. W., Krabill, W. B., Leuschen, C. J., McAdoo, D., et al. 2007. Combined airborne laser and radar altimeter measurements over the Fram Strait in May 2002. *Remote Sensing of Environment* 111(2), 182–194.

Glenn, N. F. and Carr, J. R. 2004. The effects of soil moisture on synthetic aperture radar delineation of geomorphic surfaces in the Great Basin, Nevada, USA. *Journal of Arid Environments* 56(4), 643–657.

Goel, M. and Srivastava, H. N. 1990. Monsoon trough boundary layer experiment (MONTBLEX). *Bulletin of the American Meteorological Society* 71(11), 1594–1600.

Goetz, A. F. 2009. Three decades of hyperspectral remote sensing of the Earth: A personal view. *Remote Sensing of Environment* 113, S5–S16.

Goetz, S. and Dubayah, R. 2011. Advances in remote sensing technology and implications for measuring and monitoring forest carbon stocks and change. *Carbon Management* 2(3), 231–244.

Gong, P., Wang, J., Yu, L., Zhao, Y., Zhao, Y., Liang, L., Niu, Z., et al. 2013. Finer resolution observation and monitoring of global land cover: First mapping results with Landsat TM and ETM+ data. *International Journal of Remote Sensing* 34(7), 2607–2654.

Gonzalez, R. C. and Woods, R. E. 2002. *Digital Image Processing*. Delhi, India: Pearson Education (Singapore) Ltd.

Gowen, A. A., O'Donnell, C., Cullen, P. J., Downey, G., and Frias, J. M. 2007. Hyperspectral imaging—an emerging process analytical tool for food quality and safety control. *Trends in Food Science and Technology* 18(12), 590–598.

Goyer, G. G. and Watson, R. 1963. The laser and its application to meteorology. *Bulletin of the American Meteorological Society* 44(9), 564.

Guthrie, J. and Walsh, K. 1997. Non-invasive assessment of pineapple and mango fruit quality using near infrared spectroscopy. *Animal Production Science* 37(2), 253–263.

Haack, B. and Bechdol, M. 2000. Integrating multisensor data and RADAR texture measures for land cover mapping. *Computers and Geosciences* 26(4), 411–421.

Hale, J. D., Tamblyn, C., Cash, M., and Panda, S. S. 2012. Fecal coliform and stream health analysis of Flat Creek with geo-spatial model development. In: Presented in *Southeast Lake and Watershed Management Conference*, Columbus, GA, May 13–15.

Hall, D. K., Fagre, D. B., Klasner, F., Linebaugh, G., and Liston, G. E. 2012. Analysis of ERS 1 synthetic aperture radar data of frozen lakes in northern Montana and implications for climate studies. *Journal of Geophysical Research:Oceans(1978–2012)* 99(C11), 22473–22482.

Hall, S. A., Burke, I. C., Box, D. O., Kaufmann, M. R., and Stoker, J. M. 2005. Estimating stand structure using discrete-return LiDAR: An example from low density, fire prone ponderosa pine forests. *Forest Ecology and Management* 208, 189–209.

Hodgson, M. E., Jensen, J. R., Tullis, J. A., Riordan, K. D., and Archer, C. M. 2003. Synergistic use of LiDAR and color aerial photography for mapping urban parcel imperviousness. *Photogrammetric Engineering and Remote Sensing* 69(9), 973–980.

Huisman, J. A., Hubbard, S. S., Redman, J. D., and Annan, A. P. 2003. Measuring soil water content with ground penetrating radar. *Vadose Zone Journal* 2(4), 476–491.

Jensen, J. R. 2009. *Remote Sensing of the Environment: An Earth Resource Perspective 2 /e*. Delhi, India: Pearson Education India, pp. 11–12.

Jha, M. K., Chowdhury, A., Chowdary, V. M., and Peiffer, S. 2007. Groundwater management and development by integrated remote sensing and geographic information systems: Prospects and constraints. *Water Resources Management* 21(2), 427–467.

Jhang, J., Panigrahi, S., Panda, S. S., and Borhan, M. S. 2002. Techniques for yield prediction from corn aerial images: A neural network approach. *International Journal of Agricultural and Biosystems Engineering* 3(1), 18–28.

Kasischke, E. S., Melack, J. M., and Craig Dobson, M. 1997. The use of imaging radars for ecological applications: A review. *Remote Sensing of Environment* 59(2), 141–156.

Kelly, B., Panda, S., Trettin, C., and Amatya, D. 2014. Assessment of the reach and ecological condition of freshwater tidal creeks in the lower coastal plain, Charleston County, South Carolina with advanced geospatial technology application. In: Poster Presented in *South Carolina Water Resources Conference*, Columbia, SC, October 15–16.

Kidd, R. B., Simm, R. W., and Searle, R. C. 1985. Sonar acoustic facies and sediment distribution on an area of the deep ocean floor. *Marine and Petroleum Geology* 2(3), 210–221.

Kim, M. S., Chen, Y. R., and Mehl, P. M. 2001. Hyperspectral reflectance and fluorescence imaging system for food quality and safety. *Transactions of the American Society of Agricultural Engineers* 44(3), 721–730.

Klemas, V. 2011. Remote sensing of coastal plumes and ocean fronts: Overview and case study. *Journal of Coastal Research* 28(1A), 1–7.

Kokaly, R. F., Clark, R. N., and Livo, K. E. 1998. Mapping the biology and mineralogy of Yellowstone National Park using imaging spectroscopy. In: *JPL Airborne Earth Science Workshop* (Vol. 7). Pasadena, CA, pp. 97–21.

Kokaly, R. F., Despain, D. G., Clark, R. N., and Livo, K. E. 2003. Mapping vegetation in Yellowstone National Park using spectral feature analysis of AVIRIS data. *Remote Sensing of Environment* 84(3), 437–456.

Kruse, F. A., Boardman, J. W., and Huntington, J. F. 2003. Comparison of airborne hyperspectral data and EO-1 Hyperion for mineral mapping. *IEEE Transactions on Geoscience and Remote Sensing* 41(6), 1388–1400.

Landsat Science. 2023. A new and revolutionary Landsat mission. https://landsat.gsfc.nasa.gov/satellites/landsat-next/. Accessed on November 15, 2023.

Leckie, D. G. and Ranson, K. J. 1998. Forestry applications using imaging radar. In: *Principles and Applications of Imaging Radar* (Vol. 2). New York, NY: John Wiley, pp. 435–509.

Lee, A. C. and Lucas, R. M. 2007. A LiDAR-derived canopy density model for tree stem and crown mapping in Australian forests. *Remote Sensing of Environment* 111, 493–518. doi: 10.1016/j.rse.2007.04.018.

Lees, B. G. and Ritman, K. 1991. Decision-tree and rule-induction approach to integration of remotely sensed and GIS data in mapping vegetation in disturbed or hilly environments. *Environmental Management* 15(6), 823–831.

Lefsky, M. A., Cohen, W. B., Parker, G. G., and Harding, D. J. 2002. LiDAR Remote Sensing for Ecosystem Studies LiDAR, an emerging remote sensing technology that directly measures the three-dimensional distribution of plant canopies, can accurately estimate vegetation structural attributes and should be of particular interest to forest, landscape, and global ecologists. *BioScience* 52(1), 19–30.

Lewis, A. J. and Henderson, F. M. 1998. Geomorphic and hydrologic applications of active microwave remote sensing. In: *Principles and Applications of Imaging Radar.Manual of Remote Sensing*. New York, NY: John Wiley, pp. 567–629.

Lillesand, T. M., Kiefer, R. W., and Chipman, J. W. 2004. *Remote Sensing and Image Interpretation*, 5th edn. New York, NY: John Wiley & Sons Ltd.

Loveland, T. R., Zhu, Z., Ohlen, D. O., Brown, J. F., Reed, B. C., and Yang, L. 1999. An analysis of the IGBP global land-cover characterization process. *Photogrammetric Engineering &Remote Sensing* 65(9), 1021–1032.

Lu, R. 2003. Detection of bruises on apples using near-infrared hyperspectral imaging. *Transactions of the American Society of Agricultural Engineers* 46(2), 523–530.

Martin, G. 2012. SumbandilaSat beyond repair, African defense and security news portal defence web, January 25. www.defenceweb.co.za/. Accessed on May 3, 2014.

McGlone, J. C. 2004. *Manual of Photogrammetry*, 5th edn. Bethesda, MD: ASPRS.

McKenney, B. A. and Kiesecker, J. M. 2010. Policy development for biodiversity offsets: A review of offset frameworks. *Environmental Management* 45(1), 165–176.

Metternicht, G. I. and Zinck, J. A. 2003. Remote sensing of soil salinity: Potentials and constraints. *Remote Sensing of Environment* 85(1), 1–20.

Mollicone, D., Achard, F., Federici, S., Eva, H. D., Grassi, G., Belward, A., Raes, F., et al. 2007. An incentive mechanism for reducing emissions from conversion of intact and non-intact forests. *Climatic Change* 83(4), 477–493.

Montenbruck, O. and Gill, E. 2000. *Satellite Orbits*. New York, NY: Springer.

Morgan, J. L., Gergel, S. E., and Coops, N. C. 2010. Aerial photography: A rapidly evolving tool for ecological management. *BioScience* 60(1), 47–59.

Moursund, R. A., Carlson, T. J., and Peters, R. D. 2003. A fisheries application of a dual-frequency, identification sonar, acoustic camera. *ICES Journal of Marine Science* 60, 678–683.

Murai, S. 1993. *Remote Sensing Note*. Tokyo, Japan: JARS.

NASA-Phoenix Mars Lander. 2014. www.nasa.gov/mission_pages/phoenix/news/phoenix-20080929.html. Accessed on July 2, 2014.

Nicolaï, B. M., Beullens, K., Bobelyn, E., Peirs, A., Saeys, W., Theron, K. I., and Lammertyn, J. 2007. Nondestructive measurement of fruit and vegetable quality by means of NIR spectroscopy: A review. *Postharvest Biology and Technology* 46(2), 99–118.

NSSDC. 2014. The NASA master directory held at the NASA space science data center. http://nssdc.gsfc.nasa.gov/nmc/SpacecraftQuery.jsp. Accessed on May 2, 2014.

Osborne, P. E., Alonso, J. C., and Bryant, R. G. 2001. Modelling landscape-scale habitat use using GIS and remote sensing: A case study with great bustards. *Journal of Applied Ecology* 38(2), 458–471.

Overbeck, J., Misra, D., Kinsman, N., and Panda, S. 2024. *Nearshore Bathymetry Change Detection with WorldView-2 Multispectral Satellite Data*. Manuscript to be submitted to *Remote Sensing* (MDPI).

Panda, S. S., Ames, D. P., and Panigrahi, S. 2010a. Application of vegetation indices for agricultural crop yield prediction using neural network. *Remote Sensing* 2(3), 673–696.

Panda, S. S., Andrianasolo, H., Murty, V. V. N., and Nualchawee, K. 2004. Forest management planning for soil conservation using satellite images, GIS mapping, and soil erosion modeling. *Journal of Environmental Hydrology* 12(13), 1–16.

Panda, S. S., Burry, K., and Tamblyn, C. 2012a. Wetland change and cause recognition in Georgia coastal plain. In: Presented in the *International American Society of Agricultural and Biological Engineers (ASABE) Conference 2012*, Dallas, TX, July 29—August 2. Paper # 1338205.

Panda, S. S. and Hoogenboom, G. 2013. Blueberry orchard site specific crop management with geospatial based yield modeling. In: Presented in the *International American Society of Agricultural and Biological Engineers (ASABE) Conference 2012*, Kansas City, MO, July 21–24. Paper # 1620890.

Panda, S. S., Hoogenboom, G., and Paz, J. 2009. Distinguishing blueberry bushes from mixed vegetation land-use using high resolution satellite imagery and geospatial techniques. *Computers and Electronics in Agriculture* 67(1–2), 51–59.

Panda, S. S., Hoogenboom, G., and Paz, J. 2010b. Remote sensing and geospatial technological applications for site-specific management of fruit and nut crops: A review. *Remote Sensing* 2(8), 1973–1997.

Panda, S. S., Martin, J., and Hoogenboom, G. 2011a. Blueberry crop growth analysis using climatologic factors and multi-temporal remotely sensed imageries. In: Carroll, D., ed. Published in the Peer Reviewed *Proceedings of 2011 Georgia Water Resources Conference*, Athens, GA, April 11–13. www.gawrc.org/2011proceedings.html.

Panda, S. S., Misra, D., Amatya, D., and Mukherjee, S. 2024a. Geospatial and ANN modeling approach for forest evapotranspiration estimation in varying climate. In: *Presented in the 2024 American Geophysical Union (AGU)-Chapman Conference*, Honolulu, HI, February 13–18.

Panda, S. S., Misra, D., Sahoo, D., and Jat, P. 2024b. Machine learning approach for evapotranspiration change estimation due to climate change induced permafrost thawing. In: *Presented in the 2024 American Geophysical Union (AGU)-Chapman Conference*, Honolulu, HI, February 13–18.

Panda, S. S., Nolan, J., Amatya, D., Dalton, K., Jackson, R. M., Ssegane, H., and Chescheir, G. 2012b. Stomatal conductance and leaf area index estimation using remotely sensed information and forest speciation. In: Presented in the *Third International Conference on Forests and Water in Changing Environment 2012*, Fukuoka, Japan, September 18–20.

Panda, S. S., Siddique, A., Terrill, T., Mohapatra, A., Morgan, E., Pech-Cervantes, A. A., and VanWyk, J. 2023. Precision agriculture decision support system development as a WebGIS dashboard for Lespedeza Cuneata production quality and quantity management. In: *Presented in the Machine Vision for Agricultural Robotics session of the Annual International ASABE Conference 2023*, Omaha, NE, July 09–13.

Panda, S. S., Steele, D. D., Panigrahi, S., and Ames, D. P. 2011b. Precision water management in corn using automated crop yield modeling and remotely sensed data. *International Journal of Remote Sensing Applications* 1(1), 11–21.

Panigrahi, S., Gautam, R., Gu, H., Panda, S. S., Venugopal, M., and Kizil, U. 2003. Fluorescence imaging for quality assessment of meat. ASAE Paper No. RRV03–0025, St. Joseph, MI.

Perry, M. K. and Panda, S. S. 2022. Even wetlands can burn: Developing a wildfire spatial vulnerability model for the Okefenokee Swamp for extreme geohazard mitigation decision support. In: *Annual International ASABE Conference 2022*, Houston, TX, July 17–20.

Perry, M. K., Panda, S. S., Amatya, D. M., Oyuyang, Y., Grace, J., and Jalowska, A. 2022. Comprehensive geospatial modeling for flood potential analysis in present day climate change condition: Yellowstone national park case study. *Presented in the 2022 American Geophysical Union (AGU) Conference*, Chicago, IL, December 12–16.

Petrie, G. 2009. Systematic oblique aerial photography using multiple digital cameras. *Photogrammetric Engineering & Remote Sensing* 75(2), 102–107.

Phillips, J., Tamblyn, C., Smith, A., and Panda, S. S. 2012. Impact of urbanization and point source on changes in water quality in upstream of Upper Chattahoochee River. In: Presented in *Southeast Lake and Watershed Management Conference*, Columbus, GA, May 13–15.

Poland, M., Hurwitz, S., and McCleskey, R. B. 2022. *How Might the Devastating June 2022 Floods in and Around Yellowstone National Park Influence Seismic and Hydrothermal Activity? United States Geological Survey*. United States Department of the Interior, 20 June. www.usgs.gov/observatories/yvo/news/how-might-devastating-june-2022-floods-and-around-yellowstone-national-park. Accessed on August 3, 2022.

Powell, R. L. and Matzke, N. 2004. Sources of error in accuracy assessment of thematic land-cover maps in the Brazilian Amazon. *Remote Sensing of Environment* 90(2), 221–234.

Rao, M., Awawdeh, M., and Dicks, M. 2005. Spatial allocation and environmental benefits: The impacts of the conservation reserve program in Texas County Oklahoma. In: Allen, A. W. and Vandever, M. W., eds. *The Conservation Reserve Program—Planting for the Future: Proceedings of a National Conference*, Fort Collins, CO, June 8–9.

Rao, M. and Miller, E. 2012. Mapping of serpentine soils in Lassen and Plumas National Forests. *CSU Geospatial Review* 10, 6.

Rao, M. N. 2012. Mapping snag locations in the blacks mountain experimental forest using LiDAR data. Technical Report Submitted to USFS-PSW Research Station, CA, April 19.

Rao, M. N. 2013. Mapping of serpentine soils in the Lassen and Plumas National Forest using integrated multispectral and LiDAR data. Technical Report Submitted to USFS-Mt. Hough Ranger District, CA, March 13.

Rao, M. N., Fan, G., Thomas, J., Cherian, G., Chudiwale, V., and Awawdeh, M. 2007. A web-based GIS decision support system for managing and planning USDA's Conservation Reserve Program (CRP). *Environmental Modelling and Software* 22, 1270–1280.

Renslow, M. S. 2012. Manual of airborne topographic LiDAR. ASPRS publication. ISBN 1-57083-097-5 Stock # 4587.

Riaño, D., Chuvieco, E., Condes, S., Gonzalez-Matesanz, J., and Ustin, S. L. 2004. Generation of crown bulk density for *Pinus sylvestris* L. from LiDAR. *Remote Sensing of Environment* 92, 345–352.

Riaño, D., Chuvieco, E., Ustin, S. L., Salas, J., Rodríguez-Pérez, J. R., Ribeiro, L. M., Viegas, D. X., Moreno, J. M., and Helena, F. 2007. Estimation of shrub height for fuel-type mapping combining airborne LiDAR and simultaneous color infrared ortho imaging. *International Journal of Wildland Fire* 16(3), 341–348.

Riopel, S. 2000. *The Use of RADARSAT-1 Imagery for Lithological and Structural Mapping in the Canadian High Arctic*. Ottawa, Ontario, Canada: University of Ottawa.

Ritchie, J. C., Zimba, P. V., and Everitt, J. H. 2003. Remote sensing techniques to assess water quality. *Photogrammetric Engineering and Remote Sensing* 69(6), 695–704.

Russo, J. C., Amduka, M., Gelfand, B., Pedersen, K., Lethin, R., and Springer, J. 2006. Enabling cognitive architectures for UAV mission planning. www.atl.external.lmco.com/papers/1396.pdf. Accessed on July 5, 2014.

Rylee, J., Panda, S. S., Fitzgerald, J., and Hohnhorst, D. 2012. Geospatial technology based suitability analysis for new additional reservoirs in Hall County, GA. In: Presented in *Southeast Lake and Watershed Management Conference*,, Columbus, GA, May 13–15.

Savenije, B., Geesink, G. H., Van der Palen, J. G. P., and Hemke, G. 2006. Prediction of pork quality using visible/near-infrared reflectance spectroscopy. *Meat Science* 73(1), 181–184.

Schott, J. R. 2007. *Remote Sensing: The Image Chain Approach*, 2nd edn. New York, NY: Oxford University Press, p. 1.

Schowengerdt, R. A. 2007. *Remote Sensing: Models and Methods for Image Processing*, 3rd edn. Chicago, IL: Academic Press, p. 2.

senseFly. 2014. eBee: The professional mapping drone. www.sensefly.com/drones/ebee.html. Accessed on October 10, 2014.

Sharma, J. B. and Husley, D. 2014. Integrating the UAS in undergraduate teaching and research-opportunities and challenges at the University of North Georgia. In: *Proceedings of the Joint ASPRS Pecora 2014 Conference and the ISPRS Technical Commision 1 and IAG Commision 4 Meeting*, Denver, CO, November.

Shrum, T. R., Travis, W. R., Williams, T. M., and Lih, E. 2018. Managing climate risks on the ranch with limited drought information. *Climate Risk Management* 20, 11–26.

Shaw, G. A. and Burke, H. H. K. 2003. Spectral imaging for remote sensing. *Lincoln Laboratory Journal* 14(1), 3–28.

Shibata, K. 1957. Spectroscopic studies on chlorophyll formation in intact leaves. *Journal of Biochemistry* 44(3), 147–173.

Simard, M., Pinto, N., Fisher, J. B., and Baccini, A. 2011. Mapping forest canopy height globally with spaceborne LiDAR. *Journal of Geophysical Research* 116, G04021. doi: 10.1029/2011JG001708.

Skarda, R. J., Panda, S. S., and Sharma, J. B. 2011. An assessment of the impact of retention ponds for sediment trapping in the Ada Creek and Longwood Cove using remotely sensed data and GIS analysis. In: Published in the *Proceedings of International Symposium on Erosion and Landscape Evolution Hilton Anchorage Hotel*, Anchorage, AK, September 18–21. ISELE Paper Number 11025.

Suárez, J. C., Ontiveros, C., Smith, S., and Snape, S. 2005. Use of airborne LiDAR and aerial photography in the estimation of individual tree heights in forestry. *Computers and Geosciences* 31(2), 253–262.

Swayze, G. A., Clark, R. N., Goetz, A. F. H., Livo, K. E., Breit, G. N., Kruse, F. A., Stutley, S. J., et al. 2014. Mapping advanced argillic alteration at Cuprite, Nevada using imaging spectroscopy. *Economic Geology* 109(5), 1179–1221. doi: 10.2113/econgeo.109.5.1179.

Tiffan, K. F., Rondorf, D. W., and Skalicky, J. J. 2004. Imagining fall Chinook salmon redds in the Columbia River with a dual frequency identification sonar. *North American Journal of Fisheries Management* 24, 1421–1426.

Townshend, J. R. and Justice, C. O. 1988. Selecting the spatial resolution of satellite sensors required for global monitoring of land transformations. *International Journal of Remote Sensing* 9(2), 187–236.

Townshend, J. R., Latham, J., Arino, O., Balstad, R., Belward, A., Conant, R., Elvidge, C., et al. 2008. Integrated global observation of the land: An IGOS-P theme, IGOL Report No. 8, GTOS 54, FAO, Rome, Italy, p. 74.

Trenear-Harvey, G. S. 2009. *Historical Dictionary of Air Intelligence*. Scarecrow Press.

U.S. Geological Survey. 1992a. The National Aerial Photography Program (NAPP), factsheet: Reston, VA, U.S. Geological Survey, p. 1.

U.S. Geological Survey. 1992b. NHAP and NAPP photographic enlargements, factsheet: Reston, VA, U.S. Geological Survey, p. 1.

USGS. 2024. U.S. geological survey, 2024, Landsat next (ver. 1.1, March 25, 2024): U.S. geological survey fact sheet 2024–3005, p. 2. doi: 10.3133/fs20243005. ISSN: 2327–6932 (online) ISSN: 2327–6916 (print).

Valavanis, V. D., Pierce, G. J., Zuur, A. F., Palialexis, A., Saveliev, A., Katara, I., and Wang, J. 2008. Modelling of essential fish habitat based on remote sensing, spatial analysis and GIS. *Hydrobiologia* 612(1), 5–20.

Van Diggelen, F. S. T. 2009. *A-GPS: Assisted GPS, GNSS, and SBAS*. Norwood, MA: Artech House.

Van Wagtendonk, J. W., Root, R. R., and Key, C. H. 2004. Comparison of AVIRIS and Landsat ETM+ detection capabilities for burn severity. *Remote Sensing of Environment* 92(3), 397–408.

Vergopolan, N., Chaney, N. W., Pan, M., Sheffield, J., Beck, H. E., Ferguson, C. R., . . . and Wood, E. F. 2021. SMAP-HydroBlocks, a 30-m satellite-based soil moisture dataset for the conterminous US. *Scientific Data* 8(1), 264.

Vierling, K. T., Vierling, L. A., Gould, W. A., Martinuzzi, S., and Clawges, R. M. 2008. LiDAR: Shedding new light on habitat characterization and modeling. *Frontiers in Ecology and the Environment* 6(2), 90–98.

Vuong, H. and Rao, M. 2013. Using LiDAR to estimate total aboveground biomass of redwood stands in the Jackson Demonstration State Forest, Mendocino, California, B23B-0557. In: *Proceedings of the 2013 Fall Meeting*, American Geophysical Union (AGU), San Francisco, CA, December 9–13.

Wagner, W., Bloschl, G., Pampaloni, P., Calvet, J. C., Bizzarri, B., Wigneron, J. P., and Kerr, Y. 2007. Operational readiness of microwave remote sensing of soil moisture for hydrologic applications. *Nordic Hydrology* 38(1), 1–20.

Waring, R. H., Way, J., Hunt, E. R., Morrissey, L., Ranson, K. J., Weishampel, J. F., and Franklin, S. E. 1995. Imaging radar for ecosystem studies. *BioScience* 45, 715–723.

Wilson, W. J., Yueh, S. H., Dinardo, S. J., Chazanoff, S. L., Kitiyakara, A., Li, F. K., and Rahmat-Samii, Y. 2001. Passive active L-and S-band (PALS) microwave sensor for ocean salinity and soil moisture measurements. *IEEE Transactions on Geoscience and Remote Sensing* 39(5), 1039–1048.

Withee, G. W., Smith, D. B., and Hales, M. B. 2004. Progress in multilateral Earth observation cooperation: CEOS, IGOS and the ad hoc group on earth observations. *Space Policy* 20(1), 37–43.

Yang, Z., Rao, M., Elliott, N., Kindler, S., and Popham, T. W. 2005. Using ground-based multispectral radiometry to detect stress in wheat caused by Greenbug (Homoptera: Aphididae) infestation. *Computers and Electronics in Agriculture* 47, 121–135.

Ye, B., Tian, S., Cheng, Q., and Ge, Y. 2020. Application of lithological mapping based on advanced hyperspectral imager (AHSI) imagery onboard Gaofen-5 (GF-5) satellite. *Remote Sensing* 12(23), 3990.

Yu, B., Liu, H., Wu, J., Hu, Y., and Zhang, L. 2010. Automated derivation of urban building density information using airborne LiDAR data and object-based method. *Landscape and Urban Planning* 98(3), 210–219.

Zhang, X., Han, W., Li, J., and Wang, L. 2023. Nearshore bathymetry estimation through dual-time phase satellite imagery in the absence of in-situ data. *GIScience & Remote Sensing* 60, 1. doi: 10.1080/15481603.2023.2275424.

# Part II

## Global Navigation Satellite Systems (GNSS) and Their Characteristics

# 2 Global Navigation Satellite Systems GNSS Theory and Practice

## *Evolution, State of the Art, and Future Pathways*

*Mohinder S. Grewal*

## 2.1 INTRODUCTION

Global Navigation Satellite Systems (GNSS) have become almost a household term. GNSS applications track trains, guide planes, and make the world a smaller more functional planet, as shown in Figure 2.1.

There are currently four Global Navigation Satellite Systems (GNSSs) operating or being developed. The first one is the Global Positioning System (GPS), developed by the United States (US) Department of Defense (DOD) under its Navstar program. The U.S. launched its first GPS satellite in 1978, and the system was declared with a Final Operational Capability (FOC) in April 1995. The second GNSS configuration developed was the Global Orbiting Navigation Satellite System (GLONASS), placed in orbit by the Soviet Union and now maintained by the Russian Republic. The first satellite was launched in 1982 and GLONASS was declared an operational system on September 1993. The European Union (EU) and European Space Agency (ESA 2014) is developing the Galileo System, the third GNSS. The first two operational satellites were launched in October 2011, preceded by two Galileo In-Orbit Validation Elements (GIOVE A and B) launched in 2005 and 2008. China's COMPASS, or BeiDou ("Big Dipper") is the fourth GNSS, initially developed in 2000–2003. In December 2012, the BeiDou Navigation Satellite System provided fully operational regional service. It has had global coverage since 2020.

A summary of the signal characteristics of the four GNSSs is given in Figures 2.1 and 2.2. Detailed GPS and GLONASS descriptions are given in Sections 2.2 and 2.3 respectively. Sections 2.4 and 2.5 describe Galileo and Beidou-2, respectively. Section 2.7 gives conclusions, followed by References.

To improve the accuracy and integrity of the GNSSs, various countries are developing Space Based Augmentation Systems (SBAS). SBAS uses Geostationary Earth Orbit (GEO) satellites to relay corrections and integrity information to users. SBAS is a Differential Global Navigation Satellite System (DGNSS). DGNSSs reduce the errors in GNSS-derived positions by using additional data from a reference GNSS receiver at a known location. The most common form of DGNSS involves determining the combined effects of navigation message ephemeris and satellite clock errors (including the effects of propagation) at a reference station, and transmitting corrections and integrity in real time to a user's receiver (Grewal and Andrews 2015, Grewal et al. 2020, Grewal 2023).

DOI: 10.1201/9781003541141-4

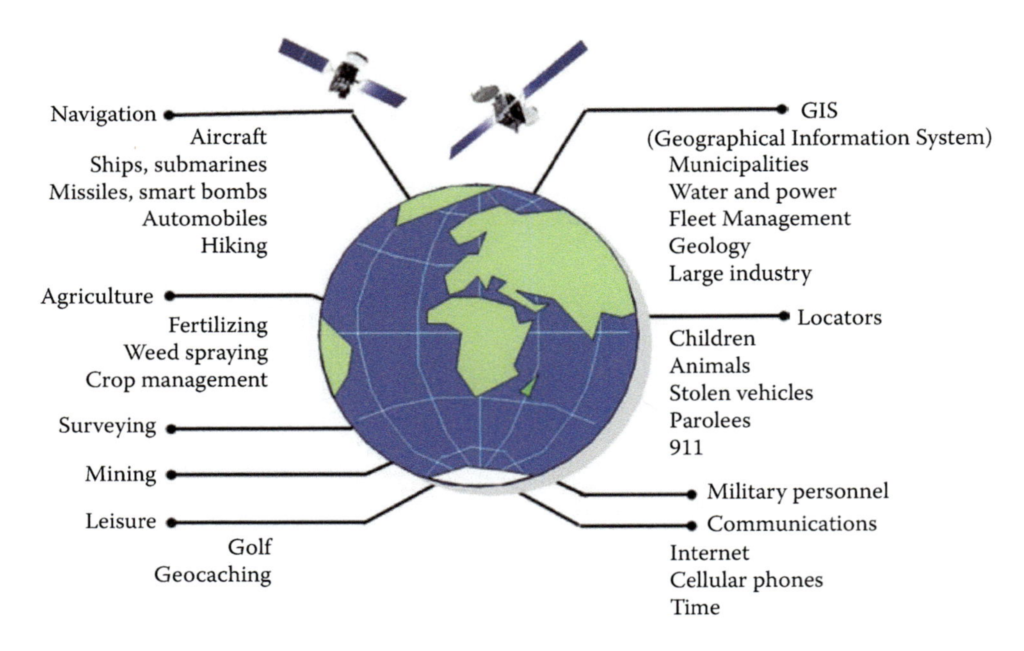


**FIGURE 2.1** Global Navigation Satellite Systems applications make the world a smaller more functional planet.

**TABLE 2.1**
**Comparison of Various Global Navigation Satellite Systems (GNSSs)**

| | GPS | GLONASS | GALILEO | COMPASS |
|---|---|---|---|---|
| Number of satellites | 31 | 24 | 30 | 56 |
| Orbit radius (km) | 26,600 | 25,510 | 29,600 | 19,100 |
| Number of orbit planes when fully developed (global) | 6 | 3 | 3 | Now regional |
| Orbit plane inclination | 55.5° | 64.8° | 54° | 55° |
| Architecture | CDMA (Code Division Multiple Access) | FDMA (Frequency Division Multiple Access) | CDMA | CDMA |
| Selective Availability (SA) | No (SA discontinued in 2000) | No | No | No |
| Service | Free | Free | Encrypted | Fee |
| Codes | C&P (Coarse Align and Precision) same for all PRN | C, P (Different codes for each PRN) | C&P (Coarse Align and Precision) same for all PRN | C&P (Coarse Align and Precision) same for all PRN |
| Carrier (MHz) | 1575.42 | 1602 | 1575.42 | 1561.098 |
| PRN (C/A) code length (chips) | 1023 | 511 | 4096 | 2046 |
| Data bit rate | 20 ms | 20 ms | 4 ms | 20 ms |

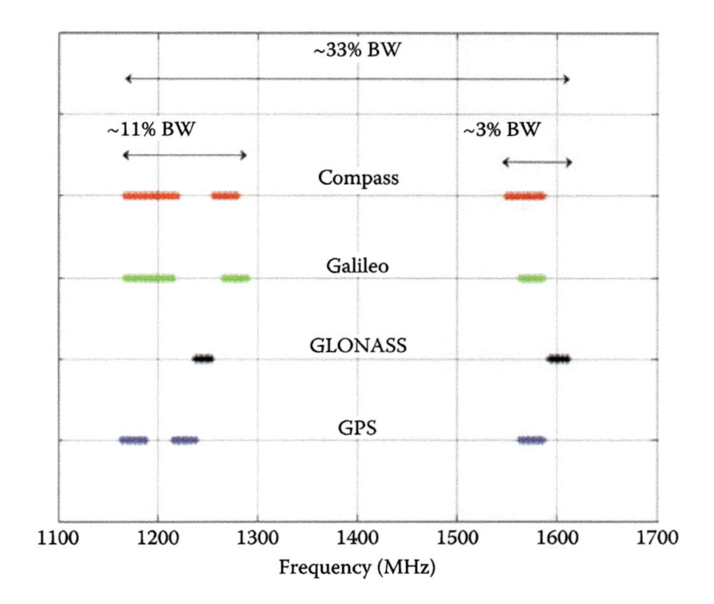


FIGURE 2.2 Comparison of frequency and bandwidth (BW) for various GNSSs.

## 2.2 GPS

### 2.2.1 GPS Orbits

The fully operational GPS includes 31 or more active satellites approximately uniformly dispersed around six circular orbits with four or more satellites each. The orbits are inclined at an angle of 55° relative to the equator and are separated from each other by multiples of 60° right ascension. The orbits are nongeostationary and approximately circular, with radii of 26,560 km and orbital periods of one-half sidereal day (≈11.967 h). Theoretically, three or more GPS satellites will always be visible from most points on the Earth's surface, and four or more GPS satellites can be used to determine an observer's position anywhere on the Earth's surface 24 hours per day.

### 2.2.2 GPS Signals

Each GPS satellite carries a cesium and/or rubidium atomic clock to provide timing information for the signals transmitted by the satellites. Internal clock correction is provided for each satellite clock. Each GPS satellite transmits two spread spectrum, L-band carrier signals—an L1 signal with carrier frequency $f_1 = 1575.42$ MHz and an L2 signal with carrier frequency $f_2 = 1227.6$ MHz. These two frequencies are integral multiples $f_1 = 1540 f_0$ and $f_2 = 1200 f_0$ of a base frequency $f_0 = 1.023$ MHz. The L1 signal from each satellite uses *binary phase-shift keying* (BPSK), modulated by two *pseudorandom noise* (PRN) codes in phase quadrature, designated as the Coarse Acquisition or C/A-code and Precise or P-code. The L2 signal from each satellite is BPSK modulated by only the P-code. A brief description of the nature of these PRN codes follows.

### 2.2.3 Selective Availability (Historical Note)

Prior to May 1, 2000, the U.S. Department of Defense deliberately derated the accuracy of GPS for "nonauthorized" (i.e., non-U.S. military) users, using a combination of methods. This selective availability (SA) included pseudorandom time dithering and truncation of the transmitted ephemerides. SA degraded the navigation solution by 100 m horizontally and 156 m vertically. The initial satellite configuration used SA with pseudorandom dithering of the onboard time reference (Janky 1997) only, but this was discontinued on May 1, 2000.

#### 2.2.3.1 Precise Positioning Service

Precise Positioning Service (PPS) is the full-accuracy (within 5 m), single-receiver GPS positioning service provided to the United States and its allied military organizations and other selected agencies. This formal, proprietary service includes access to the unencrypted P-code.

#### 2.2.3.2 Standard Positioning Service without SA

Standard Positioning Service (SPS) provides GPS single-receiver (stand-alone) positioning service to any user on a continuous, worldwide basis. SPS is intended to provide access only to the C/A-code and the L1 carrier.

#### 2.2.3.3 Standard Positioning Service with SA

The horizontal-position accuracy, as degraded by SA, advertised as 100 m when in use, the vertical-position accuracy as 156 m, and time accuracy as 334 ns—all at the 95% probability level. SPS also guaranteed the user-specified levels of coverage, availability, and reliability. This was discontinued on May 1, 2000.

### 2.2.4 Modernization of GPS

Since GPS was declared with a Final Operational Capability (FOC) in April 1995, applications and the use of GPS have evolved rapidly, especially in the civil sector. See Figure 2.1 for a sampling of applications. As a result, radically improved levels of performance have been reached in positioning, navigation, and time transfer. However, the availability of GPS has also spawned new and demanding applications that reveal certain shortcomings of the present system. Therefore, since the mid-1990s numerous governmental and civilian committees have investigated these shortcomings and requirements for a more modernized GPS.

The modernization of GPS is a difficult and complex task that requires tradeoffs in many areas. Major issues include spectrum needs and availability, military and civil performance, signal integrity and availability, financing and cost containment, and potential competing interests by other GNSSs and developing countries. Major decisions have been made for the incorporation of new civil frequencies, and new civil and military signals that will enhance the overall performance of the GPS.

GPS II F and GPS III are being designed under various contracts (Raytheon, Lockheed Martin). These will have a new L2 civil signal and new L5 signal modulated by a new code structure. These frequencies will improve the ambiguity resolution, ionospheric calculation, and C/A code positioning accuracy.

#### 2.2.4.1 Areas to Benefit from Modernization

The areas that could benefit from a modernized GPS are the following:

1. *Robust dual-frequency ionosphere correction capability for civil users:* Since only the encrypted P(Y) code appears on the L2 frequency, civil users have lacked a robust dual-frequency ionosphere. Civil users had to rely on semi-codeless tracking of the GPS L2 signal, which is not as robust as access to a full strength unencrypted signal. While civil users could employ a differential technique, this adds complexity to an ionosphere free user solution.
2. *A better civil code:* While the GPS C/A code is a good, simple spreading code, a better civil code would provide better correlation performance. Rather than just turning on the C/A code on the L2 frequency, a more advanced spreading code would provide robust ranging and ionosphere error predictions.

3. *Ability to resolve ambiguities in phase measurements needs improvement:* High-accuracy differential positioning at the centimeter level by civil users requires rapid and reliable resolution of ambiguities in phase measurements. Ambiguity resolution with single-frequency (L1) receivers generally requires a sufficient length of time for the satellite geometry to change significantly. Performance is improved with robust dual-frequency receivers. However, the effective signal to noise ratio (SNR) of the legacy P(Y) encrypted signal is dramatically reduced because the encrypted P code cannot be despread by the civil user.
4. *Dual-frequency navigation signals in the ARNS band:* The Aeronautical Radio Navigation Services (ARNS) band of frequencies are federally protected and can be used for safety-of-life applications. The GPS L1 band IS an ARNS band but the GPS L2 band is not. (The L2 band is in the Radio Navigation Satellite Services Band (RNSS) that has a substantial amount of uncontrolled signals in it.) In applications involving public safety the integrity of the current system can be improved with a robust dual-frequency capability where both GPS signals are within the ARNS bands. This is particularly true in aviation landing systems that demand the presence of an adequate number of high integrity satellite signals and functional crosschecks during precision approaches.
5. *Improvement is needed in multipath mitigation capability.* Multipath remains a dominant source of GPS positioning error and cannot be removed by differential techniques. Although certain mitigation techniques, such as multipath mitigation technology (MMT), approach theoretical performance limits for in-receiver processing, the required processing adds to receiver costs. In contrast, effective multipath rejection could be made available to all receivers by using new GPS signal designs.
6. *Military requirements in a jamming environment:* The feature of Selective Availability (SA) was suspended at 8 p.m. EDT on May 1, 2000. SA was the degradation in the autonomous positioning performance of GPS which was a concern in many civil applications requiring the full accuracy of which GPS is capable. If the GPS C/A code was ever interfered with, the accuracies that it could be afforded would not be present. Because the P(Y) code has an extremely long period (seven days), it is difficult to acquire unless some knowledge of the code timing is known. P(Y) timing information is supplied by the GPS Hand Over Word (HOW) at the beginning of every subframe. However, to read the HOW, the C/A-code must first be acquired to gain access to the navigation message. Unfortunately, the C/A-code is relatively susceptible to jamming, which would seriously impair the ability of a military receiver to acquire the P(Y) code. Direct P(Y) acquisition techniques are possible, but these techniques still require information about the satellite position, user position, and clock errors to be successful. Furthermore, an interference on the C/A coded signal, would also effect the P(Y) coded signal in the same frequency band. It would be far better if direct acquisition of a high-performance code were possible, without the need to first acquire the C/A code.
7. Additional power received from the satellite would also help military users operate more effectively. While same advantages can be gained in the user equipment, having an increased power from the satellite could provided added value.
8. *Compatibility and Operability with other GNSSs:* With the advances in other GNSSs by other nations, the requirement exists for international cooperation in the development of new GNSS signals to ensure they do not interfere with each other and potentially provide an interoperable combined GNSS service.

### 2.2.5 Elements of the Modernized GPS

Figure 2.3 illustrates the modernized GPS signal spectrum. The legacy GPS signals (L1 C/A and P(Y), and L2 P(Y)) and the additional modernized signals (L2C, L5, and GPS L1 and L2 M-code) are illustrated. The L1C signal is not shown.

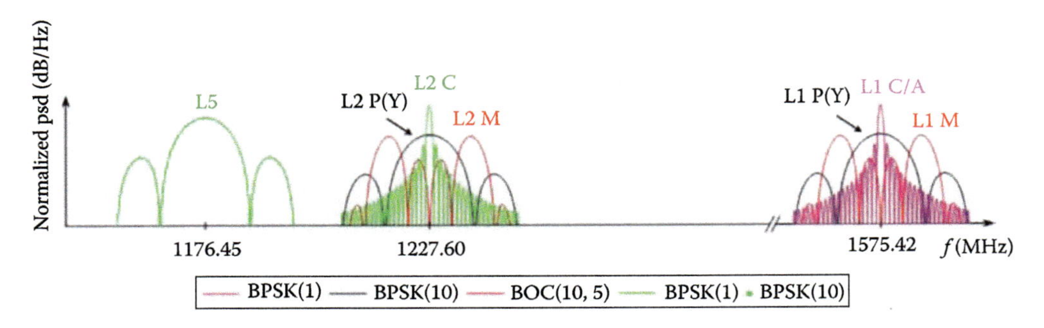


**FIGURE 2.3** A modernized GPS signal spectrum shown along with the legacy GPS signals.

The bandwidths potentially available for the modernized GPS signals is up to 24 MHz, but the compatibility and power levels relative to the other codes needs to be considered. Furthermore, assuming equal received power and filtered bandwidth, the ranging performance (with or without multipath) on a GPS signal, the performance is highly dependent upon the signal's spectral shape (or equivalently, the shape of the autocorrelation function). In this sense, the L1 C/A coded and L2 civil signals are somewhat equivalent in scope, as are the P(Y) and L5 civil signals (albeit with very different characteristics). As we will see, the military M-coded signal and other GNSS codes are different because they use different subcarrier frequencies and chipping rates. These different subcarriers, in essence, add an aspect known as frequency division multiplexing to the GPS spectrum.

The major elements of these modernized signals are discussed in the following sections.

#### 2.2.5.1 L2 Civil Signal (L2C)

This new civil signal has a new code structure that has some performance advantages over the legacy C/A code. The L2C signal offers civilian users the following improvements:

(*a*) *Robust dual-frequency ionosphere error correction.* The dispersive delay characteristic of the ionosphere proportional to $1/f^2$ can be estimated much more accurately with this new, full strength signal on the L2 frequency. Thus, civil users can choose to use a semi-codeless P(Y) L2 and C/A L1, or a new L2C and C/A L1 technique to estimate the ionosphere.

(*b*) *Carrier phase ambiguity resolution will be significantly improved.* The accessibility of the full strength L1 and L2 signals provides "widelane" measurement combinations having ambiguities that are much easier to resolve.

(*c*) *The additional L2C signal will improve robustness in acquisition and tracking.* The new spreading code identified for civil (C) users will provide more robust acquisition and tracking performance.

Originally, modernization efforts considered turning on the C/A-code at the L2 carrier frequency (1227.60 MHz) to provide the civilian community with a robust ionosphere correction capability as well as additional flexibility and robustness. However, later in the planning process it was realized that additional advantages could be obtained by replacing the planned L2 C/A signal with a new L2 civil signal (L2C). The decision was made to use this new signal, and its structure was made public early in 2001. Both the L2C and the new military M code signal (to be described) appear on the L2 carrier orthogonal to the current P(Y).

Like the C/A code, the C code is a PRN code that runs at a $1.023 \times 10^6$ cps (chips per second) rate. However, it is generated by 2:1 time-division multiplexing of two independent subcodes, each having half the chipping rate, namely $511.5 \times 10^3$ cps. Each of these subcodes is made available to the receiver by demultiplexing. These two subcodes have different periods before they repeat. The first subcode, the code-moderate (CM) has a moderate length of 10,230 chips, a 20-ms period. The moderate length of this code permits relatively easy acquisition of the signal although the 2:1 multiplexing results in a 3 dB acquisition and data demodulation loss. The second subcode, the code-long (CL) has a length of 707,250 chips, a 1.5 s period, and is data-free. The CM and CL codes are combined to provide the C code at the 1.023 Mcps rate. Navigation data can be modulated on the C code. Provisions call for no data, legacy navigation data at a 50 bps rate, or new Civil navigation (CNAV) data at a 25 bps; the CNAV data at a 25 bps rate would be encoded using a rate ½ convolutional encoding technique, to produce a 50 sps (symbols per second) data bit stream that could then be modulated onto the L2C signal. With no data, the coherent processing time can be increased substantially, thereby permitting better code and carrier tracking performance, especially at low SNR. The relatively long CL code length also generates smaller correlation sidelobes as compared to the C/A code. Details on the L2 civil signal are given by (Fontana et al. 2001).

The existing C/A code at the L1 frequency will be retained for legacy purposes.

#### 2.2.5.2 L5 Signal

Although the use of the L1 and L2C signals can satisfy most civil users, there are concerns that the L2 frequency band may be subject to unacceptable levels of interference for applications involving public safety, such as aviation. The potential for interference arises because the International Telecommunications Union (ITU) has authorized the L2 band on a co-primary basis with radiolocation services, such as high-power radars. As a result of Federal Aviation Administration (FAA) requests, the Department of Transportation and Department of Defense have called for a new civil GPS frequency, called L5 at 1176.45 MHz in the Aeronautical Radio Navigation Service (ARNS) of 960–1215 MHz. To gain maximum performance, the L5 spread spectrum codes was selected to have a higher chipping rate and longer period than do the C/A codes to allow for better accuracy measurements. Additionally, the L5 signal has two signal components in phase quadrature, one of which will not carry data modulation. The L5 signal provides the following system improvements:

1. *Ranging accuracy will improve:* Pseudorange errors due to random noise will be reduced below levels obtainable with the C/A-codes, due to the larger bandwidth of the proposed codes. As a consequence, both code-based positioning accuracy and phase ambiguity resolution performance will improve.
2. *Errors due to multipath will be reduced:* The larger bandwidth of the new codes will sharpen the peak of the code autocorrelation function, thereby reducing the shift in the peak due to multipath signal components. The eventual multipath mitigation will depend upon the final receiver design and delay of the multipath.
3. *Carrier phase tracking will improve:* Weak-signal phase tracking performance of GPS receivers is severely limited by the necessity of using a Costas (or equivalent-type) PLL to remove carrier phase reversals of the data modulation. Such loops rapidly degrade below a certain threshold (about 25–30 dB-Hz) because truly coherent integration of the carrier phase is limited to the 20 ms data bit length. In contrast, the "data-free" quadrature component of the L5 signal will permit coherent integration of the carrier for arbitrarily long periods, which will permit better phase tracking accuracy and lower tracking thresholds.
4. *Weak-signal code acquisition and tracking will be enhanced:* The "data-free" component of the L5 signal will also permit new levels of positioning capability with very weak signals. Acquisition will be improved because fully coherent integration times longer than 20 ms will be possible. Code tracking will also improve by virtue of better carrier phase tracking for the purpose of code rate aiding.
5. *The L5 signal will further support rapid and reliable carrier phase ambiguity resolution.* The L5 signal is a full strength, high chipping rate code that will provide high quality code

and carrier phase measurements. These can be used to support various code and carrier combinations for high accuracy carrier phase ambiguity resolution techniques.

6. *The codes will be better isolated from each other:* The longer length of the L5 codes will reduce the size of crosscorrelation between codes from different satellites, thus minimizing the probability of locking onto the wrong code during acquisition, even at the increased power levels of the modernized signals.
7. *Advances Navigation Messaging:* The L5 signal structure has a new civil navigation (CNAV) messaging structure that will allow for increase data integrity.

GPS modernization for the L5 signal calls for a completely new civil signal format (i.e., L5 code) at a carrier frequency of 1176.45 MHz (i.e., L5 carrier). The L5 signal is defined in a quadrature scene where the total signal power is divided equally between in-phase ($I$) and quadrature ($Q$) components. Each component is modulated with a different but synchronized 10,230 chip direct sequence L5 code transmitted at $10.23 \times 10^6$ cps (chips per second), the same rate as the P(Y) code, but with a 1 millisecond (ms) period, the same as the C/A code period. The $I$ channel is modulated with a 100 sps data stream, which is obtained by applying rate 1/2, constraint length 7, forward error correction (FEC) convolutional coding to a 50 bps navigation data message that contains a 24-bit cyclic redundancy check (CRC). The $Q$ channel is unmodulated by navigation data. However, both channels are further modulated by Neuman-Hoffman (NH) synchronization codes, which provide additional spectral spreading of narrowband interference, improve bit and symbol synchronization, and also improve crosscorrelation properties between signals from different GPS satellites. The L5 signal is shown in Figure 2.3 illustrating the Modernized GPS (and Legacy GPS) signal spectrum.

Compared to the C/A code, the ten times larger chip count of the $I$ and $Q$ channel civil L5 codes provides lower autocorrelation sidelobes, and the ten times higher chipping rate substantially improves ranging accuracy, provides better interference protection, and substantially reduces multipath errors at longer path separations (i.e., long delay multipath). Additionally, these codes were selected to reduce, as much as possible, the cross correlation between satellite signals. The absence of data modulation on the $Q$ channel permits longer coherent processing intervals in code and carrier tracking loops, with full-cycle carrier tracking in the latter. As a result, the tracking capability and phase ambiguity resolution become more robust.

Further details on the civil L5 signal can be found in references GPS Directorate (2012), Van Dierendonck and Spilker (1999), Hegarty and Van Dierendonck (1999), and Spilker and Van Dierendonck (2001).

#### 2.2.5.3 M Code

The new military (M) codes will also be transmitted on both the L1 and L2 carrier frequencies. These M codes are based on a new family of split-spectrum GNSS codes for military and new GPS civil signals (Spilker et al. 1998). The M codes will provide the following advantages to military users:

1. *Direct acquisition of the M codes will be possible:* The design of these codes will eliminate the need to first acquire the L1 C/A code with its relatively high vulnerability to jamming.
2. *Better ranging accuracy will result:* As can be seen in Figure 2.3, the M codes have significantly more energy near the edges of the bands, with a relatively small amount of energy near the band center. Since most of the C/A code power is near the band center, potential interference between the codes is mitigated. The effective bandwidth of the M codes is much larger than that of the P(Y) codes, which concentrate most of their power near the L1 or L2 carrier. Because of the modulated subcarrier, the autocorrelation function of the M codes has, not just one peak, but several peaks spaced one subcarrier period apart, with

the largest at the center. The modulated subcarrier will cause the central peak to be significantly sharpened, significantly reducing pseudorange measurement error.

3. *Error due to multipath will be reduced:* The sharp central peak of the M code autocorrelation function is less susceptible to shifting in the presence of multipath correlation function components.

The M coded signals will be transmitted on the L1 and L2 carriers, with the capability of using different codes on the two frequencies. The M codes are known as Binary Offset Carrier (BOC) encoded signals where the notation of BOC($f_{sx}$,$f_{cx}$) is used, where $f_{sx}$ represents the subcarrier multiplier, and $f_{cx}$ represents the code rate multiplier, with respect to a nominal code rate of 1.023 MHz. The M code is a BOC(10,5) code in which a 5.115 Mcps chipping sequence modulates a 10.23 MHz square wave subcarrier. Each spreading chip subtends exactly two cycles of the subcarrier, with the rising edge of the first subcarrier cycle coincident with initiation of the spreading chip. The spectrum of the BOC(10,5) code has considerably more relative power near the edges of the signal bandwidth than any of the C/A, P(Y), L2C, and L5 coded signals. As a consequence, the M coded signal has minimal spectral overlap with the other GPS transmitted signals, which permits transmission at higher power levels without mutual interference. The resulting spectrum has two lobes, one on each side of the band center, thereby producing the split-spectrum code. The M code signals are illustrated in Figure 2.3. The M code signal is transmitted in the same quadrature channel as the C/A code (i.e., with the same carrier phase), that is, in phase quadrature with the P(Y) code. The M codes are encrypted and unavailable to unauthorized users. The nominal received power level is –158 dBW at Earth. Additional details on the BOC(10,5) code can be found in a paper by Barker et al. (2000).

#### 2.2.5.4 L1C Signal

A proposed new L1 Civil (L1C) signal is planned for the next generation of GPS SVs. Although the current C/A code is planned to remain on the L1 frequency (1575.42 MHz), the additional L1C signal will add a higher performance civil signal at the L1 frequency with potential interoperability with other GNSSs. Like the L5 civil signal, the planned L1C signal will have a data-free quadrature component.

The original L1C signal structure considered a pure BOC(1,1) signal but evolved into a more complex signal that multiplexed two BOC signals. Additional complexities involved the desire to have the L1C signal interoperable with other GNSS signals, as well as potential intellectual property issues. The L1C signal contains a dataless (i.e., pilot) and data signal component transmitted in quadrature. The L1C signal for GPS that emerged from development is based upon a Time-Multiplexed BOC (TMBOC) modulation technique that synchronously time-multiplexed the BOC(1,1) and BOC(6.1) spreading codes for the pilot component (designated as $L1C_p$), and a BOC(1,1) modulated signal (designated as $L1C_D$). For both, the BOC codes are generated synchronously at rate of 1.023 MHz and based the Legendre sequence called Weil code. These codes have a period of 10 msec, so 10,230 chips are within one period. Additionally, there is an overlay code that is encoded onto the $L1C_p$ pilot channel. One bit of the overlay code has a duration, and is synchronized to the 10 msec period of the BOC code generators. The overlay code rate is 100 bps and has 1,800 bits in an 18-second period.

To generate the TMBOC signal for the $L1C_p$ channel the BOC(1,1) and BOC(6.1) spreading sequences are time-multiplexed. With 33 symbols of a BOC(1,1) sequence, four symbols are replaced with BOC(6.1) chips. These occur at symbols 0, 4, 6, and 29. Thus, with a 75% of the power planned for distribution in the pilot signal there will be 1/11th of the power in the BOC(6,1) component and 10/11th of the power in the BOC(1,1) component of the carrier.

The L1C signal also has a new navigation message structure, designated as CNAV-2, with three different subframe formats defined. Subframe 1 contains GPS time information (i.e, Time of

Interval (TOI)). Subframe 2 contains ephemeris and clock correction data. Subframe 3 is commutated over various pages and provides less time sensitive data such almanac, UTC, ionosphere, and can be expended in the future.

The split-spectrum nature of the L1C BOC(1,1) encoded signal will provide some frequency isolation from the L1 C/A encoded signal. Each spreading chip subtends exactly one cycle of the subcarrier, with the rising edge of the first subcarrier half-cycle coincident with initiation of the spreading chip. The MBOC codes provide a larger RMS bandwidth compared to pure BOC(1,1).

For many years now, cooperation at the international level has been ongoing to enable the L1C signal to be interoperable with other GNSSs. A combined interoperable signal would allow a user to ubiquitously use navigation signals from different GNSS with known and specified performances attributes.

Additional details on the L1C signal can be found in Spilker and Van Dierendonck (2001), Spilker et al. (1998), Barker et al. (2000), Issler et al. (2004), Betz et al. (2006), and GPS Directorate (2011).

### 2.2.6 GPS Satellite Blocks

The families of satellites launched prior to recent modernization efforts are referred to as Block I (1978–1985), Block II (1989–1990), and Block IIA (1990–1997); all of these satellites transmit the Legacy GPS signals (i.e, L1 C/A and P(Y) and L2 P(Y)). (The United States Naval Observatory has an up to date listing of all of the GPS satellites in use today [USNO 2012]).

In 1997 a new family, the Block IIR satellites, began to replace the older Block II/IIA family. The Block IIR satellites have several improvements, including reprogrammable processors enabling problem fixes and upgrades in flight. Eight Block IIR satellites were modernized (designated as Block IIR-M) to include the new military M code signals on both the L1 and L2 frequencies, as well as the new L2C signal on L2. The first modernized Block IIR was launched in September 2005.

To help secure the L5 frequency utilization, one of the Block IIR-M satellites (GPS IIR-20(M)), SV49, was outfitted with a special L5 payload and launched on March 24, 2009. This particular satellite had hardware configuration issues relating to the L5 payload installation and is transmitting a degraded signal. Since that time, the navigation signals have been set unhealthy in the broadcast navigation message.

The Block IIF (i.e., Follow-on) family was the next generation of GPS satellites, retaining all the capabilities of the previous blocks, but with many improvements, including an extended design life of 12 years, faster processors with more memory, and the inclusion of the new L5 signal on a third L5 frequency (1176.45 MHz). The first Block IIF satellite was launched in May 2010.

### 2.2.7 GPS III

The next block of GPS satellites planned is designated as the Block III family, which is still under development. The GPS III block of satellites and associated GPS Ground Control Segment components will represent a major advancement in capabilities for military and civil users. GPS III is planned to include all Legacy, and Modernized GPS signal components, including the new L1C and L5 signals and add specified signal integrity. The added signal integrity planned for in GPS III may be able to satisfy some of the aviation requirements (FAA 2008). Improvements for military include two high-power spot beams for the L1 and L2 military M-code signals, providing 20 dB higher received power over the earlier M code signals. However, in the fully modernized Block III satellites, the M coded signal components are planned to be radiated as physically distinct signals from a separate antenna on the same satellite. This is done in order to enable optional transmission of a spot beam for greater anti-jam resistance within a selected local region on the Earth.

The system with newer frequencies will keep growing and will be fully operational by 2027. There are currently 17 L5 equipped satellites in orbit. There need to be 24 live satellites in order to run a reliable system.

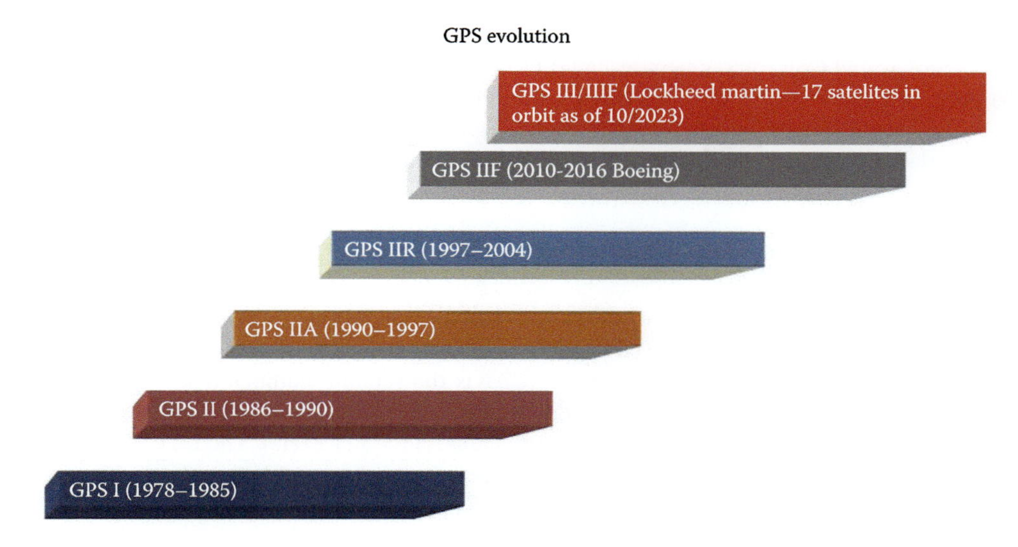


**FIGURE 2.4** GPS evolution.

The GPS continues to set the gold standard in its field. Other nations report improvements in accuracy and equivalent performance in availability. However, GPS is still the clear leader in "integrity" and is the only system accepted for international flight use. Figure 2.4 shows the evolution of GPS.

## 2.3 GLOBAL ORBITING NAVIGATION SATELLITE SYSTEM (GLONASS)

A second configuration for global positioning is the Global Orbiting Navigation Satellite System (GLONASS), placed in orbit by the former Soviet Union, and now maintained by the Russian Republic. GLONASS is the Russian GNSS. The GLONASS has similar operational requirements to GPS with some key differences in its configuration and signal structure. Like GPS, GLONASS is an all-weather, 24-hour satellite-based satellite navigation system that has a space, control, and user segment. The first GLONASS satellite was launched in 1982 and the GLONASS declared an operational system on September 24, 1993.

### 2.3.1 GLONASS Orbits

The GLONASS satellite constellation is designed to operate with 24 satellites in three orbital planes at 19,100 km altitude; (whereas GPS uses six planes at 20,180 km altitude). GLONASS calls for eight SVs equally spaced in each plane. The GLONASS orbital period is 11 hours 15 minutes, which is slightly shorter than the 11 hour 56 minutes for a GPS satellite. Because some areas of Russia are located at high latitudes, the orbital inclination of 64.8° is used as opposed to the inclination of 55° used for GPS.

GLONASS has 24 satellites, distributed approximately uniformly in three orbital planes (as opposed to six for GPS) of eight satellites each. Each orbital plane has a nominal inclination of 64.8° relative to the equator, and the three orbital planes are separated from each other by multiples of 120° right ascension. GLONASS orbits have smaller radii than GPS orbits, about 25,510 km, and a satellite period of revolution of approximately $\frac{8}{17}$ of a sidereal day. A GLONASS satellite and a GPS satellite will complete 17 and 16 revolutions, respectively, around the Earth every eight days.

Each GLONASS satellite transmits its own ephemeris and system almanac data. Via the GLONASS Ground Control Segment, each GLONASS satellite transmits its position, velocity, and lunar/solar acceleration effects in an ECEF coordinate frame (as opposed to GPS, that encodes SV positions using Keplerian orbital parameters). Glonass ECEF coordinates are referenced to the PZ-90.02 datum and time reference linked to their National Reference of Coordinated Universal Time UTC(SU) (Soviet Union, now Russia).

## 2.3.2 GLONASS Signals

The GLONASS system uses frequency-division multiplexing of independent satellite signals. Its two carrier signals corresponding to L1 and L2 have frequencies $f_1 = (1.602 + 9k/16)$ GHz and $f_2 = (1.246 + 7k/16)$ GHz, where $k = 0,1,2,\ldots,23$ is the satellite number. These frequencies lie in two bands at 1.597–1.617 GHz (L1) and 1240–1260 GHz (L2). The L1 code is modulated by a C/A code (chip rate = 0.511 MHz) and by a P code (chip rate = 5.11 MHz). The L2 code is presently modulated only by the P code. The GLONASS satellites also transmit navigational data at a rate of 50 baud. Because the satellite frequencies are distinguishable from each other, the P code and the C/A code are the same for each satellite. The methods for receiving and analyzing GLONASS signals are similar to the methods used for GPS signals (Janky 1997). GLONASS does not use any form of SA.

### 2.3.2.1 Frequency Division Multiple Access (FDMA) Signals

GLONASS uses multiple frequencies in the L-band and has used frequencies separated by a substantial distance for ionosphere mitigation (i.e., L1 and L2), but these are slightly different than the GPS L1 and L2 frequencies. One significant difference between GLONASS and GPS is that GLONASS has historically used a frequency division multiple access (FDMA) architecture as opposed to the CDMA approach used by GPS.

### 2.3.2.2 Carrier Components

The GLONASS uses two L-band frequencies, L1 and L2 as defined here. The channel numbers for GLONASS signal operation are for

$$f_{K1} = f_{01} + K\Delta f_1$$
$$f_{K2} = f_{02} + K\Delta f_2$$

Where

$$K = \text{channel number}\left(-7 \le K \le +6\right)$$
$$f_{01} = 1602M;\ \Delta f_1 = 562.5\ kHz$$
$$f_{02} = 1246M;\ \Delta f_2 = 437.5\ kHz$$

### 2.3.2.3 Spreading Codes and Modulation

With the GLONASS signals isolated in frequency, an optimum maximal-length (m-sequence) can be used as the spreading code. GLONASS utilizes two such codes, one Standard Precision Navigation Signal (SPNS) at a 0.511 Mcps rate that repeats every 1 millisecond and a second High Precision Navigation Signal (HPNS) at a 5.11 Mcps rate, that repeats every 1 second. Similar to GPS, the GLONASS signals utilized BPSK modulation and are transmitted out of a right-hand circularly polarized antenna.

### 2.3.2.4 Navigation Data Format

The format of the GLONASS navigation data is similar to the GPS navigation data format, with different names and content. The GLONASS navigation data format is organized as a Superframe that is made up of frames, where frames are made up of strings. A Superframe has a duration of 150 seconds and is made up of five frames, so each frame lasts 30 seconds. Each frame is made up of 15 strings, where a string has a duration of 2 seconds. GLONASS encodes satellite ephemeris data as Immediate Data and almanac data as Non-immediate data. There is a time mark in the GLONASS navigation data (last 0.3 second of a string) that is an encode PRN sequence.

#### 2.3.2.5 Satellite Families

While the first series of GLONASS satellites were launched from 1982 to 2003, the GLONASS-M satellites were launched beginning in 2003. These GLONASS-M satellites had improved frequency plans and the accessible signals that are known today. The new generation of GLONASS satellites is designated as GLONASS-K satellites and is considered a major modernization effort by the Russian government. These GLONASS-K satellites plan to transmit the legacy GLONASS FDMA signals as well as a new CDMA format.

Additional details of the GLONASS signal structure and GLONASS-M satellite capabilities can be found in GLONASS (2008).

### 2.3.3 Next Generation GLONASS

The satellite for the next generation of GLONASS-K launched on February 26, 2011, and continues to undergo flight tests. Twenty-four satellites consisting of four GLONASS-M satellites are all transmitting healthy signals. Recently, Russian scientists have proposed a new code-division multiple-access signal format to be broadcast on a new GLONASS L3 signal. Once implemented across the modernizing GLONASS constellation, this will facilitate interoperability with, and eventually interchangeability among, other GNSS signals. The flexible message format permits relatively easy upgrades in the navigation message (GPS World 2013).

#### 2.3.3.1 Code Division Multiple Access (CDMA) Modernization

One of the issues with a FDMA GNSS structure is the inter-channel (i.e., inter-frequency) biases that can arise within the FDMA GNSS receiver. If not properly addressed in the receiver design, these inter-channel biases can be a significant error source in the user solutions. These error sources arise because the various navigation signals pass through the components within the receiver at slightly different frequencies. The group delay thought these components are non-common, at the different frequencies, and produced different delays on the various navigation signals, coming from different satellites. This inter-frequency bias is substantially reduced (on a comparative basis) with CDMA based navigation systems, because all of the signals are transmitted at the same frequency. (The relatively small amount of Doppler received from the various CDMA navigation signals is minor when considering the group delay.)

An additional consideration with an FDMA GNSS signal structure is the amount of frequency bandwidth that is required to support the FDMA architecture. CDMA architecture typically has all the signals transmitted at the same carrier frequency, for more efficient utilization of a given bandwidth.

GLONASS has established several separate versions of its GLONASS-K satellites. The first GLONASS-K1 satellite was launched on February 26, 2011, which carried the first GLONASS CDMA signal structure and has been successfully tracked on Earth (Septentrio 2012). The GLONASS-K1 satellite is transmitted a CDMA signal at a designated L3 frequency of 1202.025 MHz (test signal), as well as the legacy GLONASS FDMA signals at L1 and L2. The CDMA signal from the GLONASS-K1 satellite is considered a test signal. The follow-on generation of satellites is designated as the GLONASS-K2 satellites (Revnivykh 2011). A full constellation of legacy and new CDMA signals are planned for these GLONASS-K2 satellites including plans to transmit its CDMA signal on or near the GPS L1 and L2 frequency bands. The GLONASS KM satellites are in the research phase and plans call for the transmission of the legacy GLONASS FDMA signals, the CDMA signals introduced in the GLONASS-K2, and a new CDMA signal on the GPS L5 frequency.

## 2.4 GALILEO

Galileo is a GNSS being developed by the European Union and the European Space Agency (ESA). Like GPS and GLONASS, it is an all-weather, 24-hour satellite-based navigation system being designed to provide various services. The program has had three development phases: (1) definition (completed), (2) development of on-orbit validation, and (3) launch of operational satellite, including additional development (ESA 2012). The first Galileo In-Orbit Validation Element (GIOVE) satellite, designated as GIOVE-A, was launched in December 28, 2005, followed by the GIOVE-B on April 27, 2008. The next two Galileo operational satellites were launched on October 21, 2011, to provide additional validation of the Galileo.

### 2.4.1 Constellation and Levels of Services

The full constellation of Galileo is planned to have 30 satellites in Medium Earth Orbits (MEOs) with an orbital radius of 23,222 km (similar to the GPS orbital radius of 20,180 km). The inclination angle of the orbital plane is 56° (GPS is 55°), with three orbital planes (GPS has six). This constellation will thus have ten satellites in each orbital plane.

Various services are planned for Galileo, including an open service (OS), commercial service (CS), public regulated, a safety of life (SoL), and a search and rescue (SAR) service. These services will be supported with different signal structures and encoding formats tailored to support the particular service.

### 2.4.2 Navigation Data and Signals

Table 2.2 lists some of the key parameters for the Galileo that will be discussed in this section. The table lists the Galileo signals, frequencies, identifiable signal component, its navigation data format, and what service the signal is intended to support. The European Union has published the Galileo Open Service (OS) Interface Control Document which contains significant detail on the Galileo signals in space (EU et al. 2010). All of the Galileo signals transmit two orthogonal signal components, where the in-phase component transmits the navigation data and the quadrature component is dataless (i.e., a pilot). These two components have a power sharing so that the dataless channel can be used to aid the receiver in acquisition and tracking of the signal. All of the individual Galileo GNSS signal components utilize phase-shift keying modulation and right-hand circular polarization (RHCP) for the navigation signals.

To support the various services for Galileo, three different navigation formats are planned for implementation: (1) A Free navigation (F/NAV) format to support OS in for the E1 and E5a signals; (2) the Integrity navigation (I/NAV) format is planned to support SoL services for the L1 and E5a signals; and (3) a Commercial navigation (C/NAV) format is to support CS.

Galileo E1 frequency (same as GPS L1) at 1574.42 MHz will have a split-spectrum-type signal around the center frequency. This signal is planned to interoperate with the GPS L1C signal, however discussions continue on the technical, political, and intellectual property aspects with its implementation. Despite these challenges, the Galileo E1 signal is planned to be a Combined BOC (CBOC) signal that is based upon two BOC signals (a BOC(1,1) and BOC(6.1) basis component) in phase quadrature. The E1 in-phase component has I/NAV data encoded on it and the quadrature phase has no data (i.e., pilot).

The Galileo E6 signal is planned to support CS at a center frequency of 1278.750 MHz, with no offset carrier, and a spreading code at a rate of 5.115 MHz (5 x 1.023 MHz). The E6 signal have

**TABLE 2.2**
**Key Galileo Signals and Parameters**

| Signal | Frequency (MHz) | Component | Data | Service |
|---|---|---|---|---|
| E1 | 1575.420 | | | |
| | | E1-B | I/NAV | OS/CS/SOL |
| | | E1-c | Pilot | |
| E6 | 1278.750 | | | |
| | | E6-B | C/NAV | CS |
| | | E6-B | Pilot | |
| E5 | 1191.795 | | | |
| E5a | 1176.450 | E5a-I | F/NAV | OS |
| | | E5a-Q | Pilot | |
| E5b | 1207.140 | E5b-I | I/NAV | OS/CS/SOL |
| | | E5b-Q | Pilot | |

two signal components in quadrature, where the C/NAV message format is encoded on the in-phase component, and no data is on the quadrature component.

The E5 signal is a unique GNSS signal that has an overall center frequency of 1191.795 MHz, with two areas of maximum power at 1176.450 MHz and 1207.140 MHz. The wideband E5 signal is generated by a modulation technique called AltBOC. The generation of the E5 signal is such that it is actually composed of two Galileo signals that can be received and processed separately, or combined by the user. The first of these two signals within the composite E5 signal, is the E5a, centered at 1176.450 MHz (same as the GPS L5). The E5a signal has again two signal components, transmitted in quadrature, where one has data (in-phase) and other does not (quadrature). The F/NAV data is encoded onto the in-phase E5a channel at a 50 sps rate. The second of these two signals within the composite E5 signal, is the E5b, centered at 1207.140 MHz. The E5b signal has two signal components, transmitted in quadrature, where one has data (in-phase) and the other does not (quadrature). The in-phase channel has the I/NAV message format to support OS, CS, and SOL applications. The data encoded within the I/NAV format will contain important integrity information necessary to support SOL applications.

### 2.4.3 Updates

The first two Galileo In-Orbit Validation (IOV) satellites were launched in October 2011 and began broadcasting in December 2011. All Galileo signals were activated on December 17, 2011 simultaneously for the first time across the European GNSS system's three spectral bands known as $E_1$(1559–1592 MHz), $E_5$(1164–1215 MHz) and $E_6$(1260–1300 MHz). The remaining two IOVs were launched in 2012.

Galileo's second generation (G2G) has entered into its orbit validation development phase. As of June 2023, the European Space Agency (ESA) awarded GMV a contract to develop the ground segment responsible for in-orbit control and validation of the Galileo Second Generation (G2G). The G2G will introduce new services, improve accuracy, and increase system security. A total of 12 satellites are expected to be launched over the next three years. As of this writing, Galileo serves more than four billion users worldwide, delivering global positioning, navigation, and clock synchronization services with positioning up to 20 cm.

The International Civil Aviation Organization (ICAO) is a United Nations agency, established to help countries share their skies to their mutual benefit. ICAO adopts international standards for Galileo and future SBASs.

Current updates at European Space Agency website (European Space Agency 2023).

## 2.5 COMPASS (BEIDOU-2)

The BeiDou Navigation Satellite System is being developed, starting with regional services, and later expanding to global services. Phase I was established in 2000 and Phase II provided for areas in China and its surrounding areas by 2012. Also known as Compass and BeiDou-2, the Chinese BDS started operations in December 2012 and has 14 active satellites in service over the Asia-Pacific region available to general users. As of May 16, 2023, BDS (Phase III) BeiDou has a constellation total of 56 satellites offering complete coverage around the globe (since 2020). China will employ the BeiDou satellite-based augmentation system (BDSBAS) to provide high-precision positioning service in railway survey and construction, according to the China Railway Siyuan Survey and Design Group Co., Ltd., per the the Xinhua News Agency. Four satellite-based and 12 ground-based observation stations will be set up along the Wufeng-Enshi railway section in central China's Hubei Province. According to the company, this is the first time that the BDSBAS will be used in the field of intelligent railway surveys (BeiDou 2014).

### 2.5.1 Compass Satellites

Compass is the Chinese developed GNSS, where Compass is an English translation for BeiDou. The Chinese government performed initial development on the Compass in the 2000–2003 timeframe with reference to BeiDou (BD)-1. BeiDou consists of 14 satellites, including five Geostationary Earth Orbit (GEO) satellites and nine Non-Geostationary Earth-Orbit (Non-GEO) satellites. The Non-GEO satellites include four Medium Earth Orbit (MEO) and five Inclined Geosynchronous Satellite Orbit (IESO) satellites.

The Compass space segment is called a Dippler Constellation that plans for an eventual 27 MEO, five GEOs, and three Inclined GEOs (IGSO) satellites. The GEOs are planned for longitude locations at 58.75°E, 80°E, 110.5°E, 140°E, and 160°E. The system uses its own datum, China Geodetic Coordinate System 2000 (CGCS2000) and time reference BeiDou Time (BDT) System, relatable to UTC. Of the BD-2 SVs, the first MEO Compass M-1 satellite was launched on April 14, 2007. The GEOs have been launched with an orbital radius of 42,164.17 km. The first IGSO SV was launched on July 31, 2010, with an inclination angle of 55° (same as GPS MEOs). In December 2012, the BeiDou Navigation Satellite System provided fully operational regional service (BeiDou 2014).

The current Interface Control Document (ICD) specifies a B1 GNSS signal at a center frequency of 1561.098 MHz. The architecture is based on a CDMA approach. The spreading codes used are Gold codes, based on 11 stages of delay, running at 2.046 Mcps, so the code will repeat after 1 millisecond. Navigation data on the MEO and IGSO SVs is at a rate of 50 bps, with a secondary code rate of 1 Kbps.

### 2.5.2 Frequency

The nominal carrier frequency is 1561.098 MHz with B1 signal is currently a quadrature phase-shift key in (QPSK) modulation. Compass now has 10 BeiDou-2 satellites operating in its constellation (BeiDou 2014).

## 2.6 SPACE-BASED AUGMENTATION SYSTEMS (SBAS)

The Space-Based Augmentation System (SBAS) uses GEO satellites to relay correction and integrity information to users. A secondary use of the GEO signal is to provide users with a GPS-like ranging source. Two independent ranging signals are generated on the ground and provided via C-band uplink to the GEO, where the navigation payload translates the uplinked signals to L1 downlink frequencies. A GEO satellite and its corresponding pair of GUS sites comprise a GEO Communication and Control Segment (GCCS). Figure 2.5 illustrates current and proposed SBASs.

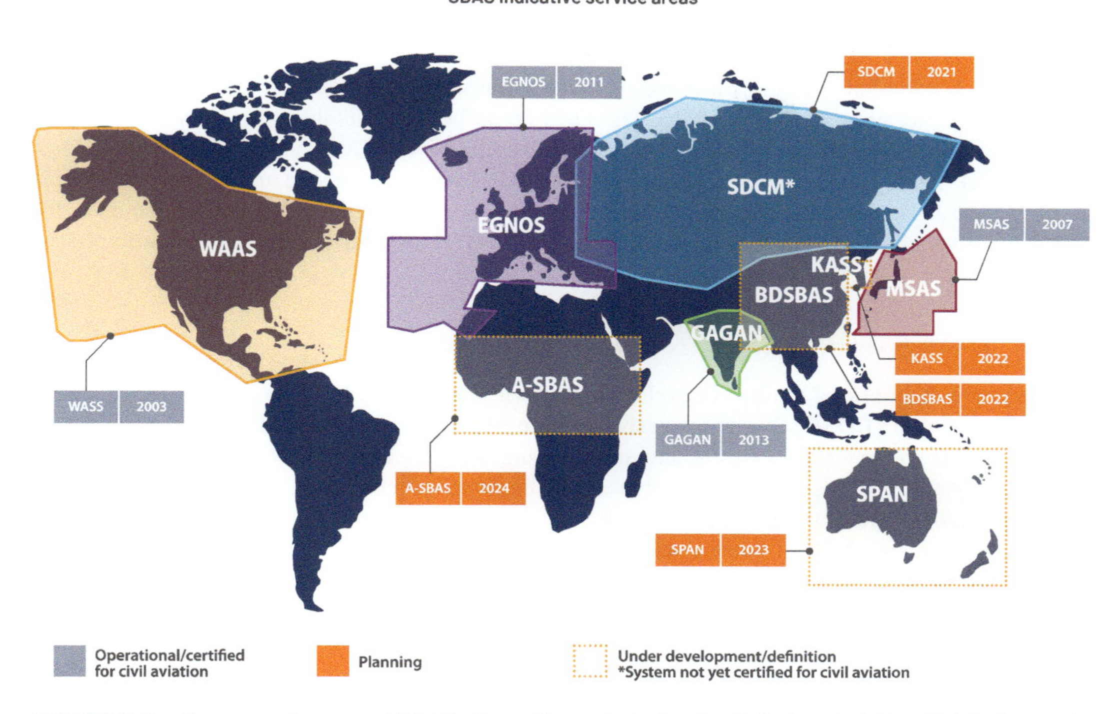


**FIGURE 2.5** Current and proposed SBASs (https://gssc.esa.int/navipedia/index.php?title=SBAS_Systems).

### 2.6.1 WAAS (Wide-Area Augmentation System)

WAAS enhances the GPS standard positioning service (SPS) and is available over a wide geographic area. The WAAS, developed by the federal Aviation Administration (FAA) together with other agencies, provides Wide-Area Differential GNSS (WASGNSS) corrections, additional ranging signals from Geostationary Earth Orbit (GEO) satellites, and integrity data on the GPS and GEO satellites. The FAA continues to make improvements to WAAS infrastructure and technical refreshes. See www.faa.gov/ato/navigation-programs/waas.

### 2.6.2 EGNOS (European Geostationary Navigation Overlay System)

EGNOS is a joint project of the European Space Agency, the European Commission, and the European Organization for the Safety of Air Navigation (Eurocontrol). Its primary service area is the European Civil Aviation Conference (ECAC) region. However, several extensions of its service area to adjacent and more remote areas are under study. See www.euspa.europa.eu/european-space/egnos/.

### 2.6.3 MSAS (MTSAT Satellite-Based Augmentation System)

The MTSAT Satellite Augmentation System (MSAS) is the Japanese Satellite Based Augmentation System (SBAS) System. The MSAS system for aviation use was declared operational in September 27, 2007, providing a service of horizontal guidance for En-route through Non-Precision Approach. The SBAS signal was first transmitted from MTSAT (Multi-functional Transport Satellites) operated by the Ministry of Land, Infrastructure, Transport and Tourism (MLIT). Since April 2020, the SBAS signal made by MLIT is transmitted from the QZS-3 GEO satellite using the QZSS SBAS transmission service. See https://gssc.esa.int/navipedia/index.php/MSAS.

### 2.6.4 QZSS (Quasi-Zenith Satellite System) (Sub-Meter Level Augmentation Service [SLAS])

The QZSS Navigation Service is a space-based positioning system being developed by the Japan Aerospace Exploration Agency (JAXA). The QZSS is to be interoperable with GPS and provides augmentation to the GPS over the regions covered by a particular Quasi-Zenith Satellite (QZS) or QZSS GEO. The Space Systems Command (SSC) has delivered the second and last payload (QZS-6, QZS-7) to Japan for launch to geosynchronous orbit. As of October 1, 2023, the launch dates for these satellites were not determined. Additional details of the QZSS can be found at https://qzss.go.jp/en/.

### 2.6.5 SNAS (Satellite Navigation Augmentation System)

The People's Republic of China is developing its own SBAS, called BeiDou satellite-based augmentation system (BDSBAS and formerly Satellite Navigation Augmentation System (SNAS)) to provide SBAS services in China and surrounding regions. In 2002, Novatel was awarded a contract to provide 12 receivers for the phase II of the development. These stations would complement the 11 ones already installed in and around Beijing for phase I of the program. BDSBAS is integrated in the BeiDou system and uses BDS-3 type satellites to broadcast SBAS L1/L5 signals, augmenting BDS and GPS. If they are recommended by ICAO SARPs, Galileo E1C/E5a and GLONASS L1/L3 will also be considered. See also https://gssc.esa.int/navipedia/index.php/SNAS.

### 2.6.6 GAGAN (GPS and GEO Augmented Navigation) on May 29, 2023

The Indian Space Research Organization (ISRO)'s Geosynchronous Satellite Launch Vehicle (GSLV) launched the first of its second generation navigation satellites. The Navigation with Indian Constellation (NavIC) system's first generation consists of seven satellites in geosynchronous orbit. This brings the NavIC satellite total to nine. See also www.isro.gov.in.

### 2.6.7 SDCM (System for Differential Corrections and Monitoring)

The System for Differential Corrections and Monitoring (SDCM) is the SBAS currently being developed in the Russian Federation by JSC (Russian Space Systems) as a component of GLONASS. The main differentiator of SDCM with respect to other SBAS systems is that it is conceived as an SBAS augmentation that would perform integrity monitoring of both GPS and GLONASS satellites. See https://gssc.esa.int/navipedia/index.php/SDCM.

### 2.6.8 KASS (Korean Augmentation Satellite System)

KASS is the future satellite-based augmentation system (SBAS) for the Republic of Korea. It is currently developed by the Korea Aerospace Research Institute (KARI) for the government of the Republic of Korea. Thales Alenia Space is the industry prime contractor for this development.

KASS will provide SBAS service compliant with ICAO SARPS Annex 10 [1] over the South Korea area with service level up to APVI. The system will comprise segments from different manufacturers or service providers. KASS system deployment began at the end of 2020 with the onsite installation of the system reference stations network. In parallel, the wide area network has been extensively tested and verified to allow continuous data transmission. Since this first deployment step, data has been collected and Thales Alenia Space has undertaken the first performance analysis of real data. See https://insidegnss.com, February 2023.

### 2.6.9 SouthPAN (Southern Positioning Augmentation Network)

SouthPAN will deliver satellite service to Australia and New Zealand, providing accurate, reliable, and instant positioning service in the Asian Pacific region. Multinational technology company Lockheed Martin Australia is contracted to develop the processing and control centers to deliver ongoing SouthPAN services.

Australia and New Zealand will be contributing to improved global coverage and interoperability for SBAS services, joining the world's countries and regions that already have their own SBAS systems.

## 2.7 CONCLUSIONS

Future developments will provide enhanced safety and services for users. Most future users are anticipated to rely on multiconstellation receivers, that is, that will receive and process GPS, GLONASS, Galileo, and Compass signals, complementing and enhancing each other. A GNSS data integrity channel (GIC) will be provided in the next generation of satellites, for example, in GPS, the GPS IIF and GPS III will provide an integrity channel. In addition, next generations will include airborne monitoring by using redundant measurements to provide signal integrity. New satellites to be built and launched will have more frequencies, higher power, and larger bandwidths. These improvements will reduce the errors such as multipath and atmospheric (ionospheric and tropospheric). The support for various services, including Open Service (OS), Commercial Service (CS), Public Regulated Service (PRS), Safety of Life Service (SOL), and Search and Rescue (SAR), will be supported with different signal structures and encoding formats, tailored to each service.

Global Navigation Satellite Systems will continue to make the world a smaller, more fully functional planet. Each of the existing systems will individually improve and seek engagement with its neighboring systems to provide a seamless overlay of safe and reliable service around the globe.

## REFERENCES


Barker, B. C., J. W. Betz, J. E. Clark, J. T. Correia, J. T. Gillis, S. Lazar, K. A. Rehborn, and J. R. Straton. 2000. *Proceedings of the 2000 National Technical Meeting of the Institute of Navigation*. Anaheim, CA. Jan: 542–549.

BeiDou Satellite Navigation System News Center. 2014. Article on first anniversary of BeiDou navigation satellite system providing full operational regional service. Available at http://en.beidou.gov.cn/ (accessed February 2014, October 2023).

Betz, J. W., M. A. Blanco, C. R. Cahn, P. A. Dafesh, C. J. Hegarty, K. W. Hudnut, V. Kasemsri, R. Keegan, K. Kovach, L. S. Lenahan, H. H. Ma, J. Rushanan, J. J. Rushanan, D. Sklar, T. A. Stansell, C. C. Wang, and S. K. Yi. 2006. Description of the L1C signal. *Proceedings of the 19th International Technical Meeting of the Satellite Division of The Institute of Navigation (ION GPS 2006)*. Fort Worth, TX. Sept: 2080–2091.

European Space Agency. 2014. www.esa.int/Our_Activities/Navigation/GNSS_Evolution/About_the_European_GNSS_Evolution_Programme (accessed February 10, 2014).

European Union, European Commission, Satellite Navigation, Galileo Open Service. Signal-in-Space Interface Control Document. 2010. OS SIS ICD. Issue 1.1:Sept. Available at http://ec.europa.eu/enterprise/policies/satnav/galileo/files/galileo-os-sis-icd-issue1-revision1_en.pdf (accessed February 10, 2014).

euspa.europa.eu/european-space/egnos/. 2023.

Fontana, R. D., W. Cheung, P. M. Novak, and T. A. Stansell. 2001. The new L2 civil signal. *Proceedings of the 14th International Technical Meeting of the Satellite Division of The Institute of Navigation (ION GPS 2001)*. Salt Lake City, UT. Sept: 617–631.

Global Navigation Satellite system (GLONASS). 2008. Interface control document L1, L2. Russian Institute of Space Device Engineering. Moscow. Version 5.1. Available at www.glonass-ianc.rsa.ru/en/ (accessed February 10, 2014).

GPS Directorate, Systems Engineering & Integration Interface Specification. 2011. Navstar GPS space segment/user segment L1C interface, IS-GPS-800B.pdf. Available at www.navcen.uscg.gov/pdf/gps/IS-GPS-800B.pdf (accessed February 10, 2014).

GPS Directorate, Systems Engineering & Integration Interface Specification, IS-GPS-705. 2012. Navstar GPS space segment/user segment L5 interface, IS-GPS-705B. Available at www.navcen.uscg.gov/pdf/gps/IS-GPS-705B.pdf (accessed February 10, 2014).

GPS World. 2013. CSR location platforms go live with China's Beidou 2tracking. Available at http://gpsworld.com/csr-location-platforms-go-live-with-chinas-beidou-2-tracking/ (accessed February 10, 2014).

Grewal, M. S. 2023. Application of Kalman filtering to GPS, INS, & navigation short course. Available at gssc.esa.int/navipedia.

Grewal, M. S. and A. P. Andrews, 2015. *Kalman Filtering: Theory & Practice Using MATLAB®*, 4th Ed. New York, John Wiley & Sons. Translated to Simplified Chinese, 2023.

Grewal, M. S., A. P. Andrews, and C. G. Bartone. 2020. *Global Navigation Satellite Systems, Inertial Navigation, and Integration*, 4th Ed. New York, John Wiley & Sons. Translated 3rd Ed. to Simplified Chinese.

Hegarty, C. and A. J. Van Dierendonck. 1999. Civil GPS/WAAS signal design and interference environment at 1176.45 MHz: Results of RTCA SC159 WG1 activities. *Proceedings of the 12th International Technical Meeting of the Satellite Division of the Institute of Navigation (ION GPS 1999).* Nashville, TN. Sept: 1727–1736.

insidegnss.com. February 2023.

Issler, J. L., L. Ries, J. M. Bourgeade, L. Lestarquit, and C. Macabiau. 2004. *Proceedings of the 17th International Technical Meeting of the Satellite Division of the Institute of Navigation (ION GPS 2004).* Long Beach, CA. Sept: 2136–2145.

Janky, J. M. May 3, 1997. Clandestine location reporting by a missing vehicle. US Patent 5629693.

Revnivykh, S. 2011. *Proceedings of the 24th International Technical Meeting of the Satellite Division of the Institute of Navigation (ION GPS 2011).* Portland, OR. Sept: 839–854.

Septentrio. April 12, 2012. Septentrio's AsteRx3 receiver tracks first GLONASS CDMA signal on L3. Available at www.insidegnss.com/node/2563 (accessed February 10, 2014).

Spilker, J. J. and A. J. Van Dierendonck. 2001. Proposed new L5 civil GPS codes. *Navigation, Journal of the Institute of Navigation* 48:3, 135–144.

Spilker, J. J., E. H. Martin, and B. W. Parkinson. 1998. A family of split spectrum GPS civil signals. *Proceedings of the 11th International Technical Meeting of the Satellite Division of the Institute of Navigation (ION GPS 1998).* Nashville, TN. Sept: 1905–1914.

Van Dierendonck, A. J. and J. J. Spilker, Jr. 1999. Proposed civil GPS signal at 1176.45 MHz: In-phase/quadrature codes at 10.23 MHz chip rate. *Proceedings of the 55th Annual Meeting of the Institute of Navigation.* Cambridge, MA. June: 761–770.

www.faa.gov/ato/navigation-programs/waas. 2023.

# 3 Global Navigation Satellite System Reflectometry for Ocean and Land Applications

*Kegen Yu, Chris Rizos, and Andrew Dempster*

## 3.1 INTRODUCTION

Investigating the use of reflected Global Navigation Satellite System (GNSS) signals for remotely sensing the Earth's surface was initiated about three decades ago (Martín-Neira 1993). Remote sensing based on processing and analyzing reflected GNSS signals is commonly termed GNSS reflectometry (GNSS-R). When using GNSS-R technique to build a remote sensing system, only the receiver needs to be designed and manufactured. The receiver platform (static or mobile; land-based, aircraft, or satellite) needs to be selected based on the specific application. In the case of an aircraft or satellite platform, the direct signal is received via a zenith-looking right-hand circularly polarized (RHCP) antenna, while the reflected signal is received through a nadir-looking left-hand circularly polarized (LHCP) antenna. The reason for such an antenna selection is that GNSS signals are designed as RHCP; however, when reflected over a ground surface, some of them are changed to be LHCP. In the case of a land-based platform, either two antennas are used to receive the direct and reflected signals separately or a single antenna is used to capture both the two signals.

The GNSS signals are always available, globally, and the signal structures are typically well-known, except for those dedicated to military use. The L-band GNSS signals are sensitive to ground surface parameters so that they can be utilized for remote sensing purposes. Recently there has been an increase in such investigations by academia and research institutions, partly because this innovative use of GNSS signals has many potential applications. In particular, space agencies such as NASA and ESA have already funded, or are going to fund, a number of projects/missions which focus on the applications of GNSS-R. The Cyclone Global Navigation Satellite System (CYGNSS) project is just one example that aims to develop a system using a constellation of eight microsatellites to improve hurricane forecasting especially with regards to the storm intensity. Another example is the ESA's Passive Relectometry and Interferometery System (PARIS) project (www.esa.int/Our_Activities/Technology/PARIS). PARIS can be used as a passive radar altimeter. Different from current radar altimeters, PARIS would measure multiple samples from different tracks and rapidly form images of mid-sized (mesoscale) phenomena such as ocean currents or tsunamis. Geophysical parameters that can be measured using a GNSS-based reflectometry system include those (but not limited to) listed in Table 3.1 (Garrison et al. 2002, Gleason 2006, Gleason et al. 2005, Font et al. 2010, Yu et al. 2012a, 2012b, Zavorotny et al. 2014, Yu 2021).

This chapter focuses on GNSS-R for ocean applications including sea surface wind speed estimation and sea surface altimetry and for land applications including soil moisture retrieval and forest change detection. In addition, past and planned satellite missions associated with GNSS-R as well as GNSS radio occultation (GNSS-RO) are summarized. Finally, some challenging issues related to GNSS remote sensing products and services are addressed.

DOI: 10.1201/9781003541141-5

**TABLE 3.1**
**Examples of Ocean and Land Applications of GNSS-R**

| Ocean Applications | Land Applications |
|---|---|
| Ocean wind speed and direction estimation | Soil moisture retrieval |
| Sea surface altimetry | Biomass density estimation |
| Tropical cyclone intensity estimation | Forest change detection |
| Sea surface salinity estimation | Land surface classification |
| Sea wave height characteristics estimation | Snow depth and snow water equivalent estimation |
| Sea surface rainfall detection | Flood monitoring |
| Oil slicks detection | Inland water detection |
| Sea ice detection | Vegetation height estimation |
| Green algae detection | |
| Storm surge detection | |

## 3.2 SATELLITE MISSIONS RELATED TO GNSS REMOTE SENSING

Useful reviews of radio occultation missions can be found in Anthes (2011) and Jin et al. (2011). The concept was first proved by the GPS/MET (GPS/Meteorology) satellite mission in 1995–1997. That satellite took few measurements but was followed by the more productive CHAMP (CHAllenging Minisatellite Payload) and SAC-C (Satellite de Aplicaciones Cientificas-C) satellites. CHAMP provided eight years of radio occultation data consisting of around 440,000 measurements from February 2001 to October 2008 (Heise et al. 2014). Those missions led to the launch in 2006 of a six-satellite constellation FORMOSAT-3 (Formosa Satel-lite mission #3)/COSMIC (Constellation Observing System for Meteorology, Ionosphere, and Climate), which provided 1,500–2,000 soundings per day, resulting in GPS-RO becoming an operational data source for weather prediction and ionospheric monitoring. Other missions that provided significant quantities of RO data are GRACE-A (Gravity Recovery and Climate Experiment), METOP-A (METeorological Operations), C/NOFS (Communications/Navigation Outage Forecasting System), TerraSAR-X, and TanDEM-X. Table 3.2 (taken from the COSMIC website www.cosmic.ucar.edu/) shows the contributions made by these and other missions to both atmospheric and ionospheric sounding. With the COSMIC constellation degrading, as it has reached its design life, a new FORMOSAT-7/COSMIC-a constellation of six satellites was launched into low-inclination orbits in 2019.

A number of satellites have been able to perform GNSS reflectometry. The first was UK-DMC from 2003–2011 (Unwin et al. 2003), the data collected by which has not been widely used, mainly due to limited amount of data. The second was UK TechDemoSat-1 (TDS-1) from 2014–2019 (Unwin et al. 2016, Tye et al. 2016), which has the main objective of sea state monitoring. The third was CYGNSS (a constellation of eight microsatellites) launched by NASA in 2016, with an initial objective of measuring the intensity of Hurricane (Ruf et al. 2013, Clarizia and Ruf 2016). Both TDS-1 and CYGNSS produced a large amount of data which have been exploited for sensing geophysical parameters. In 2017, Japan launched the microsatellite WNISAT-1R with the aim to improve the accuracy of meteorological and hydrographic forecasts (Weathernews WNISAT-1R 2023 visited). In 2019, China launched BuFeng-1 A/B twin satellites, focusing on the monitoring of sea surface wind velocity field (Jing et al. 2019). Spire Global's CubeSats launched in 2019 began to collect data to observe how GNSS signals scatter after bouncing off Earth surface features in early 2020 (Spire Global 2023 visited). In 2021, China launched Jilin-01B, the first commercial satellite, to observe a number of ocean parameters (Jilin-01B 2023 viewed). Also in 2021, China launched FY-3E Satellite, which integrated GNSS-RO and GNSS-R, jointly sensing parameters in the ionosphere, atmosphere and ocean (Zhang et al. 2021). Other missions that have been proposed are PARIS (Passive Reflectometry and Interferometry System), a dedicated mission for GNSS-R (Martin-Neira et al. 2009, 2011), and GEROS-ISS (GNSS REflectometry, Radio Occultation and Scatterometry onboard International Space Station) (GEROS 2013).

Figure 3.1 illustrates that a launched LEO satellite can receive the direct, reflected and refracted GNSS signals to enable positioning, reflectometry and occultation applications.

**TABLE 3.2**
**Contributions to Atmospheric and Ionospheric Sounding by Mission, from the COSMIC Website**

| Mission | Total Atm Occs | Total Ion Occs |
|---|---|---|
| CHAMP | 399,968 | 303,291 |
| CNOFS | 120,588 | 0 |
| COSMIC | 4,039,311 | 3,707,966 |
| GPSMET | 5,002 | 0 |
| GPSMETSA | 4,666 | 0 |
| GRACE | 273,013 | 132,817 |
| METOPA | 993,084 | 0 |
| SACC | 353,808 | 0 |
| TSX | 276,549 | 0 |
| Total | 6,465,989 | 4,144,074 |

*Source:* Anthes, R.A., *Atmos. Measure. Techn.*, 4, 1077, 2011, last updated December 7, 23:25:02 MST 2013.

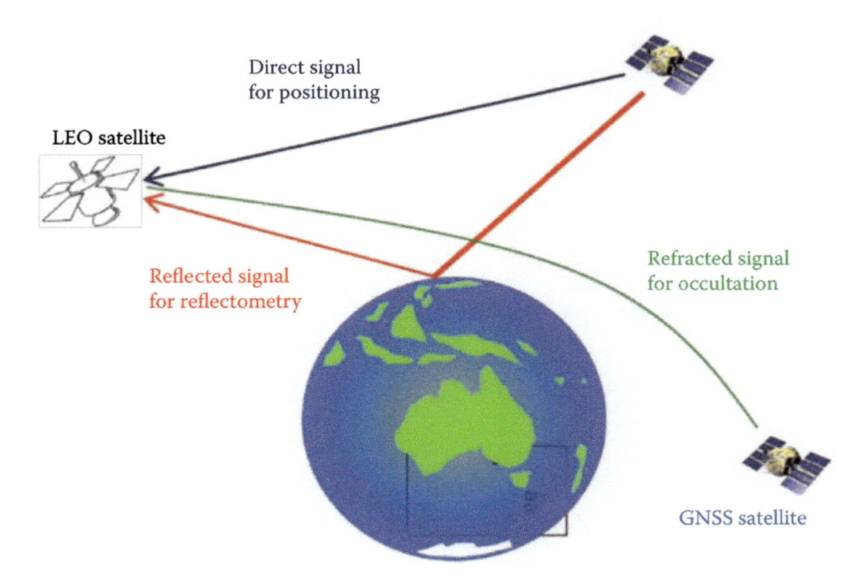


**FIGURE 3.1** Illustration of the geometry of the direct, reflected, and refracted GNSS signals and their use for different applications.

## 3.3 OCEAN OBSERVATION

This section discusses two specific GNSS-R based ocean applications: sea surface wind speed estimation and sea surface height estimation.

### 3.3.1 Sea Surface Wind Speed Estimation

Wind speed retrieval using GNSS-R is mainly based on the theoretical model of the received signal power reflected over sea surface. Since the wind speed is estimated by observing the wind-driven sea surface wave slope characteristics, the estimation is reliable only when the wind and wave interaction has reached the steady state after the wind has continuously blown the surface in one direction for at least say half an hour. The wind speed estimation accuracy can be better than 2 m/s based on processing the data observed from several GNSS-R airborne experimental campaigns. Some details of the technique are described as follows.

#### 3.3.1.1 Sea Wave Spectrum

Sea surface undulation is a complex process and sea wave heights change randomly in time and space. Sea surface roughness can be described by a number of parameters including significant wave height (SWH) and significant wave period (SWP). SWH is defined as the average height of the one-third highest waves and SWP is defined as average period of the waves used to calculate the SWH. Alternatively, wave height spectrum and wave direction spectrum can be used to describe the surface roughness. Among the wave height spectral models, the Pierson-Moskowitz model, the JONSWAP model, and the Elfouhaily model are widely studied (Pierson and Moscowitz 1964, Elfouhaily et al. 1997).

The Elfouhaily model that describes the wind-driven wave height spectrum is defined as

$$W(\kappa,\varphi)=\frac{\kappa^{-4}}{2\pi}(B_\ell(\kappa)+B_h(\kappa))(1+\Delta(\kappa)\cos(2(\varphi-\varphi_0)) \tag{3.1}$$

where $\kappa$ is the wave number, $\varphi$ is the azimuth angle, and $\varphi_0$ is the wind direction. The long wave curvature spectrum $B_\ell(\kappa)$ is defined as

$$B_\ell(\kappa)=\frac{3\times10^{-3}U_{10}}{\tilde{c}\sqrt{\Omega}}\exp\left(-\frac{\Omega}{\sqrt{10}}\left(\sqrt{\frac{\kappa}{\kappa_p}}\right)-1\right)\exp\left(-\frac{5}{4}\left(\frac{\kappa_p}{\kappa}\right)^2\right)\gamma^\Gamma \tag{3.2}$$

where $U_{10}$ is the wind speed at a height of 10 m above the sea. Note that the wind speed at a height of 19.5 m above the sea is related to $U_{10}$ by $U_{19.5}=1.026U_{10}$, showing little difference between wind speeds within the vicinity of these heights. $\Omega$ is the inverse wave age which is equal to 0.84 for a well-developed sea (driven by wind). $\kappa_p$ is the wave number of the dominant waves defined as

$$\kappa_p=\frac{g\Omega^2}{U_{10}^2} \tag{3.3}$$

where $g$ is the gravity. The other three parameters (in Equation 3.2) are defined as

$$\begin{aligned}
&\tilde{c}=\sqrt{g(1+(\kappa/\kappa_m)^2)/\kappa},\quad \kappa_m=370\\
&\Gamma=\begin{cases}1.7, & 0.83<\Omega<1\\ 1.7+6\log(\Omega), & 1<\Omega<5\end{cases}\\
&\gamma=\exp\left(-\frac{1}{2\delta^2}\left(\sqrt{\frac{\kappa}{\kappa_p}}-1\right)^2\right),\quad \delta=0.08(1+4\Omega^{-3})
\end{aligned} \tag{3.4}$$

The short-wave curvature spectrum $B_h(\kappa)$ in the Elfouhaily model is defined as

$$B_h(\kappa) = \frac{c_m \alpha_m}{2\tilde{c}} \exp\left(-\frac{1}{4}\left(\frac{\kappa}{\kappa_m} - 1\right)^2\right) \tag{3.5}$$

where $c_m = 0.23$ and the parameter $\alpha_m$ is determined by

$$\alpha_m = \begin{cases} 10^{-2}(1+\ln(u_f / c_m)), & u_f < c_m \\ 10^{-2}(1+3\ln(u_f / c_m)), & u_f \geq c_m \end{cases} \tag{3.6}$$

where $u_f$ is the friction velocity which can be iteratively computed by

$$u_f = 0.4U_{10}\left|\left(\ln\left(\frac{10}{b(u_f)}\right)\right)^{-1}\right|, \quad b(u_f) = 0.11 \times 14 \times 10^{-6} u_f^{-1} + \frac{0.48u_f^3\Omega}{gU_{10}} \tag{3.7}$$

The initial value may be chosen as $\sqrt{10^{-3}(0.81+0.065U_{10})U_{10}}$ .

The mean square slopes of the surface in the upwind direction and in the cross-wind direction are then respectively calculated by

$$mss_x = \int_0^{\kappa_*}\int_{-\pi}^{\pi} \kappa^2 \cos^2\phi\, W(\kappa,\varphi)\kappa\, d\varphi d\kappa, \qquad mss_y = \int_0^{\kappa_*}\int_{-\pi}^{\pi} \kappa^2 \sin^2\phi\, W(\kappa,\varphi)\kappa\, d\varphi d\kappa \tag{3.8}$$

where the wave number cutoff $\kappa_*$ can be calculated according to

$$\kappa_* = \frac{2\pi}{3\lambda} \tag{3.9}$$

where $\lambda$ is the wavelength (0.1904 m for the GPS L1 signal). In Garrison et al. (2002) the wave number cutoff is modified as

$$\kappa_* = \frac{2\pi\sin(\theta)}{3\lambda} \tag{3.10}$$

where $\theta$ is the incidence angle (complementary to the elevation angle of the satellite). Some details about how to simulate this model and the associated simulation codes can be found in Gleason and Gebre-Egziabher (2009).

#### 3.3.1.2 Sea Surface Scattering

As the GNSS signals arrive at the sea surface some of the signal energy is absorbed by the sea water, while the other energy is reflected. The reflected signals that travel towards the receiver will be captured by the nadir-looking antenna. Let the positions of the transmitter (on the GNSS satellite) and the receiver (on an aircraft, LEO satellite, or land-based) be $(x_t, y_t, z_t)$ and $(x_r, y_r, z_r)$ respectively. Also define the position of the scattering point on the sea surface as $(x_s, y_s, z_s)$. Then the distance from the transmitter through the scattering point to the receiver is given by

$$\begin{aligned} d_{tsr}(x_s, y_s, z_s) = & \sqrt{(x_t - x_s)^2 + (y_t - y_s)^2 + (z_t - z_s)^2} \\ & + \sqrt{(x_r - x_s)^2 + (y_r - y_s)^2 + (z_r - z_s)^2} \end{aligned} \tag{3.11}$$

The specular scattering point (SSP) is the scattering point $(x_{ssp}, y_{ssp}, z_{ssp})$ on the surface where the distance $d_{tsr}$ is minimal. With respect to the signal reflected at the SSP, the signals reflected at other scattering points arrive at the receiver with a delay given by

$$\delta\tau = d_{tsr}(x_s, y_s, z_s) / c - \tau_c \tag{3.12}$$

where $c$ is the speed of light and $\tau_c = d_{tsr}(x_{ssp}, y_{ssp}, z_{ssp})/c$. Given the transmitter and receiver positions and the delay $\delta\tau$, the scattering points define an ellipse on the surface. That is, at each specific delay the signals reflected on the ellipse will arrive at the receiver at the same time, supposing that they travel towards the nadir-looking antenna. Figure 3.2 shows an example of the iso-delay when the satellite elevation angle is 63°. The velocity vector of the aircraft is (21.166946 m/s, –52.149224 m/s, –2.527502 m/s) and the receiver position is (–33.693117°, 151.2745950°, 508.0884 m) in the WGS84 system. The satellite position is (–52.3194°, 162.2910°, 1.9903e+007 m) and its velocity vector is (–131.75275 m/s, –2727.07856 m/s, –573.77554 m/s).

Due to the relative movement between the transmitter and the receiver, Doppler frequencies are produced, resulting in the increase or decrease of the signal carrier frequency. Let the velocity vectors of the transmitter and the receiver be $\vec{V}_t$ and $\vec{V}_r$ respectively. The Doppler frequency is determined by

$$f_D = (\vec{V}_t \bullet \vec{m} - \vec{V}_r \bullet \vec{n})/\lambda \tag{3.13}$$

where $\vec{m}$ and $\vec{n}$ are the unit vectors of the incident wave and the reflected wave respectively, and "$\bullet$" denotes the vector dot product. For a given Doppler frequency, Equation 3.13 represents a hyperbola. That is, the signals reflected on such a hyperbola will have the same Doppler frequency. Figure 3.3 shows an example of the iso-Doppler map under the same circumstance as the iso-Doppler map depicted in Figure 3.2. The intersections of the iso-delay lines and the iso-Doppler lines form a network of grids that will be used to determine the power of the reflected signals arriving at the receiver for each pair of relative propagation delay and Doppler frequency.

#### 3.3.1.3 Reflected Signal Power

The reflected signals received via the nadir-looking antenna are first down-converted to IF signals. The code phase offset or delay and the Doppler frequency associated with the satellite of interest can be estimated based on processing the direct signal received via the zenith-looking antenna through code acquisition and tracking. The carrier frequency of the IF signals is then compensated for and the resulting baseband signal is correlated with a replica of the PRN code related to a specific satellite. At the central Doppler frequency the cross-correlation with a sequence of code phases produces

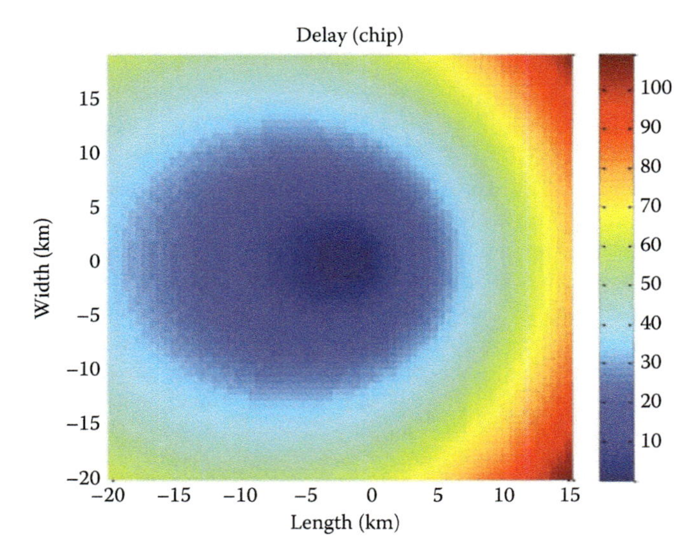


FIGURE 3.2 Example of an iso-delay map for the case where the receiver is carried by an aircraft at a speed of 202.8 km/h and an altitude of 508 m.

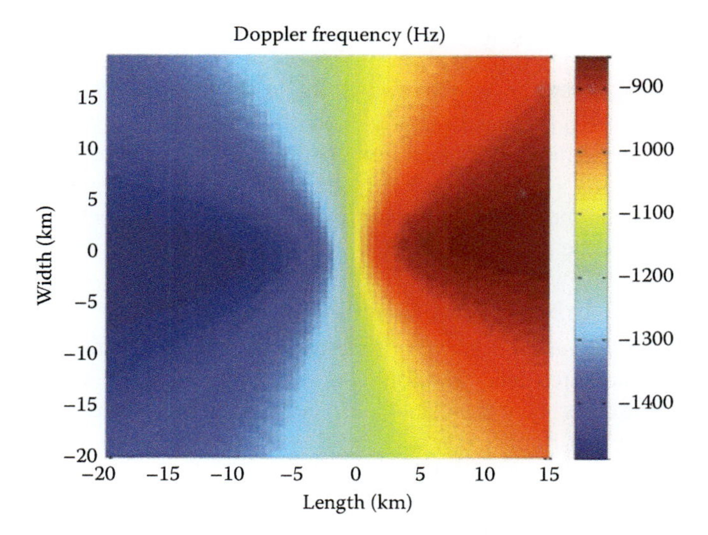


**FIGURE 3.3** Example of an iso-Doppler map for the case where the receiver is carried by an aircraft at a speed of 202.8 km/h and an altitude of 508 m.

a delay waveform (correlation power versus code phase). A delay-Doppler waveform is produced when both a sequence of Doppler frequencies and a sequence of code phases are considered.

Theoretically, the signal power with respect to a code phase and a Doppler frequency can be computed by (Zavorotny and Voronovich 2000):

$$Y(\tau, f_D) = \frac{T_i^2 \lambda^2 P_t \eta}{(4\pi)^2} \iint_A \frac{G_t G_r \sigma_0}{R_{ts}^2 R_{sr}^2} \Lambda^2(\tau - \tau_c)\ \mathrm{sinc}^2((f_D - f_c)T_i)dA \tag{3.14}$$

where $T_i$ is the coherent integration time, $P_t$ is the transmit power, $\eta$ is the atmospheric attenuation, $G_t$ and $G_r$ are the antenna gains of the transmitting and receiving antennas respectively, $R_{ts}$ and $R_{sr}$ are the distance from the transmitter to the scattering point and the distance from the scattering point to the receiver respectively, $\Lambda(\tau - \tau_c)$ is the correlation function of the PRN code with $\tau$ the delay of the replica code and $\tau_c$ the delay of the received code, sinc(·) is the sinc function representing the attenuation caused by Doppler misalignment with $f_c$ the Doppler frequency of the received signal and $f_D$ the Doppler frequency of the replica signal, $A$ is the effective scattering surface area, and the bistatic radar cross section (BRCS) $\sigma_0$ can be calculated according to

$$\sigma_0 = \frac{\pi |\rho|^2}{(\vec{q}_z \vec{q})^4} p\left(-\frac{\vec{q}_\perp}{\vec{q}_z}\right) \tag{3.15}$$

where $\rho$ is the polarization-dependent Fresnel reflection coefficient, $\vec{q}$ is the scattering unit vector that bisects the incident vector and the reflection vector, $\vec{q}_\perp$ and $\vec{q}_z$ are the horizontal and vertical components of $\vec{q}$ respectively, and $p(\cdot)$ is the probability density function (PDF) of the surface slope, which may be simply assumed as omni-directional Gaussian distribution. When performing the double integration in Equation 3.14, the size of the effective scattering area should be appropriately selected. As the flight height increases, the scattering area increases accordingly. Nevertheless, it is not necessary to make the area dimensions too large so as to reduce computational complexity. The contribution of the reflected signals beyond the effective scattering area will be negligible due to the limited antenna beam width and the fact that the power is inversely proportional to the squared distance between the GNSS satellite and the scattering point and between the scattering point and the receiver.

Figure 3.4 shows the decibel delay waveforms based on the theoretical models studied earlier. Six curves correspond to the six different wind speeds (4, 7, 10, 13, 16, and 19 m/s). Four different flight heights were tested: 0.5, 2, 5, and 10 km. In the case of 0.5 km altitude, the six curves are nearly identical and the spread versus time is rather small. Since the waveforms are insensitive to the wind variation when the flight altitude is less than 0.5 km, it would be inappropriate to use the trailing edge to perform any sea state or wind speed retrieval. As the flight altitude increases, the signal spread increases and the waveforms distinguish from each other better. Thus more accurate parameter estimates would be expected.

#### 3.3.1.4 An Example of Airborne Data–Based Approach

Figure 3.5 shows the five theoretical delay waveforms corresponding to five different wind speeds (3, 4, 5, 6, and 7 m/s) and the measured delay waveform associated with a specific satellite. The data was collected during an airborne experiment at an altitude of about 3 km. The measured waveform is produced through coherent integration of 1 millisecond IF signals and then noncoherent integration of 1,000 such 1 millisecond waveforms. It can be seen that the measured waveform has a good match with the theoretical waveform of wind speed 4 m/s which is a good estimate of the real wind speed which is about 4.3 m/s.

This is just an illustrative example to show how the wind speed is estimated. In practice, a mathematical approach will be employed to automatically produce estimation solutions. Regarding this model-matching approach, a certain number of theoretical waveforms are produced and a cost function is defined. The theoretical waveform with the minimal cost function is then selected and the corresponding wind speed is the estimate of the real wind speed. The cost function can be defined as the sum of the squared difference between the theoretical and measured waveforms, that is, least-squares fitting. However, this method requires the alignment of the two waveforms so that the difference between the two waveforms is minimized. Alternatively, a slope-based method proposed in (Zavorotny and Voronovich 2000) can be employed. However, as observed in Figure 3.5, it is

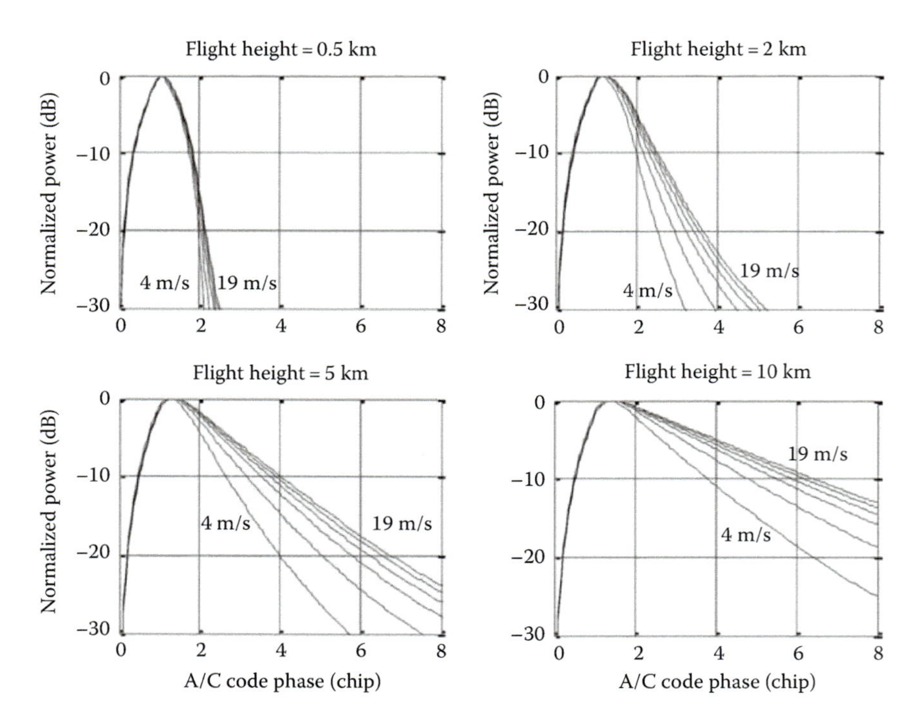


**FIGURE 3.4** Normalized correlation powers of the simulated reflected signals using the Elfouhaily wave elevation model with four different flight heights and five different wind speeds (4, 7, 10, 13, 16, and 19 m/s).

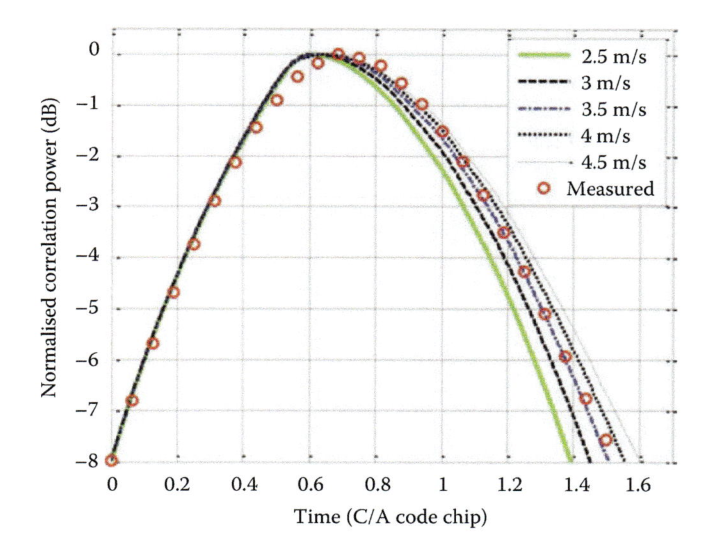


**FIGURE 3.5** Wind speed estimation through matching the measured delay waveform with a number of theoretical waveforms.

rather difficult to distinguish the slopes of neighboring curves related to different wind speeds. The area-based waveform fitting method may be a more suitable option to estimate the wind speed with the following multistep procedure (Yu et al. 2012c):

1. Interpolating both the measured and theoretical waveforms without changing the original data; when using Matlab the library function "INTERP" can be directly used to perform the interpolation.
2. Selecting the cutoff correlation power to retain the waveform above certain power lever so that the slope of the trailing edge does not change abruptly.
3. Calculating the areas of the interpolated waveforms above the cutoff power and calculating the area difference between the measured waveform and each of the theoretical waveforms.
4. Selecting the theoretical waveform that produces the minimum area difference and taking the corresponding wind speed as the estimate.

Figure 3.6 shows the average of wind speed estimates associated with eight satellites when 16 different wind directions are assumed. One observation is that although the mean wind speed estimate varies with the assumed directions, the variation is only about 0.53 m/s. That is, in this case, the variation is not significant. Also shown is the standard deviation (STD) of the wind speed estimation versus the assumed wind directions. The STD ranges from 0.52 m/s to 0.9 m/s. The true wind direction is about 110°, so this STD plot provides some useful information about the wind direction. That is, the wind direction with the smallest STDs may be treated as the actual wind direction. However, the estimation accuracy is rather low and there exists an ambiguity of 180°. More accurate wind direction estimation may be realized by using the properties of the delay Doppler map as reported in Valencia et al. (2014).

#### 3.3.1.5 An Example of Spaceborne Data–Based Approach

Comparing with airborne data, satellite data would be more useful, since satellites have much wider coverage and are in a continuous operation mode. There are two broad methods in using spaceborne GNSS-R data to retrieve sea surface wind speed. The first uses the traditional modeling method, building a mathematical model to describe the relationship between wind speed and an observation variable derived from DDM, such as normalized bistatic radar cross section (NBRCS) and leading edge slope (LES) of a delay waveform, which are closely related to sea surface roughness and hence wind speed (Bu et al. 2020). Figure 3.7 shows the relationship between wind speed data (ECMWF reanalysis data) provided by Copernicus climate change service (C3S) climate database (https://cds.climate.copernicus.eu/cdsapp#!/home) and CYGNSS NBRCS. The ECMWF data are treated as the ground-truth data or reference data, although they possess some uncertainty. From the distribution pattern of the scatter plot, two different simple models may be considered:

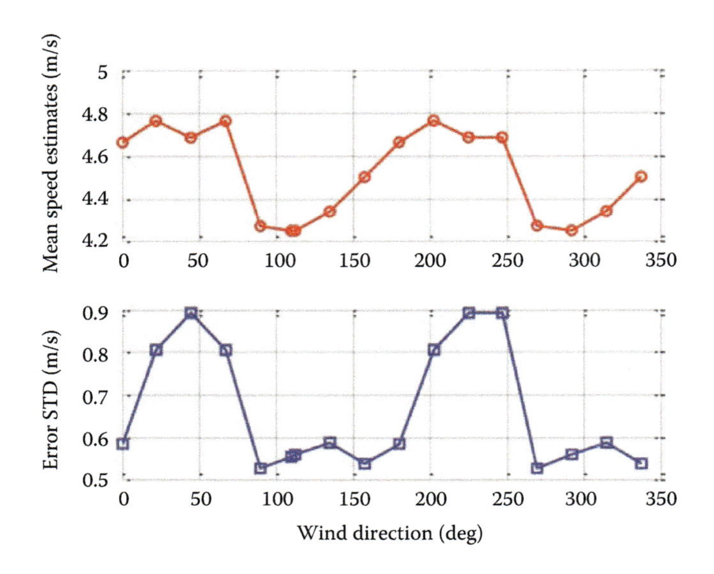


**FIGURE 3.6** Mean wind speed estimate versus assumed wind directions using the delay waveform matching method.

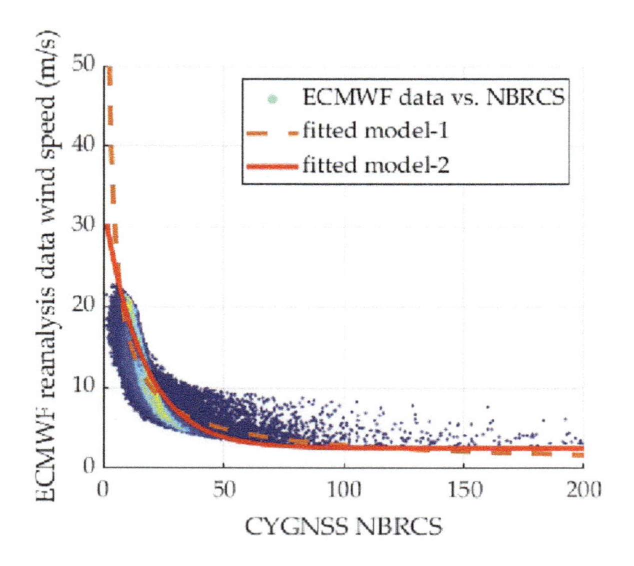


**FIGURE 3.7** Scatter plot of NBRCS and ECMWF reanalysis data and model curves generated by least-squares fitting.

Model-1:

$$s(\sigma_0) = a_1 \sigma_0^{b_1},\ 0 < \sigma_0 \leq 200, \tag{3.16}$$

Model-2:

$$s(\sigma_0) = a_2 \exp(b_2 \sigma_0) + c_2\ ,\ 0 < \sigma_0 \leq 200 \tag{3.17}$$

where $s$ is the sea surface wind speed, $\sigma_0$ is the NBRCS, and $\{a_1, b_1, a_2, b_2, c_2\}$ are the model parameters. Through least-squares fitting, the model parameters can be determined. To make the models being general, it is necessary to use a large amount of data collected in different regions and different seasons to build the models. Note that the raw satellite data are usually rather noisy and some of them can be abnormal such as due to occasional parameter anomalies of spacecraft, receiver, and antenna. Thus, denoising techniques can be used to mitigate strong noise, and quality control flags are used to label the data to avoid the use of poor quality data. That is, the model development and wind speed retrieval are based on cleaned spaceborne GNSS-R data to ensure model quality and retrieval performance. Also note that the reference wind speed data and GNSS-R observation data have different spatio-temporal resolution, so matching in time and space between the two types of data is needed in the model development.

The mathematical models generated through curve fitting like those in Figure 3.7 may not consider the effect of a range of factors such as signal-to-noise ratio, noise floor, instrument gain, incidence angle, and azimuth angle of a given specular point (Wang et al. 2022). To account for the impact of many different parameters or factors effectively, machine learning methods can be applied to build models and retrieve the sea surface wind speed. There are many machine learning methods or algorithms, including binary trees, ensembles of trees, XGBoost, LightGBM, artificial neural network, stepwise linear regression, and Gaussian support vector machine. Their complexities can be significantly different, so is their retrieval accuracy. Thus, it is useful to choose a machine learning method by considering accuracy, complexity, and reliability as well. Based on testing over one million datasets, the retrieval accuracy of the seven methods mentioned earlier, in terms of root mean square error, is between 1.4 m/s and 1.9 m/s. Figure 3.8 shows the scatter plot of the retrieved wind speed versus

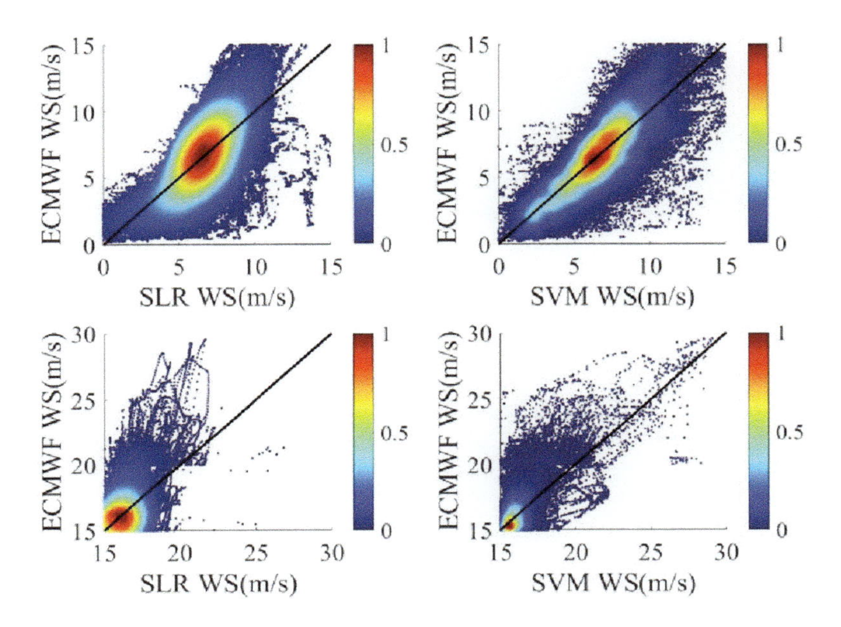


**FIGURE 3.8** Results of wind speed retrievals based on SLR and SVM.

the reference wind speed using two machine learning methods (SLR and SVM). To achieve higher accuracy, the wind speed may be divided into different ranges and different models are developed respectively. For instance, two different ranges are used in Figure 3.8, although the samples of lower wind speed (0–15 m/s) is much larger than those of higher wind speed (15–30 m/s) in this case.

## 3.3.2 Sea Surface Altimetry

### 3.3.2.1 Sea Surface Height Calculation

As shown in Figure 3.9, sea surface height (SSH) is calculated relative to the surface of the theoretical Earth ellipsoid in the WGS84 system, which has zero altitude. The SSH at a specific sea surface point is the distance from the point to the WGS84 Earth ellipsoid surface and the mean SSH is the average of a large number of these distances over an area of interest.

Figure 3.10 shows the flowchart of how to calculate the SSH using a two-loop iterative method. Specifically, for a given tentative SSH, since both the GNSS satellite position/and the receiver position estimate/are known, the SSP $(\tilde{x}_{ssp}, \tilde{y}_{ssp}, \tilde{z}_{ssp})$ on the tentative sea surface can be readily determined. The SSP estimation is realized by minimizing the total path length (TPL) from the satellite through the SSP and to the receiver, which is given by

$$\tilde{R}_{tSr} = \tilde{R}_{tS} + \tilde{R}_{Sr} \tag{3.18}$$

where

$$\begin{aligned} \tilde{R}_{tS} &= \sqrt{(x_t - \tilde{x}_{ssp})^2 + (y_t - \tilde{y}_{ssp})^2 + (z_t - \tilde{z}_{ssp})^2} \\ \tilde{R}_{Sr} &= \sqrt{(\tilde{x}_{ssp} - \hat{x}_r)^2 + (\tilde{y}_{ssp} - \hat{y}_r)^2 + (\tilde{z}_{ssp} - \hat{z}_r)^2} \end{aligned} \tag{3.19}$$

The TPL can also be estimated using the propagation time ($R_{tr}$) of the direct signal and the relative delay ($\tau_{rd}$) of the reflected signal, that is,

$$\hat{R}_{tSr} = \hat{R}_{tr} + c\hat{\tau}_{rd} \tag{3.20}$$

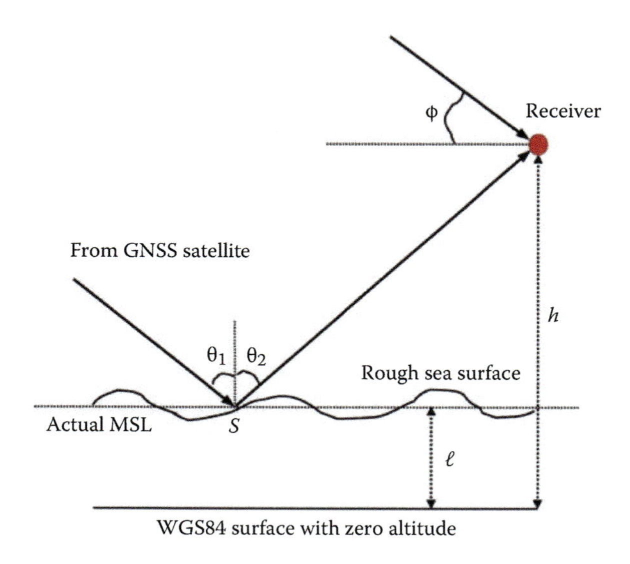


FIGURE 3.9 Geometry of the receiver, WGS84 Earth ellipsoid surface, rough sea surface, actual MS, and direct and reflected signal paths.

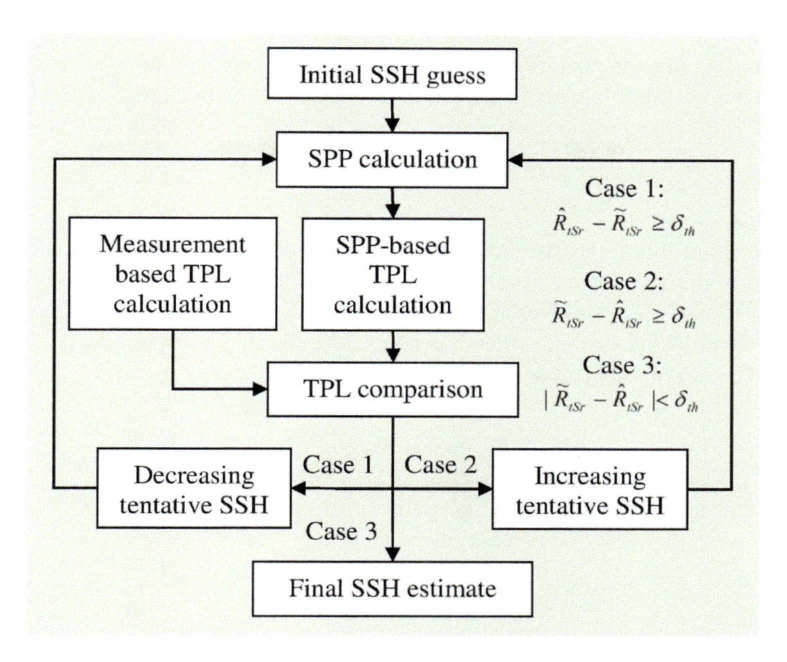


**FIGURE 3.10** Flowchart for two-loop iterative SSH calculation. $\delta_{th}$ is a small positive number.

where $c$ is the speed of light, $\hat{\tau}_{rd}$ is the estimated relative delay, and $\hat{R}_{tr}$ is the estimate of $R_{tr}$ calculated by

$$\hat{R}_{tr} = \sqrt{(x_t - \hat{x}_r)^2 + (y_t - \hat{y}_r)^2 + (z_t - \hat{z}_r)^2} \tag{3.21}$$

where the satellite position is assumed error free, while the receiver position is an estimate. How to estimate the relative delay will be discussed later. Then, as indicated in Figure 3.9, the calculated TPL by Equation 3.18 is compared with the measured TPL by Equation 3.20 to determine whether the tentative sea surface height should be increased or decreased. The process is terminated once the difference between the two TPLs is smaller than the predefined threshold.

To reduce the computational complexity, a simpler technique may be used. For instance, if $\tilde{R}_{tSr} > \hat{R}_{tSr}$, the tentative surface height is increased by a relatively larger increment such as 40 m. At the next iteration of the outer loop, if $\tilde{R}_{tSr} < \hat{R}_{tSr}$, the increment is decreased by half of the previous increment. In this way, the process will quickly converge to the steady state. Note that the specular reflection must satisfy Snell's law, that is, the two angles ($\theta_1$ and $\theta_2$ in Figure 3.9) between the incoming wave and the reflected wave, separated by the surface normal, must be equal or the difference is extremely small. Thus, the results should be tested to see if this law is satisfied.

From Equation 3.18 the partial derivatives with respect to the coordinates of the specular point can be determined as

$$\frac{\partial R_{tSr}}{\partial u_S} = \frac{u_S - x_r}{R_{rS}} + \frac{u_S - x_t}{R_{tS}}, \qquad u_S \in \{x_{ssp}, y_{ssp}, z_{ssp}\} \tag{3.22}$$

which can be rewritten in a vector form as

$$d\vec{S} = \frac{\vec{R} - \vec{S}}{R_{rS}} + \frac{\vec{T} - \vec{S}}{R_{tS}} \tag{3.23}$$

where $\vec{S}$, $\vec{R}$, and $\vec{T}$ are the position vectors of the SSP, the receiver, and the transmitter, respectively. Equation 3.23 is used to generate an iterative solution to the minimum path length. That is, at time instant $n + 1$, the SSP position is updated according to

$$\vec{S}_{n+1} = \vec{S}_n + \kappa\, d\vec{S} \tag{3.24}$$

where $\kappa$ is a constant that typically should be set as a larger value as the flight altitude increases. The initial guess of the SSP can be simply the projection of the receiver position onto the surface. At each iteration, a constraint must be applied to restrain the SSP on the surface that is $\tilde{\ell}$ m above or below the WGS84 ellipsoid surface which has a zero altitude if $\tilde{\ell}$ is a positive or negative number. That is, the SSP position is scaled according to

$$\vec{S}'_{n+1} = (r_S + \ell)\frac{\vec{S}_{n+1}}{|\vec{S}_{n+1}|} \tag{3.25}$$

where the radius of the Earth at the specular point is calculated by

$$r_S = a_{WGS84}\sqrt{\frac{1 - e_{WGS84}^2}{1 - e_{WGS84}^2(\cos\lambda_S)^2}}, \qquad \lambda_S = \arcsin\left(\frac{z_S}{|\vec{S}|}\right) \tag{3.26}$$

where $e_{WGS84} = 0.08181919084262$ and $a_{WGS84} = 6378137$ m.

Since the altitude of the WGS84 Earth's surface is zero, the WGS84 altitude of the SSP is equal to $\ell$. Clearly, the altitude of a single SSP cannot be treated as the estimate of the mean sea surface height. However, a reasonable estimate of the mean surface height will be produced through the generation and subsequent processing of the altitude estimates of many SSPs over a period of time.

#### 3.3.2.2 Calibration

Since the zenith-looking and nadir-looking antennas, the receiver, and the reference point are not in the same position, it is necessary to calibrate the relative delay measurements to remove the effect of these position differences. Note that the reference point may be set at the center of the inertial measurement unit (IMU) provided that such a device is used. Figure 3.11 illustrates the configuration of the devices. The two antennas are connected to the receiver via two cables whose lengths are $L_{CR}$ and $L_{CD}$. The actual measurement of the relative delay of the reflected signal is given by

$$L_{measured} = AS + SD + L_{DR} - (AU + L_{UR}) + \varepsilon \tag{3.27}$$

where $\varepsilon$ is the measurement error. On the other hand, using the reference point position, the relative delay is calculated as

$$L_{calculated} = AS + SC - AC \tag{3.28}$$

As described earlier, the SSH is estimated by comparing the measured and calculated relative delays. However, typically, there is an offset between the calculated and measured delays even in the absence of measurement error. That is,

$$\begin{aligned} L_{offset} &= L_{measured} - L_{calculated} \\ &= (SC - SD) + (AU - AC) + (L_{UR} - L_{DR}) \end{aligned} \tag{3.29}$$

where the measurement error is ignored.

Therefore, the measured relative delay should be calibrated by subtracting the offset from itself. The lengths of the two cables can be readily measured in advance. In case where a LNA is used to amplify the signal such as captured by the nadir-looking antenna, the path length from D to R will be the sum of the two cable lengths plus the distance between the two connection points of the LNA. The distance from U to C and that from D to C can be manually determined in advance. The positions of points U and C are estimated by the receiver and frame transformation; thus distances AC and AU can be readily calculated. Calculation of the distances SC and SD requires a knowledge of the SSH that is unknown in advance. However, initial information about the SSH or previous SSH estimation results can be exploited. The uncertainty in the SSH estimate will affect the calculation of both SC and SD in a very similar way, so that a small SSH error will have a negligible impact. Thus, distance CV can be estimated and distance SC can be calculated using elevation angle. Calculation of distance SD requires its orientation to determine the angle $\angle SDC$. In the case where the nadir-looking antenna is fixed directly beneath the reference point, the distance can be simply calculated by

$$SD = \sqrt{SC^2 + CD^2 - 2 \times SC \times CD \times \cos(90 - \varphi)} \tag{3.30}$$

Once the distances and the cable lengths in Equation 3.27 are known, the relative delay offset can be readily determined. For instance, suppose that points $U$ and $D$ are directly above and below point $C$ respectively; $AC$ = 20,000 km; satellite elevation angle is 50°; $UC$ = 0.8 m; $CD$ = 0.4 m; $CV$ = 300 m; $L_{UR}$ = 1 m; and $L_{DR}$ = 0.6 m. Then, the unknown distances are obtained from some simple calculations: $AU$ = 19999.999387 km, $SD$ = 391.32 m, $SC$ = 391.62 m. As a result, the relative relay offset is calculated to be 0.094 m. The SSH estimation error caused by this relative delay offset can be approximated as

$$\delta\ell \approx \frac{L_{offset}}{2\sin\varphi} = 6.1\text{cm} \tag{3.31}$$

Thus, when the configuration of the devices is arranged properly and the cable lengths are selected based on similar analysis, the SSH error caused by the device configuration will not be large. However, to achieve accurate altimetry, such an error must be compensated for.

#### 3.3.2.3 Relative Delay Estimation Methods

The SSH estimation accuracy is largely dependent on the performance of estimation of the total path length (TPL) of the reflected signal, from the transmitter through the SSP on the sea surface

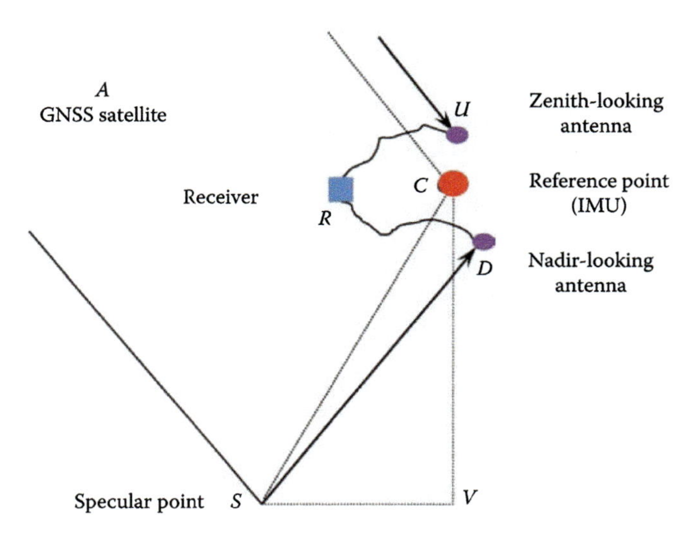


FIGURE 3.11 Geometric relationship between the devices (receiver, antennas, inertial measurement unit (IMU)) mounted on a LEO satellite or an aircraft.

and to the receiver. The TPL is equal to the sum of the path length of the direct signal and the relative delay of the reflected signal. Since the satellite position is known, the measurement error associated with the path length of the direct signal comes from the receiver position estimation error. When the receiver is given, such an error is typically not reducible, although smoothing may improve the accuracy marginally. Thus, it is important to use a GNSS receiver which can achieve satisfactory position estimation accuracy. When an accurate receiver is used, the relative delay estimation error would be dominant. Therefore, it is vital to reduce this error. The relative delay is estimated by determining the code phase of the direct signal and that of the reflected signal based on the measured correlation waveform.

There are two methods associated with the delay waveform based method. The first one makes use of the clean code (C/A code), while the second one utilises the interferometry technique. The clean code method deals with the direct signal and the reflected signal separately and only uses the C/A code to generate the delay waveform. In the interferometry technique, on the other hand, both signals are processed together. That is, the direct signal observed from a zenith-looking antenna is cross-correlated against the reflected one captured by a nadir-looking antenna. High-gain antennas are needed to achieve separability among different satellites. The interferometry technique is intended to exploit either the P(Y) code or military M code to achieve an accuracy gain at the cost of high-gain and directional antennas. In addition, a high-bandwidth front end/receiver is required.

In the case where the zenith-looking antenna is high above the ground, especially when the receiver is mounted on a satellite or on an aircraft, the code phase of the direct GNSS signal can be readily estimated by determining the location of the peak power of the correlation waveform. On the other hand, it may not be easy to obtain an accurate estimate of the desired code phase of the reflected signal forwarded from a rough sea surface. The main reason is that the location of the peak power of the reflected signal would not be the desired code phase of the reflected signal since the peak power location is shifted due to rich multipath propagation. Clearly, using the peak power location to calculate the relative delay would produce a large bias error. The time shift or offset would depend on a number of factors including the surface roughness and the receiver altitude. The peak power location of the delay waveform derivative can be used as the desired code phase of the reflected signal, but the estimate would be biased (Rius et al. 2010). It is observed that the desired code phase of the reflected signal is somewhere between the peak power location of the delay waveform and that of the waveform derivative. However, the exact location of the code phase is typically unknown.

For clarity, it is desirable to explain why the peak power location could shift when a flat sea surface is replaced with a rough sea surface. In the presence of a perfect smooth sea surface, the signal will only be reflected at the SSP and then travels to the receiver. In the presence of a rough sea surface, besides the first path signal, there will be multipath signals arriving at the receiver. Consider the idealised case where there is no noise or error so that the correlation diagram of each path is an isosceles triangle. Suppose that there are $J$ multipath signals whose delays relative to the first path are less than the GNSS code chip width (i.e., half of the correlogram triangle width). Then, the following result exists.

*The peak correlation power location of the combined multipath signals will shift from the peak correlation power location of the first path signal provided that*

$$\sum_{j=2}^{J} P_j > P_1 \tag{3.32}$$

*where $P_j$ is the peak correlation power of the $j^{th}$ path signal.*

Since both the first path signal and signals of other paths are reflected signals, the signal power of the second path and a number of following paths can be significant with respect to the first path. Thus, intuitively, Equation 3.32 would always be valid with a rough sea surface. The relationship in Equation 3.32 can be proved mathematically as in Yu et al. (2014).

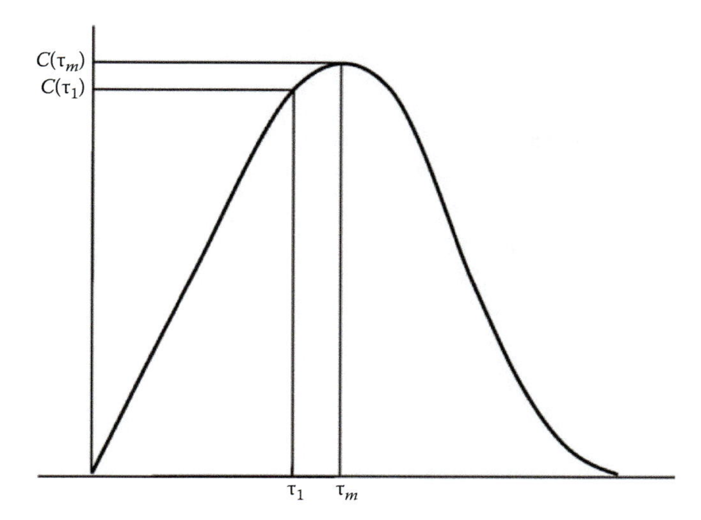


FIGURE 3.12 Illustration of code phases/delays and related correlation power (delay waveform) of reflected signal in the presence of a rough surface.

In the case where the P(Y) code or the military M code is employed, the peak power location shift can be much smaller so that the shift may be ignored. However, when using the C/A code, the shift needs to be taken into account in the presence of a rough surface. In the absence of information of accurate surface wave statistics, the power ratio–based approach to be discussed next may be employed.

#### 3.3.2.4 Power Ratio–Based Sea Surface Height Estimation—An Example

The power ratio is defined as the ratio of the correlation power at the desired code phase over the peak correlation power of the reflected signal. The desired code phase corresponds to the peak power location when the surface is perfectly smooth. Figure 3.12 is an illustration of the delay waveform of the reflected signal in the presence of a rough surface. $C(\tau_m)$ is the peak power of the reflected signal received via the down-looking antenna where $\tau_m$ is the time point at which the peak power occurs, while $C(\tau_1)$ is the correlation power at the time point $\tau_1$ where the reflected signal peak power occurs when the surface is perfectly smooth. Since $C(\tau_m)$ can be measured, the time parameter $\tau_m$ can be estimated. On the other hand, neither $C(\tau_1)$ nor $\tau_1$ can be simply measured. The power ratio is simply defined as

$$\eta = \frac{C(\tau_1)}{C(\tau_m)} \tag{3.33}$$

Clearly, in the case of a perfectly smooth sea surface, the power ratio is equal to unity. Otherwise it is less than one. Given a power ratio and the measured peak power, the power at the desired code phase ($\tau_1$) can be calculated using Equation 3.33 and then $\tau_1$ can be determined using the measured delay waveform. As a consequence, the desired delay of the reflected signal relative to the direct signal can be obtained. The relative delay is then used to calculate the SSH using the method described earlier. A sequence of SSH estimates is produced based on a time series of measured and smoothed delay waveforms. In the case of low altitude airborne altimetry, the mean SSH could be approximately the same over the surface specular reflection tracks related to several GNSS satellites with the largest elevation angles. Then, sequences of SSH estimates associated with these GNSS satellites can be jointly processed to estimate the desired power ratio and the mean SSH. One method is to define a cost function of the power ratio as

$$\psi(\eta) = \sigma(\eta) \tag{3.34}$$

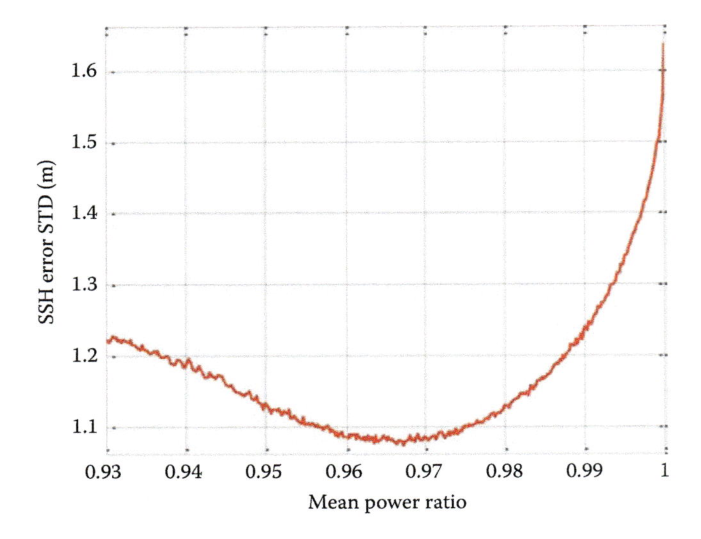


**FIGURE 3.13** Cost function (SSH error STD) versus mean power ratio for mean SSH estimation. Data related to four satellites are used.

**TABLE 3.3**
**PRN Numbers, Elevation and Azimuth Angles of Four Satellites**

| Satellite (PRN#) | 22 | 18 | 6 | 21 |
|---|---|---|---|---|
| Elevation (°) | 62.67–62.99 | 58.14–57.41 | 50.89–50.92 | 48.57–47.95 |
| Azimuth (°) | 238.19–240.29 | 148.82–149.94 | 266.29–268.22 | 78.87–79.93 |

where $\sigma$ is the standard deviation of all the sequences of SSH estimates. $N$ power ratio values correspond to $N$ cost function values. The power ratio and the sequence of the SSH estimates or the mean SSH are selected if the corresponding cost function value is the minimum. This criterion selection comes from the consideration that using the desired power ratio would yield estimates that have minimum variations. Figure 3.13 shows an example of the cost function versus the power ratio, which is basically a convex function. In this example, the mean SSH estimate is 23.36 m in the WGS84 system, while the mean SSH measured by a LiDar device is 23.44 m. Measurements related to four satellites as listed in Table 3.3 were used to produce the mean SSH estimate. The average of the mean SSH errors and the root-mean-square mean SSH errors are both 8 cm. This sea surface altimetry method is only suited for the case where the receiver is aircraft borne. In the case of the satellite-borne receiver, the theoretical model of the received reflected signal power is required to obtain the offset of the peak power location of the reflected signal in the presence of a rough sea surface. However, when the sea surface is calm or the GPS P(Y) code is used, the offset can be negligible, so that measurements associated with a single GNSS satellite can be used to retrieve the mean SSH.

## 3.4 LAND APPLICATIONS

As mentioned earlier in Table 3.2, GNSS-R technique can be used to retrieve a range of land parameters. This section briefly describes how this technique can be employed to estimate soil moisture and detect forest change.

### 3.4.1 Soil Moisture

Knowledge of soil moisture content is critical for drought and irrigation management, so as to increase crop yields and to gain a better understanding of natural processes linked to the water, energy and carbon cycles. Soil moisture can be estimated by analyzing surface scattered signals transmitted and received by active radar sensors or natural surface emission detected by microwave radiometers. Alternatively, GNSS-R signals can be employed to estimate soil moisture. Since the soil dielectric constant and reflectivity depend on soil moisture, the variation in the reflected GNSS signal power, or signal-to-noise ratio indicates changes in soil moisture (Masters et al. 2004). Using an interferometric method, the power associated with the coherent sum of the direct and reflected GNSS signals can be used to estimate soil moisture. After continuously measuring the total power for a few hours, the location of the power notch over satellite elevation can be observed, and the soil moisture may be estimated (Rodriguez-Alvarez et al. 2009). The total power received can be described as

$$\eta F(\theta) \,|\, 1 + R \cdot \exp(j\varphi) \,|^2 \tag{3.35}$$

where $\eta$ is a scaling factor, $F(\theta)$ is the antenna pattern/gain, $\theta$ is the incidence angle, $R$ is the reflectivity coefficient, and $\varphi$ is the phase difference between the incident and reflected signals. The reflectivity coefficient may be described by a three-layer reflectivity model (air and two soil layers) as

$$\zeta = \exp\left(-\left(\frac{4\pi\sigma_s}{\lambda}\right)^2\right) \frac{r_{i,i+1} + r_{i+1,i+2}\exp(\delta + j2\psi)}{1 + r_{i+1,i+2}^2 \exp(\delta + j2\psi)} \tag{3.36}$$

where $\sigma_s$ is the soil surface roughness, $\lambda$ is the signal wavelength (for GPS L1 signal, $\lambda = 19\,\text{cm}$), $r_{i,i+1}$ and $r_{i+1,i+2}$ are the Fresnel coefficients of the first and second soil layers, $\delta$ is the surface roughness correction factor, and $\psi$ is a phase term. $\eta$ may be simply set to be unity when only using the power notch position for soil moisture estimation, while it can be adjusted when using the pattern or shape of the total power. It is an advantage to use existing GPS receivers, installed primarily for geophysical and geodetic applications, to estimate soil moisture (Zavorotny et al. 2010). These receivers, if exploited effectively, could provide a global network for soil moisture monitoring. In addition to the interferometric method, the reflectometry technique also can be employed. For instance, the empirical dielectric model based method in (Wang and Schmugge 1980) may be applied to GNSS-based soil moisture retrieval. In this method, the transition moisture parameter ($\theta_t$) is defined as

$$\theta_t = 0.165 + 0.49\theta_{wt} \tag{3.37}$$

where $\theta_{wt}$ is the wilting point moisture that is a function of the percentage of the clay and sand contents. Depending on whether the soil moisture is greater or less than the transition moisture, a specific formula can be used to calculate the soil moisture. This model employs the mixing of either the dielectric constants or the refraction indices of ice, water, rock, and air, and treats the transition moisture value as an adjustable parameter. The dielectric constant of the soil can be estimated by measuring the surface reflectivity. The model is quite general since it was developed by considering many different types of soils. GNSS-based soil moisture estimation is complicated by a number of issues including surface roughness, vegetation canopy, and variation in percentage of individual soil components. To achieve reliable soil moisture estimation, these issues must be taken into account through processes such as modeling and compensation. Currently, the accuracy of GNSS-R based soil moisture estimation is largely not as good as that of some active sensors and passive microwave radiometers. Most of the GNSS-R experiments conducted for soil moisture retrieval are ground-based, but there are a good number of aircraft-borne ones. Data provided by several satellite missions are also used to infer global soil moisture information.

#### 3.4.1.1 An Example

Figure 3.14 shows an experimental setup for investigating the use of GNSS signals for soil moisture retrieval. A typical geodetic receiver and antenna was used to collect the GNSS signal. The antenna was face-up to simulate the situation where the GNSS signals are captured by a CORS (continuously operating reference stations) receiver and antenna. A network of tens of thousands of CORS has been established in the world, and the CORS data can be used to sense the environmental parameters around the CORS based on GNSS reflectometry. The ground-truth soil moisture data were obtained by a soil moisture sensor installed beneath the ground surface (Chang et al. 2019). The SNR time series recorded by the receiver is first processed by removing its low-frequency component, which is mainly contributed by the direct GNSS signal. As a result, the detrened SNR time series is mostly associated with the reflected GNSS signals. The detrended SNR time series is further denoised through filtering and fitting and the purified SNR time series is a quasi-sinusoidal signal. Then, a statistic can be defined as the mean of the absolute values of the peaks (maximums and minimums of the quasi-sinusoidal wave):

$$\bar{F} = \frac{1}{m}\sum_{i=1}^{m}|F_i| \tag{3.38}$$

where the value of $m$ is selected by considering the smallest absolute peak value $| F_i |$ should be much greater than noise level. It was observed that the statistic series and the soil moisture have a reciprocal relationship, so that the observation variable is defined as:

$$F_{reciprocal}^{(q)} = \frac{1}{k}\sum_{j=1}^{k}\frac{1}{\bar{F}_j^{(q)}} \tag{3.39}$$

where there are $k$ mean peaks of all satellites selected in a day, that is the qth day, and $\bar{F}_j^q$ is the $j$th mean peak on the $q$th day. Modeling is then carried out by using the daily time series $\{F_{reciprocal}^{(q)}\}$ and the in situ observation of the soil moisture. Using least-squares fitting, a second-order polynomial model is generated as:

$$SM = 0.0651F_{reciprocal}^2 - 0.1397F_{reciprocal} + 0.0423 \tag{3.40}$$

FIGURE 3.14 A GNSS receiver and a face-up antenna were used to collect GNSS signals in the experimental field (left); a soil moisture sensor was used to record the soil moisture about 5 cm beneath the ground surface (right).

Note that Equation 3.40 is just an empirical model that is suited to bare soil or vegetation cover is rather minor. In the presence of significant vegetation cover or significantly different type of soil, a new model should be developed.

### 3.4.2 Forest Change Detection

Forest change detection is another possible application of GNSS remote sensing. Forest change can provide useful information about global climate change impacts and carbon storage, and knowledge of forest change is vital for effective forest management. Received signal strength of the reflected GNSS signals can be employed to distinguish forest conditions from each other, since different surface cover has different reflectivity. A number of signal strength ranges may be defined to be associated with a group of surfaces such as lake/river water, typical dense forest, and cleared area due to logging (Yu et al. 2013). Figure 3.15 shows four ground specular reflection tracks of the reflected signals from four GNSS satellites received by a receiver on an aircraft. The tracks are colorized by the reflected signal power. The low signal power corresponds to the dense forest areas, while the high signal power occurs over three areas marked by A1 (cleared area), A2 (partially cleared area), and A3 (lake water). Accordingly, a surface can be classified by determining within which range its measured strength falls. In addition to signal strength, other signal characteristics such as those related to the observed correlation waveform may be used to enhance surface change detection or surface classification. The major issue related to surface change detection is the dependence of reflectivity on several factors such as soil moisture and surface roughness. Forest change may be quantified by evaluating forest biomass, which depends on the volume of both living and dead trees (leaves, branches, and trunks). The received signal power can be used to determine the scattering coefficient which, based on simulation (Ferrazzoli et al. 2011), should be a function of the biomass. However, it is a challenging problem to derive a formula that accurately describes the relationship between the scattering coefficient and the biomass.

### 3.4.3 Snow Depth

GNSS-R technology has also been investigated for snow depth measurement. The interferometric signal captured by a receiver is the sum of the direct signal and reflected signals. Considering only

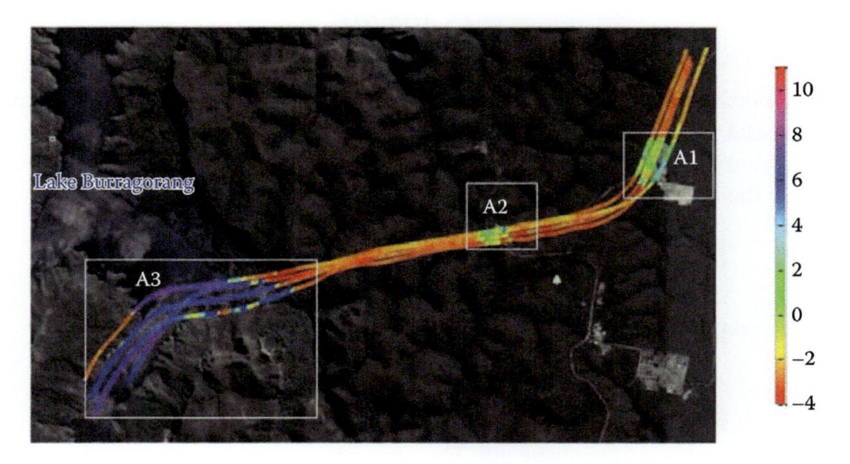


FIGURE 3.15 Ground specular reflection tracks associated with four GPS satellites, colorized by reflected signal power. The picture was generated by Google Earth and GPS Visualizer.

one reflected signal, that is, the specular reflection path signal, the propagation delay of the reflected signal relative to the direct signal can be well approximated as:

$$\Delta_m(t) = 2h\sin\theta(t) \tag{3.41}$$

where $h$ is the antenna height and $\theta(t)$ is the satellite elevation angle, which is time-varying. The resulting excess phase is given by:

$$\delta_\phi(t) = 2\pi f\frac{\Delta_m(t)}{c} = \frac{4\pi h}{\lambda}\sin\theta(t) \tag{3.42}$$

where $f$ is the carrier frequency of the satellite signal, $c$ is the speed of light, and $\lambda$ is the signal wavelength. Accordingly, the received signal can be described by

$$\begin{aligned} s(t) &= A_d(t)\sin\psi(t) + A_m(t)\sin(\psi(t) + \delta_\phi(t)) \\ &= (A_d(t) + A_m(t)\cos\delta_\phi(t))\sin\psi(t) + A_m(t)\sin\delta_\phi(t)\cos\psi(t) \\ &= A(t)\sin\tilde{\psi}(t) \end{aligned} \tag{3.43}$$

where $A_d(t)$ and $A_m(t)$ are the amplitudes of direct and reflected signal, $\psi(t)$ is the direct signal phase, and

$$\begin{aligned} A(t) &= \sqrt{A_d^2(t) + A_m^2(t) + 2A_d(t)A_m(t)\cos\delta_\phi(t)} \\ \tilde{\psi}(t) &= \psi(t) + \beta(t) \\ \beta(t) &= \tan^{-1}\frac{\alpha(t)\sin\delta_\phi(t)}{1 + \alpha(t)\cos\delta_\phi(t)} \end{aligned} \tag{3.44}$$

Here $\tilde{\psi}(t)$ is the composite phase, $\beta(t)$ is the composite excess phase (interferometric phase), and $\alpha(t) = A_m(t)/A_d(t)$ is the amplitude ratio. Clearly, the simplified received signal is a single quasi-sinusoid, which has time-varying amplitude and frequency.

#### 3.4.3.1 Signal-to-Noise Ratio

As mentioned earlier, after removing the low-frequency component of the SNR time series, which can be obtained through low-order polynomial fitting or lowpass filtering, the detrended SNR time series can be described as:

$$SNR_d(t) = \frac{2A_d(t)A_m(t)}{P_N}\cos\delta_\phi(t) \tag{3.45}$$

where $P_N$ is the noise power. It can be seen from Equations 3.42 and 3.45 that the detrended SNR is a quasi-sinusoidal signal, the instantaneous frequency of which changes with time. However, if $\sin\theta(t)$ is treated as an independent time variable, then differentiating the excess phase over $\sin\theta(t)$ produces the radian frequency as (Larson et al. 2009):

$$\frac{d(\delta_\phi(t))}{d(\sin\theta(t))} = 2\pi \times \frac{2h}{\lambda} \tag{3.46}$$

That is, the frequency of the detrended SNR series is given by

$$f = \frac{2h}{\lambda} \tag{3.47}$$

which is a constant for a given antenna height. Therefore, through spectrum analysis on the detrended SNR time series, the peak-power frequency can be determined, and hence the antenna height can be calculated by Equation 3.47 with the known signal wavelength and the frequency. The antenna height in the absence of snow on the ground can be measured in advance, so the snow depth is the antenna height difference when the ground is free of snow and when the ground is covered with certain depth of snow.

Figure 3.16 shows the power spectrum of the detrended SNR series; although it is rather different from the power spectrum of a sinusoid, a sharp pulse does occur. The significant difference comes from the fact that the detrended SNR time series includes multiple reflected signals instead of a single one and there exists noise. Note that the sampling versus the independent time variable is not uniform, so the fast Fourier transform cannot be used for the spectrum analysis. To deal with the problem of non-uniform sampling, a suited spectral analysis such as the Lomb-Scargle spectral analysis can be applied.

### 3.4.3.2 Carrier Phase and Pseudorange Combination

In addition to SNR, measurements of other parameters such as carrier phase and pseudorange, provided by ground-based geodetic receivers, can be used to measure snow depth. In the dual carrier phase combination method (Ozeki and Heki 2012), after removing the geometric effect, the phase combination is detrended to remove the effect of ionospheric delay. Then, the purified phase combination becomes:

$$M_{1,2}(t) = \lambda_1 \tilde{\psi}_1(t) - \lambda_2 \tilde{\psi}_2(t) \tag{3.48}$$

where the noise term is ignored for analytical simplicity, $\lambda_i$ and $\tilde{\psi}_i(t)$ are the wavelength and composite carrier phase of the ith frequency band signal, respectively. After some mathematical manipulations, Equation 3.48 becomes:

$$M_{1,2}(t) = \lambda_1 \beta_1(t) - \lambda_2 \beta_2(t) \tag{3.49}$$

The combination is a function of the wavelengths and interferometric phases of the dual frequency signals. As seen from Equation 3.44, the interferometric phase depends on the satellite elevation angle, surface reflectivity, and antenna gain pattern. Since the amplitude ratio $\alpha(t)$ is usually much

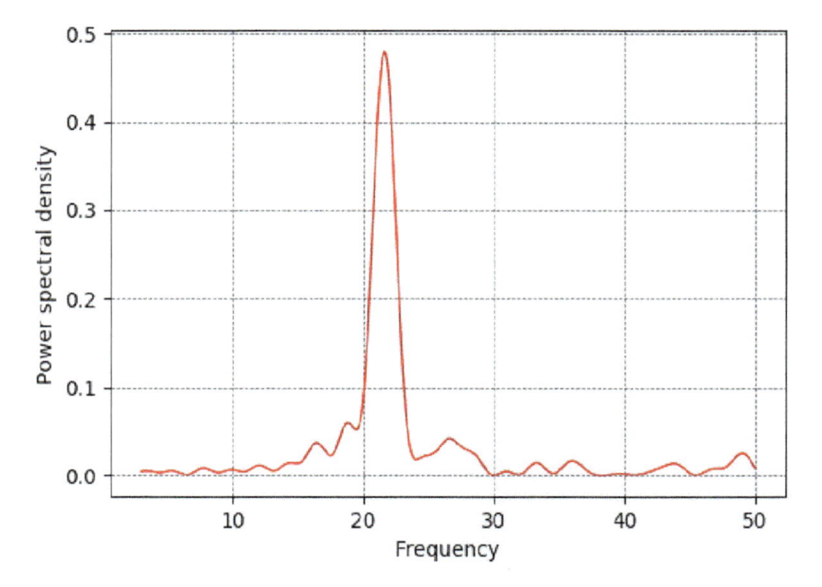


FIGURE 3.16 Example of power spectral density of detrended SNR time series with $\sin\theta(t)$ being the independent time variable.

smaller than one in the case of a face-up antenna due to reflection and that antenna gain pattern is designed to greatly suppress the reflected signal, the interferometric phase can be approximated as:

$$\begin{aligned}\beta(t) &\approx \tan^{-1}(\alpha(t)\sin\delta_{\phi}(t)) \\ &\approx \alpha(t)\sin\delta_{\phi}(t)\end{aligned} \tag{3.50}$$

where the second approximation makes use of $\tan^{-1}(x) \approx x$ if $x << 1$. Thus, the purified phase combination is approximated as the subtraction of two quasi-sinusoids. Two spectral peaks would occur in the power spectrum, since there are two different main frequencies. For such a reason and the approximations in Equation 3.50, the direct calculation of antenna height based on Equation 3.47 is not recommended; otherwise, a large error would be produced. Instead, based on Equation 3.49, a semi-empirical model can be developed to describe the relationship between antenna height and spectral peak frequency, and $\sin\theta(t)$ is still treated as the independent time variable. The amplitude ratio is a function of the surface reflectivity and the antenna gain pattern. The former may be treated as a constant for a specific reflection area, while the latter can be different for different type of antennas. Thus, a specific antenna gain pattern should be used in the model development.

It is observed from experimental data that strong noise exists in the purified phase combination. One possible reason would be that the residual ionospheric delay is still non-trivial. To deal with the issue, triple-frequency carrier phase combination can be exploited to measure snow depth (Yu et al. 2015) and the combination is given as:

$$\tilde{M}_{1,2,3}(t) = \kappa_1\hat{\psi}_1(t) + \kappa_2\hat{\psi}_2(t) + \kappa_3\hat{\psi}_3(t) \tag{3.51}$$

where

$$\kappa_1 = \lambda_1(\lambda_3^2 - \lambda_2^2);\ \kappa_2 = \lambda_2(\lambda_1^2 - \lambda_3^2);\ \kappa_3 = \lambda_3(\lambda_2^2 - \lambda_1^2) \tag{3.52}$$

Considering that the ionospheric delay is inversely proportional to the squared frequency and the constants can be dropped, it can be derived that:

$$\tilde{M}_{1,2,5}(t) = \kappa_1\beta_1(t) + \kappa_2\beta_2(t) + \kappa_5\beta_5(t) \tag{3.53}$$

As a result, the effect of geometry and ionospheric delays is removed. Figure 3.17 shows the power spectrum of the triple-frequency carrier phase combination calculated by Equation 3.53 for a given antenna height and antenna gain pattern. The spectral peak frequency of the power spectral density can be readily determined from the plot. A number of peak frequencies can be obtained for the same number of antenna heights of interest under the same antenna gain pattern. Then, a linear model can be generated through least-squares fitting, as shown in Figure 3.18. It is not a surprise that a linear model is generated, since the composite excess phase $\beta_i(t)$ can be approximated as a quasi-sinusoid and its frequency in terms of time variable is proportional to antenna height. The model is obtained for a given antenna gain pattern, so it is useful to study the effect of antenna gain pattern on the model parameters. The model would be treated as general, if the model parameters change little for a variety of antenna gain patterns.

In the absence of triple-frequency signals, dual-frequency signals can also be used to retrieve snow depth and the effect of geometry and ionospheric delay is removed. This is realized through the combination of dual-frequency carrier phases and single frequency pseudorange (Yu et al. 2019), and the combination is given by:

$$M_{1,2}(t) = \tilde{\rho}_1(t) + \kappa_1\lambda_1\tilde{\varphi}_1(t) + \kappa_2\lambda_2\tilde{\varphi}_2(t) \tag{3.54}$$

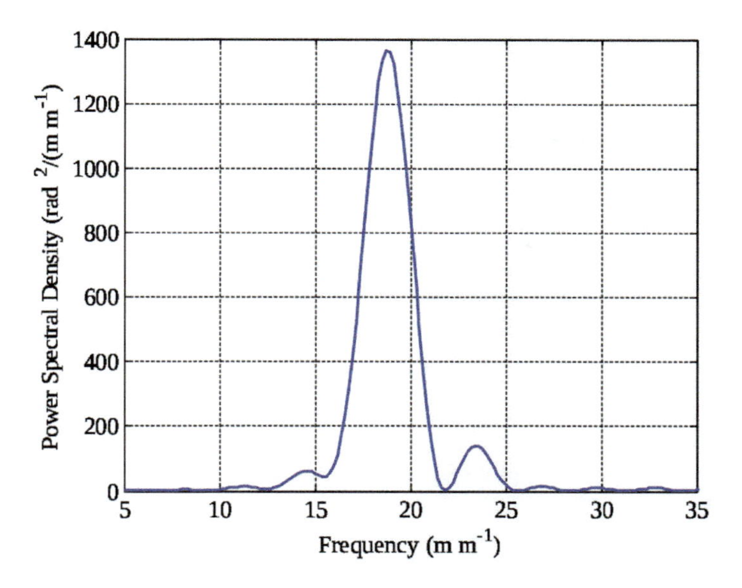


**FIGURE 3.17** Example of power spectral density of triple-frequency carrier phase combination with $\sin\theta(t)$ being the independent time variable.

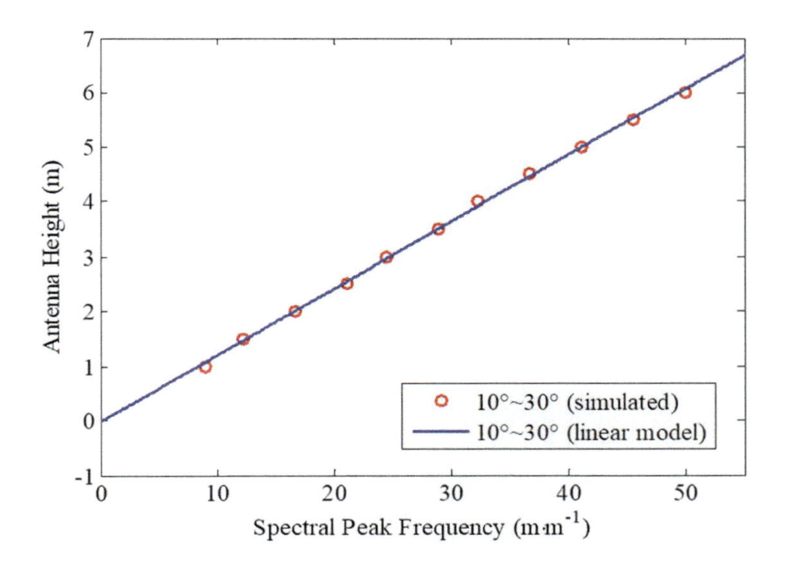


**FIGURE 3.18** Model constructed by least-squares fitting between the peak frequency and the antenna height.

where $\tilde{\rho}_1(t)$ is the pseudorange of the signal of first frequency band and

$$\kappa_1 = \frac{\lambda_1^2 + \lambda_2^2}{\lambda_1^2 - \lambda_2^2}, \qquad \kappa_2 = \frac{-2\lambda_1^2}{\lambda_1^2 - \lambda_2^2} \tag{3.55}$$

Equation 3.54 can be rewritten as:

$$M_{1,2}(t) = \ell_1(t) + \kappa_1\lambda_1\beta_1(t) + \kappa_2\lambda_2\beta_2(t) \tag{3.56}$$

where $\{\beta_i(t)\}$ are described in Equation 3.44 and $\{\ell_i(t)\}$ are the pseudorange multipath errors given by:

$$\ell(t) = \frac{2h\sin\theta(t)\cdot\alpha(t)\cdot\cos\delta_\phi(t)}{1+\alpha(t)\cos\delta_\phi(t)} \tag{3.57}$$

In the same way, a semi-empirical linear model can be developed to describe the relationship between antenna height and spectral peak frequency for a specific antenna gain pattern. Similar to the composite excess phase, the pseudorange multipath error is also approximately a quasi-sinusoid, and thus the model is linear.

## 3.5 CHALLENGING ISSUES AND FUTURE DIRECTIONS

GNSS is a global space-based positioning, navigation, and timing (PNT) capability. To come close to matching the extraordinary success of GNSS-PNT there are many technological challenges for GNSS remote sensing, especially in the case of the GNSS-R technique. Satellite-borne GNSS remote sensing must demonstrate its value as a reliable, high quality sensing technology. With the launch of multiple LEO satellites with GNSS-RO and GNSS-R capable instrumentation it seems we are at last close to a renaissance. But are more LEO missions and better algorithms for geophysical parameter extraction sufficient? In the case of GNSS-RO the answer is yes, because the meteorological/climate community now assimilates GNSS-RO products into their operational and research systems, and is eager for more. However, GNSS-R is still very much a novelty technology as far as the geosciences community is concerned.

The International Association of Geodesy (IAG) has been very successful in launching technique-specific services (see www.iag-aig.org). These services aid other geoscientists in their research, as well as supporting important applications in the wider community. Examples of the latter are the contribution of the International GNSS Service (IGS) to precise positioning (http://igs.org), and the International Earth Rotation and Reference Systems Service (IERS) to the International Terrestrial Reference Frame (ITRF) (http://itrf.ensg.ign.fr). The IGS does generate troposphere and ionosphere parameter products (www.igs.org/components/prods.html), however they are based on observations made from the global ground-based GNSS tracking network (Figure 3.19), not from satellite platforms. There are no geodetic services producing GNSS remote sensing products on a continuous, synoptic basis. From the IAG's perspective the challenging issues are related to "operationalising" GNSS remote sensing, in all its forms, so that the remotely sensed geophysical parameters (atmospheric, oceanic, wind, soil moisture, biomass, etc) are included in the suite of geodetic outputs of the Global Geodetic Observing System (GGOS) (GGOS 2023).

The nontechnical challenges therefore include the evolution of current GNSS remote sensing science missions to operational services by: *increasing* the number of LEO satellites equipped with GNSS receivers to satisfy coverage requirements in space and time; *standardizing* data formats, instrumentation, and calibration, preprocessing and analysis procedures; and establishing a *coordinating* agency or authority so as to ensure continuous, high quality product generation and dissemination. The IAG has considerable experience with geometry technique-based services (such as the IGS, ILRS, IVS, and IDS) (www.iers.org/nn_10880/IERS/EN/Organization/TechniqueCentres/TC.html?__nnn=true), however it has not yet established any "geodetic imaging" services based on technologies such as synthetic aperture radar (SAR), satellite radar altimetry, LiDAR, or GNSS remote sensing (GNSS-RO, GNSS-R). One of the future challenges is to address this shortcoming, so that GNSS remote sensing can be recognized as a geodetic technique that is making critical contributions to science and society.

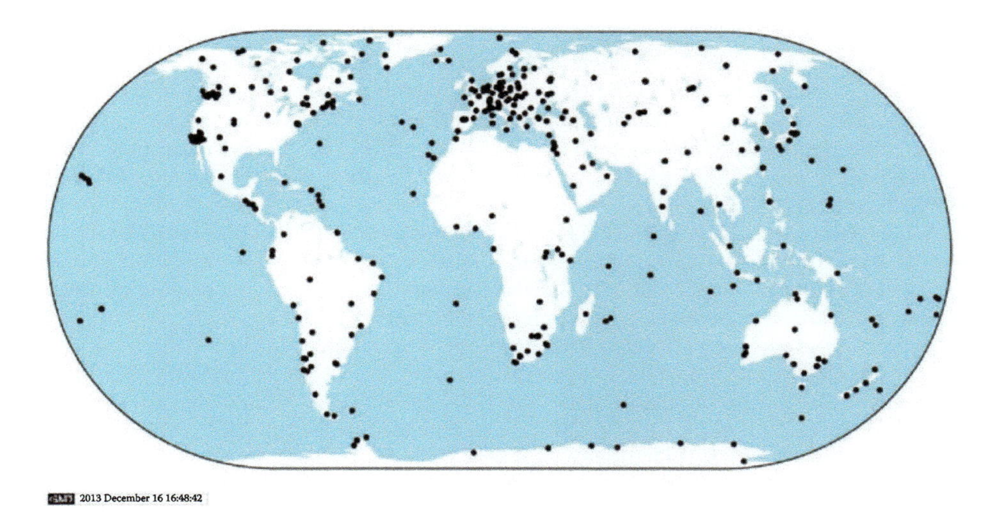


**FIGURE 3.19** One of the reasons for the success of IAG services are the permanent geodetic observing networks such as the IGS's global GNSS tracking network, providing accurate, reliable data products for science and society.

## 3.6 CONCLUSIONS

This chapter studied GNSS-based remote sensing with a focus on GNSS-R for both land and ocean applications. In particular, details about airborne and spaceborne GNSS-R based sea surface wind speed estimation and airborne sea surface altimetry were presented. The airborne wind speed estimation accuracy can be around 1 m/s, but spaceborne wind speed estimation accuracy is significantly lower; and the mean sea surface height estimation can be decimeter. Also, ground-based snow depth measurement and soil moisture measurement have been studied. In addition to the advantage of low cost, the GNSS-R technique can measure the geophysical parameters over a wide area. The major drawback of the technique may be that the accuracy can be affected considerably by modeling errors. As mentioned in the chapter, a number of dedicated satellite missions associated with GNSS remote sensing have been accomplished, several missions are scheduled, and more missions are expected to be funded. It is envisaged that as a cost-effective technique, GNSS-based remote sensing will play a significant role in a wide range of remote sensing applications.

## REFERENCES


Anthes, R.A. 2011. Exploring Earth's atmosphere with radio occultation: Contributions to weather, climate and space weather. *Atmospheric Measurement Techniques* 4:1077–1103.

Bu, J., K. Yu, Y. Zhu, N. Qian, J. Chang. 2020. Developing and testing models for sea surface wind speed estimation with GNSS-R delay Doppler maps and delay waveforms. *Remote Sensing* 12(22):1–24, 3760.

Chang, X., T. Jin, K. Yu, Y. Li, J. Li. 2019. Soil moisture estimation by GNSS multipath signal. *Remote Sensing* 11(21):1–16, 2559.

Clarizia, M.P., C.S. Ruf. 2016. Wind speed retrieval algorithm for the cyclone global navigation satellite system (CYGNSS) mission. *IEEE Transactions on Geoscience and Remote Sensing* 54(8):4419–4432.

Elfouhaily, T., B. Chapron, K. Katsaros, D. Vandemark. 1997. A unified directional spectrum for long and short wind-driven waves. *Journal of Geophysical Research* 102(C7):15781–15796.

Ferrazzoli, P., L. Guerriero, N. Pierdicca, R. Rahmoune. 2011. Forest biomass monitoring with GNSS-R: Theoretical simulations. *Advances in Space Research* 47:1823–1832.

Font, J., A. Camps, A. Borges, M. Martin-Neira, J. Boutin, N. Reul, Y.H. Kerr, A. Hahne, S. Mecklenburg. 2010. SMOS: The challenging sea surface salinity measurement from space. *Proceedings of the IEEE* 98(5):649–655.

Garrison, J.L., A. Komjathy, V.U. Zavorotny, S.J. Katzberg. 2002. Wind speed measurements using forward scattered GPS signals. *IEEE Transactions on Geoscience and Remote Sensing* 40(1):50–65.

GEROS—GNSS REflectometry, Radio Occultation and Scatterometry onboard International Space Station (GEROS-ISS), www.ice.csic.es/en/view_project.php?PID=155, viewed 9 Dec 2013.

Gleason, S.T. 2006. Land and ice sensing from low Earth orbit using GNSS bistatic radar. *Proceedings of the International Technical Meeting of the Satellite Division of the Institute of Navigation (ION GNSS)*, Fort Worth, TX, pp. 2523–2530, Sept.

Gleason, S.T., D. Gebre-Egziabher. 2009. *GNSS Applications and Methods*, Boston, Artech House.

Gleason, S.T., S. Hodgart, Y. Sun, C. Gommenginger, S. Mackin, M. Adjrad, M. Unwin. 2005. Detection and processing of bistatically reflected GPS signals from low Earth orbit for the purpose of ocean remote sensing. *IEEE Transactions on Geoscience and Remote Sensing* 43(6):1229–1241.

Global Geodetic Observing System (GGOS), www.ggos.org, viewed 17 Oct 2023.

Heise, S., J Wickert, C. Arras, G. Beyerle, A. Faber, G. Michalak, T. Schmidt, F. Zus. 2014. Reprocessing and application of GPS radio occultation data from CHAMP and GRACE, in Observation of the System Earth from Space—CHAMP, GRACE, GOCE and Future Missions, ed. By F. Flechtner, N. Sneeuw, W.-D. Schuh (GEOTECHNOLOGIEN Science Report No. 20, Springer, 2014).

Jilin-01B, www.tse.edu/tse/online/sat_jilin_1_kuanfu_1b.html, viewed 17 Oct 2023.

Jin, S., G.P. Feng, S. Gleason. 2011. Remote sensing using GNSS signals: Current status and future directions. *Advances in Space Research* 47(10):1645–1653.

Jing, C., X. Niu, C. Duan, F. Lu, G. Di, X. Yang. 2019. Sea surface wind speed retrieval from the first Chinese GNSS-R mission: Technique and preliminary results. *Remote Sensing* 11(24):1–13.

Larson, K.M., E.D. Gutmann, V.U. Zavorotny, J.J. Braun, F.G. Nievinski. 2009. Can we measure snow depth with gps receivers? *Geophysical Research Letters* 36(17):L17502.

Martín-Neira, M. 1993. A passive reflectometry and interferometry system (PARIS): Application to ocean altimetry. *ESA Journal* 17(14):331–355.

Martin-Neira, M., S. D'Addio, C. Buck, N. Floury, R. Prieto-Cerdeira. 2009. The PARIS in-orbit demonstrator. *Proceedings of IEEE IGARSS* 2:II-322–II-325.

Martin-Neira, M., S. D'Addio, C. Buck, N. Floury, R. Prieto-Cerdeira. 2011. The PARIS ocean altimeter in-orbit demonstrator. *IEEE Transactions on Geoscience and Remote Sensing* 49(6):2209–2237.

Masters, D., P. Axelrad, S.J. Katzberg. 2004. Initial results of land-reflected GPS bistatic radar measurements in SMEX02. *Remote Sensing of Environment* 92:507–520.

Ozeki, M., K. Heki. 2012. GPS snow depth meter with geometry-free linear combinations of carrier phases. *Journal of Geodesy* 86(3):209–219.

Pierson, W.J., L. Moscowitz. 1964. A proposed spectral form for fully developed wind seas based on the similarity theory of S. A. Kitaigorodskii. *Journal of Geophysical Research* 69(24):5181–5190.

Rius, A., E. Cardellach, M. Martın-Neira. 2010. Altimetric analysis of the sea-surface GPS-reflected signals. *IEEE Transactions on Geoscience and Remote Sensing* 48(4):2119–2127.

Rodriguez-Alvarez, N., X. Bosch-Lluis, A. Camps, M. Vall-Llossera, E. Valencia, J.F. Marchan-Hernandez, I. Ramos-Perez. 2009. Soil moisture retrieval using GNSS-R techniques: Experimental results over a bare soil field. *IEEE Transactions on Geoscience and Remote Sensing* 47(11):3616–3625.

Ruf, C., M. Unwin, J. Dickinson, R. Rose, M. Vincent, A. Lyons. 2013. CYGNSS: Enabling the future of hurricane prediction. *IEEE Geoscience and Remote Sensing Magazine* 1(2):52–67.

Spire Global GNSS-R satellites, https://spacenews.com/spire-gnss-reflectometry/, viewed 13 Oct 2023.

Tye, J., P. Jales, M.J. Unwin, C. Underwood. 2016. The first application of stare processing to retrieve mean square slope using the SGR-ReSI GNSS-R experiment on TDS-1. *IEEE Journal of Selected Topics in Applied Earth Observations and Remote Sensing* 9(10):4669–4677.

Unwin, M.J., P. Jales, J. Tye, C. Gommenginger, G. Foti, J. Rosello. 2016. Spaceborne GNSS-Reflectometry on TechDemoSat-1: Early mission operations and exploitation. *IEEE Journal of Selected Topics in Applied Earth Observations and Remote Sensing* 9(10):4525–4539.

Unwin, M.J., S. Gleason, M. Brennan. 2003. The space GPS reflectometry experiment on the UK disaster monitoring constellation satellite. *Proceedings of ION-GPS/GNSS*, Portland, OR, Sept.

Valencia, E., V.U. Zavorotny, D.M. Akos, A. Camps. 2014. Using DDM asymmetry metrics for wind direction retrieval from GPS ocean-scattered signals in airborne experiments. *IEEE Transactions on Geoscience and Remote Sensing* 52(7):3924–3936.

Wang, C., K. Yu, F. Qu, J. Bu, S. Han, K. Zhang. 2022. Spaceborne GNSS-R wind speed retrieval using machine learning methods. *Remote Sensing* 14(14):1–21, 3507.

Wang, J.R., T.J. Schmugge. 1980. An empirical model for the complex dielectric permittivity of soils as a function of water content. *IEEE Transactions on Geoscience and Remote Sensing* 18(4): 288–295.

Weathernews WNISAT-1R, https://global.weathernews.com/news/10564/, viewed 17 Oct 2023.

Yu, K. 2021. *Theory and Practice of GNSS Reflectometry*, Singapore, Springer.

Yu, K., C. Rizos, A.G. Dempster. 2012a. Error analysis of sea surface wind speed estimation based on GNSS airborne experiment. *Proceedings of ION GNSS*, Nashville, TE, USA, Sept.

Yu, K., C. Rizos, A.G. Dempster. 2012b Performance of GNSS-based altimetry using airborne experimental data. *Proceedings of Workshop on Reflectometry using GNSS and Other Signals of Opportunity*, West Lafayette, IN, USA, Oct.

Yu, K., C. Rizos, A.G. Dempster. 2012c. Sea surface wind speed estimation based on GNSS signal measurements. *Proceedings of International Geoscience and Remote Sensing Symposium (IGARSS)*, Munich, Germany, pp. 2587–2590, July.

Yu, K., C. Rizos, A.G. Dempster. 2013. Forest change detection using GNSS signal strength measurements. *Proceedings of International Geoscience and Remote Sensing Symposium (IGARSS)*, Melbourne, Australia, pp. 1003–1006, July.

Yu, K., C. Rizos, A.G. Dempster. 2014. GNSS-based model-free sea surface height estimation in unknown sea state scenarios. *IEEE Journal of Selected Topics in Applied Earth Observations and Remote Sensing* 7(5):1424–1435.

Yu, K., W. Ban, X. Zhang, X. Yu. 2015. Snow depth estimation based on multipath phase combination of GPS triple-frequency signals. *IEEE Transactions on Geoscience and Remote Sensing* 53(9):5100–5109.

Yu, K., Y. Li, X. Chang. 2019. Snow depth estimation based on combination of pseudorange and carrier phase of GNSS dual-frequency signals. *IEEE Transactions on Geoscience and Remote Sensing* 57(3):1817–1828.

Zavorotny, V.U., A.G. Voronovich. 2000. Scattering of GPS signals from the ocean with wind remote sensing application. *IEEE Transactions on Geoscience and Remote Sensing* 38(2):951–964.

Zavorotny, V.U., K.M. Larson, J.J. Braun, E.E. Small, E.D. Gutmann, A.L. Bilich. 2010. A physical model for GPS multipath caused by land reflection: Toward bare soil moisture retrievals. *IEEE Journal of Selected Topics in Applied Earth Observations and Remote Sensing* 3(1):100–110.

Zavorotny, V.U., S. Gleason, E. Cardellach, A. Camps. 2014. Tutorial on remote sensing using GNSS bistatic radar of opportunity. *IEEE Geoscience and Remote Sensing Magazine* 2(4):8–45.

Zhang, P., X. Hu, Q. Lu, A. Zhu, M. Lin, L. Sun, L. Chen, N. Xu. 2021. FY-3E: The first operational meteorological satellite mission in an early morning orbit. *Advances in Atmospheric Sciences* 39:1–8.

# 4 Global Navigation Satellite Systems (GNSS) for a Wide Array of Terrestrial Applications

*D. Myszor, O. Antemijczuk, M. Grygierek, M. Wierzchanowski, and K.A. Cyran*

## 4.1 INTRODUCTION

Global Navigation Satellite Systems (GNSS) provide positioning services across the globe (Kumar et al. 2023, Bhardwaj 2020, Hofmann-Wellenhof et al. 2008). Utilization of these systems gained huge popularity because of many advantages they provide, such as constant availability (although rare cases of downtimes are known), free of charge service, and good positioning accuracy (Bhardwaj 2020, Jeffrey 2010, Gleason an Gebre-Egziabher 2010, Gleason and Gebre-Egziabher 2009). In general, GNSS might be internally splat into three segments: space, control, and users (Prasad and Ruggieri 2005, Groves 2008). Space segment is comprised of satellites that are constantly orbiting around the globe. Orbits of individual satellites, within particular systems (constellation), are arranged in a way which allows for observation of at least four satellites, from almost any point of Earth surface, at the same time (Lannucci and Humphreys 2022, Petrovski and Tsujii 2012). Satellites send ceaselessly information with time stamp, precise orbit coordinates (ephemeris), description of satellites constellation (almanac), error corrections as well as health state indicating their operational state (Lannucci and Humphreys, 2022, Ward et al. 2005a,b). GNSS receiver obtains navigation signals and utilizes its internal clock in order to determine delay between time of navigation message transmission and time of message acquisition by the receiver antenna. As a result, pseudorange between transmitting satellite and receiver might be determined (calculated value is called pseudorange because various errors can influence obtained results) (Kumar et al. 2023, Grewal et al. 2007, Xu 2007, Kaplan et al. 2005). If signals from four satellites are available then based on obtained pseudoranges, and ephemeris of satellites, receiver is able to calculate coordinates in three dimensional space (Gleason et al. 2009). It is worth to mention that contemporary receivers are able to utilize many visible navigation satellites at the same time, and combined data obtained from various GNSS in order to increase the precision of obtained coordinates. System requires high time precision and synchronization of time between satellites, even small errors in time synchronization lead to significant errors at the level of the Earth's coordinates calculations. In order to achieve thiese tasks, satellites are equipped with several clocks (i.e., battery of caesium and rubidium atomic clocks) which are regularly synchronised by control segment.

Control segment is composed of installations located at the surface of the Earth. These installations form network of stations scattered around the globe. Following facilities are usually possessed by particular Global Navigation Satellite Systems: master control station as well as backup control station, set of command sites with control antennas and battery of monitoring stations. Master control station receives data from monitoring stations thus is able to control proper functioning of navigation system. In general control segment is responsible for constant monitoring of the

 DOI: 10.1201/9781003541141-6

navigation satellites' states, supervision of orbital plane drift, sustaining of constellation geometry, taking satellites out of service for maintenance tasks such as orbit modification or software update, resolution of various anomalies and malfunctions in satellite functionality. It also regularly communicates with satellites in order to synchronize theirs atomic on board clocks and adjust ephemeris of satellite orbital model. Every satellite, belonging to GNSS constellation, has a designated lifetime, therefore control segment is responsible for disposal of retired satellites, launching of new ones on early orbits as well as introduction of these satellites into navigation constellation.

User segment consists of individual, institutional (government, commercial as well as scientific) and military users which are in the possession of navigation signal receivers. Receivers might be characterized by various levels of accuracy, qualities of internal clocks, multichannel abilities (calculation of coordinates based on many satellites), and possibility of utilization of different Global Navigation Satellite Systems.

### 4.1.1 Contemporary Global Navigation Satellite Systems

Global Positioning System–Navigation Satellite Time and Ranging (GPS–NAVSTAR) (Haibo et al. 2022, El-Rabbany 2002, Letham and Letham 2008). The system is operated and controlled by the U.S. Department of Defense. Originally it was created solely for military purposes; however, later it was opened for civilian applications. With time, improvements were introduced that allowed for achievement of positioning accuracy close to that obtained with military signals (Hein 2020, Harte and Levitan 2007). Currently open signals are transmitted on two frequencies, therefore there is a possibility of autonomous mitigation of ionosphere errors when proper receiver is utilized (Xu 2007). Baseline constellation is composed of 24 slots (Figure 4.1). There are six equally spaced orbital planes, each orbit possesses four slots. Initially each slot can possess one satellite however in order to improve signal availability, especially on areas that are characterized by obstruction reach environment, three slots were expanded and can host two satellites each (fore and aft location within expanded slot are defined). Therefore 27 navigation satellites could be accommodated in baseline constellation. Constellation can possess additional spare satellites, however they do not occupy predefined slots (see Table 4.1). Devices are located in the Middle Earth Orbit at altitude of 20,200

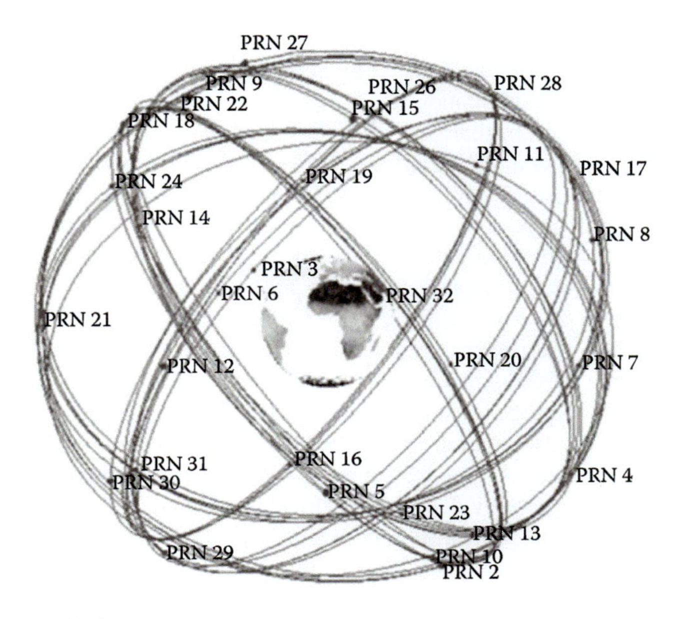


FIGURE 4.1 GPS constellation.

**TABLE 4.1**
**GPS Constellation**

| Total Number of Satellites in GPS Constellation (June 3, 2014) | 32 |
|---|---|
| Operational | 30 |
| In commissioning phase | 1 |
| In maintenance | 1 |

km and orbit the Earth approximately every 12 hours. Arrangement of satellites ensures that from almost every point of Earth at least four satellites are visible at the same time and theirs elevation angles are greater than 15° relative to the ground (Hofmann-Wellenhof et al. 2001). All satellites utilizes the same frequencies and code division multiple access (CDMA) technique. Importantly, equipment and constellation configuration are improved all the time in order to increase accuracy (see Table 4.4).

Global Navigation Satellite System (GLONASS): the system is operated by the Coordination Scientific Information Centre of the Ministry of Defence of the Russian Federation. Similarly to GPS, GLONASS was originally created for military purposes and later it was opened for civilian application. Currently, satellites transmit two types of signals open one (for civilian purposes) and the encrypted one (for military purposes). Importantly, signals for civilian purposes are broadcasted in two bands thus ionosphere effects can be mitigated if two band receiver is applied. Baseline constellation is composed of 24 satellites organized in three orbital planes, each orbit possess eight satellites (Figure 4.2). Fully operational state requires 21 satellites, additional three satellites (called active spares) are maintained for the purpose of replacement of malfunctioning devices (see Table 4.2). Orbital planes are designed in order to obtain higher accuracy than GPS in high altitudes. Devices are located in the Middle Earth Orbit at altitude of 19,100 km and orbit the Earth approximately every 11 hours. Satellites utilize 15 channel frequency division multiple access technique (Harper 2009).

GALILEO is European Union's response to military controlled GNSS systems for which civil accuracy can be degenerated or even disabled in case of conflicts. Such an action could introduce huge disturbances in many fields of industry and human lives, therefore the decision was made to create a civil controlled system that will allow for obtainment of precise positioning across the globe. Planned baseline constellation is composed of 30 satellites organized in three orbital planes, each orbit possess ten satellites (Figure 4.3). The system, in order to be fully operational, requires 27 satellites; an additional three satellites (called active spare) will be maintained (one on each orbit) for the purpose of malfunctioning devices replacement. Such numbers of satellites will allow for increase of the accuracy level in city canyons, in addition orbital planes in the constellation are designated to obtain better accuracy, at high latitudes, than GPS-based solution, thus accuracy of Northern Europe coverage will improve. Satellites are located in the Middle Earth Orbit at altitude of 23,222 km and orbit the Earth approximately every 15 hours. The system is operated by the civilian organization European Space Agency (ESA). GALILEO will provide open signal for all users (as in GPS, two frequencies are available), however there will be a possibility to purchase access to an encrypted signal, which will increase resistance to spoofing and jamming. Interestingly, an additional encrypted service will be provided for government agencies. There will also be a possibility of picking up emergency signals transmitted by beacons installed aboard ships, aircrafts, or carried by individuals. Furthermore, the system will be able to inform users about signal degradation; such ability is especially important in applications where guaranteed precision is important, for example, plane navigation (Tables 4.3 and 4.4).

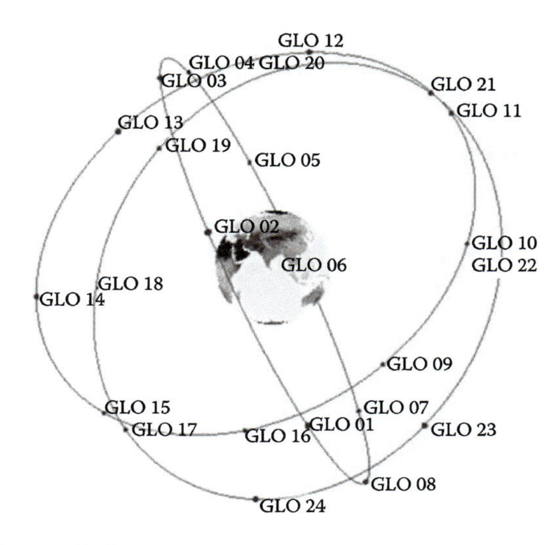


FIGURE 4.2 GLONASS constellation.

**TABLE 4.2**
**GLONASS Constellation**

| Total Number of Satellites in GLONASS Constellation (June 3, 2014) | 29 |
|---|---|
| Operational | 24 |
| In commissioning phase | 0 |
| In maintenance | 0 |
| Spares | 4 |
| In flight tests phase | 1 |

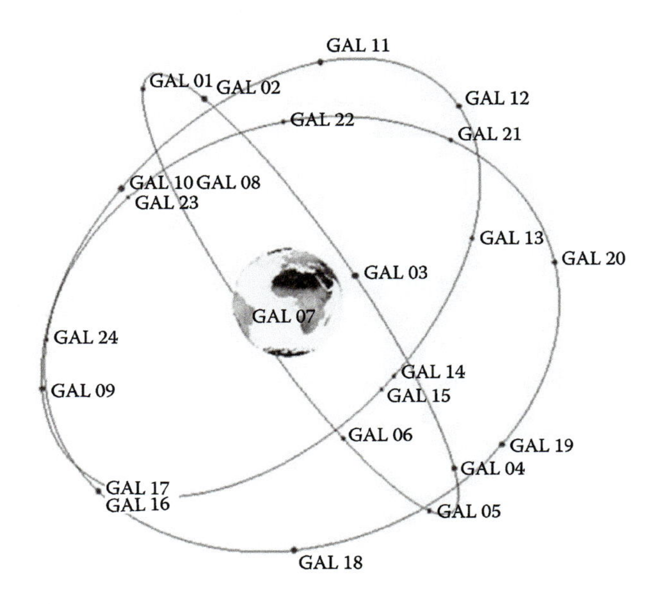


FIGURE 4.3 GALILEO constellation.

**TABLE 4.3**
**GNSS Summary**

| GNSS | No. Satellites (Base Constellation, without Spares) | Operating Country | Civilian Applications | Military Applications | Accuracy (Perfect Conditions, without Supporting Systems) |
|---|---|---|---|---|---|
| GPS | 27 | USA | True | True | Up to 7 m (Allan 1997) |
| GLONASS | 21 | Russian Federation | True | True | Up to 15 m (Miller 2000) |
| GALILEO | 27 | European Union | True | True | Constellation is not fully operational |

**TABLE 4.4**
**Historical Accuracy of GPS System**

| Year | 1993 | 1995 | 2000 | 2014 |
|---|---|---|---|---|
| Horizontal accuracy (m) | ~100 to ~300 | ~100 | ~20 m | ~7.8 m |

## 4.2 SBAS—SATELLITE-BASED AUGMENTATION SYSTEM

Precision of calculated coordinates based solely on GNSS is insufficient in many applications therefore additional Satellite Based Augmentation Systems were created that can improve the accuracy of obtained results. Among them the most popular are Wide Area Augmentation System (WAAS), European Geostationary Navigation Overlay Service (EGNOS), and Multifunctional Satellite Augmentation System (MSAS). WAAS was created by the United States; it is designated for GPS only and it is available on the area of North America. EGNOS on the other hand, was the first complex project in the field of satellite navigation systems created in cooperation between EU members; it was established in 2005 and fully operational state was achieved in 2009. The purpose of the system was to create augmentation for GPS and GLONASS over the area of Europe (ESA SP-1303 2006). Finally MSAS, was created by Japanese in order to deliver corrections to GPS signals over its area.

In order to ensure the highest positioning accuracy, these systems constantly monitor satellites constellations with net of ground monitoring stations, calculate errors, and provide correction data to the end users in real time. As a result, significant reduction of positioning errors is obtained, for example, in case of GPS, positioning accuracy can be increased from approximately 10 m when it is calculated without correction services (one frequency receiver) (DoD 2008) to approximately 2 m when EGNOS service is utilized. Interestingly, aforementioned augmentation systems are compatible; they were created in order to increase positioning precision mainly in aviation applications thus correction signals are broadcasted through satellites (Figure 4.4). It implies that corrections might be hardly obtainable in areas surrounded by high obstacles. In such a case, the internet might be utilized for correction delivery (SISNeT technology in EGNOS (Zinkiewicz et al. 2010) and ongoing work over EGNOS Data Access System [EDAS]) (Table 4.5).

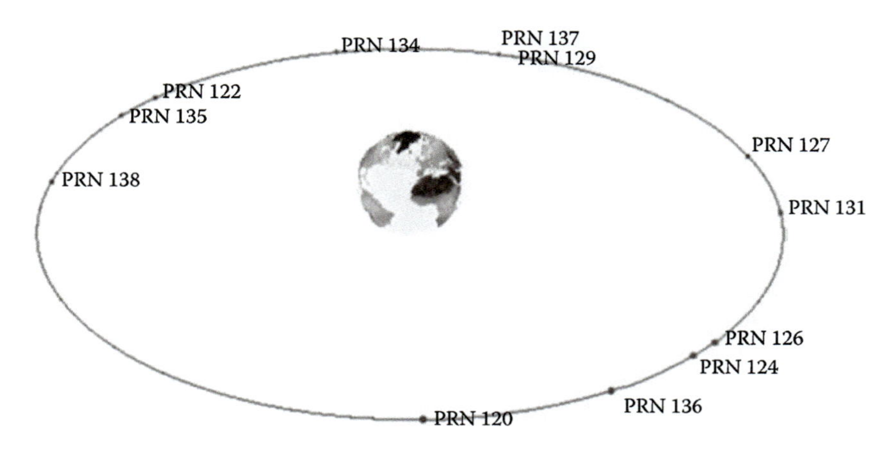


FIGURE 4.4 EGNOSS constellation.

TABLE 4.5
**GPS Accuracy When Various Augmentation Systems Are Utilized**

| GPS | WASS | NDGPS | GDGPS | AGPS | EGNOS | MSAS |
|---|---|---|---|---|---|---|
| L1 = ~7.8 m | ~3 m | ~10–15 cm | ~10 cm | ~5 m | ~3 m | ~0.8–5.3 m |

## 4.3 GNSS ERRORS

Utilization of satellite navigation systems for measuring of the object position is exposed to variety of errors that affect accuracy of designated coordinates (Lau et al. 2021, Grewal et al. 2001). Among factors that can lead to incorrect results are satellite position errors. They could be caused by various factors such as solar radiation pressure and solar flares, electromagnetic forces (Silva and Olsen 2002a, 2002b), and relativistic effects. In addition, gravity force of Earth, Sun, Moon, and other celestial bodies as well as oceanic and terrestrial tidal waves influence satellites positions thus degrade obtained coordinates. Noteworthy geometry of observed satellites in relation to the receiver antenna, also has impact on accuracy of obtained results, dilution of precision (DOP) is a geometric quantity which describes this relation (Hewlett Packard 1996). Another issue appears when change in satellites configuration occurs; jump in assessed location can be obtained and accuracy might be depleted (Lau et al. 2021, Sukkarieh et al. 1999).

Earth tides, which are small cyclical ground movements caused by gravity forces of other celestial bodies (mainly Sun and Moon) introduce issues when very precise conversion of coordinates to the location on the map is required. Similar issues are introduced by tectonic movements which cause constant displacement of reference sites on Earth's surface in relation to reference points set at the level of reference frame, for example, World Geodetic System 1984 (WGS84) (Harper 2009) or its high accuracy version International Terrestrial Reference System (ITRS). When precise measurements are required magnitude of this error cannot be underestimated. On some areas this phenomenon can introduce displacement at rate up to 120 mm per year, for example Great Britain is in constant move in reference to WGS84 and displacement is at the level of 25 mm per year. In order to mitigate this effect frames of reference which are tied to the ground and fixed in time are introduced. In Europe the most popular one is European Terrestrial Reference System 1989 (ETRS89), which can be easily converted to ITRS by simple transformation of coordinates that are published by International Earth Rotation Service (IERS). Also Earth rotator parameters are not constant, therefore an additional source of errors is introduced. As in the previous case, influence of this type of errors can be mitigated by application of corrections which could be based on data obtained from IERS. It is worthy to mention that all types of conducted calculations, for example, conversions between different reference systems are also prone to precision errors. It happens because of floating point number representation in devices' memory, which usually results in necessity of numerical rounding when extensive calculations are done.

Distortions of propagation of signals transmitted by satellites in Earth's atmosphere, which are mainly caused by ionosphere and troposphere refraction, are another cause of errors (MacGougan 2006). Signal can also be disturbed by high voltage lines and other transmitters (e.g., various GNSS signal jammers). In general, all objects that are blocking direct visual contact between receiver and transmitter (such as trees' leaves, buildings, cliff faces) are also potential source of signal degradation or even its unavailability. When signal is weakened special software algorithms might be applied, which could improve ability of position obtainment (Ziedan 2006). In addition receiver's surrounding environment might introduce phenomenon of multipath signal propagation (Lau et al. 2021, Ward et al. 2005a, 2005b); it occurs when the antenna receives the same signal reflected from various obstacles (i.e., building walls). In such case, multiple copies of the same signal reach receiver antenna in various times, which can cause the GNSS receiver to produce jumping coordinates. Although software incorporated in receivers often tries to detect such a situation and to limit influence of this phenomenon, however it still might have an effect on accuracy of obtained results. What is more requirement of visual contact between receiver and transmitter antennas rule out possibility of GNSS employment for navigation within buildings.

In another area, errors of transmitting and receiving devices are located. They could be caused by various factors such as instability of internal clocks frequency (most GNSS receivers utilize quartz-based generators, which are easily influenced by temperature fluctuations), internal noises within transmitting and receiving devices, variability of antennas phase centers. Likewise all electronic devices, navigation satellites are prone to hardware as well as software errors and malfunctions that could influence acquired positioning level accuracy or even completely disable services in given areas. In some rare cases, particular GNSS might broadcast wrong information (e.g., 11 hours of GLONASS failure in April of 2014), thus render the provided services unusable.

For a military controlled system such as GPS and GLONASS there is a possibility of temporal degradation of provided service in the selected areas. Also malicious third parties might jam the navigation signal on selected areas, although such a behavior is restricted to an extent in some countries, for example, in United States. In addition, there is a possibility to spoofing of navigation signal. The purpose of this action is to mislead receiving device so it determines its coordinates based on fake signal. Techniques applied in order to fulfil this task might include utilization of simulator of particular Global Navigation Satellite System that is connected to the physical antenna, or controlled retransmission of signals transmitted by original navigation satellites. It is worthy to mention that especially civil receivers are liable to this issue. For selected applications (military based) GPS and GLONASS provide additional encrypted services, which reduce vulnerability for spoofing and jamming. GALILEO also would provide encrypted navigation services for selected companies and government agencies such as police, search and rescue operators, humanitarian aid, fire brigades, health services, defence, coastguard, border controls, customs and civil protection units as well as critical infrastructure and networks.

In order to mitigate atmospheric errors, corrections provided by SBAS might be utilized (approximate accuracy 2 m); moreover, there is a possibility of utilization of professional two frequency (in case of GPS or GALILEO) or two band (for GLONASS) receivers, which can autonomously rule out this kind of errors (approximate accuracy 1 m) (Table 4.6).

**TABLE 4.6**
**Typical Errors, Their Magnitudes and Methods of Mitigation**

| Type of Error | Typical Magnitude of an Error (m) | Methods of Error Mitigation |
|---|---|---|
| Time base error | 3 | Utilization of signals from many satellites |
| Inaccuracy in positioning of satellite on the orbit | 2 | Utilization of correction services providing exact location of satellites on the orbit |
| Satellite orbit geometry errors | 2.5 | Utilization of correction services which provides precise orbit information |
| Ionosphere influence | 5 | Utilization of differential corrections/dual frequency receivers/models of influence of ionosphere |
| Troposphere influence | 0.5 | Utilization of differential corrections/dual frequency receivers/models of influence of troposphere |
| Propagation of electromagnetic waves | 1 | Utilization of proper algorithms |

### 4.3.1 Troposphere and Ionosphere Errors Mitigation Techniques

During research work over Research on EGNOS/Galileo in Aviation and Terrestrial Multi-sensors Mobility Applications for Emergency Prevention and Handling (EGALITE) project our team created application GPS 3D Viewer, which is designating coordinates bases on RAW GPS data. Various algorithms calculating influence of troposphere and ionosphere on acquired positions were included. Implemented model of troposphere signal propagation delay is described by the following equation:

$$TC_i = -\left(d_{hyd} + d_{wet}\right)*m\left(El_i\right) \tag{4.1}$$

where values of $[d_{hyd}, d_{wet}]$ are calculated based on the receiver antenna height above sea level and are estimated by five meteorological parameters: pressure [$P$ (mbar)], temperature [$T$ (K)], the vapor pressure of water [$e$ (mbar)], change of temperature [$\beta$ (K/m)], and change in the rate of evaporation [$\lambda$ (dimensionless)]. Value of every meteorological parameter is calculated for receiver latitude [$\varphi$] and day in astronomical year [D] (which starts on January 1); in addition average seasonal changes (presented in Table 4.8) are taken into account. Values of particular meteorological parameters, denoted for simplification as [$\xi$], are calculated with the following formula:

$$\xi\,(\varphi, D) = \xi_0\left(\varphi\right) - \Delta\xi\left(\varphi\right)*\cos\left(\frac{2\pi\left(D - D_{min}\right)}{365,25}\right) \tag{4.2}$$

where $D_{min}$ = 28 for northern latitudes, $D_{min}$ = 211 for southern latitudes, and $[\xi_0, \Delta\xi]$ represent mean seasonal fluctuation of values of parameters for latitude of receiver antenna. When latitude is lower than $|\varphi| \leq 15°$ or higher than $|\varphi| \geq 75°$, values of $[\xi_0]$ as well as $[\Delta\xi]$ are taken directly from Table 4.7. For latitudes in the range $15° < |\varphi| < 75°$ values of $[\xi_0, \Delta\xi]$ are calculated with following equations:

$$\xi_0\,(\varphi) = \xi_0(\varphi_i) + \left[\xi_0\left(\varphi_{i+1}\right) - \xi_0\left(\varphi_i\right)\right] * \frac{\varphi - \varphi_i}{\varphi_{i+1} - \varphi_i} \tag{4.3}$$

$$\Delta\xi\,(\varphi) = \Delta\xi(\varphi_i) + \left[\Delta\xi\left(\varphi_{i+1}\right) - \Delta\xi\left(\varphi_i\right)\right] * \frac{\varphi - \varphi_i}{\varphi_{i+1} - \varphi_i} \tag{4.4}$$

Mean values:

**TABLE 4.7**
**Mean Values for GPS Receiver's Troposphere Corrections**

| Latitude (°) | $P_0$ (mbar) | $T_0$ (K) | $e_0$ (mbar) | $\beta_0$ (K/m) | $\lambda_0$ |
|---|---|---|---|---|---|
| 15° or less | 1013.25 | 299.65 | 26.31 | 6.30e−3 | 2.77 |
| 30 | 1017.25 | 294.15 | 21.79 | 6.05e−3 | 3.15 |
| 45 | 1015.75 | 283.15 | 11.66 | 5.58e−3 | 2.57 |
| 60 | 1011.75 | 272.15 | 6.78 | 5.39e−3 | 1.81 |
| 75° or more | 1013.00 | 263.65 | 4.11 | 4.53e−3 | 1.55 |

Seasonal values:

**TABLE 4.8**
**Seasonal Values for GPS Receiver's Troposphere Corrections**

| Latitude (°) | $\Delta P$ (mbar) | $\Delta T$ (K) | $\Delta e$ (mbar) | $\Delta\beta$ (K/m) | $\Delta\lambda$ |
|---|---|---|---|---|---|
| 15° or less | 0.00 | 0.00 | 0.00 | 0.00e−3 | 0.00 |
| 30 | −3.75 | 7.00 | 8.85 | 0.25e−3 | 0.33 |
| 45 | −2.75 | 11.00 | 7.24 | 0.32e−3 | 0.46 |
| 60 | −1.75 | 15.00 | 5.36 | 0.81e−3 | 0.74 |
| 75° or more | −0.50 | 14.50 | 3.39 | 0.62e−3 | 0.30 |

Delays $[Z_{hyd}, Z_{wet}]$ for the level of the sea are calculated with following formulas:

$$Z_{hyd} = \frac{10^{-6} k_1 R_d P}{g_m} \tag{4.5}$$

$$Z_{wet} = \frac{10^{-6} k_2 \mathrm{R_d}}{\mathrm{g_m}(\lambda + 1) - \beta \mathrm{R_d}} * \frac{\mathrm{e}}{\mathrm{T}} \tag{4.6}$$

where $k_1 = 77.604$ K/mbar, $\mathrm{k_2} = 382{,}000\ \mathrm{K^2/mbar}$, $R_d = 287.054$ J / kg / K, and $g_m = 9.784\ \mathrm{m/s^2}$; $[\mathrm{d_{hyd}}, \mathrm{d_{wet}}]$ are calculated with:

$$\mathrm{d_{hyd}} = \left(1 - \frac{\beta \mathrm{H}}{\mathrm{T}}\right)^{\frac{\mathrm{g}}{\mathrm{R_d}\beta}} - \mathrm{Z_{hyd}} \tag{4.7}$$

$$\mathrm{d_{wet}} = \left(1 - \frac{\beta \mathrm{H}}{\mathrm{T}}\right)^{\frac{\mathrm{g_m}(\lambda+1)}{\mathrm{R_d}\beta} - 1} - \mathrm{Z_{wet}} \tag{4.8}$$

Parameter $[g]$ is equal to $9.80665\ \mathrm{m/s^2}$, height of receiver $[H]$ is measured in [m] above the sea level.

Troposphere correction function for the satellite elevation [$m\ (E_i)$] is calculated with following equation:

$$\mathrm{m(E_i)} = \frac{1.001}{\sqrt{0.002001 + \sin^2\left(\mathrm{EL_i}\right)}} \tag{4.9}$$

Importantly, the function is incorrect for elevations less than 5°. Finally $i$th satellite troposphere delay error is equal to:

$$\sigma^2_{j\,tropo} = \left(0.12 \times m\left(E\right)\right)^2 \tag{4.10}$$

and vertical troposphere error is $\sigma_{TVE} = 0.12$ m.

The influence of Earth's environment on positioning accuracy was presented on 3D graphs. Application responsible for visualization, GPS3D Viewer, obtains RAW data from Septentrio PolaRx-3 receiver and calculates antenna position with scheduled frequency. Data presented in this work were possessed with frequency equal to 1 Hz, experiment took 24 hours (Cyran et al. 2011). Gathered samples are presented on 3D graph with violet points (see Figures 4.5–4.8). Obtained results show clearly that in horizontal plane, GNSS based coordinates have deviation at the level of 1 to 2 m, however accuracy in vertical plane is worst and depending on the conditions could be at the level of a few meters. Consecutive figures show influence of taking into account ionosphere and troposphere influence on positioning accuracy. Figure 4.5 presents antenna position when troposphere and ionosphere corrections are turned off; Figures 4.6–4.8 present the influence of activation of these corrections in various combinations.

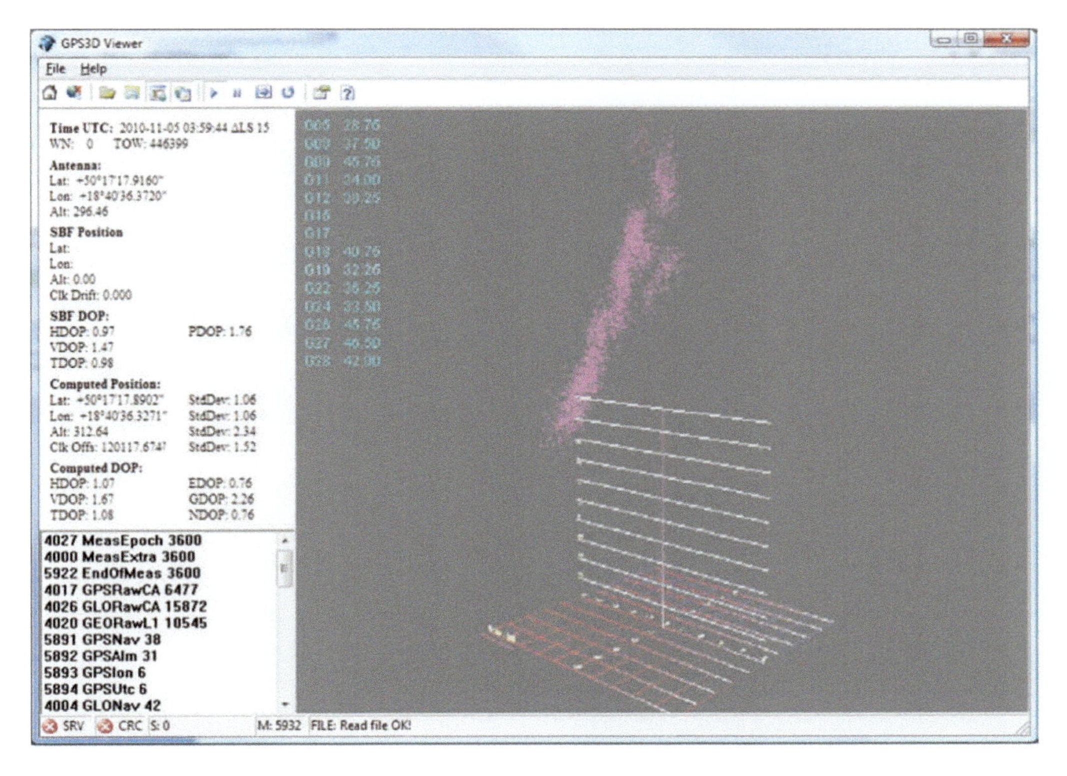


FIGURE 4.5 Influence of Earth's environment on positioning accuracy for troposphere corrections OFF, ionosphere corrections OFF.

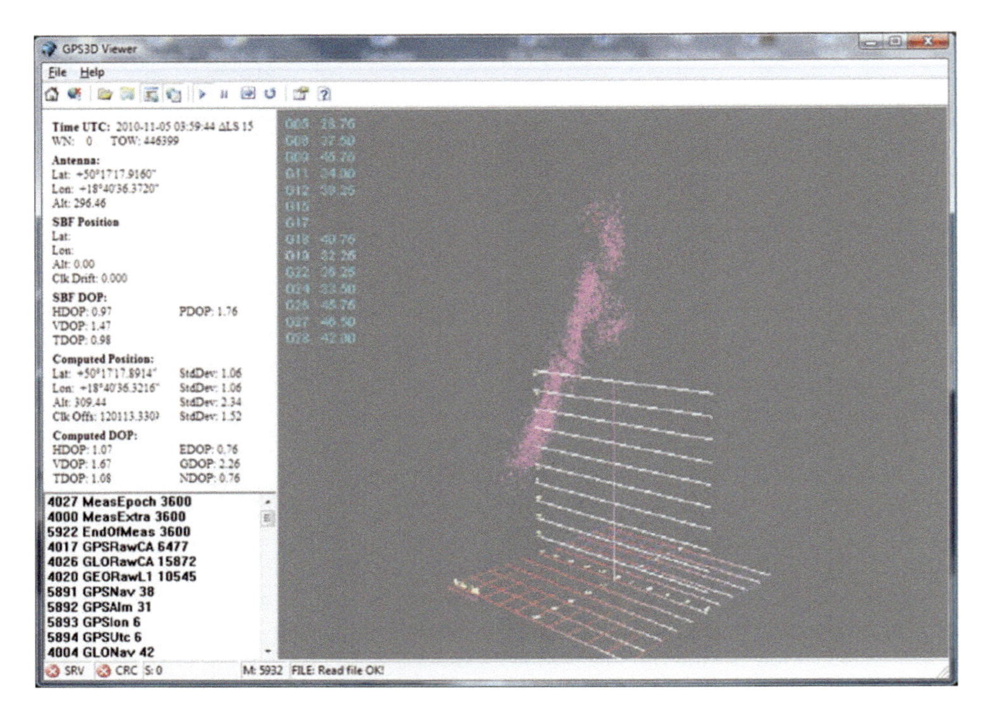


**FIGURE 4.6** Influence of Earth's environment on positioning accuracy for troposphere corrections OFF, ionosphere corrections ON.

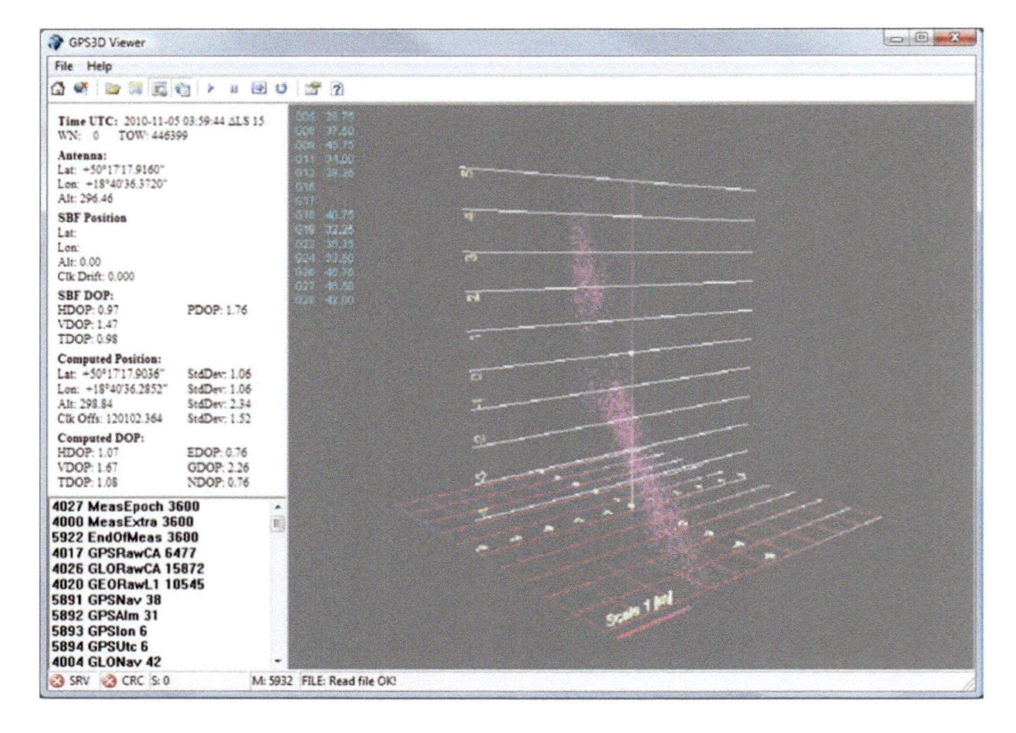


**FIGURE 4.7** Influence of Earth's environment on positioning accuracy for troposphere corrections ON, ionosphere corrections OFF.

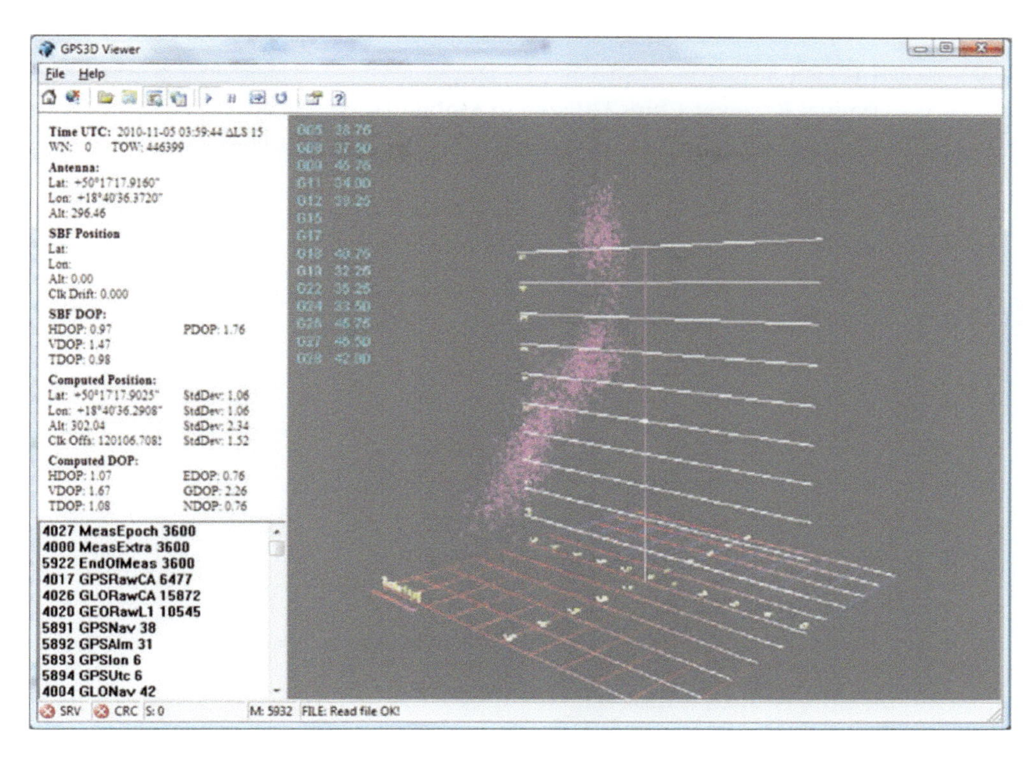


FIGURE 4.8 Influence of Earth's environment on positioning accuracy for troposphere corrections ON, ionosphere corrections ON.

## 4.4 CONSUMER GRADE GNSS POSITIONING PRECISION

Typical mobile devices (mobile phones, tablets) available on the market are equipped with multi-channel GNSS receivers, which allow for acquisition of the signal from many navigation systems at the same time (GPS, GLONASS, GALILEO); moreover, usually, acquisition of augmentation signal is also enabled. List of typical receivers is presented in Table 4.9.

Samsung Galaxy SIII, as a representative of typical mobile device, was utilized in order to present GNSS precision issues. Special application was prepared and installed on the device. Task of this application was to log coordinates of the device to internal database. Outcomes of the experiment were compared with results obtained from Septentrio PolaRx-3 receiver, which is widely utilized in aeronautics. During the experiment receivers were lying stationary in the same location, in order to provide the same conditions for both devices. Positions of receivers were logged for 24 hours, with frequency of 1 Hz. Based on positioning graphs estimation of errors was performed, according to obtained results (see Figure 4.9), Samsung Galaxy SIII is characterized by mean error at the level of 15 m, while PolaRx-3 is characterized by mean error at the level of 2 m (Figure 4.10). There are many reasons of such huge difference in precision. Receiver build in Samsung Galaxy SIII obtains positioning data based on one frequency only. In addition, it has small antenna and lower sensitivity whereas PolaRx-3 is dual frequency receiver (GPS L1 and L2 frequency) and it was equipped with lightweight high precision geodetic dual-frequency antenna PolaNt.

There is a possibility of limitation of errors when mobile device is able to utilize more than one GNSS such as Sony XPERIA S, which is equipped with dual system receiver (GPS and GLONASS). In the same conditions this device was able to achieve precision at the level of 3 m when both GPS and GLONASS data were taken into account (Figure 4.11). Interestingly, when only GLONASS was utilized then positioning precision was at the level of 18 m (Figure 4.12).

**TABLE 4.9**
**Typical Receiver Chips Utilized in Mobile Devices**

| Manufacturer | Chipset | Channel Number |
|---|---|---|
| SiRF (CSR) | SIRFstarIII | 20 |
| SiRF (CSR) | SIRFstarIV | 40 |
| MediaTek | MTK3318 | 51 |
| MediaTek | MTK3339 | 66 |
| MaxLinear | Mxl800sm | 12 |

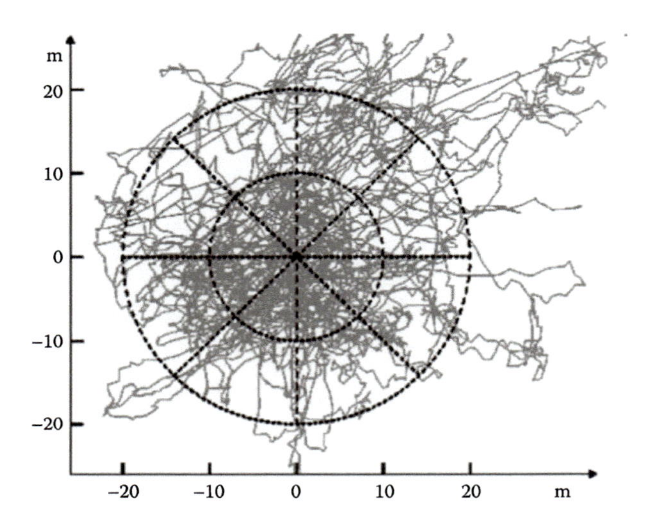


FIGURE 4.9 Graph of coordinates obtained from Samsung Galaxy III internal GNSS receiver, when GPS was utilized (24-hour period).

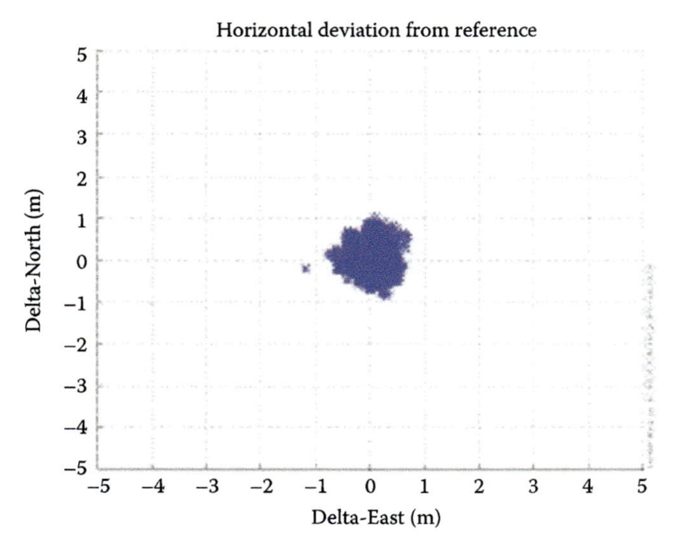


FIGURE 4.10 Graph of coordinates obtained from PolaRx-3 receiver when GPS + SBAS + two frequencies were utilized (24-hour period).

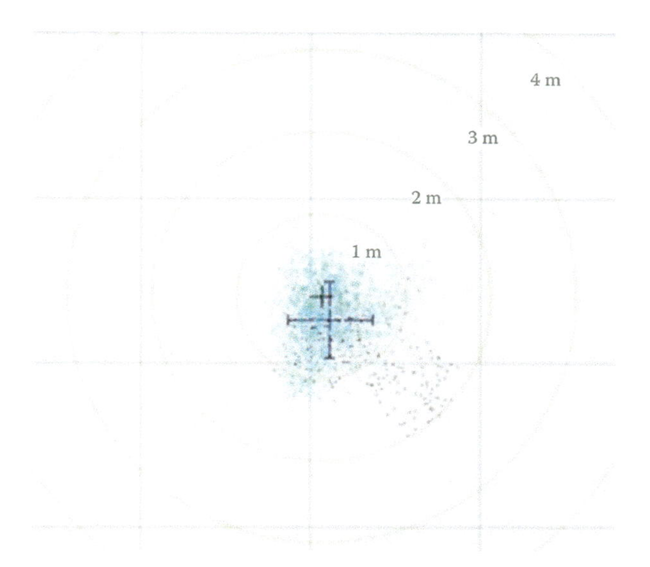


**FIGURE 4.11** Graph of coordinates obtained from Samsung XPERIA S internal GPS + GLONASS receivers, when GPS and GLONASS were active (24-hour period).

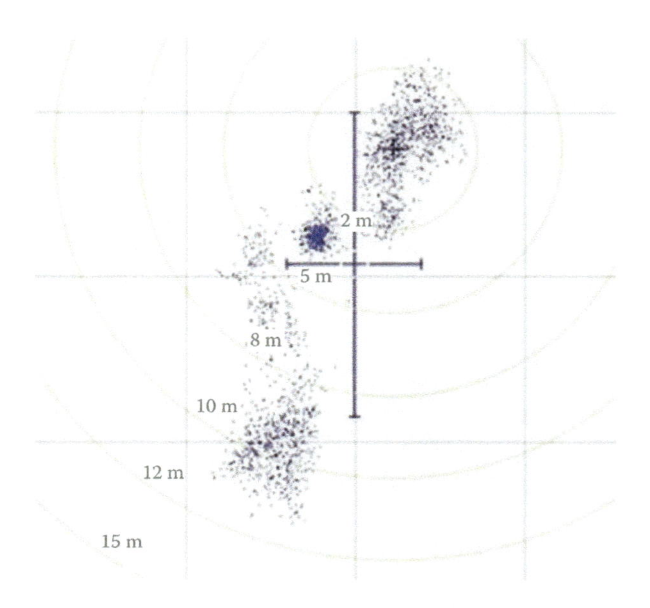


**FIGURE 4.12** Graph of coordinates obtained from Samsung XPERIA S internal GLONASS receiver when only GLONASS was active (24-hour period).

## 4.5 LOW-COST RECEIVERS

Nowadays, cheap GNSS receivers, with parameters similar to professional chips utilized by aeronautics, are available on the market. Authors of this chapter conducted studies, in which GNSS receivers, produced by U-Blox company, were employed. To test applicability of these chips custom device has been developed, schematic was presented on Figure 4.13. GNSS receiver was based on the low-cost U-Blox LEA-6S chip.

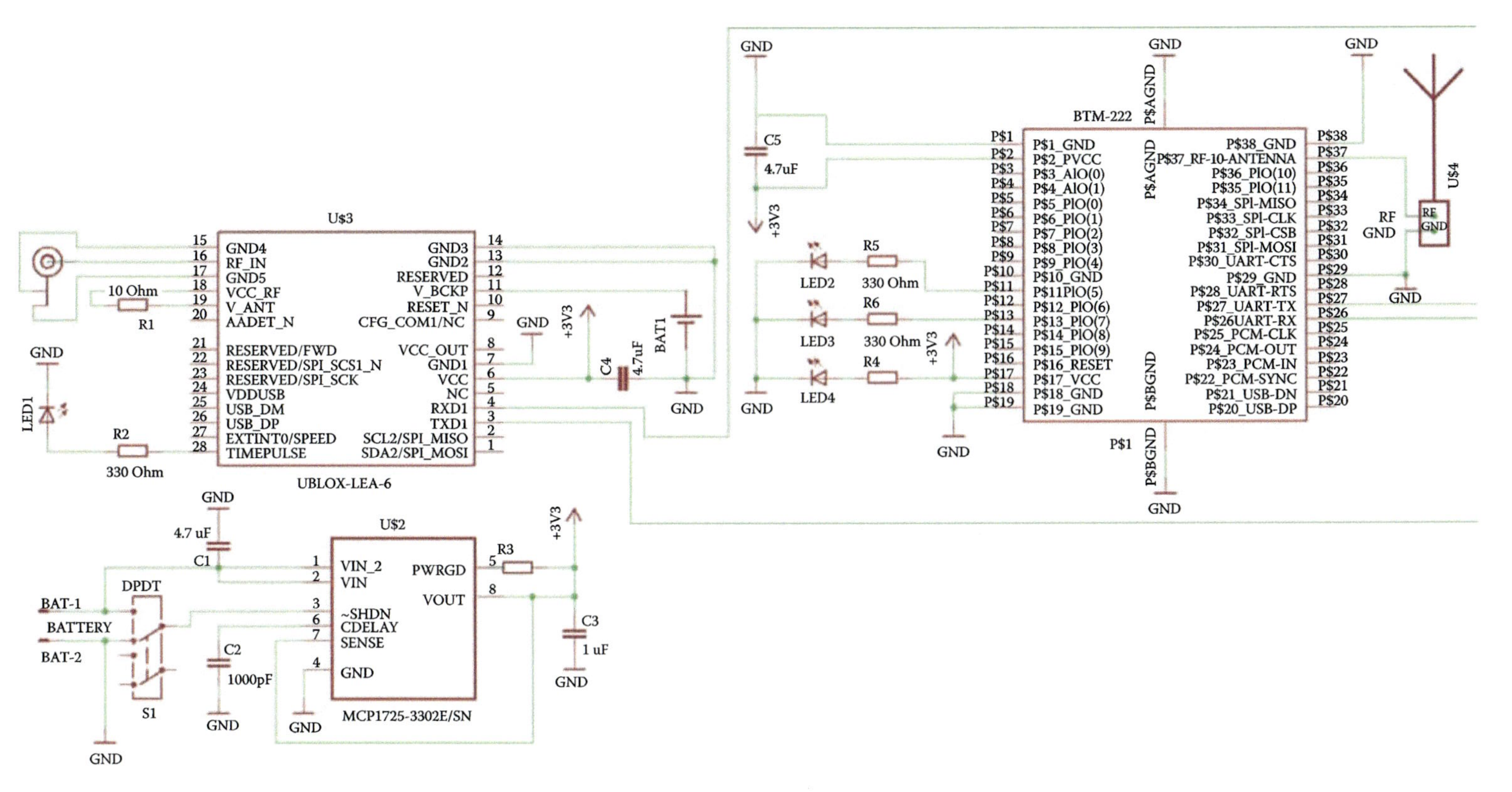


FIGURE 4.13 LEA-6S GPS receiver schematic.

During precision tests receivers were stationary and data were gathered for 24 hours. Results presented on Figure 4.14 show that LEA-6S exhibit the precision in the range 2.5 m when GPS + SBAS reference signal was utilized, it is close to the precision obtained by PolaRx-3 Septentrio receiver. In order to present precision which can be obtained at the level of different Global Navigation Satellite Systems, NEO-7 receiver, which can determine position with GPS, GLONASS, GALILEO, and QZSS was utilized. Result of survey, of two navigation systems (GPS and GLONASS), for experiments that took 24 hours and frequency refresh rate was equal to 1 Hz are presented in Figures 4.15 and 4.16 (note that scale of axis are different).

Importantly NEO-7 chip is able to work in the PPP mode (Precise Point Positioning), a method that offers enhanced positioning precision (see Figure 4.17) by utilization of the carrier phase measurements to smooth calculated pseudoranges between receiver and satellites antennas. The algorithm needs continuous carrier phase measurements to be able to smooth the pseudorange measurements effectively. Additionally, ionosphere corrections like those received from SBAS or from GPS are required. Unfortunately, technology works correctly when un-obscured sky view is available.

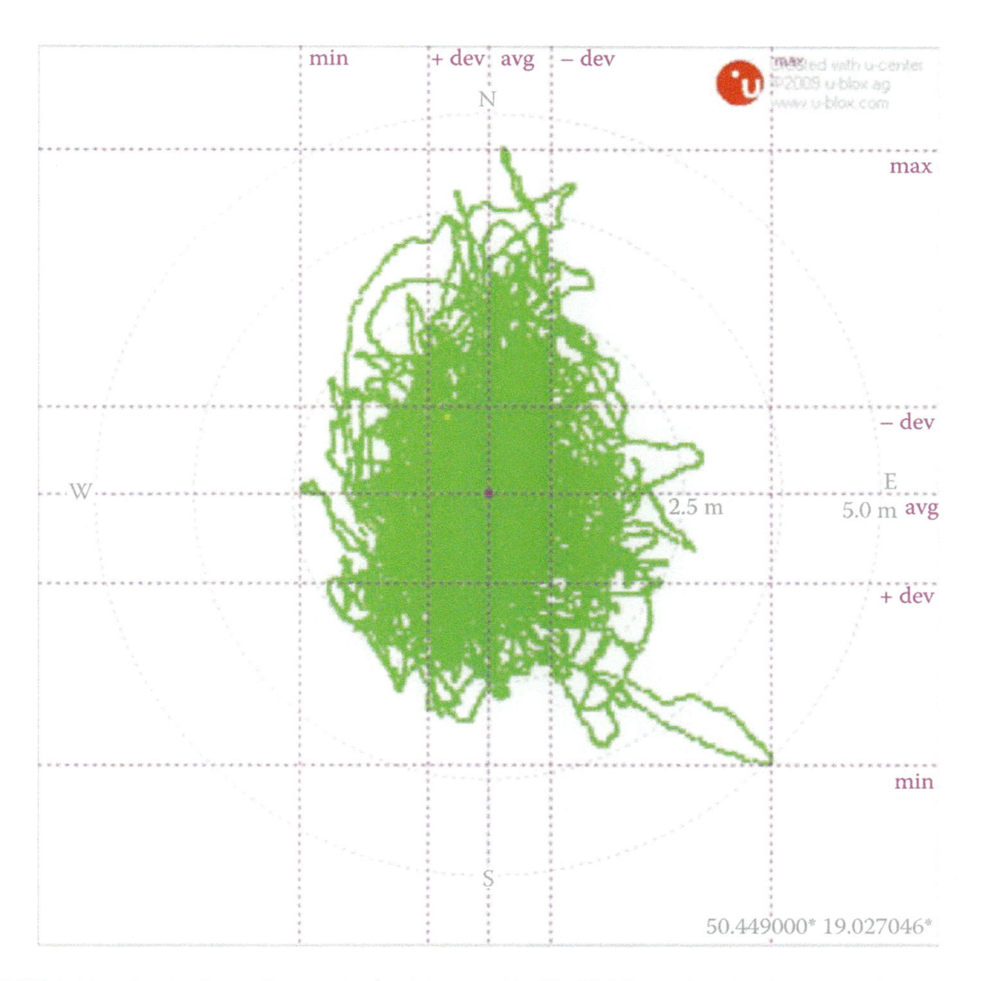


FIGURE 4.14 Graph of coordinates obtained from LEA-6S (GNSS) receiver (24-hour period).

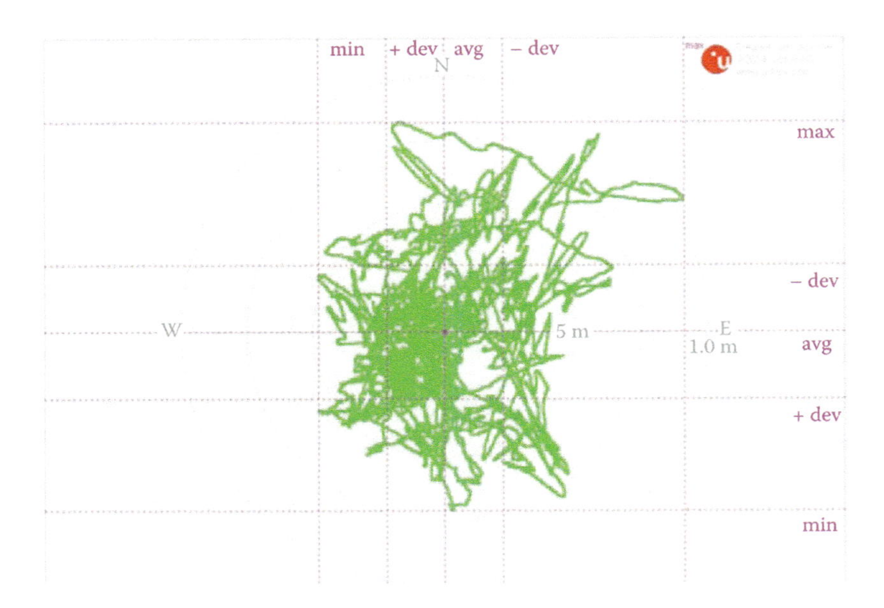


**FIGURE 4.15** Graph of coordinates obtained from NEO-7 (GNSS) receiver when GLONASS was active. Sampling frequency 1 Hz (24-hour period).

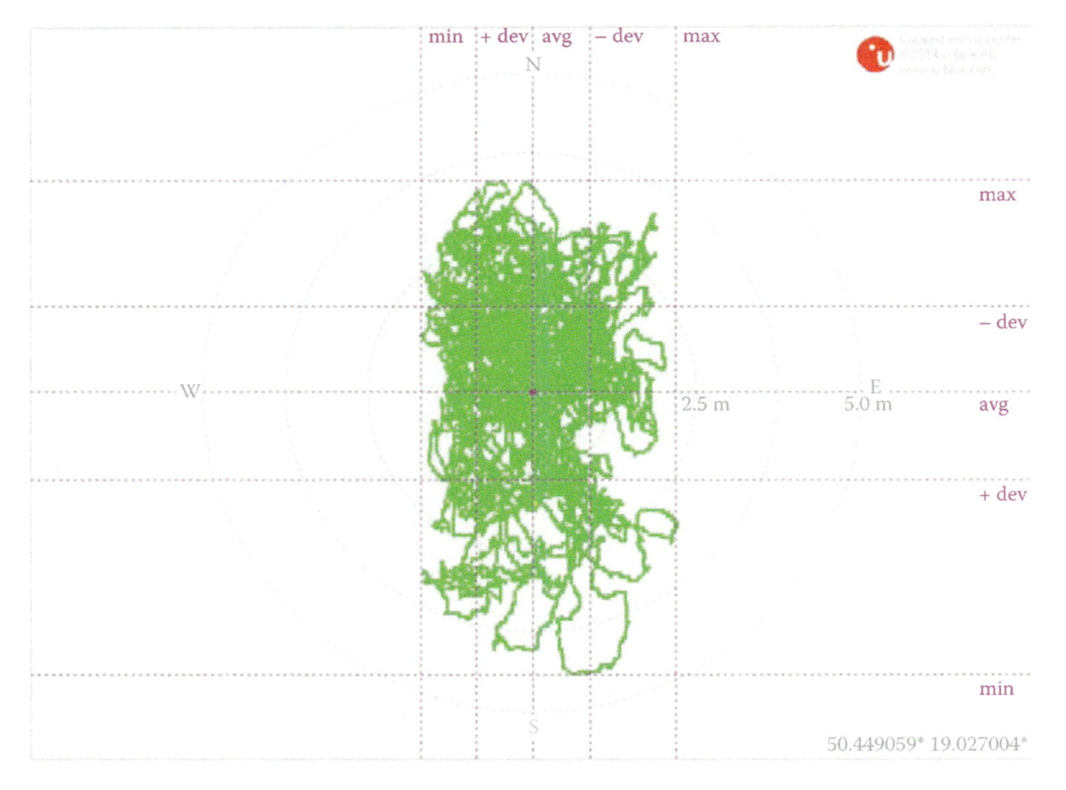


**FIGURE 4.16** Graph of coordinates obtained from NEO-7 (GNSS) receiver when GPS was active. Sampling frequency 1 Hz (24-hour period).

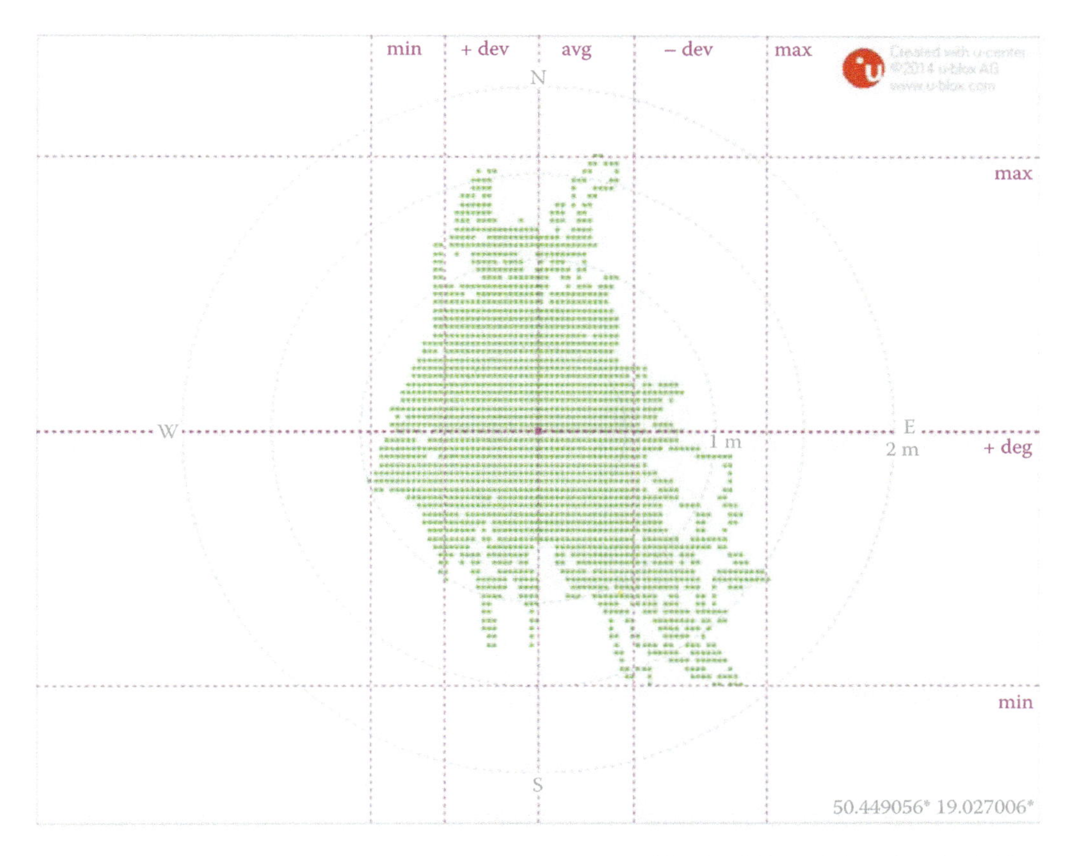


FIGURE 4.17 Graph of coordinates obtained from NEO-7 working in GPS+SBAS mode when PPP was activated (24-hour period).

## 4.6 GNSS APPLICATIONS

Global Navigation Satellite Systems (GNSS) play an important role in many various fields (see Table 4.10). Basically, two main types of applications of these services can be distinguished: civilian and military. However within these fields many common categories can be distinguished.

### 4.6.1 Consumer Applications

GNSS-based devices and applications play important role in everyday life of millions of peoples. The ability of precise determination of location allows for creation of various applications such as interactive maps with the ability of localization of the closest services in the area (such as restaurants, hotels, etc.), automatic location aware reminders (Odolinski et al. 2020, Raza et al. 2022, Hariharan et al. 2005), activities trackers which allow for visualization of people movement and creation of various statistics, automatic photo taggers that add coordinates to photos, objects trackers (Hasan et al. 2009) (e.g., cars, phones) for determination of location of stolen or lost goods, friends trackers for searching of acquaintances in the area, and many others. However, the most popular seems to be various navigation applications that can automatically calculate the best route to a destination location, and present it on the screen as 2D/3D map or in live view mode (Odolinski et al. 2020, Huang et al. 2012). What is more, if the driver misinterprets directions provided by the system, these applications can instantly work out new route. It hugely reduces time of trip preparation. Furthermore, by utilization of proper algorithms which are able to determine the best routes,

**TABLE 4.10**
**Application of GNSS**

| Categories | Applications |
|---|---|
| Consumer | Precise determination of user location |
| | Various navigation applications |
| | Localization of the closest services |
| | Automatic, location aware, reminders |
| | Activity trackers |
| | Automatic photo taggers |
| Industry | Determination of precise location of equipment and employees |
| | Location of potential clients (transportation) |
| | Monitoring of tourist flow (e.g., ski resorts) |
| | Planning of actions and processes |
| | Optimization of various processes |
| Archeology | Surveying of archaeological sites |
| | Site map creation |
| Geodesy and cartography | Preparation and update of maps |
| | On site measurements |
| | Analysis of geological and tectonic movements |
| Military | Coordination of military actions |
| | Soldiers localization |
| | Homing missiles |
| | Detailed maps creation |
| | Rescue operations |
| Aviation | Primary navigation system or support for ground-based navigation systems |

limitation of the distance that have to be traveled could be achieved. Another useful feature of navigation systems is theirs ability of utilization of database of speed limits (Bargeton et al. 2010). It could allow for fuel saving, reduction of air pollution, and increase safety on roads.

In combination with other services such as wireless internet access, navigation application can check current traffic statistics (e.g., existence of traffic jams on the route, road works) and modify the route on the fly in order to circumvent problematic areas (Bar-Gera 2007). If such data are collected from many users simultaneously then they can be utilized for traffic pattern recognition and planning of traffic issues solutions. However in such a case privacy standards should be maintained, so users which provide the data can keep anonymity. It might be hard in some cases because even if the name of the user would not be included in the package send to the data collecting center, user can be identified by many different factors for example the place of living or place of work. Therefore various methods of deeper anonymization of data were prepared for example it can obtained by simulation of system of induction loops embedded in roads, special location are designated and data are collected only when vehicle drive over such a virtual marker.

Application designated for wearable devices can also benefit from ability of current location obtainment. These applications can learn human behaviors and prediction of future user locations. It will give ability of automatic accommodation to user environment, for example, in order to increase productivity (Ashbrook and Starner 2003).

### 4.6.2 Industry Applications

GNSS-based systems are widely utilized in various workplaces and industrial applications. Employers can hugely benefit from utilization of GNSS technology especially when it is combined with wireless means of communication. Ability of precise location of equipment as well as employees allows for simplification of control processes and reallocation of resources (e.g., for soil excavation process, in order to control movement of hauling units, loaders and excavators [Odolinski et al. 2020, Raza et al. 2022, Ahn et al. 2011]), it can also reduce significantly costs.

Many fields can gain profits from utilization of this kind of systems especially those in which employees are working outside of the company headquarters, such as transportation municipal services (street sweeping, snowploughs, garbage collection, road mending) (Figure 4.18), sales departments, emergency services, agriculture, etc. GNSS based systems are successfully utilized in the area of construction works for the purpose of localization of land and marine construction sites for example in offshore oil exploration.

Application of these systems combined with logging of obtained data allow for creation of reports and introduction of various optimizations. Employers gain control because there is a possibility of validation of realization of task by employees. Routes can be checked and all deviations from the planned path, or omission of planned locations, determined. Current and historical positions of chosen unit can be obtained, detection of entering or leaving of selected area might be performed (so-called geofencing), various geo-alerts might be set. On the other hand, application of GNSS technology can support workers, that is, when salary which was paid is insufficient for the time of work, then readouts obtained from GNSS based system can confirm or reject legitimacy of workers complaint. In many cases utilization of these systems allowed for detection of law violations such as stealing of company's possession, or improper habits such as utilization of company resources for personal purposes. It is worth to mention that abidance of driving laws can easily be determined, it can reduce the amount of accidents and improve safety. Self-employed worker, such as taxi drivers, can also benefit from utilization of GNSS services. Special GNSS-based dispatching systems, which allow for easy location of passengers and determination of temporal hot spots allows for optimization of routes, reduction of passenger await time and increase of drivers incomes (Odolinski et al. 2020, Hou and Chia 2011). In addition, data possessed from systems locate in taxis are often utilized by scientists in order to conduct various analysis and researches, for example, automatic land-use classification (Pan et al. 2013), traffic flow patterns, assessment of road quality, etc.

Tourist resorts can automatic acquire information about locations that are visited by guests at various times across the day, tourist moving patterns as well as typical activities in which tourists are involved (Skov-Petersen et al. 2012). It allows for introduction of modifications in various areas such as offered services, resort layouts, travel recommendations presented to the tourists, etc. (Odolinski et al. 2020, Zheng et al. 2009). Ski resorts can monitor skiers flow and accommodate tracks accordingly, also safety might be improved. Tracking can be done through utilization of custom devices or with tourist mobile devices equipped in special application. In order to facilitate sightseeing and surveying of interesting localizations, automatic guide system based on GNSS services can be prepared, it can be combined with aforementioned statistic gathering system so data can be collected. It is worth to mention that interesting researches are also done in the field

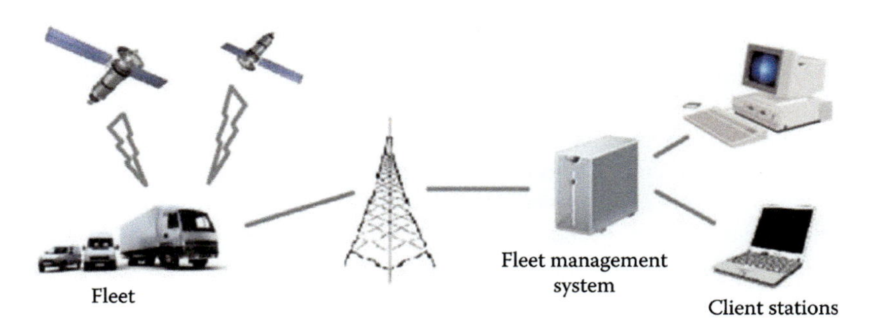


FIGURE 4.18 Fleet management system.

of geo-tagged photos utilization for the purpose of extraction of travel patterns (Raza et al. 2022, Zheng et al. 2012).

Although GNSS proved to be useful tool in company environment, there are some legal issues, dependent on the country or region in which such methods are applied, that should be taken into account at the level of planning of implementation of GNSS based solutions, that is, necessity of information of employee that system of localization of workers is utilized. Also workers' coordinates should not be logged in their private time (after work or during brakes) and special zones which are excluded from logging should be established (Inks and Loe 2005, Towns and Cobb 2012).

### 4.6.3 Transportation Applications

Public transportation benefits hugely from combination of GNSS with wireless communication abilities. Means of transportation equipped in such devices can constantly communicate their location to the control center and then information can be posted at bus/tram stops as well as at web page. It can also be distributed to dedicated mobile applications utilized by passengers. Importantly such system can be integrated with in vehicle information systems thus information about next stop can be presented to the passengers, without driver's action. Further integration, with city traffic control systems, is often implemented, so traffic lights are controlled in a way to give priority for public transportation vehicles (Raza et al. 2022). When actual vehicle position is combined with historical data and information about current state of traffic flow, predicted time of arrival of bus or tram at the line stops can be calculated. As a result not only public transport controllers can rapidly react at emerging situations and for example dispatch additional resources or redirect routes of particular lines, but also passengers can dynamically modify their travel plans and utilize different lines in order to faster achieve their destinations. Application of such systems could reduce frustration of the end users in case of delays, in addition it supports the idea of utilization of public transportation, thus counteracts formation of traffic jams and decrease pollution level. When additional sensors are incorporated, for example, automatic passengers counters, there is a possibility of creation of detailed passenger load statistics for all line's segments. Implementation of such system which ceaselessly collects data from various means of public transportation, can improve process of: schedules modification (in order to obtain better time coverage of some areas of the city), determination of the number of required vehicles, optimization of routes on particular lines and modification of timetables so they accommodate real arrival times.

It is worthy to mention that such systems installed in school buses, through automatic notifications about delays, can hugely facilitate lives of parents and increase their trust in the transportation system.

### 4.6.4 Surveying Applications

GNSS positioning systems play important role in the field of geodesy, archaeology, landscape surveys and maps creation (Ainsworth and Thomason 2003, Grenier et al. 2023, Bussios et al. 2004, Bargeton et al. 2010). It is a great tool for the purpose of mapping of large areas or difficult to access locations (Grenier et al. 2023, Bargeton et al. 2010). GNSS receivers can be carried by surveyor or mounted in the car and automatically gather coordinates data. Huge advantage of utilization of GNSS technology is the fact that analysis of obtained data can be done on site or raw data, obtained from the receiver, can be recorded and further processed in offline mode. GNSS-based measurements can be done relatively quickly when compared to the utilization of theodolite, tape, or electromagnetic distance measurement methods (Figure 4.19).

Utilization of GNSS technology in the field of archaeology has a long history, it was successfully applied to small scale project such mapping of single archaeological objects (stones/walls etc.) as well as large ones. Application of this technology significantly improves accuracy of obtained result, it allows to limit utilization of other geodetic techniques such as tapes measurements, and

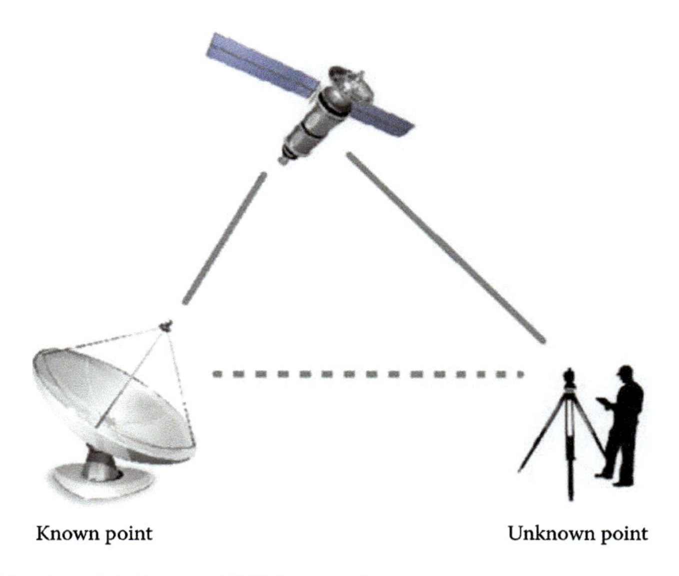


FIGURE 4.19 Utilization of dedicated GNSS-based reference station.

speed up the process of coordinate generation. As a result reduction of the number of personnel experienced in geodesy measurements involved in the realization of this kind of projects, is possible. Utilization of GNSS technology perfectly fulfil needs of the surveying processes, which usually requires determination of the location of the site in the relation to the map, followed by the determination of the location of all points of interest on the site level and determination of the relation between given point and all other points which are localized on analyzed site. In some cases accuracy provided by pure GNSS solution is insufficient therefore in order to increase precision additional techniques, that can increase accuracy of measurement to centimeters, are often employed. The most popular is utilization of dedicated GNSS-based reference stations (if such are available on selected area) and application of differential corrections (Grenier et al. 2023, Cosentino et al. 2005) or Real Time Kinematics (Wei et al. 2010). Deployment of stations is usually done at the nationality level. For every reference station very precise coordinates are determined. There are two types of stations: active ones are equipped in stationary GPS receivers, operate all the time and usually share data through dedicated websites. They are usually scattered around the area, for example, in Great Britain there were 32 active stations based on GPS (data for 2002 December); any point of Great Britain should be within 100 km from the closest active stations in addition urban areas are usually covered by more than one active station. If higher level of accuracy is required passive stations might be employed. This kind of station is usually realized as a ground maker which does not possess its own GNSS receiver, therefore user, in order to utilize this station, has to provide its own equipment which is located at passive station site for the time of survey conduction. In Great Britain there was around 900 such markers, any point of land should be within 20–30 km from the closest marker. Data for differential correction can also be obtained from SBAS.

As mentioned earlier Earth's continents are in constant move, GNSS in combination with fixed reference stations, on the surface of the ground, are great tool for measurement of such relocations, level of oceans can also be monitored (Yang and Lo 2000). In general, GNSS are valuable tool for measurement of object deformation (e.g., damns), and in engineering surveys applications (Frei et al. 1993). Furthermore, scientists investigate ways to utilize sensitivity of GNSS-based devices for measurement of various disturbances in atmosphere. Such systems can be applied in meteorology for determination of the amount of vaporised water in the atmosphere, and current researches point out that they can also be applied to volcanic ash detection, therefore improvement of the early volcanic ash alerting system on affected areas might be improved (Aranzulla et al. 2013). As a result flight safety and comfort of population living on these areas might be improved.

### 4.6.5 Accident and Disaster Recovery Applications

Studies conducted for Germany point out that in 2002 there were more than 45 private cars per 100 inhabitants and on average every vehicle was characterized by annual travel distance at the level of 12,600 km (Raza et al. 2022, Kalinowska and Hartmut 2006). These results show significance of mobility in contemporary society. Unfortunately mobility requirement entails exposure of road users to various harmful events. Statistical data gathered for UE reveal that every year over 1.5 million of people take place in accidents, to make matter worse, as a consequence of occurrence of such events more than 30,000 people annually lost their lives. It is tremendous social and economic issue. Researches for UE, done in 2000, estimated such costs at the level of 160 billion of euro per annum.

Outcomes of performed researches point out that in many cases quicker delivery of medical help could reduce the number of casualties and decrease the time of post-accident trauma recovery. There are many issues when it comes to emergency services notification, often accident victims or witnesses cannot give precise location of event occurrence, it is usually especially hard in rural areas, what is more the stated location might be incorrect. Another issue is that, in many cases victims of accident lost their consciousness, however there might be no witnesses to call for help, such a situation is common in uninhabited areas especially for accidents which occur during night hours. Up to 52% of fatalities could be avoided if victims were earlier located by medical personnel (5%), were faster transported to the hospital (12%), or advanced trauma center (32%). The EU estimates that introduction of efficient system of rapid emergency center notification, which will decreases amount of rescue actions await time and speeds up the process of medical actions delivery, thus reduces severity of injures, up to 2,500 lives can be spared and 26 billion euro can be saved per annum.

The aforementioned analysis led to the proposals of systems that are responsible for automated detection of accidents and notification of emergency center. One of such a system is developed in the course of Research on EGNOS/Galileo in Aviation and Terrestrial Multi-sensors Mobility Applications for Emergency Prevention and Handling (EGALITE). EGALITE project among other purposes, aims at the creation of efficient system of emergency handling and prevention base on mobile devices. Contemporary mobile devices (e.g., smartphones, tablets) are feature rich items equipped in many sensors (i.e., accelerometers, gyros, magnetometer) and receivers (i.e., GNSS receiver); what is more, they are easily accessible and widely available. Moreover rules introduced by U.S. Federal Communication Commission (FCC) state that in the U.S. every mobile carriers must be able to determine 911 mobile caller longitude and latitude with certain accuracy (Yang and Lo 2000). However precise localization based solely on transmitting tower was, in the time of introduction of this rule, insufficient and it required costly investments in development of infrastructure. It was one of driving force to equip mobile phones with GPS receivers, which were able introduced better positioning precision. Vast popularity of smartphones allows for utilization of implemented system by sufficient fraction of the population without incurring of any additional costs. In proposed systems readouts gathered from accelerometer build in mobile device are analysed by artificial intelligence based on feed forward artificial neural networks. Artificial intelligence was trained with real data emerged from test car crashes and non-accidents car runs. Test cases encompass various types of cars, speeds and types of roads. As a result system which is able to achieve high accuracy of accident detection (at the level of test set composed of 18,266,234 samples from accident and non-accidents runs) sensitivity equal to 0.97 and specificity equal to 1, was created. In case of accident occurrence position based on GNSS signal is determined then coordinates, time of event as well as user ID are sent to the emergency center through GSM network. Application of GNSS technology allows for quick and precise location of accident site thus contributes in significant decrease of emergency services await time. In addition time which is send to the emergency center is bases on

GNSS readouts, it provides precise time synchronization between various devices, as a result there is a possibility of determination of groups of related events and reconstruction of roads accidents in which many vehicles are involved, is facilitated.

In order to determine car environment automatically, activate/deactivate application, NFC module is utilized. An assumption is made that car should be equipped in NFC tag. When device detect presence of proper NFC card it automatically switch to active mode, in other case inactive mode is chosen. If user's device does not possess NFC capability there is a possibility of manual choosing of proper mode. Interestingly also GNSS receivers can be utilized in this area. Some creators of similar systems assumed that mobile devices must move with speed higher than some empirically determined threshold value and for the purpose of speed determination GNSS readouts were employed. However such a solution consumes a lot of power, because vehicle position has to be continuously refreshed, in addition this system will not detect accident when the car is stationary on the crossroad or in the traffic jam (Thompson et al. 2010, White et al. 2011).

In more platform-dependent solution there is a possibility of integration of mobile accident detecting application with air bags deployment controller or creation of additional hardware module which contains high precision accelerometers, is permanently attached to the car's body and communicates with mobile device through Bluetooth technology (Matthews and Adetiba 2011). Such approach can further reduce amount of required power and increase detection precision.

As mentioned earlier, an important issue of utilization of GNSS receiver, built into mobile devices, is limited battery power. Although battery power consumption of specialized GNSS receivers in perfect conditions can be as low as 10 mW (e.g., Sony CXD5600GF) typical power utilization in modern device is higher at the level above 140 mW, while operating GNSS receivers are responsible for significant fraction of battery consumption. Traffic accident detecting application can be combined with other services such as navigation or traffic jam avoidance system, then constant utilization of GNSS receiver is justified. However end users can consider continuous employment of GNSS receiver only for the purpose of accident detecting application, as a waste of energy. Researchers try to determine approaches that can decrease utilization of power through application of various time patterns of coordinates obtainment. Theoretically GNSS receiver can be turned off all the time and utilized only in case of accident detection, however the issue here is the time required for the first fix obtainment. After activation of the receiver, time of determination of location can span from single second up to a few minutes. This process is shorter (single seconds) when receiver possess current information about almanac, ephemerides and has synchronized clock with satellite time, however it extends when such information are not available, and have to be downloaded directly from satellites. For example, for GPS based system, the longest time takes collection of almanac data (at least 12.5 minutes in foster conditions, longer if GNSS signal is degraded by obstacles), however data are valid for several months. If ephemerides have to be updated (which are valid up to four to six hours [Dorsey et al. 2005]), the process can take more than 30 seconds. When only time synchronization is required then the process of fix obtainment takes a few seconds. Obtainment of the first fix can be significantly speed up if receiver is able to utilize wireless services, which provide almanac, ephemerides, accurate time and satellite status, for example, assisted GNSS (A-GNSS) (Van Diggelen 2009, Harper 2009). Importantly, then the position can be calculated within seconds even under poor satellite signal conditions. In addition utilization of such services can improve positioning accuracy, although GNSS based readouts without additional corrections, should be sufficient for the purpose of emergency site location.

Noteworthy is the fact that GNSS receivers in combination with other sensors (such as pulse oximetry sensor) can support other types of emergency situations such as heart failure (Hashmi et al. 2005).

Global navigation satellite systems play important role in disaster rescue and recovery actions support. Emergency center personnel needs fresh data about locations of rescue resources in order to made proper decision in timely manner (Ameri et al. 2009). Therefore every rescue unit should be equipped with GNSS receiver that has ability to automatically send location to an emergency center.

In addition precise interactive maps of the location, combined with automatic localization functionality allow for easy navigation on the areas on which characteristic landmarks were hindered by disaster, for example, flood (Van Westen 2000), it can also facilitate introduction of additional support units which are not familiar with terrain (e.g., rescue units from other regions/countries). Furthermore unmanned autonomic carriers, can be send at the disaster site in order to gather visual data from chosen areas. Navigation of carries and activation of the camera might be based on the location obtained from GNSS receiver. GNSS based system can also support early warning of population about disasters such as earthquakes and volcanic eruptions.

### 4.6.6 Health Researches and Monitoring Applications

Combination of Global Navigation Satellite Systems with mobile devices, which are carried ceaselessly by users, open new possibilities for various location services. Interesting application of such services is trial of assessment of contextual effects, such as environmental conditions, on human's health state, and combination of health hazards with locations (Kumar et al. 2023, Raza et al. 2022, Bhardwaj 2020, Thierry et al. 2013) (Figure 4.20).

Many studies concerning contextual effects influence on health is focused on human residential area (Thierry et al. 2013), however people spend vast amount of time outside of theirs living locations. Therefore ability of easy collection of coordinates time series and combination of these data with information about users' activities opens new possibility of research conduction. However vast amount of information which should be processed and lack of broader context in raw GNSS data are huge issues. Therefore subject of location activity is introduced and specialised algorithms are employed in order to detect activity place location, based on GNSS readouts (e.g., time distant cluster detection, kernel-based, etc.). Then data might be analysed from wider perspective and influence of series of consecutive activities locations on each other might be analysed. Such an approach might be useful for various researches, for example, in determination of causality in epidemiological (Thierry et al. 2013). In addition determination of life-space can be base for automatic estimation of older adults' activity level and their health state (Wan and Lin 2013).

### 4.6.7 Tracking Applications

Other area of GNSS-based tracking is utilized by researchers in order to find behavior of animals (Raza et al. 2022, Zagami et al. 1998). Various animals can be equipped with devices that are

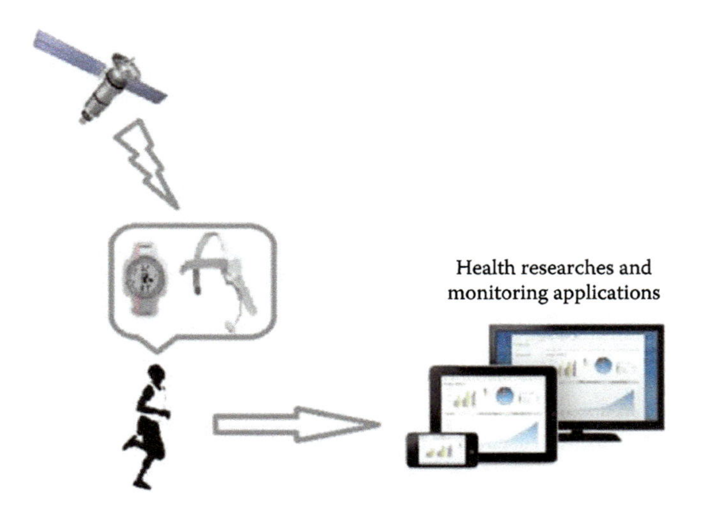


FIGURE 4.20 Application of GNSS for health researches and monitoring applications.

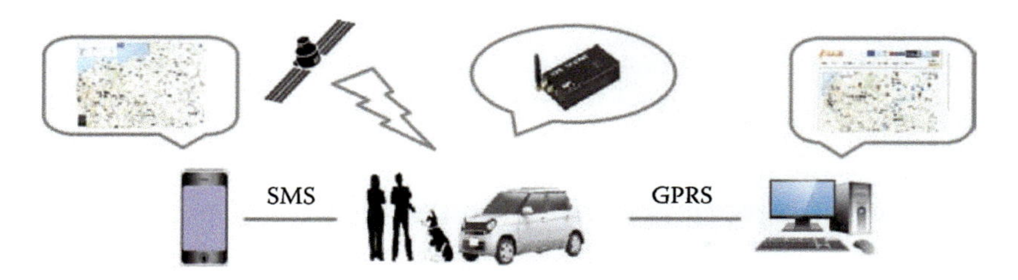


FIGURE 4.21 Global navigation satellite systems are widely utilized in tracking applications.

transmitting their locations. Such researches have many practical aspects for example lead to better understanding of causes of road collisions with animals or allow for determination of animals' grazing areas (Rutter et al. 1997). GNSS collars are often coupled with various sensors (e.g., neck movement sensor), in order to improve abilities of automatic determination of animal activity. Issue in this area of application is battery lifetime. Tracking devices should be small and light, they should be virtually unnoticeable for the tracked animal, on the other hand battery life time depends on its size and weight. When it comes to farm animals, replacement of power source is relatively easy, however often replacement of batteries in devices designed for wild animals is not possible because it introduces stress and might cause temporal modification of animal behavior. Therefore, devices utilized for this purpose are characterized by low frequency of coordinates obtainment and limited communication with central server (Kumar et al. 2023, Haibo et al. 2022, Gottardi et al. 2010) (Figure 4.21).

When it comes to person localization, GNSS-based devices are perfect base for systems that are responsible for children tracking. Parents can utilize such devices in order to check whether their descendants are in the proper location, for example, school/home. In combination with correct algorithms derogations from daily routines can be determined and caretakers can be automatically informed. Similar systems can be utilized in case of persons which dementia or similar illness, that influence ability of returning to the place of living (Landau et al. 2011). Also prison system could reduce costs because commitment of minor crimes might be punished with home detention and tracking of convicts can be done with GNSS collars.

Monitored person should be aware about carried GNSS receiver, however there is ongoing discussion whether caretakers can utilize such devices, without the knowledge or agreement of tracked person.

### 4.6.8 Unmanned Vehicles Applications

GNSS are utilized in various areas of autonomous and semi-autonomous systems. Very promising field is application of this technology in autonomous land vehicle applications, for example, for open-cast mining, agriculture applications (Kumar et al. 2023, Raza et al. 2022, Bhardwaj 2020, Heraud and Lange 2009, Prakash et al. 2012), cargo and human transportation (Manabu et al. 2006, Bevly and Cobb 2010, Ozguner et al. 2011) (especially in extreme conditions [Nagashima et al. 2013]). Also domain of autonomous robots is developing quickly. Precise control process needs ceaselessly monitoring of the state of an object, however refresh rate of typical GNSS receiver varies between 1 to 10 MHz in addition precision of particular GNSS is also limited, usually to a few meters. Furthermore, GNSS receivers are subject of errors, especially common are multipath errors and signal lost caused by occlusion of navigation signal by various obstacles, for example, high buildings in city centers. There are many problems to solve because consequences of errors in autonomous system can be tragic. In order to overcome precision issues, often signals from many satellite navigation systems, supported by SBAS, are processed at the same time. It helps to improve precision however still there is a possibility of obtainment of coordinates affected by high errors or event complete loss of GNSS signal, for example, in tunnels. Therefore advanced statistic techniques are applied in order to determine whether coordinates obtained with GNSS should be accepted and utilized. In order to be able to determine state of an object between successive accepted GNSS readouts, obtain better insight into state of controlled process, as well as increase system state sampling frequency and to be able to direct the process despite of the fact that some GNSS readouts might be rejected Internal Navigation Systems (INS) are utilized. They are composed of additional sensors which give information about state of controlled object, as a result when proper calculations are applied, location of an object can be deduced. This group includes accelerometers, gyros, magnetometers etc. therefore there is a possibility of determination of speed, and angle, of controlled object movement thus ability of calculation of its coordinates based on previously determined position, and results obtained from employed sensors.

Integrated battery of such supporting devices is referred to as Internal Measurement Unit (IMU). Main advantage of these sensors is high refresh rate (typical 100 Hz, however it can be more in case of utilization of industrial level devices) in addition most of them is jamming and deception resistant (Sukkarieh et al. 1998, Zhong et al. 2008). However huge problem of IMU application is fast accumulation of readout errors that, with time, lead to significant degradation of calculated parameters, therefore periodical calibration is required. In consequence system controlling process cannot be solely based on IMU readouts. Combination of IMU with GNSS is a great way of obtainment of high IMU frequency refresh rate with periodical calibration provided by GNSS. Usually at the heart of such systems Kalman filters are applied (Bullock et al. 2005, Groves 2008), which works in two stages: prediction—responsible for constant determination of coordinates, velocity as well as attitude based on IMU readouts, and estimation—responsible for determination of errors of aforementioned values and correction of their values when valid GNSS readout is available. In terrestrial applications, in order to minimize errors introduced by IMU, there is a possibility of live determination of values of it internal sensors biases (that are influenced by various factors, e.g., temperature) it is done every time when the controlled object is stationary. Then obtained results are taken into account in prediction stage. In addition detection of stationary state triggers reset of internally tracked velocity.

Another way of improving of GNSS positioning accuracy is utilization of magnetic markers incorporated in road lanes (Haibo et al. 2022, Hein 2020, Lau 2020, Hernandez and Kuo 2003) or application of RFID-assisted localization in combination with sensor network (Lee et al. 2009). Such systems are especially useful in emergency prevention and unmanned vehicles applications, that is, collision avoidance for Vehicular Ad hoc Network. In such a system precise location and ability of determination of ranges between cars are very important. Unfortunately systems solely base on GNSS even if supported by SBAS corrections, suffer because of precision issue which could be caused by local conditions. It happens because SBAS systems provide corrections which are valid for large areas, therefore they cannot perfectly fit to all locations because they are characterized by various environmental factors (e.g., trees, city canyons, etc.). RFID assisted concept assumes that roadsides of tracks, on which system is utilized, should be equipped with RFID tags that allow for calculation of GNSS corrections. Then vehicles could improve position precision obtained from GPS with RFID readouts and broadcast corrections through wireless radio to neighborhood cars. As a result, units that are not equipped in RFID readers, or are located on inner lane, outside of range of RFID tags, can also benefit from the system. Such an approach allows for obtainment of corrections fitted to local conditions, and enables precise determination of relation between moving objects, however it is worth to point out that it does not solve the issue of localization when GNSS signal is not available, for example, in city canyons or tunnels. Although additional equipment is required, RFID tags are already utilized on many roads, for example, for the purpose of automatic tool collection, therefore such system might base on already existing infrastructure.

### 4.6.9 Time Synchronization Applications

GNSS services require precise time synchronization between satellites and receivers clocks. Therefore another fields of applications of such systems emerged. Precise time and frequency synchronization, among different site on the globe, are important in many areas (Hewlett Packard Company 1996). High frequency trading utilizes precise time in order to time stamp transactions (Korreng 2010), mobile telecommunication requires precise time synchronization between base stations so they can efficiently share common radio spectrum (Sterzbach 1997, Chunping et al. 2003). Precise time is important in all kinds of sensor networks which collect data from the environment, for example, seismographic researches need time synchronization between sensors scattered around the globe, power companies utilize precise time for efficient energy distribution and pining down issues of disturbances in the network (Fan et al. 2005), physics can precisely measure various values (e.g., one-way light speed) (Gift 2010), and observational astronomy needs time synchronization between data collecting sites. Structural health monitoring (Kim et al. 2012) such as in pipeline transportation can precisely localise malfunction site through utilization of synchronized sensors and measurement of error signal time detection by individual sensors (e.g., acoustic ones) (Tian et al. 2010). Message encryption could also require precise time synchronization (Bahder and Golding 2004).

Utilization of GNSS services allows for achieving of time accuracy at the level of nanoseconds (HP 1996). Huge advantage of solution based on GNSS are low costs, as mentioned earlier access to GNSS signal is free of charge, therefore only maintenance of receiver infrastructure is required.

### 4.6.10 Military Applications

GNSS-based system plays important roles in military applications. It might be confirmed by the fact that GPS and GLONASS systems are controlled by army. Among many applications these systems can: be utilized by precise targeting units (Kumar et al. 2023, Raza et al. 2022, Bhardwaj 2020, Haibo et al. 2022, Hein 2020, Lau 2020, Brown et al. 1999), support troops reallocation, be incorporated into vehicles and soldiers navigation devices. It can hugely improve cooperation between troops and facilitates carrying out of activities in unfamiliar terrains. GNSS is utilized by the army not only during military operations but also in civilian supporting activities, for example, during disaster recovery. Interestingly, UE develops system which will be controlled by civilians and focused mainly on civilian applications, however services for various government agencies also will be provided (Figure 4.22).

### 4.6.11 Aeronautic Applications

Contemporary GNSS became popular tool in the area of avionic navigation (Clarke 1998). Utilization of these systems is regulated by regional legislations (Bhardwaj 2020, Haibo et al. 2022, Hein 2020, Lau 2020, Andrade 2001). In general, when pure GPS or GLONASS signals are utilized, then GNSS-based navigation equipment that is incorporated in plane is treated as supplementary system. It happens because there is lack of monitoring of positioning precision within signal broadcasted by navigation satellites, it implies that information about not complying with predetermined accuracy levels, determined solely on readouts from these systems, is missing. In addition precision of

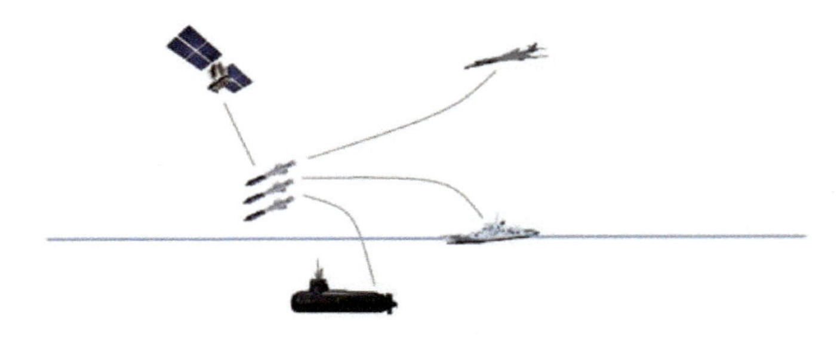

FIGURE 4.22 Various kinds of military equipment utilize Global Navigation Satellite Systems.

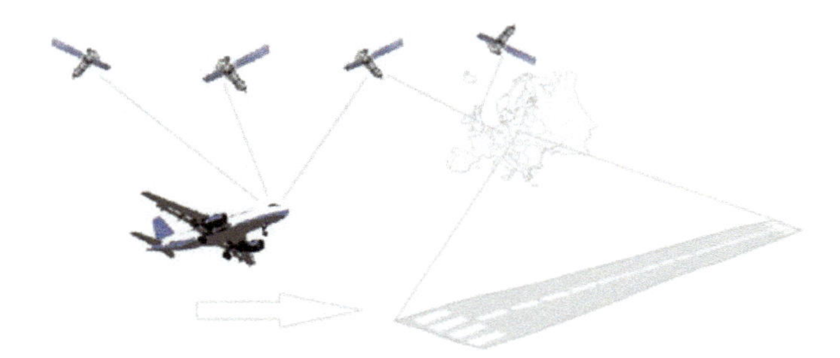

FIGURE 4.23 GNSS allow for determination of precise coordinates of planes.

obtained coordinates in vertical plane are worse than those obtained in horizontal planes. Avionics applications require high precision and reliability of GNSS-based positioning in all spatial planes. As mentioned before, in order to overcome these issues, and allow for aircraft navigation to be based primarily on GNSS, satellite-based augmentation systems were deployed, for example, in Europe, European Space Agency (ESA) together with air traffic control organizations created EGNOS. These systems provide information about GNSS failures and at the same time increase positioning accuracy. Various organizations (e.g., Institute of Computer Science in Silesian University of Technology) are involved in constant monitoring of performance correctness of GPS, EGNOS, and GLONASS systems. Positioning accuracy, accessibility, and reliability of data provided by GNSS satellites are collected and sent to EGNOS Data Collection Network (EDCN) that is controlled by EUROCONTROL, the organization that is responsible for ceaselessly monitoring of correctness of functionality of GNSS for the purpose of air navigation in European airspace (Figure 4.23).

Huge advantages of these systems are reliability and ability of infrastructure cost reduction. Therefore, all the time works are conducted should result in an increase of involvement of GNSS in this field. Application of GNSS improves work of air traffic control, ability of precise location of airplanes allows for collision avoidance and optimization of routes for reduction of travel time and fuel consumption. Promising area of GNSS-based system incorporation involves support (and in the future automation) of procedure of aircraft approach to landing so called Localizer Performance with Vertical Guidance (LPV) procedures. It allows for optimization of landing approaches paths and reduces the amount of financial resources which have to be devoted for airfield infrastructure (e.g., it can rule out necessity of investment in ILS), at the same time airplanes can approach to runways even in conditions of low visibility. Implementation of LPV procedures has many advantages among them are:

- Can be utilized on thousands of airfield across the world (more than 1,800)
- Approach path eliminates intermediate phases (dive and drive)
- Approach path is independent from airfield and airplane barometrical equipment
- Reduction of influence of low temperatures and influence of errors in flight instruments settings

Therefore, LPV procedures increase safety of air traffic. In order to perform LPV approach equipment that meets the specific requirements determined by current legislations, such as dual frequency GPS receiver compatible with SBAS, as well as special GPS receiver antenna, are required.

## 4.7 SUMMARY

Utilization of Global Navigation Satellite Systems in various terrestrial applications is characterized by increasing popularity. It is visible in number of aspects of human lives, various scientific researches are based on data gathered with GNSS; what is more, many companies rely on proper functionality of these systems. Ability of rapid determination of precise location across the globe is widely utilized in military and disaster recovery applications, even artists are trying to use GNSS data as a form of artistic expression (Lauriault and Wood 2009).

The most important factors that allowed GNSS for such a popularity are availability of navigation signals (free of costs, everywhere on Earth), good positioning precision (especially when supported by various augmentation systems), and high reliability of service (i.e., rarely cases of downtimes of GPS system functionality). In addition, after achievement of fully operational state by GALILEO system, which will be fully controlled by civilian institutions, a wave of new applications of these systems is expected.

## ACKNOWLEDGMENTS

The research leading to these results has received funding from the PEOPLE Programme (Marie Curie Actions) of the European Union's Seventh Framework Programme FP7/2007–2013 under REA grant agreement no 285462. We would also like to thank to Marcin Paszkuta for support in the figures creation.

## REFERENCES

Ahn, S.M., C. Park, J. Kang, 2011, Application of GPS fleet tracking and stochastic simulation to a lean soil excavation practice, Proceedings of the 28th ISARC, Seoul, Korea, pp. 335–336

Ainsworth, S., B. Thomason, 2003, Where on earth are we? The Global Positioning System (GPS) in archaeological field survey, Technical Paper, English Heritage Publishing

Allan, D.W., 1997, The Science of Timekeeping, Hewlett Packard

Ameri, B., D. Meger, K. Powert, 2009, Uas applications: Disaster & emergency management, ASPRS Annual Conference, Baltimore, Maryland

Andrade, A.A.L., 2001, The global navigation satellite system: Navigating into the new millennium, Ashgate, ISBN 9780754618256

Aranzulla, M., F. Cannavò, S. Scollo, G. Puglisi, G. Immè, 2013, Volcanic ash detection by GPS signal, GPS Solutions, Vol. 17(4), pp. 485–497

Ashbrook, D., T. Starner, 2003, Using GPS to learn significant locations and predict movement across, Personal and Ubiquitous Computing, Vol. 7(5), pp. 275–286

Bahder, T.B., W.M. Golding, 2004, Clock synchronization based on second-order quantum coherence of entangled photons, AIP Conf. Proc. 734, 395, Glasgow

Bar-Gera, H., 2007, Evaluation of a cellular phone-based system for measurements of traffic speeds and travel times: A case study from Israel, Transportation Research Part C, Vol. 15, pp. 380–391

Bargeton, A., F. Moutarde, F. Nashashibi, A.S. Puthon, 2010, Joint interpretation of on-board vision and static GPS cartography for determination of correct speed limit, CoRR

Bhardwaj, A. 2020. Terrestrial and satellite-based positioning and navigation systems: A review with a regional and global perspective, Engineering Proceedings, Vol. 2(1), p. 41. https://doi.org/10.3390/ecsa-7-08262.

Bevly, D.M., S. Cobb, 2010, GNSS for Vehicle Control, Artech House, ISBN 9781596933026

Brown, A.K., G. Zhang, D. Reynolds, 1999, Precision targeting using GPS/Internal-aided sensors, ION 55th Annual Meeting, Cambridge, MA

Bullock, J.B., M. Foss, G.J. Geier, M. King, 2005, Integration of GPS with Other Sensors and Network Assistance, Understanding GPS: Principles and Applications (Kaplan, E.D., Hegarty, C.J.), Artech House, ISBN 9781580538954

Bussios, N., Y. Tsolakichy, M. Tsakiri-Strati, O. Goergoula, 2004, Integrated High Resolution Satellite Image, GPS and Cartographic Data in Urban Studies, Municipality of Thessaloniki

Chunping, L., R. Shuangchen, G. Xiangdong, 2003, The programmable logic implementation of GPS/GLONASS clock synchronization ASIC'03, 5th International Conference, Vol. 2, pp. 732–735

Clarke, B., 1998, Aviator's Guide to GPS, McGraw-Hill, ISBN 9780070094932

Cosentino, R.J., D.W. Diggle, M.U. de Haag, C.J. Hegarty, D. Milbert, J. Nagle, 2005, Understanding GPS: Principles and Applications (Kaplan, E.D., Hegarty, C.J.), Artech House, ISBN 9781580538954

Cyran, K.A., D. Sokołowska, A. Zazula, B. Szady, O. Antemijczuk, 2011, Data gathering and 3D-visualization at OLEG multiconstellation station in EDCN system, Proceedings 21st International Conference on Systems Engineering, Las Vegas, USA, pp. 221–226

DoD, 2008, Global Posirioning System Standard Positioning Service Performance Standard, US Department of Defense, 4th Edition

Dorsey, A.J., W.A. Marquis, P.M. Fyfe, E.D. Kaplan, L.F. Wiederholt, 2005, GPS System Segments, Understanding GPS: Principles and Applications (Kaplan, E.D., Hegarty, C.J.), Artech House, ISBN 9781580538954

El-Rabbany, A.,2002, Introduction to GPS: The Global Positioning System, Artech House, ISBN 1-58053-183-0

ESA SP-1303 The European EGNOS Project, 2006, EGNOS The European Geostationary Navigation Overlay System: A Cornerstone of Galileo, ESA Publications Division, ISBN: 92-9092-453-5

Fan, R., I. Chakraborty, N. Lynch, 2005, Clock synchronization for wireless networks principles of distributed systems, Lecture Notes in Computer Science, Vol. 3544, pp. 400–414

Frei, E., A. Ryf, R. Scherrer, 1993, Use of the global positioning system in dam deformation and engineering surveys, SPN, Vol. 2

Gift, S.J.G., 2010, One-way light speed measurement using the synchronized clocks of the global positioning system (GPS), Physics Essays, Vol. 23(2)

Gleason, S., D. Gebre-Egziabher, 2009, GNSS Applications and Methods, Artech House, ISBN 9781596933309

Gleason, S., D. Gebre-Egziabher, 2010, Global Navigation Satellite Systems, McGraw-Hill Education (India) Pvt Limited, ISBN 9780070700291

Gottardi, E., F. Tua, B. Cargnelutti, M.L. Maublanc, J.M. Angibault, S. Said, H. Verheyden, 2010, Use of GPS activity sensors to measure active and inactive behaviours of European roe deer, Mammalia, Vol. 74(4), pp. 355–362

Grenier, A., E.S. Lohan, A. Ometov, J. Nurmi, 2023, A survey on low-power GNSS, *IEEE Communications Surveys & Tutorials*, Vol. 25(3), pp. 1482–1509, thirdquarter 2023, https://doi.org/10.1109/COMST.2023.3265841

Grewal, M.S., L.R. Weill, A.P. Andrews, 2001, Global positioning systems, inertial navigation, and integration, John Wiley and Sons, ISBN 978-0-47135-032-3

Groves, P.D., 2008, Principles of GNSS, inertial, and multi-sensor integrated navigation systems, Artech House, ISBN 9781580532556

Haibo, G., L.B. Li, S. Jia, L. Nie, T. Wu, Z. Yang, J. Shang, Y. Zheng, M. Ge, 2022, LEO Enhanced Global Navigation Satellite System (LeGNSS): Progress, opportunities, and challenges, Geo-Spatial Information Science, Vol. 25(1), pp. 1–13, https://doi.org/10.1080/10095020.2021.1978277

Hariharan, R., J. Krumm, E. Horvitz, 2005, Web-enhanced GPS, First International Conference on Location and Context Awareness LoCA'05, Berlin, pp. 95–104

Harper, N., 2009, Server-Side GPS and Assisted-GPS in Java, Artech House, Incorporated, ISBN 9781607839866

Harte, L., B. Levitan, 2007, GPS Quick Course: Technology, Systems and Operation, Althos, ISBN 9781932813708

Hasan, K.S., M. Rahman, A.L. Haque, M.A. Rahman, T. Rahman, M.M. Rasheed, 2009, Cost effective GPS-GPRS based object tracking system, Proceedings of the International MultiConference of Engineers and Computer Scientists, Vol. 1

Hashmi, N., D. Myung, M. Gaynor, S. Moulton, 2005, A sensor-based web service-enabled emergency medical response system, EESR '05, pp. 25–29

Hein, G.W., 2020, Status, perspectives and trends of satellite navigation, Satellite Navigation, Vol. 1, p. 22. https://doi.org/10.1186/s43020-020-00023-x

Heraud, J.A., A.F. Lange, 2009, Agricultural automatic vehicle guidance from horses to GPS, Agricultural Equipment Technology Conference, Louisville

Hernandez, J.I., C.Y. Kuo, 2003, Steering control of automated vehicles using absolute positioning GPS and magnetic markers, Vehicular Technology IEEE Transactions, Vol. 52(1), pp. 150–161

Hofmann-Wellenhof, B., H. Lichtenegger, E. Wasle, 2008, GNSS—Global Navigation Satellite Systems: GPS, GLONASS, Galileo, and More, Springer, ISBN 978-3-211-73012-6

Hofmann-Wellenhof, B., H. Lichtenegger, J. Collins, 2001, Global Positioning System: Theory and Practice, 5th Edition, Springer-Verlag, ISBN 978-3-211-83534-0

Hou, S.T., F. Chia, 2011, Making senses of technology: A triple contextual perspective of GPS use in the the taxi industry, Service Sciences (IJCSS), International Joint Conference on Service Sciences 2011, Taipei, Taiwan

HP Hewlett Packard Company, 1996, GPS and precision timing applications, HP Application Note, No 1272
Huang, J.Y., C.H. Tsai, S.T. Huang, 2012, The next Generation of GPs navigation systems, Communications of the ACM, Vol. 55(3), pp. 84–93
Inks, S.A., T.W. Loe, 2005, The ethical perceptions of salespeople and sales managers concerning the use of GPS tracking systems to monitor salesperson activity, Marketing Management Journal, Vol. 15(1), p. 108
Jeffrey, C., 2010, An Introduction to GNSS: GPS, GLONASS, Galileo and Other Global Navigation Satellite Systems, First Edition, NovAtel Inc., ISBN 978-0-9813754-0-3
Kalinowska, D., K. Hartmut, 2006, Motor vehicle use and travel behaviour in Germany: Determinants of car mileage, DIW-Diskussionspapiere, No. 602
Kaplan, E.D., J.L. Leva, D. Milbert, M.S. Pavloff, 2005, Fundamentals of Satellite Navigation, Understanding GPS: Principles and Applications (Kaplan, E.D., Hegarty, C.J.), Artech House, ISBN 9781580538954
Kim, R., T. Nagayama, H. Jo, B.F. Spencer, 2012, Preliminary study of low-cost GPS receivers for time synchronization of wireless sensors, Sensors and Smart Structures Technologies for Civil—Mechanical and Aerospace Systems, 83451A
Korreng, M.D., 2010, UTC time transfer for high frequency trading using IS-95 CDMA base station transmissions and IEEE-1588 precision time protocol, 42nd Annual Precise Time and Time Interval (PTTI) Meeting, EndRun Technologies, Reston, Virginia
Kumar, A., S. Kumar, P. Lal, P. Saikia, P.K. Srivastava, G.P. Petropoulos, 2023. Chapter 1: Introduction to GPS/GNSS technology (Editor(s): G.P. Petropoulos, P.K. Srivastava), GPS and GNSS Technology in Geosciences, Elsevier, 2021, pp. 3–20, ISBN 9780128186176, https://doi.org/10.1016/B978-0-12-818617-6.00001-9. (www.sciencedirect.com/science/article/pii/B9780128186176000019)
Landau, R., G.K. Auslander, S. Werner, N. Shoval, J. Heinik, 2011, Who should make the decision on the use of GPS for people with dementia?, Aging and Mental Health, Vol. 15(1), pp. 78–84
Lannucci, P.A., R.E. Humphreys, 2022, Fused low-earth-orbit GNSS, IEEE Transactions on Aerospace and Electronic Systems, https://doi.org/10.1109/TAES.2022.3180000
Lau, L., 2020, Chapter 4: GNSS multipath errors and mitigation techniques (Editor(s): G.P. Etropoulos, P.K. Srivastava), GPS and GNSS Technology in Geosciences, Elsevier, 2021, pp. 77–98, ISBN 9780128186176, https://doi.org/10.1016/B978-0-12-818617-6.00009-3. (www.sciencedirect.com/science/article/pii/B9780128186176000093)
Lauriault, T.P., J. Wood, 2009, GPS tracings: Personal cartographies, Art & Cartography Special Issue: The Cartographic Journal, Vol. 46(4), pp. 360–365
Lee, E.K., S. Yang, S.Y. Oh, M. Gerla, 2009, RF-GPS: RFID assisted localization in VANETs, Mobile Adhoc and Sensor Systems MASS '09, IEEE 6th International Conference, pp. 621–626
Letham, L., A. Letham, 2008, GPS Made Easy: Using Global Positioning Systems in the Outdoors, Rocky Mountain Books, ISBN 9781897522059
MacGougan, G.D., 2006, A Short Summary of Tropospheric Math Models By Glenn D. MacGougan, (October 11)
Manabu, O., H. Naohisa, F. Takehiko, S. Hiroshi, 2006, The application of RTK-GPS and steer-by-wire technology to the automatic driving of vehicles and an evaluation of driver behavior, IATSS Research, Vol. 30(2)
Matthews, V.M., E. Adetiba, 2011, Vehicle Accident Alert and Locator (VAAL), International Journal of Electrical & Computer Sciences IJECS-IJENS, Vol. 11(2), pp. 35–38
Miller, K.M., 2000, A review of GLONASS, The Hydrographic Journal, Vol. 98, pp. 15–22
Nagashima, K., K. Yamada, A. Tadano, 2013, Driverless antarctic tractor system, Hitachi Review, Vol. 62(3)
Odolinski, R., P.J.G. Teunissen, B. Zhang, 2020, Multi-GNSS processing, positioning and applications, Journal of Spatial Science, Vol. 65(1), pp. 3–5, https://doi.org/10.1080/14498596.2020.1687170
Ozguner, U., T. Acarman, K.A. Redmill, 2011, Autonomous Ground Vehicles, Artech House, ISBN 9781608071937
Pan, G., G. Qi, Z. Wu, D. Zhang, 2013, Land-use classification using taxi GPS traces intelligent transportation systems, IEEE Transactions, Vol. 14(1), pp. 113–123
Petrovski, I.G., T. Tsujii, 2012, Digital Satellite Navigation and Geophysics: A Practical Guide with GNSS Signal Simulator and Receiver Laboratory, Cambridge University Press, ISBN 9780521760546
Prakash, N.R., D. Kumar, K. Nandan, 2012, An autonomous vehicle for farming using GPS, International Journal of Electronics and Computer Science Engineering, Vol. 1(3), pp. 1695–1700
Prasad, R., M. Ruggieri, 2005, Applied Satellite Navigation Using GPS, GALILEO, and Augmentation Systems, Artech House, ISBN 9781580538145
Raza, S., A. Al-Kaisy, R. Teixeira, B. Meyer, 2022, The role of GNSS-RTN in transportation applications, Encyclopedia, Vol. 2(3), pp. 1237–1249. https://doi.org/10.3390/encyclopedia2030083

Rutter, S.M., N.A. Beresford, G. Roberts, 1997, Use of GPS to identify the grazing areas of hill sheep, Computers and Electronics in Agriculture, Vol. 17, pp. 177–188

Silva, J.M., R.G. Olsen, 2002a, Use of Global Positioning System (GPS) receivers under power line conductors, IEEE Power Engineering Review, Vol. 22(7), pp. 62–62

Silva, J.M., R.G. Olsen, 2002b, Use of Global Positioning System (GPS) receivers under power-line conductors, Power Delivery IEEE Transactions, Vol. 17(4), pp. 938–944

Skov-Petersen, H., R. Rupf, D. Köchli, 2012, Revealing recreational behaviour and preferences from GPS recordings, 6th International Conference on Monitoring and Management of Visitors in Recreational and Protected Areas (MMV), Stockholm, Sweden

Sterzbach, B., 1997, GPS-based Clock Synchronization in a Mobile, Real-Time Systems, Vol. 12(1), pp. 63–75

Sukkarieh, S., E.M. Nebot, H.F. Durrant-Whyte, 1998, Achieving integrity in an INS/GPS navigation loop for autonomous land vehicle applications, robotics and automation, IEEE International Conference, Vol. 4

Sukkarieh, S., E.M. Nebot, H.F. Durrant-Whyte, 1999, A high integrity IMU/GPS navigation loop for autonomous land vehicle applications, robotics and automation, IEEE Transactions, Vol. 15(3)

Thierry, B., B. Chaix, Y. Kestens, 2013, Detecting activity locations from raw GPS data: A novel kernel-based algorithm, International Journal of Health Geographics, Vol. 12, pp. 1–10

Thompson, C., J. White, B. Dougherty, A. Albright, C.D. Schmidt, 2010, Using smartphones to detect car accidents and provide situational awareness to emergency responders, Lecture Notes of the Institute for Computer Sciences, Social Informatics and Telecommunications Engineering, Vol. 48, pp. 29–42

Tian, J., J. China, H. Zhao, 2010, Pipeline damage locating based on GPS Fiducial Clock Intelligent Control and Automation (WCICA), 8th World Congress, pp. 4161–4164

Towns, D.M., L.M. Cobb, 2012, Notes on: GPS technology: Employee monitoring enters a new era, Labor Law Journal, Vol. 63(3), p. 203

Van Diggelen, F., 2009, A-GPS: Assisted GPS, GNSS, and SBAS, Artech House, ISBN-13 978-1-59693-374-3

Van Westen, C., 2000, Remote Sensing for Natural Disaster Management, International Archives of Photogrammetry and Remote Sensing. Vol. XXXIII, Part B7, Amsterdam

Wan, N., G. Lin, 2013, Life-space characterization from cellular telephone collected GPS data Computers, Environment and Urban Systems, Vol. 39, pp. 63–70

Ward, P.W., J.W. Betz, C.J. Hegarty, 2005a, GPS Satellite Signal Characteristics, Understanding GPS: Principles and Applications (Kaplan, E.D., Hegarty, C.J.), Artech House, ISBN 9781580538954

Ward, P.W., J.W. Betz, C.J. Hegarty, 2005b, Interference, Multipath and Scintillation, Understanding GPS: Principles and Applications (Kaplan, E.D., Hegarty, C.J.), Artech House, ISBN 9781580538954

Wei, W., S. Baosheng, Z. Kefa, Z. Xia, 2010, The use of GPS for the measurement of details in different areas, Electronic Commerce and Security (ISECS), Third International Symposium, pp. 59–62

White, J., C. Thompson, H. Turner, B. Dougherty, D.C. Schmidt, 2011, Automatic traffic accident detection and notification with smartphones, Mobile Networks and Applications, Vol. 16(3), pp. 285–303

Xu, G., 2007, GPS: Theory, Algorithms and Applications, Second Edition, Springer, ISBN 978-3-540-72714-9

Yang, M., C.F. Lo, 2000, Real-time kinematic GPS positioning for centimeter level ocean surface monitoring, Proceedings of the National Science Council ROC(A), Vol. 24(1), pp. 79–85

Zagami, J.M., S.A. Pari, J.J. Bussgang, K.D. Melillo, 1998, Providing universal location services using a wireless E911 location network, IEEE Communications Magazine, Vol. 36(4), pp. 66–71

Zheng, Y., L. Zhang, X. Xie, W.Y. Ma, 2009, Mining interesting locations and travel sequences from GPS trajectories, 18th International Conference on World Wide Web, Spain, pp. 791–800

Zheng, Y.T., Z.J. Zha, T.S. Chua, 2012, Mining travel patterns from geotagged photos, ACM transactions on intelligent systems and technology TIST, Vol. 3(3)

Zhong, P., et al., 2008, Adaptive wavelet transform based on cross-validation method and its application to GPS multipath mitigation, GPS Solution, Vol. 12, pp. 109–117

Ziedan, N.I., 2006, GNSS receivers for weak signals, Artech House, Incorporated, ISBN 9781596930520

Zinkiewicz, D., B. Buszke, M. Houdek, F. Toran-Marti, 2010, SISNeT as a source of EGNOS information: Overview of functionalities and applications, Satellite Navigation Technologies and European Workshop on GNSS Signals and Signal Processing (NAVITEC), 5th ESA Workshop on, Nordwijk

# Part III

## Fundamentals of Remote Sensing

### Evolution, State of the Art, and Future Possibilities

# 5 Fundamentals of Remote Sensing for Terrestrial Applications

## *Evolution, Current State of the Art, and Future Possibilities*

*Natascha Oppelt and Arnab Muhuri*

## ACRONYMS DEFINED

| | |
|---|---|
| **ADEOS** | Advanced Earth Observing Satellite |
| **AI** | Artificial Intelligence |
| **AIS** | Automated Identification System |
| **ALOS** | Advanced Land Observing Satellite |
| **API** | Application Programming Interface |
| **ARD** | Analysis Ready Data |
| **ASAR** | Advanced SAR |
| **ASTER** | Advanced Spaceborne Thermal Emission and reflection Radiometer |
| **ATLAS** | Advanced Topographic Laser Altimeter |
| **AVHRR** | Advanced Very High Resolution Radiometer |
| **AWS** | Amazon Web Services |
| **CBERS** | China/Brazil Earth Resources Satellite |
| **CCI** | Climate Change Initiative |
| **CEOS** | Committee on Earth Observation Satellites |
| **CHIME** | Copernicus Hyperspectral Imaging Mission |
| **CIR** | Color Infrared |
| **CNES** | Centre National d'Etudes Spatiales |
| **CNN** | Convolutional Neural Network |
| **DEM** | Digital Elevation Model |
| **D-InSAR** | Differential Interferometry Synthetic Aperture Radar |
| **DIAS** | Data and Information Access Services |
| **DLR** | German Aerospace Center |
| **DSM** | Digital Surface Model |
| **EnMAP** | Environmental Mapping and Analysis Program, Germany |
| **EO** | Earth Observation |
| **ERS** | European Remote Sensing |
| **ERTS** | Earth Resources Technology Satellite |
| **ESA** | European Space Agency |
| **EU** | European Union |
| **EUMETSAT** | European Organization for the Exploitation of Meteorological Satellites |
| **fAPAR** | Fraction of Absorbed Photosynthetically Active Radiation |

DOI: 10.1201/9781003541141-8

| | |
|---|---|
| **G8** | Group of Eight |
| **GeoAI** | Geospatial Artificial Intelligence |
| **GCOS** | Global Climate Observation System |
| **CGLS** | Copernicus Global Land Service |
| **GEDI** | Ecosystem Dynamics Investigation |
| **GEE** | Google Earth Engine |
| **GEO** | Group on Earth Observation |
| **GEO-CAPE** | Geostationary Coastal and Air Pollution Events |
| **GEOSS** | Global Earth Observation System of Systems |
| **GFW** | Global Forest Watch |
| **GIS** | Geographic Information System |
| **GMES** | Global Monitoring for Environment and Security |
| **HLS** | Harmonized Landsat Sentinel products |
| **HyspIRI** | Hyperspectral Infrared Imager |
| **InSAR** | Airborne Interferometric Synthetic Aperture Radar |
| **IRS** | Indian Remote Sensing |
| **ISS** | International Space Station |
| **ISRO** | Indian Space Research Organization |
| **JEM-EF** | Japanese Experiment Module-Exposed Facility |
| **JPL** | Jet Propulsion Laboratory |
| **LAI** | Leaf Area Index |
| **LiDAR** | Light Detection And Ranging |
| **LP DAAC** | Land Processes Distribute Active Archive Center |
| **LULC** | Land Use and Land Cover |
| **MERIS** | Medium Resolution Imaging Spectrometer |
| **MIR** | Mid InfraRed |
| **MODIS** | Moderate Resolution Imaging Spectrometer |
| **MPC** | Microsoft Planetary Computer |
| **MSI** | Multi-Spectral Imager |
| **MSS** | Multispectral Scanner System |
| **NASA** | National Aeronautics and Space Administration |
| **NDVI** | Normalized Difference Vegetation Index |
| **NISAR** | NASA-ISRO SAR |
| **NIR** | Near InfraRed |
| **NMS** | Non-Maximum Suppression |
| **NN** | Neural Network |
| **NOAA** | National Oceanic and Atmospheric Administration |
| **OBIA** | Object-Based Image Analysis |
| **OLCI** | Ocean and Land Colour Instrument |
| **OLI** | Operational Land Imager |
| **PALSAR** | Phased Array L-band Synthetic Aperture Radar |
| **PCA** | Principal Component Analysis |
| **PRISMA** | Hyperspectral Precursor and Application Mission, Italy |
| **RCM** | RADARSAT Constellation Mission |
| **R-CNN** | Region-based Convolutional Neural Network |
| **RF** | Random Forest |
| **RNN** | Recurrent Neural Networks |
| **RoI** | Region of Interest |
| **RPN** | Region Proposal Network |
| **SAOCOM** | Satellites for Observation and Communications, Argentina |
| **SAR** | Synthetic Aperture Radar |

| | |
|---|---|
| **SLA** | Shuttle Laser Altimeter |
| **SMAP** | Soil Moisture Active Passive |
| **SOM** | Self-Organizing Map |
| **SPOT** | Satellite Pour l'Observation de la Terre |
| **SVM** | Support Vector Machine |
| **SRTM** | Shuttle Radar Topography Mission |
| **SWIR** | Short-Wave InfraRed |
| **TIR** | Thermal or Far InfraRed |
| **TIROS** | Television Infrared Observation Satellite |
| **UAV** | Unmanned Aerial Vehicle |
| **UN** | United Nations |
| **UNFCCC** | United Nations Framework Convention on Climate Change |
| **US** | United States |
| **USDA** | United States Department of Agriculture |
| **USDOI** | United States Department of Interior |
| **USGS** | United States Geological Survey |
| **VHRR** | Very High Resolution Radiometer |
| **VIS** | Visible Light |
| **YOLO** | You Only Look Once |

## 5.1 INTRODUCTION

Remote sensing is the acquisition of information about objects from a distance without establishing a physical contact. In general, information is acquired by detecting and measuring the spectrum of the electromagnetic field emitted or reflected by the objects. However, it may also be an acoustic field or gravity/magnetic potential (Campbell & Wynne, 2011).

Today, remote sensing is applied in various fields such as astronomy and environmental science; the latter includes atmospheric, oceanic, and terrestrial applications, which is the subject of our attention. In general, terrestrial remote sensing is related to gravity, magnetic fields and electromagnetic radiation; the techniques of the latter cover the electromagnetic spectrum from the Visible Light (VIS) to the microwave region. Table 5.1 indicates the wavelength ranges of different spectral regions of the electromagnetic (EM) spectrum, which are commonly employed for Earth Observation (EO). The selection of certain ranges of the EM spectrum depends on the availability of atmospheric windows, atmospheric condition, and specific requirements for the imaging wavelength (Campbell, 2012).

Passive remote sensing in all of these regions includes sensors measuring radiation naturally reflected or emitted from the Earth's surface, atmosphere, or clouds. The VIS, Near Infrared (NIR), and Mid-Infrared (MIR) regions correspond to the reflective spectral range because the radiation is essentially solar radiation reflected from the Earth. The MIR region is also referred to as Short-Wave

**TABLE 5.1**
**Primary Spectral Regions Used in Terrestrial Remote Sensing**

| Name | Wavelength Range |
|---|---|
| Visible Light (VIS) | 0.38–0.70 μm |
| Near Infrared (NIR) | 0.70–1.30 μm |
| Mid Infrared (MIR, SWIR) | 1.30–3 μm |
| Thermal or Far Infrared (TIR) | 3–14 μm |
| Microwave | 1 mm–1 m |

**TABLE 5.2**
**Subdivisions of Active Microwave Wavelengths**

| Band Name | Wavelength Range | Band Name | Wavelength Range |
|---|---|---|---|
| Ka | 0.75–1.18 cm | C | 3.75–7.50 cm |
| Ku | 1.18–1.67 cm | S | 7.50–15.0 cm |
| K | 1.67–2.40 cm | L | 15.0–30.0 cm |
| X | 2.40–3.75 cm | P | 77–107 cm |

Infrared (SWIR). The Thermal or Far-Infrared (TIR) consists of wavelengths beyond the MIR and extends into regions that border the microwave region and cover radiation that is emitted by the Earth. Toward this end of the electromagnetic spectrum, active remote sensing, which employs an active source of radiation, is used to analyze the radar backscattering properties of the ground targets. Active imaging radars operate over a range of wavelengths indicated in Table 5.1; Table 5.2 lists the radar bands of the active microwave regions (Campbell & Wynne, 2011).

The forthcoming section discusses the history of terrestrial applications and image processing from the advent of the term *remote sensing* in the 1960s. The section "State of the Art" provides a brief overview of the vast domain of terrestrial remote sensing with focus on the present and future planned EO missions and standards in data processing, computing platforms, and policies. The readers are encouraged to refer to the other relevant chapters of the *Remote Sensing Handbook* Volume I (this volume), Volume II, and Volume III. Section 5.4 outlines the emerging technologies and challenges and provides a brief description of the ongoing discussions on data accessibility and sharing. The main conclusions are outlined in section 5.5.

## 5.2 EVOLUTION OF TERRESTRIAL APPLICATIONS

### 5.2.1 From a Qualitative Description and Visual Interpretation of the Earth Surface to Digital Data Processing

Aerial photography was the most preliminary form of remote sensing; throughout its history, proponents of aerial photography looked for advances in the technology to remotely acquire information about the Earth's resources. During the 1950s and 1960s, most analyses were visual interpretations of prints or transparencies of aerial images (Fischer et al., 1976). Back then, black and white (panchromatic, 400–700 nm) aerial photographs have been common and different patterns of the land surface appeared as a series of gray values ranging from black to white. These grayscale prints and transparencies of aerial images were interpreted manually. The development of Color InfraRed (CIR) photographs (covering green, red, and NIR wavelengths), sometimes referred to as false color, can be attributed to one of the most significant developments. Robert Colwell (1956) was one of the first who successfully used CIR image to address crop diseases. Figure 5.1 illustrates that, besides analysis of shapes, sizes, pattern, texture, shadow or spatial relationships, CIR images improved the interpretation of vegetation health since it covered the NIR region, which is sensitive to vegetation biomass and health. In CIR imagery healthy, dense vegetation appears in bright red. Lighter tones of red generally represent vegetation with a less distinct NIR plateau such as mature stands of evergreens. Agricultural fields approaching the end of growing season and dead/unhealthy plants often appear in shades of less intense red. Soils appear in white, blue-green, or tan, while water bodies are dark blue to black (USGS, 2001). Image interpretation, however, was limited to visual analysis and therefore remained qualitative and subjective.

In 1957, the former Soviet Union deployed the first artificial satellite into space: Sputnik, carrying four radio antennas to broadcast radio pulses, triggered the space race and therefore was

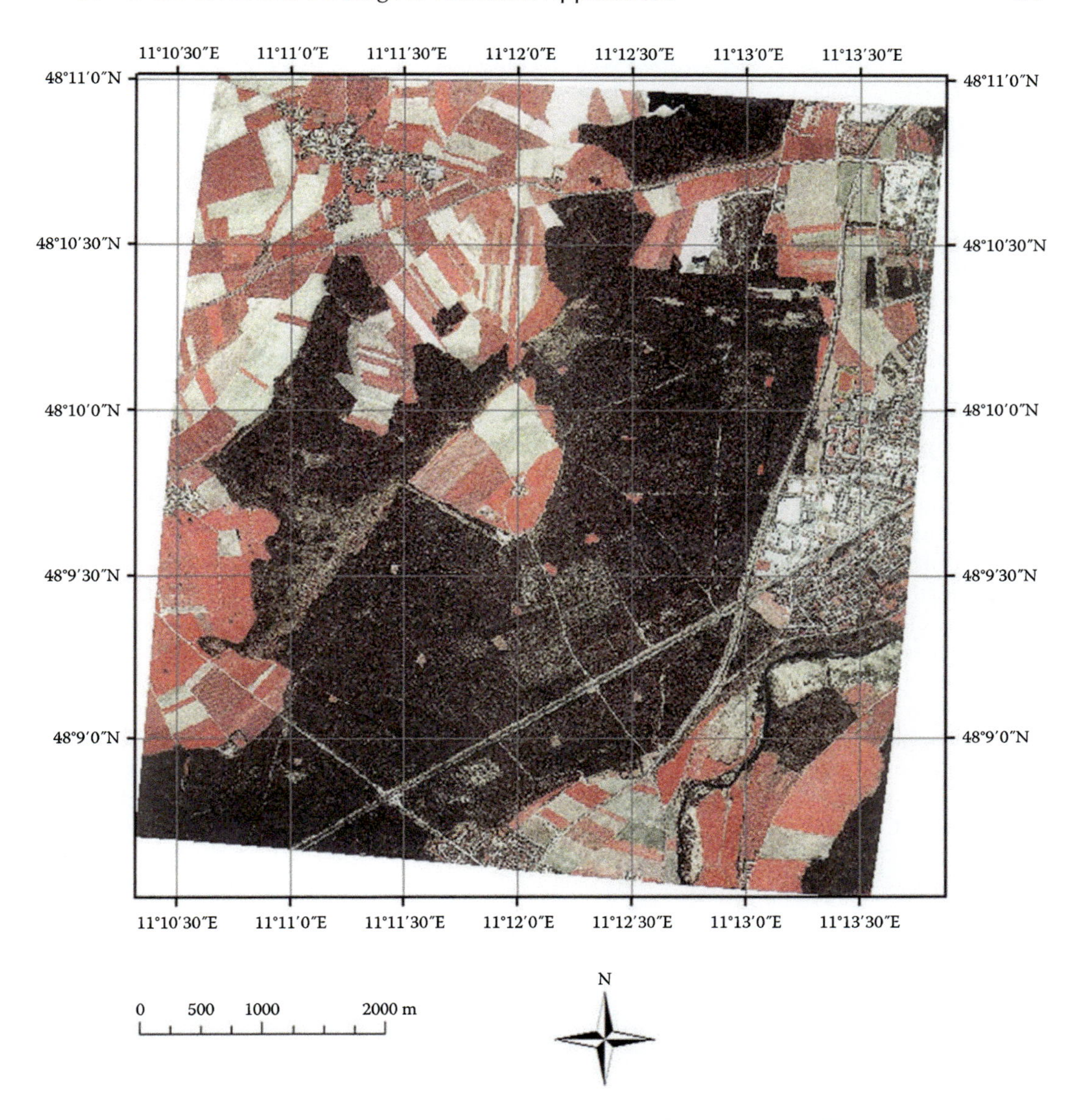


**FIGURE 5.1** Color infrared (CIR) aerial image acquired on April 14, 1983. In the East, the image covers the small town Fuerstenfeldbruck (Bavaria, Germany), a mixed forest stand crosses the image in SW–NE direction, surrounded by agricultural fields (bare soil, winter cereals, and pasture). (Geospatial data © Bavarian Surveying and Mapping Authority, 2014.)

the starting signal for the space age (Siddiqi, 2003). This was followed by the launch of NASA's (National Aeronautics and Space Administration) Television Infrared Observation Satellite (TIROS-1) in 1960. TIROS-1 provided the first systematic images of the Earth from space (Allison et al., 1962). A single television camera pointed at the Earth's surface for a limited time during each orbit and mainly collected images of North America. TIROS-1 was the first of a series of experimental weather satellites, which through TIROS-X contained television cameras and four of them also included IR sensors (Hastings & Emery, 1992).

By the late 1960s, scientists in the U.S. Departments of Agriculture (USDA) and Interior (USDOI) had proposed a space mission dedicated to acquiring synoptic, multispectral images of the Earth's surface. These datasets were aimed toward a wide range of applications such as agriculture, forestry, mineral exploration, land resource evaluation, land use planning, water resources, mapping and charting, and environmental protection. The mission was launched in 1972 as the Earth Resources Technology Satellite (ERTS), and later renamed Landsat. The launch of the Multispectral Scanner System (MSS) onboard the ERTS/Landsat was the beginning of spaceborne terrestrial remote sensing. For the first time an Earth-orbiting system provided systematic, repetitive observations of the Earth's land areas. Landsat MSS depicted large areas of the Earth's surface and, although only with four spectral bands and a pixel size of 80 m, provided routinely available data (Lillesand et al., 2008).

Also in the 1960s, the National Oceanic and Atmospheric Administration (NOAA) launched the Very High Resolution Radiometer (VHRR) onboard the NOAA-4 satellite. NOAA also introduced the direct reception of digital VHRR data free of charge to ground stations already operational in 1972. Being available in digital form, VHRR and Landsat MSS marked the beginning of the era of digital image analysis (Campbell & Wynne, 2011), which by then was limited to specialized research institutions. The technical standard of personal computers and limited availability of image analysis software limited analysis of data, which is now regarded as commonplace (Jones & Vaughan, 2010).

The availability of digital data, however, was a seeding point for the development of image analysis software and computer-based processing techniques. In 1978, with further development of VHRR, the Advanced Very High Resolution Radiometer (AVHRR) was onboard the seventh generation of TIROS-NOAA satellites. Although initially designed for observing global weather, subsequent sensors were specifically designed to measure other phenomena including terrestrial vegetation (Hastings & Emery, 1992). Therefore, the AVHRR-derived regional and global indices proved to be a very robust and useful observation (Gutman, 1991). In the 1970s, the next two Landsat satellites were launched (see Figure 5.2). During this time period, other nations decided to operate their own sensors, which led to a rapid increase of data availability. In the 1980s, SPOT-1 (Satellite Pour l'Observation de la Terre), the first of a series of multispectral satellites, was launched by the French Centre National d'Etudes Spatiales (CNES) in cooperation with Belgium and Sweden

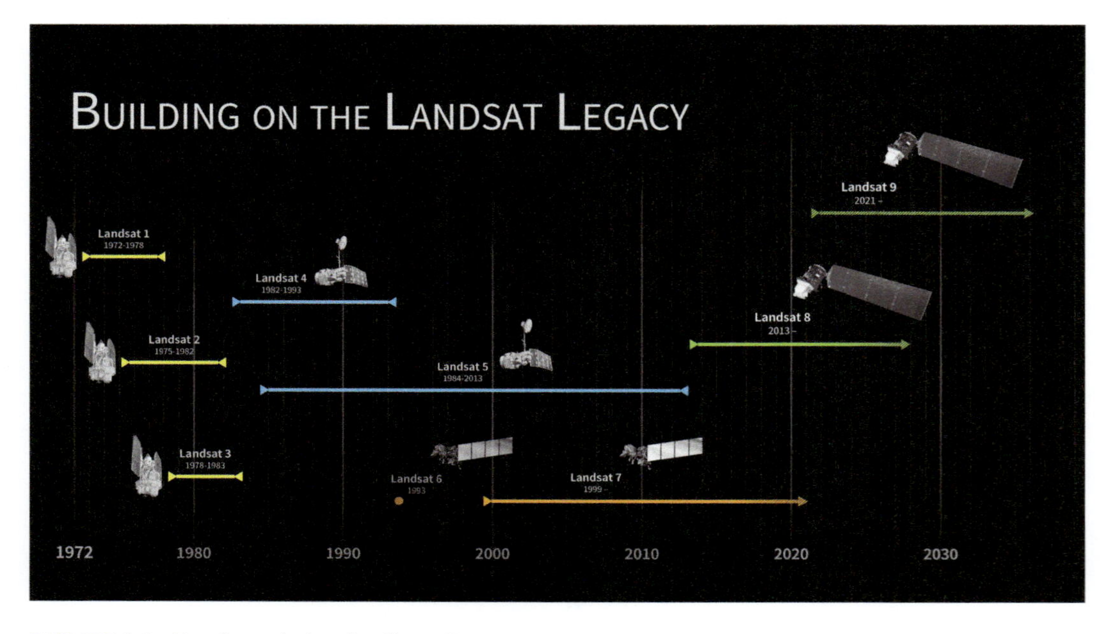


FIGURE 5.2 Landsat mission timeline. (Courtesy of NASA's Scientific Visualization Studio.)

(Chevrel et al., 1981), and India launched the first sensor of the Indian Remote Sensing (IRS) satellite program (Misty, 1998).

In the 1990s, other nations deployed their own EO satellites such as China and Brazil with their CBERS (China/Brazil Earth Resources Satellite; De Oliveira et al., 2000) or Japan with the launch of ADEOS (Advanced Earth Observing Satellite; Shimoda, 1999). This development pushed forward further progress in data processing. Dedicated software and the rapid advances in computer technology enabled users to employ essential image processing techniques (Lillesand et al., 2008).

The launch of the first civilian Synthetic Aperture Radar (SAR) in space (SeaSat in 1978) started a similar rapid development of applications as with optical imagery. Since then, a number of satellites have been deployed that carry microwave radiometers and imaging radars (Elachi, 1988; Kuschel & O'Hagan, 2010). In spite of data complexity, the parallel advances in computer techniques have popularized the applications of radar data for terrestrial applications (Jones & Vaughan, 2010). Some prominent examples are the European Remote Sensing (ERS) satellites, ERS-1 and ERS-2, which were launched in the 1990s and acquired tandem C-band SAR data for a diverse range of applications for nearly two decades (see also Section 5.3.2).

These systems offered the very first opportunity for global coverage every 3 to 22 days. With Landsat and ERS sensors in orbit, the temporal repetivity was improved. Nevertheless, true global coverage was limited by a host of issues such as cloud cover, data downlink limitations, and absence of sufficient ground stations to receive data (Goward et al., 2001). In the 1990s, a series of EO satellites were launched with a wide range of spatial, spectral, and temporal coverage capabilities, which allowed for further expansion of global coverage. In 1999, Terra, a joint spaceborne mission between US, Japan, and Canada, was launched (Running et al., 1998). Terra was the first satellite that was specifically designed to provide global coverage for monitoring of the Earth's ecosystems. Terra carried several instruments such as the Advanced Spaceborne Thermal Emission and reflection Radiometer (ASTER) and the Moderate resolution Imaging Spectrometer (MODIS; Xiong et al., 2011). For global applications, MODIS was a substantial improvement over the AVHRR sensor in spatial resolution and the diversity of spectral bands, with onboard calibration and enhanced radiometric accuracy (Running et al., 1994). In 2002, the European Space Agency (ESA) launched ENVISAT, which was one of the largest civilian EO missions. With the Advanced SAR (ASAR), ENVISAT ensured the data continuity of ERS. New instruments supplemented the mission such as atmospheric sensors to monitor trace gases and the Medium Resolution Imaging Spectrometer (MERIS). These systems were designed to support broad-scale Earth science research and allowed the monitoring of spatial patterns of environmental and climate changes (Louet & Bruzzi, 1999).

In contrast to the common approach of distributing radiance or reflectance images, the broad-scale sensors also opened a new era targeted towards the generation of standardized data products. The goal was the generation of consistent, satellite-based records of physical environmental parameters (e.g., chlorophyll, Leaf Area Index (LAI) or fraction of Absorbed Photosynthetically Active Radiation (fAPAR)). These products were successfully integrated into Earth system models to improve the understanding of connections between drivers and response variables (Sellers et al., 1997; Pielke, 2005).

During the early part of the 21st century, the internet began to facilitate public access to remote sensing imagery (Goodchild, 2007; Gould et al., 2008; see also *Remote Sensing Handbook*, Chapters 26, 27, and 28 of this volume). Over a period of time, image-based products, radiance or reflectance data were available online and free of cost such as Landsat, MODIS, and MERIS (Bontemps et al., 2011; Woodcock et al., 2008).

### 5.2.2 Development of Indices and Quantitative Assessment of Environmental Parameters

Computer-based processing of digital data directly led to the development of the quantitative assessment of environmental parameters. The development of spectral indices played a key role during this era, which aimed towards retrieval of information about the land surface by using specific spectral responses. Rouse (1974) observed that vegetation shows a typical reflectance behavior in the VIS and NIR and introduced the Normalized Difference Vegetation Index (NDVI) for MSS data in 1973:

$$\mathrm{NDVI} = \frac{(\mathrm{NIR} - \mathrm{RED})}{(\mathrm{NIR} + \mathrm{RED})} \tag{3.1}$$

where
NIR reflectance in the 0.8–1.1 μm region
RED reflectance in the 0.6–0.7 μm region

Foremost, the NDVI was used to analyze terrestrial green vegetation status and spatio-temporal dynamics of herbaceous vegetation and forests (Sellers, 1985). The normalization proved to be an important advantage for the success of the NDVI as a descriptor of vegetation variations in spite of atmospheric effects (Holben et al., 1990). Accordingly, it was used in numerous regional and global applications for studying vegetation with various sensors (e.g., Hame et al., 1997; Tucker & Sellers, 1986). Although the NDVI suffers from certain deficiencies (e.g., soil background influence) it is probably still the most well-known and common spectral index. Nowadays, there are numerous alternative forms of indices that correct for specific deficiencies of the NDVI or are adapted to specific sensors such as MODIS (Huete et al., 2002). For hyperspectral instruments a broad variety of narrow band indices has been developed to assess parameters such as LAI, fractional cover, fAPAR, or to quantify biophysical and biochemical plant parameters such as chlorophyll, accessory pigments or plant water content (Thenkabail et al., 2012). Hyperspectral indices are also commonly used for mineral identification (Sabins, 1999; van der Meer et al., 2012), monitoring of inland water constituents (Odermatt et al., 2012), and assessment of soil conditions (Pettorelli, 2013).

## 5.3 CURRENT STATE OF THE ART

At present, a large diversity of datasets is available ranging from very high spatial resolution optical data such as WorldView, GeoEye, and Quickbird (commercial sensors operated by Digital Globe) to Planet, Aster, and SPOT toward medium resolution Landsat and Sentinel-2 and coarse scale MODIS and Sentinel-3 data. Furthermore, a broad variety of satellite-borne SAR data from sensors such as ALOS, RADARSAT, TerraSAR-X, and Sentinel-1 is also available. According to their spectral and spatial characteristics, they support various terrestrial applications (Figure 5.3).

EO allows monitoring of patterns, structures, morphology, and the relationships of the built environment. In the past decades, satellite-derived data products incorporating Earth science products expanded into the mainstream environmental, meteorological and other user communities. This includes the movement of essentially science-produced research products derived from EO missions into operational observations (Brown et al., 2013). This resulted in a vast range of applications where such capabilities played a significant role. This progressed land cover monitoring applications such as mapping, change detection, modeling, and other observations of phenomena at the Earth's surface (see other contributions in *Remote Sensing Handbook*). A majority of applications is integrated in a Geographic Information System (GIS) environment (see Chapter 8 in this volume), mostly developed to provide thematic maps which may then be used by specialized groups of scientists, stakeholders or policy makers (Barrett & Curtis, 2013). For a detailed description of data quality and uncertainties, we refer to Chapter 28 in this volume). Moreover, the *Remote Sensing Handbook* covers a detailed view of a broad range of land (Vol. 2) and water (Vol. 3) applications.

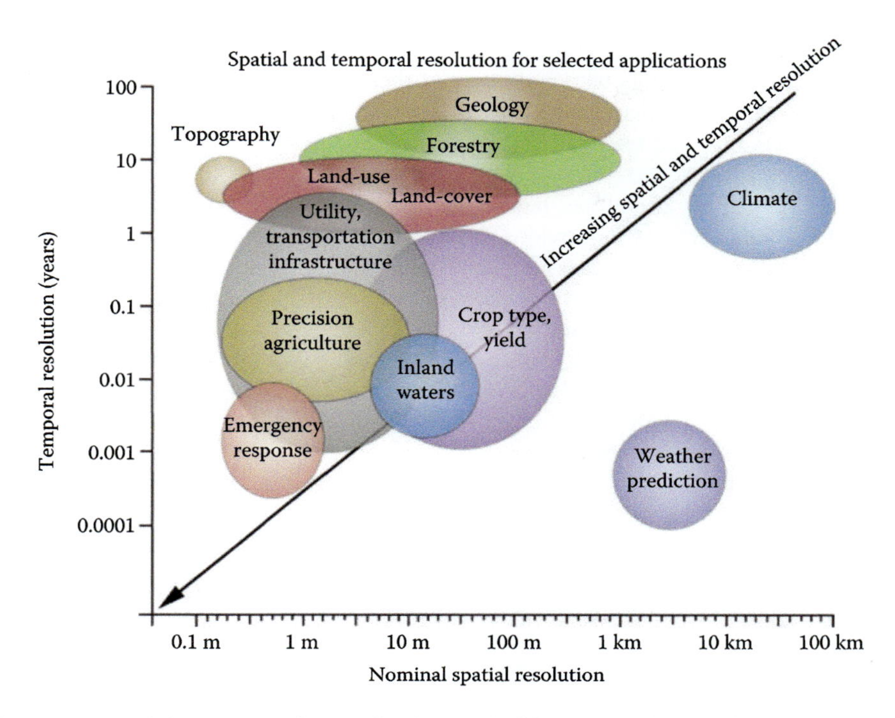


**FIGURE 5.3** Terrestrial remote sensing applications. (Modified from a figure originally published by the American Society for Photogrammetry and Remote Sensing, Bethesda, Maryland, www.asprs.org (Davis et al., 1991).)

### 5.3.1 Optical Remote Sensing

Recent developments in spaceborne remote sensing have introduced a multitude of new EO sensors and technologies. In the United States, the National Space Policy of 2020 highlighted the overarching goal of enhancing integrated predictive science, encompassing research, monitoring, assessments, and modeling of both natural and human-induced changes to Earth's land and inland water bodies. It also emphasized the establishment of a national global land surface data archive and its wide-scale distribution, emphasizing the significance of long-term observations (USA, 2020). In line with this, NASA launched Landsat 9 in 2021, providing invaluable continuity to the Landsat mission.

The U.S. also has future missions on the horizon that are suitable for land applications. One such mission is the Hyperspectral Infrared Imager (HyspIRI), currently in the study phase, which holds promise with its medium-scale hyperspectral capabilities. This instrument comprises both an imaging spectrometer covering the VIS to SWIR and a thermal imager, offering various data possibilities, particularly for carbon cycle analyses (JPL, 2018).

In Europe, ESA embarked on a new EO Strategy in 2015, marking a shift from the previous Living Planet Programme developed nearly two decades prior (Liebig et al., 2007). This updated strategy aligns with societal challenges, including food security, water resource management, energy, climate change, and civil security. It also embraces innovative computational approaches, such as cloud-based access and processing of EO data (Aschbacher, 2019). ESA introduced the Copernicus program, formerly known as Global Monitoring for Environment and Security (GMES), to support high-resolution environmental monitoring. Copernicus serves as Europe's flagship EO program, focusing on environmental and security-relevant information on a global scale. It comprises the Sentinel family of satellites (Sentinel-1 to 6) and numerous contributing missions from ESA, member states, European Organization for the Exploitation of Meteorological Satellites (EUMETSAT), and third-party mission operators who make part of their data available for Copernicus (Berger & Aschbacher, 2012). The combination of Landsat and Copernicus has ushered in a paradigm shift in EO, delivering a robust and reliable long-term operational infrastructure that fuels research and innovation.

Specifically, the Sentinel-2 satellites, launched in 2015 and 2016, offer medium spatial resolution optical imagery of land surfaces worldwide. These satellites ensure continuity in SPOT-type observations, with enhancements in spectral and temporal resolution. The Multi-Spectral Imager (MSI) instrument aboard Sentinel-2 covers the VIS, NIR and SWIR in 13 spectral bands with a spatial resolution between 10 m and 60 m, and with an additional three bands dedicated to atmospheric corrections and cloud screening. With a swath width of about 290 km and two satellites operating in tandem, Sentinel-2 achieves revisit times of five days at the equator and two to three days at mid-latitudes, facilitating monitoring of rapid changes, such as seasonal vegetation variations and improved change detection techniques.

Sentinel-3, launched in 2016, primarily supports marine services. It features the Ocean and Land Colour Instrument (OLCI), an instrument derived from the heritage of the MERIS instrument but with a broader range of wavelength bands and a spatial resolution of 300 m × 300 m in full resolution mode. OLCI also serves terrestrial vegetation monitoring, building on previous MERIS products (Bourg et al., 2023). Complementing spaceborne vegetation monitoring is ESA's forthcoming Earth Explorer mission, FLEX (Fluorescence Explorer), equipped with a high-resolution Fluorescence Imaging Spectrometer (FLORIS). FLORIS captures data in the 500–780 nm spectral range, offering global estimates of vegetation fluorescence, actual photosynthetic activity, and vegetation stress (Coppo et al., 2017).

National space missions contribute significantly to international space programs, especially for hyperspectral land monitoring. For instance, the Italian Hyperspectral Precursor and Application Mission (PRISMA), launched in 2019, offers an impressive 139 spectral bands covering a wide wavelength range from 400 to 2500 nm, opening new horizons for spaceborne imaging spectroscopy (Loizzo et al., 2018). Similarly, the German Environmental Mapping and Analysis Program (EnMAP), launched in 2022, features 244 spectral bands ranging from VIS to SWIR, setting milestones for spaceborne imaging spectroscopy with Landsat-like spatial resolution (Chabrillat et al., 2022). These sensors collect reflected radiation across the electromagnetic spectrum in narrow wavebands, enabling in-depth analysis of the Earth's surface properties (see Chapter 29 of this volume). These hyperspectral missions hold tremendous potential for applications related to ecosystem responses to environmental and anthropogenic pressures, and they lay the groundwork for multidisciplinary studies on climate change, Land Use and Land Cover (LULC) transformations, mineral exploration, hazards, and environmental pollution (Chabrillat et al., 2022; Shaik et al., 2023; Gautam et al., 2023). Due to data storage limitations, the along-track dimension is limited to 1800 km (PRISMA) and 5000 km (EnMAP) per day. Both sensors, therefore, offer on-demand technology, and priority acquisition of data for a specific date may be difficult due to a high number of requests. Therefore, an analysis may be hampered for a certain region due to the limited availability of data in the archive. Nevertheless, these hyperspectral missions enable numerous applications to decipher the response of ecosystems to natural and man-made pressure and push forward multidisciplinary applications on climate change, LULC changes, hazards, and environmental pollution (Chabrillat et al., 2022; Shaik et al., 2023; Gautam et al., 2023). Moreover, these missions are predecessors of the Copernicus Hyperspectral Imaging Mission (CHIME) starting in 2028, which will complement Sentinel-2 applications (Celesti et al., 2022).

### 5.3.2 SAR Imaging from 2D to 3D and Beyond

With the launch of Sentinel-1A in 2014, new opportunities arose with open-source data in the microwave domain. The mission initially offered repeated global acquisitions with a repeat-cycle of

12 days, which was reduced to six days with the launch of Sentinel-1B before its malfunction due to an anomaly that occurred in December 2021. The free and open data policy fueled the expansion of SAR techniques. To compensate for this loss, ESA plans the launch of Sentinel-1C in 2024.

Further SAR data are provided by the RADARSAT Constellation mission (RCM; C-band, Canada, launch in 2017) as well as the L-band missions ALOS-2 (Advanced Land Observing Satellite, Japan, launched in 2014) and SAOCOM (Satellites for Observation and Communications; Argentina, launched in 2018). Another example for an L-band SAR is the U.S. Soil Moisture mission SMAP (Soil Moisture Active Passive instrument, operated since 2015), which is being employed for measuring soil moisture for water and energy cycling analysis (NASA EOSPSO, 2019). The L-band has lower resolution than C- and X-band images but is more coherent over time, especially in vegetated regions with its better signal penetration ability (Aoki et al., 2021). Unfortunately, the active radar instrument failed 208 days after the launch of the SMAP mission leaving only its passive radar capability intact. This reduced the capability of the mission at higher resolution with active radar since the initial plan was to combine both active and passive measurements to get a better idea about the soil moisture in the top layer. However, Sentinel-1 C-band backscatter has been assisting the SMAP mission to compensate for its loss of the active radar.

Fostered by the availability of regular SAR observations from space in the 2000s, interferometric techniques were developed for several terrestrial applications, which until then had not been supported by remote sensing data (Rosen et al., 2000). Primarily dedicated to the generation of Digital Elevation Models (DEMs), culminating in the Shuttle Radar Topography Mission (SRTM), the application range quickly extended towards the measurement of displacements with centimeter accuracy using Differential Interferometry SAR (D-InSAR, Figure 5.4) techniques. Thus, mapping of co- and post-seismic displacements, volcano inflation, and glacier flow became possible over large areas supporting traditional point-wise GPS measurements. To counteract the disturbing influence of the atmosphere and temporal changes of the radar backscattering in between acquisitions, D-InSAR approaches quickly developed towards the evaluation of time series of SAR images.

The common primary objective of all these D-InSAR stacking approaches is to remove the disturbing signal component of the atmosphere (correlated in space, but random in time), and of DEM errors (using a geometric model) from the displacement signal (correlated in time and spatially limited). In this manner, D-InSAR measurements became more trustworthy and accurate, for example, the measurement of seasonal thermal dilation of buildings with accuracies down to a few millimeters was reported.

With the availability of spaceborne SAR data with resolutions in the order of 1 m (e.g., TerraSAR-X (Germany) or CosmoSkyMed (Italy)), the identification of radar backscattering centers in layover became feasible, allowing the monitoring of individual building infrastructures (Zhu & Bamler, 2014). Further applications of the D-InSAR technique are the detection of landslides, measurements of subsidence due to groundwater extraction, mining activities, gas and petroleum prospection, as well as the control of inflation due to carbon capture sequestration (Moreira et al., 2013; Kim et al., 2023).

Nowadays, RCM and NASA-ISRO SAR (NISAR) are among the most recent and discussed EO missions in the SAR community. RCM is a Canadian SAR mission and is Canada's third generation of EO satellites launched with the objective to primarily serve the operational requirements of different government departments (Thompson, 2015). The mission also serves as a continuity mission for RADARSAT-1 and RADARSAT-2 (Parashar et al., 1993; Morena et al., 2004), which aim at addressing applications for marine surveillance (ship, ice, pollution monitoring etc.), but also disaster management (earthquakes, landslides, floods etc.), and ecosystem monitoring (forest, wetland, agriculture etc.). Since it is not possible to meet the requirements with a single satellite in orbit, three identical satellites work together in the RCM to deliver daily data. Apart from the full polarimetric capability, RCM also has the capability to image the Earth in the compact polarimetric mode where a circularly polarized wave is transmitted and the backscattered signal is received in the horizontal (H) and vertical (V) polarization channels simultaneously. Another interesting capability that RCM offers is the combination of SAR and Automated Identification System (AIS), which is exploited

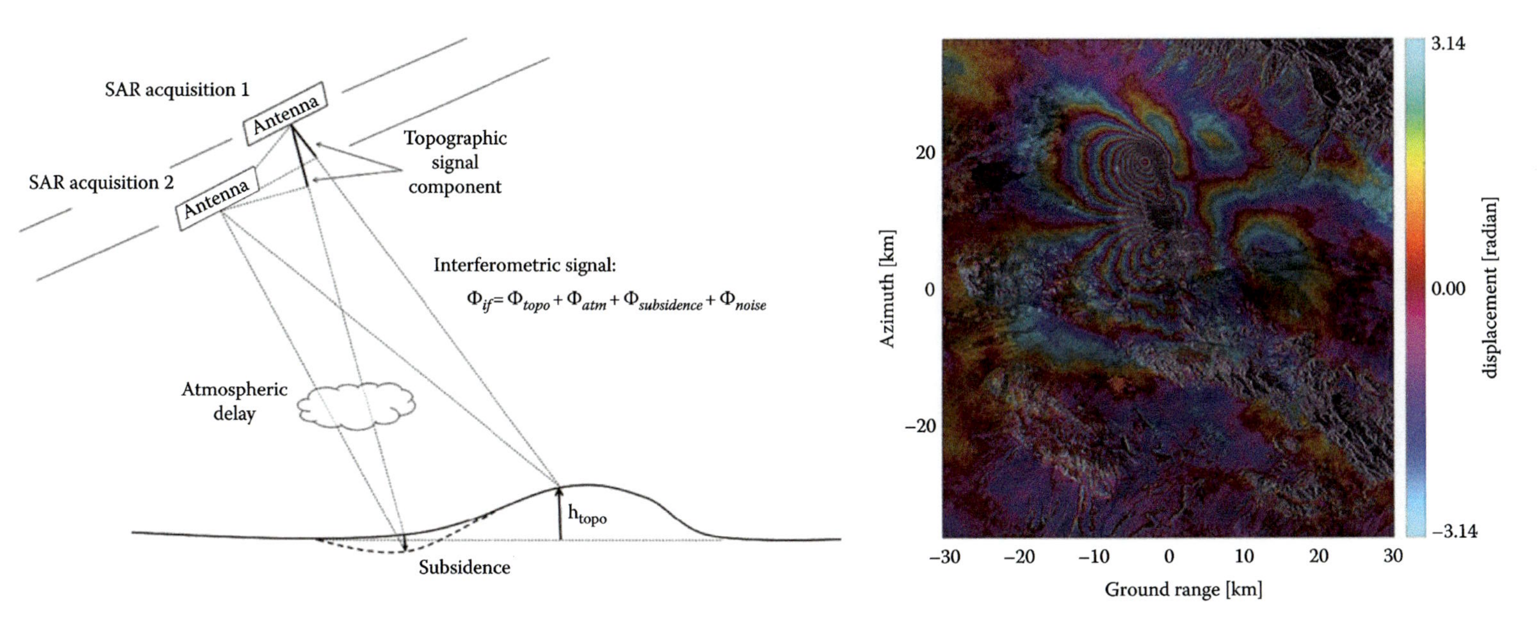


**FIGURE 5.4** Geometry for differential interferometric SAR (D-InSAR) data acquisition (left). Differential interferometric phase signature of co-seismic displacement of Bam earthquake, December 26, 2003, Iran as derived from ASAR data on ENVISAT superimposed on radar reflectivity (right). One phase/color cycle corresponds to 2.8 cm in radar line of sight. (© Data provided by ESA and processed by DLR.)

for ship detection since AIS coordinates enhance the vessel detection (Touzi & Vachon, 2015; Cote et al., 2021).

NISAR is a joint mission between India's Space Research Organization (ISRO) and NASA (Kellogg et al., 2020). The mission concept is a result of the National Academy of Science's 2007 decadal survey on future Earth observational priorities (National Research Council of the National Academies, 2007; Rosen & Kumar, 2021). NISAR is a unique satellite mission since it will use two different radar frequencies (L-band: NASA and S-band: ISRO) addressing areas like agriculture, natural disaster, crustal deformation, Alpine cryospheric processes, and coastal monitoring. NISAR will be able to perform advanced operations like simultaneous acquisition in dual frequency, high-resolution wide swath acquisition using SweepSAR technique, and has a global repetivity of nearly all of Earth's land and ice surface twice every 12 days (Kumar et al., 2016). The mission launch is scheduled in 2024 from the Satish Dhawan Space Centre located in the southern coast of India (Balakrishnan et al., 2004).

Apart from RCM and NISAR missions, ESA's Radar Observing System for Europe in L-band (ROSE-L) is planned to be launched in 2028 (Pierdicca et al., 2019). The mission will support key European policy objectives and help to fill the gaps of the presently operational Copernicus satellite constellations. Furthermore, the next generation Sentinel-1 (Sentinel-1 NG) with full polarimetric capability will complement the diversity SAR missions (Zonno et al., 2019; Davidson et al., 2022). Another example for a future European radar missions is the EarthExplorer Biomass (launch expected in 2024), which will carry a P-band SAR. These expansions to the existing Copernicus Space Component will respond to the requirements expressed by Copernicus services (Davidson et al., 2022).

### 5.3.3 LIDAR in Space

Light Detection and Ranging (LiDAR) has been shown as useful for mapping ground elevation and vegetation structure. LiDAR uses laser pulses to measure the distance from the sensor to objects on the Earth's surface. It works on the principle of time-of-flight, where the time taken for a laser pulse to bounce off an object and return to the sensor is used to calculate the distance. The data is typically collected as a point cloud, with each point representing a specific location on the ground and its associated elevation. LiDAR is limited by the availability of direct line-of-sight, and the data is also spatially limited (small-scale coverage) due to its point-based nature. However, airborne LiDAR can provide extremely high spatial resolution and help in the retrieval of very detailed information (Kostadinov et al., 2019; Deschamps-Berger et al., 2023). Therefore, airborne LiDAR has been widely employed to map vegetation properties such as forest biomass, vegetation carbon content or habitat structure (Hancock et al., 2021).

While airborne LiDAR continues to accelerate, there have been only a few non-atmospheric spaceborne LiDARs, beginning with the Shuttle Laser Altimeter (SLA) (Garvin et al., 1998). SLA was followed by the ICESat (Schutz et al., 2005) mission, which provided data until 2009. Since 2018, the Advanced Topographic Laser Altimeter (ATLAS) ia operated onboard ICESat2. In the same year, the Global Ecosystem Dynamics Investigation (GEDI) was also installed in the Japanese Experiment Module-Exposed Facility (JEM-EF) on board the International Space Station (ISS; Dubayah et al., 2020).

Current spaceborne LiDAR operate at 532 nm and can be divided into two types according to the detection principle (Wang et al., 2023b): full-waveform LiDAR and photon-counting LiDAR. Full-waveform sensors such as GEDI record the entire profile of reflected energy over time and can provide data that characterize the entire understory structure (Figure 5.5a, Wang et al., 2023b). The full-waveform echo is the result of convolution between a laser pulse and all ground targets within the spot range. In regions with large topographic relief, however, waveforms tend to broaden and overlap, aggravating interpretation of the echo (Figure 5.5b and c). Photon-counting LiDAR, such as ATLAS record the time tag of each photon received by the detector, similarly to airborne systems.

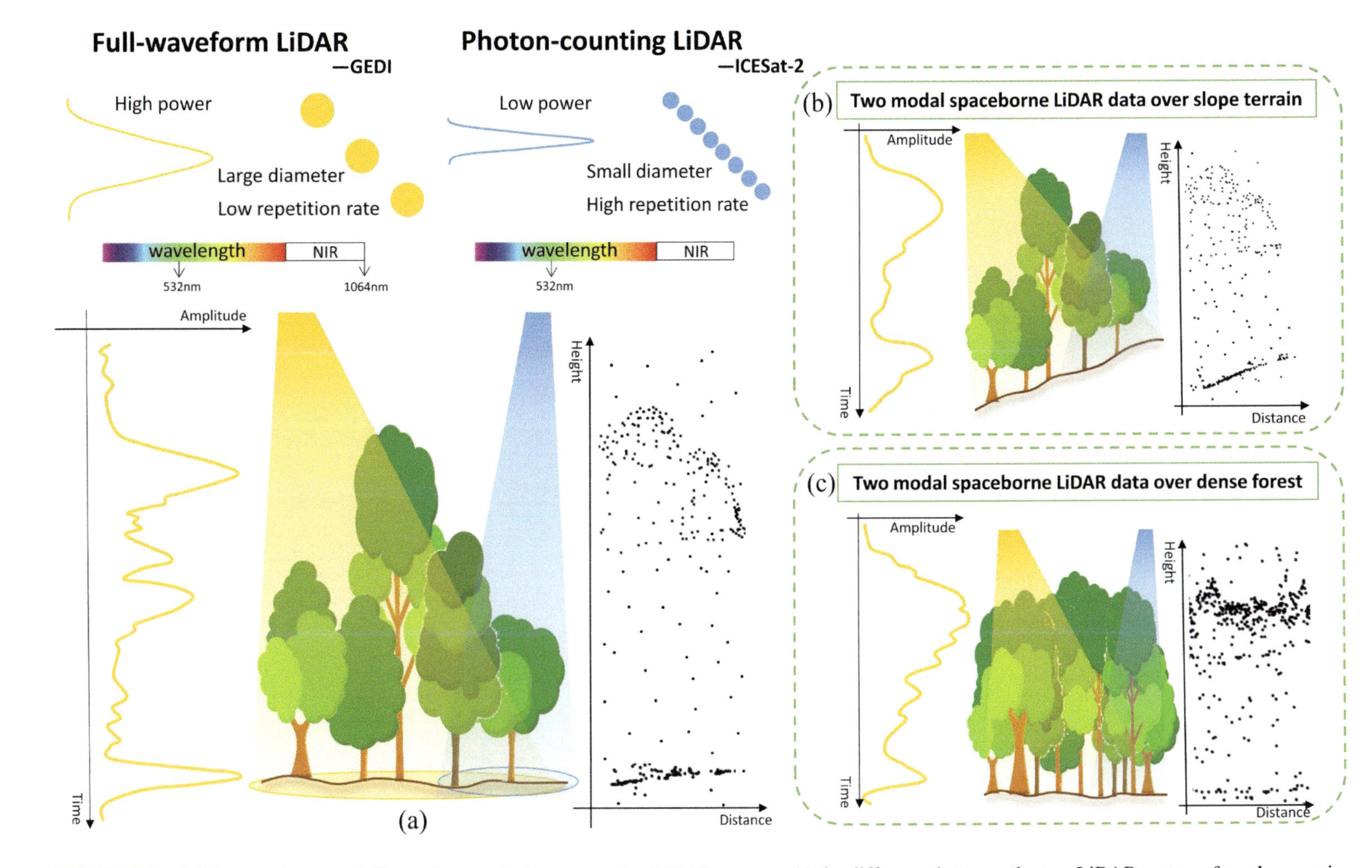


**FIGURE 5.5** Differences between full-waveform and photon-counting LiDAR systems: (a) the difference between the two LiDAR systems from laser emission to reception; (b) the variations in the LiDAR data obtained from the two systems in sloped terrain; and (c) the variations in the LiDAR data obtained from the two systems in a densely forested area. (© Wang et al., 2023a,b.)

As mentioned earlier, these systems typically operate in the Green wavelength region, where the reflectance of vegetation is low and, therefore, the photon energy is weak. Therefore, the actual density of photons received in forested area is relatively low (Neuenschwander et al., 2022). These factors currently limit the accuracy of spaceborne LiDAR-derived canopy height retrieval (Wang et al., 2023b) although the spaceborne missions in general have proven the suitability for mapping vegetation structure (Hancock et al., 2021).

### 5.3.4 Increasing Data Availability and Continuity

As already depicted, data continuity and availability are two important decisions of international data policies, which will improve EO data for a wide range of applications (Lynch et al., 2013; USA, 2020; Aschbacher, 2019). To contribute to standardized measures especially for global monitoring, continuous information of our planet and its availability are of high importance (Pereira et al., 2013). To further increase the use of EO data for new applications, the capability to develop standardized products based on automatic thematic processors for consistent knowledge generation and provision may be a prerequisite. Furthermore, the foreseeable increase in data volume will continue to induce new challenges in terms of data storage, handling, mining, processing techniques and accuracy requirements. Investment in processing is still comparatively high due to mostly not fully automation of information generation procedures due to, for example, different atmospheric conditions, land cover types, different user's requirements or the algorithms still being in experimental status. Therefore, robust methodologies are needed.

As it becomes more and more understood and accepted that single disciplines are decreasingly able to progress individually, this paradigm shift can be observed increasingly. Multidisciplinary approaches using EO-based data in combination with other data sources are developing with the idea for geographic value-adding to these spatial data beyond the sole physical measurements. Besides multidisciplinary approaches, multi-sensor applications will form a further spreading technique to enhance existing approaches, which requires temporally overlapping and well calibrated sensors. International cooperation and data sharing will also reduce instrument and launch costs and mission redundancy (Durrieu & Nelson, 2013). Joint commitments between national and international space agencies may therefore be a powerful means of stabilizing new space initiatives, and hence data availability and continuity (Morel, 2013).

#### 5.3.4.1 Data Policy as a Backbone

EO data policy refers to the framework, rules, regulations, and guidelines governing the collection, distribution, management, and usage of data acquired through EO technologies. These policies are typically established by governmental or international organizations. EO data policy underwent an alternating history, whereas the availability of EO data strongly depends on national distribution policies and pricing. NASA/U.S. Geological Survey (USGS) and ESA have a free and open data policy, following the GEO data principle, which already demonstrated a boost in research and innovation. In general, however, there co-exist three models of data distribution policy:

- Web-enabled, free: Organizations which provide data and associated products freely available for every interested user (e.g., NASA's/USGS's Landsat data or MODIS products, ESA's Copernicus data, Brazil's CBERS data)
- Commercial, at a cost: Commercial remote sensing industry, which mainly is located in the very high spatial resolution domain, began with the launch of Ikonos and Quickbird. Most commercial systems, however, operate with meter or sub-meter resolutions (e.g., WorldView, GeoEye, Pleiades (France)) and allow the detection of small-scale objects, such as elements of residential housing, commercial buildings, transportation systems, and other utilities which can assist industrial geospatial applications (Navulur et al., 2013), but may also be purchased by noncommercial users or scientists.

- Databuy: Public-private partnerships support the scientific community with cost-free data, but also established commercial EO markets, that is, assist in developing remote sensing business in a way that follow-ups may be financed by industry using the profit from selling data or products to non-scientific clients. RapidEye fleet; the TerraSAR and TanDEM-X missions as well as the SPOT sensors are successful examples for public-private partnerships.

#### 5.3.4.2 Multi-Sensor Integration—Data Fusion

The increasing availability of EO and ancillary data has drawn significant attention towards the effective use of multimodal datasets (e.g., Ghamisi et al., 2019). Researchers across diverse applications are exploring the advantages of combining multiple sensors, a practice known as sensor, image, or data fusion (Belgiu & Stein, 2019). This approach integrates temporal information with spatial and spectral/backscattering data from different sensors, providing a means to overcome limitations imposed by factors like weather conditions, time of day, or terrain-related issues.

Sensor fusion allows for the best use of each sensor's strengths, resulting in imagery with high spatial and spectral resolution (Ghassemian, 2023). Moreover, the integration of time dimension into data presentation transforms two- or three-dimensional data into four-dimensional frameworks, unlocking new possibilities for information extraction. However, the process of fusing data from various sensors requires meticulous calibration to ensure compatibility, especially when merging datasets with varying radiometric, spatial and temporal resolutions. This is crucial for applications like land cover classifications and achieving a balance between spatial and spectral resolutions (Zhu et al., 2016; Wang & Atkinson, 2018; Wang et al., 2018).

Many terrestrial applications require frequent observations at a medium spatial scale (Claverie et al., 2018; Himeur et al., 2022). To meet the demand for higher temporal resolution, especially in cloud-prone areas, data of several sensors can be combined. A prominent example is the synergistic use of Landsat-8/9 Operational Land Imager (OLI) and Sentinel-2 MSI data, which paves the way to capture near-daily data (Wulder et al., 2012) and further ensures continuity of data if one sensor fails. The data need to be processed into Analysis Ready Data (ARD), ensuring consistent spatial resolutions and geophysical parameter retrieval, regardless of the sensor used (Ergorov et al., 2018). The proliferation of freely available EO satellite data and the availability of powerful computing facilities are facilitating the processing of ARD from multiple instruments (Li & Roy, 2017). This integration of data from various sensors into a single dataset forms what is known as a "virtual constellation" (CEOS, 2019).

Enhancing spatial resolution is particularly crucial for applications like environmental monitoring, urban planning, agriculture, disaster management, and natural resource monitoring. Satellite data fusion is a common approach in this context. The primary objective of spatial resolution enhancement through data fusion is to maintain the spectral characteristics of a sensor with good spectral capabilities while improving its spatial resolution using data from a high-resolution sensor. Pan sharpening, a well-known technique in this field, enhances the spatial details of multispectral or hyperspectral images by fusing them with features extracted from a higher-resolution panchromatic image. This panchromatic data captures imagery in a broad wavelength range, resulting in higher spatial resolution. The result of image sharpening using panchromatic data is an image with improved spatial details while preserving the spectral characteristics of the original multispectral or hyperspectral data (Amro et al., 2011). Applications such as LULC classifications gain significantly from pan-sharpened data (e.g., Feng et al, 2017).

In recent years, the advent of AI and machine learning has led to significant research interest in adapting these methods for image fusion (Ghassemian, 2023). While machine learning models show great potential, they have not yet surpassed traditional methods in most cases (Wang et al., 2023a). Nevertheless, they are considered a promising avenue, especially for unsupervised learning and image generation applications (see Section 4.2).

### 5.3.5 The Development of a Commercial EO Sector

The last sections focus on government-led EO aimed at scientific research and environmental monitoring. However, the development of commercial spaceborne remote sensing has also witnessed remarkable growth and transformation over the past few decades. Over time, the commercial industry has come to realize the vast potential of spaceborne remote sensing, resulting in a growing sector dedicated to providing EO data and developing commercial applications. The progress in commercial spaceborne remote sensing can be attributed to several technological advancements such as the miniaturization of sensors and instrumentation, coupled with cost-effective satellite manufacturing. These advancements have facilitated the deployment of small satellites, including CubeSats, for example, of the PlanetScope CubeSat constellation. Another key driver of commercial remote sensing is the continuous improvement in spatial and spectral resolution. High-resolution satellites, such as those of the WorldView, Pleiades and Planet constellations have found applications in agriculture (e.g., Dakhir et al., 2021), urban planning (e.g., Warth et al., 2020), or disaster management (e.g., Adriano et al., 2021).

Online platforms and Application Programming Interfaces (APIs) allow governments, businesses, researchers, and the general public to access and analyze imagery, enabling evidence-based decision making. This development has spurred innovation and entrepreneurship in geospatial analytics. However, despite its significant expansion, the commercial spaceborne EO industry faces several challenges. High-resolution imagery, capable of capturing intricate details about individuals and properties, has given rise to privacy concerns. As a result, there are ongoing discussions regarding regulatory frameworks and ethical considerations aimed at addressing these issues. Furthermore, the competitive landscape and pricing pressures can impact the long-term viability of commercial EO ventures. In response, companies are exploring opportunities in value-added services and partnerships to ensure their sustainability (Denis et al., 2017).

### 5.3.6 Computing Platforms and the Advent of Big Data

Data analysis techniques for SAR, LiDAR, and optical data vary significantly due to the nature of the data and the applications they serve. Therefore, EO data analysis involves the use of specialized software platforms and tools tailored to different types of data. In general, powerful commercial software packages such as ENVI (NV5 Geospatial), Erdas Imagine (Hexagon Geospatial), PCI Geomatica (Catalyst) or ArcGIS (Esri) are available. They provide state-of-the-art processing of EO data for various sensors. Commercial software is often backed by a dedicated support team that can provide help and technical support when needed. This is particularly useful for organizations that require reliability and timely solutions. For users under software maintenance, providers release regular updates and patches, ensuring that the software remains up-to-date and capable of handling new sensors, data formats, and processing techniques.

One of the key benefits of open-source software lies in its typical cost-free nature. This characteristic can lead to substantial savings for individuals and organizations operating within limited budgets. Additionally, open-source software allows users to customize and extend the software to suit their specific needs. This flexibility and the access to the source code is particularly useful for researchers and developers, which can also benefit from a user community who contribute to improve the software. Table 5.3 lists some of the commonly used open-source platforms.

In most cases, these software packages are stored on a single computer and, therefore, rely on the computational resources of a local machine or a limited set of computers. The processing power is limited to the capabilities of the individual computer or a small cluster of computers, which, however, may not be sufficient for handling large remote sensing datasets, for example, time series. Moreover, there has been an increasing number of EO missions that are now generating a nearly continuous flow of data, which fostered the creation of large remote sensing data repositories that can only be exploited using adequate parallel and distributed processing techniques. To

**TABLE 5.3**
**Examples for Open-Source Software Platforms for EO Data Computing**

| Sensor Independent | Brief Description | Download |
|---|---|---|
| QGIS | GIS software useful for handling and analyzing optical remote sensing data. It has a wide range of plugins for remote sensing tasks. | https://qgis.org/en/site |
| Sentinel Application Platform SNAP | SNAP is an architecture for all Sentinel and SMOS toolboxes, allowing EO processing and analysis, for Sentinels and various other sensors. | https://step.esa.int/main/download/snap-download/ |
| Python | Popular programming language with libraries/modules for remote sensing data analysis, including raster processing, image classification, machine learning, and geospatial analysis. Several packages and libraries for processing available, for example, Rasterio, Xarray, scikit-learn, and satpy. In addition, there are toolboxes, such as Orfeo Toolbox. | Rasterio (https://rasterio.readthedocs.io/en/latest/intro.html), Xarray (https://docs.xarray.dev/en/stable/), Satpy (https://satpy.readthedocs.io/en/stable/), scikit-learn (https://scikit-learn.org/stable/), Orfeo Toolbox (www.orfeo-toolbox.org/) |
| R | Programming language for statistical computing, which also has toolboxes, such as RStoolbox. Tools for Remote Sensing Data Analysis in R providing a wide range of every-day remote sensing processing (data import, preprocessing, analysis, image classification and graphical display). Is suitable for processing large data-sets even on small workstations. | https://bleutner.github.io/RStoolbox/ |
| **Optical data** | | |
| PRISMA Toolbox | Toolbox allows to import, view and convert L1, L2B, L2C, L2D products. User can interact with hyperspectral data in HDF format and the metadata. | www.asi.it/en/earth-science/prisma/ |
| EnMAP Box | The EnMAP-Box is a free and open source plugin for QGIS. Designed to process hyperspectral data providing state-of-the-art applications and a graphical user interface for visualization and exploration of multi-band remote sensing data and spectral libraries (van der Linden et al., 2015). | https://enmap-box.readthedocs.io/en/latest/usr_section/usr_installation.html |
| **SAR** | | |
| PolSARpro | Open-source software suite for polarimetric SAR data analysis. It includes tools for data visualization, polarimetric SAR data decomposition, and classification of ALOS-1, COSMO-SkyMed, RADARSAT-2, TerraSAR-X/TanDEM-X, and Biomass data. | https://earth.esa.int/eogateway/tools/polsarpro |
| **LiDAR** | | |
| LAStools | Popular software suite for LiDAR processing, which supports operational processing of data from advanced airborne LiDAR systems and facilitates generation of Digital Surface Models (DSMs) and Digital Terrain Models (DTMs) from raw/preprocessed data. The tools provides an intuitive Graphical User Interface (GUI), a streamlined workflow and a plugin for QGIS. | https://lastools.github.io/ |
| FUSION/LDV | Specifically designed for analyzing forest vegetation characteristics (McGaughey, 2007). This analysis and visualization platform comprises two main programs, FUSION and LDV (LiDAR data viewer). The package facilitates fast, efficient, and flexible access to a range of datasets (LiDAR, IFSAR, and terrain). | http://forsys.cfr.washington.edu/fusion/fusionlatest.html |
| CloudCompare | Software platform for 3D point cloud and mesh processing, including LiDAR data analysis and visualization (Girardeau-Montaut, 2016). The tool can be used to measure the distance between trees and can be also used to convert 3D data types. | www.danielgm.net/cc/ |

give some examples for specific sensors, AVIRIS acquires almost 9 GB of data per hour. Similarly, the EO-1 Hyperion sensor acquires about 71.9 GB of data per hour (Haut et al., 2021). ESA's Copernicus programme acquires astonishing 16 TB of data per day (ESA, 2023), and commercial companies collect upwards of 100 TB per day (Mohney, 2020). The unprecedented proliferation of data has posed significant challenges in managing, processing and interpreting massive amount of EO data. Great efforts have been toward the integration of high-performance computing (HPC) in EO applications. These HPC-based approaches were seen as the most dominant yet efficient way for addressing the enormous computational requirements of big data. Nowadays, the computation capabilities available are no longer the limiting factor: however, despite of the huge processing power, the cluster-based HPC systems still remain considerably challenging (Ma et al., 2015).

A solution for big data handling provide cloud-based platforms. Early on, cloud platforms only offered storage services. These services allowed organizations to upload and store large volumes of remote sensing data securely in the cloud. Nowadays, cloud-based platforms such as Google Earth Engine (GEE; Tamiminia et al., 2020; see also Chapter 28 of this volume), Microsoft Planetary Computer (MPC; Planetary Computer, 2022), Earth on Amazon Web Services (AWS; Ferreira et al., 2022), as well as the Data and Information Access Service (DIAS providing centralized access to Copernicus data and processing tools) offer virtually unlimited processing power for data storage, computing and management. However, the handling of massive amounts of data requires high storage and computing resources, which becomes increasingly accessible to the public via cloud-based services. In most cases, cloud-based solutions follow a pay-as-you-go model, that is, users pay for the resources they use. As a result, the cost is much lower than sophisticated and expensive high-performance computers (Ferreira et al., 2022). Therefore, cloud-based computing is best means to realize global EO analysis and is expected to become the backbone of future EO computing (Guo et al., 2021). For more details it is referred to Chapter 27 of this volume.

## 5.4 EMERGING TECHNOLOGIES AND CHALLENGES

### 5.4.1 Unmanned Aerial Vehicles

In recent years, Unmanned Aerial Vehicles (UAVs), commonly known as drones, have revolutionized the field of remote sensing by offering flexible, efficient, and cost-effective platforms for data acquisition and analysis.

UAVs can be equipped with various sensors such as multi- and hyperspectral cameras, thermal sensors or even LIDAR systems to collect aerial imagery. A significant advantage of UAV-based remote sensing is the ability to obtain data at various spatial and temporal resolutions. UAVs can capture data at centimeter-scale, providing information for precise mapping and monitoring of small-scale features such as individual trees, buildings, land features or even count individual flowering plants (see Figure 5.6).

At present, UAVs are particularly useful for agricultural/forestry, cryospheric monitoring, and conversational applications (e.g., Bhardwaj et al., 2016; Gröschler et al., 2023; Inoue, 2020) as well as for the monitoring of natural hazard-related disasters (e.g., Kucharczyk & Hugenholtz, 2021). Farmers and agronomists use UAV-collected data to assess crop health, monitor plant growth patterns, and detect plant diseases or stress factors at an early stage. This information aids in optimizing resource allocation, improving crop yield, and reducing the environmental impact through efficient use of fertilizers and pesticides. Furthermore, the use of UAVs for nature conservation efforts is rapidly growing. Especially when collecting dynamic and rapidly changing features, UAVs can be deployed rapidly, allowing for timely data collection. An increasing number of studies prove the usefulness of UAV-based data for monitoring wildlife, track changes in habitats (see also Figure 5.6) and assess the anthropogenic impact on ecosystems.

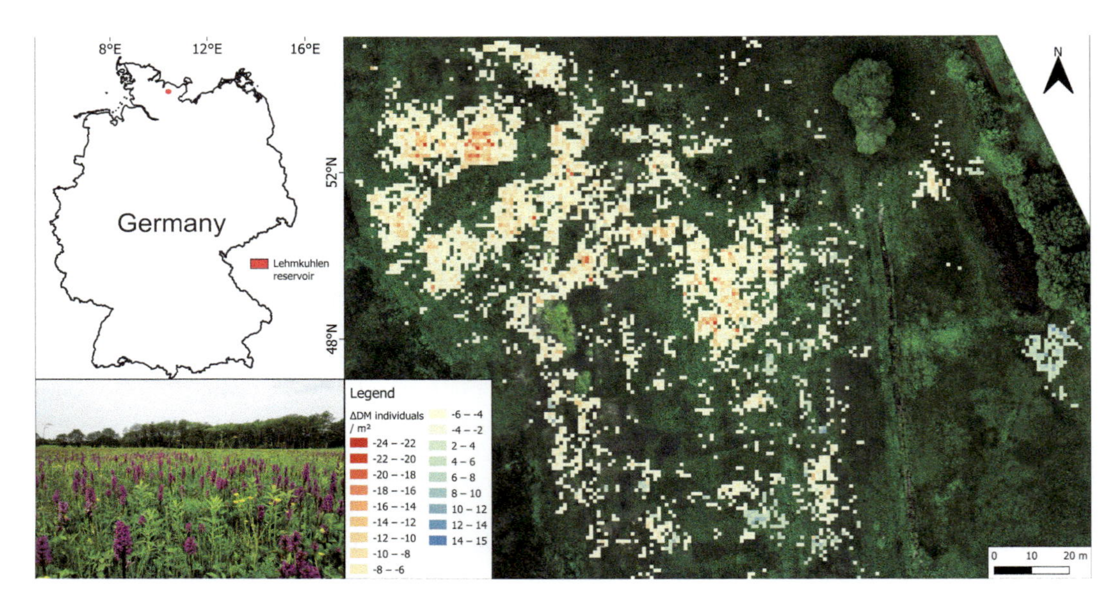


**FIGURE 5.6** Monitoring conservations measures in an area with protected plants, that is, *Dactylorhiza majalis*. Plant counts of have been conducted in consecutive years using multispectral drone surveys with a spatial resolution of about 2 cm. The map shows the net change of plants per unit area as detected during the flowering seasons of 2021 and 2022, respectively. (Modified according to Gröschler et al., 2023.)

Despite the advantages of UAV-based remote sensing, challenges exist including regulatory limitations, privacy concerns, data processing complexities, and the need for skilled operators. Addressing these challenges requires ongoing efforts to develop standardized protocols, further improvements in data processing, and enhanced training in UAV operation and data analysis. UAV technology is expected to further advance and, therefore, will play an even more important role for terrestrial applications in the future.

### 5.4.2 Integration of Machine Learning

Research in cognitive science, which studies human thought processes, has provided insights into how humans perceive, reason, and learn from data, influencing the development of machine learning algorithms. These aspects are encapsulated within mathematical and statistical constructs to build a framework that simulates human-like learning from experience. Machine learning is a subdivision of Artificial Intelligence (AI), and, at its core, aspires to construct prediction models capable of autonomously enhancing their performance by drawing insights from labeled training and validation data. These models, often composed of weighted links and activation functions (mathematical equations), strive to emulate the neural processing of information in the human brain. They incorporate concepts from interdisciplinary domains such as regression, probability theory, and hypothesis testing to mimic the intricate decision-making processes that human cognition engages in (Bishop, 2006).

The history of machine learning in EO (Lary et al., 2016) is closely tied to the development and evolution of both machine learning methods (Salcedo-Sanz et al., 2020) and EO technology (Belward & Skøien, 2015). With the exponential development of the machine learning techniques like Support Vector Machines (SVMs) and decision trees such as Random Forest (RF) in the 2000ies, their presence was realized in application areas like image classification, land cover mapping, and object recognition tasks (Mas & Flores, 2008). Today, machine learning techniques are at the forefront of terrestrial remote sensing applications (Demir et al., 2010; Hariharan et al., 2017; Chaudhuri et al., 2019; Hill et al., 2020; Rajaneesh et al., 2022; Persello et al., 2023; Joshi et al., 2023). With increasing computing capacities, this trend is expected to increase further addressing global challenges, such as climate change, natural resource management, and disaster response (Dao et al., 2019; Reichstein et al., 2019; Pan et al., 2020; Ibrahim et al., 2021; Rolnick et al., 2022; Hall et al., 2023). Nowadays, some machine learning algorithms already are established in EO computing (see Chapter 26 of this volume). The next section, therefore, describes algorithms that are promising for present and future EO applications. Due to the huge selection of machine learning algorithms, we have only provided a few examples here. The choice of algorithm always depends on the specific task, the nature of the data, the amount of available labeled data, and computational resources and, therefore, may differ from those described in the following.

#### 5.4.2.1 Supervised Learning Algorithms

The foundation of neural networks (NNs) commenced with a single-layer perceptron model that exhibited potential of supervised learning (Rosenblatt, 1958). A perceptron is essentially a single neuron with adjustable synaptic weights and bias. In the early days, the feed-forward back-propagation multi-layer perceptron (MLP) was the kind of neural network that was mainly used with remote sensing data (Lawrence, 1993; Atkinson & Tatnall, 1997). NNs can be highly complex with multiple layers and thousands or millions of parameters. They have the capacity to learn intricate patterns in data. Deeper neural networks are often considered "black box" models, making it challenging to interpret how they arrive at their predictions. The internal representations are complex and difficult to interpret directly. Hence, they can be susceptible to overfitting, especially when they have many layers and parameters. Therefore, regularization techniques are often employed to mitigate this issue. NNs can identify subtle and nonlinear patterns, which are not always achievable by statistical methods. The network does not put a requirement for normally distributed data. Consequently, data from different sources with poorly defined or unknown distributions can be integrated into the network (Notarnicola et al., 2008). Among the several types of NNs, the feed-forward MLP has been reported to perform suitably for classification and inversion. NN based inverse scattering model has been utilized for the inversion of vegetation parameters from radar backscattering coefficients (Chuah, 1993). NN trained with dense media radiative transfer model that includes the effects of multiple scattering has been employed with passive microwave measurements for inversion of snow parameters (Tsang et al., 1992). Today, advanced NN-inspired architectures such as Convolutional Neural Networks (CNNs) and Recurrent Neural Networks (RNNs) are applied to various tasks such as image classification, object detection, and semantic segmentation (Mou et al., 2017). Physics-aware algorithms with an interplay data driven and physical process-based models is perceived as the future of machine learning algorithms in EO (Eroglu et al., 2019; Camps-Valls et al., 2021). Furthermore, an amalgamation of Geocomputation and Geospatial Artificial Intelligence (GeoAI) is steering the domain of EO into a different generation, which is causing a paradigm shift in traditional geospatial perspectives (Song et al., 2023).

You Only Look Once (YOLO) is a CNN-based framework that is being employed for target detection in satellite imagery (Pham et al., 2020). YOLOs are single-shot object detection systems, which can detect and classify objects in an image in a single forward pass of the NN. YOLO uses a CNN architecture to simultaneously predict bounding boxes and class probabilities for multiple objects within an image. This technique is known for its speed (real-time capability) and utilized for applications that require quick and accurate object detection.

The technique divides the input image into a grid and assigns bounding boxes and class predictions to each grid cell. This grid-based approach enables real-time object detection and ensures that small objects in the image are not overlooked. The algorithm employs anchor boxes, which are predefined shapes that help the model predict the size and location of bounding boxes. These anchor boxes are learned during the training process. YOLO predicts class probabilities for each bounding box. This means that it cannot only detect objects but also classify them into predefined categories. The algorithm makes multi-scale predictions at different levels of the network to detect objects of various sizes (Xu & Wu, 2020). This allows the algorithm to handle both small and large objects effectively.

The region-based CNN (R-CNN) was a breakthrough in object detection (Girshick et al., 2014). Faster Region-based CNNs (Faster R-CNN) are deep learning models used for object detection and localization in EO images. Faster Region-based-CNNs are successors of R-CNNs, where the Region Proposal Networks (RPNs) replace the selective search algorithms for region proposal (Girshick, 2015). This enabled sharing of convolutional layers between the RPNs and the detection network, which essentially improved the execution speed of the algorithm. The network (Wang et al., 2018) commences with a CNN backbone, such as VGG16, ResNet, or Inception, which is pre-trained on a large dataset. Faster R-CNN features a key innovation through the integration of RPN, which serves as the extraction module in addition to the detection module (Yan et al., 2021). The RPN operates on the feature maps generated by the backbone network and is naturally implemented as a fully convolutional network and can be trained using back-propagation and stochastic gradient descent (SGD). The RPN is responsible for proposing regions of interest (RoIs) in the image where objects may be present. RPN generates high-quality region from the basic feature map and the Fast R-CNN directly detects the object in the proposed region. It uses anchor boxes, which are predefined boxes of different sizes and aspect ratios, placed at various locations on the feature maps. For each anchor box, the RPN predicts two values: an objectness score (indicating the likelihood of an object being present) and adjustments to the anchor box's coordinates to better fit an object if one is present. The RPN ranks the proposed anchor boxes based on their objectness scores (Ren et al., 2015). High-scoring anchor boxes are selected as potential RoIs. The selected RoIs are used to crop and resize the corresponding regions in the feature maps obtained from the backbone network. This process is known as RoI pooling and ensures that all RoIs have a consistent size and shape, regardless of the size of the proposed regions, that is, the RoI pooling layer helps to extract feature vectors.

The RoI feature maps are passed through fully connected layers that consist of sub branches. One of the branches called the object classification branch predicts the class label of the object within each RoI using an appropriate activation function. The bounding box regression branch refines the coordinates of the RoI's bounding box to more accurately enclose the object using a regression method. The algorithm computes losses during the training for classification and bounding box regression. The classification loss measures the difference between the predicted class labels and the ground truth labels for the objects inside the RoIs. This is typically computed using cross-entropy loss. On the other hand, the bounding box regression loss measures the difference between the predicted bounding box coordinates and the ground truth coordinates for the objects inside the RoIs. A commonly used loss function is the smooth L1 loss. The network is optimized by minimizing these two losses. During the training process, RPN's anchor boxes are also fine-tuned to generate more accurate region proposals. Certain post-processing steps, such as Non-Maximum Suppression (NMS), are applied to filter out redundant or overlapping detections and refine the final object detection results (Qiu et al., 2018).

#### 5.4.2.2 Unsupervised Learning Algorithms

Popular unsupervised machine learning algorithms are K-Means clustering (Burney & Tariq, 2014) and Principal Component Analysis (PCA), which are included in Chapter 26 of this volume. More recently, however, unsupervised algorithms such as Hierarchical Clustering and Self Organizing Maps (SOMs) have come into prominence.

Hierarchical Clustering (Sidorova, 2012) is a versatile technique used in EO, valuable for exploring the structure and relationships within remote sensing data. It is able to capture multi-scale information and reveal hierarchical relationships among data elements. This approach groups pixels or regions into hierarchies of clusters, which can be useful for hierarchical classification or object grouping. Hierarchical Clustering can be used to segment remote sensing images into regions or objects with similar spectral characteristics (Tarabalka et al., 2009). After segmenting an image into objects, hierarchical clustering can be applied to group similar objects into higher-level clusters. Large EO images with complex land cover distribution can represent the image content as a hierarchical structure with abstract and detailed levels such as categorizing urban areas into densely and sparsely built-up areas sub-classes (Yao et al., 2016). The algorithm commences with individual pixels as clusters and progressively merges them based on similarity, forming a hierarchy of clusters. This structure can capture multi-scale features in the image, leading to more detailed and context-aware segmentation results. It can be integrated into Object-Based Image Analysis (OBIA) workflows (Gonzalo-Martín et al., 2016). Hierarchical clustering can be applied to classify land cover in remote sensing data. By grouping pixels with similar spectral signatures into clusters, it is possible to identify land cover classes and their hierarchical relationships (Häme et al., 2020; Cordeiro et al., 2021). For example, clusters at higher levels of the hierarchy may represent broad land cover categories (e.g., vegetation, water), while clusters at lower levels may represent specific subtypes (e.g., types of vegetation). In hyperspectral remote sensing, where data may have a large number of spectral bands, hierarchical clustering can help reduce the dimensionality of the data by identifying clusters of similar bands. These clusters can be used as reduced sets of spectral bands for further analysis or classification (Ji et al., 2022).

The SOM, also known as the Kohonen map, consists of two layers, which are fully connected via synaptic weights (Filippi et al., 2010). The method is based upon competitive learning with an input layer having the same dimensionality as the feature vector and the Kohonen/competitive layer that serves as the output layer. The output layer is composed of a network of neurons with the adjacent neurons having a unity Euclidean distance. SOM essentially models the probability distribution of the input feature vector in the hypersphere. The approach essentially captures the topology with the weight vector modeling the probability distribution function and converts the pattern space into feature space (Tasdemir et al., 2011). This conversion essentially reduces the dimensionality. The topology-preserving feature is aimed towards conserving the neighborhood relationship of the input feature pattern. The weight vectors represent the natural tendency of a cluster of data points. SOMs have already been used for EO applications like identifying ocean current patterns, texture analysis, image deblurring, and selection of suitable window size for speckle reduction in polarimetric SAR images (Shitole et al., 2015). SOM does not assume linearity in the data (Neagoe et al., 2014); it can capture complex, nonlinear relationships between data points. A stacked architecture of SOMs has been found to be useful for processing large volume of EO data and, therefore, my play a more important role in future EO computing.

### 5.4.3 Enhanced Data Accessibility

Enhanced data accessibility and sharing of data for calibration and validation of EO algorithms is fundamental for evidence-based policy decisions and advanced scientific research. As outlined before, open data policies, user friendly download portals, the availability of sophisticated software and computing platforms pushed forward data accessibility.

Moreover, a large diversity of data is available ranging from very high to coarse-scale resolution data, which often are freely available at different processing levels from radiances to reflectances. Converting these data into valuable products for terrestrial application has already been aimed at by Bartholomé & Belward (2005), but still remains a challenge. Prominent examples are global products such as LULC classifications which provide a good global representation of general land cover classes but are neither easily comparable nor congruent to other land cover classifications (McCallum et al., 2006; Herold et al., 2008). Questions of integrity and re-usability of global LULC maps therefore will have to be further discussed. In response to this need, international bodies such as Group on Earth Observation (GEO), Global Climate Observation System (GCOS) and EO communities in Europe have been involved to establish an operational a continuous LULC observing system that includes integration, harmonization and validation of datasets (Mora et al., 2014). The Global Forest Watch (GFW, 2023) initiative may have served as a pioneer for ongoing and future initiatives providing cloud computing and crowdfunding. The commitment of many agents including space agencies and industrial partners to supporting an ongoing program was unprecedented (Baltuck et al., 2013).

GEO also promotes free and open data within its scope to build a Global Earth Observation System of Systems (GEOSS). GEO has been launched as a response to the World Summit on Sustainable Development and the Group of Eight (G8) leading industrialized countries, which called for action concerning international collaboration for exploiting the potential of EO (Nativi et al., 2013). In 2023, it includes 114 national governments and the European Commission as members; the number of 152 participating organizations includes intergovernmental, international and regional organizations such as space agencies and United Nation bodies (GEO, 2023). GEOSS is conceived as a global network of product providers aiming provision of decision-support tools to a great variety of stakeholder. Focused community portals have developed to meet the particular needs of individual communities and to develop solutions targeted to these specialized users (GEO, 2010). Together with the GFW initiative, this kind of coordinated development of methodologies, processing, and validation led to a new era of standardized data products.

#### 5.4.3.1 Harmonized EO Products and Services

The need for harmonization and standardization has been outlined before and is discussed in detail in Section 3 of this volume. A recent example of a harmonized EO product is the Harmonized Landsat and Sentinel-2 (HLS) dataset (Claverie et al., 2018). HLS data undergoes processing and standardization to ensure compatibility for cross-comparison and analysis. These datasets maintain consistency in terms of spatial resolution, spectral bands, and data formats, simplifying their concurrent use.

Optical satellite imagery quality can be impacted by cloud cover and atmospheric interference. Harmonized products often incorporate cloud masks and cloud-free composites to enhance the usability of the data. The harmonization process for Landsat and Sentinel data encompasses various aspects, including standardizing pixel resolution, map projection, tile coverage, atmospheric correction, and cloud masking using a shared radiative transfer algorithm. Additionally, it involves normalization to a common nadir view geometry and adjustments to account for their spectral response functions. A comprehensive description of the HLS processing can be found in the work by Claverie et al. (2018). Researchers and organizations aiming to combine data from both Landsat and Sentinel missions for diverse applications find these harmonized products particularly valuable since HLS serves a wide range of applications, including LULC classification, vegetation monitoring, urban planning, and environmental change detection.

The HLS project is a collaborative effort involving NASA Goddard Space Flight Center, USGS and ESA. It leverages Landsat 8/9 OLI and Sentinel-2A/B MSI data to generate harmonized reflectance products with a common pixel resolution of 30 m and a combined temporal resolution of two to three days. HLS data are accessible through EarthData Search and NASA's Land Processes Distribute Active Archive Center (LP DAAC). Furthermore, the inclusion of data from the upcoming Sentinel-2C mission, planned for launch in 2024, will further improve revisit times and enhance land monitoring capabilities (NASA EarthData, 2023).

Aligned with its EO strategy, ESA's Climate Change Initiative (CCI) strives to address the requirements outlined by the United Nations Framework Convention on Climate Change (UNFCCC). These needs center on supporting global modeling approaches that aim to comprehensively understand the intricately interconnected Earth system, encompassing physical, chemical, and biological components with complex interactions and feedback mechanisms. The UNFCCC, as emphasized in the Paris Agreement of 2016, emphasizes the necessity for systematic observation and the establishment of data archives related to the climate system (UNFCCC, 2016). These actions are deemed critical for gaining a deeper understanding of the Earth system, the climate's state, and the potential consequences of climate change.

This endeavor includes the creation of an integrated observing system that combines numerical models with long-term, meticulously calibrated, and well-documented datasets sourced from both satellites and in situ observations (Plummer et al., 2017). To achieve this, CCI was devised as a means to consolidate European EO expertise across all phases of satellite product development. It actively encourages international collaboration, both within and outside Europe, with the primary goal of providing enduring and transparent access to consistent, multi-sensor products for essential climate variables. These products are derived from global EO archives and are readily accessible online through the CCI portal (https://climate.esa.int/).

Meanwhile, CCI brings together a community of approximately 500 world-leading experts to generate comprehensive, multi-mission, and multi-decadal datasets that fulfill the requirements for the 22 essential climate variables stipulated by the UNFCCC. These datasets have fully characterized uncertainties and are validated using independent, traceable, in situ measurements. Specifically focusing on terrestrial aspects, CCI concentrates on 13 essential climate variables encompassing a range of subjects such as lakes, snow cover, glaciers and ice caps, ice sheets, permafrost, land cover, aboveground biomass, fire disturbance, soil moisture, land surface temperature, albedo, fAPAR, and LAI (ESA, 2021).

Furthermore, the Copernicus Global Land Service (CGLS) extends its offerings to encompass essential climate variables and additional biogeophysical products like burnt area, dry matter production, NDVI, vegetation condition, and global productivity. In contrast to CCI products, CGLS provides higher spatial resolution and lower update frequency, particularly tailored for regional and Pan-European scales (CGLS, 2023). CGLS processes EO data from diverse sensors to generate value-added products and land surface information. It ensures timely and consistent delivery to a diverse user base while also offering systematic near-real-time monitoring of biogeophysical parameters on a global scale using medium and coarse-scale sensors. Furthermore, CGLS offers supplementary services, such as hot spot mapping, agricultural monitoring, and ground-based validation measurements, upon request, catering to specific regions with limited geographical coverage, characterized by a lower revisit frequency and the use of high-resolution satellite data.

#### 5.4.3.2 Data Sharing and Data Repositories

With the extensive deployment of information technology, a great deal of EO data has been accumulated and stored at various organizations and departments. Sharing data and creating data repositories have become essential elements in today's scientific research landscape. These practices play a crucial role in promoting transparency, collaboration, and the efficient distribution of knowledge to support the implementation of the FAIR principles (FAIR representing Findable, Accessible, Interoperable, and Reusable). These principles outline standardized guidelines for responsible data management and governance, increasing awareness of the value of open science (Wilkinson et al., 2016).

Governmental organizations have been offering EO data at different processing levels as web services for some time, for example, NASA's Earth Explorer. More recently, funding agencies and international scientific journals promote enforcement of FAIR principles by requiring authors to publish the data, measurements, and source code underlying a publication in data repositories such as Pangaea, the GEOSS portal, NASA's EarthData, ESA's Sentinel Hub or the EarthRef Digital Archive. With the growing influence of AI and machine learning, the importance of data sharing and cross-sector repositories containing metadata, field measurements, and spectral libraries has become even more crucial to establish standardized datasets that are compatible with AI applications.

The future of data repositories and data sharing is marked by an increasing embrace of open science, technological progress, interdisciplinary cooperation, and a growing awareness of the advantages of open and easily accessible research data. Data sharing will also extend to include contributions from citizen scientists and the public, enabling large-scale data collection and collaborative efforts on a global scale. These initiatives might be part of more extensive data ecosystems that integrate data storage with data analysis tools and platforms, creating more comprehensive and user-friendly environments for data management and analysis.

However, with the rise in data sharing, there is a growing focus on data privacy and security. Data repositories will need to implement robust security measures to safeguard sensitive information while still making data available. The EU launched Gaia-X initiative aims at establishing the next generation of data infrastructure and serves as an example how to create an open, transparent, and secure federated digital ecosystem, where data and services respond to ethical rules. A key component of GAIA-X is digital sovereignty. Digital sovereignty refers to cloud services that comply with European regulation, and data sovereignty with the goal of being able to safely share data among participants in a consortium (Braud et al., 2021). The GAIA-X initiative is just one example indicating that there is a growing need to advance the data, tools, and resulting technologies, a trend that is likely to persist in the future.

#### 5.4.3.3 Ethical Constraints

Ethics, as a field of study, addresses questions concerning what is right or wrong, good or bad, and consequently offers guidance on what one should or should not do (Kochupillai & Taubenböck, 2023). The UN Principles Relating to Remote Sensing of the Earth from Space (UN, 1986), established over the period between 1970 and 1986, serve as the basis for ethical aspects of EO. These principles encompass various ethical considerations, including the equitable distribution of the benefits of remote sensing, adherence to international law, and the promotion of international cooperation. Additionally, they underscore environmental ethics by emphasizing the use of EO to safeguard the Earth's natural environment. In the broadest context, EO activities should always be oriented toward benefiting all affected parties, rather than merely being profit-driven for those conducting these activities (Wasowski, 1991).

The history of EO, however, has been largely one of science, technology, and engineering. Therefore, ethical rules and principles often were found rather abstract and inadequate to guide scientists towards ethically mindful research (Kochupillai et al., 2022). With big data, exceptionally high-resolution data, and machine learning, the immediate influence of EO on governance, policies, and the well-being of individuals, ethical constraints began to gain prominence in EO science (Harris, 2013).

Regarding the general public, Google Earth has transformed the knowledge of EO by offering comprehensive global imagery, supplemented with higher spatial resolution satellite data and aerial photographs. Google Earth images are presented in the form of straightforward color composites and are freely available. Therefore, Google Earth serves as a platform for disseminating geographic information, thereby making EO data more accessible to a broader user community. However, an ethical dilemma arises when Google Earth imagery is misused for criminal or terrorist activities, potentially granting unrestricted access to military facilities, army command centers, or nuclear plants (Chassay & Johnson, 2007; Harris, 2013).

With the increased spatial resolution of EO sensors, there is also a growing ethical concern over privacy (Fisher et al., 2021; Aranzamendi et al., 2010). Furthermore, significant emphasis has been directed toward ethics in the realm of big data (Kochupillai et al., 2022; VanValkenburg & Dufton, 2020). Studies involving geospatial information can be fraught with ethical uncertainties because professional responsibility demands recognition that the proposed methods may not translate well to other geographical and cultural contexts. In short, location is burdened with contextual specifics; for instance, when publicly accessible geospatial data could amplify the vulnerability of archaeological sites by revealing information of interest to the global antiquities market (Fisher et al., 2021).

To promote greater awareness of ethical constraints in research, it is crucial to emphasize ethics at an institutional level. At present, funding organizations are taking steps to encourage the development of ethical guidelines and protocols for data collection, processing, dissemination, and utilization, with a strong emphasis on respecting diverse cultures, genders, minority groups, and vulnerable individuals. However, there remains a need for global and regional guidelines and principles that can comprehensively address ethics and privacy within diverse cultural and political contexts and serve as substitutes for insufficient or poorly enforced legislation (Taylor, 2017). The issue of enhancing the flow of geospatial data, including EO data, while optimizing the application of cutting-edge technologies and safeguarding the rights of individuals, holds significant relevance for the future.

#### 5.4.3.4 Space Debris—A Pressing Problem

In more than 50 years of space activities, more than 4,800 launches have deployed some 5,000 satellites into orbit, of which only a minor fraction of about 1,000 remain operational today. In addition to this substantial amount of functional space hardware, numerous other objects orbit the Earth. Space debris refers to defunct satellites, satellite breakups, spent rocket stages, fragments from previous satellite collisions, and particles from the collision of other debris (Mark & Kamath, 2019). As of today, surveillance networks count about 27,000 space debris particles larger than 10 cm in size (see also Figure 5.7). These objects vary in size, ranging from small paint flecks to larger defunct satellites. They pose a significant collisions risk to active satellites, which will add even more debris into space. Beyond the trackable debris, an estimated 600,000 to 900,000 fragments sized between 5 mm and 10 cm, as well as many hundreds of thousands of pieces smaller than 5 mm, exist that cannot be monitored (David & Byvik, 2023).

The impact of these undetectable particles, however, is tremendous, as they cause damage to satellite surfaces, which can lead to malfunctions. When space debris of centimeter size collides with a satellite, the substantial kinetic energy can lead to the satellite breaking apart (Ren et al., 2023). For instance, in February 2009, a collision occurred at an altitude of 790 km

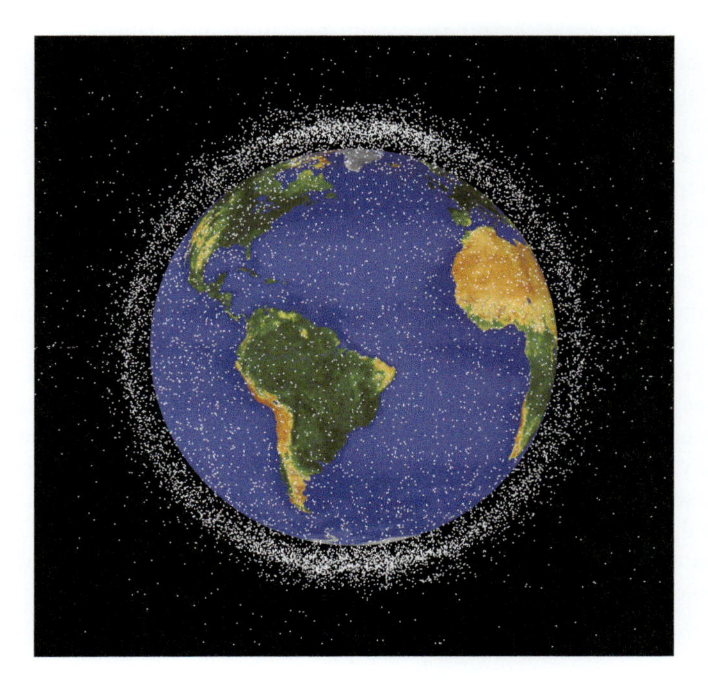

FIGURE 5.7 Computer generated image of objects in Earth orbit that are being tracked in 2019. Approximately 95% of the objects in this illustration are orbital debris, that is, not functional satellites. The orbital debris dots are scaled according to the image size of the graphic to optimize their visibility and are not scaled to Earth. (Courtesy of NASA Orbital Debris Program Office.)

between a U.S. communication satellite and a decommissioned Russian communication satellite, leading to the breakup of both satellites. This incident generated over 800 large trackable fragments and numerous small fragments that cannot be identified or cataloged (Svatina & Cherkasova, 2023).

Space services are expected to further expand as technological advancements and reduced costs remove barriers, while international collaborations in space support continue to grow. State, nonstate, and commercial actors will, therefore, increasingly gain access to EO data and services. Consequently, a wider array of entities, including governments, nongovernmental organizations, and commercial enterprises, will gain access to EO data and services. The number of EO companies and satellite service providers is expected to continue growing at least through 2025. For instance, Starlink plans to deploy up to 42,000 satellites for its satellite-based internet system (Kay, 2023). Amazon's Project Kuiper plans to establish a constellation of 3,236 small communications satellites in low-Earth orbit until 2029. The US Space Force intends to launch between 300 to 500 small communications satellites into low-Earth orbit to form the Transport Layer constellation, while China will soon complete the Hongyan constellation of 320 small communications satellites in low-Earth orbit (David & Byvik, 2023).

Today, various methods for space debris removal already exist (Mark & Kamath, 2019; Svatina & Cherkasova, 2023). However, due to the urgent worldwide concern regarding space debris, multiple countries are initiating initiatives to catalog and eliminate debris. Nonetheless, it is clear that the difficulties associated with tracking and eliminating debris will inevitably increase in the years ahead.

## 5.5 CONCLUSIONS

From the early beginning of satellite-based remote sensing in the 1970s, the use of remote sensing for terrestrial applications has experienced a remarkable increase. A multitude of applications spanning from local to global scales have been established since then. However, the needs and applications within the terrestrial EO community are highly diverse, with various user groups, sensor categories, and data requirements. Consequently, a range of distinct systems is required to cater to the specific information needs. In the past decade, the EO community has witnessed a growing synergy between missions, presenting a substantial potential to leverage sensors with varying areas of focus and data outputs. This has been notably observed in missions like Landsat, MODIS, MERIS, and the Copernicus programme. Furthermore, interdisciplinary studies that harness the increased data density to explore temporal dynamics across multiple scales indicate a trajectory toward the future development of well-calibrated, global products.

Besides affordability and sustainability, long-term scenarios are geared toward fulfilling the priorities of continuity, observation frequency, sensor (inter-)calibration, and system evolution. Continuity of operational observations drives the nature of the observations and the design of missions; the frequency of observations drives the deployment of simultaneous sensors in orbit for each class of observations; sensor calibration and (inter-)calibration play an instrumental role in ensuring data continuity, while sensor evolution is driven by the evolving needs of users, encompassing demands for new services and products, fresh observation requirements, and technological enhancements. We are on the cusp of an era characterized by an unparalleled wealth of EO data and information. However, as we embark on this journey, we must be prepared to grapple with a multitude of technological and ethical challenges that will shape the path ahead.

## REFERENCES

Adriano, B., Yokoya, N., Xia, J. & Baier, G. 2021. Big Earth Observation Data Processing for Disaster Damage Mapping. Werner, M. & Chiang, Y. Y. (Eds.). Handbook of Big Geospatial Data. Cham: Springer. DOI: 10.1007/978-3-030-55462-0_4.

Allison, L. J. (Ed.), Schnapf, A., Diesen, B. C., Martin, P. S., Schwalb, A. & Bandeen, W. R. 1962. Tiros 1 Meteorological Satellite System Final Report. NASA Technical Report. Document Number 19640007992.

Amro, I., Mateos, J., Vega, M., Molina, R. & Katsaggelos, A. K. 2011. A survey of classical methods and new trends in pansharpening of multispectral images. EURASIP Journal on Advances in Signal Processing, 1, 1–22.

Aoki, Y., Furuya, M., De Zan, F., Doin, M.-P., Eineder, M., Ohki, M. & Wright, T. 2021. L-band synthetic aperture radar: Current and future applications to Earth sciences. Earth Planets Space 73, 56. DOI: 10.1186/s40623–021–01363-x.

Aranzamendi, M. S., Sandau, R. & Schrogl, K.-U. 2010. Treaty Verification and Law Enforcement through Satellite Earth Observation. Privacy Conflicts from High Resolution Imaging. Vienna: European Space Policy Institute. p. 70.

Aschbacher, J. 2019. ESA's Earth Observation Strategy and Copernicus. Onoda, M. & Young, O. R. (Eds.). Satellite Earth Observations and Their Impact on Society and Policy. Springer Open, 81–86. DOI: 10.1007/978-981-10-3713-9.

Atkinson, P. M. & Tatnall, A. R. L. 1997. Introduction neural networks in remote sensing. International Journal of Remote Sensing, 18(4), 699–709.

Balakrishnan, S. S., Narayana Moorthi, D., Venkateswara Rao, P., Sanyal, M. K. & Radhakrishnan, D. 2004. Satish Dhawan Space Centre SHAR-Spaceport of India. 55th International Astronautical Congress of the International Astronautical Federation, the International Academy of Astronautics, and the International Institute of Space Law; 4 October–8 October, Vancouver, Canada. DOI: 10.2514/6.IAC-04-V.2.03.

Baltuck, M., Briggs, S., Loyche-Wilkie, M., McGee, A., Muchoney, D. & Skøvseth, P. E. 2013. The Global Forest Observation Initiative: Fostering the use of satellite data in forest management, reporting and verification. Carbon Management, 4(1), 17–21.

Barrett, E. C. & Curtis, L. F. 2013. Introduction to Environmental Remote Sensing. New York, USA: Routledge Taylor and Francis.

Bartholomé, E. & Belward, A. S. 2005. GLC2000: A new approach to global land cover mapping from Earth observation data. International Journal of Remote Sensing, 26(9), 1959–1977.

Belgiu, M. & Stein, A. 2019. Spatiotemporal image fusion in remote sensing. Remote Sensing, 11(7), 818. DOI: 10.3390/rs11070818.

Belward, A. S. & Skøien, J. O. 2015. Who launched what, when and why: Trends in global land-cover observation capacity from civilian earth observation satellites. ISPRS Journal of Photogrammetry and Remote Sensing, 103, 115–128.

Berger, M. & Aschbacher, J. 2012. The sentinel missions: New opportunities for science. Remote Sensing of Environment, 120, 1–276.

Bhardwaj, A., Sam, L., Martín-Torres, F. J. & Kumar, R. 2016. UAVs as remote sensing platform in glaciology: Present applications and future prospects. Remote Sensing of Environment, 175, 196–204.

Bishop, C. M. 2006. Pattern Recognition and Machine Learning. New York: Springer. p. 798. ISBN 0387310732.

Bontemps, S., Defourny, P., Bogaert, E. V., Arino, O., Kalogirou, V. & Perez, J. R. 2011. GLOBCOVER 2009. Products Description and Validation Report, UCV Louvain and ESA.

Bourg, L., Bruniquel, C., Henocq, C., Morris, H., Dash, J., Preusker, R. & Dansfeld, S. 2023. Sentinel-3 OLCI Land User Handbook. Available online at https://sentinel.esa.int/documents/247904/4598066/Sentinel-3-OLCI-Land-Handbook.pdf

Braud, A., Fromentoux, G., Radier, B. & Le Grand, O. 2021. The road to European digital sovereignty with Gaia-X and IDSA. IEEE Network, 35(2), 4–5. DOI: 10.1109/MNET.2021.9387709.

Brown, M. E., Escobar, V. M., Aschbacher, J., Milagro-Perez, M. P., Doorn, B., Macauley, M. K. & Friedl, L. 2013. Policy for robust space-based earth science, technology and applications. Space Politics, 29(1), 76–82.

Burney, S. M. A. & Tariq, H. 2014. K-means cluster analysis for image segmentation. International Journal of Computer Applications, 96(4).

Campbell, B. A. 2012. High circular polarization ratios in radar scattering from geologic targets. Journal of Geophysical Research, 177. DOI: 10.1029/2012JE004061.

Campbell, J. B. & Wynne, R. H. 2011. Introduction to Remote Sensing. New York, USA: The Guilford Press.

Camps-Valls, G., Svendsen, D. H., Cortés-Andrés, J., Mareno-Martínez, Á., Pérez-Suay, A., Adsuara, J., Martín, I., Piles, M., Muñoz-Marí, J. & Martino, L. 2021. Physics-aware machine learning for geosciences and remote sensing. IGARSS, 2086–2089.

Celesti, M., Rast, M., Adams, J., Boccia, V., Gascon, F., Isola, C. & Nieke, J. 2022. The Copernicus Hyperspectral Imaging Mission for the Environment (Chime): Status and Planning. IEEE International Geoscience and Remote Sensing Symposium. IGARSS. DOI: 10.1109/IGARSS46834.2022.9883592.

CGLS (Copernicus Global Land Service). 2023. Copernicus Global Land Service Homepage. Available online at https://land.copernicus.eu/global/

Chabrillat, S., Guanter, L., Kaufmann, H., Förster, S., Beamish, A., Brosinsky, A., Wulf, H., Asadzadeh, S., Bochow, M., Bohn, N., Bösche, N., Bracher, A., Brell, M., Buddenbaum, H., Cerra, D., Fischer, D., Hank, T., Heiden, U., Heim, B., Heldens, W., Hill, J., Hollstein, J., Hostert, P., Krasemann, H., LaPorta, S., Leitão, P. S., van der Linden, S., Mauser, W., Milewski, R., Mottus, M., Okujeni, A. Oppelt, N., Pinnel, N., Roessner, S., Röttgers, R., Schickling, A., Schneiderhan, T., Soppa, M., Staenz, K. & Segl, K. (2022). EnMAP Science Plan. Potsdam: GFZ Data Services. p. 87. DOI: 10.48440/enmap.2022.001.

Chassay, C. & Johnson, B. 2007. Google earth used to target Israel. The Guardian, October 25.

Chaudhuri, U., Banerjee, B. & Bhattacharya, A. 2019. Siamese graph convolutional network for content based remote sensing image retrieval. Computer Vision and Image Understanding, 184, 22–30.

Chevrel, M., Courtois, M. & Weill, G. 1981. The SPOT satellite remote sensing mission. Photogrammetric Engineering & Remote Sensing, 47, 1163–1171.

Chuah, H. T. 1993. An artificial neural network for inversion of vegetation parameters from radar backscatter coefficients. Journal of Electromagnetic Waves and Applications, 7(8), 1075–1092.

Claverie, M., Ju, J., Masek, J. M., Dungan, J. L., Vermote, E. F., Roger, J. C., Skakun, S. V. & Justice, C. 2018. The harmonized landsat and sentinel-2 surface reflectance data set. Remote Sensing of Environment, 219, 145–161. DOI: 10.1016/j.rse.2018.09.002.

Colwell, R.1956. Determining the prevalence of certain cereal crop diseases by means of aerial photography. Hilgardia, 26, 223–286.

Comittee on Earth Observation System of Systems (CEOS). 2019. CEOS Virtual Constellations Process Paper. Available online at https://ceos.org/document_management/Publications/Governing_Docs/Virtual-Constellations_Process%20Paper_rev1-2019.pdf (accessed October 16, 2023).

Copernicus. 2018. Copernicus DIAS Fact Sheet: The DIAS User Friendly Access to Copernicus Data and Information. Available online at www.copernicus.eu/sites/default/files/Copernicus_DIAS_Factsheet_June2018.pdf (accessed October 23, 2023).

Coppo, P., Taiti, A., Pettinato, L., Francois, M., Taccola, M. & Drusch, M. 2017. Flourescence Imaging Spectrometer (FLORIS) for ESA FLEX mission. Remote Sensing, 9(7), 649. DOI: 10.3390/rs9070649.

Cordeiro, M. C. R., Martinez, J.-M. & Peña-Luque, S. 2021. Automatic water detection from multidimensional hierarchical clustering for Sentinel-2 images and a comparison with Level 2A processors. Remote Sensing of Environment, 253, 112209. DOI: 10.1016/j.rse.2020.112209.

Cote, S., Lapointe, M., Lisle, D. D., Arsenault, E. & Wierus, M. 2021. The RADARSAT Constellation: Mission Overview and Status. EUSAR 2021, 13th European Conference on Synthetic Aperture Radar, 1–5.

Dakhir, A., Barramou, F. Z. & Omar, A. B. 2021. Optical satellite images services for precision agricultural use: A review. Advances in Science, Technology and Engineering Systems Journal, 6(3), 326–331.

Dao, D., Cang, C., Fung, C., Zhang, M., Pawlowski, N., Gonzales, R., Beglinger, N. & Zhang, C. 2019. GainForest: Scaling Climate Finance for Forest Conservation Using Interpretable Machine Learning on Satellite Imagery. ICML Climate Change AI Workshop, Long Beach, California.

David, J. E. & Byvik, C. 2023. What's Up There, Where Is It, and What's It Doing? The U.S. Space Surveillance Network. National Security Archive. 2023, March 13. Briefing Book 824.

Davidson, M., Iannini, L., Torres, R. & Geudtner, D. 2022. New perspectives for applications and services provided by future spaceborne SAR missions at the European Space Agency. IGARSS, 4720–4723.

Davis, F., Quattrochi, D., Ridd, M., Lam, M., Walsh, S., Michaelson, J., Franklin, J., Stow, D., Johanssen, G. & Johnston, C. 1991. Environmental analysis using integrated GIS and remotely sensed data: Some research needs and priorities. Photogrammetric Engineering and Remote Sensing, 57(6), 689–697.

Demir, B., Persello, C. & Bruzzone, L. 2010. Batch-mode active-learning methods for the interactive classification of remote sensing images. IEEE Transactions on Geoscience and Remote Sensing, 49(3), 1014–1031.

Denis, G., Claverie, A., Pasco, X., Darnis, J.-P., de Maupeou, B., Lafaye, M. & Morel, E. 2017. Towards disruptions in Earth observation? New Earth observation systems and markets evolution: Possible scenarios and impacts. Acta Astronautica, 137, 415–433. DOI: 10.1016/j.actaastro.2017.04.034.

De Oliveira, L. C., Lima, M. G. R. & Hubscher, G. L. 2000. CBERS: An international space cooperation program. Acta Astronautica, 47, 559–564.

Deschamps-Berger, C., Gascoin, S., Shean, D., Besso, H., Guiot, A. & López-Moreno, J. I. 2023. Evaluation of snow depth retrievals from ICESat-2 using airborne laser-scanning data. The Cryosphere, 17(7), 2779–2792.

Dubayah, R., Blair, J. B., Goetz, S., Fatoyinbo, L., Hansen, M., Healey, S., Hofton, M., Hurtt, G., Kellner, J., Luthcke, S., Armston, J., Tang, H., Duncanson, L., Hancock, S., Jantz, P., Marselis, S., Patterson, P. L., Qi, W. & Silva, C. 2020. The global ecosystem dynamics investigation: High-resolution laser ranging of the Earth's forests and topography. Science of Remote Sensing, 1, 2020, 100002. DOI: 10.1016/j.srs.2020.100002.

Durrieu, S. & Nelson, R. F. 2013. Earth observation from space: The issue of environmental sustainability. Space Politics, 29(4), 238–250.

Elachi, C.1988. Spaceborne Radar Remote Sensing: Applications and Techniques. New York, USA: IEEE Press.

Eroglu, O., Kurum, M., Boyd, D. & Gurbuz, A. C. 2019. High spatio-temporal resolution CYGNSS soil moisture estimates using artificial neural networks. Remote Sensing, 11(19), 2272.

ESA. 2021. Climate Change Initiative Reference Documentation. Status of the Cooperation between ESA's Climate Change Initiative and the EC's Copernicus Climate Change Service. ESA/PB-EO(2021)74. Available online at https://climate.esa.int/media/documents/ESA_PB-EO_2021_74_EN1.pdf (accessed October 17, 2023).

ESA (European Space Agency). 2023. 25 Times Copernicus Made the Headlines. ESA Applications Newsletter, June 7. Available online at www.esa.int/Applications/Observing_the_Earth/Copernicus/25_times_Copernicus_made_the_headlines#:~:text=Now%2C%20well%20established%20as%20the,selection%20of%2025%20Copernicus%20highlights (accessed October 17, 2023).

Ergorov, A. V., Roy, D. P., Zhang, H. K., Hansen, M. C. & Kommareddy, A. 2018. Demonstration of percent tree cover mapping using landsat Analysis Ready Data (ARD) and sensitivity with respect to landsat ARD processing level. Remote Sensing, 10, 209. DOI: 10.3390/rs10020209.

Feng, Y., Lu, Y., Moran, E. F., Dutra, L. V., Calvi, M. F. & de Oliveira, M. A. F. 2017. Examining spatial distribution and dynamic change of urban land covers in the Brazilian Amazon using multitemporal multisensor high spatial resolution satellite imagery. Remote Sensing, 9(4), 381. DOI: 10.3390/rs9040381.

Ferreira, K., Queiroz, G. R., Marujo, R. F. B. & Costa, R. W. 2022. Building Earth Observation Data Cubes on AWS. The International Archives of the Photogrammetry, Remote Sensing and Spatial Information Sciences, Volume XLIII-B3–2022, XXIV ISPRS Congress, 6–11 June, Nice, France. DOI: 10.5194/isprs-archives-XLIII-B3-2022-597-2022.

Filippi, A., Klein, A. G., Dobreva, I. & Jensen, J. R. 2010. Self-organizing map-based applications in remote sensing. Self-Organizing Maps, 432.

Fischer, W. A., Hemphill, W. R. & Kover, A. 1976. Progress in remote sensing. Photogrammetria, 32, 33–72.

Fisher, M., Fradley, M., Flohr, P., Rouhani, B. & Simi, F. 2021. Ethical considerations for remote sensing and open data in relation to the endangered archaeology in the Middle East and North Africa project. Archaeological Prospection, 28(3), 279–292.

Garvin, J., Bufton, J., Blair, J., Harding, D., Luthcke, S., Frawley, J. & Rowlands, D. 1998. Observations of the earth's topography from the Shuttle Laser Altimeter (SLA): Laser-pulse echo-recovery measurements of terrestrial surfaces. Physics and Chemistry of the Earth, 23(9–10), 1053–1068. DOI: 10.1016/S0079-1946(98)00145-1.

Gautam, R., Patel, P. N., Singh, M. K., Liu, T., Mickley, L. J., Jethva, H. & DeFries, R. S. 2023. Extreme smog challenge of India intensified by increasing lower tropospheric stability. Geophysical Research Letters, 50(11). DOI: 10.1029/2023GL103105.

GEO (Group on Earth Observations). 2010. Report on Progress Beijing Ministerial Summit-Observe, Share, Inform. Geneva, Switzerland: GEO Secretariat.

GEO (Group on Earth Observations). 2023. The GEO Website. Available online at www.earthobservations.org (accessed October 27, 2023)

GFW (Global Forest Watch). 2023. The Global Forest Watch Website. Available online at www.globalforestwatch.org/ (accessed October 17, 2023).

Ghamisi, P., Rasti, B., Yokoya, N., Wang, Q., Hofle, B., Bruzzone, L., Bovolo, F., Chi, M., Anders, K., Gloaguen, R., Atkinson, P. M. & Benediktsson, J. A. 2019. Multisource and multitemporal data fusion in remote sensing: A comprehensive review of the state of the art. IEEE Geoscience and Remote Sensing Magazine, 7(1), 6–39. DOI: 10.1109/MGRS.2018.2890023.

Ghassemian, H. 2023. A review of remote sensing fusion methods. Information Fusion, 32(A), 75–89. DOI: 10.1016/j.inffus.2016.03.003.

Girardeau-Montaut, D. 2016. Cloudcompare. CloudCompare (version 2.6.2) [GPL software]. Available online at http://www.cloudcompare.org/

Girshick, R. 2015. Fast R-CNN. IEEE International Conference on Computer Vision, 1440–1448.

Girshick, R., Donahue, J., Darrell, T. & Malik, J. 2014. Rich feature hierarchies for accurate object detection and semantic segmentation. IEEE Conference on Computer Vision and Pattern Recognition, 580–587.

Gonzalo-Martín, C., Lillo-Saavedra, M., Menasalvas, E., Fonseca-Luengo, D., García-Pedrero, A. & Costumero, R. 2016. Local optimal scale in a hierarchical segmentation method for satellite images: An OBIA approach for the agricultural landscape. Journal of Intelligent Information Systems, 46, 517–529.

Goodchild, M. F. 2007. Citizens as sensors: The world of volunteered geography. GeoJournal, 69(4), 211–221.

Gould, M., Craglia, M., Goodchild, M. F., Annoni, A., Camara, G., Kuhn, K., Mark, D., Masser, I., Liang, S. & Parsons, E. 2008. Next generation digital earth: A position paper from the Vespucci initiative for the advancement of geographic information science. International Journal of Spatial Data Infrastructures Research, 3, 146–167.

Goward, S. N., Masek, J. G., Williams, D. L., Irons, J. R. & Thompson, R. J. 2001. The Landsat 7 mission: Terrestrial research and applications for the 21st century. Remote Sensing of Environment, 78, 3–12.

Gröschler, K. C., Muhuri, A., Roy, S. K. & Oppelt, N. 2023. Monitoring the population dynamics of indicator plants in high-nature-value grassland using machine learning and drone data. Drones, 7(10), 644. DOI: 10.3390/drones7100644.

Guo, H., Fang, C., Zhongchang, S., Jie, L. & Dong, L. 2021. Big earth data: A practice of sustainability science to achieve the sustainable development goals. Science Bulletin, 66(11), 1050–1053. DOI: 10.1016/j.scib.2021.01.012.

Gutman, G. G.1991. Vegetation indices from AVHRR: An upgrade and future prospects. Remote Sensing of Environment, 35, 121–136.

Hall, O., Dompae, F., Wahab, I. & Dzanku, F. M. 2023. A review of machine learning and satellite imagery for poverty prediction: Implications for development research and applications. Journal of International Development. DOI: 10.1002/jid.3751.

Hame, T., Salli, K., Andersson, K. & Lohi, A. 1997. A new methodology for the estimation of biomass of conifer-dominated boreal forest using NOAA AVHRR data. International Journal of Remote Sensing, 18(15), 3211–3243.

Häme, T., Sirro, L., Kilpi, J., Seitsonen, L., Andersson, K. & Melkas, T. 2020. A hierarchical clustering method for land cover change detection and identification. Remote Sensing, 12, 1751. DOI: 10.3390/rs12111751.

Hancock, S., McGrath, C., Lowe, C., Davenport, I. & Woodhouse, I. 2021. Requirements for a global lidar system: Spaceborne lidar with wall-to-wall coverage. Royal Society Open Science, 8, 211166. DOI: 10.1098/rsos.211166.

Hariharan, S., Tirodkar, S., Porwal, A., Bhattacharya, A. & Joly, A. 2017. Random forest-based prospectivity modelling of greenfield terrains using sparse deposit data: An example from the Tanami Region, Western Australia. Natural Resources Research, 26, 489–507.

Harris, R. 2013. Reflections on the value of ethics in relation to Earth observation. International Journal of Remote Sensing, 34(4), 1207–1219.

Hastings, D. A. & Emery, W. 1992. The advanced very high resolution radiometer (AVHRR): A brief reference guide. Photogrammetric Engineering & Remote Sensing, 58(8), 1183–1188.

Haut, J. M., Paoletti, M. E., Moreno-Álvarez, S., Plaza, J., Rico-Gallego, J.-A. & Plaza, A. 2021. Distributed deep learning for remote Sensing data interpretation. Proceedings of the IEEE, 109(8), 1320–1349, August. DOI: 10.1109/JPROC.2021.3063258.

Herold, M., Mayaux, P., Woodcock, C. E., Baccini, A. & Schmullius, C. 2008. Some challenges in global land cover mapping: An assessment of agreement and accuracy in existing 1 km datasets. Remote Sensing of Environment, 112(5), 2538–2556.

Hill, P. R., Kumar, A., Temimi, M. & Bull, D. R. 2020. HABNet: Machine learning, remote sensing-based detection of harmful algal blooms. IEEE Journal of Selected Topics in Applied Earth Observations and Remote Sensing, 13, 3229–3239.

Himeur, Y., Rimal, B., Tiwary, A. & Abbes, A. 2022. Using artificial intelligence and data fusion for environmental monitoring: A review and future perspectives. Information Fusion, 86–87, 44–75. DOI: 10.1016/j.inffus.2022.06.003.

Holben, B. N., Kaufman, Y. J. & Kenall, J. D. 1990. NOAA-11 AVHRR visible and near-IR inflight calibration. International Journal of Remote Sensing, 11(8), 1511–1519.

Huete, A., Didan, D., Miura, T., Rodriguez, E. P. & Gao, X. 2002. Overview of the radiometric and biophysical performance of MODIS vegetation indices. Remote Sensing of Environment, 83(1–2), 195–213.

Ibrahim, S. K., Ziedan, I. E. & Ahmed, A. 2021. Study of climate change detection in North-East Africa using machine learning and satellite data. IEEE Journal of Selected Topics in Applied Earth Observations and Remote Sensing, 14, 11080–11094.

Inoue, Y. 2020. Satellite- and drone-based remote sensing of crops and soils for smart farming: A review. Soil Science and Plant Nutrition, 66, 798–810. DOI: 10.1080/00380768.2020.1738899.

Ji, H., Zuo, Z. & Han, Q.-L. 2022. A divisive hierarchical clustering approach to hyperspectral band selection. IEEE Transactions on Instrumentation and Measurement, 71, 5014312.

Jones, H. G. & Vaughan, R. A. 2010. Remote Sensing of Vegetation: Principles, Techniques, and Applications. New York, USA: Oxford University Press.

Joshi, G., Natsuaki, R. & Hirose, A. 2023. Application of inverse mapping for automated determination of normalized indices useful for land surface classification. IEEE Journal of Selected Topics in Applied Earth Observations and Remote Sensing, 16, 7804–7818. DOI: 10.1109/JSTARS.2023.3308049.

JPL (Jet Propulsion Laboratory, California Institute of Technology). 2018. HyspIRI Final Report. Available online at http://hyspiri.jpl.nasa.gov/ (accessed October 6, 2023).

Kay, G. 2023. Everything we know about Elon Musk's Starlink satellites and future internet plans. Business Insider, August 9.

Kellogg, K., Hoffman, P., Standley, S., Shaffer, S., Rosen, P., Edelstein, W., Dunn, C., Baker, C., Barela, P., Shen, Y., Guerrero, A. M., Xaypraseuth, P., Sagi, V. R., Harinath, N., Kumar, R., Bhan, R. & Sarma, C. V. H. S. 2020. NASA-ISRO Synthetic Aperture Radar (NISAR) Mission. IEEE Aerospace Conference, Big Sky, MT, USA, 1–21.

Kim, J., Lin, S. Y., Singh, T. & Singh, R. P. 2023. InSAR time series analysis to evaluate subsidence risk of monumental Chandigarh City (India) and surroundings. IEEE Transactions on Geoscience and Remote Sensing. DOI: 10.1109/TGRS.2023.3305863.

Kochupillai, M., Kahl, M., Schmitt, M., Taubenböck, H. & Zhu, X. X. 2022. Earth observation and artificial intelligence: Understanding emerging ethical issues and opportunities. IEEE Geoscience and Remote Sensing Magazine, 10(4), 90–124. DOI: 10.1109/MGRS.2022.3208357.

Kochupillai, M. & Taubenböck, H. 2023. Conducting Ethically Mindful Earth Observation Research: The Case of Slum Mapping. IGARSS 2023–2023 IEEE International Geoscience and Remote Sensing Symposium, Pasadena (USA), 1937–1940. DOI: 10.1109/IGARSS52108.2023.10281725.

Kostadinov, T. S., Schumer, R., Hausner, M., Bormann, K. J., Gaffney, R., McGwire, K., Painter, T. H., Tyler, S. & Harpold, A. A. 2019. Watershed-scale mapping of fractional snow cover under conifer forest canopy using lidar. Remote Sensing of Environment, 222, 34–49.

Kucharczyk, M. & Hugenholtz, C. H. 2021. Remote sensing of natural hazard-related disasters with small drones: Global trends, biases, and research opportunities. Remote Sensing of Environment, 264, 112577. DOI: 10.1016/j.rse.2021.112577.

Kumar, R., Rosen, P. & Misra, T. 2016. NASA-ISRO synthetic aperture radar: Science and applications. In Earth Observing Missions and Sensors: Development, Implementation, and Characterization IV, 9881, 988103.

Kuschel, H. & O'Hagan, D. 2010. Passive Radar from History to Future. Proceeding of the 11th International Radar Symposium, 16–18 June, Vilnuis (Lithuania), 1–4. ISBN 978-1-4244-5613-0.

Lary, D. J., Alavi, A. H., Gandomi, A. H. & Walker, A. L. 2016. Machine learning in geosciences and remote sensing. Geoscience Frontiers, 7(1), 3–10.

Lawrence, J. 1993. Introduction to neural networks. California Scientific Software, 324.

Li, J. & Roy, D. P. 2017. A global analysis of Sentinel-2A, Sentinel-2B and Landsat-8 data revisit intervals and implications for terrestrial monitoring. Remote Sensing, 9, 902. DOI: 10.3390/rs9090902.

Liebig, V., Briggs, S. & Grassl, H. 2007. The changing Earth. ESA Bulletin, 129, 8–17.

Lillesand, T., Kiefer, R. W. & Chipma, J. 2008. Remote Sensing and Image Interpretation. New York: Wiley & Sons.

Loizzo, R., Guarini, R., Longo, F., Scopa, T., Formaro, R., Facchinetti, C. & Varacalli, G. 2018. Prisma: The Italian Hyperspectral Mission. IEEE International Geoscience and Remote Sensing Symposium. IGARSS. DOI: 10.1109/IGARSS.2018.8518512.

Louet, J. & Bruzzi, S. 1999. ENVISAT mission and system. IEEE International Geoscience and Remote Sensing Symposium. IGARSS '99 Proc, 3, 1680–1682.
Lynch, M., Maslin, H., Balzter, H. & Sweeting, M. 2013. Sustainability: Choose satellites to monitor deforestation. Nature, 496, 293–294.
Ma, Y., Wu, H., Wang, L., Huang, B., Ranjan, R., Zomaya, A. & Jie, W. 2015. Remote sensing big data computing: Challenges and opportunities. Future Generation Computer Systems, 51, 47–60. DOI: 10.1016/j.future.2014.10.029.
Mark, C. P. & Kamath, S. 2019. Review of active space debris removal methods. Space Policy, 47, 194–206. DOI: 10.1016/j.spacepol.2018.12.005.
Mas, J. F. & Flores, J. J. 2008. The application of artificial neural networks to the analysis of remotely sensed data. International Journal of Remote Sensing, 29(3), 617–663. DOI: 10.1080/01431160701352154.
McCallum, I., Obersteiner, M., Nilsson, S. & Shvidenko, A. 2006. A spatial comparison of four satellite derived 1km global land cover datasets. International Journal of Applied Earth Observation and Geoinformation, 8(4), 246–255.
McGaughey, R. 2007. Fusion/LDV: Software for Lidar Data Analysis and Visualization. USDA Forest Service, Pacific Northwest Research Station.
Misty, D.1998. India's emerging space program. Pacific Affairs, 71, 151–174.
Mohney, J. 2020. Terabytes from space: Satellite imaging is filling data centers. Data Center Frontier, April 28. Available online at www.datacenterfrontier.com/ (accessed October 20, 2023).
Mora, B., Tsendbazar, N. E., Herold., M. & Arino, O. 2014. Global land cover mapping: Current status and future trends: Land use land cover. Remote Sensing and Digital Image Processing, 18, 11–30.
Moreira, A., Prats-Iraola, P., Yonis, M., Krieger, G., Hajnsek, I. & Papathanassiou, K. P. 2013. A tutorial on synthetic aperture radar. IEEE Geoscience and Remote Sensing Magazine, 43.
Morel, P. 2013. Advancing earth observation from space: A global challenge. Space Policy, 29, 175–180.
Morena, L. C, James, K. V. & Beck, J. 2004. An introduction to the RADARSAT-2 mission. Canadian Journal of Remote Sensing, 30(3).
Mou, L., Ghamisi, P. & Zhu, X. X. 2017. Deep recurrent neural networks for hyperspectral image classification. IEEE Transactions on Geoscience and Remote Sensing, 55(7), 3639–3655.
NASA EarthData (NASA's Earth Science Data Systems Program). 2023. Harmonized Landsat and Sentinel-2 (HLS). Available online at www.earthdata.nasa.gov/esds/harmonized-landsat-sentinel-2 (accessed October 16, 2023).
NASA EOSPSO (NASA's Earth Observing System Project Science Office). 2019. SMAP Mission. Available online at http://eospso.nasa.gov/missions/soil-moisture-active-passive (last updated on October 10, 2019, accessed April 4, 2024).
National Research Council of the National Academies. 2007. Earth Science and Applications from Space: National Imperatives for the Next Decade and Beyond. Washington, DC: The National Academies Press. DOI: 10.17226/11820.
Nativi, S., Mazzetti, P., Craglia, M. & Pirrone, N. 2013. The GEOSS solution for enabling data interoperability and integrative research. Environmental Science and Pollution Research, 21, 4177–4192.
Navulur, K., Pacifici, F. & Baugh, B. 2013. Trends in optical commercial remote sensing industry. IEEE Geoscience and Remote Sensing Magazine, December, 57–64.
Neagoe, V.-E., Stoica, R.-M., Ciurea, A.-I., Bruzzone, L. & Bovolo, F. 2014. Concurrent self-organizing maps for supervised/unsupervised change detection in remote sensing images. IEEE Journal of Selected Topics in Applied Earth Observations and Remote Sensing, 7(8), 3525–3533.
Neuenschwander, A., Magruder, L., Guenther, E., Hancock, S., Purslow, M. 2022. Radiometric assessment of ICESat-2 over vegetated surfaces. Remote Sensing, 14, 787. DOI: 10.3390/rs14030787.
Notarnicola, C., Angiulli, M. & Posa, F. 2008. Soil moisture retrieval from remotely sensed data: Neural network approach versus Bayesian method. IEEE Transactions on Geoscience and Remote Sensing, 46(2), 547–557.
Odermatt, D., Gitelson, A., Brando, V. E. & Schaepman, M. 2012. Review of constituent retrieval in optically deep and complex waters from satellite imagery. Remote Sensing of Environment, 118, 116–126.
Pan, Z., Xu, J., Guo, Y., Hu, Y. & Wang, G. 2020. Deep learning segmentation and classification for urban village using a worldview satellite image based on U-Net. Remote Sensing, 12(10), 1574.
Parashar, S., Langham, E., McNally, J. & Ahmed, S. 1993. RADARSAT mission requirements and concept. Canadian Journal of Remote Sensing, 19(4).

Pereira, M., Ferrier, S., Walters, M., Geller, G. N., Jongman, R. H. G., Scholes, R. J., Bruford, M. W., Brummitt, N., Butchart, S. H. M., Cardoso, A. C., Coops, N. C., Dulloo, E., Faith, D. P., Freyhof, J., Gregory, R. D., Heip, C., Höft, R., Hurtt, G., Jetz, W., Karp, D., McGeoch, M. A., Obura, D., Onoda, Y., Pettorelli, N., Reyers, B., Sayre, R., Scharlemann, J. P. W., Stuart, S. N., Turak, E., Walpole, M. & Wegmann, M. 2013. Essential biodiversity variables. Science, 339, 277–278.

Persello, C., Grift, J., Fan, X., Paris, C., Hänsch, R., Koeva, M. & Nelson, A. 2023. AI4SmallFarms: A data set for crop field delineation in Southeast Asian smallholder farms. IEEE Geoscience and Remote Sensing Letters, 20, 1–5. DOI: 10.1109/LGRS.2023.3323095.

Pettorelli, N. 2013. The Normalized Difference Vegetation Index. New York: Oxford University Press.

Pham, M.-T., Courtrai, L., Friguet, C., Lefèvre, S. & Baussard, A. 2020. YOLO-Fine: One-stage detector of small objects under various backgrounds in remote sensing images. Remote Sensing, 12(15), 2501.

Pielke, R. A. 2005. Land use and climate change. Science, 310, 1625–1626.

Pierdicca, N., Davidson, M., Chini, M., Dierking, W., Djavidnia, S., Haarpaintner, J., Hajduch, G., Laurin, G. V., Lavalle, M., López-Martínez, C., Nagler, T. & Su, B. 2019. The Copernicus L-band SAR mission ROSE-L (radar observing system for Europe). In Active and Passive Microwave Remote Sensing for Environmental Monitoring III, 11154, 111540E.

Planetary Computer. 2022. Available online at https://planetarycomputer.microsoft.com/ (accessed October 23, 2023).

Plummer, S., Lecomte, P. & Doherty, M. 2017. The ESA Climate Change Initiative (CCI): A European contribution to the generation of the global climate observing system. Remote Sensing of Environment, 203, 2–8.

Qiu, S., Wen, G., Deng, Z., Liu, J. & Fan, Y. 2018. Accurate non-maximum suppression for object detection in high-resolution remote sensing images. Remote Sensing Letters, 9(3), 237–246.

Rajaneesh, A., Vishnu, C. L., Oommen, T., Rajesh, V. J. & Sajinkumar, K. S. 2022. Machine learning as a tool to classify extra-terrestrial landslides: A dossier from Valles Marineris, Mars. Icarus, 376, 114886.

Reichstein, M., Camps-Valls, G., Stevens, B., Jung, M., Denzler, J., Carvalhais, N. & Prabhat, P. 2019. Deep learning and process understanding for data-driven Earth system science. Nature, 566(7743), 195–204.

Ren, S., Gong, Z., Wu, Q., Song, G., Zhang, Q., Zhang, P., Chen, C. & Cao, Y. 2023. Satellite breakup behaviors and model under the hypervelocity impact and explosion: A review. Defense Technology, 27, 284–307. DOI: 10.1016/j.dt.2022.08.004.

Ren, S., He, K., Girshick, R. & Sun, J. 2015. Faster R-CNN: Towards real-time object detection with region proposal networks. Advances in Neural Information Processing Systems, 28.

Rolnick, D., Donti, P. L., Kaack, L. H., Kochanski, K., Lacoste, A., Sankaran, K., Ross, A. S., Milojevic-Dupont, N., Jaques, N., Waldman-Brown, A., Luccioni, A. S., Maharaj, T., Sherwin, E. D., Mukkavilli, S. K., Kording, K. P., Gomes, C. P., Ng, A. Y., Hassabis, D., Platt, J. C., Creutzig, F., Chayes, J. & Bengio, Y. 2022. Tackling climate change with machine learning. ACM Computing Surveys (CSUR), 55(2), 1–96.

Rosen, P. A., Hensley, S., Joughin, I. R., Li, F. K., Madsen, S. N., Rodriguez, E. & Goldstein, R. M. 2000. Synthetic aperture radar interferometry. Proceedings of the IEEE, 88, 333–382.

Rosen, P. A. & Kumar, R. 2021. NASA-ISRO SAR (NISAR) mission status. IEEE Radar Conference (RadarConf21), 1–6.

Rosenblatt, F. 1958. The perceptron: A probabilistic model for information storage and organization in the brain. Psychological Review, 65(6), 386.

Rouse, J. W. 1974. Monitoring the Vernal Advancement and Retrogradation of Natural Vegetation. NASA/GSFCT Type II Report, Greenbelt, MD, USA.

Running, S. W., Collatz, G. J., Diner, D. J., Kahle, A. B. & Salomonson, V. V. (Eds.). 1998. Special issue on EOS AM-1 platform, instruments, and scientific data. IEEE Transactions on Geoscience and Remote Sensing, 36.

Running, S. W., Justice, C. O., Salomonson, V., Hall, D., Barker, J., Kaufmann, Y. J., Strahler, A. H., Huete, A. R., Muller, J. P., Vanderbilt, V., Wan, Z. M., Teillet, P. & Carneggie, D. 1994. Terrestrial remote sensing science and algorithms planned for EOS/MODIS. International Journal of Remote Sensing, 15, 3587–3620.

Sabins, F. F. 1999. Remote sensing for mineral exploitation. Ore Geology Reviews, 14, 157–183.

Salcedo-Sanz, S., Ghamisi, P., Piles, M., Werner, M., Cuadra, L., Moreno-Martínez, A., Izquierdo-Verdiguier, E., Muñoz-Marí, J., Mosavi, A. & Camps-Valls, G. 2020. Machine learning information fusion in Earth observation: A comprehensive review of methods, applications and data sources. Information Fusion, 63, 256–272.

Schutz, B. E., Zwally, H. J., Shuman, C. A., Hancock, D. & DiMarzio, J. P. 2005. Overview of the ICESat mission. Geophysical Research Letters, 32, L21S01. DOI: 10.1029/2005GL024009.

Sellers, P. J. 1985. Canopy reflectance, photosynthesis and transpiration. International Journal of Remote Sensing, 6, 1335–1372.

Sellers, P. J., Dickinson, R. E., Randall, D. A., Betts, A. K., Hall, F. G., Berry, J. A., Collatz, G. J., Denning, A. S., Mooney, H. A., Nobre, C. A., Sato, N., Field, C. B. & Henderson-Sellers, A. 1997. Modeling the exchanges of energy, water, and carbon between continents and the atmosphere. Science, 275, 502–509.

Shaik, R. U., Periasamy, S., & Zeng, W. 2023. Potential assessment of PRISMA hyperspectral imagery for remote sensing applications. Remote Sensing, 15, 1378. DOI: 10.3390/rs15051378.

Shimoda, H. 1999. ADEOS overview. IEEE Transactions on Geoscience and Remote Sensing, 37, 1465–1471.

Shitole, S., De, S., Rao, Y. S., Krishna Mohan, B. & Das, A. 2015. Selection of suitable window size for speckle reduction and deblurring using SOFM in polarimetric SAR images. Journal of the Indian Society of Remote Sensing, 43, 739–750.

Siddiqi, A. A. 2003. Sputnik and the Soviet Space Challenge. Gainesville, FL: University Florida Press.

Sidorova, V. S. 2012. Hierarchical cluster algorithm for remote sensing data of earth. Pattern Recognition and Image Analysis, 22, 373–379.

Song, Y., Kalacska, M., Gašparović, M., Yao, J. & Najibi, N. 2023. Advances in geocomputation and geospatial artificial intelligence (GeoAI) for mapping. International Journal of Applied Earth Observation and Geoinformation, 103300.

Svatina, V. V. & Cherkasova, M. V. 2023. Space debris removal: Review of technologies and techniques: Flexible or virtual connection between space debris and service spacecraft. Acta Astronautica, 203, 840–853.

Tamiminia, H., Bahram, S., Mahdianpari, M., Quackenbush, L., Adeli, S. & Brisco, B. 2020. Google Earth Engine for geo-big data applications: A meta-analysis and systematic review. ISPRS Journal of Photogrammetry and Remote Sensing, 164, 152–170. DOI: 10.1016/j.isprsjprs.

Tarabalka, Y., Benediktsson, J. A. & Chanussot, J. 2009. Spectral-spatial classification of hyperspectral imagery based on partitional clustering techniques. IEEE Transactions on Geoscience and Remote Sensing, 47(8), 2973–2987.

Tasdemir, K., Milenov, P. & Tapsall, B. 2011. Topology-based hierarchical clustering of self-organizing maps. IEEE Transactions on Neural Networks, 22(3), 474–485.020.04.001.

Taylor, L. 2017. What is data justice? The case for connecting digital rights and freedoms globally. Big Data & Society, 4(2), 2053951717736335.

Thenkabail, P. S., Lyon, J. G. & Huete, A. 2012. Hyperspectral Remote Sensing of Vegetation. Boca Raton: CRC Press.

Thompson, A. A. 2015. Overview of the RADARSAT constellation mission. Canadian Journal of Remote Sensing, 41(5), 401–407.

Touzi, R. & Vachon, P. W. 2015. RCM polarimetric SAR for enhanced ship detection and classification. Canadian Journal of Remote Sensing, 41(5), 473–484.

Tsang, L., Chen, Z., Oh, S., Marks, R. J., & Chang, A. T. C. 1992. Inversion of snow parameters from passive microwave remote sensing measurements by a neural network trained with a multiple scattering model. IEEE Transactions on Geoscience and Remote Sensing, 30(5), 1015–1024.

Tucker, C. J. & Sellers, P. J. 1986. Satellite remote sensing of primary production. International Journal of Remote Sensing, 7, 1395–1416.

UN (United Nations). 1986. Principles Relating to Remote Sensing of the Earth from Space, UN General Assembly, A/RES/41/65, 95th Plenary Meeting.

UNFCCC (United Nations Framework Convention on Climate Change). 2016. Annex to Report of the Conference of the Parties on Its Twenty-First Session, held in Paris from 30 November to 13 December 2015, Addendum. Part Two: Action taken by the Conference of the Parties at Its Twenty-First Session, FCCC/CP/2015/10/Add.1. Available online at http://unfccc.int/paris_agreement/items/9485.php.

USA. 2020. National Space Policy of the United States of America. Washington, DC: United States Federal Government.

USGS (US Geological Survey). 2001. Understanding color-infrared photography. USGS Fact Sheet, 129(1).

Van der Linden, S., Rabe, A., Held, M., Jakimow, B., Leitão, P. J., Okujeni, A., Schwieder, M., Suess, S. & Hostert, P. 2015. The EnMAP-Box: A toolbox and application programming interface for EnMAP data processing. Remote Sensing, 7, 11249–11266. DOI: 10.3390/rs70911249.

Van der Meer, F. D., van der Werff, H., van Ruitenbeek, F. J., Hecker, C. A., Bakker, W. H., Noomen, M. F., van der Meijde, M., Carranza, E. J. M., de Smeth, J. B. & Woldai, T. 2012. Multi-and hyperspectral geologic remote sensing: A review. International Journal of Applied Earth Observations and Remote Sensing, 14, 112–128.

Van Valkenburg, P. & Dufton, J. A. 2020. Big archaeology: Horizons and blindspots. Journal of Field Archaeology, 45(1), 1–7.

Wang, Q. & Atkinson, P. M. 2018. Spatio-temporal fusion for daily Sentinel-2 images. Remote Sensing of Environment, 204, 31–42.

Wang, Q., Zhang, X., Chen, G., Dai, F., Gong, Y. & Zhu, K. 2018. Change detection based on Faster R-CNN for high-resolution remote sensing images. Remote Sensing Letters, 9(10), 923–932.

Wang, S., Liu, C., Li, W., Jia, S. & Yue, H. 2023b. Hybrid model for estimating forest canopy heights using fused multimodal spaceborne LiDAR data and optical imagery. International Journal of Applied Earth Observation and Geoinformation, 122, 103431. DOI: 10.1016/j.jag.2023.103431.

Wang, Z., Ma, Y. & Zhang, Y. 2023a. Review of pixel-based remote sensing image fusion based on deep learning. Information Fusion, 90, 36–58. DOI: 10.1016/j.inffus.2022.09.008.

Warth, G., Braun, A., Assmann, O., Fleckenstein, K. & Hochschild, V. 2020. Prediction of socio-economic indicators for urban planning using VHR satellite imagery and spatial analysis. Remote Sensing, 12, 1730. DOI: 10.3390/rs12111730.

Wasowski, R. J. 1991. Some ethical aspects of international satellite remote sensing. Photogrammetric Engineering and Remote Sensing, 57, 41–48.

Wilkinson, M. D, Dumontier, M., Aalbersberg, U. J., Appleton, G., Axton, M., Baak, A., Blomberg, N., Boiten, J.-W., da Silva Santos, L. B., Bourne, P. E., et al. 2016. The FAIR guiding principles for scientific data management and stewardship. Scientific Data, 3, 160018. DOI: 10.1038/sdata.2016.18.

Woodcock, C. E., Allen, R., Anderson, M., Belward, A., Bindschadler, R., Cohen, W., Gao, F., Goward, S. N., Helder, D., Helmer, E., Nemani, R., Oreopoulos, L., Schott, J., Thenkabail, P. S., Vermote, E. F., Vogelmann, J., Wulder, M. A. & Whynne, R. 2008. Free access to Landsat imagery. Science, 320(5879), 1011. DOI: 10.1126/science.320.5879.1011a.

Wulder, M. A., Masek, J. G., Cohen, W. B., Loveland, T. R. & Woodcock, C. E. 2012. Opening the archive: How free Landsat data has enables the science and monitoring promise of Landsat. Remote Sensing of Environment, 122, 2–10.

Xiong, X., Wenny, B. N. & Barnes, W. L. 2011. Overview of NASA Earth Observing Systems Terra and Aqua moderate resolution spectroradiometer instrument calibration algorithms and on-orbit performance. Journal of Applied Remote Sensing, 3(1), 032501. DOI: 10.1117/1.3180864.

Xu, D. & Wu, Y. 2020. Improved YOLO-V3 with DenseNet for multi-scale remote sensing target detection. Sensors, 20(15), 4276.

Yan, D., Li, G., Li, X., Zhang, H., Lei, H., Lu, K., Cheng, M. & Zhu, F. 2021. An improved faster R-CNN method to detect tailings ponds from high-resolution remote sensing images. Remote Sensing, 13(11), 2052.

Yao, W., Loffeld, O. & Datcu, M. 2016. Application and evaluation of a hierarchical patch clustering method for remote sensing images. IEEE Journal of Selected Topics in Applied Earth Observations and Remote Sensing, 9(6), 2279–2289.

Zhu, E., Helmer, E. Gao, F., Liu, D., Chen, J. & Lefsky, M. A. 2016. A flexible spatiotemporal method for fusing satellite images with different resolutions. Remote Sensing of Environment, 172: 165–177.

Zhu, X. & Bamler, R. 2014. Super-resolving SAR tomography for multi-dimensional imaging of urban areas. IEEE Signal Processing Magazine, ISSN, 1053–5888.

Zonno, M., Matar, J., Queiroz de Almeida, F., Younis, M., Reimann, J., Rodriguez-Cassola, M., Krieger, G., Perrera, A., De Giorgi, M., Torrini, A., Cucinella, C. & Tossaint, M. 2019. System and Mission Trade-Offs for the Sentinel-1 Next Generation Phase-0 Study. Microwaves and Radar Institute, German Aerospace Center (DLR), Oberpfaffenhofen.

# Part IV

## Data Normalization, Harmonization, and Inter-Sensor Calibration

# 6 Overview of Satellite Image Radiometry in the Solar-Reflective Optical Domain

*Philippe M. Teillet*

## LIST OF ACRONYMS

| ACRONYM | Definition |
|---|---|
| **AVHRR** | Advanced Very High Resolution Radiometer |
| **BRDF** | Bidirectional Reflectance Distribution Function |
| **CEOS** | Committee on Earth Observation Satellites |
| **CLARREO** | Climate Absolute Radiance and Refractivity Observatory |
| **DEM** | Digital Elevation Model |
| **DTM** | Digital Terrain Model |
| **ETM+** | Enhanced Thematic Mapper Plus |
| **GEO** | Group on Earth Observations |
| **GEOSS** | Global Earth Observation System of Systems |
| **GIFOV** | Ground Instantaneous Field of View |
| **GIS** | Geographic Information System |
| **IFOV** | Instantaneous Field of View |
| **IVOS** | Infrared Visible Optical Subgroup |
| **LAI** | Leaf Area Index |
| **MODIS** | MODerate resolution Imaging Spectroradiometer |
| **MODTRAN** | MODerate resolution atmospheric TRANsmission |
| **NASA** | National Aeronautics and Space Administration |
| **NIST** | National Institute of Standards and Technology |
| **NOAA** | National Oceanic and Atmospheric Administration |
| **NOMAD** | Networked Online Mapping of Atmospheric Data |
| **NRC** | National Research Council |
| **PSF** | Point Spread Function |
| **QA4EO** | Quality Assurance Framework for Earth Observation |
| **SI** | International System of Units |
| **SRBC** | Solar-Radiation-Based Calibration |
| **TM** | Thematic Mapper |
| **TOA** | Top of Atmosphere |
| **TRUTHS** | Traceable Radiometry Underpinning Terrestrial- and Helio-Studies |
| **UTM** | Universal Transverse Mercator |
| **WGCV** | Working Group on Calibration and Validation |

DOI: 10.1201/9781003541141-10

## 6.1 INTRODUCTION

Modern-day remote sensing satellite systems yield high-quality digital images that provide both synoptic and detailed observations of the Earth from space. The steps that have led to this unprecedented geospatial technology are many and even a summary of that development is beyond the scope of this chapter. Details of the key historical elements of Earth observation remote sensing, such as aviation, rockets, space travel, orbiting satellites, imaging, and lunar and planetary exploration, can be found in numerous books (e.g., Panda et al., 2024, Masek et al., 2020, Burrows, 1999, Kramer, 2001, Jensen, 2006). Nonetheless, it is worth noting that it has long been known that seeing our planet from above brings significant advantages.[1]

Countless applications of satellite imagery have been and continue to be developed, the vast majority for qualitative, every-day uses that benefit nevertheless from the laboratory geometric and radiometric quality of the digital images available. That said, there is also tremendous interest in the quantitative use of satellite images to retrieve and monitor information about the current and changing states of geophysical and biophysical variables, particularly those involved in climate, Earth resources, and environment. Applications include, but are not limited to, vegetation analysis (e.g., agriculture, forestry, precision farming), disaster monitoring, environmental monitoring, watershed management, urban growth analysis, bathymetry, geological mapping, mineral exploration, and intelligence data gathering. In that light, this chapter outlines the key considerations that need to be addressed to ensure that terrestrial variables derived from satellite sensor systems operating in the solar-reflective optical domain are calibrated radiometrically to a common physical scale.

There are 78 space agencies in the world and at least 50 of them are operating currently or planning to launch over the coming decade, 327 Earth observation satellite missions carrying 845 sensors (400+ different sensors, some being repeats in a series).[2] The data quality varies significantly between sensor systems due to differences in mission requirements, sensor characteristics and the different calibration methodologies utilized.

In this chapter, emphasis is placed on the post-launch methodologies used to convert digital image data in the solar reflective domain to radiometrically-calibrated products for the user community. For most targets of interest, the retrieval of land, atmosphere and ocean data and information from satellite image data requires sensor radiometric calibration, retrieval of surface reflectance (involving correction for atmospheric effects), allowance for geometric effects on image radiometry, and an understanding of scene spectral reflectance behavior. The primary terrestrial variable that serves as an example and a common thread in this short treatment is that of reflectance, an Earth surface parameter of wide interest.

## 6.2 THE NEED FOR DATA STANDARDIZATION

Environmental monitoring requirements and responsibilities spanning from local communities to planetary scales continue to multiply. Accordingly, building to a significant extent on military technologies, government agencies in many countries have developed geostrategic technologies and applications that make use of space-based observations of the Earth. Much of the focus of Earth sciences that make use of these satellite sensor systems is on improving predictions of Earth system changes, both short-term and long-term, primarily with respect to climate, population growth, land transformation, pollution, and biodiversity. As such, remote sensing systems help provide more solid underpinnings for decisions that impact us all in terms of quality of life and economic consequences.

The extent to which new Earth observation technologies can contribute information of significant economic, social, environmental, strategic, and political value depends critically on developments in data standardization and quality assurance (MacDonald, 1997, Teillet et al., 1997a). Thus, remote sensing calibration and validation are essential aspects of Earth observation measurements and methods used to estimate terrestrial variables to ensure that they are not compromised by sensor effects and data processing artefacts. Accordingly, research progress in remote sensing calibration has been featured periodically in review articles (e.g., Naethe et al., 2024, Shen et al., 2023, Masek et al., 2022, 2021, Page et al., 2019, Claverie et al., 2018, Yang et al., 2017, Slater, 1984, 1985, 1988a, Teillet, 1986; Fraser and Kaufman, 1986, Duggin, 1986, 1987, Price, 1987a, Ahern et al., 1989, Duggin and Robinove, 1990, Che and Price, 1992, Slater et al., 1995, 2001, Teillet, 1997a, Teillet et al, 1997a, Secker et al., 2001, Teillet, 2005, Chander et al., 2013a), special journal issues and conference proceedings (e.g., Markham and Barker, 1985, Price, 1987b, Slater, 1988b, Jackson, 1990, Guenther, 1991, Connolly and Tolar, 1994, Bruegge and Butler, 1996, Teillet, 1997b, Chander et al., 2013b), as well as book chapters and reports (e.g., Muench, 1981, Malila and Anderson, 1986, Butler et al., 2005). The challenge is to ensure that measurements and methods yield self-consistent and accurate geophysical and biophysical data, even though the measurements are made with a variety of different satellite sensors under different observational conditions and the parameter retrieval methodologies vary.

The most stringent satellite data requirements are driven by the need for long-term monitoring of terrestrial variables to determine condition and detect change, as well as to provide inputs to regional and global carbon, energy, and water process models. These requirements have been well documented and include climate and land surface variables such as surface temperature, albedo, fraction of absorbed photosynthetically active radiation, and net primary productivity (e.g., Claverie et al., 2018, Yang et al., 2027, Guenther et al., 1996, 1997, Ohring et al., 2007). Similar user requirements have been documented for oceans (e.g., sea surface temperature and ocean color) (e.g., Gordon, 1987) and atmospheres (e.g., temperature, water vapor, precipitation, ozone, clouds and aerosols, radiation, and trace species) (e.g., Page et al., 2029, Claverie et al., 2018, Yang et al., 2027, Guenther et al., 1996, 1997, Edwards et al., 2004).

Calibration is defined by the international Committee on Earth Observation Satellites (CEOS) Working Group on Calibration and Validation as the process of quantitatively defining the system response to known, controlled signal inputs (www.ceos.org). Generally, "calibration-validation" or "cal-val" refers to the entire suite of processing algorithms used to convert raw data into accurate and useful Earth science data that are verified to be self-consistent (Teillet, 1997a). Calibration can include radiometric, geometric, spectral, temporal, and polarimetric aspects. This chapter focuses on radiometric considerations in the solar-reflective optical domain, but it is clear that all these different aspects are important considerations for any measurement device regardless of what part of the electromagnetic spectrum it covers (optical, microwave, or other) if it is to yield useful data and information (Teillet et al., 2004a). The present chapter together with the next four chapters of this volume provide readers with a comprehensive overview of remote sensing calibration.

The Global Earth Observation System of Systems (GEOSS) of the Group on Earth Observations (GEO) aims to deliver timely and comprehensive "knowledge information products" to meet the needs of its nine "Societal Benefit Areas," of which the most demanding, in terms of accuracy, is climate (Ohring et al., 2005, 2007). This vision builds on the synergistic use of a system of disparate sensing systems that were or are being built for a multitude of applications, and requires the establishment of a worldwide coordinated operational framework to facilitate interoperability and harmonization. CEOS, considered to be the space arm of GEO, has led the development of a Quality Assurance Framework for Earth Observation (QA4EO),[3] which is based on the adoption of key guidelines. These guidelines have been derived from "best practises" for implementation by the community under the auspices of GEO. The guidelines define the generic processes and activities needed to put in place an operational QA4EO. Their use will facilitate the assignment of quality indicators to the output of every step in Earth observation data processing chains to demonstrate the level of traceability to internationally agreed reference standards (SI where possible). The QA4EO was endorsed by the CEOS Working Group for Calibration and Validation (WGCV) at its 29th plenary in September 2008 and endorsed by the CEOS Plenary in November 2008.

The inadequacy of proper calibration and validation in diverse Earth observation applications has been identified at many workshops and in a variety of user need and market studies (e.g., Naethe et al., 2024, Panda et al., 2024, Shen et al., 2023, Masek et al., 2022, 2021, 2020, Page et al., 2019, Claverie et al., 2018, Yang et al., 2017, Sweet et al., 1992, Sellers et al., 1995, Hall et al., 1991, Horler, 1996, Horler and Teillet, 1996, Teillet, 1997c, 1998, Hegyi, 2004). Raw or uncorrected imagery cannot be used to provide meaningful information for natural resource management, environmental monitoring, and climate studies. As Schowengerdt (2007) puts it, raw sensor digital counts are "simply numbers, without physical units." Thus, quantitative uses of Earth observation rely on conversion of the data to physical units, certifiably traceable to the International System of Units (SI) standards, in order to enable the comparison of Earth data from different sensors or from any given sensor over time, whether it be from days to decades (e.g., Fox, 1999, Helder et al., 2012). Moreover, even for qualitative applications, remote sensing remains in some measure untapped[4] and it will only become a mainstream information technology when it provides reliability of supply, consistent data quality, and plug-and-play capability (Teillet et al., 1997a).

## 6.3 OVERVIEW

Radiometry and radiation propagation in the remote sensing setting are well described in textbooks (e.g., Naethe, et al., 2024, Panda et al., 2024, Shen et al., 2023, Masek et al., 2022, 2021, 2020, Slater, 1980, Chen, 1996, Wyatt et al., 1998, Morain and Budge, 2004, Schott, 2007). It is not the role of this chapter to cover fundamental radiometric terms and concepts but rather to provide a brief overview of the key steps involved in converting digital image data from satellite optical systems to calibrated surface quantities, spectral reflectance in particular, in the Earth science context. To begin, it is instructive to look at the paths photons take from their source, the Sun, to the entrance aperture of any given satellite sensor in Earth's orbit. Figure 6.1 portrays the processes involved, with five main pathways and associated interactions, from a phenomenological perspective.

1. Photons from the Sun propagate through the Earth's atmosphere, are reflected by the target of interest, and then propagate back through the atmosphere to the satellite sensor (direct sunlight path).
2. Photons from the Sun are scattered by the Earth's atmosphere (thus generating diffuse sky illumination), are reflected by the target of interest, and then propagate back through the atmosphere to the satellite sensor (diffuse skylight path).
3. Photons from the Sun are scattered by the Earth's atmosphere back to the sensor without reaching the surface of the Earth (i.e., path radiance, which can be as much as 10% of the total signal at visible wavelengths on hazy days).
4. Photons from the Sun propagate through the Earth's atmosphere, are reflected from background objects and subsequently from the target of interest, and then propagate back through the atmosphere to the satellite sensor (multiple reflections).
5. Photons from the Sun propagate through the Earth's atmosphere, are reflected from surrounding surfaces, and then scattered into the line of sight of the sensor (the so-called adjacency effect).

Atmospheric radiative transfer codes used for satellite image correction typically take 1 through 3 into account. Some codes take 5 into account, but those that include 4 are less common.

Teillet (2005) lists over 60 considerations and factors that affect the radiometry of Earth observation image data, grouped into seven categories: sensor characteristics, sensor regime, illumination conditions, observation domain, path medium, target characteristics, and product generation domain. While some of the factors are straightforward, such as the variation in Earth–Sun distance, others merit extensive treatments in their own right, such as atmospheric scattering and absorption. This chapter touches on the key post-launch factors, but it is beyond the scope of the chapter to provide details on the remaining factors and considerations.

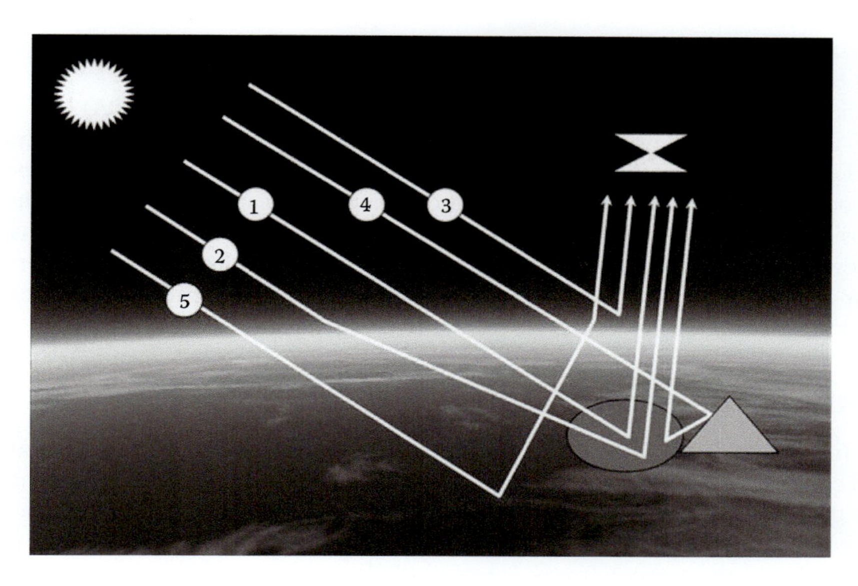


**FIGURE 6.1** Schematic depicting photon pathways from the Sun to the entrance aperture of a satellite sensor: (1) direct solar illumination, (2) diffuse sky illumination, (3) atmospheric path radiance, (4) background object reflections, and (5) adjacency effects.

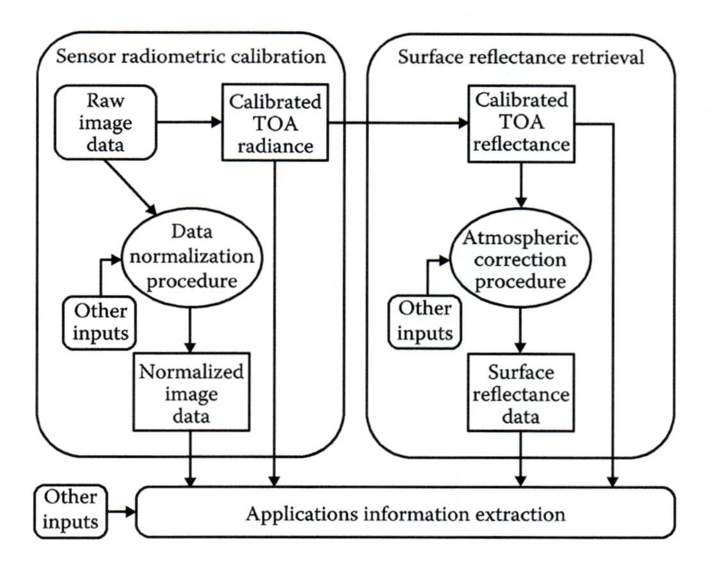


**FIGURE 6.2** Radiometric data flow options in preparation for information extraction. The additional dimensions of spectral characterization and georadiometric effects on image radiometry (not shown, but discussed in the text) enter into all of the elements portrayed.

Figure 6.2 presents a data flow scheme that emphasizes the main radiometric preprocessing steps prior to information extraction. The main elements portrayed are sensor radiometric calibration, surface reflectance retrieval (i.e., atmospheric correction), radiometric normalization, and information extraction. Information extraction is a separate subject that is not treated in this chapter, and normalization is touched on briefly only. However, the additional dimensions of spectral characterization and georadiometric effects on image radiometry, not easily shown in the data flow in Figure 6.2 because they enter into all of the elements, will be discussed.

Raw image data and calibrated radiances at the top of the atmosphere (TOA) (Figure 6.2) are referred to as Level 0 and Level 1 data, respectively. Level 2 data consisting of radiometrically calibrated and geolocated physical variables such as surface reflectance, emittance, and temperature are preferred or required in applications dependent on quantitative analyses. Still higher data product levels consist of information products that are spectral, spatial or temporal integrations or aggregates of lower-level data. Few remote sensing data product generation systems and services offer Level 2 or higher-level products. Therefore, users must typically undertake this processing work themselves using image analysis software or in collaboration with private industry.

## 6.4 SENSOR RADIOMETRIC CALIBRATION

The most fundamental part of the calibration-validation process is sensor radiometric calibration, a broad and complex field that imposes the greatest limitations on quantitative applications of remote sensing (e.g., Teillet, 1997a, Teillet et al., 1997a). The methodologies and instrumentation involved can be grouped into three domains (e.g., Dinguirard and Slater, 1999): (1) on the ground before launch, (2) onboard the spacecraft post-launch, including reference to lamp sources and/or solar illumination (e.g., Markham et al., 1997), and (3) vicarious approaches using suitable Earth scenes or extra-terrestrial targets imaged in-flight (cf., later in this chapter and the next chapter (Teillet and Chander, 2014)).

As for any scientific instrument, extensive and expensive efforts are devoted to building stable optical imaging sensors and to characterizing them to the fullest extent possible before launch (e.g., Guenther et al., 1996). Pre-launch calibrations in a laboratory are easier to control and perform than methods used after launch. Sensor calibration coefficients are determined before launch using radiation sources, whose calibrations are traceable to national laboratory standards.

After launch, sensor radiometric response usually degrades over time such that it must be monitored continually and recalibrated as necessary throughout the life of the system to ensure the maintenance of high data quality. In many instances, considerable time and effort go into the development and use of onboard systems such as standard lamps (blackbody sources for thermal bands), and/or solar diffuser panels to monitor the post-launch calibration performance of satellite sensors. Nevertheless, even the status of the onboard calibration systems must be verified over time via independent means. Vicarious methods (described in Section 4.4) provide these independent data and yield calibration information over the mission lifetime. Further discussion of pre-launch and onboard calibration systems is beyond the scope of this chapter.

Whereas pre-launch methods encompass a vast array of painstaking sensor characterizations in the laboratory (and on very rare occasions outdoors, so-called solar-radiation-based calibration (SRBC) (e.g., Naethe et al., 2024, Masek et al., 2022, 2021, Biggar et al., 1993, Mueller et al., 1996, Dinguirard et al., 1997), post-launch radiometric calibration based on onboard systems or vicarious methods is devoted primarily to the monitoring of the radiometric responsivities or gain coefficients for each sensor in each spectral band over time. Bias coefficients are determined generally on the basis of deep space or dark-shutter readings. In recent years, unified approaches have been pursued to link the three domains (pre-launch, onboard, and vicarious calibrations), plus the field instruments deployed at the surface in support of vicarious calibration (e.g., Slater et al., 2001, Butler et al., 2005). Error analyses of such approaches indicate that uncertainties in the 2–3% (1 $\sigma$) range with respect to exo-atmospheric solar irradiance are attainable in each domain and can be accurately related to national laboratory standards (Dinguirard and Slater, 1999). In all cases, the objective is traceability of radiometric calibration accuracies to the International System of Units (SI) for science users (e.g., Pollock et al., 2003) and data products with consistent quality for the broader user community.

As an example, the pre-launch coefficients of the Landsat-5 Thematic Mapper (TM) were revised after launch because the percent changes in radiometric gain calibration coefficients in the blue, green and red spectral bands attained −1.0%, −0.85%, and −0.55% per annum, respectively, and the coefficients in the near-infrared and two shortwave infrared spectral bands changed by 0%, +0.1%, and −0.2% per annum, respectively (Teillet et al., 2004a, 2004b). These changes are significant given the almost 30-year lifetime of the Landsat-5 mission. The impact of these radiometric gain degradations is exemplified in one study by the underestimation of leaf area index (LAI) over the course of a decade by 4% and 1% for rangeland and grassland, respectively, whereas there would be negligible impact on the LAI of black spruce (Teillet et al., 2004a, 2004b).

### 6.4.1 Dynamic Range

Radiance describes the flux of energy impinging on a given area from a specified direction in a given spectral band. Conventional units for radiance, used in this document, are Watts per square meter per steradian per micrometer ($W \cdot m^{-2} \cdot sr^{-1} \cdot \mu m^{-1}$), although some data suppliers use slightly different units. The dynamic range in radiance units is set by sensor design and calibration specialists to cover the full radiance range of scenes to be imaged in that satellite mission. For sensor systems with linear response in the radiance range of interest, the dynamic range is specified in terms of the minimum radiance, $L^*_{min}$, which corresponds to a digital image level of $Q_{cal,\,min}$ (counts), and the maximum radiance, $L^*_{max}$, which corresponds to a digital image level of $Q_{cal,\,max}$ (counts) (e.g., Markham and Barker, 1987). The asterisk indicates a TOA quantity. The four parameters, $Q_{cal,\,min}$, $Q_{cal,\,max}$, $L^*_{min}$, and $L^*_{max}$, establish a known linear relationship between calibrated digital data, $Q_{cal}$, and radiance at the sensor, $L^*$ (cf., next section).

### 6.4.2 Converting Digital Counts to At-Sensor Radiance

In each spectral band, the outputs of most optical satellite remote sensing systems are quantized raw image data, $Q$, which are transmitted to the ground (or recorded onboard and then subsequently transmitted to the ground). After housekeeping corrections and relative radiometric calibration, the raw data are archived on the ground as Level-0 data in units of digital counts. Relative radiometric calibration of satellite imagery primarily involves characterizing and correcting the differences between detector gain and bias levels in sensor systems that utilize multiple detectors in a given spectral band.

To derive radiometrically calibrated (Level 1) data requires knowing the relationship between $Q$ and the TOA at-satellite radiance input to the sensor (absolute radiometric calibration). Rather than computing radiance directly, image product generation systems often transform raw data, $Q$, to scaled calibrated data, $Q_{cal}$, taking care of changes in sensor calibration performance in the process. For such $Q_{cal}$ products, the user can then use a set of time-invariant calibration coefficients to convert calibrated digital counts, $Q_{cal}$, to TOA radiance, $L^*$. Thus, product generation systems require time-dependent information about the sensor's radiometric calibration performance, a requirement that remains one of the most challenging aspects of the production and quantitative use of satellite image data (e.g., Naethe et al., 2024, Shen et al., 2023, Slater and Biggar, 1996, Slater et al., 2001, Teillet and Chander, 2014).

The linear radiometric calibration equation is

$$L^* = \frac{1}{G}[Q_{cal} - Q_o] \tag{6.1}$$

The gain and bias calibration parameters G and $Q_o$ are specified as follows:

$$G = \frac{Q_{cal,\max} - Q_{cal,\min}}{L^*_{\max} - L^*_{\min}} \text{ (counts per unit radiance)} \tag{6.2}$$

and

$$Q_o = Q_{cal,\min} - (Q_{cal,\max} - Q_{cal,\min}) \frac{L^*_{\min}}{L^*_{\max} - L^*_{\min}} \text{ (counts)} \tag{6.3}$$

If a particular application involves the utilization of images from multiple sensors, it is necessary to transform the image data from the different sensors to common physical units such as radiance before they can be compared. If a particular application involves the relative inter-comparison of several images from a single sensor, the data can be used in the form of calibrated counts, $Q_{cal}$. In the latter case, it is not necessary to transform to radiances.

It is worth noting that, while the sensor radiometric calibration equation is as given in Equation 6.1, thus relating the input to the sensor (at-sensor radiance) to the output (digital counts) in proper engineering terms, users are more interested in the inverse computation that converts digital counts in their image data to at-sensor radiance. Unfortunately, the coefficients for this inverse calculation are at times also called "gain" and "offset," which can lead to confusion. These inverse computation "gain" and "offset" coefficients have units that differ from those of the sensor radiometric calibration gain and offset coefficients defined in Equations 6.2 and 6.3, that is, the inverse computation "gain" and "offset" coefficients are in units of "radiance per unit count" and "radiance," respectively.

### 6.4.3 Converting At-Sensor Radiance to At-Sensor Reflectance

A full description of reflectance, whether it be TOA or at the surface, is given by the bidirectional reflectance distribution function (BRDF), which is a function of all incident ($i$) and reflected ($r$) zenith angles ($\theta$) and azimuthal angles ($\varphi$) (Nicodemus, 1965, 1970):

$$\rho_{BRDF} = \frac{L(\theta_r, \varphi_r)}{E(\theta_i, \varphi_i)} \qquad [sr^{-1}] \tag{6.4}$$

Units for radiance, $L$, are as already defined ($W \cdot m^{-2} \cdot sr^{-1} \cdot \mu m^{-1}$) and units for irradiance, $E$, are Watts per square meter per micrometer ($W \cdot m^{-2} \cdot \mu m^{-1}$). Note that the units of BRDF are per steradian ($sr^{-1}$). Since full BRDF characterizations of terrestrial surfaces are nontrivial, remote sensing most often makes use of simplified representations such as the Lambertian reflectance. The radiance from the idealized surface known as a Lambertian surface is the same in all directions, that is, the decreasing intensity with angle follows a cosine law and is exactly compensated by a decrease in projected area. More advanced treatments are needed to deal with surfaces that are, to any significant extent, non-Lambertian and/or sloped with respect to the horizontal.

The TOA or at-sensor reflectance, $\rho^*$, can be defined as a function of the upwelling radiance, $L^*$ ($W \cdot m^{-2} \cdot sr^{-1} \cdot \mu m^{-1}$), observed by the satellite sensor divided by the exo-atmospheric solar irradiance, $E_0$ ($W \cdot m^{-2} \cdot \mu m^{-1}$), as follows (Schott, 2007):

$$\rho^*(\bar{\lambda}) = \frac{\pi \, d_s^2 \, L^*(\bar{\lambda})}{E_0(\bar{\lambda}) \cos(\theta_z)} \tag{6.5}$$

where $\lambda$ indicates quantities that are wavelength-dependent and $\bar{\lambda}$ is used to indicate that these variables are integrated over a given spectral band. As formulated earlier, the reflectance $\rho^*$ is a dimensionless factor between 0 and 1. Equation 6.5 is exact only for a Lambertian surface, that is, one for which the radiance is constant as a function of angle (hence the geometric factor $\pi$). $E_0$ is the exo-atmospheric solar irradiance (Watts per square meter per micrometer), $q_z$ is the solar zenith angle, and $d_s$ is the Earth–Sun distance in dimensionless Astronomical Units. Exo-atmospheric solar irradiance is a difficult quantity to measure and it does vary a few percent over time. Hence, an average model for $E_0$ is always used and several such models have been developed (e.g., Thuillier et al., 2003). Therefore, care must be taken to ensure that comparisons between different calibration methods and results are made on the basis of the same $E_0$ data.

The use of TOA reflectance as opposed to TOA radiance corrects for at least some of the sources of variation that affect satellite data, in particular variations in solar illumination caused by the diurnal cycle and by cyclical changes in the Earth–Sun distance, as well as differences in solar irradiance due to differences between similar spectral bands from sensor to sensor (discussed later in this chapter).

### 6.4.4 Vicarious Calibration

Earth surfaces with appropriate characteristics have long been used to provide post-launch updates of the radiometric calibration of satellite sensors, a methodology often referred to as vicarious or ground-look calibration (e.g., Slater et al., 1987, Teillet et al., 1990, Biggar et al., 1994, Thome, 2001). A comprehensive summary of this topic is given in the next chapter of this volume (Teillet and Chander, 2014). Accordingly, only a few aspects are mentioned here.

Reflectance-based or radiance-based methods use surface and atmospheric measurements to estimate TOA radiance at the entrance aperture of a given satellite sensor in order to monitor and, if needed, provide updates of sensor radiometric calibration. Initially, field measurement campaigns by specialised teams at calibration reference test sites targeted only one sensor per sortie. In subsequent years, with the increase in the number of sensors that passed over a given test site on a given day, it became possible with careful planning to undertake vicarious calibrations for several sensors per sortie (e.g., Thome et al., 1998). Efforts such as these are resource intensive and take time to complete. Hence, it has been of considerable interest to develop less expensive complementary approaches that can provide more frequent calibration updates, even if they are less accurate individually. The use of standard reference test sites to compare or transfer radiometric calibration between satellite sensors, that is, cross-calibration, has also been investigated (e.g., Masek et al., 2022, 2021, Claverie et al., 2018, Yang et al., 2027, Teillet et al., 2001a, Thome et al., 2003, Chander et al., 2013a, Helder et al., 2013), with and without near-simultaneous or coincident surface measurements. With greater timeliness and reduced costs in mind, new methods are being explored that yield useful results without near-simultaneous or coincident measurements by field teams (e.g., Teillet et al., 2001b, Rao et al., 2003, Thome, 2005, Thome et al., 2008, McCorkel et al., 2013).

Vicarious calibration can also be undertaken with respect to lunar views (e.g., Kieffer and Wildey, 1985, 1996, Kieffer et al., 2002, Barnes et al., 2004, Stone et al., 2005, 2013, Stone, 2008, Xiong et al., 2008) and/or bright stars (e.g., Chang et al., 2012). Although such ultra-stable targets are imaged without atmospheric effects, they are relatively low-radiance sources and they require high-risk spacecraft platform manoeuvres to achieve. Some sensors are designed to acquire space views to provide dark target calibration data.

Other vicarious calibration or cross-calibration methods take advantage of non-land Earth targets, including atmospheric Rayleigh scattering (e.g., Naethe et al., 2024, Shen et al., 2023, Masek et al., 2022, Claverie et al., 2018, Yang et al., 2017, Vermote et al., 1992; Kaufman and Holben, 1993, Dilligeard et al., 1997, Meygret et al., 2000), ocean sunglint (e.g., Kaufman and Holben, 1993, Vermote and Kaufman, 1995, Luderer et al., 2005), snow/ice fields (e.g., Loeb, 1997, Tahnk and Coakley, 2001, Nieke et al, 2003, Six et al., 2004), and clouds (e.g., Vermote and Kaufman, 1995, Iwabuchi, 2003, Doelling et al., 2004, 2010, Hu et al., 2004, Fougnie and Bach, 2009).

Early and ongoing work worth mentioning in the context of vicarious calibration is the case of the series of Advanced Very High Resolution Radiometer (AVHRR) sensors operating on National Oceanic and Atmospheric Administration (NOAA) satellites since 1978. The AVHRR program did not include onboard post-launch radiometric calibration for the solar-reflective spectral bands. As a result, a wide variety of vicarious methodologies were developed to generate updates of prelaunch radiometric calibrations during the course of the mission on-orbit. AVHRR radiometric gain degradation proved to be significant post-launch and the degradations differed considerably from sensor to sensor in the AVHRR series (e.g., Naethe, et al., 2024, Shen et al., 2023, Masek et al., 2022, 2021, Frouin and Gautier, 1987, Smith et al., 1988, Holben et al., 1990, Staylor, 1990, Teillet et al., 1990, Whitlock et al., 1990, Brest and Rossow, 1992, Che and Price, 1992, Mitchell et al., 1992, Teillet, 1992, Vermote et al., 1992, Abel et al., 1993, Kaufman and Holben, 1993, Vermote and Kaufman, 1995, Loeb, 1997, Cabot et al., 2000, O'Brien and Mitchell, 2001, Cao and Heidinger, 2002, Iwabuchi, 2003, Wu et al., 2003, Doelling et al., 2004, Vermote and Saleous, 2006). Quantitative Earth observation applications taking advantage of AVHRR synoptic daily coverage found it difficult to keep up with radiometric calibration changes until various researchers and, eventually, NOAA developed and implemented appropriate approaches and communication tools (e.g., Naethe et al., 2024, Shen et al., 2023, Masek et al., 2022, 2021, 2020, Page et al., 2029, Claverie et al., 2018, Yang et al., 2027, Teillet and Holben, 1994, Cihlar and Teillet, 1995, Rao and Chen, 1995, 1996, 1999, Brest et al., 1997, Tahnk and Coakley, 2001).

Compared to onboard radiometric calibration techniques, the main disadvantages of vicarious calibration are that the methods suffer from lower precision and lower temporal sampling frequency. The main advantage of vicarious calibration is that the sensor typically acquires calibration data in the same modality as it acquires Earth image data, that is, with the same source spectrum, similar illumination conditions, and the same spectral bands.

## 6.5 SURFACE REFLECTANCE RETRIEVAL

Surface reflectance in the solar reflective part of the electromagnetic spectrum is a primary geophysical variable of interest in many applications, either for its own sake or to generate other geophysical or biophysical quantities. The Lambertian surface spectral reflectance, $\rho_{surf}$ (dimensionless between 0 and 1), is defined as $\pi$ times the ratio of the radiance, $L$ ($W \cdot m^{-2} \cdot sr^{-1} \cdot \mu m^{-1}$), upwelling from the target (as measured just above the target) divided by the downwelling irradiance, $E$ ($W \cdot m^{-2} \cdot \mu m^{-1}$), that illuminates the target (Schott, 2007):

$$\rho_{surf}(\bar{\lambda}) = \frac{\pi L(\bar{\lambda})}{E(\bar{\lambda})} \tag{6.6}$$

As already noted (Figure 6.1), the downwelling irradiance signal at the surface, E, is made up of several components in addition to the direct solar and diffuse sky contributions, a consideration that must also be taken into account when making ground-based radiance measurements. The sky irradiance distribution is anisotropic, but it can be modeled reasonably well to take into account the increase in diffuse radiation in the circumsolar region and towards the horizon (Dave, 1077, Temps and Coulson, 1977, Steven, 1977, Klucher, 1979, Hooper and Brunger, 1980, Kirchner et al., 1982, Hooper et al., 1987, Brunger and Hooper, 1993). This refinement is especially helpful where topography obscures parts of the sky from a given point on the ground (Hay and McKay, 1985, Hay et al., 1986, Duguay, 1993).

### 6.5.1 ATMOSPHERIC CORRECTION

Scattering and absorption due to aerosols and gases in the atmosphere modify radiation making its way from outside the atmosphere down to the target, as well as the surface reflectance propagated through the atmosphere to the satellite sensor (e.g., Chahine, 1983, Schott, 2007). In the solar-reflective spectral domain, molecular (Rayleigh) scattering ($\lambda^{-4}$) is strongest in the blue (e.g., contributing ~0.07 in reflectance for 1,013 mbar), aerosol scattering ($\lambda^{-2}$ to $\lambda^{+0.6}$) is strongest in the visible (e.g., contributing ~0.04 in reflectance for 10 km visibility), ozone gas absorption is strongest in the green (e.g., ~10% effect on atmospheric transmission for 0.35 cm-atm), and water vapor absorption is strongest in the near-infrared and shortwave infrared (e.g., ~12% effect on transmission for 3 g/cm$^2$). Interestingly, atmospheric source and loss effects roughly cancel out for Earth surface reflectances around 0.25; hence, this is a good reflectance for vicarious calibration reference sites since atmospheric effects are minimized.

Given that the optical properties of the Earth's atmosphere are not uniform spatially or temporally, image corrections for these effects in the solar-reflective spectral bands are needed to put satellite data on the same radiometric scale for investigations intended to monitor terrestrial surfaces quantitatively over time and space. A compact form of the relationship between surface and TOA reflectance is given as (Tanré et al., 1990):

$$\rho^* = \tau_g \left( \frac{\tau_s \, \tau_v \, \rho_{surf}}{1 - \bar{\rho} \, s} + \rho_a \right) \tag{6.7}$$

All quantities are wavelength dependent. $\tau_g$ is atmospheric gas transmittance, $\tau_s$ is downward scattering transmittance in the solar illumination direction, $\tau_v$ is upward scattering transmittance in the sensor view direction, $\rho_a$ is atmospheric reflectance, $\bar{\rho}$ is the average reflectance of surrounding surfaces, and $s$ is the spherical albedo (i.e., the reflectance) of the atmosphere. All reflectances and transmittances in Equation 6.6 are dimensionless factors between 0 and 1. The atmospheric quantities are computed using radiative transfer codes.

The detailed textbook treatment by Schott (2007) is also instructive and worth citing. It explains all of the contributions to the upwelling spectral radiance reaching the satellite sensor, including the thermal energy paths not presented in this chapter. The result is the so-called Big Equation for the TOA spectral radiance (cf., Schott (2007) for details).

The need for and various approaches to atmospheric correction have long been examined in the literature (e.g., Masek et al., 2022, 2021, Page et al., 2019, Claverie et al., 2018, Yang et al., 2027, Turner and Spencer, 1972, O'Neill et al., 1978, Ahern et al., 1979, Slater and Jackson, 1982, Richardson, 1982, Tanré et al., 1983, Moran et al., 1992, Kaufman et al., 1997) and atmospheric radiative transfer codes are available in modern remote sensing image analysis systems. Some solutions may involve the use of a radiative transfer code but many make use of pre-computed look-up table results to save computation time. A widely accepted atmospheric correction code for research purposes is MODTRAN5 (MODerate resolution atmospheric TRANsmission code version 5) (Berk et al., 2006). Most of the predominantly used codes tend to disagree significantly only for very large aerosol optical depths and/or large off-nadir illumination or viewing geometries (>60°). Thus, the choice of code is not an important factor except for the correction of high spectral resolution (hyperspectral) data (e.g., Staenz et al., 1994, 2002). Monochromatic computations should not be used (Teillet, 1989) and bandpass calculations based on relative spectral response profiles with 0.0025-micrometer grid spacing or finer are recommended. Another caution is that Rayleigh scattering, a long-understood phenomenon, is not always computed accurately in atmospheric correction codes (Teillet, 1990a, He et al., 2006).

Operational atmospheric correction for surface reflectance retrieval depends on ready access to timely and accurate information on atmospheric variables such as aerosol optical depth (e.g., Teillet et al., 1994) and water vapor content. Although the computational tools for image correction are available, the user is still left with the problem of obtaining the necessary atmospheric variables for input to the image correction. The main possibilities in this respect are as follows.

1. Measure the required parameters in the field at the same time as image acquisition.
2. Query an online source of radiosonde and/or Sun photometer network data that may have acquired parameters near the time and location of the image acquisition (e.g., Holben et al., 1998).
3. Assume fixed standard values for the atmospheric parameters for the geographic region of interest (e.g., Fedosejevs et al., 2000).
4. Use climatological values developed over time (e.g., Bokoye et al., 2001).

5. Estimate the required parameters from the image data themselves, using techniques such as the dark target approach (e.g., Chavez, 1988, 1996, Teillet and Fedosejevs, 1995, Liang et al., 1997, Song et al., 2001) and atmospheric absorption feature extraction (e.g., Staenz et al., 2002).
6. Use data assimilation results based on dynamic models driven by analyzed meteorological data (e.g., O'Neill et al., 2002).

The challenge is to make optimum use of available ground-based and satellite-based atmospheric optical measurements and to ensure they are consistent and easily accessible. A prototype atmospheric optical parameter estimation system was development to provide optical parameters online for any time and place across extended regions such as Canada or North America based on available climate to meteorological scale data and models (O'Neill et al., 2002, Freemantle et al., 2002). Aside from making atmospheric corrections more operational, this concept of Networked Online Mapping of Atmospheric Data (NOMAD) transfers the responsibility of making available quality atmospheric parameters from the user to the scientific and technical specialists maintaining the atmospheric parameter server. Nevertheless, at the end of the day, the significant uncertainties that can arise from the atmospheric correction process with uncertain inputs (e.g., Kaufman et al., 1997, Teillet et al., 2004a), too varied to characterize in a summary way, are such that users often prefer to seek out imagery acquired on relatively haze-free days, or to apply normalization techniques (e.g., Schroeder et al., 2006) rather than undertake atmospheric correction.

## 6.6 GEOMETRIC EFFECTS ON IMAGE RADIOMETRY

### 6.6.1 The Pixel

The nature of the remote sensing image pixel has been given some but, arguably, not enough attention (Townshend, 1981, Duggin, 1986, Fisher, 1997, Cracknell, 1998, Townshend et al., 2000). Clearly, a pixel is not a true geographical object, nor does it correspond exactly to the spatial resolution of the imaging sensor. The pixel is a sampling of the non-uniform point spread function (PSF) of a given instantaneous field-of-view (IFOV) of the sensor. Space does not allow elaboration on the subject of PSFs of satellite imaging sensors. A good treatment of this topic can be found in Schowengerdt (2007).

It is easy to forget that a substantial proportion of the signal apparently coming from the surface area represented by a given pixel comes from surrounding areas and from the atmospheric path. A significant proportion of any given satellite image consists of mixed pixels. Moreover, in most imaging systems, pixels at large off-nadir view angles each encompass significantly larger surface areas than do pixels closer to nadir because of the projection effect. Large field-of-view whisk broom sensors such as the MODerate-resolution Imaging Spectroradiometer (MODIS) are subject to the panoramic "bow tie" effect, yielding scans that partially overlap at off-nadir angles (Souri and Azizi, 2013). Further still, cross-track ground sampling intervals increase with increasing view angle because of Earth curvature; the flat-Earth approximation is good to within 4% for off-nadir view angles less than 23° (Schowengerdt, 2007). These considerations have important ramifications on the per-pixel characterization of regions of interest and on the integration of remote sensing data into geographical information systems (GIS). Thus, contrary to common practice, land cover properties should be reported at spatial resolutions coarser than the individual remote sensing pixels.

The case of the AVHRR sensor series illustrates this point. Nominally, the nadir-view size of AVHRR pixels is 1.1 km by 1.1 km. With the large off-nadir angles that constitute a significant portion of each AVHRR image, the majority of pixels are greater than 1.1 km in size, reaching as much as ten times the area of nadir pixels at the largest off-nadir angles (55° at ground level). Moreover, it has been estimated that the nadir pixel sampling at 1.1 km represents approximately only 27% of the radiation captured by the ground IFOV (GIFOV) (personal communication, NASA Langley

Research Center). It has also been found that, in the case of the Landsat-5 TM, the cross-track GIFOV is on the order of 40–45 m instead of the nominal resolution of 30 m (Schowengerdt et al., 1985).

### 6.6.2 Illumination and Viewing Geometries through the Atmosphere

Except for large angles, the influence of varying illumination and viewing paths through the atmosphere is generally well handled by most atmospheric correction algorithms. However, image data acquired for high-latitude regions, where Sun angles are typically far from the zenith, are more difficult to correct for atmospheric effects. In addition, while some atmospheric correction codes incorporate simple BRDF models to allow for multiple surface-atmosphere scatterings (e.g., the 6S code;[5] Vermote et al., 2006, Kotchenova et al., 2006, Kotchenova and Vermote, 2007), computationally intensive multiple scattering in the radiative transfer is not always a component of atmospheric correction algorithms and, hence, is not always taken into consideration when deriving reflectance information about a surface from satellite image data. The inclusion of multiple scattering is a greater concern for ocean color parameter retrieval (e.g., Gordon and Wang, 1994) than for land surface reflectance retrieval.

### 6.6.3 Atmospheric Refraction

For typical air densities, the atmospheric refraction correction for the entire atmosphere down to sea level is less 0.1° for zenith angles less than 80° (Scarpace and Wolf, 1973, McCartney, 1976, Chu, 1983, Egan, 1985). Accordingly, atmospheric refraction only becomes important for satellite sensors designed to study the Earth's atmosphere at near-grazing angles.

### 6.6.4 Reflectance Anisotropy

Reflectance anisotropy as a function of illumination and viewing geometries is a fundamental property of any terrestrial surface and is best described in terms of BRDF (Equation 6.4). Some terrestrial surfaces have highly anisotropic reflectance properties and this has an important influence on image radiometric characteristics. Forest canopies are particularly subject to these effects (cf., Figure 6.3). Methodologies to model and deal with BRDF effects are well documented in the literature (e.g., Li and Strahler, 1992, Chen and Cihlar, 1997, White et al., 2002a, Pinty et al., 2004).

FIGURE 6.3 Digital imagery showing the same black spruce canopy taken at the indicated angles from a truck-based platform. The effect of reflectance anisotropy is clearly evident (solar illumination is from the right). Because these anisotropy effects are at the scale of trees and their components, they will give rise to sub-pixel variations in satellite sensor data of any spatial resolution.

Nevertheless, although anisotropic reflectance effects have been studied extensively, they remain challenging to deal with in an operational setting and there are many other geometric effects to consider apart from the BRDF.

### 6.6.5 Adjacency Effects

As already noted, some of the photons reaching the satellite sensor will have propagated through the Earth's atmosphere, reflected from surfaces surrounding any given instantaneous field-of-view (IFOV), and then scattered into the line of sight of the sensor (Otterman and Fraser, 1979, Dave, 1980, Dana, 1982, Meckler and Kaufman, 1982, Deschamps et al., 1983, Otterman et al., 1983, Kaufman and Fraser, 1984). This adjacency effect can give rise to blurring and contrast reduction at the boundaries of IFOVs. While the preference here is to include it as a georadiometric effect, the adjacency effect is commonly discussed as part of atmospheric correction. However, more often than not, the adjacency effect is ignored or, at best, taken into account in an approximate way (Tanré et al., 1990, Vermote et al., 2006).

### 6.6.6 Topographic Effects

A number of approaches have been developed to address the influence of surface topography on the radiometric properties of pixels in remotely sensed image data. Terrain elevation variations cause variable atmospheric path lengths that image correction algorithms have to address in the contexts of remote sensing from both airborne platforms in the atmosphere and satellite platforms above the atmosphere (e.g., Naethe, et al., 2024, Shen et al., 2023, Claverie et al., 2018, Yang et al., 2027, Teillet and Santer, 1991, 1992). Terrain elevation variations also cause shadow effects that introduce errors in image understanding in general (Woodham and Lee, 1985, Woodham and Gray, 1987) and in surface reflectance retrieval in particular (Giles, 2001, Adeline et al., 2013).

Image corrections for terrain slope and aspect variations are more difficult to carry out. Some approaches have used normalization techniques (e.g., Naethe, et al., 2024, Shen et al., 2023, Masek et al., 2022, 2021, Holben and Justice, 1981, Allen, 2000), with mixed results (Richter, 1997, 1998). Instead, most radiometric corrections for topographic effects use digital terrain elevation, slope, aspect and other derivatives (i.e., a digital terrain model (DTM)) to describe explicitly the surface topography. Two main categories of these approaches are (1) slope-aspect corrections that use some function of the cosine of the incident solar angle (Smith et al., 1980, Teillet et al., 1982, Cavayas, 1987, Meyer et al., 1993, Ekstrand, 1996, Riaño et al., 2003, Soenen et al., 2005, 2008) and (2) model-based corrections that deal with solar radiation interactions with surface targets (Hugli and Frei, 1983, Proy et al., 1989, Gu and Gillespie, 1998, Dymond and Shepherd, 1999, Shepherd and Dymond, 2003, Li et al., 2012) and with atmospheric radiative transfer codes (Richter, 1997, 1998).

In practice, slope-aspect corrections may involve the prior rectification of an image to a map-based coordinate system, typically the Universal Transverse Mercator (UTM) projection, which matches the map coordinate system of the DTM. Although this simplifies image processing and facilitates image classification studies involving various images in a common map projection, there are disadvantages to the image rotation and resampling that occur during the map rectification process. Significant radiometric errors can be introduced as a result of resampling (Forster and Trinder, 1984), even using an optimum parametric cubic convolution algorithm (Schowengerdt et al., 1983). Additionally, the slope-aspect correction procedure has to keep track of the scan direction to specify view angles properly. A better approach is to not apply the map transformation to the image before slope-aspect correction and, instead, make use of the map transformation equations in reverse to apply the slope-aspect correction in image space.

A rule of thumb to keep in mind is that the spatial resolution of the digital elevation model (DEM) used to define the slope and aspect data in the DTM should be finer, by a factor of nine, than the spatial resolution of the image data to be corrected for slope-aspect effects. For example, terrain elevation data with horizontal spatial resolutions of 33 m by 33 m should be used to generate the slope and aspect angles to be used to correct imagery whose spatial resolution is 100 m by 100 m.

In principle, topographic slope-aspect corrections should also involve atmospheric and surface reflectance models to allow for the proper treatment of the effects due to terrain elevation variations, BRDF, irradiance from adjacent slopes, and diffuse sky illumination distribution (Kimes and Kirchner, 1981, Sjoberg and Horn, 1983, Bruhl and Zdunkowski, 1983, Cavayas et al., 1983, Woodham and Lee, 1985).

### 6.6.7 Position of the Sun

Uncertainties in knowledge of the position of the Sun can give rise to significant errors in shadow area estimation in rugged terrain (Teillet et al., 1986). The largest error by far is the use of a single set of solar zenith and azimuth angles for the radiometric processing and correction of an entire satellite image. For example, the lack of specification of within-scene location for a Landsat image leads to uncertainties in solar position of 4.5° (zenith angle) by 2.5° (azimuth angle). It is not uncommon to use a single set of solar angles for Landsat images, whereas, for radiometric corrections applied to wide-area coverage satellite imaging sensors such as AVHRR or MODIS, it is routine to specify solar angles on a sub-scene basis. Other contributions to the uncertainty in solar position include, in decreasing order of importance: finite size of the solar disk (30 minutes), lack of specification of within-scene time (15 minutes (azimuth angle) by eight minutes (zenith angle)), use of solar ephemeris from an inappropriate epoch, atmospheric refraction effects for large solar zenith angles, and horizontal parallax (target at Earth surface not Earth center). The potential cumulative uncertainty effect in the Landsat case is 5° (zenith angle) by 3.5° (azimuth angle).

## 6.7 SPECTRAL CHARACTERIZATION

Although it has received less attention than other aspects of satellite image radiometry, spectral characterization is an important consideration for proper surface reflectance retrieval, regardless of how wide or narrow the spectral bands may be. Spectral bands designed for specific applications and data products are susceptible to post-launch variations in spectral bandpasses and they usually differ between different sensors and missions. If spectral bands have changed in wavelength position or bandwidth post-launch, or if there are uncertainties as to their characteristics, there is a direct impact on radiometric and atmospheric processing, as well as on derived information products (Naethe, et al., 2024, Shen et al., 2023, Masek et al., 2022, 2021, Suits et al., 1988, Teillet, 1990b, Teillet and Irons, 1990, Flittner and Slater, 1991, Goetz et al., 1995, Teillet et al., 1997b, Trishchenko, 2009, Steven et al., 2003). In practice, there is little that users can do to take into account post-launch changes in spectral band performance. Even when spectral bands perform as designed, users should be aware that similar information products derived from different sensors with analogous spectral bands that do not match exactly are not directly comparable (Teillet et al., 2007, Teillet and Ren, 2008). These spectral band difference effects are scene-dependent and, hence, not corrected easily.

A limited number of investigations have been undertaken to assess radiometric calibration errors due to differences in spectral response functions between satellite sensors when attempting cross-calibration based on near-simultaneous imaging of common ground look targets in analogous spectral bands (Teillet et al., 2001c, 2004b, 2007, Trishchenko et al., 2002, Rao et al., 2003, Doelling et al., 2012, Chander et al., 2013c, Henry et al., 2013). A specific example is the radiometric cross-calibration of the Landsat-7 Enhanced Thematic Mapper Plus (ETM+) and Landsat-5 Thematic Mapper (TM) sensors based on early-mission tandem-orbit datasets (Teillet et al., 2001c, 2004b), which included adjustments for spectral band differences between the two Landsat sensors. Spectral band difference effects were shown to be significant (an effect that can be 5% or more), despite the close similarity in the spectral filters and response functions of the Landsat sensors. A variety of terrestrial surfaces were assessed regarding their suitability for Landsat radiometric cross-calibration in the absence of surface reflectance spectra. This line of inquiry was extended to Earth observation sensors on several satellite platforms, indicating that large differences can arise (Teillet et al., 2007, Teillet and Ren, 2008).

## 6.8 NORMALIZATION APPROACHES

Many applications are not dependent on securing radiometric calibration on an absolute scale. Only relative changes between images are of interest such that normalization procedures can be used to bypass steps in Figure 6.2. Spectral band transformations (e.g., band ratios or principal component analyses) have long been used to mitigate atmospheric and/or topographic effects (e.g., Crist, 1985). A variety of empirical radiometric normalization methods have been developed (Naethe, et al., 2024, Shen et al., 2023, Masek et al., 2022, 2021, Schott et al., 1988, Hall et al., 1991, Yuan and Elvidge, 1996, e.g., Schroeder et al., 2006).

## 6.9 PROCESSING CONSIDERATIONS

This chapter has so far provided an overview of the principal physical effects affecting satellite image radiometry. Additional concerns arise in the domain of image processing that, under certain circumstances, can affect image radiometry significantly.

Nominally, image processing software will first calibrate digital counts to TOA radiance and, in some cases, convert TOA radiance to TOA reflectance. In any case, the image data are then in physical units such that atmospheric correction to estimate surface reflectance can be done. Adjustments for bidirectional and/or topographic effects should be included in the latter step. It is important to note that these radiometry-related image processing steps involve equations that are linear or nonlinear. The associated software must be constrained to prevent user application of the various calibration/correction steps in different sequences because linear and nonlinear transformations are not commutative (Teillet, 1986). In principle, all radiometric corrections should be completed prior to any geometric processing.

Moreover, it is important to understand how the sequencing of spectral band integration can affect image radiometry, especially in processing for surface reflectance retrieval (i.e., atmospheric correction). Detailed atmospheric computations can be done monochromatically and the results integrated over the spectral band involved, or band-integrated quantities can be used in the atmospheric computations. Depending on the surface reflectance spectrum and the relative spectral response profile of the spectral band involved, the two approaches can yield results that differ significantly, and the proper choice of approach depends mainly on whether the image data are multispectral or hyperspectral in character (Teillet, 1989, Teillet and Irons, 1990).

## 6.10 DISCUSSION OF FUTURE TRENDS

Although radiometric calibration is a very specialised aspect of remote sensing, it constitutes an essential component of Earth observation systems and their utilization. It is crucial for the generation of useful long-term data records, as well as for independent data quality control. In that light, this chapter outlines the key post-launch methodologies used to convert digital image data to calibrated data products. Nevertheless, post-launch calibration and the associated quality assurance infrastructures, which should be behind-the-scenes activities as with any mature technology, remain semi-operational at best. The CEOS QA4EO effort, noted in Section 2 in the context of GEOSS, is an important step in the direction of operational best practices for traceability and interoperability. Two other calibration-related initiatives for the future are noted here, both representing challenging space-based undertakings.

The Traceable Radiometry Underpinning Terrestrial- and Helio-Studies (TRUTHS) mission concept (Fox et al., 2002, 2011) offers a novel approach to the provision of key scientific data with unprecedented radiometric accuracy for Earth observation and solar studies. TRUTHS will calibrate its instrumentation directly to SI on orbit, overcoming the usual uncertainties associated with drifts of sensor calibration by using an electrical rather than an optical standard as the basis of its radiometric calibration. A space-based cryogenic radiometer together with its associated calibration chain to a terrestrial primary standard will provide a space-based standard reference for measurements of both the Sun and the Earth. The TRUTHS mission has the potential to improve the performance and accuracy of Earth observation missions by an order of magnitude (on the order of 0.3% in the solar reflective domain). As a result, TRUTHS will provide the necessary advances to ensure that Earth system datasets have sufficient radiometric long-term precision for the reliable detection and evaluation of global change.

One of the missions recommended for earliest possible implementation by the U.S. National Research Council (NRC) Decadal Survey report (NRC, 2007) is the Climate Absolute Radiance and Refractivity Observatory (CLARREO), a joint NASA/NOAA mission. NOAA's contribution is to be sensors for total and spectral solar irradiance measurements and Earth energy budget climate data records. NASA's contribution is to be sensors for the measurement of spectrally resolved thermal infrared and reflected solar radiation with high absolute accuracy. The CLARREO mission objective pertinent to the topic of this chapter is the provision of a space-based high-accuracy calibration standard to enable calibration and intercalibration of Earth observation satellite sensors. To accomplish this, CLARREO will include a Reflected Solar Suite consisting of two push-broom hyperspectral imagers covering 320–2,300 nanometres in a single instrument package. Reflectance will be obtained via ratios of Earth-view data to solar-view data. High accuracy will be achieved by precisely calibrating the instruments via the solar view combined with a sensor model to transfer National Institute of Standards and Technology (NIST) laboratory standards to orbit (Anderson et al., 2008, Thome et al., 2010, Sandford et al., 2010, Lukashin et al., 2013). As for TRUTHS, the CLARREO mission has the potential to improve the performance and accuracy of Earth observation missions by an order of magnitude (on the order of 0.3% in the solar reflective domain).

The success of QA4EO is far from guaranteed, and the TRUTHS and CLARREO missions have yet to be approved. In the early 21st century, innovation in Earth observation as well as the subfield of data standardization requires making judicious choices. As the geostrategic technologies of Earth observation seek adoption by mainstream information society, it is hoped that young people pursuing optical engineering, radiometry, metrology, and remote sensing physics will strive to ensure the data quality and product validation needed for Earth observation to underpin sound decision making.

## 6.11 CONCLUDING REMARKS

This chapter has provided a broad overview of the key considerations involved in the radiometric calibration and correction of image data from satellite sensor systems operating in the solar-reflective optical domain. Relevant pioneering research over the past four decades has been featured, with emphasis on sensor radiometric calibration, retrieval of surface reflectance (involving correction for atmospheric effects), allowance for geometric effects on image radiometry, and an understanding of target spectral reflectance behavior. Surface reflectance served as an important Earth surface variable of wide interest. The need for radiometric calibration and correction is well documented in the literature and growing in importance as Earth science data products are being derived increasingly from a multiplicity of different satellite sensor systems, and data and information demands for the purposes of sound decision making require increasingly greater precision and accuracy.

Sensor radiometric calibration is the most fundamental consideration given that uncertainties in radiometric calibration translate directly into the same amount of radiometric uncertainty in all products derived downstream. Uncertainties within 2–3% (1 $\sigma$) with respect to exo-atmospheric solar irradiance are attainable routinely, relative to national laboratory standards, and missions such as TRUTHS and CLARREO promise an order-of-magnitude improvement. Recommendations for future research on vicarious calibration using terrestrial reference standard sites can be found in the next chapter (Teillet and Chander, 2014).

However, the retrieval of terrestrial surface variables is subject to many other considerations, as outlined extensively in this chapter. Significant spatial and temporal variations in the optical properties of the Earth's atmosphere necessitate image corrections for these effects in the solar-reflective spectral bands to put satellite data on the same radiometric scale for studies intended to monitor terrestrial surfaces over time and space. Although the computational tools for image correction for atmospheric effects are available, they require user inputs reliant on ready access to timely and accurate data on atmospheric variables such as aerosol optical depth and water vapor content. The net result is that surface reflectance retrieval from remotely sensed image data is operational for larger programs (e.g., MODIS, Landsat, Sentinel (Masek et al., 2022, 2021, Vermote et al., 1997, 2002, Schaaf et al., 2002)), but remains laborious for individual user studies. It would be very helpful if space agencies and data suppliers more often took the extra steps to generate and offer surface reflectance and other higher-level products to the user community at reasonable or no cost. Progress can also come about if an agency with the pertinent scientific and technical resources and specialists maintained a dynamic open-access atmospheric parameter server along the lines of NOMAD.

The chapter also described numerous geometric factors and phenomena that give rise to radiometric effects in satellite image data. Most of these geometric considerations are tractable, but a few are more challenging to address, especially anisotropic reflectance properties of surfaces and topographic effects. A progressive perspective on BRDF is to view angular signatures of spectral reflectance as an information source as opposed to something that has to be corrected (e.g., White et al., 2001, 2002b). Indeed, a greater number of satellite sensor systems than can perform angular remote sensing would yield substantial information returns on investment (M. Verstraete, personal communication). Slope-aspect corrections will benefit from the highest-resolution DEMs, which are becoming increasingly available for the entire surface of the Earth.[6]

Although spectral considerations have received relatively less attention than other aspects of satellite image radiometry, the literature has grown considerable over the past decade. The increased attention is partly because of the advent of hyperspectral remote sensing (beyond the scope of this chapter to treat), which demands numerous preprocessing corrections. Also, efforts devoted to radiometric calibration, a first-order consideration, have been well advanced by the research community such that increased attention is now being devoted, belatedly, to spectral considerations. Proper surface reflectance retrieval depends on a good understanding of the state of sensor spectral band characteristics post-launch, given that they have a direct impact on radiometric and atmospheric processing, as well as on the spectral properties of the scene (the atmosphere and the surface targets of interest).

Based on best practices and recommended for implementation and use throughout the GEO community, QA4EO is the way forward for sensor radiometric calibration. It is hoped that QA4EO will grow to encompass the other important elements of satellite image radiometry, including atmospheric, geometric and spectral influences on Earth science data.

## ACKNOWLEDGMENTS

The author gratefully acknowledges guidance over the years from Philip N. Slater and many substantive discussions on remote sensing radiometry with Kurtis J. Thome, Nigel P. Fox, Robert P. Gauthier, and Gunar Fedosejevs. The author also wishes to thank Gyanesh Chander for taking the time to read and provide detailed comments on drafts of this chapter.

## NOTES

1 "Man must rise above the atmosphere and beyond to understand fully the world in which he lives." Socrates, 700 BC.

2 These numbers are according to *The Earth Observation Handbook* (2014) (www.eohandbook.com/) of the international Committee on Earth Observation Satellites (CEOS), which consists of the majority of the space agencies worldwide. The numbers exclude missions by military agencies and commercial companies.

3 http://qa4eo.org/index.html.

4 A recent estimate indicates that 99.5% of newly created digital data of all kinds are never analyzed (Regalado, 2013).

5 6S: Second Simulation of the Satellite Signal in the Solar Spectrum.

6 E.g., WorldDEM: www.astrium-geo.com/worlddem/ and OpenTopography: www.opentopography.org/index.php

## REFERENCES


Abel, P., B. Guenther, R.N. Galimore, and J.W. Cooper, 1993, Calibration results for NOAA-11 AVHRR channels 1 and 2 from congruent path aircraft observations, *Journal of Atmospheric and Oceanic Technology*, 10(4): 493–508.

Adeline, K., M. Chen, X. Briottet, S. Pang, and N. Paparoditis, 2013, Shadow detection in very high spatial resolution aerial images: A comparative study, *ISPRS Journal of Photogrammetry and Remote Sensing*, 80: 21–38.

Ahern, F.J., R.J. Brown, J. Cihlar, R. Gauthier, J. Murphy, R.A. Neville, and P.M. Teillet, 1989, Radiometric correction of visible and infrared remote sensing data at the Canada Centre for Remote Sensing, In *Remote Sensing Yearbook 1988/89*, Editors: A. Cracknell and L. Hayes, Taylor and Francis, Philadelphia, PA, pp. 101–127.

Ahern, F.J., P.M. Teillet, and D.G. Goodenough, 1979, Transformation of atmospheric and solar illumination conditions on the CCRS image analysis system, *Proceedings of the 5th International Symposium on Machine Processing of Remotely Sensed Data*, West Lafayette, IN, pp. 34–51.

Allen, T.R., 2000, Topographic normalization of Landsat Thematic Mapper data in three mountain environments, *Geocarto International*, 15: 13–19.

Anderson, D., K.W. Jucks, and D.F. Young, 2008, The NRC decadal survey climate absolute radiance and refractivity observatory: NASA implementation, *Proceedings of the IEEE International Geoscience and Remote Sensing Symposium (IGARSS)*, Boston, MA, pp. 9–11.

Barnes, R.A., R.E. Eplee, Jr., F.S. Patt, H.H. Kieffer, T.C. Stone, G. Meister, J.J. Butler, and C.R. McClain, 2004, Comparison of SeaWiFS measurements of the moon with the U.S. geological survey lunar model, *Applied Optics*, 43(31): 5838–5854.

Berk, A., G.P. Anderson, P.K. Acharya, L.S. Bernstein, L. Muratov, J. Lee, M. Fox, S.M. Adler-Golden, J.H. Chetwynd, M.L. Hoke, R.B. Lockwood, J.A. Gardner, T.W. Cooley, C.C. Borel, P.E. Lewis, and E.P. Shettle, 2006, MODTRAN5: 2006 update, *Proc. SPIE Conference 6233 on Algorithms and Technologies for Multispectral, Hyperspectral, and Ultraspectral Imagery* XII, 62331F.

Biggar, S.F., P.N. Slater, and D.I. Gellman, 1994, Uncertainties in the in-flight calibration of sensors with reference to measured ground sites in the 0.4 to 1.1 μm range, *Remote Sensing of Environment*, 48: 245–252.

Biggar, S.F., P.N. Slater, K.J. Thome, A.W. Holmes, and R.A. Barnes, 1993, Preflight solar-based calibration of SeaWiFS, *Proceedings of SPIE Conference 1939*, pp. 233–242.

Bokoye, A.I., A. Royer, N.T. O'Neill, P. Cliche, G. Fedosejevs, P.M. Teillet, and L.J.B. McArthur, 2001, Characterization of Atmospheric Aerosols across Canada from a ground-based Sunphotometer network: AEROCAN, *Atmosphere-Ocean*, 39(4): 429–456.

Brest, C.L., and W.B. Rossow, 1992, Radiometric calibration and monitoring of NOAA AVHRR data for ISCCP, *International Journal of Remote Sensing*, 13(2): 235–273.

Brest, C.L., W.B. Rossow, and M.D. Roiter, 1997, Update of radiance calibrations for ISCCP, *Journal of Atmospheric and Oceanic Technology*, 14(5): 1091–1109.

Bruegge, C., and J. Butler, Editors, 1996, Special issue on earth observing system calibration, *Journal of Atmospheric and Oceanographic Technology*, 13(2): 273–544.

Bruhl, C., and W. Zdunkowski, 1983, An approximate calculation method for parallel and diffuse irradiances on inclined surfaces in the presence of obstructing mountains or buildings, *Archives for Meteorology,Geophysics, and Bioclimatology (Series B)*, 32(2–3): 111–129.

Brunger, A.P., and F.C. Hooper, 1993, Anisotropic sky radiance model based on narrow field of view measurements of shortwave radiance, *Solar Energy*, 51(1): 53–64, Erratum: 51(6): 523.

Burrows, W.E., 1999, *This New Ocean: The Story of the First Space Age*, Modern Library Paperbacks, 752 pages.

Butler, J.J., B.C. Johnson, and R.A. Barnes, 2005, The calibration and characterization of Earth remote sensing and environmental monitoring instruments. In *Optical Radiometry, Volume 41, Experimental Methods in the Physical Sciences*. Editors: A.C. Parr, R.U. Datla and J.L. Gardner. Treatise Editors R. Celotta and T. Lucatorto. Elsevier/Academic Press, pp. 453–534.

Cabot, F., O. Hagolle, and P. Henry, 2000, Relative and multitemporal calibration of AVHRR, SeaWiFS, and VEGETATION using POLDER characterization of desert sites, *Proceedings of International Geoscience and Remote Sensing Symposium 2000*, Honolulu, Hawaii, pp. 2188–2190.

Cao, C., and A.K. Heidinger, 2002, Inter-comparison of the longwave infrared channels of MODIS and AVHRR/NOAA-16 using simultaneous nadir observations at orbit intersections, *Proceedings of SPIE Conference 4814 on Earth Observing Systems VII*, Seattle, Washington, Editor: W.L. Barnes. SPIE, Bellingham, Washington, pp. 306–316.

Cavayas, F., 1987, Modelling and correction of topographic effect using multi-temporal satellite images, *Canadian Journal of Remote Sensing*, 13(2): 49–67.

Cavayas, F., G. Rochon, and P. Teillet, 1983, Estimation des reflectances bidirectionnelles par analyse des images Landsat: problèmes et possibilités de solutions, Comptes rendus du 8ème Symposium canadien sur la télédétection et 4ème Congres de l'Association québécoise de télédétection, Montréal, Québec, p. 645.

Chahine, M.T., 1983, Interaction mechanisms within the atmosphere, Chapter 5 in *Manual of Remote Sensing*, second edition, Editor: R.N. Colwell, American Society of Photogrammetry, Falls Church, VA, The Sheridan Press.

Chander, G., T.J. Hewison, N. Fox, X. Wu, X. Xiong, and W.J. Blackwell, 2013a, Overview of intercalibration of satellite instruments, *IEEE Transactions on Geoscience and Remote Sensing*, 51(3 SI): 1056–1080.

Chander, G., T.J. Hewison, N. Fox, X. Wu, X. Xiong, and W.J. Blackwell, Guest Editors, 2013b, Special issue on intercalibration of satellite instruments, *IEEE Transactions on Geoscience and Remote Sensing*, 51(3 SI): 491 pages.

Chander, G., N. Mishra, D.L. Helder, D. Aaron, A. Angal, T. Choi, X. Xiong, and D. Doelling, 2013c, Applications of Spectral Band Adjustment Factors (SBAF) for cross-calibration, *IEEE Transactions on Geoscience and Remote Sensing*, 51(3 SI): 1267–1281.

Chang, I.L., C. Dean, Z. Li, M. Weinreb, X. Wu, and P.A.V.B. Swamy, 2012, Refined algorithms for star-based monitoring of GOES imager visible channel responsivities, *Proceedings of SPIE Earth Observing Systems XVII*, San Diego, CA, pp. 851 00R.

Chavez, P.S., 1988, An improved dark-object subtraction technique for atmospheric scattering correction of multispectral data, *Remote Sensing of Environment*, 24: 459–479.

Chavez, P.S., 1996, Image-based atmospheric correction—Revisited and improved, *Photogrammetric Engineering and Remote Sensing*, 62(9): 1025–1036.

Che, N. and J.C. Price, 1992, Survey of radiometric calibration results and methods for visible and near-infrared channels of NOAA-7, NOAA-9, and NOAA-11 AVHRRs, *Remote Sensing of Environment*, 41(1): 19–27.

Chen, J.M., and J. Cihlar, 1997, A hotspot function in a simple bidirectional reflectance model for satellite applications, *Journal of Geophysical Research—Atmospheres*, 102(D22): 25907–25913.

Chen, H.S., 1996, *Remote Sensing Calibration Systems—An Introduction*. ISBN 0-937194-38-7, A. Deepak Publishing, Hampton, VI, USA.

Chu, W.P., 1983, Calculations of atmospheric refraction for spacecraft remote sensing applications, *Applied Optics*, 22(5): 721–725.

Cihlar, J., and P.M. Teillet, 1995, Forward piecewise linear calibration model for quasi-real-time processing of AVHRR data, *Canadian Journal of Remote Sensing*, 21(1): 22–27.

Claverie, M., J. Ju, J.G. Masek, J.L. Dungan, E.F. Vermote, J.C. Roger, S.V. Skakun, and C. Justice, 2018, The harmonized landsat and sentinel-2 surface reflectance data set, *Remote Sensing of Environment*, 219: 145–161, ISSN 0034–4257, https://doi.org/10.1016/j.rse.2018.09.002. (www.sciencedirect.com/science/article/pii/S0034425718304139)

Connolly, J.I., and B. Tolar, 1994, Standards and calibration, workshop II, *Proceedings of the First International Symposium on Spectral Sensing Research (ISSSR)*, San Diego, CA, pp. 87–111.

Cracknell, A.P., 1998, Synergy in remote sensing—what's in a pixel?, *International Journal of Remote Sensing*, 19(11): 2025–2047.

Crist, E.P., 1985, A TM Tasseled Cap equivalent transformation for reflectance factor data, *Remote Sensing of Environment*, 17: 301–306.

Dana, R.W., 1982, Background reflectance effects in Landsat data, *Applied Optics*, 21(22): 4106–4111.

Dave, J.V., 1980, Effect of atmospheric conditions on remote sensing of a surface nonhomogeneity, *Photogrammetric Engineering and Remote Sensing*, 46: 1173–1180.

Deschamps, P.Y., M. Herman, and D. Tanré, 1983, Definitions of atmospheric radiance and transmittances in remote sensing, *Remote Sensing of Environment*, 13(1): 89–92.

Dilligeard, E., X. Briottet, J.L. Deuze, and R.P. Santer, 1997, SPOT calibration of blue and green channels using Rayleigh scattering over clear oceans, *Proceedings of SPIE Conference on Advanced Next-Generation Satellites II*, Taormina, Italy, pp. 373–379.

Dinguirard, M., J. Mueller, F. Sirou, and T. Tremas, 1997, Comparison of ScaRaB ground calibration in the short wave and long wave domains. Special Issue on NEWRAD '97, Tucson, Arizona. *Metrologia*, 35: 597–601.

Dinguirard, M., and P.N. Slater, 1999, Calibration of space-multispectral imaging sensors: A review, *Remote Sensing of Environment*, 68(3): 194–205.

Doelling, D.R., G. Hong, D. Morstad, R. Bhatt, A. Gopalan, and X. Xiong, 2010, The characterization of deep convective cloud albedo as a calibration target using MODIS reflectances, *Proceedings of SPIE Conference 7862 on Earth Observing Missions Sensors: Development, Implementation, and Characterization*, Editors: X. Xiong, C. Kim, and H. Shimoda, Incheon, Republic of Korea, p. 786 20I.

Doelling, D.R., C. Lukashin, P. Minnis, B. Scarino, and D. Morstad, 2012, Spectral reflectance corrections for satellite intercalibrations using SCHIAMACHY data. *IEEE Geoscience and Remote Sensing Letters*, 9(1): 119–123.

Doelling, D.R., L. Nguyen, and P. Minnis, 2004, On the use of deep convective clouds to calibrate AVHRR data, *Proceedings of SPIE Conference 5542 on Earth Observing Systems IX*, Editors: W.L. Barnes and J.J. Butler, Denverr, CO, pp. 281–289.

Duggin, M.J., 1986, Variance in radiance recorded from heterogeneous targets in the optical-reflective, middle-infrared, and thermal-infrared regions, *Applied Optics*, 25(23): 4246–4252.

Duggin, M.J., 1987, Impact of radiance variations on satellite sensor calibration. *Applied Optics*, 26(7): 1264–1271.

Duggin, M.J., and C.J. Robinove 1990, Assumptions implicit in remote sensing data acquisition and analysis. *International Journal of Remote Sensing*, 11(10): 1669–1694.

Duguay, C.R. 1993, Radiation modeling in mountainous terrain review and status, *Mountain Research and Development*, 13(4): 339–357.

Dymond, J.R., and J.D. Shepherd, 1999, Correction of the topographic effect in remote sensing, *IEEE Transactions on Geoscience and Remote Sensing*, 37: 2618–2620.

Edwards, D.P., L.K. Emmons, D.A. Hauglustaine, D.A. Chu, J.C. Gille, Y.J. Kaufman, G. Petron, L.N. Yurganov, L. Giglio, M.N. Deeter, V. Yudin, D.C. Ziskin, J. Warner, J.-F. Lamarque, G.L. Francis, S.P. Ho, D. Mao, J. Chen, E.I. Grechko, and J.R. Drummond, 2004, Observations of carbon monoxide and aerosols from the Terra satellite: Northern Hemisphere variability, *Journal of Geophysical Research*, 109: D24202, doi: 10.1029/2004JD004727.

Egan, W.G., 1985, *Photometry and Polarization in Remote Sensing*, Elsevier, New York, 503 pages.

Ekstrand, S., 1996, Landsat TM-based forest damage assessment: Correction for topographic effects, *Photogrammetric Engineering and Remote Sensing*, 62(2): 151–161.

Fedosejevs, G., N.T. O'Neill, A. Royer, P.M. Teillet, A.I. Bokoye, and B. McArthur, 2000, Aerosol optical depth for atmospheric correction of AVHRR composite data, *Canadian Journal of Remote Sensing*, 26(4): 273–284.

Fisher, P., 1997, The pixel: A snare and a delusion, *International Journal of Remote Sensing*, 18(3): 679–685.

Flittner, D.E., and P.N. Slater, 1991, Stability of narrow-band filter radiometers in the solar-reflective range, *Photogrammetric Engineering and Remote Sensing*, 57(2): 165–171.

Forster, B.C., and J.C. Trinder, 1984, An examination of the effects of resampling on classification accuracy, *Proceedings of the 3rd Australasian Remote Sensing Conference*, Queensland, Australia, pp. 106–115.

Fougnie, B., and R. Bach, 2009, Monitoring of radiometric sensitivity changes of space sensors using deep convective clouds: Operational application to PARASOL, *IEEE Transactions on Geoscience and Remote Sensing*, 47(3): 851–861.

Fox, N.P., 1999, Improving the accuracy and traceability of radiometric measurements to SI for remote sensing instrumentation, *Proceedings of the 4thInternational Airborne Remote Sensing Conference and Exhibition/21st Canadian Symposium on Remote Sensing*, ERIM International, Ottawa, ON, 1: 304–311.

Fox, N., J. Aiken, J.J. Barnett, X. Briottet, R. Carvell, C. Frohlich, S.B. Groom, O. Hagolle, J.D. Haigh, H.H. Kieffer, J. Lean, D.B. Pollock, T. Quinn, M.C.W. Sandford, M. Schaepman, K.P. Shine, W.K. Schmutz, P.M. Teillet, K.J. Thome, M.M. Verstraete, and E. Zalewski, 2002, Traceable Radiometry Underpinning Terrestrial- and Helio-Studies (TRUTHS), *Proceedings of SPIE Conference 4881 on Sensors, Systems, and Next-Generation Satellites VIII*, Heraklion, Crete, Greece, 12 pages.

Fox, N., A. Kaiser-Weiss, W. Schmutz, K. Thome, D. Young, B. Wielicki, R. Winkler, and E. Woolliams, 2011, Accurate radiometry from space: An essential tool for climate studies, *Philosophical Transactions of the Royal Society A Mathematical, Physical, and Engineering Sciences*, 369(1953): 4028–4063.

Fraser, R.S., and Y.J. Kaufman, 1986, Calibration of satellite sensors after launch, *Applied Optics*, 25(7): 1177–1185.

Freemantle, J., N.T. O'Neill, P.M. Teillet, A. Royer, J.-P. Blanchet, M. Aubé, S. Thulasiraman, F. Vachon, S. Gong, and M. Versi, 2002, Using web services for atmospheric correction of remote sensing data, *Proceedings of the 2002International Geoscience and Remote Sensing Symposium (IGARSS'02) and the Twenty-Fourth Canadian Symposium on Remote Sensing*, Toronto, ON, 5: 2939–2941, also on CD-ROM.

Frouin, R., and C. Gautier, 1987, Calibration of NOAA-7 AVHRR, GOES-5, and GOES-6 VISSR/VAS solar channels, *Remote Sensing of Environment*, 22(1): 73–101.

Giles, P. 2001, Remote sensing and cast shadows in mountainous terrain, *Photogrammetric Engineering and Remote Sensing*, 67(7): 833–840.

Goetz, A.F.H., K.B. Heidebrecht, and T.G. Chrien, 1995, High accuracy in-flight wavelength calibration of imaging spectrometry data, *Summaries of the Fifth Annual JPL Airborne Earth Science Workshop, Volume 1, AVIRIS Workshop*, Pasadena, CA, pp. 67–69.

Gordon, H.R., 1987, Calibration requirements and methodology for remote sensors viewing the ocean in the visible, *Remote Sensing of Environment*, 22: 103–126.

Gordon, H.R., and M. Wang, 1994, Retrieval of water-leaving radiance and aerosol optical thickness over the oceans with SeaWiFS: A preliminary algorithm, *Applied Optics*, 33(3): 443–452.

Gu, D., and A. Gillespie, 1998, Topographic normalization of Landsat TM images of forest based on sub-pixel sun-canopy-sensor geometry, *Remote Sensing of Environment*, 64: 166–175.

Guenther, B.W., 1991, Calibration of passive remote observing optical and microwave instrumentation, *SPIE Volume 1493*, Orlando, FL, ISBN 0-8194-0602-3, SPIE—The International Society for Optical Engineering, Bellingham, Washington 98227–0010, USA, 304 pages.

Guenther, B.W., W. Barnes, E. Knight, J. Barker, J. Harnden, R. Weber, M. Roberto, G. Godden, H. Montgomery, and P. Abel, 1996, MODIS calibration: A brief review of the strategy for the at-launch calibration approach, *Journal of Atmospheric and Oceanographic Technology*, 13(2): 274–285.

Guenther, B.W., J. Butler, and P. Ardanuy, Editors, 1997, *Workshop on Strategies for Calibration and Validation of Global Change Measurements:May 10– 12, 1995,*NASA Reference Publication 1397, NASA/GSFC, Greenbelt, MD 20771, USA, 125 pages.

Hall, F.G., D.E. Strebel, J.E. Nickerson, and S.J. Goetz, 1991, Radiometric rectification: Toward a common radiometric response among multidate, multisensor images, *Remote Sensing of Environment*, 35: 11–27.

Hay, J.E., and D.C. McKay, 1985, Estimating solar irradiance on inclined surfaces: A review and assessment of methodologies, *International Journal of Solar Energy*, 3(4–5): 203–240.

Hay, J.E., R. Perez, D.C. and McKay, 1986, Addendum and Errata to the Paper "Estimating solar irradiance on inclined surfaces: A review and assessment of methodologies", *International Journal of Solar Energy*, 4(1): 321–324.

He, X., D. Pan, Y. Bai, and F. Gong, 2006, A general purpose exact Rayleigh scattering look-up table for ocean color remote sensing, *Acta Oceanologica Sinica*, 25(1): 48–56.

Hegyi, F., 2004, *Alignment of Earth Observation Calibration and Validation Activities*, Report delivered by Hegyi Geomatics International Inc (HGI) to the Canada Centre for Remote Sensing, 588 Booth Street, Ottawa, Ontario, K1A 0Y7, Canada, March 2004, 47 pages.

Helder, D., K. Thome, D. Aaron, L. Leigh, J. Czapla-Myers, N. Leisso, S. Biggar, and N. Anderson, 2012, Recent surface reflectance measurement campaigns with emphasis on best practices, SI traceability and uncertainty estimation, *Metrologia*, 49(2): S21–S28.

Helder, D., K. Thome, N. Mishra, G. Chander, X. Xiong, A. Angal, and T. Choi, 2013, Absolute radiometric calibration of landsat using a pseudo invariant calibration site, *IEEE Transactions on Geoscience and Remote Sensing*, 51(3 SI): 1360–1369.

Henry, P., G. Chander, B. Fougnie, C. Thomas, and X. Xiong, 2013, Assessment of spectral band impact on inter-calibration over desert sites using simulation based on EO-1 Hyperion data, *IEEE Transactions on Geoscience and Remote Sensing*, 51(3 SI): 1297–1308.

Holben, B.N., T.F. Eck, I. Slutsker, D. Tanré, J.P. Buis, A. Setzer, E. Vermote, J.A. Reagan, Y. Kaufman, T. Nakajima, F. Lavenu, I. Jankowiak, and A. Smirnov, 1998, AERONET—A federated instrument network and data archive for aerosol characterization, *Remote Sensing of Environment* 66: 1–16.

Holben, B.N., and C.O. Justice, 1981, An examination of spectral band ratioing to reduce the topographic effect on remotely sensed data, *International Journal of Remote Sensing* 2(2): 115–133.

Holben, B.N., Y.J. Kaufman, and J.D. Kendall, 1990, NOAA-11 AVHRR visible and near-IR inflight calibration, *International Journal of Remote Sensing* 11(8): 1511–1519.

Hooper, F.C., A.P. Brunger, and C.S. Chan, 1987, A clear sky model of diffuse sky radiance, *Journal of Solar Energy Engineering*, 109: 9–14.

Hooper, F.C., and A.P. Brunger, 1980, A model for the angular distribution of sky radiance, *Journal of Solar Energy Engineering*, 102: 196–202.

Horler, D.N.H., 1996, *Framework Study on Calibration/Validation User Requirements as Part of the Ground Infrastructure Program of the Canadian Long-Term Space Plan*, Final Report to the Canada Centre for Remote Sensing, Ottawa, Ontario, K1A 0Y7, Canada, by Horler Information Inc. for Contract 23413–5-E178/01-SQ, 53 pages.

Horler, D.N.H., and P.M. Teillet, 1996, *Workshop on Canadian Earth Observation Calibration and Validation, Workshop Report*, Canada Centre for Remote Sensing, Ottawa, ON, K1A 0Y7, Canada, 59 pages.

Hu, Y., B.A. Wielicki, P. Yang, P.W. Stackhouse, Jr., B. Lin, and D.F. Young, 2004, Application of deep convective cloud albedo observation to satellite-based study of the terrestrial atmosphere: Monitoring the stability of spaceborne measurements and assessing absorption anomaly, *IEEE Transactions on Geoscience and Remote Sensing*, 42(11): 2594–2599.

Hugli, H., and W. Frei, 1983, Understanding anisotropic reflectance in mountainous terrain, *Photogrammetric Engineering and Remote Sensing*, 49(5): 671–683.

Iwabuchi, H., 2003, Calibration of the visible and near-infrared channels of NOAA-11 and-14 AVHRRs by using reflections from molecular atmosphere and stratus cloud, *International Journal of Remote Sensing*, 24(24): 5367–5378.

Jackson, R.D., Editor, 1990, Special issue on coincident satellite, aircraft, and field measurements at the Maricopa Agricultural Center (MAC), *Remote Sensing of Environment*, 32(2&3): 77–228.

Jensen, J.R., 2006, *Remote Sensing of the Environment: An Earth Resources Perspective*, second edition, Prentice Hall, Denver, CO, 608 pages.

Kaufman, Y.J., and R.S. Fraser, 1984, Atmospheric effects on classification of finite fields, *Remote Sensing of Environment*, 15: 95–118.

Kaufman, Y.J. and B.N. Holben, 1993, Calibration of the AVHRR visible and near-IR bands by atmospheric scattering, ocean glint and desert reflection, *International Journal of Remote Sensing*, 14(1): 21–52.

Kaufman, Y.J., D. Tanré, H.R. Gordon, T. Nakajima, J. Lenoble, R. Frouin, H. Grassl, B.M., Herman, M.D. King, and P.M. Teillet, 1997, Passive remote sensing of tropospheric aerosol and atmospheric correction for the aerosol effect, *Journal of Geophysical Research*, 102(D14): 16815–16830.

Kieffer, H.H., T.C. Stone, R.A. Barnes, S. Bender, R.E. Eplee, Jr., J. Mendenhall, and L. Ong, 2002, On-orbit radiometric calibration over time and between spacecraft using the moon, *Proceedings of SPIE Conference on Sensors, Systems, and Next-Generation Satellites VI*, Crete, Greece, pp. 287–298.

Kieffer, H.H., and R.L. Wildey, 1985, Absolute calibration of Landsat instruments using the moon, *Photogrammetric Engineering and Remote Sensing*, 51(9): 1391–1393.

Kieffer, H.H., and R.L. Wildey, 1996, Establishing the moon as a spectral radiance standard, *Journal of Atmospheric and Oceanic Technology*, 13(2): 360–375.

Kimes, D.S., and J.A. Kirchner, 1981, Modeling the effects of various radiant transfers in mountainous terrain on sensor response, *IEEE Transactions on Geoscience and Remote Sensing*, 19: 100–108.

Kirchner, J.A., S. Youkhana, and J.A. Smith, 1982, Influence of sky radiance distribution on the ratio technique for estimating bidirectional reflectance, *Photogrammetric Engineering and Remote Sensing*, 48: 955–959.

Klucher, T.M., 1979, Evaluation of models to predict insolation on tilted surfaces, *Solar Energy*, 23(2): 111–114.

Kotchenova, S.Y., and E.F. Vermote, 2007, Validation of a vector version of the 6S radiative transfer code for atmospheric correction of satellite data. Part II: Homogeneous Lambertian and anisotropic surfaces, *Applied Optics*, 46(20): 4455–4464.

Kotchenova, S.Y., E.F. Vermote, R. Matarrese, and F. Klemm, 2006, Validation of a vector version of the 6S radiative transfer code for atmospheric correction of satellite data. Part I: Path radiance, *Applied Optics*, 45: 6762–6774.

Kramer, H.J., 2001, *Observation of the Earth and Its Environment: Survey of Missions and Sensors*, fourth edition, Springer, New York, 1510 pages.

Li, F., D.L.B. Jupp, M. Thankappan, L. Lymburner, N. Mueller, A. Lewis, and A. Held, 2012, A physics-based atmospheric and BRDF correction for Landsat data over mountainous terrain, *Remote Sensing of Environment*, 124: 756–770.

Li, X., and A.H. Strahler, 1992, Geometric-optical bidirectional reflectance modeling of the discrete crown vegetation canopy: Effect of crown shape and mutual shadowing, *IEEE Transactions on Geoscience and Remote Sensing*, 30(2): 276–292.

Liang, S., H. Fallah-Adl, S. Kalluri, J. JaJa, Y.J. Kaufman, and J.R.G. Townshend, 1997, An operational atmospheric correction algorithmfor Landsat Thematic Mapper imagery over the land, *Journal of Geophysical Research*, 102: 17173–17186.

Loeb, N.G., 1997, In-flight calibration of NOAA AVHRR visible and near-IR bands over Greenland and Antarctica, *International Journal of Remote Sensing*, 18(3): 477–490.

Luderer, G., J.A. Coakley, Jr., and W.R. Tahnk, 2005, Using sun glint to check the relative calibration of reflected spectral radiances, *Journal of Atmospheric and Oceanic Technology*, 22(10): 1480–1493.

Lukashin, C., B. Wielicki, D. Young, K. Thome, Z. Jin, and W. Sun, 2013, Uncertainty estimates for imager reference inter-calibration with CLARREO reflected solar spectrometer, *IEEE Transactions on Geoscience and Remote Sensing*, 51(3 SI): 1425–1436.

MacDonald, J.S., 1997, From space data to information, *Proceedings of the ISPRS Joint Workshop on Sensors and Mapping From Space*, Hannover, Germany, Editor: G. Konecny, Institute of Photogrammetry and Engineering Surveys, University of Hannover, Nienburger Str. 1, D-30167, Hannover, Germany, pp. 233–240.

Malila, W.A., and D.M. Anderson, 1986, *Satellite Data Availability and Calibration Documentation for Land Surface Climatology Studies*, Environmental Research Institute of Masek, J.G., Ju, J., Claverie, M., Skakun, S., Roger, J.C., Vermote, E., Franch, B., Yin, Z., Dungan. J.L. 2022. Harmonized Landsat Sentinel-2 (HLS) Product User Guide Product Version 2.0.

Masek, J., J. Ju, J. Roger, S. Skakun, E. Vermote, M. Claverie, J. Dungan, Z. Yin, B. Freitag, and C. Justice, 2021, HLS Sentinel-2 MSI surface reflectance daily global 30m v2.0., distributed by NASA EOSDIS Land Processes DAAC, https://doi.org/10.5067/HLS/HLSS30.002.

Masek, J.G., J. Ju, M. Claverie, S. Skakun, J.C. Roger, E. Vermote, B. Franch, Z. Yin, and J.L. Dungan, 2022, Harmonized Landsat Sentinel-2 (HLS) Product User Guide Product Version 2.0.

Masek, J.G., M.A. Wulder, B. Markham, J. McCorkel, C.J. Crawford, J. Storey, and D.T. Jenstrom, 2020, Landsat 9: Empowering open science and applications through continuity, *Remote Sensing of Environment*, 248: 111968, ISSN 0034–4257, https://doi.org/10.1016/j.rse.2020.111968. (www.sciencedirect.com/science/article/pii/S0034425720303382)

Markham, B.L., and J.L. Barker, 1987, Radiometric properties of U.S. processed landsat MSS data, *Remote Sensing of Environment*, 22: 39–71.

Markham, B.L., and J.L. Barker, Editors, 1985, Special issue on landsat image data quality assessment, *Photogrammetric Engineering and Remote Sensing*, 51: 1245–1493.

Markham, B.L., W.C. Boncyk, D.L. Helder, and J.L. Barker, 1997, Landsat-7 enhanced thematic mapper plus radiometric calibration, *Canadian Journal of Remote Sensing*, 23(4): 318–332.

McCartney, E.J., 1976, *Optics of the Atmosphere*, John Wiley, Hoboken, NJ, 408 pages.

McCorkel, J., K. Thome, and R. Lockwood, 2013, Absolute radiometric calibration of narrow-swath imaging sensors with reference to non-coincident wide-swath sensors, *IEEE Transactions on Geoscience and Remote Sensing*, 51(3 SI): 1309–1318.

Meckler, Y., and Y.J. Kaufman, 1982, Contrast reduction by the atmosphere and retrieval of nonuniform surface reflectance, *Applied Optics*, 21: 310–316.

Meyer, P., K.I. Itten, T. Kellenberger, S. Sandmeier, and R. Sandmeier, 1993, Radiometric corrections of topographically induced effects on Landsat TM data in an alpine environment, *ISPRS Journal of Photogrammetry and Remote Sensing*, 48(4): 17–28.

Meygret, A., X. Briottet, P.J. Henry, and O. Hagolle, 2000, Calibration of SPOT4 HRVIR and vegetation cameras over Rayleigh scattering. *Proceedings of SPIE Conference 4135 on Earth Observing Systems V*, Editor: W.L. Barnes. SPIE, Bellingham, Washington, San Diego, CA, pp. 302–313.

Mitchell, R.M., D.M. O'Brien, and B.W. Forgan, 1992, Calibration of the NOAA AVHRR Shortwave channels using split pass imagery: I. Pilot study, *Remote Sensing of Environment*, 40(1): 57–65.

Morain, S.A. and A.M. Budge, Editors, 2004, *Postlaunch Calibration of Satellite Sensors*, ISPRS Book Series Volume 2, *Proceedings of the International Workshop on Radiometric and Geometric Calibration*, A.A. Balkema Publishers, Gulfport, MS, 193 pages.

Moran, M.S., R.D. Jackson, P.N. Slater, and P.M. Teillet, 1992, Evaluation of atmospheric correction procedures for visible and near-infrared satellite sensor output, *Remote Sensing of Environment*, 41: 169–184.

Mueller, J., R. Stulhmann, R. Becker, E. Raschke, J.L. Monge, and P. Burkert, 1996. Ground based calibration facility for the scanner for radiation budget instrument in the solar spectral domain. *Metrologia*, 32: 657–660.

Muench, H.S., 1981, *Calibration of Geosynchronous Satellite Video Sensors*, Report No. AFGL-TR-81–0050, Air Force Geophysical Laboratory, Hanscom, MA.

Naethe, P., A.D. Sanctis, A. Burkart, P.K.E. Campbell, R. Colombo, B.D. Mauro, A. Damm, T. El-Madany, F. Fava, J.A. Gamon, K.F. Huemmrich, M. Migliavacca, E. Paul-Limoges, U. Rascher, M. Rossini, D. Schüttemeyer, G. Tagliabue, Y. Zhang, and T. Julitta, 2024, Towards a standardized, ground-based network of hyperspectral measurements: Combining time series from autonomous field spectrometers with Sentinel-2, *Remote Sensing of Environment*, 303: 114013, ISSN 0034–4257, https://doi.org/10.1016/j.rse.2024.114013. (www.sciencedirect.com/science/article/pii/S0034425724000245)

Nicodemus, F.E., 1965, Directional reflectance and emissivity of an opaque surface, *Applied Optics*, 4(7): 767–775.

Nicodemus, F.E., 1970, Reflectance nomenclature and directional reflectance and emissivity, *Applied Optics*, 9(6): 1474–1475.

Nieke, J., T. Aoki, T. Tanikawa, H. Motoyoshi, M. Hori, and Y. Nakajima, 2003, Cross-calibration of satellite sensors over snow fields, *Proceedings of SPIE Conference 5151 on Earth Observing Systems VIII*, Editor: W. L. Barnes, SPIE, San Diego, CA, Volume 5151, pp. 406–414.

NRC, 2007, *Earth Science and Applications from Space: National Imperatives for the Next Decade and Beyond*, National Research Council, The National Academies, 500 Fifth St. N.W., Washington, DC, 20001, 428 pages.

O'Brien, D.M. and Mitchell, R.M., 2001, An error budget for cross-calibration of AVHRR shortwave channels against ATSR-2, *Remote Sensing of Environment*, 75(2): 216–229.

Ohring, G., B. Wielicki, R. Spencer, B. Emery, and R. Datla. 2005. Satellite instrument calibration for measuring global climate change: Report of a workshop. *Bulletin of the American Meteorological Society*, 86: 1303–1313.

Ohring, G., J. Tansock, W. Emery, J. Butler, F. L, F. Weng, K.S. Germain, B. Wielicki, C. Cao, M. Goldberg, J. Xiong, G. Fraser, D. Kunkee, D. Winker, L. Miller, S. Ungar, D. Tobin, J.G. Anderson, D. Pollock, S. Shipley, A. Thurgood, G. Kopp, P. Ardanuy, and T. Stone, 2007, Achieving satellite instrument calibration for climate change, *EOS, Transactions American Geophysical Union*, 88(11): 136.

O'Neill, N.T., J.R. Miller, and F.J. Ahern, 1978, Radiative transfer calculations for remote sensing applications, *Proceedings of the 5th Canadian Symposium on Remote Sensing*, Victoria, BC, p. 572.

O'Neill, N.T., P.M. Teillet, A. Royer, J.-P. Blanchet, M. Aubé, J. Freemantle, S. Gong, D. Stanley, S. Thulasiraman, and F. Vachon, 2002, Concept of a central optical parameter server for atmospheric corrections of remote sensing data, *Proceedings of the 2002International Geoscience and Remote Sensing Symposium (IGARSS'02) and the Twenty-Fourth Canadian Symposium on Remote Sensing*, Toronto, ON, Volume V, pp. 2951–2953, also on CD-ROM.

Otterman, J., M. Dishon, and S. Rehavi, 1983, Point spread functions in imaging a Lambert surface from zenith through a thin scattering layer, *International Journal of Remote Sensing*, 4: 583.

Otterman, J., and R.S. Fraser, 1979, Adjacency effects on imaging by surface reflection and atmospheric scattering: Cross radiance to zenith, *Applied Optics*, 18: 2852.

Page, B.P., L.G. Olmanson, and D.R. Mishra, 2019, A harmonized image processing workflow using Sentinel-2/MSI and Landsat-8/OLI for mapping water clarity in optically variable lake systems, *Remote Sensing of Environment*, 231: 111284, ISSN 0034–4257, https://doi.org/10.1016/j.rse.2019.111284. (www.sciencedirect.com/science/article/pii/S0034425719303037)

Panda, S.S., M.N. Rao, P.S. Thenkabail, and J.E. Fitzgerald, 2024, Remote sensing systems—platforms and sensors: Aerial, satellites, UAVs, optical, radar, and LiDAR, Chapter 1. In P.S. Thenkabail, Editor-in-Chief. *Remote Sensing Handbook, Volume 1, Second Edition: Remotely sensed data characterization, classification, and accuracies*. Taylor and Francis Inc./CRC Press, Boca Raton, London, New York.

Pinty, B., J.-L. Widlowski, M. Taberner, N. Gobron, M.M. Verstraete, M. Disney, F. Gascon, J.-P. Gastellu, L. Jiang, A. Kuusk, P. Lewis, X. Li, W. Ni-Meister, T. Nilson, P. North, W. Qin, L. Su, S. Tang, R. Thompson, W. Verhoef, H. Wang, J. Wang, G. Yan, and H. Zang, 2004, Radiation Transfer Model Intercomparison (RAMI) exercise: Results from the second phase, *Journal of Geophysical Research*, V109(D06).

Pollock, D.B., T.L. Murdock, R.U. Datla, and A. Thompson, 2003, Data uncertainty traced to SI units. Results reported in the international system of units, *International Journal of Remote Sensing*, 24(2): 225–235.

Price, J.C., 1987a, Radiometric calibration of satellite sensors in the visible and near infrared: History and outlook, *Remote Sensing of Environment*, 22: 3–9.

Price, J.C., Editor, 1987b, Special issue on radiometric calibration of satellite data, *Remote Sensing of Environment*, 22(1): 1–158.

Proy, C., D. Tanré, and P.Y. Deschamps, 1989, Evaluation of topographic effects in remotely sensed data, *Remote Sensing of Environment*, 30(1): 21–32.

Rao, C.R.N., and Chen, J., 1995, Intersatellite calibration linkages for the visible and near-infared channels of the advanced very high-resolution radiometer on the NOAA-7, NOAA-9, and NOAA-11 spacecraft, *International Journal of Remote Sensing*, 16(11): 1931–1942.

Rao, C.R.N., and J.H. Chen, 1996, Postlaunch calibration of the visible and near-infrared channels of the advanced very high-resolution radiometer on the NOAA-14 spacecraft, *International Journal of Remote Sensing*, 17(14): 2743–2747.

Rao, C.R.N., and J.H. Chen, 1999, Revised postlaunch calibration of the visible and near-infrared channels of the Advanced Very High-Resolution Radiometer (AVHRR) on the NOAA-14 spacecraft, *International Journal of Remote Sensing*, 20(18): 3485–3491.

Rao, C.R.N., C. Cao, and N. Zhang, 2003, Inter-calibration of the moderate-resolution imaging spectroradiometer and the along-track scanning radiometer-2, *International Journal of Remote Sensing*, 24(9): 1913–1924.

Regalado, A., 2013, The data made me do it, *MIT Technology Review*, 116(4): 63–64.

Riaño, D., E. Chuvieco, J. Salas, and I. Aguado, 2003, Assessment of different topographic corrections in landsat-TM data for mapping vegetation types (2003), *IEEE Transactions on Geoscience and Remote Sensing*, 41(5): 1056–1061.

Richardson, A.J., 1982, Relating Landsat digital count values to ground reflectances for optical thin atmospheric conditions, *Applied Optics*, 21: 1457.

Richter, R., 1997, Correction of atmospheric and topographic effects for high spatial resolution satellite images, *International Journal of Remote Sensing*, 18: 1099–1111.

Richter, R. 1998, Correction of satellite imagery of mountainous terrain, *Applied Optics*, 37(18): 4004–4014.

Sandford, S.P., D.F. Young, J.M. Corliss, B.A. Wielicki, M.J. Gazarik, M.G. Mlynczak, A.D. Little, C.D. Jones, P.W. Speth, D.E. Shick, K.E. Brown, K.J. Thome, and J.H. Hair, 2010, CLARREO: Cornerstone of the climate observing system measuring decadal change through accurate emitted infrared and reflected solar spectra and radio occultation, *Proceedings of SPIE Conference 7826 on Sensors, Systems, and Next-Generation Satellites XIV*, Editors: R. Meynart, S.P. Neeck, and H. Shimoda, Toulouse, France, 782611.

Scarpace, F.L. and P.R. Wolf, 1973, Atmospheric refraction, *Photogrammetric Engineering*, 39: 521.

Schaaf, C.B., F. Gao, A.H. Strahler, W. Lucht, X. Li, T. Tsang, N.C. Strugnell, X. Zhang, Y. Jin, J.-P. Muller, P. Lewis, M. Barnsley, P. Hobson, M. Disney, G. Roberts, M. Dunderdale, C. Doll, R.P. d'Entremont, B. Hu, S. Liang, J.L. Privette, and D. Roy, 2002, First operational BRDF, albedo nadir reflectance products from MODIS, *Remote Sensing of Environment*, 83(1–2): 135–148.

Schott, J.R., 2007, *Remote Sensing, The Image Chain Approach*, Second Edition, ISBN-10: 0195178173, ISBN-13: 978–0195178173, Oxford University Press, New York.

Schott, J.R., C. Salvaggio, and W.J. Volchok, 1988, Radiometric scene normalization using pseudo-invariant features, *Remote Sensing of Environment*, 26: 1–16.

Schowengerdt, R.A. 2007. *Remote Sensing: Models and Methods for Image Processing*, Third Edition, Academic Press, Elsevier, 515 pages.

Schowengerdt, R.A., C. Archwamety, and R.C. Wrigley, 1985, Landsat thematic mapper image-derived MTF, *Photogrammetric Engineering and Remote Sensing of Environment*, 51(9): 1395–1406.

Schowengerdt, R.A., S.K. Park, and R.T. Gray, 1983, An optimized cubic interpolator for image resampling, *Proceedings of the 17th International Symposium on Remote Sensing of Environment*, Ann Arbor, MI, p. 1291.

Schroeder, T.A., W.B. Cohen, C. Song, M.J. Canty, and Z. Yang, 2006, Radiometric correction of multi-temporal Landsat data for characterization of early successional forest patterns in western Oregon, *Remote Sensing of Environment*, 103: 16–26.

Secker, J., K. Staenz, R.P. Gauthier, and B. Budkewitsch, 2001, Vicarious calibration of hyperspectral sensors in operational environments, *Remote Sensing of Environment*, 26: 81–92.

Sellers, P.J., B.W. Meeson, F.G. Hall, G. Asrar, R.E. Murphy, R.A. Schiffer, F.P. Bretherton, R.E. Dickinson, R.G. Ellingson, C.B. Field, K.F. Huemmrich, C.O. Justice, J.M. Melack, N.T. Roulet, D.S. Schimel, and P.D. Try, 1995, Remote sensing of the land surface for studies of global change: Models—algorithms—experiments, *Remote Sensing of Environment*, 51: 3–26.

Shen, Y., X. Zhang, Z. Yang, Y. Ye, J. Wang, S. Gao, Y. Liu, W. Wang, K.H. Tran, and J. Ju, 2023, Developing an operational algorithm for near-real-time monitoring of crop progress at field scales by fusing harmonized Landsat and Sentinel-2 time series with geostationary satellite observations, *Remote Sensing of Environment*, 296: 113729, ISSN 0034–4257, https://doi.org/10.1016/j.rse.2023.113729. (www.sciencedirect.com/science/article/pii/S0034425723002808)

Shepherd, J.D., and J.R. Dymond, 2003, Correcting satellite imagery for the variance of reflectance and illumination with topography, *International Journal of Remote Sensing*, 24(17): 3503–3514.

Six, D., M. Fily, S. Alvain, P. Henry, and J.P. Benoist, 2004, Surface characterisation of the Dome Concordia area (Antarctica) as a potential satellite calibration site, using SPOT4/VEGETATION instrument, *Remote Sensing of Environment*, 89(1): 83–94.

Sjoberg, R.W., and B.K.P. Horn, 1983, Atmospheric effects in satellite imaging of mountainous terrain, *Applied Optics*, 22: 1702.

Slater, P.N., 1980, *Remote Sensing, Optics and Optical Systems*, ISBN 0-201-07250-5, Addison-Wesley Publishing Company, Reading, MA.

Slater, P.N., 1984, The importance and attainment of absolute radiometric calibration, *Proceedings of SPIE Critical Rev . Remote Sensing, SPIE Volume 475*, pp. 34–40.

Slater, P.N., 1985, Radiometric considerations in remote sensing, *Proceedings of the Institute of Electrical and Electronics Engineers*, 73(6): 997–1011.

Slater, P.N., 1988a, Review of the calibration of radiometric measurements from satellite to ground level, *International Archives of Photogrammetry and Remote Sensing*, 27(B11): 726–724.

Slater, P.N., 1988b, Recent advances in sensors, radiometry, and data processing for remote sensing, *SPIE Volume 924*, Orlando, FL, ISBN 0-89252-959-8, SPIE—The International Society for Optical Engineering, Bellingham, Washington 98227–0010, USA, 341 pages.

Slater, P.N., and S.F. Biggar, 1996, Suggestions for radiometric calibration coefficient generation, *Journal of Atmospheric and Oceanographic Technology*, 13(2): 376–382.

Slater, P.N., S.F. Biggar, R.G. Holm, R.D. Jackson, Y. Mao, M.S. Moran, J.M. Palmer, and B. Yuan, 1987, Reflectance- and radiance-based methods for the in-flight absolute calibration of multispectral sensors, *Remote Sensing of Environment*, 22: 11–37.

Slater, P.N., S.F. Biggar, J.M. Palmer, and K.J. Thome, 1995, Unified approach to pre- and in-flight satellite-sensor absolute radiometric calibration, *Proceedings of the SPIE Europto Symposium*, SPIE Vol. 2583, Paris, France, pp. 130–141.

Slater, P.N., S.F. Biggar, J.M. Palmer, and K.J. Thome, 2001, Unified approach to absolute radiometric calibration in the solar-reflective range, *Remote Sensing of Environment*, 77: 293–303.

Slater, P.N., and R.D. Jackson, 1982, Atmospheric effects on radiation reflected from soil and vegetation as measured by orbital sensors using various scanning directions, *Applied Optics*, 21: 3923.

Smith, G.R., R.H. Levin, P. Abel, and H. Jacobowitz, 1988, Calibration of the solar channels and NOAA-9 AVHRR using high altitude aircraft measurements, *Journal of Atmospheric and Oceanic Technology*, 5: 631–639.

Smith, J.A., T.L. Lin, and K.J. Ranson, 1980, The lambertian assumption and landsat data, *Photogrammetric Engineering and Remote Sensing*, 46: 1183–1189.

Soenen, S.A., D.R. Peddle, and C.A. Coburn, 2005, A modified sun-canopy-sensor topographic correction in forested terrain, *IEEE Transactions on Geoscience and Remote Sensing*, 43(9): 2148–2159.

Soenen, S., D. Peddle, C. Coburn, R. Hall, and F. Hall, 2008, Improved topographic correction of forest image data using a 3-D canopy reflectance model in multiple forward mode. *International Journal of Remote Sensing*, 29(4): 1107–1027.

Song, C., C.E. Woodcock, K.C. Seto, M. Pax-Lenney, and S.A. Macomber, 2001, Classification and change detection using landsat TM data: When and how to correct atmospheric effects, *Remote Sensing of Environment*, 75: 230–244.

Staenz, K., J. Secker, B.C. Gao, C. Davis, and C. Nadeau, 2002, Radiative transfer codes applied to hyperspectral data for the retrieval of surface reflectance, *ISPRS Journal of Photogrammetry and Remote Sensing*, 57(3): 194–203.

Staenz, K., D.J. Williams, G. Fedosejevs, and P.M. Teillet, 1994, Surface reflectance retrieval from imaging spectrometer data using three atmospheric codes, *Proceedings of SPIE EUROPTO ' 94,* Rome, Italy, SPIE Vol.2318, pp. 17–28.

Staylor, W.F., 1990, Degradation rates of the AVHRR visible channel for the NOAA 6, 7, and 9 spacecraft, *Journal of Atmospheric and Oceanic Technology*, 7(3): 411–423.

Steven, M.D., 1977, Standard distribution of clear sky radiance, *The Quarterly Journal of the Royal Meteorological Society*, 106: 57.

Steven, M.D., T.J. Malthus, F. Baret, H. Xu, and M.J. Chopping, 2003, Intercalibration of vegetation indices from different sensor systems, *Remote Sensing Environment* 88(4): 412–422.

Stone, T.C., 2008, Radiometric calibration stability and inter-calibration of solar-band instruments in orbit using the moon, *Proceedings of SPIE Conference on Earth Observing Systems XIII*, San Diego, CA, p. 708 10X.

Stone, T.C., H.H. Kieffer, and I.F. Grant, 2005, Potential for calibration of geostationary meteorological satellite imagers using the moon, *Proceedings of SPIE Conference on Earth Observing Systems X*, San Diego, CA, pp. 1–9.

Stone, T.C., W.B. Rossow, J. Ferrier, and L.M. Hinkelmann, 2013, Evaluation of ISCCP multi-satellite radiance calibration for geostationary imager visible channels using the moon, *IEEE Transactions on Geoscience and Remote Sensing*, 51(3): 1255–1266.

Souri, A.H., and A. Azizi, 2013, Removing bowtie phenomenon by correction of panoramic effect in MODIS imagery, *International Journal of Computer Applications*, 68(3): 12–16.

Suits, G.H., W.A. Malila, and T.M. Weller, 1988, The prospects for detecting spectral shifts due to satellite sensor ageing, *Remote Sensing of Environment*, 26: 17–29.

Sweet, R.J.M, J.C. Elliott, and J.R. Beasley, 1992, Research needs to encourage the growth of the earth observation applications market, *Proceedings of the 18th Annual Conference of the Remote Sensing Society: Remote Sensing From Research to Operation*, Editors: P.A. Cracknell and R.A. Vaughan, University of Dundee, UK, 15–17 September, pp. 399–407.

Tahnk, W.R., and J.A. Coakley, 2001, Updated calibration coefficients for NOAA-14 AVHRR channels 1 and 2, *International Journal of Remote Sensing*, 22(15): 3053–3057.

Tanré, D., C. Deroo, P. Duhaut, M. Herman, J.J. Morcrette, and J. Perbos, 1990, Description of a computer code to simulate the satellite signal in the solar spectrum: The 5S code, *International Journal of Remote Sensing*, 11: 659–668.

Tanré, R.C., M. Herman, and P.Y. Deschamps, 1983, Influence of the atmosphere on space measurements of directional properties, *Applied Optics*, 22: 733.

Teillet, P.M., 1986, Image correction for radiometric effects in remote sensing, *International Journal of Remote Sensing*, 7: 1637–1651.

Teillet, P.M., 1989, Surface reflectance retrieval using atmospheric correction algorithms, *Proceedings of the 1989International Geoscience and Remote Sensing Symposium (IGARSS'89) and the Twelfth Canadian Symposium on Remote Sensing*, Vancouver, BC, pp. 864–867.

Teillet, P.M., 1990a, Rayleigh optical depth comparisons from various sources, *Applied Optics*, 29: 1897–1900.

Teillet, P.M., 1990b, Effects of spectral shifts on sensor response, *Proceedings of the ISPRS Commission VII Symposium*, Victoria, BC, Canada, pp. 59–65.

Teillet, P.M., 1992, An algorithm for the radiometric and atmospheric correction of AVHRR data in the solar reflective channels, *Remote Sensing of Environment*, 41: 185–195.

Teillet, P.M., 1997a, A status overview of earth observation calibration/validation for terrestrial applications, *Canadian Journal of Remote Sensing*, 23(4): 291–298.

Teillet, P.M., Editor, 1997b, Special issue on calibration/validation, *Canadian Journal of Remote Sensing*, 23(4): 289–423.

Teillet, P.M., 1997c, *Report on the Second Canadian Workshop on Earth Observation Calibration and Validation*, Canada Centre for Remote Sensing, Ottawa, Ontario, K1A 0Y7, Canada, 66 pages.

Teillet, P.M., 1998, *Report on the Third Canadian Workshop on Earth Observation Calibration and Validation*, Canada Centre for Remote Sensing, Ottawa, Ontario, K1A 0Y7, Canada, 74 pages.

Teillet, P.M., 2005, Satellite image radiometry: From photons to calibrated earth science data, *Physics in Canada*, September/October, pp. 301–310.

Teillet, P.M., and G. Chander, 2014, Post-launch radiometric calibration of satellite-based optical sensors with emphasis on terrestrial reference standard sites, Volume 1, Chapter 4, In *The Remote Sensing Handbook*, Editor: Prasad S. Thenkabail, Taylor and Francis Inc./CRC Press, Boca Raton, FL.

Teillet, P.M., and G. Fedosejevs, 1995, On the dark target approach to atmospheric correction of remotely sensed data, *Canadian Journal of Remote Sensing*, 21(4): 374–387.

Teillet, P.M., and B.N. Holben, 1994, Towards operational radiometric calibration of NOAA AVHRR imagery in the visible and near-infrared channels, *Canadian Journal of Remote Sensing*, 20(1): 1–10.

Teillet, P.M., and J.R. Irons, 1990, Spectral variability effects on the atmospheric correction of imaging spectrometer data for surface reflectance retrieval, *Proceedings of the ISPRS Commission VII Symposium*, Victoria, BC, pp. 579–583.

Teillet, P.M., and X. Ren, 2008, Spectral band difference effects on vegetation indices derived from multiple satellite sensor data, *Canadian Journal of Remote Sensing*, 34(3): 159–173.

Teillet, P.M., and R.P. Santer, 1991, Terrain elevation and sensor altitude dependence in a semi-analytical atmospheric code, *Canadian Journal of Remote Sensing*, 17: 3644.

Teillet, P.M., and K. Staenz, 1992, Atmospheric effects due to topography on MODIS vegetation index data simulated from AVIRIS imagery over mountainous terrain, *Canadian Journal of Remote Sensing*, 18(4): 283–291.

Teillet, P.M., B. Guindon, and D.G. Goodenough, 1982, On the slope aspect correction of multi-spectral scanner data, *Canadian Journal of Remote Sensing*, 8: 84–106.

Teillet, P.M., M. Lasserre, and C.G. Vigneault, 1986, An evaluation of sun angle computation algorithms, *Proceedings of the Tenth Canadian Symposium on Remote Sensing*, Edmonton, Alberta, pp. 91–100.

Teillet, P.M., P.N. Slater, Y. Ding, R.P. Santer, R.D. Jackson and M.S. Moran, 1990, Three methods for the absolute calibration of the NOAA AVHRR sensors in-flight, *Remote Sensing of Environment*, 31: 105–120.

Teillet, P.M., G. Fedosejevs, F.J. Ahern and R.P. Gauthier, 1994, Sensitivity of surface reflectance retrieval to uncertainties in aerosol optical properties, *Applied Optics*, 33(18): 3933–3940.

Teillet, P.M., D.N.H. Horler, and N.T. O'Neill, 1997a, Calibration, validation, and quality assurance in remote sensing: A new paradigm, *Canadian Journal of Remote Sensing*, 23(4): 401–414.

Teillet, P.M., K. Staenz, and D.J. Williams, 1997b, Effects of spectral, spatial, and radiometric characteristics on remote sensing vegetation indices for forested regions, *Remote Sensing of Environment*, 61: 139–149.

Teillet, P.M., G. Fedosejevs, R.P. Gauthier, N.T. O'Neill, K.J. Thome, S.F. Biggar, H. Ripley, and A. Meygret, 2001a, A generalized approach to the vicarious calibration of multiple earth observation sensors using hyperspectral data, *Remote Sensing of Environment*, 77(3): 304–327.

Teillet, P.M., K.J. Thome, N. Fox, and J.T. Morisette, 2001b, Earth observation sensor calibration using A Global Instrumented and Automated Network of Test Sites (GIANTS), *Proceedings of SPIE Conference 4550 on Sensors, Systems, and Next-Generation Satellites V*, Editors: H. Fujisada, J.B. Lurie, and K. Weber, SPIE, Toulouse, France, Volume 4550, pp. 246–254.

Teillet, P.M., J.L. Barker, B.L. Markham, R.R. Irish, G. Fedosejevs, and J.C. Storey, 2001c, Radiometric cross-calibration of the Landsat-7 ETM+ and Landsat-5 TM sensors based on tandem data sets, *Remote Sensing of Environment*, 78(1–2): 39–54.

Teillet, P.M., G. Fedosejevs, R.K. Hawkins, T.I. Lukowski, R.A. Neville, K. Staenz, R. Touzi, J. van der Sanden, and J. Wolfe, 2004a, *Importance of Data Standardization for Generating High Quality Earth Observation Products for Natural Resource Management*, Earth Sciences Sector, Natural Resources Canada, 588 Booth Street, Ottawa, ON, K1A 0Y7, 42 pages.

Teillet, P.M., G. Fedosejevs, and K.J. Thome, 2004b, Spectral band difference effects on radiometric cross-calibration between multiple satellite sensors in the landsat solar-reflective spectral domain, workshop on inter-comparison of large-scale optical and infrared sensors, *Proceedings of SPIE Conference 5570 on Sensors, Systems, and Next-Generation Satellites VIII*, Editors: R. Meynart, S.P. Neeck, and H. Shimoda, Maspalomas, Canary Islands, Spain, pp. 307–316.

Teillet, P.M., G. Fedosejevs, K.J. Thome, and J.L. Barker, 2007, Impacts of spectral band difference effects on radiometric cross-calibration between satellite sensors in the solar-reflective spectral domain, *Remote Sensing of Environment*, 110(3): 393–409.

Temps, R.C., and K.L. Coulson, 1977, Solar radiation incident upon slopes of different orientations, *Solar Energy*, 19: 179.

Thome, K.J., 2001, Absolute radiometric calibration of landsat 7 ETM+ using the reflectance-based method, *Remote Sensing of Environment*, 78(1–2): 27–38.

Thome, K., 2005, Sampling and uncertainty issues in trending reflectance-based vicarious calibration results, *Proceedings of SPIE Conference 5882 on Earth Observing Systems X*, Editor: J.J. Butler, SPIE, Bellingham, Washington, San Diego, CA, pp. 1–11.

Thome, K., S. Schiller, J. Conel, K. Arai, and S. Tsuchida, 1998, Results of the 1996 earth observing system vicarious calibration joint campaign at Lunar Lake Playa, *Metrologia*, 35: 631–638.

Thome, K.J, S.F. Biggar, and W. Wisniewski, 2003, Cross comparison of EO-1 sensors and other earth resources sensors to landsat-7 ETM+ using railroad valley playa, *IEEE Transactions on Geoscience and Remote Sensing*, 41: 1180–1188.

Thome, K., J. Czapla-Myers, N. Leisso, J. McCorkel, and J. Buchanan, 2008, Intercomparison of imaging sensors using automated ground measurements, *Proceedings of IEEE Conference on Remote Sensing: The Next Generation*, IEEE, Piscataway, New Jersey, Boston, MA, pp. 1332–1335.

Thome, K., R. Barnes, R. Baize, J. O'Connell, and J. Hair, 2010, Calibration of the reflected solar instrument for the climate absolute radiance and refractivity observatory, *Proceedings of the 2010IEEE International Geoscience and Remote Sensing Symposium (IGARSS)*, Honolulu, Hawaii, pp. 2275–2278.

Thuillier, G., M. Hersé, D. Labs, T. Foujols, W. Peetermans, D. Gillotay, P.C. Simon, and H. Mandel, 2003, The solar spectral irradiance from 200 to 2400 nm as measured by the SOLSPEC spectrometer from the atlas and eureca missions, *Solar Physics*, 214(1): 1–22.

Townshend, J.R.G., 1981, The spatial resolving power of earth resources satellites, *Progress in Physical Geography*, 5(1): 32–55.

Townshend, J.R.G., C. Huang, S.N.V. Kalluri, R.S. Defries, and S. Liang, 2000, Beware of per-pixel characterization of land cover, *International Journal of Remote Sensing*, 21(4): 839–843.

Trishchenko, A.P., J. Cihlar, and Z. Li, 2002, Effects of spectral response function on surface reflectance and NDVI measured with moderate resolution satellite sensors, *Remote Sensing of Environment*, 81(1): 1–18.

Trishchenko, A.P., 2009, Effects of spectral response function on surface reflectance and NDVI measured with moderate resolution satellite sensors: Extension to AVHRR NOAA-17, 18 and METOP-A, *Remote Sensing of Environment*, 113(2): 335–341.

Turner, R.E., and Spencer, M.M., 1972, Atmospheric model for correction of spacecraft data, *Proceedings of the 8th International Symposium on Remote Sensing of Environment*, Ann Arbor, MI, p. 895.

Vermote, E., and Y.J. Kaufman, 1995, Absolute calibration of AVHRR visible and near-infrared channels using ocean and cloud views, *International Journal of Remote Sensing*, 16(13): 2317–2340.

Vermote, E.F., and Saleous, N.Z., 2006, Calibration of NOAA16 AVHRR over a desert site using MODIS data, *Remote Sensing of Environment*, 105(3): 214–220.

Vermote, E.F., R. Santer, P.Y. Deschamps, and M. Herman, 1992, In-flight calibration of large field of view sensors at short wavelengths using Rayleigh scattering, *International Journal of Remote Sensing*, 13(18): 3409–3429.

Vermote, E.F., N. El Saleous, C.O. Justice, Y.J. Kaufman, J.L. Privette, L. Remer, J.C. Roger and D. Tanre, 1997, Atmospheric correction of visible to middle-infrared EOS-MODIS data over land surfaces: Background, operational algorithm and validation, *Journal of Geophysical Research*, 102: 17131–17141.

Vermote, E.F., N.Z. El Saleous, and C.O. Justice, 2002, Atmospheric correction of MODIS data in the visible to middle infrared: First results, *Remote Sensing of Environment*, 83(1–2): 97–111.

Vermote, E.F., D. Tanré, J.L. Deuzé, M. Herman, J.J. Morcrette, S.Y. Kotchenova, and T. Miura, 2006, *Second Simulation of the Satellite Signal in the Solar Spectrum (6S), 6S User Guide Version 3 (November,2006)*, www.6s.ltdri.org.

White, H.P., J.R. Miller, and J.M. Chen, 2001, Four-Scale Linear Model for Anisotropic Reflectance (FLAIR) for plant canopies. I: Model description and partial validation, *IEEE Transactions on Geoscience and Remote Sensing*, 39(5): 1072–1083.

White, H.P., L. Sun, K. Staenz, R.A. Fernandes, and C. Champagne, 2002a, Determining the contribution of shaded elements of a canopy to remotely sensed hyperspectral signatures, *Proceedings of the 1st International Symposium on Recent Advances on Quantitative Remote Sensing*, Torrent, Valencia, Spain.

White, HP., J.R. Miller, and J.M. Chen, 2002b, Four-Scale Linear Model for Anisotropic Reflectance (FLAIR) for plant canopies II: Partial validation and inversion using field measurements, *IEEE Transactions on Geoscience and Remote Sensing*, 40(5): 1038–1046.

Whitlock, C.H., W.F. Staylor, J.T. Suttles, G. Smith, R. Levin, R. Frouin, C. Gautier, P.M. Teillet, P.N. Slater, Y.J. Kaufman, B.N. Holben, W.B. Rossow, B.C., and S.R. LeCroy, 1990, AVHRR and VISSR satellite instrument calibration results for both cirrus and marine stratocumulus IFO periods, *Proceedings of FIRE Science Meeting*, NASA Langley Research Center, Hampton, Virginia, Vail, CO, pp. 141–146.

Woodham, R.J., and T.K. Lee, 1985, Photometric method for radiometric correction of multispectral scanner data, *Canadian Journal of Remote Sensing*, 11: 132.

Woodham, R.J., and M.H. Gray, 1987. Analytic method for radiometric correction of satellite multispectral scanner data, *IEEE Transactions on Geoscience and Remote Sensing*, GE-25: 258–271.

Wu, A., C. Cao, and X. Xiong, 2003, Intercomparison of the 11- and 12-μm bands of Terra and Aqua MODIS using NOAA-17 AVHRR, *Proceedings of SPIE Conference 5151 on Earth Observing Systems VIII*, Editor: W.L. Barnes. SPIE, Bellingham, Washington, San Diego, CA, pp. 384–394.

Wyatt, C.L., V. Privalsky, and R. Datla, 1998, *Recommended Practice: Symbols, Terms, Units and Uncertainty Analysis for Radiometric Sensor Calibration*, NIST Handbook 152, National Institute of Standards and Technology, Gaithersburg, MD 20899–0001, USA, 120 pages.

Xiong, X., J. Sun, and W. Barnes, 2008, Intercomparison of on-orbit calibration consistency between Terra and Aqua MODIS reflective solar bands using the moon, *IEEE Geoscience and Remote Sensing Letters*, 5(4): 778–782.

Yang, H., S. Li, J. Chen, X. Zhang, and S. Xu. 2017, The standardization and harmonization of land cover classification systems towards harmonized datasets: A review. *ISPRS International Journal of Geo-Information*, 6(5): 154. https://doi.org/10.3390/ijgi6050154

Yuan, D., and C.D. Elvidge, 1996, Comparison of relative radiometric normalization techniques, *ISPRS Journal of Photogrammetry and Remote Sensing*, 51: 117–126.

# 7 Post-Launch Radiometric Calibration of Satellite-Based Optical Sensors with Emphasis on Terrestrial Reference Standard Sites

*Philippe M. Teillet and Gyanesh Chander*

## 7.1 INTRODUCTION

Scientists and decision makers addressing local, regional, and/or global issues rely increasingly on operational use of data and information obtained from a multiplicity of satellite-based Earth Observation (EO) sensor systems. It is imperative that they be able to rely on the accuracy of EO data and information products. Accordingly, the characterization and calibration of these sensors are critical elements of EO programs.

Chapter 6 in this olume provided an overview of satellite image radiometry in the solar-reflective domain (Teillet, 2014). The present chapter delves more deeply into post-launch sensor radiometric calibration methodologies, encompassing optical sensors operating in the visible, near-infrared, and shortwave infrared spectral regions, with particular emphasis on the use of terrestrial reference standard sites. Although a lot of time and effort are devoted to pre-launch radiometric calibration, post-launch changes to satellite sensors and their radiometric calibration performance necessitate ongoing calibration monitoring and maintenance (e.g., Román et al., 2024; Dong et al., 2023; Russell et al., 2023; Voskanian et al., 2023; Niro et al., 2021; Sterckx et al., 2020; Rodrigo et al., 2019; Shrestha et al., 2019, Slater et al., 2001; Butler et al., 2005).

Calibration of satellite sensor radiometric response is generally classified into relative calibration and absolute calibration (e.g., Román et al., 2024; Dong et al., 2023; Russell et al., 2023; Voskanian et al., 2023; Niro et al., 2021; Sterckx et al., 2020; Rodrigo et al., 2019; Shrestha et al., 2019; Kastner and Slater, 1982; Slater, 1984; Teillet, 1986). Relative calibration mainly deals with compensation for detector-to-detector differences. Ideally, detectors constructed from the same materials should respond identically to the same incident energy. However, typically, detectors do not respond identically, resulting in differences in detector gain and bias levels that lead to "striping" in the image data. This striping can be corrected by selecting a stable reference detector, then scaling and shifting the other detector responses to identical targets to the reference detector's gain and bias. This process is called detector-to-detector relative radiometric calibration. Absolute radiometric calibration enables the conversion of image digital counts (DCs) to physical units in the International System of Units (SI) such as at-sensor spectral radiance (W $m^{-2}$ $sr^{-1}$ $\mu m^{-1}$). Because DCs from one sensor bear no relation to DCs from a different sensor, conversion to at-sensor spectral radiance is a fundamental step that enables the comparison of similar products from different sensors. The additional step of converting the at-sensor radiance to top-of-atmosphere (TOA) reflectance is also important, as outlined in Chapter 6 in this Volume (Teillet, 2014). Absolute radiometric calibration procedures (Slater, 1984) include (1) pre-launch calibration, with only focal plane calibration after

 DOI: 10.1201/9781003541141-11

launch; (2) post-launch reference to onboard standard sources or to the Sun via a diffuser, usually not illuminating the full aperture, and (3) post-launch reference to an Earth target of known reflectance, to the Moon, or to bright stars. Procedure category (3) is usually called vicarious calibration. Vicarious calibration can also be used for relative calibration to compensate for temporal trends in sensor response.

Onboard radiometric calibration systems can provide good temporal sampling with high precision that allows trending of system responses. A disadvantage of onboard calibration approaches is that they add significantly to the complexity and cost of satellite missions. Vicarious calibration techniques involving Earth targets provide independent, full-aperture calibrations with relatively high accuracy. While reference to an Earth target requires an accurate determination of atmospheric optical properties (mainly aerosol optical depth) at the time of satellite overpass, it has the important advantage of replicating actual conditions of image acquisition, with a full irradiation of the entrance aperture. The sensor acquires data in the same modality as it acquires Earth image data, that is, in the same spectral bands, with the same source spectrum and under typical illumination conditions. The main disadvantages of vicarious calibration are that the methods yield lower precision and have lower temporal sampling frequencies compared to onboard radiometric calibration approaches.

Terrestrial surfaces with suitable characteristics have long served as benchmarks or reference standard sites, via either vicarious calibration or cross-calibration, to assess the post-launch radiometric calibration performance of satellite optical sensors (e.g., Román et al., 2024; Russell et al., 2023; Shrestha et al., 2019; Teillet and Chander, 2010a; Chander et al., 2013a, 2013b). The use of such sites is the only practical way of evaluating radiometric calibration biases between sensors, and it also provides a means of bridging gaps in measurement continuity. Accordingly, after a brief review of onboard calibration and other vicarious calibration methodologies, this chapter focuses on the use of terrestrial reference standard sites for post-launch sensor radiometric calibration of optical sensors operating in the visible, near-infrared, and shortwave infrared spectral regions.

## 7.2 POST-LAUNCH SENSOR RADIOMETRIC CALIBRATION METHODOLOGIES

As noted in Chapter 6 in this volume (Teillet, 2014), the establishment of worldwide coordinated and operational calibration efforts is critical to achieving the goals of quantitative EO programs in general, and the goals of endeavors such as the Global Earth Observation System of Systems (GEOSS)[1] of the Group on Earth Observations (GEO), as well as the Global Space-Based Inter-Calibration System (GSICS).[2] Accordingly, the international Committee on Earth Observation Satellites (CEOS), considered to be the space arm of GEO, has led the development of a Quality Assurance Framework for Earth Observation (QA4EO),[3] which is based on the adoption of key guidelines to help address difficult issues of traceability and interoperability, especially in the post-launch environment.

Whereas pre-launch calibration methodologies typically include radiometric, geometric, spectral, and polarization characterizations, post-launch calibration typically involves primarily radiometric and geometric calibration. One exception is NASA's Terra and Aqua MODerate-resolution Imaging Spectroradiometers (MODIS), which include an onboard Spectro-Radiometric Calibration Assembly (SRCA) designed to monitor the visible, near-infrared, and short-wave infrared spectral bands on MODIS (e.g., Román et al., 2024; Xiong et al., 2006a; 2010a; Choi et al., 2013). Post-launch radiometric calibration can be categorized into onboard calibration and vicarious calibration, the topics of the following Sections.

### 7.2.1 ONBOARD RADIOMETRIC CALIBRATION

Some satellite sensor systems utilize onboard devices to help monitor and, if necessary, update radiometric calibration coefficients. Partial aperture calibrators include standard lamps and/or solar diffuser panels. If done well, such approaches can capture sensor drift, instability, and sensor and electronic ageing problems. Shutters or deep space views provide dark readings.

Although there are numerous satellite sensor systems that make use of onboard radiometric calibration systems, the Landsat series of whisk-broom sensors is a notable example. Tremendous efforts have been dedicated to understanding and using the onboard calibration lamps (Russell et al., 2023; Voskanian et al., 2023; Niro et al., 2021; Sterckx et al., 2020; Rodrigo et al., 2019; Shrestha et al., 2019; Barker, 1985; Markham and Barker, 1985) and partial-aperture or full-aperture diffuse reflectance panels (Markham et al., 2003). These efforts are a measure of the importance of the Landsat series of sensors that have provided a readily accessible, global, and seasonal archive of relatively high spatial resolution data spanning over four decades (Voskanian et al., 2023; Niro et al., 2021; Sterckx et al., 2020; Rodrigo et al., 2019; Shrestha et al., 2019; Goward and Masek, 2001; Goward et al., 2006; Markham and Helder, 2012; Wulder and Masek, 2012). It should also be noted that the Landsat radiometric calibration experience was a critical contribution to the challenging radiometric calibration of the Earth Observing System (EOS) MODIS whisk-broom sensors.

Key characteristics of the Multi-Spectral Scanners (MSS) sensors on Landsat missions 1 through 5 are documented mainly in the so-called gray literature (e.g., Bartolucci and Davis, 1983). For the purposes of this chapter, the main MSS onboard calibration systems were internal calibrators (IC) consisting of a shutter wheel with a mirror and a neutral density filter, and two redundant tungsten-filament lamps (e.g., Murphy, 1984, 1986; Markham and Barker, 1987; Thome et al., 1997b). The mirror reflected light from the IC lamps through the neutral density filter and onto the focal plane. The neutral density filter was wedge-shaped, yielding a variable attenuation as it rotated with the shutter wheel. The IC did not test the full optical path of the MSS sensors.

The Thematic Mapper (TM) sensors on Landsat-4 (L4) and Landsat-5 (L5), and the Landsat-7 (L7) Enhanced TM Plus (ETM+) have been well documented (e.g., Markham et al., 1998; Markham et al., 2012). Briefly, the TMs and the ETM+ have an un-cooled primary focal plane containing 16 silicon (Si) detectors per band for the four visible and near-infrared bands. The L7 ETM+ also has a panchromatic band with 32 detectors. The TMs and the ETM+ have a cold focal plane containing 16 indium antimonide (InSb) detectors for each of the two short-wave infrared bands, plus four mercury cadmium telluride (HgCdTe) detectors (eight for the ETM+) for the thermal emissive band.

The Landsat TM and ETM+ sensors also incorporate IC systems for onboard radiometric calibration monitoring for the solar-reflective spectral bands (Mika, 1997).[4] These IC devices have a shutter arm that oscillates back and forth directly in front of the primary focal plane. On the TMs, the IC features three lamps that cycle through an eight-lamp state sequence over the course of a 24-second "scene" of data, whereas the ETM+ has two lamps (four lamp states). A complex, synchronized system involving the shutter, a scan mirror and photodiodes record IC lamp signals, dark shutter signals, and Earth reflected light for each scan line. Algorithms to use the IC data for radiometric calibration were developed and implemented in operational Landsat product generation systems around the world. Techniques were developed to analyze the IC data, mainly for the long lifetime of the L5 TM (Russell et al., 2023; Voskanian et al., 2023; Niro et al., 2021; Sterckx et al., 2020; Rodrigo et al., 2019; Shrestha et al., 2019; Metzler and Malila, 1985; Helder, 1996, Helder et al., 1996, 1997; Markham et al., 1998; Helder et al., 1998a, 1998b), but also for the L4 TM (Niro et al., 2021; Sterckx et al., 2020; Rodrigo et al., 2019; Shrestha et al., 2019; Chander et al., 2007a; Markham and Helder, 2012; Helder et al., 2013). Over time, these trend analyses led to a detailed understanding of several image artefacts introduced by various TM sensor characteristics. Within-scene relative calibration algorithms were developed and implemented in product generation systems to remove most of these artefacts, which include striping, scan-correlated shift, memory effect, coherent noise, and oscillations due to cold focal plane icing (e.g., Teillet et al., 2004a; Helder and Micijevic, 2004; Helder and Ruggles, 2004). Landsat processing systems make corrections for all artefacts except the coherent noise effect, which is typically on the order of 0.15 digital counts or less.

Historically, the TM radiometric calibration procedure used the instrument's response to the IC lamps on a scene-by-scene basis to determine the gain and offset of each detector. Since May 2003, the L5 TM data processed and distributed by the U.S. Geological Survey (USGS) Earth Resources Observation and Science (EROS) Center were updated to use a lifetime gain look-up-table (LUT) model to radiometrically calibrate the L5 TM data instead of using scene-specific IC lamp based gains (Chander and Markham, 2003). The new procedure for the reflective bands (1–5, 7) is based on a lifetime radiometric calibration curve for the instrument derived from the instrument's IC, cross-calibration with the L7 ETM+, and vicarious measurements. This change has improved absolute calibration accuracy, consistency over time, and consistency with L7 ETM+ data (Chander et al., 2004a). With this change, the radiometric scaling coefficients were updated as well (Chander et al., 2007b, 2009; Helder et al., 2008). Users need to apply the new L5 TM rescaling factors to convert calibrated digital counts to TOA at-sensor spectral radiance.

The L7 ETM+ has a Full Aperture Solar Calibrator (FASC). This capability features a near-Lambertian YB71 panel of known reflectance that is deployed periodically (approximately monthly) in front of the ETM+ aperture and diffusely reflects solar radiation into the full aperture of the sensor. With knowledge of the solar irradiance and geometric conditions, the FASC serves as an independent, full aperture calibrator. The ETM+ also has a Partial Aperture Solar Calibrator (PASC) that consists of a small passive device with a set of optics that allow the ETM+ to image the Sun (daily) through small holes. On-orbit performance of the ETM+ calibrator systems can be found in Markham et al. (2003, 2004a, 2004b, 2012).

While the IC data from the onboard systems played a central role in the radiometric calibration of each Landsat mission, the vicarious calibration techniques described in the following Sections made it possible to establish and validate radiometric calibration across the full family of Landsat sensors going back to 1972, including the MSS sensors (Helder et al., 2012a). The overall effort was undertaken over the past 15 years, approximately, and is summarized by Markham and Helder (2012). The key result from the tremendous amount of work involved is a consistently calibrated Landsat data archive that spans more than four decades with total uncertainties on the order of 10% or less for most sensors and bands.

In particular (Markham and Helder, 2012), the L1 through L5 MSS sensors are within 10% absolute radiometric accuracy in all but near-infrared spectral band 4, and with the other exceptions of L2 red spectral band 3 (11%) and L1 visible spectral bands 1–3 (11%, 11%, and 12%, respectively). MSS band 4 is very broad spectrally and includes large atmospheric absorption features. Thus, the uncertainty for this band in the MSS series grows much more rapidly to the point where the absolute radiometric calibration uncertainty for L1 MSS band 4 approaches 25%. For the solar reflective spectral bands of the L4 TM, L5 TM, and L7 ETM+ sensors, Markham and Helder (2012) report absolute radiometric calibration uncertainties of 5%, 7%, and 9%, respectively.

The Optical Land Imager (OLI) on Landsat-8 (L8) uses push-broom scanning technology. As a result, OLI radiometric calibration benefits from the long history of calibration experience with the push-broom sensors on the French Satellite Pour l'Observation de la Terre (SPOT) satellites (e.g., Valorge et al., 2004), as well as from the ALI (Advanced Land Imager) technology demonstration sensor flown on NASA's Earth Observing-1 (EO-1) satellite (e.g., Mendenhall et al., 2003, 2005). The spectral bands of the L8 OLI sensor[5] include the L7 ETM+ sensor bands augmented by a deep blue spectral band designed for water resources and coastal zone investigation, and an infrared spectral band at 1.37 micrometers for cirrus cloud detection. A quality assurance band is also included with each data product to indicate the presence of clouds, water, and/or snow.

The L8 OLI radiometric calibration system[6] consists of two solar diffusers and a shutter, and two stimulation lamp assemblies (Markham et al., 2010; Reuter et al., 2011). The diffusers provide a full-system full-aperture calibration when the shutter is open and a dark reference when the shutter is closed. Each lamp assembly contains three tungsten lamps, operated at constant current and monitored by a silicon photodiode, which illuminate the OLI detectors through the full optical system. One "working" lamp is used daily for intra-orbit calibration, a "reference" lamp is used approximately monthly, and a "pristine" lamp is used approximately twice a year. L8 sensor characterization and calibration performance is reviewed in a special journal issue edited by Markham et al. (2015).

### 7.2.2 Vicarious Calibration Methodologies

The predominant vicarious calibration approach is the measurement, on satellite overpass days, of pertinent surface and atmospheric optical properties at terrestrial sites with suitable characteristics to estimate at-sensor radiance or TOA reflectance. Comparisons of these estimates with image-based values computed using the canonical calibration coefficients provide post-launch monitoring of the radiometric calibration of satellite sensors, either individually or by cross-calibration between sensors (e.g., Román et al., 2024; Dong et al., 2023; Russell et al., 2023; Voskanian et al., 2023; Niro et al., 2021; Sterckx et al., 2020; Rodrigo et al., 2019; Shrestha et al., 2019; Slater et al., 1987; Teillet et al., 1990; Biggar et al., 1994; Thome, 2001; Teillet, 2014). Well-understood or instrumented sites serve as reference standard sites, whereas other terrestrial targets that change very little in surface reflectance serve as pseudo-invariant features (PIFs), also known as pseudo-invariant calibration sites (PICS) in the present context. A comprehensive review of terrestrial reference standard sites used for post-launch radiometric calibration is given in the next section of this chapter.

Error analyses of vicarious calibration approaches indicate that uncertainties in the 2–3% (1 $\sigma$) range with respect to exo-atmospheric solar irradiance (W $m^{-2}$ $\mu m^{-1}$) are attainable and can be accurately related to national laboratory standards (Dinguirard and Slater, 1999). The objective is traceability of radiometric calibration accuracies to SI units for science users (e.g., Pollock et al., 2003) and data products with consistent quality for the broader user community.

Although different satellite sensors image the Earth in analogous spectral bands, the spectral bands almost never match exactly such that data acquired over identical targets and derived information products are not directly comparable (Teillet et al., 2007a; Teillet and Ren, 2008). Relatively few investigations have been undertaken to assess radiometric calibration errors due to differences in spectral band response functions between satellite sensors when attempting cross-calibration based on near-simultaneous imaging of common ground look targets in analogous spectral bands (Sterckx et al., 2020; Rodrigo et al., 2019; Shrestha et al., 2019; Teillet et al., 2001a, 2004b, 2007a; Trishchenko et al., 2002; Rao et al., 2003; Teillet et al., 2007a; Teillet and Ren, 2008; Doelling et al., 2012; Chander et al., 2013c; Henry et al., 2013). Such spectral band difference effects (SBDEs) can be significant. For instance, Teillet et al. (2007a) simulated SBDEs affecting vicarious-calibration-based TOA reflectance comparisons between many satellite sensors. The following summary examples are for Railroad Valley Playa in Nevada, a frequently used vicarious calibration test site, and for an atmospheric aerosol optical depth of 0.05 and a solar zenith angle of 60°. Comparisons in visible spectral bands were generally within 3%, some were within 5%, and worst cases were around 7% between Terra-MISR[7] and various other sensors. Comparisons in near-infrared spectral bands were generally within 3%, some were within 10%, and worst cases were around 21% between AVHRR[8] and various other sensors. Comparisons in shortwave infrared spectral bands around 1.65 micrometres were generally within 3%, and worst cases were around 8% between TM or CBERS-IRMSS[9] and various other sensors. Comparisons in shortwave infrared spectral bands around 2.2 micrometres were generally within 6%, and worst cases were around 21% between MODIS or ADEOS-GLI[10] and various other sensors.

## 7.3 VICARIOUS CALIBRATION VIA TERRESTRIAL REFERENCE STANDARD SITES

Publications in the form of research articles, special journal issues, reports, and books have periodically provided overviews or reviews of satellite sensor radiometric calibration with some mention of vicarious calibration (Román et al., 2024; Dong et al., 2023; Russell et al., 2023; Voskanian et al.,

2023; Niro et al., 2021; Sterckx et al., 2020; Rodrigo et al., 2019; Shrestha et al., 2019; Slater, 1980, 1984, 1985; Markham and Barker, 1985; Malila and Anderson, 1986; Price, 1987a, 1987b; Ahern et al., 1988; Nithianandam et al., 1993; Bruegge and Butler, 1996; Chen, 1996; Slater and Biggar, 1996; Teillet, 1997a, 1997b; Dinguirard and Slater, 1999; Slater et al., 2001; Morain and Budge, 2004; Valorge et al., 2004; Butler et al., 2005; Chander et al., 2013a, 2013b). This section provides an overview of the use of terrestrial reference standard sites for post-launch sensor radiometric calibration from historical, current, and future perspectives. Emphasis is placed on optical sensors operating in the visible, near-infrared, and shortwave infrared spectral regions.

### 7.3.1 Historical Perspective

Reflectance and other properties of land surfaces and atmospheric phenomena now used by many for vicarious calibration were first examined decades ago using surface and airborne measurements in the contexts of the micrometeorology of arid zones (Ashburn and Weldon, 1956), investigations of natural landing areas for aircraft (Molineux et al., 1971), and other analyses (Davis and Cox, 1982; Smith, 1984; Bowker and Davis, 1992; Warren et al., 1998). In scene understanding studies in support of the first meteorological satellite observations, Salomonson and Marlatt (1968) examined the directional reflectance properties of cloud tops, snow, and white gypsum sand (hydrous calcium sulphate) at *White Sands*, New Mexico (WSNM), in the United States, which are three surface target types that have since been used extensively for vicarious calibration of satellite imaging sensors.

The potential use of suitable terrestrial targets on or near the Earth's surface for the post-launch radiometric calibration of satellite sensors was first examined by a number of investigators for various satellite sensor systems and for several surface types, including WSNM, deserts, playas, salt flats, ocean, cloud tops, snow or ice fields, pseudo-invariant features, and uniformly vegetated areas (Román et al., 2024; Dong et al., 2023; Russell et al., 2023; Voskanian et al., 2023; Niro et al., 2021; Sterckx et al., 2020; Rodrigo et al., 2019; Shrestha et al., 2019; Coulson and Jacobowitz, 1972; Muench, 1981; Kastner and Slater, 1982; Koepke, 1982; Hovis et al., 1985; Mueller, 1985; Fraser and Kaufman, 1986; Staylor, 1986; Duggin, 1987; Schott et al., 1988; Biggar et al., 1994; Teillet et al., 1998a; Le Marshall et al., 1999; Smith et al., 2002; Kamstrup and Hansen, 2003; Nieke et al., 2004; Martiny et al., 2005; Anderson and Milton, 2006; Wu et al., 2008a, 2008c; Teillet et al., 2010b). Desiderata for terrestrial reference site selection have been well documented (e.g., Scott et al., 1996):

- High spatial uniformity over a large area (within 3%)
- Surface reflectance greater than 0.3 across solar-reflective wavelengths
- Flat spectral reflectance across solar-reflective wavelengths
- Temporally invariant surface properties (within 2%)
- Horizontal surface with nearly Lambertian reflectance
- High altitude, far from ocean, urban, and industrial areas
- Arid region with low probability of cloud cover or airborne particles

It is also worth noting that the characterization of any given calibration reference site usually involves "nested" sites: the specific surface measurement site and the overall calibration reference site to be imaged by satellite sensors, which must be well represented by the measurement site's spectral reflectance properties.

Early research on surface measurement methodologies for land surfaces focused to a considerable extent on the alkali flats region of WSNM (cf. Table 7.1 for selected literature references). These vicarious calibration methodologies were used to provide post-launch radiometric calibration updates for satellite sensor systems such as the L4 and L5 TM sensors (Slater, 1986; Slater et al., 1987). A notable focus of attention was the provision of post-launch radiometric calibration for the Advanced Very High Resolution Radiometer (AVHRR) sensors on the series of National Oceanic

**TABLE 7.1**
**Selected Literature for vVcarious Calibration Targets Mentioned in the Text**

| Test Site | Literature References |
|---|---|
| *White Sands*, New Mexico, USA | Castle et al., 1984; Smith and Levin, 1985; Smith et al., 1985; Whitlock et al., 1990a, 1990b; Wheeler et al., 1994 |
| *The Moon* | Kieffer and Wildey, 1985, 1996; Kieffer et al., 2002; Barnes et al., 2004; Stone et al., 2005, 2013; Stone, 2008; Xiong et al., 2008 |
| *La Crau*, France | Xing-Fa et al., 1990; Santer et al., 1992, 1997; Gu et al., 1992; Moran et al., 1995; Richter, 1997; Rondeaux et al., 1998; Six, 2002 |
| *Saharan and Arabian Deserts*, North Africa | Henry et al., 1993; Cosnefroy et al., 1994, 1996, 1997; Cabot, 1997; Cabot et al., 1998, 1999, 2000; Miesch et al., 2003; Rao et al., 2003; Markham et al., 2006; Vermote and Saleous, 2006; Wu et al., 2008b; Lachérade et al., 2013 |
| *Atmospheric molecular (Rayleigh) scattering* | Vermote et al., 1992; Kaufman and Holben, 1993; Dilligeard et al., 1997; Meygret et al., 2000 |
| *Saharan and Arabian Deserts*, North Africa | Henry et al., 1993; Cosnefroy et al., 1994, 1996, 1997; Cabot, 1997; Cabot et al., 1998, 1999, 2000; Miesch et al., 2003; Rao et al., 2003; Markham et al., 2006; Vermote and Saleous, 2006; Wu et al., 2008b |
| *Ocean sunglint* | Kaufman and Holben, 1993; Vermote and Kaufman, 1995; Hagolle et al., 2004; Luderer et al., 2005 |
| *Clouds* | Desormeaux et al., 1993; Vermote and Kaufman, 1995; Iwabuchi, 2003; Doelling et al., 2004, 2010, 2013; Hu et al., 2004; Fougnie and Bach, 2009 |
| *Sites in Australia* | Graetz et al., 1994; Prata et al., 1996; Mitchell et al., 1997; O'Brien and Mitchell, 2001; de Vries et al., 2007 |
| *Dunhuang*, China | Min and Zhu, 1995; Wu et al., 1994, 1997; Xiao et al., 2001; Hu et al., 2001, 2009, 2010; Liu et al., 2004; Zhang et al., 2001, 2004, 2008, 2009; Li et al., 2009 |
| *Railroad Valley Playa*, USA | Scott et al., 1996; Snyder et al., 1997; Thome, 2001, 2005; Matsunaga et al., 2001; Thome et al., 2000, 2001, 2003a, 2003b, 2004a, 2006, 2007, 2008; Biggar et al., 2003; McCorkel et al., 2006; D'Amico et al., 2006; Czapla-Myers et al., 2007, 2008, 2010; Angal et al., 2008; Kerola and Bruegge, 2009 |
| *Ivanpah Playa*, USA | Thome et al., 1996, 2004a; Villa-Aleman et al., 2003a, 2003b; Thome, 2005; Rodger et al., 2005 |
| *Snow/ice fields* | Loeb, 1997; Tahnk and Coakley, 2001; Nieke et al, 2003; Six et al., 2004 |
| *Negev Desert*, Israel | Bushlin et al., 1997; Gilead and Karnieli, 2004; Faiman et al., 2004; Gilead, 2005 |
| *Lunar Lake Playa*, USA | Thome et al., 1998; Bannari et al., 2005 |
| *Dome-C*, Antarctica | Six et al., 2004, 2005; Tomasi et al., 2008; Wu et al., 2008b; Wenny and Xiong, 2008; Wenny et al., 2009; Xiong et al., 2009a, 2009b; Bouvet and Ramoino, 2010; Cao et al., 2010; Potts et al., 2013 |
| *Frenchman Flat*, USA | Gross et al., 2007; Helmlinger et al., 2007; Kerola et al., 2009 |
| *Tuz Golu*, Turkey | Gurol et al., 2008, 2010 |
| *Sonoran Desert*, Mexico, aka Yuma, Arizona, USA | Angal et al., 2010a, 2010b, 2011; Kim and Lee, 2013 |

and Atmospheric Administration (NOAA) satellites, because of the widespread utilization of AVHRR image data and because AVHRRs do not have onboard calibration systems (Russell et al., 2023; Voskanian et al., 2023; Niro et al., 2021; Sterckx et al., 2020; Rodrigo et al., 2019; Shrestha et al., 2019; Duggin, 1987; Frouin and Gautier, 1987; Smith et al., 1988; Teillet et al., 1990; Holben et al., 1990; Staylor, 1990; Mitchell et al., 1992; Brest and Rossow, 1992; Che and Price, 1992; Kaufman and Holben, 1993; Abel et al., 1993; Teillet and Holben, 1994; Cihlar and Teillet, 1995; Rao and Chen, 1995, 1996, 1999; Brest et al., 1997; Tahnk and Coakley, 2001; Cao and Heidinger, 2002; Wu et al., 2003, 2008a, 2008c; Cao et al., 2005; Wu et al., 2010).

Other deserts and playas used as reference standard sites include (cf. Table 7.1 for selected literature references) *La Crau* in France, several *sites in Australia, Dunhuang* in China, *Railroad Valley Playa* in the United States, *Ivanpah Playa* in the United States, the *Negev Desert* in Israel, *Lunar Lake Playa* in the United States, *Frenchman Flat* in the United States, the *Sonoran Desert* in Mexico (also known as Yuma, Arizona), and *Tuz Golu* in Turkey. About 20 large areas in the *Saharan and Arabian Deserts* of North Africa have been used without surface observations to enable cross-calibration and/or uniformity calibration for a variety of sensor systems for several decades. These large uniform sites are important for the calibration of push-broom sensor detector arrays, such as those on the SPOT satellite systems, given that it is difficult to avoid detector-to-detector differences if the whole array cannot be exposed easily to a uniform radiance field. Because significant episodes of airborne particles can occur in these regions, only the clearest image acquisitions are used for radiometric calibration purposes.

Additionally, other vicarious calibration or cross-calibration methods take advantage of non-land Earth targets, including (cf. Table 7.1 for selected literature references) atmospheric molecular (Rayleigh) scattering, ocean sunglint, snow/ice fields, and clouds. A snowfield of particular interest is the *Dome-C* site in Antarctica, where image data from many overpasses are available during the summer months. A key facility for the vicarious calibration of ocean color satellite sensors is the Marine Optical BuoY (MOBY), an autonomous optically instrumented buoy moored off the island of Lanai, Hawaii, in the United States (Clark et al., 1997; Brown et al., 2007).

The Remote Sensing Group at the University of Arizona pioneered vicarious calibration methodologies that involved not only WSNM but also the playas at Edwards Air Force Base, Railroad Valley, Lunar Lake, and Ivanpah. Many of the key surface measurement methodologies central to vicarious calibration were developed as a result of close collaborations between the University of Arizona and the U.S. Department of Agriculture (USDA) (Phoenix), focused to a considerable extent on field campaigns at the Maricopa Agricultural Center (e.g., Jackson, 1990), but also involving many field campaigns at WSNM and elsewhere. These pioneering efforts, together with leading-edge work by NASA scientists and collaborations with scientists at the National Institute for Standards Technology (NIST), played an important role in the calibration programs of NASA's EOS (e.g., Bruegge and Butler, 1996; Slater et al., 1996; Xiong et al., 2010b). Similar methodologies and efforts have also been key elements of other EO systems (e.g., Voskanian et al., 2023; Niro et al., 2021; Sterckx et al., 2020; Rodrigo et al., 2019; Shrestha et al., 2019; Guenther et al., 1996; Markham et al., 1997; Schroeder et al., 2001; Secker et al., 2001; Teillet et al., 2001a, 2006, 2007a; Black et al., 2003; Schiller, 2003; Chander et al., 2007a, 2007c; Biggar et al., 2003; Thome et al., 1997a, 1997b, 2003a, 2003b; Chander et al., 2004b; Xiong et al., 2006b; Chander et al., 2013b; Henry et al., 2013). There have also been systematic and sustained vicarious calibration efforts over many years by the Centre National d'Etudes Spatiales (CNES), the European Space Agency (ESA), the UK's National Physical Laboratory (NPL), the Japan Aerospace Exploration Agency (JAXA), and the U.S. Geological Survey (USGS) Earth Resources Observation and Science (EROS) Center.

Vicarious calibration can also be undertaken by imaging the Moon, a stable target whose dynamic illumination variations can be computed (cf. Table 7.1 for selected literature references), and/or by imaging bright stars (e.g., Chang et al., 2012). Although ultra-stable stellar targets are imaged without atmospheric effects, they are relatively low radiance sources and, as with imaging the Moon, they require high-risk spacecraft platform maneuvers to achieve. Satellite sensors often use data acquisitions of deep space views without spacecraft platform maneuvers to provide dark target calibration data.

As the importance of remote sensing calibration was increasingly acknowledged and as more terrestrial sites were investigated for use as reference standard test sites for post-launch sensor calibration, specialists in the international community endeavored to put together databases of worldwide calibration facilities, including test sites and instruments. Early pilot efforts were those of the international CEOS in 1993 (spearheaded by Barton, Guenther, and others) and NASA's EOS in 1995 (spearheaded by Butler, Starr, Reber, and others). Calls for contributions led to editions of such databases (e.g., Butler et al., 2001), but these early databases were never fully populated and they no longer appear to be active. Current efforts in this regard by CEOS, ESA, and the USGS are outlined in the next Section.

During the 1990s, the role of Earth monitoring systems of all kinds was increasingly cast in the context of climate observations. For example, the Global Climate Observing System (GCOS) was established in 1992 to focus on satellite and in situ observations for climate in the atmospheric, oceanic, and terrestrial domains.[11] As part of the process of identifying climate observation requirements in 1999, GCOS established a set of Climate Monitoring Principles[12] that were adopted in 2003 by the World Meteorological Organization (WMO) and CEOS. The GCOS Climate Monitoring Principles refer specifically to the critical role of calibrated satellite sensor observations.

A generalized approach to the vicarious calibration of multiple EO sensors using hyperspectral data was demonstrated by Teillet et al. (1998a, 1998b, 2001b) in a project on quality assurance and stability reference (QUASAR) monitoring. The approach uses spatially-extensive airborne or satellite hyperspectral imagery of a terrestrial reference standard site as spectroradiometric reference data to carry out vicarious radiometric calibrations for all optical satellite sensors that image the site on the same day.

The establishment of a global instrumented and automated network of test sites (GIANTS) for post-launch radiometric calibration of Earth observation sensors was outlined by Teillet et al. (2001c). The GIANTS concept proposes that a small number of well-characterized benchmark test sites and datasets be supported for the calibration of all space-based optical sensors imaging the Earth. A core set of surface sensors, measurements, and protocols should be standardized across all participating test sites and measurement datasets should undergo identical processing at a central secretariat. GIANTS is intended to supplement calibration information already available from other calibration systems and efforts, reduce the resources required by individual agencies, and provide greater consistency for terrestrial monitoring studies based on multiple sensor systems. The network approach is also intended to explore of the use of automation, communication, coordination, visibility, and education, all of which can be facilitated by greater use of advanced ground-based sensor and telecommunication technologies.

### 7.3.2 Current Developments

More recent initiatives in the domain of reference standard sites include ongoing methodology research (e.g., Román et al., 2024; Dong et al., 2023; Russell et al., 2023; Voskanian et al., 2023; Niro et al., 2021; Sterckx et al., 2020; Rodrigo et al., 2019; Shrestha et al., 2019; Wang et al., 2011; Thome, 2012; Anderson et al., 2013; Helder et al., 2012b, 2013; Govaerts et al., 2013; Thome et al., 2013), as well as the GEOSS, CEOS, and QA4EO efforts mentioned in the Introduction. Notably and significantly, the CEOS WGCV subgroup on Infrared Visible Optical Sensors (IVOS) has worked in recent years with collaborators around the world to establish a core set of CEOS-endorsed, globally distributed, reference standard test sites for the post-launch calibration of space-based optical imaging sensors. There are now eight CEOS reference instrumented sites (Table 7.2, Figure 7.1) and six CEOS reference PICS (Table 7.3, Figure 7.2). The instrumented sites are mainly used for field campaigns to obtain or update radiometric gain coefficients and they serve as a focus for international efforts, facilitating traceability and cross-comparison to evaluate biases between current and future sensors and methods in a harmonized manner. The pseudo-invariant reference standard test sites are desert areas characterised by high surface reflectance, some sand dunes, little or no vegetation, and low atmospheric aerosol loadings. Such sites can be used to evaluate the long-term stability of a given sensor and to facilitate cross-comparisons between multiple sensors (e.g., Helder et al., 2010; Chander et al., 2013b). They can also be used for the validation of selected higher-order climate variable products. As a precursor to the selection of the CEOS-endorsed core sites, a list of potential candidates was developed (Teillet et al., 2007b) and an online catalog of worldwide test sites used for sensor characterization was implemented by the USGS (Chander et al., 2007d).[13] The catalog provides ready access to this vital information by the global community.

## TABLE 7.2
## Core Set of CEOS-Endorsed Instrumented Reference Standard Test Sites

| # | Site | WRS-2 Path/ Row | Centre Latitude (°), Longitude (°), and Altitude ASL (m) | Point of Contact |
|---|---|---|---|---|
| 1 | Dome C, Antarctica | 88–89–90/113 | −74.50, +123.00, 3215 | Stephen Warren, University of Washington, USA |
| 2 | Dunhuang, China | 137/32 | +40.13, +94.34, 1220 | Xiuqing Hu, National Satellite Meteorological Center, China |
| 3 | Frenchman Flat, USA | 40/34 | +36.81, −115.93, 940 | Carol J. Bruegge, NASA Jet Propulsion Laboratory, USA |
| 4 | Ivanpah Playa, USA | 39/35 | +35.5692, −115.3976, 813 | Kurtis J. Thome, NASA Goddard Space Flight Center, USA |
| 5 | La Crau, France | 196/30 | +43.47, +4.97, 28 | Patrice Henry, Centre National d'Etudes Spatiales, France |
| 6 | Negev Desert, Israel | 174/39 | +30.11, +35.01, 334 | Arnon Karnieli, Ben Gurion University, Israel |
| 7 | Railroad Valley Playa, USA | 40/33 | +38.50, −115.69, 1435 | Kurtis J. Thome, NASA Goddard Space Flight Center, USA |
| 8 | Tuz Golu, Turkey | 177/33 | +38.83, +33.33, 905 | Selime Gurol, Tubitak Uzay (Space Technologies Research Institute), Turkey |

*Note*: WRS is the Landsat Worldwide Reference System and ASL = Above Sea Level. Other characteristics of some of these test sites are given in Teillet et al. (2007b).

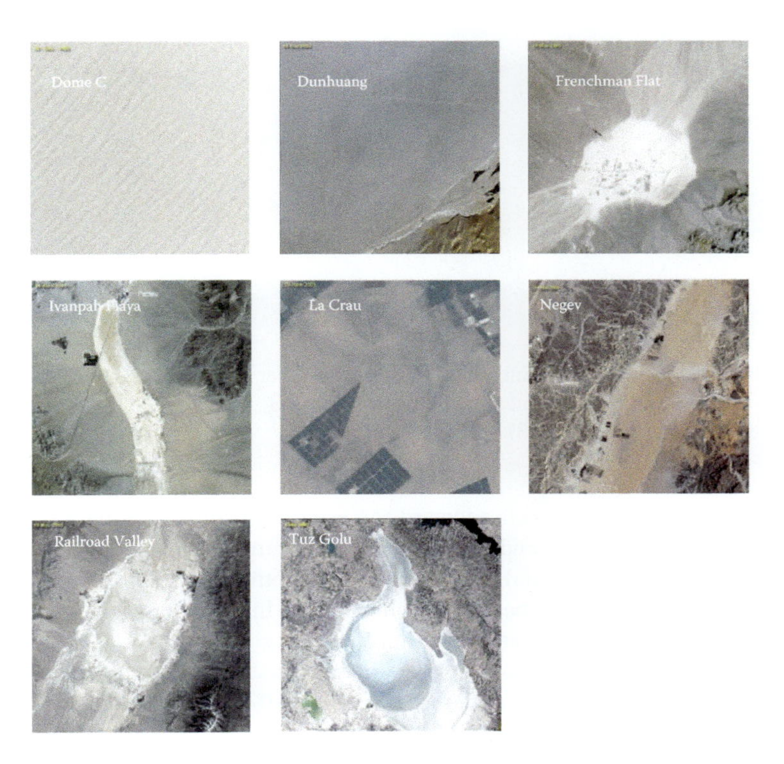


**FIGURE 7.1** Examples of Landsat-7 Enhanced Thematic Mapper Plus (ETM+) imagery of the core set of CEOS-endorsed instrumented reference standard test sites (N.B., scales differ, North is up in all cases, and site coordinates can be found in Table 7.2).

**TABLE 7.3**
**Core Set of CEOS-Endorsed Pseudo-Invariant Reference Standard Test Sites**

| # | Site | WRS-2 Path/Row | Center Latitude (°), Longitude (°), and Altitude ASL (m) |
|---|---|---|---|
| 1 | Libya 4 | 181/40 | +28.55, +23.39, 118 |
| 2 | Mauritania 1 | 201/47 | +19.40, −9.30, 392 |
| 3 | Mauritania 2 | 201/46 | +20.85, −8.78, 384 |
| 4 | Algeria 3 | 192/39 | +30.32, +7.66, 245 |
| 5 | Libya 1 | 187/43 | +24.42, +13.35, 648 |
| 6 | Algeria 5 | 195/39 | +31.02, +2.23, 530 |

*Notes:* WRS is the Landsat Worldwide Reference System and ASL = Above Sea Level. Note that Mauritania 1 and 2 are considered to be one site. The point of contact for these sites is Patrice Henry, Centre National d'Etudes Spatiales (CNES), France. Other characteristics of these test sites are given in Teillet et al. (2007b).

**FIGURE 7.2** Examples of Landsat-7 Enhanced Thematic Mapper Plus (ETM+) imagery of the core set of CEOS-endorsed pseudo-invariant reference standard test sites (N.B., scales differ, North is up in all cases, and site coordinates can be found in Table 7.3).

CEOS/WGCV/IVOS is also working toward the establishment of optimum methodologies for the characterization and use of the endorsed reference instrumented sites. The principal criteria in selecting these sites, other than spatial uniformity and brightness, etc., was that they were all fully and regularly calibrated by ground-based instrumentation. In some cases, instrumentation is permanently deployed; in others, it is transported to the site for specific characterization campaigns. However, in all cases, the basis for assigning a value to the surface reflectance and its subsequent propagation to the TOA as radiance (for comparison to satellite imager radiances) is derived from measurements made by ground survey teams. Cost is a significant issue for the instrumentation (and associated performance) needed for test-sites for establishing strategies for the long-term maintenance of CEOS test-sites. In addition, having a limited number of sites provides an opportunity for the owners/operators of Earth-viewing optical imagers (including commercial operators) to routinely collect data over the CEOS-endorsed reference sites.

In principle, radiometric calibration coefficients should not depend on the geographic locations of reference standard sites. However, the methodologies involved in vicarious calibration via instrumented sites involve characterization of surface properties and atmospheric conditions by different teams and, pending the adoption of standardized approaches, different instrumentation. Similarly, vicarious calibration via PICS involves methods for atmospheric compensation. These diversities are deemed beneficial in that they can be used to yield an improved understanding of the role of atmospheric effects and different modeling approaches on calibration results.

In any given year, there are typically several field campaigns undertaken by various agencies to carry out vicarious calibrations for a number of satellite optical sensor systems. Thus, a logical next step is to formalise and test the GIANTS concept via the concatenation of CEOS-endorsed core sites (Teillet and Fox, 2009). The concatenation concept proposes the deployment of as many field campaigns at as many of the core sites as possible during a given time period (e.g., one month) to generate updates for as many satellite optical sensors as possible. The pilot project would make a concrete initial test of the GIANTS approach (at this stage excluding an "automated" aspect in most if not all instances).

The trade-off between automation and the accuracy demands of radiometric calibration is an important consideration that has yet to be investigated to a significant extent. Several organizations have begun to explore automated radiometric systems for vicarious calibration (Niro et al., 2021; Sterckx et al., 2020; Rodrigo et al., 2019; Shrestha et al., 2019; Thome et al., 2004b; Czapla-Myers et al., 2007, 2008; Gross et al., 2007; Helmlinger et al., 2007; Kerola et al., 2009; Meygret et al., 2011). Another topic that has received insufficient attention by the calibration research community is the use of data at wavelengths outside the solar-reflective spectrum to help characterise reference standard sites used for optical sensor calibration (e.g., Teillet et al., 1995; Blumberg and Freilikher, 2001; Floury et al., 2002; Brogioni et al., 2006).

The UK's NPL has contributed significantly in a number of ways to new methods and systems, as well as to international collaboration, for radiometric calibration of satellite sensors (Fox, 1999), including the development of bold new space mission concepts for advanced cryogenic-based radiometric calibration capabilities (Fox et al., 2002). NPL has also coordinated several CEOS campaigns in the context of GEO tasks to compare satellite calibration approaches, recently using the Dome-C site and also leading major comparison exercises focused on the Tuz Golu site in Turkey.[14]

CEOS is providing data and information for calibration and validation (cal/val) of EO data through its Cal/Val Portal.[15] The original Cal/Val Portal started in 2007 and a new version is now moving to a completely operational system providing new tools, datasets, results, and upcoming campaign information on a regular basis. Among other initiatives, the enhanced Cal/Val Portal supports measurement and comparison campaigns by means of a prototype database with data and information on reference standard sites.

### 7.3.3 Next Steps

Advocacy of the use of CEOS-endorsed reference standard sites and the establishment and use of some sort of certification process as part of "best practice" methodologies should lead to improved measurement consistency between satellite sensors, reduce costs, and help underpin the accurate monitoring of planetary change.

With respect to the CEOS-endorsed reference standard instrumented sites, the following steps are recommended for the near term: (1) increase the number of sites from eight to ten; (2) gather more complete site characterization data and information, especially with respect to their reflectance anisotropy properties and their temporal stabilities; (3) define a recommended standard set of core measurements; (4) formalize the set of sites into an operational network; (5) organize local, regional, national, and international field campaigns; (6) acquire and archive imagery of all of the sites on an ongoing basis; (7) develop online calibration data access infrastructures; and (8) continue to improve vicarious calibration methodologies. The current set of endorsed sites is distributed unevenly across the globe, which should be kept in mind to the extent possible when selecting additional instrumented sites. However, the adoption and use of new sites will depend not only on the suitability of site characteristics, but also on the interests, capabilities, and funding of research groups, as well as logistical and political accessibility.

Still with respect to the CEOS-endorsed reference standard instrumented sites, the following steps are recommended for the longer term: (1) continue to acquire and archive imagery on an ongoing basis for all of the sites; (2) continue to improve vicarious calibration methodologies; (3) establish traceability chains for site measurement data; (4) develop recommended guidelines and best practices for using the network of sites; and (5) endorse and advocate compliance with recommended guidelines and best practices. ISO standards for radiometric calibration of remote sensing data have been proposed (Di, 2004), making a high-level start toward certification.

## 7.4 CONCLUDING REMARKS

This chapter has provided a broad overview of the post-launch radiometric calibration of image data from satellite sensor systems operating in the solar-reflective optical domain. Even though satellite sensors are very well characterized radiometrically prior to launch, post-launch changes to satellite sensors and their radiometric calibration performance necessitate ongoing calibration monitoring and adjustment.

Onboard radiometric calibration systems have been implemented since the early days of satellite imaging sensors. This chapter highlighted the rich history of Landsat onboard sensor calibration and the significant efforts that were required by the absence of such capability for the AVHRR sensors, which provided data for countless broad-scale studies of the Earth (e.g., Román et al., 2024; Dong et al., 2023; Russell et al., 2023; Voskanian et al., 2023; Niro et al., 2021; Sterckx et al., 2020; Rodrigo et al., 2019; Shrestha et al., 2019; Townshend, 1994; Townshend et al., 1994; Teillet et al., 2000). The chapter focused extensively on vicarious calibration methodologies, important complements to the use of complex and expensive onboard calibration systems that seldom test the full optics of sensors. Terrestrial reference standard sites for vicarious calibration were featured and the relevant pioneering research over many decades was documented.

Today's mainstream information society routinely makes extensive qualitative use of satellite imagery. With respect to quantitative use, the significant improvement in the knowledge of the radiometric performance of satellite sensor systems, as described in this chapter, is indicative of the advancement in the state of the art of post-launch radiometric calibration methods. It is also indicative of the need for persistent calibration of satellite sensors over mission lifetimes, as well as the need for a variety of calibration methodologies in order to assess the true radiometric performance of any given sensor as accurately as possible. Accordingly, the use of terrestrial reference standard sites for post-launch monitoring of satellite sensor radiometric performance increasingly constitutes a key component of current and future satellite sensor calibration strategies. With continued international cooperation on test site calibration methodologies among remote sensing agencies and organizations, remotely sensed EO data can be made more accurate and useful to help develop an improved and sound understanding of our planet and its climate.

The success of climate monitoring missions will depend critically on significant advances in post-launch calibration. The planned Climate Absolute Radiance and Refractivity Observatory (CLARREO) (Anderson et al., 2008; Thome et al., 2010; Sandford et al., 2010; Lukashin et al., 2013) and the proposed Traceable Radiometry Underpinning Terrestrial- and Helio-Studies (TRUTHS) mission (Fox et al., 2002, 2011) offer novel approaches to the provision of key scientific data with unprecedented radiometric accuracy for terrestrial and solar studies. Missions such as CLARREO and TRUTHS will establish well-calibrated reference targets and standards to support other EO missions as well as solar and lunar observations. Once launched, these sensor systems would in effect provide the de-facto primary reference standards for the global ensemble of EO systems. CLARREO and TRUTHS promise to improve the performance and accuracy of Earth observation missions by an order of magnitude (on the order of 0.3% in the solar reflective domain). Post-launch radiometric calibration based on metrology-quality observatories in space lies at the heart of these proposed missions, which, as a result, will be integral and indispensable tools to help maximize the benefits of all EO systems.

## ACKNOWLEDGMENTS

The post-launch radiometric calibration methodologies documented in this chapter are the culmination of decades of effort on the part of many individuals and agencies. The authors thank CEOS/WGCV/IVOS members in particular for their stewardship and contributions toward progress in EO calibration and validation.

## NOTES

1. www.earthobservations.org/geoss.shtml
2. http://gsics.wmo.int/ GSICS is part of the World Meteorological Organization (WMO). It is an international collaboration to monitor, improve and harmonize data quality from operational environmental satellites for climate monitoring and weather forecasting.
3. http://qa4eo.org/index.html.
4. An assessment of the Landsat-5 thermal band calibration has been carried out by Barsi et al. (2003, 2007) and Padula et al. (2010).
5. http://landsat.usgs.gov/band_designations_landsat_satellites.php
6. https://directory.eoportal.org/web/eoportal/satellite-missions/l/landsat-8-ldcm
7. Terra-MISR: Multi-angle Imaging Spectro-Radiometer on NASA's Terra satellite.
8. AVHRR: Advanced Very High Resolution Radiometer.
9. CBERS-IRMSS: China–Brazil Earth Resources Satellite Infrared Multi-Spectral Scanner.
10. ADEOS-GLI: Japan's Advanced Earth Observing Satellite Global Imager.
11. www.wmo.int/pages/prog/gcos/index.php
12. www.wmo.int/pages/prog/gcos/index.php?name=ClimateMonitoringPrinciples
13. http://calval.cr.usgs.gov/sites_catalog_map.php
14. http://calvalportal.ceos.org/cvp/web/guest/tuz-golu-campaign
15. http://calvalportal.ceos.org

## REFERENCES

Abel, P., Guenther, B., Galimore, R.N., and Cooper, J.W. 1993. Calibration results for NOAA-11 AVHRR channels 1 and 2 from congruent path aircraft observations. *Journal of Atmospheric and Oceanic Technology*, 10(4): 493–508.

Ahern, F.J., Brown, R.J., Cihlar, J., Gauthier, R., Murphy, J., Neville, R.A., and Teillet, P.M. 1988. Radiometric correction of visible and infrared remote sensing data at the Canada centre for remote sensing. In *Remote Sensing Yearbook 1988*. Edited by A. Cracknell and L. Hayes. Taylor and Francis, Philadelphia, PA.

Anderson, D., Jucks, K.W., and Young, D.F. 2008. The NRC decadal survey climate absolute radiance and refractivity observatory: NASA implementation, *Proceedings of the IEEE International Geoscience and Remote Sensing Symposium (IGARSS)*, Boston, Massachusetts, pp. 9–11.

Anderson, K., and Milton, E.J. 2006. On the temporal stability of ground calibration targets: Implications for the reproducibility of remote sensing methodologies. *International Journal of Remote Sensing*, 27(16): 3365–3374.

Anderson, N., Czapla-Myers, J., Leisso, N., Biggar, S., Burkhart, C., Kingston, R., and Thome, K. 2013. Design and calibration of field deployable ground-viewing radiometers. *Applied Optics*, 52(2): 231–240.

Angal, A., Chander, G., Choi, T., Wu, A., and Xiong, X. 2010a. The use of the Sonoran Desert as a pseudo-invariant site for optical sensor cross-calibration and long-term stability monitoring. *Proceedings of the 2010 International Geoscience and Remote Sensing Symposium (IGARSS)*, Honolulu, Hawaii, pp. 1656–1659.

Angal, A., Chander, G., Xiong, X., Choi, T., and Wu, A. 2011. Characterization of the Sonoran desert as a radiometric calibration target for Earth observing sensors. *Journal of Applied Remote Sensing*, 5(1): 059502.

Angal, A., Choi, T.J., Chander, G., and Xiong, X. 2008. Monitoring on-orbit stability of Terra MODIS and Landsat 7 ETM+ reflective solar bands using Railroad Valley Playa, Nevada (RVPN) test site. *Proceedings of IEEE Conference on Remote Sensing: The Next Generation*, Boston, Massachusetts. IEEE, Piscataway, New Jersey, IV, pp. 1364–1367.

Angal, A., Xiong, X., Choi, T.-Y., Chander, G., and Wu, A. 2010b. Using the Sonoran and Libyan Desert test sites to monitor the temporal stability of reflective solar bands for Landsat 7 Enhanced Thematic Mapper Plus and Terra MODerate-resolution Imaging Spectroradiometer sensors. *Journal of Applied Remote Sensing*, 4(1): 043525-12.

Ashburn, E.V., and Weldon, R.G. 1956. Spectral diffuse reflectance of desert surfaces. *Journal of the Optical Society of America*, 46(8): 583–586.

Bannari, A., Omari, K., Teillet, P.M., and Fedosejevs, G. 2005. Potential of Getis statistics to characterize the radiometric uniformity and stability of test sites used for the calibration of Earth observation sensors. *IEEE Transactions on Geoscience and Remote Sensing*, 43(12): 2918–2926.

Barker, J.L., Ed. 1985. *Proceedings of the Landsat-4 Science Characterization Early Results Symposium*, 22–24 February 1983, NASA Goddard Space Flight Center, Greenbelt, Maryland, NASA Conference Publication 2355.

Barnes, R.A., Eplee, R.E., Jr., Patt, F.S., Kieffer, H.H., Stone, T.C., Meister, G., Butler, J.J., and McClain, C.R. 2004. Comparison of SeaWiFS measurements of the moon with the U.S. geological survey lunar model. *Applied Optics*, 43(31): 5838–5854.

Barsi, J.A., Hook, S.J., Schott, J.R., Raqueno, N.G., and Markham, B.L. 2007. Landsat-5 thematic mapper thermal band calibration update. *Geoscience and Remote Sensing Letters*, 4(4): 552–555.

Barsi, J.A., Schott, J.R., Palluconi, F.D., Helder, D.L., Hook, S.J., Markham, B.L., Chander, G., and O'Donnell, E.M. 2003. Landsat TM and ETM+ thermal band calibration, *Canadian Journal of Remote Sensing*, 29(2): 141–153.

Bartolucci, L.A., and S.M. Davis. 1983. The calibration of Landsat MSS data as an analysis tool. *LARS Technical Reports Paper 71, LARS Technical Report 062283, Proceedings of the 1983 Machine Processing of Remotely Sensed Data Symposium*, Laboratory for Applications of Remote Sensing, West Lafayette, Indiana, pp. 279–287.

Biggar, S.F., Slater, P.N., and Gellman, D.I. 1994. Uncertainties in the in-flight calibration of sensors with reference to measured ground sites in the 0.4 to 1.1 μm range. *Remote Sensing of Environment*, 48: 245–252.

Biggar, S.F., Thome, K.J., and Wisniewski, W. 2003. Vicarious radiometric calibration of EO-1 sensors by reference to high-reflectance ground targets. *IEEE Transactions on Geoscience and Remote Sensing*, 41(6): 1174–1179.

Black, S.E., Helder, D.L., and Schiller, S.J. 2003. Irradiance-based cross-calibration of Landsat-5 and Landsat-7 thematic mapper sensors. *International Journal of Remote Sensing*, 24(2): 287–304.

Blumberg, D.G., and Freilikher, V. 2001. Soil water-content and surface roughness retrieval using ERS-2 SAR data in the Negev Desert, Israel. *Journal of Arid Environments*, 49(3): 449–464.

Bouvet, M., and Ramoino, F. 2010. Radiometric intercomparison of AATSR, MERIS, and Aqua MODIS over Dome Concordia (Antarctica). *Canadian Journal of Remote Sensing*, 36(5): 464–473.

Bowker, D.E., and Davis, R.E. 1992. Influence of atmospheric aerosols and desert reflectance properties on satellite radiance measurements. *International Journal of Remote Sensing*, 13(16): 3105–3126.

Brest, C.L., and Rossow, W.B. 1992. Radiometric calibration and monitoring of NOAA AVHRR data for ISCCP. *International Journal of Remote Sensing*, 13(2): 235–273.

Brest, C.L., Rossow, W.B., and Roiter, M.D. 1997. Update of radiance calibrations for ISCCP. *Journal of Atmospheric and Oceanic Technology*, 14(5): 1091–1109.

Brogioni, M., Macelloni, G., and Pampaloni, P. 2006. Temporal and spatial variability of multi-frequency microwave emission from the east Antarctic plateau. *Proceedings of IEEE Conference on Remote Sensing: A Natural Global Partnership*, Denver, Colorado. IEEE, Piscataway, New Jersey, pp. 3820–3823.

Brown, S.W., Flora, S.J., Feinholz, M.E., Yarbrough, M.A., Houlihan, T., Peters, D., Kim, Y.S., Mueller, J., Johnson, B.C., and Clark, D.K. 2007. The Marine Optical BuoY (MOBY) radiometric calibration and uncertainty budget for ocean color satellite sensor vicarious calibration. *Proceedings of SPIE Conference 6744 on Sensors, Systems, and Next-Generation Satellites XI*, Florence, Italy. Edited by R. Meynart, S.P. Neeck, H. Shimoda and S. Habib. SPIE, Bellingham, Washington, pp. 67441M.

Bruegge, C., and Butler, J., Eds. 1996. Special issue on earth observing system calibration. *Journal of Atmospheric and Oceanographic Technology*, 13(2): 273–544.

Bushlin, Y., Ben-Shalom, A., Sheffer, D., Steinman, A., Dimmeler, A., Clement, D., and Strobel, R. 1997. Background properties in arid climates: Measurements and analysis. *Proceedings of SPIE Conference 3062 on Targets and Backgrounds: Characterization and Representation III*, Orlando, Florida. Edited by W.R. Watkins and D. Clement. SPIE, Bellingham, Washington, pp. 311–321.

Butler, J.J., Johnson., B.C., and Barnes, R.A. 2005. The calibration and characterization of Earth remote sensing and environmental monitoring instruments. In *Optical Radiometry, Volume 41, Experimental Methods in the Physical Sciences*. Edited by A.C. Parr, R.U. Datla and J.L. Gardner. Treatise Editors R. Celotta and T. Lucatorto. Elsevier/Academic Press, Cambridge, MA, pp. 453–534.

Butler, J.J., Wanchoo, L., and Truong, L. 2001. CEOS database of worldwide calibration facilities and validation test sites. *Proceedings of SPIE Conference 4169 on Sensors, Systems, and Next-Generation Satellites IV*, Barcelona, Spain. Edited by H. Fujisada, J.B. Lurie, A. Ropertz and K. Weber. SPIE, Bellingham, Washington, pp. 202–208.

Cabot, F. 1997. Proposal for the development of a repository for in-flight calibration of optical sensors over terrestrial targets. *Proceedings of SPIE Conference 3117 on Earth Observing Systems II*, San Diego, Califonia. Edited by W.L. Barnes. SPIE, Bellingham, Washington, pp. 148–155.

Cabot, F., Hagolle, O., Cosnefroy, H., and Briottet, X. 1998. Intercalibration using desertic sites as a reference target. *Proceedings of International Geoscience and Remote Sensing Symposium, 1998*, Seattle, Washington, Vol. 5, pp. 2713–2715.

Cabot, F., Hagolle, O., and Henry, P. 2000. Relative and multitemporal calibration of AVHRR, SeaWiFS, and VEGETATION using POLDER characterization of desert sites. *Proceedings of International Geoscience and Remote Sensing Symposium, 2000*, Honolulu, Hawaii, IEEE, pp. 2188–2190.

Cabot, F., Hagolle, O., Ruffel, C., and Henry, P.J. 1999. Remote sensing data respository for in-flight calibration of optical sensors over terrestrial targets. *Proceedings of SPIE Conference 3750 on Earth Observing Systems IV*, Denver, Colorado. Edited by W.L. Barnes. SPIE, Bellingham, Washington, pp. 514–523.

Cao, C., and Heidinger, A.K. 2002. Inter-comparison of the longwave infrared channels of MODIS and AVHRR/NOAA-16 using simultaneous nadir observations at orbit intersections. *Proceedings of SPIE Conference 4814 on Earth Observing Systems VII*, Seattle, Washington. Edited by W.L. Barnes. SPIE, Bellingham, Washington, pp. 306–316.

Cao, C., Uprety, S., Xiong, X., Wu, A., Jing, P., Smith, D., Chander, G., Fox, N., and Ungar, S. 2010. Establishing the Antarctic Dome C community reference standard site towards consistent measurements from Earth observation satellites. *Canadian Journal of Remote Sensing*, 36(5): 498–513.

Cao, C., Xu, H., Sullivan, J., McMillin, L., Ciren, P., and Hou, Y. 2005. Intersatellite radiance biases for the High Resolution Infrared Radiation Sounders (HIRS) onboard NOAA-15, -16, and -17 from simultaneous nadir observations. *Journal of Atmospheric and Oceanic Technology*, 22(4): 381–395.

Castle, K.R., Holm, R.G., Kastner, C.J., Palmer, J.M., Slater, P.N., Dinguirard, M., Ezra, C.E., Jackson, R.D., and Savage, R.K. 1984. In-flight absolute radiometric calibration of the thematic mapper. *IEEE Transactions on Geoscience and Remote Sensing*, 22(3): 251–255.

Chander, G., Angal, A., Choi, T.J., Meyer, D.J., Xiong, X.J., and Teillet, P.M. 2007c. Cross-calibration of the Terra MODIS, Landsat 7 ETM+ and EO-1 ALI sensors using near-simultaneous surface observation over the Railroad Valley Playa, Nevada, test site. *Proceedings of SPIE Conference 6677 on Earth Observing Systems XII*, San Diego, California. Edited by J.J. Butler and J. Xiong. SPIE, Bellingham, Washington, pp. 66770Y.

Chander, G., Christopherson, J.B., Stensaas, G.L., and Teillet, P.M. 2007d. Online catalogue of worldwide test sites for the postlaunch characterization and calibration of optical sensors. *Proceedings of the 58th International Astronautical Congress*, Hyderabad, India. International Astronautical Federation, pp. 2043–2051.

Chander, G., Helder, D.L., Malla, R., Micijevic, E., and Mettler, C.J. 2007a. Consistency of L4 TM absolute calibration with respect to the L5 TM sensor based on near-simultaneous image acquisition. *Proceedings of SPIE Conference 6677 on Earth Observing Systems XII*, San Diego, California. Edited by J.J. Butler and J. Xiong. SPIE, Bellingham, Washington, pp. 66770F.

Chander, G., Helder, D.L., Markham, B.L., Dewald, J.D., Kaita, E., Thome, K.J., Micijevic, E., and Ruggles, T.A. 2004a. Landsat-5 TM reflective-band absolute radiometric calibration. *IEEE Transactions on Geoscience and Remote Sensing*, 42: 2747–2760.

Chander, G., Hewison, T.J., Fox, N., Wu, X., Xiong, X., and Blackwell, W.J., Guest Editors. 2013a. Special issue on intercalibration of satellite instruments. *IEEE Transactions on Geoscience and Remote Sensing*, 51(3 SI): 491.

Chander, G., Hewison, T.J., Fox, N., Wu, X., Xiong, X., and Blackwell, W.J. 2013b. Overview of intercalibration of satellite instruments. *IEEE Transactions on Geoscience and Remote Sensing*, 51(3 SI): 1056–1080.

Chander, G., and Markham, B.L. 2003. Revised Landsat-5 TM radiometric calibration procedures and postcalibration dynamic ranges. *IEEE Transactions on Geoscience and Remote Sensing*, 41: 2674–2677.

Chander, G., Markham, B.L., and Barsi, J.A. 2007b. Revised Landsat-5 Thematic Mapper radiometric calibration. *IEEE Geoscience and Remote Sensing Letters*, 4: 490–494.

Chander, G., Markham, B.L., and Helder, D.L. 2009. Summary of current radiometric calibration coefficients for Landsat MSS, TM, ETM+, and EO-1 ALI sensors. *Remote Sensing of Environment*, 113: 893–903.

Chander, G., Meyer, D.J., and Helder, D.L. 2004b. Cross calibration of the Landsat-7 ETM+ and EO-1 ALI sensors. *IEEE Transactions on Geoscience and Remote Sensing*, 42(12): 2821–2831.

Chander, G., Mishra, N., Helder, D.L., Aaron, D., Angal, A., Choi, T., Xiong, X., and Doelling, D. 2013c. Applications of Spectral Band Adjustment Factors (SBAF) for cross-calibration. *IEEE Transactions on Geoscience and Remote Sensing*, 51(3 SI): 1267–1281.

Chang, I.L., C. Dean, Z. Li, M. Weinreb, X. Wu, and P.A.V.B. Swamy. 2012. Refined algorithms for star-based monitoring of GOES imager visible channel responsivities. *Proceedings of SPIE Earth Observing Systems XVII*, San Diego, California, pp. 851 00R.

Che, N., and Price, J.C. 1992. Survey of radiometric calibration results and methods for visible and near-infrared channels of NOAA-7, NOAA-9, and NOAA-11 AVHRRs. *Remote Sensing of Environment*, 41(1): 19–27.

Chen, H.S. 1996. *Remote Sensing Calibration Systems: An Introduction*. ISBN 0-937194-38-7, A. Deepak Publishing, Hampton, VA.

Choi, T., Xiong, X., Wang, Z., and D. Link. 2013. Terra and Aqua MODIS on-orbit spectral characterization for reflective solar bands, *Proceedings of SPIE Conference 8724, Ocean Sensing and Monitoring V*, 87240Y, doi:10.1117/12.2016640.

Cihlar, J., and Teillet, P.M. 1995. Forward piecewise linear calibration model for quasi-real-time processing of AVHRR data. *Canadian Journal of Remote Sensing*, 21(1): 22–27.

Clark, D., Gordon, H., Voss, K., Ge, Y., Broenkow, W., and Trees, C. 1997. Validation of atmospheric correction over the oceans. *Journal of Geophysical Research*, 102(D14): 17209–17217.

Cosnefroy, H., Briottet, X., Leroy, M., Lecomte, P., and Santer, R. 1994. In field characterization of Saharan sites reflectances for the calibration of optical satellite sensors. *Proceedings of IEEE Conference on Surface and Atmospheric Remote Sensing: Technologies, Data Analysis and Interpretatio*, Pasadena, California, IEEE, Piscataway, New Jersey, pp. 1500–1502.

Cosnefroy, H., Briottet, X., Leroy, M., Lecomte, P., and Santer, R. 1997. A field experiment in Saharan Algeria for the calibration of optical satellite sensors. *International Journal of Remote Sensing*, 18(16): 3337–3359.

Cosnefroy, H., Leroy, M., and Briottet, X. 1996. Selection and characterization of Saharan and Arabian desert sites for the calibration of optical satellite sensors. *Remote Sensing of Environment*, 58(1): 101–114.

Coulson, K.L., and Jacobowitz, H. 1972. *Proposed Calibration Target for the Visible Channel of a Satellite Radiometer*. NOAA Technical Report NESS 62, UDC 551.507.362.2:551.508.21, National Oceanic and Atmospheric Administration, Washington, D.C., p. 31.

Czapla-Myers, J.S., Thome, K.J., and Buchanan, J.H. 2007. Implication of spatial uniformity on vicarious calibration using automated test sites. *Proceedings of SPIE Conference 6677 on Earth Observing Systems XII*, San Diego, California. Edited by J.J. Butler and J. Xiong. SPIE, Bellingham, Washington, pp. 66770U.

Czapla-Myers, J.S., Thome, K.J., Cocilovo, B.R., McCorkel, J.T., and Buchanan, J.H. 2008. Temporal, spectral, and spatial study of the automated vicarious calibration test site at Railroad Valley, Nevada. *Proceedings of SPIE Conference 7081 on Earth Observing Systems XIII*, San Diego, California. Edited by J.J. Butler and J. Xiong. SPIE, Bellingham, Washington, pp. 70810I.

Czapla-Myers, J.S., Thome, K.J., and Leisso, N.P. 2010. Radiometric calibration of earth-observing sensors using an automated test site at Railroad Valley, Nevada. *Canadian Journal of Remote Sensing*, 36(5): 474–487.

D'Amico, J., Thome, K., and Czapla-Myers, J. 2006. Validation of large-footprint reflectance-based calibration using coincident MODIS and ASTER data. *Proceedings of SPIE Conference 6296 on Earth Observing Systems XI*, San Diego, California. Edited by J.J. Butler and J. Xiong. SPIE, Bellingham, Washington, pp. 629612–629618.

Davis, J.M., and Cox, S.K. 1982. Reflected solar radiances from regional scale scenes. *Journal of Applied Meteorology*, 21(11): 1698–1712.

Desormeaux, Y., Rossow, W.B., Brest, C.L., and Campbell, G.G. 1993. Normalization and calibration of geostationary satellite radiances for the International Satellite Cloud Climatology Project. *Journal of Atmospheric and Oceanic Technology*, 10(3): 304–325.

De Vries, C., Danaher, T., Denham, R., Scarth, P., and Phinn, S. 2007. An operational radiometric calibration procedure for the Landsat sensors based on pseudo-invariant target sites. *Remote Sensing of Environment*, 107(3): 414–429.

Di, L. 2004. A proposed ISO/TC 211 standards project on radiometric calibration of remote sensing data. *Proceedings of the International Workshop on Radiometric and Geometric Calibration: ISPRS Book Series Volume 2, Postlaunch Calibration of Satellite Sensors*, Gulfport, Mississippi. Edited by S.A. Morain and A.M. Budge. Taylor and Francis, London, pp. 53–56.

Dilligeard, E., Briottet, X., Deuze, J.L., and Santer, R.P. 1997. SPOT calibration of blue and green channels using Rayleigh scattering over clear oceans. *Proceedings of SPIE Conference on Advanced Next-Generation Satellites II*, Taormina, Italy, pp. 373–379.

Dinguirard, M., and Slater, P.N. 1999. Calibration of space-multispectral imaging sensors: A review. *Remote Sensing of Environment*, 68(3): 194–205.

Doelling, D.R., Hong, G., Morstad, D., Bhatt, R., Gopalan, A., and Xiong, X. 2010. The characterization of deep convective cloud albedo as a calibration target using MODIS reflectances. *Proceedings of SPIE Conference 7862 on Earth Observing Missions Sensors: Development, Implementation, and Characterization*. Edited by X. Xiong, C. Kim and H. Shimoda. Incheon, Republic of Korea, p. 78620I.

Doelling, D.R., Lukashin, C., Minnis, P., Scarino, B., and Morstad, D. 2012. Spectral reflectance corrections for satellite intercalibrations using SCHIAMACHY data. *IEEE Geoscience and Remote Sensing Letters*, 9(1): 119–123.

Doelling, D.R., Morstad, D., Scarino, B.R., Bhatt, R., and Gopalan, A. 2013. The characterization of deep convective clouds as an invariant calibration target and as a visible calibration technique. *IEEE Transactions on Geoscience and Remote Sensing*, 51(3 SI): 1147–1159.

Doelling, D.R., Nguyen, L., and Minnis, P. 2004. On the use of deep convective clouds to calibrate AVHRR data. *Proceedings of SPIE Conference 5542 on Earth Observing Systems IX*, Denverr, Colorado. Edited by W.L. Barnes and J.J. Butler, pp. 281–289.

Dong, J., Chen, Y., Chen, X., and Xu, Q. 2023. Radiometric cross-calibration of wide-field-of-view cameras based on Gaofen-1/6 satellite synergistic observations using Landsat-8 Operational Land Imager images: A solution for off-nadir wide-field-of-view associated problems. *Remote Sensing*, 15(15): 3851. https://doi.org/10.3390/rs15153851

Duggin, M.J. 1987. Impact of radiance variations on satellite sensor calibration. *Applied Optics*, 26(7): 1264–1271.

Faiman, D., Feuermann, D., Ibbetson, P., Medwed, B., Zemel, A., Ianetz, A., Liubansky, V., Setter, I., and Suraqui, S. 2004. The negev radiation survey. *Journal of Solar Energy Engineering-Transactions of the ASME*, 126(3): 906–914.

Floury, N., Drinkwater, M., and Witasse, O. 2002. L-band brightness temperature of ice sheets in Antarctica: Emission modelling, ionospheric contribution and temporal stability. *Proceedings of IEEE Conference on Remote Sensing: Integrating Our View of the Planet*, Toronto, Ontario. IEEE, Piscataway, New Jersey, pp. 2103–2105.

Fougnie, B., and Bach, R. 2009. Monitoring of radiometric sensitivity changes of space sensors using deep convective clouds: Operational application to PARASOL. *IEEE Transactions on Geoscience and Remote Sensing*, 47(3): 851–861.

Fox, N.P. 1999. Improving the accuracy and traceability of radiometric measurements to SI for remote sensing instrumentation. *Proceedings of the 4th International Airborne Remote Sensing Conference and Exhibition/21st Canadian Symposium on Remote Sensing*. Ottawa, Ontario. ERIM International, I, pp. 304–311.

Fox, N.P., Aiken, J., Barnett, J.J., Briottet, X., Carvell, R., Froehlich, C., Groom, S.B., Hagolle, O., Haigh, J.D., Kieffer, H.H., Lean, J., Pollock, D.B., Quinn, T.J., Sandford, M.C.W., Schaepman, M.E., Shine, K.P., Schmutz, W.K., Teillet, P.M., Thome, K.J., Verstraete, M.M., and Zalewski, E.F. 2002. Traceable radiometry underpinning terrestrial- and Helio-studies (TRUTHS). *Advances in Space Research*, 32(11): 2253–2261.

Fox, N.P., Kaiser-Weiss, A., Schmutz, W., Thome, K., Young, D., Wielicki, B., Winkler, R., and Woolliams, E. 2011. Accurate radiometry from space: An essential tool for climate studies. *Philosophical Transaction of the Royal Society A: Mathematical, Physical, and Engineering Sciences*, 369(1953): 4028–4063.

Fraser, R.S., and Kaufman, Y.J. 1986. Calibration of satellite sensors after launch. *Applied Optics*, 25(7): 1177–1185.

Frouin, R., and Gautier, C. 1987. Calibration of NOAA-7 AVHRR, GOES-5, and GOES-6 VISSR/VAS solar channels. *Remote Sensing of Environment*, 22(1): 73–101.

Gilead, U. 2005. *Locating and Examining Potential Sites for Vicarious Radiometric Calibration of Space Multi-Spectral Imaging Sensors in the Negev Desert*. Ben Gurion University, Beer Sheva, Israel, p. 105.

Gilead, U., and Karnieli, A. 2004. Locating potential vicarious calibration sites for high-spectral resolution sensors in the Israeli Negev Desert by GIS analysis. In *Postlaunch Calibration of Satellite Sensors, ISPRS Book Series Volume 2, Proceedings of the International Workshop on Radiometric and Geometric Calibration*, Gulfport, Mississippi. Edited by S.A. Morain and A.M. Budge. A.A. Balkema Publishers, pp. 181–187.

Govaerts, Y., Sterckx, S., and Adriaensen, S. 2013. Use of simulated reflectances over bright desert target as an absolute calibration reference. *Remote Sensing Letters*, 4(6): 523–531.

Goward, S.N., Arvidson, T., Williams, D.L., Faundeen, J.L., Irons, J.R., and Franks, S. 2006. Historical record of Landsat global coverage-mission operations, NSLRSDA, and international cooperator stations. *Photogrammetric Engineering and Remote Sensing*, 72(10): 1155–1169.

Goward, S.N., and Masek, J.G. Eds. 2001. Special issue on "Landsat 7". *Remote Sensing of Environment*, 78(1–2): 222.

Graetz, R.D., Wilson, M.A., Prata, A.J., Barton, I.J., and Mitchell, R.M. 1994. A Continental Instrumented Ground Site Network (CIGSN, Australia): A prerequisite for the detection, interpretation and quantification of global change. *Proceedings of 1994 International Geoscience and Remote Sensing Symposium (IGARSS'94)*, Pasadena, California. IEEE, Piscataway, New Jersey, pp. 1254–1256.

Gross, H.N., Bruegge, C.J., and Helmlinger, M.C. 2007. Unattended vicarious calibration of a low Earth orbit visible-near infrared sensor. *Proceedings of AIAA Space 2007 Conference*, Long Beach, California. AIAA, Reston, Virginia, pp. 786–793.

Gu, X.F., Guyot, G., and Verbrugghe, M. 1992. Evaluation of measurement errors in ground surface reflectance for satellite calibration. *International Journal of Remote Sensing*, 13(14): 2531–2546.

Guenther, B., Barnes, W., Knight, E., Barker, J., Harnden, J., Weber, R., Roberto, M., Godden, G., Montgomery, H., and Abel, P. 1996. MODIS calibration: A brief review of the strategy for the at-launch calibration approach. *Journal of Atmospheric and Oceanic Technology*, 13(2): 274–285.

Gürol, S., Behnert, I., Özen, H., Deadman, A., Fox, N., and Leloğlu, U.M. 2010. Tuz Gölü: New CEOS reference standard test site for infrared visible optical sensors. *Canadian Journal of Remote Sensing*, 36(5): 553–565.

Gurol, S., Ozen, H., Leloglu, U.M., and Tunali, E. 2008. Tuz Golu: New absolute radiometric calibration test site. *Proceedings of the XXI ISPRS Congress: Silk Road for Information from Imagery*, Beijing, China, pp. 35–40.

Hagolle, O., Nicolas, J.M., Fougnie, B., Cabot, F., and Henry, P. 2004. Absolute calibration of VEGETATION derived from an interband method based on sunglint over ocean sites. *IEEE Transactions on Geoscience and Remote Sensing*, 42(7): 1472–1481.

Helder, D.L. 1996. A radiometric calibration archive for Landsat TM. *Proceedings of the SPIE Conference 2758 on Algorithms for Multispectral and Hyperspectral Imagery*, Orlando, Florida, pp. 273–284.

Helder, D.L., Barker, J., Boncyk, W., and Markham, B.L. 1996. Short term calibration of Landsat TM: Recent findings and suggested techniques. *Proceedings of 1996International Geoscience and Remote Sensing Symposium (IGARSS'96)*, Lincoln, Nebraska, pp. 1286–1289.

Helder, D.L., Basnet, B., and Morstad, D.L. 2010. Optimized identification of worldwide radiometric pseudo-invariant calibration sites. *Canadian Journal of Remote Sensing*, 36(5): 527–539.

Helder, D.L., Boncyk, W., and Morfitt, R. 1997. Landsat TM memory effect characterization and correction. *Canadian Journal of Remote Sensing*, 23(4): 299–308.

Helder, D.L., Boncyk, W., and Morfitt, R. 1998a. Absolute calibration of the Landsat thematic mapper using the internal calibrator. *Proceedings of 1998 International Geoscience and Remote Sensing Symposium (IGARSS'98)*, Seattle, Washington, pp. 2716–2718.

Helder, D.L., Boncyk, W., and Morfitt, R. 1998b. Landsat TM memory effect characterization and correction. *Canadian Journal of Remote Sensing*, 23(4): 299–301.

Helder, D.L., Karki, S., Bhatt, R., Micijevic, E., Aaron, D., and Jasinski, B. 2012a. Radiometric calibration of the Landsat MSS sensor series. *IEEE Transactions on Geoscience and Remote Sensing*, 50(6): 2380–2399.

Helder, D.L., Markham, B.L., Thome, K.J., Barsi, J.A., Chander, G., and Malla, R. 2008. Updated radiometric calibration for the Landsat-5 Thematic Mapper reflective bands. *IEEE Transactions on Geoscience and Remote Sensing*, 46: 3309–3325.

Helder, D.L., and Micijevic, E. 2004. Landsat-5 thematic mapper outgassing effects. *IEEE Transactions on Geoscience and Remote Sensing*, 42: 2717–2729.

Helder, D.L., and Ruggles, T.A. 2004. Landsat thematic mapper reflective-band radiometric artifacts. *IEEE Transactions on Geoscience and Remote Sensing*, 42: 2704–2716.

Helder, D.L., Thome, K., Aaron, D., Leigh, L., Czapla-Myers, J., Leisso, N., Biggar, S., and Anderson, N. 2012b. Recent surface reflectance measurement campaigns with emphasis on best practices: SI traceability and uncertainty estimation. *Metrologia*, 49(2): S21–S28.

Helder, D.L., Thome, K., Mishra, N., Chander, G., Xiong, X., Angal, A., and Choi, T. 2013. Absolute radiometric calibration of Landsat using a Pseudo Invariant Calibration Site. *IEEE Transactions on Geoscience and Remote Sensing*, 51(3 SI): 1360–1369.

Helmlinger, M.C., Bruegge, C.J., Lubka, E.H., and Gross, H.N. 2007. LED spectrometer (LSpec) autonomous vicarious calibration facility. *Proceedings of SPIE Conference 6677 on Earth Observing Systems XII*, San Diego, California. Edited by J.J. Butler and J. Xiong. SPIE, Bellingham, Washington, pp. 66770V.

Henry, P., Chander, G., Fougnie, B., Thomas, C., and Xiong, X. 2013. Assessment of spectral band impact on inter-calibration over desert sites using simulation based on EO-1 Hyperion data. *IEEE Transactions on Geoscience and Remote Sensing*, 51(3 SI): 1297–1308.

Henry, P., Dinguirard, M., and Bidilis, M. 1993. SPOT multitemporal calibration over stable desert areas. *Proceedings of SPIE International Symposium on Aerospace Remote Sensing*, Orlando, Florida, pp. 67–76.

Holben, B.N., Kaufman, Y.J., and Kendall, J.D. 1990. NOAA-11 AVHRR visible and near-IR inflight calibration. *International Journal of Remote Sensing*, 11(8): 1511–1519.

Hovis, W.A., Knoll, J.S., and Smith, G.R. 1985. Aircraft measurements for calibration of an orbiting spacecraft sensor. *Applied Optics*, 24: 407–410.

Hu, X., Liu, J., Sun, L., Rong, Z., Li, Y., Zhang, Y., Zheng, Z., Wu, R., Zhang, L., and Gu, X. 2010. Characterization of CRCS Dunhuang test site and vicarious calibration utilization for Fengyun (FY) series sensors. *Canadian Journal of Remote Sensing*, 36(5): 566–582.

Hu, X., Zhang, Y., Liu, Z., Zhang, G., Huang, Y., Qiu, K., Wang, Y., Zhang, L., Zhu, X., and Rong, Z. 2001. Optical characteristics of China Radiometric Calibration Site for Remote Sensing Satellite Sensors (CRCSRSSS). *Proceeding of SPIE Conference 4151 on Hyperspectral Remote Sensing of the Land and Atmosphere*, Sendai, Japan. Edited by W.L. Smith and Y. Yasuoka. SPIE, Bellingham, Washington, pp. 77–86.

Hu, X.Q., Liu, J.J., Qiu, K.M., Fan, T.X., Zhang, Y.X., Rong, Z.G., and Zhang, L.J. 2009. New method study of sites vicarious calibration for SZ-3/CMODIS. *Spectroscopy and Spectral Analysis*, 29(5): 1153–1159.

Hu, Y., Wielicki, B.A., Yang, P., Stackhouse, P.W., Jr., Lin, B., and Young, D.F. 2004. Application of deep convective cloud albedo observation to satellite-based study of the terrestrial atmosphere: Monitoring the stability of spaceborne measurements and assessing absorption anomaly. *IEEE Transactions on Geoscience and Remote Sensing*, 42(11): 2594–2599.

Iwabuchi, H. 2003. Calibration of the visible and near-infrared channels of NOAA-11 and-14 AVHRRs by using reflections from molecular atmosphere and stratus cloud. *International Journal of Remote Sensing*, 24(24): 5367–5378.

Jackson, R.D. Ed. 1990. Special issue on coincident satellite, aircraft, and field measurements at the Maricopa Agricultural Center (MAC). *Remote Sensing of Environment*, 32(2–3): 77–228.

Kamstrup, N., and Hansen, L.B. 2003. Improved calibration of Landsat-5 TM applicable for high-latitude and dark areas. *International Journal of Remote Sensing*, 24(24): 5345–5365.

Kastner, C.J., and Slater, P.N. 1982. In-flight radiometric calibration of advanced remote sensing systems. *Proceeding of Field Measurement and Calibration Using Electro-Optical Equipment: Issues and Requirements*, San Diego, California. Edited by F.M. Zweibaum and H. Register. SPIE, Bellingham, Washington, pp. 1–8.

Kaufman, Y.J., and Holben, B.N. 1993. Calibration of the AVHRR visible and near-IR bands by atmospheric scattering, ocean glint and desert reflection. *International Journal of Remote Sensing*, 14(1): 21–52.

Kerola, D.X., and Bruegge, C.J. 2009. Desert test site uniformity analysis. *Proceedings of SPIE Conference 7452 on Earth Observing Systems XIV*, San Diego, California. Edited by J.J. Butler, X. Xiong and X. Gu. SPIE, Bellingham, Washington, pp. 74520C–74528.

Kerola, D.X., Bruegge, C.J., Gross, H.N., and Helmlinger, M.C. 2009. On-orbit calibration of the EO-1 Hyperion and Advanced Land Imager (ALI) sensors using the LED spectrometer (LSpec) automated facility. *IEEE Transactions on Geoscience and Remote Sensing*, 47(4): 1244–1255.

Kieffer, H.H., and Wildey, R.L. 1985. Absolute calibration of Landsat instruments using the moon. *Photogrammetric Engineering and Remote Sensing*, 51(9): 1391–1393.

Kieffer, H.H., Stone, T.C., Barnes, R.A., Bender, S., Eplee, R.E., Jr., Mendenhall, J., and Ong, L. 2002. On-orbit radiometric calibration over time and between spacecraft using the moon. *Proceedings of SPIE Conference on Sensors, Systems, and Next-Generation Satellites VI*, Crete, Greece, pp. 287–298.

Kieffer, H.H., and Wildey, R.L. 1996. Establishing the moon as a spectral radiance standard. *Journal of Atmospheric and Oceanic Technology*, 13(2): 360–375.

Kim, W., and Lee, S. 2013. Study on radiometric variability of the Sonoran Desert for vicarious calibration of satellite sensors. *Korean Journal of Remote Sensing*, 29(2): 209–218.

Koepke, P. 1982. Vicarious satellite calibration in the solar spectral range by means of calculated radiances and its application to Meteosat. *Applied Optics*, 21(15): 2845–2854.

Lachérade, S., Fougnie, B., Henry, P., and Gamet, P. 2013. Cross calibration over desert sites: Description, methodology, and operational implementation. *IEEE Transactions on Geoscience and Remote Sensing*, 51(3 SI): 1098–1113.

Le Marshall, J.F., Simpson, J.J., and Jin, Z.H. 1999. Satellite calibration using a collocated nadir observation technique: Theoretical basis and application to the GMS-5 pathfinder benchmark period. *IEEE Transactions on Geoscience and Remote Sensing*, 37(1): 499–507.

Li, Y., Zhang, Y., Liu, J., Rong, Z.G., and Zhang, L.J. 2009. Calibration of the visible and near-infrared channels of the FY-2C/FY-2D GEO meteorological satellite at radiometric site. *Guangxue Xuebao /Acta Optica Sinica*, 29(1): 41–46.

Liu, J.J., Li, Z., Qiao, Y.L., Liu, Y.J., and Zhang, Y.X. 2004. A new method for cross-calibration of two satellite sensors. *International Journal of Remote Sensing*, 25(23): 5267–5281.

Loeb, N.G. 1997. In-flight calibration of NOAA AVHRR visible and near-IR bands over Greenland and Antarctica. *International Journal of Remote Sensing*, 18(3): 477–490.

Luderer, G., Coakley, J.A., Jr., and Tahnk, W.R. 2005. Using sun glint to check the relative calibration of reflected spectral radiances. *Journal of Atmospheric and Oceanic Technology*, 22(10): 1480–1493.

Lukashin, C., Wielicki, B., Young, D., Thome, K., Jin, Z., and Sun, W. 2013. Uncertainty estimates for imager reference inter-calibration with CLARREO reflected solar spectrometer. *IEEE Transactions on Geoscience and Remote Sensing*, 51(3 SI): 1425–1436.

Malila, W.A., and Anderson, D.M. 1986. *Satellite Data Availability and Calibration Documentation for Land Surface Climatology Studies*. NASA/GSFC, Greenbelt, Maryland, Report No. 180300–1-F, p. 214.

Markham, B., Storey, J., and Morfitt, R. 2015. Landsat-8 sensor characterization and calibration. *Remote Sensing*, 7: 2279–2282. https://doi.org/10.3390/rs70302279

Markham, B.L., and Barker, J.L. Eds. 1985. Special issue on Landsat Image Data Quality Assessment (LIDQA). *Photogrammetric Engineering and Remote Sensing*, 51(9): 1245–1493.

Markham, B.L., and Barker, J.L. 1987. Radiometric properties of U.S. processed landsat MSS data. *Remote Sensing of Environment*, 22(1): 39–71. https://doi.org/10.1016/0034-4257(87)90027-7

Markham, B.L., Barker, J.L., Kaita, E., Seiferth, J., and Morfitt, R. 2003. On-orbit performance of the landsat-7 ETM+ radiometric calibrators. *International Journal of Remote Sen sing*, 24(2): 265–286.

Markham, B.L., Barsi, J.A., Helder, D.L., Thome, K.J., and Barker, J.L. 2006. Evaluation of the Landsat-5 TM radiometric calibration history using desert test sites. *Proceedings of SPIE Conference 6361 on Sensors, Systems, and Next-Generation Satellites X*, Stockholm, Sweden. Edited by R. Meynart, S.P. Neeck and H. Shimoda. SPIE, Bellingham, Washington, pp. 63610.

Markham, B.L., Boncyk, W.C., Helder, D.L., and Barker, J.L. 1997. Landsat-7 Enhanced Thematic Mapper Plus radiometric calibration. *Canadian Journal of Remote Sensing*, 23(4): 318–332.

Markham, B.L., Dabney, P.W., Knight, E.J., Kvaran, G., Barsi, J.A., Murphy-Morris, J.E., and Pedelty, J.A. 2010. The Landsat Data Continuity Mission Operational Land Imager (OLI) radiometric calibration. *Proc eedings of IEEE International Geoscience and Remote Sensing Symposium (IGARSS) 2010*, Honolulu, Hawaii, p. 4.

Markham, B.L., Haque, O., Barsi, J.A., Micijevic, E., Helder, D.L., Thome, K., Aaron, D., and Czapla-Myers, J.S. 2012. Landsat-7 ETM+: 12 years on-orbit reflective-band radiometric performance. *IEEE Transactions on Geoscience and Remote Sensing*, 50(5): 2056–2062.

Markham, B.L., and Helder, D.L. 2012. Forty-year calibrated record of earth-reflected radiance from Landsat: A review. *Remote Sensing of Environment*, 122: 30–40.

Markham, B.L., Seiferth, J.C., Smid, J., and Barker, J.L. 1998. Lifetime responsivity behavior of the Landsat-5 Thematic Mapper. *Proceedings of SPIE Conference 3427*, San Diego, California, pp. 420–431.

Markham, B.L., Storey, J.C., Williams, D., and Irons, J. 2004a. Landsat sensor performance: History and current status. *IEEE Transactions on Geoscience and Remote Sensing*, 42(12): 2691–2694.

Markham, B.L., Thome, K., Barsi, J., Kaita, E., Helder, D., Barker, J., and Scaramuzza, P. 2004b. Landsat-7 ETM+ on-orbit reflective-band radiometric stability and absolute calibration. *IEEE Transactions on Geoscience and Remote Sensing*, 42(12): 2810–2820.

Martiny, N., Santer, R., and Smolskaia, I. 2005. Vicarious calibration of MERIS over dark waters in the near infrared. *Remote Sensing of Environment*, 94(4): 475–490.

Matsunaga, T., Nonaka, T., Sawabe, Y., Moriyama, M., Tonooka, H., and Fukasawa, H. 2001. Vicarious and cross calibration methods for satellite thermal infrared sensors using hot ground targets. *Proceedings of IEEE Conference on Scanning the Present and Resolving the Future*, Sydney, Australia. IEEE, Piscataway, New Jersey, pp. 1841–1843.

McCorkel, J., Thome, K., Biggar, S., and Kuester, M. 2006. Radiometric calibration of Advanced Land Imager using reflectance-based results between 2001 and 2005. *Proceedings of SPIE Conference 6296 on Earth Observing Systems XI*, San Diego, California. Edited by J.J. Butler and J. Xiong. SPIE, Bellingham, Washington, p. 62960G.

Mendenhall, J.A., Hearn, D.R., Lencioni, D.E., Digenis, C.J., and Ong, L. 2003. Summary of the EO-1 ALI performance for the first 2.5 years on-orbit. *Proceedings of SPIE Conference 5151 on Earth Observing Systems VIII*, Edited by W.L. Barnes. San Diego, California, pp. 574–585.

Mendenhall, J.A., Lencioni, D.E., and Evans, J.B. 2005. Spectral and radiometric calibration of the Advanced Land Imager. *Lincoln Laboratory Journal*, 15(2): 207–224.

Metzler, M.D., and Malila, W.A. 1985. Characterization and comparison of Landsat-4 and Landsat-5 Thematic Mapper data. *Photogrammetric Engineering and Remote Sensing*, 51(9): 1315–1330.

Meygret, A., Briottet, X., Henry, P.J., and Hagolle, O. 2000. Calibration of SPOT4 HRVIR and vegetation cameras over Rayleigh scattering. *Proceedings of SPIE Conference 4135 on Earth Observing Systems V*, San Diego, California. Edited by W.L. Barnes. SPIE, Bellingham, Washington, pp. 302–313.

Meygret, A., Santer, R., and Berthelot, B. 2011. ROSAS: A robotic station for atmosphere and surface characterization dedicated to on-orbit calibration. *Proceedings of SPIE Conference 8153 on Earth Observing Systems XVI*, 815311: 12.

Miesch, C., Cabot, F., Briottet, X., and Henry, P. 2003. Assimilation method to derive spectral ground reflectance of desert sites from satellite datasets. *Remote Sensing of Environment*, 87(2–3): 359–370.

Mika, A.M. 1997. Three decades of Landsat instruments. *Photogrammetric Engineering and Remote Sensing*, 63(7): 839–852.

Min, X., and Zhu, Y. 1995. Properties of atmospheric aerosol extinction for the satellite radiometric calibration site of Dunhuang, China. *Proceedings of IEEE Conference on Quantitative Remote Sensing for Science and Applications*, Firenze, Italy. IEEE, Piscataway, New Jersey, pp. 1858–1860.

Mitchell, R.M., O'Brien, D.M., Edwards, M., Elsum, C.C., and Graetz, R.D. 1997. Selection and initial characterization of a bright calibration site in the Strzelecki Desert, South Australia. *Canadian Journal of Remote Sensing*, 23(4): 342–353.

Mitchell, R.M., O'Brien, D.M., and Forgan, B.W. 1992. Calibration of the NOAA AVHRR Shortwave channels using split pass imagery: I. Pilot study. *Remote Sensing of Environment*, 40(1): 57–65.

Molineux, C.E., Bliamptis, E.E., and Neal, J.T. 1971. *A Remote-Sensing Investigation of Four Mojave Playas*. Environmental Research Papers, No. 352, AFCRL-71–0235, 16 April 1971. Air Force Cambridge Research Laboratories, Hanscom Field, Bedford, Massachusetts 01730, USA, p. 70.

Morain, S.A., and Budge, A.M., Eds. 2004. *Postlaunch calibration of satellite sensors*, ISPRS Book Series Volume 2. *Proceedings of the International Workshop on Radiometric and Geometric Calibration*, Gulfport, Mississippi. A.A. Balkema Publishers, p. 193.

Moran, M.S., Jackson, R.D., Clarke, T.R., Qi, J., Cabot, F., Thome, K.J., and Markham, B.L. 1995. Reflectance factor retrieval from Landsat TM and SPOT HRV data for bright and dark targets. *Remote Sensing of Environment*, 52(3): 218–230.

Mueller, J.L. 1985. NIMBUS-7 CZCS: Confirmation of its radiometric sensitivity decay-rate through 1982. *Applied Optics*, 24(7): 1043–1047.

Muench, H.S. 1981. *Calibration of geosynchronous satellite video sensors*. Report No. AFGL-TR-81–0050, Air Force Geophysical Laboratory, Hanscom, Massachusetts.

Murphy, J.M. 1984. *Radiometric Correction of Landsat Thematic Mapper Data, CCRS Digital Methods Division Technical Memorandum DMD-TM#84–368*. Canada Centre for Remote Sensing, 588 Booth Street, Ottawa, Ontario, K1A 0Y7, Canada.

Murphy, J.M. 1986. Within-scene radiometric correction of Landsat Thematic Mapper data in Canadian production systems. *Proceedings of SPIE Conference 660 on Earth Remote Sensing Using Landsat Thematic Mapper and SPOT Sensor Systems*, Innsbruck, Austria, pp. 25–31.

Nieke, J., Aoki, T., Tanikawa, T., Motoyoshi, H., and Hori, M. 2004. A satellite cross-calibration experiment. *IEEE Geoscience and Remote Sensing Letters*, 1(3): 215–219.

Nieke, J., Aoki, T., Tanikawa, T., Motoyoshi, H., Hori, M., and Nakajima, Y. 2003. Cross-calibration of satellite sensors over snow fields. *Proceedings of SPIE Conference 5151 on Earth Observing Systems VIII*, San Diego, California. Edited by W. L. Barnes. SPIE Volume 5151, pp. 406–414.

Niro, F., Goryl, P., Dransfeld, S., Boccia, V., Gascon, F., Adams, J., Themann, B., Scifoni, S., and Doxani, G. 2021. European Space Agency (ESA) calibration/validation strategy for Optical Land-Imaging satellites and pathway towards interoperability. *Remote Sensing*, 13(15): 3003. https://doi.org/10.3390/rs13153003

Nithianandam, J., Guenther, B.W., and Allison, L.J. 1993. An anecdotal review of NASA Earth observing satellite remote sensors and radiometric calibration methods. *Metrologia*, 30(4): 207–212.

O'Brien, D.M., and Mitchell, R.M. 2001. An error budget for cross-calibration of AVHRR shortwave channels against ATSR-2. *Remote Sensing of Environment*, 75(2): 216–229.

Padula, F.P., Schott, J.R., Barsi, J.A., Raqueno, N.G., and Hook, S.J. 2010. Calibration of Landsat 5 thermal infrared channel: Updated calibration history and assessment of the errors associated with the methodology. *Canadian Journal of Remote Sensing*, 36(5): 617–630.

Pollock, D.B., Murdock, T.L., Datla, R.U., and Thompson, A. 2003. Data uncertainty traced to SI units: Results reported in the International System of Units. *International Journal of Remote Sensing*, 24(2): 225–235.

Potts, D.R., Mackin, S., Muller, J.-P., and Fox, N. 2013. Sensor intercalibration over Dome C for the ESA GlobAlbedo project. *IEEE Transactions on Geoscience and Remote Sensing*, 51(3 SI): 1139–1146.

Prata, A.J., Cechet, R.P., Grant, I.F., and Rutter, G.F. 1996. The Australian Continental Integrated Ground-truth Site Network (CIGSN): Satellite data calibration and validation, and first results. *Proceedings of the Second SEIKEN Symposium on Global Environmental Monitoring from Space*, Tokyo, Japan. SEIKEN Symposium, pp. 245–261.

Price, J.C. 1987a. Radiometric calibration of satellite sensors in the visible and near-infrared: History and outlook. *Remote Sensing of Environment*, 22(1): 3–9.

Price, J.C., Ed. 1987b. Special issue on radiometric calibration of satellite data. *Remote Sensing of Environment*, 22(1): 1–158.

Rao, C.R.N., and Chen, J. 1995. Intersatellite calibration linkages for the visible and near-infared channels of the Advanced Very High-Resolution Radiometer on the NOAA-7, NOAA-9, and NOAA-11 spacecraft. *International Journal of Remote Sensing*, 16(11): 1931–1942.

Rao, C.R.N., and Chen, J.H. 1996. Postlaunch calibration of the visible and near-infrared channels of the Advanced Very High-Resolution Radiometer on the NOAA-14 spacecraft. *International Journal of Remote Sensing*, 17(14): 2743–2747.

Rao, C.R.N., Cao, C., and Zhang, N. 2003. Inter-calibration of the MODerate-resolution imaging spectroradiometer and the along-track scanning radiometer-2. *International Journal of Remote Sensing*, 24(9): 1913–1924.

Rao, C.R.N., and Chen, J.H. 1999. Revised postlaunch calibration of the visible and near-infrared channels of the Advanced Very High-Resolution Radiometer (AVHRR) on the NOAA-14 spacecraft. *International Journal of Remote Sensing*, 20(18): 3485–3491.

Reuter, D., Irons, J., Lunsford, A., Montanaro, M., Pellerano, F., Richardson, C., Smith, R., Tesfaye, Z., and Thome, K. 2011. Operational Land Imager (OLI) and the Thermal Infrared Sensor (TIRS) on the Landsat Data Continuity Mission (LDCM). *Proceedings of SPIE Conference 8048 on Algorithms and Technologies for Multispectral, Hyperspectral, and Ultraspectral Imagery XVII*. Edited by S.S. Shen and P.E. Lewis. Orlando, Florida, p. 804819.

Richter, R. 1997. On the in-flight absolute calibration of high spatial resolution spaceborne sensors using small ground targets. *International Journal of Remote Sensing*, 18(13): 2827–2833.

Rodger, A.P., Balick, L.K., and Clodius, W.B. 2005. The performance of the Multispectral Thermal Imager (MTI) surface temperature retrieval algorithm at three sites. *IEEE Transactions on Geoscience and Remote Sensing*, 43(3): 658–665.

Rodrigo, J.F., Gil, J., Salvador, P., Gómez, D., Sanz, J., and Casanova, J.L. 2019. Analysis of spatial and temporal variability in Libya-4 with Landsat 8 and Sentinel-2 data for optimized ground target location. *Remote Sensing*, 11(24): 2909. https://doi.org/10.3390/rs11242909

Román, M.O., Justice, C., Paynter, I., Boucher, P.B., Devadiga, S., Endsley, A., Erb, A., Friedl, M., Gao, H., Giglio, L., Gray, J.M., Hall, D., Hulley, G., Kimball, J., Knyazikhin, Y., Lyapustin, A., Myneni, R.B., Noojipady, P., Pu, J., Riggs, G., Sarkar, S., Schaaf, C., Shah, D., Tran, K.H., Vermote, E., Wang, D., Wang, Z., Wu, A., Ye, Y., Shen, Y., Zhang, S., Zhang, S., Zhang, X., Zhao, M., Davidson, C., and Wolfe, R. 2024. Continuity between NASA MODIS collection 6.1 and VIIRS collection 2 land products. *Remote Sensing of Environment*, 302(2024): 113963, ISSN 0034-4257. https://doi.org/10.1016/j.rse.2023.113963. (www.sciencedirect.com/science/article/pii/S0034425723005151)

Rondeaux, G., Steven, M.D., Clark, J.A., and Mackay, G. 1998. La Crau: A European test site for remote sensing validation. *International Journal of Remote Sensing*, 19(14): 2775–2788.

Russell, B.J., Soffer, R.J., Ientilucci, E.J., Kuester, M.A., Conran, D.N., Arroyo-Mora, J.P., Ochoa, T., Durell, C., and Holt, J. 2023. The ground to space CALibration experiment (G-SCALE): Simultaneous validation of UAV, airborne, and satellite imagers for earth observation using specular targets. *Remote Sensing*, 15(2): 294. https://doi.org/10.3390/rs15020294

Salomonson, V.V., and Marlatt, W.E. 1968. Anisotropic solar reflectance over white sand, snow, and stratus clouds. *Journal of Applied Meteorology* 7: 475–483.

Sandford, S.P., Young, D.F., Corliss, J.M., Wielicki, B.A., Gazarik, M.J., Mlynczak, M.G., Little, A.D., Jones, C.D., Speth, P.W., Shick, D.E., Brown, K.E., Thome, K.J., and Hair, J.H. 2010. CLARREO: Cornerstone of the climate observing system measuring decadal change through accurate emitted infrared and reflected solar spectra and radio occultation. *Proceedings of SPIE Conference 7826 on Sensors, Systems, and Next-Generation Satellites XIV*. Edited by R. Meynart, S.P. Neeck and H. Shimoda. Toulouse, France, p. 782611.

Santer, R., Gu, X.F., Guyot, G., Deuze, J.L., Devaux, C., Vermote, E., and Verbrugghe, M. 1992. SPOT calibration at the La-Crau test site (France). *Remote Sensing of Environment*, 41(2–3): 227–237.

Santer, R., Schmectig, C., and Thome, K.J. 1997. BRDF and surface-surround effects on SPOT-HRV vicarious calibration. *Proceedings of SPIE Conference 2957 on Advanced and Next-Generation Satellites II*, Taormina, Italy. Edited by H. Fujisada, G. Calamai and M.N. Sweeting. SPIE, Bellingham, Washington, pp. 344–354.

Schiller, S.J. 2003. Technique for estimating uncertainties in top-of-atmosphere radiances derived by vicarious calibration. *Proceedings of SPIE Conference 5151 on Earth Observing Systems VIII*, San Diego, California. Edited by W.L. Barnes. SPIE, Bellingham, Washington, pp. 502–516.

Schott, J.R., Salvaggio, C., and Volchok, W.J. 1988. Radiometric scene normalization using pseudo-invariant features. *Remote Sensing of Environment*, 26: 1–16.

Schroeder, M., Poutier, L., Muller, R., Dinguirard, M., Reinartz, P., and Briottet, X. 2001. Intercalibration of optical satellites: A case study with MOMS and SPOT. *Aerospace Science and Technology*, 5(4): 305–315.

Scott, K.P., Thome, K.J., and Brownlee, M.R. 1996. Evaluation of Railroad Valley Playa for use in vicarious calibration. *Proceedings of SPIE Conference 2818 on Multispectral Imaging for Terrestrial Applications*, Denver, Colorado. Edited by B. Huberty, J.B. Lurie, J.A. Caylor, P. Coppin and P.C. Robert. SPIE, Bellingham, Washington, pp. 158–166.

Secker, J., Staenz, K., Gauthier, R.P., and Budkewitsch, P. 2001. Vicarious calibration of airborne hyperspectral sensors in operational environments. *Remote Sensing of Environment*, 76(1): 81–92.

Shrestha, M., Leigh, L., and Helder, D. 2019. Classification of North Africa for use as an Extended Pseudo Invariant Calibration Sites (EPICS) for radiometric calibration and stability monitoring of optical satellite sensors. *Remote Sensing*, 11(7): 875. https://doi.org/10.3390/rs11070875

Six, C. 2002. *Automatic Station for the In-Flight Calibration of the Satellite Sensors: Application to the SPOT /HRV on the La Crau test site*. Universite du Littoral-Cote d'Opale, Dunkerque, France, p. 147.

Six, D., Fily, M., Alvain, S., Henry, P., and Benoist, J.P. 2004. Surface characterisation of the Dome Concordia area (Antarctica) as a potential satellite calibration site, using SPOT4/VEGETATION instrument. *Remote Sensing of Environment*, 89(1): 83–94.

Six, D., Fily, M., Blarel, L., and Goloub, P. 2005. First aerosol optical thickness measurements at Dome C (East Antarctica), summer season 2003–2004. *Atmospheric Environment*, 39(28): 5041–5050.

Slater, P.N. 1980. *Remote Sensing, Optics and Optical Systems*. Addison-Wesley Publishing Company, Reading, MA.

Slater, P.N. 1984. The importance and attainment of accurate absolute radiometric calibration. *Proceedings of SPIE Conference 475 on Remote Sensing*, Arlington, Virginia. Edited by P.N. Slater. SPIE, Bellingham, Washington, pp. 34–40.

Slater, P.N. 1985. Radiometric considerations in remote-sensing. *Proceedings of the IEEE*, 73(6): 997–1011.

Slater, P.N. 1986. Variations in in-flight absolute radiometric performance. In *International Satellite Land-Surface Climatology Project (ISLSCP) Conference*, Rome, Italy. European Space Agency, Paris, France, pp. 357–363.

Slater, P.N., and Biggar, S.F. 1996. Suggestions for radiometric calibration coefficient generation. *Journal of Atmospheric and Oceanic Technology*, 13(2): 376–382.

Slater, P.N., Biggar, S.F., Holm, R.G., Jackson, R.D., Mao, Y., Moran, M.S., Palmer, J.M., and Yuan, B. 1987. Reflectance-based and radiance-based methods for the in-flight absolute calibration of multispectral sensors. *Remote Sensing of Environment*, 22(1): 11–37.

Slater, P.N., Biggar, S.F., Palmer, J.M., and Thome, K.J. 2001. Unified approach to absolute radiometric calibration in the solar-reflective range. *Remote Sensing of Environment*, 77(3): 293–303.

Slater, P.N., Biggar, S.F., Thome, K.J., Gellman, D.I., and Spyak, P.R. 1996. Vicarious radiometric calibration of EOS sensors. *Journal of Atmospheric and Oceanographic Technology*, 13(2): 349–359.

Smith, D.L., Mutlow, C.T., and Rao, C.R.N. 2002. Calibration monitoring of the visible and near-infrared channels of the along-track scanning radiometer-2 by use of stable terrestrial sites. *Applied Optics*, 41(3): 515–523.

Smith, G.R. 1984. *Surface Soil Moisture Measurements of the White Sands, New Mexico*. NOAA Technical Report NESDIS 7, PB85–135754, National Oceanic and Atmospheric Administration, Washington, D.C., p. 12.

Smith, G.R., and Levin, R.H. 1985. *High Altitude Measured Radiance of White Sands, New Mexico, in the 400–2000 nm Band Using a Filter Wedge Spectrometer*. NOAA Technical Report NESDIS 21, PB85–206084, National Oceanic and Atmospheric Administration, Washington, D.C., p. 17.

Smith, G.R., Levin, R.H., Abel, P., and Jacobowitz., H. 1988. Calibration of the solar channels and NOAA-9 AVHRR using high altitude aircraft measurements. *Journal of Atmospheric and Oceanic Technology*, 5: 631–639.

Smith, G.R., Levin, R.H., and Knoll, J.S. 1985. *An Atlas of High Altitude Aircraft Measured Radiance of White Sands, New Mexico, in the 450–1050 nm Band*. NOAA Technical Report NESDIS 20, PB85–204501, National Oceanic and Atmospheric Administration, Washington, D.C., p. 29.

Snyder, W.C., Wan, Z.M., Zhang, Y.L., and Feng, Y.Z. 1997. Requirements for satellite land surface temperature validation using a silt playa. *Remote Sensing of Environment*, 61(2): 279–289.

Staylor, W.F. 1986. *Site selection and directional models of deserts used for ERBE validation targets*. NASA Technical Paper 2540, NASA, Washington, D.C.

Staylor, W.F. 1990. Degradation rates of the AVHRR visible channel for the NOAA 6, 7, and 9 spacecraft. *Journal of Atmospheric and Oceanic Technology*, 7(3): 411–423.

Sterckx, S., Brown, I., Kääb, A., Krol, M., Morrow, R., Veefkind, P., Boersma, K.F., De Mazière, M., Fox, N., and Thorne, P. 2020. Towards a European Cal/Val service for earth observation. *International Journal of Remote Sensing*, 41:12, 4496–4511. doi:10.1080/01431161.2020.1718240

Stone, T.C. 2008. Radiometric calibration stability and inter-calibration of solar-band instruments in orbit using the moon. *Proceedings of SPIE Conference on Earth Observing Systems XIII*, San Diego, California, p. 70810X.

Stone, T.C., Kieffer, H.H., and Grant, I.F. 2005. Potential for calibration of geostationary meteorological satellite imagers using the moon, *Proceedings of SPIE Conference on Earth Observing Systems X*, San Diego, California, pp. 1–9.

Stone, T.C., Rossow, W.B., Ferrier, J., and Hinkelmann, L.M. 2013. Evaluation of ISCCP multi-satellite radiance calibration for geostationary imager visible channels using the moon. *IEEE Transactions on Geoscience and Remote Sensing*, 51(3): 1255–1266.

Tahnk, W.R., and Coakley, J.A. 2001. Updated calibration coefficients for NOAA-14 AVHRR channels 1 and 2. *International Journal of Remote Sensing*, 22(15): 3053–3057.

Teillet, P.M. 1986. Image correction for radiometric effects in remote sensing. *International Journal of Remote Sensing*, 7: 1637–1651.

Teillet, P.M. 1997a. A status overview of earth observation calibration/validation for terrestrial applications. *Canadian Journal of Remote Sensing: Special Issue on Calibration/Validation*, 23(4): 291–298.

Teillet, P.M. Ed. 1997b. Special issue on calibration/validation. *Canadian Journal of Remote Sensing*, 23(4): 289–423.

Teillet, P.M. 2014. Overview of satellite image radiometry in the solar-reflective optical domain, Volume 1, Chapter 3. In *The Remote Sensing Handbook* (First Edition). Edited by P.S. Thenkabail. Boca Raton, FL, p. 2304.

Teillet, P.M., Barker, J.L., Markham, B.L., Irish, R.R., Fedosejevs, G., and Storey, J.C. 2001a. Radiometric cross-calibration of the Landsat-7 ETM+ and Landsat-5 TM sensors based on tandem data sets. *Remote Sensing of Environment*, 78(1–2): 39–54.

Teillet, P.M., Barsi, J.A., Chander, G., and Thome, K.J. 2007b. Prime candidate earth targets for the postlaunch radiometric calibration of space-based optical imaging instruments. *Proceedings of the SPIE Conference* 6677 *on Earth Observing Systems XII*, San Diego, CA, USA. Edited by J.J. Butler and J. Xiong. SPIE, Bellingham, Washington, pp. 66770S1–12.

Teillet, P.M., and Chander, G. 2010a. Terrestrial reference standard sites for postlaunch sensor calibration. *Canadian Journal of Remote Sensing*, 36(5): 437–450.

Teillet, P.M., El Saleous, N., Hansen, M.C., Eidenshink, J.C., Justice, C.O., and Townshend, J.R.G. 2000. An evaluation of the global 1-km AVHRR land data set. *International Journal of Remote Sensing*, 21(10): 1987–2021.

Teillet, P.M., Fedosejevs, G., and Gauthier, R.P. 1998a. Operational radiometric calibration of broadscale satellite sensors using hyperspectral airborne remote sensing of prairie rangeland: First trials. *Metrologia*, 35(4): 639–641.

Teillet, P.M., Fedosejevs, G., Gauthier, D., D'Iorio, M.A., Rivard, B., and Budkewitsch, P. 1995. Initial examination of Radar imagery of optical radiometric calibration sites. *Proceedings of SPIE Conference 2583 on Advanced and Next-Generation Satellites*, Paris, France. Edited by H. Fujisada and M.N. Sweeting. SPIE, Bellingham, Washington, pp. 154–165.

Teillet, P.M., Fedosejevs, G., Gauthier, R.P., and Schowengerdt, R.A. 1998b. Uniformity characterization of land test sites used for radiometric calibration of Earth observation sensors. *Proceedings of the Twentieth Canadian Symposium on Remote Sensing*, Calgary, Alberta. Canadian Remote Sensing Society, Ottawa, Canada, pp. 1–4.

Teillet, P.M., Fedosejevs, G., Gauthier, R.P., O'Neill, N.T., Thome, K.J., Biggar, S.F., Ripley, H., and Meygret, A. 2001b. A generalized approach to the vicarious calibration of multiple Earth observation sensors using hyperspectral data. *Remote Sensing of Environment*, 77(3): 304–327.

Teillet, P.M., Fedosejevs, G., and Thome, K.J. 2004b. Spectral band difference effects on radiometric cross-calibration between multiple satellite sensors in the Landsat solar-reflective spectral domain, workshop on inter-comparison of large-scale optical and infrared sensors. *Proceedings of SPIE Conference 5570 on Sensors, Systems, and Next-Generation Satellites VIII*, Edited by R. Meynart, S.P. Neeck, and H. Shimoda, Maspalomas, Canary Islands, Spain, pp. 307–316.

Teillet, P.M., Fedosejevs, G., Thome, K.J., and Barker, J.L. 2007a. Impacts of spectral band difference effects on radiometric cross-calibration between satellite sensors in the solar-reflective spectral domain. *Remote Sensing of Environment*, 110(3): 393–409.

Teillet, P.M., and Fox, N.P. 2009. Concatenation of terrestrial reference standard sites for systematic postlaunch calibration monitoring of multiple space-based imaging sensors. *Proceedings of SPIE Conference 7474 on Sensors, Systems, and Next-Generation Satellites XIII*, Berlin, Germany. Edited by R. Meynart, S.P. Neeck and H. Shimoda. SPIE, Bellingham, Washington, pp. 747410.

Teillet, P.M., Helder, D.L., Ruggles, T.A., Landry, R., Ahern, F.J., Higgs, N.J., Barsi, J., Chander, G., Markham, B.L., Barker, J.L., Thome, K.J., Schott, J.R., and Palluconi, F.D. 2004a. A definitive calibration record for the Landsat-5 Thematic Mapper anchored to the Landsat-7 radiometric scale, *Canadian Journal of Remote Sensing*, 30(4): 631–643.

Teillet, P.M., and Holben, B.N. 1994. Towards operational radiometric calibration of NOAA AVHRR imagery in the visible and near-infrared channels. *Canadian Journal of Remote Sensing*, 20(1): 1–10.

Teillet, P.M., Markham, B.L., and Irish, R.R. 2006. Landsat cross-calibration based on near simultaneous imaging of common ground targets. *Remote Sensing of Environment*, 102(3–4): 264–270.

Teillet, P.M., and Ren, X. 2008. Spectral band difference effects on vegetation indices derived from multiple satellite sensor data. *Canadian Journal of Remote Sensing*, 34(3): 159–173.

Teillet, P.M., Ren, X., and Smith, A.M. 2010b. Suitability of rangeland terrain for satellite remote sensing calibration. *Canadian Journal of Remote Sensing*, 36(5): 451–463.

Teillet, P.M., Slater, P.N., Ding, Y., Santer, R.P., Jackson, R.D., and Moran, M.S. 1990. Three methods for the absolute calibration of the NOAA AVHRR sensors in-flight. *Remote Sensing of Environment*, 31(2): 105–120.

Teillet, P.M., Thome, K.J., Fox, N.P., and Morisette, J.T. 2001c. Earth observation sensor calibration using a global instrumented and automated network of test sites (GIANTS). *Proceedings of SPIE Conference 4550 on Sensors, Systems, and Next-Generation Satellites V*, Toulouse, France. Edited by H. Fujisada, J.B. Lurie and K. Weber. SPIE, Bellingham, Washington, pp. 246–254.

Thome, K.J. 2001. Absolute radiometric calibration of Landsat 7 ETM+ using the reflectance-based method. *Remote Sensing of Environment*, 78(1–2): 27–38.

Thome, K.J. 2005. Sampling and uncertainty issues in trending reflectance-based vicarious calibration results. *Proceedings of SPIE Conference 5882 on Earth Observing Systems X*, San Diego, California. Edited by J.J. Butler. SPIE, Bellingham, Washington, pp. 1–11.

Thome, K.J. 2012. Characterization approaches to place invariant sites on traceable scales. *Proceedings of IEEE International Geoscience and Remote Sensing Symposium (IGARSS) 2012*, pp. 7019–7022.

Thome, K.J., Barnes, R., Baize, R., O'Connell, J., and Hair, J. 2010. Calibration of the reflected solar instrument for the climate absolute radiance and refractivity observatory. *Proceedings of the 2010IEEE International Geoscience and Remote Sensing Symposium (IGARSS)*, Honolulu, Hawaii, pp. 2275–2278.

Thome, K.J., Biggar, S.F., and Wisniewski, W. 2003a. Cross comparison of EO-1 sensors and other earth resources sensors to Landsat-7 ETM+ using Railroad Valley Playa. *IEEE Transactions on Geoscience and Remote Sensing*, 41(6): 1180–1188.

Thome, K.J., Crowther, B.G., and Biggar, S.F. 1997a. Reflectance- and irradiance-based calibration of Landsat-5 thematic mapper. *Canadian Journal of Remote Sensing*, 23(4): 309–317.

Thome, K.J., Czapla-Myers, J., and Biggar, S. 2003b. Vicarious calibration of Aqua and Terra MODIS. *Proceedings of SPIE Conference 5151 on Earth Observing Systems VIII*, San Diego, California. Edited by W.L. Barnes. SPIE, Bellingham, Washington, pp. 395–405.

Thome, K.J., Czapla-Myers, J., and Biggar, J. 2004b. Ground-monitor radiometer system for vicarious calibration. *Proceedings of SPIE Conference 5546 on Imaging Spectrometry X*, Denver, Colorado. Edited by S.S. Shen and P.E. Lewis. SPIE, Bellingham, Washington, pp. 223–232.

Thome, K.J., Czapla-Myers, J., Leisso, N., McCorkel, J., and Buchanan, J. 2008. Intercomparison of imaging sensors using automated ground measurements. *Proceedings of IEEE Conference on Remote Sensing: The Next Generation*, Boston, Massachusetts. IEEE, Piscataway, New Jersey. pp. 1332–1335.

Thome, K.J., Czapla-Myers, J., and McCorkel, J. 2007. Retrieval of surface BRDF for reflectance-based calibration. *Proceedings of SPIE Conference 6677 on Earth Observing Systems XII*, San Diego, CA, USA. Edited by J.J. Butler and J. Xiong. SPIE, Bellingham, Washington, pp. 66770T-66711.

Thome, K.J., D'Amico, J., and Hugon, C. 2006. Intercomparison of Terra ASTER, MISR, and MODIS, and Landsat-7 ETM+. *Proceedings of IEEE Conference on Remote Sensing: A Natural Global Partnership*, Denver, Colorado. IEEE, Piscataway, New Jersey, pp. 1772–1775.

Thome, K.J., Gustafson-Bold, C., Slater, P.N., and Farrand, W.H. 1996. In-flight radiometric calibration of HYDICE using a reflectance-based approach. *Proceedings of SPIE Conference 2821 on Hyperspectral Remote Sensing and Applications*, Denver, Colorado. Edited by S.S. Shen. SPIE, Bellingham, Washington, pp. 311–319.

Thome, K.J., Helder, D.L., Aaron, D., and Dewald, J.D. 2004a. Landsat-5 TM and Landsat-7 ETM+ absolute radiometric calibration using the reflectance-based method. *IEEE Transactions on Geoscience and Remote Sensing*, 42(12): 2777–2785.

Thome, K.J., Markham, B., Barker, J., Slater, P., and Biggar, S. 1997b. Radiometric calibration of Landsat. *Photogrammetric Engineering and Remote Sensing*, 63(7): 853–858.

Thome, K.J., McCorkel, J., and Czapla-Myers, J. 2013. In-situ transfer standard and coincident-view intercomparisons for sensor cross-calibration. *IEEE Transactions on Geoscience and Remote Sensing*, 51 (3 SI): 1088–1097.

Thome, K.J., Schiller, S., Conel, J., Arai, K., and Tsuchida, S. 1998. Results of the 1996 earth observing system vicarious calibration joint campaign at Lunar Lake Playa, Nevada. *Metrologia*, 35(4): 631–638.

Thome, K.J., Smith, N., and Scott, K. 2001. Vicarious calibration of MODIS using Railroad Valley Playa. *Proceedings of IEEE Conference on Scanning the Present and Resolving the Future*, Sydney, Australia. IEEE, Piscataway, New Jersey, pp. 1209–1211.

Thome, K.J., Whittington, E.E., Smith, N., Nandy, P., and Zalewski, E.F. 2000. Ground-reference techniques for the absolute radiometric calibration of MODIS. *Proceedings of SPIE Conference 4135 on Earth Observing Systems V*, San Diego, CA, USA. Edited by W.L. Barnes. SPIE, Bellingham, Washington, pp. 51–59.

Tomasi, C., Petkov, B., Benedetti, E., Valenziano, L., Lupi, A., Vitale, V., and Bonafe, U. 2008. A refined calibration procedure of two-channel sun photometers to measure atmospheric precipitable water at various Antarctic sites. *Journal of Atmospheric and Oceanic Technology*, 25(2): 213–229.

Townshend, J.R.G. 1994. Global data sets for land applications from the advanced very high resolution radiometer: An introduction. *International Journal of Remote Sensing*, 15(17): 3319–3332.

Townshend, J.R.G., Justice, C.O., Skole, D., Malingreau, J.-P., Cihlar, J., Teillet, P., Sadowski, F., and Ruttenberg, S. 1994. The 1-km resolution global data set: Needs of the international geosphere biosphere programme. *International Journal of Remote Sensing*, 15(17): 3417–3441.

Trishchenko, A.P., Cihlar, J., and Li, Z. 2002. Effects of spectral response function on surface reflectance and NDVI measured with moderate resolution satellite sensors. *Remote Sensing of Environment*, 81(1): 1–18.

Valorge, C., Meygret, A., Lebègue, L., Henry, P., Bouillon, A., Gachet, R., Breton, E., Léger, D., and Viallefont, F. 2004. 40 years of experience with SPOT in-flight calibration. *Proceedings of the International Workshop on Radiometric and Geometric Calibration: ISPRS Book Series Volume 2, Postlaunch Calibration of Satellite Sensors*, Gulfport, Mississippi. Edited by S.A. Morain and A.M. Budge. A.A. Balkema Publishers, pp. 119–133.

Vermote, E.F., and Kaufman, Y.J. 1995. Absolute calibration of AVHRR visible and near-infrared channels using ocean and cloud views. *International Journal of Remote Sensing*, 16(13): 2317–2340.

Vermote, E.F., and Saleous, N.Z. 2006. Calibration of NOAA16 AVHRR over a desert site using MODIS data. *Remote Sensing of Environment*, 105(3): 214–220.

Vermote, E.F., Santer, R., Deschamps, P.Y., and Herman, M. 1992. In-flight calibration of large field of view sensors at short wavelengths using Rayleigh scattering. *International Journal of Remote Sensing*, 13(18): 3409–3429.

Villa-Aleman, E., Kurzeja, R.J., and Pendergast, M.M. 2003a. Assessment of Ivanpah Playa as a site for thermal vicarious calibration for the MTI satellite. *Proceedings of SPIE Conference 5093 on Algorithms and Technologies for Multispectral, Hyperspectral, and Ultraspectral Imagery IX*, Orlando, Florida. Edited by S.S. Shen and P.E. Lewis. SPIE, Bellingham, Washington, pp. 331–342.

Villa-Aleman, E., Kurzeja, R.J., and Pendergast, M.M. 2003b. Temporal, spatial, and spectral variability at the Ivanpah Playa vicarious calibration site. *Proceedings of SPIE Conference 5093 on Algorithms and Technologies for Multispectral, Hyperspectral, and Ultraspectral Imagery IX*, Orlando, Florida. Edited by S.S. Shen and P.E. Lewis. SPIE, Bellingham, Washington, pp. 320–330.

Voskanian, N., Thome, K., Wenny, B.N., Tahersima, M.H. and Yarahmadi, M. 2023. Combining RadCalNet sites for radiometric cross calibration of Landsat 9 and Landsat 8 operational land imagers (OLIs). *Remote Sensing*, 15(24): 5752. https://doi.org/10.3390/rs15245752

Wang, Y., Czapla-Myers, J., Lyapustin, A., Thome, K., and Dutton, E. 2011. AERONET-based surface reflectance validation network (ASRVN) data evaluation: Case study for Railroad Valley calibration site. *Remote Sensing of Environment*, 115(10): 2710–2717.

Warren, S.G., Brandt, R.E., and Hinton, P.O. 1998. Effect of surface roughness on bidirectional reflectance of Antarctic snow. *Journal of Geophysical Research-Planets*, 103(E11): 25789–25807.

Wenny, B.N., and Xiong, X. 2008. Using a cold earth surface target to characterize long-term stability of the MODIS thermal emissive bands. *IEEE Geoscience and Remote Sensing Letters*, 5(2): 162–165.

Wenny, B.N., Xiong, X., and Dodd, J. 2009. MODIS thermal emissive band calibration stability derived from surface targets. *Proceedings of SPIE Conference 7474 on Sensors, Systems, and Next-Generation Satellites XIII*, Berlin, Germany. Edited by R. Meynart, S.P. Neeck and H. Shimoda. SPIE, Bellingham, Washington, pp. 74740W.

Wheeler, R.J., Lecroy, S.R., Whitlock, C.H., Purgold, G.C., and Swanson, J.S. 1994. Surface characteristics for the alkali flats and dunes regions at White-Sands-Missile-Range, New-Mexico. *Remote Sensing of Environment*, 48(2): 181–190.

Whitlock, C.H., Staylor, W.F., Darnell, W.L., Chou, M.-D., Dedieu, G., Deschamps, P.Y., Ellis, J., Gautier, C., Frouin, R., Pinker, R.T., Laslo, I., Rossow, W.B., and Tarpley, D. 1990a. Comparison of surface radiation budget satellite algorithms for downwelled shortwave irradiance with wisconsin FIRE/SRB surface truth data. *Proceedings of the 7th AMS Conference on Atmospheric Radiation*, San Francisco, California, pp. 237–242.

Whitlock, C.H., Staylor, W.F., Suttles, J.T., Smith, G., Levin, R., Frouin, R., Gautier, C., Teillet, P.M., Slater, P.N., Kaufman, Y.J., Holben, B.N., Rossow, W.B., C., B., and LeCroy, S.R. 1990b. AVHRR and VISSR satellite instrument calibration results for both cirrus and marine stratocumulus IFO periods. *Proceedings of FIRE Science Meeting*, Vail, Colorado. NASA Langley Research Center, Hampton, Virginia, pp. 141–146.

Wu, A., Cao, C., and Xiong, X. 2003. Intercomparison of the 11- and 12-μm bands of Terra and Aqua MODIS using NOAA-17 AVHRR. *Proceedings of SPIE Conference 5151 on Earth Observing Systems VIII*, San Diego, California. Edited by W.L. Barnes. SPIE, Bellingham, Washington, pp. 384–394.

Wu, A., Xiong, X., and Cao, C. 2008b. Examination of calibration performance of multiple POS sensors using measurements over the Dome C site in Antarctica. *Proceedings of SPIE Conference 7106 on Sensors, Systems, and Next-Generation Satellites XII*, Cardiff, Wales, United Kingdom. Edited by R. Meynart, S.P. Neeck, H. Shimoda and S. Habib. SPIE, Bellingham, Washington, pp. 71060W.

Wu, A., Xiong, X., and Cao, C. 2008c. Terra and Aqua MODIS intercomparison of three reflective solar bands using AVHRR onboard the NOAA-KLM satellites. *International Journal of Remote Sensing*, 29(7): 1997–2010.

Wu, A., Xiong, X., Cao, C., and Angal, A. 2008a. Monitoring MODIS calibration stability of visible and near-IR bands from observed top-of-atmosphere BRDF-normalized reflectances over Libyan Desert and Antarctic surfaces. *Proceedings of SPIE Conference 7081 on Earth Observing Systems XIII*, San Diego, Califonia. Edited by J.J. Butler and J. Xiong. SPIE, Bellingham, Washington, pp. 708113–708119.

Wu, D., Yin, Y., Wang, Z., Gu, X., Verbrugghe, M., and Guyot, G. 1997. Radiometric Characterisation of Dunhuang Satellite Calibration Test Site (China). *Proceedings of the Seventh International Symposium on Physical Measurements and Signatures in Remote Sensing*, Courchevel, France. Edited by G. Guyot and T. Phulpin. Taylor and Francis, Rotterdam, Balkema. pp. 151–160.

Wu, D., Zhu, Y., Wang, Z., Ge, B., and Yin, Y. 1994. The building of radiometric calibration test site for satellite sensors in China. *Proceedings of the Sixth International Symposium on Physical Measurements and Signatures in Remote Sensing*, Val D'Isere, France. ISPRS, pp. 167–171.

Wu, X., Sullivan, J.T., and Heidinger, A.K. 2010. Operational calibration of the advanced very high resolution radiometer (AVHRR) visible and near-infrared channels. *Canadian Journal of Remote Sensing*, 36(5): 602–616.

Wulder, M.A., and Masek, J.G. Ed. 2012. Landsat legacy special issue. *Remote Sensing of Environment*, 122: 202.

Xiao, Q., Liu, J., Yu, H., and Zhang, H. 2001. Analysis and evaluation of optical uniformity for Dunhuang calibration site by airborne spectrum survey data. *Proceedings of China Remote Sensing Sensors Radiometric Calibration*, Beijing, China. Ocean Press, pp. 136–142.

Xing-Fa, G., Guyot, G., and Verbrugghe, M. 1990. Evaluation of measurement errors on the reflectance of "La Crau," the French SPOT calibration area. *Porceedings of the 10th EARSeL Symposium on New European Systems, Sensors and Applications*, Toulouse, France. Edited by G. Konecny. European Association of Remote Sensing Laboratories, Boulogne-Billancourt, France, pp. 121–133.

Xiong, X., Che, N., Xie, Y., Moyer, D., Barnes, W., Guenther, B., and Salomonson, V. 2006a. Four-years of on-orbit spectral characterization results for Aqua MODIS reflective solar bands. *Proceedings of SPIE Conference 6361, Sensors, Systems, and Next-Generation Satellites X*, Edited by R. Meynart, S.P. Neeck, and H. Shimoda, Stockholm, Sweden, pp. 63610S. doi:10.1117/12.687163

Xiong, X., Choi, T., Che, N., Wang, Z., Dodd, J., Xie, Y., and Barnes, W. 2010a. Results and lessons from a decade of Terra MODIS on-orbit spectral characterization, *Proceedings of SPIE Conference 7862, Earth Observing Missions and Sensors: Development, Implementation, and Characterization*, Edited by X. Xiong, C. Kim, and H. Shimoda, Incheon, Republic of Korea, pp. 78620M. doi:10.1117/12.868930

Xiong, X., Sun, J., and Barnes, W. 2008. Intercomparison of on-orbit calibration consistency between Terra and Aqua MODIS reflective solar bands using the moon. *IEEE Geoscience and Remote Sensing Letters*, 5(4): 778–782.

Xiong, X., Wu, A., Angal, A., and Wenny, B. 2009a. Recent progress on cross-comparison of Terra and Aqua MODIS calibration using Dome C. *Proceddings of SPIE Conference 7474 on Sensors, Systems, and Next-Generation Satellites XIII*, Berlin, Germany. Edited by R. Meynart, S.P. Neeck and H. Shimoda. SPIE, Bellingham, Washington, p. 747411.

Xiong, X., Wu, A., Sun, J., and Wenny, B. 2006b. An overview of intercomparison methodologies for Terra and Aqua MODIS calibration. *Proceedings of SPIE Conference 6296 on Earth Observing Systems XI*, San Diego, California. Edited by J.J. Butler and J. Xiong. SPIE, Bellingham, Washington, pp. 62960C.

Xiong, X.X., Wu, A.S., and Wenny, B.N. 2009b. Using dome C for moderate resolution imaging spectroradiometer calibration stability and consistency. *Journal of Applied Remote Sensing*, 3(1): 033520.

Xiong, X., Wu, A., Wenny, B.N., Choi, J., and Angal, A. 2010b. Progress and lessons from MODIS calibration intercomparison using ground test sites. *Canadian Journal of Remote Sensing*, 36(5): 540–552.

Zhang, Y., Li, Y., Rong, Z.G., Hu, X.Q., Zhang, L.J., and Liu, J.J. 2009. Field measurement of Gobi surface emissivity spectrum at Dunhuang calibration site of China. *Spectroscopy and Spectral Analysis*, 29(5): 1213–1217.

Zhang, Y., Qiu, K., Hu, X., Rong, Z., and Zhang, L. 2004. Vicarious radiometric calibration of satellite FY-1D sensors at visible and near infrared channels. *Acta Meteorologica Sinica*, 18(4): 505–516.

Zhang, Y., Rong, Z., Hu, X., Liu, J., Zhang, L., Li, Y., and Zhang, X. 2008. Field measurement of Gobi surface emissivity using CE312 and infragold board at Dunhuang calibration site of China. *Proceedings of IEEE Conference on Remote Sensing: The Next Generation*, Boston, Massachusetts. IEEE, Piscataway, New Jersey, pp. 358–360.

Zhang, Y., Zhang, G., Liu, Z., Zhang, L., Zhu, S., Rong, Z., and Qiu, K. 2001. Spectral reflectance measurements at the China radiometric calibration test site for the remote sensing satellite sensor. *Acta Meteorologica Sinica*, 15(3): 377–382.

# 8 Remote Sensing Data Normalization

*Rudiger Gens and Jordi Cristóbal Rosselló*

## ACRONYMS AND ABBREVIATIONS

| | |
|---|---|
| 6SV | Second Simulation of a Satellite Signal in the Solar Spectrum Vector |
| ALOS | Advanced Land Observing Satellite |
| ARD | Analysis Ready Data |
| BRDF | Bidirectional Reflectance Distribution Function |
| CEOS | Committee on Earth Observation Satellites |
| DN | Digital number |
| GEE | Google Earth Engine |
| InSAR | Interferometric SAR |
| JAXA | Japan Aerospace Exploration Agency |
| LSE | Land Surface Emissivity |
| LST | Land Surface Temperature |
| MintPy | Miami InSAR time-series software in Python |
| NDVI | Normalized Difference Vegetation Index |
| SAR | Synthetic Aperture Radar |
| TIR | Thermal Infrared |
| TOA | Top of Atmosphere |

## 8.1 INTRODUCTION

The increasing access to remote sensing data from different platforms, acquired at different spatial, spectral, and temporal resolutions, is continuously widening the scope of applications of these datasets. Parallel advancements in computational science and technology have led to the development of more sophisticated data processing and analysis tools. While early remote sensing studies focused on detecting a feature or phenomenon, the current practice is to conduct multitemporal studies and time series analyses based on multiple data sources, including optical, microwave, and thermal imagery. There are several calibration and normalization issues that need to be resolved before these more complex monitoring and change detection studies can be accomplished.

In this chapter, the terms remote sensing data and remote sensing images are used interchangeably. To be able to use data in any combined fashion, each individual dataset needs to have the same reference that allows quantitative comparisons. The most generic term for this is *data normalization* and for remote sensing data is known as *radiometric normalization*. Depending on the data source, the processing and necessary corrections differ. Although the terms and definitions used for remote sensing data from different sources might vary, there are two main categories of radiometric normalization (Bao *et al.*, 2012): absolute normalization (methods based on radiative transfer methods that account for atmospheric, illumination, and sensor differences) and relative normalization (techniques that minimize the effects of changing atmospheric and solar conditions in one or a series of images, relative to a standard image). For optical and thermal imagery, *radiometric corrections* need to be applied to account for atmospheric conditions, solar angle, or sensor view angle


DOI: 10.1201/9781003541141-12

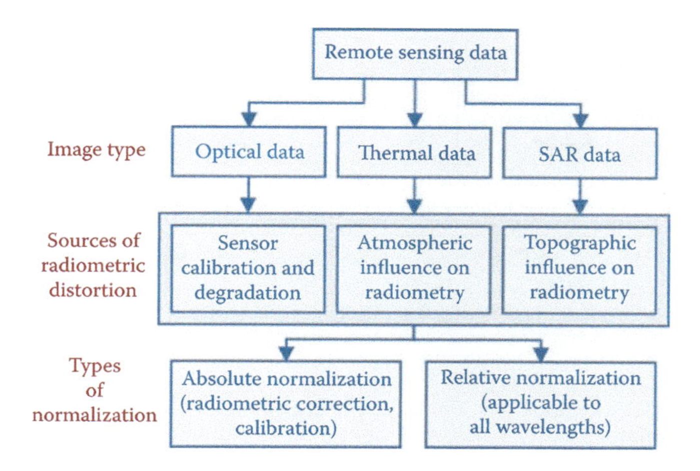


**FIGURE 8.1** Remote sensing data normalization overview for various types of imagery.

(Chen *et al.*, 2005; Du *et al.*, 2002), in addition to the sensor pre-launch and post-launch calibration. These corrections help to convert the raw signal recorded at the sensor to physically meaningful and measurable values, such as ground reflectance or ground temperatures. The quantitative use of Synthetic Aperture Radar (SAR) data requires *calibrated imagery* (Freeman, 1992). Specifically, the SAR processor used for image generation needs to be calibrated, and the calibration parameters then need to be applied to the data to generate calibrated images.

When combining image datasets, it is assumed that the images are corrected for imaging geometry, that is, optical and thermal imagery have been *orthorectified* (Lillesand *et al.*, 2007) and SAR imagery has been *terrain corrected* (Small, 2011). In the case of SAR data, it should be noted the radiometric terrain correction also corrects the pixel brightness due to geometric distortions (Small, 2011). These geometric corrections are prerequisites for the co-registration of any multitemporal imagery. Image co-registration and radiometric corrections are considered the most important steps in monitoring activities such as change detection (Hussain *et al.*, 2013).

Figure 8.1 summarizes the elements associated with the topic of data normalization covered in this book chapter. The chapter focuses on three image types in the optical, thermal, and microwave spectrum and the approaches for their absolute and relative normalization. Several sources of radiometric distortion, such as sensor calibration and degradation, as well as the atmospheric and topographic influence on the radiometry, need to be considered. Depending on the wavelength region, absolute normalization may involve radiometric correction or calibration, whereas relative normalization applies to all wavelengths.

## 8.2 REMOTE SENSING DATA

Remote sensing data, as addressed in this chapter, refers to images acquired anywhere from the visible to the microwave region of the electromagnetic spectrum (Figure 8.2).

The visible and infrared regions generally range from 0.4 μm through 14 μm in wavelength, with the visible region occupying the shorter wavelength end from 400 nm to 700 nm. In remote sensing literature, there is considerable discrepancy and no general consensus on how to classify and name the infrared portion of the spectrum and where to set the boundaries (Gupta, 2003; Quattrochi *et al.*, 2009). The disagreement stems from the fact that the dominant process operating in the shorter wavelength end of the infrared region can either be reflection (at temperatures close to the ambient temperature of Earth), or it can be emission (when the target is at much higher temperatures) with the amount of energy emitted is dependent on the temperature of the target as guided

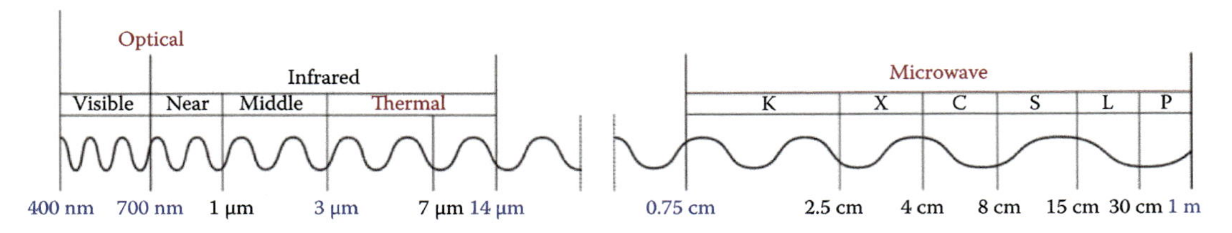


**FIGURE 8.2** Optical, thermal, and SAR data and their respective wavelengths in the electromagnetic spectrum. (Adapted from Gens, 2009.)

by the Planck's function (Prakash and Gens, 2010; Prakash and Gupta, 1999). For data normalization, the logical way to categorize the data is not by the absolute wavelengths but by the dominant physical processes of reflection and emission. Therefore, we classify the data as optical and thermal, where optical data occupy the shorter wavelength portion and the thermal data occupy the longer wavelength portion of this range.

Optical and thermal remote sensing depend on incoming radiation from the Sun and emitted radiation from the Earth's surface, respectively. It does not depend on any external energy source and, therefore, is also classified as passive remote sensing. Passive remote sensing data can be acquired in broad spectral bands, for example, panchromatic visible imagery, as well as in narrower bands as multispectral (spectral bandwidth in the order of 100 nm) and hyperspectral data (spectral bandwidth in the order of about 10 nm) (Gens, 2009).

SAR data are acquired in the microwave region of the electromagnetic spectrum between 0.75 cm and 1 m in wavelength. As the energy is transmitted as a single frequency from the sensor itself, this technique is classified as active remote sensing. The SAR sensor not only transmits energy but also records the energy backscattered from the target. SAR data can be acquired day and night and in all weather conditions (Gens, 2009). Unless used for SAR interferometric processing, the imagery is not affected by atmospheric effects and does not need to be corrected for its influence. SAR signals can be transmitted and received with horizontal and vertical polarization. The various combinations of polarizations have different backscatter behaviors and, therefore, provide additional information complementary to the spectral information. Data acquired in passive mode within the microwave region are not addressed in this chapter. However, readers should refer to Chapter 1 of this volume to get a greater understanding of various sensors and their characteristics.

## 8.3 SOURCES OF RADIOMETRIC DISTORTION

The main sources of radiometric distortions stem from issues related to sensor calibration and sensor degradation, atmospheric interactions, and influences of topographic variations on image radiometry.

### 8.3.1 Sensor Calibration and Degradation

The Committee on Earth Observation Satellites (CEOS) defines sensor calibration as quantitatively defining the system response to known, controlled signal inputs. The purpose of calibration activities is to ensure that the user can retrieve as accurate and meaningful quantitative information from the remote sensing images as possible. All sensors and onboard calibration devices undergo a rigorous pre-launch calibration but need to be routinely recalibrated due to degradation over time (Chander *et al.*, 2009; Wang *et al.*, 2012). Post-launch calibration activities help to correct the onboard calibrators and to verify that the signal response has not drifted away from the original response for the controlled signal inputs. Any error in sensor calibration will propagate and cause errors in quantitative retrievals from remote sensing data. Excellent reviews of pre- and post-launch calibration techniques for optical and thermal systems are available in the literature (e.g., Datla *et al.*, 2011; Schott *et al.*, 2012; Xiong *et al.*, 2009).

Typically, the agency responsible for the satellite launch and operation undertakes the calibration tasks and provides calibration parameters, also called calibration constants, as part of the metadata associated with the image data. An end user then applies the correct calibration constants in the respective radiometric correction algorithms to convert the image digital numbers (DNs) to derived physical parameters such as ground reflectances, ground temperatures, etc. For the sake of brevity, a further discussion on optical and sensor calibration is not presented in this chapter. SAR calibration details are presented in Subsection 8.4.1.2 under absolute radiometric correction.

### 8.3.2 Atmospheric Influence on Radiometry

The atmosphere plays a large role in attenuating the signal recorded by the sensor. Atmospheric particles cause selective scattering, absorption, and emission influencing the signal from the target. The atmosphere intervenes twice in optical images: once when the electromagnetic radiations travel from the source (Sun) to the Earth and the second time when the radiations travel after reflection from the Earth to the sensor. In thermal images, the atmospheric influence comes into play only once as the emitted signal from the Earth travels up to the satellite sensor. The atmosphere is largely transmissive in the microwave region, and for most application purposes, with the exception of applications that rely on SAR interferometric (InSAR) processing; the atmosphere has an insignificant influence on SAR data.

To reduce atmospheric effects in interferometric time series analysis, a processing flow has been implemented in the Miami InSAR time-series software in Python (MintPy) that formulates the time series as a weighted least square inversion (Yunjun *et al.*, 2019). The deterministic phase component is then corrected for the tropospheric phase delay using global atmospheric models (Jolivet *et al.*, 2011, 2014; Li *et al.*, 2009; Onn and Zebker, 2006; Yu *et al.*, 2018) or the delay-elevation ratio (Bekaert *et al.*, 2015; Doin *et al.*, 2009; Lin *et al.*, 2010), as well as the topographic residual and/or the phase ramp. This leads to a noise-reduced displacement time series (Yunjun *et al.*, 2019).

Teasing out the pure signal from the target from a mixed signal response coming from the target and atmosphere then becomes important. Common methods for atmospheric correction in the optical and thermal regions are discussed in Subsection 8.4.1.1.

### 8.3.3 Topographic Influence on Radiometry

Surface topographic variations influence the radiometric response of optical, thermal, and SAR data by influencing the source-target-sensor geometry. The influence is most pronounced in high altitude and high latitude areas, showing high topographic variations.

#### 8.3.3.1 Effect of Topography on Optical Data

Differences in illumination conditions due to solar position at satellite overpass with respect to surface slope and aspect or elevation can produce similar reflectance responses for similar terrain features (Vanonckelen *et al.*, 2013). Accounting for a topographic correction, is then important to calculate surface reflectances accurately, especially in high-relief areas (Hantson and Chuvieco, 2011). According to Pons *et al.* (2014), there are several methodologies that account for topographic effects based on the phenological stage (Riaño *et al.*, 2003; Hantson and Chuvieco, 2011; Meyer *et al.*, 1993; Vincini and Reeder, 2000), surface anisotropy typically described as a function of illumination and observation geometry by the bidirectional reflectance distribution function (BRDF) (Schaepman-Strub *et al.*, 2006), or the cosine topographic correction model (Teillet *et al.*, 1982). The cosine topographic correction is most convenient for automated radiometric correction procedures. Methodologies accounting for the phenological stage require knowledge on the different

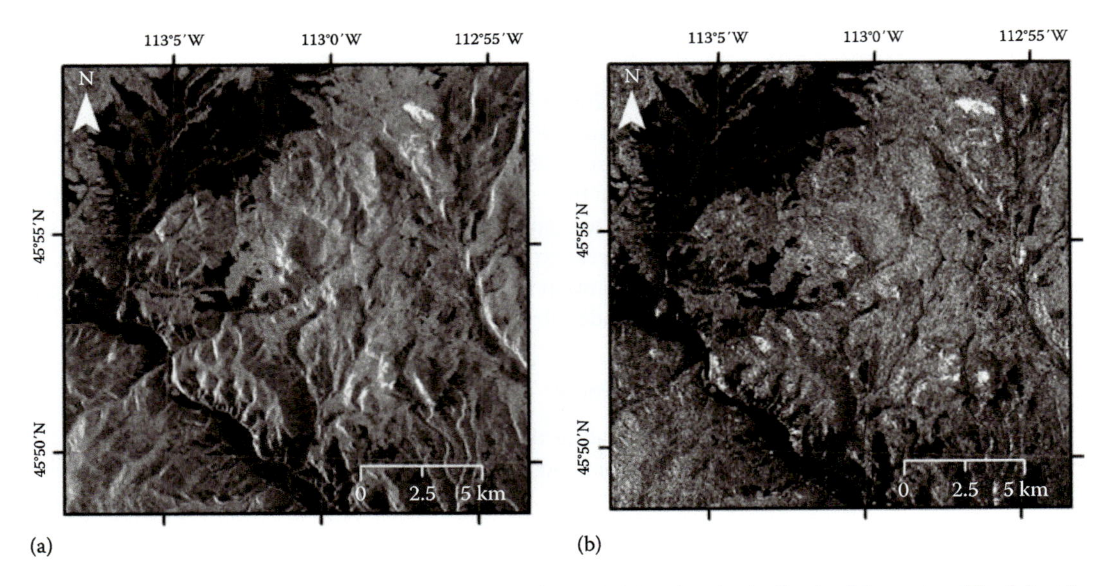


FIGURE 8.3 Example of terrain correction of ALOS PALSAR data in the Rocky Mountains. The left side (a) is the uncorrected version, clearly showing the topographic structure of the mountains. In the image on the right (b) the geometric distortions have been removed, so the topography appears flattened. (Imagery copyright Japan Aerospace Exploration Agency (JAXA) (2006).)

land cover classes or ground reference information, while BRDF models have limitations, as they at times fail to remove angular effects in spectral bands sensitive to water vapor absorption and caused by large seasonal oscillation (Kim *et al.*, 2012). Although, they also need information that is difficult to obtain for regional or long-term studies (Goslee, 2012), it is recommended applying anisotropy compensation for mosaicking, mineral mapping, or the retrieval of plant traits while it not be convenient for physical-based information retrievals (Vögtli *et al.*, 2024).

#### 8.3.3.2 Effect of Topography on SAR Data

SAR systems have a side-looking geometry. This viewing geometry leads to geometric distortions, especially in areas with high topography. The information from slopes facing the sensor is compressed, resulting in brighter pixels; while slopes facing away from the sensor in shadow regions, not covered by any signal at all, appear dark.

This geometric distortion can be corrected by using a digital elevation model. While this technique, called terrain correction, shifts the pixels into the correct geolocation, it is not able to completely recover the radiometric information (as shown in Figure 8.3). By calculating the area that was covered by the signal, the correction factors, later multiplied by the geometrically corrected SAR image, can be determined (Small, 2011). However, the distribution of the individual scatterers and their contribution to the received backscattered signal within this area remains unknown.

## 8.4 RADIOMETRIC NORMALIZATION

There are two types of radiometric normalization: absolute and relative. The absolute radiometric normalization, also known as radiometric correction, helps to derive the absolute reflectance of targets at the Earth's surface. The relative radiometric normalization adjusts the radiometric properties of targets within an image to match a reference image (Janzen *et al.*, 2006; Yuan and Elvidge, 1996). Depending on the application needs, an absolute radiometric correction may not be required

**TABLE 8.1**
**Summary of Absolute and Relative Normalization for the Various Data Types**

| | | Absolute Normalization | | Relative Normalization | |
|---|---|---|---|---|---|
| | | **Technique** | **Effect** | **Technique** | **Effect** |
| Data | Optical<br>Thermal | Radiative transfer methods | Accounts for changes in satellite sensor calibration over time, difference among in-band solar spectral irradiance, solar angle, and atmospheric interferences | Statistical methods using histograms, linear regression, etc. | Minimizes effects of changing atmospheric and solar conditions |
| | SAR | Analysis in homogeneous area with known backscatter | Adjusts the intensity to reference backscatter level | Statistical methods using information from common area | Preserves dominating scattering mechanism |

in all cases. A relative radiometric normalization based on radiometric information in the imagery is sufficient for change detection studies or supervised land cover classifications (Canty *et al.*, 2004). In general, absolute radiometric corrections are required to derive quantitative parameters such as biophysical variables from remote sensing images. For other applications, such as studies focused on relative change from one time to another, relative radiometric normalization may be sufficient. The characteristics of the absolute and relative normalization are summarized in Table 8.1.

### 8.4.1 Absolute Radiometric Correction

Techniques for absolute radiometric correction use in situ measurements or model data to convert from DN to an absolute scale. Residual effects might still require a relative radiometric normalization.

#### 8.4.1.1 Correction of Optical and Thermal Data

Absolute radiometric correction is a two-step process. In a first step, the sensor-specific calibration parameters (gain and offset), determined usually prior to launch, are applied to convert the DNs (dimensionless) into spectral radiance $L_{sat}$ (in W $m^{-2}$ $sr^{-1}$ $\mu m^{-1}$) using the following equation (Chen *et al.*, 2005):

$$L_{sat} = \text{DN} \times \text{Gain} + \text{Offset}$$

$L_{sat}$ can then be converted to top of atmosphere (TOA) reflectances or temperatures. Applying a radiometric correction to retrieve surface reflectance or land surface temperature (kinetic temperature) is needed to be able to analyze long-time series of remote sensing data.

Surface reflectance $\rho_{surface}$ can be determined by:

$$\rho_{surface} = \frac{\left(L_{sat} - L_{path}\right)\pi}{E\,\tau}$$

where $L_{path}$ is the path radiance, $E$ is the exoatmopsheric irradiance on the ground target, and $\tau$ is the transmission of the atmosphere (Lillesand *et al.*, 2007).

Other methodologies to retrieve surface reflectance can be found in Vicente-Serrano (2008). It is important to note that an absolute method should account for changes in satellite sensor calibration over time, differences among in-band solar spectral irradiance, solar angle, and variability in Earth–Sun distance and atmospheric interferences (Carvalho *et al.*, 2013). To remove the atmospheric effects, detailed information on atmospheric parameters such as aerosols, water vapor, or ozone is often required. Local atmospheric measurements or re-analysis data are a common source of atmospheric information. However, atmospheric radiosondes are usually not available at the time of satellite pass and a single atmospheric radiosonde might not be representative of the atmospheric conditions of wide-swath satellite images such as the one provided by Landsat, NOAA-AVHRR or TERRA/AQUA-MODIS sensors, especially in areas with highly variable relief (Cristóbal *et al.*, 2009). AERONET network is another important source of atmospheric data, but as in the case of radiosonde data, ground network distribution might not be wide enough to provide atmospheric parameters over large areas (Themistocleous *et al.*, 2012). Re-analysis data can also provide atmospheric inputs for atmospheric correction; however, its current spatial resolution is still too coarse to be applied to medium or coarse-resolution imagery.

An example of atmospheric correction in the thermal infrared (TIR) using a single-channel algorithm proposed by Pons and Solé-Sugrañes (1994) and Pons *et al.* (2014) using Landsat-5 TM data is shown in Figure 8.4. In remote areas such as Alaska (US), the collection of atmospheric data is challenging because of remoteness, winter conditions, and the high costs of maintaining ground-based measurement sensors (Cristóbal *et al.*, 2012). Therefore, there is a real need for methods requiring few atmospheric inputs. The radiometric correction shown in Figure 8.4, based on the dark object subtraction (extracted from the image histogram) and the cosine topographic correction model (computed through a digital elevation model), allows for the reduction of the number of undesired artifacts due to the atmospheric effects or differential illumination that are results of time of day, location on Earth and relief (zones being more illuminated than others, shadows, etc.), minimizing the effect of these factors on the image data.

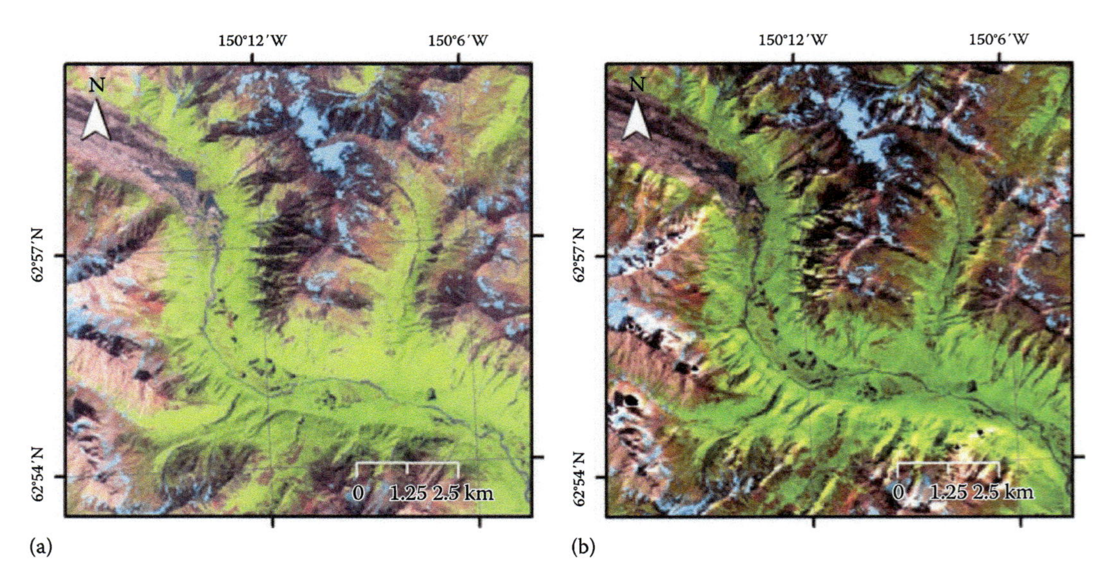


**FIGURE 8.4** Example of radiometric correction of optical data in Alaska (U.S.) from a Landsat-5 TM image of July 7, 2009 (standard false color composite). The left panel (a) is the uncorrected version, clearly showing the topographic structure of the mountains. The right panel (b) is the radiometrically corrected image where topographic and atmospheric effects have been removed. For example, areas with a similar land cover that occupy different sides of a hilly terrain show self-cast shadows on the left panel but present similar reflectivity on the right panel. (Images are courtesy of the U.S. Geological Survey.)

In the case of the TIR region, a suitable and popular procedure to retrieve land surface temperature (LST) is by the inversion of the radiative transfer equation. The following expression can be then applied to a certain sensor channel (or wavelength interval).

$$L_{sensor,\,\lambda} = \left[\varepsilon_\lambda\, B_\lambda\, T_s + \left(1-\varepsilon_\lambda\right) L^{\downarrow}_{atm,\,\lambda}\right] \tau_\lambda + L^{\uparrow}_{atm,\lambda}$$

where $L_{sensor}$ is at-sensor radiance (in W m$^{-2}$ sr$^{-1}$ µm$^{-1}$), ε is land surface emissivity (dimensionless), λ is the wavelength (in µm), $T_s$ is the LST (in K), $L_{atm,}\lambda\downarrow$ is the downwelling atmospheric radiance (hemispherical flux divided by pi and in W m$^{-2}$ sr$^{-1}$ µm$^{-1}$), $L_{atm,}\lambda\uparrow$ is the upwelling atmospheric radiance (path radiance at λ wavelength and in W m$^{-2}$ sr$^{-1}$ µm$^{-1}$), and τ is the atmospheric transmissivity (dimensionless). *B* term is Planck's law, expressed as follows:

$$B_\lambda\left(T_s\right) = -\frac{2\pi hc^2}{\lambda^5\left[\exp\left(\frac{hc}{k\lambda T_s}\right)-1\right]}$$

where *c* is the speed of light (2.998 × 10$^8$ ms$^{-1}$), *h* is Planck's constant (6.626076 × 10$^{-34}$ Js), and *k* is the Boltzmann constant (1.3806 × 10$^{-23}$ JK$^{-1}$). Usually, LST retrieval methods are developed based on the available thermal sensor bands and can be classified in single-channel, split-window or temperature and emissivity separation algorithms. A single-channel algorithm is applied to thermal sensors with only one band in the TIR (Cristóbal *et al.*, 2009, Jiménez-Muñoz and Sobrino, 2003, Cristóbal *et al.*, 2018), such as Landsat-5, Landsat-7, or Landsat-8 (band 10) missions. In the case when two thermal bands are available, such as in NOAA-AVHRR, Landsat-8 or Landsat-9 TIRS missions, split-window algorithms can be applied (Wan and Dozier, 1996; Jimenez-Muñoz *et al.*, 2014). The temperature and emissivity separation algorithm can be applied when three or more bands are available (Gillespie *et al.*, 1998), as in the Terra ASTER or Suomi NPP VIIRS missions. Single-channel and split-window methods also require knowledge of land surface emissivity and water vapor. Land surface emissivity estimates for bare soil and vegetation covers can be retrieved for operational processing using methods based on the normalized difference vegetation index (NDVI) (Valor and Caselles, 2005; Sobrino and Raissouni, 2000; Sobrino *et al.*, 2008). Water vapor can be obtained by means of local radiosounding or at a regional scale by means of remote sensing data (Sobrino *et al.*, 1999), or via image-based water vapor products such as the TERRA/AQUA MODIS product (MOD/MYD_05).

An example of atmospheric correction in the TIR using a single-channel algorithm proposed by Cristóbal *et al.* (2009) using Landsat-5 TM data, water vapor estimates from the MODIS water vapor product, and emissivity retrieved using Sobrino and Raissouni (2000) is shown the in the middle panel of Figure 8.5 (left panel shows a radiometric correction of optical bands to assist the interpretation of the LST image). This method works well in areas where only remote sensing estimates of water vapor products and LSE are available as input for LST estimation. The magnitude to which brightness temperature (TOA temperatures not corrected by atmospheric or emissivity effects) are corrected is shown in the left panel of Figure 8.5. This magnitude difference is computed by subtracting the LST corrected by atmospheric and emissivity effects (mid-panel) and the brightness temperature. In this case, removing atmospheric and emissivity effects leads to a correction around 2 K and 4 K, meaning that brightness temperature is clearly underestimating LST and that absolute radiometric correction methodologies are needed for accurate LST retrievals.

#### 8.4.1.2 SAR Calibration

To calibrate a SAR image, the SAR processor needs to be calibrated, that is, calibration coefficients have been determined. Applying these coefficients to the SAR data calibrates the SAR image. The digital numbers of the original image get converted into a power scale; a ratio of the power that is

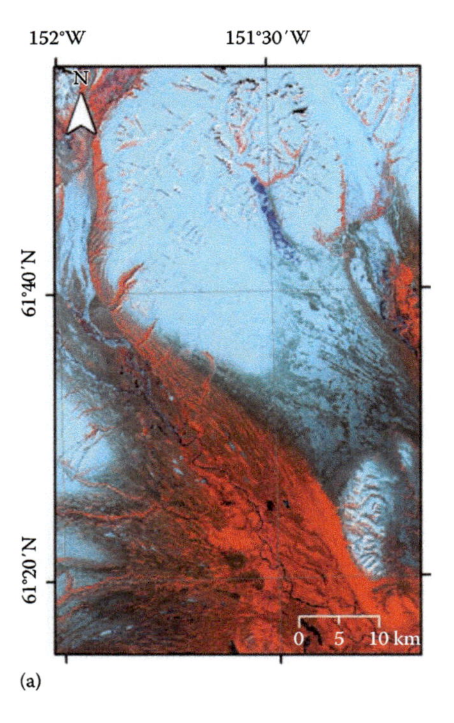


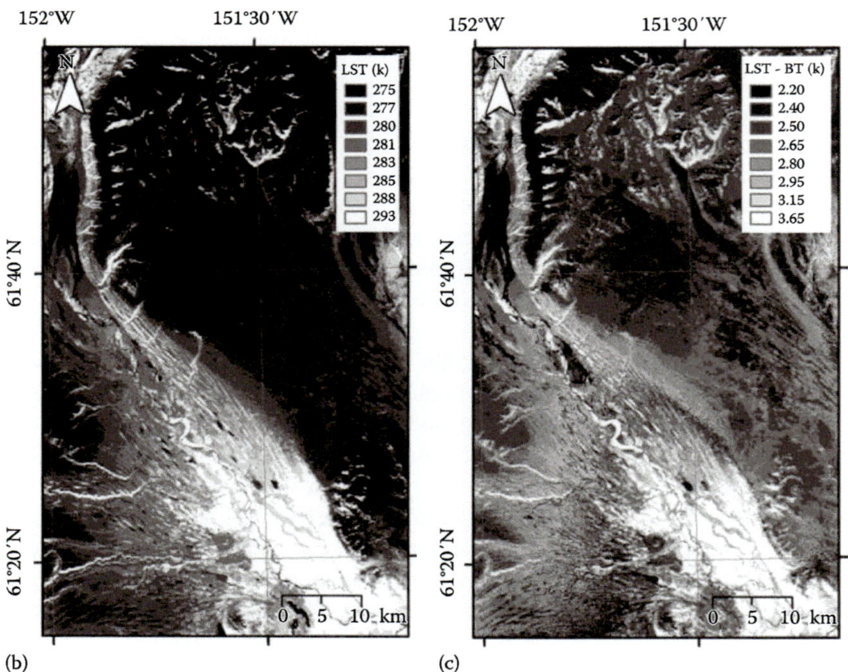


FIGURE 8.5 Example of atmospheric correction of thermal data in Alaska (U.S.) from a Landsat-5 TM image of May 4, 2009 (standard false color composite). The left panel (a) is the optical image after radiometric correction; the center panel (b) is the LST derived after correcting for atmospheric effects and emissivity; and the right panel (c) is the difference image generated by subtracting the LST corrected for atmospheric effects and emissivity and the brightness temperature, BT (not corrected for emissivity or atmospheric effects). (Images are courtesy of the U.S. Geological Survey.)

backscattered to the power sent within a resolution cell. Calibrated images are often transformed from the power scale into the logarithmic dB scale.

To derive geophysical parameters, the calibration needs to meet certain requirements. The absolute calibration needs to be ±1 dB, the long-term relative calibration ±0.5 dB, and the short-term relative calibration better than 0.5 dB (Freeman, 1992).

Another technique that can provide radiometric calibration uses so-called permanent scatterers. These are natural targets in a stack of SAR images that are stable over time and act as corner reflectors with an unknown radar cross-section and a quality that can be estimated by repeated observations (D'Aria *et al.*, 2010). These are also exploited in permanent scatterer interferometry that allows long-term time series analysis (Ferretti *et al.*, 2001).

### 8.4.2 Relative Radiometric Correction

A relative correction transforms the digital numbers to a common scale, adjusting the radiometric properties of an image to match a reference image (de Carvalho *et al.*, 2013).

#### 8.4.2.1 Correction of Optical and Thermal Data

The goal of image normalization for the relative radiometric correction of optical and thermal data is to reduce the radiometric influence of non-surface factors so that the differences in DN between satellite images from different dates will reflect actual changes on the surface of the Earth (Heo and Fitz-Hugh, 2000). Several techniques for relative radiometric correction have been developed in the last decades such as robust and linear regression (El Hajj *et al.*, 2008; Olsson, 1993; Wessman, 1987), histogram matching (Chavez and MacKinnon, 1994; Liang, 2002), the use of invariant areas (Eckhardt *et al.*, 1990; Jensen *et al.*, 1995; Michener and Houhoulis, 1997), the use of pseudo-invariant areas (Bao *et al.*, 2012; Pons *et al.*, 2014; Schott *et al.*, 1988; Zhou *et al.*, 2011), Gaussian method (Singh, 1989), among others. Linear regression, invariant, and pseudo-invariant methods also require an appropriate selection of stable or quasi-stable radiometric areas, areas that can be selected using methods such as multivariate alteration detection (Nielsen *et al.*, 1998; Scheidt *et al.*, 2008). Although there is currently no consensus on what method is more suitable, more objective, and automatic methods with less manual intervention and that take advantage of long time series of remote sensing data, such as those based on the pseudo-invariant areas, are preferred for relative radiometric normalization of optical and thermal imagery.

#### 8.4.2.2 Relative Correction of SAR Data

Most often absolute radiometric correction of SAR data is sufficient for analysis. Sometimes a relative radiometric normalization is merited, especially when adjacent scenes have significant seasonal differences. The normalization of SAR polarimetric data is slightly more complicated, as the correction is not supposed to change the scattering mechanism. For relative radiometric normalization, Shimada and Ohtaki (2010) used a polygonal curve approximation to suppress differences in intensity between neighboring image strips. Antropov *et al.* (2012) extended the relative correlation approach of Shimada and Ohtaki by using the span of the covariance matrix to calculate the corrective gain. There are various ways to apply the radiometric correction in this case. Lee *et al.* (2004) used only those pixels from the overlapping areas, where a dominating scattering mechanism is preserved. However, for several applications where the purpose is to generate a thematically classified product based on relative clustering of backscatter values within an image, simpler radiometric normalization techniques, such as color balance or histogram matching, used popularly in optical remote sensing can prove to be equally efficient.

Figure 8.6a shows an example mosaic from the northern foothills of the Alaska Range generated using two adjacent SAR polarimetric images from different seasons. The images were terrain corrected prior to generating the mosaic. A general brightness contrast is visible across the image boundary (the red dashed line in Figure 8.6b). Note that the boundary is not straight because of

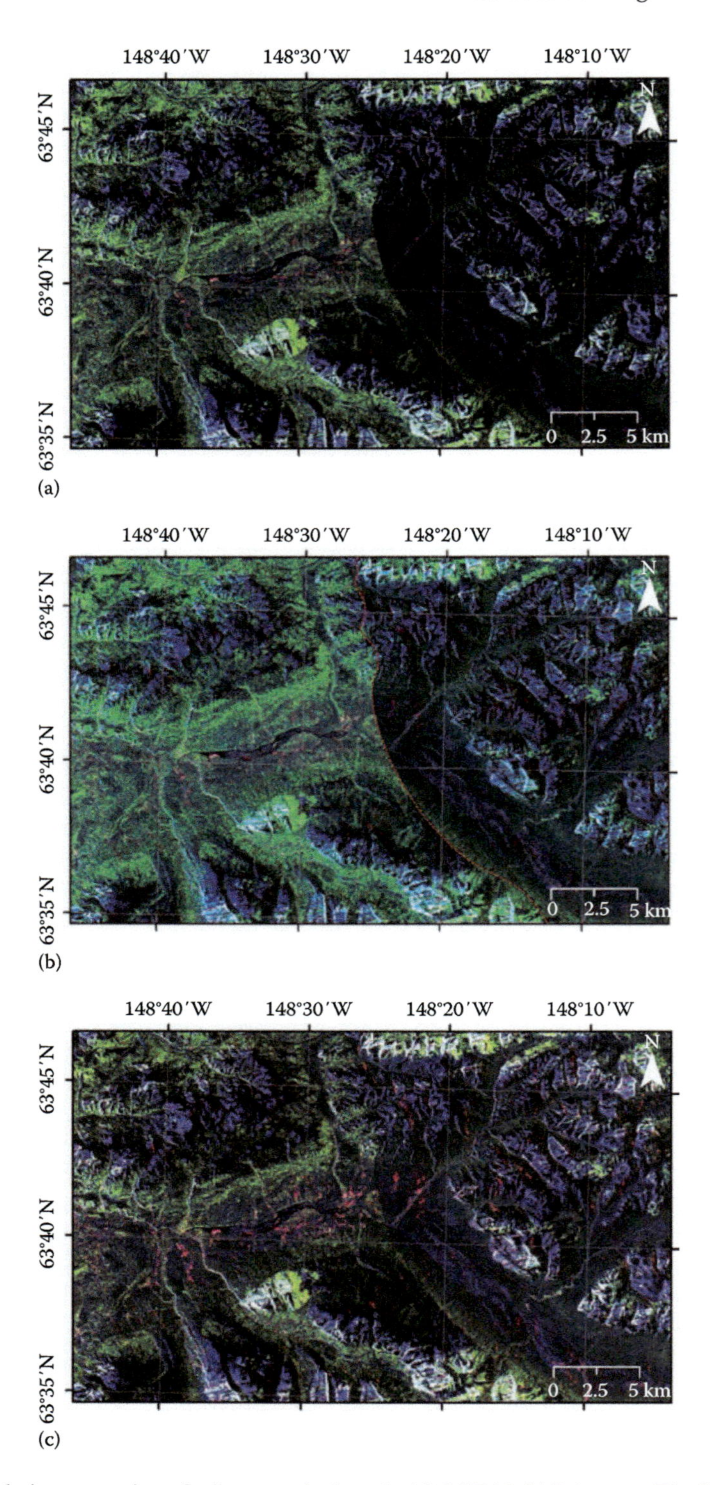


**FIGURE 8.6** Relative correction of adjacent polarimetric ALOS PALSAR images. The Yamaguchi decompositions have been terrain corrected. The first image, (a), is the uncorrected mosaic. In the second image, (b), the boundary between the two images is indicated (not a straight line because of the terrain correction). The last mosaic, (c), shows the relative corrected mosaic. (Imagery copyright Japan Aerospace Exploration Agency (JAXA) (2007).)

terrain correction. Relative radiometric normalization using a histogram match of the overlapping area was applied to the terrain corrected image pair. The mosaic generated using these normalized images yielded a superior product (Figure 8.6c) for mapping and classification applications.

## 8.5 ANALYSIS READY RADIOMETRIC DATASETS

In the last decade, several initiatives started to establish more homogeneous, harmonized, and accurate radiometric datasets based on standard corrections or by application-specific processing use radiative transfer models together with atmospheric characterization data and image-based aerosol retrieval methods (Doxani *et al.*, 2023). Moreover, currently there is a trend of generating Analysis Ready Data (ARD) including both high accurate geometric and radiometric datasets (Dwyer *et al.*, 2018, Frantz, 2019, Potapov *et al.*, 2020).

An ARD example is the freely available Landsat Collection 2 Level 2 surface reflectance and products since 1982 for United States (https://earthexplorer.usgs.gov) that provides both greater radiometric consistency and improved geolocation needed for time series analyses. In this product, surface reflectance is derived with global coverage for daytime images (solar zenith <76°) using algorithms based on the Second Simulation of a Satellite Signal in the Solar Spectrum Vector (6SV) radiative transfer code (Crawford *et al.*, 2023). Another example is the Normalised Radar Backscatter product for Land for Copernicus Sentinel-1 that is designed to be a high-quality ARD compliant with the CEOS ARD for land (https://ceos.org/ard/files/PFS/NRB/v5.5/CARD4L-PFS_NRB_v5.5.pdf).

Finally, ARD products can be also easily accessed through cloud computing platform such as Google Earth Engine (GEE) that effectively address the challenges of big data analysis data over large areas and monitoring the environment for long periods of time (Ghorbanian *et al.*, 2020).

## 8.6 CONCLUSIONS

In general, absolute radiometric correction methods are required to derive quantitative parameters such as biophysical variables from remote sensing images. However, relative radiometric normalization may be sufficient for other applications, such as studies focused on relative change from one time to another. In the case of absolute radiometric correction methods, removing atmospheric effects often requires detailed information on atmospheric parameters such as aerosols, water vapor, or ozone, which are not often available at regional scales. Therefore, absolute radiometric correction methods that minimize input variables are preferred for optical and thermal imagery.

For their study on the radiometric cross-calibration of Landsat sensors, Teillet *et al.* (2001) concluded that the most limiting factor of their approach is the need to adjust for spectral band differences between the two sensors, introducing a dependency on knowledge mainly about the surface reflectance spectrum of the scene. Particularly, the cross-calibration of the shortwave infrared bands remains an issue without these spectra (Teillet *et al.*, 2001).

## REFERENCES

Antropov, O., Rauste, Y., Lonnqvist, A. & Hame, T. 2012. PolSAR mosaic normalization for improved land-cover mapping. IEEE Geoscience and Remote Sensing Letters, 9, 1074–1078. Doi: 10.1109/LGRS.2012.2190263

Bao, N. S., Lechner, A. M., Fletcher, A., Mellor, A., Mulligan, D. & Bai, Z. K. 2012. Comparison of relative radiometric normalization methods using pseudo-invariant features for change detection studies in rural and urban landscapes. Journal of Applied Remote Sensing, 6, 063578–063571. Doi: 10.1117/1.JRS.6.063578

Bekaert, D. P. S., Hooper, A. & Wright, T. J. 2015. A spatially-variable power-law tropospheric correction technique for InSAR data. Journal of Geophysical Research—Solid Earth, 120, 1345–1356. Doi: 10.1002/2014JB011558

Canty, M. J., Nielsen, A. A. & Schmidt, M. 2004. Automatic radiometric normalization of multitemporal satellite imagery. Remote Sensing of Environment, 91, 441–451. Doi: 10.1016/j.rse.2003.10.024

Chander, G., Markham, B. L. & Helder, D. L. 2009. Summary of current radiometric calibration coefficients for Landsat MSS, TM ETM+ and EO-1 ALI sensors. Remote Sensing of Environment, 113, 893–903. Doi: 10.1016/j.rse.2009.01.007

Chavez, P. S. & MacKinnon, D. L. 1994. Automatic detection of vegetation changes in the southwestern United States using remotely sensed images. Photogrammetric Engineering and Remote Sensing, 60, 571–583.

Chen, X. X., Vierling, L. & Deering, D. 2005. A simple and effective radiometric correction method to improve landscape change detection across sensors and across time. Remote Sensing of Environment, 98, 63–79. Doi: 10.1016/j.rse.2005.05.021

Crawford, C. J., Roy, D. P., Arab, S., Barnes, C., Vermote, E., Hulley, G., Gerace, A., Choate, M., Engebretson, C., Micijevic, E., Schmidt, G., Anderson, C., Anderson, M., Bouchard, M., Cook, B., Dittmeier, R., Howard, D., Jenkerson, C., Kim, M., Kleyians, T., Maiersperger, T., Mueller, C., Neigh, C., Owen, L., Page, B., Pahlevan, N, Rengarajan, R., Roger, J-C., Sayler, K., Scaramuzza, P., Skakun, S., Yan, L., Zhang, H. K., Zhu, Z. & Zahn, S. 2023. The 50-year Landsat collection 2 archive. Science of Remote Sensing, 8, 100103. Doi: 10.1016/j.srs.2023.100103

Cristóbal, J., Jiménez-Muñoz, J. C., Prakash, A., Mattar, C., Skoković, D. & Sobrino, J. A. 2018. An improved single-channel method to retrieve land surface temperature from the Landsat-8 thermal band. Remote Sensing, 10, 431. Doi: 10.3390/rs10030431

Cristóbal, J., Jiménez-Muñoz, J. C., Sobrino, J. A., Ninyerola, M. & Pons, X. 2009. Improvements in land surface temperature retrieval from the Landsat series thermal band using water vapor and air temperature. Journal of Geophysical Research—Atmospheres, 114. Doi: 10.1029/2008JD010616

Cristóbal, J., Prakash, A., Starkenburg, D, Fochesatto, J., Anderson, M. A., Kustas, W. P., Alfieri, J. G., Gens, R. & Kane, D. 2012. Energy fluxes retrieval on an Alaskan Arctic and Sub-Arctic vegetation by means MODIS imagery and the DTD method. AGU Fall meeting. December 3–7, San Francisco, USA.

D'Aria, D., Ferretti, A., Guarnieri, A. M. & Tebaldini, S. 2010. SAR calibration aided by permanent scatterers. IEEE Transactions on Geoscience and Remote Sensing, 48, 2076–2086. Doi: 10.1109/TGRS.2009.2033672

Datla, R. U., Rice, J. P., Lykke, K. R., Johnson, B. C., Butler, J. J. & Xiong, X. 2011. Best practice guidelines for pre-launch characterization and calibration of instruments for passive optical remote sensing. Journal of Research of the National Institute of Standards and Technology, 116, 621–646. Doi: 10.6028/jres.116.009

De Carvalho, O. A., Guimaraes, R. F., Silva, N. C., Gillespie, A. R., Gomes, R. A. T., Silva, C. R. & De Carvalho, A. P. F. 2013. Radiometric normalization of temporal images combining automatic detection of pseudo-invariant features from the distance and similarity spectral measures, density scatterplot analysis, and robust regression. Remote Sensing, 5, 2763–2794. Doi: 10.3390/rs5062763

Doin, M. P., Lasserre, C., Peltzer, G., Cavalié, O. & Doubre, C. 2009. Corrections of stratified tropospheric delays in SAR interferometry: Validation with global atmospheric models. Journal of Applied Geophysics, 69, 35–50. Doi:10.1016/j.jappgeo.2009.03.010

Doxani, G., Vermote, E. F., Roger, J-C, Skakun, S., Gascon, F., Collison, A., De Keukelaere, L., Desjardins, C., Frantz, D., Hagolle, O., Kim, M., Louis, J., Pacifici, F., Pflug, B., Poilvé, H., Ramon, D., Richter, R. & Yin, F. 2023. Atmospheric correction inter-comparison exercise, ACIX-II land: An assessment of atmospheric correction processors for Landsat 8 and Sentinel-2 over land. Remote Sensing of Environment, 285, 113412. Doi: 10.1016/j.rse.2022.113412

Du, Y., Teillet, P. M. & Cihlar, J. 2002. Radiometric normalization of multitemporal high-resolution satellite images with quality control for land cover change detection. Remote Sensing of Environment, 82, 123–134. Doi: 10.1016/S0034–4257(02)00029–9

Dwyer, J. L., Roy, D. P., Sauer, B., Jenkerson, C. B., Zhang, H. K. & Lymburner, L. 2018. Analysis ready data: Enabling analysis of the Landsat archive. Remote Sensing, 10(9), 1363. Doi: 10.3390/rs10091363

Eckhardt, D. W., Verdin, J. P. & Lyford, G. R. 1990. Automated update of an irrigated lands GIS using SPOT HRV imagery. Photogrammetric Engineering and Remote Sensing, 56, 1515–1522.

El Hajj, M., Bégué, A., Lafrance, B., Hagolle, O., Dedieu, G. & Rumeau, M. 2008. Relative radiometric normalization and atmospheric correction of a SPOT 5 time series. Sensors, 8, 2774–2791. Doi: 10.3390/s8042774

Ferretti, A., Prati, C. & Rocca, F. 2001. Permanent scatterers in SAR interferometry. IEEE Transactions on Geoscience and Remote Sensing, 39, 8–20. Doi: 10.1109/36.898661

Frantz, D. 2019. FORCE—Landsat+ Sentinel-2 analysis ready data and beyond. Remote Sensing, 11(9), 1124. Doi: 10.3390/rs11091124

Freeman, A. 1992. SAR calibration—an overview. IEEE Transactions on Geoscience and Remote Sensing, 30, 1107–1121. Doi: 10.1109/36.193786

Gens, R. 2009. Spectral Information Content of Remote Sensing Imagery. In: Li, D., Shan J. & Gong, J. (eds.) Geospatial Technology for Earth Observation. Springer.

Ghorbanian, A., Ahmadi, S. A., Kakooei, M., Moghimi, A., Mirmazloumi, S. M., Moghaddam, A. H. A., Mahdavi, S., Ghahremanloo, M., Parsian, S., Wu, Q. & Brisco, B. 2020. Google Earth Engine cloud computing platform for remote sensing big data applications: A comprehensive review. IEEE Journal of Selected Topics in Applied Earth Observations and Remote Sensing, 13, 5326–5350. Doi: 10.1109/JSTARS.2020.3021052

Gillespie, A., Rokugawa, S., Matsunaga, T., Cothern, J. S., Hook, S. & Kahle, A. B. 1998. A temperature and emissivity separation algorithm for advanced spaceborne thermal emission and reflection radiometer (ASTER) images. IEEE Transactions on Geoscience and Remote Sensing, 36, 1113–1126. Doi: 10.1109/36.700995

Goslee, S. C. 2012. Topographic corrections of satellite data for regional monitoring. Photogrammetric Engineering and Remote Sensing, 78, 973–981.

Gupta, R. P. 2003. Remote Sensing Geology, 2nd Edition. Springer. 656p., ISBN-13: 978–3540431855.

Hantson, S. & Chuvieco, E. 2011. Evaluation of different topographic correction methods for Landsat imagery. International Journal of Applied Earth Observation and Geoinformation, 13, 691–700. Doi: 10.1016/j.jag.2011.05.001

Heo, J. & FitzHugh, T. W. 2000. A standardized radiometric normalization method for change detection using remotely sensed imagery. Photogrammetric Engineering and Remote Sensing, 66, 173–181.

Hussain, M., Chen, D. M., Cheng, A., Wei, H. & Stanley, D. 2013. Change detection from remotely sensed images: From pixel-based to object-based approaches. ISPRS Journal of Photogrammetry and Remote Sensing, 80, 91–106. Doi: 10.1016/j.isprsjprs.2013.03.006

Janzen, D. T., Fredeen, A. L. & Wheate, R. D. 2006. Radiometric correction techniques and accuracy assessment for Landsat TM data in remote forested regions. Canadian Journal of Remote Sensing, 32, 330–340. Doi: 10.5589/m06–028

Jensen, J. R., Rutchey, K., Koch, M. S. & Narumalani, S. 1995. Inland wetland change detection in the Everglades water conservation area 2A using a time series of normalized remotely sensed data. Photogrammetric Engineering and Remote Sensing, 61, 199–209.

Jiménez-Muñoz, J. C. & Sobrino, J. A. 2003. A generalized single-channel method for retrieving land surface temperature from remote sensing data. Journal of Geophysical Research—Atmospheres, 108. Doi: 10.1029/2003JD003480

Jimenez-Munoz, J. C., Sobrino, J. A., Skokovic, D., Mattar, C. & Cristobal, J. 2014. Land surface temperature retrieval methods from Landsat-8 hermal infrared sensor data. IEEE Geoscience and Remote Sensing Letters, 11, 1840–1843. Doi: 10.1109/LGRS.2014.2312032

Jolivet, R., Agram, P. S., Lin, N. Y., Simons, M., Doin, M. P., Peltzer, G. & Li, Z. 2014. Improving InSAR geodesy using global atmospheric models. Journal of Geophysical Research—Solid Earth, 119, 2324–2341. Doi: 10.1002/2013JB010588

Jolivet, R., Grandin, R., Lasserre, C., Doin, M. P. & Peltzer, G. 2011. Systematic InSAR tropospheric phase delay corrections from global meteorological reanalysis data. Geophysical Research Letters, 38, L17311. Doi:10.1029/2011GL048757

Kim, D. S., Pyeon, M. W., Eo, Y. D., Byun, Y. G. & Kim, Y. I. 2012. Automatic pseudo-invariant feature extraction for the relative radiometric normalization of Hyperion hyperspectral images. GIScience & Remote Sensing, 49, 755–773. Doi: 10.2747/1548–1603.49.5.755

Lee, J. S., Grunes, M. R., Pottier, E. & Ferro-Famil, L. 2004. Unsupervised terrain classification preserving polarimetric scattering characteristics. IEEE Transactions on Geoscience and Remote Sensing, 42, 722–731. Doi: 10.1109/TGRS.2003.819883

Li, Z., Fielding, E., Cross, P. & Preusker, R. 2009. Advanced InSAR atmospheric correction: MERIS/MODIS combination and stacked water vapour models. International Journal of Remote Sensing, 30, 3343–3363. Doi: 10.1080/01431160802562172

Liang, S. 2002. Estimation of Land Surface Biophysical Variables. In: Kong, J. A. (ed.) Quantitative Remote Sensing of Land Surfaces. Wiley. pp. 247–264.

Lillesand, T., Kiefer, R. W. & Chipman, A. 2007. Remote Sensing and Image Interpretation. Wiley.
Lin, Y. N. N., Simons, M., Hetland, E. A., Muse, P. & DiCaprio, C. 2010. A multiscale approach to estimating topographically correlated propagation delays in radar interferograms. Geochemistry, Geophysics, Geosystems, 11, Doi: 10.1029/2010GC003228
Meyer, P., Itten, K. I., Kellenbenberger, T., Sandmeier, S. & Sandmeier, R. 1993. Radiometric corrections of topographically induced effects on Landsat TM data in an alpine environment. ISPRS Journal of Photogrammetry and Remote Sensing, 48, 17–28. Doi: 10.1016/0924–2716(93)90028-L
Michener, W. K. & Houhoulis, P. F. 1997. Detection of vegetation hanges associated with extensive flooding in a forested ecosystem. Photogrammetric Engineering and Remote Sensing, 63, 173–181.
Nielsen, A. A., Conradsen, K. & Simpson, J. J. (1998). Multivariate alteration detection (MAD) and MAF post-processing in multispectral, bitemporal image data: New approaches to change detection studies. Remote Sensing of Environment, 64, 1–19. Doi: 10.1016/S0034–4257(97)00162–4
Olsson, H. 1993. Regression functions for multi-temporal relative calibration of thematic mapper data over Boreal forest. Remote Sensing of Environment, 46, 89–102. Doi: 10.1016/0034–4257(93)90034-U
Onn, F. & Zebker, H. A. 2006. Correction for interferometric synthetic aperture radar atmospheric phase artifacts using time series of zenith wet delay observations from a GPS network. Journal of Geophysical Research—Solid Earth, 111. Doi: 10.1029/2005JB004012
Pons, X., Pesquer, L., Cristóbal, J. & González-Guerrero, O. 2014. Automatic and improved radiometric correction of Landsat imagery using reference values from MODIS surface reflectance images. International Journal of Applied Earth Observation and Geoinformation, 243–254. Doi: 10.1016/j.jag.2014.06.002
Pons, X. & Solé-Sugrañes, L. 1994. A simple radiometric correction model to improve automatic mapping of vegetation from multispectral satellite data. Remote Sensing of Environment, 48, 191–204. Doi: 10.1016/0034–4257(94)90141–4
Potapov, P., Hansen, M. C., Kommareddy, I., Kommareddy, A., Turubanova, S., Pickens, A., Adusei, B., Tyukavina, A. & Ying, Q. 2020. Landsat analysis ready data for global land cover and land cover change mapping. Remote Sensing, 12, 426. Doi: 10.3390/rs12030426
Prakash, A. & Gens, R. 2010. Remote Sensing of Coal Fires. In: Stracher, G. B., Prakash, A. & Sokol, E. V. (eds.) Coal and Peat Fires: A Global Perspective. Volume 1, Coal—Combustion and Geology. Elsevier.
Prakash, A. & Gupta, R. P. 1999. Surface fires in Jharia Coalfield, India—their distribution and estimation of area and temperature from TM data. International Journal of Remote Sensing, 20, 1935–1946. Doi: 10.1080/014311699212281
Quattrochi, D. A., Prakash, A., Evena, M., Wright, R., Hall, D. K., Anderson, M., Kustas, W. P., Allen, R. G., Pagano, T. & Coolbaugh, M. F. 2009. Thermal Remote Sensing: Theory, Sensors, and Applications. In: Jackson, M. (ed) Manual of Remote Sensing 1.1: Earth Observing Platforms & Sensors. ASPRS, 550 p.
Riaño, D., Chuvieco, E., Salas, J. & Aguado, I. 2003. Assessment of different topographic corrections in Landsat-TM data for mapping vegetation types. IEEE Transactions on Geoscience and Remote Sensing, 41, 1056–1061. Doi: 10.1109/TGRS.2003.811693
Schaepman-Strub, G., Schaepman, M., Painter, T., Dangel, S. & Martonchik, J. 2006. Reflectance quantities in optical remote sensing—definitions and case studies. Remote Sensing of Environment, 103(1). 27–42. Doi: 10.1016/j.rse.2006.03.002
Scheidt, S., Ramsey, M. & Lancaster, N. 2008. Radiometric normalization and image mosaic generation of ASTER thermal infrared data: An application to extensive sand sheets and dune fields. Remote Sensing of Environment, 112, 920–933. Doi: 10.1016/j.rse.2007.06.020
Schott, J. R., Hook, S. J., Barsi, J. A., Markham, B. L., Miller, J., Padula, F. P. & Raqueno, N. G. 2012. Thermal infrared radiometric calibration of the entire Landsat 4, 5, and 7 archive (1982–2010). Remote Sensing of Environment, 122, 41–49. Doi: 10.1016/j.rse.2011.07.022
Schott, J. R., Salvaggio, C. & Vochok, W. J. 1988. Radiometric scene normalization using pseudo-invariant features. Remote Sensing of Environment, 26, 1–16. Doi: 10.1016/0034–4257(88)90116–2
Shimada, M. & Ohtaki, T. 2010. Generating large-scale high-quality SAR mosaic datasets: Application to PALSAR data for global monitoring. IEEE Journal of Selected Topics in Applied Earth Observations and Remote Sensing, 3, 637–656. Doi: 10.1109/JSTARS.2010.2077619
Singh, A. 1989. Digital change detection techniques using remotely sensed data. International Journal of Remote Sensing, 10, 989–1103. Doi: 10.1080/01431168908903939
Small, D. 2011. Flattening gamma: Radiometric terrain correction for SAR imagery. IEEE Transactions on Geoscience and Remote Sensing, 49, 3081–3093. Doi: 10.1109/TGRS.2011.2120616

Sobrino, J. A. & Raissouni, N. 2000. Toward remote sensing methods for land cover dynamic monitoring: Application to Morocco. International Journal of Remote Sensing, 21, 353–366. Doi: 10.1080/014311600210876

Sobrino, J. A., Jiménez-Muñoz, J. C., Sòria, G., Romaguera, M., Guanter, L., Moreno, J., Plaza, A. & Martínez, P. 2008. Land surface emissivity retrieval from different VNIR and TIR sensors. IEEE Transactions on Geoscience and Remote Sensing, 46, 316–327. Doi: 10.1109/TGRS.2007.904834

Sobrino, J. A., Raissouni, N., Simarro, J., Nerry, F. & François, P. 1999. Atmospheric water vapour content over land surfaces derived from the AVHRR data. Application to the Iberian Peninsula, IEEE Transactions and Geoscience and Remote Sensing, 37, 1425–1434. Doi: 10.1109/36.763306

Teillet, P. M., Barker, J. L., Markham, B. L., Irish, R. R., Fedosjevs, G. & Storey, J. C. 2001. Radiometric cross-calibration of the Landsat-7 ETM+ and Landsat-5 TM sensors based on tandem data sets. Remote Sensing of Environment, 78, 39–54. Doi: 10.1016/S0034–4257(01)00248–6

Teillet, P. M., Guindon, B. & Goodeonugh, D. G. 1982. On the slope-aspect correction of multispectral scanner data. Canadian Journal of Remote Sensing, 8, 84–106. Doi: 10.1080/07038992.1982.10855028

Themistocleous, K., Hadjimitsis, D. G., Retalis, A. & Chrysoulakis, N. 2012. Development of a new image based atmospheric correction algorithm for aerosol optical thickness retrieval using the darkest pixel method. Journal of Applied Remote Sensing, 6, 063538. Doi: 10.1117/1.JRS.6.063538

Valor, E & Caselles, V. 2005. Validation of the Vegetation Cover Method for Land Surface Emissivity Estimation. In: Caselles, V., Valor, E. & Coll, C. (eds.) Recent Research Developments in Thermal Remote Sensing. Research Signpost, pp. 1–20.

Vanonckelen, S., Lhermitte, S. & Van Rompaey, A. 2013. The effect of atmospheric and topographic correction methods on land cover classification accuracy. International Journal of Applied Earth Observation and Geoinformation, 24, 9–21. Doi: 10.1016/j.jag.2013.02.003

Vicente-Serrano, S. M., Pérez-Cabello, F. & Lasanta, T. 2008. Assessment of radiometric correction techniques in analyzing vegetation variability and change using time series of Landsat images. Remote Sensing of Environment, 112, 3916–3934. Doi: 10.1016/j.rse.2008.06.011

Vincini, M. & Reeder, D. 2000. Minnaert Topographic Normalization of Landsat TM Imagery in Rugged Forest Areas. In: International Archives of Photogrammetry and Remote Sensing. Vol. XXXIII, Part B7. International Society for Photogrammetry and Remote Sensing.

Vögtli, M., Schläpfer, D., Schuman, M. C., Schaepman, M. E., Kneubühler, M. & Damm, A. 2024. Effects of atmospheric, topographic, and BRDF correction on imaging spectroscopy-derived data products. IEEE Journal of Selected Topics in Applied Earth Observations and Remote Sensing, 17, 109–126. Doi: 10.1109/JSTARS.2023.3325926

Wan, Z. & Dozier, J. 1996. A generalized split-window algorithm for retrieving land-surface temperature from space. IEEE Transactions on Geoscience and Remote Sensing, 34, 892–905. Doi: 10.1109/36.508406

Wang, D. D., Morton, D., Masek, J., Wu, A. S., Nagol, J., Xiong, X. X., Levy, R., Vermote, E. & Wolfe, R. 2012. Impact of sensor degradation on the MODIS NDVI time series. Remote Sensing of Environment, 119, 55–61. Doi: 10.1016/j.rse.2011.12.001

Wessman, C. A. 1987. Analysis of Landsat Thematic Mapper imagery over UW arboretum and Blackhawk Island, Ph.D. Dissertation, University of Wisconsin-Madison.

Xiong, X. X., Wenny, B. N. & Barnes, W. L. 2009. Overview of NASA earth observing systems Terra and Aqua moderate resolution imaging spectroradiometer instrument calibration algorithms and on-orbit performance. Journal of Applied Remote Sensing, 3, 032501. Doi: 10.1117/1.3180864

Yu, C., Li, Z. & Penna, N. T. 2018. Interferometric synthetic aperture radar atmospheric correction using a GPS-based iterative tropospheric decomposition model. Remote Sensing of the Environment, 204, 109–121. Doi: 10.1016/j.rse.2017.10.038

Yuan, D. & Elvidge, C. D. 1996. Comparison of relative radiometric normalization techniques. ISPRS Journal of Photogrammetry and Remote Sensing, 51, 117–126. Doi: 10.1016/0924–2716(96)00018–4

Yunjun, Z., Fattahi, H. & Amelung, F. 2019. Small baseline InSAR time series analysis: Unwrapping error correction and noise reduction. Computers & Geosciences, 133, 104331, Doi:10.1016/j.cageo.2019.104331

Zhou, Q., Li, B. & Chen, Y. 2011. Remote sensing change detection and process analysis of long-term land use change and human impacts. Ambio, 40, 807–818. Doi: 10.1007/s13280-011-0157-1

# 9 Satellite Data Degradations and Their Impacts on High-Level Products

*Aolin Jia and Dongdong Wang*

## 9.1 INTRODUCTION

It has been half a century since the launch of the first generation Earth observation satellites, such as former Soviet Union's Sputnik 1 and the U.S.'s Television Infrared Observation Satellite Program (TIROS)-1 (Tatem et al. 2008). Compared with traditional observation approaches, one important advantage of satellite remote sensing is its global mapping capability. Satellite data have thus become an unprecedented source to study global environmental changes (e.g., Kaufman et al. 2002; Nemani et al. 2003). Thanks to the continuity of satellite missions, the length of remote sensing data record is continuously increasing. Currently, we have 40+ years archive of NOAA AVHRR at spatial resolution suitable for continental or global studies (Beck et al. 2011) and Landsat TM/ETM+ data at ecosystem scales (Markham and Helder 2012). Entering the new century, data records from new generation remote sensors such as the Satellite Pour l'Observation de la Terre (SPOT) VEGETATION are also lengthening. More than two decades of MODIS data become available (2000–present) and serve as valuable resource for the atmospheric (King et al. 2003), terrestrial (Justice et al. 2002), and oceanic (Esaias et al. 1998) research community with well-designed spectral configuration and improved radiometric and geometric accuracy (Justice et al. 1998).

With progress in sensor design and manufacturing techniques, reliability of radiometric calibration has improved substantially. Detection of subtle trends in environmental variables is extremely sensitive to calibration drifts of satellite sensors (Datla et al. 2004). Errors in geometric coregistration will also affect the analysis of multitemporal data and thus detection of changes and trends (Townshend et al. 1992). Stable, consistent, and reliable time series data with extended temporal coverage and accurate geolocation information are greatly needed to study interannual variability and long-term trends of global environments.

Substantial efforts have been devoted to ensure the absolute and relative accuracy of radiative quantities measured by remote sensors. Particularly, various calibration approaches (e.g., pre-launch, onboard, vicarious, cross-platform) are employed to convert sensors' signature (e.g., digital number) to accurate values of energy emitted and scattered by the Earth system (Dinguirard and Slater 1999). Before launch of satellites, sensors' characteristics of spectral, radiometric, and spatial responses are measured in labs at various levels to produce pre-launch calibration metrics. The actual calibration coefficients hardly remain constant because of the impacts from the launching process and the sensor degradation during the operation in space (Mekler and Kaufman 1995). To monitor sensitivity changes of sensors in radiometric response, many satellite sensors are equipped with onboard calibration units, such as internal lamp, solar diffuser, blackbody, and so on. For sensors without onboard calibrators, observations over pseudo-invariant targets are typically used to adjust calibration coefficients so that sensor degradation can be taken into account. Such technique of vicarious calibration is also applied to sensors with onboard calibration devices


DOI: 10.1201/9781003541141-13

as an additional source to verify performance of satellite detectors and their calibration systems (Wu et al. 2008).

Despite the various efforts, it is still a challenge to maintain the accuracy and precision of radiometric calibration due to various reasons discussed here. The harsh space environment where satellite sensors operate makes it extremely hard to keep the onboard calibration devices function ideally as designed. For example, the performance of internal illuminating sources and diffusers may degrade over time as well (Helder et al. 1998). Vicarious calibration is independent on degradation of onboard instruments, but it has its own uncertainties and limitations. Stable atmospheric and surface conditions are usually assumed at the calibration sites, although both tend to change. In addition, undetected cloud or cloud shadow and effects of bidirectional reflectance distribution function (BRDF) are among other major sources of uncertainty. As a result, satellite data records are prone to contain time-dependent radiometric drifts of some levels even after such sophisticated calibration efforts have been made. Substantial degradations have been found to exist in both sensors without onboard calibrators such as AVHRR (Wu and Zhong 1994) and those with onboard calibration devices as Landsat/TM (Helder et al. 1998) and Terra/MODIS (Wang et al. 2012).

Degradation in satellite signature (i.e., top-of-atmosphere (TOA) observations) will eventually translate into errors and bias in high level products through the satellite data processing chain, if the issue of data degradation is not well addressed in the design of retrieval algorithms. Due to the time-dependence nature of data degradations, their impacts will be especially evident in analysis of time series data, such as trend detection and investigation of interannual variability. In such cases, subtle changes of environmental parameters of interest will be obscured by artificial trends embedded in the time series of degraded data. For instance, analysis of four AVHRR NDVI datasets revealed inconsistent trends over Europe, Africa and the Sahel (Beck et al. 2011). Difference in handling sensor degradation can be an important factor of such discrepancies. Comparison of NDVI trends derived from Aqua/MODIS and Terra/MODIS showed contradictory results over boreal North America (Wang et al. 2012). Further data analysis suggested this be caused by the degradation issue in Terra/MODIS Collection 5 data. Similarly, inconsistent trends were observed in aerosol (Levy et al. 2010) and ocean color (Djavidnia et al. 2010) when comparing products derived from Aqua/MODIS and Terra/MODIS.

In addition to radiometric responses, other characteristics of satellites and sensors may also change with time. For example, orbits of Sun-synchronous satellites may drift and cause shift in solar zenith angle (SZA) (Ignatov et al. 2004). This is called orbital drift, which is a common issue over the course of the NOAA series' satellite lifetime because earlier satellites did not have active control to maintain a consistent daily overpass time over a long period. Thus, the Equator Crossing Time also drifts from local noon time to later afternoon for the Afternoon Constellation and from morning to earlier morning for the Morning Constellation This can lead to artifact drifts in time series of some satellite products, similar to effects of sensor degradation. Thus, we here treat orbital drift as one type of data degradations as well. The issues of sensor degradation exist in not only optical sensors (visible and NIR channels) but also thermal infrared (TIR) sensors and products due to the substantially diurnal temperature changes (Jia et al. 2022a; Jia et al. 2023a).

In this chapter, we will mainly focus on the degradation of optical and TIR data, introduce some common degradation issues, discuss their impacts on high level products, and summarize approaches, methods, and techniques to address these degradation issues.

## 9.2 COMMON ISSUES OF DATA DEGRADATION

This section briefly summarizes three types (Table 9.1) of common degradation issues of satellite data that consist of progressive changes in radiometric response and drifts in satellite orbital parameters, and two types of radiometric degradations (one with onboard calibration and one without onboard calibration).

**TABLE 9.1**
**Major Types of Data Degradation and Their Impacts on Satellite Products**

| Types | Summary | Impacts | Solution | Example | Reference |
|---|---|---|---|---|---|
| Sensor degradation without onboard calibration unit | Response of detectors changes with time due to the harsh environment where satellites operate. | Inaccurate radiance and time-dependent artifacts in time series | Using pseudo-invariant targets on the ground to adjust calibration coefficients (vicarious calibration) | NOAA AVHRR | (Molling et al. 2010) |
| Degradation of onboard calibration unit | Characteristics of onboard calibration device themselves degrade with time. | Inaccurate radiance and time-dependent artifacts in time series | Combining calibration with additional sources, such as lunar observation and pseudo-invariant targets | MODIS/Terra | (Wu et al. 2013) |
| Orbital drift | Orbits of Sun-synchronous satellites drift and cause changes in Equator Crossing Time. | (1) Gradual shift in solar zenith angle and variations in surface reflectance due to BRDF<br>(2) Artificially cooling effect in thermal infrared record due to diurnal temperature change | (1) Post-correction of time series data using statistical or BRDF models.<br>(2) Post correction: data assimilation method using long-term simulation | NOAA AVHRR | (Privette et al. 1995) |

### 9.2.1 Radiometric Degradation without Onboard Calibration Systems

Unlike ground-based measuring instruments, spaceborne sensors are usually physically inaccessible after satellites are launched. It is thus impossible to monitor the stability of their responses by periodically calibrating them with traditional approaches of laboratory measurement. To achieve reliable performance, extremely stable materials are typically used to manufacture satellite detectors. Calibration coefficients which convert satellite signature (e.g., digital number) to radiance or other radiative variables are measured in laboratory before launching. The sensors without onboard calibration units don't have mechanisms to update the coefficients on orbit. However, the characteristics of satellite detectors tend to change since the moment when the pre-launch coefficients are measured. The radiometric response of sensors will degrade due to the impacts from the launching process and the harsh space environment (Rao and Chen 1995).

Visible and near infrared (NIR) bands of AVHRR don't have onboard calibration systems. Staylor (1990) evaluated the calibration stability of AVHRR visible bands onboard NOAA 6, 7, and 9 satellites by monitoring their performance over the Libyan Desert. By assuming a fixed degradation rate, an exponential degradation model was used to fit AVHRR time series (Staylor 1990). The annual degradation rates of Channel 1 of NOAA 6, 7, and 9 were found 0%, 4%, and 6%, respectively (Staylor 1990). Masonis and Warren (2001) used data of the same bands onboard NOAA 9, 10, and 11 over Greenland to estimate their degradation rates. They found the sensor "degradation" is not only the reduced sensitivity to EM signals. NOAA 11 had an annual increased rate of 2.3% in radiometric gain (Masonis and Warren 2001). Similarly, MODIS also are found to have increased gains in Band 1 and 2 (Wu et al. 2013).

Due to the existence of radiometric degradation, measurements from such sensors cannot be directly used in quantitative studies (Gorman and McGregor 1994; Gutman 1999b), which require accurate values of radiance to quantify variables of interest. For such sensors, in practice, vicarious calibration is usually employed to account for the time-variant drifts of radiometric response (Dinguirard and Slater 1999). After launching, calibration coefficients are updated by benchmarking observations over various types of stable surfaces, such as desert (Rao and Chen 1995), ocean and high cloud (Vermote and Kaufman 1995), snow and ice (Loeb 1997). In addition, the observation can also be adjusted by comparing with other well-calibrated satellite images such as MODIS (Heidinger et al. 2002) or airborne data (Smith et al. 1988). Besides the calibration of individual bands, variable of interest (e.g., NDVI) can also be directly adjusted by a similar procedure of vicarious calibration (Los 1998).

Different vicarious calibration practice chooses different calibration sites and uses various forms of degradation equations. Lots of factors contribute to uncertainties in such methods. Reflectivity of pseudo-invariant targets is hardly a constant. Though the selected target is relatively isotropic in terms of its reflectivity, BRDF effects are a major source of uncertainties. Moreover, changes in atmospheric conditions can cause errors, especially aerosol from volcano eruption, water vapor effects for NIR bands. Issues of sub-pixel or undetected cloud cannot be negligible either. As a result, the derived calibration coefficients may contain large errors. Molling et al. (2010) compared AVHRR reflectances of a desert region calculated from ten groups of calibration coefficients and found the difference among them can be as large as 20% [Figure 1 of (Molling et al. 2010)].

Ideally, vicarious calibration is expected to account for all the effects of sensor degradation, so that the satellite data after vicarious calibration can be free of problems of sensor degradation. However, as we mentioned, the vicarious calibration has its disadvantages and limitations. An alternative strategy of onboard calibration is in great need.

### 9.2.2 Radiometric Degradation with Onboard Calibration Device

Given the aforementioned problems of vicarious calibration, an alternative way is to continuously monitor and correct sensor degradation on orbit. Actually, many sensors have units designed for onboard calibration, for example, Solar Diffuser (SD) for MODIS (Xiong and Barnes 2006); internal calibrator for TM (Chander and Markham 2003) and two additional devices (Full Aperture Solar Calibrator, Partial Aperture Solar Calibrator) for ETM+ (Markham et al. 2004). Theoretically, a well-designed onboard calibration system should be able to account for the sensor degradation and correct it by using the updated calibration coefficients.

For instance, MODIS is equipped with a sophisticated onboard calibration system for its visible, NIR and shortwave infrared bands (also known as reflective solar bands, RSB) (Xiong et al. 2010). SD and solar diffuser stability monitor (SDSM) are two key devices of this system (Xiong and Barnes 2006). The radiometric response of the MODIS detector is calibrated by viewing SD, which has known reflecting characteristics. The degradation of SD is monitored by SDSM. SDSM is actually a type of radiometer, which determines SD degradation by observing the Sun and SD alternately (Xiong et al. 2007). In this process, spectral, angular, and mirror-side dependency of reflectance and radiometric response has been well taken into account. Moreover, periodic view of the moon's surface is used as an additional calibration source to independently check the stability of radiometric calibration (Xiong et al. 2010). Normally, such well-designed onboard calibration system is able to update the calibration coefficients as needed to compensate the degradation of the detector, so that reliable calibration results can be generated during the lifetime of the sensor. As a matter of fact, MODIS/Aqua is proved to achieve a very high level of calibration stability. Combination of four lines of evidence indicated that SRB reflectance derived from MODIS/Aqua has changed less than 1% during its decadal operation (Wu et al. 2013).

However, the other sensor of twin MODIS, MODIS/Terra, was found to have a serious issue of calibration drifts in some SRB bands (Franz et al. 2008; Wu et al. 2013). According to the original strategy, the SD door opens only when onboard calibration is in progress to reduce the exposure of SD to the Sun. Due to an SD anomaly of Terra/MODIS, the door remains open after 2003 (Xiong and Barnes 2006). Terra/MODIS has thus experienced much worse degradation problems than Aqua/MODIS since then. Due to the difference of viewing geometry between onboard calibration and observing the Earth, stability of SD BRDF is essential for SDSM to track its degradation (Wu et al. 2013). As a result, bi-directionality of SD reflectance and changes in response versus scan angle (RVS) are major sources of uncertainties in calibrating MODIS in flight (Xiong et al. 2007). Because of its excessive exposure, the changes in BRDF of SD is believed to be the major cause of the calibration drift of MODIS/Terra (Wu et al. 2013). This degradation is found to be dependent on view zenith angle, spectral band and mirror side. The shorter wavelength bands have the most series degradation. The stability of Terra/MODIS blue band reflectance changed as much as 7% for nadir observations (Figure 9.1.) (Wang et al. 2012). With the lesson learned, the C6 reprocessing of MODIS/Terra data will use stable observations of the Moon and desert to correct the errors in radiometric calibration (Wenny et al. 2010). The calibration drift issue of MODIS/Terra is expected to be reduced substantially through reprocessing (Wu et al. 2013).

Landsat5/TM is another example of onboard calibration drift. Landsat 5 initially used onboard units to provide scene-by-scene calibration coefficients. However, the performance of the internal tungsten lamps of Landsat 5 failed to remain constant (Helder et al. 1998). The vicarious calibration and onboard calibration suggested that there existed trends in sensor response (Thome et al. 1997). Scene-by-scene calibration coefficients derived from the onboard calibrator are not reliable anymore (Chander and Markham 2003). As a result, a new calibration method based on monitoring desert and cross-calibration with Landsat7/ETM+ is developed (Chander et al. 2007). Compared to Landsat5/TM, ETM+ is rather stable. Annual degradation rates of visible and NIR bands of ETM+ are smaller than 0.4% (Markham et al. 2004).

### 9.2.3 Orbit Drift of Sun-Synchronous Satellites

Sun-synchronous is a feature of many polar orbiting Earth observation missions. Such satellites will always cross the equator at the same local time to minimize effects of diurnal changes in

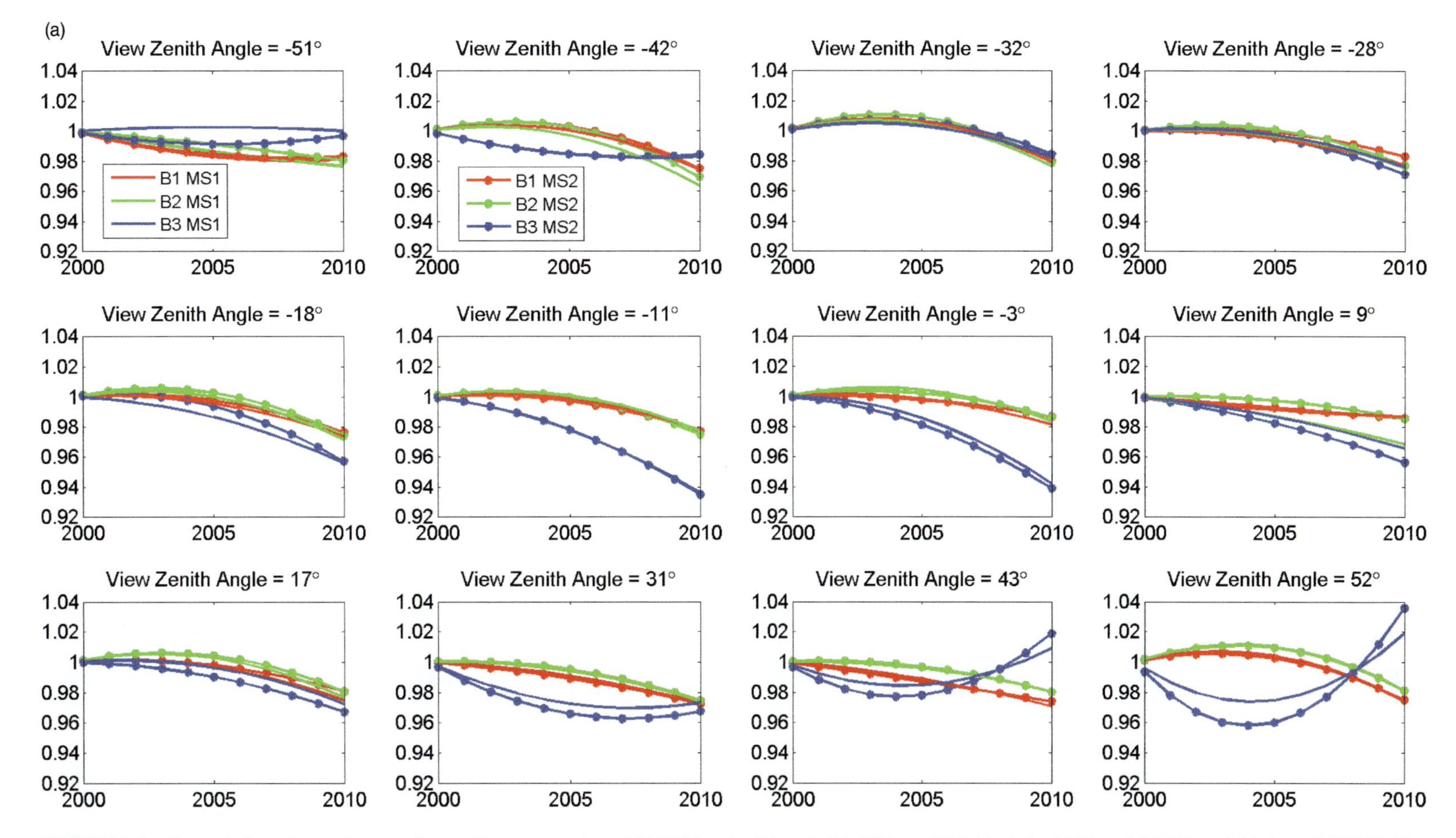


**FIGURE 9.1** Degradation of top-of-atmosphere reflectance at three MODIS bands (B1: red, B2: NIR, and B3: blue) for (a) Terra/MODIS and (b) Aqua/MODIS. The degradation rates are dependent on view zenith angle, spectral band, and mirror side (MS). [Figure S1 from (Wang et al. 2012).]

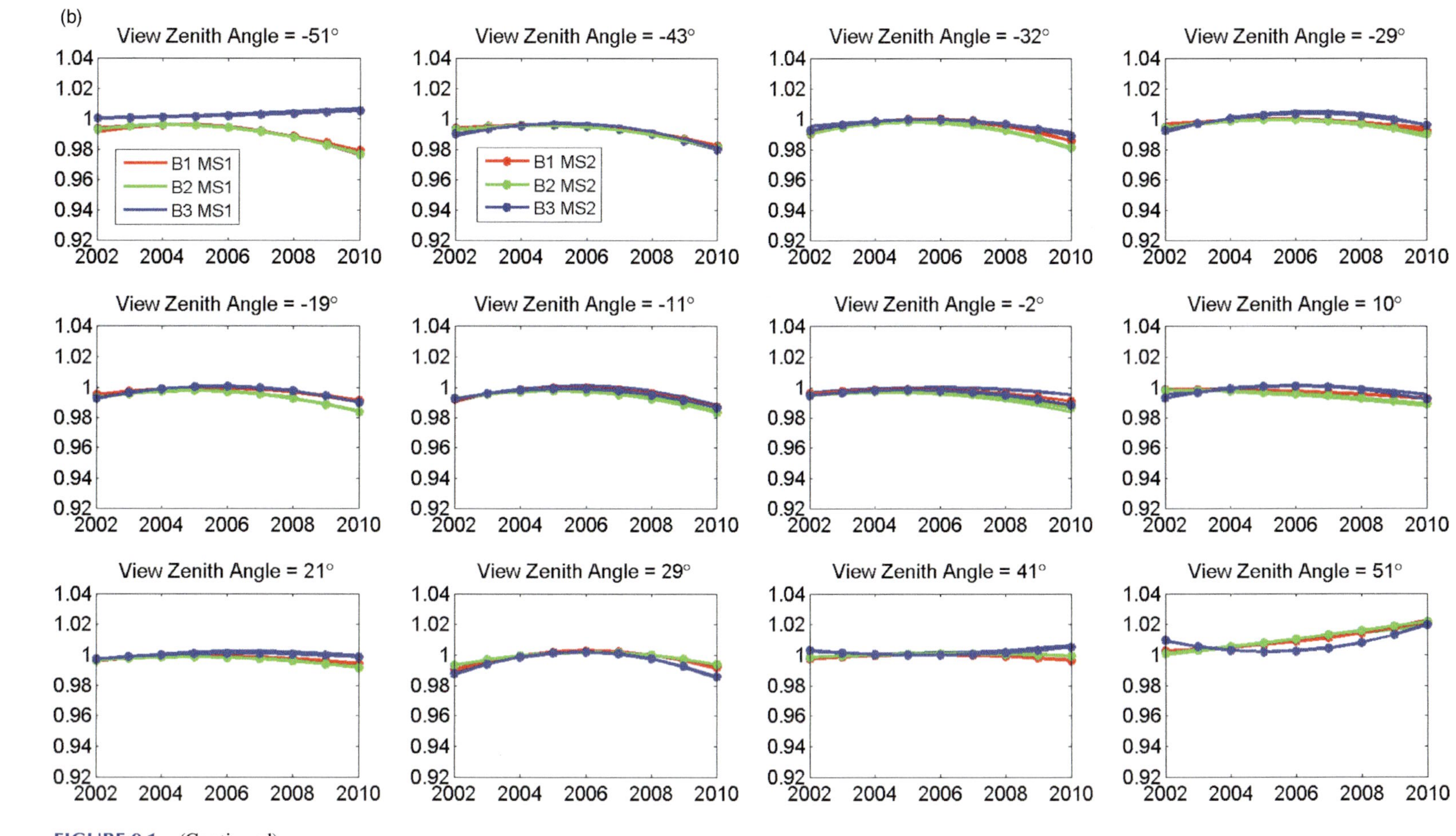


**FIGURE 9.1** (Continued)

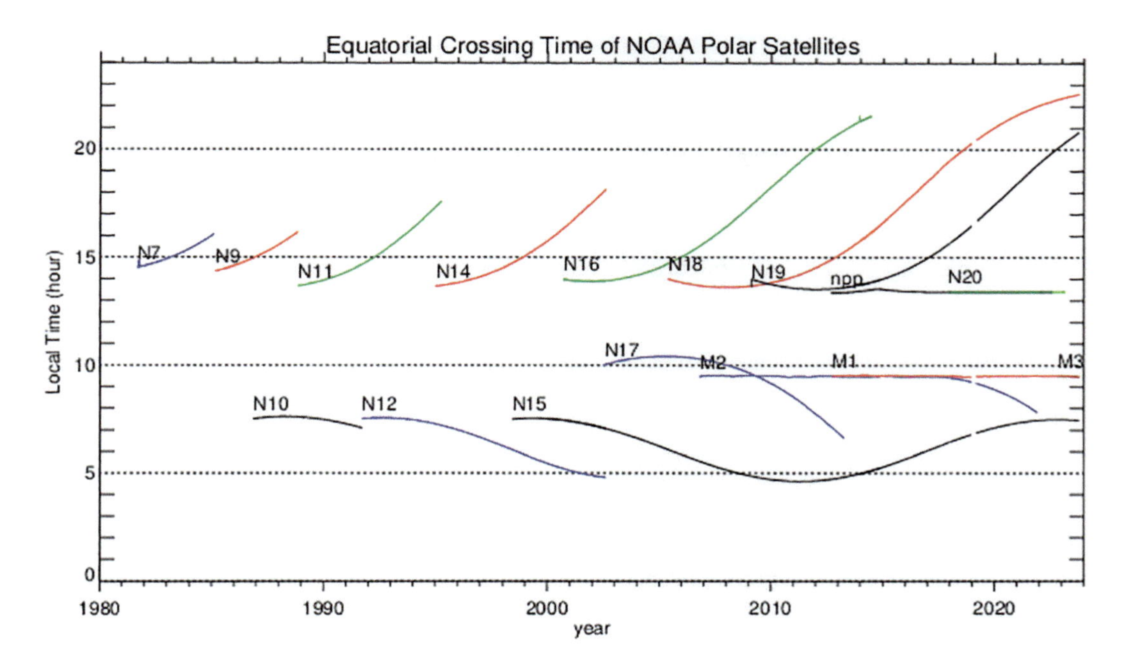


**FIGURE 9.2** Changes in equator crossing time for NOAA satellites. [data from NOAA (www.star.nesdis.noaa.gov/smcd/emb/vci/VH/vh_avhrr_ect.php)].

variables of interest. Besides, this also assures that the observations have relatively similar illumination conditions. Due to BRDF effects, fixed illumination geometry will make reflectance comparable with each other. However, the designated orbits will change during the operation due to various perturbations (Ignatov et al. 2004). One straightforward way to address the issue is to accordingly adjust the satellite orbit from time to time. However, it will be a problem for satellites without the capability of adjusting flight on orbit, for example, NOAA heritage satellites. For NOAA morning satellites, equator crossing time (ECT) becomes gradually earlier, and for afternoon satellites, the overpassing time moves towards later hours. ECT can change as long as several hours during the lifetime of NOAA satellites (Figure 9.2). In addition, Terra satellites have been operating for more than 20 years, and it maintains the ECT at 10:30 a.m. mean local time to allow five sensors onboard to collect consistent, simultaneous data. In 2020, Terra implemented its final inclination maneuver, and now it has drifted to earlier ECT. By fall 2022, its ECT will be at 10:15 a.m. and the drift will continue.

The change in crossing time will translate into shift of solar zenith angle. The angular changes are dependent on latitudes. Higher latitudes see greater seasonal variability but relative smaller shift of SZA, whereas equatorial area experiences substantial change of SZA (Privette et al. 1995). The progressive orbit drift will add a layer of noise into the temporal signals for some applications. The impacts mainly are in two aspects. ECT changes will affect features with diurnal variations, for example, cloud (Devasthale et al. 2012) and surface temperature (Li et al. 2013). The other issue is the changes in viewing geometry. Correction of atmospheric effects, for example, scattering of molecule and aerosol and absorption of water vapor, is dependent on the length of path and viewing geometry. In addition, surface reflectance is also dependent on viewing geometry because of surface BRDF effects. In Section 9.3.2, we use AVHRR data as example to illustrate how orbit shifts may affect NDVI and land surface temperature (LST) products.

## 9.3 IMPACTS OF DATA DEGRADATIONS ON HIGH-LEVEL PRODUCTS

### 9.3.1 MODIS/Terra Degradation

As we learned in Section 9.2.2, SRB of MODIS/Terra has experienced noticeable issues of calibration drifts, mainly caused by changes of bidirectional reflectance of SD. The actual degradation rates are dependent on spectral bands, view zenith angle and mirror side. Generally speaking, bands of shorter wavelength and near nadir observations have the worst problem of degradation. Blue bands are important inputs to algorithms to generate products of ocean color and aerosol loadings. Large discrepancies are observed for the two products derived from Terra/MODIS and Aqua/MODIS. Sensor degradation of Terra/MODIS is likely to be a major of such difference. Besides, some products such as NDVI may also be impacted by the degradation issue of blue bands through the processing chain of MODIS products, although they don't directly use blue bands as inputs.

In a simulation study, Wang et al. (2012) analyzed how degradation of Terra/MODIS blue band affects its NDVI product and how the embedded errors in NDVI impact data analysis such as trend detection. As is shown in Figure 9.1, TOA reflectance of Terra/MODIS red and NIR bands doesn't have serious issue of degradation. However, the MODIS NDVI algorithm needs surface reflectance of red and NIR as inputs. MODIS uses the Dense Dark Vegetation algorithm to atmospherically correct reflectance of red and NIR bands (Kaufman et al. 1997), which heavily depends on AOD information derived from the blue band. Using the actual degradation rates of blue, red and NIR bands at TOA as inputs, the induced errors in AOD and NDVI are simulated for various combination of view zenith angle, aerosol loadings and land surface types (Figure 9.3). The simulated results are consistent with the observed difference between Aqua/MODIS and Terra/MODIS corresponding products.

Although the annual degradation rate in the high level NDVI products is only with the magnitude of $10^{-3}$, it has great implication for the study of global vegetation change with the decadal time series of MODIS NDVI time series. Because this is close to natural changing rates of vegetation without disturbance such as fire, insect burst, and logging. Under such circumstances, Monte Carlo

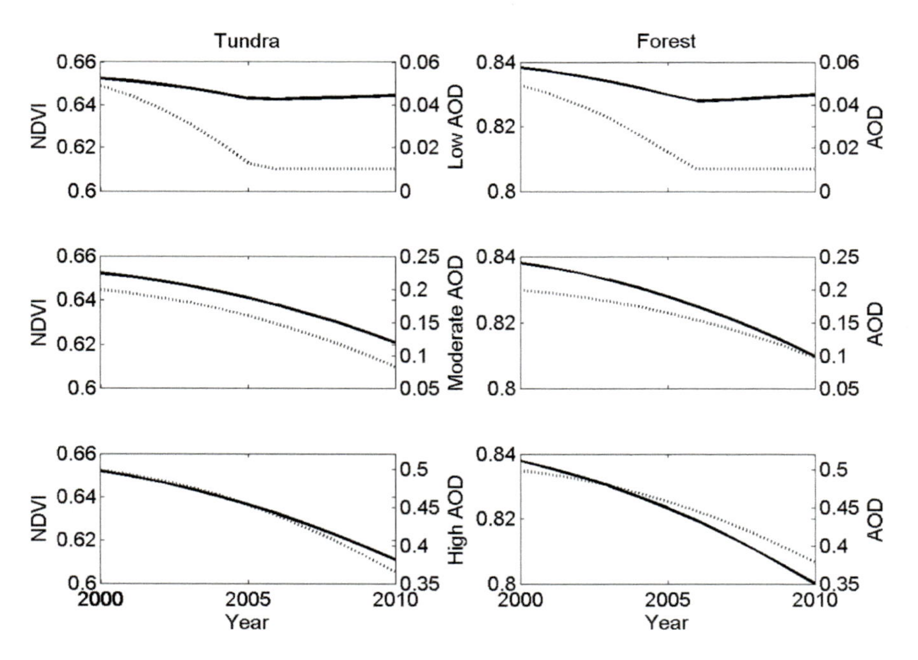


FIGURE 9.3 Temporal changes of NDVI (solid lines) and AOD (dashed lines) from 2000 to 2010, simulated with the degradation of Terra/MODIS as inputs. Two biomes (tundra (left) and forest (right)) and three levels of aerosol loadings (0.05, 0.20, and 0.50) are used in simulation. [From (Wang et al. 2012).]

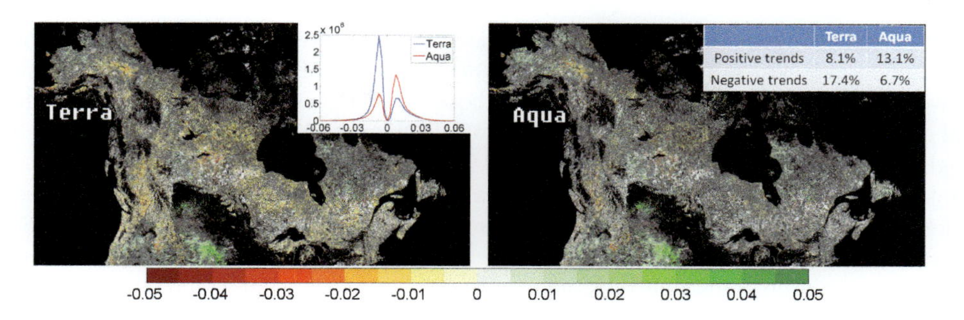


**FIGURE 9.4** Boreal North America maps of NDVI trends between 2002 to 2010 calculated from the two actual MODIS products (Terra: left and Aqua: right). White pixels are those recently disturbed by fire, wood harvest, etc. Black is non-tundra or forest area. [From (Wang et al. 2012).]

simulation suggested that it is extremely difficult to reliably detect trends of NDVI from the decadal MODIS data. Figure 9.4 shows the boreal North America maps of NDVI trends between 2002 to 2010 derived from Aqua/MODIS and Terra/MODIS respectively. Historical disturbance datasets were used to exclude pixels disturbed by various natural and anthropogenic factors. The two maps display similar patterns over some area where the change rates are large. However, there exists systematic difference between the trends derived from the two datasets. As a result, Aqua data suggests two thirds of significant trends are positive trends where Terra data show opposite statistics (Wang et al. 2012). Due to the potential impact of Terra/MODIS C5 data on trend analysis, it is suggested that user should use Aqua/MODIS data in analyzing interannual variability before reprocessed MODIS data are available.

## 9.3.2 AVHRR Orbital Drift and Correction Methods

### 9.3.2.1 NDVI

Orbit drift won't always be a problem for data users. If the users' models can explicitly handle data's variations in viewing geometry, their results will not be affected by the effects of orbit drift. For example, in calculation of NDVI, if BRDF-adjusted reflectance can be used, orbit drift will not necessarily lead to artificial changes in NDVI time series. However, due to the sparse temporal sampling of AVHRR, it is hard to model BRDF from AVHRR observations. The NDVI products typically directly use directional reflectance at the viewing geometry as inputs. As a ratio of two bands, NDVI is less affected by the variations in viewing geometry. Nevertheless, orbit drift still produces spurious variations on top of vegetation dynamics (Latifovic et al. 2012). The impacts of orbit drift on NDVI are dependent on latitude, season and land cover type. Privette et al. (1995) simulated the changes of TOA NDVI caused by the orbit drift of the NOAA 11 satellite. The drops in NDVI are most significant in boreal winter and the equator. After five years on orbit, the decrease of NDVI caused by orbit drift can be as great as 10% at the equator [Figure 5 of (Privette et al. 1995)].

To remove effects of orbit drift, various kinds of algorithms have been developed. Some are based on statistical analysis to isolate signals caused by changes in SZA (Sobrino et al. 2008). Some utilize prior knowledge of vegetation BRDF to estimate NDVI difference induced by angular variations (Bacour et al. 2006). Los et al. (2005) developed a BRDF-based correction approach without assumption of biome-dependent BRDF shapes. Taken a first order approximation, the NDVI equation in terms of Ross-Li kernels ($K_L$ and $K_R$) is reduced to (Los et al. 2005):

$$NDVI(\theta_v, \theta_s, \Delta\phi) = k_i' + k_g' K_L + k_v' K_R$$

Even in this reduced form, it is difficult to estimate the kernel parameters ($k_i'$, $k_g'$, and $k_v'$) because of the limited temporal sampling of AVHRR data. The authors further divided NDVI variability into two parts: those caused by phenology and those by BRDF effects. Using both simulated and actual data, the authors demonstrated the effects of BRDF on NDVI can be reduced by 50% to 85% through their method.

Impacts of NOAA satellite orbit drift are usually coupled with other issues such as sensor degradation of AVHRR (Staylor 1990), insufficient atmospheric correction (Nagol et al. 2009). So, data quality of AVHRR time series is heavily dependent on approaches of data processing and analysis. Large difference in both absolute value and temporal change exist among analysis using different versions of AVHRR NDVI datasets (Beck et al. 2011). Since the aim of data correction is to isolate vegetation variability from noise and artifacts in NDVI time series, an alternative way is to treat all the artifacts originated from data degradation together and correct them all at once (Jiang et al. 2008; Latifovic et al. 2012).

#### 9.3.2.2 LST

LST has a clear diurnal temperature cycle (DTC) mainly heated by incoming solar radiation (Jia et al. 2020), and the DTCs vary at different land cover types, and it can be as large as 35 K at barren land in summer (Figure 9.5). Moreover, the DTC magnitude is also determined by the thermal inertia of the underlying surface that is controlled by cover types and moisture conditions (Hu et al. 2020).

However, due to the orbital drift issue, the instantaneous recording time of afternoon NOAA satellites on each day changed from noon to late afternoon, thus an artificial "cooling effect" is reflected in the raw TIR BTs and retrieved instantaneous LST series. Figure 9.6 illustrates that

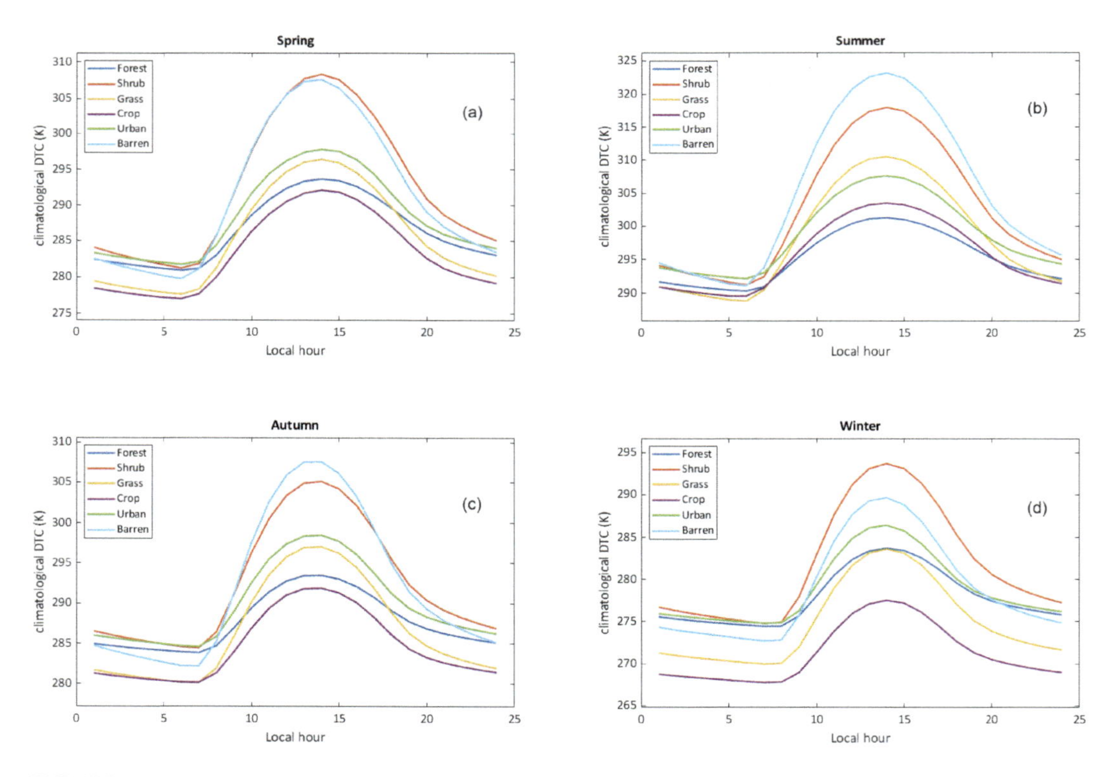


FIGURE 9.5 Typical DTCs for different land cover types: (a) forest, (b) shrubland, (c) grassland, (d) cropland, (e) urban, and (f) barren area. [Images from Figure 15 of Jia et al. (2022a)]

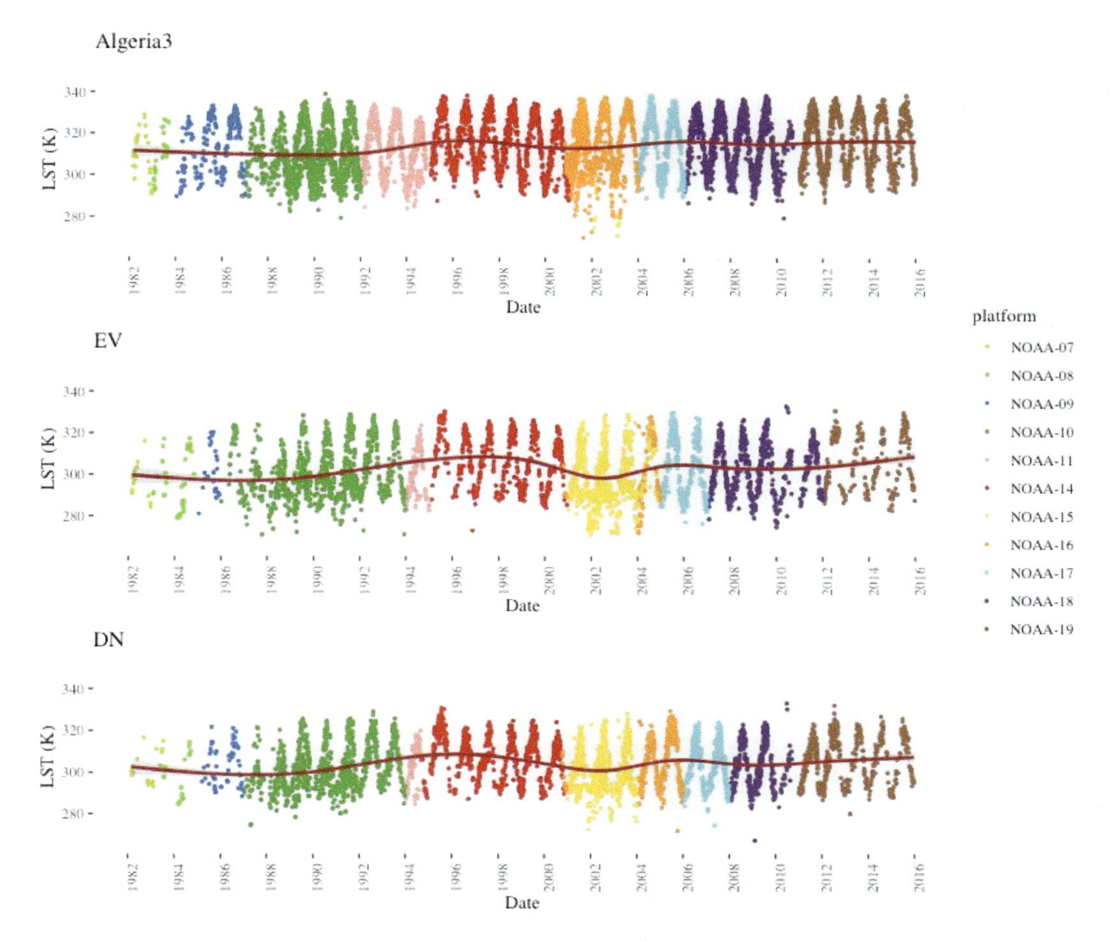


**FIGURE 9.6** LST time series at three locations: Algeria3, Évora (EV), and Doñana (DN). [Image from Figure 12 of Reiners et al. (2021).]

the LST series shows a "cooling" trend during the course of each AVHRR sensor. Arid areas were mostly affected due to the larger DTC. Moreover, clear data discontinuity is also reflected due to the sensor change every four to five years. Such discontinuity of satellite products will introduce significant uncertainty and even spurious conclusions in surface energy budget analysis (Jia et al. 2018). Accordingly, in order to obtain long-term, stable, and continuous LST products, methodologies have been developed for orbital drift issue correction. Overall, they can be classified into two categories: SZA-based temperature correction and trigonometric function-based DTC modeling.

SZA-based methods build the statistical relationship between SZA and temperature anomalies, and then by setting the SZA at a mean reference time, the LST will be corrected accordingly. This method originates from Gutman (1999a), and the relationship between cosine SZA and temperature adjustment value was regressed at six land cover types globally. Climatology is required for anomaly calculation of SZA and temperature. Gleason et al. (2002) tested this method which was found to suffer from greater variability, and the mean reference time was modified to a standard reference time. Sobrino et al. (2008) followed the SZA-based scheme and revised the method to pixel-based regression thus no land cover information was needed. In addition, the revised method can also implement an intercalibration within sensors. Julien and Sobrino (2012) developed a double-fitting model in which temperature anomalies were fitted with both viewing time and the second order of polynomial fit of SZA anomalies. By doing so, the method would not affect the natural trend of the temperature variation. However, the temperature variability in a day is not only related to SZA and land cover, and such simple empirical models cannot perform accurate correction and only provided a possible "guess" of the LST series.

The second method is to correct the drift by fitting a DTC model. A typical DTC model is mainly represented by a trigonometric function at the daytime, and researchers usually solve the DTC model equation using multiple clear-sky observations in a day (Duan et al. 2014; Quan et al. 2014); however, it won't work well for early AVHRR LSTs as only single sensor was operated at a certain period (Figure 9.2). In addition, LST can be only retrieved at clear-sky, whereas nearly half time and space over the globe, especially at high latitudes (Jia et al. 2023b). Jin and Treadon (2003) started to correct the AVHRR LST drift using DTC models for typical land cover types, seasons, and latitude bands, of which parameters were pre-defined in a look-up table and simulated from the National Center for Atmospheric Research (NCAR) Climate Community Model (CCM) 3 coupled with the land surface model, Biosphere—Atmosphere Transfer Scheme (BATS) (Bonan et al. 2002). Global long-term LST trend before 2000 was then quantified by AVHRR historical data (Jin 2004). However, due to the discontinuity of the look-up table, the LST maps also show clear spatiotemporal discontinuity among latitude bands and seasons. Liu et al. (2019) proposed a new method to fit DTC model parameters using more than four spatially neighboring temperature values of which viewing times are generally different. Ma et al. (2020) followed this idea and generated the first drifted corrected AVHRR LST product, of which the DTC model was revised by separating vegetation and soil temperatures. Nevertheless, such fitting solution for DTC models is not stable because the viewing times of spatially closing pixels are very similar, thus abnormal values may appear in the results.

Moreover, previous studies focused on correcting the drifted LSTs to a certain reference time; as a matter of fact, the long-term daily-mean LST series, rather than instantaneous values, are more useful to the research community, such as obtaining monthly value for analyzing climate warming (Chen et al. 2017) and determining the long-term surface radiation budget (Gallego-Elvira et al. 2016). Researchers have been working on daily-mean LST estimation by using limited instantaneous observations at noon and midnight times, whereas considerable biases were found based on site validation (Jia et al. 2022a). Therefore, rather than converting drifted temperature values to a certain fixed time in a day, daily mean temperature estimation shows high priority in corresponding studies.

In this chapter, we developed an innovative Two-step Correction method to correct the orbital drift issue of AVHRR LST and converted instantaneous retrievals to the daily-mean scale. Global DTC climatology from multiple geostationary satellite products and historical reanalysis of skin temperature from fifth-generation European Centre for Medium-Range Weather Forecasts (ECMWF) atmospheric reanalysis were also employed.

Clear-sky AVHRR LSTs were retrieved from AVHRR Level-1b Global Area Coverage (GAC) historical data (Bas et al. 2021). The spatial resolution is about 4 km at the nadir, and each sensor can pass a location at least twice per day (daytime and nighttime) (Di and Rundquist 1994). To implement the split window retrieval method, two thermal infrared bands were utilized. The NOAA-07, 09, 11, 12, 14, 15, 16, 18, 19, M1, and M2 (Figure 9.2) were used to make sure that generally there are at least four times recordings at each location in a clear-sky day from two platforms. The data coverage of one sensor was used as short as possible to minimize the impact of orbital drift. Essentially this method can include all historical records from more than two AVHRR sensors in a day, whereas based on the analysis, 4 observations in a day are enough for retrieving daily mean LSTs with satisfactory accuracy. Figure 9.2 also reveals that the viewing time after 2000 is relatively stable if the LST resource selection is always switched if a new sensor has been launched. The early period in the 1980s may not satisfy such requirement while our analysis suggest that by only including one single sensor it won't bring considerable uncertainty to the results. Detailed calibration and reprojection for GAC L1B data were processed by two Python packages: Pyresample and Pygac (Li et al. 2022). The clear-sky AVHRR LST retrieval method mainly followed the GLASS LST algorithm (Ma et al. 2020). Records with a VZA larger than 40° are not included to maintain high geolocation accuracy in data applications based on Wu et al. (2020). The ground validation networks are mainly from Surface Radiation Budget (SURFRAD).

Overall, the two-step method considers the hourly ERA5 reanalysis data as the target and corrects the reanalysis LST series by assimilating discontinuous remote sensing products. In other words, we consider the official ERA5 hourly LST mainly has two error sources: systemic bias and random errors in history. In the first step, the simulated hourly LSTs from ERA5 were initially calibrated by hourly satellite LST climatology data generated by Jia et al. (2023a), thus the systematic biases (Nogueira et al. 2021) were removed over the globe; in the next step, the calibrated DTCs of ERA5 were further corrected from 1982 to 2021 using AVHRR instantaneous LSTs on the corresponding passing time day by day. The fusion method employed Kalman Filtering (KF). The assimilated DTCs in a day were averaged to obtain daily mean LSTs and the drift correction was finished. In the end, the surface energy balance (SEB) correction was performed to the LST series (Jia et al. 2021), and the cloud effect was estimated using long-term satellite radiation products (Xu et al. 2022). Even though all-sky LST estimation has been combined with the two-step correction method, the all-sky LST estimation is beyond the discussion in this chapter. In fact, the two-step correction for the orbital drift can be utilized as an independent technique by fusing AVHRR LSTs with skin temperatures from any historical reanalysis data.

First, the systematical bias of the ERA5 LST over the globe is determined by comparing the hourly LST climatology of ERA5 ($LST_{clim}^{ERA5}$) from 2011 to 2021 and satellites ( $LST_{clim}^{sat}$ ).

$$Bias = LST_{clim}^{ERA5} - LST_{clim}^{sat},$$

$$LST_c = LST^{ERA5} - Bias,$$

where climatology is the average of the hourly LSTs at each hour in a day and each day of the year from 2011 to 2021, which represents the general magnitude and variability at the climate scale (Jia et al. 2022b) (data length of the climatology at each location: 365 × 24), and such calibration was implemented pixel by pixel. The systematical bias of ERA5 was minimized by the calibration.

The second step correction used KF to fuse the AVHRR LSTs with the calibrated ERA5 LSTs. The calibrated LST series was used to generate an annual dynamic model for data assimilation, and the annual dynamic model can be represented as follows:

$$LST_t = F_t \times LST_{t-1}$$

$$F_t = 1 + \frac{1}{Z_t + \delta} \times \frac{dZ_t}{dt}$$

$Z_t$ is considered as the difference in the LST series at day $t$ and day $t$—1, which is the temporal profile from the LST dynamic model over one year, and δ = 0.01 avoids a null denominator. And then as long as clear-sky AVHRR LST is available, the annual dynamic modeling can be corrected at the corresponding time based on KF (Bishop and Welch 2001):

$$z_k = x_k + v_k$$

$$\hat{x}_k^- = A\hat{x}_{k-1} + \omega_{k-1}$$

$$P^-{}_k = AP_{k-1}A^T + Q$$

where, $z_k$, the AVHRR LST at day $k$, is represented by the retrieved value, $x_k$, with uncertainty, $v_k$ (covariance is $R$), $\hat{x}_k^-$, is the prior estimate of the LST directly from the annual dynamic model, $A$, and $\omega_{k-1}$ is the model error with a covariance of $Q$. $R$ is determined by the site validation in advance as was set to 9; the initial value of $Q$ was equal to the squared ERA RMSE, where ERA5 RMSE is the RMSE of ERA5 LST, calculated by comparing samples on clear days with AVHRR retrievals in this year. $Q$ will be updated based on Kalman Gain ($K_k$) accordingly. To obtain the Kalman gain:

$$\hat{x}_k = \hat{x}_k^- + K_k\left(z_k - \hat{x}_k^-\right)$$

$$P_k = \left(I - K_k\right)P^-{}_k$$

$$K_k = P^-{}_k\left(P^-{}_k + R\right)^{-1}$$

$P^-{}_k$ is the prior covariance estimate after the prediction. The correction part included the final LST estimation ($\hat{x}_k$) corrected from the $\hat{x}_k^-$ via the $P_k$. The output error covariance was $P_k$. $I$ is the unit matrix. Therefore, after assimilating historical AVHRR retrievals, the ERA5 series was further corrected in detail. After temporal aggregation, the daily-mean LST can be obtained.

Long-term temperature continuity analyzed is implemented here to demonstrate that the two-step correction method well mitigated the orbital drift problem. The monthly temperature anomaly of AVHRR LST is compared to corresponding ground measurements (Figure 9.7) from 1990s to 2021.

The temporal analysis indicates that the orbital drift issue and the discontinuity caused by sensor change have been well mitigated and. The annual anomalies of AVHRR and sites matched with each other very well, no matter the general trend or the monthly variability. Therefore, the ground assessment reveals the reliable feasibility of the Two-step Correction method in AVHRR LST orbital drift issue.

## 9.4 CONCLUDING REMARKS

Analysis of long-term satellite data record has become an important tool to investigate how the Earth system changes over time and how its components interact with each other. Detection of subtle changes of global environment prefers consistent and stable time series data without artifact drifts. Unfortunately, spaceborne remote sensing data suffer from various issues of degradation during satellite's operation in the harsh space environment. This chapter summarizes some common phenomena of data degradations and discusses their impacts on high level satellite products.

Changes in radiometric response of spaceborne sensors cause time-variant drifts in TOA radiance, which translate into artificial trends in high level products of environmental variables. For sensors without onboard calibration devices, vicarious calibration is a key step to account for such artifacts in the remote sensing data. However, lots of environmental factors make it a challenge to eliminate the effects of data degradation only through vicarious calibration. So, onboard calibration systems are designed to address such problems. However, reliability of onboard calibration is also affected by its own degradation. The onboard calibration system needs collaboration with other calibration methods. Combination of onboard and vicarious calibration, utilization of multiple independent data sources, such as lunar view (Barnes et al. 2001; Cao et al. 2009), is a promising pathway to improve reliability of long-term data archive.

Any measurements have uncertainties, including remote sensing data. Users should always keep uncertainties of satellite data in mind when using them in quantitative studies. Inaccurate absolute calibration leads to bias or errors in satellite retrievals. Degradation of satellite data causes time-variant drift in long-term time series of satellite products. Caution should be exercised when drawing conclusion regarding long-term trends from data with possible artificial drifts. Meanwhile, efforts to evaluate accuracy of absolute calibration as well as long-term stability of calibration are greatly needed for climate change study.

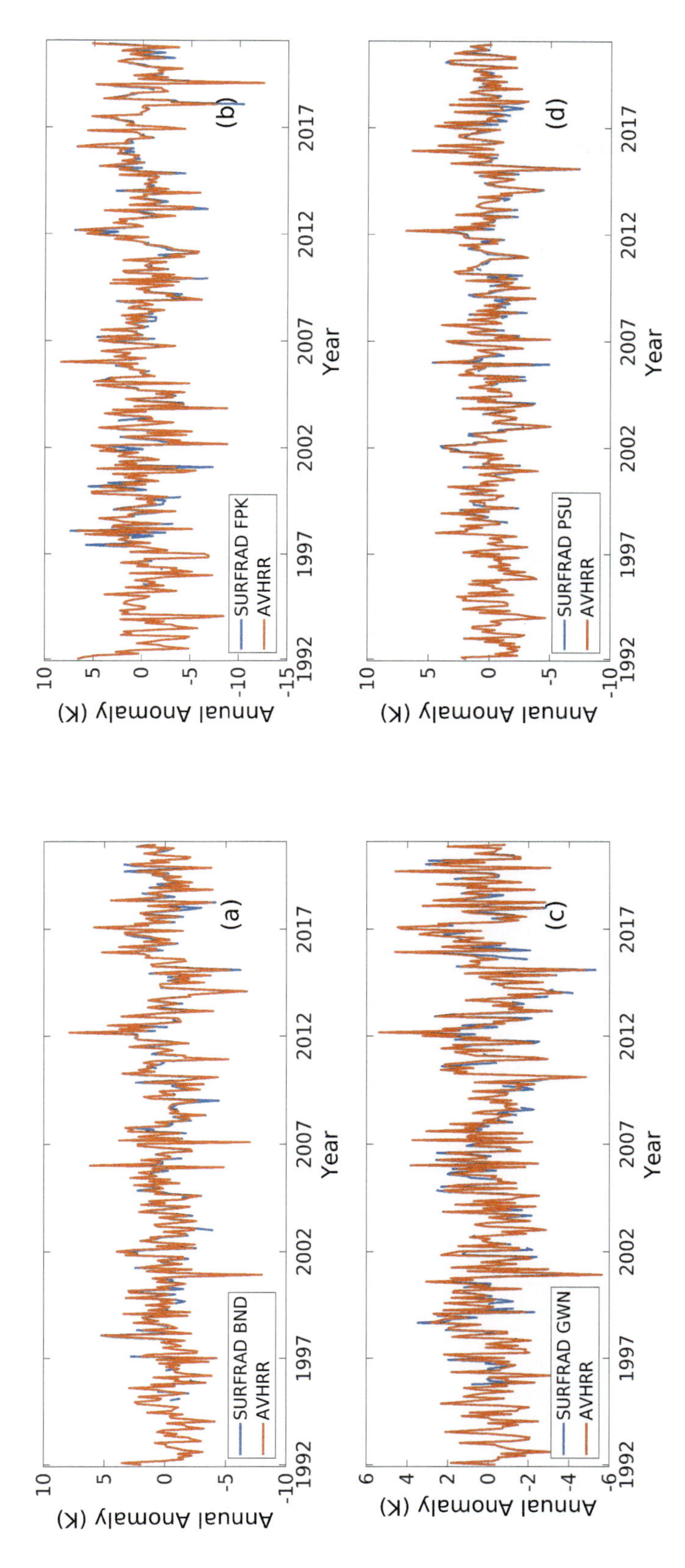


**FIGURE 9.7** Temperature monthly variability comparison between AVHRR LST and SURFRAD ground measurements at (a) Bondville (BND) site, (b) Fort Peck (FPK) site, (c) Goodwin Creek (GWN) site, and (d) Penn State University (PSU) site.

## REFERENCES

Bacour, C., Breon, F.M., & Maignan, F. (2006). Normalization of the directional effects in NOAA-AVHRR reflectance measurements for an improved monitoring of vegetation cycles. *Remote Sensing of Environment, 102,* 402–413

Barnes, R.A., Eplee, R.E., Schmidt, G.M., Patt, F.S., & McClain, C.R. (2001). Calibration of SeaWiFS. I. Direct techniques. *Applied Optics, 40,* 6682–6700

Bas, S., Debaecker, V., Kocaman, S., Saunier, S., Garcia, K., & Just, D. (2021). Investigations on the geometric quality of AVHRR level 1B imagery aboard MetOp-A. *PFG—Journal of Photogrammetry, Remote Sensing and Geoinformation Science,* 1–16

Beck, H.E., McVicar, T.R., van Dijk, A.I.J.M., Schellekens, J., de Jeu, R.A.M., & Bruijnzeel, L.A. (2011). Global evaluation of four AVHRR-NDVI data sets: Intercomparison and assessment against Landsat imagery. *Remote Sensing of Environment, 115,* 2547–2563

Bishop, G., & Welch, G. (2001). An introduction to the Kalman filter. *Proc of SIGGRAPH, Course, 8,* 41

Bonan, G.B., Oleson, K.W., Vertenstein, M., Levis, S., Zeng, X., Dai, Y., Dickinson, R.E., & Yang, Z.-L. (2002). The land surface climatology of the community land model coupled to the NCAR community climate model. *Journal of Climate, 15,* 3123–3149

Cao, C.Y., Vermote, E., & Xiong, X.X. (2009). Using AVHRR lunar observations for NDVI long-term climate change detection. *Journal of Geophysical Research-Atmospheres, 114*

Chander, G., & Markham, B. (2003). Revised Landsat-5 TM radiometric calibration procedures and postcalibration dynamic ranges. *IEEE Transactions on Geoscience and Remote Sensing, 41,* 2674–2677

Chander, G., Markham, B.L., & Barsi, J.A. (2007). Revised Landsat-5 thematic mapper radiometric calibration. *IEEE Geoscience and Remote Sensing Letters, 4,* 490–494

Chen, X., Su, Z., Ma, Y., Cleverly, J., & Liddell, M. (2017). An accurate estimate of monthly mean land surface temperatures from MODIS clear-sky retrievals. *Journal of Hydrometeorology, 18,* 2827–2847

Datla, R., Emery, B., Ohring, G., Spencer, R., & Wielicki, B. (2004). Stability and Accuracy Requirements for Satellite Remote Sensing Instrumentation for Global Climate Change Monitoring. In S.A. Morain, & A.M. Budge (Eds.), *Post-Launch Calibration of Satellite Sensors.* CRC Press, Boca Raton, FL

Devasthale, A., Karlsson, K.G., Quaas, J., & Grassl, H. (2012). Correcting orbital drift signal in the time series of AVHRR derived convective cloud fraction using rotated empirical orthogonal function. *Atmospheric Measurement Techniques, 5,* 267–273

Di, L., & Rundquist, D.C. (1994). A one-step algorithm for correction and calibration of AVHRR level 1b data. *Photogrammetric Engineering and Remote Sensing, 60,* 165–171

Dinguirard, M., & Slater, P.N. (1999). Calibration of space-multispectral imaging sensors: A review. *Remote Sensing of Environment, 68,* 194–205

Djavidnia, S., Melin, F., & Hoepffner, N. (2010). Comparison of global ocean colour data records. *Ocean Science, 6,* 61–76

Duan, S.-B., Li, Z.-L., Tang, B.-H., Wu, H., Tang, R., Bi, Y., & Zhou, G. (2014). Estimation of diurnal cycle of land surface temperature at high temporal and spatial resolution from clear-sky MODIS data. *Remote Sensing, 6,* 3247–3262

Esaias, W.E., Abbott, M.R., Barton, I., Brown, O.B., Campbell, J.W., Carder, K.L., Clark, D.K., Evans, R.H., Hoge, F.E., Gordon, H.R., Balch, W.M., Letelier, R., & Minnett, P.J. (1998). An overview of MODIS capabilities for ocean science observations. *IEEE Transactions on Geoscience and Remote Sensing, 36,* 1250–1265

Franz, B.A., Kwiatkowska, E.J., Meister, G., & McClain, C.R. (2008). Moderate resolution imaging spectroradiometer on terra: Limitations for ocean color applications. *Journal of Applied Remote Sensing, 2*

Gallego-Elvira, B., Taylor, C.M., Harris, P.P., Ghent, D., Veal, K.L., & Folwell, S.S. (2016). Global observational diagnosis of soil moisture control on the land surface energy balance. *Geophysical Research Letters, 43,* 2623–2631

Gleason, A.C., Prince, S.D., Goetz, S.J., & Small, J. (2002). Effects of orbital drift on land surface temperature measured by AVHRR thermal sensors. *Remote Sensing of Environment, 79,*147–165

Gorman, A.J., & McGregor, J. (1994). Some considerations for using AVHRR data in climatological studies 2. Instrument performance. *International Journal of Remote Sensing, 15,* 549–565

Gutman, G. (1999a). On the monitoring of land surface temperatures with the NOAA/AVHRR: Removing the effect of satellite orbit drift. *International Journal of Remote Sensing, 20,* 3407–3413

Gutman, G.G. (1999b). On the use of long-term global data of land reflectances and vegetation indices derived from the advanced very high resolution radiometer. *Journal of Geophysical Research-Atmospheres, 104,* 6241–6255

Heidinger, A.K., Cao, C.Y., & Sullivan, J.T. (2002). Using moderate resolution imaging spectrometer (MODIS) to calibrate advanced very high resolution radiometer reflectance channels. *Journal of Geophysical Research-Atmospheres, 107*

Helder, D., Boucyk, W., & Morfitt, R. (1998). Absolute calibration of the Landsat Thematic Mapper using the internal calibrator. In, *1998 IEEE International Geoscience and Remote Sensing Symposium Proceedings* (pp. 2716–2718)

Hu, T., Renzullo, L.J., van Dijk, A.I., He, J., Tian, S., Xu, Z., Zhou, J., Liu, T., & Liu, Q. (2020). Monitoring agricultural drought in Australia using MTSAT-2 land surface temperature retrievals. *Remote Sensing of Environment, 236,*111419

Ignatov, A., Laszlo, I., Harrod, E.D., Kidwell, K.B., & Goodrum, G.P. (2004). Equator crossing times for NOAA, ERS and EOS sun-synchronous satellites. *International Journal of Remote Sensing, 25,* 5255–5266

Jia, A., Liang, S., Jiang, B., Zhang, X., & Wang, G. (2018). Comprehensive assessment of global surface net radiation products and uncertainty analysis. *Journal of Geophysical Research: Atmospheres, 123,* 1970–1989

Jia, A., Liang, S., & Wang, D. (2022a). Generating a 2-km, all-sky, hourly land surface temperature product from Advanced Baseline Imager data. *Remote Sensing of Environment, 278,* 113105

Jia, A., Liang, S., Wang, D., Jiang, B., & Zhang, X. (2020). Air pollution slows down surface warming over the Tibetan Plateau. *Atmos pheric Chemistry and Physics, 20,* 881–899

Jia, A., Liang, S., Wang, D., Ma, L., Wang, Z., & Xu, S. (2023a). Global hourly, 5 km, all-sky land surface temperature data from 2011 to 2021 based on integrating geostationary and polar-orbiting satellite data. *Earth System Science Data, 15,* 869–895

Jia, A., Ma, H., Liang, S., & Wang, D. (2021). Cloudy-sky land surface temperature from VIIRS and MODIS satellite data using a surface energy balance-based method. *Remote Sensing of Environment, 263,* 112566

Jia, A., Wang, D., Liang, S., Peng, J., & Yu, Y. (2022b). Global daily actual and snow-free blue-sky land surface albedo climatology from 20-year MODIS products. *Journal of Geophysical Research: Atmospheres,* e2021JD035987

Jia, A., Wang, D., Liang, S., Peng, J., & Yu, Y. (2023b). Improved cloudy-sky snow albedo estimates using passive microwave and VIIRS data. *ISPRS Journal of Photogrammetry and Remote Sensing, 196,* 340–355

Jiang, L., Tarpley, J.D., Mitchell, K.E., Zhou, S., Kogan, F.N., & Guo, W. (2008). Adjusting for long-term anomalous trends in NOAA's global vegetation index data sets. *IEEE Transactions on Geoscience and Remote Sensing, 46,* 409–422

Jin, M. (2004). Analysis of land skin temperature using AVHRR observations. *Bulletin of the American Meteorological Society, 85,* 587–600

Jin, M., & Treadon, R. (2003). Correcting the orbit drift effect on AVHRR land surface skin temperature measurements. *International Journal of Remote Sensing, 24,* 4543–4558

Julien, Y., & Sobrino, J.A. (2012). Correcting AVHRR long term data record V3 estimated LST from orbital drift effects. *Remote Sensing of Environment, 123,* 207–219

Justice, C.O., Townshend, J.R.G., Vermote, E.F., Masuoka, E., Wolfe, R.E., Saleous, N., Roy, D.P., & Morisette, J.T. (2002). An overview of MODIS land data processing and product status. *Remote Sensing of Environment, 83,* 3–15

Justice, C.O., Vermote, E., Townshend, J.R.G., Defries, R., Roy, D.P., Hall, D.K., Salomonson, V.V., Privette, J.L., Riggs, G., Strahler, A., Lucht, W., Myneni, R.B., Knyazikhin, Y., Running, S.W., Nemani, R.R., Wan, Z.M., Huete, A.R., van Leeuwen, W., Wolfe, R.E., Giglio, L., Muller, J.P., Lewis, P., & Barnsley, M.J. (1998). The moderate resolution imaging spectroradiometer (MODIS): Land remote sensing for global change research. *IEEE Transactions on Geoscience and Remote Sensing, 36,* 1228–1249

Kaufman, Y.J., Tanre, D., & Boucher, O. (2002). A satellite view of aerosols in the climate system. *Nature, 419,* 215–223

Kaufman, Y.J., Tanre, D., Remer, L.A., Vermote, E.F., Chu, A., & Holben, B.N. (1997). Operational remote sensing of tropospheric aerosol over land from EOS moderate resolution imaging spectroradiometer. *Journal of Geophysical Research-Atmospheres, 102,* 17051–17067

King, M.D., Menzel, W.P., Kaufman, Y.J., Tanre, D., Gao, B.C., Platnick, S., Ackerman, S.A., Remer, L.A., Pincus, R., & Hubanks, P.A. (2003). Cloud and aerosol properties, precipitable water, and profiles of temperature and water vapor from MODIS. *IEEE Transactions on Geoscience and Remote Sensing, 41,* 442–458

Latifovic, R., Pouliot, D., & Dillabaugh, C. (2012). Identification and correction of systematic error in NOAA AVHRR long-term satellite data record. *Remote Sensing of Environment, 127*, 84–97

Levy, R.C., Remer, L.A., Kleidman, R.G., Mattoo, S., Ichoku, C., Kahn, R., & Eck, T.F. (2010). Global evaluation of the collection 5 MODIS dark-target aerosol products over land. *Atmospheric Chemistry and Physics, 10*, 10399–10420

Li, J.-H., Li, Z.-L., Liu, X., & Duan, S.-B. (2022). A global historical twice-daily (daytime and nighttime) land surface temperature dataset produced by AVHRR observations from 1981 to 2005. *Earth System Science Data Discussions,* 1–38

Li, Z.-L., Tang, B.-H., Wu, H., Ren, H., Yan, G., Wan, Z., Trigo, I.F., & Sobrino, J.A. (2013). Satellite-derived land surface temperature: Current status and perspectives. *Remote Sensing of Environment, 131*, 14–37

Liu, X., Tang, B.-H., Yan, G., Li, Z.-L., & Liang, S. (2019). Retrieval of global orbit drift corrected land surface temperature from long-term AVHRR data. *Remote Sensing, 11*, 2843

Loeb, N.G. (1997). In-flight calibration of NOAA AVHRR visible and near-IR bands over Greenland and Antarctica. *International Journal of Remote Sensing, 18*, 477–490

Los, S.O. (1998). Estimation of the ratio of sensor degradation between NOAA AVHRR channels 1 and 2 from monthly NDVI composites. *IEEE Transactions on Geoscience and Remote Sensing, 36*, 206–213

Los, S.O., North, P.R.J., Grey, W.M.F., & Barnsley, M.J. (2005). A method to convert AVHRR normalized difference vegetation index time series to a standard viewing and illumination geometry. *Remote Sensing of Environment, 99*, 400–411

Ma, J., Zhou, J., Göttsche, F.-M., Liang, S., Wang, S., & Li, M. (2020). A global long-term (1981–2000) land surface temperature product for NOAA AVHRR. *Earth System Science Data, 12*, 3247–3268

Markham, B.L., & Helder, D.L. (2012). Forty-year calibrated record of earth-reflected radiance from Landsat: A review. *Remote Sensing of Environment, 122*, 30–40

Markham, B.L., Thome, K.J., Barsi, J.A., Kaita, E., Helder, D.L., Barker, J.L., & Scaramuzza, P.L. (2004). Landsat-7 ETM+ on-orbit reflective-band radiometric stability and absolute calibration. *IEEE Transactions on Geoscience and Remote Sensing, 42*, 2810–2820

Masonis, S.J., & Warren, S.G. (2001). Gain of the AVHRR visible channel as tracked using bidirectional reflectance of Antarctic and Greenland snow. *International Journal of Remote Sensing, 22*, 1495–1520

Mekler, Y., & Kaufman, Y.J. (1995). Possible causes of calibration degradation of the advanced very high-resolution radiometer visible and near-infrared channels. *Applied Optics, 34*, 1059–1062

Molling, C.C., Heidinger, A.K., Straka, W.C., & Wu, X.Q. (2010). Calibrations for AVHRR channels 1 and 2: Review and path towards consensus. *International Journal of Remote Sensing, 31*, 6519–6540

Nagol, J.R., Vermote, E.F., & Prince, S.D. (2009). Effects of atmospheric variation on AVHRR NDVI data. *Remote Sensing of Environment, 113*, 392–397

Nemani, R.R., Keeling, C.D., Hashimoto, H., Jolly, W.M., Piper, S.C., Tucker, C.J., Myneni, R.B., & Running, S.W. (2003). Climate-driven increases in global terrestrial net primary production from 1982 to 1999. *Science, 300*, 1560–1563

Nogueira, M., Boussetta, S., Balsamo, G., Albergel, C., Trigo, I.F., Johannsen, F., Miralles, D.G., & Dutra, E. (2021). Upgrading land-cover and vegetation seasonality in the ECMWF coupled system: Verification with FLUXNET sites, METEOSAT satellite land surface temperatures, and ERA5 atmospheric reanalysis. *Journal of Geophysical Research: Atmospheres, 126*, e2020JD034163

Privette, J.L., Fowler, C., Wick, G.A., Baldwin, D., & Emery, W.J. (1995). Effects of orbital drift on advanced very high-resolution radiometer products—normalized difference vegetation index and sea surface temperature. *Remote Sensing of Environment, 53*, 164–171

Quan, J., Chen, Y., Zhan, W., Wang, J., Voogt, J., & Li, J. (2014). A hybrid method combining neighborhood information from satellite data with modeled diurnal temperature cycles over consecutive days. *Remote Sensing of Environment, 155*, 257–274

Rao, C.R.N., & Chen, J. (1995). Intersatellite calibration linkages for the visible and near-infared channels of the advanced very high-resolution radiometer on the NOAA-7, NOAA -9, and NOAA -11 spacecraft. *International Journal of Remote Sensing, 16*, 1931–1942

Reiners, P., Asam, S., Frey, C., Holzwarth, S., Bachmann, M., Sobrino, J., Göttsche, F.-M., Bendix, J., & Kuenzer, C. (2021). Validation of AVHRR land surface temperature with MODIS and in situ LST—a timeline thematic processor. *Remote Sensing, 13*, 3473

Smith, G.R., Levin, R.H., Abel, P., & Jacobowitz, H. (1988). Calibration of the solar channels of the NOAA-9 AVHRR using high altitude aircraft measurements. *Journal of Atmospheric and Oceanic Technology, 5*, 631–639

Sobrino, J.A., Julien, Y., Atitar, M., & Nerry, F. (2008). NOAA-AVHRR orbital drift correction from solar zenithal angle data. *IEEE Transactions on Geoscience and Remote Sensing, 46*, 4014–4019

Staylor, W.F. (1990). Degradation rates of the AVHRR visible channel for the NOAA-6, NOAA-7, and NOAA-9 spacecraft. *Journal of Atmospheric and Oceanic Technology, 7*, 411–423

Tatem, A.J., Goetz, S.J., & Hay, S.I. (2008). Fifty years of earth-observation satellites—views from space have led to countless advances on the ground in both scientific knowledge and daily life. *American Scientist, 96*, 390–398

Thome, K., Markham, B., Barker, J., Slater, P., & Biggar, S. (1997). Radiometric calibration of Landsat. *Photogrammetric Engineering and Remote Sensing, 63*, 853–858

Townshend, J.R.G., Justice, C.O., Gurney, C., & McManus, J. (1992). The impact of misregistration on change detection. *IEEE Transactions on Geoscience and Remote Sensing, 30*, 1054–1060

Vermote, E., & Kaufman, Y.J. (1995). Absolute calibration of AVHRR visible and near-infrared channels using ocean and cloud views. *International Journal of Remote Sensing, 16*, 2317–2340

Wang, D.D., Morton, D., Masek, J., Wu, A.S., Nagol, J., Xiong, X.X., Levy, R., Vermote, E., & Wolfe, R. (2012). Impact of sensor degradation on the MODIS NDVI time series. *Remote Sensing of Environment, 119*, 55–61

Wenny, B.N., Sun, J., Xiong, X., Wu, A., Chen, H., Angal, A., Choi, T., Chen, N., Madhavan, S., Geng, X., Kuyper, J., & Tan, L. (2010). MODIS calibration algorithm improvements developed for Collection 6 Level-1B. In, *SPIE Proceddings* (Vol. 7807, pp. 331–339). SPIE

Wu, A., Angal, A., Xiong, X., & Cao, C. (2008). Monitoring MODIS calibration stability of visible and near-IR bands from observed top-of-atmosphere BRDF-normalized reflectances over Libyan Desert and Antarctic surfaces. *Proceedings of SPIE 7081*, 708113

Wu, A., & Zhong, Q. (1994). A method for determining the sensor degradation rates of NOAA AVHRR channels 1 and 2. *Journal of Applied Meteorology, 33*, 118–122

Wu, A.S., Xiong, X.X., Doelling, D.R., Morstad, D., Angal, A., & Bhatt, R. (2013). Characterization of Terra and Aqua MODIS VIS, NIR, and SWIR spectral bands' calibration stability. *IEEE Transactions on Geoscience and Remote Sensing, 51*, 4330–4338

Wu, X., Naegeli, K., & Wunderle, S. (2020). Geometric accuracy assessment of coarse-resolution satellite datasets: A study based on AVHRR GAC data at the sub-pixel level. *Earth System Science Data, 12*, 539–553

Xiong, X., Sun, J., Barnes, W., Salomonson, V., Esposito, J., Erives, H., & Guenther, B. (2007). Multiyear on-orbit calibration and performance of Terra MODIS reflective solar bands. *IEEE Transactions on Geoscience and Remote Sensing, 45*, 879–889

Xiong, X.X., & Barnes, W. (2006). An overview of MODIS radiometric calibration and characterization. *Advances in Atmospheric Sciences, 23*, 69–79

Xiong, X.X., Sun, J.Q., Xie, X.B., Barnes, W.L., & Salomonson, V.V. (2010). On-Orbit calibration and performance of Aqua MODIS reflective solar bands. *IEEE Transactions on Geoscience and Remote Sensing, 48*, 535–546

Xu, J., Liang, S., & Jiang, B. (2022). A global long-term (1981–2019) daily land surface radiation budget product from AVHRR satellite data using a residual convolutional neural network. *Earth System Science Data, 14*, 2315–2341

# Part V

## Vegetation Index Standardization and Cross-Calibration of Data from Multiple Sensors

# 10 Inter- and Intra-Sensor Spectral Compatibility and Calibration of the Enhanced Vegetation Indices

*Tomoaki Miura, Kenta Obata, Hiroki Yoshioka, and Alfredo Huete*

## ACRONYMS

| | |
|---|---|
| AERONET | Aerosol Robotic Network |
| ABI | Advanced Baseline Imager |
| AHI | Advanced Himawari Imager |
| AOT | Aerosol Optical Thickness |
| ARVI | Atmospherically Resistance Vegetation Index |
| ASTER | Advanced Spaceborne Thermal Emission and Reflection Radiometer |
| AVHRR | Advanced Very High Resolution Radiometer |
| BRDF | Bidirectional Reflectance Distribution Function |
| CAI | Cloud and Aerosol Imager |
| ccL | Continuity-corrected Landsat |
| DSCOVR | Deep Space Climate Observatory |
| EO-1 | Earth Observing-1 |
| EOS | Earth Observing System |
| EPIC | Earth Polychromatic Imaging Camera |
| ESA | European Space Agency |
| ETM+ | Enhanced Thematic Mapper Plus |
| EVI | Enhanced Vegetation Index |
| EVI2 | Two-band Enhanced Vegetation Index (without a blue band) |
| GF-1 | Geofen-1 |
| GLI | Global Imager |
| GMFR | Geometric Mean Functional Regression |
| GPP | Gross Primary Productivity |
| GOES-R | Geostationary Operational Environmental Satellites—R Series |
| HJ-1 | Huanjing-1 |
| HRVIR | Haute Résolution dans le Visible et l'Infra-Rouge |
| JPSS | Joint Polar Satellite System |
| LAI | Leaf Area Index |
| MccL | MODIS continuity-corrected Landsat |
| MERIS | Medium Resolution Imaging Spectrometer |
| MISR | Multi-angle Imaging SpectroRadiometer |
| MODIS | Moderate Resolution Imaging Spectroradiometer |
| MSI | MultiSpectral Instrument |

DOI: 10.1201/9781003541141-15

| | |
|---|---|
| NASA | National Aeronautics and Space Administration |
| NIR | Near-infrared |
| NOAA | National Oceanic and Atmospheric Administration |
| NPP | National Polar-orbiting Partnership |
| NDVI | Normalized Difference Vegetation Index |
| OLI | Operational Land Imager |
| OLI-2 | Operational Land Imager-2 |
| PAC | Partial Atmosphere Correction |
| RMSD | Root Mean Square Difference |
| SAVI | Soil Adjusted Vegetation Index |
| SeaWiFS | Sea-Viewing Wide Field-of-View Sensor |
| SGLI | Second-generation Global Imager |
| SPOT | Système Pour l'Observation de la Terre |
| sRMPD | Systematic Square Root of Mean Product Difference |
| TM | Thematic Mapper |
| TOA | Top-of-the-Atmosphere |
| TOC | Top-of-Canopy |
| uRMPD | Unsystematic Square Root of Mean Product Difference |
| VGT | VEGETATION |
| VIIRS | Visible Infrared Imaging Radiometer Suite |
| WVC | Wide-View Charge-Coupled Device Camera |

## 10.1 INTRODUCTION

The enhanced vegetation index (EVI), an index developed for the National Aeronautics and Space Administration (NASA) Earth Observing System (EOS) Moderate Resolution Imaging Spectroradiometer (MODIS) mission (Huete et al., 2002), has been shown useful across a wide range of terrestrial vegetation studies. These include climate-vegetation interactions (Ponce Campos et al., 2013; Saleska et al., 2007; Seddon et al., 2016), gross primary productivity (GPP) (Chen et al., 2011; Guanter et al., 2014; Ichii et al., 2017; Shi et al., 2017; Sims et al., 2008), land-atmosphere energy fluxes (Jung et al., 2019), drought impact assessment (Song et al., 2013), land cover classification (Friedl et al., 2010), and phenology (Zhang et al., 2006).

The EVI has also been derived from sensors other than MODIS and used in characterizing terrestrial vegetation conditions and dynamics. These include Système Pour l'Observation de la Terre (SPOT) VEGETATION (VGT) for GPP estimation (e.g., Xiao et al., 2004; Xiao et al., 2003), Landsat Thematic Mapper (TM) for floristic diversity assessment (Cabacinha & de Castro, 2009), and Landsat Enhanced Thematic Mapper Plus (ETM+) and IKONOS for leaf area index (LAI) estimation (Soudani et al., 2006).

In the last decade, a number of new satellite sensors were launched and have been placed in orbit as data continuity and/or new-generation missions, all of which have spectral bands suitable for the EVI. Suomi National Polar-orbiting Partnership (NPP) and Joint Polar Satellite System (JPSS) Visible Infrared Imaging Radiometer Suite (VIIRS) is a moderate resolution sensor to continue the data streams from the National Oceanic and Atmospheric Administration (NOAA) Advanced Very High Resolution Radiometer (AVHRR) sensor series and from EOS MODIS (Cao et al., 2014). PROBA-V is a European Space Agency (ESA)-owned satellite mission launched to continue the moderate resolution data stream of SPOT VGT (Sterckx et al., 2013). Landsat 8 Operational Land Imager (OLI) and Landsat 9 OLI-2, and Sentinel-2 MultiSpectral Instrument (MSI) provide the continuation of and the enhancement to Landsat type, medium spatial resolution data with a greater number of spectral bands and higher radiometric capabilities (Irons et al., 2012). In particular, the EVI is included as one of the VIIRS standard products (Justice et al., 2013; Vargas et al., 2013) and used to derive VIIRS phenology products (Zhang et al., 2018). The potential of the EVI hypertemporal data derived from new-generation geostationary satellite sensors, such as Himawari-8 Advanced Himawari Imager (AHI) and Geostationary Operational Environmental Satellites–R Series (GOES-R) Advanced Baseline Imager (ABI), as well as those from Deep Space Climate Observatory (DSCOVR) Earth Polychromatic Imaging Camera (EPIC), a satellite in the Lagrange Point 1 orbit, for phenology monitoring and characterization have been explored (Tran et al., 2020; Weber et al., 2020; Yan et al., 2019; Zhao et al., 2022).

It is thus of great importance to develop a good understanding of EVI spectral compatibilities across sensors and also of significant interest to investigate whether the EVI data record begun with EOS MODIS can be temporally and spatially extended with other satellite sensors for improved monitoring capabilities and climate science studies. This chapter focuses on the EVI and discusses inter-sensor spectral compatibility of the EVI. We also discuss inter-sensor spectral compatibility of a two-band version of the EVI without a blue band, or EVI2 (Jiang et al., 2008), and its intra-sensor compatibility with the EVI. Specific objectives of this chapter are to:

1. Present a comprehensive review of inter- and intra-sensor spectral compatibility of the EVI and EVI2
2. Evaluate the atmospheric impact on spectral compatibility of the EVI and EVI2 across sensors
3. Discuss cross-sensor calibration methodologies for EVI and EVI2 spectral compatibility

## 10.2 THE ENHANCED VEGETATION INDICES

The EVI is a three-band index, requiring reflectances of the near-infrared (NIR) ($\rho_{\text{NIR}}$), red ($\rho_{\text{red}}$), and blue ($\rho_{\text{blue}}$) bands. The index was designed to optimize the vegetation signal through a de-coupling of the canopy background signal, with a reduction in atmospheric aerosol influences, and with improved sensitivity in high biomass region, which complements the conventional, normalized difference vegetation index (NDVI) (Huete et al., 2002). The EVI (dimensionless) takes the form

$$EVI = G\frac{\rho_{NIR} - \rho_{red}}{\rho_{NIR} + C_1\rho_{red} - C_2\rho_{blue} + L} \tag{10.1}$$

where $L$ (reflectance unit, dimensionless) is the canopy background brightness adjustment factor that came from the Soil Adjusted Vegetation Index (SAVI) (Huete, 1988), $C_1$ (dimensionless) and $C_2$ (dimensionless) are the coefficients of the aerosol resistance term adapted from the Atmospherically Resistant Vegetation Index (ARVI) (Kaufman & Tanré, 1992), and $G$ (dimensionless) is the gain factor which adjusts the EVI dynamic range to a comparable one to that of the NDVI.

The input reflectances in Equation 10.1 need to be corrected for a partial atmosphere (i.e., molecular scattering and gaseous absorption effects) or the total atmospheric effects including aerosols. Miura et al. (2001) demonstrated the effectiveness of the EVI in reducing residual aerosol contaminations in the total-atmosphere corrected reflectances that arise due to the highly spatially and temporally variable nature of aerosol loadings and properties, and uncertainties in the aerosol loading estimations. The coefficients adopted in the MODIS EVI algorithm are $L = 1$, $C_1 = 6$, $C_2 = 7.5$, and $G = 2.5$ (Huete et al., 1997), which can be used for all of the preceding input reflectance types.

While many studies found that the EVI was advantageous in vegetation productivity assessments and characterization of vegetation biophysical properties [e.g., an improved surrogate measure of GPP (Rahman et al., 2005)], several potential issues have been reported for the EVI. First, by its very design, the EVI is limited to a sensor system with a blue band in addition to a red and NIR bands (Equation 10.1). The EVI is not applicable to, for example, SPOT Haute Résolution dans le Visible et l'Infra-Rouge (HRVIR), EOS Advanced Spaceborne Thermal Emission and Reflection Radiometer (ASTER), or AVHRR. Second, the EVI algorithm can produce faulty values for snow/ice-contaminated observations that have higher blue and red reflectances than NIR reflectance

(Didan & Huete, 2006; Huete et al., 2002; Vargas et al., 2013). Finally, Fensholt et al. (2006) suggested that the consistency of EVI values across different sensors may be more problematic than that of the NDVI due to variable and more difficult atmospheric correction schemes of the blue reflectance as described later (Section 10.4).

In order to address the preceding issues, a two-band version of the EVI without a blue band, or EVI2, was developed, which is functionally equivalent to the EVI although slightly more prone to aerosol noise (Jiang et al., 2008). Jiang et al. (2008) noted that the blue band in the EVI is rather aimed at reducing noise and uncertainties associated with highly variable atmospheric aerosols than at providing additional biophysical information on vegetation.

The EVI2 was derived by optimization of the linear vegetation index (LVI) to attain the best similarity with the EVI, particularly when atmospheric effects are insignificant and data quality is good (Jiang et al., 2008). The LVI was a newly-proposed index for the EVI2 derivation that incorporated the soil-adjustment factor of SAVI with a linearity-adjustment factor. The derived EVI2 (dimensionless) is of the form

$$EVI2 = G \cdot \frac{\rho_{NIR} - \rho_{red}}{\rho_{NIR} + C\rho_{red} + L} \tag{10.2}$$

where $G = 2.5$ (dimensionless), $C = 2.4$ (dimensionless), and $L = 1.0$ (reflectance unit, dimensionless) for the MODIS spectral bands (Jiang et al., 2008). Jiang et al. (2008) used MODIS reflectance data extracted from 40 globally distributed sites to obtain these coefficients.

The consistency between EVI and EVI2 across various land cover types demonstrated that their similarity was independent of land cover. Comparisons of EVI and EVI2 time series further revealed that their similarity was seasonally independent. The EVI2 (Equation 10.2) has been adopted as the EVI backup algorithm, which is used as a substitute for the EVI for snow/ice-contaminated pixels, since the Collection 6 MODIS VI Product suite.

## 10.3 MULTI-SENSOR COMPATIBILITY OF EVI AND EVI2: A REVIEW

As in the case of the NDVI, the EVI and EVI2 from different sensors are likely subject to systematic differences due to differences in sensor/platform characteristics and product generation algorithms. The spectral bandpass is one key sensor characteristic that varies widely among sensors (Figure 10.1). Although focused on the NDVI, Swinnen and Veroustraete (2008) provide a comprehensive list of factors that would potentially impact inter-sensor compatibility between SPOT VGT and NOAA AVHRR.

Unlike the NDVI, only a small number of studies have addressed the issue of multi-sensor EVI and/or EVI2 compatibility (Table 10.1). Most of these studies used MODIS EVI or EVI2 as a reference in examining inter-sensor compatibility, as these indices were originally designed for MODIS. Several recent studies focused on inter-sensor EVI and/or EVI2 compatibility among medium resolution sensors, such as ETM+, OLI, and MSI (e.g., Sulla-Menashe et al., 2016; Arekhi et al., 2019). Whereas early studies examined spectral compatibility across different sensor bandpasses using hyperspectral data (e.g., Miura & Yoshioka, 2011), recent studies used actual sensor data to evaluate multi-sensor EVI and/or EVI2 compatibility for specific applications and for additional newer sensors (Nouri et al., 2020; Weber et al., 2020).

While all these studies investigated primarily compatibility of the EVI across sensors or of the EVI2 among different sensors using atmospherically corrected, surface reflectance data, several studies evaluated two additional details. First, Kim et al. (2010), Nouri et al. (2020), and Zhen et al. (2023) examined intra-sensor EVI-to-EVI2 compatibility for VGT, SeaWiFS, ETM+, OLI, and MSI. Kim et al. (2010), Miura et al. (2008), and Yamamoto et al. (2012) also examined inter-sensor EVI-to-EVI2 compatibility for AVHRR and MODIS, and for ASTER and MODIS. Second, Miura and Yoshioka (2011) assessed the impact of different atmospheric correction schemes on cross-sensor

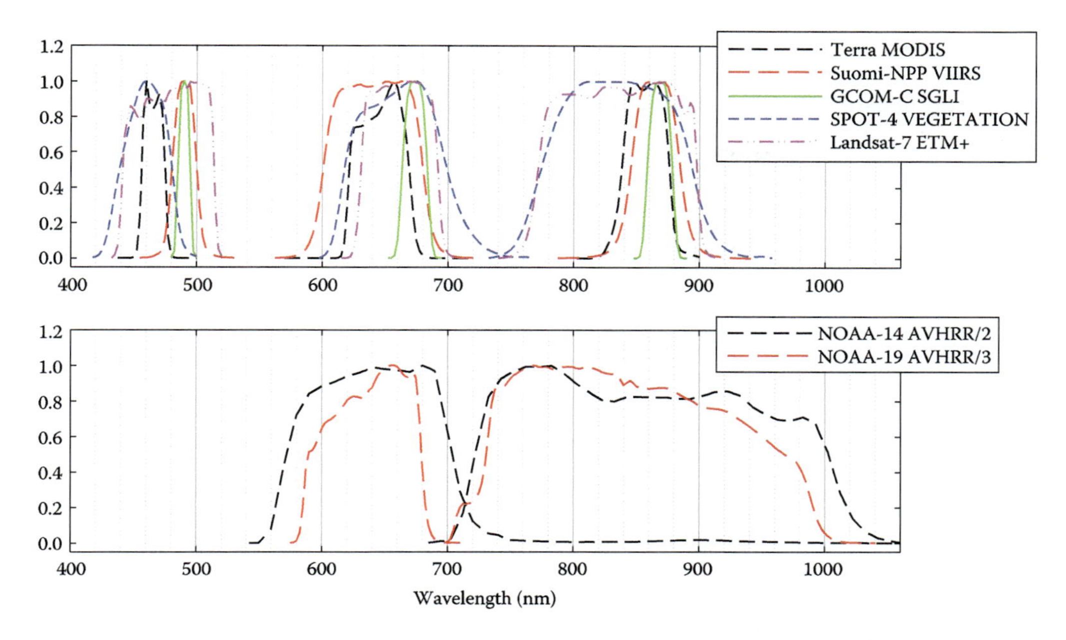


**FIGURE 10.1** Normalized spectral response curves of blue, red, and NIR bands of selected satellite sensors. The AVHRR sensors do not have a blue band.

EVI2 compatibility using a scenario where total atmosphere-corrected EVI2 from MODIS and VGT were compared with partial atmosphere-corrected EVI2 from AVHRR and VIIRS. Later, we highlight key findings from some of the early studies.

Ferreira et al. (2003) investigated the utility of spectral vegetation indices in monitoring seasonal dynamics of Brazilian savanna vegetation formations. Airborne hyperspectral data were acquired and convolved to the MODIS and ETM+ bandpasses and converted to the EVI. Inter-sensor comparisons of seasonal dynamics, based on spectral bandpass properties, showed that the simulated ETM+ EVI had better seasonal discrimination capability than that of MODIS. They attributed this finding to the closer proximity between the ETM+ red and NIR band-centers.

Fensholt et al. (2006) evaluated the quality of MODIS EVI, Medium Resolution Imaging Spectrometer (MERIS) EVI, and VGT EVI using in situ measurements and their inter-sensor compatibility using actual satellite products at a grassland site in Senegal. A good agreement between the EVI from satellite and from in situ measured MODIS data was found, indicating an accurate atmospheric correction of the MODIS red, NIR, and blue spectral bands. On the other hand, the consistency of the EVI from MERIS and VGT with MODIS EVI was not as good as was found among the NDVI from these sensors. Fensholt et al. (2006) attributed this reduced EVI consistency to different, sensor-specific atmospheric correction schemes used in blue band reflectance retrievals.

Miura et al. (2008) examined the compatibility of ASTER EVI2 and MODIS EVI using actual satellite data extracted from randomly located, globally distributed locations. A robust linear relationship was found between the two satellite indices with a mean difference of 0.012 EVI units (ASTER minus MODIS).

Kim et al. (2010) performed a band decomposition analysis in which bandpass contributions to observed cross-sensor VI difference were identified. They found that disparities in blue bandpasses were the primary cause of EVI differences between MODIS and other coarse resolution sensors. The highest compatibility was between VIIRS and MODIS EVI2 while AVHRR EVI2 was the least compatible to MODIS.

**TABLE 10.1**
**List of Multi-Sensor EVI/EVI2 Compatibility Studies**

| Reference | Sensor/VI Pair | | | Data Analyzed |
|---|---|---|---|---|
| Ferreira et al. (2003) | MODIS EVI | vs. | ETM+ EVI | Airborne Hyperspectral Data |
| Fensholt et al. (2006) | MODIS EVI | vs. | MERIS EVI | Actual Sensor Data |
| | MODIS EVI | vs. | VGT EVI | Actual Sensor Data |
| Miura et al. (2008) | MODIS EVI | vs. | ASTER EVI2 | Actual Sensor Data |
| Kim et al. (2010) | MODIS EVI | vs. | VIIRS EVI | EO-1[a] Hyperion Data |
| | MODIS EVI | vs. | VGT EVI | |
| | MODIS EVI | vs. | SeaWiFS[b] EVI | |
| | MODIS EVI2 | vs. | VIIRS EVI2 | EO-1 Hyperion Data |
| | MODIS EVI2 | vs. | VGT EVI2 | |
| | MODIS EVI2 | vs. | SeaWiFS EVI2 | |
| | MODIS EVI2 | vs. | AVHRR/2 EVI2 | |
| | MODIS EVI | vs. | MODIS EVI2 | EO-1 Hyperion Data |
| | VGT EVI | vs. | VGT EVI2 | |
| | SeaWiFS EVI | vs. | SeaWiFS EVI2 | |
| | MODIS EVI | vs. | AVHRR/2 EVI2 | |
| Miura and Yoshioka (2011) | MODIS EVI | vs, | VGT EVI | EO-1 Hyperion Data |
| | MODIS EVI | vs. | GLI[c] EVI | |
| | MODIS EVI | vs. | VIIRS EVI | |
| | MODIS EVI | vs. | SeaWiFS EVI | |
| | MODIS EVI | vs. | SGLI[d] EVI | |
| | MODIS EVI2 | vs. | AVHRR/2 EVI2 | EO-1 Hyperion Data |
| | MODIS EVI2 | vs. | AVHRR/3 EVI2 | |
| | MODIS EVI2 | vs. | VGT EVI2 | |
| | MODIS EVI2 | vs. | GLI EVI2 | |
| | MODIS EVI2 | vs. | VIIRS EVI2 | |
| | MODIS EVI2 | vs. | SeaWiFS EVI2 | |
| | MODIS EVI2 | vs. | CAI[e] EVI2 | |
| | MODIS EVI2 | vs. | SGLI EVI2 | |
| | MODIS EVI2 | vs. | AVHRR/2 EVI2 | EO-1 Hyperion Data |
| | MODIS EVI2 | vs. | VIIRS EVI2 | |
| | VGT EVI2 | vs. | AVHRR/2 EVI2 | |
| | VGT EVI2 | vs. | VIIRS EVI2 | |
| Yamamoto et al. (2012) | MODIS EVI | vs. | ASTER EVI2 | Actual Sensor Data |
| | MODIS EVI | vs. | MODIS/ASTER EVI | |
| Obata et al. (2013) | MODIS EVI | vs. | VIIRS EVI | Actual Sensor Data |
| She et al. (2015) | OLI EVI | vs. | ETM+ EVI | Ground Hyperspectral Data<br>Actual Sensor Data |
| Sulla-Menashe et al. (2016) | ETM+ EVI | vs. | TM EVI | Actual Sensor Data |
| Obata et al. (2016) | MODIS EVI | vs. | VIIRS EVI | Actual Sensor Data |
| Jarchow et al. (2018) | MODIS EVI | vs. | VIIRS EVI | Actual Sensor Data |
| | MODIS EVI | vs. | TM EVI | |
| | MODIS EVI | vs. | OLI EVI | |
| Miura et al. (2018) | MODIS EVI | vs. | VIIRS EVI | Actual Sensor Data |
| | MODIS EVI2 | vs. | VIIRS EVI2 | |
| Arekhi et al. (2019) | OLI EVI | vs. | MSI EVI | Actual Sensor Data |
| Mancino et al. (2020) | OLI EVI | vs. | ETM+ EVI | Actual Sensor Data |
| Mansaray et al. (2020) | HJ-1 WVC[f] EVI | vs. | GF-1[g] EVI | Actual Sensor Data |
| | HJ-1 WVC EVI | vs. | OLI EVI | |
| | HJ-1 WVC EVI | vs. | MSI EVI | |

**TABLE 10.1** *(Continued)*
**List of Multi-Sensor EVI/EVI2 Compatibility Studies**

| Reference | Sensor/VI Pair | | | Data Analyzed |
|---|---|---|---|---|
| Nouri et al. (2020) | MODIS EVI | vs. | MODIS EVI2 | Actual Sensor Data |
| | OLI EVI | vs. | OLI EVI2 | |
| | ETM+ EVI | vs. | ETM+ EVI2 | |
| Abbasi et al. (2021) | MODIS EVI | vs. | OLI EVI | Actual Sensor Data |
| | MODIS EVI | vs. | ETM+ EVI | |
| | MODIS EVI | vs. | TM EVI | |
| | MODIS EVI2 | vs. | OLI EVI2 | Actual Sensor Data |
| | MODIS EVI2 | vs. | ETM+ EVI2 | |
| | MODIS EVI2 | vs. | TM EVI2 | |
| | ccL[h] EVI | vs. | ccL EVI2 | Actual Sensor Data |
| | MccL[i] EVI | vs. | MccL EVI2 | |
| Miura et al. (2021) | MODIS EVI | vs. | VIIRS EVI | Actual Sensor Data |
| | MODIS EVI2 | vs. | VIIRS EVI2 | |
| | MODIS EVI | vs. | MODIS EVI2 | Actual Sensor Data |
| | VIIRS EVI | vs. | VIIRS EVI2 | |
| | MODIS EVI | vs. | VIIRS EVI2 | Actual Sensor Data |
| | MODIS EVI2 | vs. | VIIRS EVI | |
| Zhen et al. (2023) | MODIS EVI | vs. | MODIS EVI2 | USGS Spectral Library |
| | OLI EVI | vs. | OLI EVI2 | Actual Sensor Data |
| | MSI EVI | vs. | MSI EVI2 | |

[a] *EO-1:* Earth Observing-1

[b] *SeaWiFS:* Sea-Viewing Wide Field-of-View Sensor

[c] *GLI:* Global Imager

[d] *SGLI:* Second-Generation Global Imager

[e] *CAI:* Cloud and Aerosol Imager

[f] *HJ-1 WVC:* Huanjing-1 Wide-View Charge-Coupled Device Camera

[g] *GF-1:* Geofen-1

[h] *ccL:* continuity-corrected Landsat

[i] *MccL:* MODIS continuity-corrected Landsat

Miura and Yoshioka (2011) also empirically examined inter-sensor EVI and EVI2 relationships using MODIS as a reference, but for a larger number of sensors. They also found that spectral bandpass differences resulted in systematic differences on inter-sensor EVI2 relationships with MODIS, but the relationships were very linear. Inter-sensor EVI relationships between MODIS and other sensors were linear for most of sensors, although several sensors had either curve-linear or incompatible relationships with MODIS EVI, all of which have spectrally very different blue band from the MODIS blue bandpass.

## 10.4 ATMOSPHERIC IMPACT ON INTER- AND INTRA-SENSOR SPECTRAL COMPATIBILITY OF EVI AND EVI2

In this section, we present a spectral compatibility analysis conducted to fill in some of the knowledge gaps already identified, namely, the impact of operational atmospheric corrections on inter- and intra-sensor spectral compatibility of the EVI and EVI2. We specifically evaluated: (1) the effects of residual aerosols on inter-sensor EVI-to-EVI, inter-sensor EVI2-to-EVI2, and intra-sensor EVI2-to-EVI relationships, and (2) the effects of different atmospheric correction schemes on inter-sensor EVI2 compatibility between AVHRR and other sensors. The analysis was conducted by bandpass simulations with a global set of Hyperion hyperspectral data (Miura et al., 2013).

### 10.4.1 Materials and Methods

#### 10.4.1.1 Hyperion Data and Preprocessing

The dataset used in Miura et al. (2013) and adapted here consisted of 37 Level 1R EO-1 Hyperion hyperspectral scenes obtained over 15 globally distributed Aerosol Robotic Network (AERONET) sites (Holben et al., 2001; Middleton et al., 2010; Pearlman et al., 2003). Hyperion is a pushbroom sensor with 256 pixels in the cross-track direction (~7.7 km swath at 30 m spatial resolution), acquiring Earth-reflected radiation from 400 nm to 2,500 nm at 10 nm nominal sampling intervals (Pearlman et al., 2003). Level 1R Hyperion scenes provide radiometrically corrected and radiometrically calibrated radiance values, and are corrected for the VNIR/SWIR inter-spectrometer misregistration (USGS, 2011).

The sites/scenes represented 13 out of the 17 International Geosphere-Biosphere Programme land cover types, with the four cover types not represented being: evergreen broadleaf forests, permanent wetlands, snow and ice, and water bodies. Coincident Level 2 AERONET data (±1 hour of the Hyperion acquisition time) were available for all of the scenes. Aerosol optical thickness (AOT) at 550 nm ranged from 0.02 (clean) to 0.53 (turbid) for this dataset (Figure 10.2). See Table 10.1 of Miura et al. (2013) for additional details of this Hyperion dataset.

The Hyperion scenes were spectrally and spatially convolved to simulate top-of-the-atmosphere (TOA) reflectances of the VIIRS, MODIS, SGLI, VGT, AVHRR/3, AVHRR/2, and ETM+ bands at 1 km spatial resolution. The spectral response functions of these spectral bands (Figure 10.1) were interpolated to Hyperion band center wavelengths for each Hyperion pixel (Pearlman et al., 2003). The MODIS point spread function, which is, to the first order, triangular in the scan direction and rectangular in the track direction, was assumed for the spatial convolution (Wolfe et al., 2002).

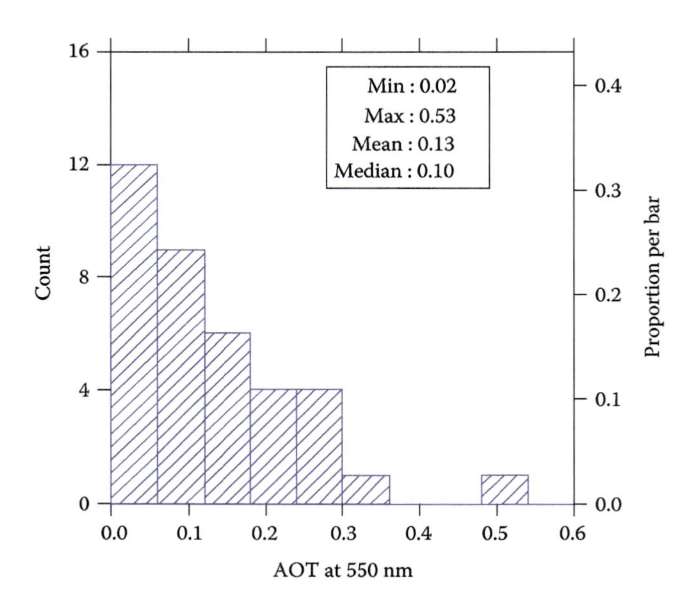


FIGURE 10.2 Histogram and univariate statistics of atmospheric aerosol optical thickness (AOT) at 550 nm of Hyperion scenes used in this study. (Adapted and reprinted with permission from Miura et al. (2013).)

#### 10.4.1.2 Simulation of Atmospheric Correction

The “6S” radiative transfer code was used to atmospherically correct the convolved Hyperion scenes (Kotchenova & Vermote, 2007; Vermote et al., 2006). For each scene, the 6S code was constrained with the corresponding in situ AERONET atmospheric measurements, view and solar zenith, and relative azimuth angles at the time of the image acquisition, and the site elevation obtained from GTOPO30.

Three atmospheric corrections were applied. First, all the scenes were corrected for total atmospheric effects including aerosols, that is, TOA reflectances were reduced to surface or top-of-canopy (TOC) reflectances. Second, all the scenes were corrected for total atmospheric effects, but with incorrect AOT values (AERONET-measured AOT values ±0.05 AOT at 550 nm) to simulate residual aerosol effects in the retrieved surface reflectances. Finally, we applied a partial atmosphere correction (PAC) on the AVHRR bandpass reflectances where all atmospheric effects except for aerosols were accounted for.

The EVI and EVI2 were computed from each of these retrieved surface reflectances (i.e., TOC reflectances, TOC reflectances subject to residual aerosol effects, and PAC reflectances). Approximately, 60 1-km pixels were randomly selected and extracted from each scene, totaling ~2,000 pixels over the 37 scenes. This random sample dataset was used for the analyses described next.

#### 10.4.1.3 Statistical Difference Analysis

Three statistical measures of difference were employed to quantitatively evaluate and compare inter- and intra-sensor compatibility of the EVI and EVI2: the root mean square difference (RMSD), systematic square root of mean product difference (sRMPD), and unsystematic square root of mean product difference (uRMPD) (Ji & Gallo, 2006):

$$RMSD = \sqrt{\frac{1}{n}\sum_{i=1}^{n}\left(X_{1,i} - X_{2,i}\right)^2} \tag{10.3}$$

$$sRMPD = \sqrt{\frac{1}{n}\sum_{i=1}^{n}\left[\left(X_{1,i} - X_{2,i}\right)^2 - \left|\hat{X}_{1,i} - X_{1,i}\right|\left|\hat{X}_{2,i} - X_{2,i}\right|\right]} \tag{10.4}$$

$$uRMPD = \sqrt{\frac{1}{n}\sum_{i=1}^{n}\left(\left|\hat{X}_{1,i} - X_{1,i}\right|\left|\hat{X}_{2,i} - X_{2,i}\right|\right)} \tag{10.5}$$

where $X_1$ and $X_2$ are the EVI or EVI2 for sensor-1 and -2, respectively, and $\hat{X}_1$ and $\hat{X}_2$ are the corresponding points of $X_1$ and $X_2$ on the regression line that is the best estimate of the primary trend of the $X_1$ vs. $X_2$ scatter. sRMPD and uRMPD decouple RMSD into its systematic and unsystematic components, respectively; sRMPD measures how far the scatter (trend or regression line) is from the line of $X_1 = X_2$, or the bias errors, and uRMPD measures the magnitude of data scattering (secondary variation) about the regression line (Ji & Gallo, 2006). Following the protocol developed by Ji and Gallo (2006), regression lines were estimated using the geometric mean functional regression (GMFR), which considers that both variables of interest are subject to error. Unlike the ordinary least squares regression, the derived regression line is symmetric about $X_1$ and $X_2$, or invertible.

### 10.4.2 Results

#### 10.4.2.1 Inter-Sensor EVI

EVI differences for all possible combinations among the five sensors of VIIRS, MODIS, SGLI, VGT4, and ETM+ for both atmospheric corrections with and without AOT errors are plotted in Figure 10.3 and summarized in Table 10.2 (see Appendix 10.A.1 for GMFR equations). All sensor pairs examined here were subject to systematic differences of which magnitudes varied among sensor

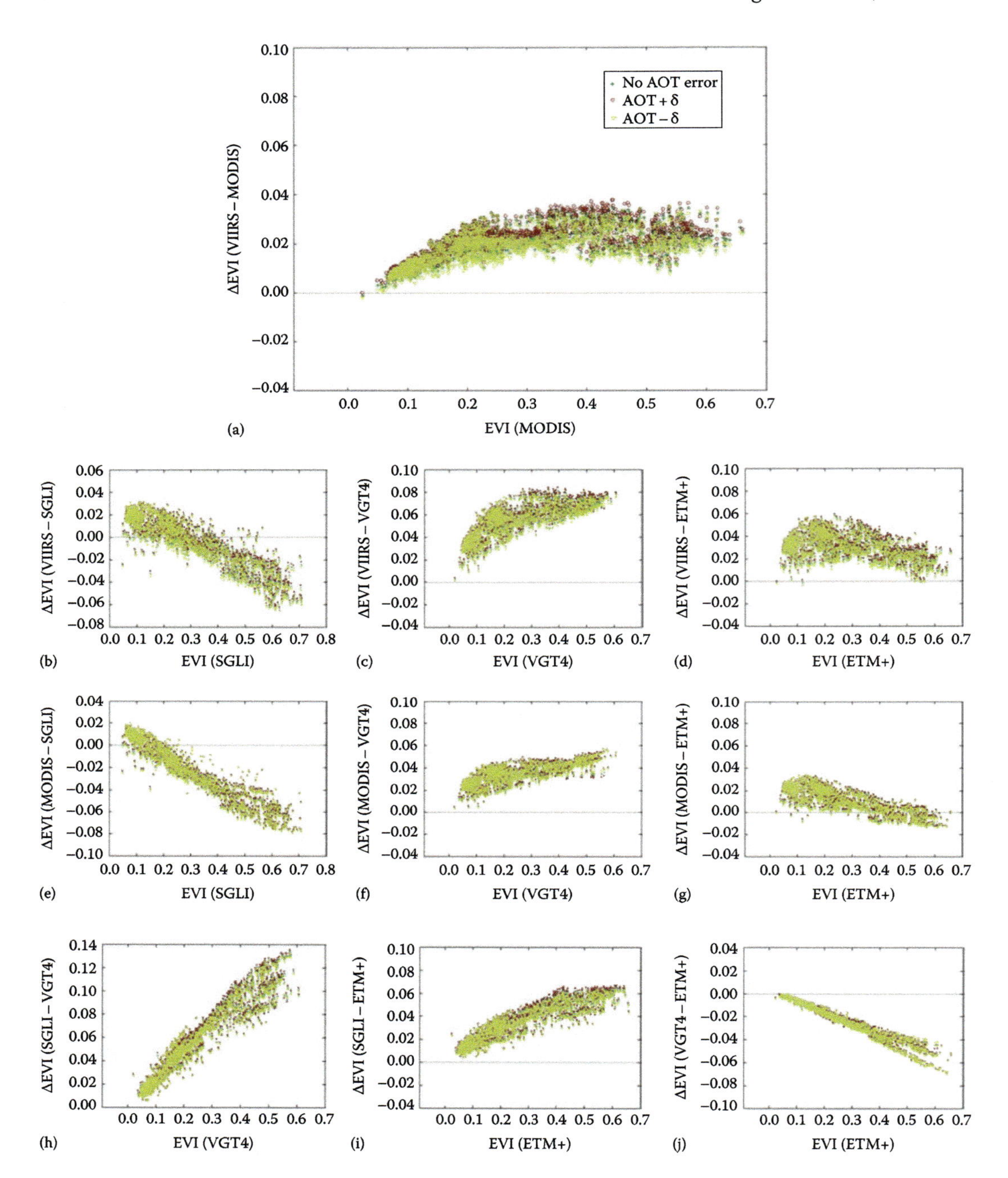


**FIGURE 10.3** Inter-sensor EVI difference across VIIRS, MODIS, SGLI, VGT4, and ETM+ bandpasses subject to atmospheric aerosol correction error.

pairs. The systematic differences basically linearly increased or decreased with EVI values although some curve-linearity was seen for those between VIIRS and MODIS, VGT4, or ETM+ (Figure 10.3a, c, and d, respectively). The smallest and largest systematic differences (in terms of sRMPD) were observed for MODIS vs. ETM+ and SGLI vs. VGT4, respectively (Table 10.2). The magnitudes of unsystematic differences also changed among sensor pairs, and uRMPD were the smallest for VGT4 vs. ETM+ and the largest for VIIRS vs. SGLI and ETM+ (Figure 10.3 and Table 10.2).

Overlapped symbols in Figure 10.3 indicate that inter-sensor EVI differences remained nearly the same when the input surface reflectances were subject to residual aerosol contaminations. For all the sensor pairs examined here, RMSD, sRMPD, and uRMPD of inter-sensor EVI were the same whether the input surface reflectances were subject to AOT errors or not (Table 10.2). These results indicate that the atmospheric resistance of the EVI functions well for inter-sensor EVI.

#### 10.4.2.2 Inter-Sensor EVI2

Inter-sensor EVI2 relationships were similar to those of EVI for the sensor pairs examined here (Figure 10.4). They were, however, very different for VIIRS vs. MODIS in that the inter-sensor EVI2 relationship was very linear and their biases were, on average, negative and smaller than those of EVI (Figure 10.4a).

For the sensor pairs studied here, inter-sensor EVI2 differences were basically smaller than those of EVI when the input reflectances were not subject to errors. RMSD, sRMPD, and uRMPD for inter-sensor EVI2 relationships ranged for 0.003–0.035, 0.002–0.034, and 0.002–0.008, respectively (Table 10.3) (see Appendix 10.A.2 for GMFR equations), whereas the corresponding values for inter-sensor EVI relationships were 0.02–0.06 (RMSD), 0.019–0.06 (sRMPD), and 0.003–0.009 (uRMPD) (Table 10.3). Table 10.4 lists the ratios of these three different statistics for every sensor pairs. One can observe that, except for MODIS vs. VGT4 and MODIS vs. ETM+, all the ratios were less than 1.0, indicating smaller systematic and unsystematic, and overall differences for inter-sensor EVI2 relationships.

On the other hand, unlike the EVI, inter-sensor EVI2 differences varied when the input reflectances were subject to residual aerosol contaminations due to atmospheric correction errors. In Figure 10.4, it can be seen that EVI2 differences formed three separate trends, corresponding to the three correction error scenarios for all the sensor pairs. It can also be seen as a 20% or more increase in uRMPD upon the contaminations (the smallest uRMPD ratio in Table 10.3 being 1.2). The resultant uRMPD of inter-sensor EVI2 relationships were nearly the same or larger (0.9–1.7 times) than the corresponding uRMPD of inter-sensor EVI relationships which were found insensitive to residual aerosol contaminations (the most right column in Table 10.4). The residual aerosol contaminations did not impact sRMPD because we used symmetric errors to perturb EVI2. These results indicate that the atmospheric resistance of the EVI with a blue band is advantageous in deriving inter-sensor EVI relationships.

**TABLE 10.2**
**Statistics for Inter-Sensor Spectral Difference: EVI**

| | TOC with No AOT Error | | | TOC with AOT Error of +/−0.05 | | | Ratio (TOC Error/TOC No Error) | | |
|---|---|---|---|---|---|---|---|---|---|
| | RMSD | sRMPD | uRMPD | RMSD | sRMPD | uRMPD | RMSD | sRMPD | uRMPD |
| VIIRS vs. MODIS | 0.019 | 0.019 | 0.005 | 0.020 | 0.019 | 0.005 | 1.0 | 1.0 | 1.0 |
| VIIRS vs. SGLI | 0.021 | 0.019 | 0.009 | 0.021 | 0.019 | 0.009 | 1.0 | 1.0 | 1.0 |
| VIIRS vs. VGT4 | 0.054 | 0.053 | 0.008 | 0.054 | 0.053 | 0.008 | 1.0 | 1.0 | 1.0 |
| VIIRS vs. ETM+ | 0.033 | 0.032 | 0.009 | 0.034 | 0.032 | 0.009 | 1.0 | 1.0 | 1.0 |
| MODIS vs. SGLI | 0.031 | 0.030 | 0.008 | 0.031 | 0.030 | 0.008 | 1.0 | 1.0 | 1.0 |
| MODIS vs. VGT4 | 0.035 | 0.034 | 0.005 | 0.035 | 0.034 | 0.005 | 1.0 | 1.0 | 1.0 |
| MODIS vs. ETM+ | 0.018 | 0.017 | 0.006 | 0.018 | 0.017 | 0.006 | 1.0 | 1.0 | 1.0 |
| SGLI vs. VGT4 | 0.060 | 0.060 | 0.008 | 0.060 | 0.060 | 0.008 | 1.0 | 1.0 | 1.0 |
| SGLI vs. ETM+ | 0.035 | 0.035 | 0.006 | 0.035 | 0.035 | 0.006 | 1.0 | 1.0 | 1.0 |
| VGT4 vs. ETM+ | 0.026 | 0.025 | 0.003 | 0.026 | 0.026 | 0.003 | 1.0 | 1.0 | 1.0 |

In order to evaluate and confirm the atmospheric resistance of the EVI against the EVI2, differences between EVI with and without residual contaminations as well as those between EVI2 with and without residual contaminations are plotted in Figure 10.5 and their RMSD summarized in Table 10.5. For the five sensor bandpasses, the EVI successfully reduced residual aerosol errors to less than ±0.005 (Figure 10.5) with RMSD of 0.002 or less (Table 10.5). EVI2 errors exceeded 0.015 with RMSD of 0.005 or more for the AOT estimation errors of ±0.05.

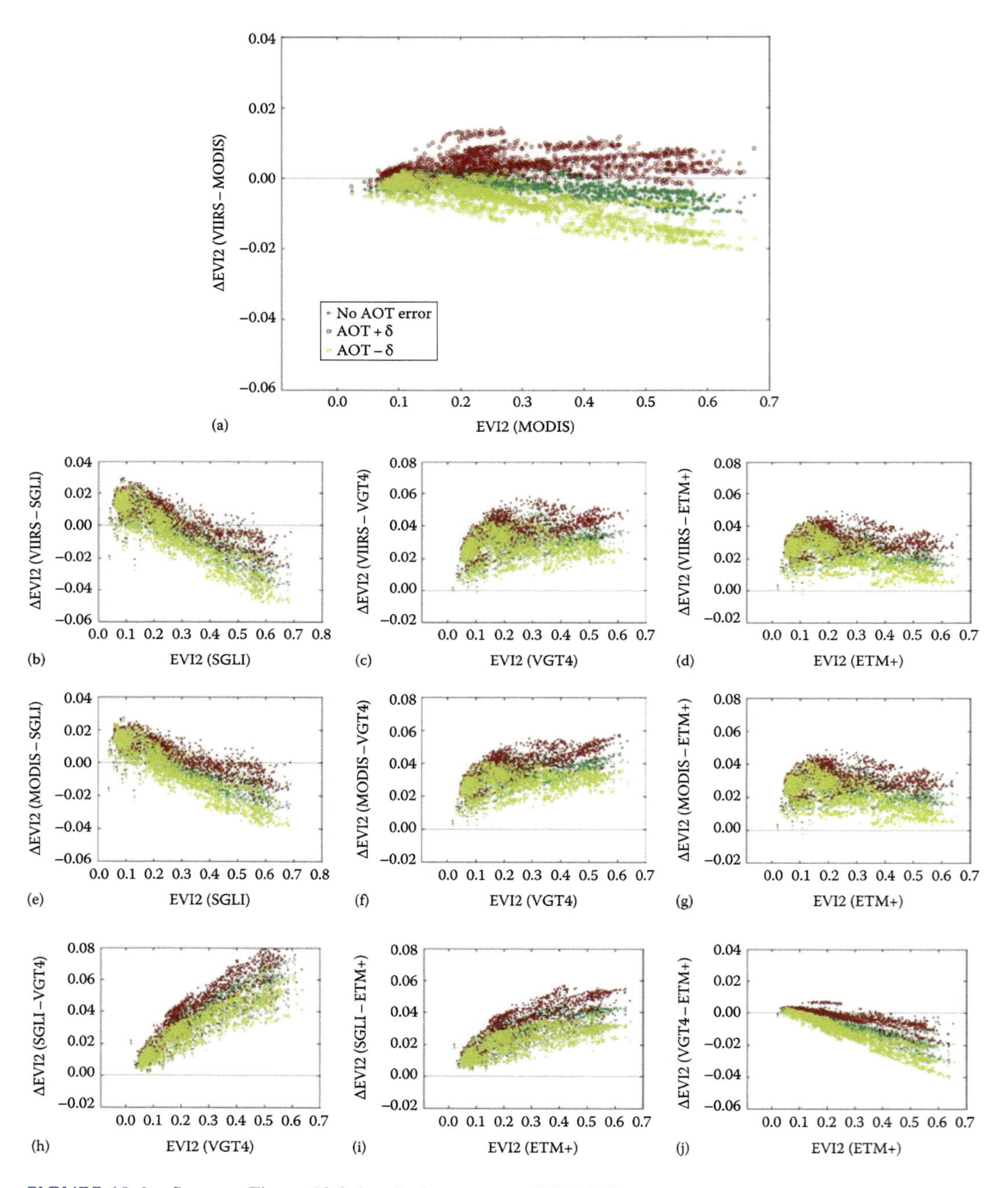


**FIGURE 10.4** Same as Figure 10.3, but for inter-sensor EVI2 difference.

## TABLE 10.3

### Statistics for Inter-Sensor Spectral Difference: EVI2

| | TOC with No AOT Error | | | TOC with AOT Error of +/−0.05 | | | Ratio (TOC Error/TOC No Error) | | |
|---|---|---|---|---|---|---|---|---|---|
| | RMSD | sRMPD | uRMPD | RMSD | sRMPD | uRMPD | RMSD | sRMPD | uRMPD |
| VIIRS vs. MODIS | 0.003 | 0.002 | 0.002 | 0.006 | 0.002 | 0.006 | 2.1 | 0.8 | 3.2 |
| VIIRS vs. SGLI | 0.015 | 0.014 | 0.007 | 0.016 | 0.014 | 0.009 | 1.1 | 1.0 | 1.3 |
| VIIRS vs. VGT4 | 0.031 | 0.031 | 0.007 | 0.032 | 0.031 | 0.009 | 1.0 | 1.0 | 1.3 |
| VIIRS vs. ETM+ | 0.026 | 0.025 | 0.008 | 0.027 | 0.025 | 0.010 | 1.0 | 1.0 | 1.2 |
| MODIS vs. SGLI | 0.013 | 0.012 | 0.005 | 0.014 | 0.012 | 0.008 | 1.1 | 1.0 | 1.5 |
| MODIS vs. VGT4 | 0.033 | 0.032 | 0.006 | 0.034 | 0.033 | 0.008 | 1.0 | 1.0 | 1.4 |
| MODIS vs. ETM+ | 0.027 | 0.026 | 0.006 | 0.028 | 0.027 | 0.008 | 1.0 | 1.0 | 1.3 |
| SGLI vs. VGT4 | 0.035 | 0.034 | 0.006 | 0.036 | 0.035 | 0.008 | 1.0 | 1.0 | 1.3 |
| SGLI vs. ETM+ | 0.026 | 0.026 | 0.005 | 0.027 | 0.026 | 0.008 | 1.0 | 1.0 | 1.4 |
| VGT4 vs. ETM+ | 0.010 | 0.009 | 0.002 | 0.011 | 0.009 | 0.006 | 1.1 | 1.0 | 2.6 |

## TABLE 10.4

### Statistics for Inter-Sensor Spectral Difference: EVI2 to EVI Ratio

| | TOC with No AOT Error | | | TOC with AOT Error of +/−0.05 | | |
|---|---|---|---|---|---|---|
| | RMSD | sRMPD | uRMPD | RMSD | sRMPD | uRMPD |
| VIIRS vs. MODIS | 0.2 | 0.1 | 0.4 | 0.3 | 0.1 | 1.2 |
| VIIRS vs. SGLI | 0.7 | 0.7 | 0.7 | 0.8 | 0.7 | 0.9 |
| VIIRS vs. VGT4 | 0.6 | 0.6 | 0.8 | 0.6 | 0.6 | 1.0 |
| VIIRS vs. ETM+ | 0.8 | 0.8 | 0.8 | 0.8 | 0.8 | 1.0 |
| MODIS vs. SGLI | 0.4 | 0.4 | 0.7 | 0.5 | 0.4 | 1.0 |
| MODIS vs. VGT4 | 1.0 | 0.9 | 1.1 | 1.0 | 1.0 | 1.5 |
| MODIS vs. ETM+ | 1.5 | 1.6 | 1.0 | 1.6 | 1.6 | 1.4 |
| SGLI vs. VGT4 | 0.6 | 0.6 | 0.8 | 0.6 | 0.6 | 1.0 |
| SGLI vs. ETM+ | 0.7 | 0.7 | 0.9 | 0.8 | 0.8 | 1.2 |
| VGT4 vs. ETM+ | 0.4 | 0.4 | 0.7 | 0.4 | 0.4 | 1.7 |

## TABLE 10.5

### Comparison between EVI and EVI2 Resistance to Residual Aerosol Effects

| RMSD | EVI | EVI2 | EVI2-to-EVI Ratio |
|---|---|---|---|
| VIIRS | 0.002 | 0.006 | 3.5 |
| MODIS | 0.001 | 0.006 | 6.8 |
| SGLI | 0.001 | 0.006 | 4.5 |
| VGT4 | 0.001 | 0.005 | 7.6 |
| ETM+ | 0.001 | 0.005 | 4.8 |

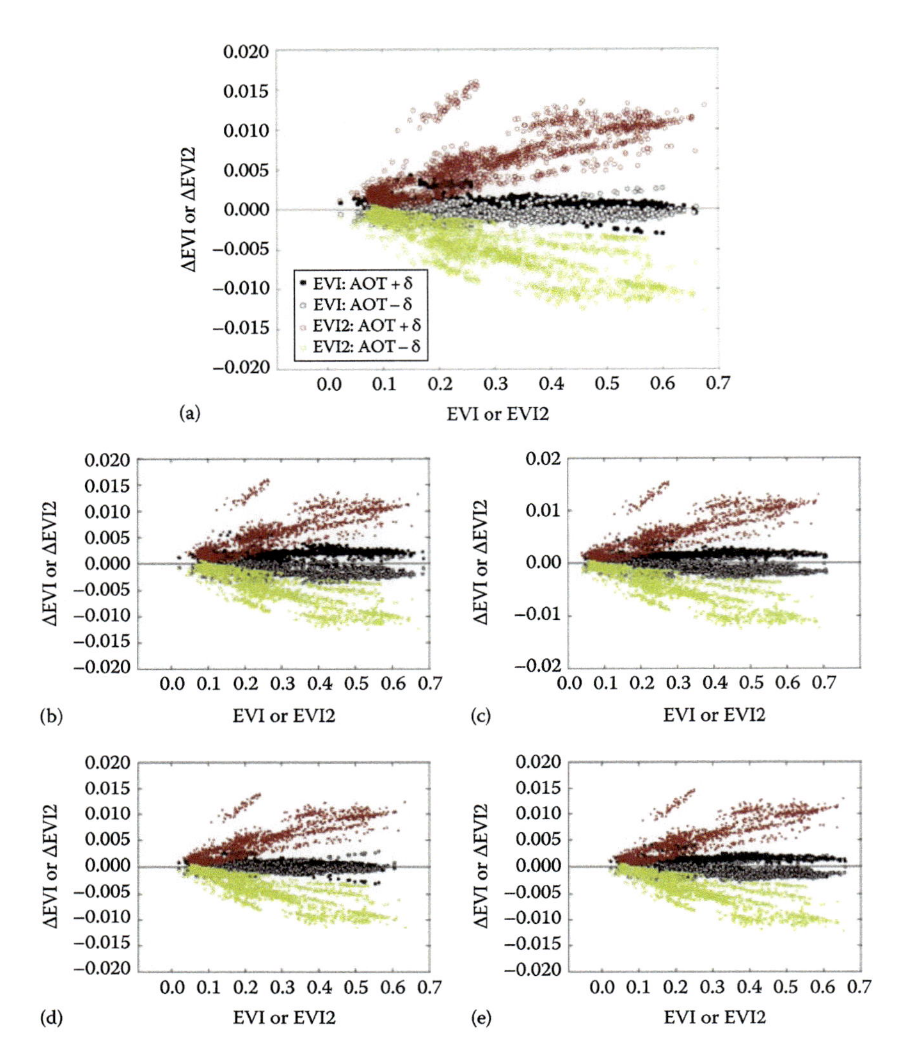


**FIGURE 10.5** Comparison of EVI and EVI2 resistance to atmospheric aerosol correction error for five sensor bandpasses.

#### 10.4.2.3 Intra-Sensor EVI2 versus EVI

EVI2-to-EVI differences with and without residual aerosol errors are plotted as a function of EVI in Figure 10.6 and their statistics summarized in Table 10.6 (see Appendix 10.A.3 for GMFR equations) for the purpose of analyzing intra-sensor EVI2-to-EVI compatibilities for the sensors other than MODIS. Intra-sensor EVI2 versus EVI relationships of all sensor bandpasses were subject to systematic differences although their magnitudes changed from sensor to sensor. They were the largest for SGLI with sRMPD of 0.015–0.016 and the smallest for ETM+ with sRMPD of 0.006–0.007 (Table 10.6). These sRMPD values were basically all less than those observed for inter-sensor EVI or EVI2 relationships, except for that of VIIRS EVI2 versus MODIS EVI2 that had the smallest sRMPD and that of VGT4 EVI2 versus ETM+ EVI2.

Unsystematic variations of these intra-sensor relationships were about the same across all the sensors with uRMPD ranging from 0.007 to 0.008 without residual aerosols and from 0.008 to 0.01 with residual aerosols (Table 10.6). The magnitudes of uRMPD were comparable to those of inter-sensor EVI2 relationships (Table 10.3).

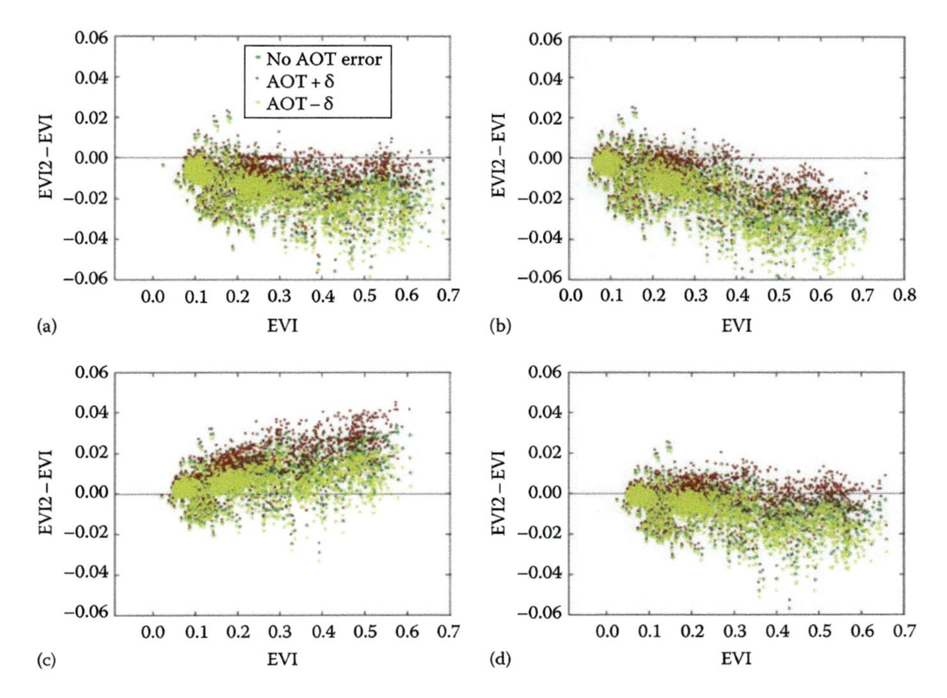


FIGURE 10.6 Intra-sensor EVI versus EVI2 differences subject to atmospheric aerosol correction error for four sensor bandpasses.

TABLE 10.6
**Difference Statistics for Intra-Sensor EVI2 and EVI Compatibility**

| | TOC with No AOT Error | | | TOC with AOT Error of +/−0.05 | | | Ratio (TOC Error/TOC No Error) | | |
|---|---|---|---|---|---|---|---|---|---|
| | RMSD | sRMPD | uRMPD | RMSD | sRMPD | uRMPD | RMSD | sRMPD | uRMPD |
| **VIIRS** | 0.016 | 0.014 | 0.008 | 0.016 | 0.013 | 0.010 | 1.02 | 0.96 | 1.14 |
| **SGLI** | 0.018 | 0.016 | 0.008 | 0.018 | 0.015 | 0.010 | 1.01 | 0.97 | 1.16 |
| **VGT4** | 0.012 | 0.010 | 0.007 | 0.013 | 0.010 | 0.008 | 1.11 | 1.05 | 1.22 |
| **ETM+** | 0.010 | 0.007 | 0.007 | 0.011 | 0.006 | 0.009 | 1.08 | 0.93 | 1.18 |

### 10.4.2.4 Inter-Sensor Compatibility with AVHRR EVI2

Lastly, we present inter-sensor compatibility between TOC EVI2 and partial-atmosphere corrected (PAC) AVHRR EVI2. No operational aerosol correction of AVHRR data has been implemented and, thus, this provides a realistic scenario on EVI2 compatibility/continuity with AVHRR. For comparisons, AVHRR TOC EVI2 was also analyzed.

Figure 10.7a and b show example plots for VIIRS TOC EVI2 versus AVHRR/3 TOC EVI2 and for VIIRS TOC EVI2 versus AVHRR/3 PAC EVI2 differences, respectively. Whether corrected for total or partial atmosphere, AVHRR/3 EVI2 had a linear relationship with VIIRS EVI2. However, AVHRR/3 PAC EVI2 had larger systematic and unsystematic differences than the TOC counterpart against VIIRS EVI2. sRMPD and uRMPD for the former were 0.028 and 0.007, respectively, whereas those for the latter were 0.041 and 0.011, respectively (Table 10.7). The same trends were observed for all the other sensors when paired with AVHRR/3 or AVHRR/2 (Table 10.7). AVHRR EVI2 had the largest differences with SGLI and the smallest differences with VGT4 (Table 10.7).

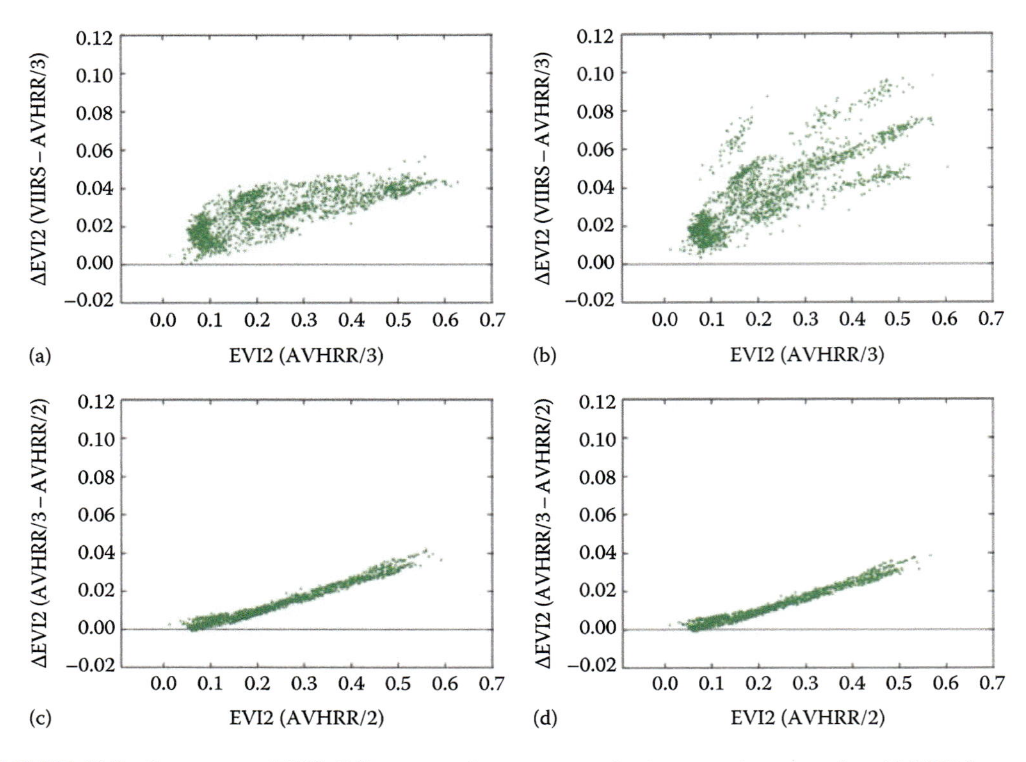


**FIGURE 10.7** Inter-sensor EVI2 difference under two atmospheric correction scenarios: (a) VIIRS versus AVHRR/3 for TOC, (b) VIIRS versus AVHRR/3 for PAC, (c) AVHRR/3 versus AVHRR/2 for TOC, and (d) AVHRR/3 versus AVHRR/2 for PAC.

**TABLE 10.7**

**Difference Statistics of Inter-Sensor AVHRR EVI2 Compatibility (See Appendix 10.A for Corresponding GMFR Equations)**

| EVI2 vs. AVHRR EVI2 | TOC vs. TOC | | | TOC vs. PAC | | | Ratio (PAC/TOC) | | |
|---|---|---|---|---|---|---|---|---|---|
| | RMSD | sRMPD | uRMPD | RMSD | sRMPD | uRMPD | RMSD | sRMPD | uRMPD |
| **VIIRS vs. AVHRR/3** | 0.029 | 0.028 | 0.007 | 0.042 | 0.041 | 0.011 | 1.5 | 1.5 | 1.7 |
| **MODIS vs. AVHRR/3** | 0.031 | 0.030 | 0.006 | 0.044 | 0.043 | 0.011 | 1.4 | 1.4 | 1.8 |
| **SGLI vs. AVHRR/3** | 0.035 | 0.034 | 0.008 | 0.049 | 0.047 | 0.013 | 1.4 | 1.4 | 1.6 |
| **VGT4 vs. AVHRR/3** | 0.008 | 0.007 | 0.004 | 0.018 | 0.015 | 0.010 | 2.2 | 2.2 | 2.2 |
| **ETM+ vs. AVHRR/3** | 0.014 | 0.013 | 0.005 | 0.027 | 0.025 | 0.010 | 1.9 | 1.9 | 1.9 |
| **VIIRS vs. AVHRR/2** | 0.043 | 0.042 | 0.006 | 0.055 | 0.054 | 0.010 | 1.3 | 1.3 | 1.8 |
| **MODIS vs. AVHRR/2** | 0.045 | 0.045 | 0.006 | 0.058 | 0.057 | 0.010 | 1.3 | 1.3 | 1.8 |
| **SGLI vs. AVHRR/2** | 0.050 | 0.050 | 0.008 | 0.063 | 0.062 | 0.012 | 1.3 | 1.2 | 1.5 |
| **VGT4 vs. AVHRR/2** | 0.018 | 0.017 | 0.005 | 0.030 | 0.029 | 0.010 | 1.7 | 1.7 | 1.9 |
| **ETM+ vs. AVHRR/2** | 0.027 | 0.027 | 0.006 | 0.040 | 0.038 | 0.010 | 1.4 | 1.4 | 1.7 |
| **AVHRR/3 vs. AVHRR/2** | 0.015 | 0.015 | 0.002 | 0.014* | 0.014* | 0.001* | 0.9 | 0.9 | 1.0 |

* *These statistics are for AVHRR/3 PAC EVI2 versus AVHRR/2 PAC EVI2.*

AVHRR/3 and AVHRR/2 EVI2 showed very high compatibility regardless of the level of atmospheric correction as far as both sensor data had the same levels of atmospheric correction (Figure 10.7c and d). Although they were subject to some systematic differences (sRMPD ~0.015), they were subject to very small unsystematic variations (uRMPD ~0.002) and their relationships were nearly independent of atmospheres, indicating the compatible sensitivity of their spectral bandpasses to atmospheric (aerosol) effects.

In summary, inter-sensor EVI relationships were insensitive to residual aerosol contaminations due to EVI's atmospheric resistance. The EVI2 showed higher multi-sensor compatibility than the EVI, with smaller bias and unsystematic errors, which was however only when the input reflectances were not subject to residual aerosol contaminations. Intra-sensor EVI versus EVI2 relationships were subject to the same magnitudes of unsystematic errors as inter-sensor EVI2 relationships. Intra-sensor EVI2-to-EVI relationships were subject to significant biases for some sensors. This indicated the need of EVI and/or EVI2 coefficient adjustments, which were optimized for the MODIS bandpasses by design. Backward compatibility with AVHRR EVI2 would be a feasible option only when aerosol source of variability are reduced in AVHRR EVI2 either by atmospheric correction or by temporal compositing to select cleanest observations.

## 10.5 INTER-SENSOR CALIBRATION OF EVI AND EVI2

Some of the studies listed in Table 10.1 have proposed inter-sensor calibration methods for the EVI and EVI2. These cross-calibration methods can be divided into empirical (Abbasi et al., 2021; Kim et al., 2010) and theoretical (Obata et al., 2013, 2016; Yoshioka et al., 2012) approaches.

Kim et al. (2010) used a simple linear model and derived EVI-to-EVI and EVI2-to-EVI2 spectral translation equations between MODIS and select medium resolution sensors

$$X_{Sensor} = a \cdot X_{MODIS} + b + \varepsilon \tag{10.6}$$

where $X$ is the EVI or EVI2, the subscript, *Sensor*, refers to either VIIRS, VGT4, SeaWiFS, or AVHRR/2, and $\varepsilon$ is the unexplained residual error.

Kim et al. (2010) also developed intra-sensor EVI-to-EVI2 translation equations with the simple linear model for the VIIRS, VGT4, and SeaWiFS bandpasses:

$$EVI2_{Sensor} = a \cdot EVI_{Sensor} + b + \varepsilon \tag{10.7}$$

Mean absolute deviations/differences (MAD) were used to evaluate translation results:

$$MAD = \frac{1}{n}\sum_{i=1}^{n}\left|X_{1,i} - X_{2,i}\right| \tag{10.8}$$

where $X$'s are the EVI, EVI2, translated (predicted) EVI, or translated (predicted) EVI2 using the derived simple linear equations and two of the four variables are selected and inserted into Equation 10.8 depending on the quantity of interest. Hyperion data collected along an tropical forest-savanna eco-gradient in Brazil were used in all of these derivations and evaluations of cross-calibration equations.

The derived, simple linear spectral translation equations performed well on the Hyperion-simulated bandpass data (Table 10.8). For both the EVI and EVI2, MAD were reduced to 0.005 or less (in EVI/EVI2 units) upon the translations for most cases, which corresponded to 31–87% reductions in MAD (Table 10.8a, b). For the EVI-to-EVI2 spectral calibration, the equations did not perform as good as those for the inter-sensor calibration. Reductions in MAD were 35% or less and the largest MAD of 0.0154 was observed for VGT4 (Table 10.8c).

Yoshioka et al. (2012) and Obata et al. (2013) analytically derived EVI and/or EVI2 spectral translation equations. Both studies used the vegetation isoline equations in the derivations, which are based upon the physics of vegetation-photon interactions (Yoshioka, 2004).

**TABLE 10.8**

**Comparison of Mean Absolute Deviation (MAD) before and after Spectral Translation Using Simple Linear Cross-Calibration Equation**

(a) EVI inter-sensor spectral compatibility with MODIS

| | MAD for $EVI_{Sensor}$ vs. $EVI_{MODIS}$ | MAD for $EVI_{Sensor}$ vs. pred. $EVI_{Sensor}$ | % Reduction |
|---|---|---|---|
| VIIRS | 0.0118 | 0.0052 | 56% |
| VGT4 | 0.0207 | 0.0042 | 80% |
| SeaWiFS | 0.0108 | 0.0070 | 35% |

(b) EVI2 inter-sensor spectral compatibility with MODIS

| | MAD for $EVI2_{Sensor}$ vs. $EVI2_{MODIS}$ | MAD for $EVI2_{Sensor}$ vs. pred. $EVI2_{Sensor}$ | % Reduction |
|---|---|---|---|
| VIIRS | 0.0027 | 0.0018 | 33% |
| VGT4 | 0.0255 | 0.0033 | 87% |
| SeaWiFS | 0.0048 | 0.0033 | 31% |
| AVHRR/2 | 0.0411 | 0.0059 | 86% |

(c) EVI-to-EVI2 intra-sensor compatibility

| | MAD for $EVI2_{Sensor}$ vs. $EVI_{Sensor}$ | MAD for $EVI2_{Sensor}$ vs. pred. $EVI2_{Sensor}$ | % Reduction |
|---|---|---|---|
| VIIRS | 0.0108 | 0.0077 | 29% |
| VGT4 | 0.0236 | 0.0154 | 35% |
| SeaWiFS | 0.0103 | 0.0091 | 12% |

*Note*: Adapted from Kim et al., (2010).

Yoshioka et al. (2012) used the vegetation isoline equations first to eliminate NIR and blue reflectances from the EVI equation or NIR reflectance from the EVI2 equation, and then to relate two EVI or EVI2 using the red reflectances. Derived is a spectral transformation equation that relates two EVI or two EVI2:

$$v_a = G \cdot \frac{h_1 v_b - h_2}{h_3 v_b - h_4} \tag{10.9}$$

where $v_a$ and $v_b$ are the EVI or EVI2 for sensors *a* and *b*, respectively. The four coefficients, $h_i$ ($i = 1, \ldots, 4$), are actually functions and their values change with vegetation amount, soil, and atmosphere.

In Obata et al. (2013), MODIS red, NIR, and blue reflectances were expressed as a function of VIIRS red, NIR, and blue reflectances, respectively, using the vegetation isoline equations. Substituting MODIS reflectances with the functions resulted in a MODIS-compatible EVI, $\hat{v}_m$ :

$$\hat{v}_m = G \cdot \frac{\rho_{NIR} - K_1 \rho_{red} + K_2}{\rho_{NIR} + K_1 C_1 \rho_{red} - K_3 C_2 \rho_{blue} + K_4} \tag{10.10}$$

where $K_i$ ($i = 1, \ldots, 4$) are the spectral transformation coefficients. As $h_i$ in Equation 10.9, $K_i$ are functions and change with vegetation, soil, and atmosphere conditions.

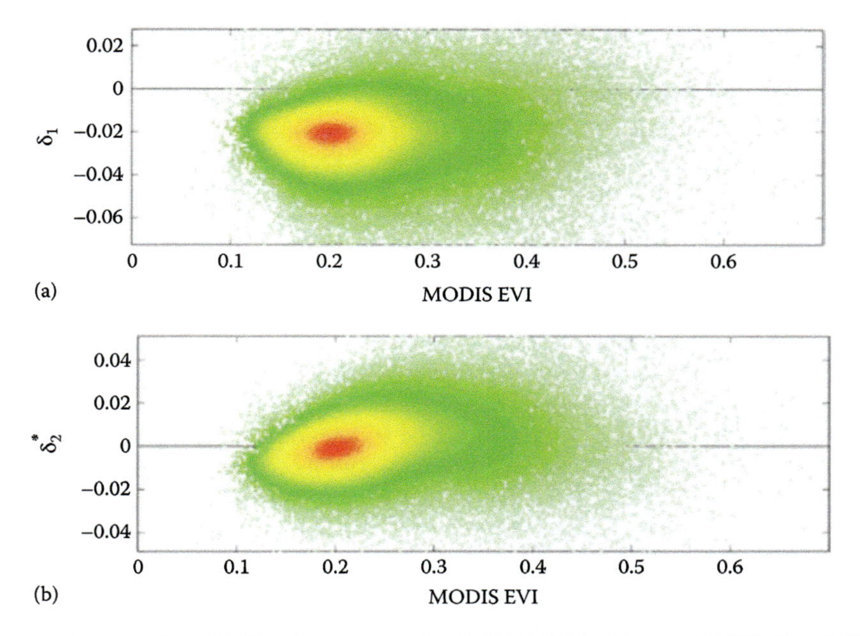


**FIGURE 10.8** Density plot of EVI differences against MODIS EVI with actual MODIS and VIIRS data. (a) MODIS EVI minus VIIRS EVI ($\delta_1$) against MODIS EVI ($v_m$); (b) MODIS EVI minus MODIS-compatible VIIRS EVI using optimized parameters ($\delta_2$*) against MODIS EVI ($v_m$). Adapted from Obata et al. (2013).

Obata et al. (2013), however, found via optimization for both model-simulated and actual satellite datasets, that a single set of coefficient values can be derived with a reasonable level of accuracy. Their simulation dataset was generated with the PROSAIL canopy reflectance model (Feret et al., 2008) and the 6S atmospheric radiative transfer model. The actual MODIS-VIIRS dataset was populated by extracting pairs of Aqua MODIS and VIIRS daily nadir-view surface reflectance spectra on the same date (in August 2013) at the same coordinates over North America. Refer to Obata et al. (2013) for further details on these datasets.

For the simulated dataset, the mean of $\delta_1$ (MODIS EVI minus VIIRS EVI) and $\delta_2$* (MODIS EVI minus MODIS-compatible VIIRS EVI) were −0.008 and 0.0001, respectively, and variability of $\delta_2$* was much smaller than that of $\delta_1$ (RMSE changed from 0.01 to 0.0017; an 83% reduction) (Obata et al., 2013). For the actual dataset, the mean and RMSE of $\delta_2$* were 0.001 and 0.023, a 95% and 41% reduction, respectively, from those of $\delta_1$ (−0.022 and 0.039) (Figure 10.8).

## 10.6 DISCUSSIONS AND FUTURE DIRECTIONS

In this chapter, inter- and intra-sensor compatibilities of the EVI and EVI2 were reviewed with a focus on spectral, atmospheric, and calibration issues. Overall, the EVI2 showed higher inter-sensor spectral compatibilities than the EVI. The highest spectral compatibility was observed between MODIS EVI2 and VIIRS EVI2, and AVHRR/2 EVI2 and AVHRR/3 EVI2 although both inter-sensor EVI2 relationships were subject to systematic differences. When the residual aerosol effects were considered, however, inter-sensor spectral compatibilities of the EVI2 and EVI were at comparable levels. The blue band in the EVI is advantageous in reducing residual aerosol effects, but brings an added complication in inter-sensor compatibilities. Likewise, the higher inter-sensor EVI2 compatibilities do not necessarily translate into higher absolute accuracies of the EVI2.

Among the limited number of studies and available cross-calibration methods, the approach of Obata et al. (2013) appears to be the most promising. The most significant advantage of the approach would be that the translation technique simultaneously optimizes for bandpass differences and for the EVI coefficients, which can not be accomplished with the simple linear regression approach. The potential pitfall in the approach is a nonexistence of the unique set of coefficient values, as it requires nonlinear regression to find the optimum coefficient values.

Inter-sensor spectral compatibility of AVHRR EVI2 with other sensors is challenging, in particular with MODIS, VIIRS, and SGLI of which red and NIR spectral bands are much narrower than those of AVHRR (i.e., Figure 10.1). Backward compatibility of these sensors with AVHRR EVI2 would be a feasible option only when aerosol and water vapor sources of variability are reduced in AVHRR data either by atmospheric correction or by temporal compositing to select cleanest observations. Another approach would be to generate more than one EVI/EVI2 long-term data record in parallel, for example, one solely from the AVHRR sensor series and another by merging MODIS and VIIRS EVI products. Inter-comparisons of the two data records over the current overlapping period can be used to develop a methodology for analyzing historical AVHRR EVI2 time series.

Empirical investigations with actual satellite data should still continue as new sensors are continuously developed and launched. For cross-sensor compatibility and calibration of actual satellite data, additional factors need to be taken into consideration, including differences in spatial resolution, overpass time (morning vs. afternoon), and Sun-target-viewing geometry (Sun and view zenith angle differences due to different sensor platform orbits) (e.g., Morton et al., 2014; Sims et al., 2011; Wang et al., 2022).

## APPENDIX 10.A INTER- AND INTRA-SENSOR SPECTRAL RELATIONSHIPS FOR EVI AND EVI2

The following equations were estimated for the Hyperion dataset (without AOT errors) described in Section 10.4.1 by GMFR.

**TABLE 10.A.1**
**Inter-Sensor EVI vsersus EVI Spectral Relationships**

| *Y* vs. *X* | Intercept [a] | Slope [a] | $R^2$ |
|---|---|---|---|
| **VIIRS vs. MODIS** | 0.010 | 1.033 | 0.999 |
| **VIIRS vs. SGLI** | 0.029 | 0.893 | 0.997 |
| **VIIRS vs. VGT4** | 0.033 | 1.083 | 0.997 |
| **VIIRS vs. ETM+** | 0.038 | 0.975 | 0.997 |
| **MODIS vs. SGLI** | 0.019 | 0.864 | 0.998 |
| **MODIS vs. VGT4** | 0.023 | 1.048 | 0.999 |
| **MODIS vs. ETM+** | 0.027 | 0.944 | 0.999 |
| **SGLI vs. VGT4** | 0.005 | 1.213 | 0.998 |
| **SGLI vs. ETM+** | 0.010 | 1.092 | 0.999 |
| **VGT4 vs. ETM+** | 0.004 | 0.900 | 1.000 |

[a] Useful for trend analysis; use with caution on actual satellite data.

### TABLE 10.A.2
**Inter-Sensor EVI2 versus EVI2 Spectral Relationships**

| *Y* vs. *X* | Intercept [a] | Slope [a] | $R^2$ |
|---|---|---|---|
| VIIRS vs. MODIS | 0.001 | 0.989 | 1.000 |
| VIIRS vs. SGLI | 0.021 | 0.918 | 0.998 |
| VIIRS vs. VGT4 | 0.026 | 1.018 | 0.998 |
| VIIRS vs. ETM+ | 0.031 | 0.971 | 0.998 |
| MODIS vs. SGLI | 0.020 | 0.929 | 0.999 |
| MODIS vs. VGT4 | 0.025 | 1.030 | 0.999 |
| MODIS vs. ETM+ | 0.030 | 0.982 | 0.998 |
| SGLI vs. VGT4 | 0.006 | 1.108 | 0.999 |
| SGLI vs. ETM+ | 0.011 | 1.057 | 0.999 |
| VGT4 vs. ETM+ | 0.005 | 0.954 | 1.000 |

[a] Useful for trend analysis; use with caution on actual satellite data.

### TABLE 10.A.3
**Intra-Sensor EVI2 versus EVI Relationships**

| *EVI2* vs. *EVI* | Intercept [a] | Slope [a] | $R^2$ |
|---|---|---|---|
| VIIRS | −0.006 | 0.973 | 0.997 |
| SGLI | 0.002 | 0.946 | 0.998 |
| VGT4 | 0.000 | 1.036 | 0.998 |
| ETM+ | 0.000 | 0.977 | 0.998 |

[a] Useful for trend analysis; use with caution on actual satellite data.

### TABLE 10.A.4
**Inter-Sensor Relationships for AVHRR EVI2**

| | TOC vs. TOC | | | TOC vs. PAC | | |
|---|---|---|---|---|---|---|
| *Y* vs. *X* | Intercept [a] | Slope [a] | $R^2$ | Intercept [b] | Slope [b] | $R^2$ |
| VIIRS vs. AVHRR/3 | 0.013 | 1.058 | 0.998 | 0.008 | 1.148 | 0.999 |
| MODIS vs. AVHRR/3 | 0.012 | 1.070 | 0.998 | −0.013 | 1.236 | 0.997 |
| SGLI vs. AVHRR/3 | −0.009 | 1.152 | 0.997 | −0.017 | 1.115 | 0.999 |
| VGT4 vs. AVHRR/3 | −0.013 | 1.039 | 0.999 | −0.023 | 1.169 | 0.998 |
| ETM+ vs. AVHRR/3 | −0.018 | 1.089 | 0.999 | 0.007 | 1.199 | 0.995 |
| VIIRS vs. AVHRR/2 | 0.011 | 1.119 | 0.994 | 0.006 | 1.212 | 0.995 |
| MODIS vs. AVHRR/2 | 0.010 | 1.132 | 0.994 | −0.016 | 1.305 | 0.993 |
| SGLI vs. AVHRR/2 | −0.011 | 1.219 | 0.993 | −0.019 | 1.178 | 0.995 |
| VGT4 vs. AVHRR/2 | −0.015 | 1.099 | 0.995 | −0.025 | 1.234 | 0.995 |
| ETM+ vs. AVHRR/2 | −0.021 | 1.152 | 0.995 | −0.004 | 1.073 | 1.000 |
| AVHRR/3 vs. AVHRR/2 | 0.009 | 1.135 | 0.998 | −0.004* | 1.071* | 1.000 |

[a] Useful for trend analysis; use with caution on actual satellite data.
[b] Useful for trend analysis; do not use on actual satellite data.
* The intercept and slope values are for AVHRR/3 PAC EVI2 versus AVHRR/2 PAC EVI2.

## REFERENCES

Abbasi, N., Nouri, H., Didan, K., Barreto-Muñoz, A., Chavoshi Borujeni, S., Salemi, H., . . . Nagler, P. (2021). Estimating actual evapotranspiration over croplands using vegetation index methods and dynamic harvested area. *Remote Sensing*, *13*(24). doi:10.3390/rs13245167

Arekhi, M., Goksel, C., Balik Sanli, F., & Senel, G. (2019). Comparative evaluation of the spectral and spatial consistency of Sentinel-2 and Landsat-8 OLI data for Igneada longos forest. *ISPRS International Journal of Geo-Information*, *8*(2). doi:10.3390/ijgi8020056

Cabacinha, C. D., & de Castro, S. S. (2009). Relationships between floristic diversity and vegetation indices, forest structure and landscape metrics of fragments in Brazilian Cerrado. *Forest Ecology and Management*, *257*(10), 2157–2165. doi:10.1016/j.foreco.2009.02.030

Cao, C., DeLuccia, F. J., Xiong, X., Wolfe, R., & Weng, F. (2014). Early on-orbit performance of the visible infrared imaging radiometer suite onboard the Suomi National Polar-Orbiting Partnership (S-NPP) satellite. *IEEE Transactions on Geoscience and Remote Sensing*, *52*(2), 1142–1156. doi:10.1109/TGRS.2013.2247768

Chen, M., Zhuang, Q., Cook, D. R., Coulter, R., Pekour, M., Scott, R. L., . . . Bible, K. (2011). Quantification of terrestrial ecosystem carbon dynamics in the conterminous United States combining a process-based biogeochemical model and MODIS and AmeriFlux data. *Biogeosciences*, *8*(9), 2665–2688. doi:10.5194/bg-8-2665-2011

Didan, K., & Huete, A. (2006). *MODIS Vegetation Index Product Series Collection 5 Change Summary*, last accessed on 13 July 2014. Retrieved from http://landweb.nascom.nasa.gov/QA_WWW/forPage/MOD13_VI_C5_Changes_Document_06_28_06.pdf

Fensholt, R., Sandholt, I., & Stisen, S. (2006). Evaluating MODIS, MERIS, and VEGETATION—Vegetation indices using in situ measurements in a semiarid environment. *IEEE Transactions on Geoscience and Remote Sensing*, *44*(7), 1774–1786. doi:10.1109/TGRS.2006.875940

Feret, J.-B., FranÃ§ois, C., Asner, G. P., Gitelson, A. A., Martin, R. E., Bidel, L. P. R., . . . Jacquemoud, S. (2008). PROSPECT-4 and 5: Advances in the leaf optical properties model separating photosynthetic pigments. *Remote Sensing of Environment*, *112*(6), 3030–3043. doi:10.1016/j.rse.2008.02.012

Ferreira, L. G., Yoshioka, H., Huete, A., & Sano, E. E. (2003). Seasonal landscape and spectral vegetation index dynamics in the Brazilian Cerrado: An analysis within the Large-Scale Biosphere-Atmosphere experiment in AmazÙnia (LBA). *Remote Sensing of Environment*, *87*(4), 534–550. doi:10.1016/j.rse.2002.09.003

Friedl, M. A., Sulla-Menashe, D., Tan, B., Schneider, A., Ramankutty, N., Sibley, A., & Huang, X. (2010). MODIS collection 5 global land cover: Algorithm refinements and characterization of new datasets. *Remote Sensing of Environment*, *114*(1), 168–182. doi:10.1016/j.rse.2009.08.016

Guanter, L., Zhang, Y., Jung, M., Joiner, J., Voigt, M., Berry, J. A., . . . Griffis, T. J. (2014). Global and time-resolved monitoring of crop photosynthesis with chlorophyll fluorescence. *Proceedings of the National Academy of Sciences*, *111*(14), E1327–E1333. doi:10.1073/pnas.1320008111

Holben, B. N., Tanré, D., Smirnov, A., Eck, T. F., Slutsker, I., Abuhassan, N., . . . Zibordi, G. (2001). An emerging ground-based aerosol climatology: Aerosol optical depth from AERONET. *Journal of Geophysical Research*, *106*, 12067–12097.

Huete, A., Didan, K., Miura, T., Rodriguez, E. P., Gao, X., & Ferreira, L. G. (2002). Overview of the radiometric and biophysical performance of the MODIS vegetation indices. *Remote Sensing of Environment*, *83*(1–2), 195–213. doi:10.1016/S0034-4257(02)00096-2

Huete, A. R. (1988). A Soil-Adjusted Vegetation Index (SAVI). *Remote Sensing of Environment*, *25*(3), 295–309. doi:10.1016/0034-4257(88)90106-X

Huete, A. R., Liu, H. Q., Batchily, K., & van, L., W. (1997). A comparison of vegetation indices over a global set of TM images for EOS-MODIS. *Remote Sensing of Environment*, *59*(3), 440–451. doi:10.1016/S0034-4257(96)00112-5

Ichii, K., Ueyama, M., Kondo, M., Saigusa, N., Kim, J., Alberto, M. C., . . . Zhao, F. (2017). New data-driven estimation of terrestrial CO2 fluxes in Asia using a standardized database of eddy covariance measurements, remote sensing data, and support vector regression. *J. Geophys. Res .-Biogeo.*, *122*(4), 767–795. doi:10.1002/2016JG003640

Irons, J. R., Dwyer, J. L., & Barsi, J. A. (2012). The next Landsat satellite: The Landsat data continuity mission. *Remote Sensing of Environment*, *122*, 11–21. Retrieved from www.sciencedirect.com/science/article/pii/S0034425712000363

Jarchow, C. J., Didan, K., Barreto-Muñoz, A., Nagler, P. L., & Glenn, E. P. (2018). Application and comparison of the MODIS-derived enhanced vegetation index to VIIRS, Landsat 5 TM and Landsat 8 OLI platforms: A case study in the arid Colorado River Delta, Mexico. *Sensors*, *18*(5). doi:10.3390/s18051546

Ji, L., & Gallo, K. (2006). An agreement coefficient for image comparison. *Photogrammetric Engineering and Remote Sensing*, *72*(7), 823–833.

Jiang, Z., Huete, A. R., Didan, K., & Miura, T. (2008). Development of a two-band enhanced vegetation index without a blue band. *Remote Sensing of Environment*, *112*(10), 3833–3845. doi:10.1016/j.rse.2008.06.006

Jung, M., Koirala, S., Weber, U., Ichii, K., Gans, F., Camps-Valls, G., . . . Reichstein, M. (2019). The FLUXCOM ensemble of global land-atmosphere energy fluxes. *Scientific Data*, *6*(1), 74. doi:10.1038/s41597-019-0076-8

Justice, C. O., Román, M. O., Csiszar, I., Vermote, E. F., Wolfe, R. E., Hook, S. J., . . . Masuoka, E. J. (2013). Land and cryosphere products from Suomi NPP VIIRS: Overview and status. *J. Geophys. Res .-Atmos.*, *118*(17), 9753–9765. doi:10.1002/jgrd.50771

Kaufman, Y. J., & Tanré, D. (1992). Atmospherically Resistant Vegetation Index (ARVI) for EOS-MODIS. *IEEE Transactions on Geoscience and Remote Sensing*, *30*(2), 261–270.

Kim, Y., Huete, A. R., Miura, T., & Jiang, Z. (2010). Spectral compatibility of vegetation indices across sensors: A band decomposition analysis with Hyperion data. *Journal of Applied Remote Sensing*, *4*, 043520. doi:10.1117/1.3400635

Kotchenova, S. Y., & Vermote, E. F. (2007). Validation of a vector version of the 6S radiative transfer code for atmospheric correction of satellite data: Part II. Homogeneous Lambertian and anisotropic surfaces. *Applied Optics*, *46*(20), 4455–4464. Retrieved from http://ao.osa.org/abstract.cfm?URI=ao-46-20-4455

Mancino, G., Ferrara, A., Padula, A., & Nolè, A. (2020). Cross-comparison between Landsat 8 (OLI) and Landsat 7 (ETM+) derived vegetation indices in a Mediterranean environment. *Remote Sensing*, *12*(2). doi:10.3390/rs12020291

Mansaray, L. R., Kanu, A. S., Yang, L., Huang, J., & Wang, F. (2020). Evaluation of machine learning models for rice dry biomass estimation and mapping using quad-source optical imagery. *GIScience & Remote Sensing*, *57*(6), 785–796. doi:10.1080/15481603.2020.1799546

Middleton, E. M., Campbell, P. K. E., Ungar, S. G., Ong, L., Zhang, Q., Huemmrich, K. F., . . . Frye, S. W. (2010). Using EO-1 Hyperion images to prototype environmental products for HyspIRI. *2010IEEE International Geoscience and Remote Sensing Symposium (IGARSS)*, 4256–4259. doi:10.1109/IGARSS.2010.5648946

Miura, T., Huete, A. R., Yoshioka, H., & Holben, B. N. (2001). An error and sensitivity analysis of atmospheric resistant vegetation indices derived from dark target-based atmospheric correction. *Remote Sensing of Environment*, *78*(3), 284–298. doi:10.1016/S0034-4257(01)00223-1

Miura, T., Muratsuchi, J., & Vargas, M. (2018). Assessment of cross-sensor vegetation index compatibility between VIIRS and MODIS using near-coincident observations. *Journal of Applied Remote Sensing*, *12*(4), 045004–045012. doi:10.1117/1.JRS.12.045004

Miura, T., Smith, C. Z., & Yoshioka, H. (2021). Validation and analysis of Terra and Aqua MODIS, and SNPP VIIRS vegetation indices under zero vegetation conditions: A case study using Railroad Valley Playa. *Remote Sensing of Environment*, *257*, 112344. doi:10.1016/j.rse.2021.112344

Miura, T., Turner, J. P., & Huete, A. R. (2013). Spectral compatibility of the NDVI across VIIRS, MODIS, and AVHRR: An analysis of atmospheric effects using EO-1 Hyperion. *IEEE Transactions on Geoscience and Remote Sensing*, *51*(3), 1349–1359. doi:10.1109/TGRS.2012.2224118

Miura, T., & Yoshioka, H. (2011). Hyperspectral data in long-term, cross-sensor continuity studies. In P. S. Thenkabail, J. G. Lyon, & A. Huete (Eds.), *Hyperspectral Remote Sensing of Terrestrial Vegetation* (pp. 611–613). CRC Press, Taylor and Francis Group.

Miura, T., Yoshioka, H., Fujiwara, K., & Yamamoto, H. (2008). Inter-comparison of ASTER and MODIS surface reflectance and vegetation index products for synergistic applications to natural resource and environmental monitoring. *Sensors*, *8*(4), 2480–2499. doi:10.3390/s8042480

Morton, D. C., Nagol, J., Carabajal, C. C., Rosette, J., Palace, M., Cook, B. D., . . . North, P. R. J. (2014). Amazon forests maintain consistent canopy structure and greenness during the dry season. *Nature*, *506*(7487), 221–224. doi:10.1038/nature13006

Nouri, H., Nagler, P., Chavoshi Borujeni, S., Barreto Munez, A., Alaghmand, S., Noori, B., . . . Didan, K. (2020). Effect of spatial resolution of satellite images on estimating the greenness and evapotranspiration of urban green spaces. *Hydrological Processes*, *34*, 3183–3199. doi:10.1002/hyp.13790

Obata, K., Miura, T., Yoshioka, H., Huete, A., & Vargas, M. (2016). Spectral cross-calibration of VIIRS enhanced vegetation index with MODIS: A case study using year-long global data. *Remote Sensing*, *8*(1), 34. doi:10.3390/rs8010034

Obata, K., Miura, T., Yoshioka, H., & Huete, A. R. (2013). Derivation of a MODIS-compatible enhanced vegetation index from visible infrared imaging radiometer suite spectral reflectances using vegetation isoline equations. *Journal of Applied Remote Sensing*, *7*(1), 073467. doi:10.1117/1.JRS.7.073467

Pearlman, J. S., Barry, P. S., Segal, C. C., Shepanski, J., Beiso, D., & Carman, S. L. (2003). Hyperion, a space-based imaging spectrometer. *IEEE Transactions on Geoscience and Remote Sensing*, *41*(6), 1160–1173.

Ponce Campos, G. E., Moran, M. S., Huete, A., Zhang, Y., Bresloff, C., Huxman, T. E., . . . Starks, P. J. (2013). Ecosystem resilience despite large-scale altered hydroclimatic conditions. *Nature*, *494*(7437), 349–352. doi:10.1038/nature11836

Rahman, A. F., Sims, D. A., Cordova, V. D., & El-Masri, B. Z. (2005). Potential of MODIS EVI and surface temperature for directly estimating per-pixel ecosystem C fluxes. *Geophysical Research Letters*, *32*(19), L19404. doi:10.1029/2005GL024127

Saleska, S. R., Didan, K., Huete, A. R., & da Rocha, H. R. (2007). Amazon forests green-up during 2005 drought. *Science*, *318*(5850), 612. doi:10.1126/science.1146663

Seddon, A. W. R., Macias-Fauria, M., Long, P. R., Benz, D., & Willis, K. J. (2016). Sensitivity of global terrestrial ecosystems to climate variability. *Nature*, *531*(7593), 229–232. doi:10.1038/nature16986

She, X., Zhang, L., Cen, Y., Wu, T., Huang, C., & Baig, M. H. (2015). Comparison of the continuity of vegetation indices derived from Landsat 8 OLI and Landsat 7 ETM+ data among different vegetation types. *Remote Sensing*, *7*(10), 13485–13506. doi:10.3390/rs71013485

Shi, H., Li, L., Eamus, D., Huete, A., Cleverly, J., Tian, X., . . . Carrara, A. (2017). Assessing the ability of MODIS EVI to estimate terrestrial ecosystem gross primary production of multiple land cover types. *Ecological Indicators*, *72*, 153–164. doi:10.1016/j.ecolind.2016.08.022

Sims, D. A., Rahman, A. F., Cordova, V. D., El-Masri, B. Z., Baldocchi, D. D., Bolstad, P. V., . . . Xu, L. (2008). A new model of gross primary productivity for North American ecosystems based solely on the enhanced vegetation index and land surface temperature from MODIS. *Remote Sensing of Environment*, *112*(4), 1633–1646. doi:10.1016/j.rse.2007.08.004

Sims, D. A., Rahman, A. F., Vermote, E. F., & Jiang, Z. (2011). Seasonal and inter-annual variation in view angle effects on MODIS vegetation indices at three forest sites. *Remote Sensing of Environment*, *115*(12), 3112–3120. doi:10.1016/j.rse.2011.06.018

Song, Y., Njoroge, J. B., & Morimoto, Y. (2013). Drought impact assessment from monitoring the seasonality of vegetation condition using long-term time-series satellite images: A case study of Mt. Kenya region. *Environmental Monitoring and Assessment*, *185*(5), 4117–4124. doi:10.1007/s10661-012-2854-z

Soudani, K., Francois, C., le Maire, G., Le Dantec, V., & Dufrene, E. (2006). Comparative analysis of IKONOS, SPOT, and ETM+ data for leaf area index estimation in temperate coniferous and deciduous forest stands. *Remote Sensing of Environment*, *102*(1–2), 161–175. doi:10.1016/j.rse.2006.02.004

Sterckx, S., Livens, S., & Adriaensen, S. (2013). Rayleigh, deep convective clouds, and cross-sensor desert vicarious calibration validation for the PROBA-V mission. *IEEE Transactions on Geoscience and Remote Sensing*, *51*(3), 1437–1452. doi:10.1109/TGRS.2012.2236682

Sulla-Menashe, D., Friedl, M. A., & Woodcock, C. E. (2016). Sources of bias and variability in long-term Landsat time series over Canadian boreal forests. *Remote Sensing of Environment*, *177*, 206–219. doi:10.1016/j.rse.2016.02.041

Swinnen, E., & Veroustraete, F. (2008). Extending the SPOT-VEGETATION NDVI time series (1998–2006) back in time with NOAA-AVHRR data (1985–1998) for Southern Africa. *IEEE Transactions on Geoscience and Remote Sensing*, *46*(2), 558–572. doi:10.1109/TGRS.2007.909948

Tran, N. N., Huete, A., Nguyen, H., Grant, I., Miura, T., Ma, X., . . . Ebert, E. (2020). Seasonal comparisons of Himawari-8 AHI and MODIS vegetation indices over latitudinal Australian grassland sites. *Remote Sensing*, *12*(15), 2494. doi:10.3390/rs12152494

USGS. (2011). *EO-1 (Earth Observing-1)*, last accessed on 31 January 2012. Retrieved from http://eros.usgs.gov/#/Find_Data/Products_and_Data_Available/ALI

Vargas, M., Miura, T., Shabanov, N., & Kato, A. (2013). An initial assessment of Suomi NPP VIIRS vegetation index EDR. *J. Geophys. Res .-Atmos.*, *118*(22), 12, 301–312,316. doi:10.1002/2013JD020439

Vermote, E., Tanré, D., Deuzé, J. L., Herman, M., Morcrette, J. J., & Kotchenova, S. Y. (2006). *Second Simulation of a Satellite Signal in the Solar Spectrum—Vector (6SV) User Guide Version 3* (p. 243).

Wang, W., Wang, Y., Lyapustin, A., Hashimoto, H., Park, T., Michaelis, A., & Nemani, R. (2022). A novel atmospheric correction algorithm to exploit the diurnal variability in hypertemporal geostationary observations. *Remote Sensing*, *14*(4), 964. doi:10.3390/rs14040964

Weber, M., Hao, D., Asrar, G. R., Zhou, Y., Li, X., & Chen, M. (2020). Exploring the use of DSCOVR/EPIC satellite observations to monitor vegetation phenology. *Remote Sensing*, *12*(15), 2384. doi:10.3390/rs12152384

Wolfe, R. E., Nishihama, M., Fleig, A. J., Kuyper, J. A., Roy, D. P., Storey, J. C., & Patt, F. S. (2002). Achieving sub-pixel geolocation accuracy in support of MODIS land science. *Remote Sensing of Environment*, *83*(1–2), 31–49. doi:10.1016/S0034-4257(02)00085-8

Xiao, X., Braswell, B., Zhang, Q., Boles, S., Frolking, S., & Moore, B. I. I. I. (2003). Sensitivity of vegetation indices to atmospheric aerosols: Continental-scale observations in Northern Asia. *Remote Sensing of Environment*, *84*, 385–392.

Xiao, X. M., Hollinger, D., Aber, J., Goltz, M., Davidson, E. A., Zhang, Q. Y., & Moore, B. (2004). Satellite-based modeling of gross primary production in an evergreen needleleaf forest. *Remote Sensing of Environment*, *89*(4), 519–534. doi:10.1016/j.rse.2003.11.008

Yamamoto, H., Miura, T., & Tsuchida, S. (2012). Advanced Spaceborne Thermal Emission and Reflection Radometer (ASTER) Enhanced Vegetation Index (EVI) products from Global Earth Observation (GEO) Grid: An assessment using Moderate Resolution Imaging Spectroradiometer (MODIS) for synergistic applications. *Remote Sensing*, *4*(8), 2277–2293. doi:10.3390/rs4082277

Yan, D., Zhang, X., Nagai, S., Yu, Y., Akitsu, T., Nasahara, K. N., . . . Maeda, T. (2019). Evaluating land surface phenology from the Advanced Himawari Imager using observations from MODIS and the phenological eyes network. *International Journal of Applied Earth Observation and Geoinformation*, *79*, 71–83. doi:10.1016/j.jag.2019.02.011

Yoshioka, H. (2004). Vegetation isoline equations for an atmosphere-canopy-soil system. *IEEE Transactions on Geoscience and Remote Sensing*, *42*(1), 166–175.

Yoshioka, H., Miura, T., & Obata, K. (2012). Derivation of relationships between spectral vegetation indices from multiple sensors based on vegetation isolines. *Remote Sensing*, *4*(3), 583–597. doi:10.3390/rs4030583

Zhang, X. Y., Friedl, M. A., & Schaaf, C. B. (2006). Global vegetation phenology from Moderate Resolution Imaging Spectroradiometer (MODIS): Evaluation of global patterns and comparison with in situ measurements. *Journal of Geophysical Research*, *111*(G4), G04017. doi:10.1029/2006JG000217

Zhang, X. F., Jayavelu, S., Liu, L., Friedl, M. A., Henebry, G. M., Liu, Y., . . . Gray, J. (2018). Evaluation of land surface phenology from VIIRS data using time series of PhenoCam imagery. *Agricultural and Forest Meteorology*, *256– 257*, 137–149. doi:10.1016/j.agrformet.2018.03.003

Zhao, Y., Wang, M., Zhao, T., Luo, Y., Li, Y., Yan, K., . . . Ma, X. (2022). Evaluating the potential of H8/AHI geostationary observations for monitoring vegetation phenology over different ecosystem types in northern China. *International Journal of Applied Earth Observation and Geoinformation*, *112*, 102933. doi:10.1016/j.jag.2022.102933

Zhen, Z., Chen, S., Yin, T., & Gastellu-Etchegorry, J.-P. (2023). Globally quantitative analysis of the impact of atmosphere and spectral response function on 2-band enhanced vegetation index (EVI2) over Sentinel-2 and Landsat-8. *ISPRS Journal of Photogrammetry and Remote Sensing*, *205*, 206–226. doi:10.1016/j.isprsjprs.2023.09.024

# 11 Toward Standardization of Vegetation Indices

*Michael D. Steven, Timothy J. Malthus, and Frédéric Baret*

## ABBREVIATIONS/ACRONYMS

| | |
|---|---|
| **6S** | Second Simulation of a Satellite Signal in the Solar Spectrum—an atmospheric correction model |
| **AATSR** | Advanced Along-Track Scanning Radiometer |
| **ASTER** | Advanced Spaceborne Thermal Emission and Reflection Radiometer |
| **ATSR2** | Along-Track Scanning Radiometer—2 |
| **AVHRR** | Advanced Very High Resolution Radiometer |
| **BRDF** | Bidirectional Reflectance Distribution Function |
| **CEOS** | Committee on Earth Observation Satellites |
| **CHRIS** | Compact High Resolution Imaging Spectrometer on the PROBA satellite |
| **DMC** | Disaster Monitoring Constellation |
| **DN** | Digital Number |
| **Etm+** | Enhanced Thematic Mapper (Landsat) |
| **EVI** | Enhanced Vegetation Index |
| **fAPAR** | The fraction of Photosynthetically Active Radiation Absorbed by a vegetation canopy |
| **Formosat** | Name of a commercial high-resolution satellite |
| **IKONOS** | Name of a commercial high-resolution satellite |
| **IRS** | Indian Remote Sensing Satellite |
| **ISO** | International Organization for Standardization |
| **Kompsat** | Korea Multi-Purpose Satellite |
| **LAI** | Leaf Area Index |
| **Landsat** | A system of Earth observation satellites in operation since 1972 |
| **LISS** | Linear Imaging Self Scanning Sensor (IRS) |
| **MERIS** | MEdium Resolution Imaging Spectrometer |
| **MISR** | Multi-angle Imaging SpectroRadiometer |
| **MODIS** | MOderate Resolution Imaging Spectrometer |
| **MODIS-TIP** | Two-stream Inversion Package applied to MODIS products by the European Joint Research Centre |
| **MODTRAN** | MODerate resolution atmospheric TRANsmission—an atmospheric correction model |
| **MSAVI** | Modified Soil-Adjusted Vegetation Index |
| **MSG/SEVIRI** | Meteosat Second Generation/Spinning Enhanced Visible and Infrared Imager |
| **MSS** | Multi-Spectral Scanner (Landsat) |
| **NASA** | National Aeronautics and Space Administration |
| **NDVI** | Normalised Difference Vegetation Index |
| **NOAA** | National Oceanic and Atmospheric Administration |
| **OrbView** | Name of a commercial high-resolution satellite |
| **OSAVI** | Optimised Soil-Adjusted Vegetation Index |


DOI: 10.1201/9781003541141-16

| | |
|---|---|
| **PAR** | Photosynthetically Active Radiation |
| **POLDER** | POLarization and Directionality of the Earth's Reflectances—an optical imaging radiometer |
| **PROSAIL** | A combination of the PROSPECT leaf optical properties model with the SAIL canopy bidirectional reflectance model |
| **PROSPECT** | A model of leaf optical properties |
| **QuickBird** | Name of a commercial high-resolution satellite |
| **SAIL** | Scattering by Arbitrarily Inclined Leaves—a canopy bidirectional reflectance model |
| **SAVI** | Soil-Adjusted Vegetation Index |
| **SeaWIFS** | Sea-Viewing Wide Field-of-View Sensor |
| **SPOT** | Système Probatoire d'Observation de la Terre |
| **SPOT-VEG** | The VEGETATION sensor on the SPOT satellite |
| **TM** | Thematic Mapper (Landsat) |
| **TSAVI** | Transformed Soil-Adjusted Vegetation Index |
| **VEGETATION** | 1 km resolution monitoring instrument on the SPOT satellite |
| **Venμs** | Vegetation and Environment monitoring on a New MicroSatellite |
| **VI** | Vegetation Index |

## 11.1 INTRODUCTION—VEGETATION INDICES AND THEIR USES

The concept of the vegetation index is one of the lasting success stories of terrestrial remote sensing. The physiological and anatomical characteristics of vegetation give rise to distinctive spectral features that allow its presence to be detected in any environment and with suitable precautions permit the properties of the vegetation canopy to be inferred from the reflected spectrum. Healthy vegetation absorbs visible (especially red) light via chlorophyll and other pigments. In the near-infrared (NIR) where no absorbers are active, light is strongly reflected by foliage because the juxtaposition of cells, essentially containing water, with air spaces between, creates a strongly scattering medium (Gates et al. 1965; Gausman and Allen 1973). The resulting reflectance spectrum (Figure 11.1) is highly characteristic and is recognisable as the spectral signature of vegetation even when distorted by other environmental variables. When used to monitor growing vegetation the spectral signal of vegetation is mixed with that of soil or other backgrounds, which tend to show a much flatter response across this spectral region. The result is that the NIR tends to increase with vegetation cover, for example as a crop grows, while the red reflectance decreases, ending at 100% cover with a spectrum close (but not identical) to that of a single leaf.

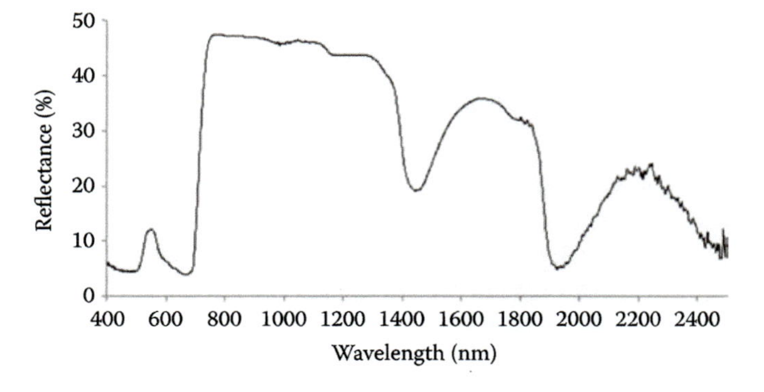


FIGURE 11.1 Laboratory-measured reflectance spectrum of a bean leaf. The jitters at longer wavelengths are instrumental noise. (Source: the author.)

The key factor in the development of vegetation indices is the increasing contrast between the reflectance in the two bands. To encapsulate this contrast, early work identified a variety of combinations of the near-infrared and visible bands, of which the best known by far (although by far from the best) is the Normalized Difference Vegetation Index NDVI, defined as:

$$\text{NDVI} = \frac{\rho_{\text{NIR}} - \rho_{\text{red}}}{\rho_{\text{NIR}} + \rho_{\text{red}}} \tag{11.1}$$

where ρ is the spectral reflectance (dimensionless) in the near-infrared (NIR) or red spectral band. The NDVI is functionally equivalent to the simple ratio of the two bands (Perry and Lautenshlager 1984). However its formulation ensures that the value of the NDVI ranges strictly from –1 to +1, which is computationally more convenient than the ratio, which has no upper bound. A wide variety of alternative formulations have been used, for example the Soil Adjusted Vegetation Index SAVI, (Huete 1988), defined as:

$$\text{SAVI} = (1+L)\frac{(\rho_{\text{NIR}} - \rho_{\text{red}})}{(\rho_{\text{NIR}} + \rho_{\text{red}} + L)} \tag{11.2}$$

The formulation of SAVI differs from NDVI by including a "soil calibration factor *L*," which adjusts for variability in the index introduced by soil reflectance characteristics. Although Huete (1988) found that the optimal value of *L* varied with vegetation density, a mid-range value of 0.5 was found to provide effective correction for variations due to soil background across the full range of densities. Later variants on this approach varied the value of *L*, with Rondeaux, Steven and Baret (1996) proposing Optimized SAVI (OSAVI) with $L = 0.16$ on the basis of an optimization across a range of agricultural soils. Others include MSAVI, the Modified Soil-Adjusted Vegetation Index (Qi et al. 1994), which employs a self-adjusted value of *L* based on the spectral reflectance data themselves, and TSAVI, the Transformed Soil-Adjusted Vegetation Index (Baret and Guyot 1991), where *L* is based on information about the soil characteristics. Most VIs are reformulations based on the same two spectral bands, but some, such as the Enhanced Vegetation Index, EVI (Huete et al. 1997), introduce additional spectral information, usually in an attempt to reduce atmospheric sensitivity. Each index has its advocates and many have particular merits, but key features are that they are intrinsically dimensionless and that they all ultimately share the characteristic of using the contrast between NIR and red reflectance as the primary measure of vegetation. For most of this discussion, the term *vegetation index* (VI) will be used generically to apply to all of them (Almeida-Ñauñay et al. 2022; Ferchichi et al. 2022; Rehman et al. 2022; Swoish et al. 2022; Zeng et al. 2022; Carneiro et al. 2020; Mancino et al. 2020; Jarchow et al. 2018; Macfarlane et al. 2017; Psomiadis et al. 2017; Robinson et al. 2017; Xue and Su 2017).

### 11.1.1 Vegetation Index Applications

The applications of vegetation indices are based on their ability to measure foliage density in a consistent manner across a wide range of vegetation types. In early field studies VIs were successfully related to leaf area index, LAI defined as the total (single-sided) area of leaf per unit area of ground ($m^2/m^2$, i.e., dimensionless), canopy chlorophyll ($g\ m^{-2}$), wet and dry biomass ($g\ m^{-2}$), the fraction of ground covered by leaves (dimensionless), primary productivity ($g\ m^{-2}\ d^{-1}$), the fraction of photosynthetically active radiation absorbed by the canopy fAPAR (dimensionless) and other variables (Tucker 1977; Tucker, Elgin and McMurtrey 1979; Holben, Tucker and Fan 1980; Steven, Biscoe and Jaggard 1983). As all of these factors are measures of foliage density, they tend to be highly inter-correlated in any individual study. However, these variables are hierarchically linked by the process of canopy photosynthesis which converts absorbed PAR to fixed energy in biomass (Figure 11.2). The reflection of light by a plant canopy is largely determined by the total area of leaf and the projections of that area towards both the source of illumination and the detection device. This introduces a dependence on leaf angle distribution (Verhoef 1985) and to a lesser extent other factors such as clumping (Gower, Kucharik and Norman 1999). As they involve similar projections, this is closely connected to the way in which leaf area and angle combine to determine leaf cover fraction or PAR absorption (Steven et al 1986), so that as indicated in Figure 11.2 the response of a vegetation index relates most directly to these variables. An illustration of this link is that in spite of variations with time of day, Pinter (1993) found that the relationship of fAPAR with various VIs was independent of illumination angle. Figure 11.3 shows the modeled relationship between OSAVI (representative here of VIs generally) and LAI for canopies with different leaf angle distributions, while Figure 11.4 shows that the relationships essentially collapse onto a single curve when expressed as functions of leaf cover fraction (Steven 1998). The same behavior can be found with NDVI, where the merger of curves for different leaf angle distributions is slightly less tight, although not enough to introduce serious variations. The implication of this merger is that there is redundancy in the biophysical parameters and that VIs can be used to measure foliage density in a way that is largely independent of canopy structure and to some extent, species. Moreover as leaf cover fraction is itself closely related, through similar dependence on leaf area and structure, to the interception of solar radiation by plant canopies (Steven et al. 1986), VIs lend themselves to the estimation of light capture by vegetation (or fAPAR). Following Monteith (1977) who established the model outlined in Figure 11.2, showing that light capture was the key determinant of the conversion of solar energy to biomass, remote sensing of this variable leads to direct estimation of productivity in crops (Steven, Biscoe and Jaggard 1983; Wiegand et al. 1991). The same argument allows VIs to be applied in large scale monitoring of vegetation dynamics (Goward et al. 1993), with a range of applications in agriculture (Maselli et al. 2000) or ecology, (Pettorelli et al. 2005). Direct estimation of fAPAR on a global scale is now the focus of several major space programs, using the MODIS, MERIS, SeaWIFS, MODIS-TIP, SPOT-VEG, and AVHRR systems (Picket-Heaps et al. 2014).

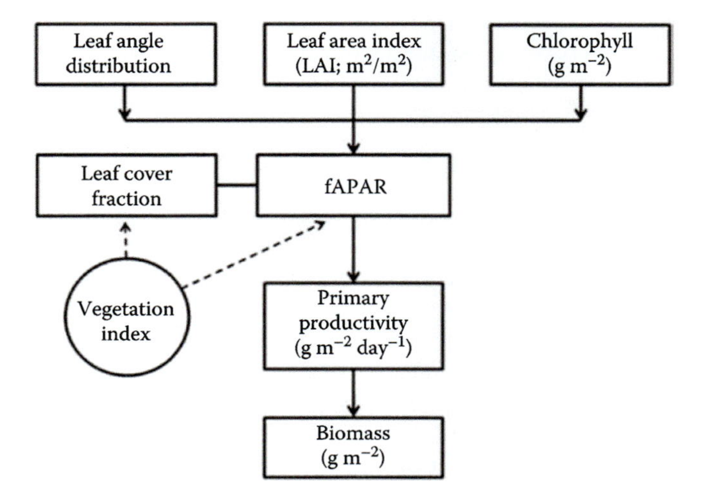


**FIGURE 11.2** Hierarchical diagram showing the relationships between various measures of canopy density and their link with vegetation indices.

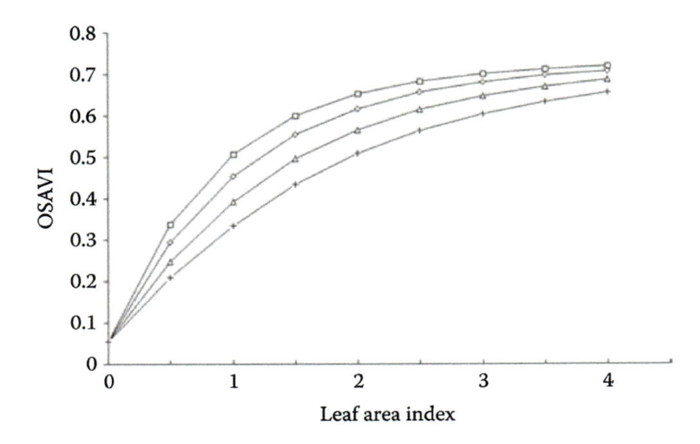


**FIGURE 11.3** Sensitivity of OSAVI to leaf angle distribution for a range of leaf area indices, for ellipsoidal distributions with mean leaf angles of 30°, 45°, 57.3°, and 65°+: all the parameters are dimensionless. (Steven 1998).

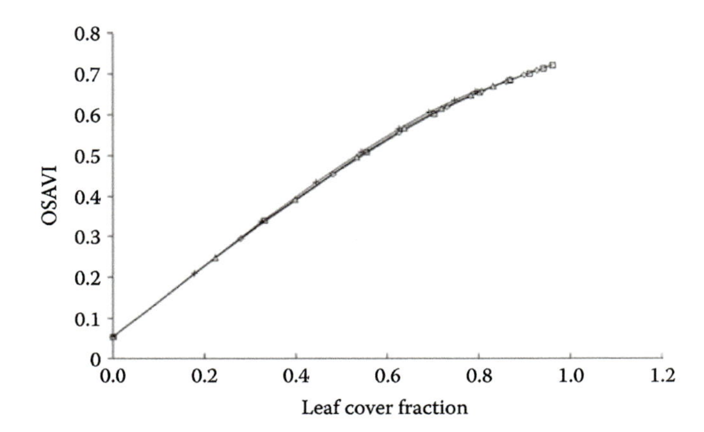


**FIGURE 11.4** Sensitivity of OSAVI cover estimates to leaf angle distribution: parameters as for Figure 11.3. (Steven 1998).

## 11.2 THE NEED FOR STANDARDS

All sciences commence with a lengthy period of exploration, with a diversity of idiosyncratic approaches until the difficulties of rationalising different methodologies leads to a movement for standardization. The benefits of standardization that are particularly pertinent to remote sensing are interoperability of systems and data continuity. The use of standards also helps to generate authority for standardized products and with increasing use, a greater familiarity with their capabilities and limits.

A persistent issue in vegetation monitoring is the acquisition of sufficient data to capture the dynamics of plant growth. Plant growth requires water, usually supplied by rainfall, so the more productive vegetated regions are frequently cloudy, obstructing the view of satellite sensors (Heller 1961). The data acquisition problem for satellite systems has broadly been resolved in two ways: at higher resolutions by the development of systems such as SPOT and WorldView with pointable cameras and PlanetScope a virtual constellation of similar sensors that can target particular sites several times within the satellite repeat cycle; and at lower resolutions with more frequent data acquisition by selectively compositing over periods of ten days or more (Gutman 1991; van Leeuwen et al. 1999). Alternatively, vegetation monitoring can be performed by aircraft that can fly beneath the cloud deck (Jaggard and Clark 1990), or by monitoring systems mounted on mobile farm machinery for precision agriculture. All these solutions introduce their own problems. In particular, both pointing and compositing tend to increase the range of viewing angles, and to a lesser extent solar angles, used in the vegetation index product. Monitoring under cloud requires additional data to normalise for incident solar irradiance and measures different reflectance characteristics corresponding to the multi-directional diffuse illumination of the target. Corrections can be made for angular effects using a model of the bidirectional reflectance distribution function (BRDF), but require *a priori* knowledge of the vegetation type (Steven 1998; Bacour Bréon and Maignan 2006).

A complementary approach is to combine data from more than one system, sometimes referred to as the use of a virtual constellation (CEOS 2006; Martínez-Beltrán et al. 2009). Long term environmental analysis may also require meta-analysis of data from a range of systems (Boyd and Foody 2011). Key to these approaches is the adoption of a set of operating standards for the systems to be combined. The increasing focus on long-term continuity of vegetation observations, particularly for monitoring environmental change at larger scales, has led to considerable interest in back-calibrating data from earlier systems such as Landsat (operational since 1972) and AVHRR (since 1978), as near as possible to current standards, to establish the long-term baseline (Almeida-Ñauñay et al. 2022; Ferchichi et al. 2022; Rehman et al. 2022; Swoish et al. 2022; Zeng et al. 2022; Carneiro et al. 2020; Mancino et al. 2020; Jarchow et al. 2018; Macfarlane et al. 2017; Psomiadis et al. 2017; Robinson et al. 2017; Xue and Su 2017; Brown et al. 2006; Samain, Geiger and Roujean 2006). Precise calibration of the instruments is required and attention to variations in BRDF associated with different orbital characteristics (Teillet, Markham and Irish 2006; Röder, Kuemmerle and Hill 2005; Martínez-Beltrán et al. 2009). Thus, standardization is required firstly for individual sensors, to account for variability in calibration and other observational parameters over time. Secondly standardization is required between sensors to allow the interoperability of different systems. Coupled with these instrumental issues is the need to specify standard observing conditions, with unambiguous correction procedures to account for deviations from the standard. Achievement of these goals would in principle allow consistent monitoring of vegetation across time and across spatial scales. However, although standards for many activities within the field of remote sensing are being promoted by CEOS, ISO, NASA and other agencies, no concerted attempt has yet been made to standardize vegetation indices.

### 11.2.1 Vegetation Index Formula

At present a wide range of VIs are generated by different Earth observation systems and data users. Le Maire, François and Dufrêne (2004) evaluated 61 indices for their sensitivity to chlorophyll across different types of leaves and Agapiou, Hadjimitsis and Alexakis (2012) evaluated 71 established indices for distinguishing archaeological crop marks. Neither list is exhaustive. As well as highlighting the inadequacies of NDVI, the enormous proliferation of indices indicates both the need for a standard and the difficulty of achieving one. Although a number of indices have been developed that are claimed to represent specific properties of vegetation—water content, chlorophyll, carotenoids and radiation use efficiency (Peñuelas and Filella 1998)—these usually require high spectral resolution and are not the concern here. In general, the different broadband vegetation indices are highly interrelated and are usually used to represent the same vegetation parameters. In most cases, their differences lie in the generality of the data that have been used to test them and the degree to which they can suppress factors extraneous to the estimation of vegetation characteristics, such as soil or atmospheric effects. However different indices do have distinct responses and may yield differing estimates of derived parameters: Boyd et al. (2011) found that estimates of phenological event dates such as the start and end of the growing season differed by several days when different indices were used, with uncertainties from these differences "as large as those arising from climatic perturbations."

### 11.2.2 Alternatives to Standard Vegetation Indices

While a standard vegetation index would therefore seem desirable, it is nonetheless worth discussing the alternatives. If standards are developed prematurely, there is a danger of lock-in to a standard that proves to be inadequate. To some extent, this is the case with the NDVI. Many papers have pointed out its excessive sensitivity to soil characteristics (Huete 1988; Rondeaux, Steven and Baret 1996), but having been adopted as an operational product from the early 1980s (Townshend and Justice 1986), it has become the *de facto* standard for later systems; a recent web search (June 2014) reported over 5,000 academic papers on this index. One alternative to vegetation indices is to estimate vegetation parameters directly by inverting a vegetation canopy reflectance model with Earth observation data. As the basis for productivity estimation, fAPAR has been recognised as an essential climate variable for global modeling and is already a routine product of the MODIS, MERIS, VEGETATION, and MSG/SEVIRI systems. Gobron et al. (2008) evaluated the effects of radiometric uncertainties on the MERIS product and estimated that errors in fAPAR estimation should be ≤0.1 but D'Odorico et al. (2014) found inconsistencies, particular in forest, when comparing three fAPAR algorithms over Europe. Similar issues were found by Picket-Heaps et al. (2014) who tested six alternative fAPAR products over Australia. They suggest that current fAPAR products are not reliable enough to be fed into biogeochemical process models or used in data fusion approaches. A further difficulty with these products is that they rely on different types of data input including, in most cases, multi-angular reflectance data. While greater reliability may be achievable with future systems or better models, it would not be realistic to generate the equivalent product from historic data where key inputs are missing. However, a vegetation index, although less directly applicable as an input to process models (it remains just an index), can within limits be standardized.

## 11.3 SOURCES OF VARIATION IN VEGETATION INDICES

Although VIs respond primarily to foliage density, however expressed, they are critically affected by a range of other factors (Almeida-Ñauñay et al. 2022; Ferchichi et al. 2022; Rehman et al. 2022; Swoish et al. 2022; Zeng et al. 2022; Carneiro et al. 2020; Mancino et al. 2020; Jarchow et al. 2018; Macfarlane et al. 2017; Psomiadis et al. 2017; Robinson et al. 2017; Xue and Su 2017). Van Leeuwen (2006) distinguishes between uncertainties related to input parameters, VI formulation and product generation issues such as compositing rules. Most of the discussion here relates to the first of these categories. Table 11.1 classifies the sources of extraneous variation into environmental, observational and instrumental categories, as discussed further later.

**TABLE 11.1**
**Sources of Extraneous Variation in Vegetation Indices**

| Source of Variation | Approach to Overcome | Key References |
|---|---|---|
| **Environmental factors** | | |
| Soil background | Soil adjusted indices | (Huete 1988; Rondeaux, Steven and Baret 1996). |
| Atmosphere | Atmospheric modeling | (Vermote et al. 1997; Berk et al. 1998). |
| **Observation parameters** | | |
| Solar angle | Canopy BRDF modeling | (Jacquemoud et al. 2009; Vermote, Justice and Bréon 2009). |
| Viewing angle | | |
| **Instrumental factors** | | |
| Pixel size | Aggregation, scale modeling | (Martínez-Beltrán et al. 2009; Obata, Miura and Yoshioka 2012). |
| Spectral bands | VI intercalibration | (Steven et al. 2003); this chapter |

### 11.3.1 Soil Background

Dependence on soil reflectance is inherent in the formulation of VIs as measures of vegetation-soil contrast. With the NDVI, darker soils will tend to amplify the vegetation component of the signal while brighter soils will tend to suppress it (Huete 1988; Rondeaux, Steven and Baret 1996). Field studies that relate NDVI to measures of foliage density are almost invariably conducted on a single soil type and almost inevitably achieve relationships with high correlations; but subsequent attempts to transfer the relationships to other environments are often disappointing due to changes in the background effect of soil type. Individual soils also decrease in brightness with wetting (Bowers and Hanks 1965; Rondeaux, Steven and Baret 1996). Soil effects in NDVI can be as high as 50% of the dynamic range, but are considerably reduced in the SAVI range of indices (Rondeaux, Steven and Baret 1996), corresponding to a maximum error of about 0.05 in the estimation of leaf cover fraction.

### 11.3.2 Atmospheric Effects

Atmospheric effects change the radiances measured so that the top-of-atmosphere reflectances generate a different VI from that observed at the surface. It is essential to point out here that to standardize a vegetation index requires the index to be based on surface reflectances as defined in Equation 11.1 or 11.2. Any other measurement, such as top-of-atmosphere reflectance, represents a combination of signals from both vegetation and atmosphere, but using equivalent surface reflectance data eliminates many of the errors and allows comparison of indices (Guyot and Gu 1994). Measurements must also be calibrated according to best practice (Price 1987). Many studies in the past have applied the vegetation index concept rather loosely, some even using uncalibrated digital numbers (DN) to compute the index. Such formulations can provide strong correlations with vegetation density measures in individual studies, but cannot easily be compared with studies using different instruments or formulations. Zhou et al. (2009) found large sensor-dependent differences between NDVIs for various systems depending on whether they were DN based, radiance based or reflectance based. Hadjimitsis et al. (2010) found a mean difference of 18% between uncorrected and atmospherically-corrected NDVI values and more modest, but still troublesome, differences in a range of other indices. Miura, Turner and Huete (2013) evaluated atmospheric effects on inter-sensor compatibility and showed that atmospheric correction to top of canopy led to the greatest consistency between systems. Corrections for atmospheric effects to retrieve the surface reflectance are therefore a strict necessity. A variety of empirical and model-based approaches are available (Mahiny and Turner 2007). The 6S (Vermote et al. 1997) and MODTRAN (Berk et al. 1998) atmospheric models are widely used, but difficulties can arise in acquisition of the atmospheric parameters required to make the correction, particularly in remote locations, with uncertainties associated mainly with aerosol content (Nagol, Vermote and Prince 2009). Fortunately, atmospheric effects in VIs are somewhat limited by the ratio construction of most indices coupled with the fact that the visible and near-infrared bands are to a large extent affected in similar ways by atmospheric aerosol; Steven (1998) found that OSAVI estimates of leaf cover fraction for mid-latitude summer monitoring conditions were relatively insensitive to quite large errors in modeled atmospheric parameters, resulting in no more than 4–5% error in the cover estimate.

### 11.3.3 Directional Effects

The complex structure of vegetation canopies as assemblages of leaves and other components suspended above the surface gives rise to a more or less complex bidirectional reflectance distribution function (BRDF) that leads to a VI dependence on solar angle and viewing angle. The BRDF of vegetation is largely controlled by the relative fractions of illuminated soil, foliage, and shadow visible in a given direction (Guyot 1990; Schluessel et al. 1994). At large incidence angles the scene will appear to be fully vegetated, even when LAI is small, whereas the nadir view will show a greater proportion of soil and maximises sensitivity to leaf area and canopy structure. In general, variations of reflected radiance with either solar or viewing angle are greatest for erectophile leaf angle distributions and least for planophile distributions. Tucker et al. (2005) showed that vegetation indices showed large variations with viewing angle (and, by the principle of reciprocity, solar angle) that are dependent on both index and vegetation type. Sims et al. (2011) and Moura et al. (2012) also found that directional effects were index-sensitive, being greater for EVI than for NDVI. If uncorrected, solar and view angle effects may vary with orbital drift of long-term satellites (Tucker et al. 2005; Brown et al. 2006). Angular effects also occur as a result of the compositing approaches used to generate cloud-free vegetation products from global datasets (van Leeuwen, Huete and Laing 1999). Where the vegetation characteristics are known, angular effects on vegetation indices can be estimated by canopy models (Shultis 1991; Steven 1998; Tucker et al. 2005; Jacquemoud et al. 2009) and as the solar and viewing angles are known precisely for any observation, the residual errors are not large. For global scale vegetation monitoring, where it is impractical to apply individual models to specific vegetation types, Vermote, Justice and Bréon (2009) demonstrated that BRDF corrections could be made on the basis of an assumption that the shape of the BRDF varies more slowly than the magnitude of the reflectance. The shape can then be quantified by two parameters R and V (related in broad terms to vegetation and roughness) that can be derived by inversion of the time series.

BRDF effects can also occur with cloudiness, topography and atmospheric state. These factors change the relative contribution of diffuse solar irradiance which comprises a distribution of incident angles (Steven 1977; Steven and Unsworth 1980) interacting with different parts of the BRDF and generating spectral and angular reflectances different to those from the unidirectional direct solar irradiance. By comparing canopy measurements made under full sunlight and simulated cloud (entirely diffuse illumination), Steven (2004) found differences up to 0.15 in NDVI between canopies in sunlight and shade, but demonstrated that the canopy spectra under standard conditions could be reconstructed to a precision of 10–15% from the shaded measurements. Overall, these studies indicate that although angular responses are problematic and complex, they are susceptible to correction by modeling the BRDF.

### 11.3.4 Pixel-Size Effects

A number of studies have commented on the effects of pixel aggregation when integrating data from multiple sources (van Leeuwen 2006; Tarnavsky, Garrigues and Brown 2008; Martínez-Beltrán et al. 2009). A fundamental issue is that VIs are not linear functions of reflectance, so that spatial averages of the VI are not precisely equivalent to VI values calculated on the basis of averaging the radiances or reflectances (Figure 11.5). The difference is not large in the example shown, but would in principle increase with greater heterogeneity of the surface and with the number of pixels to be aggregated. Obata, Miura and Yoshioka (2012) developed a theoretical basis for dealing with scaling effects based on monotonic behavior of the effect as a function of spatial resolution. Martínez-Beltrán et al. (2009) indicate that in practice this nonlinearity has not been found to be the major issue and that geolocation uncertainty is a more serious source of error in inter-sensor comparison. However Munyati and Mboweni found that aggregation in areas of sparse, patchy vegetation could lead to underestimation of productivity.

### 11.3.5 Spectral Band Effects

In addition to variability of the VI itself, any standardization protocol must also account for differences between measurement systems. Even when instruments are precisely calibrated and all the proper corrections are applied for BRDF and atmospheric effects, indices from the various measurement systems differ systematically due to differences in the position, width, and shape

| | Red | | | | NIR | |
|---|---|---|---|---|---|---|
| | 0.1 | 0.15 | 0.2 | 0.6 | 0.55 | 0.5 |
| | 0.15 | 0.2 | 0.25 | 0.6 | 0.55 | 0.5 |
| | 0.2 | 0.25 | 0.3 | 0.6 | 0.55 | 0.5 |
| Averages | | 0.200 | | | 0.550 | |

NDVI from average = 0.467

| NDVI | | |
|---|---|---|
| 0.71 | 0.57 | 0.43 |
| 0.60 | 0.47 | 0.33 |
| 0.50 | 0.38 | 0.25 |

Average NDVI = 0.471

FIGURE 11.5 Hypothetical illustration on the effect of NDVI nonlinearity on aggregation of nine pixels into one. The mean of the NDVIs across a 3 × 3 pixel area (lower box) is not identical to the NDVI calculated from the mean reflectance values (upper two boxes).

of the wavebands used (Gallo and Daughtry 1987; Guyot and Gu 1994). Band placement does not critically affect the behavior or strength of relationships with vegetation density, but can give rise to large differences in the fitted coefficients (Lee et al. 2004; Soudani et al. 2006). Differences in index are inevitable because different sensors measure different parts of the vegetation–soil reflectance spectrum and consequently respond in differing degrees to the biophysical variables concerned. In addition, Teillet and Ren (2008) found that spectral differences are generated by the spectral dependence of atmospheric gas transmittance. However, it transpires that in most cases these differences are more quantitative than qualitative, with almost perfect correlations between indices recorded by different systems (Steven et al. 2003; Steven, Malthus and Baret 2007). This study proposes the application of the relationships established in these papers as a key step in the standardization of vegetation indices.

## 11.4 VEGETATION INDEX INTERCALIBRATION APPROACHES

Inter-comparison of vegetation indices can be performed by direct comparison of measurements made by different instruments, by simulation from model spectra or by partial simulation (of band responses) from field reflectance spectra (Swoish et al. 2022; Zeng et al. 2022; Carneiro et al. 2020; Mancino et al. 2020; Xue and Su 2017). Direct comparisons are limited to particular instruments and are subject to errors due to non-simultaneity, inexact image co-registration and atmospheric effects; they also fail to distinguish spectral band effects from sensor calibration errors. Although model spectra are highly versatile, particularly for sensitivity tests, their direct application requires supreme confidence that the model parameterization captures all the important sources of variation. Simulation from data has the advantage in this context in that the spectral data are realistic, the simulated results are not limited to particular instruments, and the inter-comparisons are all made on a common dataset.

### 11.4.1 Intercalibration of Vegetation Indices after Steven et al. (2003)

The analysis by Steven et al. (2003) used a database of high resolution spectra of vegetation canopies to simulate near-infrared and visible band responses of particular instruments and then compared vegetation indices as measured by different simulated observing systems. The database, described in more detail in Steven et al. (2003) comprised a set of 166 nadir-viewing bidirectional reflectance spectra measured over canopies of contrasting architecture (sugar beet and maize) in the UK and France in experiments conducted in 1989 and 1990. The plant canopies had a full range of leaf area indices and soil backgrounds, while contrasting leaf colors were achieved by treatments with disease or diluted herbicide. Spectral band responses were simulated by convolving the top-of-canopy spectral radiance data with the full spectral response function of each of the sensors tested and normalising with the corresponding convolved data for the reference panel used in the field, adjusted for its true reflectance. The spectral response functions were found in the literature, or from the web, or obtained by personal communication and were digitised every 1 nm to match the spectral data (Figure 11.6). The operators of the OrbView-2 and OrbView-3 systems were unwilling to release the spectral response functions of their instruments, so Steven, Malthus and Baret (2007) tested two alternative models: a box function across the nominal wavelength range and a Gaussian fitted so that the nominal waveband limits were the half-power points. The wavebands for Venμs, which had not been precisely defined, were also modeled with a Gaussian on the basis of the developer's advice. The simulated band reflectances were then applied to compute the vegetation indices NDVI, SAVI and OSAVI for the different systems as well as for a hypothetically ideal narrow-band sensor pair based on narrow bands at 670 and 815 nm, proposed as a standard. With narrow bands, the reflectance values are no longer sensitive to band width or spectral response function. However, to reduce the effects of instrumental noise in the original database, bandwidths of 20 nm centered on the nominal wavelengths were used to determine the reflectances for the standard bands. The VIs from the different simulated systems were then compared (Ferchichi et al. 2022; Rehman et al. 2022; Swoish et al. 2022; Zeng et al. 2022).

Figures 11.7 and 11.8 show NDVI values for two systems compared with the corresponding NDVIs for the proposed standard bands at 670 and 815 nm. Vegetation indices from all systems were highly linearly correlated. NOAA8 and CHRIS as shown here represent the extremes of slope, with NDVI for NOAA8 being as much 19% lower than the standard on the same target. The strength and linearity of the correlations (minimum $r^2 = 0.984$) means that NDVIs recorded by different observation systems can be intercalibrated to a degree of precision of about ±1%. Steven et al. (2003) tabulated two-way intersystem conversion coefficients for 15 operational systems as well as the standard narrow-band sensor pair. In a later update, Steven, Malthus and Baret (2007) extended the intercalibration of vegetation indices forwards to include conversion coefficients for orbiting sensor systems launched since the 2003 paper, and backwards to include historical variations in the NOAA AVHRR system, a total of 41 systems (Table 11.2). Two-way inter-conversion tables are no longer practical for the large number of systems involved, so for simplicity the conversions are given relative to an NDVI based on the standard pair of narrow bands at 670 and 815 nm. These bands were originally chosen to be close to the optimum, maximising the NDVI value, although in fact

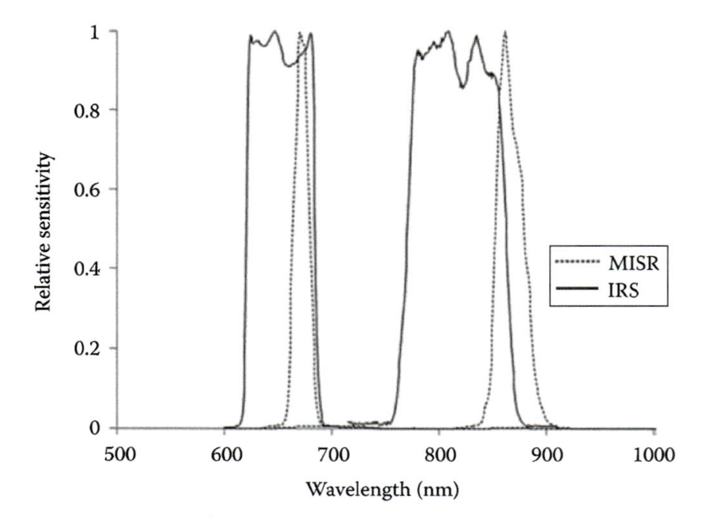


FIGURE 11.6 Example of spectral response functions for two of the systems simulated. (Steven et al. 2003).

the NDVI for the CHRIS system using the L14 near-infrared band (Figure 11.8) does have a slightly greater dynamic range.

Linear regressions for SAVI, OSAVI, and NDVI differ in slope and intercept by no more than about 0.0008 and 0.013, respectively, across the whole range of indices. As these differences are considerably less than typical errors of measurement, a single conversion table is adequate for the range of vegetation index formulations considered. It is also possible to convert from one operational system to another using Table 11.2 to convert first to the standard as an intermediate stage and then from the standard to the second system. On examples tested, the error in this two-stage process, as compared with direct conversion between the systems, was up to 0.01 in slope and 0.007 in intercept.

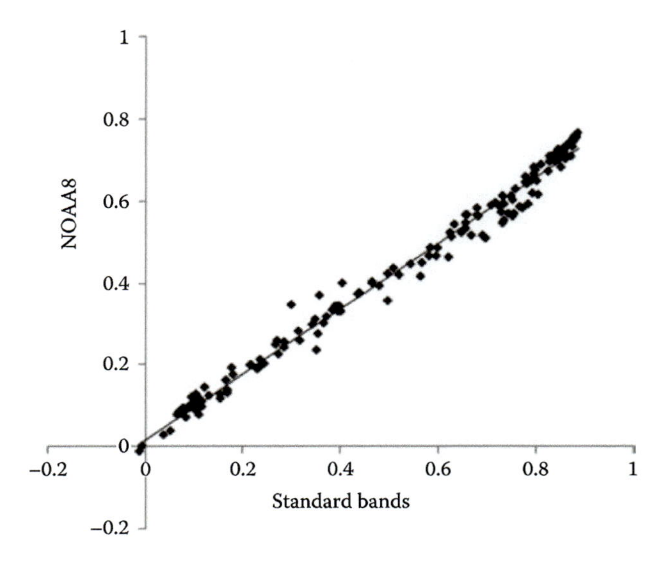


FIGURE 11.7 Regression of NDVI based on NOAA8 bands against the standard bands 670 ± 10 and 815 ± 10 nm. (Steven, Malthus and Baret 2007).

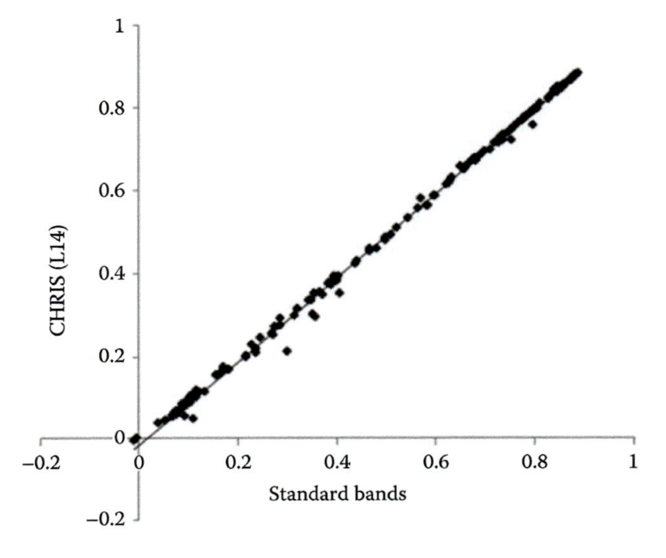


FIGURE 11.8 Regression of NDVI based on the CHRIS near-infrared (L14) and red bands against the standard bands. (Steven, Malthus and Baret 2007).

**TABLE 11.2**
**Conversion Coefficients for VIs (NDVI, SAVI, and OSAVI) from Different Systems Based on Simulations from the Database of the Full NIR and Visible Spectral Responses of Each System (Where Alternative Options for a System Exist, Both Are Shown)**

| | A | | B | |
|---|---|---|---|---|
| | Sensor vs. Standard | | Standard vs. Sensor | |
| Satellite sensor | Intercept | Slope | Intercept | Slope |
| ALI | –0.005 | 0.965 | 0.006 | 1.034 |
| ASTER, using band 3B | –0.001 | 0.933 | 0.003 | 1.068 |
| ASTER, using band 3N | 0.000 | 0.933 | 0.002 | 1.068 |
| ATSR2/AATSR | 0.008 | 0.968 | –0.006 | 1.030 |
| CHRIS, using band L14 | –0.015 | 1.009 | 0.016 | 0.989 |
| CHRIS, using band L15 | 0.005 | 0.991 | –0.004 | 1.007 |
| DMC | 0.006 | 0.954 | –0.005 | 1.046 |
| Formosat | 0.002 | 0.936 | 0.000 | 1.065 |
| Ikonos | –0.010 | 0.870 | 0.015 | 1.144 |
| IRS | 0.005 | 0.950 | –0.004 | 1.050 |
| Kompsat | 0.004 | 0.942 | –0.003 | 1.058 |
| Landsat 5 TM | 0.005 | 0.938 | –0.003 | 1.063 |
| Landsat 7 ETM+ | 0.003 | 0.957 | –0.002 | 1.041 |
| Landsat MSS | 0.029 | 0.883 | –0.024 | 1.115 |
| MERIS | 0.008 | 0.983 | –0.008 | 1.016 |
| MISR | 0.005 | 0.985 | –0.005 | 1.014 |
| MODIS | 0.017 | 0.935 | –0.015 | 1.065 |
| NOAA10 | 0.003 | 0.854 | 0.001 | 1.160 |
| NOAA11 | 0.015 | 0.831 | –0.011 | 1.188 |
| NOAA12 | 0.015 | 0.844 | –0.012 | 1.173 |
| NOAA13 | 0.017 | 0.835 | –0.014 | 1.184 |
| NOAA14 | 0.016 | 0.837 | –0.013 | 1.180 |
| NOAA15 | 0.016 | 0.902 | –0.014 | 1.100 |
| NOAA16 | 0.017 | 0.897 | –0.015 | 1.107 |
| NOAA17 | 0.016 | 0.904 | –0.014 | 1.098 |
| NOAA18 | 0.017 | 0.905 | –0.014 | 1.097 |
| NOAA6 | 0.021 | 0.850 | –0.018 | 1.163 |
| NOAA7 | 0.015 | 0.857 | –0.012 | 1.155 |
| NOAA8 | 0.015 | 0.807 | –0.012 | 1.226 |
| NOAA9 | 0.015 | 0.839 | –0.012 | 1.179 |
| OrbView-2, using block fcn | 0.005 | 0.989 | –0.004 | 1.009 |
| OrbView-2, using Gaussian | 0.005 | 0.982 | –0.005 | 1.016 |
| OrbView-3, using block fcn | 0.002 | 0.937 | 0.000 | 1.063 |
| OrbView-3, using Gaussian | 0.002 | 0.857 | 0.001 | 1.159 |
| POLDER | 0.005 | 0.985 | –0.005 | 1.014 |
| QuickBird | 0.000 | 0.909 | 0.002 | 1.096 |
| Seawifs | 0.005 | 0.982 | –0.004 | 1.016 |
| Severi MSG | 0.012 | 0.926 | –0.010 | 1.076 |
| Spot2 Hrv2 | 0.012 | 0.921 | –0.011 | 1.081 |
| Spot4 Hrv2 | 0.010 | 0.917 | –0.008 | 1.085 |
| SPOT5 | 0.010 | 0.928 | –0.008 | 1.073 |
| Venus, using band B10 with Gaussian | –0.012 | 0.984 | 0.013 | 1.015 |
| Venus, using band B11 with Gaussian | 0.007 | 0.967 | –0.006 | 1.032 |

*Source:* Steven, Malthus and Baret (2007).

Table 11.2 allows VI data from any of the listed systems to be corrected to the corresponding VI for the standard pair of spectral bands. To convert a vegetation index from an operational system $VI_{op}$ to the standard, $VI_{std}$, Equation 11.3 is applied, using the slope and intercept values from column B. To convert from the standard to a particular operational system, Equation 11.4 is applied, using the slope and intercept values from column A.

$$VI_{std} = VI_{op} \times [slope]_B + [intercept]_B \tag{11.3}$$

$$VI_{op} = VI_{std} \times [slope]_A + [intercept]_A \tag{11.4}$$

For the OrbView systems where the detailed spectral response functions were unavailable, there are two results in Table 11.2, corresponding to the alternative assumptions applied. For OrbView-2 both methods give comparable results so that adjustment to the standard can be made to better than 1% precision. For OrbView-3 however, the difference is about 8%, indicating that in the absence of further information, this system is unsuitable for applications requiring intercalibration with others.

### 11.4.2 Validation of Cross-Sensor Conversion

Although direct intercomparisons of sensors suffer from greater errors and limitations than the simulation approach applied in Section 11.4.1, they are valuable in validating the findings. Steven et al. (2003) reported image-based comparisons between SPOT HRV and Landsat TM data and between ATSR-2 and AVHRR. In both studies, pixels were aggregated across quasi-uniform targets to minimise co-registration errors and the empirical results were in reasonable agreement (±0.03) with the simulated cross-sensor calibrations. Studies by Guyot and Gu (1994) also generated coefficients that supported the findings, but previous simulations by Gallo and Daughtry (1987) were significantly different, although similarly based on simulation from field data. More recent studies have included both direct comparisons of image data and various forms of simulation. The quality of image comparisons in the literature is however quite variable, with many suffering from substantial differences in image acquisition time, different solar or viewing angles, insufficient aggregation to overcome co-registration errors or other problems. Such studies may show differences between sensors but have insufficient precision to test conversion factors. Selected results are discussed later.

Martínez-Beltrán et al. (2009) compared ETM+, TM, LISS, ASTER, QuickBird and AVHRR data on selected sites in south eastern Spain covering a wide range of surface types (Figure 11.9). All images were near-nadir viewing and the image pairs compared were no more than a few hours apart. Although comparisons were of top-of-atmosphere NDVI, without atmospheric correction, they found that with sufficient aggregation to reduce noise, their cross-sensor relationships were linear, reasonably precise and in good agreement with the results of Steven et al. (2003). However, Figure 11.9 shows the difficulty of establishing reliable cross-sensor relationships by direct comparison. In spite of the use of aggregation and homogenous areas for comparison, substantial differences in coefficients remain. Further discussions of vegetation indices within and between sensors can be found in these papers (Zeng et al. 2022; Carneiro et al. 2020; Mancino et al. 2020; Jarchow et al. 2018; Macfarlane et al. 2017; Psomiadis et al. 2017; Robinson et al. 2017; Xue and Su 2017).

Ji et al. (2008) evaluated differences between AVHRR and MODIS in two years of data over the conterminous United States. In addition to their own findings that the differences are substantial and 20% systematic, they compare their results with 17 previous cross-sensor studies. Gallo et al. (2005) compared NDVI values for MODIS and AVHRR over the United States for identical 16-day compositing periods and found linear relationships between NDVI values from different sensors. Their regression slopes differ from Steven et al. (2003) by no more than 0.02 indicating good agreement within the limits of the data; although the compositing process can introduce a systematic upward bias in NDVI (Goward et al. 1993), this would probably be similar for both systems. Fensholt, Sandholt, and Stisen (2006) compared MERIS, MODIS, and VEGETATION products on grass savannah in Senegal using wide angle in situ measurements with band radiometers designed

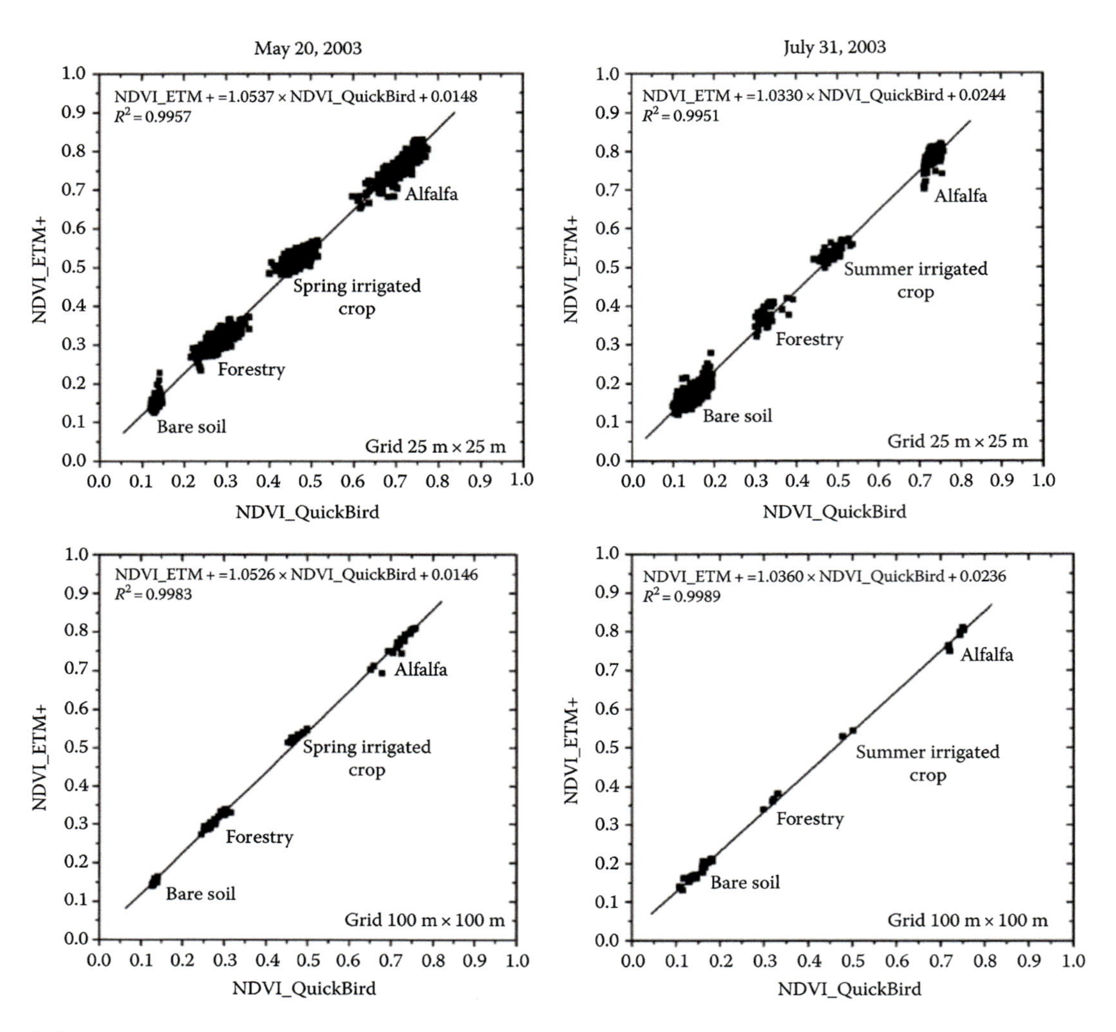


**FIGURE 11.9** Intercomparisons of ETM+ and QuickBird NDVI in homogeneous zones at 25 m and 100 m scales. The data for July 31, 2003 include a prior conversion from TM to ETM+ NDVI. (Martínez-Beltrán et al. 2009).

to approximate the relevant bands. They report generally good agreement with Steven et al. (2003) but with higher MERIS sensitivity to vegetation than predicted. However, the accuracy of their comparisons depends on the degree to which the in situ sensor bands match those of the satellite instruments. In addition as they noted in their study, wide angle measurements would exaggerate vegetation indices, particularly in the middle of the range, although, given the extensive overlap of the spectral bands, the angular response effect is likely to be very similar for the different systems compared.

Trishchenko, Cihlar and Li (2002) and d'Odorico et al. (2013) have applied quadratic correction factors for inter-sensor corrections. There is no doubt that a quadratic function will provide a better statistical fit to any given set of data, even data which appear strongly linear as in Figures 11.7 and 11.8, but fitting a curve to such data makes the coefficients relatively unstable as the additional coefficient tends to counteract the previous ones and it is unclear whether the extra coefficient is justified by a general gain in accuracy when applied to independent data. Steven et al. (2003) compared the effect of corrections and the polynomial corrections of Trishchenko, Cihlar and Li

(2002). In general the corrections by Steven et al. (2003) gave the greater reduction of error, except for open or sparse grassland with low NDVI, where values were overcorrected, while the method of Trishchenko, Cihlar and Li (2002) provided a more consistent correction across all land cover classes.

Song, Ma and Veroustraete (2010) found linear conversions between AVHRR GIMMS and SPOT-VGT NDVI values, but the coefficients varied regionally across China. Figure 11.10 shows a map of average differences between the GIMMS and VGT products. Mountainous regions (C) show both positive and negative differences, while the borderline between regions A and B, where the differences are larger, and regions C and D, where they are lower, correspond to a major climatic and geological boundary. When rates of change of seasonally integrated NDVI were computed, there were substantial differences between the two systems (Figure 11.11).

Miura, Huete and Yoshioka (2006) compared NDVI values for a number of systems on Hyperion hyperspectral image data over tropical forest and savannah in Brazil. They applied a similar approach to Steven et al. (2003); Steven Malthus and Baret (2007), combining the atmospherically corrected data with spectral response functions to simulate surface measured radiance in various bands. Although relationships between simulated radiances in paired bands were found to be land cover dependent, the relationships for NDVI were independent of land cover. However, they were sufficiently nonlinear to require fitting a quadratic function for an adequate conversion between systems.

Inter-sensor conversions have also been evaluated by modeling approaches (Jarchow et al. 2018; Xue and Su 2017). Van Leeuwen et al. (2006) simulated NDVI from AVHRR, MODIS, and VIIRS using the SAIL model with a wide range of LAI. The model was parameterised with data inputs from spectral libraries of vegetation (two spectra), soil (two spectra) and snow (one spectrum). Their result for NOAA16 versus MODIS is within 0.01 of values predicted from Table 11.1, but

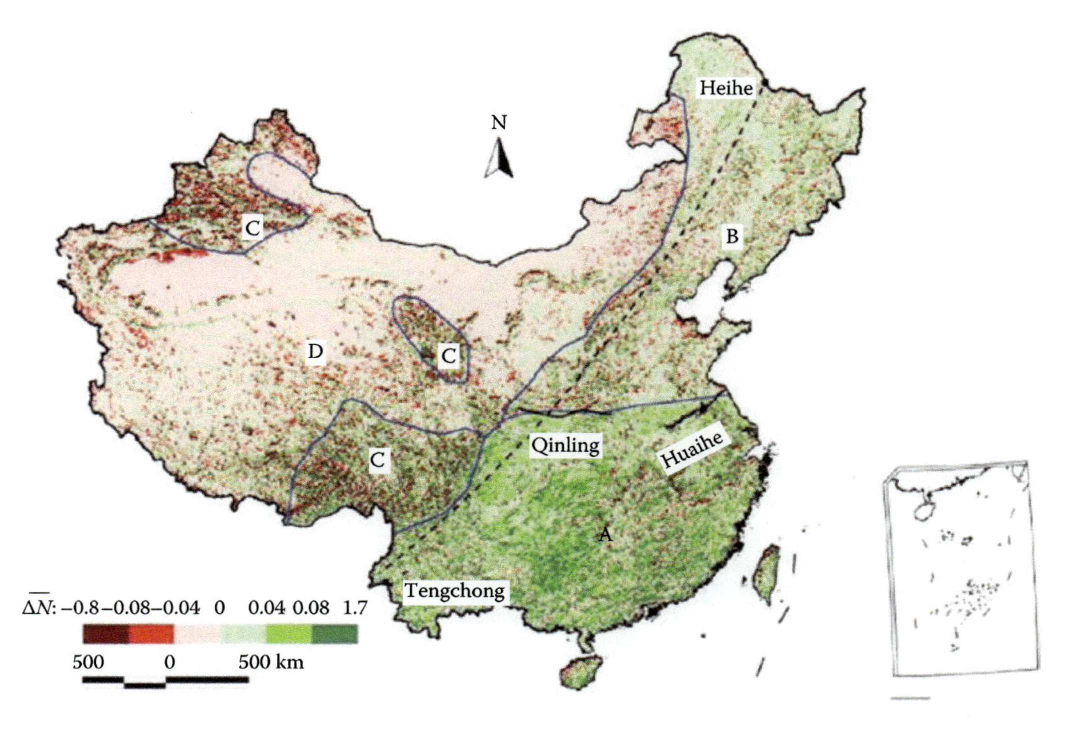


FIGURE 11.10 Map of mean annual differences of VGT and GIMMS NDVI for 1998–2003 in China. (Song, Ma and Veroustraete 2010).

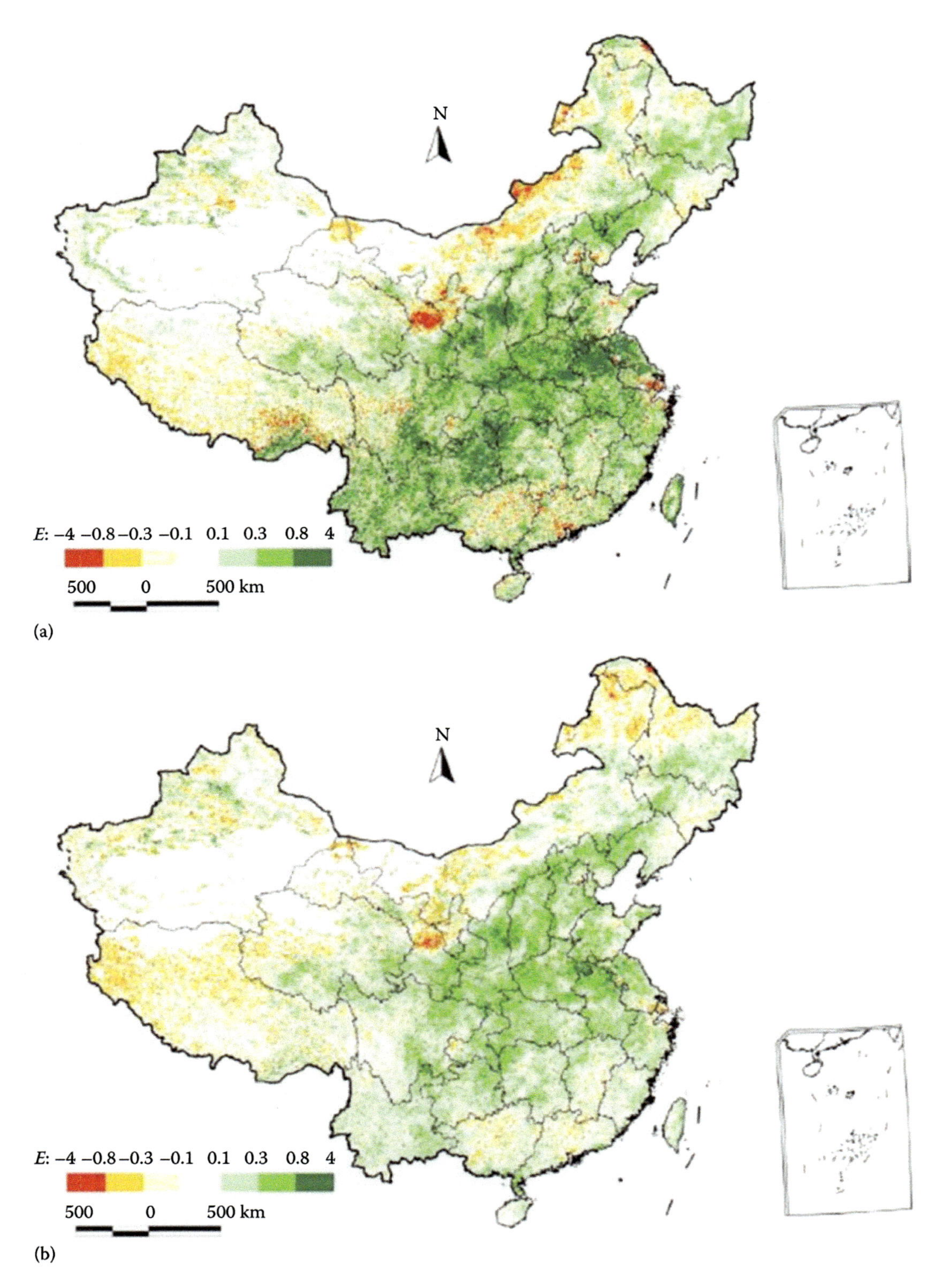


**FIGURE 11.11** Rate of change ($y^{-1}$) of seasonally integrated NDVI in China from 1998 to 2006 for (a) VGT and (b) GIMMS systems. (Song, Ma and Veroustraete 2010).

their prediction for NOAA14 has a slope 0.03 higher. Gonsamo and Chen (2013) used a coupled PROSPECT + SAIL (or PROSAIL) and 6S model to generate synthetic data corresponding to 21 satellite sensors. Their simulations included a range of leaf characteristics, LAI from 0 to 6 with a fixed spherical leaf angle distribution, a range of solar and viewing angles and eight distinct backgrounds. They found good agreement with both Steven et al. (2003) and van Leeuwen et al. (2006), but less so with other studies that had used quadratic fitting. Soudani et al. (2006) also found good agreement with Steven et al. (2003) for PROSAIL simulations of IKONOS, ETM+, and SPOT NDVIs from forest canopies; the maximum differences in slope and intercept were 0.383 and 0.0183 respectively. D'Odorico et al. (2013) assessed the general effectiveness of synthetic calibration data generated by radiative transfer modeling for cross-sensor calibration. In general, airborne spectral measurements gave better results, but model-generated data were found to be a good substitute for regional or global monitoring.

## 11.5 VEGETATION INDEX STANDARDIZATION

In proposing a standard for vegetation indices, some general principles should be considered (Ferchichi et al. 2022; Swoish et al. 2022; Zeng et al. 2022; Carneiro et al. 2020; Mancino et al. 2020; Psomiadis et al. 2017; Robinson et al. 2017; Xue and Su 2017).

**Principle 1—Standardization should be applicable to vegetation indices from all systems**, including the very earliest, to meet the aim of providing a long term baseline of global vegetation measurements. For this reason the index chosen should be based on just two bands, the red and the near-infrared, in spite of improvements in performance reported with incorporation of additional band information.

**Principle 2—Standards should be capable of evolving over time**. A vegetation index standard must be traceable and, if possible, reversible, as new knowledge or improved modeling may require readjustment of past data. With an evolving standard, complete metadata are essential to allow later adjustments, and adopters of the standard should specify precisely what corrections and procedures have been applied in adjusting measured data.

**Principle 3—Standardization procedures should be modular**, so that corrections for spectral bands, soil background, BRDF and other factors are performed separately. Modularity improves the ability to trace (and if necessary reverse) processes and errors in the system. The sequence of correction procedures should also be considered as not all the operations are strictly commutative. Radiometric corrections—calibration and atmospheric correction—should come first, followed by corrections for illumination and viewing geometry as these may vary to some extent with the particular bands of the Earth observation system. Finally, the vegetation index should be calculated and adjustment made for the spectral bands as described in Section 11.4.1 (Equation 11.3).

### 11.5.1 Proposals for Vegetation Index Standards

In practice, atmospheric corrections are usually combined in a single model with calibration. The 6S (Vermote et al. 1997) and MODTRAN (Berk et al. (1998) models are both widely used and appropriate here. Callieco and Dell'Acqua (2011) investigated differences between these models, but found that significant effects were confined to wavelengths shorter than 500 nm, so the choice of model should have minimal impact on vegetation indices. Such models also account for variations in Earth–Sun distance and the effects of solar and viewing angle on the atmospheric effects. Variations in solar output are not modeled, but are typically of the order of 0.1% and can safely be ignored.

A BRDF model is needed to account for effects of viewing and illumination directions. The PROSAIL model (Jacquemoud et al. 2009) is widely respected and sufficiently versatile for this purpose. To standardize for bidirectional effects it is also necessary to adopt standard solar and viewing angles. Bacour, Bréon and Maignan (2006) proposed standardizing on viewing at nadir with a solar zenith angle of 40°. Conveniently, these angles approximate to the average measurement conditions that apply to the canopy spectral database used by Steven et al. (2003); Steven, Malthus and Baret (2007), so adoption of these angles as standard would ensure a self-consistent system. The solar angle proposed is also a reasonable mid-value for summer viewing conditions at mid-latitudes. The bidirectional effects depend on vegetation type which can be modeled by PROSAIL on the basis of leaf angle distribution. If this information is unavailable, or impractical to implement such as in global scale applications, corrections based on a spherical leaf angle distribution are recommended here.

The instrument-specific vegetation index can next be calculated from the corrected surface reflectance values. Adjustment should then be made to determine the corresponding index for the standard bands—670 and 815 nm are proposed here—by applying Equation 11.3 with the appropriate conversion coefficients from Table 11.2.

Multiple standards for indices may be necessary (Zeng et al. 2022; Carneiro et al. 2020; Mancino et al. 2020; Jarchow et al. 2018; Macfarlane et al. 2017; Psomiadis et al. 2017; Robinson et al. 2017; Xue and Su 2017), at least in the short term. The NDVI is so well established that its continuance is unavoidable, in spite of its deficiencies. An index adjusted to minimize sensitivity to variable soil should also be included, but there are many claimants. The simplest is the Soil Adjusted Vegetation index, SAVI defined in Equation 11.2 (Huete 1988), with a value of 0.5 for the adjustment factor L. The advantage of the SAVI formulation is that the adjustment factor eliminates the need for specific calibration to different soils (Huete and Liu 1994). Re-evaluation by Rondeaux, Steven and Baret (1996) of a range of $L$ values found that a value of 0.16 gave a slightly better performance, particularly when the soil types in the analysis were restricted to those likely to be found in agriculture and the Optimized Soil Adjusted Vegetation Index OSAVI was proposed incorporating this value. In the analysis by Rondeaux, Steven and Baret (1996) the variance in the index due to soil type was reduced from 7.5% for NDVI to 1.1% and 1.7% for SAVI and OSAVI, respectively. The case for OSAVI was largely based on its performance with a restricted set of agricultural soil types where it reduced the soil signal to 0.06% with the residual error distributed evenly across the range of foliage cover. With the wider dataset that might be more applicable for global application, SAVI and MSAVI were slightly better performers. SAVI (with $L = 0.5$) is recommended here over MSAVI for its greater simplicity and its known compatibility with the conversion coefficients in Table 11.2. Both OSAVI and the original SAVI are based on analyses of relatively narrow datasets and further studies over a wider range of soil types may lead to better understanding of the errors and an improved general index, but a perfect soil adjustment is not attainable and residual errors of the order of 5% can be expected in the estimation of fractional vegetation cover (Rondeaux, Steven and Baret 1996).

A summary of these proposals is shown in Table 11.3.

### 11.5.2 Limits to Standardization

Finally, it must be recognised that there are limits to the ability to correct and standardize vegetation indices (Rehman et al. 2022; Swoish et al. 2022; Zeng et al. 2022; Carneiro et al. 2020; Mancino et al. 2020; Jarchow et al. 2018; Robinson et al. 2017; Xue and Su 2017). The potential pitfalls of failing to apply limits to the operational parameters are illustrated in Figure 11.12, which represents the average of MODIS EVI observations over two month periods around the solstices. On close examination there appears to be a slight increase in EVI along the Arctic Circle in the middle of the northern winter. This is emphatically not a result of increased vegetation! At this time of year in this location, the Sun is grazing the horizon and as a result of its low altitude is severely depleted in shorter wavelengths by Rayleigh scattering in the atmosphere. The effect of this selective depletion is that although the Sun would appear red to a ground observer, it is actually relatively weak in the red compared to the near-infrared. Sunlight reflected from snow therefore has an exaggerated NIR: red ratio that enhances the vegetation index. We recommend this image as a puzzle and a warning

**TABLE 11.3**

**Summary of Specific Proposals Made in This Chapter for Vegetation Index Standards**

| Standard Parameters | Value | Remarks |
|---|---|---|
| VI formula | SAVI | Preferred |
| | NDVI | Tolerated |
| Spectral bands | 815 nm | Nominal band width 20 nm, centered on these wavelengths |
| Near-infrared | 670 nm | |
| Red | | |
| View zenith angle | 0° | After Bacour, Bréon and Maignan (2006) |
| Solar zenith angle | 40° | |
| Atmospheric correction | To surface reflectance | 6S or MODTRAN routines recommended* |
| BRDF corrections | To standard angles | PROSAIL recommended with spherical leaf angle distribution by default* |

* These are interim recommendations and do not preclude alternative or improved procedures.

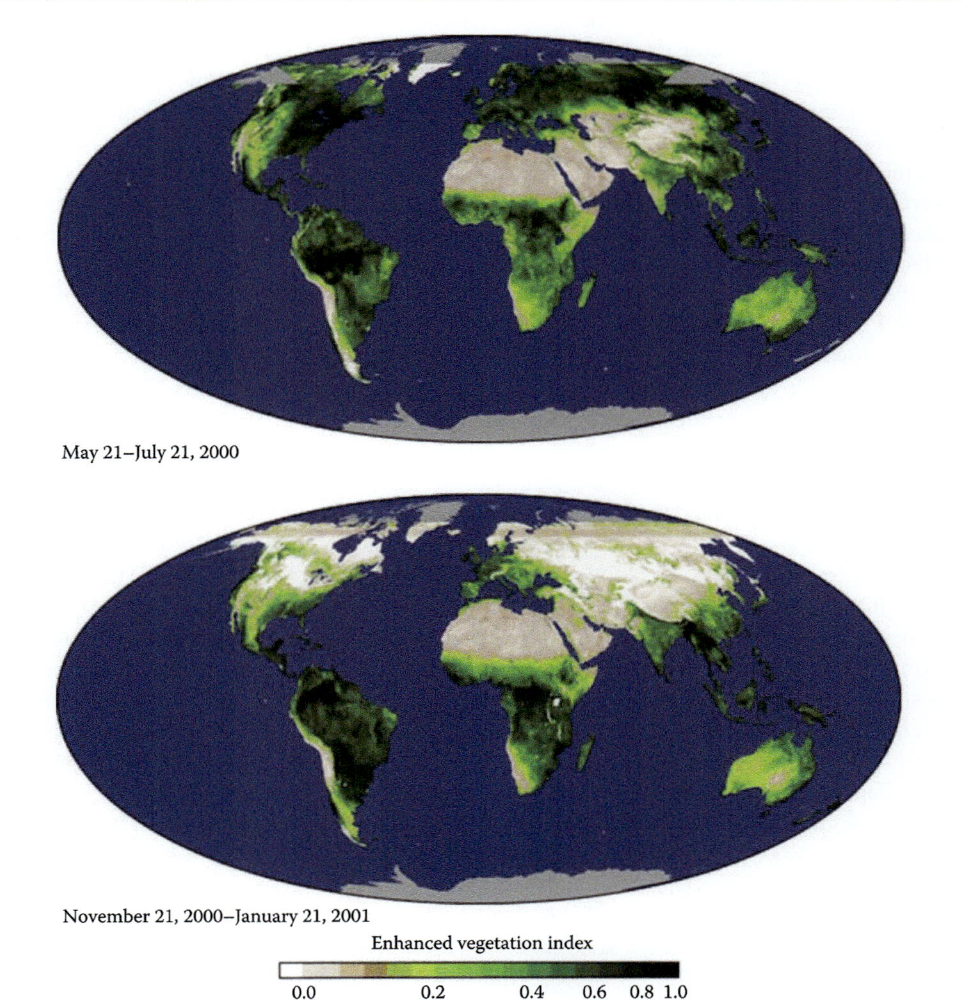


**FIGURE 11.12** Composited seasonal MODIS EVI observations illustrating the effects of extreme solar angles on the index. (http://earthobservatory.nasa.gov/IOTD/view.php?id=2033).

for students of remote sensing! The problem is easily avoided: the spectral balance of solar irradiation is conservative once solar elevation exceeds 10° (Monteith and Unsworth 1990). Variations in spectral irradiance can probably be modeled with reasonable accuracy within a few degrees of the horizon, but until this can be reliably demonstrated, observations beyond a solar zenith angle of 80° should be excluded. The same limit applies to observation angles, as selective Rayleigh scattering works in exactly the same way on the upwelling reflected radiance.

## 11.6 DISCUSSION

There is a general difficulty with both atmospheric correction and BRDF modeling that the models require input data for parameterization. Atmospheric correction parameters are difficult to obtain for remote areas, but Steven (1998) found that OSAVI estimates were affected to a very minor extent by errors in the model assumptions. In a global monitoring context it would also be possible to incorporate satellite aerosol observations into a routine correction procedure. For BRDF modeling, the leaf angle distribution is critical. In a simulation of the effect of variable SPOT viewing angles, Steven (1998) found variations in OSAVI up to 4% relative to vertical viewing. This difference corresponded to 30° leftwards tilt of the SPOT camera and larger differences will occur at greater angles. However, as the off-nadir effect can be estimated by the model, it should not of itself be a major source of error, while errors associated with the assumed leaf angle distribution can be expected to be second order relative to the overall magnitude of the off-nadir effect.

Since the paper by Steven et al. (2003), a number of studies have applied the coefficients provided to convert for cross-sensor effects (Almeida-Ñauñay et al. 2022; Ferchichi et al. 2022; Rehman et al. 2022; Swoish et al. 2022; Zeng et al. 2022; Carneiro et al. 2020; Mancino et al. 2020; Jarchow et al. 2018; Macfarlane et al. 2017; Psomiadis et al. 2017; Robinson et al. 2017; Xue and Su 2017; Sun et al. 2007; Fisher, Tu and Baldocchi 2008; Pouliot, Latifovic and Olthof 2009; Tang et al. 2009; Zhang et al. 2009; Propastin and Erasmi 2010; Ouyang et al. 2010, 2012). Others have tested the conversion equations against data or modeling approaches as described in Section 11.4.2. The validation studies discussed provide general support for the idea of intercalibrating vegetation indices and in many cases, support for the linear coefficients provided by Steven et al. (2003) and given in extended form in Table 11.2. The accuracy suggested by these studies is of the order of ±0.03 in the slope. This is probably about the limit of accuracy for a validation study in this context. More pertinent here is the relative precision of the conversions, which is about ±0.01.

Despite this support, the generality of the inter-sensor relationships should be considered. Song, Ma and Veroustraete (2010) found that differences between AVHRR GIMMS and SPOT VEGETATION NDVI were both land cover and seasonally dependent. However it is difficult to assess whether such variations are fundamentally linked to the basic observations, or an artefact of the compositing and processing procedures applied to generate one or other of the datasets. It is also possible that these dependencies, and the earlier finding by Swinnen and Veroustraete (2008) that the equations of Steven et al. (2003) overcorrected low values, may be related to the high variability of NDVI with respect to soil. D'Odorico et al. (2013) referred to earlier studies that mention land cover dependence, but these studies related to differences in VIs rather than their ratio. Miura, Huete and Yoshioka (2006), as well as the study by D'Odorico et al. (2013) itself, showed that conversion to NDVI takes out most of the land cover dependence that is found in the individual bands.

A further concern is that the cross-sensor relationships have been established on a relatively narrow base of data (Ferchichi et al. 2022; Swoish et al. 2022' Zeng et al. 2022; Carneiro et al. 2020; Mancino et al. 2020; Jarchow et al. 2018; Macfarlane et al. 2017; Psomiadis et al. 2017). The database used by Steven et al. (2003); Steven, Malthus and Baret (2007) comprised 166 independent measurements of spectral reflectance of strictly agricultural canopies. Other studies have even fewer measurements and while model simulations may generate large synthetic datasets, their empirical support-base is often very narrow. In addition, many of the studies discussed in Section 11.4.2

have applied the same models, so similar findings are to be expected. Forest cover classes, where deep shadow may be a significant component of the target, are not well represented in the datasets, although the study by Soudani et al. (2006) suggests that the coefficients of Table 11.2 still provide a good correction. As noted by Steven et al. (2003), previous studies have also found senescent biomass, or litter to be a significant factor. The results of Miura, Huete and Yoshioka (2006), which required a quadratic correction between sensors, suggest that the relations established in Table 11.2 are not universal. A broader database might help to resolve these issues, but as Miura, Huete and Yoshioka pointed out, observations with different spectral bands are inherently different and may introduce bias into downstream products. The same caveat applies to correction for BRDF effects. Nevertheless, standardization is imperative for a range of issues and does not preclude use of the data in original format for specific purposes. Steven et al. (2003) argued that restriction of the intercalibration database to cultivated vegetation may actually have allowed a tighter statistical relationship to be established than would have occurred with a broader, more representative dataset. Differences between indices are in most cases quite small and applying conversions based on agricultural datasets may help to maintain relationships with fAPAR and related parameters that have been well validated in such environments.

## 11.7 CONCLUSIONS

The outcome of this study is that corrections are possible, to an acceptable degree of precision, for the main sources of variation in vegetation indices: soil background, atmospheric, BRDF, and spectral band effects. The uncertainty of these corrections is about 5% for soil, when using a soil adjusted vegetation index (Rondeaux, Steven and Baret 1996), 3% for poorly characterised atmospheric aerosol amount (Steven 1998), 4% for off-nadir view angle effects up to 30°, if poorly characterized (Steven 1998) and about 1% for spectral band effects. The corresponding uncertainties in downstream products such as fAPAR will be slightly larger. Greater errors can also be expected at larger angles of view or solar zenith angles, and if errors combine malignly, but the overall effect of atmospheric and BRDF errors can be reduced substantially by application of simple well-established models and appropriate input data. Further studies are needed to extend the database for comparison of sensors and assessment of BRDF effects, to assess the errors of modeling and to validate the procedures and their effect on VIs and downstream products. It is anticipated that the errors will become better characterized as procedures evolve. Nevertheless, there is sufficient evidence to proceed now. Standardization will have immense benefits for vegetation monitoring (Almeida-Ñauñay et al. 2022; Ferchichi et al. 2022; Rehman et al. 2022; Swoish et al. 2022; Zeng et al. 2022; Carneiro et al. 2020; Mancino et al. 2020; Jarchow et al. 2018; Macfarlane et al. 2017; Psomiadis et al. 2017; Robinson et al. 2017; Xue and Su 2017), both on short-timescale regional studies where the constellation approach is applied to monitor the dynamics of specific land covers and in large scale studies of environmental history where records from early satellites are merged with modern data.

## REFERENCES

Agapiou, A., D. G. Hadjimitsis, and D. D. Alexakis. 2012. Evaluation of broadband and narrowband vegetation indices for the identification of archaeological crop marks. *Rem Sens* 4:3892–3919.

Almeida-Ñauñay, A. F., M. Villeta, M. Quemada and A. M. Tarquis. 2022. Assessment of drought indexes on different time scales: A case in semiarid mediterranean grasslands. *Rem Sens* 14, no. 3:565. https://doi.org/10.3390/rs14030565

Bacour, C., F. M. Bréon and F. Maignan. 2006. Normalization of the directional effects in NOAA—AVHRR reflectance measurements for an improved monitoring of vegetation cycles. *Rem Sens Environ* 102:402–413.

Baret, F. and G. Guyot. 1991. Potentials and limits of vegetation indices for LAI and APAR assessment. *Rem Sens Environ* 35:161–173.

Berk, A., L. S. Bernstein, G. P. Anderson, P. K. Acharya, D. C. Robertson, J. H. Chetwynd, and S. M. Adler-Golden. 1998. MODTRAN cloud and multiple scattering upgrades with application to AVIRIS. *Rem Sens Environ* 65:367–375.

Bowers, S. A. and R. J. Hanks. 1965. Reflection of radiant energy from soil. *Soil Sci* 100:130–138.

Boyd, D. S., S. Almond, J. Dash, P. J. Curran and R. A. Hill. 2011. Phenology of vegetation in Southern England from Envisat MERIS terrestrial chlorophyll index (MTCI) data. *Int J Rem Sens* 32:8421–8447.

Boyd, D. S. and G. M. Foody. 2011. An overview of recent remote sensing and GIS based research in ecological informatics. *Ecol Informatics* 6:25–36.

Brown, M., J. E. Pinzón, K. Didan, J. T. Morisette and C. J. Tucker. 2006. Evaluation of the consistency of long-term NDVI time series derived from AVHRR, SPOT-Vegetation, SeaWiFS, MODIS, and Landsat ETM+ sensors. *IEEE Trans Geosci Rem Sens* 44:1787–1793.

Callieco, F. and F. Dell'Acqua. 2011. A comparison between two radiative transfer models for atmospheric correction over a wide range of wavelengths. *Int J Rem Sens* 32:1357–1370.

Carneiro, M., Angeli Furlani, F., Zerbato, C. E., et al. 2020. Comparison between vegetation indices for detecting spatial and temporal variabilities in soybean crop using canopy sensors. *Precision Agric* 21:979–1007. https://doi.org/10.1007/s11119-019-09704-3

CEOS. 2006. *The CEOS Virtual Constellation Concept V0.4,Committee on Earth Observation Satellites*. http://igos-cryosphere.org/docs/CEOS_constellations.doc (accessed July 25, 2014).

D'Odorico, P., A. Gonsamo, A. Damm and M. E. Schaepman. 2013. Experimental evaluation of Sentinel-2 spectral response functions for NDVI time-series continuity. *IEEE Trans Geosci Rem Sens* 51:1336–1348.

D'Odorico, P., A. Gonsamo, B. Pinty, N. Gobron, N. Coops, E. Mendez and M. E. Schaepman. 2014. Intercomparison of fraction of absorbed photosynthetically active radiation products derived from satellite data over Europe. *Rem Sens Environ* 142:141–154.

Fensholt, R., I. Sandholt and S. Stisen. 2006. Evaluating MODIS, MERIS, and VEGETATION vegetation indices using in situ measurements in a semiarid environment. *IEEE Trans Geosci Rem Sens* 44:1774–1786.

Ferchichi, A., Abbes, A. B., Barra, V., Farah, I. R. 2022. Forecasting vegetation indices from spatio-temporal remotely sensed data using deep learning-based approaches: A systematic literature review. Ecol Inform 68:101552, ISSN 1574–9541. https://doi.org/10.1016/j.ecoinf.2022.101552. (www.sciencedirect.com/science/article/pii/S1574954122000012)

Fisher, J. B., K. P. Tu and D. D. Baldocchi. 2008. Global estimates of the land-atmosphere water flux based on monthly AVHRR and ISLSCP-II data, validated at 16 FLUXNET sites. *Rem Sens Environ* 112:901–919.

Gallo, K. P. and C. S. T. Daughtry. 1987. Differences in vegetation indices for simulated Landsat-5 MSS and TM, NOAA-9 AVHRR and SPOT-1 sensor systems. *Rem Sens Environ* 23:439–452.

Gallo, K., L. Ji, B. Reed, J. Eidenshink and J. Dwyer. 2005. Multi-platform comparisons of MODIS and AVHRR normalized difference vegetation index data. *Rem Sens Environ* 99:221–231.

Gates, D. M., H. J. Keegan, J. C. Schleter and V. R. Weidner. 1965. Spectral properties of plants. *App Opt* 4:11–20.

Gausman, H. W. and W. A. Allen. 1973. Optical parameters of leaves of 30 plant species. *Plant Physiol* 52:57–62.

Gobron, N., B. Pinty, O. Aussedat, M. Taberner, O. Faber, F. Mélin, T. Lavergne, M. Robustelli and P. Snoeij. 2008. Uncertainty estimates for the FAPAR operational products derived from MERIS—Impact of top-of-atmosphere radiance uncertainties and validation with field data. *Rem Sens Environ* 112:1871–1883.

Gonsamo, A. and J. M. Chen. 2013. Spectral response function comparability among 21 satellite sensors for vegetation monitoring. *IEEE Trans Geosci Rem Sens* 51:1319–1335.

Goward, S. N., D. G. Dye, S. Turner and J. Yang. 1993. Objective assessment of the NOAA global vegetation index data product. *Int J Rem Sens* 14:3365–3394.

Gower, S. T., C. J. Kucharik and J. K. M. Norman. 1999. Direct and indirect estimation of leaf area index, $f_{APAR}$, and net primary production of terrestrial ecosystems. *Rem Sens Environ* 70:29–51.

Gutman, G. 1991. Vegetation indices from AVHRR: An update and future prospects. *Rem Sens Environ* 35:121–136.

Guyot, G. 1990. Optical properties of vegetation canopies. In M. D. Steven and J. A. Clark (eds), *Applications of Remote Sensing in Agriculture*, (London, Butterworths) pp 19–43.

Guyot, G. and X. F. Gu. 1994. Effect of radiometric corrections on NDVI—Determined from SPOT HRV and Landsat TM data. *Rem Sens Environ* 49:169–180.

Hadjimitsis, D. G., G. Papdavid, A. Agapiou, K. Themistocleous, M. G. Hadjimitsis, A. Retalis, S. Michaelides, N. Chrysoulakis, L. Toulios and C. R. I. Clayton. 2010. Atmospheric correction for satellite remotely sensed data intended for agricultural applications: Impact on vegetation indices. *Nat Haz Earth Syst Sci* 10:89–95.

Heller, J. 1961. *Catch 22*. (New York, Simon and Schuster).

Holben, B. N., C. J. Tucker and C. J. Fan. 1980. Spectral assessment of soybean leaf area and leaf biomass. *Photogramm Eng Rem Sens* 46:651–656.

Huete, A. R. 1988. A Soil Adjusted Vegetation Index (SAVI). *Int J Rem Sens* 9:295–309.

Huete, A. R. and H. Q. Liu. 1994. An error and sensitivity analysis of the atmospheric- and soil-correcting variants of the NDVI for the MODIS-EOS. *IEEE Trans Geosci Rem Sens* 32:897–905.

Huete, A. R., H. Q. Liu, K. Batchily and W. van Leeuwen. 1997. A comparison of vegetation indices over a global set of TM images for EOSMODIS. *Rem Sens Environ* 59:440–451.

Jacquemoud, S., W. Verhoef, F. Baret, C. Bacour, P. J. Zarco-Tejada, G. P. Asner, C. François and S. L. Ustin. 2009. PROSPECT + SAIL models: A review of use for vegetation characterization. *Rem Sens Environ* 113:S56–S66.

Jaggard, K. W. and C. J. Clark. 1990. Remote sensing to predict the yield of sugar beet in England. In M.D. Steven and J.A. Clark (eds), *Applications of Remote Sensing in Agriculture*, (London, Butterworths) pp 201–208.

Jarchow, C. J., K. Didan, A. Barreto-Muñoz, P. L. Nagler and E. P. Glenn. 2018. Application and comparison of the MODIS-derived enhanced vegetation Index to VIIRS, Landsat 5 TM and Landsat 8 OLI platforms: A case study in the Arid Colorado River Delta, Mexico. *Sens* 18, no. 5:1546. https://doi.org/10.3390/s18051546

Ji, L., K. Gallo, J. C. Eidenshink and J. Dwyer. 2008. Agreement evaluation of AVHRR and MODIS 16-day composite NDVI data sets. *Int J Rem Sens* 29:4839–4861.

Lee, K.-S., W. B. Cohen, R. E. Kennedy, T. K. Maiersperger and S. T. Gower. 2004. Hyperspectral versus multispectral data for estimating leaf area index in four different biomes. *Rem Sens Environ* 91:508–520.

Le Maire, G., C. François and E. Dufrêne. 2004. Towards universal broad leaf chlorophyll indices using PROSPECT simulated database and hyperspectral reflectance measurements. *Rem Sens Environ* 89:1–28.

Macfarlane, C., Grigg, A. H. and Daws, M. I. 2017. A standardised Landsat time series (1973–2016) of forest leaf area index using pseudoinvariant features and spectral vegetation index isolines and a catchment hydrology application. Remote Sens Appl Soc Environ 6:1–14, ISSN 2352–9385, https://doi.org/10.1016/j.rsase.2017.01.006. (www.sciencedirect.com/science/article/pii/S2352938516300921)

Mahiny, A. S. and B. Turner. 2007. A comparison of four common atmospheric correction methods. *Photogramm Eng Rem Sens* 73:361–368.

Mancino, G., A. Ferrara, A. Padula and A. Nolè. 2020. Cross-Comparison between Landsat 8 (OLI) and Landsat 7 (ETM+) derived vegetation indices in a mediterranean environment. *Remote Sens* 12, no. 2:291. https://doi.org/10.3390/rs12020291

Martínez-Beltrán, C., M. A. O. Jochum, A. Calera and J. Meliá. 2009. Multisensor comparison of NDVI for a semi-arid environment in Spain. *Int J Rem Sens* 30:1355–1384.

Maselli, F., S. Romanelli, L. Bottai and G. Maracchi. 2000. Processing of GAC NDVI data for yield forecasting in the Sahelian region. *Int J Rem Sens* 21:3509–3523.

Miura, T., A. Huete and H. Yoshioka. 2006. An empirical investigation of cross-sensor relationships of NDVI and red/near-infrared reflectance using EO-1 Hyperion data. *Rem Sens Environ* 100:223–236.

Miura, T., J. P. Turner and A. R. Huete. 2013. Spectral compatibility of the NDVI across VIIRS, MODIS and AVHRR: An analysis of atmospheric effects using EO-1 Hyperion. *IEEE Trans Geosci Rem Sens* 51:1349–1359.

Monteith, J. L. 1977. Climate and the efficiency of crop production in Britain. *Phil Trans Roy Soc, London, Ser B* 281:277–294.

Monteith, J. L. and M. H. Unsworth. 1990. *Principles of Environmental Physics*, 2nd Edition, (London, Edward Arnold).

Moreno, J. S. Kimball, D. E. Naugle, T. A. Erickson and A. D. Richardson. 2017. A dynamic landsat derived Normalized Difference Vegetation Index (NDVI) product for the conterminous United States. *Remote Sens* 9, no. 8:863. https://doi.org/10.3390/rs9080863

Moura, Y. M., L. S. Galvão, J. R. dos Santos, D. A. Roberts and F. M. Breunig. 2012. Use of MISR/Terra data to study intra- and inter-annual EVI variations in the dry season of tropical forest. *Rem Sens Environ* 127:260–270.

Nagol, J., E. F. Vermote and S. D. Prince. 2009. Effects of atmospheric variation on AVHRR NDVI data. *Rem Sens Environ* 113:392–397.

Obata, K., T. Miura and H. Yoshioka. 2012. Analysis of the scaling effects in the area-averaged fraction of vegetation cover retrieved using an NDVI-isoline-based linear mixture model. *Sens* 4:2156–2180.

Ouyang, W., F. H. Hao, C. Zhao and C. Lin. 2010. Vegetation response to 30 years hydropower cascade exploitation in upper stream of Yellow River. *Comms Nonlin Sci Num Simul* 15:128–1941.

Ouyang, W., F. H. Hao, A. K. Skidmore, T. A. Groen, A. G. Toxopeus and T. Wang. 2012. Integration of multisensor data to assess grassland dynamics in a Yellow River sub-watershed. *Ecol Indic* 18:163–170.

Peñuelas, J. and I. Filella. 1998. Visible and near-infrared techniques for diagnosing plant physiological status. *Trends Plant Sci* 3:151–156.

Perry, C. R. and L. F. Lautenshlager. 1984. Functional equivalence of spectral vegetation indices. *Rem Sens Environ* 14:169–182.

Pettorelli, N., J. O. Vik, A. Mysterud, J.-M. Gaillard, C. J. Tucker and N. C. Stenseth. 2005. Using the satellite-derived NDVI to assess ecological responses to environmental change. *Trends Ecol Evol* 20:503–510.

Picket-Heaps, C. A., J. G. Canadell, P. R. Briggs, N. Gobron, V. Haverd, M. J. Paget, B. Pinty and M. R. Raupach. 2014. Evaluation of six satellite-derived Fraction of Absorbed Photosynthetic Active Radiation (FAPAR) products across the Australian continent. *Rem Sens Environ* 140:241–256.

Pinter, P. J. 1993. Solar angle independence in the relationship between absorbed PAR and remotely sensed data for alfalfa. *Rem Sens Environ* 46:19–25.

Psomiadis, E., Dercas, N., Dalezios, N. R. and Spyropoulos, N. V. 2017. Evaluation and cross-comparison of vegetation indices for crop monitoring from sentinel-2 and worldview-2 images. Proc. SPIE 10421, Remote Sens Agric, Ecosyst, and Hydrol XIX, 104211B (2 November). https://doi.org/10.1117/12.2278217

Pouliot, D., R. Latifovic and I. Olthof. 2009. Trends in vegetation NDVI from 1 km AVHRR data over Canada for the period 1985–2006. *Int J Rem Sens* 30:149–168.

Price, J. C. 1987. Calibration of satellite radiometers and the comparison of vegetation indices. *Rem Sens Environ* 21:15–27.

Propastin, P. and S. Erasmi. 2010. A physically based approach to model LAI from MODIS 250 m data in a tropical region. *Int J App Earth Obs Geoinf* 12:47–59.

Qi, J., A. S. Chehbouni, A. R. Huete, Y. H. Kerr and S. Sorooshian. 1994. A modified soil adjusted vegetation index. *Rem Sens Environ* 48:119–126.

Rehman, T. H., M. E. Lundy and B. A. Linquist. 2022. Comparative sensitivity of vegetation indices measured via proximal and aerial sensors for assessing N status and predicting grain yield in rice cropping systems. *Remote Sens* 14, no. 12:2770. https://doi.org/10.3390/rs14122770

Röder, A., T. Kuemmerle and J. Hill. 2005. Extension of retrospective datasets using multiple sensors: An approach to radiometric intercalibration of Landsat TM and MSS data. *Rem Sens Environ* 95:195–210.

Rondeaux, G., M. D. Steven and F. Baret. 1996. Optimization of soil adjusted vegetation indices. *Rem Sens Environ* 55:95–107.

Samain, O., B. Geiger and J.-L. Roujean. 2006. Spectral normalisation and fusion of optical sensors for the retrieval of BRDF and Albedo: Application to VEGETATION, MODIS, and MERIS data sets. *IEEE Trans Geosci Rem Sens* 44:3166–3179.

Schluessel, G., R. E. Dickinson, J. L. Privette, W. J. Emery and R. Kokaly. 1994. Modeling the bidirectional reflectance distribution function of mixed finite plant canopies and soil. *J Geophys Res-Atmos* 99(D5):10577–10600.

Shultis, J. K. 1991. Calculated sensitivities of several optical radiometric indices for vegetation canopies. *Rem Sens Environ* 38:211–228.

Sims, D. A., A. F. Rahman, E. F. Vermote and Z. Jiang. 2011. Seasonal and inter-annual variation in view angle effects on MODIS vegetation indices at three forest sites. *Rem Sens Environ* 115:3112–3120.

Song, Y., M. Ma, and F. Veroustraete. 2010. Comparison and conversion of AVHRR GIMMS and SPOT VEGETATION NDVI data in China. *Int J Rem Sens* 31:2377–2392.

Soudani, K., C. François, G. le Maire, V. Le Dantec and E. Dufrêne. 2006. Comparative analysis of IKONOS, SPOT, and ETM+ data for leaf area index estimation in temperate coniferous and deciduous forest stands. *Rem Sens Environ* 102:161–175.

Steven, M. D. 1977. Standard distributions of clear sky radiance. *Quart J Roy Met Soc* 103:457–465.

Steven, M. D. 1998. The sensitivity of the OSAVI vegetation index to observational parameters. *Rem Sens Environ* 63:49–60.

Steven, M. D. 2004. Correcting the effects of field of view and varying illumination in spectral measurements of crops. *Precis Agric* 5:51–68.

Steven, M. D., P. V. Biscoe and K. W. Jaggard. 1983. Estimation of sugar beet productivity from reflection in the red and infrared spectral bands. *Int J Rem Sens* 4:325334.

Steven, M. D., P. V. Biscoe, K. W. Jaggard and J. Paruntu. 1986. Foliage cover and radiation interception. *Field Crops Res* 13:75–87.

Steven, M. D. and M. H. Unsworth. 1980. The angular distribution and interception of diffuse solar radiation below overcast skies. *Quart J Roy Met Soc* 106:5761.

Steven, M. D., T. J. Malthus, F. Baret, H. Xu and M. Chopping. 2003. Intercalibration of vegetation Indices from different Sensor Systems. *Rem Sens Environ* 88:412422.

Steven, M. D., T. J. Malthus and F. Baret. 2007. Intercalibration of vegetation indices—An update. In M. E. Schaepman, S. Liang, N. Groot and M. Kneubühler (eds), *10th International Symposium on Physical Measurements and Spectral Signatures in Remote Sensing International Archives of the Photogrammetry, Remote Sensing and Spatial Information Sciences*, Vol. XXXVI, Part 7/C50, ISPRS, Davos (CH). pp 1682–1777.

Sun, Z., Q. Wang, Z. Ouyang, M. Watanabe, B. Matsushita and T. Fukushima. 2007. Evaluation of MOD16 algorithm using MODIS and ground observational data in winter wheat field in North China plain. *Hydrol Proc* 21:1196–1206.

Swinnen, E. and F. Veroustraete. 2008. Extending the SPOT-VEGETATION NDVI time series (1998–2006) back in time with NOAA-AVHRR data (1985–1998) for Southern Africa. *IEEE Trans Geosci Rem Sens* 46:558–572.

Swoish, M., J. F. Da Cunha Leme Filho, M. S. Reiter, J. B. Campbell and W. E. Thomason. 2022. Comparing satellites and vegetation indices for cover crop biomass estimation. Comput Electron Agric 196:106900, ISSN 0168–1699. https://doi.org/10.1016/j.compag.2022.106900. (www.sciencedirect.com/science/article/pii/S0168169922002174)

Tang, Q., S. Peterson, R. H. Cuenca, Y. Hagimoto and D. P. Lettenmaier. 2009. Satellite-based near-real-time estimation of irrigated crop water consumption. *J Geophys Res* 114:D05114. https://doi.org/10.1029/2008JD010854

Tarnavsky, E., S. Garrigues and M. Brown. 2008. Multiscale geostatistical analysis of AVHRR, SPOT-VGT, and MODIS global NDVI products. *Rem Sens Environ* 112:535–549.

Teillet, P. M., B. L. Markham and R. R. Irish. 2006. Landsat cross-calibration based on near simultaneous imaging of common ground targets. *Rem Sens Environ* 102:264–270.

Teillet, P. M. and X. Ren. 2008. Spectral band difference effects on vegetation indices derived from multiple satellite sensor data. *Can J Rem Sens* 34:159–173.

Townshend, J. R. G. and C. O. Justice. 1986. Analysis of the dynamics of African vegetation using the normalised difference vegetation index. *Int J Rem Sens* 7:1435–1445.

Trishchenko, A. P., J. Cihlar and Z. Li. 2002. Effects of spectral response function on surface reflectance and NDVI measured with moderate resolution satellite sensors. *Rem Sens Environ* 81:1–18.

Tucker, C. J. 1977. Spectral estimation of grass canopy variables. *Rem Sens Environ* 6:11–26.

Tucker, C. J., J. H. Elgin and J. E. McMurtrey. 1979. Temporal spectral measurements of corn and soybean crops. *Photogramm Eng Rem Sens* 45:643–653.

Tucker, C. J., J. E. Pinzon, M. E. Brown, D. A. Slayback, E. W. Pak, R. Mahoney, E. F. Vermote and N. El Saleous. 2005. An extended AVHRR 8-km NDVI dataset compatible with MODIS and SPOT vegetation NDVI data. *Int J Rem Sens* 26:4485–4498.

Van Leeuwen, W. J. D. 2006. Spectral vegetation indices and uncertainty: Insights from a user's perspective. *IEEE Trans Geosci Rem Sens* 44:1931–1933.

Van Leeuwen, W. J. D., A. R. Huete and T. W. Laing. 1999. MODIS vegetation index compositing approach: A prototype with AVHRR data. *Rem Sens Environ* 69:264–280.

Van Leeuwen, W. J. D., B. J. Orr, S. E. Marsh and S. M. Herrmann. 2006. Multi-sensor data continuity: Uncertainties and implications for vegetation monitoring applications. *Rem Sens Environ* 100:67–81.

Verhoef, W. 1985. Light scattering by leaf layers with application to canopy reflectance modeling: The SAIL model. *Rem Sens Environ* 16:125–141.

Vermote, E. F., D. Tanré, J. L. Deuze, M. Herman and J. J. Morcrette. 1997. Second simulation of the satellite signal in the solar spectrum: An overview. *IEEE Trans Geosci Rem Sens* 35:675–686.

Vermote, E., C. O. Justice and F.-M. Bréon. 2009. Towards a generalised approach for correction of the BRDF effect in MODIS directional reflectances. *IEEE Trans Geosci Rem Sens* 47:898–908.

Wiegand, C. L., A. J. Richardson, D. E. Escobar and A. H. Gerbermann. 1991. Vegetation indices in crop assessments. *Rem Sens Environ* 35:105–119.

Xue, J. and B. Su. 2017. Significant remote sensing vegetation indices: A review of developments and applications. J Sens 2017, Article ID 1353691:17. https://doi.org/10.1155/2017/1353691

Zeng, Y., Hao, D., Huete, A. et al. 2022. Optical vegetation indices for monitoring terrestrial ecosystems globally. *Nat Rev Earth Environ* 3:477–493. https://doi.org/10.1038/s43017-022-00298-5

Zhang, Y., M. Xu, J. Adams and X. Wang. 2009. Can Landsat imagery detect tree line dynamics. *Int J Rem Sens* 30:1327–1340.

Zhou, X., H. Guan, H. Xie and J. L. Wilson. 2009. Analysis and optimization of NDVI definitions and areal fraction models in remote sensing of vegetation. *Int J Rem Sens* 30:721–751.

# Part VI

## Crowdsourcing and Remote Sensing Data

# 12 Crowdsourcing and Remote Sensing
## *Combining Two Views of Planet Earth*

*Fabio Dell'Acqua and Silvio Dell'Acqua*

## LIST OF ACRONYMS AND THEIR MEANINGS

| | |
|---|---|
| **AI** | Artificial Intelligence |
| **AR** | Augmented Reality |
| **BLE** | Bluetooth Low Energy |
| **CIIM** | Community Internet Intensity Map |
| **DeFi** | Decentralized Finance |
| **DYFI** | Did You Feel It |
| **EO** | Earth Observation |
| **GNU** | GNU's Not Unix |
| **GPS** | Global Positioning System |
| **Hi-Res** | High-Resolution |
| **MMI** | Modified Mercalli Intensity |
| **NFT** | Non-Fungible Token |
| **OSM** | OpenStreetMap |
| **TCP/IP** | Transmission Control Protocol/Internet Protocol |
| **UGC** | User Generated Content |
| **USGS** | United States Geological Survey |
| **VGI** | Volunteered Geographic Information |
| **XYO** | XY Oracle Network |

## 12.1 INTRODUCTION

Although the term *crowdsourcing*, according to the official records, was used for the first time in 2006 in a *Wired* journal paper by Howe (2006), the concept of "on-the-fly" recruiting of a generally large set of contributors to solving a specific problem dates back by centuries. For instance, it is well known that in 1714 the British government offered a money prize (the Longitude Prize) to anyone who could devise a reliable method to compute the longitude a ship was located at (Sobel, 1998); this was an example of entrusting a nonformalized pool of individuals—potentially possessing unexplored intelligent capabilities—with the task of solving a specific problem in information retrieval. Probably less widely known is the fact that Toyota, the Japanese car manufacturer, started a competition in 1936 to define its logo, selected among 27,000 candidate entries (DesignCrowd, 2010). The winning logo has been in use to the present day.

Several examples of such calls can be found across history, but it was not until the advent of the World Wide Web, and especially the Web 2.0, that crowdsourcing found the optimal environment for being widely applied in practice. The internet, indeed, allows recruiting contributors

DOI: 10.1201/9781003541141-18

from virtually any place on the Earth, sharing resources easily, and transmitting results immediately. This is why the internet will play an important role in this book chapter. The internet itself is based on the shared contribution of the infrastructures that physically compose it, yet the term *crowdsourcing* refers to applications requiring intelligence or otherwise capacity of a human operator to solve problems that machines are still unable to deal with. An example is the so-called "wiki" philosophy, marked in 2001 by the birth of the free encyclopaedia named "Wikipedia" (2001) maintained by millions of volunteers, and later extended to tens of similarly conceived projects. Although Wikipedia writers work for free, this is not necessary a rule for all crowdsourced work: in 2005, for example, the Amazon Mechanical Turk was launched, featuring paid, online crowdsourcing services. The name of the service was chosen in reference with the famous mechanical Turkish chess player built in 1770 by Wolfgang Von Kempelen (1734–1804) to shock the Austrian empress Maria Theresa: faking a mechanical chess player, it actually contained a disguised human operator—somehow like the modern realization of crowdsourcing, "hiding" human intelligence behind an electronic interface and infrastructure. Despite advances at a fast pace in Artificial Intelligence (AI), the Amazon Mechanical Turk is still in place today, testifying that human beings are still irreplaceable even for some of the "mechanical" labor out there.

This chapter will describe and analyze how the concept of recruiting a large pool of casual contributors to generate information is applied to the domain of remote sensing and Earth Observation, where crowdsourcing and the "citizen sensor" may represent an effective way to fill in the geospatial information gaps left behind by "specialized" remote sensing operation using conventional infrastructure like spaceborne and airborne sensors. The chapter is organized as follows: Section 12.2 introduces the concept of crowdsourcing; Section 12.3 discusses the concepts of User Generated Content (UGC), Web 2.0, social media, and the emergence of Web 3.0. Section 12.4 moves closer to the aim of the chapter by presenting the Citizen Sensor and Volunteered Geographic Information (VGI). Section 12.5 presents some examples of implementations of the crowdsourcing concept in a geospatial and remote sensing environment. Section 12.6 closes the chapter while drawing some conclusions.

## 12.2 WHAT IS CROWDSOURCING

As already mentioned, the term *crowdsourcing* was coined by Howe (2006) as a neologism generated on the model of *outsourcing* while implying the meaning of *sourcing* to the *crowd*; that is, entrusting an external provider for the provision of a service, which external provider happens to be a group of people not determined a priori rather than a specific, single agent.

After Howe's invention of the term *crowdsourcing*, several derived terms have been born. Some of the most notable ones are:

1. *Crowd-voting*: collecting opinions (possibly forced to fit into a pre-defined set of options) on a given topic or product
2. *Crowd-funding*: collective funding of a project by many small investors/donors
3. *Creative crowdsourcing*: co-working on a collective artwork
4. *Wisdom of the crowd*: merging contributions by individuals, based on the idea that "collective ideas" are better that "individual ideas," that is, asking many people increases chances of finding the right solution to a problem
5. *Microwork*: splitting a big task into micro-tasks entrusted to several human agents (see Amazon Mechanical Turk, 2005)

As we will see in the following, crowdsourcing as a support to Earth Observation mainly falls into the categories of crowd-voting and wisdom of the crowd, although other specific concepts like the "Citizen Sensor" (Sheth, 2009) need to be considered for a complete picture.

Crowdsourcing can thus be conceived as a production model, or problem-solving model, exploiting a sort of "distributed intelligence" or—as we will see for our topic of interest—"distributed data collection," typically geospatial information for our case. In its most classical conception, the solving of a given problem, or the development of a project, are entrusted to a group of individuals—a *crowd*—not necessarily defined *a priori*, which contribute, individually and/or through mutual interactions, to finding or defining a solution to the proposed problem, possibly subject to the proposer's approval. The solution typically ends up belonging to the company or organization that initially proposed the problem, while the individuals that contributed to the solution are compensated for their work through small money payments, prizes, or even simply moral or intellectual satisfaction.

The advantages of such a model with respect to traditional commercial models are obvious. First of all, the solution may be found at a reasonable cost—even zero, in terms of raw labor cost, if the remuneration consists of personal satisfaction as in the case of Wikipedia. Then, the pool of considered, possible solutions—or, more relevant for our *remote sensing* case, of collectable information—is much broader than any single problem solver or information collector can reach.

On the other hand, the process of crowdsourcing often raises concerns about the quality of the information produced (Allahbakhsh et al., 2013), as contributors might have different levels of skills and expertise, possibly insufficient for performing certain tasks—even simple data collection in many cases requires a minimum degree of knowledge about the observed targets; contributors might also have various and even biased interests and incentives; even malicious activities can be set up to bias the output of a crowdsourcing operation. The usual countermeasures act on the sides of assessing workers' profiles and reputation (Jøsang et al., 2007; Alfaro et al., 2011) and of smartly designing both the task (Dow et al., 2012) and the compensation schemes (Scekic et al., 2012).

## 12.3 GENERATIONS OF "THE WEB"

### 12.3.1 From Web 1.0 to Web 2.0 and Social Media

The way the internet looked, when it started becoming popular in the second half of the 1990s of the 20th century, was termed *Web 1.0* in retrospect. The term was coined at the beginning of the new century to mark the fact that new features were emerging within the net and somehow a new kind of internet was developing, the Web 2.0 (O'Reilly, 2007).

"Web 2.0" does not mean a new revision of the World Wide Web: it is still based on TCP/IP protocols, and all the elements that characterized the web since its beginning, like hypertext and linking among contents. Web 2.0 is rather the arriving point of a seamless evolution in the approach to online information.

Web 1.0 was indeed based on static websites, where users mostly access information (or data) generated by a limited set of other agents. A clear distinction is in place between (comparatively few) information contributors on the one side, and (many more) information consumers on the other side, with few exceptions of mixed behaviors. A diverse array of search engines was designed and established to enable the efficient retrieval of pertinent information amidst a vast flood of extraneous or irrelevant data. Figure 12.1 depicts the concept of Web 1.0.

The Web 2.0 is instead based on a high level of interaction between the website and the user, which frequently is, at the same time, also a generator of the information contained in the websites, as highlighted in the scheme on Figure 12.2. Web 2.0 allows users to "publish," that is, make available to anyone requesting to access the website, some self-generated pieces of information. This is the so-called User Generated Content, or UGC (OECD, 2007), which entails creative effort and nonprofessional, nonstandard labor.

The means through which this can be done are various and include the so-called social networks, photo and video sharing websites, blogs, forums, wikis, and other web services. The information can be made completely public, or restrictions can be placed on who can access what, through, for example, registration and/or "linking" among specific individuals or groups.

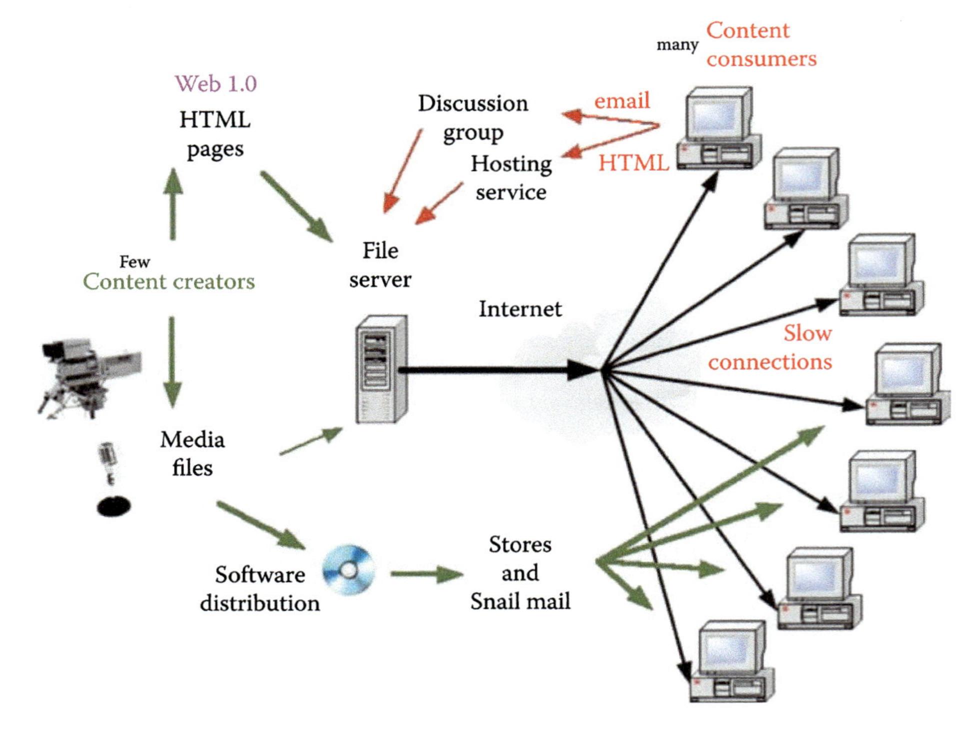


FIGURE 12.1 Schematic concept of Web 1.0. (From WikiVersity, GNU license.)

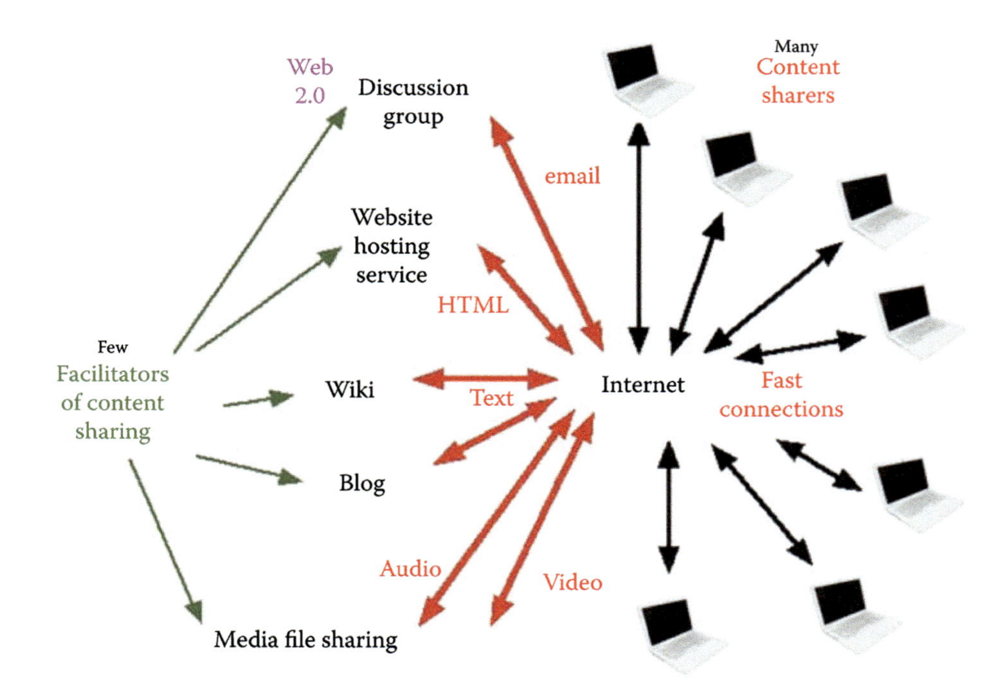


FIGURE 12.2 Schematic concept of Web 2.0. (From WikiVersity, GNU license.)

By uploading material on such sites, commenting material posted by others, and interacting with other individuals, the users generate a large amount of information, which is highly dynamic in that it is continuously amended, topped, and made to evolve including through closed-loop feedback. The phenomenon was spontaneous, still fostered and sped up by the availability of user interfaces, software, and online storage tools, perfectly integrated in the web, which increasingly steered towards a smarter platform enabling different applications including remote GIS (WebGIS) ones.

Not only the huge amount of shared information, but especially its evolution and the possibility of live interaction, build up a sort of "collective intelligence" (Levy, 1999) or "global brain" (Heylighen, 1997). This also casts a spotlight on Web 2.0 as a possible tool to practically implement crowdsourcing as discussed later.

The *social media*, as defined by Kaplan and Haenlein (2010), are the expression of massive UGC generation in the framework of the Web 2.0 as an enabling technology. Kaplan and Haenlein (2010) define different types of social media, but the most relevant for our purposes are the following:

- Collaborative projects—wikis, social bookmarking
- Content communities—sharing of content among users
- Social networking sites—much less interesting as the contents are not public by default

Social media's UGC include geographic tags and may be instrumental in building the so-called Citizen Sensor Networks, as discussed in Section 12.4.

### 12.3.2 The Emergence of Web 3.0

In the second decade of 21st century, some critical aspects of Web 2.0 started to become apparent, including the following:

- **Centralization:** Web 2.0 is dominated by a few large technology companies controlling the vast majority of user data and online flows.
- **Poor privacy level:** Said companies often collect user data without their consent, and possibly also sell them to advertisers, once more without consent from the actual "owners" of the traded information.
- **Censorship:** Interactive platforms generate an illusion of open discussion, but the reality is Web 2.0 platforms can censor content and ban users at their discretion.

The preceding factors together led to increasing concerns about privacy, covert surveillance, and the preservation of free speech and net neutrality.

As a reaction, efforts started to be placed in creating a more decentralized, privacy-preserving, open, neutral internet. Their goal was summarized into the concept of Web 3.0, a term commonly attributed to Gavin Wood, a computer scientist, and the co-founder of Ethereum. The term was used earlier by Hendler (2009), but in a different sense involving the rise of the "Semantic Web" (Berners-Lee et al., 2001). The current definition of Web 3.0 captures a novel stage in the evolution of the internet which is still in the making. Some of the concepts supposedly enabling the advent of Web 3.0 include:

- **Blockchain**, a distributed ledger technology capable of creating secure and transparent records of transactions without the need for a centralized recording service. Blockchain can thus power decentralized applications and services such as
    a. **Non-Fungible Tokens (NFTs)**, meaning a unique digital identifier recorded on a blockchain, and used to certify ownership and authenticity. NFTs cannot be copied, substituted, or subdivided.

b. **Decentralized Finance (DeFi)**, using blockchain technology to remove third parties and centralized institutions from financial transactions.

- **Artificial Intelligence (AI)**, which can be used to improve the user experience of Web 3.0 applications and services by, for example, generating ad hoc creative content.
- **Augmented/virtual reality (AR/VR)**. These items are specially relevant as they enable the creation of a virtual world where novel, *virtual geospatial information* is created that envisages a new geospatial ecosystem.

At the time of writing, the internet is still dominated by Web 2.0, with Web 3.0 slowly carving its space and expanding. Although Web 3.0 is still junior, it deserves attention as an enabler and/or multiplier of geospatial data generation as in the examples presented in Section 12.5.

## 12.4 THE CITIZEN SENSOR

Today, the Earth supports about 8.1 billion people, according to the UN Population Division Data Portal (UN-PDDR, 2023); every year those people, according to statistics (IDC, 2023), purchase millions of *smartphones* with positioning, picture-taking, video- and sound-recording capabilities. Practically every inhabited area of the Earth hosts at least one device with capabilities of acquiring environmental data—in a wide sense—and of transmitting it to the "cyberspace." The device plus the user fit the definition of "citizen sensor" given by Sheth (2009); whether in an organized or informal way, a distributed set of Citizen Sensors forms a "Citizen Sensor Network," that is, a distributed network of potential, mobile, connected sensing instruments. The concept of Citizen Sensor, however, finds its natural place, and gains even more sense, in the framework of the Volunteered Geographic Information (VGI) concept (Goodchild, 2007), that is, a special case of the more general web phenomenon of UGC. VGI is defined as "the widespread engagement of large numbers of private citizens . . . in the creation of geographic information" and it can be seen as the vehicle through which sparsely collected information is conveyed to a single, virtual place where it becomes usable, including by those who created it.

Amongst the most notable examples of VGI collectors, we find Wikimapia (2007) and OpenStreetMap (2004) websites. The former encourages participants to post comments about geo-referenced locations; the latter is an international effort to create a free source of map data through volunteers' efforts.

In Figure 12.3 a comparison between the Google Maps© version and the OpenStreetMap (OSM) version of a map on the same area highlights the amount of information that the "Citizen Sensor" is willing to contribute to a public VGI repository. Note that this version of the OSM map does not display the 3D building information here, which is however frequently contributed (Goetz and Zipf, 2012; Over et al., 2010).

It is however to be noted that the comparison represented in Figure 12.3 refers to a small-sized town in a developed country, where a large fraction of the population has access to electronic navigation devices and fast internet connections. One may wonder what is the real coverage to expect when the sight is expanded to a global level, including areas where such favorable conditions are not found at all.

The level of coverage offered by a crowdsourced mapping service like OSM is an interesting issue, but it is difficult to assess for various reasons:

1. No reference data ensuring full coverage of the entire globe is available, against which to measure OSM's coverage.
2. No standard definition of items to be included in a map is provided, so a definite response on what is actually mapped versus what *should have been mapped* cannot be given. For instance, an unpaved road could be considered unimportant if the focus is on residential

area accessibility, while if the map is to be used in an agricultural context, the same road becomes important; still on the unpaved road, in a developing country it may be as important as a paved one in a residential neighborhood.

3. Assessing a global coverage of an ever-changing dataset, in diverse environments across the globe is a tricky task. Buildings appear so well delineated and separated from each other in Figure 12.3; other sorts of "buildings" may not be so easy to tell from each other in a context such as that of, for example, informal settlements in urban areas with social problems.

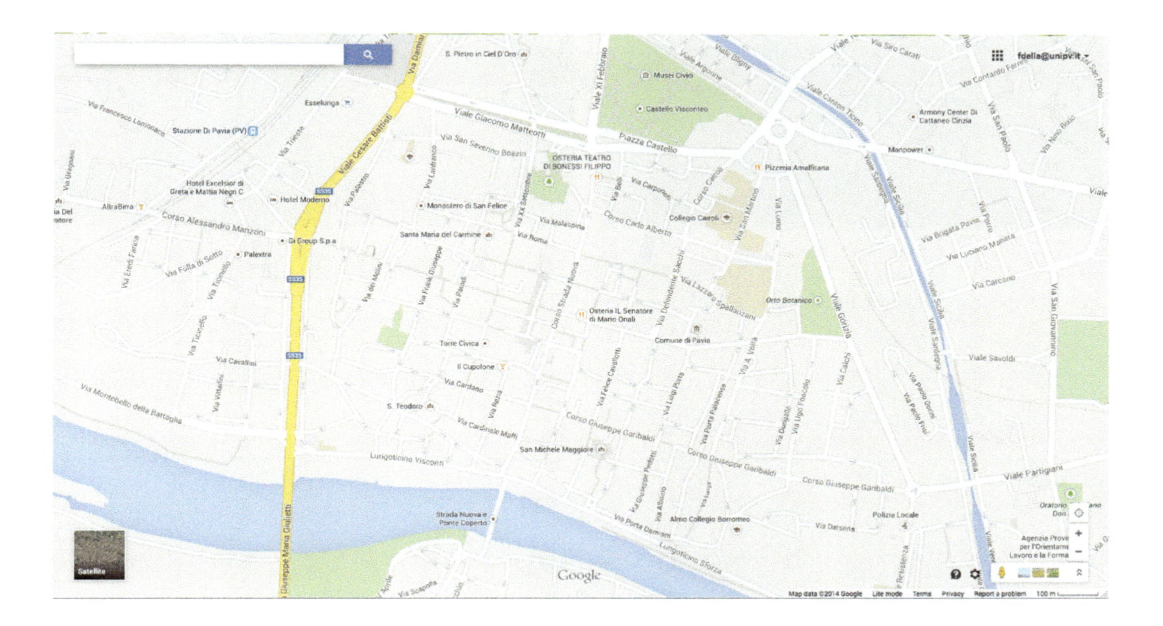

FIGURE 12.3 The city center of Pavia, Italy, as mapped on (a) the Google Map© website and (b) the OpenStreetMap website.

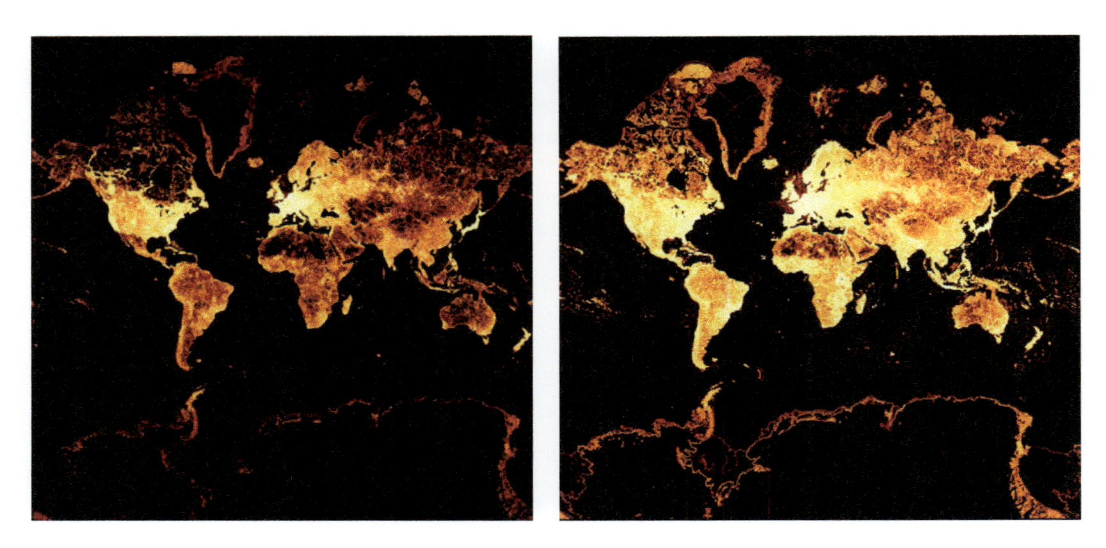

FIGURE 12.4 OSM Node Density 2014 (a), OSM Node Density 2023 (b). (© Martin Raifar, CC-BY—source data © OpenStreetMap contributors, ODbL.)

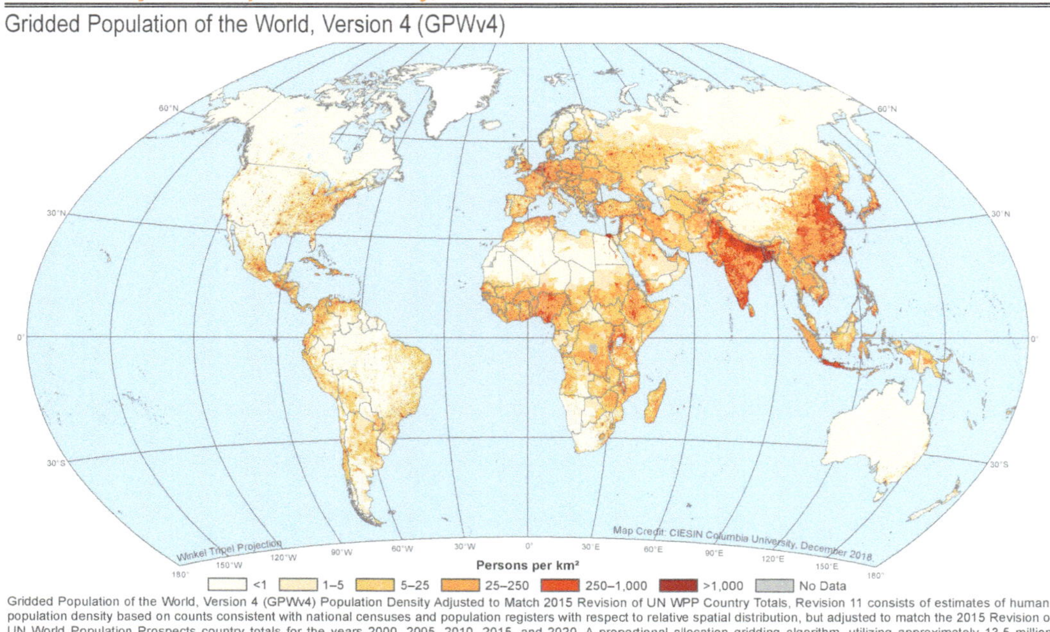


FIGURE 12.5 Global population density in year 2020 according to CIESIN as available from NASA SEDAC. (Copyright CIESIN of Columbia University, license: CC-BY-4.0.)

Notwithstanding these issues, local studies provide encouraging results (Yeboah et al., 2021); at a global level, qualitative assessments also report good coverage (Barrington-Leigh and Millard-Ball, 2017). Interesting clues can be derived from the "OSM node density map" published yearly by Raifar (2014) up to the present day. As can be seen in Figure 12.4a, the color-coded density correlates well with the population distribution depicted in Figure 12.5, and so does node growth, which can be estimated by comparison with the same node density map nine years later (Figure 12.4b). Lower coverage rates are reported in noninhabited (or nearly so) areas such as Saharan Africa, inland Greenland, or the Australian Outback, although the coverage densified in these areas as well during the considered nine-year span. All in all, the maps suggest that where one expects to find settlements and thus roads, they are actually reported on OSM.

Similarly, the accuracy of crowdsourced maps is also difficult to assess; yet, some studies have been carried out, highlighting that especially local knowledge—of the sort you can expect from OSM contributors—may help achieve high accuracy levels (Atwal et al., 2022).

In this framework, the deployment of Web 3.0 provides additional opportunities thanks to its characteristics of decentralization, transparency, immutability, and scalability. Blockchain technology makes it more difficult for governments or other organizations to censor or control data; distributed ledgers allow tracking the flow of data and ensuring that it is not tampered with. Scalability ensures that many geographic-labeled transactions can be handled from an extensive pool of contributors without straining the system.

## 12.5 CURRENT IMPLEMENTATIONS

In this section we will see how the concepts illustrated so far find their application in the real world. Various ways of using VGI mechanisms are illustrated here in different fields of applications, including disaster management, land cover mapping, and real-time, crowd-driven location information collection.

### 12.5.1 Seismic Risk: The "Did You Feel It" Service

Although crowdsourcing is potentially very useful to assess damage to buildings and structures in the post-disaster phase (Kerle, 2011), the earliest examples in the field of seismic risk focus on mapping the phenomenon itself rather than its consequences.

The "Did You Feel It" (DYFI; Wald et al., 2011) program was developed by the United States Geological Survey (USGS) in the early 1990s, long before the concept of crowdsourcing came to be formalized by Howe (2006), and can be considered the first notable example of crowdsourcing applied to natural hazard management. DIFY consists of a website (DYFI, 1991) onto which individuals (Citizen Sensors) can report their time-stamped, georeferenced observations regarding a seismic phenomenon they have experienced. The idea underpinning DYFI is to best exploit the public engagement (an informal Citizen Sensor Network) to generate a map of the event which be useful, in case of a relevant event, to get a prompt input to the emergency procedures, but also to build a historical record of minor quakes to be used as an input to earthquake hazard models. The map is called "Community Internet Intensity Map" (CIIM), and it is effectively a consolidated output of a crowdsourcing mechanism.

A short online questionnaire is proposed to the candidate contributor, including three different sections on context, experience, effects. The questionnaire was designed to generate reports fitting the Modified Mercalli Intensity (MMI) through suitable mapping of responses into MMI grades (Dengler and Moley, 1994) (Dengler and Dewey, 1998).

Every filled questionnaire will thus result into an estimated MMI grade, linked to the micro-region to which the geolocation of the contributor corresponds. Micro-regions are defined conventionally to correspond with U.S. zip codes as a compromise between statistical significance of the aggregates and spatial resolution. An average value of the linked, estimated MMI grades is computed on every

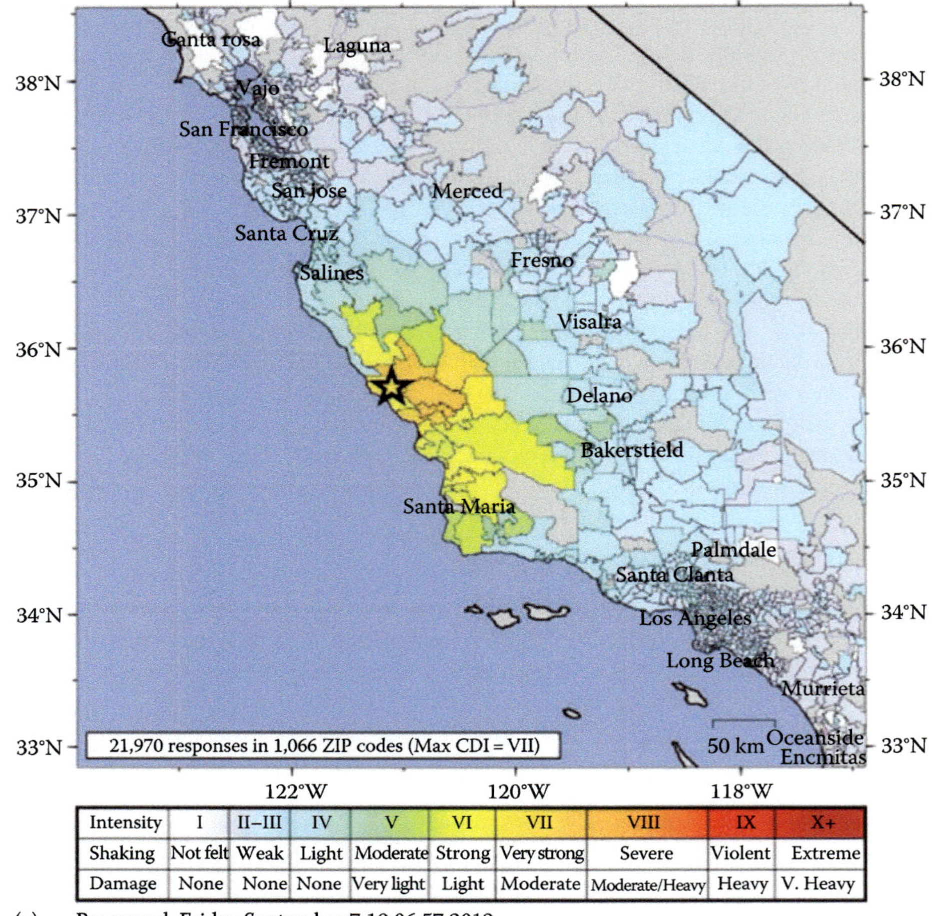


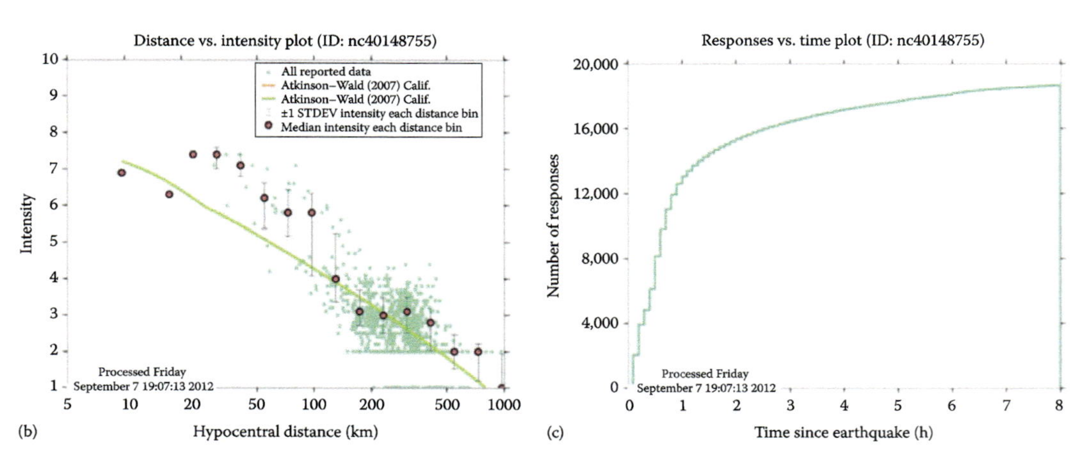


**FIGURE 12.6** DYFI products for the Central California event on December 22, 2003. (a) Color-coded Community Internet Intensity Map. (b) Estimated intensity versus distance plotting of aggregated responses. (c) Cumulated responses versus time elapsed. (All images from the DYFI website.)

micro-region, and attached to it. Visually, the average is turned into a color through a conventional color palette ranging from white (nonperceived) to dark red (extreme). Within a few minutes from the earthquake occurrence a CIIM starts building on the website, often collecting thousands of contributions for clearly felt quakes (from grade IV upward). Together with the CIIM map, a graph depicting how the average intensity decays with distance is also constructed; examples are shown in Figure 12.6.

The DYFI program has recorded an increasing success. Nowadays, the total input questionnaire count exceeds five million units (Quitoriano and Wald, 2020), and the service has been scaled up to worldwide coverage, allowing global contributors to feed their observation through the dedicated "outside U.S." section of the website. In Figure 12.7, the spatial distribution of questionnaires filed in 2022 is translated into a map with color-coded earthquake intensity, providing a compelling view of the seismic activity across the North American continent. This view is very interesting to analyze seismicity on a continental scale.

A few words are in order on the accuracy of DYFI data, because they make an interesting case of how crowdsourcing can be useful not only for qualitative but also for quantitative purposes. As Atkinson and Wald (2007) stated, DYFI data "make up in quantity what they may lack in quality." The authors focused on the 2004 Parkfield, CA, USA, earthquake, comparing MMI data collected through the DYFI system across the affected ZIP codes with the measured earthquake ground motions from local instruments, as cataloged by the USGS's ShakeMap web-based database, and estimated using consolidated models. The two data series match very well, not only in terms of general MMI versus distance trend, but also on more subtle features such as a flattening of attenuation in the distance range from 70 to 150 km. The uncertainty bars of DYFI data appeared a bit wider than those associated with the model-based estimation, but this is a reasonable price to be paid in exchange for an incredibly dense network of "virtual sensors."

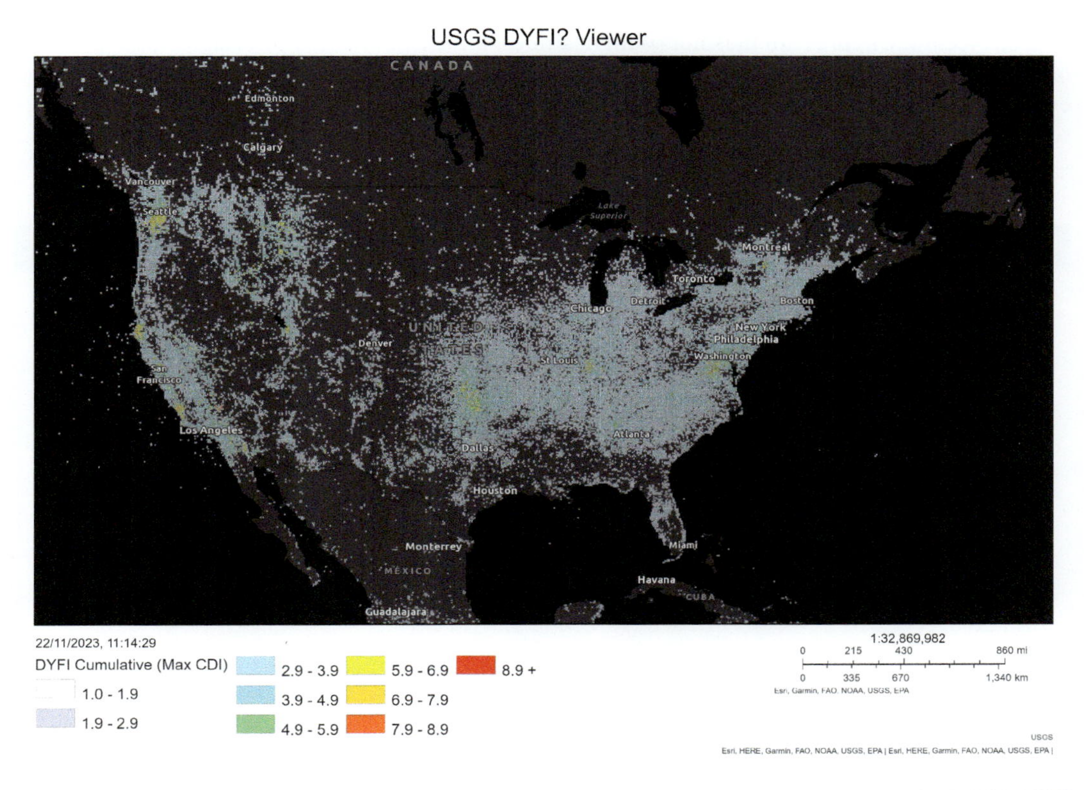


**FIGURE 12.7** Spatial coverage of questionnaires filed in 2022 on the DYFI system. (From the USGS DYFI website.)

### 12.5.2 Land Cover Updating

Land cover mapping from multispectral spaceborne data dates back to the beginnings of the LANDSAT satellite series in 1972 (Anderson et al., 1976). It soon became evident that land cover classification from remotely sensed data is far from perfect, even in ideal conditions (Congalton, 1991). Also in this case, crowdsourcing can offer some help, with the Geo-Wiki Project (Fritz et al., 2009). This project aggregates a global network of volunteers willing to help improve the quality of global land cover maps. Each single contributor is requested to review hotspot maps of global land cover and determine, based on what they actually see in Google Earth and their local knowledge, if the land cover maps are correct or incorrect at the spot under review. Georeferenced ground pictures can be uploaded and support the proposed land cover class. Each input is recorded in a database, along with (possible) uploaded photos, to be used in the future for the creation of a new and improved global land cover map. Figure 12.8 shows the interface of GEO-Wiki, where the familiar Google Earth Hi-Res mosaic is overlaid with a much spatially coarser land cover map from the GlobCover dataset.

A note on accuracy should also be made here. Accuracy assessment of crowdsourced land cover data encounters the same issues that have been pointed out in Section 12.4 for completeness assessment of crowdsourced maps. The issues are made even worse by the higher variability of land cover with respect to, for example, settlement location or road geometry. However, assessments have been attempted, but obviously on datasets whose size is very small compared to the global world coverage. Foody et al. (2013) reported a kappa coefficient around 0.7 for a 10-class Geo-Wiki classification experiments based on labeling of about 300 samples carried out by 65 volunteers among which the ten most productive were selected for the final assessment. Similar figures in terms of accuracy (60–70%) were reported in another Geo-Wiki experiment (See et al., 2013) where the difference between expert and nonexpert contributors proved narrower than one may expect. Accuracy figures are thus not so high, still even the combination of disagreeing inputs can bring valuable information (Foody et al., 2013). The situation has been improving since then, to the extent that Geo-Wiki is now considered one among the reference land cover databases for large-scale applications (Tsendbazar et al., 2021).

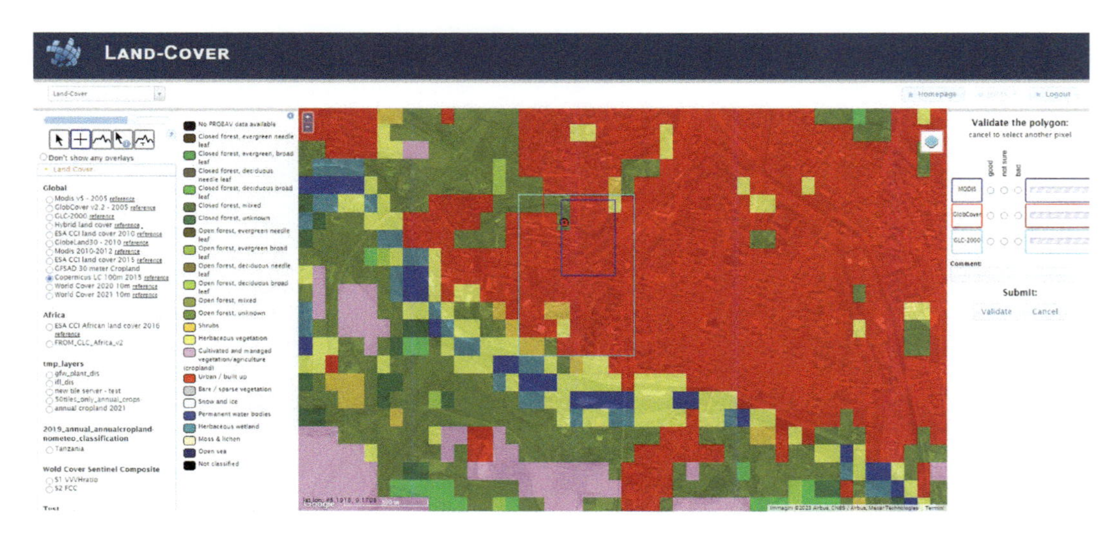


FIGURE 12.8 The Geo-Wiki Interface after connecting as a guest user. The Copernicus GLC land cover map is shown overlaid on the Google Earth© image the concerned area, and the user can report on the agreement or lack thereof among the recorded and the actual land cover class by clicking on the appropriate radio buttons on the right bar.

### 12.5.3 Merging into the Blockchain Technology and Web 3.0

The previous examples required cooperation by the user in providing data items, but models exist that can do without any involvement of the human operator, except a simple initial setting ("set and forget"). The Bluetooth Low Energy (BLE) wireless technology (Gomez et al., 2012), available since the late 2000s, has provided the ability to automate the collection of many types of data, with extremely little or no human intervention.

BLE devices called "beacons," positioned in specific places or aboard moving vehicles, are equipped with sensors capable of automatically detecting the required data and storing them. Uninterrupted internet access is not needed, because the beacon can be connected via BLE to a smartphone, possibly even an unknown "by-passer smartphone," which will take care of receiving the data packets and retransmitting them through the network to their processing centers, in a completely transparent manner. Perhaps the most popular of these applications consists of the so-called "anti-lost tags" or "findable," that is, small BLE devices used as anti-loss trackers in the form of, for example, key rings or suitcase labels. They do not include an internal GNSS system but rather rely on that of the device to which they are connected, for example to memorize the last position seen or to activate an alarm when communication with the device is interrupted (the object was left behind).

Although these devices individually can only operate within 15–60 m, that is, the BLE range of action, they evolved to allow the localization of an object "at a distance"; this is enabled by the construction of a network of smartphones whose owners agreed joining the system. Each smartphone can transparently scan the environment for BLE devices, exchange information with them, and transmit the retrieved data, all without the need for human cooperation and guaranteeing privacy at the same time.

The more crowded an area is, the higher the probability that the object will be located by a passing smartphone, which will automatically report—through the ad hoc virtual network—the GPS coordinates of the place where the beacon was "spotted." If the beacon includes a device for sensing of environmental parameters, a VGI-based environmental monitoring system is established.

One of these BLE-crowdsourcing networks, for example, named XY Oracle Network or XYO Network (Trouw et al., 2018) in 2017 combined BLE and geolocation with the potential of blockchain technology to guarantee the integrity of the data collected. Special beacons called "Sentinels" have also been introduced with the task of "certifying" the truthfulness of the data transmitted from the device to the network, thus providing a sort of "geographical signature."

Furthermore, a native cryptocurrency has been implemented on the same blockchain which allows both to reward those who make their device available—to different extents—and to purchase the services offered by the network. In a sense, this is the next step, or the "Web 3.0 step" of the "Citizen Sensor" evolution.

## 12.6 CONCLUSIONS

It is a fact that, after the initial enthusiasm for the apparently pervasive reach of space-based remote sensing, reality kicked in and disappointment followed. The foundations of detecting any type of material on the Earth surface by observing it from above were laid early (Clark and Roush, 1984), but translating these principles into something operational turned out to be tougher than expected. A part of the problem was the inherent difficulty in translating noisy measures of faraway physical quantities into estimates that could be relied upon. Yet, another part was instead connected with the limitations in coverage (information on point $x$ at time $y$ can't be acquired because no sensor is there at that time) and visibility of observed features (information on feature $z$ is not accessible simply because it is not visible from above).

Both these latter issues can in principle be addressed through integration of crowdsourcing into the production chain of geospatial information. Crowdsourcing, seen as an immersive, distributed network of countless in situ sensors, carries a potential to:

- Substantially increase the *a priori* likelihood of "being there" when "it matters"
- Gain accessibility to features that are invisible to the average, faraway nadir or quasi-nadir sensors, like, for example, small details, or reflectance of a vertical surface

In addition to the preceding points, the introduction of human beings in the loop scales up the observation paradigm from purely quantitative to semantic—even after including and weighing all the implied risks of data pollution, misunderstanding, misinterpretation, this is still a big step up.

This is not however the only power of crowdsourcing that comes from the active participation of intelligent humans in a task assigned to them. People are social beings and as such are also oriented towards sharing information and helping one another; crowdsourcing capitalises on this orientation, allowing to gather more information (both in terms of quantity and depth), from many more "sensors" and much faster.

## REFERENCES


Alfaro, L.D., et al. (2011). "Reputation Systems for Open Collaboration," Community ACM, vol. 54, no. 8, pp. 81–87.

Allahbakhsh, M., Benatallah, B., Ignjatovic, A., Motahari-Nezhad, H.R., Bertino, E., Dustdar, S. (2013). "Quality Control in Crowdsourcing Systems: Issues and Directions," Internet Computing, IEEE, vol. 17, no. 2, pp. 76,81, March–April. doi: 10.1109/MIC.2013.20

Amazon Mechanical Turk. (2005). Online. Available: www.mturk.com/

Anderson, J.R., Hardy, E.E., Roach, J.T., Witmer, R.E. (1976). A Land Use and Land Cover Classification System for Use with Remote Sensor Data. Government Printing Office, Washington, DC, US Geological Survey, Professional Paper 964.

Atkinson, G.M., Wald, D.J. (2007). "Did You Feel It?," Intensity Data: A Surprisingly Good Measure of Earthquake Ground Motion: Seismological Research Letters, vol. 78, pp. 362–368, May/June. doi: 10.1785/gssrl.78.3.362

Atwal, K.S., Anderson, T., Pfoser, D., et al. (2022). "Predicting Building Types Using OpenStreetMap," Science Reports, vol. 12, p. 19976. doi: 10.1038/s41598-022-24263-w

Barrington-Leigh, C., Millard-Ball, A. (2017). "The World's User-Generated Road Map Is More Than 80% Complete," PLoS ONE, vol. 12, no. 8, p. e0180698.

Berners-Lee, T., Hendler, J., Lassila, O. (2001). "The Semantic Web," Scientific American, vol. 284, no. 5, pp. 35–43.

Clark, R.N., Roush, T.L. (1984). "Reflectance Spectroscopy: Quantitative Analysis Techniques for Remote Sensing Applications," Journal of Geophysical Research: Solid Earth (1978–2012), vol. 89, no. B7, pp 6329–6340, 10 July.

Congalton, R.G. (1991). "A Review of Assessing the Accuracy of Classifications of Remotely Sensed Data," Remote Sensing of Environment, vol. 37, no. 1, pp. 35–46, July.

Dengler, L.A., Dewey, J.W. (1998). "An Intensity Survey of Households Affected by the Northridge, California, Earthquake of 17 January 1994," Bulletin of the Seismological Society of America, vol. 88, pp. 441–462.

Dengler, L.A., Moley, K. (1994). "Toward a Quantitative, Rapid Response Estimation of Intensities," Seismological Research Letters, vol. 65, p. 48.

DesignCrowd. (2010). "5 Famous Logo Contests-Toyota, Google, Wikipedia & More!," Online. Available: http://blog.designcrowd.com/article/218/5-famous-logo-contests-toyota-google-wikipedia-more

Dow, S.P., Kulkarni, A., Klemmer, S.R., Hartmann, B. (2012). "Shepherding the Crowd Yields Better Work," Proceedings 2012 ACM Conference Computer Supported Cooperative Work (CSCW 12), ACM, pp. 1013–1022.

DYFI. (1991). "Did You Feel It," Earthquake Hazards Program. Online. Available: http://earthquake.usgs.gov/earthquakes/dyfi/

Foody, G.M., See, L., Fritz, S., Van der Velde, M., Perger, C., Schill, C., Boyd, D.S. (2013). "Assessing the Accuracy of Volunteered Geographic Information Arising from Multiple Contributors to an Internet Based Collaborative Project," Transactions in GIS, vol. 17, no. 6, pp. 847–860.

Fritz, S., McCallum, I., Schill, C., Perger, C., Grillmayer, R., Achard, F., Kraxner, F., Obersteiner, M. (2009). "Geo-Wiki.Org: The Use of Crowdsourcing to Improve Global Land Cover," Remote Sensing, vol. 1, no. 3, pp. 345–354. doi: 10.3390/rs1030345. Open access.

Goetz, M., Zipf, A. (2012). "Towards Defining a Framework for the Automatic Derivation of 3D CityGML Models from Volunteered Geographic Information," International Journal of 3-D Information Modeling, vol. 1, pp. 496–507.

Gomez, C., Oller, J., Paradells, J. (2012). "Overview and Evaluation of Bluetooth Low Energy: An Emerging Low-Power Wireless Technology," Sensors, vol. 12, pp. 11734–11753. doi: 10.3390/s120911734

Goodchild, M.F. (2007). "Citizens as Sensors: The World of Volunteered Geography," GeoJournal, vol. 69, pp. 211–221. doi: 10.1007/s10708-007-9111-y

Hendler, J. (2009). "Web 3.0 Emerging," Computer, vol. 42, no. 1, pp. 111, 113, January. doi: 10.1109/MC.2009.30

Heylighen, F. (1997). "Towards a Global Brain: Integrating Individuals into the World-Wide Electronic Network," in: D.S. der Sinne, U. Brandes & C. Neumann (Eds.), Göttingen: Steidl Verlag.

Howe, J. (2006). "The Rise of Crowdsourcing," Wired Magazine—Issue 14.06—June 2006. Available: www.wired.com/wired/archive/14.06/crowds_pr.html

IDC. (2023). "International Data Corporation (IDC) Worldwide Quarterly Mobile Phone Tracker," Online. Available: www.idc.com/getdoc.jsp?containerId=IDC_P8397

Jøsang, A., Ismail, R., and Boyd, C. (2007). "A Survey of Trust and Reputation Systems for Online Service Provision," Decision Support Systems, vol. 43, no. 2, pp. 618–644.

Kaplan, A.M., Haenlein, M. (2010). "Users of the World, Unite! The Challenges and Opportunities of Social Media," Business Horizons, vol. 53, pp. 59–68.

Kerle, N. (2011). "Remote Sensing Based Post-Disaster Damage Mapping: Ready for a Collaborative Approach?," IEEE Earthzine. Posted on March 23rd, 2011 in Articles, Disaster Management Theme, Earth Observation.

Levy, P. (1999). "Collective Intelligence: Mankind's Emerging World in Cyberspace," New York: Basic Books. ISBN: 9780738202617

O'Reilly, T. (2007). "What Is Web 2.0: Design Patterns and Business Models for the Next Generation of Software," Journal of Communications & Strategies, no. 65.

OECD. (2007). "Participative Web and User-Created Content: Web 2.0, Wikis, and Social Networking," Paris: Organisation for Economic Co-Operation and Development.

Openstreetmap. (2004). "Collaborative Mapping Web Site," Online. Available: www.openstreetmap.org/

Over, M., Schilling, A., Neubauer, S., Zipf, A. (2010). "Generating Web-Based 3D City Models from OpenStreetMap: The Current Situation in Germany," Computers Environment and Urban Systems, vol. 34, pp. 496–507.

Quitoriano, V., Wald, D.J. (2020). "USGS 'Did You Feel It?': Science and Lessons from 20 Years of Citizen Science-Based Macroseismology," Frontiers in Earth Science, vol. 8, p. 120.

Raifar, M. (2023). "OSM Node Density 2023," Map. Available: http://tyrasd.github.io/osm-node-density/; explanation Available: www.openstreetmap.org/user/tyr_asd/diary/22363. Accessed on 18th November, 2023.

Scekic, O., Truong, H., Dustdar, S. (2012). "Modeling Rewards and Incentive Mechanisms for Social BPM," in: A. Barros et al. (Eds.), Business Process Management, vol. 7481, Berlin and Heidelberg: Springer, pp. 150–155.

See, L., Comber, A., Salk, C., Fritz, S., van der Velde, M., Perger, C., Schill, C., McCallum, I., Kraxner, F., Obersteiner, M. (2013). "Comparing the Quality of Crowdsourced Data Contributed by Expert and Non-Experts," PLoS ONE, vol. 8, no. 7, p. e69958. doi: 10.1371/journal.pone.0069958

Sheth, A. (2009). "Citizen Sensing, Social Signals, and Enriching Human Experience," Internet Computing, IEEE, vol. 13, no. 4, pp. 87,92, July–August. doi: 10.1109/MIC.2009.77

Sobel, D. (1998). "Longitude: The True Story of a Lone Genius Who Solved the Greatest Scientific Problem of His Time," London: Fourth Estate Ltd., p. 6, ISBN 1-85702-571-7

Trouw, A., Levin, M., Scheper, S. (2018). "The XY Oracle Network: The Proof-of-Origin Based Cryptographic Location Network," San Diego, CA, USA: XYO Network. White paper. Available: https://social-feed-media.s3.ap-southeast-1.amazonaws.com/pdfs/1674033247474-XYO.pdf Accessed on 21st November, 2023.

Tsendbazar, N., Herold, M., Li, L., Tarko, A., De Bruin, S., Masiliunas, D., Lesiv, M., Fritz, S., Buchhorn, M., Smets, B., Van De Kerchove, R., Duerauer, M. (2021). "Towards Operational Validation of Annual Global Land Cover Maps," Remote Sensing of Environment, vol. 266, p. 112686.

UN-PDDR. (2023). "UN Population Division Data Portal." Available: https://population.un.org/dataportal/home

Wald, D.J., Quitoriano, V., Worden, B., Hopper, M., Dewey, J.W. (2011). "USGS 'Did You Feel It?' Internet-Based Macroseismic Intensity Maps," Annals of Geophysics, vol. 54, no. 6. doi: 10.4401/ag-5354

Wikimapia. (2007). "Collaborative Mapping Web Site: Not a Part of Non-Profit Wikimedia Foundation," Online. Available: http://wikimapia.org/. Accessed on 20th November 2023.

Wikipedia. (2001). "Wikipedia: The Free On-Line Encyclopaedia." Online. Available: www.wikipedia.org/

Yeboah, G., Porto de Albuquerque, J., Troilo, R., Tregonning, G., Perera, S., Ahmed, S.A.K.S., Ajisola, M., Alam, O., Aujla, N., Azam, S.I., et al. (2021). Analysis of OpenStreetMap Data Quality at Different Stages of a Participatory Mapping Process: Evidence from Slums in Africa and Asia," ISPRS International Journal of Geo-Information, vol. 10, p. 265. doi: 10.3390/ijgi10040265

# *Part VII*

## *Cloud Computing and Remote Sensing*

# 13 Processing Remote Sensing Data in Cloud-Computing Environments

*Ramanathan Sugumaran, James W. Hegeman, Vivek B. Sardeshmukh, and Marc P. Armstrong*

## ACRONYMS AND ABBREVIATIONS

| | |
|---|---|
| **AWS** | Amazon Web Services |
| **CLiPS** | Cloud-based LiDAR Processing System |
| **CPU** | Central Processing Unit |
| **EC2** | Elastic Compute Cloud |
| **EOSDIS** | Earth Observing System Data and Information System |
| **GIS** | Geographic Information System |
| **GPGPU** | General-Purpose Computing on Graphics Processing Units |
| **GPU** | Graphics Processing Unit |
| **HPC** | High-Performance Computing |
| **HTC** | High-Throughput Computing |
| **IaaS** | Infrastructure as a Service |
| **MODIS** | Moderate Resolution Imaging Spectroradiometer |
| **OCC** | Open Cloud Consortium |
| **OGC** | Open Geospatial Compliant |
| **PaaS** | Platform as a Service |
| **Saas** | Software as a Service |
| **TIN** | Triangulated Irregular Network |
| **UAV** | Unpiloted Aerial Vehicles |
| **USGS** | United States Geological Survey |
| **VPN** | Virtual Private Network |
| **WMS** | Web Map Service |

During the past four decades, scientific communities around the world have regularly accumulated massive collections of remotely sensed data from ground, aerial, and satellite platforms. In the United States, these collections include the U.S. Geological Survey's (USGS) 37-year record of Landsat satellite images (comprising petabytes of data) (USGS, 2011), the NASA Earth Observing System Data and Information System (EOSDIS)—having multiple data centers and more than 7.5 petabytes of archived imagery (Hyspeed Computing, 2013)—as well as current NASA systems that record approximately 5 TB of remote sensing-related data per day (Vatsavai et al., 2012). In addition, new data-capture technologies such as LiDAR are used routinely to produce multiple petabytes of three-dimensional (3D) remotely sensed data representing topographic information (Sugumaran et al., 2011). These technologies have galvanized changes in the way remotely sensed data are collected, managed, and analyzed (Pérez-Cutillas et al., 2023, Liu et al., 2023, Thenkabail et al., 2021, Amani et al., 2020, Phalke et al., 2020, Antunes et al., 2019, Li et al., 2019, Oliphant et al., 2019,

DOI: 10.1201/9781003541141-20

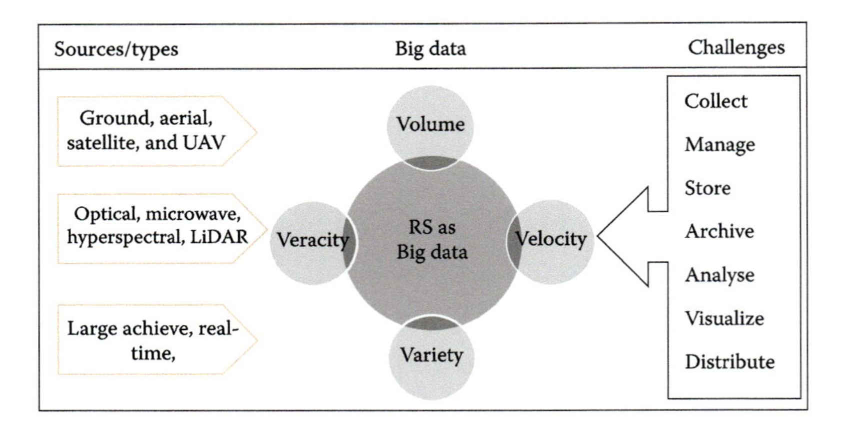


FIGURE 13.1 Remote sensing: big-data sources and challenges.

Teluguntla et al., 2018, Yan et al., 2018, Yao et al., 2020, Wang et al., 2018, Gorelick et al., 2017, Teluguntla et al., 2017, Xiong et al., 2017a, 2017b, Yang et al., 2017a, 2017b, Teluguntla et al., 2016). On the sensor side, great progress has been made in optical, microwave, and hyperspectral remote sensing with (1) spatial resolutions extending from kilometers to sub-meters; (2) temporal resolutions ranging from weeks to 30 minutes; (3) spectral resolutions ranging from single bands to hundreds of bands; and (4) radiometric resolutions ranging from 8 to 16 bits. The platform side has also seen rapid development during the past three decades. Satellite and aerial platforms have continued to mature and are producing large quantities of remote sensing data. Moreover, sensors deployed on unpiloted aerial vehicles (UAVs) have recently begun to produce massive quantities of very high resolution data.

The technological nexus of continuously increasing spatial, temporal, spectral, and radiometric resolutions of inexpensive sensors, on a range of platforms, along with internet data accessibility is creating a flood of remote sensing data that can easily be included in what is commonly referred to as "big data." This term refers to datasets that have grown sufficiently large that they have become difficult to store, manage, share, and analyse using conventional software tools (White, 2012). Big data are often thought to span four dimensions: volume (data quantity), velocity (real-time processing), variety (source multiplicity), and veracity (data accuracy) (IBM, 2012). Operating hand-in-glove with Moore's law, the growth of big data is largely a consequence of advances in acquisition technology and increases in storage capacity (Pérez-Cutillas et al., 2023, Liu et al., 2023, Thenkabail et al., 2021, Amani et al., 2020, Phalke et al., 2020, Antunes et al., 2019, Li et al., 2019, Oliphant et al., 2019, Teluguntla et al., 2018, Yan et al., 2018, Yao et al., 2020, Wang et al., 2018, Gorelick et al., 2017, Teluguntla et al., 2017, Xiong et al., 2017a, 2017b, Yang et al., 2017a, 2017b, Teluguntla et al., 2016). Figure 13.1 summarizes the overall sources and challenges presented by big remote sensing data.

## 13.1 BIG-DATA PROCESSING CHALLENGES

As the pace of imaging technology has continued to advance, the provision of affordable technology for dealing with issues such as storing, processing, managing, archiving, disseminating, and analyzing large volumes of remote sensing information has lagged. One major challenge is related to the computational power required to process these massive data sources. Traditionally, desktop computers with single or multiple cores have been used to process remote sensing data for small areas. In contrast, large- or macro-scale remote-sensing applications may require high-performance computing (HPC) technologies; general-purpose computing using graphics processing units (GPGPU), as well as parallel, cluster, and distributed-computing approaches are gaining broad acceptance (Gorelick et al., 2017, Xiong et al., 2017a, 2017b, Yang et al., 2017a, 2017b, Teluguntla et al., 2016, Simmhan and Ramakrishnan, 2010, Plaza et al., 2006, Shekhar et al., 2012, González et al., 2009). Given these architectural advances, the analysis of big data presents new challenges to both cluster-infrastructure software and parallel-application design, and it requires the development of new computational methods. These methods and several articles about the importance of HPC in remote sensing are featured in special journal issues, books, and conferences devoted to this topic (Thenkabail et al., 2021, Gorelick et al., 2017, Thenkabail, 2013, Plaza and Chang, 2008, Lee et al., 2011, Plaza and Chang, 2007).

Graphics processing units (GPUs) have been widely used (in GPGPU applications) to address remote-sensing problems (Pérez-Cutillas et al., 2023, Liu et al., 2023, Thenkabail et al., 2021, Wang et al., 2018, Gorelick et al., 2017, Teluguntla et al., 2017, Xiong et al., 2017a, 2017b, Christophe et al., 2011, Chang et al., 2011, Song et al., 2011, Yang et al., 2011). Oryspayev and others (2012) developed an approach to LiDAR processing that used data-mining algorithms coupled with parallel computing technology. A specific comparison was made between the use of multiple CPUs (Intel Xeon Nehalem chipsets), and GPUs (Intel i7 Core CPUs using the NVIDIA Tesla s1070 GPU cards). The experimental results demonstrated that the GPU option was up to 35 times faster than the CPU option. In a similar vein, distributed parallel approaches have also been developed. Haifang (2003) implemented various algorithms using a heterogeneous grid computing environment, and Liu et al. (2010) analyzed the efficiency improved by grid computing for the Maximum Likelihood Classification method. Yue et al. (2010) used cluster computing to solve remote-sensing-image fusion, filtering, and segmentation. Commodity cluster-based parallel processing of various multispectral and hyperspectral imagery has also been used by various authors (e.g., Plaza et al., 2006).

While high-performance computing (HPC) environments such as clusters, grid computing, and supercomputers (Pérez-Cutillas et al., 2023, Liu et al., 2023, Thenkabail et al., 2021, Amani et al., 2020, Phalke et al., 2020, Gorelick et al., 2017, Simmhan and Ramakrishnan, 2010) can be used, these platforms require significant investments in equipment and maintenance (Ostermann et al., 2010), and individual researchers and many government agencies do not have routine access to these resources. In addition to data volume, the variety and update rate of datasets often exceed the capacity of commonly used computing and database technologies (Antunes et al., 2019, Li et al., 2019, Yan et al., 2018, Yao et al., 2020, Teluguntla et al., 2017, Xiong et al., 2017a, 2017b, Shekhar et al., 2012, Yang et al., 2011, Wang et al., 2009). As a result of these limitations, users have begun to search for less-expensive solutions for the development of large-scale data-intensive remote sensing applications. Cloud computing provides a potential solution to this challenge due to its scalability advantages in data storage and processing, and its relatively low cost as compared to user-owned, high-power compute clusters (Kumar et al., 2013). The goal of this chapter is to provide a short review of remote-sensing applications using cloud-computing environments, as well as a case study that provides greater implementation detail. An introduction to cloud computing is provided in Section 13.2, and then in Section 13.3 various applications, including a detailed case study, are described to illustrate the advantages of cloud-computing environments.

## 13.2 INTRODUCTION TO CLOUD COMPUTING

### 13.2.1 Definitions

*Cloud computing* is a vague term, as nebulous as its eponym. *The NIST Definition of Cloud Computing* (Mell and Grance, 2011) defines cloud computing as "a model for enabling ubiquitous, convenient, on-demand network access to a shared pool of configurable computing resources." In research, "cloud computing" is a popular idiom for distributed computing, encompassing the same fundamental concepts—multiplicity, parallelism, and fault tolerance. Distributed computing has come to the forefront as an area of research over the past two decades as dataset sizes have outstripped traditional sequential processing power, even of modern high-performance processors, and as the bottlenecks of large-scale computation have moved outside the CPU (to storage I/O, for example). To employ distributed or cloud computing means to leverage the additional hardware and computing throughput available from large networks of machines. Because of the challenges inherent in optimizing resource scale and utilization for a problem, two central focuses of cloud computing in practice have been (1) the elasticity of resource provisioning; and (2) abstraction layers capable of simplifying these challenges for users. Indeed, as exemplified in an eScience Institute document (2012), it is these two qualities of cloud computing—elasticity and abstraction—that are often most important from the end-user's perspective. Thus, while research with cloud computing generally focuses on the distributed scalability of the cloud, the characteristic feature of cloud computing, in practice, is flexibility. There are several related terms associated with cloud computing as explained next.

*Cluster computing* is a similar, but more limited, model of parallel/distributed computing, in which the machines comprised by the cluster are usually assumed to be tightly connected by a low-latency, private network. This generally implies physical locality of the system itself. The concept of cluster computing was a precursor to today's notion of *high-performance computing* (Gorelick et al., 2017, Lee et al., 2011). Cluster computing differs from cloud computing in its emphasis—a compute cluster is often a localized system dedicated to one particular problem (or class of problems) at a time.

*Grid computing* can also be thought of as a subset of cloud computing having a slightly different emphasis (Buyya et al., 2009). A compute grid is a distributed system, often encompassing machines physically separated by a large distance, together with a *job scheduler* that allows a user to easily and simultaneously run a certain small set of programs on many different data inputs—the action of a user's program on one such input constitutes a *task*, or *job*. Grid computing is thus the use of a parallel/distributed system for solving a large problem or accomplishing a large collection of tasks which would take too long to complete on a single machine. In grid computing, the problem to be solved can generally be broken down into many nearly identical tasks which can run concurrently and independently on the nodes of the system (see, for example, Wang and Armstrong, 2003). Once complete, the solutions to, or results of, these tasks are aggregated.

The type of computational problem-solving approach exemplified by grid computing is also known as *high-throughput computing* (HTC), in which similar computations are done independently by a (large) number of processors, and the network interconnect is used primarily for data distribution and results aggregation. Another, distinct, distributed computational paradigm is *high-performance computing* (HPC), which emphasizes slightly different aspects of the system. Whereas high-throughput computing emphasizes the problem division, high-performance computing refers to the use of many high-power servers connected by a fast (usually >10 Gbps) network, under any algorithmic paradigm. An HPC algorithm may also qualify as HTC, or it may require much more intermediate communication between processes in order to accomplish the goal. In other words, in high-performance computing, individual processes/processors may need to communicate very rarely, or very often, during the intermediate stages of a computation. The concepts of HTC and HPC thus focus on slightly different qualities of a whole system, and are neither identical nor mutually exclusive. In Figure 13.2, an example HTC workflow is depicted. The problem input is divided into many segments, each of which is sent to a node of the system; depending on processing capacities, a single node may receive multiple segments. Several tasks (in Figure 13.2, three) may be required to complete the processing of each segment. Thus, in Figure 13.2, any two tasks $T_{i,j}$ with the same first (i) index are distinct but operate (in series) on the same segment of the input, whereas any two tasks with the same second (j) index are identical but operate on different portions of the input. After processing of each segment, the individual results are merged in some manner to form the output. Unlike HTC, there is no canonical diagram for HPC, since HPC is defined more by the power of the computational resources and admits many algorithmic paradigms.

Presently, there are several large-scale commercial options for cloud computing, and many larger institutions have their own distributed systems which users may use as a "cloud." Amazon's Elastic

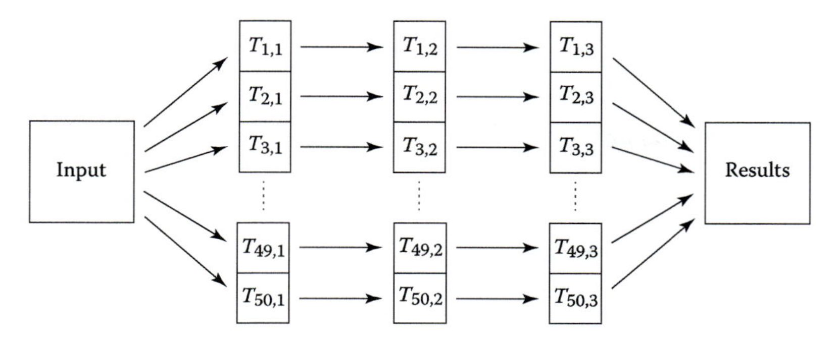


**FIGURE 13.2** A typical HTC workflow model.

Compute Cloud (EC2) was the first public cloud to be available to large numbers of users over the internet. Subsequently, Google and Microsoft, as well as several other vendors, have also begun to offer large-scale public cloud services. Since the distributed nature of a cloud is opposite of the traditional mainframe server, most cloud-computing systems in practice run some version of the GNU/Linux operating system. However, Microsoft Windows Server is also an option for cloud systems, and indeed the availability of GIS-specific software packages makes Windows Server relevant in this domain.

### 13.2.2 Cloud Paradigms

While a variety of common terms have arisen—*public cloud, private cloud, hybrid cloud*—these models of computing are fundamentally the same, differing only in ancillary issues such as security and usability.

A *public cloud* is a cloud system available to the public (or some subset of the public) over the World Wide Web. A public cloud may even rely on internet infrastructure for "internal" network connectivity. In practice, compute access to most public clouds is available for rent to the general public. Amazon's Elastic Compute Cloud (EC2) is the quintessential public cloud—Amazon Web Services was an early leader in providing computing power as a commercial service. Others in the commercial sector have followed suit, such as Google with their Google Cloud Platform (Thenkabail et al., 2021, Gorelick et al., 2017). The emergence of public clouds is a good example of an economy of scale—powerful servers and high-performance networks can be expensive, and require specialized expertise to administer and support. Many institutions possess problems for which cloud computing is [a part of] an ideal solution, but only the largest have a sufficient quantity of such problems that it makes financial sense to own and manage their own servers. Because of the expense, cloud-computing systems of scale are often only financially practical under high load—smaller organizations without a sufficient volume of computational problems may find themselves only utilizing a private system, say, 50% of the time. In this scenario, the costly overhead looms even larger compared to the cost-benefit analysis of problems solved versus power consumption.

A *private cloud* is just that—a private network of computers owned by the using entity, usually segregated from the public internet. Examples of private clouds include any compute clusters to which access is limited, such as the research clusters commonly operated by and within large universities. Private cloud computing makes sense under many scenarios: (1) For research and development, it may be necessary to have more control over the system than is afforded by the typical public cloud. Often times, complete control of all aspects of a system can only be achieved when the system is private. (2) When it is financially feasible to own a system whose response time would be sufficient for the users, and if the users have a sufficient quantity of computational tasks that the system would be highly utilized, it could make sense to use a private cloud—thus cutting out the middle man from the services.

Finally, the term *hybrid cloud* is sometimes used to refer to an aggregate system of which some portion is owned, or fully controlled, by the user and some portion is available over a public network. A hybrid cloud can make sense when different parts of the cloud are being used for distinctly different tasks and a different cloud-computing model would be ideal for each. For example, an institution with critical data it needs to process may opt to store the data permanently on a smaller, private cloud, where members have full control over data security, but send some portion of the data to a public cloud, on demand, for processing. While a private cloud offers more control to its owners/users, a public cloud can be easier to use, and can usually provide greater computational power, with little or no overhead cost. As such, a hybrid cloud model may make sense in practice for many businesses and smaller, short-to-medium-term research endeavors.

### 13.2.3 Cloud Service Models

*Infrastructure as a Service:* The fundamental concept of outsourcing and commercialization of compute time is referred to as *Infrastructure as a Service* (IaaS). In this approach, a cloud service provider (for instance, Amazon) owns and operates a collection of networked servers available for rent. At a base level, the cloud service provider provisions rental machines (often virtual machines) with a client's desired operating system, as well as related facilities and tools. For instance, a client might rent compute time from a cloud service provider and request that the rental machines run Fedora Linux. The physical machines, together with the network, Linux kernel, and Fedora distribution constitute the infrastructure, and this system in its entirety is the product provided to the client.

*Platform as a Service*: The concept of a *Platform as a Service* (PaaS) lies on top of IaaS. For many clients, a compute infrastructure alone is not sufficient for their goals—there is a large gulf between the presence of computing infrastructure and the desired end result. For end users who may have a higher-level abstraction of their computational task(s), an intermediate platform—a computational service that provides more than just an operating system and network and is managed at that higher level—may be appropriate. In PaaS, a software platform is provisioned by the cloud service provider, and can be comprised of one or more different software layers. One common example is Hadoop, by Apache. Hadoop is an open-source implementation of Google's MapReduce framework that provides the MapReduce computational model along with an underlying distributed file system. In addition, there are a variety of additional layers compatible with Hadoop, which can exist on top of the Hadoop-MapReduce platform—a common such layer is the data warehouse Hive, also developed by Apache. For more information on the MapReduce computing framework, see Dean and Ghemawat (2004).

*Software as a Service:* At a higher level than PaaS is *Software as a Service* (SaaS), in which the cloud service provider furnishes a complete software package designed for a particular domain. The end user need only perform configuration-level tasks on the system before it is ready for use. This model of computational services is well suited for organizations that must perform ubiquitous tasks, such as vehicle location tracking or legal document preparation, and who have no desire for, or commitment to, computational research and development. SaaS is becoming increasingly popular as an operational business model.

### 13.2.4 Advantages and Limitations of Cloud Computing

On a practical level, the cloud-computing paradigm has many advantages: access to high-performance computing systems; "pay-as-you-go" payment schemes, with few overhead costs; on-demand provisioning of resources; highly scalable/elastic compute and storage resources; and automated data reliability (Pérez-Cutillas et al., 2023, Liu et al., 2023, Thenkabail et al., 2021, Amani et al., 2020, Phalke et al., 2020, Antunes et al., 2019, Li et al., 2019, Oliphant et al., 2019, Teluguntla et al., 2018, Yan et al., 2018, Yao et al., 2020, Wang et al., 2018, Gorelick et al., 2017, Teluguntla et al., 2017, Xiong et al., 2017a, 2017b, Yang et al., 2017a, 2017b, Teluguntla et al., 2016, Cui et al., 2010, Huang et al., 2010, McEvoy and Schulze, 2008, Rezgui et al., 2013, Watson et al., 2008, Yang and Huang,

2013). Note that this last improvement, high data reliability, is different from data security—one of the major challenges in cloud computing. By its very nature of high accessibility/availability, the most scalable type of cloud computing—the use of a public cloud—is inherently less secure than a computing model based on private control of the entire system. The security challenges of using a public cloud are many-fold: depending on privacy requirements, it may not be acceptable that data resides on the cloud service provider's machines; users are reliant on the internal security controls of the service provider to prevent unauthorized data access, and the public network infrastructure that must be traversed for data and compute access may be compromised. More elaborate (and from a performance perspective, costly) security and encryption measures must commonly be taken when using a public cloud, such as the use of a *Virtual Private Network* (VPN).

On a fundamental level, the advantages of cloud computing lie in the distributed paradigm of the hardware systems and the prospect of lowered barriers to access through the commoditization of computing power itself. The cloud-computing paradigm provides flexibility, scalability, robustness, and, when managed by a dedicated provider, ease-of-use. In order to make best use of these new computational opportunities, however, software design must take the distributed/parallel nature of cloud computing into account—the power of cloud computing lies not in any single machine, but in the capacities of the system as a whole. Problems of a high-throughput-computing (HTC) nature are natural candidates for solving in the cloud because of the intrinsically scalable aspects of HTC solutions. More generally, for all problem paradigms, the communication requirements of the problem play a role in selecting the appropriate service model when seeking a cloud-computing solution—if substantial communication is required between compute nodes in a particular algorithm, a lower-level service model (IaaS) may be necessary. For example, it may be difficult to find an appropriate, off-the-shelf SaaS-level software implementation for an algorithmic approach that requires compute nodes to communicate heavily. In contrast, in HTC, the focus of the software developer can be narrowed to the proper provisioning of resources; network bandwidth is often a bottleneck only during the dissemination of problem input and gathering of output. Thus, an analysis akin to Amdahl's Law plays an important role in determining the practicality of a particular scale of cloud computing for running a specific algorithm or solving a particular problem.

Finally, on issues of cloud security and liability: it should be emphasized that, in part because cloud computing is not an established commodity, many contractual aspects (e.g., various liabilities) are left to the user and provider to agree upon.

## 13.3 CLOUD-COMPUTING-BASED REMOTE-SENSING-RELATED APPLICATIONS

This section provides a short summary of remote sensing applications that use cloud-computing environments as well as a more detailed case study. High-performance computing (HPC) frameworks (for example, supercomputers at various research organizations) can potentially be widely adapted for remote-sensing applications (Thenkabail et al., 2021, Oliphant et al., 2019, Teluguntla et al., 2018, Gorelick et al., 2017, Teluguntla et al., 2017, Xiong et al., 2017a, 2017b, Yang et al., 2017a, 2017b, Teluguntla et al., 2016, Lee et al., 2011, Plaza et al., 2011, Simmhan and Ramakrishnan, 2010, Parulekar, 1994). The main limitations associated with the use of HPC are (1) HPC resources are not readily available for large user communities and (2) HPC resources are expensive to acquire and maintain (Ostermann et al., 2010).

Krishnan et al. (2010) evaluated a MapReduce approach for LiDAR gridding on a small private cluster consisting of around eight to ten commodity computers. They investigated the effects of several parameters, including grid resolution and dataset size, on performance. For their software implementation using Hadoop, the authors experimented with Hadoop-specific factors, such as the number of reducers allocated for a problem and the inherent concurrency therein. In their particular study, using quad-core machines with 8 GB of main memory and connected by gigabit Ethernet, they found that doubling the size of their Hadoop cluster from four to eight nodes had little effect on their experimental runtimes. They thus showed that their solution to the task was not strictly compute-bound. On the other hand, the authors did note a substantial degradation in performance for their single-node, nondistributed algorithm control (implemented in C++) when the problem size grew larger than could fit in main memory. Krishnan et al. (2010) concluded that Hadoop could be a useful framework for algorithms that process large-scale spatial data (roughly 150 million points), but that certain serial elements, such as output generation, could still be rate-limiting, especially on commodity hardware. Their work also (1) motivates the study of systems with larger memories (this stems from their experience with HPC resources); and (2) demonstrated that the task of designing optimal systems for the processing of massive spatial data is complex.

Project *Matsu* is an open source project for processing satellite imaginary using a community cloud (Bennett and Grossman, 2012). This project, a collaboration between NASA and the Open Cloud Consortium (OCC), has been developed to process data from NASA's EO-1 satellite and to develop open source technology for public cloud-based processing of satellite imagery. Most computations were completed using a Hadoop framework running on nine compute nodes with 54 compute cores and 352GB of RAM. The stated goal of this project (http://matsu.opensciencedatacloud.org/) is (1) to use an open source cloud-based infrastructure to make high quality satellite image data accessible through an Open-Geospatial-Consortium-compliant (OGC-compliant) Web Map Service (WMS); (2) to develop an open-source cloud-based analytic framework for analyzing individual images and collections of images; and (3) to generalize this framework to manage and analyze other types of spatial-temporal data. This project also features an on-demand cloud-based disaster assessment capability through satellite image comparisons. The image comparisons are done via a MapReduce job using a Hadoop-streaming interface. As an example, this project hosts a website that provides real-time information about flood prediction and assessment in Namibia. The final data is served to end-users using a standard OGC Web Map Service and Web Coverage Processing Service tools.

Oryspayev et al. (2012) studied LiDAR data reduction algorithms that were implemented using the GPGPU and multicore CPU architectures available on the Amazon Web Services (AWS) Elastic Compute Cloud (EC2). This paper tests the veracity of a vertex-decimation algorithm for reducing LiDAR data size/density and analyses the performance of this approach on multicore-CPU and GPU technologies, to better understand processing time and efficiency. The paper documents the performance of various GPGPU and multicore CPU machines including Tesla family GPUs and the Intel's multicore i-CPU series for the data reduction problem using large-scale LiDAR data. The study raises several questions about implementation of spatial-data processing algorithms on GPGPU machines, such as how to reduce overhead during the initialization of devices and how to optimize algorithms to minimize data transfer between CPUs/GPUs.

Eldawy and Mokbel (2013) developed an open-source framework, *SpatialHadoop*, that extends Hadoop by providing native support for spatial data. As an extension of Hadoop, the framework operates similarly—programs are written in terms of Map and Reduce functions, though the system is optimized to exploit underlying properties and characteristics of spatial data. As case studies, SpatialHadoop has three spatial operations, range queries, *k*-nearest-neighbor queries, and spatial join.

Cary et al. (2009) studied the performance of the MapReduce framework for bulk-construction of *R*-trees and aerial image quality computation on both vector and raster data. They deployed their MapReduce implementations using the Hadoop framework on the Google and IBM clouds. The authors presented results that demonstrate the scalability of MapReduce, and the effect of parallelism on the quality of the results. This paper also studied various metrics to compare the performance of their implemented algorithms including execution time, correctness, and tile quality. Their results indicate that the appropriate application of MapReduce could dramatically improve task completion times and also provide close to linear scalability. This study motivates further investigation of the MapReduce framework for other spatial-data-handling problems.

Li et al. (2010) studied the integration of data from ground-based sensors with the Moderate-Resolution Imaging Spectroradiometer (MODIS) satellite data using the Windows Azure cloud platform. Specifically, the authors provide a novel approach to re-project input data into timeframe- and resolution-aligned geographically formatted data and also develop a novel reduction technique to derive important new environmental data through the integration of satellite and ground-based data. Slightly modified Windows Azure abstractions and APIs were used to accomplish the re-projection and reduction steps. They suggest that cloud computing has a great potential for efficiently processing satellite data. It should be noted that their current framework doesn't fit into the MapReduce framework since it uses Azure's general queue-based task model.

Berriman et al. (2010) compared various toolkits to create image mosaics and to manage their provenance using both the Amazon EC2 cloud and the Abe high performance cluster at NCSA, UIUC. They conducted a series of experiments to study performance and costs associated with different types of tasks (I/O-bound, CPU-bound, and memory-bound) in these two environments. Their experiments show that for I/O-bound applications, the most expensive resources are not necessarily the most cost-effective, that data transfer costs can exceed the processing costs for I/O-bound applications on Amazon EC2, and that the resources offered by Amazon EC2 are generally less powerful than those available in the Abe high-performance cluster and consequently do not offer the same levels of performance. They concluded from their results that cloud computing offers a powerful and cost-effective new resource for compute and memory intensive remote sensing applications.

### 13.3.1 A Case Study: Cloud-Based LiDAR Processing System (CLiPS)

To more completely illustrate an approach to cloud computing of remote sensing information, in this section we describe a LiDAR application. In this Cloud-based LiDAR Processing System (CLiPS) project, we use a state-wide (Iowa) LiDAR data repository (Iowa DNR, 2009) in which data is distributed to the public as a collection of 34,000 tiles, each covering 4 km$^2$; this comprises roughly 7 terabytes of data. In order to process this massive data, CLiPS was designed (Figure 13.3) as a web portal implemented using Adobe's Flex framework along with ESRI's ArcGIS API for Flex (ESRI, 2012, Sugumaran et al., 2014), OpenLayers, and the Amazon EC2 cloud environment. CLiPS use a three-tier client-server model (Figure 13.3). The top tier supports user interaction with the system, the second tier provides process management services, such as monitoring, and analysis, and the third tier is dedicated to data and file services. The client-side interface was developed using Flex and the server-side uses custom-built tools, constructed from open source products.

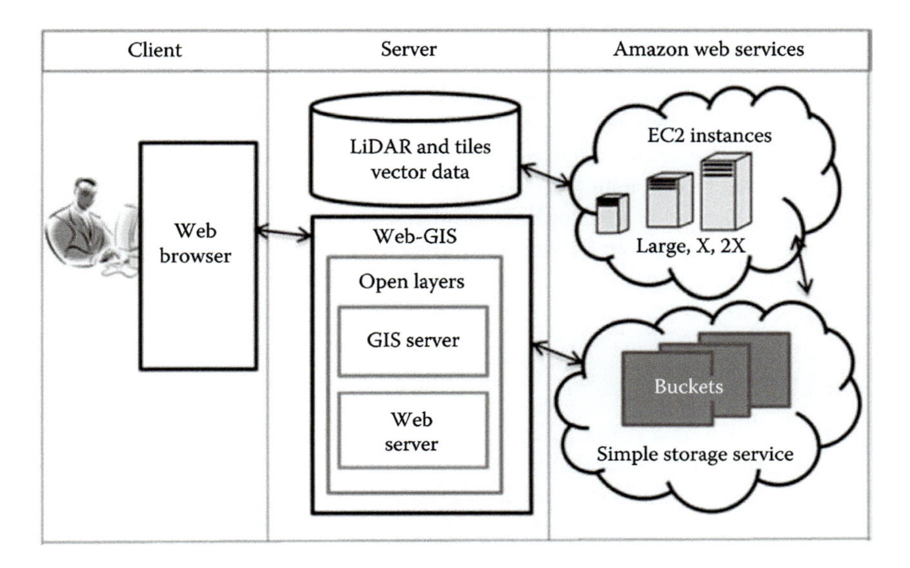


**FIGURE 13.3** Overall architecture developed for CLiPS.

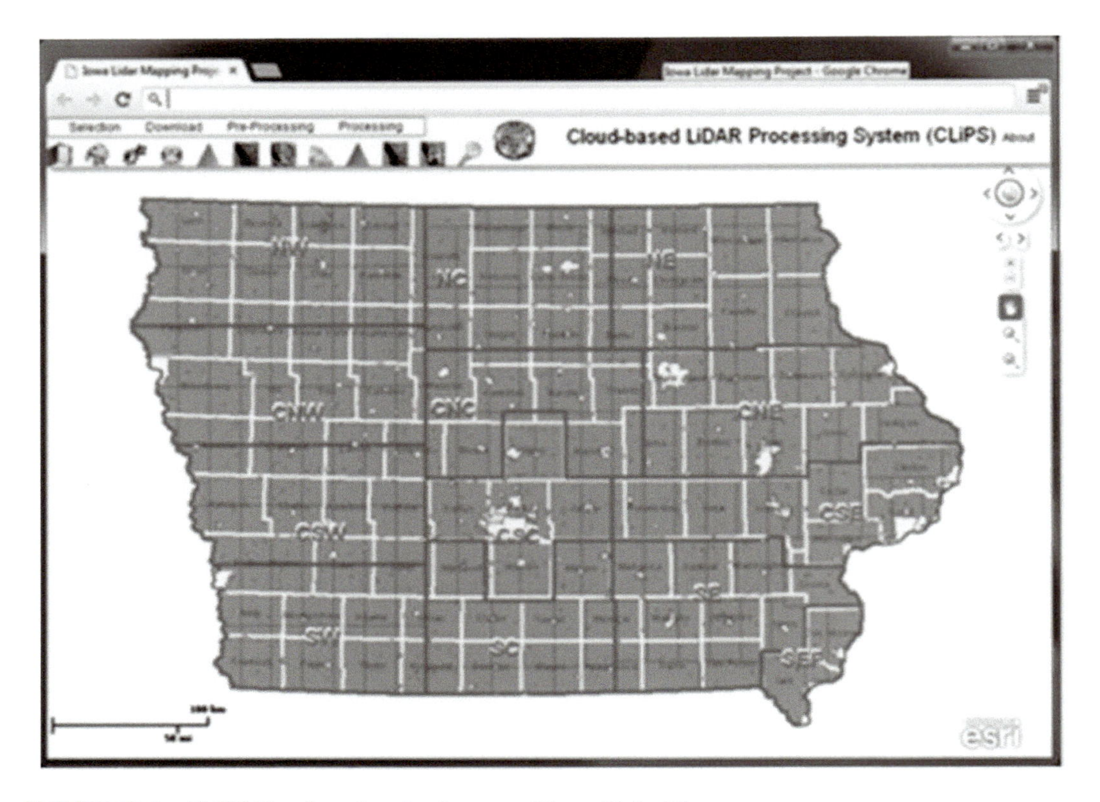


FIGURE 13.4 CLiPS User Interface for the state of Iowa, United States.

Figure 13.4 shows the user interface developed for this study. The interactive user interface requires a user to, first, select a region of interest on a map and then select Amazon web services credentials and computing resources as well as a location to which results will be sent for downloading (this is typically an email address). The system was tested by creating a TIN from a LiDAR point cloud using 18 dataset and processor configurations: three terrain data (flat, undulating, urban), two tile sizes (9 and 25) and three Amazon EC2 processing configurations (large, Xlarge, and double Xlarge). The undulating terrain dataset took more time than the other terrain types for 5 × 5 tile groups, while the urban terrain was the most computationally intensive for the 3 × 3 tile groups used in this study (Sugumaran et al., 2014). The results clearly show that as computer power increases, processing times decrease for all three types of LiDAR terrain data. The various combinations in our evaluations showed that even with up to 25 tiles with varying processing configuration types, each request required less than an hour and cost less than a dollar for data processing (e.g., TIN creation). Moreover, the cost of uploading data from our server and data storage on the cloud was less than $50. Thus, the overall cost for our test using the Amazon cloud was less than $100, an amount that is affordable by most users.

## 13.4 CONCLUSIONS

It is abundantly clear that as a consequence of technological and sensor licensing improvements, the spatial, spectral, temporal, and radiometric resolution of remote sensing imagery will continue to increase. This translates into massive quantities of data that must be processed to glean meaningful information that can be used in a variety of decision support and visualization applications. Such data quantities quickly overwhelm the capabilities of even the most powerful desktop systems available now and in the foreseeable future. As a consequence, researchers continue to explore cost-effective, yet powerful, computing environments that can be harnessed for remote sensing applications. Dedicated HPC systems are expensive to acquire and maintain and have relatively short half-lives. This significantly diminishes this alternative as a practical solution. Instead, the availability of high-speed communication technologies now makes the use of distributed, pay-as-you-go resources an attractive choice for researchers and government agencies, as well as users in the private sector. One term that is used to describe these distributed resources is cloud computing. The cloud provides new patterns for deploying remote-sensing-data processing and provides easy, inexpensive access to servers, elastic scalability, managed infrastructure, and low complexity of deployment (Pérez-Cutillas et al., 2023, Liu et al., 2023, Thenkabail et al., 2021, Amani et al., 2020, Phalke et al., 2020, Antunes et al., 2019, Li et al., 2019, Oliphant et al., 2019, Teluguntla et al., 2018, Yan et al., 2018, Yao et al., 2020, Wang et al., 2018, Gorelick et al., 2017, Teluguntla et al., 2017, Xiong et al., 2017a, 2017b, Yang et al., 2017a, 2017b, Teluguntla et al., 2016).

Despite these significant advantages, the use of cloud computing does have some limitations (Gorelick et al., 2017). First, since data is distributed across networks to distributed, but unknown, locations, security can become problematic. Thus, public cloud resources cannot be used exclusively for applications that require the processing of many types of individual-level information. This limitation, however, is not normally significant for remote sensing applications. Second, different types of spatial algorithms may require considerable amounts of inter-processor communication, particularly in big data applications, and certain elements of cloud infrastructure may therefore induce processing latencies that may be unacceptable to users.

## ACKNOWLEDGMENTS

This research was conducted with support from an Amazon Research Grant and USGS—AmericaView projects.

## REFERENCES

Amani, M. et al. 2020. Google Earth Engine cloud computing platform for remote sensing big data applications: A comprehensive review, IEEE Journal of Selected Topics in Applied Earth Observations and Remote Sensing, 13, 5326–5350, doi: 10.1109/JSTARS.2020.3021052.

Antunes, R.R., Blaschke, T., Tiede, D., Bias, E.S., Costa, G.A.O.P., & Happ, P.N. 2019. Proof of concept of a novel cloud computing approach for object-based remote sensing data analysis and classification, GIScience and Remote Sensing, 56(4), 536–553, doi: 10.1080/15481603.2018.1538621.

Bennett, C., & Grossman, R. 2012. OCC Project Matsu: An Open-Source Project for Cloud-Based Processing of Satellite Imagery to Support the Earth Sciences. http://matsu.opensciencedatacloud.org (Accessed March 4, 2014).

Berriman, G.B., Deelman, E., Groth, P., & Juve, G. 2010. The application of cloud computing to the creation of image mosaics and management of their provenance. In SPIE Astronomical Telescopes + Instrumentation, San Diego, CA: International Society for Optics and Photonics, vol. 7740F.

Buyya, R., Yeo, C.S., Venugopal, S., Broberg, J., & Brandic, I. 2009. Cloud computing and emerging IT platforms: Vision, hype, and reality for delivering computing as the 5th utility, Future Generation Computer Systems, 25, 599–616.

Cary, A., Sun, Z., Hristidis, V., & Rishe, N. 2009. Experiences on processing spatial data with MapReduce. In Scientific and Statistical Database Management, Berlin and Heidelberg: Springer, pp. 302–319.

Chang, C.C., Chang, Y.L., Huang, M.Y., & Huang, B. 2011. Accelerating regular LDPC code decoders on GPUs, IEEE Journal of Selected Topics in Applied Earth Observations and Remote Sensing (JSTARS), 4(3), 653–659.

Christophe, E., Michel, J., & Inglada, J. 2011. Remote sensing processing: From multicore to GPU, IEEE Journal of Selected Topics in Applied Earth Observations and Remote Sensing (JSTARS), 4(3), 643–652.

Cui, D., Wu, Y., & Zhang, Q. 2010. Massive spatial data processing model based on cloud computing model. In Computational Science and Optimization CSO, 2010 Third International Joint Conference on Computational Science and Optimization, Huangshan, China, vol. 2, pp. 347–350.

Dean, J., & Ghemawat, S. 2004. MapReduce: Simplified data processing on large clusters. In Proc. of the 6th Conference on Symposium on Operating Systems Design and Implementation, San Francisco, CA, vol. 6.
Eldawy, A., & Mokbel, M.F. 2013. A demonstration of SpatialHadoop: An efficient MapReduce framework for spatial data, Proceedings of the VLDB Endowment, 6(12), 1230–1233.
eScience Institute, University of Washington. 2012. Understanding Cloud Computing for Research and Teaching. http://escience.washington.edu/get-help-now/understanding-cloud-computing-research-and-teaching (Accessed March 4, 2014).
ESRI. 2012. ArcGIS Server. www.esri.com/software/arcgis/arcgisserver (Accessed March 23, 2012).
González, J.F., Rodríguez, M.C., & Nistal, M.L. 2009. Enhancing reusability in learning management systems through the integration of third-party tools. In Frontiers in Education Conference (FIE), 2009 39th IEEE, San Antonio, TX, pp. 1–6.
Gorelick, N., Hancher, M., Dixon, M., Ilyushchenko, S., Thau, D., & Moore, R. 2017. Google Earth Engine: Planetary-scale geospatial analysis for everyone, Remote Sensing of Environment, 202, 18–27, ISSN 0034-4257, doi: 10.1016/j.rse.2017.06.031. (www.sciencedirect.com/science/article/pii/S0034425717302900)
Haifang, Z. 2003. Study and Implementation of Parallel Algorithms for Remote Sensing Image Processing, National University of Defence Technology.
Huang, Q., Yang, C., Nebert, D., Liu, K., & Wu, H. 2010. Cloud computing for geosciences: Deployment of GEOSS clearinghouse on Amazon's EC2. In HPDGIS '10: Proceedings of the ACM SIGSPATIAL International Workshop on High Performance and Distributed Geographic Information Systems, San Jose, CA, pp. 35–38.
Hyspeed Computing. 2013. Big Data and Remote Sensing—Where Does All This Imagery Fit into the Picture? http://hyspeedblog.wordpress.com/2013/03/22/big-data-and-remote-sensing-where-does-all-this-imagery-fit-into-the-picture (Accessed March 2014).
IBM. 2012. Bringing Big Data to the Enterprise. http://www-01.ibm.com/software/data/bigdata (Accessed April 4, 2013).
Iowa DNR. 2009. State of Iowa. Retrieved April 29, 2009, from: www.iowadnr.gov/mapping/lidar/index.html (Accessed April 4, 2013).
Krishnan, S., Bary, C., & Crosby, C. 2010. Evaluation of MapReduce for gridding LIDAR data, *CloudCom*. In IEEE Second International Conference on Cloud Computing Technology and Science, Indianapolis, IN, pp. 33–40.
Kumar, N., Lester, D., Marchetti, A., Hammann, G., & Longmont, A. 2013. Demystifying Cloud Computing for Remote Sensing Application. http://eijournal.com/newsite/wp-content/uploads/2013/06/cloudcomputing.pdf.
Lee, C.A., Gasster, S.D., Plaza, A., Chang, C.I., & Huang, B. 2011. Recent developments in high performance computing for remote sensing—a review, IEEE Journal of Selected Topics in Applied Earth Observations and Remote Sensing, 4(3), 508–527.
Li, J., Humphrey, M., Agarwal, D., Jackson, K., Ingen, C., & Ryu, Y. 2010. eScience in the cloud: A MODIS satellite data reprojection and reduction pipeline in the windows Azure platform. In IEEE International Parallel and Distributed Processing Symposium (IPDPS), Atlanta, GA, pp. 1–10.
Li, Z., Shen, H., Cheng, Q., Liu, Y., You, S., & He, Z. 2019. Deep learning based cloud detection for medium and high resolution remote sensing images of different sensors, ISPRS Journal of Photogrammetry and Remote Sensing, 150, 197–212, ISSN 0924-2716, doi: 10.1016/j.isprsjprs.2019.02.017. (www.sciencedirect.com/science/article/pii/S0924271619300565)
Liu, T. et al. 2010. Remote sensing image classification techniques based on the maximum likelihood method, FuJian Computer, 001, 7–8.
Liu, Z., Chen, Y., & Chen, C. 2023. Analysis of the spatiotemporal characteristics and influencing factors of the NDVI based on the GEE cloud platform and Landsat images, Remote Sensing, 15(20), 4980, doi: 10.3390/rs15204980.
McEvoy, G.V., & Schulze, B. 2008. Using clouds to address grid limitations. In Proc. of the 6th International Workshop on Middleware for Grid Computing Leuven, Belgium: MGC, December 1–5.
Mell, P., & Grance, T. 2011. The NIST definition of cloud computing. National Institute of Standards and Technology, Special Publication 800-145. https://nvlpubs.nist.gov/nistpubs/legacy/sp/nistspecialpublication800-145.pdf
Oliphant, A., Thenkabail, P.S., Teluguntla, P., Xiong, J., Gumma, M.K., Congalton, R., & Yadav, K. 2019. Mapping cropland extent of Southeast and Northeast Asia using multi-year time-series Landsat 30m data using random forest classifier on Google Earth Engine, International Journal of Applied Earth Observation and Geoinformation, 81, 110–124, doi: 10.1016/j.jag.2018.11.014.

Oryspayev, D., Sugumaran, R., DeGroote, J., & Gray, P. 2012. LiDAR data reduction using vertex decimation and processing with GPGPU and multicore CPU technology, Computers and Geosciences, 43, 118–125.

Ostermann, S., Iosup, A., Yigitbasi, N., Prodan, R., Fahringer, T., & Epema, D. 2010. A performance analysis of EC2 cloud computing services for scientific computing. In Cloud Computing, Berlin and Heidelberg: Springer, pp. 115–131.

Parulekar, R. et al. 1994. High performance computing for land cover dynamics. In Pattern Recognition, 1994. Vol. 2—Conference B: Computer Vision & Image Processing, Proceedings of the 12th IAPR International Conference on Pattern Recognition, Jerusalem, Israel: IEEE.

Pérez-Cutillas, P., Pérez-Navarro, A., Conesa-García, C., Antonio Zema, D., & Amado-Álvarez, J.P. 2023. What is going on within Google Earth Engine? A systematic review and meta-analysis, Remote Sensing Applications: Society and Environment, 29, 100907, ISSN 2352-9385, doi: 10.1016/j.rsase.2022.100907. (www.sciencedirect.com/science/article/pii/S2352938522002154)

Phalke, A.R., Özdoğan, M., Thenkabail, P.S., Erickson, T., Gorelick, N., Yadav, K., & Congalton, R.G. 2020. Mapping croplands of Europe, Middle East, Russia, and Central Asia using Landsat, Random Forest, and Google Earth Engine, ISPRS Journal of Photogrammetry and Remote Sensing, 167, 104–122, ISSN 0924-2716, doi: 10.1016/j.isprsjprs.2020.06.022. (www.sciencedirect.com/science/article/pii/S0924271620301805) IP-116983.

Plaza, A., & Chang, C.-I. 2008. High Performance Computing in Remote Sensing. Boca Raton, FL: CRC Press, 2007.

Plaza, A., & Chang, C.-I. 2008. Special issue on high performance computing for hyperspectral imaging, International Journal of High Performance Computing Applications, 22(4), 363–365.

Plaza, A., Du, Q., Chang, Y.-L., & King, R.L. 2011. High performance computing for hyperspectral remote sensing, IEEE Journal of Selected Topics in Applied Earth Observations and Remote Sensing (JSTARS), 4(3), 528–544.

Plaza, A., Plaza, J., Paz, A., & Sanchez, S. 2011. Parallel hyperspectral image and signal processing, IEEE Signal Processing Magazine, 28(3), 119–126.

Plaza, A., Valencia, D., Plaza, J., & Martinez, P. 2006. Commodity cluster-based parallel processing of hyperspectral imagery, Journal of Parallel and Distributed Computing, 66(3), 345–358.

Rezgui, A., Malik, Z., & Yang, C. 2013. High-resolution spatial interpolation on cloud platforms. In Proceedings of the 28th Annual ACM Symposium on Applied Computing, New York, NY: ACM, pp. 377–382.

Shekhar, S., Gunturi, V., Evans, M.R., & Yang, K. 2012. Spatial big-data challenges intersecting mobility and cloud computing. In Proceedings of the Eleventh ACM International Workshop on Data Engineering for Wireless and Mobile Access, New York, NY, pp. 1–6.

Simmhan, Y., & Ramakrishnan, L. 2010. Comparison of resource platform selection approaches for scientific workflows. In Proceedings of the 19th ACM International Symposium on High Performance Distributed Computing, New York, NY: ACM, pp. 445–450.

Song, C., Li, Y., & Huang, B. 2011. A GPU-accelerated wavelet decompression system with SPIHT and Reed-Solomon decoding for satellite images, IEEE Journal of Selected Topics in Applied Earth Observations and Remote Sensing (JSTARS), 4(3), 683–690.

Sugumaran, R., Burnett, J., & Armstrong, M.P. 2014. Using a cloud computing environment to process large 3D spatial datasets. H. Karimi (ed.) Big Data: Techniques and Technologies in Geoinformatics. Boca Raton, FL: CRC Press, pp. 53–65.

Sugumaran, R., Oryspayev, D., & Gray, P. 2011. GPU-based cloud performance for LiDAR data processing. In COM.Geo 2011: 2nd International Conference and Exhibition on Computing for Geospatial Research and Applications, Washington DC, USA, May 23–25.

Teluguntla, P., Thenkabail, P.S., Oliphant, A., Xiong, J., & Gumma, M.K. 2018. A 30m Landsat-derived cropland extent product of Australia and China using random forest machine learning algorithm on Google Earth Engine cloud computing platform, ISPRS Journal of Photogrammetry and Remote Sensing, 144, 325–340, ISSN 0924-2716, doi: 10.1016/j.isprsjprs.2018.07.017.

Teluguntla, P., Thenkabail, P.S., Xiong, J., Gumma, M.K., Congalton, R.G., Oliphant, A., Poehnelt, J., Yadav, K., Rao, M., & Massey, R. 2017. Spectral Matching Techniques (SMTs) and Automated Cropland Classification Algorithms (ACCAs) for mapping croplands of Australia using MODIS 250-m time-series (2000–2015) data, International Journal of Digital Earth, 10(9), doi: 10.1080/17538947.2016.1267269. IP-074181.

Teluguntla, P., Thenkabail, P.S., Xiong, J., Gumma, M.K., Giri, C., Milesi, C., Ozdogan, M., Congalton, R., Tilton, J., Sankey, T., Massey, R., Phalke, A., & Yadav, K. 2016. NASA Making Earth System Data Records for Use in Research Environments (MEaSUREs) Global Food Security Support Analysis Data (GFSAD) Crop Mask 2010 Global 1 km V001 [Data set]. NASA EOSDIS Land Processes DAAC (Accessed 24, 2022), doi: 10.5067/MEaSUREs/GFSAD/GFSAD1KCM.001.

Thenkabail, P. 2013. Special issue on high performance computing in remote sensing, Remote Sensing.

Thenkabail, P.S., Teluguntla, P.G., Xiong, J., Oliphant, A., Congalton, R.G., Ozdogan, M., Gumma, M.K., Tilton, J.C., Giri, C., Milesi, C., Phalke, A., Massey, R., Yadav, K., Sankey, T., Zhong, Y., Aneece, I., & Foley, D. 2021. Global cropland-extent product at 30m resolution (GCEP30) derived from Landsat satellite time-series data for the year 2015 using multiple machine-learning algorithms on Google Earth Engine cloud: U.S. Geological Survey Professional Paper 1868, p. 63, doi: 10.3133/pp1868. https://lpdaac.usgs.gov/news/release-of-gfsad-30meter-cropland-extent-products/. IP-119164.

USGS. 2011. Landsat. http://landsat.usgs.gov, (accessed April 7, 2013).

Vatsavai, R.R., Ganguly, A., Chandola, V., Stefanidis, A., Klasky, S., & Shekhar, S. 2012. Spatiotemporal data mining in the era of big spatial data: Algorithms and applications. In Proceedings of the 1st ACM SIGSPATIAL International Workshop on Analytics for Big Geospatial Data, New York, NY, pp. 1–10.

Wang, L., Ma, Y., Yan, J., Chang, V., & Zomaya, A.Y. 2018. pipsCloud: High performance cloud computing for remote sensing big data management and processing, Future Generation Computer Systems, 78(Part 1), 353–368, ISSN 0167-739X, doi: 10.1016/j.future.2016.06.009. (www.sciencedirect.com/science/article/pii/S0167739X16301923)

Wang, S., & Armstrong, M.P. 2003. A quadtree approach to domain decomposition for spatial interpolation in grid computing environments, Parallel Computing, 29(10), 1481–1504.

Wang, Y., Wang, S., & Zhou, D. 2009. Retrieving and indexing spatial data in the cloud computing environment. In Proc. of the 1st International Conference on Cloud Computing, Beijing, China, December 1–4, *Lecture Notes in Computer Sciences*, vol. 5931, pp. 322–331.

Watson, P., Lord, P., Gibson, F., Periorellis, P., & Pitsilis, G. 2008. Cloud computing for e-Science with CARMEN. In 2nd Iberian Grid Infrastructure Conference Proceedings, Porto, Portugal, pp. 3–14.

White, T. 2012. Hadoop: The Definitive Guide. Sebastopol, CA: O'Reilly Media, Inc.

Xiong, J., Thenkabail, P.S., Gumma, M., Teluguntla, P., Poehnelt, J., Congalton, R., & Yadav, K. 2017b. Automated cropland mapping of continental Africa using Google Earth Engine cloud computing, ISPRS Journal of Photogrammetry and Remote Sensing, 126, 225–244, doi: 10.1016/j.isprsjprs.2017.01.019.

Xiong, J., Thenkabail, P.S., Tilton, J.C., Gumma, M.K., Teluguntla, P., Oliphant, A., Congalton, R.G., Yadav, K., & Gorelick, N. 2017a. Nominal 30m cropland extent map of continental Africa by integrating pixel-based and object-based algorithms using Sentinel-2 and Landsat-8 data on Google Earth Engine, Remote Sensing, 9(10), 1065, doi: 10.3390/rs9101065. (www.mdpi.com/2072-4292/9/10/1065) IP-088538.

Yan, J., Ma, Y., Wang, L., Raymond Choo, K., & Jie, W. 2018. A cloud-based remote sensing data production system, Future Generation Computer Systems, 86, 1154–1166, ISSN 0167-739X, doi: 10.1016/j.future.2017.02.044. (www.sciencedirect.com/science/article/pii/S0167739X17303035)

Yang, C., & Huang, Q. 2013. Spatial Cloud Computing: A Practical Approach. Boca Raton, FL: CRC Press.

Yang, C., Huang, Q., Li, Z., Liu, K., & Hu, F. 2017b. Big data and cloud computing: Innovation opportunities and challenges, International Journal of Digital Earth, 10(1), 13–53, doi: 10.1080/17538947.2016.1239771.

Yang, C., Yu, M., Hu, F., Jiang, Y., & Li, Y. 2017a. Utilizing cloud computing to address big geospatial data challenges, Computers, Environment and Urban Systems, 61(Part B), 120–128, ISSN 0198-9715, doi: 10.1016/j.compenvurbsys.2016.10.010. (www.sciencedirect.com/science/article/pii/S0198971516303106)

Yang, H., Du, Q., & Chen, G. 2011. Unsupervised hyperspectral band selection using graphics processing units, IEEE Journal of Selected Topics in Applied Earth Observations and Remote Sensing (JSTARS), 4(3), 660–668.

Yao, X., Li, G., Xia, J., Ben, J., Cao, Q., Zhao, L., Ma, Y., Zhang, L., & Zhu, D. 2020. Enabling the big earth observation data via cloud computing and DGGS: Opportunities and challenges, Remote Sensing, 12(1), 62, doi: 10.3390/rs12010062.

Yue, P., Gong, J., Di, L., Yuan, J., Sun, L., Sun, Z., & Wang, Q. 2010. GeoPW: Laying blocks for the geospatial processing web, Transactions in GIS, 14(6), 755–772.

# 14 Cloud Computing in Remote Sensing

## *A Comprehensive Assessment of State of the Arts*

*Lizhe Wang, Jining Yan, Yan Ma, Xiaohui Huang, Jiabao Li, Sheng Wang, Haixu He, Ao Long, and Xiaohan Zhang*

## ACRONYMS AND DEFINITIONS

**CCRS** Cloud Computing in Remote Sensing DaaS Data as a Service
**SaaS** Software as a Service PaaS Platform as a Service IaaS Infrastructure as a Service
**NASA** National Aeronautics and Space Administration EOSDIS Earth Observing System Data and Information System
**ESA** European Space Agency
**USGS** United States Geological Survey GEE Google Earth Engine
**PIE** Pixel Information Expert OGE Open Geospatial Engine ARD Analysis Ready Data UMM Unified Metadata Model ODC Open Data Cube
**SDG** Sustainable Development Goals AI Artificial Intelligence
**HPC** High Performance Computing DAG Directed Acyclic Graph

## 14.1 INTRODUCTION TO CLOUD COMPUTING IN REMOTE SENSING

### 14.1.1 Big Earth Observation Data

The rapid progress in sensor technologies has catalyzed the remarkable growth of Earth observation devices, encompassing satellites, space shuttles, unmanned aerial vehicles, and more. A pertinent illustration of this growth is the presence of 1,192 satellites worldwide in orbit as of the conclusion of 2021 [1]. Earth observation devices, equipped with diverse sensors, including multi-polarized, multi-angle, active, and passive sensors, have been instrumental in generating substantial volumes of remote sensing data characterized by high-spectral, high-spatial, and high-temporal resolutions [2]. Furthermore, the maturation of ubiquitous perception technologies, such as backpack-mounted LiDAR systems, street view data collection vehicles, and traffic cameras, has significantly enhanced the comprehensiveness of the stereoscopic observation network across the realms of space, airborne, maritime, ground, and human environments [3].

As the annual summary of the National Aeronautics and Space Administration's (NASA) Earth Observing System Data and Information System (EOSDIS) for the period from October 1, 2021, to September 30, 2022, it reveals a comprehensive Earth science data archive encompassing data from satellites, aircraft, in situ measurements, and various sources, amounting to 71.64 petabytes (PB). Projections indicate that by 2030, the EOSDIS archive is poised to exceed 320 PB [4]. Furthermore, the China Center for Resource Satellite Data and Application (CRESDA) has amassed a total archive volume of over 19.79 million satellite scenes as of October 15, 2023 [5]. These staggering figures

DOI 10.1201/9781003541141-21

underscore the categorization of remote sensing data as "big data" due to their vast quantity, rapid generation, diverse typology, and substantial significance. Indeed, the era of big Earth observation data has already arrived.

A large volume of remote sensing data has been utilized in various applications, such as global change and response, land and resource management, environmental pollution monitoring, human activity footprint detection, etc. The expansion of remote sensing data applications also poses the need to provide real-time or near-real-time, and personalized services. For example, since the beginning of Copernicus operations in February 8, 2023, the European Space Agency (ESA) Sentinel data dashboard has published 58,265,412 remote sensing products, 650,956 registered users, and a total of 512.23 PB of user downloaded data [6]. The Jilin-1 Mofang-01A(R) satellite is equipped with a built-in Atlas2-NPU edge computing device, which provides it with the ability to identify forest fire points onboard and also provides users with the ability to download results within 13 seconds [7]. In addition, Google Earth Engine (GEE) library provide over 800 remote sensing processing-related functions, thereby aiming to satisfy the personalized data processing needs of different users [8].

### 14.1.2 Challenges of Big Earth Observation Data

The explosive growth of such a huge amount of remote sensing data and the personalized, real-time, or near-real-time needs of diversified applications pose the following new challenges for the management, processing, analysis, and sharing of remote sensing data [9].

- The escalating volume and diversity of remote sensing data present a substantial resource for advancing remote sensing applications. However, this abundance also introduces novel challenges in data management. Initially, the shift from centralized to distributed storage architectures has been necessitated by the sheer volume of remote sensing data. Institutions leverage evolving distributed systems to manage expanding data resources. Consequently, the central concern revolves around the efficient organization of these vast data repositories to facilitate scalable storage and high-performance processing. Furthermore, the heterogeneity in semantic descriptions of multisource remote sensing data, stored by various stakeholders, complicates the retrieval of relevant information. Thus, expediting the discovery of remote sensing data emerges as an additional imperative challenge.
- Remote sensing data processing and analysis. Efficient and reliable processing is critical to promote the usage of remote sensing data in diversified applications following the "data-information-knowledge" transformation pattern. Processing a huge amount of data in a reasonable manner is difficult because remote sensing data processing is data-intensive, thereby implying that remote sensing data processing technologies are constantly evolving. Moreover, remote sensing-based applications are becoming increasingly complex due to the analysis of multisource remote sensing data, thereby causing the processing procedures of remote sensing data to also be complicated [10]. The trend of fine division of labor and collaborative integration in remote sensing data processing, as well as the real-time processing requirements on the edge side, pose challenges to the processing and analysis of remote sensing data.
- Remote sensing data distribution and sharing. As an important aspect of the remote sensing data life cycle, data distribution and sharing face challenges in various respects such as data standard formulation, data quality, permission control, and data security, etc. Initially, it is necessary to establish comprehensive data standards and service standards to facilitate the flow and sharing of data, thereby breaking down data flow

barriers among institutions and ensuring data availability and interoperability. Then, effective permission control mechanisms need to be established to prevent unauthorized data access and abuse. Next, metadata is vital to help users understand the source and accuracy of remote sensing data. Finally, delivering remote sensing data in an accurate and quick manner with high quality to meet the different data needs of users is also a challenge [11].

Cloud computing provides users with heterogeneous resources following the pay-as-you-go pattern to satisfy various application requirements. In recent years, cloud computing has assisted the development of remote sensing in terms of large-scale data storage and management, processing, and analysis surrounding complex remote sensing data analysis tasks [12]. Representative platforms, such as the GEE, PIE-Engine Studio, etc., are adopted by users for their massive data storage and high-performance computing capabilities and have played a significant role in global change, environmental protection, and other areas [13].

### 14.1.3 The Basic Idea of Cloud Computing in Remote Sensing

Cloud computing in remote sensing (CCRS) refers to providing on-demand, easily scalable remote sensing data storage, processing, analysis, visualization, and other services through the network to meet users' diverse needs for remote sensing applications [14]. CCRS is an emerging pattern that combines remote sensing technology with cloud computing. It has brought disruptive changes in remote sensing data acquisition, processing, analysis, and sharing. Moreover, CCRS aims to improve the management convenience, processing efficiency, and the extent of opening and sharing of remote sensing data, by applying cloud computing technologies.

According to the user's control over resources and service differences, the service models of CCRS can be divided into four categories: infrastructure as a service (IaaS), platform as a service (PaaS), software as a service (SaaS), and data as a service (DaaS) [15]. IaaS provides hardware infrastructure deployment services, such as computing, storage, network, and virtualization resources. PaaS is a running environment of cloud computing applications, providing application deployment and management services. SaaS provides the entire application stack, thereby making it accessible and usable for complete cloud applications delivered to users. DaaS provides users with browsing, retrieval, downloading, and sharing of remote sensing data and value-added products of all levels.

In recent years, as cloud computing services have transformed from resource delivery to cloud-native value empowerment, remote sensing cloud computing has also undergone a new evolution and development direction. First, remote sensing cloud computing will promote the development of more efficient data storage and management solutions, including data compression, tiered storage, and automated data cleaning. Second, the popularity of remote sensing applications will push cloud computing providers to develop services specifically for remote sensing data processing to meet users' diverse, one-stop service needs, such as data acquisition, calculation analysis, and result visualization. In the future, the focus of CCRS technology will continue to shift upward, and modern application capabilities that are loosely coupled, assembleable, and easy to operate will become the focus [16].

### 14.1.4 Cloud-Computing Service Patterns in Remote Sensing

Different from the traditional single-computer or cluster computing, CCRS, a new computing and service framework, has demonstrated a new type of service pattern in the storage and management, processing and analysis, and distribution and sharing of remote sensing data (Figure 14.1).

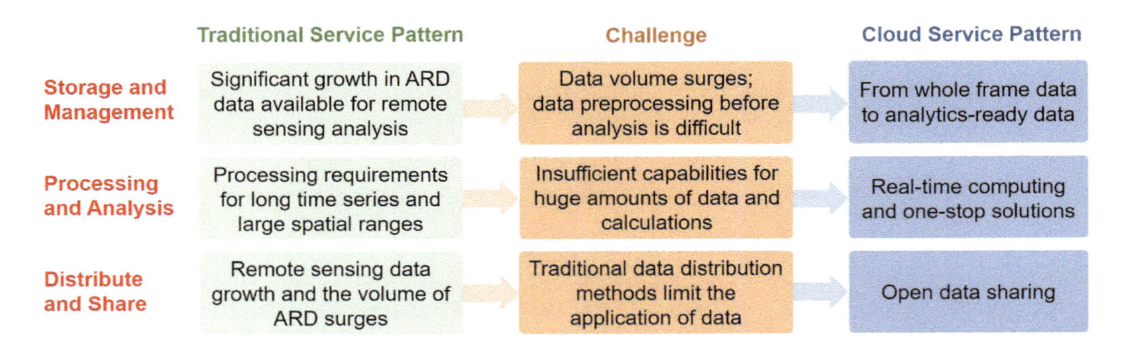


**FIGURE 14.1** Cloud computing service patterns in remote sensing.

1. *Scene File to Analysis Ready Data*: Traditional remote sensing data storage and management is often based on the satellite orbit strip or view organization pattern, with the image frame as the basic unit. Therefore, remote sensing data requires significant effort in calibration, geometric correction, and atmospheric correction before it can be utilized for analysis.

With the rising popularity of the DataCube cloud computing platform in Australia [17], the analysis ready data (ARD) pattern of data organization and management has been widely accepted, thereby marking a seminal milestone in the field of Earth observation. The concept of remote sensing ARD aims to provide preprocessed and calibrated remote sensing imageries to facilitate immediate analysis with minimal user effort [18]. In 2017, the USGS released the first Landsat analysis-ready datasets. In addition, the FORCE system [19] even produces highly analysis-ready data (hARD) to immediately fit machine learning-based high-level analysis workflows.

Summarily, the advent of remote sensing ARD greatly changed the manner in which data is harnessed, thereby enabling researchers to bypass time-consuming data preprocessing and directly focus on advanced analysis [19]. Furthermore, the convenient ease of use of analysis-ready datasets and the interoperability among them has also opened doors for many global-scale scientific applications through time series that were not possible a few decades ago. A few typical examples including Alexey's terrestrial monitoring of the United States using Landsat 4,5, and 7 ARD sets through a 30-year time series [20] and Peter Potapov's global land cover and change mapping adopting Landsat analysis-ready datasets [21].

2. *Local Customized-Processing to On-the-Fly Computing*: The conventional local customized-processing paradigm, which follows a "download-preprocess-store-analyze" cycle is significantly challenged with tremendous data volume and huge computational needs [22]. In the context of large-scale processing, it is almost infeasible for the conventional processing cycle to download all the necessary data "in place" analysis. Meanwhile, despite the high-performance computing employed, the tremendous computational demands for efficient processing of massive imageries are still beyond the capacity of traditional batch processing-based analysis infrastructure, which is characterized by limited computing capacity, network bandwith, and storage capacity. Therefore, the demand for timely and efficient processing of such vast imageries has become paramount and is further fuelling a transformative computing paradigm evolution. Fortunately, the expansion of cloud computing, characterized by elastic resource provisioning and serverless computing, has brought on-the-fly cloud processing to light and made it one of the most promising techniques to tackle these challenges. As defined by Wu Jin, on-the-fly computing [23] is a simplified cloud computing model which offers abundant rich-datatype

analysis-ready remote sensing data and significantly easier programming with delicate analysis operators to achieve on-demand processing that is on maps as an immediate response. Leveraging the scalability and computational power of the state-of-the-art cloud infrastructure to accommodate elastic serverless computing, the novel paradigm of on-the-fly computing is capable of furnishing on-demand data analysis online with instant responses. By shipping code to data that persists in the cloud storage, the on-the-fly processing could work with vast amounts of analysis-ready datasets "in place." With the trending of on-the-fly cloud computing paradigm, several typical examples spring up, such as AWS Lambda platform [24], GEE, and PIE-Engine Studio [25]. GEE is the most influential and widely esteemed instance, as it takes a breakthough step forward toward on-the-fly processing.

3. *Open Data Sharing*: The sheer volume of remote sensing data, particularly the boom of ARD, has triggered a paradigm shift in remote sensing data sharing [26]. Led by USGS's free and open data policy enacted in 2008, the European Copernicus program followed to provide free Sentinel datasets. This also spurred other agencies to globally embrace open data practices. Benefiting from open data initiatives, the humungous remote sensing image archives worldwide have opened up for free access and there has been an almost exponential increase in data use [27]. The annual data downloads from USGS in 2017 have undergone a 20-fold increase compared with those in 2009. Inevitably, the sweeping open data policy gives rise to the shifting paradigm in data sharing from a traditional data distribution manner to an open data sharing model. Open data sharing is mostly enabled by data sharing platforms that host vast collections of data to provide simplified data searching and accessing through web portals at no cost. A few typical showcases are NASA's distributed EOSDIS system that offers free data search and downloading as well as the ESA's Sentinel Hub platform that allows limited access to Sentinel data through its free plan. Overall, commercial Cloud platforms, like AWS Cloud and GEE Cloud, are leveraged to host large volumes of Landsat and Sentinel ARDs through collaborations between space agencies and commercial entities in a public-private partnership.

## 14.2 TECHNOLOGIES OF CCRS

The technology system of cloud computing in remote sensing, as illustrated in Figure 14.2, encompasses various essential components, including ubiquitous remote sensing, data integration, data organization and management, data processing, data sharing and distribution, as well as data mining, among others.

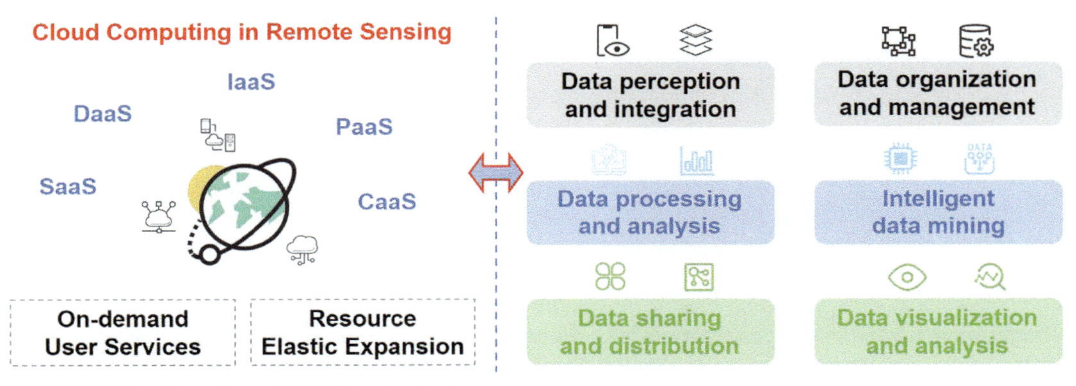


FIGURE14.2 Overview of CCRS technologies.

### 14.2.1 Cloud-Based Ubiquitous Remote Sensing and Data Integration

1. *Ubiquitous Sensing of Remote Sensing Data*: Ubiquitous sensing of remote sensing data refers to the use of sensors and communication devices that are widely present in the environment to perceive people and the environment. It is divided into two methods: active sensing and passive sensing [28]. Common active sensing methods include Earth observation sensor networks, drone sensing, Lidar, etc. Passive sensing mainly involves wired or wireless communication networks, cameras, the internet, etc. The key technologies for ubiquitous remote sensing include spatio-temporal benchmark alignment, high-precision matching of spatial information, time series analysis, spatial distribution pattern mining, machine learning, and deep learning [29].
2. *Multisource Remote Sensing Data Integration*: Multisource remote sensing data integration refers to the logical and physical organic concentration of ubiquitous sensing data from different sources, formats, and characteristics, and fully considers the attributes of remote sensing data to perform corresponding data cleaning, format conversion, and metadata extraction, etc. [30]. Cleaning of remote sensing data mainly eliminates images with thick cloud layers through built-in cloud detection algorithms in central servers or edge computing devices, such as the typical cloud detection algorithm Fmask [31]. Further, data format conversion is performed to uniformly convert remote sensing data from different sources into the internal format of the cloud computing platform to facilitate subsequent data processing and analysis. For example, in the Open DataCube (ODC) cloud computing system, remote sensing data are uniformly converted into NetCDF format to save storage space [17]. Metadata extraction uses a unified metadata model (UMM) to describe integrated multisource data and support rapid data discovery and retrieval [32]. The typical metadata repository is Common Metadata Repository (CMR) in NASA EarthData [33], which can translate various formats of metadata, such as Directory Interchange Format(DIF) [34], EOS Clearing House (ECHO) [35], Service Entry Resource Format (SERF) [36], ISO 19115–1 and ISO 19115–2 [14], to UMM [37] without requiring individual mapping between each pair of formats.

### 14.2.2 Cloud-Based Remote Sensing Data Organization and Management

The data organization and management model in the cloud computing environment can be mainly divided into two methods: scene-file-based data organization and multidimensional-index-based data cube organization.

1. *Scene-File-Based Data Organization*: Each scene is referenced to imagery files and its metadata for self description which has a corresponding maximum bounding rectangle (MRB) to define the boundary of data [38]. Conventionally, the path and row index of the orbital track when scenes are snapshotted are usually utilized by the satellite data provider for the data catalog. However, in large-scale applications, the efficient organization of substantial volumes of remote sensing imagery scenes to facilitate rapid data indexing and retrieval is always a crucial yet intricate task. To cope with this challenge, a number of solutions incorporating R-tree or B-tree indexing structures have historically been adopted to enhance the performance of the retrieval and organization of global remote sensing data, such as the Hilbert-*R*+-tree-based data organization 3 fostered by pipsCloud [10] and the logical segmentation indexing (LSI) model [39].

Within the context of scene-based file organization, the data is naturally structured into two-dimensional scene-based files, representing an acquisition-oriented approach that primarily considers spatial dimensions. However, when dealing with extensive temporal analysis involving time series, this scene-file oriented organizational method may prove to be inefficient. The process of

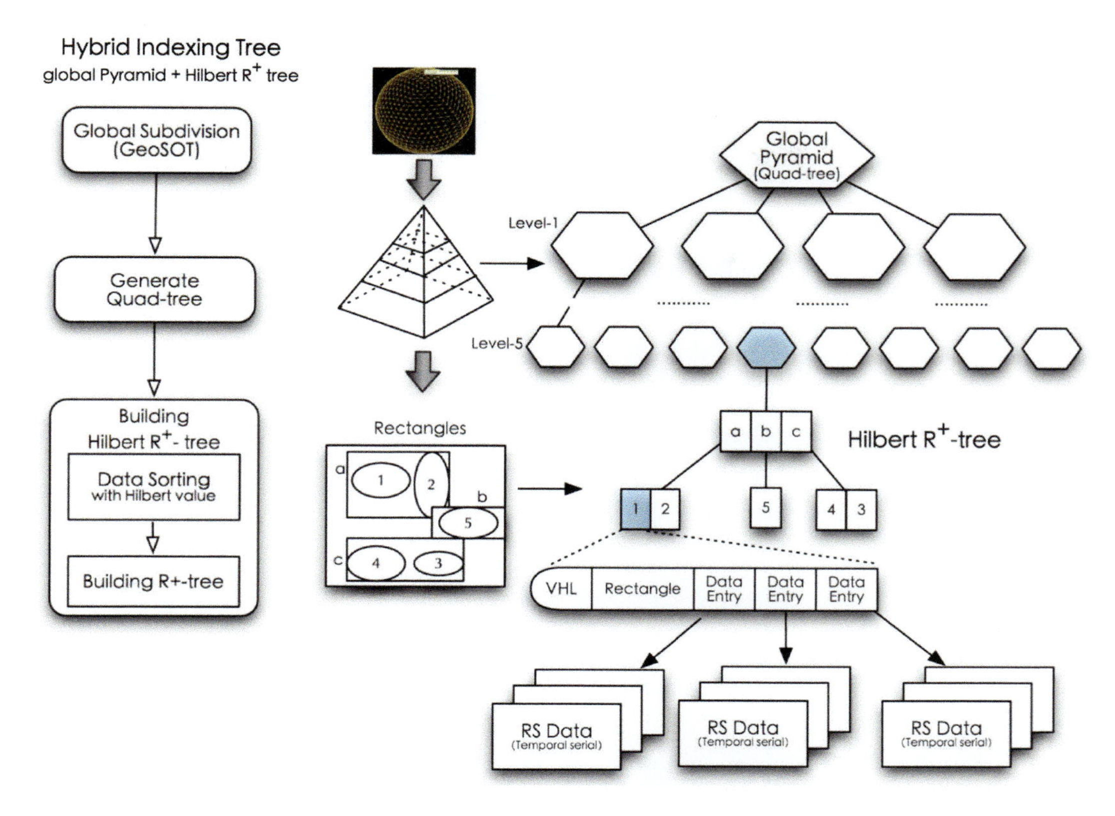


FIGURE 14.3 Hilbert-$R^+$-tree-based data organization.

accessing massive data across time series may entail time-consuming searches and the extraction of time-series data from a vast repository of scene-based files. This can result in a significant and frequent pattern of accessing temporal data across multiple scene files. Such data access patterns can lead to a substantial increase in input/output (I/O) operations and an overall reduction in processing efficiency. Consequently, the conventional scene-file-based data organization hinders the comprehensive analysis of large-scale Earth observation data.

2. *Multidimensional-Index-Based Data Cube Organization*: The Data Cube Organization, as is depicted in Figure 14.4, typically organizes remote sensing ARD as massive gridded multidimensional data with indexing along different dimensions or facts. It spatially partitions analysis-ready datasets into regular and nonoverlapping tiles of gridded data through the data ingestion process [17]. According to the timestamp of the data capture, these gridded data tiles are stacked with a visual of a "cube" and then packaged into multidimensional data files (e.g., netCDF files) or stored in array databases. The extra index information is also generated through data ingestion and added into the data cube database (e.g., Postgresql, HBase [40]) for direct data operation on these native multidimensional data along different facts. Following this method, groups of data cubes could be then organized into user-tailored data collections for analysis for different facets, such as satellites and thematic information.

In practical terms, the utilization of N-dimensional array data structures and associated interfaces serves to enhance the accessibility of multidimensional data within data cubes, as exemplified by frameworks like Xarray. In response to the burgeoning popularity of data cubes, multiple

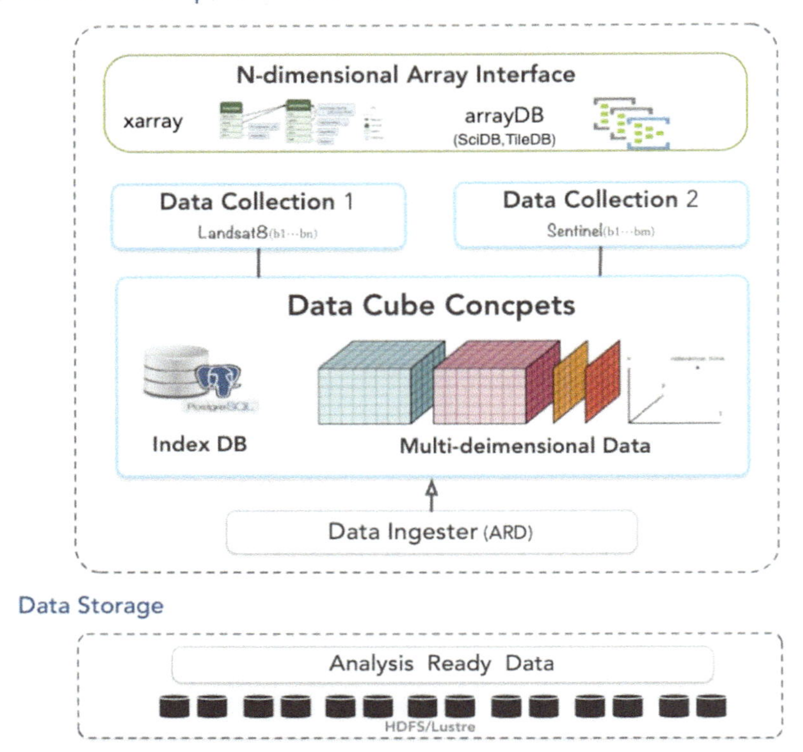


**FIGURE 14.4** Data cube concept and infrastructure.

canonical infrastructure implementations designed to embody the concept of data cubes have emerged. Prominent among these implementations are platforms such as ODC [17], EarthServer [41], SciDB [42], and GEE.

Compared with traditional scene-file-based data archives in which data are flattened and organized as messes of files with ad hoc cumbersome metadata encoding in file names or directories, the data cube approach tackles big data challenges by facilitating simplified intuitive data accessing stacks of spatially aligned pixels as regular multidimensional arrays for temporal or even multisensor analyses as well as straightforward scalability of parallelism over smaller sub-cubes for better performance. However, it should be pointed out that the significant work of building data cubes through data retiling during ingestion is not trivial. Moreover, the various initiatives of data cube concepts with diverse technical solutions have also caused difficulites in interoperability [43].

### 14.2.3 Cloud-Based Remote Sensing Data Processing

1. *High Performance Computing in the Cloud Center*: High performance computing (HPC) in the cloud center adopts a distributed computing model, in which components of a software system are shared among multiple computers or nodes. Even though the software components may be spread out across multiple computers in multiple locations, they are run as a single system [44].

In the context of distributed computing technologies, it generally encompasses two types: data parallel computing and model parallel computing. Data parallelism involves dividing the data into multiple parts, with each computing node processing a portion of the data. Sun et al. [45] applied data parallelism to train deep neural networks with extremely large batches. Model parallelism divides the deep learning model into multiple sub-models that are concurrently processed on different computing nodes [46]. Figure 14.5 (b) illustrates model parallelism, often employed when dealing with large-scale models that cannot be accommodated within a single computing node. In such cases, the model is fragmented and computations are distributed across multiple nodes. For example, Hong et al. [47] adopted model parallelism in training Transformer-structured large models for hyperspectral classification.

Further, Message Passing Interface (MPI) [48], Apache Hadoop MapReduce [49], and Spark [50] stand out as renowned distributed/parallel computing frameworks extensively employed in HPC in remote sensing cloud data centers. MPI, Apache Hadoop MapReduce, and Spark all decompose large-scale remote sensing data processing tasks into numerous small subtasks, which can be executed in parallel on different processors or servers, thereby greatly reducing the time required to complete complex computing tasks. What needs to be added is that Spark achieves high-performance in-memory computing by transforming the dataset involved in computations into a distributed resilient distributed dataset (RDD) and employing directed acyclic graphs (DAG) for job scheduling. Presently, the integration of the Spark platform with remote sensing big data is becoming increasingly integral. This integration has given rise to a series of distributed computing engines tailored for remote sensing data processing and analysis, including GeoTrellis, Geomesa, Sedona [51], etc. These engines empower developers to effortlessly manage large-scale remote sensing data within Spark cluster systems.

2. *Edge Computing*: Edge computing is a distributed computing architecture that can move applications, data, and computing services from network central nodes to edge nodes for processing. This computing mode is initiated on the edge devices to produce faster network service responses and meet the basic needs in various fields, including real-time response, intelligent applications, security and privacy protection [52]. The concept and technical system of CCRS are constantly evolving, among which "cloud-edge-device" collaborative computing is a prominent development direction, which has had a profound impact on CCRS. A typical edge computing case is depicted in Figure 14.6.

The application of edge computing in CCRS has gradually become a trend. Based on the ground center cloud, space-based edge devices, satellite sensors, and user devices, distributed collaborative tasks (such as sensing, computing, navigation, positioning and communication, etc.) can be effectively completed. Under the space-based "cloud-edge-device" intelligent service architecture, it can provide spatio-temporal continuous information services for typical applications and situation awareness scenarios such as disaster emergency rescue, public health or security events, smart cities, and personalized services [53]. For example, in the scenario of satellite Earth observation, the low-level processing for autonomous target detection and reasoning based on remote sensing data can be transferred to edge devices with reasoning capabilities to achieve cloud-edge collaborative applications [54]. In order to achieve real-time intelligent interpretation on orbiting satellites, Zhang et al. proposed a distributed device-cloud collaborative network model for remote sensing image classification [55]. In addition, edge computing can also provide efficient and reliable new data processing solutions for multiple agricultural scenarios and complex tasks. Moreover, edge computing is combined with deep learning and applied in many aspects, such as crop disease monitoring, growth environment monitoring, and unmanned farm management.

The core advantage of edge intelligent remote sensing is to efficiently solve complex computing tsks that cannot be solved in the traditional remote sensing data processing pattern through the integration of "cloud-edge-device" collaborative technology and artificial intelligence, thereby reducing the data communication between space-based remote sensing payloads and ground stations, and enhancing system security and reliability [56]. In short, edge computing will have a profound impact on the development and application of remote sensing cloud computing, particularly

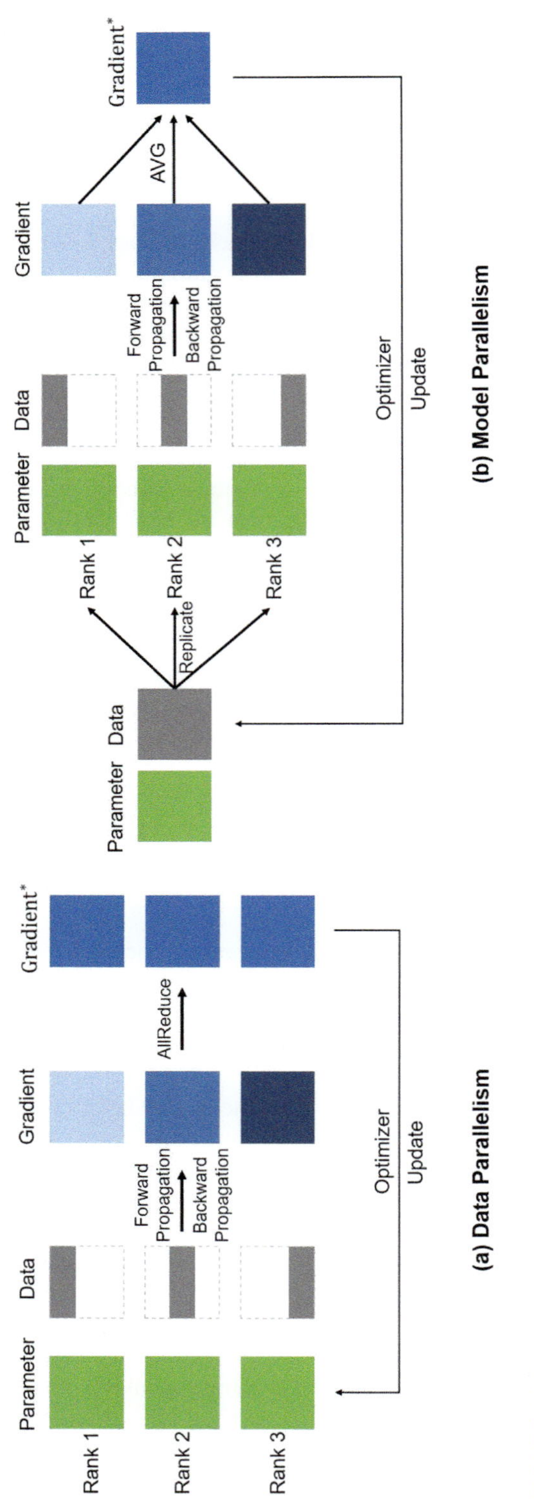


FIGURE 14.5 Two common models of HPC in cloud centers.

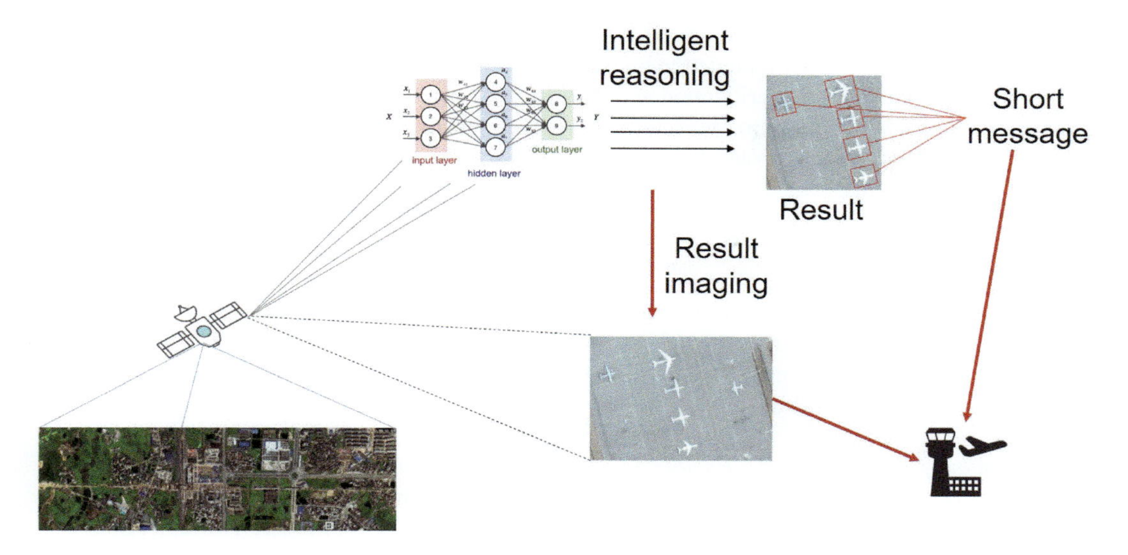


**FIGURE 14.6** Schematic diagram of edge computing in CCRS.

in terms of achieving fast response with high real-time performance, thereby effectively reducing transmission and storage requirements and improving privacy protection and data security [57]. In the future, with the leapfrog development of the space-based Internet of Things (IoT), users will be provided with fast, accurate, safe, and intelligent services as well as provide solid technical support for the construction of intelligent space-based systems.

### 14.2.4 Cloud-Based Remote Sensing Data Sharing and Distribution

In cloud-based remote sensing platforms, remote sensing data and information products are generally stored in file systems. The query and retrieval capabilities are provided through metadata technology. Among these, large remote sensing scientific data centers have developed standard rules for dataset naming to ensure consistency and uniqueness of dataset names, and utilize a relatively stable standardized directory structure to store data for each dataset. With the advancement of cloud computing technologies, the sharing and information services provided by remote sensing data is gradually shifting to cloud-based storage, transfer, and sharing.

In 2023, European Space Agency (ESA) and NASA performed a successful large-scale data transfer using a cloud infrastructure to transfer Sentinel-2 data between the two agencies [58]. ESA was able to provide access to NASA for the S2L1C archive between November 28, 2015, and September 30, 2020. This archive has 10,916,055 products, with a total size transfer of 5.7 PB. These products are divided into 89 buckets. NASA initiated this archive to be transferred from OVHCloud-hosted cold storage (OpenStack Swift based) to NASA-controlled AWS S3 storage in the U.S. West region. The final solution is based on open source Apache Airflow and Apache Airavata MFT. The transfer was optimized to the network bandwidth. NASA ran 20 parallel agents to pull the data at a rate of 20 Gbps. The 5.7 PB archive was transferred in 27 calendar days [58].

In additon, authenticity and non-repudiation are two core concepts in information security regarding the integrity and legitimacy of data transmission [59]. Because data is managed and transferred every day, it is important to verify the sender's origin (authentication) and ensure that during transmission or any data duplication, the data was not intercepted or altered in any manner (integrity). When you have both authenticity and integrity, the legitimacy of the data cannot be denied;

therefore, all parties can be confident in their data security (nonrepudiation). Approaches to manage data replicas on cloud mainly include the human and technical approaches. In the human approach, data filename inspection and verification and validation by the authoritative source, may be the two effective methods. With regard to the technical approach, persistent identifiers, Hash code, cryptography, digital signature (e.g., Watermark) and blockchain mechanism (e.g., KSI Blockchain) [60], can effectively grant the data integrity and authenticity.

### 14.2.5 Cloud-Based Remote Sensing Data Mining and Analysis

Cloud-based remote sensing data mining methods are in a continuous phase of evolution and expansion to adapt to the increasingly complex and diverse data and task requirements. These methods can be categorized into machine-learning-based, deep-learning-based, and large-model-based remote sensing data mining and analysis.

1. *Machine-Learning-Based Remote Sensing Data Mining and Analysis*: Machine learning has been successfully applied across various remote sensing applications, including agriculture, forestry, urban planning, environmental protection, and more. Through feature engineering, it is possible to extract characteristics such as pixel values, textures, shapes, and more from remote sensing images, followed by the application of various machine learning algorithms for tasks encompassing classification, regression, or clustering. For example, the GEE platform integrates various machine learning data mining techniques for different task scenarios [12], such as random forests, SVMs, etc. Relying on machine learning technologies, task objectives are achieved through data mining and feature extraction, thereby providing a more professional and efficient approach for the processing and analysis of remote sensing data. This further propels research and applications in the field of remote sensing. In-depth research into knowledge discovery methods based on machine learning allows us to categorize these methods into three typical types, based on their developmental stages: (1) rule-based methods, (2) data-driven methods, and (3) integrated methods [61], as depicted in Table 14.1.

**TABLE 14.1**
**Machine Learning-Based Knowledge Discovery Methods and Their Commonly Used Algorithms**

| Category | Common Algorithms | References |
|---|---|---|
| **Rule-based approaches** | Expert system | Matsuyama et al. [62] |
| | Decision tree | Min et al. [63]; Ye et al. [64] |
| | Association rule | Han et al. [65] |
| **Data-driven approaches** | K-NN | Kotsiantis et al. [66]; Jiang et al. [67] |
| | DTW | Romani et al. [68]; Guan et al. [69] |
| | Naïve Bayes classifier | Suthaharan et al. [70]; Singha et al. [71] |
| | Bayesian networks | Yan et al. [72]; Liao et al. [73] |
| | SVM and SVR | Wale et al. [74] |
| | Heuristic search | Morgan et al. [75]; Georganos et al. [76] |
| | Genetic algorithm | Tso et al. [77] |
| **Ensemble methods** | Random forest | Babbar et al. [78]; Yan et al. [79] |
| | AdaBoost | Dou et al. [80]; Liu et al. [81] |

**TABLE 14.2**
**Deep Learning-Based Knowledge Discovery Methods and Their Commonly Used Algorithms**

| Category | Common Algorithms | References |
|---|---|---|
| **Convolutional neural network** | VGG | Chaib et al. [82] |
| | FCN | Fu et al. [83] |
| | UNet | Wang et al. [84] |
| | TCN | Yan et al. [85] |
| | Fast-RCNN | Yao et al. [86] |
| **Recurrent neural network** | LSTM | Crisostomo et al. [87] |
| | GRU | Zhao et al. [88] |
| **Graph neural network** | GCN | Zhang et al. [89] |
| | GAT | Sha et al. [90] |
| **Transformer-based** | Vit | Deng et al. [91] |
| | ChangeFormer | Bandara et al. [92] |
| | Segformer | Tang et al. [93] |
| **Deep reinforcement learning** | Deep Q-Learning | Franois et al. [94]; Cheng et al. [95] |
| | Deep Q-Network | Fu et al. [96] |

**TABLE 14.3**
**Application Examples of Remote Sensing Data Mining and Analysis Based on Large Models**

| Task | Description | References |
|---|---|---|
| **Remote sensing image captioning** | Utilize machine comprehension to interpret the content of remote sensing images and describe it in natural language. | Shi et al. [97]; Lu et al. [98]; Wei et al. [99] |
| **Text-based remote sensing image generation** | Text-based image generation combines natural language processing and computer vision to create realistic images from textual descriptions. | Bejiga et al. [100]; Xu et al. [101] |
| **Text-based remote sensing image retrieval** | The primary objective of image retrieval is to extract specific images from large datasets. The fundamental idea is to narrow down the search scope for the target image and retrieve images that match a particular query. | Yuan et al. [102]; Yuan et al. [103]; Rahhal et al. [104] |
| **Remote sensing visual question answering, RSVQA** | RSVQA systems enable non-expert users to interact with remote sensing images using natural language questions as queries, facilitating user-friendly and advanced understanding of the images. | Chappuis et al. [105]; Yuan et al. [106] |
| **Remote sensing visual grounding** | By employing natural language guidance to locate objects in remote sensing scenarios, remote sensing visual localization offers object-level comprehension and convenience for end users. | Zhan et al. [107] |
| **Zero-Shot remote sensing scene classification** | The objective of zero-shot remote sensing scene classification is to identify pre-viously unseen scene concepts by referencing the semantic relationships between visual features and semantic categories. | Sumbul et al. [108]; Li et al.[109] |
| **Few-Shot remote sensing object detection** | Few-shot remote sensing object detection aims to detect points of interest in remote sensing images using only a limited number of annotated samples. | Kim et al. [110]; Lu et al. [111] |

*(Continued)*

**TABLE 14.3** *(Continued)*
**Application Examples of Remote Sensing Data Mining and Analysis Based on Large Models**

| Task | Description | References |
|---|---|---|
| **Scene-level remote sensing data processing and analysis using the PanGu large model** | The PanGu large model can be widely applied in agriculture, urban planning, environmental monitoring, and more. For instance, the PanGu meteorological model can be directly used in meteorological scenarios like typhoons, providing accurate predictions for variables such as geopotential, wind speed, and temperature. | Bi et al. [112] |

2. *Deep-Learning-Based Remote Sensing Data Mining and Analysis*: Deep-learning-based data mining and analysis, compared to traditional machine learning methods, can fully leverage the advantages of remote sensing big data. By employing automatic feature extraction, they effectively mitigate uncertainties within vast datasets and comprehensively reveal concealed relationships among various types of big data. This, in turn, facilitates a comprehensive analysis of Earth science phenomena and developmental patterns, thereby enabling a detailed examination of existing information and precise forecasting of future trends. Leveraging large-scale time series data from satellite, remote sensing, meteorological, and other Earth science data sources along with high-performance computing, large-scale data storage, and distributed computing, individuals can rapidly engage in various Earth science research endeavors. High-performance remote sensing cloud computing platforms, exemplified by GEE, PIE-Engine Studio, and SDG Big Data Platform (https://sdg.casearth.cn/en), empower users to effortlessly access and utilize massive remote sensing data stored on the cloud. This facilitates large-scale land cover classification, change detection, and other data mining research based on deep learning models [113]. Through an in-depth examination of contemporary knowledge discovery methods based on deep learning, we categorize them into several typical types based on model architectures and applications: (1) convolutional neural network, (2) recurrent neural network, (3) graph neural networks, (4) transformer-based approaches, and (5) deep reinforcement learning, as illustrated in Table 14.2.
3. *Large-Model-Based Remote Sensing Data Mining and Analysis*: With the emergence of vision-based foundational models utilizing Transformer networks (e.g., CLIP [114], Florence [115]) and large language models (e.g., GPT-3 [116], OPT [117]), showcasing zero-shot capabilities and generalization in visual and language understanding tasks, there has been substantial scholarly attention directed towards pretrained foundational models. Owing to their extensive parameter count, these pre-trained foundational models are often referred to as "large-scale remote sensing models." In the remote sensing domain, many researchers have embarked on the path of increasing model parameters and optimizing these models using large volumes of data. Remote sensing pretrained foundational models have expanded from models like JointSAREO with 20 million parameters [118] to the Scale-MAE model with 300 million parameters [119]. The volume of training data has also grown from hundreds of thousands to millions. This expansion has resulted in substantial improvements in accuracy across various downstream remote sensing visual tasks.

Amidst the trend of large models, in 2023, PIESAT Information Technology Co. Ltd. introduced the "Tianquan" visual large model [120], which is aimed at multimodel remote sensing data, addressing the limitations of sample annotation and model generalization in the existing "AI + remote sensing" business model. In the same year, SenseTime Technologies developed a remote sensing large model with a scale of 350 million, introducing the SenseEarth platform [121], endowed with large-scale artificial intelligence (AI) processing capabilities, including fully automated road extraction based on satellite images, ship detection, land use classification, and change detection, among other artificial intelligence interpretation functions. Alibaba Damo Academy developed the universal segmentation interpretation model AIE-SEG for AI Earth [122], intelligently interpreting remote sensing image data and supporting single-object extraction, comprehensive surface element extraction, and two-phase image change detection among other artificial intelligence interpretation functions, thereby achieving rapid extraction in the remote sensing field without the need for samples.

In recent years, several early endeavors have sought to explore the use of vision-language models and remote sensing large models for various remote sensing data analysis tasks, including remote sensing image captioning, text-based remote sensing image generation, text-based remote sensing image retrieval, visual question-answering, scene classification, semantic segmentation, object detection, and more. For specific details, refer to Table 14.3.

## 14.3 CLOUD-COMPUTING-BASED REMOTE SENSING SYSTEMS

Relying on powerful physical or virtual storage, computing and network resources, as well as integrating data sensing and integration, data organization and management, high-performance processing, data sharing and distribution, data mining and distribution technologies, cloud-computing-based remote sensing system provides users with on-demand, elastic, and scalable DaaS, SaaS, PaaS, and IaaS services.

According to the main service types provided by each cloud systems, they can be divided into DaaS-oriented CCRS system, SaaS-oriented CCRS system, PaaS-oriented CCRS system, and IaaS-orientedCCRS system.

### 14.3.1 The DaaS-Oriented CCRS System

The DaaS-oriented CCRS system is the earliest and most widely used remote sensing cloud service model. It is usually built on multiple satellite data centers and relies on the data storage and computing capabilities offered by distributed computing clusters to provide users with massive data distribution and sharing services. Typical representatives include NASA EarthData, ESA Sentinel-Hub, EarthServer, AI Earth, etc.

1. *NASA EarthData*: With the launch of new, high-data-volume Earth observation missions, NASA's ability to effectively ingest, process, and archive large amounts of data requires the most cost-effective, flexible, and scalable data-management architectures and technologies. EOSDIS provides end-to-end capabilities for managing NASA Earth science data from satellites, aircraft, in situ measurements, and other sources. The migration of EOSDIS data into the Earthdata Cloud benefits users by providing them with new ways to access NASA's collection of Earth science datasets, improving the efficiency of data systems operations, increasing user autonomy, maximizing flexibility, and offering shared services and controls. And the EarthData Cloud is a key component of the ESDS Transform to Open Science (TOPS) mission, which provides the visibility, advocacy, and community resources to support and enable the shift to open science. The Earthdata Cloud architecture became operational in July 2019 and key EOSDIS services, such as NASA's Common Metadata Repository (CMR) and EarthData Search, were deployed in it. Currently, the platform provides over 59 PB of data sharing and is expected to archive over 320 PB of data by 2030.

2. *ESA Sentinel Hub*: Sentinel Hub is a data service platform developed by the ESA for the application of remote sensing data. Its primary mission is to support users in easily accessing, processing, analyzing, and visualizing Earth observation data from satellites such as Sentinel, among others [123]. It is used to facilitate various applications, including scientific research, big data analysis, and mapping services.

The ESA Sentinel Hub platform primarily provides data storage, data processing, and data distribution platforms. This platform utilizes a highly optimized cloud computing architecture, thereby enabling the rapid acquisition of large-scale remote sensing data and exhibiting strong overall performance. Currently, the platform is applied in diverse fields, including land cover and land use monitoring, meteorology and climate research, disaster monitoring and emergency response, ocean and coastline monitoring, natural ecosystems, and agricultural management.

3. *EarthServer*: The EarthServer project is coordinated by Jacobs University of Bremen, and in partial cooperation with European research centers, NASA, and several private companies under the support of the European Community's Seventh Framework Programme from 2011 to 2014 [124]. The goal of EarthServer is to provide researchers with a platform supporting open access and ad hoc analytics of analysis-ready remote sensing data based on the browser-server architecture. EarthServer comprises the front end and the back end from the perspective of system architecture. The back end mainly includes a Rasdaman, an array DBMS used for the storage of analysis-ready remote sensing data. The front end includes two main components, a browser-based web interface, and data services. The browser-based web interface is designed to provide the ability of data discovery, data accessing, data processing, etc, which are built on top of data services. Moreover, the communications between the front end and the back end are realized based on standardized services such as HTTP, WCS, WCPS, etc.

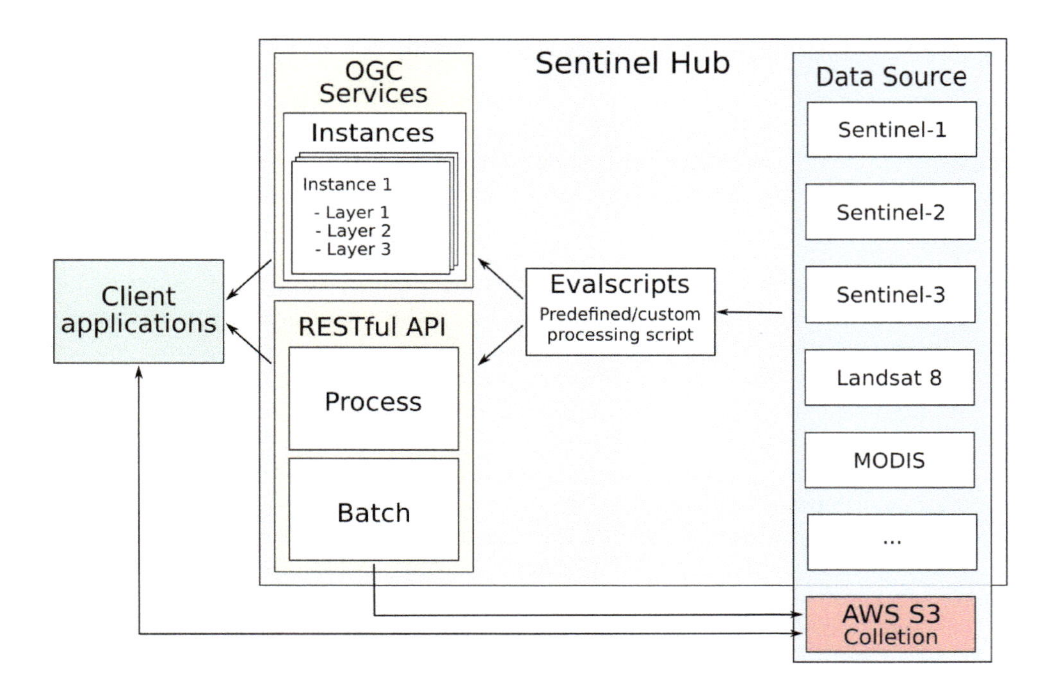


**FIGURE 14.7** Sentinel Hub platform system architecture diagram.

### 14.3.2 The SaaS-Oriented CCRS System

The SaaS-oriented CCRS system, mainly relies on the powerful computing and storage resources of the cloud center and uses software or tool sets provided by system developers or shared by users for remote sensing data processing and analysis, to provide users with online data processing services over the internet on a pay-as-you-go basis. Normally, consumers access software using a thin client via a web browser. Typical representatives include NASA Earth Exchange (NEX), pipsCloud, Huawei Pangu Large Model, ENVI Services Engine, etc.

1. *NASA Earth Exchange*: The NASA Earth Exchange (NEX) is an open, integrated, and centralized research platform designed by NASA, which aims to provide researchers with the ability to explore and analyze satellite data archived in NASA's physically distributed data centers by combining NASA's powerful data resources, software resources and high-performance computing resources [125].

NEX is developed in a distributed manner, satellite data archived in multiple data centers can be discovered and accessed by researchers via NEX. In addition, NEX also provides a high-performance solution to process satellite data in a distributed manner without data transmission between different data centers. Researchers can design their processing procedures with the help of tools provided by NEX to satisfy the requirements of their applications. NEX then executes the defined processing procedures on the underlying high-performance clusters.

2. *pipsCloud*: pipsCloud is a cloud-enabled high-performance remote sensing data processing system for large-scale remote sensing applications [10]. It is built on a high-performance virtual cluster generated by the OpenStack cloud computing platform and helps actualize the integration and management [14] as well as online processing and product production [126] of massive remote sensing data through remote sensing data organization and indexing [40], parallel processing [72], and other technical means (Figure 14.8). pipsCloud integrates over 100 common remote sensing data processing and product production algorithms, combines China Centre for Resources Satellite Data and Application (CRESDA), National Ocean Satellite Application Center (NSOAS), National Satellite Meteorological Center (NSMC), Computer Network Information Center (CNIC), Twenty First Century Aerospace Technology Co. Ltd (21AT), and Institute of Remote Sensing and Digital Earth (RADI), and realizes the production of over 40 remote sensing common products [127].
3. *Huawei Pangu Large Model*: Released on July 7, 2023, Huawei Pangu large model focuses on the basic tasks of natural language processing, computer vision, and scientific computing, encompassing text analysis, image processing, numerical simulation, model training, and decision making and evaluation. It has been widely used in the fields of government affairs, finance, manufacturing, pharmaceuticals, mines, railroads, and meteorology.

The release of the Pangu large model will have a multifaceted impact on the field of remote sensing cloud computing. First, its powerful processing capability will enhance the accuracy and efficiency of the analysis and processing of cloud-based remote sensing data. Second, in terms of automation and intelligence, the Pangu large model is expected to reduce the need for manual intervention and accelerate remote sensing data analysis. Furthermore, the support of the Pangu large model for multimodel remote sensing data will fully utilize the application value of remote sensing big data and broaden the application scope.

Finally, with the integration of the Pangu large model and remote sensing cloud technology, more application areas will be expanded, including agriculture, urban planning, and environmental monitoring.

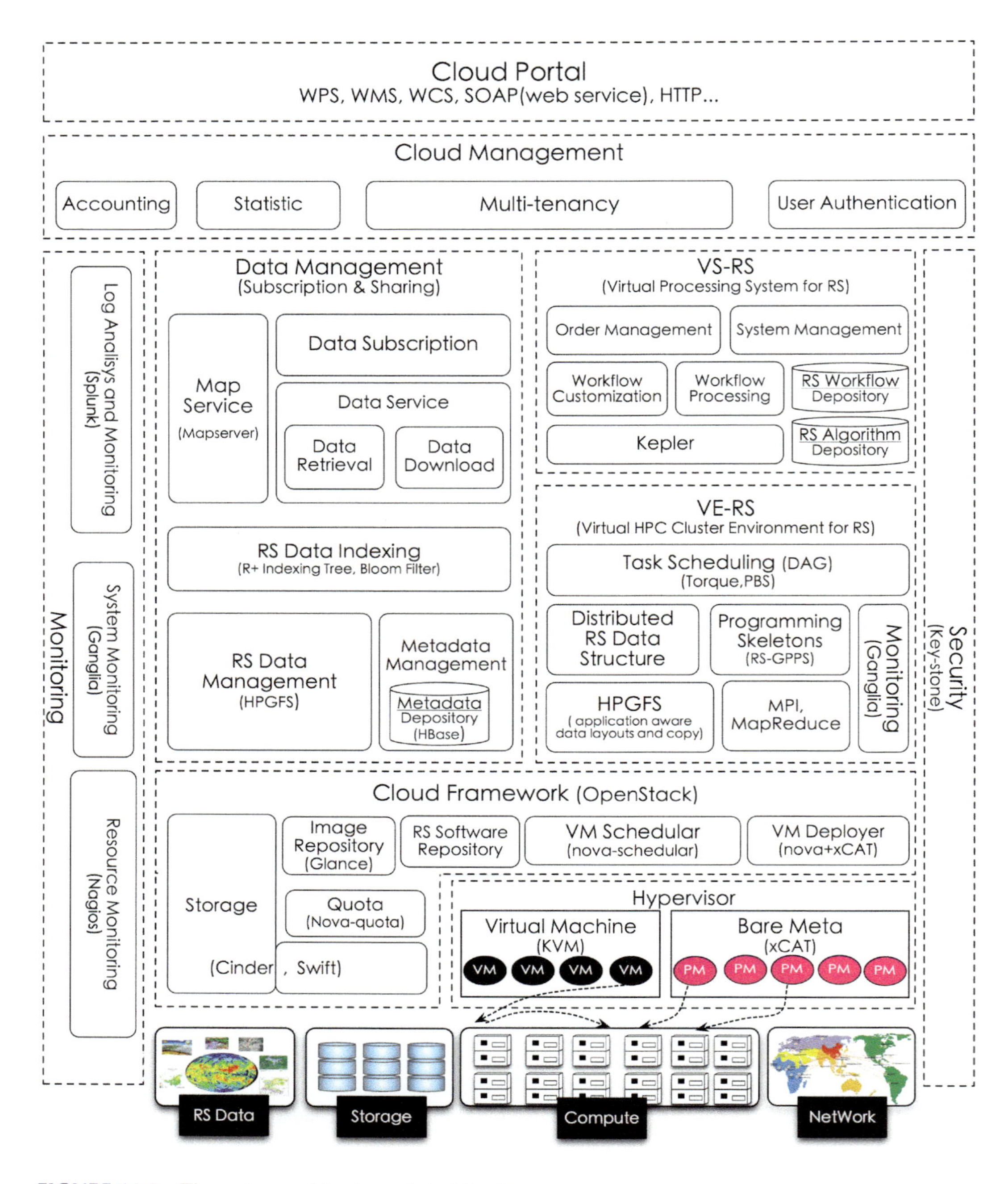


FIGURE 14.8 The system architecture of pipsCloud.

### 14.3.3 The PaaS-Oriented CCRS System

The PaaS-oriented CCRS system, mainly relies on the powerful computing resources of the cloud center and uses tool sets or function libraries provided by system developers or shared by users for remote sensing data processing and analysis in order to provide users with online programming processing services. Consumers adopt PaaS to create and deploy applications without considering the expense and complexity of buying and managing Integrated Development Environment

(IDE) licenses, the underlying application infrastructure and middleware, or development tools and other resources. Typical representatives include GEE, PIE-Engine Studio, Open Geospatial Engine (OGE), Open Data Cube (ODC), AI Earth, etc.

1. *GEE*: The GEE is a remote sensing cloud computing platform introduced by Google in 2010 [128]. It was designed for the storage and analysis of remote sensing data, including satellite imagery, meteorological data, geospatial information, and more. It provides a convenient means to access, process, and analyze extensive Earth observation data, thereby providing solutions to Earth-related environmental challenges. GEE leverages Google's core infrastructure and cutting-edge technologies like artificial intelligence to enable online computation and analysis of global-scale Earth science data, thereby ensuring efficient and secure cloud services for users (Figure 14.9) [129].

The GEE platform integrates massive geospatial data and allows users to upload various data types, including vector and raster data. Further, the platform combines a catalog of petabytes of geospatial data with global-scale analytical capabilities. With its data visualization and computational analysis capabilities, it can address tasks such as remote sensing data processing, feature extraction, image classification, and time series analysis. Moreover, the platform provides callable application programming interfaces (APIs) that enable users to develop algorithms online using JavaScript API, generate specialized data products, or deploy interactive applications supported by GEE platform resources [130].

GEE enables access to vast satellite imagery and other Earth observation data worldwide, thereby offering sufficient computational power for data processing. It particularly excels in long-term time series analysis of remote sensing big data, particularly in areas such as change detection, trend mapping, and quantifying differences on the Earth's surface.

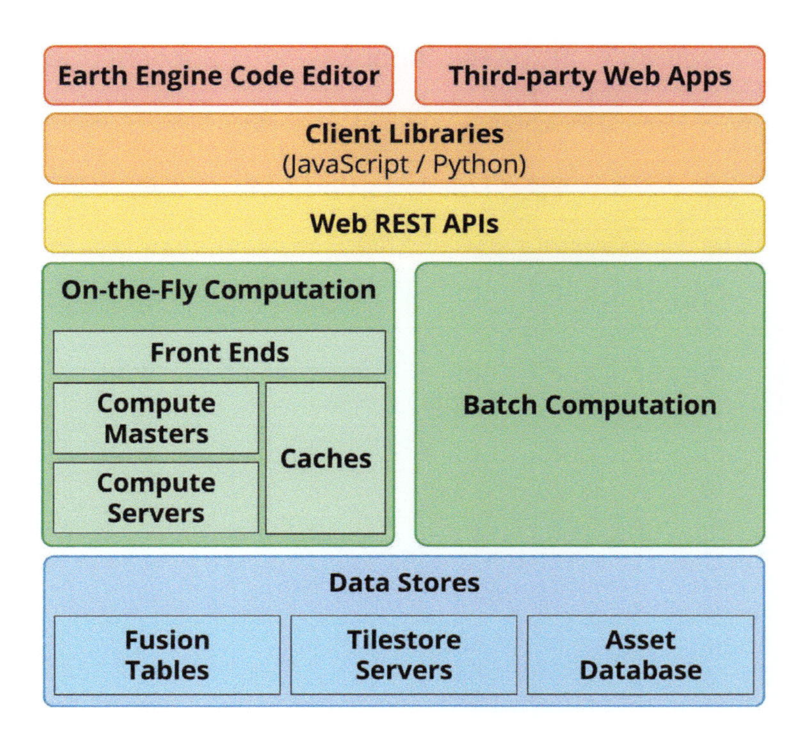


FIGURE 14.9 Google Earth Engine platform system architecture diagram.

2. *PIE-Engine Studio*: PIE-Engine Studio is a spatiotemporal remote sensing cloud computing platform developed by Space Macrograph Company. It is based on container cloud technology and is tailored to the Earth science domain. This platform seamlessly integrates a vast repository of geospatial data, interactive programming analytics, real-time distributed computing, and data visualization capabilities, thereby providing users with a comprehensive solution for the processing, analysis, visualization, and result export of multiscale remote sensing data and products [25]. The platform encompasses a range of technologies, including multisource remote sensing data processing, distributed resource scheduling, real-time computation, batch computation, and deep learning frameworks.

Currently, the remote sensing cloud platform offers access to over 160 public datasets, with a total data volume exceeding 7 petabytes and comprising over ten million scene images. Users can seamlessly import and analyze these data in the platform's code editor, thereby supporting both research and industrial applications. The system architecture of PIE-Engine Studio, depicted in Figure 14.10, is underpinned by bottom-layer data services and operator libraries. The platform is capable of delivering distributed data computing services, encompassing real-time and asynchronous computation, and also offers JavaScript and Python versions of APIs. This enables users to edit code online, thereby facilitating the processing and analysis of multisource remote sensing imagery and vector data. Currently, the platform has found applications in various Earth science domains, including surface change monitoring, agricultural resource management, natural disaster monitoring, urban planning, and more [131].

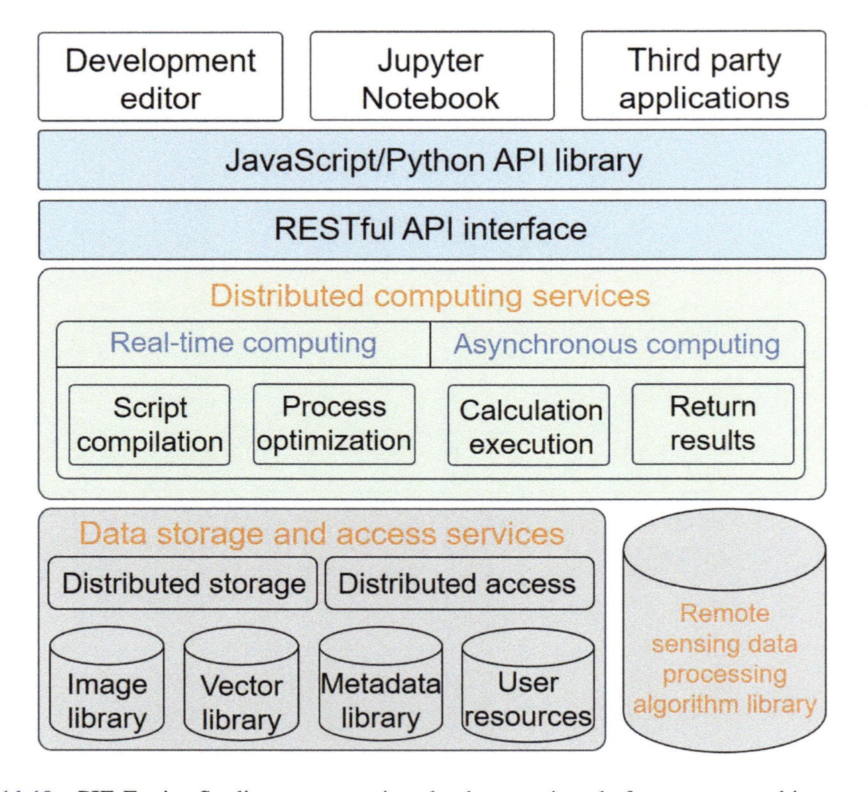


**FIGURE 14.10** PIE-Engine Studio remote sensing cloud computing platform system architecture.

3. *OGE*: OGE is an open Earth engine service platform developed by Wuhan University. It is committed to establishing a digital Earth engine system with deep coupling and open sharing of computing, data, and algorithms. This platform aims at the national strategic needs of "Digital China." It studies the organization and management methods of global spatiotemporal observation data, builds high-performance geographical analysis and AI analysis models, and serves the construction of digital twin cities in various provinces and cities with autonomous and controllable Twin Earth engines. It also provides an Earth spatio-temporal information base for multiple applications oriented to remote sensing data processing. Furthermore, the OGE platform is used to process and analyze Earth science and geographic information data. It can perform various Earth data operations, including data processing, spatial analysis, map drawing, model building, etc, to meet the needs of the geographic information field. The platform mainly has the following functions: data processing and format conversion; spatial analysis, including spatial query, map overlay, and buffer analysis; map generation and data visualization; and provision of open interfaces and plug-in support. The platform relies on remote sensing intelligent cloud computing services to serve scientific researchers in the field of geosciences on the one hand and to assist the construction of digital twin cities and smart Earth systems on the other.
4. *ODC*: ODC is an analytical framework consisting of a set of data structures and tools dedicated to organizing and analyzing massive amounts of Earth Observation (EO) data and accessing and manipulating them through a set of command-line tools and Python APIs.

Figure 14.11 demonstrates the architecture of the ODC. Data acquisition and inflow represents the process of acquisition and preparation of EO data before indexing by the ODC. The Data Cube Infrastructure illustrates the main core of the ODC, where EO data are indexed, stored, and delivered to the user through Python API. Moreover, the data and application platforms comprise auxiliary application modules, such as job management and authentication [132].

For data access, ODC provides implementations of OGC web services such as WCS, WMS, and WMTS. In addition, ODC provides tools to facilitate ODC deployment through containers. Among these are an implementation of the ODC Python API exposed through a REST API, an application for extracting statistics from indexed data, and an implementation of a web portal [133].

In this web portal, it is possible to discover available products, select and retrieve datasets (GeoTIFF and NetCDF), and perform analyses through the available applications. Currently, as the most typical example of ODC, AGDC uses a cluster of computers with the Lustre distributed file system to store over 300,000 Landsat images covering Australia.

5. *AI Earth*: AI Earth [122] is an Earth science cloud platform developed by the Alibaba DAMO Academy team. It focuses on innovative research on intelligent computational analysis in Earth science, with the mission of solving fundamental, cutting-edge, and operational problems in the field of Earth science computing. It is built on Alibaba Cloud and can provide high-performance computing capabilities and storage services on the cloud. This platform can provide a low-threshold, interfaced cloud GIS workspace for the online processing of multisource Earth observation data. AI Earth platform can integrate PB-level open-source satellite remote sensing data (covering the industry's most mainstream Landsat 8, Landsat 9, Sentinel-1, and Sentinel-2) and more than ten kinds of remote sensing AI algorithms (involving terrain classification, change detection, and SAR water extraction). With those data and algorithms, AI Earth can provide efficient data storage and management capabilities to accommodate large-scale geoscience data. Simultaneously, AI Earth is equipped with powerful AI algorithms for automated data processing, feature extraction, model training, and result analysis. AI Earth also provides

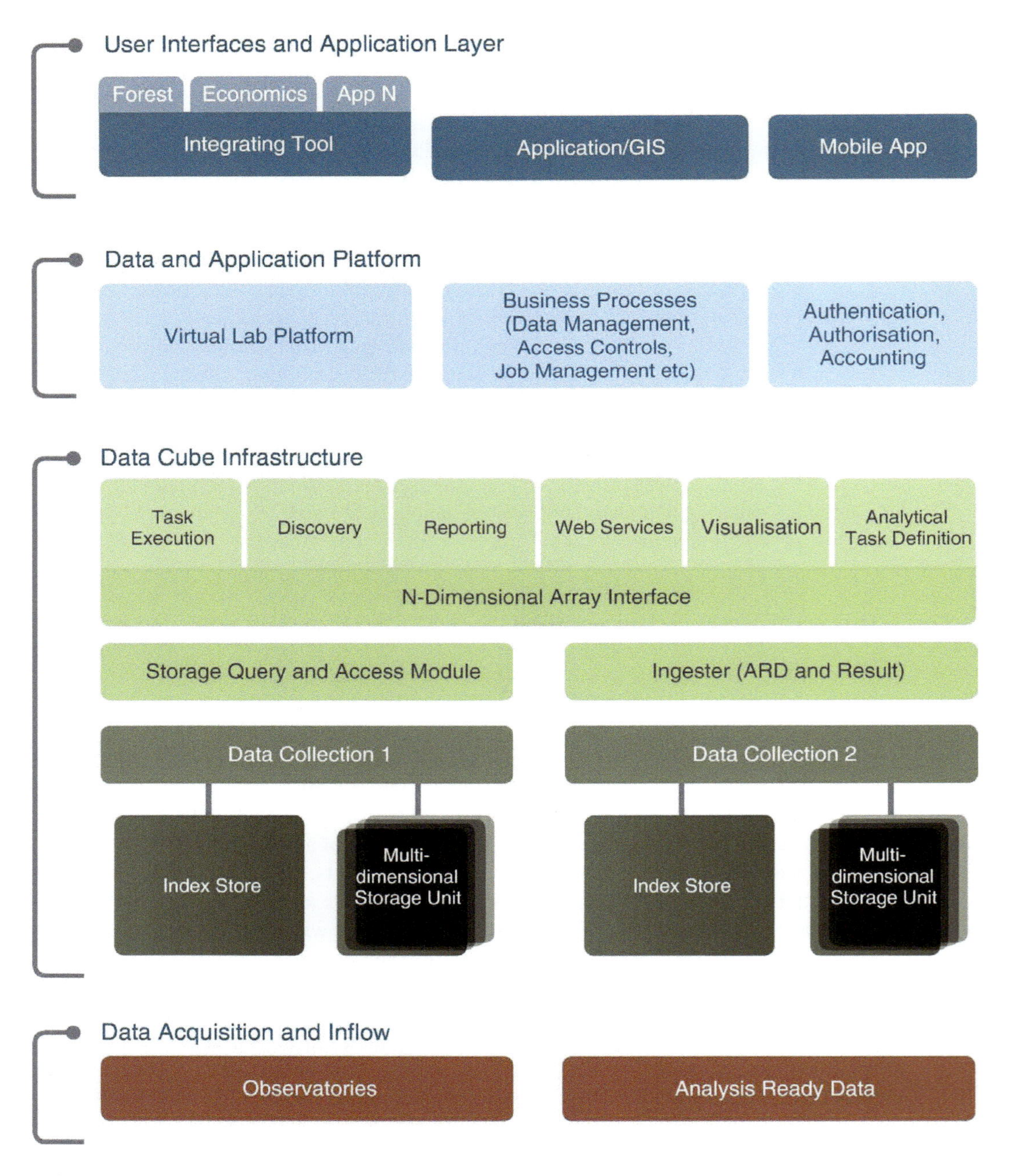


**FIGURE 14.11** Architecture diagram of the ODC platform.

data visualization tools to help users understand geoscience data more intuitively. As the AI Earth released, provides researchers with the efficiency of satellite remote sensing data processing such as multisource data retrieval, online data processing, cloud GIS workspace, comprehensive data management, AI model training, developer mode, weather AI forecast service.

### 14.3.4 The IaaS-Oriented Remote Sensing Cloud System

The IaaS-oriented CCRS system adopts virtualization technologies to provide consumers with on-demand provisioning of infrastructural resources (e.g., networks, storages, virtual servers etc.) on a pay-as-you-go basis. IaaS helps consumers avoid the expense and complexity of buying and managing physical servers and other data center infrastructure. Consumers are able to quickly scale up or scale down infrastructural resources on demand and only pay for what they use. IaaS providers need to hold massive computing and storage clusters and IaaS providers with high market share include Amazon, Google, Rackspace, Microsoft, etc., and other public cloud service providers.

In addition, relying on the Big Earth Data Science Engineering Program (CASEarth) and International Research Center of Big Data for Sustainable Development Goals (CBAS), Chinese scientists have developed the SDG Big Data Platform [134], which is a comprehensive cloud service platform integrating Daas, SaaS, PaaS, and IaaS.

The SDG Big Data Platform aims to integrate big Earth data for the monitoring and prediction of SDGs and provide decision support for the implementation of SDGs. The services provided by SDG Big Data Platform include:

- Iaas. It includes virtual computers, high-performance computing resources, and object storage resources.
- DaaS. The CASEarth DataBank Remote Sensing Data Engine forms a long-time sequence of ready-to-use (RTU) remote sensing data through the integrated processing and standardized processing of multisource Earth observation data.
- SaaS. DeepLearning Cloud System is a one-stop AI-based cloud service for industrial applications, such as remote sensing image analysis and supports domestic IT infrastructure.
- PaaS. EarthDataMiner provides a cloud-based online integrated development environment, which supports scientists to implements Big Earth data mining analysis and visualization through code development, Jupyter Notebook, and other tools.

### 14.3.5 Summary

Through the preceding detailed description of the four types of typical CCRS systems, we refer to the work of Vitor et al. [141] and use 12 capabilities as criteria for evaluating each platform. In the subsequent discussion, we introduce each capability name and its description, with comprehensive elaboration provided in Table 14.4 for reference.

- **Servers Type**: the CCRS system categories corresponding to the platform include Daas, SaaS, PaaS, and IaaS
- **Programming language**: programming languages supported by the platform
- **Data abstraction**: the capacity to provide data abstraction, hiding from scientists details about how data is stored without limiting its access mode
- **Processing abstraction**: the capacity to provide data processing abstraction, hiding from scientists details about where and how data is processed, without limiting its processing power
- **Physical infrastructure abstraction**: the capacity to hide from scientists aspects regarding the number of servers, hardware, and software resources
- **Open governance**: the capacity of the scientific community to participate in the governance and development of the platform
- **Reproducibility of science**: the capacity to provide means that allow scientist to share their analysis and/or reproduce the results among other researchers
- **Infrastructure replicability**: the capacity to replicate the software stack, processes, and data on own infrastructure

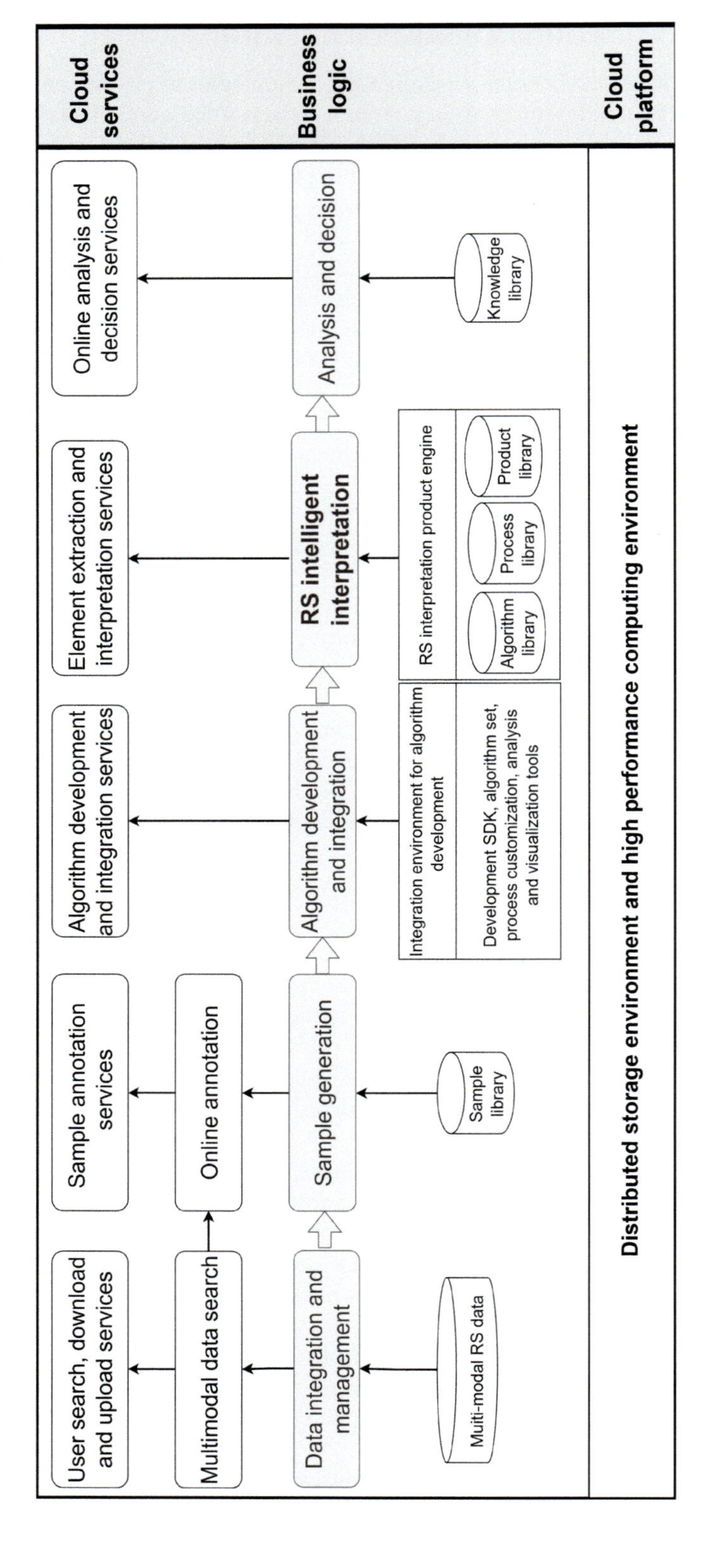


FIGURE 14.12 The technology route of cloud computing supporting global surface intelligence interpretation.

**TABLE 14.4**
**Comparison of Typical Remote Sensing Cloud Computing Platforms**

| Platform | Servers Type | Programming Language | Data Abstraction | Processing Abstraction | Physical Infrastructure Abstraction | Open Governance | Reproducibility of Science | Infrastructure Replicability | Processing Scalability | Storage Scalability | Data Access Interoperability | Extensibility |
|---|---|---|---|---|---|---|---|---|---|---|---|---|
| **NASA EarthData [135]** | DaaS | Python, R, Matlab | **High:** Multisource Dataset | **Medium:** Developable and shareable code from the Earthdata Code Collaborative | **Medium:** Both data storage and processing infrastructure | **High:** Defined governance process | **Medium:** Code and datasets shareable without guarantee to be reproducible | **Low:** Proprietary closed source software | **High:** Closed solution | **High:** Closed solution | **High:** OGC Services | **Low:** Proprietary closed source software |
| **ESA Sentinel Hub [136]** | DaaS | Python, Java | **High:** Data Source, Instances and Layers | **Medium:** Custom scripts (Evalscripts) layers perform pixel-wise processing | **High**: Both data storage and processing infrastructure | **Low:** Proprietary closed source software | **Low:** Without any ease | **Low:** Proprietary closed source software | **High:** Closed solution | **High:** Closed solution | **High:** OGC Services | **Low:** Proprietary closed source software |
| **EarthServer [124]** | DaaS | Python | **High:** DataCube-based analysis-ready data | **Low:** User runs his own code | **Medium:** Only data storage infrastructure | **Medium**: Only open source repository | **Low:** Without any ease | **Low:** Proprietary closed source software | **Low:** User runs his own code | **High:** Rasdaman array database | **High:** OGC WCS and WCPS standards | **Low:** Proprietary closed source software |
| **NEX [137]** | SaaS | Matlab | **Low**: Direct file handling | **Medium:** Computing-intensive analyses | **High**: Both data storage and processing infrastructure | **Medium**: Only open source repository | **Low:** without any ease | **Low:** Proprietary closed source software | **Medium:** Data pro-cessing API | **High**: File System and Spectra Logic tape storage | **High:** OGC Services | **Low:** Proprietary closed source software |

*(Continued)*

**TABLE 14.4** *(Continued)*

**Comparison of Typical Remote Sensing Cloud Computing Platforms**

| Platform | Servers Type | Programming Language | Data Abstraction | Processing Abstraction | Physical Infrastructure Abstraction | Open Governance | Reproducibility of Science | Infrastructure Replicability | Processing Scalability | Storage Scalability | Data Access Interoperability | Extensibility |
|---|---|---|---|---|---|---|---|---|---|---|---|---|
| **pipsCloud [10]** | SaaS | JavaScript | **Low**: Direct file handling | **Low:** User runs his own code | **Medium:** Only data storage infrastructure | **Low:** Proprietary closed source software | **Low:** Without any ease | **Low:** Proprietary closed source software | **Medium:** A template application available | **High:** Distributed File System | **Low:** Without any ease | **Low:** Proprietary closed source software |
| **Huawei Pangu [112]** | SaaS | Python | **Low**: Direct file handling | **High:** Proprietary Processes | **High**: Both data storage and processing infrastructure | **Low:** Proprietary closed source software | **Low:** Without any ease | **Low:** Proprietary closed source software | **High:** Closed solution | **High:** Closed solution | **Low:** Without any ease | **Low:** Proprietary closed source software |
| **GEE [128]** | DaaS, PaaS | JavaScript, Python | **High**: Image, image collection, feature and feature Collection | **High:** Predefined pixel-wise functions | **High**: Both data storage and processing infrastructure | **Low:** Proprietary closed source software | **Medium:** Data links and scripts shareable without guarantee to be reproducible | **Low:** Proprietary closed source software | **High**: Code automatically executed in parallel using a Map Reduce approach | **High:** Google storage services | **Medium:** Tile service | **Low:** Proprietary closed source software |
| **PIE-Engine Studio[138]** | Daas, PaaS | JavaScript, Python | **High:** Product and dataset | **High:** Predefined pixel-wise functions | **High**: Both data storage and processing infrastructure | **Low:** Proprietary closed source software | **Low:** Without any ease | **Low:** Proprietary closed source software | **Medium:** A template application available | **High:** Distributed File System | **Medium:** Tile service | **Low:** Proprietary closed source software |
| **OGE [139]** | DaaS, PaaS | Python | **High**: Data product, sample data for deep learning | **Medium:** Predefined pixel-wise functions | **High**: Both data storage and processing infrastructure | **Medium**: Only open source repository | **Low:** Without any ease | **Medium:** Open source code with basic documentation available | **Medium:** A template application available | **High:** Distributed File System | **Low:** Without any ease | **Medium:** Open source software integrated with proprietary |

| | | | | | | | | | | | | |
|---|---|---|---|---|---|---|---|---|---|---|---|---|
| **ODC [132]** | DaaS, PaaS | Python | **High:** Product and dataset | **Medium:** Xarray and celery | **Medium:** Only data storage infrastructure | **High:** Defined governance process | **Low:** Without any ease | **High:** Open source code docker containers and documentation available | **Medium:** A template application available(Python and Celery) | **High:** Distributed File System, S3 and HTTP | **High:** OGC Services | **High:** Open source and modular code |
| **Candela [140]** | DaaS, PaaS | Python | **High:** data products, S3 | **Low:** User runs his own code | **High**: Both data storage and processing infrastructure | **Low:** Proprietary closed source repository | **Low:** without any ease | **Low:** Proprietary closed source software | **High:** OpenStack API | **High**: Cloud computing, object storage | **High:** OGC Services | **Low:** Proprietary closed source software |
| **AI Earth [122]** | DaaS, PaaS | Python | **High:** Product and Dataset | **Medium:** Predefined processing functions | **High**: Both data storage and processing infrastructure | **Low:** Proprietary software, closed source software | **Low:** Scripts shareable without guarantee to be reproducible | **Low:** Proprietary closed source software | **High**: Aliyun computing services | **High:** Aliyun storage services | **High:** OGC Services | **Low:** Proprietary closed source software |
| **SDG Big Data Platform** [134] | DaaS, SaaS, PaaS, IaaS | Python, Java, Scala | **High:** Product and Dataset | **Low:** User runs his own code | **High**: Both data storage and processing infrastructure | **Low:** Proprietary software | **Low:** Without any ease | **Low:** Proprietary closed source software | **High:** Elastic expansion based on cloud computing | **High:** Elastic expansion based on cloud computing | **High:** OGC Services | **Low:** Proprietary closed source software |

- **Processing scalability**: the capacity to scale processing performance by adding more resources without a direct impact on the way scientists conducts their analysis
- **Storage scalability**: the capability to scale storage space by adding more resources without a direct impact on how scientists access data
- **Data access interoperability**: the capacity to provide means, based on standardized interfaces, that allow other applications to access analysis results or datasets available in the platform
- **Extensibility**: the capacity to add new software tools that utilize the storage and processing modules available internally in the platform

## 14.4 APPLICATIONS OF CCRS SYSTEM IN GLOBAL SURFACE INTELLIGENT INTERPRETATION

### 14.4.1 Challenges of Global Surface Intelligence Interpretation

With the rapid advancement of Earth observation satellites, the utilization of large-scale satellite images, aerial images, and other remote sensing data for automated or semi-automated tasks such as land cover classification, change monitoring, and object identification has become the mainstream approach for global surface environmental monitoring. However, the current intelligent interpretation of global surface remote sensing still faces the following challenges [142].

Primarily, the volume of remote sensing image is immense, thereby posing challenges in data storage and management. Furthermore, traditional manual processing methods are increasingly inadequate to meet the demands of efficiency and fast processing. Finally, the interpretation of global-scale surface objects often involves the utilization of remote sensing data from different sensors and satellites. The fusion and processing of multi-modal remote sensing data present challenges that need to be addressed.

### 14.4.2 Cloud-Based Solution for Global Surface Intelligent Interpretation

Cloud computing technology has been introduced to address the demands for the storage, management, and intelligent processing of large-scale remote sensing data. As illustrated in Figure 14.12, supported by cloud computing platforms, the business workflow of large-scale remote sensing intelligent interpretation primarily includes data storage and management, sample generation, algorithm development, remote sensing intelligent interpretation, online analysis, and decision making. First, remote sensing clouds offer extensive capabilities for large-scale data storage, management, and computation through a distributed storage environment and high-performance computing platforms. This encompasses cloud-based storage, online access, and parallel computing services for multisource data such as spectral, spatial, and time-series. In addition, remote sensing clouds provides support for online sample annotation and generation, as well as the development, training, and inference of machine learning models, which greatly broadens the application scope and spatial and temporal scales of remote sensing data [193]. Finally, cloud services for online analysis, collaboration, and sharing further streamline user interaction and cooperation.

Currently, with the extensive support of remote sensing clouds for frameworks such as deep learning, the integration of remote sensing cloud computing platforms with big data–driven AI methods has become deeply coupled. This integration significantly broadens the value and efficiency of massive remote sensing data in serving Earth science research. It presents new opportunities for the intelligent interpretation and rapid analysis of global-scale long time-series remote sensing surface data [194].

**TABLE 14.5**
**Cases of Cloud Computing Supported Global Surface Intelligent Interpretation Application**

| Related Fields | Contents | References |
|---|---|---|
| **Land use/cover** | Global/regional mapping of forest change | Bastin et al. [143]; Ceccherimi et al. [144]; Fine et al. [145]; Hansen et al. [146]; Qin et al. [147] |
| | Clobal/regional surface water changes | Donchyts et al. [148]; Pekel et al. [149]; Wang et al. [150]; Zou et al. [151] |
| | Farmland mapping | |
| | Oil palm mapping | Zeng et al. [152] |
| | City mapping | Ordway et al. [153] |
| | Lake phytoplankton mapping | Liu et al. [154] |
| | Global tidal flat changes | Ho et al. [155] |
| | Delta change mapping | Murray et al. [156] |
| | Changes in global river ice cover | Nienhuis et al. [157] |
| | Coastal erosion | Yang et al. [158] |
| | Landform recognition | Overeem et al. [159] |
| | | Ielpi et al. [160]; Valenza et al. [161] |
| **Vegetation changes** | Forest restoration potential | Bastin et al. [143] |
| | Arctic changes | Myers-Smith et al. [162] |
| | Plant invasion | Venter et al. [163] |
| | Carbon cycle | Badgleyet et al. [164]; Bogard et al. [165]; Maxwell et al. [166]; Roopsind et al. [167]; Stocker et al. [168] |
| | Plant transpiration | |
| | Phenological changes | |
| | Production estimate | Liu et al. [169] |
| | Biodiversity | Laskin et al. [170] |
| | | Burke et al. [171] |
| | | Betts et al. [172]; Dethier et al. [173]; Jung et al. [174] |
| | | Pfeifer et al. [175] |
| **Animal activity** | Animal habitat | Joshi et al. [176] |
| | Animal evolution | Miller et al. [177] |
| | Changes in animal communities | Giezendanner et al. [178]; Hendershot et al. [179] |
| **Climate environment** | Natural disaster | Walter et al. [180] |
| | Climate change | Tuckett et al. [181]; Venter et al. [182] |
| | Cryosphere | Chudley et al. [183]; Kraaijenbrink et al. [184]; Ryan et al. [185] |
| **Human social and economic activities** | Global urban accessibility | Weiss et al. [186] |
| | Human health | MacDonald et al. [187]; Wu et al. [188] |
| | Socioeconomic applications | Muller et al. [189]; Watmough et al. [190]; Yeh et al. [191] |
| | Archeology | |
| | | Orengo et al. [192] |

### 14.4.3 GEE Supporting Global Surface Intelligent Interpretation

From an application perspective, the utilization of remote sensing cloud computing in the intelligent interpretation of the global surface environment has encompassed various domains, including natural resource monitoring, land use, urban planning, environmental monitoring, and agricultural

**TABLE 14.6**
**Global Surface Intelligent Interpretation Visualization Products**

| Type | Content | Websites | References |
|---|---|---|---|
| **Land use/cover** | Global forest monitoring | https://www.globalforestwatch.org | Rates et al. [198] |
| | Global surface water changes (1985–2016) | http://aqua-monitor.deltares.nl https://global-surfacewater.appspot.com | Donchyts et al. [199]; Pekel et al. [149] |
| | Environmental monitoring | | Beyet et al. [200] |
| | Online land cover maps | http://www.openforis.org https://remap-app.org | Murray et al. [201] |
| **Biology** | Evapotranspiration estimation | http://eeflux-training.appspot.com | Allen et al. [202] |
| | Biodiversity mapping | http://www.earthenv.org | – |
| | Life mapping | https://mol.org | Jetz et al. [203] |
| **Disaster** | Atmospheric simulation calculation | https://clim-engine.appspot.com/climateEngine | Huntington et al. [204] |
| | Historical environment and climate analysis | https://earthmap.org | Morales et al. [205] |
| **Human activity** | Malaria distribution map | https://www.disarm.io | – |
| | Global fisheries monitor | http://globalfishingwatch.org | Merten et al. [206] |
| | Global accessibility mapping | https://accessmapper.appspot.com | Weiss et al. [186] |

management. For example, Hansen et al. [195] achieved a continuous 12-year global forest distribution mapping, thereby providing insights into the processes of forest depletion and augmentation worldwide. Gong et al. [196] produced a distribution map of global impervious surfaces over 34 consecutive years, which is vital for investigating urban expansion and land changes. Dong et al. [197] released a 30-m spatial mapping product of rice fields in the three northeastern provinces of China, thereby offering foundational references for food security assessment, water resource management, and agricultural cultivation. Taking GEE as an example, the application cases of global surface intelligent interpretation on GEE are summarized in Table 14.5.

In addition, while GEE serves large-scale remote sensing intelligent interpretation tasks, it has also formed a series of visualization products or application platforms, as presented in Table 14.6.

In summary, the introduction of cloud computing helps to solve key problems encountered in the intelligent interpretation of global surface remote sensing, such as data storage and management, high-performance computing, distributed processing, computational resources, and collaboration and resource sharing. In addition, remote sensing clouds have moved into the research fields of the ecological environment, land use, and human socioeconomic activities at the global scale, thereby promoting the in-depth understanding of human beings among the dynamic changes in the global surface environment.

## 14.5 CONCLUSIONS

CCRS, distinguished by its substantial storage capacity, robust high-performance computing capabilities, and the flexible, on-demand elastic expansion service model, offers a dependable infrastructure to facilitate the comprehensive analysis and extraction of valuable insights from large-scale remote sensing datasets. In the present landscape, characterized by the widespread adoption of remote sensing applications and the escalating need for tailored solutions, CCRS technology is currently undergoing significant maturation, with a concurrent and notable increase in their adoption and utilization within the public domain.

In light of the swift evolution of significant technological domains such as big data, IoT, and AI, alongside other information technologies, the landscape of CCRS is concurrently unveiling novel trajectories of progress. These emergent trends encompass a spectrum of transformative dimensions as follows:

- Integration with other advanced technologies. The degree of coupling and correlation between CCRS and big data, AI, and other technologies has been significantly enhanced. The threshold for using CCRS platforms will be lowered, and the scope of applications will be further expanded.
- Business refinement. As the "cloud-edge-device" process continues to advance, resource deployment will become more global and distributed, and the CCRS platform will provide refined application services for various scenarios.
- Public-oriented service. As the public's personalized needs are gradually activated, a public-centered Earth observation data sharing process will evolve accordingly. The CCRS platform is moving from industry application to mass application at an unprecedented speed.
- More intelligence application. The CCRS platforms should have the capabilities of autonomous learning, data processing, model training and reasoning, and support joint development and deployment of cloud and edge devices. It would be preferable to present a variety of algorithm frameworks and model libraries to meet the differentiated needs of users and developers in industrial scenarios and improve the efficiency of AI model training and application.

## ACKNOWLEDGMENTS

This research was supported by the National Natural Science Foundation of China under Grant 41925007 and Grant U21A2013.

## REFERENCES

[1] B. Zhang, Y. Wu, B. Zhao, J. Chanussot, D. Hong, J. Yao, L. Gao, Progress and challenges in intelligent remote sensing satellite systems, IEEE Journal of Selected Topics in Applied Earth Observations and Remote Sensing 15 (2022) 1814–1822.

[2] D. Li, M. Wang, F. Yang, R. Dai, Internet intelligent remote sensing scientific experimental satellite LuoJia3–01, Geo-spatial Information Science (2023) 1–5.

[3] Y. Ma, H. Wu, L. Wang, B. Huang, R. Ranjan, A. Zomaya, W. Jie, Remote sensing big data computing: Challenges and opportunities, Future Generation Computer Systems 51 (2015) 47–60.

[4] NASA, Earth observing system data and information system (EOSDIS), Earthdata Cloud Evolution, www.earthdata.nasa.gov/eosdis/cloud-evolution, Last accessed on 2023–10–15 (2023).

[5] CRESDA, China center for resource satallite data and application, https://data.cresda.cn/#/home, Last accessed on 2023–10–15 (2023).

[6] M. Albani, Esa agency report, in: Proceedings of the 55th Meeting of the Working Group on Information Systems & Services, Cordoba Argentina, 18–20, April 2023.

[7] Everyday Astronaut, Hyperbola-1 jilin-1 mofang-01a(r), https://everydayastronaut.com/hyperbola-1-jilin-1-mofang-01ar/, Last accessed on 2023–10–15 (2023).

[8] N. Gorelick, M. Hancher, M. Dixon, S. Ilyushchenko, D. Thau, R. Moore, Google Earth Engine: Planetary-scale geospatial analysis for everyone, Remote sensing of Environment 202 (2017) 18–27.

[9] X. Zhang, Y. Zhou, J. Luo, Deep learning for processing and analysis of remote sensing big data: A technical review, Big Earth Data 6 (4) (2022) 527–560.

[10] L. Wang, Y. Ma, J. Yan, V. Chang, A. Y. Zomaya, Pipscloud: High performance cloud computing for remote sensing big data management and processing, Future Generation Computer Systems 78 (2018) 353–368.

[11] P. Liu, L. Di, Q. Du, L. Wang, Remote sensing big data: Theory, methods and applications, Remote Sensing 10 (5) (2018) 711.

[12] M. Amani, A. Ghorbanian, S. A. Ahmadi, M. Kakooei, A. Moghimi, S. M. Mirmazloumi, S. H. A. Moghaddam, S. Mahdavi, M. Ghahremanloo, S. Parsian, et al., Google Earth Engine cloud computing platform for remote sensing big data applications: A comprehensive review, IEEE Journal of Selected Topics in Applied Earth Observations and Remote Sensing 13 (2020) 5326–5350.
[13] C. Wang, Y. Liu, L. Guo, W. Guo, F. Liao, H. Zhu, A smart agricultural service platform for crop planting, monitoring and management-pie-engine landscape, in: 2021 9th International Conference on Agro-Geoinformatics (Agro-Geoinformatics), IEEE, 2021, pp. 1–6.
[14] L. Wang, J. Yan, Y. Ma, Cloud computing in remote sensing, CRC Press, 2019.
[15] P. Srivastava, R. Khan, A review paper on cloud computing, International Journal of Advanced Research in Computer Science and Software Engineering 8 (6) (2018) 17–20.
[16] M. Maray, J. Shuja, Computation offloading in mobile cloud computing and mobile edge computing: Survey, taxonomy, and open issues, Mobile Information Systems 2022 (2022).
[17] A. Lewis, S. Oliver, L. Lymburner, B. Evans, L. Wyborn, N. Mueller, G. Raevksi, J. Hooke, R. Woodcock, J. Sixsmith, et al., The Australian geoscience data cube: Foundations and lessons learned, Remote Sensing of Environment 202 (2017) 276–292.
[18] Z. Zhu, Science of landsat analysis ready data, Remote Sensing, 11 (18) (2019) 2166.
[19] D. Frantz, Force-Landsat+ sentinel-2 analysis ready data and beyond, Remote Sensing 11 (9) (2019) 1124.
[20] A. V. Egorov, D. P. Roy, H. K. Zhang, Z. Li, L. Yan, H. Huang, Landsat 4, 5 and 7 (1982 to 2017) analysis ready data (ARD) observation coverage over the conterminous united states and implications for terrestrial monitoring, Remote Sensing 11 (4) (2019) 447.
[21] P. Potapov, X. Li, A. Hernandez-Serna, A. Tyukavina, M. C. Hansen, A. Kommareddy, A. Pickens, S. Turubanova, H. Tang, C. E. Silva, et al., Mapping global forest canopy height through integration of GEDI and Landsat data, Remote Sensing of Environment 253 (2021) 112165.
[22] J. Sun, Y. Zhang, Z. Wu, Y. Zhu, X. Yin, Z. Ding, Z. Wei, J. Plaza, A. Plaza, An efficient and scalable framework for processing remotely sensed big data in cloud computing environments, IEEE Transactions on Geoscience and Remote Sensing 57 (7) (2019) 4294–4308.
[23] J. Wu, M. Wu, H. Li, L. Li, L. Li, A serverless-based, on-the-fly computing framework for remote sensing image collection, Remote Sensing 14 (7) (2022) 1728.
[24] V. Giménez-Alventosa, G. Moltó, M. Caballer, A framework and a performance assessment for serverless mapreduce on AWS lambda, Future Generation Computer Systems 97 (2019) 259–274.
[25] D. Chang, Z. Wang, X. Ning, Z. Li, L. Zhang, X. Liu, Vegetation changes in yellow river delta wetlands from 2018 to 2020 using pie-engine and short time series sentinel-2 images, Frontiers in Marine Science 9 (2022) 977050.
[26] P. Baumann, D. Misev, V. Merticariu, B. P. Huu, Datacubes: Towards space/time analysis-ready data, Service-Oriented Mapping: Changing Paradigm in Map Production and Geoinformation Management (2019) 269–299.
[27] E. Alvarez-Vanhard, T. Corpetti, T. Houet, Uav & satellite synergies for optical remote sensing applications: A literature review, Science of remote sensing 3 (2021) 100019.
[28] G. Sagl, B. Resch, Mobile phones as ubiquitous social and environmental geo-sensors, in: Encyclopedia of mobile phone behavior, IGI Global, 2015, pp. 1194–1213.
[29] Y. Yuanxi, W. Jianrong, Ubiquitous perception and space mapping, Acta Geodaetica et Cartographica Sinica 52 (1) (2023) 1.
[30] J. Li, C. Zhou, Studies on multi-scale geo-spatial data integration, Advances in Earth Science 15 (1) (2000) 48.
[31] Z. Zhu, S. Wang, C. E. Woodcock, Improvement and expansion of the Fmask algorithm: Cloud, cloud shadow, and snow detection for Landsats 4–7, 8, and sentinel 2 images, Remote Sensing of Environment 159 (2015) 269–277. doi:10.1016/j.rse.2014.12.014, www.sciencedirect.com/science/article/pii/S0034425714005069
[32] A. Aizawa, K. Oyama, A fast linkage detection scheme for multi-source information integration, in: International Workshop on Challenges in Web Information Retrieval and Integration, IEEE, 2005, pp. 30–39.
[33] V. M. Escobar, M. Srinivasan, S. D. Arias, Improving NASA's earth observation systems and data programs through the engagement of mission early adopters, Springer International Publishing, Cham, 2016, pp. 223–267. doi:10.1007/978-3-319-33438-7_9, doi:10.1007/978-3-319-33438-79

[34] G. S. Barton, Directory interchange format: A metadata tool for the NOAA earth system data directory, in: The Role of Metadata in Managing Large Environmental Science Datasets, Pacific Northwest Laboratory, Richland, Washington, USA, 1995, pp. 19–23.
[35] M. Burnett, B. Weinstein, A. Mitchell, Echo-enabling interoperability with nasa earth science data and services, in: 2007 IEEE International Geoscience and Remote Sensing Symposium, IEEE, 2007, pp. 4012–4015.
[36] J. Behnke, A. Mitchell, H. Ramapriyan, Nasa's earth observing data and information system-near-term challenges, Data Science Journal 18 (2019) 40–40.
[37] H. Ramapriyan, The role and evolution of Nasa's earth science data systems, in: Institute of Electrical and Electronic Engineers (IEEE) EDS/CAS Chapter Meeting, no. GSFC-E-DAA-TN24713, 2015.
[38] S. Kopp, P. Becker, A. Doshi, D. J. Wright, K. Zhang, H. Xu, Achieving the full vision of earth observation data cubes, Data 4 (3) (2019). doi:10.3390/data4030094, www.mdpi.com/2306-5729/4/3/94
[39] J. Fan, J. Yan, Y. Ma, L. Wang, Big data integration in remote sensing across a distributed metadata-based spatial infrastructure, Remote Sensing 10 (1) (2017) 7.
[40] J. Yan, Y. Liu, L. Wang, Z. Wang, X. Huang, H. Liu, An efficient organization method for large-scale and long time-series remote sensing data in a cloud computing environment, IEEE Journal of Selected Topics in Applied Earth Observations and Remote Sensing 14 (2021) 9350–9363. doi:10.1109/JSTARS.2021.3110900
[41] P. Baumann, P. Mazzetti, J. Ungar, R. Barbera, Big data analytics for earth sciences: The earthserver approach, International Journal of Digital Earth 9 (1) (2016) 3–29. doi:10.1080/17538947.2014.1003106
[42] Z. Tan, P. Yue, J. Gong, An array database approach for earth observation data management and processing, ISPRS International Journal of Geo-Information 6 (7) (2017). doi:10.3390/ijgi6070220, www.mdpi.com/2220-9964/6/7/220
[43] M. D. Mahecha, F. Gans, G. Brandt, R. Christiansen, S. E. Cornell, N. Fomferra, G. Kraemer, J. Peters, P. Bodesheim, G. Camps-Valls, J. F. Donges, W. Dorigo, L. M. Estupinan-Suarez, V. H. Gutierrez-Velez, M. Gutwin, M. Jung, M. C. Londoño, D. G. Miralles, P. Papastefanou, M. Reichstein, Earth system data cubes unravel global multivariate dynamics, Earth System Dynamics 11 (1) (2020) 201–234. doi:10.5194/esd-11-201-2020, https://esd.copernicus.org/articles/11/201/2020/
[44] C. A. Lee, S. D. Gasster, A. Plaza, C.-I. Chang, B. Huang, Recent developments in high performance computing for remote sensing: A review, IEEE Journal of Selected Topics in Applied Earth Observations and Remote Sensing 4 (3) (2011) 508–527.
[45] P. Sun, Y. Wen, R. Han, W. Feng, S. Yan, Gradientflow: Optimizing network performance for large-scale distributed DNN training, IEEE Transactions Big Data 8 (2) (2022) 495–507.
[46] L. Liu, Y. Liu, J. Yan, H. Liu, M. Li, J. Wang, K. Zhou, Object detection in large-scale remote sensing images with a distributed deep learning framework, IEEE Journal of Selected Topics in Applied Earth Observations and Remote Sensing 15 (2022) 8142–8154. doi:10.1109/JSTARS.2022.3206085.
[47] D. Wang, Q. Zhang, Y. Xu, J. Zhang, B. Du, D. Tao, L. Zhang, Advancing plain vision transformer toward remote sensing foundation model, IEEE Transaction Geoscience Remote Sensing 61 (2023) 1–15.
[48] C. A. Lee, S. D. Gasster, A. Plaza, C. I. Chang, B. Huang, Recent developments in high performance computing for remote sensing: A review, IEEE Journal of Selected Topics in Applied Earth Observations and Remote Sensing 4 (3) (2011).
[49] I. Polato, R. Re´, A. Goldman, F. Kon, A comprehensive view of Hadoop research: A systematic literature review, Journal of Network and Computer Applications 46 (2014) 1–25.
[50] S. Salloum, R. Dautov, X. Chen, P. X. Peng, J. Z. Huang, Big data analytics on apache spark, International Journal of Data Science and Analytics 1 (2016) 145–164.
[51] N. Wang, F. Chen, B. Yu, Y. Qin, Segmentation of large-scale remotely sensed images on a spark platform: A strategy for handling massive image tiles with the mapreduce model, ISPRS Journal of Photogrammetry and Remote Sensing 162 (2020) 137–147.
[52] W. Z. Khan, E. Ahmed, S. Hakak, I. Yaqoob, A. Ahmed, Edge computing: A survey, Future Generation Computer Systems 97 (2019) 219–235.
[53] A. A. Abdellatif, A. Mohamed, C. F. Chiasserini, M. Tlili, A. Erbad, Edge computing for smart health: Context-aware approaches, opportunities, and challenges, IEEE Network 33 (3) (2019) 196–203.
[54] V. K. Prasad, M. D. Bhavsar, S. Tanwar, Influence of monitoring: Fog and edge computing, Scalable Computing: Practice and Experience 20 (2) (2019) 365–376.

[55] T. Zhang, Z. Wang, P. Cheng, G. Xu, X. Sun, Dcnnet: A distributed convolutional neural network for remote sensing image classification, IEEE Transactions on Geoscience and Remote Sensing 61 (2023) 1–18.

[56] N. Abbas, Y. Zhang, A. Taherkordi, T. Skeie, Mobile edge computing: A survey, IEEE Internet of Things Journal 5 (1) (2017) 450–465.

[57] R. Mahmud, R. Kotagiri, R. Buyya, Fog computing: A taxonomy, survey and future directions, Internet of Everything: Algorithms, Methodologies, Technologies and Perspectives (2018) 103–130.

[58] S. Marru, B. Freitag, D. Wannipurage, U. K. Bommala, P. Pradier, C. Demange, N. Pantha, T. Mukherjee, B. Rosich, E. Monjoux, R. Ramachandran, Blaze: A high-performance, scalable, and efficient data transfer framework with configurable and extensible features: Principles, implementation, and evaluation of a transatlantic inter-cloud data transfer case study, in: 2023 IEEE 16th International Conference on Cloud Computing (CLOUD), 2023, pp. 58–68. doi:10.1109/CLOUD60044.2023.00016.

[59] D. Newman, M. Morahan, WGISS-54–09: Authenticating data replicas in the cloud, in: The 55th Meeting of the Working Group on Information Systems and Services, 2023.

[60] J. Yan, L. Wang, F. Zhang, X. Chen, X. Huang, J. Li, Blockchain application in remote sensing big data management and production, in: Blockchains for Network Security: Principles, Technologies and Applications, 2020.

[61] Y.-T. Zhuang, F. Wu, C. Chen, Y.-H. Pan, Challenges and opportunities: From big data to knowledge in ai 2.0, Frontiers of Information Technology & Electronic Engineering 18 (2017) 3–14.

[62] T. Matsuyama, Knowledge-based aerial image understanding systems and expert systems for image processing, Geoscience and Remote Sensing, IEEE Transactions on GE-25 (3) (1987) 305–316.

[63] M. Xu, P. Watanachaturaporn, P. K. Varshney, M. K. Arora, Decision tree regression for soft classification of remote sensing data, Remote Sensing of Environment 97 (3) (2005) 322–336.

[64] J. Ye, N. Wang, M. Sun, Q. Liu, N. Ding, M. Li, A new method for the rapid determination of fire disturbance events using gee and the VCT algorithm: A case study in southwestern and northeastern China, Remote Sensing 15 (2) (2023) 413.

[65] J. Han, J. Pei, Y. Yin, Mining frequent patterns without candidate generation, ACM Sigmod Record 29 (2) (2000) 1–12.

[66] S. B. Kotsiantis, I. Zaharakis, P. Pintelas, Supervised machine learning: A review of classification techniques, Emerging Artificial Intelligence Applications in Computer Engineering 160 (1) (2007) 3–24.

[67] F. Jiang, M. Deng, J. Tang, L. Fu, H. Sun, Integrating spaceborne lidar and sentinel-2 images to estimate forest aboveground biomass in northern China, Carbon Balance and Management 17 (1) (2022) 1–13.

[68] L. A. S. Romani, R. R. V. Goncalves, J. Zullo, C. Traina, A. J. M. Traina, New DTW-based method to similarity search in sugar cane regions represented by climate and remote sensing time series, in: Geoscience and Remote Sensing Symposium (IGARSS), 2010 IEEE International, 2010.

[69] G. Xudong, H. Chong, L. Gaohuan, M. Xuelian, L. Qingsheng, Mapping rice cropping systems in vietnam using an NDVI-based time-series similarity measurement based on DTW distance, Remote Sensing 8 (1) (2016) 19.

[70] S. Shan, Big data analytics: Machine learning and Bayesian learning perspectives: What is done? What is not?, Wiley Interdisciplinary Reviews: Data Mining and Knowledge Discovery 9 (2018) e1283.

[71] C. Singha, K. C. Swain, Spatial analyses of cyclone Amphan induced flood inundation mapping using sentinel-1a SAR images through gee cloud, in: Computer Vision and Robotics: Proceedings of CVR 2021, Springer, 2022, pp. 65–83.

[72] Y. Ma, L. Wang, Y. Albert, D. Chen, R. Ranjan, Task-tree based large-scale mosaicking for massive remote sensed imageries with dynamic DAG scheduling, IEEE Transactions on Parallel & Distributed Systems 25 (8) (2013) 2126–2137.

[73] Y. Liao, G. Liu, H. Luan, M. Zheng, G. Deng, Comparative study on remote sensing image classifier of Jiulong river basin, in: International Conference on Geoinformatics and Data Analysis, Springer, 2022, pp. 88–94.

[74] P. B. Wale, V. Dhaigude, S. Mishra, Comparative analysis of image classification capabilities of support vector machine (SVM) and random forest (RF) with Google Earth Engine (GEE) platform: A case study of Sangamner, Maharashtra, Intercontinental Geoinformation Days 6 (2023) 113–116.

[75] A. J. Morgan, Simulated annealing approach to temperature-emissivity separation in thermal remote sensing, Journal of Applied Remote Sensing 10 (4) (2016) 040501.

[76] S. Georganos, T. Grippa, S. Vanhuysse, M. Lennert, M. Shimoni, S. Kalogirou, E. Wolff, Less is more: Optimizing classification performance through feature selection in a very-high-resolution remote sensing object-based urban application, Taylor & Francis, 2018.
[77] B. C. K. Tso, P. M. Mather, Classification of multisource remote sensing imagery using a genetic algorithm and Markov random fields, IEEE Transactions on Geoscience & Remote Sensing 37 (3) (1999) 1255–1260.
[78] S. Babbar, Review-mastering the game of go with deep neural networks and tree search, Nature 529 (7587) (2016) 484–489.
[79] X. Yan, J. Li, D. Yang, J. Li, T. Ma, Y. Su, J. Shao, R. Zhang, A random forest algorithm for landsat image chromatic aberration restoration based on gee cloud platform: A case study of Yucatán peninsula, Mexico, Remote Sensing 14 (20) (2022) 5154.
[80] C. Dou, Remote sensing imagery classification using AdaBoost with a weight vector (WV AdaBoost), Remote Sensing Letters (2017).
[81] H. Liu, M. Chen, H. Chen, Y. Li, C. Xie, B. Tian, C. Wang, P. Ge, Remote sensing extraction of agricultural land in Shandong province, China, from 2016 to 2020 based on Google Earth Engine, Remote Sensing 14 (22) (2022) 5672.
[82] S. Chaib, H. Liu, Y. Gu, H. Yao, Deep feature fusion for VHR remote sensing scene classification, IEEE Transactions on Geoscience and Remote Sensing 55 (8) (2017) 4775–4784.
[83] G. Fu, C. Liu, R. Zhou, T. Sun, Q. Zhang, Classification for high resolution remote sensing imagery using a fully convolutional network, Remote Sensing 9 (5) (2017) 498.
[84] X. Wang, Z. Hu, S. Shi, M. Hou, L. Xu, X. Zhang, A deep learning method for optimizing semantic segmentation accuracy of remote sensing images based on improved UNet, Scientific Reports 13 (1) (2023) 7600.
[85] J. Yan, X. Chen, Y. Chen, D. Liang, Multistep prediction of land cover from dense time series remote sensing images with temporal convolutional networks, IEEE Journal of Selected Topics in Applied Earth Observations and Remote Sensing 13 (2020) 5149–5161.
[86] X. Yao, X. Feng, J. Han, G. Cheng, L. Guo, Automatic weakly supervised object detection from high spatial resolution remote sensing images via dynamic curriculum learning, IEEE Transactions on Geoscience and Remote Sensing 59 (1) (2020) 675–685.
[87] H. Crisóstomo de Castro Filho, O. Ab´ılio de Carvalho Ju´nior, O. L. Ferreira de Carvalho, P. Pozzobon de Bem, R. dos Santos de Moura, A. Olino de Albuquerque, C. Rosa Silva, P. H. Guimaraes Ferreira, R. Fontes Guimara˜es, R. A. Trancoso Gomes, Rice crop detection using lstm, bi-lstm, and machine learning models from sentinel-1 time series, Remote Sensing 12 (16) (2020) 2655.
[88] Y. Zhao, Y. Ban, Goes-r time series for early detection of wildfires with deep gru-network, Remote Sensing 14 (17) (2022) 4347.
[89] X. Zhang, X. Tan, G. Chen, K. Zhu, P. Liao, T. Wang, Object-based classification framework of remote sensing images with graph convolutional networks, IEEE Geoscience and Remote Sensing Letters 19 (2021) 1–5.
[90] A. Sha, B. Wang, X. Wu, L. Zhang, Semisupervised classification for hyperspectral images using graph attention networks, IEEE Geoscience and Remote Sensing Letters 18 (1) (2020) 157–161.
[91] P. Deng, K. Xu, H. Huang, When CNNs meet vision transformer: A joint framework for remote sensing scene classification, IEEE Geoscience and Remote Sensing Letters 19 (2021) 1–5.
[92] W. G. C. Bandara, V. M. Patel, A transformer-based Siamese network for change detection, in: IGARSS 2022–2022 IEEE International Geo-science and Remote Sensing Symposium, IEEE, 2022, pp. 207–210.
[93] X. Tang, Z. Tu, Y. Wang, M. Liu, D. Li, X. Fan, Automatic detection of coseismic landslides using a new transformer method, Remote Sensing 14 (12) (2022) 2884.
[94] V. Franois-Lavet, P. Henderson, R. Islam, M. G. Bellemare, J. Pineau, An introduction to deep reinforcement learning, Foundations and Trends® in Machine Learning 11 (3–4) (2018) 219–354.
[95] G. Cheng, J. Han, A survey on object detection in optical remote sensing images, ISPRS Journal of Photogrammetry & Remote Sensing 117 (2016) 11–28.
[96] K. Fu, Y. Li, H. Sun, X. Yang, G. Xu, Y. Li, X. Sun, A ship rotation detection model in remote sensing images based on feature fusion pyramid network and deep reinforcement learning, Remote Sensing 10 (12) (2018).
[97] Z. Shi, Z. Zou, Can a machine generate humanlike language descriptions for a remote sensing image?, IEEE Transactions on Geoscience and Remote Sensing 55 (6) (2017) 3623–3634.

[98] X. Lu, B. Wang, X. Zheng, X. Li, Exploring models and data for remote sensing image caption generation, IEEE Transactions on Geoscience Remote Sensing 56 (4) (2018) 2183–2195.
[99] T. Wei, W. Yuan, J. Luo, W. Zhang, L. Lu, VLCA: Vision-language aligning model with cross-modal attention for bilingual remote sensing image captioning, Journal of Systems Engineering and Electronics 34 (1) (2023) 9–18.
[100] M. B. Bejiga, F. Melgani, A. Vascotto, Retro-remote sensing: Generating images from ancient texts, IEEE Journal of Selected Topics in Applied Earth Observations Remote Sensing 12 (3) (2019) 950–960.
[101] Y. Xu, W. Yu, P. Ghamisi, M. Kopp, S. Hochreiter, Txt2Img-MHN: Remote sensing image generation from text using modern Hopfield networks, IEEE Transactions on Image Processing 32 (2023) 5737–5750.
[102] Z. Yuan, W. Zhang, C. Tian, X. Rong, Z. Zhang, H. Wang, K. Fu, X. Sun, Remote sensing cross-modal text-image retrieval based on global and local information, IEEE Transactions on Geoscience and Remote Sensing 60 (2022) 1–16.
[103] Z. Yuan, W. Zhang, X. Rong, X. Li, J. Chen, H. Wang, K. Fu, X. Sun, A lightweight multi-scale cross-modal text-image retrieval method in remote sensing, IEEE Transactions on Geoscience and Remote Sensing 60 (2022) 1–19.
[104] M. M. A. Rahhal, M. A. Bencherif, Y. Bazi, A. Alharbi, M. L. Mekhalfi, Contrasting dual transformer architectures for multi-modal remote sensing image retrieval, Applied Sciences 13 (1) (2022) 282.
[105] C. Chappuis, V. Zermatten, S. Lobry, B. L. Saux, D. Tuia, Promptrsvqa: Prompting visual context to a language model for remote sensing visual question answering, in: IEEE/CVF Conference on Computer Vision and Pattern Recognition Workshops, CVPR Workshops 2022, New Orleans, LA, June 19–20, IEEE, 2022, pp. 1371–1380.
[106] Z. Yuan, L. Mou, Q. Wang, X. X. Zhu, From easy to hard: Learning language-guided curriculum for visual question answering on remote sensing data, IEEE Transactions on Geoscience Remote Sensing 60 (2022) 1–11.
[107] Y. Zhan, Z. Xiong, Y. Yuan, RSVG: Exploring data and models for visual grounding on remote sensing data, IEEE Transactions Geoscience Remote Sensing 61 (2023) 1–13.
[108] G. Sumbul, R. G. Cinbis, S. Aksoy, Fine-grained object recognition and zero-shot learning in remote sensing imagery, IEEE Transactions on Geoscience and Remote Sensing 56 (2) (2017) 770–779.
[109] Y. Li, Z. Zhu, J.-G. Yu, Y. Zhang, Learning deep cross-modal embedding networks for zero-shot remote sensing image scene classification, IEEE Transactions on Geoscience and Remote Sensing 59 (12) (2021) 10590–10603.
[110] G. Kim, H.-G. Jung, S.-W. Lee, Few-shot object detection via knowledge transfer, in: 2020 IEEE International Conference on Systems, Man, and Cybernetics (SMC), 2020, pp. 3564–3569.
[111] X. Lu, X. Sun, W. Diao, Y. Mao, J. Li, Y. Zhang, P. Wang, K. Fu, Few-shot object detection in aerial imagery guided by text-modal knowledge, IEEE Transactions on Geoscience and Remote Sensing 61 (2023) 1–19.
[112] K. Bi, L. Xie, H. Zhang, X. Chen, X. Gu, Q. Tian, Accurate medium-range global weather forecasting with 3d neural networks, Nature (2023) 1–6.
[113] J. Yao, J. Wu, C. Xiao, Z. Zhang, J. Li, The classification method study of crops remote sensing with deep learning, machine learning, and Google Earth Engine, Remote Sensing 14 (12) (2022) 2758.
[114] A. Radford, J. W. Kim, C. Hallacy, A. Ramesh, G. Goh, S. Agarwal, G. Sastry, A. Askell, P. Mishkin, J. Clark, et al., Learning transferable visual models from natural language supervision, In International Conference on Machine Learning, PMLR, 2021, pp. 8748–8763.
[115] L. Yuan, D. Chen, Y. Chen, N. Codella, X. Dai, J. Gao, H. Hu, X. Huang, B. Li, C. Li, et al., Florence: A new foundation model for computer vision, arXiv preprint arXiv:2111.11432, 2021.
[116] T. Brown, B. Mann, N. Ryder, M. Subbiah, J. D. Kaplan, P. Dhariwal, A. Neelakantan, P. Shyam, G. Sastry, A. Askell, et al., Language models are few-shot learners, Advances in Neural Information Processing Systems 33 (2020) 1877–1901.
[117] S. Zhang, S. Roller, N. Goyal, M. Artetxe, M. Chen, S. Chen, C. Dewan, M. Diab, X. Li, X. Lin, et al., Opt: Open pre-trained transformer language models, arXiv preprint arXiv:2205.01068, 2022.
[118] Y. Wang, C. M. Albrecht, X. X. Zhu, Self-supervised vision transformers for joint sar-optical representation learning, in: IGARSS 2022–2022 IEEE International Geoscience and Remote Sensing Symposium, IEEE, 2022, pp. 139–142.
[119] C. J. Reed, R. Gupta, S. Li, S. Brockman, C. Funk, B. Clipp, K. Keutzer, S. Candido, M. Uyttendaele, T. Darrell, Scale-mae: A scale-aware masked autoencoder for multiscale geospatial representation learning, in: Proceedings of the IEEE/CVF International Conference on Computer Vision, 2023, pp. 4088–4099.

[120] PIESAT, Piesat information technology co., ltd., www.piesat.cn/website/en-us/src/home.html, Last accessed on 2023–10–15 (2023).
[121] Sensetime, Senseearth ai-powered geospatial analytics and application platform, www.sensetime.com/en/, Last accessed on 2023–10–15 (2023).
[122] A. D. Academy, Ai earth, analytical insight of earth, https://engine-aiearth.aliyun.com/#/portal/open-data, Last accessed on 2023–10–15 (2023).
[123] S. S. D. Hub, US geological survey distribution of European space agency's sentinel-2 data, US Geological Survey. Technical report (2017).
[124] P. Baumann, P. Mazzetti, J. Ungar, R. Barbera, D. Barboni, A. Beccati, L. Bigagli, E. Boldrini, R. Bruno, A. Calanducci, et al., Big data analytics for earth sciences: The earthserver approach, International Journal of Digital Earth 9 (1) (2016) 3–29.
[125] W. Wang, S. Li, H. Hashimoto, H. Takenaka, A. Higuchi, S. Kalluri, R. Nemani, An introduction to the Geostationary-Nasa Earth Exchange (GEONEX) products: 1. top-of-atmosphere reflectance and brightness temperature, Remote Sensing 12 (8) (2020) 1267.
[126] J. Yan, Y. Ma, L. Wang, K.-K. R. Choo, W. Jie, A cloud-based remote sensing data production system, Future Generation Computer Systems 86 (2018) 1154–1166.
[127] J. Zhang, J. Yan, Y. Ma, D. Xu, W. Jie, Infrastructures and services for remote sensing data production management across multiple satellite data centers, Cluster Computing 19 (3) (2016) 1–18.
[128] R. Moore, M. Hansen, Google Earth Engine: A new cloud-computing platform for global-scale earth observation data and analysis, in: AGU Fall Meeting Abstracts, vol. 2011, 2011, pp. IN43C–02.
[129] O. Mutanga, L. Kumar, Google Earth Engine applications, Remote Sensing 11 (5) (2019) 591.
[130] H. Tamiminia, B. Salehi, M. Mahdianpari, L. Quackenbush, S. Adeli, B. Brisco, Google Earth Engine for geo-big data applications: A meta-analysis and systematic review, ISPRS Journal of Photogrammetry and Remote Sensing 164 (2020) 152–170.
[131] Y. Liu, C. Wang, L. Guo, W. Guo, H. Zhu, C. Li, Fast and high-accuracy cotton planting distribution identification from sentinel-2 images based on pie-engine studio, in: 2021 International Conference on Electronic Information Technology and Smart Agriculture (ICEITSA), IEEE, 2021, pp. 556–560.
[132] B. Killough, Overview of the open data cube initiative, in: IGARSS 2018–2018 IEEE International Geoscience and Remote Sensing Symposium, IEEE, 2018, pp. 8629–8632.
[133] S. Bansal, S. Patel, I. Shah, P. Patel, P. Makwana, D. R. Thakker, et al., AGDC: Automatic garbage detection and collection, arXiv preprint arXiv:1908.05849, 2019.
[134] CBAS, Sdg big data platform, https://sdg.casearth.cn/SDGDashboard, Last accessed on 2023–10–17 (2023).
[135] B. Kobler, J. Berbert, P. Caulk, P. Hariharan, Architecture and design of storage and data management for the NASA earth observing system data and information system (eosdis), in: Proceedings of IEEE 14th Symposium on Mass Storage Systems, IEEE, 1995, pp. 65–76.
[136] S. Hub, Cloud API for satellite imagery, www.sentinel-hub.com, Last accessed on 2023–10–17 (2023).
[137] R. Nemani, Nasa earth exchange: Next generation earth science collaborative, ISPRS-International Archives of the Photogrammetry, Remote Sensing and Spatial Information Sciences 3820 (2011) 17–17.
[138] PIE-Engine, Pixel information expert engine, https://engine.piesat.cn, Last accessed on 2023–10–17 (2023).
[139] OGE, Open geospatial engine, http://oge.whu.edu.cn, Last accessed on 2023–10–17 (2023).
[140] J.-F. ois Rolland, F. Castel, A. Haugommard, M. Aubrun, W. Yao, O. Dumitru, M. Datcu, M. Bylicki, B.-H. Tran, N. Aussenac-Gillles, et al., Candela: A cloud platform for Copernicus earth observation data analytics, in: IGARSS 2020–2020 IEEE International Geoscience and Remote Sensing Symposium, IEEE, 2020, pp. 3104–3107.
[141] V. C. Gomes, G. R. Queiroz, K. R. Ferreira, An overview of platforms for big earth observation data management and analysis, Remote Sensing 12 (8) (2020) 1253.
[142] W. Han, X. Zhang, Y. Wang, L. Wang, X. Huang, J. Li, S. Wang, W. Chen, X. Li, R. Feng, et al., A survey of machine learning and deep learning in remote sensing of geological environment: Challenges, advances, and opportunities, ISPRS Journal of Photogrammetry and Remote Sensing 202 (2023) 87–113.
[143] J.-F. Bastin, Y. Finegold, C. Garcia, D. Mollicone, M. Rezende, D. Routh, C. M. Zohner, T. W. Crowther, The global tree restoration potential, Science 365 (6448) (2019) 76–79.
[144] G. Ceccherini, G. Duveiller, G. Grassi, G. Lemoine, V. Avitabile, R. Pilli, A. Cescatti, Abrupt increase in harvested forest area over Europe after 2015, Nature 583 (7814) (2020) 72–77.

[145] M. Finer, S. Novoa, M. J. Weisse, R. Petersen, J. Mascaro, T. Souto, F. Stearns, R. G. Martinez, Combating deforestation: From satellite to intervention, Science 360 (6395) (2018) 1303–1305.

[146] A. J. Hansen, P. Burns, J. Ervin, S. J. Goetz, M. Hansen, O. Venter, J. E. Watson, P. A. Jantz, A. L. Virnig, K. Barnett, et al., A policy-driven framework for conserving the best of Earth's remaining moist tropical forests, Nature Ecology & Evolution 4 (10) (2020) 1377–1384.

[147] Y. Qin, X. Xiao, J. Dong, Y. Zhang, X. Wu, Y. Shimabukuro, E. Arai, C. Biradar, J. Wang, Z. Zou, et al., Improved estimates of forest cover and loss in the Brazilian Amazon in 2000–2017, Nature Sustainability 2 (8) (2019) 764–772.

[148] G. Donchyts, F. Baart, H. Winsemius, N. Gorelick, J. Kwadijk, N. Van De Giesen, Earth's surface water change over the past 30 years, Nature Climate Change 6 (9) (2016) 810–813.

[149] J.-F. Pekel, A. Cottam, N. Gorelick, A. S. Belward, High-resolution mapping of global surface water and its long-term changes, Nature 540 (7633) (2016) 418–422.

[150] X. Wang, X. Xiao, Z. Zou, J. Dong, Y. Qin, R. B. Doughty, M. A. Menarguez, B. Chen, J. Wang, H. Ye, et al., Gainers and losers of surface and terrestrial water resources in China during 1989–2016, Nature Communications 11 (1) (2020) 3471.

[151] Z. Zou, X. Xiao, J. Dong, Y. Qin, R. B. Doughty, M. A. Menarguez, G. Zhang, J. Wang, Divergent trends of open-surface water body area in the contiguous United States from 1984 to 2016, Proceedings of the National Academy of Sciences 115 (15) (2018) 3810–3815.

[152] Z. Zeng, L. Estes, A. D. Ziegler, A. Chen, T. Searchinger, F. Hua, K. Guan, A. Jintrawet, E. F. Wood, Highland cropland expansion and forest loss in southeast Asia in the twenty-first century, Nature Geoscience 11 (8) (2018) 556–562.

[153] E. M. Ordway, R. L. Naylor, R. N. Nkongho, E. F. Lambin, Oil palm expansion and deforestation in southwest Cameroon associated with proliferation of informal mills, Nature Communications 10 (1) (2019) 114.

[154] X. Liu, Y. Huang, X. Xu, X. Li, X. Li, P. Ciais, P. Lin, K. Gong, A. D. Ziegler, A. Chen, et al., High-spatiotemporal-resolution mapping of global urban change from 1985 to 2015, Nature Sustainability 3 (7) (2020) 564–570.

[155] J. C. Ho, A. M. Michalak, N. Pahlevan, Widespread global increase in intense lake phytoplankton blooms since the 1980s, Nature 574 (7780) (2019) 667–670.

[156] N. J. Murray, S. R. Phinn, M. DeWitt, R. Ferrari, R. Johnston, M. B. Lyons, N. Clinton, D. Thau, R. A. Fuller, The global distribution and trajectory of tidal flats, Nature 565 (7738) (2019) 222–225.

[157] J. H. Nienhuis, A. D. Ashton, D. A. Edmonds, A. Hoitink, A. J. Kettner, J. C. Rowland, T. E. Törnqvist, Global-scale human impact on delta morphology has led to net land area gain, Nature 577 (7791) (2020) 514–518.

[158] X. Yang, T. M. Pavelsky, G. H. Allen, The past and future of global river ice, Nature 577 (7788) (2020) 69–73.

[159] I. Overeem, B. D. Hudson, J. P. Syvitski, A. B. Mikkelsen, B. Hasholt, M. Van Den Broeke, B. Noe¨l, M. Morlighem, Substantial export of suspended sediment to the global oceans from glacial erosion in Greenland, Nature Geoscience 10 (11) (2017) 859–863.

[160] A. Ielpi, M. G. Lapoˆtre, A tenfold slowdown in river meander migration driven by plant life, Nature Geoscience 13 (1) (2020) 82–86.

[161] J. Valenza, D. Edmonds, T. Hwang, S. Roy, Downstream changes in river avulsion style are related to channel morphology, Nature Communications 11 (1) (2020) 2116.

[162] I. H. Myers-Smith, J. T. Kerby, G. K. Phoenix, J. W. Bjerke, H. E. Epstein, J. J. Assmann, C. John, L. Andreu-Hayles, S. Angers-Blondin, P. S. Beck, et al., Complexity revealed in the greening of the Arctic, Nature Climate Change 10 (2) (2020) 106–117.

[163] Z. Venter, M. Cramer, H.-J. Hawkins, Drivers of woody plant encroachment over Africa, Nature Communications 9 (1) (2018) 2272.

[164] G. Badgley, C. B. Field, J. A. Berry, Canopy near-infrared reflectance and terrestrial photosynthesis, Science Advances 3 (3) (2017) e1602244.

[165] M. J. Bogard, C. D. Kuhn, S. E. Johnston, R. G. Striegl, G. W. Holtgrieve, M. M. Dornblaser, R. G. Spencer, K. P. Wickland, D. E. Butman, Negligible cycling of terrestrial carbon in many lakes of the arid circumpolar landscape, Nature Geoscience 12 (3) (2019) 180–185.

[166] S. L. Maxwell, T. Evans, J. E. Watson, A. Morel, H. Grantham, A. Duncan, N. Harris, P. Potapov, R. K. Runting, O. Venter, et al., Degradation and forgone removals increase the carbon impact of intact forest loss by 626%, Science Advances 5 (10) (2019) eaax2546.

[167] A. Roopsind, B. Sohngen, J. Brandt, Evidence that a national REDD+ program reduces tree cover loss and carbon emissions in a high forest cover, low deforestation country, Proceedings of the National Academy of Sciences 116 (49) (2019) 24492–24499.

[168] B. D. Stocker, J. Zscheischler, T. F. Keenan, I. C. Prentice, S. I. Seneviratne, J. Pen˜uelas, Drought impacts on terrestrial primary production underestimated by satellite monitoring, Nature Geoscience 12 (4) (2019) 264–270.

[169] Y. Liu, M. Kumar, G. G. Katul, X. Feng, A. G. Konings, Plant hydraulics accentuates the effect of atmospheric moisture stress on transpiration, Nature Climate Change 10 (7) (2020) 691–695.

[170] D. N. Laskin, G. J. McDermid, S. E. Nielsen, S. J. Marshall, D. R. Roberts, A. Montaghi, Advances in phenology are conserved across scale in present and future climates, Nature Climate Change 9 (5) (2019) 419–425.

[171] M. Burke, D. B. Lobell, Satellite-based assessment of yield variation and its determinants in smallholder African systems, Proceedings of the National Academy of Sciences 114 (9) (2017) 2189–2194.

[172] M. G. Betts, C. Wolf, W. J. Ripple, B. Phalan, K. A. Millers, A. Duarte, S. H. Butchart, T. Levi, Global forest loss disproportionately erodes biodiversity in intact landscapes, Nature 547 (7664) (2017) 441–444.

[173] E. N. Dethier, S. L. Sartain, D. A. Lutz, Heightened levels and seasonal inversion of riverine suspended sediment in a tropical biodiversity hot spot due to artisanal gold mining, Proceedings of the National Academy of Sciences 116 (48) (2019) 23936–23941.

[174] M. Jung, P. Rowhani, J. P. Scharlemann, Impacts of past abrupt land change on local biodiversity globally, Nature Communications 10 (1) (2019) 5474.

[175] M. Pfeifer, V. Lefebvre, C. Peres, C. Banks-Leite, O. Wearn, C. Marsh, S. Butchart, V. Arroyo-Rodríguez, J. Barlow, A. Cerezo, et al., Creation of forest edges has a global impact on forest vertebrates, Nature 551 (7679) (2017) 187–191.

[176] A. R. Joshi, E. Dinerstein, E. Wikramanayake, M. L. Anderson, D. Olson, B. S. Jones, J. Seidensticker, S. Lumpkin, M. C. Hansen, N. C. Sizer, et al., Tracking changes and preventing loss in critical tiger habitat, Science Advances 2 (4) (2016) e1501675.

[177] E. T. Miller, G. M. Leighton, B. G. Freeman, A. C. Lees, R. A. Ligon, Ecological and geographical overlap drive plumage evolution and mimicry in woodpeckers, Nature Communications 10 (1) (2019) 1602.

[178] J. Giezendanner, D. Pasetto, J. Perez-Saez, C. Cerrato, R. Viterbi, S. Terzago, E. Palazzi, A. Rinaldo, Earth and field observations underpin metapopulation dynamics in complex landscapes: Near-term study on carabids, Proceedings of the National Academy of Sciences 117 (23) (2020) 12877–12884.

[179] J. N. Hendershot, J. R. Smith, C. B. Anderson, A. D. Letten, L. O. Frishkoff, J. R. Zook, T. Fukami, G. C. Daily, Intensive farming drives long-term shifts in avian community composition, Nature 579 (7799) (2020) 393–396.

[180] T. R. Walter, M. Haghshenas Haghighi, F. M. Schneider, D. Coppola, M. Motagh, J. Saul, A. Babeyko, T. Dahm, V. R. Troll, F. Tilmann, et al., Complex hazard cascade culminating in the Anak Krakatau sector collapse, Nature Communications 10 (1) (2019) 4339.

[181] P. Tuckett, J. Ely, A. Sole, S. Livingstone, B. Davison, J. van Wessem, J. Howard, Rapid accelerations of Antarctic peninsula outlet glaciers driven by surface melt, Nature Communications 10 (2019) 1–8.

[182] Z. S. Venter, K. Aunan, S. Chowdhury, J. Lelieveld, Covid-19 lock-downs cause global air pollution declines, Proceedings of the National Academy of Sciences 117 (32) (2020) 18984–18990.

[183] T. R. Chudley, P. Christoffersen, S. H. Doyle, M. Bougamont, C. M. Schoonman, B. Hubbard, M. R. James, Supraglacial lake drainage at a fast-flowing Greenlandic outlet glacier, Proceedings of the National Academy of Sciences 116 (51) (2019) 25468–25477.

[184] P. D. Kraaijenbrink, M. F. Bierkens, A. F. Lutz, W. Immerzeel, Impact of a global temperature rise of 1.5 degrees celsius on Asia's glaciers, Nature 549 (7671) (2017) 257–260.

[185] J. Ryan, L. Smith, D. Van As, S. Cooley, M. Cooper, L. Pitcher, A. Hubbard, Greenland ice sheet surface melt amplified by snowline migration and bare ice exposure, Science Advances 5 (3) (2019) eaav3738.

[186] D. J. Weiss, A. Nelson, H. Gibson, W. Temperley, S. Peedell, A. Lieber, M. Hancher, E. Poyart, S. Belchior, N. Fullman, et al., A global map of travel time to cities to assess inequalities in accessibility in 2015, Nature 553 (7688) (2018) 333–336.

[187] A. J. MacDonald, E. A. Mordecai, Amazon deforestation drives malaria transmission, and malaria burden reduces forest clearing, Proceedings of the National Academy of Sciences 116 (44) (2019) 22212–22218.

[188] X. Wu, D. Braun, J. Schwartz, M. Kioumourtzoglou, F. Dominici, Evaluating the impact of long-term exposure to fine particulate matter on mortality among the elderly, Science Advances 6 (29) (2020) eaba5692.

[189] M. F. Müller, J. Yoon, S. M. Gorelick, N. Avisse, A. Tilmant, Impact of the Syrian refugee crisis on land use and transboundary freshwater resources, Proceedings of the National Academy of Sciences 113 (52) (2016) 14932–14937.

[190] G. R. Watmough, C. L. Marcinko, C. Sullivan, K. Tschirhart, P. K. Mutuo, C. A. Palm, J.-C. Svenning, Socioecologically informed use of remote sensing data to predict rural household poverty, Proceedings of the National Academy of Sciences 116 (4) (2019) 1213–1218.

[191] C. Yeh, A. Perez, A. Driscoll, G. Azzari, Z. Tang, D. Lobell, S. Ermon, M. Burke, Using publicly available satellite imagery and deep learning to understand economic well-being in Africa, Nature Communications 11 (1) (2020) 2583.

[192] H. A. Orengo, F. C. Conesa, A. Garcia-Molsosa, A. Lobo, A. S. Green, M. Madella, C. A. Petrie, Automated detection of archaeological mounds using machine-learning classification of multisensor and multitemporal satellite data, Proceedings of the National Academy of Sciences 117 (31) (2020) 18240–18250.

[193] H. Guo, Big data, big science, big discovery: Review of codata workshop on big data for international scientific programmes, Bulletin of Chinese Academy of Sciences 29 (4) (2014) 500–506.

[194] E. R. DeLancey, J. F. Simms, M. Mahdianpari, B. Brisco, C. Mahoney, J. Kariyeva, Comparing deep learning and shallow learning for large-scale wetland classification in Alberta, Canada, Remote Sensing 12 (1) (2019) 2.

[195] M. C. Hansen, P. V. Potapov, R. Moore, M. Hancher, S. A. Turubanova, A. Tyukavina, D. Thau, S. V. Stehman, S. J. Goetz, T. R. Loveland, et al., High-resolution global maps of 21st-century forest cover change, Science 342 (6160) (2013) 850–853.

[196] Annual maps of global artificial impervious area (GAIA) between 1985 and 2018, Remote Sensing of Environment 236 (2020) 111510. doi:10.1016/j.rse.2019.111510.

[197] J. Dong, X. Xiao, M. A. Menarguez, G. Zhang, Y. Qin, D. Thau, C. Biradar, B. Moore III, Mapping paddy rice planting area in northeastern Asia with landsat 8 images, phenology-based algorithm and Google Earth Engine, Remote Sensing of Environment 185 (2016) 142–154.

[198] G. F. W. I. D. Rates, G. Statistics, www.glob-alforestwatch.org/dashboards/country,IDN/, Last accessed on 2020–11–19.

[199] G. Donchyts, F. Baart, H. Winsemius, N. Gorelick, Planetary-scale surface water detection from space, in: AGU Fall Meeting Abstracts, vol. 2017, 2017, pp. IN44A-01.

[200] A. Bey, A. Sánchez-Paus Díaz, D. Maniatis, G. Marchi, D. Mollicone, S. Ricci, J.-F. Bastin, R. Moore, S. Federici, M. Rezende, et al., Collect earth: Land use and land cover assessment through augmented visual interpretation, Remote Sensing 8 (10) (2016) 807.

[201] N. J. Murray, D. A. Keith, D. Simpson, J. H. Wilshire, R. M. Lucas, Remap: An online remote sensing application for land cover classification and monitoring, Methods in Ecology and Evolution 9 (9) (2018) 2019–2027.

[202] R. G. Allen, C. Morton, B. Kamble, A. Kilic, J. Huntington, D. Thau, N. Gorelick, T. Erickson, R. Moore, R. Trezza, et al., EEFlux: A landsat-based evapotranspiration mapping tool on the Google Earth Engine, in: 2015 ASABE/IA Irrigation Symposium: Emerging Technologies for Sustainable Irrigation-A Tribute to the Career of Terry Howell, Sr. Conference Proceedings, American Society of Agricultural and Biological Engineers, 2015, pp. 1–11.

[203] W. Jetz, J. M. McPherson, R. P. Guralnick, Integrating biodiversity distribution knowledge: Toward a global map of life, Trends in Ecology & Evolution 27 (3) (2012) 151–159.

[204] J. L. Huntington, K. C. Hegewisch, B. Daudert, C. G. Morton, J. T. Abatzoglou, D. J. McEvoy, T. Erickson, Climate engine: Cloud computing and visualization of climate and remote sensing data for advanced natural resource monitoring and process understanding, Bulletin of the American Meteorological Society 98 (11) (2017) 2397–2410.

[205] C. Morales, A. S.-P. Díaz, D. Dionisio, L. Guarnieri, G. Marchi, D. Maniatis, D. Mollicone, Earth map: A novel tool for fast performance of advanced land monitoring and climate assessment, Journal of Remote Sensing 3 (2023) 0003.

[206] W. Merten, A. Reyer, J. Savitz, J. Amos, P. Woods, B. Sullivan, Global Fishing Watch: Bringing transparency to global commercial fisheries, arXiv preprint arXiv:1609.08756, 2016.

# Part VIII

## Google Earth for Remote Sensing

# 15 The Legacy of Google Earth in Remote Sensing

*John E. Bailey and Josh Williams*

## ACRONYMS AND ABBREVIATIONS

| | |
|---|---|
| **API** | Application programming Interface |
| **CPU** | Central Processing Units |
| **DEM** | Digital Elevation Model |
| **GE** | Google Earth |
| **GEE** | Google Earth Engine |
| **GEC** | Google Earth Community |
| **GUI** | Graphical User Interface |
| **KML** | Keyhole Markup Language |
| **LiDAR** | Light Detection and Ranging |
| **NAIL** | Neighbors Against Irresponsible Logging |
| **NDVI** | Normalized Difference Vegetation Index |
| **NGA** | National Geospatial-Intelligence Agency |
| **NGO** | Nongovernmental Organization |
| **NOAA** | National Oceanic and Atmospheric Administration |
| **OGC** | Open Geospatial Consortium |
| **SRTM** | Shuttle Radar Topographic Mapping Mission |
| **WYSIWYG** | "What you see is what you get" |
| **XML** | Extensible Markup Language |

## 15.1 INTRODUCTION

Almost two decades have now passed since the release of Google Earth (GE). It is hard to fully measure the impact it has had on all areas of society, as it opened possibilities in the world of remote sensing to everyone from the youngest student looking at satellite images for the first time to an experienced professor who was now able to access imagery at a scale and speed never before possible. At the same time its detractors will point to limitations in the accuracy of GE's representation of the globe and the application's lack of ability as an analytical tool, to dismiss its importance. Depending on your perspective it can be easy to either view GE as a novelty application, or inversely, overstate the role it has played in the development of remote sensing applications. Judgment of its true impact requires consideration of the nuances of both viewpoints.

When GE was launched in 2005 (Google LLC. 2005), and subsequently rapidly gained popularity (Lubick 2005), though many scientists, analysts, and other remote sensing experts initially dismissed it as mostly a novelty that had some uses but was not a particularly ground-breaking application due to its limitations as an analytical platform and inherent inaccuracies associated with being a 2D projection of an assumed perfectly spherical Earth. These critics saw the wave of everyday users accessing Google Earth to "view my house" and many dismissed it as a fad.

While some criticisms of Google Earth's technical limitations are fair, they overlook the importance it has played in demonstrating and developing technology that allows rapid "3D" visualization

DOI: 10.1201/9781003541141-23

of geolocated imagery for the whole planet, so-called virtual globes (Bailey 2010). Google Earth wasn't the first technology to demonstrate this ability, e.g., Microsoft Encarta Virtual Globe 1998 Edition (Microsoft News 1997), or the first to be widely publicized, e.g., NASA WorldWind (Hogan and Kim 2004), and new applications continue to be developed, e.g., Mapbox Globe view (Anton and van Dongen 2022). But GE was the application that popularized this type of technology as a global phenomena, and this will always be the application's legacy in the history of remote sensing. The idea that knowing what your town, neighborhood, house looks like from space has gone from being specialist knowledge to an everyday concept that you can access through the phone in your pocket.

At the same time Google Earth is not the all seeing entity some believe it is. In the 1992 sci-fi novel *Snow Crash* (Stephenson 1992) the Central Intelligence Corporation (CIC) developed a piece of panoptic software called "Earth" that is able to zoom into real-time, high-resolution views of anywhere on the planet. Snow Crash is one of several influences that have been discussed as part of the origin story behind Google Earth (Bar-Zeev 2007). Google Earth does not offer this kind of data feed, although the modern era of high-altitude drone surveillance does suggest that something close to the CIC Earth could now be technologically feasible.

In 2004, the technology that was to later become GE was called "EarthViewer" (Figure 15.1) and was the proprietary property of Keyhole Inc, a Silicon Valley start-up venture formed in 2001, acquired by Google that year (Wall Street Journal 2004). It was one of several computer applications creating a 3D model of Earth or other planets that evolved in the early 2000s as technology caught up with ideas (Foerch 2017). For many they were seen as realization of a vision that had been laid out in different forms by many from futurists (Fuller 1962) to academics (Chen and Bailey 2011; Whitmeyer et al. 2012), to artists (Eames and Eames 1977), and to science fiction writers

FIGURE 15.1 Screenshot of Keyhole's EarthViewer 3D showing the Port of Tokyo near Hamazakibashi Junction.

(Stephenson 1992). Some credit Al Gore's 1998 address to the California Science Center (Gore 1998) as succinctly verbalizing the concept of a "Digital Earth," of which GE ultimately became the most popular actualization.

This chapter provides a brief overview of the evolution of the GE technology (Section 15.2) and explores how it has become the go-to application for free and accessible viewing of high-resolution satellite imagery (Section 15.3). It also traces the timeline of GE's development, by different user groups, as a canvas for geospatial visualization and storytelling (Section 15.4), and concludes by considering the legacy of GE in remote sensing (Section 15.5).

## 15.2 VERSIONS AND EVOLUTION OF GOOGLE EARTH

GE is currently available for a range of operating systems and platforms (Google LLC. 2023a). These can primarily be differentiated into three categories: (1) desktop, (2) mobile, and (3) web (Figure 15.2). The desktop version (Google Earth Pro) is a downloadable binary that is currently supported on PC, Mac, and Linux systems. The web version (earth.google.com) is accessible through all common web browsers. The mobile version is available as Android and iOS Apps are similar to the web version. Over its two decades of development, GE has evolved in terms of the availability and features present on the different platforms. Different functionality has come and gone, and sometimes returned, but the core of the technology is the rendering of a 3D globe combination of terrain data overlain by imagery.

The original base Digital Elevation Model (DEM) used to model terrain was created by the Shuttle Radar Topography Mission (SRTM), Using data collected in February 2000 by modified radar instruments on the 11-day STS-99 mission, SRTM generated the most complete high-resolution

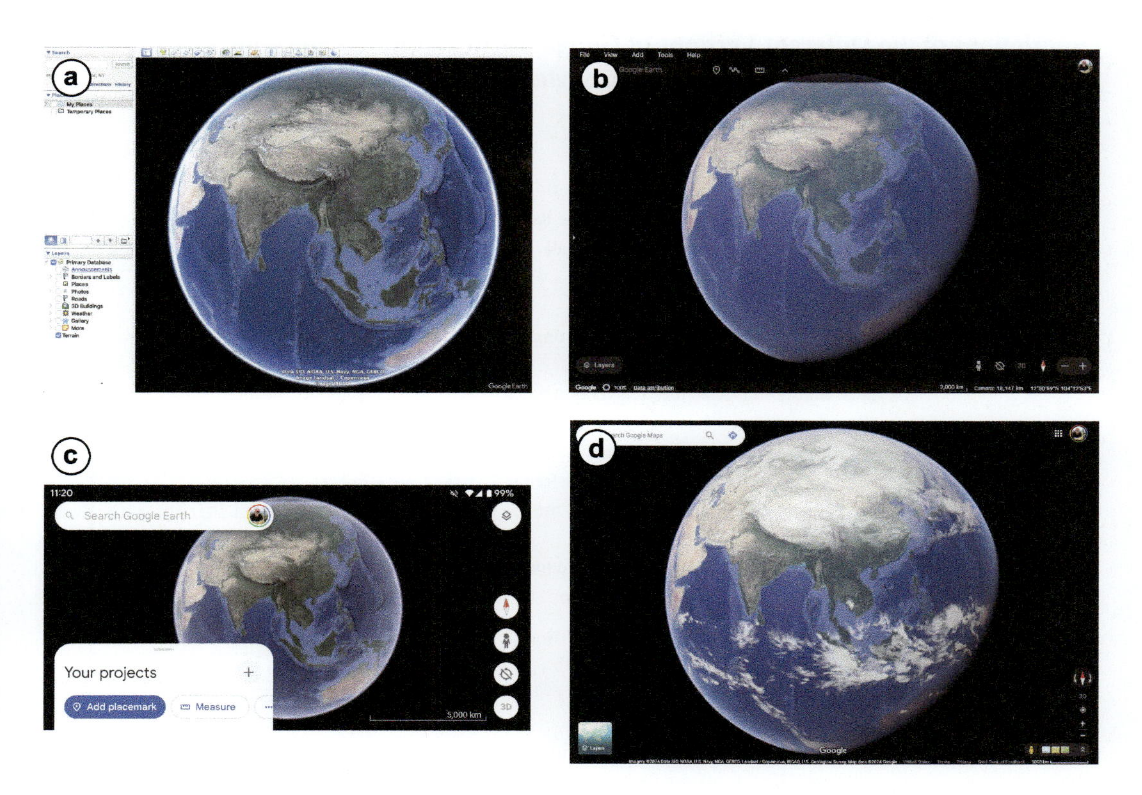


FIGURE 15.2 Screenshots of Google Earth on different platforms. (a) GE Pro for Desktop (64-bit build for Mac: 7.3.6.9345), (b) web-based Google Earth (build: 10.43.0.2), (c) Mobile Google Earth for Android on a Pixel 4A-5G screen (build: 10.41.0.7), (d) the Earth "Globe view" in Google Maps.

digital topographic database of Earth available at that time (Farr et al. 2007). It was supplemented by other datasets for high latitudes and mountainous regions that required higher-resolution data to show the true topography. There are many areas, for example, western United States, The Swiss Alps, and Canary Islands, where high-resolution light detection and ranging (LiDAR) datasets have been acquired and supplant the SRTM imagery. The original base imagery for GE was Landsat 30 m multispectral data, pan sharpened using Landsat's 15 m panchromatic imagery. These data are now supplemented by images with comparable resolution from the European Union's Copernicus mission, notably Sentinel-2's MultiSpectral Instrument (MSI). However, for many regions and most urban centers, very high spatial resolution imagery (VHRI; sub-meter to 10 m resolution) has been acquired from DigitalGlobe, Système Pour l'Observation de la Terre (SPOT) Image, and other commercial satellite providers. High resolution aerial photography is also integrated as many of these flights provide both imagery and derived terrain data. This collection method is primarily used for urban centers to create photorealistic 3D buildings (Dennis 2017).

Ironically, when many users first "found my house" they may have done so not using GE at all or perhaps did so through a dataset (Google Street View) that is part of GE and Google Maps, and also once had its own application. The Google Geo landscape is constantly changing and evolving, and confusion between Earth, Maps and Street View is understandable, especially since the products themselves are intertwined across different versions of GE (Table 15.1). Google Maps is Google's core web-mapping platform, with web, mobile, and API versions. Web Maps can be placed in "Satellite map type" and "Globe view," providing GE-type viewing of satellite and aerial imagery overlain on a DEM that can be navigated with the correct keyboard and mouse combination.

**TABLE 15.1**
**Descriptions of the Different Versions of Google Earth Supported Since the Original Keyhole Technology Was Repackaged by Google**

| GE Version | Description | Timeline 2005 2007 2009 2011 2013 2015 2017 2019 2021 2023 |
|---|---|---|
| **Google Earth Basic** | Free, desktop application with the same base imagery and geodata used to support all versions. Windows, Mac and Linux. | |
| Google Earth Plus | Subscription version ($20/year) with additional features focused on supporting GPS integration and data import. This functionality moved to the free version when it was deprecated. | |
| Google Earth Pro | Subscription version with additional functionality ($400/year) that became free in Jan. 2015 and replaced the "basic" as the only desktop version in 2017. | |
| Google Earth Enterprise | High cost annual subscription version with dedicated server and technical support. | |
| Google Earth API | The GE Plug-in and JavaScript API allows for Google Earth to be integrated into web pages, but without all the functionality of desktop GE. | |
| Google Earth—Web | After 2+ years of development to rewrite the codebase, GE was released as a web-based application. | |
| Google Earth—mobile app | Currently available for both Android and iOS. Early Apps were redesigned in 2017 to use architecture similar to the web version. | |

**TABLE 15.1** *(Continued)*
**Descriptions of the Different Versions of Google Earth Supported Since the Original Keyhole Technology Was Repackaged by Google**

| GE Version | Description | Timeline 2005 2007 2009 2011 2013 2015 2017 2019 2021 2023 |
|---|---|---|
| Google Earth VR | Available through Valve's Steam computer gaming platform and supports Oculus Rift and HTC Vive virtual reality headsets and controllers. | |
| Google Earth Engine (GEE) | Despite the name, and although inspired by GE, GEE uses the Google Maps API not GE. | |
| Google Maps—Globe | Google's mapping platform leverages the same geodatabases and imagery servers as GE. Inclusion of GE-like views directly in Google Maps has varied over time. Currently the web version allows viewing of a 3D "Globe view," with some limitations on tilting. | |

*Note:* The timeline gray boxes show the years a specific version was supported. Darker gray indicates features were integrated from deprecated versions. Dates represent when at least one platform was supported for that version of the application.

The mobile-based version of Maps can also display satellite data and offers limited tilted viewing. Unlike the web version the underlying DEM is not present in Mobile Maps. Similarly there are 3D tiling options available for the Maps API, but they do not provide the full 3D terrain experience. There was previously a Google Earth API but it was retired in 2014 (Table 15.1). All of the Google Maps platforms provide access to the Street View 360 imagery, where users can see ground-level photography of their houses (or other places). Street View can also be accessed through all versions of Google Earth. Maps and Earth provide the geolocational context for the Street View imagery.

## 15.3 GOOGLE EARTH'S FREE AND ACCESSIBLE HIGH-RESOLUTION IMAGERY

The most enduring legacy of GE is the paradigm shift it created for remote sensing, specifically in terms of imagery accessibility. GE has integrated high and very high spatial resolution imagery seamlessly for the entire world, and demonstrated that the technology infrastructure exists to make all data viewable with just the processing power available on a simple laptop or phone.

While the computing revolution starting in the 1960s and 1970s, ultimately created a path for the creation of Keyhole's EarthViewer 3D and other virtual globes, the timing of this development was a consequence of the collision of technological advances in three other key areas. The 1980s–1990s gaming industry primarily drove a need for high-performance graphics cards that provided computer performance suitable for displaying a virtual globe. The 2000s global expansion of broadband internet connectivity provided bandwidth that allowed these applications to function. Meanwhile the modern era of commercially available high-resolution satellite data began with the launches of GeoEye's Ikonos (in 1999) and DigitalGlobe's Quickbird (in 2001) satellites, and for the first time almost global coverage of Earth at centimeters to meters resolution was available to all who could afford it. These combined to create an evolutionary step in the world of remote sensing. For hundreds of millions of people the concept that satellites can be used to observe from Earth moved from the realm of science fiction and something scientists do, to any everyday viewpoint for anyone with a computer and an internet connection.

### 15.3.1 Google Earth's First Major Application: Hurricane Katrina

An opportunity for GE to display its potential occurred just weeks after the Google announcement of the reworked Keyhole technology (Google LLC. 2005). On August 29, 2005, Hurricane Katrina made landfall in Louisiana, causing devastation across the region (Knabb et al. 2005). Nowhere was the effect felt more than in the city of New Orleans. Much of the city lies below sea level and the high rainfall produced by the hurricane led to protective riverbank levees breaking, and the city was flooded. Thousands of people were trapped in their houses and as the floodwaters rose, people escaped to the roofs of their houses but often without any form of communication.

In order to find those stranded, plan logistics, and work out access routes, the National Oceanic and Atmospheric Administration (NOAA) captured thousands of high-resolution aerial images (NOAA 2005). This information was invaluable for disaster response efforts, with more than five million photos downloaded from the NOAA website each day during the first week (Nourbakhsh et al. 2006). However, processing and navigating this information in a useful manner was a colossal effort and one for which GE was to provide the backbone. The GE engineers worked around the clock to serve the images through the GE client application as KML overlays. These geolocated, stitched together layers of data were searchable and easily accessible by federal agencies, disaster response groups, and the general public. For example, this imagery was used to identify intact churches in flood-free suburbs of New Orleans that could be used as outlet centers for aid donations (Kilday 2018). The National Geospatial-Intelligence Agency (NGA) later recognized Google's contribution with a Hurricane Katrina Recognition Award, but more importantly it demonstrated GE's facility to be a tool with real-world applications.

### 15.3.2 Google Earth Engine: Analysis of Imagery

Despite the obvious benefits for efforts such as crisis response, the ability of every user to become armchair remote sensing analysts also raised concerns. Some governments raised security concerns (Finkelberg 2007), meanwhile conspiracy theorists now had evidence to back up their claims of discovering Atlantis or a hidden alien base (Geens 2009; Google LLC. 2009).

Even Google's Katrina efforts came under fire when the base imagery reverted back to pre-Katrina views (Sullivan 2007), which was later explained by the fact this had been a "rush job" that needed to be properly integrated into the core geodatabases.

But for many communities, the positives far outweighed any negative connotations as these tools opened up exploration of the whole planet for professionals and curious public alike. In particular, it has developed as an observation tool for structural geologists, geomorphologists, volcanologists, hydrologists, and other scientists who require easy access to detailed imagery. In these cases, GE has enabled studies at a scale that, while technically possible, was not practical even with extensive, time-consuming and well-funded field campaigns (Figure 15.3).

Yet, a lingering limitation and criticism was that while GE provided users with the ability to view an immense archive of imagery, and with the release of Google Earth ver. 5.0 (Lowensohn 2009) could even scroll back to older images for a given area, the utility was limited. GE did not provide analysis tools to make useful observations on current and especially the archive of imagery. There were power users doing creative things through content created using Keyhole Markup Language (KML; see Chapter 15.4) but specialists who wished to directly analyze the images had limited options.

These needs were recognized early on in the development of GE and inspired creation of a parallel project—Google Earth Engine (GEE)—whose name is a misnomer in the sense that the application is built on the Google Maps API, not the Google Earth rendering technology. As such, GEE is worthy of its own topic of discussion but given that its development was inspired by the needs of Google Earth's users it is appropriate to provide an overview here. Google Earth Engine (GEE) is a cloud computing platform for hosting and processing satellite imagery and other Earth observation data (Gorelick et al. 2017). It leverages Google's large distributed server capabilities to analyze imagery. It uses the same imagery and GIS databases created to support GE, and makes 50 years of satellite imagery available to allow users to map changes and trends, and quantify differences for the Earth's surface. Example applications include, but are not limited to, detecting deforestation (Hansen et al. 2013), mapping crop yields (Lobell et al. 2015), delineation of forest transition zones

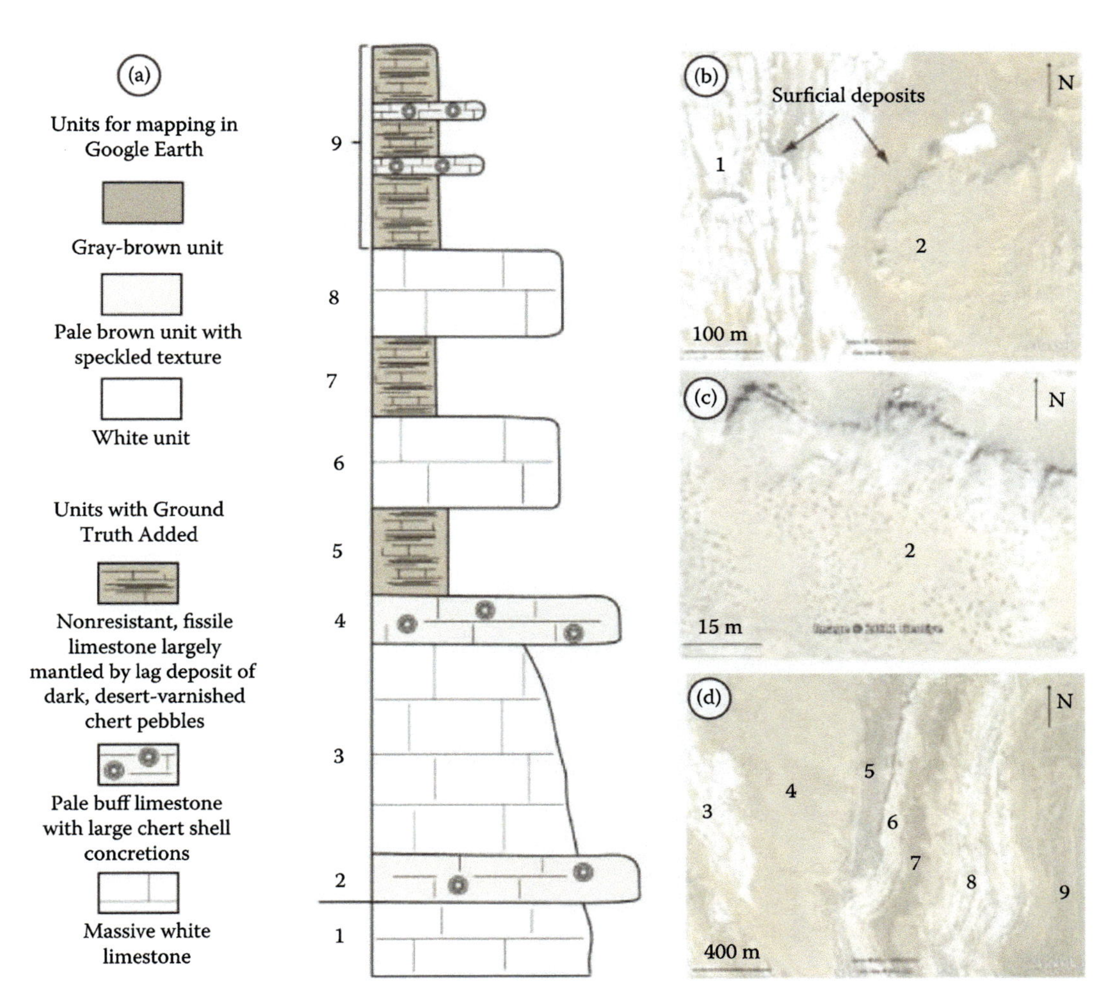


FIGURE 15.3 Tewksbury et al. (2012) used Google Earth to study structural geology in remote areas of Egypt. Using Google Earth imagery they established a stratigraphy consisting of one subunit in the El-Rufuf Formation (Unit 1) and eight subunits in the Drunka Formation (Units 2–9).

(Wei et al. 2020), monitoring changes to tiger habitat (Joshi et al. 2016), identifying the distribution of mangrove forests (Giri et al. 2014), global mapping of surface water (Pekel et al. 2016), and settlement and population mapping (Patel et al. 2015).

What makes Earth Engine different as a remote sensing tool is the ability to rapidly run large computations by distributing the processing across Google's large network of servers (Gandhi 2023). Efficient use of this cloud-based infrastructure is facilitated by the Earth Engine API through which the users specify the computations to be performed. However, the API is designed so that users do not need to be concerned with how this computation is then distributed across the server network, only that the results are assembled for them, greatly simplifying the code the user needs to supply (Gandhi 2023).

Consequently Earth Engine is very approachable for users who are not familiar with writing code. The Earth Engine API is designed to be language agnostic, and Google provides official client libraries to use the API from both JavaScript and Python. While scripting in Python or R might be more common for scientists, the Earth Engine JavaScript API is the most mature and the easiest entry point. The web-based Earth Engine platform includes a code editor (Figure 15.4) that allows users to access the JavaScript API without any installation (Google LLC. 2023b). It also provides

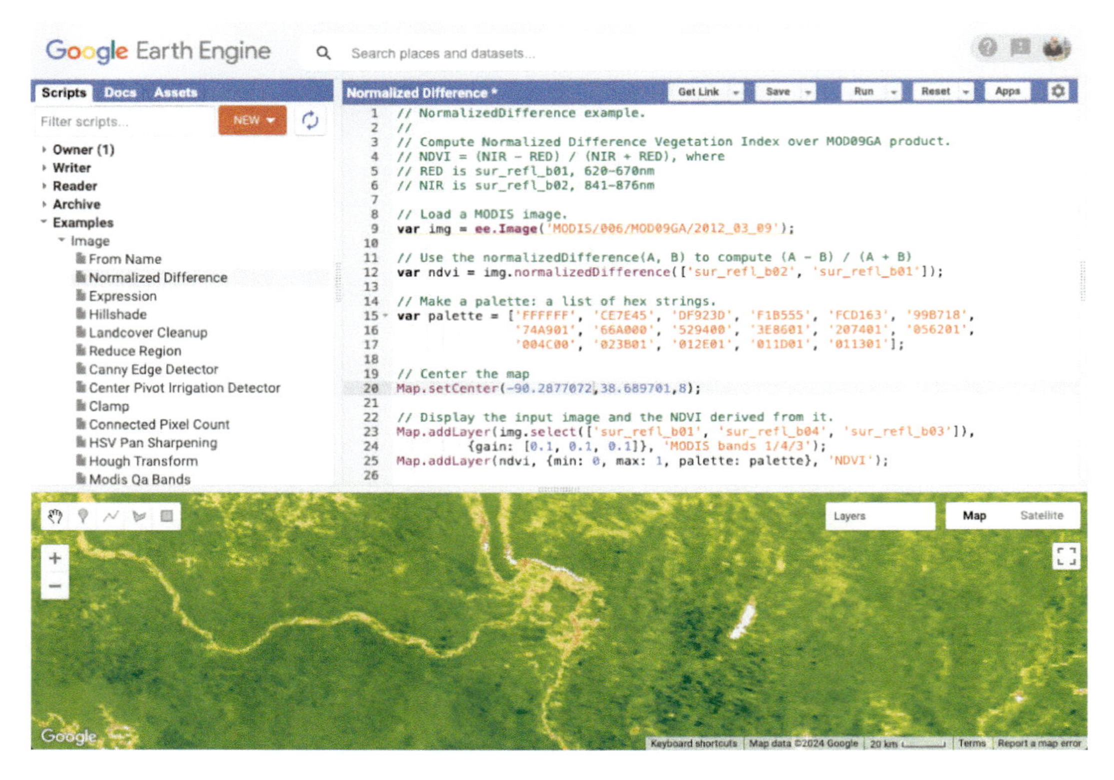


**FIGURE 15.4** The Google Earth Engine code editor interface (https://code.earthengine.google.com/). A few lines of GEE code generates a NDVI image highlighting the confluence of the Missouri and Mississippi Rivers at St. Louis, MO. Users can resize and pan around the map view and the generated view will be recalculated on-the-fly.

additional functionality to display results on a map, save scripts, manage tasks, share code, and access documentation (Gandhi 2023).

Supported by Google Developer Advocates, a strong GEE user community has developed and together these groups have collaborated to create documentation and tutorials to support both new and experienced users (Cardille et al. 2023). Similarly, user-generated documentation and remote sensing tutorials using GEE can also be found on github (Kochenour 2020) and other open source sites.

### 15.3.3 Timelapse Project

An example of the power of GEE is provided by the Timelapse project, which leverages the massive archive of Landsat data, a program managed by the USGS that has been acquiring images of the Earth's surface since 1972. Using GEE a global mosaic was constructed with each frame of the Timelapse animation constructed from a year of Landsat imagery, creating an annual 1.7 terapixel snapshot of the Earth at 30 m resolution. Through a combination of image stacking and pixel matching, the composite images were created free from clouds and without the data loss created by the scan line corrector failure on Landsat 7.

The initial version covered 29 years (from 1984 to 2012), and processed 2,068,467 scenes, a total of 909 TB of data. This required 2 million CPU (central processing unit) hours and processing was distributed over 66,000 CPUs allowing the mosaics to be computed in one and a half days. Since then, several updates have multiplied the number of images used, and by the 2020 version the animated Timelapse video was created from more than 24 million satellite images, equaling 20 petabytes of data. The quadrillions of pixels were combined into a single 4.4 terapixel-sized video mosaic—the equivalent of 530,000 videos in 4K resolution—yet, the distributed processing using Google's Cloud (and the last decade of server improvements) meant that computing time was comparable to the initial version (Moore 2021).

The most recent update added 2021–2022 data, and the latest updates also now incorporate Sentinel data, along with the Landsat archive (Herwig 2023). This provides a four decade record of change on the Earth's surface, with five themes emerging (Moore 2021): forest change, urban growth, warming temperatures, sources of energy, and our world's fragile beauty. River system changes (Figure 15.5), receding glaciers, and an expansion of mining operations are especially vividly highlighted by Timelapse (Klugger and Walsh 2013).

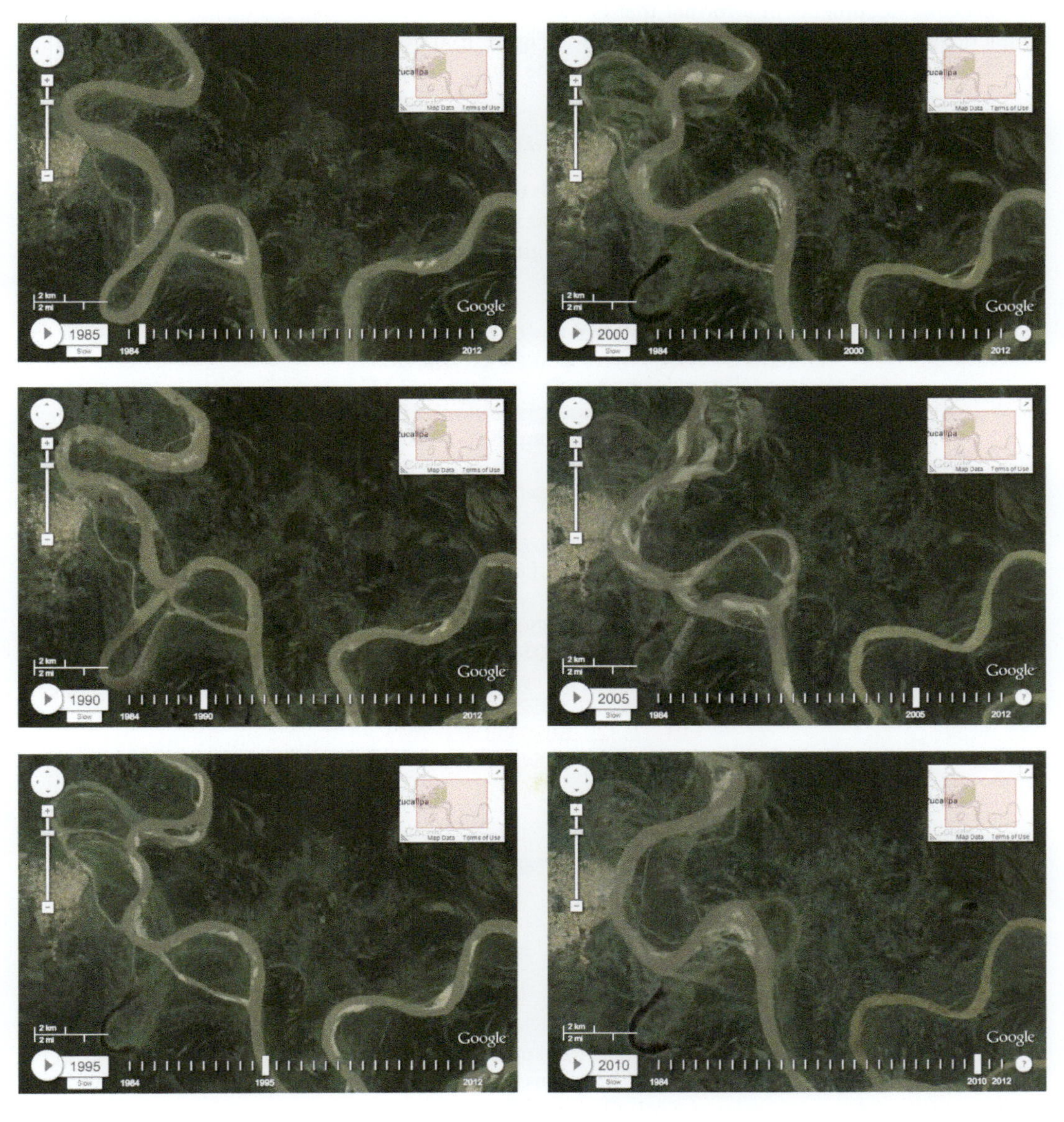


**FIGURE 15.5** Timelapse, powered by Google Earth Engine, uses four decades of Landsat imagery to show the meandering changes of the Ucayali River, near Pucallpa, Peru.

## 15.4 EVOLUTION OF GOOGLE EARTH AS A GEOSPATIAL VISUALIZATION TOOL

Aside from the accessibility of imagery, GE's other significant contribution has been the democratization of the ability for all users to create visualizations from geospatial data. Upon launch this capability was mentioned but the emphasis was on GE being a searchable 3D map rather than as a visualization tool. The ability for users to craft their own content was referenced as "easy creation and sharing of annotations among users" (Google LLC. 2005). However, as scientists and remote sensing professionals began to explore with GE, they realized the potential offered by Keyhole Markup Language (KML), the code supported by the application, to visualize user generated content in the form of both vector (placemakers, lines, polygons) and raster (overlays) features (Wernecke 2009). KML is an Extensible Markup Language (XML), a computer code that defines a set of rules for encoding documents in a format that is human readable but is also annotated in a way that is syntactically distinguishable from regular text (Figure 15.6). Originally developed by Keyhole Inc., in 2008 it became an international standard of the Open Geospatial Consortium (OGC).

### 15.4.1 Google Earth for Telling Geospatial Stories of Societal Importance

Through KML, GE became the canvas on which users can display geospatial information and add a narrative about the landscape that they live on, study, and/or care about. The trailblazer for this use was a citizen action group's response to a logging plan for their neighborhood in Los Gatos, CA.

```
<?xml version="1.0" encoding="UTF-8"?>
<kml xmlns="http://www.opengis.net/kml/2.2"
xmlns:gx="http://www.google.com/kml/ext/2.2"
xmlns:kml="http://www.opengis.net/kml/2.2"
xmlns:atom="http://www.w3.org/2005/Atom">
<Document>
    <name>Tanzania</name>
    <Placemark>
        <name>Mount Kilimanjaro</name>
        <description>Africa's highest mountain</description>
        <LookAt>
            <longitude>37.35008</longitude>
            <latitude>3.08901</latitude>
            <altitude>0</altitude>
            <heading>14.60520065400174</heading>
            <tilt>66.32916850433209</tilt>
            <range>9499.909842132638</range>
        </LookAt>
        <Point>
            <coordinates>
                37.35562729999999,-3.0674247,0
            </coordinates>
        </Point>
    </Placemark>
</Document>
</kml>
```

FIGURE 15.6 A simple example of a Keyhole Markup Language (KML) file. Different components of the text have been colored (in this example) for the purposes of identification. The gray text is namespace information, which references the rules of KML file types. The black text creates a container for KML features. The blue text creates a placemark, with a name, at the given coordinates. The green text gives a description that appears in a balloon associated with the location. The red text defines the default "fly-to" viewpoint for this placemark.

The Neighbors Against Irresponsible Logging (NAIL) was a nonprofit organization formed in 2005 by Rebecca Moore, a software engineer who had just joined the GE team. After receiving a low-quality black and white leaflet in her mail outlining a company's logging plan for the Santa Cruz forest she lived in, Moore used GE to create powerful visualizations of the true scale of the plan that helped rally her neighbors into action (Figure 15.7). The press picked up on these visualizations and

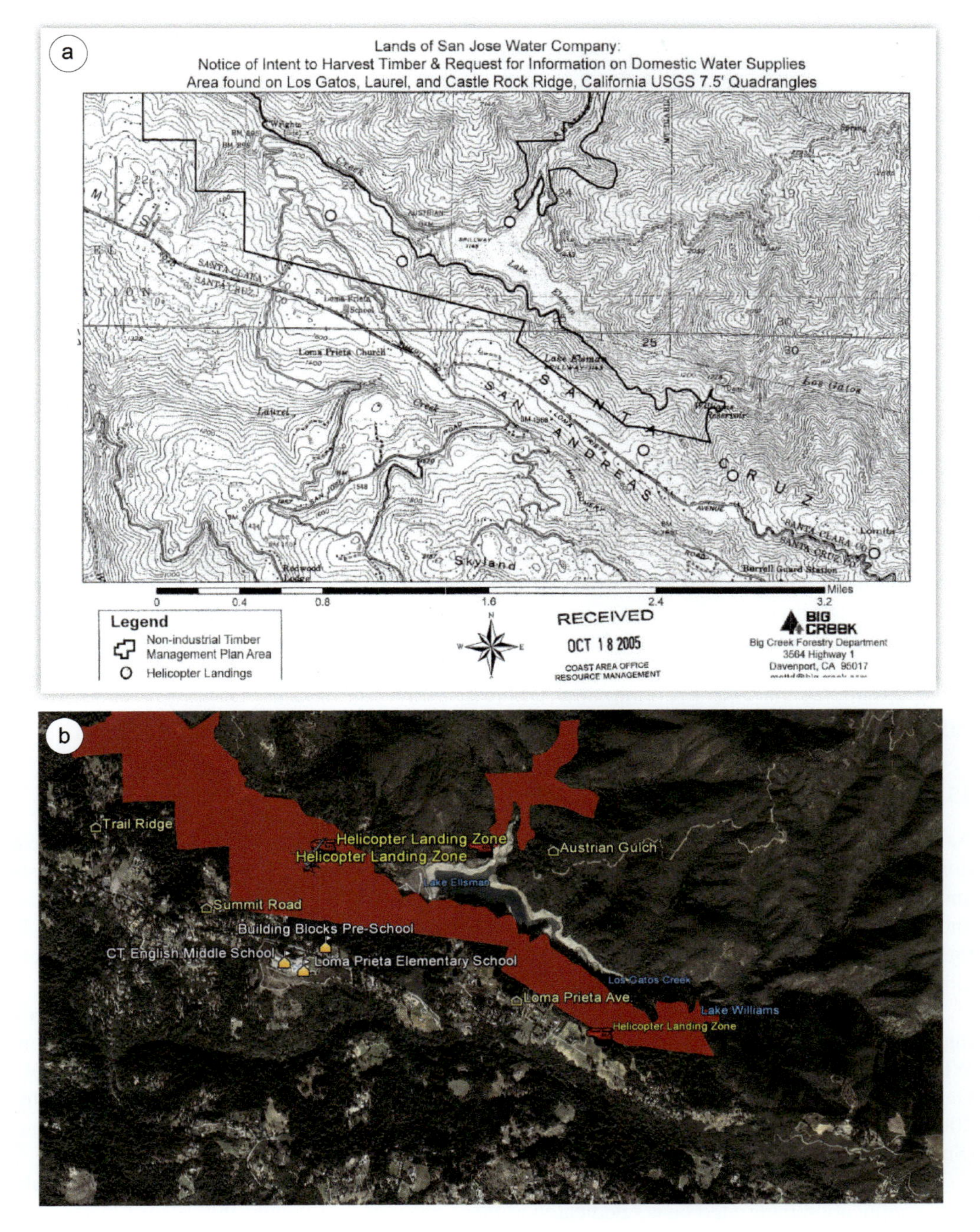


FIGURE 15.7 Information on proposed logging given to a neighborhood in Los Gatos, CA. (a) The original map that was circulated to neighbors by the logging company. (b) A Google Earth visualization of the same data created by a Googler who lived in the neighborhood.

the subsequent coverage led to the proposed logging plan being ruled ineligible by the California Department of Forestry (Moore n.d.).

While not the original intent, the success of NAIL led to conservation and nonprofit groups contacting Moore for help with telling their stories. NGOs such as the Sierra Club, the Jane Goodall Institute, and Appalachian Voices began to use GE as an educational platform for their messages (Google Earth Outreach 2023; Figure 15.8). The interest in GE as an outreach tool led Moore to form the GE Outreach team, which focuses on empowering nonprofit organizations and outreach groups to use "Google Geo for Good."

### 15.4.2 Early Adopters: Bloggers and Journalists

Other early adoption users included bloggers and journalists who also grasped GE's potential for bringing their stories to life (Lubick 2005; Butler 2006a). Where this broke new ground was rather than GE just being another way to create colorful illustrations, it gave the journalists themselves the power to great dynamic representation of the data. This is what Declan Butler, a Nature reporter and early champion for GE abilities, did with data points showing the spread of avian flu (Butler 2006b, 2006c; Figure 15.9). Although wider acceptance was slower at first, the scientific community also embraced the possibilities offered by GE for both research and education (Gramling 2007; Simonite 2007; Bailey and Chen 2011; Yu and Gong 2012).

### 15.4.3 Early Science Application: Volcanology

One of the first fields to demonstrate GE's capabilities to dynamically visualize Earth observations was volcanology (Figure 15.10). The Smithsonian Global Volcanology Program worked with the GE team to develop a volcano layer (Figure 15.10a), which became one of the most frequently accessed datasets in the application (Venzke et al. 2006). Meanwhile researchers in the Alaska Volcano Observatory remote sensing group developed the use of KML for observing real-time thermal satellite imagery (Bailey and Dehn 2006; Figure 15.10b) and modeling volcanic ash clouds (Webley et al. 2009a, 2009b; Figure 15.10c and d). Further development included locating volcanic earthquakes (De Paor 2008), assessment of volcanic hazards to aviation (Williams and Thomas 2011), visualizing volcanic gas plumes (Wright et al. 2009) and mapping of volcanic geomorphology (de Silva and Bailey 2017).

### 15.4.4 Power Users: Scientists and Educators

As GE evolved as an application, so did its target audience. During the first few years innovation was driven by the professional power users (remote sensing users, bloggers and journalists) and enthusiastic hobbyists who were posting to the Google Earth Community (GEC) BBS/forum, another legacy of the Keyhole acquisition where the community first began. Wider acceptance by the scientific and the education community followed, and by the end of its first decade, these users, along with the non-profit organizations were driving the innovation.

Aside from volcanology, scientific disciplines that quickly accepted GE as a useful remote sensing tool for a specific discipline included structural geology (Lisle 2006; De Paor 2008; De Paor and Whitmeyer 2011; Blenkinsop 2012, Figure 15.11a), cryospheric studies (Ballagh et al. 2007, 2011; Gergely et al. 2008, Figure 15.11b), hydraulic studies and management (Silberbauer and Geldenhuys 2008; Chien and Tan 2011; USGS 2014), and extreme weather tracking, particularly hurricanes (Smith and Lakshmanan 2006, 2011; Turk et al. 2011; Liu et al. 2015).

Weather and storm tracking was also a favorite topic for bloggers who discussed GE, especially as the application integrated layers dedicated to these phenomena (Taylor 2008 Figure 15.11c). The integration of hurricane tracking is especially poignant given the use of GE to serve post-Katrina imagery (see Section 15.3.1) and ultimately highlights its ability to provide "before and after"

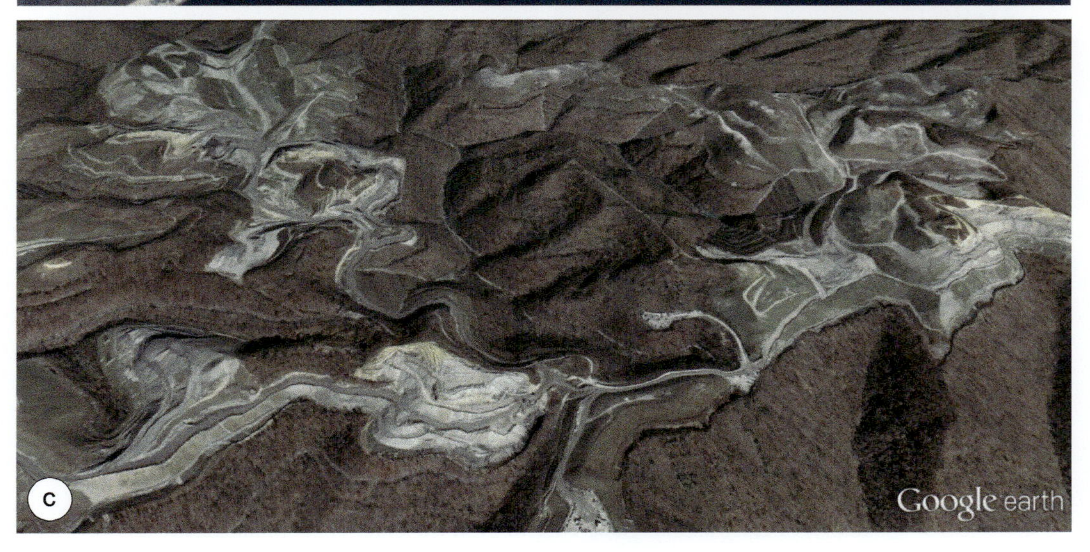


**FIGURE 15.8** Examples of NGOs' uses of Google Earth. (a) The Sierra Club visualized the impacts of sea level rise on Vancouver, Canada (Sheppard and Cizek 2008). (b) The Jane Goodall Institute created a geolocated blog about the Chimpanzees' activity (Google Earth 2007). (c) Appalachian Voices used Google Earth imagery and KML overlays to show mountaintop removal due to open-pit mining (Appalachian Voices 2014).

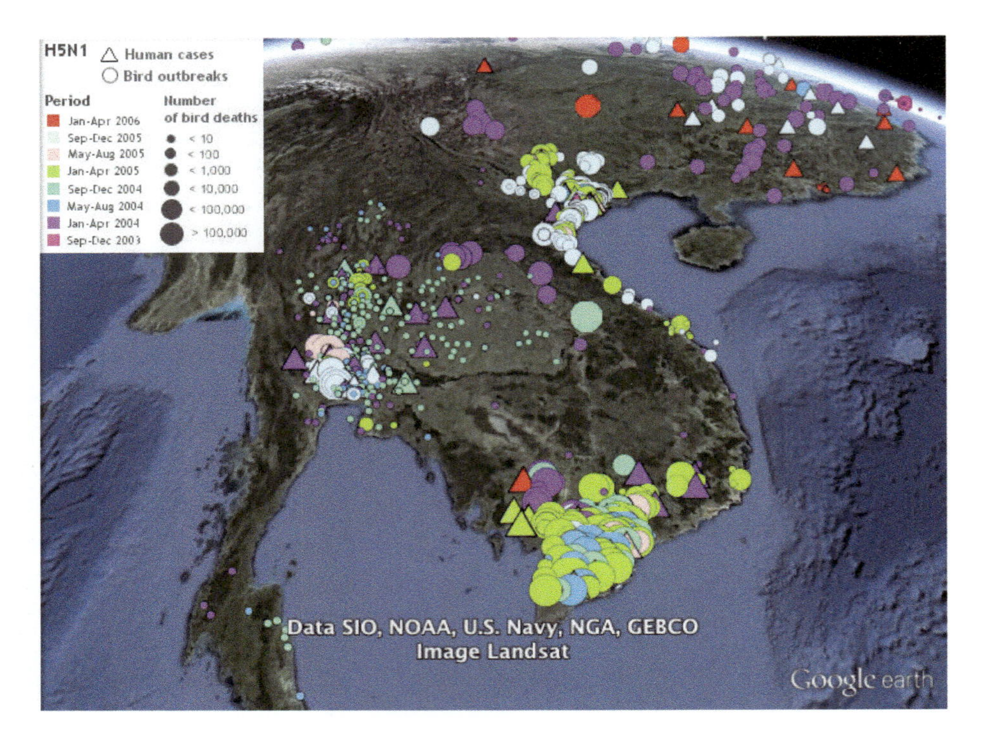


**FIGURE 15.9** Visualizing the spread of the N5H1 virus (aka avian flu) using Google Earth and KML (Butler 2006b, 2006c).

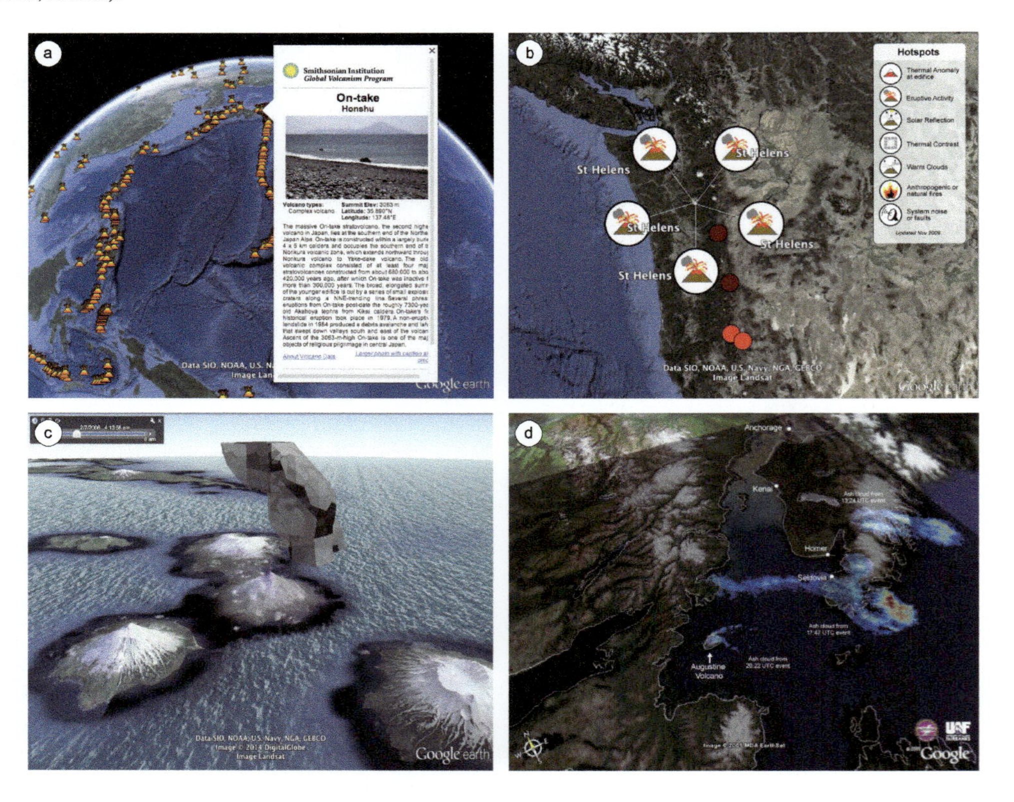


**FIGURE 15.10** Examples of early uses of Google Earth by volcanologists. (a) Smithsonian GVP's Google Earth layer of Holocene volcanoes (Venzke et al. 2006). (b) Locations of thermal hotspots shown using KML derived from thermal IR images geolocated in GE. (c) 3D visualization of an ash plume erupted from Cleveland Volcano in the Aleutian Islands. (d) Ash cloud density identified using AVHRR Thermal IR imagery and visualized as translucent overlays in GE. Images (b) to (d) created by the Alaska Volcano Observatory Remote Sensing Group.

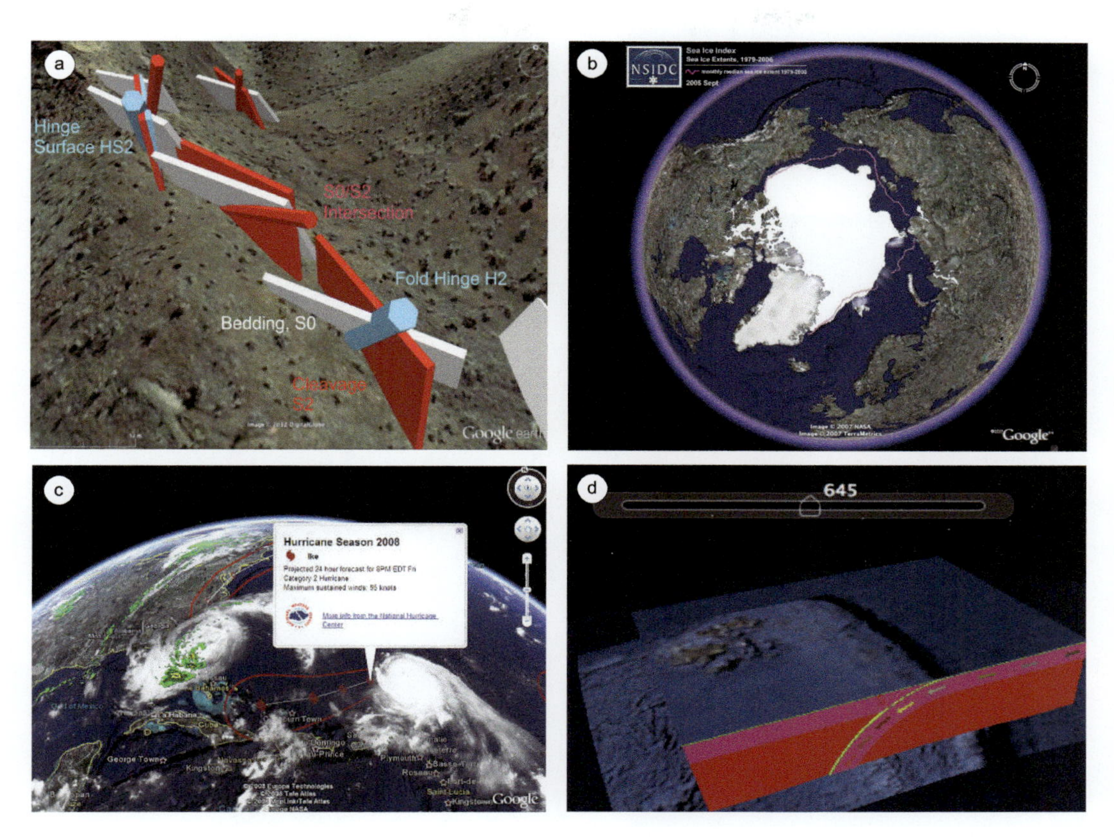


FIGURE 15.11 Examples of science and education uses of KML in Google Earth. (a) 3D planes and lines representing the hinge surfaces and hinges of small-scale folds on the western limb of Mary Kathleen synform, Queensland, Australia (Blenkinsop 2012.). (b) Arctic and Antarctic sea ice minimum and maximum extents from 1979 through present visualized using KML by National Snow and Ice Data Center (Ballagh et al. 2007; Gergely et al. 2008). (c) Hurricane Ike being tracked in the 2008 hurricane season Google Earth layer (Taylor 2008). (d) Animated COLLADA model of the Tonga Trench (De Paor et al. 2012).

perspectives of the landscape. Which ultimately is what these scientific fields (and many others) are using GE for, because they want to use Earth imagery to tell a story about the surface's appearance and any change.

From an education perspective these stories most naturally took the form of virtual field trips. The obvious attraction of GE for teachers was an ability to travel places they otherwise couldn't hope to go. Initially these were simple fly-throughs of the 3D landscape (Lamb and Johnson 2010; Krakowka 2012; Lang et al. 2012), that then became KML-enhanced tours (Rueger and Beck 2012; Treves and Bailey 2012) and eventually complex interactive experiences using the Earth API (Dordevic and Wild 2012; De Paor et al. 2012, 2016, Figure 15.11d). However, deprecation of the Earth API led to a return of focus on in-application, KML-powered experiences, and more simple use of the imagery as a canvas for student exploration and skills assessment (Patterson 2007; Schultz et al. 2008; Johnson et al. 2011), Monet and Greene 2012; Kulo and Bodzin 2013; Gold et al. 2015; Blank et al. 2016). This highlighted the need for GE development to pivot to a web-based experience in order to become a more ubiquitous tool for education users, and in turn was part of the motivation behind the redevelopment of GE into a web-based application.

### 15.4.5 User-Generated Storytelling

The role of Google Earth in remote sensing has shifted as the technology has evolved. The functionality of the web-based version has packaged an ability to display legacy KML with a new interface to create "*Projects*" (Figure 15.12a). *Projects* still allow users to add vector data, but raster overlays

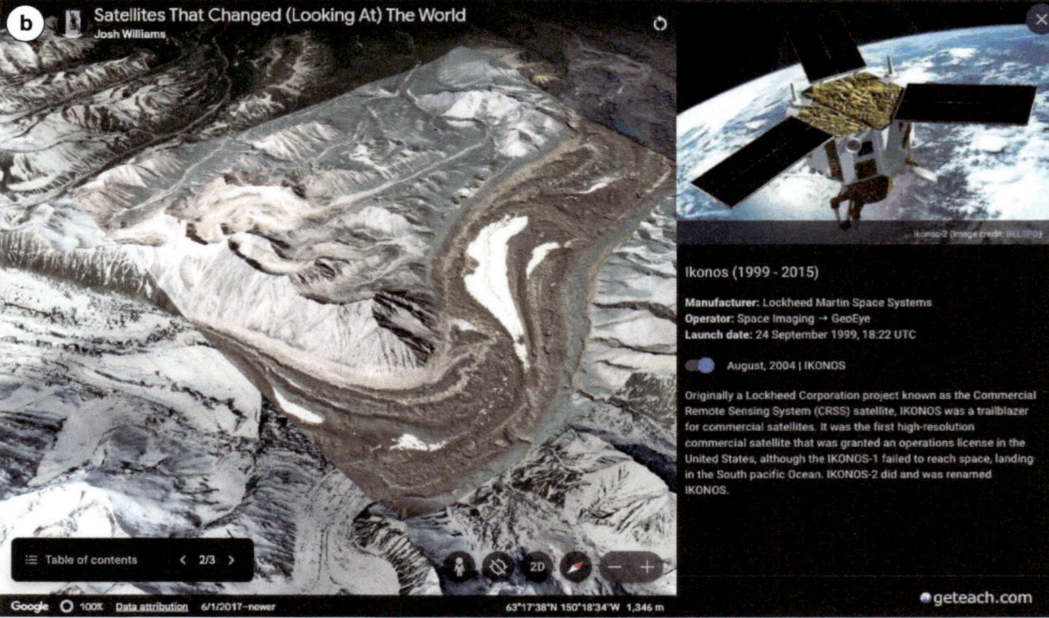


FIGURE 15.12 Projects allow creation of and viewing of content in the web and mobile versions of Google Earth. (a) The landing page for a Project highlighting important satellites in the history of Earth remote sensing. (b) Ikonos-2 image captured August 3, 2004, showing a glacier flowing down Denali in Alaska, into McKinley River Valley. The full project can be viewed at https://geteach.com/remoteearth.

can only be added as an image tile pyramid created through other applications, limiting functionality for nonexpert users. In many ways the content visualization development in Google Earth has come full circle because the primary goal of *Projects* is to provide a framework for users to create geospatial stories, similar to the focus for early adopters (Figure 15.12b). The advantage of the *Projects* is this is now integrated with Google Drive, which allows for collaboration and easy sharing. The down side is the loss of customization that KML provides because access to the underlying code is mostly lost when the content is cloud-hosted. *Projects* do provide for direct editing of the HTML code defining the format of information panels and windows.

This web-based GE also lacks other features that are favorites of scientist and educators users, such as the topographic profile tool, and access to historical high-resolution imagery. Although the Timelapse data are now an integrated layer providing opportunities to view annually averaged, cloud-free, landscape changes over time. Even with all the changes the usability of GE continues to inspire use by those who explore and teach about our planet (Sawaguchi 2018).

## 15.5 STRENGTHS, LIMITATIONS, AND LEGACY OF GOOGLE EARTH

The use of GE is now woven into the fabric of the geoscience and remote sensing communities. For many it is an easy way to generate location maps or provide a "quick view" of target sites, yet the fact that is a default action for so many is evidence of its permanent influence. The unprecedented access GE provided demonstrated it is possible to make a fast, user-friendly portal for viewing satellite imagery. Subsequently online databases of imagery have had to improve their own capabilities as users now expect better functionality and quality graphical displays.

When users choose GE over other applications that could provide similar (or potentially better) remote sensing capabilities, they primarily chose to do so for three reasons:

1. Cost: Google Earth, and more importantly access to the imagery, is free.
2. Accessibility: The process of continually making imagery and terrain data available at a global scale, for a variety of platforms and operating systems, requires considerable computing infrastructure. Google's resources allow them to do this at a level few others are capable of.
3. Usability: The user-friendly interface and associated what you see is what you get (WYSIWYG) graphical user interfaces (GUIs) allows anyone to explore imagery and create custom annotations.

However, despite these advantages some have always viewed GE as just a *fun way* to look at imagery, rather than an actual remote sensing tool or GIS platform. From the beginning they did not consider it a "real" or "full" GIS (Avraam 2009), though some disagreed (Turner 2008). For others the idea that GE was a type of *GIS for the masses* (Bader and Glennon 2007) was acceptable as it became the go-to technology for presentations and illustrations using Earth imagery.

Bailey (2015) offers a reflection on the achievements of GE in remote sensing, a decade after its first release, and focuses on its democratization of access to, and viewing of, satellite imagery. At the same time there was a sense of untapped potential especially as Google had just acquired Skybox Imaging, a company building and launching its own satellites to acquire high-resolution satellite imagery and high-definition video. Another decade on, that landscape has changed significantly. In 2017, Alphabet (the parent company that now owns Google) announced it was selling Terra Bella (the new rebrand name of Skybox) to another imagery startup company, Planet Labs.

This apparent de-emphasis of the acquisition of satellite imagery created some interesting times for Google Earth. Its subsequent re-development and relaunch as a web-based platform was crucial for its continued existence, but has ultimately evolved into a very different application. Its future role seems uncertain given the still significant differences between the web-based version and Google Earth Pro, and the loyalty of users to the latter. It is perhaps telling that despite a gradual push over the last five years to de-emphasize and eventually deprecate the stand-alone application, GE Pro remains supported.

Despite this, the legacy of GE remains undeniable. The name "Google Earth" is recognized around the world, and its influence has extended through science, education, media, and popular culture. The movie "Lion," based on the true story of a man who became lost as a boy in India but used Google Earth to discover a way back to his birth town and family 25 years later, was nominated for multiple academy awards (Shah 2016). It was the ultimate example of the impact that making satellite imagery available to all can have on lives.

Even if the application itself is one day consigned to Silicon Valley history, Google Maps remains a core part of Google's services, and while not on a globe, Google Earth Engine continues to develop as a power imagery analysis platform that leverages the vast archive satellite data initially acquired to support GE. The development and use of Google Earth has been an important chapter in the realm of remote sensing, providing an everyday tool for scientists, educators, government, and the general public alike, and has fundamentally changed how we view our planet (and house).

## ACKNOWLEDGMENTS

True appreciation of the evolution of Google Earth involves acknowledging countless Keyhole and Google software engineers, product managers and program managers, who developed and continue to transform the technology. In addition, the role of researchers, teachers, bloggers, and citizen scientists cannot be understated either. It is fair to say that GE would not have become what it has without the active Google Earth Community (GEC—a user forum formerly known as the Keyhole BBS) and blogging community. GE has impacted the lives of countless groups and individuals. Special thanks is given by the authors to former colleagues and collaborators including GE team members past and present; Tina Ornduff, Josie Wernecke, Micheal Ashbridge, Bent Hagemark, Mano Marks, Michael Weiss-Malik, Nick Clinton, Dusty Reid, Emily Henderson, and the teachers who formed part of the *Google Earth Educators group*, all of whom played critical roles in shaping the authors' own journeys with Google Earth.

## REFERENCES

Anton, M., and I. van Dongen. 2022, June 28. 3D Globe Map: A new way to experience the world. *Mapbox Blog*, www.mapbox.com/blog/globe-view (accessed 27-Sep-2023).

Appalachian Voices. 2014. Mountaintop removal in Google Earth. http://ilovemountains.org/google_earth_tutorial (accessed 5-Jan-2024).

Avraam, M. 2009. Google Maps and GIS. https://web.archive.org/web/20091220123311/http://michalisavraam.org/2009/10/google-maps-and-gis (accessed 30-Dec-2023).

Bader, J., and A. Glennon 2007. Virtual Globes: GIS for the masses? *Association of American Geographers Annual Meeting*, Paper Session 4510, 17–21 April, San Francisco, CA, USA.

Bailey, J.E. 2010. Entry for "Virtual Globe". In Warf, B. (ed.) *Encyclopedia of Geography*, Sage Publications, Los Angeles, CA, p. 3528.

Bailey, J.E. 2015. Google Earth for remote sensing. In Thenkabail, P. (ed.) *Remotely Sensed Data Characterization, Classification, and Accuracies*, Taylor & Francis, Section XI, Chapter 28, p. 565–582. doi:10.1201/b19355.

Bailey, J.E., and A. Chen. 2011. The role of Virtual Globes in geoscience. *Computers & Geosciences*. 37(1):1–2.

Bailey, J.E., and J. Dehn. 2006. Volcano monitoring using Google Earth. Geoinformatics 2006-Abstracts, *USGS Scientific Investigations* Report 2006-5201, p. 25., 10–12 May, Reston, VI, USA.

Ballagh, L.M., B.H. Raup, R.E. Duerr, S.J.S. Khalsa, C. Helm, D. Flower, and A. Gupte. 2011. Representing scientific data sets in KML: Methods and challenges. *Computers & Geosciences* . 37(1):57–64. doi:10.1016/j.cageo.2010.05.004.

Ballagh, L.M., M.A. Parsons, and R. Swick. 2007. Visualising cryospheric images in a virtual environment: Present challenges and future implications. *Polar Record* 43(4):305–310. doi:10.1017/S0032247407006523.

Bar-Zeev, A. 2007, September 27. The word on snow crash and Google Earth. *Reality Prime*, https://web.archive.org/web/20190311141042/www.realityprime.com/blog/2007/09/the-word-on-snow-crash-and-google-earth/ (accessed 30-Sep-2023).

Blank, L.M., H. Almquist, J. Estrada, and J. Crews. 2016. Factors affecting student success with a Google Earth-based Earth Science curriculum. *Journal of Science Education and Technology*. 25(1):77–90.

Blenkinsop, T.G. 2012. Visualizing structural geology: From Excel to Google Earth. *Computers & Geoscience*. 45:52–56. doi:10.1016/j.cageo.2012.03.007.

Butler, D. 2006a. The Web-Wide world. *Nature* 439:776–778.

Butler, D. 2006b. Avian flu maps in Google Earth, https://web.archive.org/web/20160315125549/; http://declanbutler.info/blog/?p=16 (accessed 30-Sep-2023).

Butler, D. 2006c. The spread of avian flu with time: New maps exploiting Google Earth's time series function, https://web.archive.org/web/20160304125910/; http://declanbutler.info/blog/?p=58 (accessed 30-Sep-2023).

Cardille, J., M. Crowley, D. Saah, and N. Clinton. 2023. *Cloud-Based Remote Sensing with Google Earth Engine: Fundamentals and Applications*. SpringerOpen, London, www.eefabook.org/.

Chen, A., and J.E. Bailey. 2011. Virtual globes in science. *Computers & Geoscience*. 37(1):1–110.

Chien, N.Q., and S.K. Tan. 2011. Google Earth as a tool in 2-D hydrodynamic modeling. *Computers & Geoscience*. 37(1):38–46. doi:10.1016/j.cageo.2010.03.006.

Dennis, N. [Nat and Friends]. 2017. Google Earth's incredible 3D imagery, explained [Video]. YouTube, www.youtube.com/watch?v=suo_aUTUpps.

De Paor, D.G. 2008. Enhanced visualization of seismic focal mechanisms and centroid moment tensors using solid models, surface bump-outs, and Google Earth. *J. Virtual Explorer* 29, paper 2. In De Paor D. (ed.) *Google Earth Science*, https://virtualexplorer.com.au/system/files/papers/00195/assets/seismic-model-visualization.pdf

De Paor, D.G., F. Coba, and S. Burgin. 2016. A Google Earth grand tour of the terrestrial planets. *Journal of Geoscience Education*. 64(4):292–302. doi:10.5408/15-116.1.

De Paor, D.G., S.C. Wild, and M.M. Dordevic. 2012. Emergent and animated COLLADA models of the Tonga Trench and Samoa Archipelago: Implications for geoscience modeling, education, and research. *Geosphere*. 8(2):491–506. doi:10. 1130/GES00758.1.

De Paor, D.G., and S.J. Whitmeyer. 2011. Geological and geophysical modeling on virtual globes using KML, COLLADA, and Javascript. *Computers & Geoscience*. 37(1):100–110. doi:10.1016/j.cageo.2010.05.003.

de Silva, S., and J. Bailey. 2017. Some unique surface patterns on ignimbrites on Earth: A "birds eye" view as a guide for planetary mappers, in Wilson L, de Silva S, Kerber L, Patrick Whelley P (eds.), Pattern to process: Remotely sensed observations of volcanic deposits and their implications for surface processes. *Journal of Volcanology and Geothermal Research*. 342:47–60. doi:10.1016/j.jvolgeores.2017.06.009.s

Dordevic, M.M., and S.C. Wild. 2012. Avatars and multi-student interactions in Google Earth-based virtual field experiences. In Whitmeyer, S.J., J.E. Bailey, D.G. De Paor, and T. Ornduff, eds., Google Earth and virtual visualizations in geoscience education and research. *Geological Society of America Special Paper*. 492:315–321. doi:10.1130/2012.2492(22).

Eames, C., and R. Eames 1977. *Powers of Ten: A Film Dealing with the Relative Size of Things in the Universe and the Effect of Adding Another Zero*, IBM, New York.

Farr, T.G., P.A. Rosen, E. Caro, R. Crippen, R. Duren, S. Hensley, M. Kobrick, M. Paller, E. Rodriguez, L. Roth, D. Seal, S. Shaffer, J. Shimada, J. Umland, M. Werner, M. Oskin, D. Burbank, and D. Alsdorf. 2007. The shuttle radar topography mission. *Reviews of Geophysics*. 45(2). doi:10.1029/2005RG000183.

Finkelberg, A.S. 2007. Space, place, and database: Layers of digital cartography. [Master's thesis] Massachusetts Institute of Technology, Dept. of Comparative Media Studies.

Foerch, A. 2017, November 1. The genesis of Google Earth. *Trajectory Magazine*, https://trajectorymagazine.com/genesis-google-earth/ (accessed 17-Nov-2023).

Fuller, B. 1962. *Education Automation, Freeing the Scholar to Return to His Studies*, Southern Illinois University Press, Carbondale and Edwardsville, Feffer & Simons Inc., London and Amsterdam, ISBN 0-8093-0137-7.

Gandhi, U. 2023. JavaScript and the Earth engine API. In Cardille J., M. Crowley, D. Saah, and N. Clinton (eds.) *Cloud-Based Remote Sensing with Google Earth Engine: Fundamentals and Applications*, SpringerOpen, London, p. 3–18.

Geens, S. 2009. Media stupidity watch: No, It's not Atlantis. *Ogle Earth Blog*, http://ogleearth.com/2009/02/media-stupidity-watch-no-its-not-atlantis/ (accessed 1-Dec-2023).

Gergely, K.L., T.M. Haran, and B. Billingsley. 2008. Virtual globe visualizations of cryospheric data at the national snow and ice data center. *EOS, Transactions, American Geophysical Union* 89(53).

Giri, C., J. Long, S. Abbas, R.M. Murali, F.M. Qamer, and D. Thau. 2014. Current status and dynamics of mangrove forests of South Asia. *Journal of Environmental Management*. 148:101–111. doi:10.1016/j.jenvman.2014.01.020.

Gold, A.U., K. Kirk, D. Morrison, S. Lynds, S.B. Sullivan, A. Grachev, and O. Persson. 2015. Arctic climate connections curriculum: A model for bringing authentic data into the classroom. *Journal of Geoscience Education*. 63(3):185–197. doi:10.5408/14-030.1.

Google Earth. 2007, June 23. Jane Goodall Gombe chimpanzee blog in Google Earth [Video]. *Youtube*, www.youtube.com/watch?v=_N805-TyUiY.

Google LLC. 2005. Google launches free 3D mapping and search product. *Google Blog*, http://googlepress.blogspot.com/2005/06/google-launches-free-3d-mapping-and_28.html (accessed 21-Sep-2023).

Google LLC. 2009. Atlantis? No, it Atlant-isn't. *Google Blog*, https://googleblog.blogspot.com/2009/02/atlantis-no-it-atlant-isnt.html (accessed 1-Dec-2023).

Google LLC. 2023a. Google Earth versions, www.google.com/earth/about/versions/ (accessed 1-Dec-2023).

Google LLC. 2023b. Google Earth Engine, https://earthengine.google.org (accessed 1-Dec-2023).

Google Earth Outreach. 2023. Get inspired by organizations who have used Google mapping tools for good, www.google.com/earth/outreach/success-stories (accessed 1-Dec-2023).

Gore, A. 1998. The Digital Earth: Understanding our planet in the 21st century. Speech given at the California Science Center, Los Angeles, CA, on 31 Jan 1998, https://web.archive.org/web/20060901182138/www.digitalearth.gov/VP19980131.html (accessed 30-Sep-2023).

Gorelick, N., M. Hancher, M. Dixon, S. Ilyushchenko, D. Thau, and R. Moore. 2017. Google Earth Engine: Planetary-scale geospatial analysis for everyone. *Remote Sensing of the Environment*. 202:18–27. doi:10.1016/j.rse.2017.06.031.

Gramling, C. 2007, February. Google Planet-with virtual globes, Earth scientists see a new world. *Geotimes*:38–40, www.geotimes.org/feb07/trends.html.

Hansen, M.C., P.V. Potapov, R. Moore, M. Hancher, S.A. Turubanova, A. Tyukavina, D. Thau, S.V. Stehman, S.J. Goetz, T.R. Loveland, A. Kommareddy, A. Egorov, L. Chini, C.O. Justice, and J.R.G. Townshend. 2013. High-resolution global maps of 21st-century forest cover change. *Science*. 342:850–853. doi:10.1126/science.1244693.

Herwig, C. 2023, April 4. See the planet change with new imagery in Google Earth Timelapse. *The Keyword*, https://blog.google/products/earth/new-imagery-google-earth-timelapse-videos (accessed 20-Dec-2023).

Hogan, P., and R. Kim. 2004. NASA planetary visualization tool. *EOS, Transactions, American Geophysical Union*. 85(47).

Johnson, N.D., N.P. Lang, and K.T. Zophy. 2011. Overcoming assessment problems in Google Earth-based assignments. *Journal of Geoscience Education*. 59:99–105.

Joshi, A.R., E. Dinerstein, E. Wikramanayake, M.L. Anderson, D. Olson, B.S. Jones, J. Seidensticker, S. Lumpkin, M.C. Hansen, N.C. Sizer, C.L. Davis, S. Palminteri, and N.R. Hahn. 2016. Tracking changes and preventing loss in critical tiger habitat. *Science Advances*. 2(4), e1501675. doi:10.1126/sciadv.1501675.

Kilday, B. 2018. *Never Lost Again: The Google Mapping Revolution That Sparked New Industries and Augmented Our Reality Hardcover*, Harper Business, New York, p. 368.

Klugger, J., and B. Walsh. 2013. Timelapse. *TIME*, http://world.time.com/timelapse/ (accessed 5-Jan-2024).

Knabb, R.D., J.R. Rhome, and D.P. Brown. 2005. Hurricane Katrina: August 23–30, 2005 (Tropical Cyclone Report). *National Oceanic and Atmospheric Administration—Hurricane Research Division*, www.nhc.noaa.gov/data/tcr/AL122005_Katrina.pdf (accessed 10-Dec-2023).

Kochenour, C. 2020. Remote sensing with Google Earth Engine. *GitHub*, https://calekochenour.github.io/remote-sensing-textbook (accessed 2-Jan-2023).

Krakowka, A.R. 2012. Field trips as valuable learning experiences in geography courses. *Journal of Geography*. 111(6):236–244. doi:10.1080/00221341.2012.707674.

Kulo, V., and A. Bodzin. 2013. The impact of a geospatial technology-supported energy curriculum on middle school students' science achievement. *Journal of Science Education and Technology*. 22(1):25–36. doi:10.1007/s10956-012-9373-0.

Lamb, A., and L. Johnson 2010. Virtual expeditions: Google Earth, GIS, and geovisualization technologies in teaching and learning. *Teacher Librarian* 37(3):81–85.

Lang, N.P., K.T. Lang, and B.M. Camodeca. 2012. A geology-focused virtual field trip to Tenerife, Spain. In Whitmeyer, S.J., J.E. Bailey, D.G. De Paor, and T. Ornduff (eds.) Google Earth and virtual visualizations in geoscience education and research. *Geological Society of America Special Paper* 492:323–334. doi:10.1130/2012.2492(23).

Lisle, R.J. 2006. Google Earth: A new geological resource. *Geology* 22(1):29–32. doi:10.1111/j.1365–2451.2006.00546.x.

Liu, P., J. Gong, M. Yu. 2015. Visualizing and analyzing dynamic meteorological data with virtual globes: A case study of tropical cyclones. *Environmental Modelling & Software* 64:80–93. doi:10.1016/j.envsoft.2014.11.014.

Lobell, D.B., D. Thau, C. Seifert, E. Engle, and B. Little. 2015. A scalable satellite-based crop yield mapper. *Remote Sensing of the Environment*. 164:324–333. doi:10.1016/j.rse.2015.04.021.

Lowensohn, J. 2009, February 2. Images: Google Earth 5.0 travels oceans, time, and space. *CNET*, www.cnet.com/pictures/images-google-earth-5-0-travels-oceans-time-and-space (accessed 20-Dec-2023).

Lubick, N. 2005, December. Spinning around the globe online. *Geotimes*, www.geotimes.org/dec05/geomedia.html.

Microsoft News. 1997, November 20. Now a Virtual Globe, Not just a world atlas, https://news.microsoft.com/1997/11/20/now-a-virtual-globe-not-just-a-world-atlas/.

Monet, J., and T Greene. 2012. Using Google Earth and satellite imagery to foster place-based teaching in an introductory physical geology course. *Journal of Geoscience Education*. 60(1):10–20. doi:10.5408/10-203.1.

Moore, R. n.d. Neighbors Against Irresponsible Logging (NAIL), www.google.com/earth/outreach/success-stories/neighbors-against-irresponsible-logging-nail (accessed 1-Dec-2023).

Moore, R. 2021, April 15. Time flies in Google Earth's biggest update in years. *The Keyword*, https://blog.google/products/earth/timelapse-in-google-earth/ (accessed 1-Dec-2023).

National Oceanic and Atmospheric Association. 2005. Hurricane Katrina images, www.aoml.noaa.gov/hrd/Storm_pages/katrina2005/sat.html (accessed 10-Dec-2023).

Nourbakhsh, I., R. Sargent, A. Wright, K. Cramer, B. McClendon, and M. Jones. 2006. Mapping Disaster Zones. *Nature* 439:787–788. doi:10.1038/439787a.

Patel, N.N., E. Angiuli, P. Gamba, A. Gaughan, G. Lisini, F.R. Stevens, A.J. Tatem, and G. Trianni. 2015. Multitemporal settlement and population mapping from Landsat using Google Earth Engine. *International Journal of Applied Earth Observations and Geoinformation*. 35(Part B):199–208. doi:10.1016/j.jag.2014.09.005.

Patterson, T.C. 2007. Google Earth as a (not just) geography education tool. *Journal of Geography*. 106(4):145–152. doi:10.1080/00221340701678032.

Pekel, J.F., A. Cottam, N. Gorelick, and A.S. Belward 2016. High-resolution mapping of global surface water and its long-term changes. *Nature* 540:418–422. doi:10.1038/nature20584.

Rueger, B.F., and E.N. Beck. 2012. Benedict Arnold's march to Quebec in 1775: An historical characterization using Google Earth. In Whitmeyer, S.J., J.E. Bailey, D.G. De Paor, and T. Ornduff (eds.) Google Earth and virtual visualizations in geoscience education and research. *Geological Society of America Special Paper* 492:347–354. doi:10.1130/2012.2492(25).

Sawaguchi, T. 2018. Geoscience education using a brand-new Google Earth. *Terrae Didatica*. 14(4):415–416. doi:10.20396/td.v14i4.8654165.

Schultz, R.B., J.J. Kerski, and T.C. Patterson. 2008. The use of virtual globes as a spatial teaching tool with suggestions for metadata standards. *Journal of Geography*. 107(1):27–34. doi:10.1080/00221340802049844.

Shah, G. 2016, November 22. Google Earth: The 25-year search. *The Keyword*, https://blog.google/products/maps/google-earth-25-year-search (accessed 1-Dec-2023).

Sheppard, S.R.J., and P. Cizek. 2008. The ethics of Google Earth: Crossing thresholds from spatial data to landscape visualization. *Journal of Environmental Management*. 90(6):2102–2117. doi:10.1016/j.jenvman.2007.09.012.

Silberbauer, M.J., and W. Geldenhuys. 2008. Using keyhole markup language to create a spatial interface to South African water resource data through Google Earth. In *FOSS4G 2008 Free and Open Source Software for Geospatial*, OSGeo, Cape Town, South Africa.

Simonite, T. 2007, May 2. Virtual Earths let researchers "mash up" data. *New Scientist*, www.newscientist.com/article/dn11773 (accessed 5-Jan-2024).

Smith, T.M., and V. Lakshmanan. 2006. Utilizing Google Earth as a GIS platform for weather applications. *American Meteorological Society*. 22nd International Conference on Interactive Information Processing Systems (IIPS) for Meteorology, Oceanography, and Hydrology, 27 January 27–3 February, Atlanta, GA, USA.

Smith, T.M., and V. Lakshmanan. 2011. Real-time, rapidly updating severe weather products for virtual globes. *Computers & Geoscience*. 37(1):3–12. doi:10.1016/j.cageo.2010.03.023.

Stephenson, N. 1992. *Snow Crash*. Bantam Dell, Random House Inc., New York, NY, p. 470. ISBN 0-553-38095-8.

Sullivan, D. 2007, March 30. Google under fire for showing Pre-Katrina New Orleans images; Others doing the same? *Search Engine Land*, https://searchengineland.com/google-under-fire-for-showing-pre-katrina-new-orleans-images-others-doing-the-same-10858 (accessed 5-Jan-2024).

Taylor, F. 2008. Google Earth Hurricane tracking layer. *Google Earth Blog*, www.gearthblog.com/blog/archives/2008/09/google_earth_hurricane_tracking_lay.html (accessed 12-Jan-2024).

Tewksbury, B.J., A.A.K. Dokmak, E.A. Tarabees, and A.S. Mansour. 2012. Google Earth and geologic research in remote regions of the developing world: An example from the Western Desert of Egypt. In Whitmeyer, S.J., J.E. Bailey, D.G. De Paor, and T. Ornduff (eds.) Google Earth and virtual visualizations in geoscience education and research. *Geological Society of America Special Paper* 492:23–36. doi:10.1130/2012.2492(02).

Treves, R., and J.E. Bailey. 2012. Best practices on how to design Google Earth tours for education. In Whitmeyer, S.J., J.E. Bailey, D.G. De Paor, and T. Ornduff (eds.) Google Earth and virtual visualizations in geoscience education and research. *Geological Society of America Special Paper* 492:383–394. doi:10.1130/2012.2492(28).

Turk, J.F., J. Hawkins, K. Richardson, and M. Surratt. 2011. A tropical cyclone application for virtual globes. *Computers & Geoscience*. 37(1):13–24. doi:10.1016/j.cageo.2010.05.001.

Turner, A. 2008, June 12. Is GoogleMaps GIS? *High Earth Orbit*, https://web.archive.org/web/20160326061418/http://highearthorbit.com/is-googlemaps-gis (accessed 30-Sep-2023).

USGS. 2014. *WaterWatch*, http://waterwatch.usgs.gov/kml.html (accessed 22-Sep-2023).

Venzke, E., L. Siebert, and J.F. Luhr. 2006. Smithsonian volcano data on Google Earth. *EOS, Transactions, American Geophysical Union*. 87(52).

Wall Street Journal. 2004, October 27. Google acquires Keyhole, www.wsj.com/articles/SB109888284313557107 (accessed 17-Dec-2023).

Webley, P.W., J. Dehn, J. Lovick, K.G. Dean, J.E. Bailey, and L. Valcic. 2009a. Near-real-time volcanic ash cloud detection: Experiences from the Alaska Volcano observatory. In Mastin L, and P. Webley. eds. Volcanic ash clouds, *Journal of Volcanology and Geothermal Research*. Special Issue 186:79–90. doi:10.1016/j.jvolgeores.2009.02.010.

Webley, P.W., K. Dean, J.E. Bailey, J. Dehn, and R. Peterson. 2009b. Automated forecasting of volcanic ash dispersion utilizing Virtual Globes. *Natural Hazards* 51(2):345–361. doi:10.1007/s11069-008-9246-2.

Wei C., D.N. Karger, and A.M. Wilson. 2020. Spatial detection of alpine treeline ecotones in the Western United States. *Remote Sensing of the Environment*. 240, 111672. doi:10.1016/j.rse.2020.111672.

Wernecke, J. 2009. *The KML Handbook: Geographic Visualization for the Web*, Addison-Wesley, Upper Saddle river, NJ, p. 368. ISBN-13: 978-0-321-52559-8.

Whitmeyer, S.J., J.E. Bailey, D.G. De Paor, and T. Ornduff. 2012. Google Earth and virtual visualizations in geoscience education and research. *Geological Society of America Special Paper*. 492, 468. ISBN 978-0-8137-2492-8.

Williams, D.B., and H.E. Thomas. 2011. An assessment of volcanic hazards to aviation: A case study from the 2009 Sarychev Peak eruption. *Geomatics, Natural Hazards and Risk*. 2(3):233–246. doi:10.1080/19475705.2011.558117.

Wright, T.E., M. Burton, D.M. Pyle, and T. Caltabiano. 2009. Visualising volcanic gas plumes with virtual globes. *Computers & Geoscience*. 35(9):1837–1842. doi:10.1016/j.cageo.2009.02.005.

Yu, L., and P. Gong. 2012. Google Earth as a virtual globe tool for Earth science applications at the global scale: Progress and perspectives. *International Journal of Remote Sensing*. 33(12):3966–3986. doi:10.1080/01431161.2011.636081.

# Part IX

## Accuracies, Errors, and Uncertainties of Remote Sensing Derived Products

# 16 Assessing Positional and Thematic Accuracies of Maps Generated from Remotely Sensed Data

*Russell G. Congalton*

## ABBREVIATIONS

| | |
|---|---|
| **ASPRS** | American Society for Photogrammetry and Remote Sensing |
| **CMAS** | Circular Map Accuracy Standard |
| **FEMA** | Federal Emergency Management Agency |
| **FGDC** | Federal Geographic Data Committee |
| **GPS** | Global Positioning System |
| **MAS** | Map Accuracy Standard |
| **MMU** | Minimum Mapping Unit |
| **MSS** | MultiSpectral Scanner |
| **NIR** | Near Infrared |
| **NMAS** | National Map Accuracy Standards |
| **NSSDA** | National Standard for Spatial Data Accuracy |
| **OBIA** | Object-Based Image Analysis |
| **RMSE** | Root Mean Square Error |
| **USDA** | United States Department of Agriculture |

## 16.1 INTRODUCTION

This chapter is devoted to assessing the accuracy of maps generated from remotely sensed data. In the ideal world, it would be great to list the ten simple steps that the reader must follow in order to conduct such an accuracy assessment. Unfortunately, map accuracy assessment does not follow such simple procedures. Instead, there are many significant factors that must be carefully considered from the beginning of the project as well as several statistical and practical methodologies that must be balanced to achieve a successful and valid assessment. Therefore, instead of a few simple steps, this chapter presents these considerations and methodologies in a flow chart (Figure 16.1) to help the reader begin to see all the components that must be thought out and planned for to conduct a valid accuracy assessment. The rest of this chapter deals with each of the parts of this flow chart.

## 16.2 ASSESSING MAP ACCURACY

Today, assessing the accuracy of a map generated from remotely sensed imagery is a routine component of most mapping projects. However, this was not always the case. By the end of World War II, the use of aerial photography and photo interpretation was a well-established means of mapping and monitoring the Earth's resources. Maps generated from aerial photos for such uses as agricultural

DOI: 10.1201/9781003541141-25 

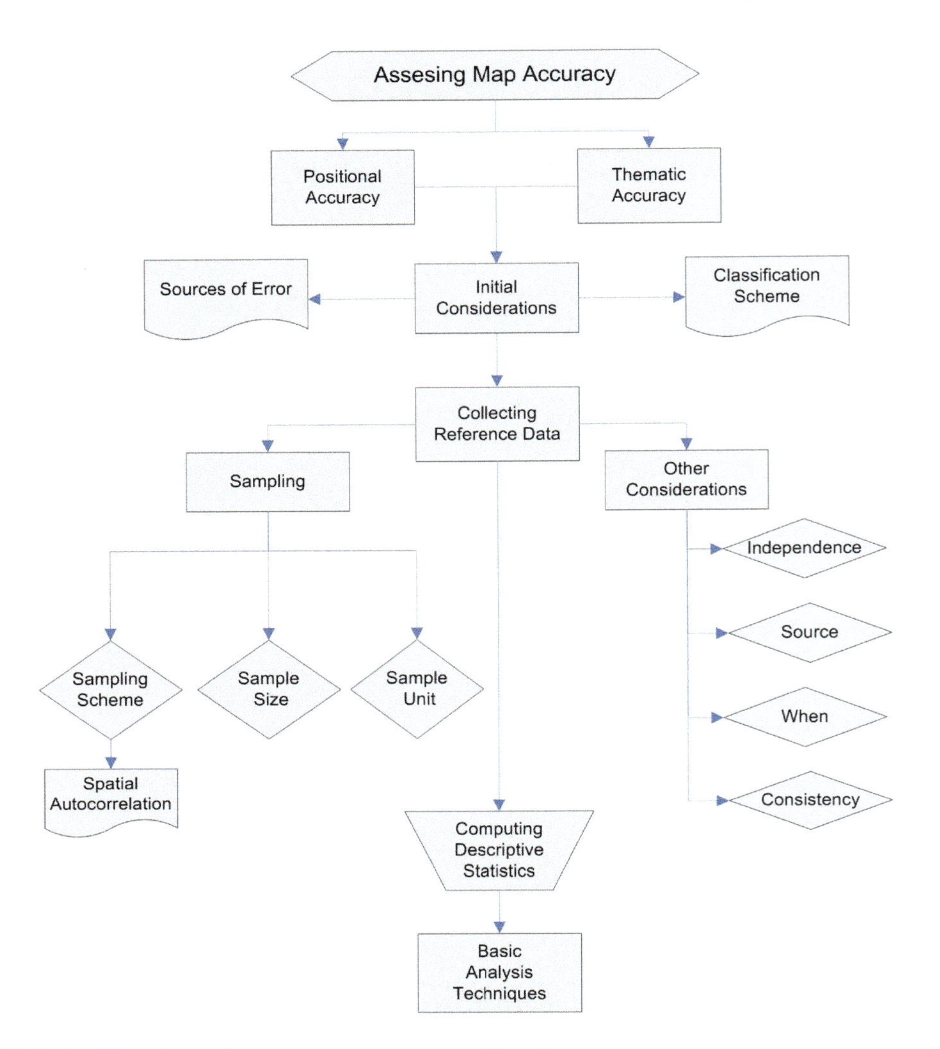


**FIGURE 16.1** Flow chart of accuracy assessment considerations.

monitoring, forest inventory, geologic exploration, and many others became commonplace. A key component of every photo interpretation project included the necessary field visits to train the interpreter to recognize the objects of interest on the ground and on the aerial photos. It was generally recognized that the human interpreter drew lines on the photo (i.e., created polygons) around areas that seemed to be distinct from each other (i.e., had more variation between polygons than within a polygon) and then did their best to appropriately label the polygons. Those areas that were difficult to label were checked in the field during another field visit. Very little thought was given to any type of quantitative evaluation or accuracy assessment of the resulting photo interpreted maps.

A notable exception to this lack of interest in photo interpretation accuracies occurred in the early 1950s. Some researchers wishing to promote the field of photo interpretation as a science recognized the need to evaluate the accuracy of their work and published a number of papers on the topic (Sammi 1950, Katz 1952, Young 1955, Colwell 1955). A panel discussion entitled "Reliability of Measured Values" was held at the 18th Annual Meeting of the American Society of Photogrammetry in 1952. At this discussion, Mr. Amrom Katz (1952), the panel chair, made a strong case for the use of statistics in photogrammetry. The results of all these papers and discussions culminated with a paper by Young and Stoeckler (1956). In this paper, these authors proposed techniques for a quantitative evaluation of photo interpretation, including the use of an error matrix to compare field and photo classifications, and went on to present a discussion of the boundary error problem when labeling polygons. Unfortunately, nothing really became of these forward-thinking ideas for about the next 25 years.

In 1972, the first Landsat satellite was launched thus strongly promoting the field of digital image processing. That first Landsat multispectral scanner (MSS) had only four bands and sensed in only three wavelengths (green, red, and 2 NIR), but it completely changed the way that we looked at creating maps from remotely sensed imagery. Instead of using the human being as the interpreter, a computer could now be used to manipulate the digital values in the imagery and produce a map. Since the computer and more specifically, some mathematical/statistical algorithm implemented by the computer, was now the entity producing the map from the digital remotely sensed imagery, questions about the accuracy (i.e., how good the map was) soon followed.

Actually, in any new technology, there is an initial exuberance regarding the use of that technology without much thought to the quality of what is being produced. This is quite natural as the excitement of the new technology dominates and many new uses of the technology are investigated. However, as technology begins to mature, questions about quality and limitations are inevitable. Such was the case for digital image processing of the Landsat imagery. During the first five to ten years after the launch of Landsat, many exaggerated claims about the quality and detail that could be produced from the imagery prevailed such that there was a great overselling of the technology. Fortunately, by the early 1980s researchers began to question some of these claims and initiated the development of methods for assessing the accuracy of maps derived from remotely sensed data. Of particular importance was a working conference held at the U.S. Geological Survey EROS Data Center in Sioux Falls, South Dakota, in November of 1980 dedicated to exploring methods to assess the accuracy of Landsat imagery (Mead and Szajgin 1981).

It must be noted at this point that map accuracy really has two components: positional accuracy and thematic accuracy. Positional accuracy deals with the accuracy of the location of map features, and measures how far a spatial feature on a map is from its true or reference location on the ground (Bolstad and Manson 2022). Thematic accuracy deals with the labels or attributes of the features of a map, and measures whether the mapped feature labels are different from the true or reference feature label. If you are in the wrong place, but label the map correctly, there will be an error in the map accuracy. If you are in the correct place, but label the map incorrectly, there will be an error in the map accuracy. Both types of assessment are critical. Therefore, this chapter deals with both types of accuracy and the flow chart in Figure 16.1 is appropriate for both types also.

## 16.3 POSITIONAL MAP ACCURACY ASSESSMENT

In order to make a map, you must know where you are. Therefore, it is critical in any mapping project that the determination of the exact same location on both the image or map and the reference data (often the ground) be assessed (see Figure 16.2). If this correspondence is not attained, then being in the wrong location may result in a thematic error. For example, it is possible to be in the right place and mislabel (incorrectly measure, observe, or label) the attribute. This error would be a thematic error. However, it is also possible to correctly label the attribute, but be in the wrong place. This locational error could then also lead to a thematic error. Positional error/accuracy and thematic error/accuracy are not independent of each, and it is critical that every possible effort be made to control both of them in order to effectively assess the accuracy of any map.

The history of assessing the positional accuracy of maps begins with the National Map Accuracy Standards (NMAS) from the U.S. Bureau of the Budget in 1947. NMAS is a very simple and straightforward standard that states:

- "for horizontal accuracy, not more than 10% of the points tested may be in error by more than 1/30th of an inch (at map scale) for maps larger than 1:20,000 scale, or by more than 1/50th of an inch for maps of 1:20,000 scale or smaller", and
- "for vertical accuracy, not more than 10% of the elevation tested may be in error by more than one half the contour interval."

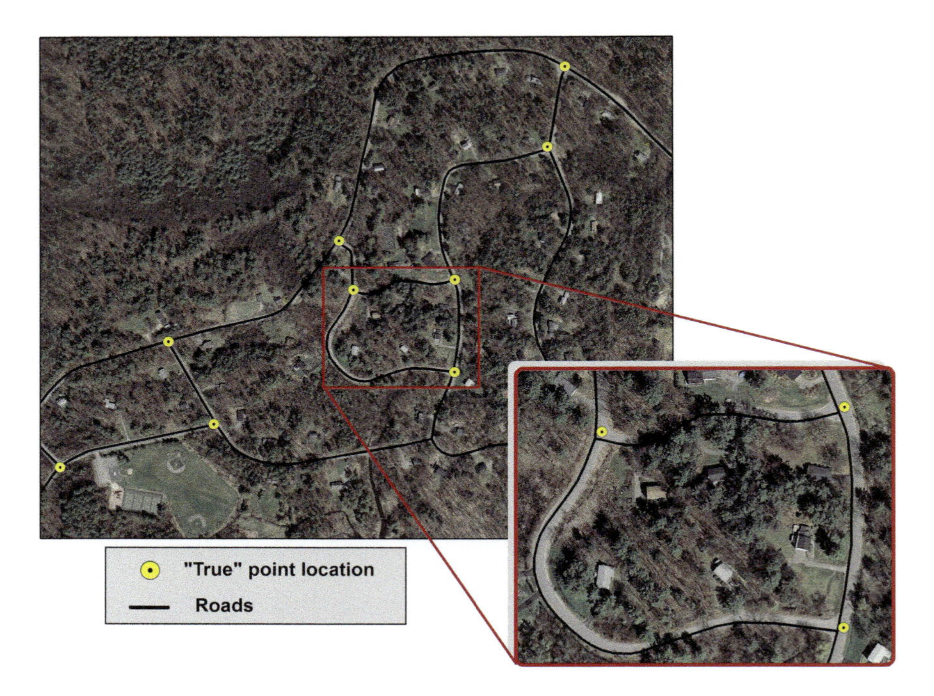


FIGURE 16.2 Example of positional accuracy showing that the road intersections as indicated by the yellow dots (reference points) and the roads are not in the same location.

While easy to apply, NMAS provides no information about determining statistical bounds around the error, but rather just uses the "percentile method" for accepting or rejecting the map as accurate. Therefore, NMAS is not commonly used today.

The next step in developing methods for positional accuracy assessment resulted from the work of Greenwalt and Schultz (1962, 1968). Their report (The Principles of Error Theory and Cartographic Applications, Greenwalt and Schultz 1962, 1968) proposed equations by which to estimate the maximum error interval for a given probability. They computed a one-dimensional map accuracy standard (MAS) used for elevation/vertical (z) data assessment and a two-dimensional circular map accuracy standard (CMAS) for horizontal (x & y) data assessment. Greenwalt and Schultz assumed that map errors are normally distributed and used the equations for MAS to estimate the interval around the mean vertical error and the equations for CMAS to estimate the interval around the mean horizontal error within which 90% of the error should occur. These equations can also be used to estimate the distribution of errors at other probabilities other than 90% and offers more flexibility than the previous NMAS.

In 1989, the American Society for Photogrammetry and Remote Sensing (ASPRS) released the ASPRS Interim Accuracy Standards for Large Scale Maps (ASPRS 1989). Instead of stipulating that no more than 10% of the errors may exceed a given maximum value (the NMAS approach), the ASPRS standard computes a mean error from a set of samples and then provides a threshold value that this mean error cannot exceed. This standard is expressed in ground units and not map units as NMAS had used. The document also reviews and confirms the work of Greenwalt and Schultz, but does not make any recommendations about the use of these equations.

The National Standard for Spatial Data Accuracy (NSSDA) was released by the U.S. Federal Geographic Data Committee (FGDC) in 1998. This standard represents the currently approved method for assessing positional accuracy and establishes some much needed guidelines for measuring, analyzing, and reporting positional accuracy for maps as well as other geo-referenced imagery (FGDC 1998). These guidelines have been widely adopted by those in the federal, state, and local government and also in the private sector. The full version of the NSSDA standards (FGDC 1998.) can be downloaded at www.fgdc.gov/standards/projects/FGDC-standards-projects/accuracy/part3/chapter3.

NSSDA does not use a maximum allowable error as the decision point if a map has acceptable positional error at any scale. Instead, it recommends determining an allowable error threshold as needed such that accuracy is then reported in ground distances at the 95% confidence level. This 95% level is stricter than previous standards which tended to use the 90% level. NSSDA incorporates the work of Greenwalt and Schultz and used these equations to determine accuracy as a maximum threshold error at a given or specified probability. However, there are issues with the calculations of the NSSDA as there is a mistake in the calculations that will be discussed later in this section.

At the turn of the century, three new documents/guidelines were developed that deal with positional accuracy, but especially with regard to LiDAR data. These documents include (1) Guidelines and Specifications for Flood Hazard Mapping Partners (FEMA 2003), which specifies that there must be 20 samples taken in each of the major vegetation cover types (at least three cover types); (2) the ASPRS Guidelines for Reporting Vertical Accuracy of LiDAR Data (ASPRS 2004), which confirms the FEMA standard, but uses a slightly different vegetation cover type classification; and (3) Guidelines for Digital Elevation Data from the National Digital Elevation Program (NDEP 2004), which agrees with the various accuracy measures specified in the ASPRS (2004) guidelines.

In 2014, ASPRS published the ASPRS Positional Accuracy Standards for Digital Geospatial Data (Edition 1, Version 1.0) (ASPRS 2014). These standards replace previous work from ASPRS and are designed to fully account for digital geospatial data. Most previous standards were still dealing with issues with hardcopy maps while this standard is fully revised for digital (softcopy) geospatial data. A follow-on to this standard has recently been published by ASPRS as Edition 2, Version 1.0 (ASPRS 2023). This revised standard has some significant changes from Edition 1. These changes include (1) the elimination of references to a 95% confidence level as an accuracy measure since RMSE (root mean square error) is considered a more reliable and understood measure, (2) a relaxation of the ground control and checkpoint accuracies because of the cost of collection and improvement in data collection, (3) modification of assessing the accuracy of Vegetated Vertical Accuracy (VVA) for LiDAR data since the technology has not reached the level of direct comparison with nonvegetated areas, (4) an increase in the minimum number of checkpoints for product accuracy assessment from 20 to 30 because doing so increases the statistical validity of the assessment, and (5) the use of a new term called Three-Dimensional Positional Accuracy which combines the horizontal and vertical accuracies together (ASPRS 2023).

### 16.3.1 Initial Considerations

Important initial considerations when assessing positional accuracy include both sources of error and appropriate classification scheme. In fact, in positional accuracy assessment, these two considerations are intimately linked together. Historically, guidelines such as NMAS and even NSSDA did not specify anything about where the samples should be taken to conduct the assessment. However, with the advent of LiDAR data it was quickly recognized that land cover or vegetation type at the location the samples were collected significantly impacts the accuracy of the results. Therefore, vegetation or land cover type is considered an important source of error in positional accuracy assessment and has been mitigated by taking a minimum number of samples (usually 20) in each of the major vegetation or land cover types regardless of the size of the area (FEMA 2003, ASPRS 2004). As previously mentioned, the exact vegetation/land cover classes (i.e., the classification scheme) varies between the FEMA and ASPRS guidelines, but both agree that samples must be acquired in a variety of cover types in order to have a valid assessment. Recently, this standard has been modified recognizing that the accuracy achieved in vegetative areas will not be as good as that in nonvegetated ones (ASPRS 2023).

### 16.3.2 Collecting Reference Data

There are many factors that must be considered when collecting the reference data used to assess the positional accuracy of the map (see Figure 16.1). If the reference data are not collected properly then the entire assessment may be invalid. Many of these considerations involve sampling while others deal with issues such as independence, source, timing of collection, and consistency.

Independence is a key component of the assessment process. The data used must be independent from any data used in the registration of the map or other spatial data. This fact is commonly understood in statistical analysis, but there seems to still be many examples in positional accuracy where statements about the goodness of the position are still expressed using the data that was employed to register the map to the ground. These data are clearly not an independent dataset and represent a most optimistic and invalid estimate of the actual positional accuracy.

Obviously, the source of the reference data must be more accurate than the map that is being assessed. Sometimes, a map of larger scale is sufficient. In other situations, a survey using GPS or other equipment is required. NSSDA recommends data be "of the highest accuracy that is feasible and practical" (FGDC 1998). Others have suggested that the positional reference data be from one to four times more accurate than the anticipated accuracy of the map being tested (e.g., Ager 2004, NDEP 2004, ASPRS 2004, ASPRS 2014). Given recent improvements in data acquisition and processing, the most recent ASPRS standard (ASPRS 2023) reduced the accuracy standard to three times the anticipated accuracy of the map being tested.

When and who collects the reference data can also be important although typically these factors are more important in thematic accuracy assessment. It is important that major changes have not taken place due to earthquakes or other natural phenomena that could alter the positions of objects (again, this is a rare situation). Also, the collection of the data must be consistent between multiple collectors. In other words, if more than one individual is collecting the data, it is important to establish objective procedures to ensure that the data are collecting similarly by all collectors.

Since the assessment of the positional accuracy of a map is not performed at every place on the map, but rather at a series of points, it is critical that the sampling be valid to achieve an appropriate assessment. Sampling involves determining the number of samples, the sample unit (identification), and the sampling scheme or distribution. Again, failure to properly plan for these factors will result in an invalid assessment. The NSSDA (FGDC 1998), states that a minimum of 20 samples must be used in the assessment. Some of the new guidelines, which require selecting samples from a number of vegetation/land cover classes, require 20 samples per class while some have increased that number to 30 per class (NDEP 2004, ASPRS 2004). From a statistical standpoint, taking 30 samples is often recommended and it is a positive step that the new ASPRS standard (ASPRS 2023) recommends this amount. Unfortunately, many assessments have been conducted with either non-independent samples or too few samples resulting in statistically invalid results.

The sample unit is positional accuracy assessment is actually a point. In fact, it is critical that these points be well-defined and "represent a feature for which the horizontal position is known to a high degree of accuracy and position with respect to the geodetic datum" (FGDC 1998). Any ambiguity about where a point is located will disqualify it as a sample point for use in the positional accuracy assessment.

Finally, the sampling scheme determines how the samples are collected throughout the map (i.e., the sample distribution). It is important that the samples are distributed throughout the map so that the entire map is assessed. The full range of variation in the map should be considered including topography, vegetation/land cover, and important features. Figure 16.3 demonstrates a stratified approach proposed by ASPRS (1989) that guarantees that the samples are distributed throughout the map by dividing the map into quadrants and forcing a minimum of 20% of the samples into each quadrant. This method also minimizes spatial autocorrelation in the sampling by setting a minimum spacing between sample points such that no two points can be closer together than $d/10$ where $d$ is the diagonal dimension of the map. This method seems to be a very effective way to insure that the sampling considerations for positional accuracy assessment are not ignored.

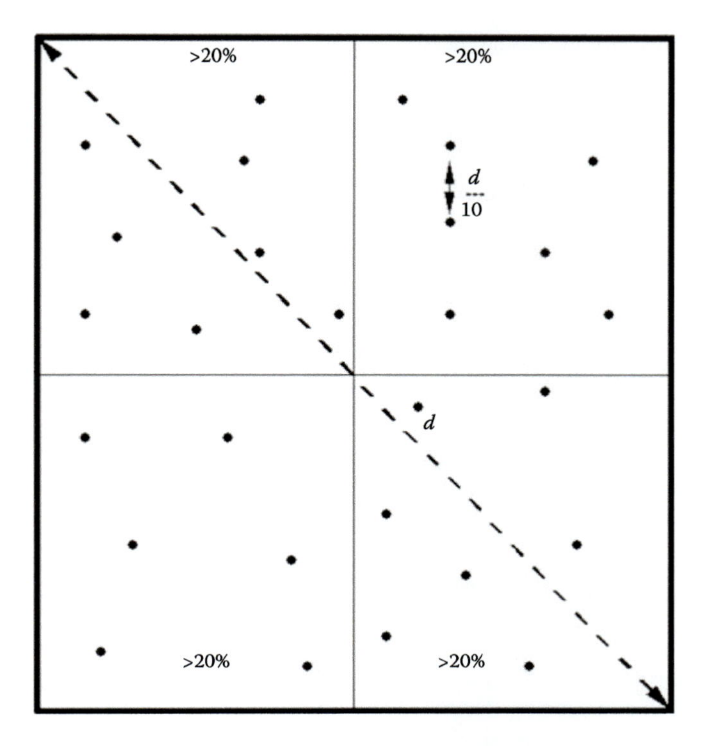


FIGURE 16.3 Sampling scheme showing samples distributed throughout the map.

### 16.3.3 Computing Descriptive Statistics

As previously discussed, assessing positional accuracy is performed using a sample of data to estimate the agreement (i.e., fit) between the map or other geospatial data and a reference dataset that is assumed to be correct. This analysis is done through the computation of a number of statistics. The key statistic that is computed in positional accuracy assessment is the root mean square error (RMSE). RMSE is simply the square root of the mean of squared differences between the samples on the map and those same samples on the reference data. The differences between the map and reference sample points are squared to insure positive values since the simple arithmetic difference between the map and reference sample points can be either negative or positive.

As alluded to previously, positional accuracy can be measured in the vertical dimension ($z$) or in the horizontal dimension which involves both $x$ and $y$. The analysis is similar in each case, but it is little more complicated for the horizontal dimension. The equation for RMSE for vertical accuracy is:

$$\text{RMSE}_v = \sqrt{\frac{\sum_i^n (e_{vi})^2}{n}} \tag{16.1}$$

where

$$e_{vi} = v_{ri} - v_{mi} \text{ and} \tag{16.2}$$

$v_{ri}$ equals the reference elevation at the $i$th sample point,
$v_{mi}$ equals the map elevation at the $i$th sample point, and
$n$ is the number of samples.

The equation for RMSE for horizontal accuracy is:

$$\text{RMSE}_h = \sqrt{\frac{\sum_i^n e_{hi}^2}{n}} \quad (16.3)$$

where

$$e_{hi}^2 = (x_{ri} - x_{mi})^2 + (y_{ri} - y_{mi})^2 \quad (16.4)$$

$x_{ri}$ and $y_{ri}$ are the reference coordinates, $x_{mi}$ and $y_{mi}$ are the map coordinates for the $i_{th}$ sample point, and $n$ is the number of samples.

The NSSDA states that positional accuracy be reported at the 95% level defined as "95% of the locations in the data set will have an error with respect to the reference position that is equal to or less than the computed statistic" (FGDC 1998). The equation for computing NSSDA for vertical accuracy is given in the guidelines as:

$$\text{NSSDA Vertical Accuracy}_v = 1.96\ (\text{RMSE}_v) \quad (16.5)$$

The equation for computing the NSSDA for horizontal accuracy is a little more complicated because of the two dimensions ($x$ and $y$). It is possible to have a different distribution of errors in the $x$ direction than in the y direction causing the errors around the position to be oblong in shape rather than circular as one would expect if the errors were equally distributed. Most use the simplified equation ignoring the distribution of errors and compute NSSDA for horizontal accuracy as:

$$\text{NSSDA Horizontal Accuracy} = 1.7308 * \text{RMSE}_h \quad (16.6)$$

Table 16.1 shows the computation of NSSDA and RMSE for a small horizontal dataset. Careful study of this table shows why the differences between the reference position and map position are squared to eliminate the positive and negative values. These computations can easily be executed in an Excel Spreadsheet or other software package to quickly compute the currently required statistics for assessing the positional accuracy of any map or geospatial dataset.

It should be noted here that there has been significant confusion in the mapping community over some of the computations used in assessing positional accuracy. This confusion results because the term RMSE as used by mapping professionals differs from the term RMSE used in statistics. In addition, the term "standard error" is unfortunately used to depict different parameters in different professions. While most statistics textbooks define the standard error as the square root of the variance of the population of means, $\sigma_{\overline{X}}$, many mapping texts define the standard error as the square root of the variance of the population signified by $\sigma$. Statisticians call this term the standard deviation. As a result some of the equations that are correct in the work by Greenwalt and Schultz have been misapplied or at least misinterpreted by the FGDC for use in developing the NSSDA. While it is not possible in this chapter to provide a full explanation of this confusion and the statistical ramifications that have resulted from this issue, a full discussion is provided in Congalton and Green (2019). Currently, NSSDA is the accepted standard and most positional accuracy assessments report this statistic along with RMSE. Newer guidelines as reviewed previously have suggested refinements including sampling in the major vegetation/land cover types as well as other ways to assess positional accuracy. The latest standards set by ASPRS are especially relevant for dealing with digital geospatial data (ASPRS 2023).

**TABLE 16.1** Example of Computing Positional Accuracy (RMSE and NSSDA)

| Sample no. | *x* (ref) | *x* (map) | *x* diff | (*x* diff)$^2$ | *y* (ref) | *y* (map) | *y* diff | (*y* diff)$^2$ | Σ *x* diff$^2$ + *y* diff$^2$ |
|---|---|---|---|---|---|---|---|---|---|
| 1 | 36.5 | 37.2 | −0.7 | 0.49 | 842.5 | 843.7 | −1.2 | 1.44 | 1.93 |
| 2 | 12.0 | 11.1 | 0.9 | 0.81 | 900.0 | 896.2 | 3.8 | 14.44 | 15.25 |
| 3 | 66.8 | 66.7 | 0.1 | 0.01 | 1010.1 | 1009.1 | 1.0 | 1.00 | 1.01 |
| 4 | 4.1 | 4.2 | −0.1 | 0.01 | 786.5 | 782.4 | 4.1 | 16.81 | 16.82 |
| 5 | 9.1 | 8.5 | 0.6 | 0.36 | 655.8 | 658.2 | −2.4 | 5.76 | 6.12 |
| 6 | 77.0 | 79.0 | −2.0 | 4.00 | 676.0 | 672.1 | 3.9 | 15.21 | 19.21 |
| 7 | 112.1 | 112.1 | 0.0 | 0.00 | 655.1 | 666.2 | −11.1 | 123.21 | 123.21 |
| 8 | 99.0 | 99.8 | −0.8 | 0.64 | 688.1 | 689.2 | −1.1 | 1.21 | 1.85 |
| 9 | 55.5 | 54.9 | 0.6 | 0.36 | 744.3 | 737.1 | 7.2 | 51.84 | 52.20 |
| 10 | 102.1 | 110.5 | −8.4 | 70.56 | 810.7 | 811.2 | −0.5 | 0.25 | 70.81 |
| 11 | 6.8 | 6.7 | 0.1 | 0.01 | 845.6 | 847.2 | −1.6 | 2.56 | 2.57 |
| 12 | 200.5 | 198.2 | 2.3 | 5.29 | 902.1 | 902.4 | −0.3 | 0.09 | 5.38 |
| 13 | 252.1 | 250.9 | 1.2 | 1.44 | 937.4 | 939.2 | −1.8 | 3.24 | 4.68 |
| 14 | 260.8 | 262.0 | −1.2 | 1.44 | 945.5 | 940.1 | 5.4 | 29.16 | 30.60 |
| 15 | 300.9 | 299.5 | 1.4 | 1.96 | 960.8 | 962.0 | −1.2 | 1.44 | 3.40 |
| 16 | 266.8 | 267.0 | 1.8 | 3.24 | 971.2 | 971.1 | 0.1 | 0.01 | 3.25 |
| 17 | 142.1 | 135.8 | 6.3 | 39.69 | 1000.3 | 999.1 | 1.2 | 1.44 | 41.13 |
| 18 | 96.7 | 96.5 | 0.2 | 0.04 | 1006.8 | 1008.2 | −1.4 | 1.96 | 2.00 |
| 19 | 42.0 | 36.9 | 5.1 | 26.01 | 1101.3 | 1110.0 | −8.7 | 75.69 | 101.70 |
| 20 | 12.2 | 15.1 | −2.9 | 8.41 | 999.9 | 1002.0 | −2.1 | 4.41 | 12.82 |
| | | | | | | | | Sum | 515.94 |
| | | | | | | | | Ave. | 25.78 |
| | | | | | | | | RMSE | 5.08 |
| | | | | | | | | NSSDA | 8.79 |

## 16.4 THEMATIC MAP ACCURACY ASSESSMENT

The history of assessing the thematic accuracy of maps derived from remotely sensed data is shorter than that of positional accuracy. Except for a brief effort to quantitatively assess the accuracy of photo interpretation in the 1950s, the history of thematic accuracy assessment began shortly after the advent of digital imagery (i.e., the launch of the first Landsat). Researchers, notably Hord and Brooner (1976), van Genderen and Lock (1977), and Ginevan (1979), proposed criteria and basic techniques for testing overall thematic map accuracy. In the early 1980s, more in-depth studies were conducted, and new techniques proposed (Mead and Szajgin 1981, Rosenfield et al. 1982, Congalton et al. 1983, Aronoff 1985). Finally, from the late 1980s up to the present time, a great deal of work has been conducted on thematic accuracy assessment. As we develop increasingly high spatial and spectral resolution sensors, more complex machine learning classification algorithms including object-based image analysis (OBIA), and faster computer hardware with easier to use software including cloud computing, it is vital that our methods for effectively assessing thematic maps continues to improve as well.

The history of thematic accuracy assessment has progressed through a number of stages as it developed and matured. As already discussed, the initial excitement of any new technology typically results in a time of intensive use of the technology without much, if any, attention paid to quality of the results. Such was the case for digital image analysis. This stage quickly moved into the next stage in which the map "looks good." There are a number of ways to qualitatively assess that a map "looks good" including visually comparison of the thematic map to the original imagery (Congalton 2001) or perhaps to other image sources (e.g., Google Earth or the like). The map could be shown to those especially familiar with the area or the map could be visually compared to other thematic maps of the area. Each of these methods could convince the analyst that the map "looks good." It is important that a thematic map "look good." If the analyst is not convinced the map is of sufficient quality from some qualitative assessment, then there really is no sense in conducting a quantitative accuracy assessment. Therefore, it "looks good" is a necessary, but not a sufficient characteristic of any thematic map. Unfortunately, some are still stuck in the it "looks good" stage of thematic map accuracy assessment and never go any further to effectively assess their maps.

Quickly following the "it looks good" stage of thematic map assessment was the stage called non-site-specific assessment. In this stage, some effort was made to obtain quantitative estimates of the accuracy of the thematic map. However, these estimates represent results of the map as a whole but do not say anything about the accuracy at any specific location on the map. Instead, estimates such as the total hectares of forest on the map are compared to some reference estimate of the total forest hectares in that same area. While non-site-specific assessments do begin to introduce some quantitative assessment into the process, the results are left lacking in that nothing can be said about any specific location on the map. Therefore, this method rapidly gave way to the current stage of thematic map accuracy assessment called site-specific accuracy assessment.

Site-specific thematic accuracy assessment is represented in the form of an error matrix (Figure 16.4), also called a contingency table in statistics. The process of creating an error matrix is demonstrated in Figure 16.5. An error matrix is a square array of numbers set out in rows and columns which expresses the number of sample units assigned to a particular category (i.e., land cover class) in one classification as compared to the number of sample units assigned to a particular category (i.e., land cover class) in another classification. Typically, one of the classifications is considered to be correct (or at least of significantly higher accuracy than the other) and is called the reference data. Unfortunately, it has become common practice to call these reference data the "ground truth." While the reference data should have high accuracy, there is no guarantee that it is 100% correct and using the term "ground truth" is not appropriate. This author hopes that the geospatial community will abandon this term for ones that better represent the actual data such as reference data or ground data or field data or the like. The columns of the error matrix typically represent the reference data while the rows indicate the map derived from remotely sensed data or other geospatial dataset. It is possible to switch the headings for the row and columns so the analyst must take care to notice how the axes of the matrix are labeled before any further analysis is undertaken. Because it is important

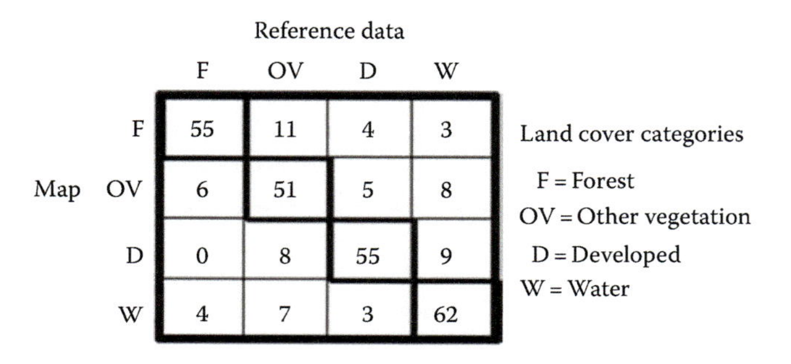


| Map \ Reference data | F | OV | D | W |
|---|---|---|---|---|
| F | 55 | 11 | 4 | 3 |
| OV | 6 | 51 | 5 | 8 |
| D | 0 | 8 | 55 | 9 |
| W | 4 | 7 | 3 | 62 |

**FIGURE 16.4** An example error matrix.

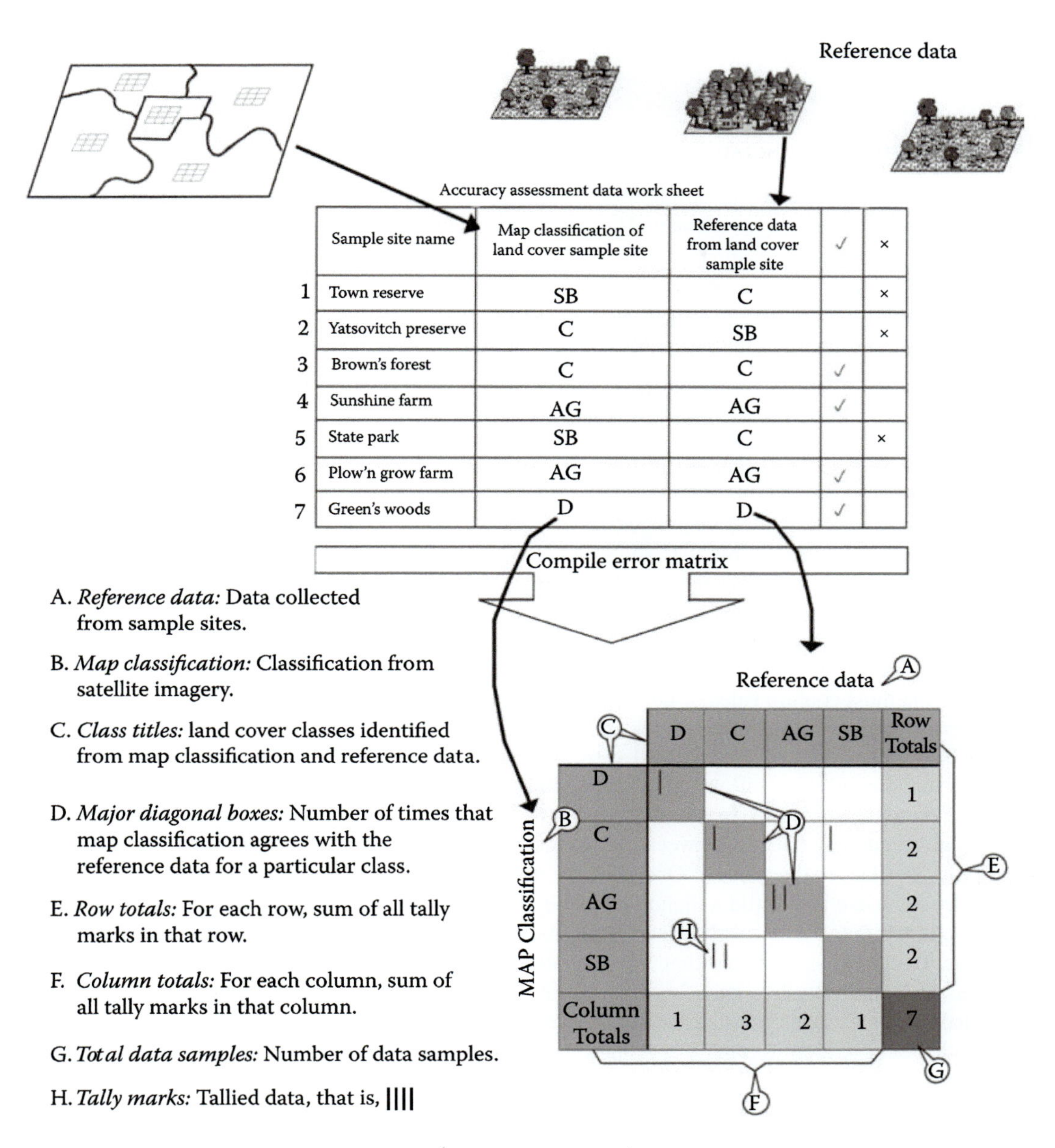


Accuracy assessment

Overall accuracy indicates how well the map identifies the land cover type on the ground.

Producer's accuracy indicates how many times a land cover type on the ground was identified as the land cover type on the map. It expresses how well the map producer identified a land cover type on the map from the satellite imagery data.

User's accuracy indicates how many times a land cover type on the map is really the land cover type on the ground. It expresses how well a person using the map will find that land cover type on the ground.

**FIGURE 16.5** A diagram showing how an error matrix is generated from a series of samples. (From Landcover Protocols of the GLOBE Teacher's Manual (www.globe.gov).)

to confirm which axis of the error matrix is which, it is imperative that the axes are properly labeled to avoid confusion.

### 16.4.1 Initial Considerations

There are two very important initial considerations that must be taken into account before beginning a thematic accuracy assessment. These are classification scheme and sources of error.

#### 16.4.1.1 Classification Scheme

Thematic maps are an abstraction of the real world and as such use some type of classification scheme to group together and simplify reality. Determining which classification scheme to use or developing your own must be completed at the beginning of the mapping project. Failure to select an efficient and effective scheme from as early in the project as possible dooms the project to significant inefficiencies and potential failure. Whether the thematic map is of land cover or vegetation or soil types or some other thematic information, it is critical that the scheme represent the information that is necessary for the end user. Therefore, it is important to bring the analyst(s) producing the map and the user(s) of the map together to agree on a scheme. However, it is also important that this information be discernible with the remotely sensed data selected for the mapping. Therefore, selecting the appropriate scheme at the beginning of the project facilitates choosing suitable imagery and allows reference data to be collected for use in assessing the accuracy of the resulting thematic map simultaneously.

Any classification scheme selected should have the following four characteristics: (1) **Definition:** The scheme must contain rules or definitions that explicitly define each of the map classes. It is not sufficient to simply list the map classes and assume that everyone agrees on exactly what each class is. For example, a forest may be defined as trees greater that 5 m tall. Using this definition, a grouping of trees that are only 3 m tall would not be defined as a forest. (2) **Mutually Exclusive:** The scheme must be mutually exclusive. In other words, every effort must be made using the class definitions to clearly eliminate any overlap between classes so that every area on the map or ground falls into one and only one class in the scheme. One effective method for ensuring that the classification scheme is mutually exclusive is to build a dichotomous key to aid in labeling the map classes. At a minimum, well thought out and extensively tested definitions are a must! (3) **Totally Exhaustive:** The scheme must be totally exhaustive. In other words, every area on the map or ground must fall into one of the classes in the classification scheme. An effective means of insuring that your scheme is totally exhaustive is to include an "Other" class. However, if a large portion of the thematic map ends up in the "Other" class, then it may be appropriate to rethink the classification scheme as perhaps some important information on the map has been neglected. (4) **Hierarchical:** Finally, it is useful if the classification scheme is hierarchical. That is having multiple levels of detail (a hierarchy) such that the map can be generated and especially assessed for accuracy at different levels of detail. For example, the forest class may be further divided in conifer and deciduous classes that add more detail. The conifer class may be further divided into a pine class and a non-pine class. As we will see shortly, it may not be possible to assess the accuracy of the map at the most detailed level because of costs of collecting the reference data. Therefore, it may be possible with a hierarchical classification scheme to map to one level of the hierarchy but to assess the accuracy to a lesser level of detail.

In determining the appropriate classification scheme for a thematic mapping project, the analyst must also consider the concept of **minimum mapping unit (mmu)**. The mmu is the smallest area that is uniquely delineated on a thematic map. This concept is widely known and used in photo/image interpretation where the maps are created from aerial photographs/imagery by a human interpreter at a given scale. However, the concept has not been as widely adopted (or perhaps forgotten) in digital imagery as the imagery can be rendered at any scale and pixel size seems to be more of the limiting factor. As will be seen in the discussion of sample unit, the pixel should not be selected as the minimum area to be mapped on digital data and therefore, consideration of an mmu even on digital mapping projects is essential.

#### 16.4.1.2 Sources of Error

Creating thematic maps from remotely sensed imagery is possible because there is a strong linkage between what can be sensed on the imagery and what is actually happening on the ground. However, this correlation is not perfect and every situation where this correlation breaks down is a source of potential error between the map and the ground. Lunetta et al. (1991) presented a discussion of these sources of error which included errors from image acquisition, errors from data processing, errors that occur within the analysis including data conversion, errors that can occur in the accuracy assessment process itself, and finally error in decision making and implementation.

While it is clear that error can enter a thematic mapping project from a great variety of sources, little has been done to evaluate these errors and prioritize methods for dealing with them. Congalton and Brennan (1999) and Congalton (2009) proposed an error budget analysis method that not only clearly lists the sources of error in any mapping project, but also documents a method for evaluating the error contribution, implementation difficulty, and implementation priority for each error. Table 16.2 shows an example of this error budgeting.

**TABLE 16.2** Error Budget Analysis for a Remote Sensing Project

| Error Source | Error Contribution Potential | Implementation Difficulty | Implementation Priority | Error Assessment Technique |
|---|---|---|---|---|
| **Systematic** | | | | |
| Sensor | Low | 5 | 21 | Instrumentation & Analysis |
| **Natural** | | | | |
| Atmosphere | Medium | 4 | 20 | Instrumentation & Analysis |
| **Preprocessing** | | | | |
| Geometric Registration | Low | 2 | | Positional Accuracy Assessment |
| Image Masking | Medium | 3 | 7 | Single Date Error Matrix |
| **Derivative Data** | | | | |
| Band Ratios | Low | 1 | 18 | Data Exploration |
| NDVI | Low | 1 | 18 | Data Exploration |
| PCA | Low | 1 | | Data Exploration |
| TCA | Low | 1 | 8 | Data Exploration |
| Other Ancillary Data | Medium | 3 | 9 | Data Exploration |
| **Classification** | | | | |
| Classification Scheme | Medium | 2 | 14 | Single Date Error Matrix |
| Training Data Collection | Medium | 3 | 15 | Single Date Error Matrix |
| Classification Algorithm | Medium | 3 | 11 | Single Date Error Matrix |
| **Post-Processing** | | | | |
| Data Conversion | High | 2 | 12 | Single Date Error Matrix |
| **Accuracy Assessment** | | | | |
| Sample Unit | Low | 2 | 13 | Single Date Error Matrix |
| Sample Size | Low | 2 | 16 | Single Date Error Matrix |
| Sampling Scheme | Medium | 3 | 17 | Single Date Error Matrix |
| Spatial Autocorrelation | Low | 1 | 3 | GeoStatistical Analysis |
| Positional Accuracy | High | 2 | 2 | RMSE/NSSDA |
| **Final Product** | | | | |
| Decision making | Medium | 2 | 1 | Sensitivity Analysis |
| Implementation | Medium | 2 | 6 | Sensitivity Analysis |

**Error Contribution Potential**—ranked from low to medium to high; **Implementation Difficulty**—ranked from 1: not very difficult to 5: extremely difficult; **Implementation Priority**—ranked from 1 to *n* showing the order in which to implement improvements

### 16.4.2 Collecting Reference Data

There are many factors that must be considered when collecting the reference data used to assess the thematic accuracy of the map. Collection of reference data is more complicated for thematic accuracy assessment than positional accuracy assessment and therefore, there is even more risk that the entire assessment may be invalid if the reference data are poorly or improperly collected. Many of the factors that must be considered involve the sampling process while others deal with issues such as independence, source, timing of collection, and consistency.

#### 16.4.2.1 Independence

It is absolutely critical that the reference data used to assetss the thematic map accuracy be independent of any other data used for training or any other purpose during the mapping project. This independence can be achieved in two ways. First, the collection of the reference data can be done at a separate time and/or by different personnel from the collection of the training data. While effective, this method is inefficient. A second approach involves collecting the training and reference data simultaneously and then splitting the data into two sets whereby the reference dataset is put aside and not viewed by any project personnel until it is time to conduct the accuracy assessment.

#### 16.4.2.2 Sources of Reference Data

The source of the reference data in any thematic map accuracy assessment is a function of the complexity of the classification scheme (i.e., level of map detail) and the budget of the project. It is fortuitous if existing maps (e.g., USDA Cropland data layer) or ground data (e.g., U.S. Forest Service Inventory Data) can be used as the reference data. However, differences in classification scheme and most often size of the sample unit typically limit using existing reference data, no matter how tempting it is. Often, the existing data are too old, the positional accuracy too poor, or the sampling plot size (sample unit) is too small to be used as reference data. Therefore, more commonly, the reference data are newly collected information assumed to be more accurate than the map that is being assessed. Manual image/photo interpretation has often been used to assess thematic maps generated from moderate resolution digital imagery, ground/field visits are used to assess maps made from high resolution imagery, and manual image interpretation has been used to assess maps created from automated classification techniques. No matter what source is selected as the reference data, it is imperative that the data are effectively and appropriately collected so as to provide a viable dataset from which to validly assess the accuracy of the thematic map. Therefore, it is necessary that an explanation and justification of the data used for reference data must be part of the accuracy assessment report.

#### 16.4.2.3 When Should the Reference Data Be Collected

Timing for collecting reference data for assessing the accuracy of thematic maps clearly depends on the type of information being mapped. Some information is just more timely than others. For example, if mapping agricultural crops, it is very important that the reference data be collected as near to the day the imagery was collected as possible (in some cases, same day collection is important). In other examples where change occurs on a much slower time scale, collection of the reference data within a year or two or five, may be sufficient. Clearly, if a change occurs between the date of the remotely sensed image capture and the data of the reference data collection, then the reference data will be of little or no use in effectively assessing the accuracy of the map.

#### 16.4.2.4 Consistency in Reference Data Collection

A good classification scheme consisting of labels with clear rules or definitions goes a long way to ensuring consistency in reference data collection. A guidebook documenting the procedures used to collect the data is also important. Finally, a field form, whether on paper or using some sort of data logger, completes the process. Having these three components facilitates training of all field personnel whether it is a single collector or a large group being disbursed throughout the study area. Objectivity in the data collection process is key and having a good reference data collection form will help enforce this need. Regardless of the classification scheme, the form will have certain common components such as the names of the collectors, date of collection, location (GPS) information, a key to the map classes (dichotomous keys are excellent), the actual map class determined by the collection, and a place to describe any issues, special findings, or problems that occurred at the collection site.

#### 16.4.2.5 Sampling

As in positional accuracy assessment, thematic accuracy assessment relies on a statistical valid sample of reference data to compare to the map to generate the error matrix and compute the required statistics. It is not possible to assess the entire map (a total enumeration) because of time, money, and effort constraints. Therefore, sampling (a subset of the map) is used to facilitate the accuracy assessment. The sampling process for collecting reference data for assessing the accuracy of thematic maps derived from remotely sensed data can be divided into 3 components: the sample unit, the sample size, and the sampling scheme including spatial autocorrelation. Each of these is very important to insure proper and valid reference data collection.

##### *16.4.2.5.1 The Sample Unit*

The sample unit is the size in area of each sample. This size is typically measured in pixels or measurement units such as hectares or meters on a side. Historically, a single pixel has often been used as the sample unit for assessing thematic map accuracy. Unfortunately, a single pixel is a very poor choice for the sample unit because a pixel is arbitrary, difficult to locate, and typically smaller than the minimum mapping unit defined by the project. Even with all the latest advances in GPS, terrain correction, and geometric registration, all reference data sample units have some positional errors. Mid-resolution sensors such as Landsat consider positional errors of one half a pixel (10–15 m) to be reasonable. Higher spatial resolution sensors have smaller pixel sizes but increased positional errors if expressed in terms of pixels (i.e., a sensor with 1 m pixels could have a positional error of 10–15 m—the same amount as Landsat, but what is half of a Landsat pixel represents a 10–15 pixel error with this sensor). This issue is especially relevant for imagery taken off-nadir (not vertical to the ground). If a Landsat pixel is registered to the ground with 10–15 m accuracy and the GPS unit used to locate the center of the pixel on the ground has an accuracy of 10–15 m, then it is totally impossible to use that single pixel as the sample unit for assessing the thematic accuracy of the map because the pixel simply can not be located accurately. These days there is lots of attention paid to citizen science and the positional accuracy of data collected by citizen scientists can be even worse than 15 m. The problem is only exacerbated more with increased spatial resolution imagery. In addition, depending on the classification scheme, the selection of a single pixel will not represent the map classes in the scheme. For example, if there is a mixed conifer/hardwood forest class, it is likely that a single pixel would be either a conifer or a hardwood but not both. Therefore, the map class would be falsely shown as inaccurate in the error matrix.

Therefore, the appropriate sample units are either a cluster of pixels or a polygon that adequately represent the classification scheme of the project. Considering positional accuracy limitations, a cluster of pixels, typically 3 × 3 pixels for moderate resolution imagery, has been a good choice. A homogeneous cluster of pixels minimizes positional error problems while maintaining a constant sized sample unit. If assessing the accuracy of a thematic map made from course resolution imagery such as MODIS or AVHRR, it is unlikely to find many homogeneous single pixels let alone a 3 × 3 cluster. In this case, it may be necessary to document the proportions of the sample unit in each land cover class and set rules for labeling the reference data accordingly. While the complexity of conducting such an assessment is beyond the scope of this chapter, recent work by Gu and Congalton (2020, 2021) may provide some additional insight for the interested reader.

If assessing the accuracy of a thematic map made from fine resolution, it is important to select appropriate sampling units that account for the positional error. For example, if the imagery has 1 m pixels and a positional accuracy of 10 m, then a 3 × 3 cluster will not compensate for positional error. Instead, a sample unit of perhaps 20 × 20 pixels is needed to compensate for a 10 m registration error.

It is important to remember that if a cluster of pixels is selected as the sample unit, then the cluster (sample unit) represents a single sample and is tallied as only one sample unit in the error matrix. There are far too many examples in the literature where an analyst has selected a cluster of pixels and then considered each pixel in the cluster as a separate sample unit. Collecting reference data in this way is incorrect and results in an invalid assessment. It should also be noted that typically the clusters selected for sampling are homogeneous (this may not be possible with coarse resolution imagery). Selecting homogenous clusters have the great advantage of minimizing positional error, but may result in a biased or inflated assessment as borders between classes (i.e., heterogeneous areas) are avoided and therefore, these edge errors may not be assessed. Therefore, again, it is imperative that the method used for the assessment be fully described. Finally, remember that while a 3 × 3 pixel cluster should work well for moderate resolution imagery, much larger clusters (in terms of number of pixels, not meters) must be collected for higher spatial resolution imagery in order to account for positional error.

Thematic maps generated from manual interpretation and more recently from object-based image analysis (OBIA) are vector maps where polygons are delineated on the imagery instead of the typical rasters (i.e., pixels). In these situations, the polygon should be used as the sample unit for collecting the reference data. There are a few new issues that arise when a polygon is used for accuracy assessment. First, the polygon may be much larger that a 3 × 3 pixel cluster and therefore, more effort may need to go into accurately labeling that polygon on the reference data (MacLean et al. 2013). Second, the polygons are no longer of the same area as was the case for a 3 × 3 pixel cluster. Therefore, a more area-based error matrix may be appropriate instead of just tallying the sample units (Maclean and Congalton 2012, MacLean and Congalton 2013, Olofsson et al. 2013). This very important consideration will be covered more in the section on advanced analysis techniques later in this chapter.

#### 16.4.2.5.2 The Sample Size

Collecting enough samples to obtain a statistically valid reference dataset is perhaps the most challenging component of assessing the accuracy of a thematic map. In almost every situation, there is a balance between what is required to be statistically valid and what is affordable within the given budget. Early on, many researchers including Hord and Brooner (1976), van Genderen and Lock (1977), Hay (1979), Ginevan (1979), Rosenfield et al. (1982), and Congalton (1988b), published equations and guidelines for choosing the appropriate sample size. Initial efforts used the binomial equation to estimate the number of samples required to report the overall accuracy of a map. However, sampling to complete an error matrix is not a binomial situation, but rather a multinomial one in which there is one correct answer and $n - 1$ wrong answers (where $n$ = the number of thematic classes) for each sample (Congalton 1988b, Congalton and Green 2009). Tortora (1978) presents the equations needed to compute the required sample size using the multinomial sampling approach. The number of samples depends on the number of thematic classes and the allowable error and the desired confidence level. Congalton (1988b) determined, using an extensive series of Monte Carlo simulations, that most maps could be assessed using 50 samples per thematic map class (maps less than about 500,000 hectares and 12 or less thematic map classes). More complex maps with more than 12 thematic map classes and covering larger areas should use 75–100 samples per thematic map class. Assessing the thematic accuracy of global and/or continental maps require partitioning the area into some type of ecological zones with the requisite number of samples per zone. Therefore, assessing these maps requires some minimum number of samples per zone times the number of zones.

It should be noted that practical considerations are a key component for determining sample size. There is a trade-off between the budget and the number of samples needed. There is a point that is reached when the samples per thematic class are under 30 per class that the assessment loses statistical validity completely and it may be better not to even attempt the accuracy assessment.

#### *16.4.2.5.3 The Sampling Scheme*

A number of sampling schemes (how to collect the samples) have been suggested for collecting the reference data including simple random sampling, systematic sampling, stratified sampling, stratified random sampling, cluster sampling, and others. While random sampling has valuable statistical properties, it is often impractical to implement because of the costs of getting to every random location and because thematic classes consisting of small areas are left undersampled. Systematic sampling ensures samples are distributed over the entire study area and is very effective if higher resolution imagery is the source used to obtain the reference data. Systematic sampling has similar problems of access when collecting ground reference data. Cluster sampling, that is collecting a number of sample units within close proximity to each other, offers certain efficiencies, but care must be taken to not take too many samples too close together because of spatial autocorrelation (samples are not independent of each other). Therefore, most reference data collection efforts use stratified sampling in order to insure some minimum number of samples in each stratum (thematic map class).

Many accuracy assessments use the rule just expressed and include 50 or so samples in each thematic map class. In some situations, it may be appropriate to have more samples in certain thematic classes than in others. These situations include when some classes are more important or when some classes occupy more of the map area than the others. Care is necessary here to ensure some minimum number of samples in each thematic class so as to be able to determine the accuracy of every thematic map class. For example, imagine an area that is 95% sand and water (see Figure 16.6) and the other 5% is divided among commercial, industrial, and residential areas. If a stratified sampling approach was not employed where a minimum number of samples were taken from each map class and instead a random sample of 250 samples were selected, it is possible to achieve a high map accuracy and yet poorly map 3 of the 5 map classes. Probability tells us that if 250 samples were randomly selected from a map that was 95% sand and water then 95% or 238 of

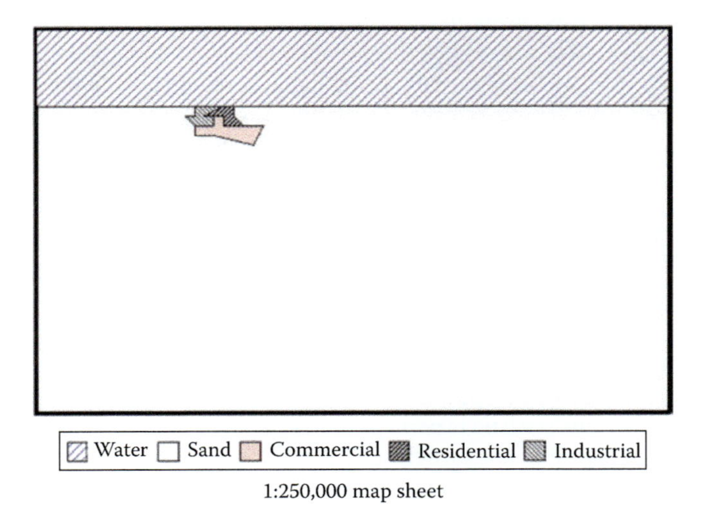


**FIGURE 16.6** A map showing an area of 95% sand and water with a small portion (5%) of the area commercial, residential, and industrial land cover types.

these samples would fall in the sand and water. Only 12 samples would fall within the other three map classes (commercial, industrial, and residential). If the sand and water were correctly mapped, but the other classes were incorrect all the time, we would still determine that this map was 95% correct based on this sampling scheme. This is a statistically valid and yet completely misleading result of our random sampling strategy. Therefore, selection of the appropriate sampling scheme is as important as selecting the correct sample unit and number of samples.

A final concept that influences the choice of sampling scheme and how the samples are selected is called spatial autocorrelation. Spatial autocorrelation occurs when the presence, absence or degree of a certain characteristic affects the presence, absence of degree of that same characteristic in neighboring units (Cliff and Ord 1973). In other words, the concept involves independence between neighboring samples. If an error is made at a certain location and positive spatial autocorrelation exists, then it is more likely that the same error will occur in nearby locations. Congalton (1988a) and Pugh and Congalton (2001) have demonstrated that spatial autocorrelation is an important consideration in the collection of reference data and therefore, samples should be taken so as to maximize the separation between samples as practical. Additionally, if cluster sampling is necessary, the number of sample clusters taken at a location should be as few as possible and they should be as far apart as possible.

Given the complexities, considerations, and choices that are involved in collecting proper reference data to assess the thematic accuracy of a map generated from remotely sensed data, it is absolutely required that a report be produced to document the process. Simply reporting an error matrix or, even worse, some summary accuracy measures without describing the choices made and the justification for these choices results in an incomplete assessment. Such a document is necessary for the map user to thoroughly understand the accuracy assessment process.

### 16.4.3 Computing Descriptive Statistics

Assuming that the error matrix has been created using a valid reference data collection approach as described earlier, the matrix is then the starting point for a number of descriptive statistics including overall accuracy, producer's accuracy, and user's accuracy. Figure 16.7 contains the same error matrix as was presented in Figure 16.4, but now shows the computation of some descriptive statistics. Remember that an error matrix is a very effective way to represent map accuracy because the individual accuracies of each category are easily discerned along with both the errors of inclusion (commission errors) and errors of exclusion (omission errors) present in the map. Just like a coin has two sides (heads and tails), map errors have two components (omission error and commission error). A commission error can be defined as including an area into a thematic class when it doesn't belong to that class, while an omission error is excluding that area from the thematic class in which it truly does belong. Each and every error is an omission from the correct thematic map class and a commission to a wrong thematic map class.

Figure 16.7 shows that six areas (sample units) were mapped as Other Vegetation while the reference data shows that they were actually Forest. In other words, six areas were omitted from the Forest class and committed to the incorrect Other Vegetation class. In addition to showing the omission and commission errors, the error matrix in Figure 16.7 also shows the computation of three other accuracy measures: (1) overall accuracy, (2) producer's accuracy, and (3) user's accuracy (Story and Congalton 1986). Overall accuracy is computed by summing up the major diagonal of the error matrix (i.e., the correctly mapped sample units) and dividing by the total number of sample units in the matrix. While this statistic is perhaps the most commonly reported measure of thematic map accuracy, this value alone is not sufficient. It is vital that the error matrix be reported so that other measures of accuracy can be computed and that the errors in the map can be clearly seen and understood.

Overall accuracy represents the accuracy of the entire map. However, in addition, we may want to know the accuracy of an individual thematic map class. Computation of individual class

| | | Reference data F | OV | D | W | Row total |
|---|---|---|---|---|---|---|
| Map | F | 55 | 11 | 4 | 3 | 73 |
| | OV | 6 | 51 | 5 | 8 | 70 |
| | D | 0 | 8 | 55 | 9 | 72 |
| | W | 4 | 7 | 3 | 62 | 76 |
| Column total | | 65 | 77 | 67 | 82 | 291 |

Land cover categories
F = Forest
OV = Other vegetation
D = Developed
W = Water

Overall accuracy = (55 + 51 + 55 + 62 )/291 = 123/291 = 77%

| Producer's accuracy | User's accuracy |
|---|---|
| F = 55/65 = 85% | F = 55/73 = 75% |
| OV = 51/77 = 66% | OV = 51/70 = 73% |
| D = 55/67 = 82% | D = 55/72 = 76% |
| W = 62/82 = 75% | W = 62/76 = 82% |

**FIGURE 16.7** Error matrix showing the calculations for overall, producer's, and user's accuracies. Note: (1) As the map producer, 85% of the forest sample units (in the reference data) are correctly classified as forest on the map; (2) As a map user, 75% of the sample units classified as forest on the map are indeed forests.

accuracies is a little more complicated than overall accuracy and requires two measures: producer's accuracy and user's accuracy (Story and Congalton 1986). The producer of the map may want to know how well they mapped a certain thematic map class, the producer's accuracy. This value is computed by dividing the value from the major diagonal (the agreement) for that class by the total number of samples in that map class as indicated by the sum of the reference data for that class. Looking at Figure 16.7 shows that the map producer called 55 areas Forest while the reference data indicates that there were a total of 65 Forest areas. Ten areas were omitted from the Forest and of these six were committed to Other Vegetation and four were committed to Water. So, 55/65 samples were correctly called Forest for a Forest Producer's accuracy of 85%. However, this is only half the story. If you now observe the user's perspective of the map you see once again that 55 samples were called Forest on the map that were actually Forest, but in addition the map called 11 samples Forest that were actually Other Vegetation and four samples Forest that were actually Developed and three samples Forest that were actually Water. The map, therefore, called 73 samples Forest, but only 55 are actually Forest. There was commission error of 18 samples into the Forest that were not Forest. The Forest User's accuracy is then computed by dividing the major diagonal value for the Forest class by the total number of samples mapped as Forest, 55/73 = 75%. In evaluating the accuracy of an individual map class, it is important to consider both the producer's and the user's accuracies.

Overall, producer's, and user's accuracies can also be explained with the following equations. Begin with n samples which are distributed into $k^2$ cells where each sample is assigned to one of $k$ thematic map classes in the map (usually the rows) and, independently, to one of the same $k$ thematic map classes in the reference dataset (usually the columns). Then, let $n^{ij}$ denote the number of samples mapped into thematic map class $i$ ($i = 1, 2, \ldots, k$) in the map and thematic map class $j$ ($j = 1, 2, \ldots, k$) in the reference dataset.

Let

$$n_{i+} = \sum_{j=1}^{k} n_{ij} \tag{16.7}$$

be the number of samples classified into thematic map class "i" in the map, and

$$n_{+j} = \sum_{i=1}^{k} n_{ij} \tag{16.8}$$

be the number of samples classified into thematic map class "j" in the reference dataset.

Then, the overall accuracy between map and the reference data can then be computed as follows:

$$\text{Overall accuracy} = \frac{\sum_{i=1}^{k} n_{ii}}{n} \tag{16.9}$$

Producer's accuracy can be computed by

$$\text{Producer's accuracy } j = \frac{n_{jj}}{n_{+j}} \tag{16.10}$$

and the user's accuracy can be computed by

$$\text{User's accuracy}_i = \frac{n_{ii}}{n_{i+}} \tag{16.11}$$

### 16.4.4 Basic Analysis Techniques

After computing the descriptive statistics from the error matrix, some additional analysis techniques can also be implemented to learn even more about the results of the accuracy assessment. The first of these techniques is called "Margfit" and is used to normalize the error matrix to remove the effect of sample size for comparison with other error matrices. Margfit uses a standard iterative proportional fitting algorithm which normalizes the matrix by forcing each row and column total (i.e., the marginal) to sum to one. The result is a normalized matrix in which differences in sample sizes used to generate different matrices are eliminated and the individual cell values in the matrix are directly comparable with another matrix. Figure 16.8 shows the results of the Margfit analysis on Figure 16.7. Notice that each value in the matrix is now a percentage of 1 and multiplying each value by 100 would result in the accuracy for each cell. The major diagonal values represent the accuracy of each thematic map class and are a combination of the individual producer's and user's accuracies for each class. For example, the accuracy of the Forest class is shown as 0.7908 or 78%. Finally, as in computing overall accuracy, the normalized accuracy of the error matrix can be computed by summing the values in the major diagonal and dividing by four (the number of map classes and also the sum all values in the matrix. The normalized accuracy of Figure 16.8 is 76% while the overall accuracy is 77% (see Figure 16.7).

In order to fully appreciate the usefulness of the Margfit analysis it is necessary to have a second error matrix to compare to the first. Figure 16.9 presents an error matrix for the same study area as was assessed in Figure 16.7. However, the analyst collected more accuracy assessment sample units from which to assess the quality of the map. Overall, Producer's, and User's accuracies were computed as described and shown in Figure 16.9. Note that 356 sample units were used for this assessment, while only 291 sample units were selected for the assessment in Figure 16.7. The overall accuracy of the error matrix in Figure 16.9 is 84%.

The results of the Margfit analysis on Figure 16.9 are shown in Figure 16.10. Again, the individual cell values represent thematic class accuracies and the normalized accuracy is 83%.

Because the original error matrices (Figures 16.7 and 16.9) have been normalized (Figures 16.8 and 16.10), the values in the normalized error matrices can be directly compared. For example,

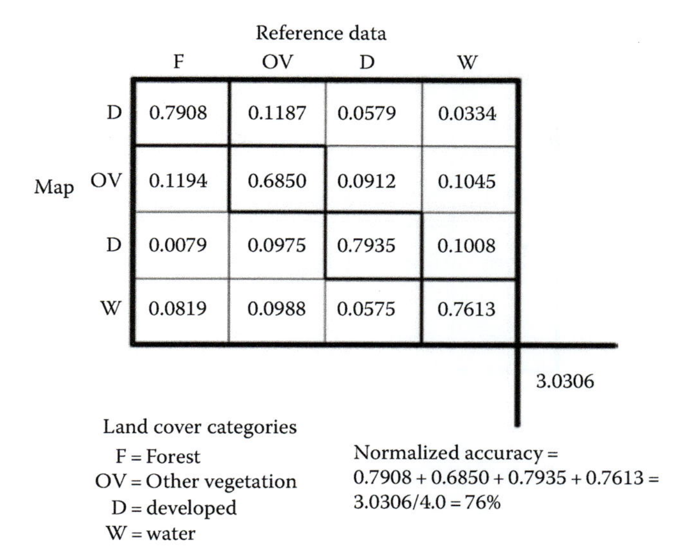


FIGURE 16.8 Error matrix showing the results of the Margfit analysis on Figure 16.7.

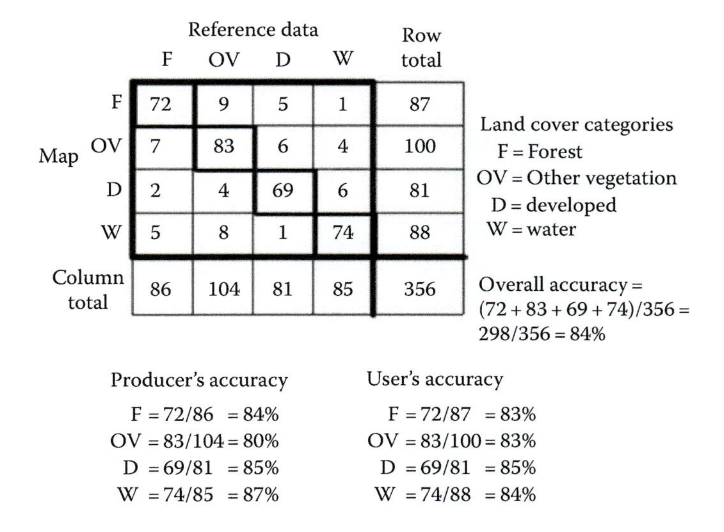


FIGURE 16.9 An error matrix from the same study area as Figure 16.7, but the analyst selected more sampling units for accuracy assessment.

comparing the 55 correct sample units in Figure 16.7 for the Forest class with the 72 correct sample units in Figure 16.9 for the Forest class would not have much meaning since the matrices were generated using different total sample sizes (291 and 356, respectively). However, with Margfit the matrices are normalized so that these values are now directly comparable with Figure 16.8 having a Forest accuracy of 79% and Figure 16.10 having a Forest accuracy of 83%. It seems that the analyst who created the map assessed using the error matrix in Figure 16.9 used a method that produced better Forest class accuracy. This normalization process produces a single accuracy value for each of the map classes that is a combination of the producer's and user's accuracy for that class. In this way, a single number rather than two separate values can be used to represent a class accuracy.

Reference data

| | | F | OV | D | W |
|---|---|---|---|---|---|
| Map | D | 0.8265 | 0.0909 | 0.0641 | 0.0180 |
| | OV | 0.0843 | 0.7872 | 0.0747 | 0.0532 |
| | D | 0.0297 | 0.0449 | 0.8446 | 0.0813 |
| | W | 0.0595 | 0.0771 | 0.0166 | 0.8475 |

3.3058

Land cover categories

F = Forest
OV = Other vegetation
D = Developed
W = Water

Normalized accuracy =
0.8265 + 0.7872 + 0.8446 + 0..8475 =
3..3058/4.0 = 83%

**FIGURE 16.10** Error matrix showing the results of the Margfit analysis on Figure 16.9.

A second technique that has often been used in accuracy assessment is called "Kappa" (Cohen 1960). The results of performing a Kappa analysis is a KHAT statistic (an estimate of Kappa) which like overall accuracy and normalized accuracy can be used as another measure of agreement or accuracy. The KHAT statistic is computed as:

$$\hat{K} = \frac{N\sum_{i=1}^{r} x_{ii} - \sum_{i=1}^{r}\left(x_{i+} * x_{+i}\right)}{N^2 - \sum_{i=1}^{r}\left(x_{i+} * x_{+i}\right)} \quad (16.12)$$

where $r$ is the number of rows in the matrix, is the number of observations in row $i$ and column i, and are the marginal totals of row $i$ and column $i$, respectively, and $N$ is the total number of observations (Bishop et al. 1975).

The equations necessary for computing the variance of the KHAT statistic and the standard normal deviate can be found in many publications including Congalton et al. (1983), Rosenfield and Fitzpatrick-Lins (1986), Hudson and Ramm (1987), Congalton (1991), and Congalton and Green (1999, 2009, 2019). Table 16.3 shows three accuracy measures that can be computed easily from the error matrix. These measures include overall accuracy computed from the original matrix, normalized accuracy computed from the normalized matrix, and kappa. Which measure or measures to use for expressing thematic map accuracy is open for discussion. Each brings a different amount of information into the calculation. Overall accuracy simply sums the major diagonal and divides by the total number of sampling units. Normalized accuracy directly incorporates the off-diagonal elements of the matrix (i.e., the errors) through the normalization (i.e., iterative proportional fitting) process. Kappa indirectly incorporates the errors by using the sums of the row and column totals in the computation of the statistic. Therefore, each measure represents different information and should be evaluated accordingly. Landis and Koch (1977) proposed certain range values of the Kappa statistic to represent different levels of agreement. A value greater than 80% represents high agreement while a range between 60% and 80% represents moderate agreement and values below 40% represent weak agreement. Presenting the full error matrix allows the analyst to compute any or all of these three measures of thematic accuracy.

**TABLE 16.3** Three Measures of Thematic Map Accuracy

| | Matrix #1 | Matrix #2 |
|---|---|---|
| **Overall accuracy** | 77% | 84% |
| **Normalized Accuracy** | 76% | 83% |
| **Kappa** | 69% | 78% |

While the Kappa statistic (KHAT) has been used extensively as a measure of accuracy, a number of researchers have pointed out the shortcomings of using it as such (Zhenkui and Redmond 1995, Naesset 1996, Pontius and Millones 2011, Foody 2020). However, the major contribution of the Kappa analysis is not as a measure of accuracy, but rather as a technique used to determine if one error matrix is statistically significant than another. Such a test using a matrix derived from one classification approach could then determine if that matrix is statistically significantly different than a matrix derived from a second classification approach. In other words, is technique A statistically significantly better than technique B. Actually, two tests are possible with the Kappa analysis. The first is to test if a single error matrix is significantly better than random. In other words, is the map better than a random assignment of pixels? (Hopefully it is.) The KHAT statistic is asymptotically normally distributed and therefore, confidence intervals can be calculated using the approximate large sample variance (Congalton and Green 2019). The test to determine if a classification is better than random uses the $Z$ test as follows:

$$Z = \frac{\text{KHAT}}{\sqrt{\hat{\text{var}}(\text{KHAT})}} \tag{16.13}$$

where $Z$ is standardized and normally distributed.

The test to determine if two error matrices are significantly different from one another uses the following $Z$ test:

$$Z = \frac{|\text{KHAT} - \text{KHAT}_2|}{\sqrt{\hat{\text{var}}(\text{KHAT}_1) + \hat{\text{var}}(\text{KHAT}_2)}} \tag{16.14}$$

where $\text{KHAT}_1$ represents the KHAT statistic from one error matrix, and $\text{KHAT}_2$ is the statistic from the other matrix.

Running the Kappa test of significance on the matrix in Figure 16.7 produces a $Z$ value of 20.8. At the 95% confidence level, at $Z$ value above 1.96 is considered significant so it can be concluded that the process that went into making this map is significantly better than a random distribution of thematic labels to the pixels. The same can be said for the matrix in Figure 16.9 with a $Z$ value of 29.9. However, the most valuable question is whether these two matrices are statistically significantly different than the other. The resulting test statistic for this Kappa analysis is 2.2 again indicating statistical significance. Therefore, the matrices are significantly different and looking at the accuracy values from Table 16.3 shows that matrix 2 (Figure 16.9) is the better of the two. If matrix 2 (i.e., the map assessed in Figure 16.9) was produced with a different classification algorithm then matrix 1, then it can be said that the algorithm is better. This ability to determine if one matrix and therefore, map is statistically significantly better than another is the real power of the Kappa analysis.

### 16.4.5 Advanced Analysis Techniques

In addition to the basic analysis techniques for thematic map accuracy assessment described earlier, there are two other more advanced techniques that must be seriously considered depending on the methodologies used to make the map. These techniques include: fuzzy accuracy assessment and area-based error matrix generation.

#### 16.4.5.1 Fuzzy Accuracy Assessment

One of the four characteristics of a good classification scheme is that it is mutually exclusive. That is, any area on the map should fall into one and only one map class. While this is a valid goal for any scheme, in reality there are always issues with achieving this goal. Gopal and Woodcock (1994) proposed the use of fuzzy set theory in thematic accuracy assessment to recognize the possibility that ambiguity exists in determining the appropriate map class. They further state that it is rare to find the situation where one map class is exactly right and all the other map classes are equally wrong.

There are many places that fuzziness can enter into a thematic map assessment. The classification scheme itself is a large source as many map classes are more a continuum that has been grouped into a single class. For example, mapping hardwood forest vs. conifer forest is easy if the forest is 100% hardwoods or 100% conifer. However, as the percentages of each get closer to 50%, it becomes harder to decide which type of forest it is. Human interpretation also is an important source of variation. In many cases, training data used in the classification process and accuracy assessment data used in the validation process are generated from manual interpretation of high-resolution imagery. Even visiting field sample sites on the ground is not without variation as one observer may determine the boundary between a class is in one place and another observer put it in another. Therefore, depending on the classification scheme used and the methods employed to collect the training and reference data, some acknowledgement of the potential fuzziness in the map is justified.

Green and Congalton (2004) and more fully described in Congalton and Green (2009, 2019) have proposed the modification of the standard error matrix to incorporate fuzziness into the assessment process. In this case, the matrix consists of the correct values, the acceptable values, and the unacceptable values. Use of the fuzzy error matrix is extremely powerful as it incorporates all the benefits of the standard error matrix while allowing for situations were classification scheme breaks are artificial points along a continuum and/or where interpreter variability is difficult to control. The best way to understand the fuzzy error matrix is to provide an example as shown in Figure 16.11.

| | | Reference data | | | | User's accuracies | | | |
|---|---|---|---|---|---|---|---|---|---|
| | | F | OV | D | W | Deterministic totals | Deterministic accuracies | Fuzzy totals | Fuzzy accuracies |
| Map | F | 55 | 4, 7 | 0,4 | 0,3 | 55/73 | 75% | 59/73 | 81% |
| | OV | 2,4 | 51 | 0,5 | 2,6 | 51/70 | 73% | 55/70 | 79% |
| | D | 0,0 | 0,8 | 55 | 0,9 | 55/72 | 76% | 55/72 | 76% |
| | W | 1,3 | 0,7 | 0,3 | 62 | 62/76 | 82% | 63/76 | 83% |

| Producer's accuracies | | | | |
|---|---|---|---|---|
| Deterministic totals | 55/65 | 51/77 | 55/67 | 62/82 |
| Deterministic accuracies | 85% | 66% | 82% | 75% |
| Fuzzy totals | 58/65 | 55/77 | 55/67 | 64/82 |
| Fuzzy accuracies | 89% | 71% | 82% | 78% |

Overall accuracies

Deterministic 223/291 = 77%

Fuzzy 232/291 = 80%

Land cover categories F = Forest OV= Other vegetation D= developed W= water

FIGURE 16.11 Deterministic and fuzzy error matrix example.

A quick examination of Figure 16.11 reveals that it is the same error matrix as in Figure 16.7 with the exception that there are now two values in the off-diagonal elements of the matrix. The values in the major diagonal still represent the best determination of the correct classification. For example, there are 55 sample units that the map said were the forest class and the reference data agreed. However, the off-diagonal elements now have two values; the first value represents acceptable matches while the second value represents the errors. Looking at Figure 16.11 shows that the map labeled an area as forest when it was really other vegetation a total of 11 times. However, four times this label is considered acceptable because the area, when visited to collect the reference data, was actually young trees that had not yet grown tall enough to meet the definition of forest which stated that the trees must be 3 m tall to be considered a forest. The other seven times, the other vegetation was actually other vegetation (i.e., not trees) and the map is in error. Fuzziness only comes into play in the matrix where appropriate. It is not an excuse to accept errors, but rather a mechanism to account for variability and is especially helpful with the reference data. Note that no confusion between the forest class and the developed class is considered acceptable. Instead, the first value in the off-diagonal is zero and the second value shows the errors (4). So, what about the other vegetation/water confusion? Again looking at the fuzzy error matrix shows that twice the map said the sample unit was other vegetation when the reference data said that it was water. The fuzziness here can be explained in that there were Lilly pads and other green vegetation growing on the surface of a pond. The pond is clearly water, but it is acceptable in this case to call it other vegetation because of what was floating on the surface.

This new error matrix (Figure 16.11) is really two error matrices in one. It is the original error matrix (now called the deterministic error matrix) and it is also a fuzzy error matrix. The difference is simply how one handles the two values in the off-diagonals. In the case of the deterministic error matrix, only the values in the cells of the major diagonal are considered correct. Therefore, the deterministic overall accuracy, the deterministic row and column totals, and the deterministic producer's and user's accuracies are identical in every way to the calculations shown in Figure 16.7. However, the fuzzy error matrix calculations are a little bit different as they include the acceptable values (represented by the first value in the off-diagonal cells) along with the correct values (represented by the cells on the major diagonal). Therefore, the fuzzy producer's accuracy for the Forest class is 58/65 = 89% where the 58 is equal to the 55 correct values plus two acceptables from the Other Vegetation class and one acceptable from the Water class. The fuzzy overall accuracy is 232/291 = 80% which includes summing the values on the major diagonal plus all the acceptables (first values in the off-diagonal cells). That is adding 55 + 51 + 55 + 62 + 2 + 1 + 4 + 2 = 232.

In some cases, the difference between the deterministic error matrix and the fuzzy error matrix can be quite small. This situation would occur when the classification scheme is rather general and the differences between the thematic map classes are large and distinct. In other situations, the differences in the matrices can be quite large especially if the classification scheme is quite complex with great variability among the classes. Another situation in which the fuzzy error matrix can be very important is when the only reference data available is difficult to interpret causing great variation in the reference data. For example, using medium scale aerial photographs (i.e., 1:25,000) as the reference data source for assessing a map generated from Landsat Thematic Mapper imagery using a classification scheme that has over 20 land cover types. In this case and because only the photos could be used as the reference data without any ground verification, the interpreters would likely select a best class label and then perhaps one to three acceptable labels for each of the reference data sample units.

An effective way to implement this fuzzy accuracy assessment method is through the use of a reference data collection form. All the thematic map classes can be listed on the form and the reference data collector can then indicate which map class they believe is the correct one. The collector would also have the option then of indicating if any of the other map classes would be an acceptable label for that particular sampling unit and also which would be not acceptable. This approach would work equally well if the collector was in the field making measurements or observations from the center of the sampling unit or if they were interpreting some imagery to label the reference data. Clearly, making measurements in the field to label a certain sample unit should result in less fuzziness (i.e., fewer acceptable labels included) than reference data collected by interpretation of some medium scale imagery. Regardless of the reference data collection approach, producing a fuzzy error matrix with its counterpart the deterministic error matrix, does provide an effective way of compensating for fuzziness in the classification process while also giving the information contained within the traditional error matrix.

#### 16.4.5.2 Area-Based Error Matrix Generation

Historically, most land cover/vegetation/thematic maps created from digital imagery have analyzed the individual image pixels. More recently, Object-based Image Analysis (OBIA) has become increasingly popular. In OBIA, pixels are grouped together into segments or objects and classified together instead of as individual pixels. This approach more closely mimics human interpretation and incorporates significant additional information about the object than was possible for the individual pixel (Blaschke 2010). With a pixel-based map, it is very appropriate to take a cluster of pixels as the sample unit for the reference data collection used in the error matrix generation. As previously described in this chapter, for medium-resolution imagery such as Landsat, a 3 × 3 homogeneous cluster of pixels is used as a single reference sample unit. Therefore, since all the sample units are exactly the same size, the error matrix is generated by simply tallying each sample unit as either correct (on the major diagonal of the error matrix) or incorrect (in the appropriate off-diagonal cell). However, for assessing the accuracy of maps created using OBIA, objects of various sizes have been created to replace the equal-area sample units. It is possible then to use these objects as the reference sampling unit (i.e., polygons). If this is done, the sampling units are no longer equal areas and therefore can not simply be tallied in the error matrix (Radoux et al. 2010, MacLean and Congalton 2012, Olofsson et al. 2013). Instead, the actual area of each of the reference sampling units must be incorporated into the error matrix and the matrix becomes an area-based error matrix. In the traditional error matrix, each reference sample unit has equal weight since each has the same area. In the area-based error matrix, each reference sample unit is weighted by the area of the sample. Creating this area-based error matrix is no more difficult than the tally method used to create the traditional matrix accept that the areas are input into the matrix instead of just a tally. MacLean and Congalton (2012) present a comparison of these methods. Radoux et al. (2010) not only documents these methods, but also shows that some sampling efficiencies can be gained by knowing the sizes of all the objects in the map and using this information to reduce the number of sample units needed to create the matrix.

It should be noted that area-weighted error matrices are possible even with pixel-based classifications if different sized clusters of pixels are selected for the reference sampling units. Also, it is possible to use a cluster-based reference sampling unit for OBIA-based maps. However, the unequal area of the objects in an OBIA-based map more readily leads one to conduct an area-based accuracy assessment and therefore, this technique is described in this part of the chapter.

## 16.5 CONCLUSIONS

This chapter has presented a discussion of the many considerations necessary to conduct a valid accuracy assessment. Both positional error and thematic error must be carefully considered as an error in one can result in an error in the other. While there is no definitive, step by step method for conducting a valid assessment, it is clear that the assessment must be carefully thought-out and planned from the very beginning of the mapping project. All decisions regarding how the assessment is conducted must be thoroughly documented so that the process can be easily understood and reproduced by anyone interested in the map.

Accuracy assessment is an essential component of any mapping project. However, accuracy assessment is also an extremely expensive component. Therefore, every effort must be made to balance statistical validity with what is practically attainable. Improper assessments can cost the entire budget for the project. Poorly designed assessments can turn out to be statistically invalid causing the assessment to not be worth the effort. However, with care, proper consideration, and wise planning accuracy assessment can be an extremely satisfying part of the project that not only shows the accuracy of the map, but also indicates where the map can be improved in the future.

## 16.6 SOME FINAL ENCOURAGEMENT AND WARNINGS

Our ability to assess the accuracy of maps made from remotely sensed imagery, both positionally and thematically, has progressed steadily since the late 1970s. This chapter has presented a balanced approach showing the various considerations and decisions to be made by the analyst wishing to conduct an assessment of their map. Given the many choices that must be made, the analyst is strongly encouraged to document their process so that everyone can see and understand what has been done.

As any technology develops, processes become more automated. Such is true for the accuracy assessment process. While having this process automated can be very useful, it also should come with a warning. It is easy to allow the software to conduct the assessment without the analyst thinking through the considerations and decisions outlined in this chapter. Relying on software without understanding exactly what it is doing is not always a good idea. The analyst is encouraged to dig into the software user's guide to fully understand the assumptions being made and look for where the user can provide input. Regardless, the analyst should still document the process that was performed to conduct the assessment. This author has read far too many papers that provide no details about how the assessment was performed, the number of samples taken, the sample size, and even those with no error matrices provided, but only some summary values that provide little information to judge if the assessment is valid.

It is a fact in the academic world that in order to publish a paper, some new contribution needs to be made. This need has both positive and negative results. It is possible to find papers that present very interesting academic but extremely complicated and esoteric methods for assessing map accuracy. A recent paper has even suggested that a complete modeling approach could be used to perform the accuracy assessment without the need for any reference data (Foody 2020). While this approach is viable, this author believes that most practitioners that use maps generated from remotely sensed data would like to see an assessment conducted with valid reference data rather than using a model to assess another model (the map).

## REFERENCES

Ager, T. 2004. An Analysis of Metric Accuracy Definitions and Methods of Computation. Unpublished memo prepared for the National Geospatial-Intelligence Agency. InnoVision. March.

Aronoff, S. 1985. The minimum accuracy value as an index of classification accuracy. *Photogrammetric Engineering and Remote Sensing*. Vol. 51, No. 1, pp. 99–111.

ASPRS. 1989. ASPRS interim accuracy standards for large-scale maps: Photogrammetric engineering and remote sensing. *Photogrammetric Engineering & Remote Sensing*. Vol. 54, No. 7, pp. 1038–1041.

ASPRS. 2004. *ASPRS Guidelines, Vertical Accuracy Reporting for LiDAR Data*. American Society for Photogrammetry and Remote Sensing, Bethesda, MD.

ASPRS. 2014. ASPRS positional accuracy standards for digital geospatial data, Edition 1, Version 1.0. *Photogrammetric Engineering and Remote Sensing*. Vol. 81, No. 3, pp. A1–A26.

ASPRS. 2023. ASPRS positional accuracy standards for digital geospatial data, Edition 2, Version 1.0. https://publicdocuments.asprs.org/PositionalAccuracyStd-Ed2-V1.

Bishop, Y., S. Fienberg, and P. Holland. 1975. *Discrete Multivariate Analysis: Theory and Practice*. MIT Press, Cambridge, MA, p. 575.

Blaschke, T. 2010. Object based image analysis for remote sensing. *ISPRS Journal of Photogrammetry and Remote Sensing*. Vol. 65, pp. 2–16.

Bolstad, P., and S. Manson. 2022. *GIS Fundamentals*. 7th edition. Eider Press, White Bear Lake, MN. p. 543.

Cliff, A. D., and J. K. Ord. 1973. *Spatial Autocorrelation*. Pion Limited. London, England. p. 178.

Cohen, J. 1960. A coefficient of agreement for nominal scales. *Educational and Psychological Measurement*. Vol. 20, No. 1, pp. 37–40.

Colwell, R. N. 1955. The PI picture in 1955. *Photogrammetric Engineering*. Vol. 21, No. 5, pp. 720–724.

Congalton, R. G. 1988a. Using spatial autocorrelation analysis to explore errors in maps generated from remotely sensed data. *Photogrammetric Engineering and Remote Sensing*. Vol. 54, No. 5, pp. 587–592.

Congalton, R. G. 1988b. A comparison of sampling schemes used in generating error matrices for assessing the accuracy of maps generated from remotely sensed data. *Photogrammetric Engineering and Remote Sensing*. Vol. 54, No. 5, pp. 593–600.

Congalton, R. G. 1991. A review of assessing the accuracy of classifications of remotely sensed data. *Remote Sensing of Environment*. Vol. 37, pp. 35–46.

Congalton, R. G. 2001. Accuracy assessment and validation of remotely sensed and other spatial information. *The International Journal of Wildland Fire*. Vol. 10, pp. 321–328.

Congalton, R. G. 2009. Accuracy and error analysis of global and local maps: Lessons learned and future considerations. IN: Thenkabail, P., Lyon, J., Turral, H., and Biradar, C. (Eds.), *Remote Sensing of Global Croplands for Food Security*. CRC/Taylor & Francis, Boca Raton, FL, pp. 441–458.

Congalton, R. G., and K. Green. 1999. *Assessing the Accuracy of Remotely Sensed Data: Principles and Practices*. CRC/Lewis Press, Boca Raton, FL, p. 137.

Congalton, R. G., and K. Green. 2009. *Assessing the Accuracy of Remotely Sensed Data: Principles and Practices*. 2nd Edition. CRC/Taylor & Francis, Boca Raton, FL, p. 183.

Congalton, R. G., and K. Green. 2019. *Assessing the Accuracy of Remotely Sensed Data: Principles and Practices*. 3rd Edition. CRC/Taylor & Francis, Boca Raton, FL, p. 328.

Congalton, R. G., and M. Brennan. 1999. Error in remotely sensed data analysis: Evaluation and reduction. Proceedings of the Sixty Fifth Annual Meeting of the American Society of Photogrammetry and Remote Sensing, Portland, OR. pp. 729–732 (CD-ROM).

Congalton, R. G., R. G. Oderwald, and R. A. Mead. 1983. Assessing Landsat classification accuracy using discrete multivariate statistical techniques. *Photogrammetric Engineering and Remote Sensing*. Vol. 49, No. 12, pp. 1671–1678.

Federal Emergency Management Agency (FEMA). 2003. *Guidelines and Specifications for Flood Hazard Mapping Partners*. US Government.

Federal Geographic Data Committee, Subcommittee for Base Cartographic Data. 1998. *Geospatial Positioning Accuracy Standards. Part 3: National Standard for Spatial Data Accuracy*. FGDC-STD-007.3–1998. Federal Geographic Data Committee, Washington, DC, p. 24.

Foody, G. M. 2020. Explaining the unsuitability of the Kappa coefficient in the assessment and comparison of the accuracy of thematic maps obtained by image classification. *Remote Sensing of Environment*. Vol. 239. https://doi.org/10.1016/j.rse.2019.111630.

Ginevan, M. E. 1979. Testing land-use map accuracy: Another look. *Photogrammetric Engineering and Remote Sensing*. Vol. 45, No. 10, pp. 1371–1377.

Gopal, S., and C. Woodcock. 1994. Theory and methods for accuracy assessment of thematic maps using fuzzy sets. *Photogrammetric Engineering and Remote Sensing*. Vol. 60, No. 2, pp. 181–188.

Green, K., and R. Congalton. 2004. An error matrix approach to fuzzy accuracy assessment: The NIMA Geocover project: A peer-reviewed chapter. IN: Lunetta, R. S. and Lyon, J. G. (Eds.), *Remote Sensing and GIS Accuracy Assessment*. CRC Press, Boca Raton, FL, p. 304.

Greenwalt, C., and M. Schultz. 1962 & 1968. *Principles of Error Theory and Cartographic Applications*. United States Air Force. Aeronautical Chart and Information Center. ACIC Technical Report Number 96. St. Louis, Missouri. 60 pages plus appendices. This report is cited in the ASPRS standards as ACIC, 1962.

Gu, J., and R. G. Congalton. 2020. Analysis of the impact of positional accuracy when using a single pixel for thematic accuracy assessment. *Remote Sensing*. Vol. 12, p. 4093. https://doi.org/10.3390/rs12244093.

Gu, J., and R. G. Congalton. 2021. Analysis of the impact of positional accuracy when using a block of pixels for thematic accuracy assessment. *Geographies*. Vol. 1, pp. 143–165. https://doi.org/10.3390/geographies1020009.

Hay, A. M. 1979. Sampling designs to test land-use map accuracy. *Photogrammetric Engineering and Remote Sensing*. Vol. 45, No. 4, pp. 529–533.

Hord, R. M., and W. Brooner. 1976. Land-use map accuracy criteria. *Photogrammetric Engineering and Remote Sensing*. Vol. 42, No. 5, pp. 671–677.

Hudson, W., and C. Ramm. 1987. Correct formulation of the kappa coefficient of agreement. *Photogrammetric Engineering and Remote Sensing*. Vol. 53, No. 4, pp. 421–422.

Katz, A. H. 1952. Photogrammetry needs statistics. *Photogrammetric Engineering*. Vol. 18, No. 3, pp. 536–542.

Landis, J., and G. Koch. 1977. The measurement of observer agreement for categorical data. *Biometrics*. Vol. 33, pp. 159–174.

Lunetta, R., R. Congalton, L. Fenstermaker, J. Jensen, K. McGwire, and L. Tinney. 1991. Remote sensing and geographic information system data integration: Error sources and research issues. *Photogrammetric Engineering and Remote Sensing*. Vol. 57, No. 6, pp. 677–687.

MacLean, M., M. Campbell, D. Maynard, M. Ducey, and R. Congalton. 2013. Requirements for labeling forest polygons in an object-based image analysis classification. *International Journal of Remote Sensing*. Vol. 34, No. 7, pp. 2531–2547.

MacLean, M., and R. Congalton. 2012. Map accuracy assessment issues when using an object-oriented approach. Proceedings of the Annual Meeting of the American Society of Photogrammetry and Remote Sensing, Sacramento, CA, p. 5. (CD-ROM).

MacLean, M., and R. Congalton. 2013. Applicability of multi-date land cover mapping using Landsat 5 TM imagery in the Northeastern US. *Photogrammetric Engineering and Remote Sensing*. Vol. 79, No. 4, pp. 359–368.

Mead, R. A., and J. Szajgin. 1981. Landsat classification accuracy assessment procedures: An account of a national working conference. *LARS Symposia*. Paper 426. http://docs.lib.purdue.edu/lars_symp/426 (accessed on September 18, 2023).

Naesset, E. 1996. Conditional tau coefficient for assessment of producer's accuracy of classified remotely sensed data. *ISPRS Journal of Photogrammetry and Remote Sensing*. Vol. 51, No. 2, pp. 91–98.

NDEP. 2004. *Guidelines for Digital Elevation Data*. Vertion 1.0. National Digital Elevation Program, May 10.

Olofsson, P., G. Foody, S. Stehman, and C. Woodcock. 2013. Making better use of accuracy data in land change studies: Estimating accuracy and area and quantifying uncertainty using stratified estimation. *Remote Sensing of Environment*. Vol. 129, pp. 122–131, https://doi.org/10.1016/j.rse.2012.10.031.

Pontius, G., and M. Millones. 2011. Death to kappa: Birth of quantity disagreement and allocation disagreement for accuracy assessment. *International Journal of Remote Sensing*. Vol. 32, No. 15, pp. 4407–4429.

Pugh, S., and R. Congalton. 2001. Applying spatial autocorrelation analysis to evaluate error in New England forest cover type maps derived from Landsat Thematic Mapper Data. *Photogrammetric Engineering and Remote Sensing*. Vol. 67, No. 5, pp. 613–620.

Radoux, J., R. Bogaert, D. Fasbender, and P. Defourny, 2010. Thematic accuracy assessment of geographic object-based image classification. *International Journal of Geographical Information Science*, Vol. 25, No. 6, pp. 895–911.

Rosenfield, G. H., and K. Fitzpatrick-Lins. 1986. A coefficient of agreement as a measure of thematic classification accuracy. *Photogrammetric Engineering and Remote Sensing*. Vol. 52, No. 2, pp. 223–227.

Rosenfield, G. H., K. Fitzpatrick-Lins, and H. Ling. 1982. Sampling for thematic map accuracy testing. *Photogrammetric Engineering and Remote Sensing*. Vol. 48, No. 1, pp. 131–137.

Sammi, J. C. 1950. The application of statistics to photogrammetry. *Photogrammetric Engineering* .Vol. 16, No. 5, pp. 681–685.

Story, M., and R. Congalton. 1986. Accuracy assessment: A user's perspective. *Photogrammetric Engineering and Remote Sensing*. Vol. 52, No. 3, pp. 397–399.

Tortora, R. 1978. A note on sample size estimation for multinomial populations. *The American Statistician*. Vol. 32, No. 3, pp. 100–102.

Van Genderen, J. L., and B. F. Lock. 1977. Testing land use map accuracy. *Photogrammetric Engineering and Remote Sensing*. Vol. 43, No. 9, pp. 1135–1137.

Young, H. E. 1955. The need for quantitative evaluation of the photo interpretation system. *Photogrammetric Engineering*. Vol. 21, No. 5, pp. 712–714.

Young, H. E., and E. G. Stoeckler. 1956. Quantitative evaluation of photo interpretation mapping. *Photogrammetric Engineering*. Vol. 22, No. 1, pp. 137–143.

Zhenkui, M., and R. Redmond. 1995. Tau coefficients for accuracy assessment of remote sensing data. *Photogrammetric Engineering and Remote Sensing*. Vol. 61, No. 4, pp. 435–439.

# Part X

## Remote Sensing Law

# 17 Remote Sensing Law

## *An Overview of Its Development and Its Trajectory in the Global Context*

*P.J. Blount*

## ACRONYM AND DEFINITION

| | |
|---|---|
| **CD** | Conference on Disarmament |
| **EO** | Earth Observation |
| **GNSS** | Global Navigation Satellite System |
| **INPE** | National Institute for Space Research (Brazil) |
| **NTM** | National Technical Means |
| **UNCOPUOS** | United Nations Committee on the Peaceful Uses of OuterSpace |
| **UNGA** | United Nations General Assembly |

## 17.1 INTRODUCTION

The law relating to remote sensing is a complex mix of international and national laws, regulations, policies, and agreements. This body of remote sensing law has numerous facets that regulate a number of uses of remote sensing satellites, related technology, and collected data. This chapter will serve as a survey of the laws and regulations that govern remote sensing activities across a variety applications and a variety of legal spaces.

This overview of remote sensing law will assert that there are essentially four bodies of law that govern remote sensing and define the its regulatory regime. These are not distinct bodies; they overlap and intersect with each other, and this typology is not intended to obscure these interactions but rather to serve as an explanatory framework. This chapter will address each of these bodies of law chronologically in the order in which each emerged, though it will not attempt to identify the specific temporal locations due to the fluidity and subjectivity of such assignments. The chapter begins with a brief examination of relevant international space law as developed in the space treaty regime. Next, it will turn to remote sensing law as developed in relation to military uses with a particular focus on disarmament treaty verification. Third, it will address international remote sensing law applicable to civil remote sensing satellites as articulated in the United Nations Principles on Remote Sensing.[1] Finally, it will argue that remote sensing law is increasingly being subsumed into the developing field of Geospatial Law, which results from processes of commercialization, globalization, and technological convergence. The purpose of this exercise will be to trace a trend of increasing adoption of domestic law regimes that intersect with international governance systems.[2] These laws are in a constant state of flux as a result of rapid advancements in technology and different perceptions on how we manage space data gathering that is agreeable to all.

The thrust of this chapter will be to address laws and regulations specific to remote sensing activities. Many of the provisions of the Outer Space Treaty apply to all space activities including remote

DOI: 10.1201/9781003541141-27

sensing activities, and reference will be made to these provisions as a whole, but in this chapter we will endeavor to focus our attention on legal provisions that are specific to remote sensing activities. The goal will be to highlight the regulations that are most relevant to remote sensing activities, and the end uses of remotely sensed data. Finally, it should be noted that while this survey of remote sensing law seeks to give a broad overview of the law, it is not by any means comprehensive, and the reader's attention is drawn to the footnotes for further readings and elaboration.

## 17.2 FOUR PILLARS OF REMOTE SENSING LAW

Remote Sensing Law is a multifaceted legal regime that draws on a number of regulatory systems. This chapter will attempt to address the four core regulatory systems, namely traditional space law, disarmament law, international remote sensing law, and geospatial law. These regimes should be understood as separate regulatory spaces that interact with each other in a variety of ways. Indeed, depending on the issue being addressed they can reorient themselves to the others, and it would be folly to suggest that they are always oriented in a specified way. These regimes shift from discrete and separate to hierarchical to overlapping as the context of their application changes. The best description would be that their relationship is discursive in nature. It is a system of regulatory feedback loops across different governance spaces (see Figure 17.1). For example, international law creates obligations for states, but state implementation of these obligations often leads to influencing the content of the obligation.[3]

This chapter will proceed by addressing these varying regimes as discrete, but will endeavor to highlight the areas wherein we can observe their interactions. It should not be assumed that at any given moment, one regime maintains primacy over another or that the delineations among them are by any means clear. Instead, this serves as a framework for understanding the overall governance system for remote sensing technologies.

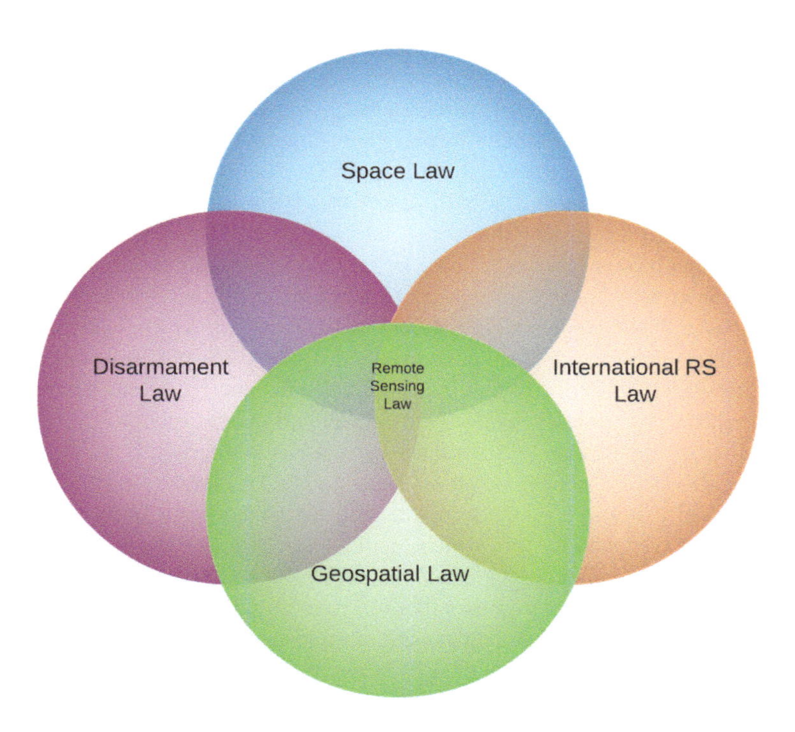


**FIGURE 17.1** Remote sensing law is the confluence of multiple legal regimes.

### 17.2.1 The Space Treaty Regime

The core body of international space law is found in four multilateral treaties.[4] These treaties set the parameters for lawful conduct by states when engaging in outer space activities. There are several introductions to these core treaties, therefore this chapter will refrain from a full evaluation of these and instead focus on provisions of special concern to remote sensing satellites.[5]

Article I of the Outer Space Treaty grants all states free access to outer space, without interference by other states. By coupling freedom of access with the principle of non-appropriation found in Article II, which places space outside the sovereign territory of all states, the negotiators of the Outer Space Treaty enshrined the principle of freedom of overflight by space objects over other state's territories.[6] This is of critical importance to remote sensing activities, since it allows states to observe another state from the non-sovereign vantage point of space. Since no state has ever protested the overflight of its territory by another state's space object, overflight is a customary norm that preexists the Outer Space Treaty.[7]

Article VI of the Outer Space Treaty places a duty on states to "authorize and supervise" nongovernmental actors, and gives states "international responsibility" for nongovernmental actors. This means that many states have adopted full regulatory regimes in order to manage the risk of commercial actors. These regimes, which take the form of licensing regimes, are meant to ensure government oversight over nongovernmental actors. A primary concern with many of these licensing regimes is to ensure that the nongovernmental satellite operators comply with minimum standards of conduct in its activities, so that the state avoids situations of international responsibility. These regimes emphasize "cradle to grave" operations plans that are meant to reduce liability issues as well mitigate space debris.[8] These licenses are often attached to the act of launching.[9] Some states, however, have licensing regimes specifically for remote sensing activities that ensure compliance with national security and global data policies.[10] These regimes will be addressed in Section 17.2.3.1.

There are numerous information sharing clauses in the Outer Space Treaty, which read together lead to an overall duty to warn other states of situations of impending disaster. Most relevant is Article IX's requirement that states engaging in space activities be guided by "the principles of cooperation and mutual assistance."[11] While the information sharing provisions are not directly related to remote sensing, they build strong evidence that there is a customary norm of information sharing in a wide variety of contexts as a way to increase cooperation in order to use space "for the benefit of all mankind." This norm is further articulated in the Remote Sensing Principles (Section 17.2.3).

Finally, reference should be made to the liability provision found in Article VII of the Outer Space Treaty and further articulated in the Liability Convention. Whether damage caused by remote sensing data might represent "damage caused by [a state's] space object" is a contested interpretation of the treaty.[12] However, this author would argue that it is an increasingly irrelevant debate. While parsing the language of the treaty is an excellent exercise, state practice is a better touchstone for determining the practicality of the argument. State reluctance to apply the Liability Convention in clear cases such as *Cosmos 954* and the *Cosmos-Iridium* collision, points to the likelihood that damage as a result of data presents a weak case for the application of the Liability Convention.[13] In this case, state practice indicates that such damage would most likely sit outside the scope of damage defined in the Outer Space Treaty and the Liability Convention, though this is admittedly an argument based less on strict legal analysis and more on a constructivist view of norm creation and definition, in which practice identifies the content of the norm.[14]

### 17.2.2 Disarmament and Verification Law

The space race and the resulting international law of outer space is a product of the Cold War, the law's early development was linked to both technological diplomacy as well as military uses.[15] The launch systems represented ballistic missile technology; the communications applications could support the emerging concept of a globally projected military; and remote sensing could open up intelligence secrets about an adversary's capabilities.[16] While one of the core guiding principles and central legal obligations in space law is the use of space for peaceful purposes, it is well established that this allows states to use space technologies in accordance with "peaceful" as set out in the UN Charter as a *de minimis* threshold.[17] As a result, states are considered to be engaging in peaceful activities when they deploy and use defensive technologies in a non-aggressive way, this is supported by the fact that until the launch of *Landsat-1*, "remote sensing was exclusively developed and used for military purposes."[18]

This use of remote sensing was critical in the negotiation of disarmament and arms control agreements between the United States and the Soviet Union. "[F]reedom of space" was "perhaps the principal aim of U.S. space policy throughout the first ten years of the space age."[19] The technology allowed the two states to overcome the problem of verification. The Soviets rejected verification by on-site visits or overflights, and the United States rejected any agreement without reliable verification measures.[20] Once it was clear that both sides were using satellites for remote sensing of the other's territory, they were able to agree on satellite technology as the core method of verification.[21] While some states have challenged the military and intelligence remote sensing regimes as non-peaceful the "international acceptance of the legitimacy of intelligence gathering from outer space was effectively secured with the enshrinement of space-based photoreconnaissance in the Cold War strategic arms control regime."[22] This was made possible by a legal structure, articulated in the Outer Space Treaty and embodied in customary international law, that ensured a nearly complete freedom of access and use outside of the Earth's atmosphere. As a result, there was no way either state could object to the use of satellites to sense its territory, because such activity was "consistent with international law."[23] These technologies were indicated in disarmament agreements under the term National Technical Means (NTM),[24] a phrase that would remain officially undefined until its meaning was revealed in a speech by Jimmy Carter.[25]

The use of satellites as the central method of verification has implications for space security, since treaties that use NTM often require that parties to the treaty not interfere with the other's NTM.[26] Arguably, this historically helped to solidify space security by creating a *de facto* draw-down of space weaponry development, due to the risk that any interference with space objects could interfere with NTM and thereby offset the strategic stability these agreements provided.[27] It can be argued that this limitation is becoming less relevant as space is populated by an increasingly diverse set of actors, and as this set of bilateral treaties has been collapsing as key states, namely Russia and the United States, have been withdrawing from them.

While there has been scholarship on expanding the verification possibilities of remote sensing satellites to other areas of enforcement, little movement has been made. France made a proposal at the Conference on Disarmament (CD) in 1978 for an "international satellite body for monitoring disarmament treaties."[28] This measure was rejected by the CD, and there has been no progress in the CD on such a system. It is unlikely, that in the near term, under current political conditions, that sufficient political will for such a system to be implemented. There is however a great deal of potential for remote sensing technologies to play a role in the international lawmaking toolkit for its verification and monitoring capabilities in a variety of areas outside disarmament such as human rights and the environment.[29]

### 17.2.3 Remote Sensing Principles

The core of international legal principles dealing with remote sensing activities are contained in UN General Assembly (UNGA) Resolution 41/65, Principles Relating to Remote Sensing of the Earth from Outer Space. These principles are the "first and foundational source of policy guidance for remote sensing activities."[30]

The Principles are contained in a UNGA resolution, and as such, The Principles themselves are not binding as legal rules.[31] However, to the extent that they represent customary international law, these principles are binding. The existence of customary international law it is evinced by two criteria: *opinio juris* and state practice. *Opinio juris* represents a subjective belief by states that they are legally bound by a rule. In the case of the Remote Sensing Principles, *opinio juris* is found in the consensus process of negotiation at the United Nations Committee on the Peaceful Uses of Outer Space (UNCOPUOS) and by the unanimous adoption by the General Assembly.[32] In light of the fact that there are no denunciations of The Principles, that states often appeal to The Principles in their domestic laws, and that states make claims of compliance with The Principles, a sufficient level of acceptance may have been achieved to constitute *opinio juris*.[33]

The second evidentiary parameter that a rule constitutes a binding customary norm requires that states indeed follow the rule they believe to be binding. State practice, essentially, helps define the content of the norm.[34] If states are found to be claiming compliance and are doing so in good faith, then their compliance becomes an epistemic unit informing that norm's interpretation. In other words, when states think they are complying with a norm, their actions construct the meaning and content of that norm.[35] Since states may disagree as to the exact content of a norm's meaning, customary norms are often better defined along a spectrum of acceptable state practice.[36] In this analysis the key question is not if states are complying, but instead, how states are complying. This is because "[c]ustom as a source of international law, leads to the recognition of the legality of the existing practice if there is general consent . . . to the observable rule of conduct."[37]

The inquiry that follows will address the Remote Sensing Principles[38] in two different groupings. First, it will identify Principles that are substantively based in preexisting binding legal principles. It will then address the "legal innovations" of the UN Principles, which articulated new content into the international law governing remote sensing. These innovations will be evaluated by examining the spectrum of state practice to illustrate acceptable state conduct that defines the content of these norms.

#### 17.2.3.1 Principles Based on Preexisting Legal Obligations

Numerous of The Principles appeal to preexisting legal obligations that can be found in general international law or in the *lex specialis* of international space law. Essentially, these principles make explicit the application of more general legal standards, that arguably would apply regardless of the acknowledgment of such in The Principles. There are three different groupings of these: those that implement general international law, those that implement international space law, and those that seek to increase international cooperation.

##### *17.2.3.1.1 General International Law: Principles III, IV, XV*

Principles III makes explicit the application of general international law to remote sensing activities.[39] This principle makes specific reference to three specific treaty instruments: the UN Charter, The Outer Space Treaty, and International Telecommunications Union Instruments.[40] This is explicit acknowledgment that these treaties represent preexisting law and that the principles should not be read in such a way as to conflict with those instruments. Principle III also acknowledges that other general international law principles apply, so for instance, the law of armed conflict, private international law principles, and environmental law principles could all be applied to space in the proper context. While this acknowledgment is important, it is hardly necessary, since The principles are embodied in a document defined under international law, and the form reifies the notion that its content exists within those parameters.

While referencing a preexisting right to the freedom of use of outer space, Principle IV specifically notes the that remote sensing activities shall be

> conducted on the basis of respect for the principle of full and permanent sovereignty of all States and peoples over their own wealth and natural resources, with due regard to the rights and interests, in accordance with international law, of other States and entities under their jurisdiction. Such activities shall not be conducted in a manner detrimental to the legitimate rights and interests of the sensed State.[41]

This principle seeks to reaffirm "the freedom of data collection and distribution," while at the same time ensuring "respect of the principle of full and permanent sovereignty of all States and peoples over their own wealth and natural resources."[42] The content of this norm is consistent with the principles of nonintervention and sovereign equality that sit at the heart of the international legal system.[43] The formulation in The Principles reflects the concern of developing nations that developed nations would use satellite technology to exploit the resources of the developing world and advance economic imperialism.[44] However, as already addressed, the right to overflight[45] had long been established, and "such protestations have never adversely affected operational progress."[46] Once deployed, technology is difficult to withdraw, and as a result the developing nations settled for a compromise in the requirements for data distribution (see Section 17.2.3.2). This principle cannot be read as extending or elaborating on the concept of sovereignty as already established in international law, and in no way requires a state to "seek consent or authorization" from a state it intends on sensing.[47]

It is worth noting that the intersection of remote sensing and sovereignty could become more relevant in the future if the principle of responsibility to protect gains traction as a justification for military intervention for gross human rights violation.[48] In such scenarios, Earth observation data could be used to expose the violations that give rise to and legitimate international action. In this capacity, remote sensing as a global technology can continue to play a role in the way in which the international community constructs the idea of sovereignty. Indeed, there have been efforts to leverage remote sensing data to expose gross human rights violations such as in Sudan,[49] Darfur,[50] and North Korea.[51]

Finally, Principle XV states that disputes will be settled in accordance with the norm of the peaceful settlement of disputes, which is previously enshrined in Article 1 of the UN Charter.[52]

#### *17.2.3.1.2 General Space Law: Principles IX and XIV*

As noted earlier, the Remote Sensing Principles acknowledge the applicability of the Outer Space Treaty to remote sensing activities, and The Principles reference the treaty throughout. Two provisions, though, are concerned explicitly with general space law provisions.

Principle IX affirms the registration obligation found in the Registration Convention[53] as well as the Article IX obligation to provide information at the request of affected states.[54] It lowers the threshold for requesting such information from the Article IX standard of "harmful interference" to "affected" states. However, the scope of the obligation is likely the same in practice, where relevant information will likely be readily available for civil systems and scarce for military systems (see Section 17.2.3.2).

Principle XIV incorporates Article VI of the Outer Space Treaty, which requires supervision and authorization of remote sensing activities and has already been discussed. It also acknowledges that in no way does the Article VI obligation preclude the application of the law of state responsibility to remote sensing activities.[55] Though not explicitly stated in the Outer Space Treaty, the customary law of state responsibility is certainly applicable to states as they engage in space activities, so this can hardly be seen as an innovation.

#### *17.2.3.1.3 International Cooperation: Principles II, V, VI, VII, VIII, and XIII*

A large portion on international space law is dedicated to promoting the soft obligation of international cooperation. It is soft because the obligation is always defined by terms to be negotiated by the states engaging in the cooperation. Additionally, while it can be said that there is an obligation to engage in international cooperation, there is no corresponding, affirmative right to international cooperation. The following provisions are continuations of this cooperation framework.

Principle II states that remote sensing should be done for the benefit of all nations and "taking into particular consideration the needs of the developing countries."[56] This principle is an extension of Article I of the Outer Space Treaty that states that space activities "shall be carried out for the benefit and in the interests of all countries, irrespective of their degree of economic or scientific development."[57] The Remote Sensing Principles add an additional obligation to take into consideration the special needs of developing countries. At best though, this obligation is weak. The legal regime, puts strong emphasis on cooperative activities in space, but the terms and extent of the cooperation are, by design, entirely up to the states involved. As a result, the addition of this clause on developing countries represents the extension of an obligation that is more akin to an international policy statement. However, it is an important statement, since "[i]n developing nations . . . these technologies have proven essential for developing public policies on issues such as deforestation assessment and management, urban planning, agricultural production, and environmental assessment."[58]

That is not to say that states willfully ignore this obligation. On the contrary, there is abundant evidence that states actively engage in such cooperation.[59] The need to extend an extra obligation to developing countries, which are burdened by technological lag, has been adopted as a set of principles in UNGA Resolution 51/122, The Declaration on International Cooperation in the Exploration and Use of Outer Space for the Benefit and in the Interest of All States, Taking into Particular Account the Needs of Developing Countries.[60] This resolution represents international consensus that developing countries should have access to space technologies, but the principles enunciated give states broad leeway to determine the extent to which they will engage in such cooperation. This certainly makes the terms of this principle very aspirational.

Despite the aspirational nature, such principles of conduct have led to an expansion of space technologies to developing nations, and "*remote sensing is no longer a technology, which only superpowers can enjoy; it is now getting diffused among states around the globe, including emerging and developing ones.*"[61] Specifically, in the field of remote sensing, states have sought to assist developing nations through their data policies, which will be addressed in depth later (Section 17.2.3.2.3). Japan, for instance, seeks to make a "safe, secure, and affluent society through the establishment of an effective space-based data use policy."[62] Another mode of assistance, is allowing direct access to the satellite data via ground stations. A great example of this is the highly successful *Landsat* program that the United States implemented, which allowed any state to build and receive the raw data from the *Landsat* system.[63]

Principles V, VI, VII, VIII, and XIII all incorporate the general obligation for international cooperation in space activities and can arguably be read more as a list of possible cooperative activities. Principle V makes a general appeal for cooperation, which echoes the "principle of cooperation" found in the Outer Space Treaty's Article IX.[64] Principle VI puts forth a specific example of a type of cooperation in the form of data receiving and processing facilities within the framework of regional agreements,[65] and Principle VII encourages states to exchange technical assistance on "mutually agreed" terms. Principle VIII requires the UN to promote international cooperation. Article XIII gives sensed states a right to request consultations to seek cooperative opportunities with the sensing state.[66] Each of these principles sets forth cooperation in relatively soft terms under which states can pursue international cooperation as they see fit.

### 17.2.3.2 Legal Innovations

The following principles represent legal innovations, the content of which is defined by state practice. These principles added new substance to the legal framework governing state remote sensing activities, and as such represent legal innovations

#### *17.2.3.2.1 Definitions: Principle I*

The Principles begin with a definitions section which serves to scope the application of the principles. The principles begin by defining remote sensing as "the sensing of the Earth's surface from space by making use of the properties of electromagnetic waves emitted, reflected or diffracted by the sensed objects, for the purpose of improving natural resources management, land use and the protection of the environment."[67] This is a twofold definition. It incorporates both a technical and intent parameters. The first element describes in a technical sense what remote sensing activities will be governed by The Principles (so for instance, GNSS is excluded). The second element, though, limits the scope of application by making the principles only applicable to activities of a specific nature, which is the political dimension of the definition. Specifically, this definition purposely places military remote sensing beyond the pale of The Principle's applicability.[68]

Next, The Principles define three types of information, making distinction among primary data, processed data, and analyzed information. These definitions are consistent with technical usage. Primary data is the "raw data that are acquired by the remote sensors."[69] Processed data is value added data that has been processed into a usable form, and analyzed information is the knowledge gained when remote sensing data is interpreted along with other inputs.[70]

Finally, Principle I defines remote sensing activities as "the operation of remote sensing space systems, primary data collection and storage stations, and activities in processing, interpreting and disseminating the processed data."[71] The significance of this definition is its expansiveness in including not only ground-link stations, but also the data analysis and dissemination. This holistic definition is important as it scopes these activities sufficiently broad to allow the principles the competency to articulate obligations in regards to data access.

##### *17.2.3.2.2 Environmental and Disaster Provisions: Principles X and XI*

Principles X and XI are environmental provisions. Principle X states that remote sensing should be used to protect the Earth's environment, and that data collected that can help to avert environmental degradation. This acknowledges that "remote sensing data can be used to assess and locate damage, monitor the progression and effect of corrective measures, verify the application of environmental treaties and assist in the response to man-made and natural disasters."[72] Article XI on the other hand seeks to encourage states to exchange data in the wake of natural disasters as a way to alleviate suffering occurring in affected areas. Both of these provisions are data sharing provisions and reflect core values found in Article IX of the Outer Space Treaty.[73] These principles support a "general duty to inform" other states of environmental damage.[74]

Principle X finds support in the "general principles of environmental and human rights law."[75] It first requires that states use remote sensing to "*promote*" the protection of the environment, which is a fairly broad, yet soft, requirement.[76] Then it requires states to share "information . . . that is capable of averting any phenomena harmful to the Earth's natural environment." The second obligation is surprisingly narrow in application. It only requires that this data should be shared when the information is "capable of averting" the environmental harm. Post-incident information as well as cumulative data that might give indications as to environmental trends are not included in this sharing regime.[77] Since compliance with this provision is difficult to monitor, it seems that the narrowness is likely unimportant. With the amount of available data and expanding access, states are increasingly able to get the information they need to manage the domestic environmental concerns.

Indeed, state practice indicates that states are actively engaging in environmental monitoring through remote sensing and increasingly making that data openly available. Japan has adopted the policy that remote sensing satellites should be used as "The Guardian of the Environment."[78] Korea uses remote sensing for "environmental protection purposes,"[79] and China uses satellites for "environmental monitoring."[80] An EU directive mandates that "all information held by public authorities relating to imminent threats to human health or the environment is immediately disseminated to the public likely to be affected."[81] State practice will continue to evolve in relation to this principle as the international community seeks to deal with global environmental problems.[82] In particular, as the climate change debate intensifies remote sensing technologies will likely be critical in both engineering and monitoring responses to climate change. These technologies will gather the evidence that will underlay the political debates surrounding the existence of, scope of, and responses to the problem. Additionally, it is likely that environmental treaties negotiated in the future will need to rely on satellite technology for verification purposes.[83]

Principle XI is worded similarly to Principle X and also has a two pronged obligation. First, it invokes a broad humanitarian obligation to "promote" the use of remote sensing technology to protect humans from natural disasters.[84] The second prong is that states should promptly share "processed data and analyzed information" with states that have been or will soon be affected by a natural disaster. This is a much broader secondary obligation than that articulated under Principle X, and is certainly geared towards the immediacy of the threat to life that disasters cause.

Principle XI has seen a great deal of state action. Several international systems have been established to facilitate the sharing of satellite data for disaster response, including the Disasters Charter,[85] UN-SPIDER,[86] and Asia Sentinel.[87] One of the most prominent of these initiatives is The Charter on Space and Disaster Cooperation, the purpose of which is to "provide EO data at times of natural or technological disasters . . . to all authorised users free of charge."[88] This agreement was established in 2000 and includes "a broad range of participants beyond Nation-States to enable pragmatic responses to a disaster."[89] China has launched satellites for disaster management and is involved in both UN-SPIDER and Asia Sentinel.[90] While there have been a number of successful programs, Ito argues that specific shortcomings in the principles themselves inhibit effective use of remote sensing technologies for disaster warning, response, and management.[91]

#### *17.2.3.2.3 Nondiscriminatory Access: Principle XII*

The most significant legal innovation in The Principles is nondiscriminatory access. Principle XII requires that sensed states be given "primary data and the processed data concerning the territory under its jurisdiction . . . on a non-discriminatory basis and on reasonable cost terms."[92] Additionally, sensed states are given a right of access, on the same terms, to analyzed data that any state engaged in remote sensing activities may have. There are two keys to this obligation: nondiscriminatory and reasonable cost terms. Nondiscriminatory has a "common interpretation . . . that the sensing States have an obligation to provide the data to the sensed States under the same conditions as other States that wish to access the data."[93] Reasonable cost basis, on the other hand, is "ambiguous and open to different interpretations," and "it in no way serves as a general guideline for price settings."[94] Principle XII is often seen as a compromise between the developed and developing nations in the negotiating process, wherein data access is given in exchange for freedom to sense.[95]

Nondiscriminatory access has been implemented by a critical mass of states, and "[g]enerally speaking, all national data policies and laws contain the same fundamental principles."[96] Gabrynowicz notes that these principles are "making data available for scientific, social, and economic benefit and restricting access to some data for national security reasons," and that "[d]ifferences occur in variables."[97] Significantly, data provisions (both the nondiscriminatory access mandates and the national security exceptions) are found in the licenses and agreements between states and private or commercial actors.[98]

The policy of nondiscriminatory access was "driven by Cold War foreign policy" and finds its roots in United States national policy, which had the "goal of influencing allies and nonaligned Nations by demonstrating technological superiority and encouraging them to use the data."[99] The United States law and policy grants nondiscriminatory access to all data from government funded civil satellites, and provides that commercial satellites systems can use "reasonable commercial terms and conditions."[100] It does give the governments of sensed states increased rights to data from commercial actors, but that access is still on "reasonable terms and conditions."[101] Nondiscriminatory access policies include Japan's *GOSAT* data policy;[102] Brazil's National Institute for Space Research's (INPE) policy, which is "to give out free on the Internet all remote sensing data received by INPE, the resulting maps, and the software for image processing and GIS";[103] *Landsat* data which is available free on the internet;[104] and, similarly, GeoScience Australia's policy to make "data free on the Internet."[105] Korea and Japan both have graduated system wherein data for different types of users is priced at different points.[106] India grants nondiscriminatory access for data above a certain resolution.[107] The Canadian law provides that the government of a sensed state has a right to obtain data from a commercial actor on "reasonable terms."[108] The French civilian remote sensing policy is "based on the promotion of a space imagery global market where data could be acquired on a nondiscriminatory basis."[109] The German Act on Satellite Data Security "favourable to commercial dissemination, create[s] de facto a wide database accessible to all third persons on a non-discriminatory basis."[110] The EU's Copernicus program uses a Creative Commons license to ensure that the data is "full, free, and open."[111] Other examples include Argentina, Malaysia, South Africa, and Thailand.[112]

In addition to actions at the state level, the Group on Earth Observations has developed data sharing principles that endorse "the full and open access to data and . . . Information."[113] This is based on an underlying philosophy that this data is a public good, and data should be distributed as such.[114] The EU has adopted a directive to "ensure[] that environmental information is systematically available and disseminated to the public,"[115] as well as the INSPIRE Directive, which "promote[s] unrestricted access to spatial data."[116] China and Brazil distribute *CBERS* data free of charge to many users and has established a specific program to distribute this data to African states.[117]

Full and open access is limited by national security concerns, but *"[d]ata denial is the exception, not the rule."*[118] Not only do The Principles not apply to military activities, they do not apply to civil or commercial imagery with national security implications.[119] As a result, when a state determines that a civil or commercial remote sensing satellite has taken imagery that affects national security concerns, that state does not have to distribute that data on a nondiscriminatory basis. Proper analysis of such provisions begs the question of "where on the spectrum is the point passed that . . . required data access moves into the reasonable national security constraints."[120] National security exceptions are also often implemented via the licensing process that states have implemented in accordance with Principle XIV and Article VI of the Outer Space Treaty.[121] National Security restrictions come in two different categories: transactional restrictions on the distribution of data and the use of "shutter control" regulations that allow a government to restrict, for national security reasons, a nongovernmental satellite company's "collection or distribution of data."[122]

The current trend is for states to restrict data on a transactional basis in which the "specifics of each request" are examined.[123] One of the most developed of these regimes is in the German satellite data law which implements "a two phase procedure: the sensitivity check undertaken by the data provider and the granting of the permit by the responsibility authority."[124] Under this regime, data dissemination is undertaken only after that data has been cleared through a process that determines "any potential endangerment of security."[125] France's 2003–2008 Military Program Law seeks to "[ensure] that certain civil programs comply with defense requirements."[126] While *CBERS* has some open access, Zhao argues that "remote sensing data are strictly controlled by the Chinese government on the grounds of protecting 'state secrets.'"[127] Additionally, Italy, India, Israel, Japan, and Korea have all implemented national security exceptions.[128]

Shutter control provisions are found in a handful of state regulations, but likely exists *de facto* in many nations. Canada[129] and the United States[130] have shutter control provisions that they can use to interrupt a satellite operator from distributing or collecting data that may harm national security. Canada's legal regime was adopted explicitly to ensure that remote sensing providers would comply with national security issues.[131] The regime allows the government to "interrupt normal service of a satellite" in cases of "the most serious of national security, national defence, and foreign policy/international obligations concerns."[132] The United States provisions require licensees to agree to "[s]pecific limitations on operational performance, including, but not limited to, limitations on data collection and dissemination" in their licenses on a case by case basis.[133]

### 17.2.4 Geospatial Law

Remote sensing data, primarily as a result of the internet, has become increasingly available to individuals and more prominent in society. Products such as Google Earth and extensive media usage of satellite imagery have brought remote sensing data into mainstream consciousness. Along with other technological advances, such as GPS, an array of geo-location and mapping applications are increasingly available to individuals globally through devices like smartphones and tablets, and geo-technologies are being embedded directly into these devices. This means that an array of legal issues have arisen, primarily in the domestic law context, which situates remote sensing activities within a broader complex of geo-location technologies. This has given rise to the area of geospatial law, which is loosely defined as the body of law that governs the collection, use, and distribution of geospatial information.

This new trend is directly related to three phenomenon: commercialization, globalization, and technological convergence. The first is commercialization. This process occurred in a number of countries and has driven the development of the domestic law in those countries.[134] Canada adopted The Remote Sensing Space Systems Act of 2005 after "advances in satellite remote sensing technology in the private sector started to drive the development of commercial space systems."[135] Similarly, Germany's data security law was triggered by the development of Public-Private Partnerships and commercial use of data.[136] Commercialization does not just refer to the operation of the satellite, it also cuts across issues such as "whether or not data should be distributed by the public sector or the private sector."[137] For instance, France's SPOT Image is a commercial data distributor of public data,[138] and India has made data from government remote sensing satellites commercially available.[139] Commercialization creates legal issues because the UN principles "were agreed [to] at a time when the commercialization of remote sensing activities were still not envisaged."[140] Commercialization is not a completed process (and will most likely remain that way indefinitely), and an important lesson of its limitations can be found in the failed *Landsat* commercialization in the United States.[141]

Next, is the process of globalization,[142] the modern version of which has cultural and conceptual roots in the space age and the re-imagination of the world as a global space.[143] Held argues that globalization is the "widening, deepening and speeding up of worldwide interconnectedness in all aspects of contemporary social life, from the cultural to the criminal, the financial to the spiritual."[144] It leads to a combining of the "local and the global" to create "distant proximities."[145] Essentially, globalization changes the space that society exists in, and importantly, "satellite imagery, digital maps, and associated information have transformed our ability for understanding the forces that shape geographical space."[146] As such globalization is an information driven process, and one of the effects of globalization has been increased access to global technologies.[147] This is certainly true in the field of remote sensing with numerous states now engaging in remote sensing activities including new entrants such as Nigeria, Algeria, Morocco, Egypt, Kenya, South Africa, Colombia, and Turkey.[148] More important, though, is the fact that remote sensing is a global technology, and as such it facilitates the conceptualization of the world as a global space.[149]

Finally, the process of technological convergence has changed the way in which we interact with this type of data. Technological convergence is the process through which technologies that were once significantly distinct (e.g., a map and a phone) are integrated and become indistinguishable. Current convergence trends are connected to the way in which information has been redefined as data in a technical sense. This is not to imply that information is data, but instead, that technologically they are indistinguishable. For example, when a user downloads a remote sensing image, s/he receives information when s/he look at the image. However, as that image was transferred over the internet, it was simply data broken down into manageable packets, and the application on the end users device reassembled it into information.[150] This means that data is now highly integrateble (depending on creativity and innovation in applications) and highly portable (depending on bandwidth). This can cause an array of issues since integrated data packages that pull from a variety of sources. For example, mapping software on mobile devices may pull on geo-location information provided by the device, map databases, remote sensing data, and crowd sourced information, among other sources. When this information becomes data, it becomes indistinguishable as separate sets of knowledge and becomes a single data point for the user. This is one of the features that has allowed "ICT technologies [to reduce] geography and physical distance in politics, economics, and war . . . render[ing] territory less important."[151] While territory has become less important, location has become increasingly important,[152] meaning that *"[g]lobal politics no longer reflects . . . neat, exclusive territorial boxes,"* creating a need to "rethink . . . and design a wide variety of maps that take into account the many forms of political activity."[153] Remote sensing data has become part of technological convergence, and there is a trend towards "increas[ing] the use of data."[154] Additionally, convergence has made the internet the core tool for improving transparency, and geospatial technologies play an important role in transparency.[155]

These three processes have significantly changed the place of remote sensing data in society and is impacting how operators engage in these activities. Market trends show that new remote sensing companies are seeking to create their own data products rather than simply selling imagery This naturally results in legal tensions as the law is recalibrated to cope with emerging technologies. Commercialization challenges remote sensing law by introducing new types of stakeholders that do not fit into the international law regime, such as companies engaged in radio-frequency remote sensing instead of imagery collection. Globalization challenges the regime because "the overlap of authority and governance among subnational, national, transnational, and international polities is fraught with difficulties."[156] And, technological convergence challenges the regime by reorienting how and why remote sensing data is used.[157] The resulting complexity situates remote sensing law within the bounds of geospatial law, at least as far as data and information issues are involved. This chapter will not attempt to give a full account of geospatial law, which is a broad category encompassing a variety of legal issues, instead it will attempt to focus on prominent issues that are directly related to remote sensing technologies.

#### 17.2.4.1 Privacy and Security

The 9/11 terrorists attacks shattered the notion that globalization was necessarily a sign of positive progress. Not only did globalization enable global public goods and mass communication, it also enabled global threats like terrorism. In this context privacy has become an increasingly important issue This results from both of the faces of globalization. The increasing access to and use of remote sensing imagery by the general population has generated significant privacy issues based on global connectivity and content coupled with global threats. The UN Principles "offer only fairly minimal guidance on the topic of privacy of individual persons and entities," and as a result privacy protections "are consequently to be found only at national level,"[158] but at the national level, "[p]rivacy is a broad area of law, and relevant legislation is unclear in concept and application."[159]

In the wake of the 9/11 attack there were "sweeping organizational changes" in the United States defense and intelligence organizations which "g[ave] rise to myriad potential statutory and regulatory impediments."[160] This creates specific issue about the use of U.S. military satellites for use of surveillance of United States citizens, which raises "particularly vexing and controversial" issues.[161] While the freedom of use of remote sensing satellites is established at the international law level, these technologies are often treated differently under domestic law.[162] In the American context, this represents a cultural value that has been embedded in the law.[163] The Reconstruction era Posse Comitatus Act restricts the use of the military in law enforcement, and the values found in the Posse Commitatus Act have been furthered formalized through laws, regulations, policies, and procedures.[164] The application of Posse Comitatus and its progeny to remote sensing technologies is less clear in practice, though there are clear exceptions.[165] However, the core value of separating these institutions from domestic law enforcement can be found in the law and policies governing remote sensing technologies.

United States law requires defense collection of information on U.S. persons to be gathered using the "least intrusive means," which requires the use of a four step methodology.[166] Petras points out that satellite data presents an issue since it is unclear whether that data represents a "publicly available" source of information, in light of commercial imagery available on the internet (the first tier of the methodology), or whether it such information retrieval requires "a warrant or approval by the Attorney General" (the fourth tier of the methodology).[167] National Geospatial-Intelligence Agency procedures require that imagery gathered by domestic satellites must be for a "foreign intelligence mission" and only if commercial imagery is "unsuitable or unavailable."[168] Any other military request must be for "authorized activities of the armed forces."[169] Consistent with this theme there are further procedures to limit the use of such technologies in the American context. The collection can not be "focused on a specific U.S. person" and must be in connection to activities that are not "domestic" in nature.[170]

Since satellite technology is not entirely out of the picture for law enforcement, the restrictions on government observation of citizen's is primarily contained in search and seizure law, and as remote sensing technology continues to "become a viable law enforcement tool," issues of proper use by governments will become more important.[171] In the United States, this is governed by a constitutional restriction found in the Fourth Amendment, which forbids the government from engaging in unreasonable searches without a proper warrant.[172] The standards involved come from a line of cases that hold that searches that breach a "reasonable expectation of privacy" violate constitutional protections.[173] However, courts have been challenged by technologies that make observations of individuals by remotely sensing them or their property. While the Supreme Court has never ruled directly on the use of satellite remote sensing data, there is a line of cases that can be traced, which gives some indication of how satellites might be treated under United States law.

Several cases address the use of aerial surveillance and hold that the government has fairly wide discretion to use aerial surveillance without a warrant.[174] Of particular note, though, is the dicta in *Dow Chemical* in which the Court posits that certain technologies not available to the general public, like satellite technology, might violate the Fourth Amendment.[175] This theme is taken up in the *Kyllo v. United States* case. In this case police used a thermal imager to determine whether grow lights were being used inside a residence to grow marijuana.[176] The court determined that the technology that was used was not available to the general public and as a result it violated a reasonable expectation of privacy.[177] Of course, remote sensing technology is now arguably available to the general public as a result of technological convergence.[178] A final and more recent case is *Jones v. United States*, which addressed the use of a GPS device on a car. The opinion was split, but one of the concurring opinions posited that such searches created a mosaic of the life of an individual and that such mosaics breach a reasonable expectation of privacy.[179] If such a standard were to be adopted by the court in the future, it would certainly have implications for uses of remote sensing data in law enforcement.

In general, states have a variety of approaches to privacy issues. Korea, in Article 17 of its Space Development Promotion Act, has legislated that the "Government shall endeavor to avoid any invasion of privacy during the course of utilizing satellite information."[180] Lee notes that privacy is a "fundamental right" provided for in the Korean Constitution. Legal regimes create a variety of contexts in which privacy must be considered; for example, remote sensing data issues have been identified in the Health Insurance Portability and Accounting Act in the United States.[181] Other states do not have fully formed privacy protections. Zhao notes that Hong Kong has "no comprehensive data protection law."[182] In the Canadian remote sensing regime, "there are no provisions . . . dealing with privacy, and no privacy conditions have been incorporated in the first remote sensing satellite system license," therefore privacy concerns must be addressed under Canada's Constitution and Privacy Act.[183] Additionally, commercial use of data has caused numerous concerns. In particular the Google suite of geo-technologies have come under extensive fire for privacy violations,[184] and as content providers increasingly incorporate geospatial data, privacy breaches are becoming more common.[185] This creates cross jurisdictional problems since content providers operate in a multiplicity of jurisdictions at any given time.

Possibly the most significant recent development in this area is the adoption of the General Data Protection Regulation by the European Union.[186] This law imposes a strict regime on entities that process the personally identifiable information of EU residents and has had broad ranging impacts globally. To the extent that remote sensing data creates or is intermingled with personal data, GDPR can have significant impacts on a satellite operator.[187]

#### 17.2.4.2 Use of Remote Sensing Data as Evidence

Data derived from remote sensing activities has become increasingly important in judicial proceedings, this is related to both the integration of geolocation technologies into everyday lives and to large scale problems such as environmental damage. The ability to introduce this type of data is governed by rules of evidence for the particular jurisdiction meaning that geospatial and remote sensing data must be fit into a preexisting evidentiary categories.[188] One of the core issues has been authentication of remote sensing evidence in courts due to its digital nature.[189] This problem has been addressed through technical means such as "digital signatures,"[190] and complete authentication procedures, as in the case of USGS/EROS data.[191]

Korea,[192] Queensland, Australia,[193] Singapore,[194] and the United States[195] use satellite technology for investigating and prosecuting of breaches of environmental laws. Satellite data has also found its way into evidence in a variety of international Institutions.[196] The International Court of Justice has used satellite data in numerous cases, which often involve boundary disputes.[197] The Permanent Court of Arbitration and the International Criminal Court has both made use of Earth observation data as evidence.[198] The use of this data in both civil and criminal courts will increase across a number of areas as data becomes increasingly embedded into society.

#### 17.2.4.3 Torts

Misuse of geospatial data could result in torts. Navigation systems providers have been repeatedly sued for incorrect data resulting in incorrect directions.[199] A U.S. court held a map provider liable for incorrect data that it purchased from a third party,[200] and Nigerian law can hold "GIS professionals . . . legally accountable for the accuracy and reliability of information stored in their databases, sold, or issued to the public."[201] Ito states that "[d]ata suppliers bear liability risks in cases where the population is affected by disasters or aid workers are injured as a result of inappropriate instructions."[202] While Ito is arguing for a more complete international legal regime, data torts will likely continue to be handled *ad hoc* at the domestic judicial level.

#### 17.2.4.4 Intellectual Property

The UN Principles are "completely silent" as to intellectual property rights in data generators.[203] This is likely because, at the time of the negotiations, these activities had not yet seen a sufficient amount of commercial or private activity to warrant addressing the issue. Commercialization has resulted in a variety of users making intellectual property rights a growing concern for remote sensing data providers and users. Generally, "current practice is that the ownership of data stays with data generators and the copyright is claimed by the majority of data generators," but it is not clear to what extent these claims can be maintained for raw data especially in jurisdictions where a "certain degree of human intellectual intervention" is required for information to be copyright protected.[204] The nature of remote sensing data means that "[c]opyright may not be the most appropriate regime to protect EO data."[205]

State regulations vary significantly. Goh notes a Singaporean case in which a data provider was held to violate copyright protections by distributing maps created with unlicensed data.[206] The EU protects intellectual property rights in "databases to secure remuneration to the maker of the database," which Harris argues could have implications for remote sensing technologies.[207] He notes that problems could arise when database protections are involved with public data.[208] Nigeria explicitly grants intellectual property rights to data producers via its National Geoinformation Policy,[209] and JAXA maintains copyright to data it creates in Japan.[210] The Korean regime can protect remote sensing imagery via copyright or database protections.[211] Additionally, Germany has denied copyright of *Landsat* data, while France has allowed it.[212]

## 17.3 CONCLUSION

Remote sensing will remain an important technology in the foreseeable future. Based on the international principles examined in this chapter, remote sensing is regulated within a stable international legal framework that is permissive in nature. Emerging challenges will come from the increasing use of remote sensing data in a data driven society. National regulations will need to balance among competing interests of national security, individual liberty, and commercial growth in an increasingly complex technological infrastructure.

## NOTES

1 The Principles Relating to Remote Sensing of the Earth from Outer Space, U.N.G.A. Res. 41/65 (Dec. 3, 1986).
2 Joanne Irene Gabrynowicz, *The Land Remote Sensing Laws and Policies of National Governments: A Global Survey* (University, MS: National Center for Remote Sensing, Air, and Space Law 2007) 7.
3 P. J. Blount, "Renovating Space: The Future of International Space Law," *Denver Journal Internation Law & Policy*, v. 40 (2012) 515–686.
4 The Treaty on Principles Governing the Activities of States in the Exploration and Use of Outer Space, including the Moon and Other Celestial Bodies (the "Outer Space Treaty") (entered into force on 10 October 1967); The Agreement on the Rescue of Astronauts, the Return of Astronauts and the Return of Objects Launched into Outer Space (the "Rescue Agreement") (entered into force on 3 December 1968); The Convention on International Liability for Damage Caused by Space Objects (the "Liability Convention") (entered into force on 1 September 1972); The Convention on Registration of Objects Launched into Outer Space (the "Registration Convention") (entered into force on 15 September 1976). On the space law making process see Paul G. Dembling and Daniel M. Arons, "The Evolution of the Outer Space Treaty," *Journal of Air Law & Commerce*, v. 33 (1967) 419 and Sergio Marchisio, "The Evolutionary Stages of the Legal Subcommittee of the United Nations Committee on the Peaceful Uses of Outer Space (COPUOS)," *Journal of Space Law*, v. 31 (2005) 21.
5 For a variety of approaches see I.H.Ph. Diederiks-Verschoor, *An Introduction to Space Law* (The Hague: Kluwer 1999); Francis Lyall and Paul B. Larsen, *Space Law: A Treatise* (Aldershot, UK: Ashgate 2013); Glenn H. Reynolds and Robert P. Merges, *Outer Space: Problems of Law and Policy* (Boulder, CO: Westview Press 1997); and Manfred Lachs, *The Law of Outer Space: An Experience in Contemporary Law-Making* (Leiden: Martinus Nijhoff 1972, 2010).
6 Ram Jahku, "International Law Governing the Acquisition and Dissemination of Satellite Imagery," *Journal of Space Law*, v. 29 (2003) 65, 73–77; M. Seara Vásquez, *Cosmic International Law* (Detroit: Wayne State University Press 1965) 164.
7 Jahku, "International Law," 73–74. *See also* P.J. Blount, "Legal Issues Related to Satellite Orbits," in *Oxford Encyclopedia of Planetary Science* (Oxford: Oxford University Press).
8 For instance, one of the primary goals of the Canadian licensing regime is to regulate "the operation of the satellite itself." Thomas Gillon, "Regulating Remote Sensing Space Systems in Canada—New Legislation for a New Era," *Journal of Space Law*, v. 34 (2008) 19, 26–27. See also Bruce Mann, "First License Issued Under Canada's Remote Sensing Satellite Legislation," *The Journal of Space Law*, v. 34 (2008) 67, 69–71.
9 For instance, the Chinese regime is based on launch licenses. "Chinese Law: Registration, Launching and Licensing Space Objects," *Journal of Space Law*, v. 33 (2007) 437.
10 Canada, Remote Sensing Space Systems Act, Gillon, "Regulating Remote Sensing," 26 (noting a second concern of the Canadian licensing regime is "the distribution of raw data and remote sensing products produced by satellites"); Germany, "Satellite Data Security Act," *Journal of Space Law*, v. 34 (2008) 115; and the United States, Licensing of Private Remote Sensing Systems, 15 C.F.R. 960 (2013).
11 Outer Space Treaty, Art. IX; see also Articles III, V, VIII, X, XI, and XII.
12 Ito notes that the Outer Space Treaty and the Liability Convention "are silent as to the types of damage associated with satellite remote sensing." Atsuyo Ito, "Improvement to the Legal Regime for the Effective Use of Satellite Remote Sensing Data for Disaster Management and Protection of the Environment," *Journal of Space Law*, v. 34 (2008) 45, 51.
13 See generally, Michael Listner, "Iridium 33 and Cosmos 2251 Three Years Later: Where Are We Now?," *The Space Review*, Feb. 13, 2012, www.thespacereview.com/article/2023/1.
14 The author has developed and applied this type of analysis to other areas of space law. See generally, P.J. Blount, "Developments in Space Security and Their Legal Implications," *Law/Technology*, 44/2 (2011) 18 and P.J. Blount, "Limits on Space Weapons: Incorporating the Law of War into the Corpus Juris Spatialis," in *Proceedings of the 51st Colloquium on the Law of Outer Space* (AIAA 2009).

15 See Joanne Irene Gabrynowicz, "Space Law: Its Cold War Origins and Challenges in the Era of Globalization," *Suffolk University Law Review* v. 37 (2004) 1043.

16 Legally these opportunities were facilitated by a "regime of complete freedom." Olusoji Nester John, Eguaroje Ezekiel, and S.O. Mohammed, "Legal Regime of Remote Sensing and Geographic Information Systems in Nigeria," in *Proceedings of the International Institute of Space Law* (The Hague: Eleven 2011) 260.

17 Blount, "Limits on Space Weapons" and Christopher M. Petras, "'Eyes' On Freedom—A View of the Law Governing Military Uses of Satellite Reconnaissance in U.S. Homeland Defense," *The Journal of Space Law*, v. 31 (2005) 81, 88–90.

18 Jahku, "International Law," 70.

19 Petras, "Eyes," 86.

20 Vásquez, *Cosmic*, 163–164.

21 For a history of such a systems, see generally, Jeffrey T. Richelson, *America's Space Sentinels: DSP Satellites and National Security* (Lawrence, KA: University Press of Kansas 1999).

22 Petras, "Eyes," 89–90.

23 *Id.* 87.

24 For example SALT I & SALT II, Diederiks-Verschoor, *Introduction*, 84–85; ABM Treaty, Petras, "Eyes," 90; CFE Treaty, id. 92–93; and Treaty between the United States of America and the Russian Federation on Measures for the Further Reduction and Limitation of Strategic Offensive Arms (signed April 10, 2010) Art. X(b) and (c).

25 Jimmy Carter, "Remarks at the Congressional Space Medal of Honor Awards Ceremony, Kennedy Space Center, FL," *Weekly Compilation of Presidential Documents*, v. 14/40 (Oct. 8, 1978) 1684. Importantly, Petras notes that NTM includes a wider range of technologies than just remote sensing satellites. Petras, "Eyes," 91.

26 Petras, "Eyes," 91 and Diederiks-Verschoor, *Introduction*, 85.

27 Roger G. Harrison, *Space and Verification, Vol. 1: Policy Implications* (Colorado Springs, CO: Eisenhower Center for Space and Defense Studies) 9.

28 Philippe Achilleas, "French Remote Sensing Law," *The Journal of Space Law*, v. 34 (2008) 1, 3; Diederiks-Verschoor, *Introduction*, 84.

29 See, for example, Masami Onada, "Satellite Earth Observation as 'Systematic Observation' in Multilateral Environmental Treaties," *The Journal of Space Law*, v. 31 (2005) 339 and Sandra Cabrera Alvarado, "The Analysis of existing International Space Cooperation Initiatives for UNESCO's World Heritage Sites," IAC-10.E3.1A.14 (2010).

30 Paul F. Uhlir, et al., "Toward Implementation of the Global Earth Observation System of Systems Data Sharing Principles," *The Journal of Space Law*, v. 35 (2009) 201, 209.

31 On the legal status of the UN Principles, see generally, Andrei Terekov, "UN General Assembly Resolutions and Outer Space Law," *Proceedings of the International Institute of Space Law* (1997) 97, 100–101 and Vladimir Kopal, "The Role of United Nations Declarations of Principles in the Progressive Development of Space Law," *The Journal of Space Law*, v. 16 (1988) 5, 14–20.

32 See UNGA Meeting Record no. A/41/PV.95 (Dec. 3, 1986). Some commentators argue that this rises to the level of "instant custom." For instance, Susan M. Jackson, "Cultural Lag and the International Law of Remote Sensing," *Brooklyn Journal of International Law*, v. 23 (1997) 853, 873. Additionally, Hosenball notes that the language evinces that states made a "commitment to live by these *Principles.*" Joanne Irene Gabrynowicz, ed., *The UN Principles Relating to Remote Sensing of the Earth from Space: A Legislative History—Interviews of Members of the United States Delegation* (University, MS: National Center for Remote Sensing and Space Law 2002) 35.

33 For instance, the UN Principles are the "guiding principle for the practice of Earth remote sensing and was, accordingly, considered in the preparation" of the German satellite data law. Bernhard Schmidt-Tedd and Max Kroyman, "Current Status and Recent Developments in German Remote Sensing Law," *The Journal of Space Law*, v. 34 (2008) 97, 105. For United States and Japan as evidence of custom see Gabrynowicz, *UN Principles*, 42. The nondiscriminatory access principle has repeatedly been incorporated in United States law and in international agreements. Gabrynowicz, *Land Remote Sensing*, 6. Jahku notes that there is, a divide on the issue, but that Principle XII at least, creates a legal obligation. Jahku, "International Law," 86–88.

34 See note 14, *supra*.

35 So, for instance, nondiscriminatory access is being "more narrowly construed" as a norm due to state practice that has expanded national security exceptions over data. Gabrynowicz, *Land Remote Sensing*, 11.
36 For instance, while all jurisdictions recognize a crime of "murder," the elements of the crime changes as jurisdictional borders are crossed. As a result, we can say that the prohibition against murder is a general principle, but that the specific content of that principle exists in a spectrum of meaning. Customary norms suffer from a similar lack of clarity and requires a spectrum analysis (if you'll allow me the pun). Furthermore, this type of analysis is particularly apt in high technology cases where there is unequal distribution of access coupled with wide acceptance of negotiated rules.
37 Vladen S. Vereschetin and Gennady M. Danilenko, "Custom as a Source of International Law of Outer Space," *The Journal of Space Law*, v. 13 (1985) 22, 30.
38 When referring specifically to the UN Principles, this chapter will capitalize (i.e. The Principles), but when used a as a generic term, "principles" will be lower case.
39 Compare to the Outer Space Treaty, Article III.
40 UN Principles, III. The Outer Space Treaty is also referenced in Principle IV.
41 UN Principles, IV
42 Achilleas, "French," 2.
43 UN Charter, Art. 1 & 2.
44 Phetole Patrick Sekhula, "The Right to Remote Sense Data: Impact of Multilateral Cooperation on International Space Law [sic]," *Proceedings of the International Institute of Space Law* (The Hague: Eleven 2011) 228, 233. Such contentions reflect current critiques of neoliberalism, for instance, see David Harvey, *A Brief History of Neoliberalism* (Oxford: Oxford University Press 2005) ("Information technology is the privileged technology of neoliberalism").
45 Yan Ling, "Remote Sensing Data Distribution and Application in the Environmental Protection, Disaster Prevention, and Urban Planning in China," *The Journal of Space Law*, v. 36 (2010) 435, 436–437.
46 Diederiks-Verschoor, *Introduction*, 83.
47 Jahku, "International Law," 79.
48 U.N.G.A. Res. 60/1 2005 World Summit Outcome (Oct. 24, 2005). R2P is an emerging norm that supports broader ability of the international community to intervene when states are committing gross human rights violations against their own citizens, such as genocide. *Id.* paras. 138–140.
49 Satellite Sentinel Project, www.satsentinel.org/ (last visited Feb. 22, 2014).
50 Eyes on Darfur, www.eyesondarfur.org/ (last visited Feb. 22, 2014).
51 United Nations High Commission on Human Rights, *Report of the Commission of Inquiry on Human Rights in the Democratic People's Republic of Korea*, A/HRC/25/63 (Feb. 17, 2014) (relying on satellite data as evidence).
52 UN Charter, Art. 1.
53 Outer Space Treaty, Art. VII; Registration Convention, Art. II; and UNGA Res. 62/102, Recommendations on Enhancing the Practice of States and International Intergovernmental Organizations in Registering Space Objects (Dec. 17, 2007). See generally, Kai-Uwe Schrogl and Niklas Hedman, "The U.N. General Assembly Resolution 62/102 of 17 December 2007 on 'Recommendations on Enhancing the Practice of States and International Intergovernmental Organizations in Registering Space Objects,'" *The Journal of Space Law*, v. 34 (2008) 141. schrogl and hedman. It should not be assumed that all states necessarily comply with registration requirements, for a remote sensing specific case see Jean-Francois Mayence, "Belgian Legal Framework for Earth Observation Activities," *The Journal of Space Law*, v. 34 (2008) 89, 92–94.
54 Outer Space Treaty, Art. IX.
55 See generally, International Law Commission, *Draft Articles on Responsibility of States for Internationally Wrongful Acts, with commentaries* (2001).
56 UN Principles, II.
57 Outer Space Treaty, Art. I.
58 Hilcea Santos Ferreira and Gilberto Camara, "Current Status and Recent Developments in Brazilian Remote Sensing Law," *The Journal of Space Law*, v. 34 (2008) 11.
59 Diederiks-Verschoor, *Introduction*, 83 (noting a trend towards increased cooperation).
60 UNGA Res. 51/122, The Declaration on International Cooperation in the Exploration and Use of Outer Space for the Benefit and in the Interest of All States, Taking into Particular Account the Needs

of Developing Countries (Dec. 13, 1996). See also, Elena Carpanelli and Brendan Cohen, "A Legal Assessment of the 1996 Declaration on Space Benefits on the Occasion of Its Fifteenth Anniversary," *The Journal of Space Law*, v. 38 (2012) 1.

61 Ikuko Kuriyama, "Environmental Monitoring Cooperation Paves the Way for Common Rules on Remote Sensing," *The Journal of Space Law*, v. 36 (2010) 567.

62 Setsuko Aoki, "Japanese Law and Regulations Concerning Remote Sensing Activities," *The Journal of Space Law*, v. 36 (2010) 335, 337.

63 For instance Iran maintains a receiving station for Landsat data. Parviz Tarikhi, "Mahdasht Satellite Receiving Station Verging into a Space Center," *Res Communis*, Oct. 13, 2008, http://rescommunis.olemiss.edu/2008/10/13/guest-blogger-parviz-tarikhi-mahdasht-satellite-receiving-station-verging-into-a-space-center/.

64 Outer Space Treaty, Art. IX;. Diederiks-Vershoor, *Intorduction*, 82 ("Principle V strengthens the other provisions of the Space Treaty's Article IX").

65 An example of such cooperation would be the Landsat Ground Station Operators' Working Group, which "attempt[s] to address . . . how data policies should be formulated." John F. Graham and Joanne Irene Gabrynowicz, eds., *Landsat 7: Past Present and Future* (University, MS: National Remote Sensing and Space Law Center 2002) 294–5. On ground station policy and operations, see generally, Joanne Irene Gabrynowicz, "Earth Observation: The View from the Ground Up," *Space Policy* (Aug. 1997) 229.

66 It should be noted that this consultation request is different in scope than the consultations afforded by Article IX of the Outer Space Treaty. The Outer Space Treaty's consultation provision is triggered by harmful interference, whereas Principle XIII is triggered by remote sensing activities (as defined in the principles) and an opportunity for cooperation between the two states. While arguably, this expansion could have been included in the innovations section that follows, it is included here due to both its softness as an obligation and its strong connection to the use of space for international cooperation. Additionally, tracking compliance and use of this principle is difficult due to the informal nature of "consultations."

67 UN Principles, I(a). This definition is legal in nature, and is lacking as a technical definition. Vasquez and Lara suggest—as a definition to help bridge the gap between law and policy makers and scientists—that remote sensing is "a way of obtaining information about an object by analyzing the data acquired through a device that is not in physical contact with said object." Fermin Romero Vásquez and Sergio Camacho Lara, "What Lawyers Need to Know about Science to Effectively Make and Address Laws for Remote Sensing and Environmental Monitoring," *The Journal of Space Law* v. 36 (2010) 365, 367.

68 See Graham and Gabrynowicz, *Landsat-7*, 18–19 (quoting Neil Hosenball as stating that the United Nations Principles "do not apply to domestic military systems" and that the issue was never raised during the negotiations); Achilleas, "French," 2; Diederiks-Verschoor, *Introduction*, 86.

69 Remote Sensing Principles, I.

70 *Id*. For more on the different definitions of data, see Jahku, "International Law," 66–67.

71 Remote Sensing Principles, I(e).

72 Elena Carpanelli and Melissa K. Force, "The Protection of the Earth Natural Environment *through* Space Activities: A General Overview of Some Legal Issues," *Proceedings of the International Institute of Space Law* (The Hague: Eleven 2011) 31.

73 Outer Space Treaty, Art. IX.

74 Carpanelli and Force, "Protection," 33. Carpanelli and Force argue that this particular provision is part of a dynamic process of customary international law development and that the extent to which it represents custom is still being defined. *Id*. 34. Onoda also argues that there is "general obligation of international environmental law for States to cooperate in promotion of global Earth observation to protect the environment." Onoda, "Satellite," 341.

75 Carpanelli and Force, "Protection," 32.

76 Remote Sensing Principles, X.

77 It should be noted that meteorological data is "considered a public good used for the benefit of all." Jahku, "International Law," 84.

78 Aoki, "Japanese," 345, 348. For example, see Japan's *Greenhouse Gases Observation Satellite* (*GOSAT*). *Id*. 338–340.

79 Jae Gon Lee, "Remote Sensing Issues as They Relate to Korea," *The Journal of Space Law*, v. 36 (2010) 415, 419, 423–424.

80 Ling, "Remote Sensing," 439, 451–452.

81 EU Directive 2003/4/EC, On public access to environmental information and repealing Council Directive 90/313/EEC (Jan. 28, 2003). See Ray Harris, "Current Status and Recent Development in UK and European Remote Sensing Law and Policy," *The Journal of Space Law*, v. 34 (2008) 33, 37.
82 "Global cooperation in environmental data is on the rise." Angeline Asangire Oprong and Vincent Rwehumbiza, "A Glance at the Earth Observation Policies and Regulations and the Impact on Developing Countries: Focusing on the African Continent," *Proceedings of the International Institute of Space Law* (The Hague: Eleven 2011) 251, 252.
83 See, for example, Vasquez and Lara, "What Lawyers," 371–377; Annette Froehlich, "Space Related Data: From Justice to Development," *Proceedings of the International Institute of Space Law* (The Hague: Eleven 2011) 221, 221–222; Onoda, "Satellite," 339.
84 Remote Sensing Principles, XI.
85 Charter On Cooperation To Achieve The Coordinated Use Of Space Facilities In The Event Of Natural Or Technological Disasters Rev.3 (25/4/2000), www.disasterscharter.org/web/charter/charter.
86 UNOOSA, "About UN-SPIDER," www.unoosa.org/oosa/en/unspider/index.html (last visited Feb. 22, 2014) (UN-SPIDER is a "United Nations programme, with the following mission statement: 'Ensure that all countries and international and regional organizations have access to and develop the capacity to use all types of space-based information to support the full disaster management cycle'").
87 Sentinel Asia, "Sentinel Asia: Disaster Management Support System in the Asia-Pacific Region," www.aprsaf.org/initiatives/sentinel_asia/pdf/Sentinel-Asia.pdf (last visited Feb. 22, 2014).
88 Ray Harris, "Science, Policy, and Evidence in EO," in Ray Purdy and Denise Leung, eds., *Evidence from Earth Observation Satellites* (Leiden: Martinus Nijhoff Publishers 2013) 43, 47; Josie Beets, "The International Charter on Space and Major Disasters and International Disaster Law: The Need for Collaboration and Coordination," *Air & Space Lawyer*, v. 22/4 (2010) 12–15.
89 Uhlir et al., "Toward Implementation," 211.
90 Ling, "Remote Sensing," 439, 453.
91 See generally, Ito, "Improvement." These critiques are echoed by Jackson, "Cultural Lag," 872–873.
92 Remote Sensing Principles, XII.
93 Ito, "Improvement," 49–50.
94 Ito, "Improvement," 50; see also, Oprong and Rwehumbiza, "Glance," 257; Gabrynowicz, *UN Principles*, 31.
95 Schmidt-Tedd and Kroyman, "Current Status," 105; Harris, "Science," 45.
96 Gabrynowicz, *Land Remote Sensing*, 3.
97 *Id.* There is at least one exception to this rule found in Israel's ImageSat, which "openly promotes exclusivity and secrecy." *Id. at* 21. For more on ImageSat, see Jason Crook, "Corporate Sovereign Symbiosis: *Wilson v. ImageSat International*, Shareholders' Actions, and the Dualistic Nature of State-Owned Corporations," *The Journal of Space Law*, v. 33 (2007) 411.
98 For instance, France in the SPOT Image commercial policy. Achilleas, "French," 5. Gabrynowicz notes that policies are still more common than formal legal regulation. Gabrynowicz, *Land Remote Sensing*, 7–10.
99 Gabrynowicz, *Land Remote Sensing*, 5. For more on law and policy connected to *Landsat* see generally Graham and Gabrynowicz, *Landsat-7* and Joanne Irene Gabrynowicz, "The Perils of *Landsat* from Grassroots to Globalization: A Comprehensive Review of US Remote Sensing Law with a Few Thoughts for the Future," *Chicago Journal of International Law*, v. 6/1 (2005) 45.
100 Jahku, "International Law," 88–89. For an extensive, though dated, overview of data availability under US law see Joanne Irene Gabrynowicz, "Defining Data Availability For Commercial Remote Sensing Systems Under United States Federal Law," *Annals of Air and Space Law*, v. 23 (1999) 93. *Landsat* data is now available free on the internet. NASA Landsat Science, "The Numbers Behind Landsat," http://landsat.gsfc.nasa.gov/?page_id=9 (last visited Feb. 22, 2014).
101 Jahku, "International Law," 89; 15 C.F.R. § 960.11(b)(10); 15 C.F.R. § 960.12.
102 Aoki, "Japanese," 339.
103 Ferreira and Camara, "Current Status," 15.
104 USGS, "Landsat," http://landsat.usgs.gov/ (accessed May, 15, 2014).
105 Gabrynowicz, *Land Remote Sensing*, 16.
106 Lee, "Remote Sensing," 429–430; Gabrynowicz, *Land Remote Sensing*, 18–19.
107 Gabrynowicz, *Land Remote Sensing*, 20.

108 Remote Sensing Space Systems Act (S.C. 2005, c. 45), 8(4)(c). See also Mann, "First License," 78.
109 Achilleas, "French," 2. Achilleas further notes that in order to "secure the market from a legal point of view, France has supported the adoption" of the Remote Sensing Principles. *Id.*
110 Schmidt-Tedd and Kroyman, "Current Status," 105. Nondiscriminatory in this context means that "it is impossible for a customer to prevent a third person from accessing data about a specific region." *Id.*
111 Sandra Cabrera Alvarado, "The Regulation of the 'Open Data' Policy and its Elements: The Legal Perspective of the EU Copernicus Programme," in P.J. Blount and Mahulena Hofmann, eds., *Space Law in a Networked World* (Leiden: Brill 2023) 256–272.
112 Gabrynowizc, *Land Remote Sensing*, i–xvii.
113 Uhlir et al., "Toward Implementation," 220.
114 *Id.*
115 Harris, "Current Status," 37. Harris notes that the term "reasonable cost" is similar to, but less definite than "cost of fulfilling a user request." *Id.*
116 Catherine Doldirina, "The Impact of Copyright Protection and Public Sector Information Regulations on the Availability of Remote Sensing Data," in Ray Purdy and Denise Leung, eds., *Evidence from Earth Observation Satellites* (Leiden: Martinus Nijhoff Publishers 2013) 293, 303.
117 Ling, "Remote Sensing," 446–447; Ferreira and Camara, "Current Status," 15. Users outside of this regime can gain access from "licensed representatives." CBERS Data Policy, *The Journal of Space Law*, v. 31 (2005) 281. See also Jose Monserrat Filho and Alvaro Frabrico dos Santos, "Chinese-Brazilian Accord on Distribution of CBERS Products," *The Journal of Space Law*, v. 31 (2005) 271.
118 Gabrynowicz, *Land Remote Sensing*, 11; Uhlir et al., "Toward Implementation," 230–231. Jahku argues that such laws and policies are inconsistent with Principle XII. Jahku, "International Law," 89–90. However, it should be noted that these exceptions have become prevalent enough that they can, "be considered part of the acceptable state practice under the norm." This is especially so in light of the prevalence of national security exceptions across a wide range of international law regimes. See also Sekhula, "The Right," 229 (arguing that the nondiscriminatory access rule to "have proved unworkable in practice and [to] require new interpretation consistent with contemporary practice").
119 Gabrynowicz, *Land Remote Sensing*, 13 ("regardless of actual practice, no Nation or data supplier wants to appear to denounce the nondiscriminatory access policy").
120 Gabrynowicz, *Land Remote Sensing*, 22.
121 But see Jahku, "International Law," 83–84 (noting a contractual agreement between India and Space Imaging restricting imagery distributed in India).
122 Petras, "Eyes," 95. Gabrynowicz notes "[a]ll current and pending national legislation and policy provide for some sort of 'shutter control,'" which correctly classifies all national security exceptions to nondiscriminatory access as forms of shutter control. Gabrynowicz, *Land Remote Sensing*, 13. This is a classification made on the outcome. The distinction made in this paper is at the point of government intervention. The transactional controls are made at the data level, whereas shutter control, though including distribution limitations, also gives a state authority over collection activities as well This division is meant to be explicatory, and it hardly represents two distinct categories.
123 Gabrynowicz, *Land Remote Sensing*, 3.
124 Schmidt-Tedd & Kroyman, "Current Status," 109.
125 *Id.* This sensitivity check takes into account all aspects of the transfer including the identity of the customer. *Id.*
126 Achilleas, "French," 3. According to Achilleas, though France lacks a specific remote sensing law "governmental control is is imposed on the SPOT Image commercial policy." *Id.* at 5. The state can restrict: "data when hostile entities might use data representing protected and sensitive French areas . . ., location of French troops abroad . . ., or the location of allied troops abroad." *Id.* at 6.
127 Yun Zhao, "Regulation of Remote Sensing Activities in Hong Kong: Privacy, Access, Security, Copyright and the Case of Google," *Journal of Space Law*, v. 36 (2010) 547, 563. Kuriyama delineates between a U.S. approach to data distribution and a Chinese approach. Kuriyama, "Environmental Monitoring," 569.
128 Gabrynowicz, *Land Remote Sensing*, 11–12 and Lee, "Remote Sensing," 431–432; Aoki, "Japanese," 338.
129 Gillon, "Regulating," 29–30; Mann, "First License," 80–81.
130 15 C.F.R. 960 (2013).
131 Gillon, "Regulating," 25. Mann adds three additional "public interest" justifications for the law: "the environment," "public health," and "safety of persons and property." Mann, "First License," 69.

132 Gillon, "Regulating," 29–30. Shutter control restrictions in Canada can apply to spatial, temporal, or resolution elements of the remote sensing activities. *Id.* at 30. The regime also gives the government priority access to data in "cases of emergency response . . ., in support of requests for aid of a civil power, or in support of Canadian Forces," which moves such requests "to the front of the order queue." *Id.* at 30–31; Mann, "First License," 81.

133 15 C.F.R. 960.11 (2013).

134 Gabrynowicz notes that "[t]he distinction between 'public' and 'private' in the remote sensing space segment is disappearing worldwide. What constitutes 'commercial' operations varies among nations." Gabrynowicz, *Land Remote Sensing*, 3, 15–16. While these definitional elements are important, the purpose here is to acknowledge the general trend as opposed to engaging with the competing definitions. In the context of this paper, commercialization represents the processes along this spectrum that result in an increased number of private, nongovernmental actors in the field. Additionally, Gabrynowicz notes that there is a rising trend of new government entities "organized like private corporations." *Id.* at 16–18. See generally, Jackson, "Cultural Lag" (arguing that the UN Principles Regime is inadequate to cope with increased commercialization). For an early view of the of the remote sensing industry see generally, John Graham and Joanne Irene Gabrynowicz, *The Remote Sensing Industry: A CEO Forum* (University, MS: National Remote Sensing and Space Law Center 2002).

135 Gillon, "Regulating," 19.

136 Schmidt-Tedd and Kroyman, "Current Status," 100–102.

137 Graham and Gabrynowicz, *Landsat-7*, 291.

138 Jackson, "Cultural Lag," 859.

139 Gabrynowicz, *Land Remote Sensing*, 6.

140 John, "Legal Regime," 261.

141 Gabrynowicz, *Land Remote Sensing*, 5–6; Gabrynowicz, "The Perils," 50–59; and, generally, Graham and Gabrynowicz, *Landsat-7*.

142 This author ascribes to the view that globalization is a process that can retract and recede. For a fuller account of different theories of globalization and its meaning see generally, Yale H. Ferguson and Richard W. Mansbach, *Globalization: The Return of Borders to a Borderless World* (London: Routledge 2012).

143 Ferguson and Mansbach note that one of the features of globalization is the non-exclusiveness of "territoriality" and note that outer space is one of the environments that have allowed territoriality to be reimagined as nonexclusive. *Id.* at 21. For a more in depth argument that the space age is an important historical factor in modern globalization see P.J. Blount and Jake Fussell, "Musical Counter Narratives: Space, Skepticism, and Religion in American Music," AIAA 52nd Aerospace Sciences Meeting (January 2014).

144 David Held referenced in Ferguson and Mansbach, *Globalization*, 17.

145 *Id.*

146 Ferreira and Camara, "Current Status," 11.

147 Ferguson and Mansbach, *Globalization*, 22, 167 ("Globalization advocates also reject critics' concerns about modern technology. . . . More information creates an informed citizenry. . . . And, although a digital divide exists new technologies are reaching more and more people and accelerating economic development in poor countries.") In relation to space technologies see generally Joanne Irene Gabrynowicz, "The International Space Treaty Regime in the Era of Globalization." *Ad Astra* (Fall 2005) 30.

148 Oprong and Rwehumbiza, "A Glance," 255; Gabrynowicz, *Land Remote Sensing*, 7; Jahku, "International Law," 68–69.

149 For example, Gabrynowicz argues that Landsat is a "national program with an inherently global function." Gabrynowicz, "The Perils," 47.

150 The is a result of the network design of the internet. David C. Clark and Susan Landau, "Untangling Attribution," in Committee on Deterring Cyberattacks, *Proceedings of a Workshop on Deterring Cyberattacks: Informing Strategies and Developing Options for U.S. Policy* (Washington, DC: National Academies Press 2010) 25, 27. This design is critical to the functionality of the internet as we know it.

151 Ferguson and Mansbach, *Globalization*, 112.

152 Location is identity. Clark and Landau, "Untangling Attribution," 25.

153 Ferguson and Mansbach, *Globalization*, 112.

154 Gabrynowicz, *Land Remote Sensing*, 14. As an example of convergence specific to remote sensing, India has seen moves to regulate "web-based image suppliers." *Id.* 21.

155 Gabrynowicz, *Land Remote Sensing*, 11.
156 Ferguson and Mansbach, *Globalization*, 134.
157 Ferreira and Camara link the internet to change in how remote sensing data is used. Ferreira and Camara, "Current Status," 17.
158 Frans G. von der Dunk, "Outer Space Law Principles and Privacy," in Ray Purdy and Denise Leung, eds., *Evidence from Earth Observation Satellites* (Leiden: Martinus Nijhoff Publishers 2013) 243, 257.
159 George Cho, "Privacy and EO: An Overview of Legal Issues," in Ray Purdy and Denise Leung, eds., *Evidence from Earth Observation Satellites* (Leiden: Martinus Nijhoff Publishers 2013) 259, 261. Despite the concept's lack of definition, Cho does note that privacy is included as a fundamental human right in the Universal Declaration of Human Rights. *Id.* 263.
160 Petras, "Eyes," 81. Petras notes specifically that 9/11 caused the United States national security regime to turn inward. *Id.* 82–83.
161 *Id.* 81. The securitization phenomenon will be addressed in the American context since it is locus of this trend.
162 *Id.* 94.
163 *Id.* 102.
164 While this is a value, it should be noted it is not a constitutional value, though it reflects core American Values. Congress has the ability to waive the prohibition in certain circumstances, which in a securitization context means that political will can shift this value. See Petras, "Eyes," 104–105 (noting the Congressional waiver for military participation in anti-drug trafficking activities near U.S. Borders).
165 Petras, "Eyes," 110–113. For a common line of cases addressing the Posse Commitatus Act, see U.S. v. Yunis, 924 F.2d 1086 (1991); Wrynn v. U.S., 200 F.Supp. 457 (1961); U.S. v. Red Feather, 392 F.Supp. (1975); Bissonette v. Haig, 776 F.2d 1384 (1985); U.S. v. Rasheed, 802 F.Supp. 312 (1992); and U.S. v. Kahn, 35 F.3d 426 (1994).
166 Petras, "Eyes," 97.
167 *Id.* 97–98.
168 *Id.* 100, 108–109.
169 *Id.* 100–101.
170 *Id.* 101. Domestic activities are those "that take place within the the United States that do not involve a significant connection with a foreign power, organization, or person." *Id.* Petras specifically argues that international terrorist organizations operations within the borders of the United States would sit outside this definition. *Id.* 110.
171 Surya Gablin Gunasekara, "The March of Science: Fourth AMendment Implications on Remote Sensing in Criminal Law," *Journal of Space Law*, v. 36 (2010) 115.
172 U.S. Constitution. 4th Amendment.
173 Gunasekara, "The March," 118–119.
174 Gunasekara identifies *Ciraolo, Dow Chemical*, and *Riley*. Gunasekara, "The March," 120–124.
175 Petras, "Eyes," 99.
176 Gunasekara, "The March," 124–126.
177 *Id.*
178 *Id.* 135–136; Petras, "Eyes," 99.
179 U.S. v. Jones, No. 10–1259, concurring Opinion of Justice Sotomayor (Jan. 23, 2012).
180 Lee, "Remote Sensing," 425.
181 Paul M. Secunda, "A Mosquito in the Ointment: Adverse HIPPAA Implications for Health-Related Remote Sensing Research and a 'Reasonable' Solution," *Journal of Space Law*, v. 30 (2004) 251.
182 Zhao, "Regulation," 551.
183 Mann, "First License," 85–87. Mann argues that in Canada "[p]rivacy rights are adequately protected." *Id.* 87. It should also be noted that the Canadian Supreme Court in a case very similar to the United States' *Kyllo* case addressed earlier, split from the US rational. *Id.* 86–87; Gunasekara, "The March," 130–134.
184 "Google faces lawsuits over Gmail, Street View privacy," *CBC News*, Oct. 2, 2013, www.cbc.ca/news/business/google-faces-lawsuits-over-gmail-street-view-privacy-1.1876594; Scott Foresman, "Google Faces $50 Million Lawsuit over Android Location Tracking," *Ars Technica*, April 30, 2011, http://arstechnica.com/tech-policy/2011/04/google-faces-50-million-lawsuit-over-android-location-tracking/.
185 For example, Sam Frizell, "Tinder Security Flaw Exposed Users' Locations," *Time*, Feb. 19, 2014, http://techland.time.com/2014/02/19/tinder-app-user-location-security-flaw/; Brian X. Chen, "iPhone Tracks

Your Every Move, and There's a Map for That," *Wired*, Apr. 20, 2011, www.wired.com/gadgetlab/2011/04/iphone-tracks/.

186 "General Data Protection Regulation," EU Regulation 2016/679 § (2016).

187 Laura Keogh, "EU Data Protection Considerations for the Space Sector," in P.J. Blount and Mahulena Hofmann, eds., *Space Law in a Networked World* (Brill 2023) 230–255.

188 For analysis of these rules in different jurisdictions see, England and Wales, Germany, The Netherlands, Belgium, Sa'id Mosteshar, "EO in the European Union: Legal Considerations," in Ray Purdy and Denise Leung, eds., *Evidence from Earth Observation Satellites* (Leiden: Martinus Nijhoff Publishers 2013) 147–175; Nigeria, John, "Legal Regime," 262–263; Singapore, Geradine Goh Escolar, "The Use of EO Data as Evidence in the Courts of Singapore," in Ray Purdy and Denise Leung, eds., *Evidence from Earth Observation Satellites* (Leiden: Martinus Nijhoff Publishers 2013) 93, 98–108; United States, Kris Dighe et al., "The Use of Satellite Imagery in Environmental Crimes Prosecutions in the United States: A Developing Area," in Ray Purdy and Denise Leung, eds., *Evidence from Earth Observation Satellites* (Leiden: Martinus Nijhoff Publishers 2013) 65, 81–90 and Merideth Wright, "The Use of Remote Sensing Evidence at Trial in the United States—One State Court Judge's Observations," in Ray Purdy and Denise Leung, eds., *Evidence from Earth Observation Satellites* (Leiden: Martinus Nijhoff Publishers 2013) 313–320.

189 Alan Shipman, "Authentification of Images," in Ray Purdy and Denise Leung, eds., *Evidence from Earth Observation Satellites* (Leiden: Martinus Nijhoff Publishers 2013) 359–377.

190 Willibald Croi, Frederic-Michael Foeteler, and Harold Linke, "Introducing Digital Signatures and Time-Stamps in the EO Data Processing Chain," in Ray Purdy and Denise Leung, eds., *Evidence from Earth Observation Satellites* (Leiden: Martinus Nijhoff Publishers 2013) 379–398.

191 Ronald J. Rychlak, Joanne Irene Gabrynowicz, and Rick Crowsey, "Legal Certification of Digital Data: The Earth Resources Observation and Science Data Center Project," *Journal of Space Law*, v. 33 (2007) 195, 212–218.

192 Lee, "Remote Sensing," 423–424.

193 Bruce Goulevitch, "Ten Years of Using Earth Observation Data in Support of Queensland's Vegetation Management Framework," in Ray Purdy and Denise Leung, eds., *Evidence from Earth Observation Satellites* (Leiden: Martinus Nijhoff Publishers 2013) 113.

194 Escolar, "Use of EO," 108–110.

195 Dighe et al., "The Use," 65; Rychlak et al., "Legal Certification," 206–210.

196 See generally Maureen Williams, "Satellite Evidence in International Institutions," in Ray Purdy and Denise Leung, eds., *Evidence from Earth Observation Satellites* (Leiden: Martinus Nijhoff Publishers 2013) 195–216.

197 See generally Froehlich, "Space Related," 222–225.

198 PCA, Eya David Macauley, "The Use of EO Technologies in Court by the Office of the Prosecutor of the International Criminal Court," in Ray Purdy and Denise Leung, eds., *Evidence from Earth Observation Satellites* (Leiden: Martinus Nijhoff Publishers 2013) 217, 224–225; ICC, *Id.* at 226–238.

199 For example, Liz Gannes, "GPS App Strava Sued Over Cyclist's Death," All Things D, June 19, 2012, http://allthingsd.com/20120619/gps-app-strava-sued-over-cyclists-death/; Adam Hadhazy, "Bad Directions from Google Maps Lead to Lawsuit," *Tech News Daily*, June 3, 2010, www.technewsdaily.com/556-bad-directions-from-google-maps-lead-to-lawsuit.html.

200 Brocklesby v. U.S. 767 F.2d 1288 (9th Cir. 1985).

201 John, "Legal Regime," 263.

202 Ito, "Improvement," 57.

203 *Id.* 50.

204 *Id.* 55. For an overview of copyright implications for EO data see Doldirina, "Impact," 293–310. For an example of such a regime see Zhao on Hong Kong, Zhao, "Regulation," 552–554.

205 Doldirina, "Impact," 298. Cromer presents a full analysis of the conflicts between the copyright regime and the space law regime. Julie D. Cromer, "How on Earth Terrestrial Laws Can Protect Geospatial Data," *Journal if Space Law*, v. 32 (2006) 253.

206 Escolar, "Use of EO," 111–112.

207 Harris, "Current Status," 37–38.

208 *Id.* 38.

209 John, "Legal Regime," 262. See also Tare Birisbe, "Outer Space Activities and Intellectual Property Protection in Nigeria," *Journal of Space Law*, v. 32 (2006) 229, 241–245.

210 Gabrynowicz, *Land Remote Sensing*, 18.
211 Lee, "Remote Sensing," 426–428.
212 Martha Mejia-Kaiser, "Copyright Claims for Meteosat and Landsat Images Under Court Challenge," *Journal of Space Law*, v. 32 (2006) 293.

## REFERENCES


Achilleas, Philippe, "French Remote Sensing Law," *The Journal of Space Law*, v. 34 (2008) 1.

Agreement on the Rescue of Astronauts, the Return of Astronauts and the Return of Objects Launched into Outer Space (the "Rescue Agreement") (entered into force on 3 December 1968).

Alvarado, Sandra Cabrera, "The Analysis of existing International Space Cooperation Initiatives for UNESCO's World Heritage Sites," IAC-10.E3.1A.14 (2010).

Aoki, Setsuko, "Japanese Law and Regulations Concerning Remote Sensing Activities," *The Journal of Space Law*, v. 36 (2010) 335.

Beets, Josie, "The International Charter on Space and Major Disasters and International Disaster Law: The Need for Collaboration and Coordination," *Air & Space Lawyer*, v. 22/4 (2010) 12–15.

Birisbe, Tare, "Outer Space Activities and Intellectual Poperty Protection in Nigeria," *The Journal of Space Law*, v. 32 (2006) 229.

Bissonette v. Haig, 776 F.2d 1384 (1985).

Blount, P.J., "Developments in Space Security and Their Legal Implications," *Law /Technology*, 44/2 (2011) 18.

Blount, P.J., "Limits on Space Weapons: Incorporating the Law of War into the Corpus Juris Spatialis," in *Proceedings of the 51 st Colloquium on the Law of Outer Space* (AIAA 2009)

Blount, P.J. and Jake Fussell, "Musical Counter Narratives: Space, Skepticism, and Religion in American Music," AIAA 52nd Aerospace Sciences Meeting (January 2014).

Brocklesby v. U.S. 767 F.2d 1288 (9th Cir. 1985).

Carpanelli, Elena and Brendan Cohen, "A Legal Assessment of the 1996 Declaration on Space Benefits on the Occasion of Its Fifteenth Anniversary," *The Journal of Space Law*, v. 38 (2012) 1.

Carpanelli, Elena and Melissa K. Force, "The Protection of the Earth Natural Environment *through* Space Activities: A General Overview of Some Legal Issues," *Proceedings of the International Institute of Space Law* (The Hague: Eleven 2011) 31.

Carter, Jimmy, "Remarks at the Congressional Space Medal of Honor Awards Ceremony, Kennedy Space Center, FL," *Weekly Compilation of Presidential Documents*, v. 14/40 (Oct. 8, 1978) 1684.

CBERS Data Policy, *The Journal of Space Law*, v. 31 (2005) 281.

Charter on Cooperation to Achieve the Coordinated Use of Space Facilities in the Event of Natural or Technological Disasters Rev. 3, Apr. 25, 2000, www.disasterscharter.org/web/charter/charter.

Chen, Brian X., "iPhone Tracks Your Every Move, and There's a Map for That," *WIred*, Apr. 20, 2011, www.wired.com/gadgetlab/2011/04/iphone-tracks/.

"Chinese Law: Registration, Launching and Licensing Space Objects," *Journal of Space Law*, v. 33 (2007) 437.

Cho, George, "Privacy and EO: An Overview of Legal Issues," in Ray Purdy and Denise Leung, eds., *Evidence from Earth Observation Satellites* (Leiden: Martinus Nijhoff Publishers 2013) 259.

Clark, David C. and Susan Landau, "Untangling Attribution," in Committee on Deterring Cyberattacks, *Proceedings of a Workshop on Deterring Cyberattacks: Informing Strategies and Developing Options for U.S. Policy* (Washington, DC: National Academies Press 2010) 25.

Convention on International Liability for Damage Caused by Space Objects (the "Liability Convention") (entered into force on 1 September 1972).

Convention on Registration of Objects Launched into Outer Space (the "Registration Convention") (entered into force on 15 September 1976).

Croi, Willibald, Frederic-Michael Foeteler, and Harold Linke, "Introducing Digital Signatures and Time-Stamps in the EO Data Processing Chain," in Ray Purdy and Denise Leung, eds., *Evidence from Earth Observation Satellites* (Leiden: Martinus Nijhoff Publishers 2013) 379.

Cromer, Julie D., "How on Earth Terrestrial Laws Can Protect Geospatial Data," *The Journal of Space Law,* v. 32 (2006) 253.

Crook, Jason, "Corporate Sovereign Symbiosis: *Wilson v. ImageSat International*, Shareholders' Actions, and the Dualistic Nature of State-Owned Corporations," *The Journal of Space Law*, v. 33 (2007) 411.

Dembling, Paul G. and Daniel M. Arons, "The Evolution of the Outer Space Treaty," *Journal of Air Law & Commerce*, v. 33 (1967) 419–456.

Diederiks-Verschoor, I.H.P., *An Introduction to Space Law* (The Hague: Kluwer 1999).

Dighe, Kris, Todd Mikolop, Raymond W. Mushal, and David O'Connell, "The Use of Satellite Imagery in Environmental Crimes Prosecutions in the United States: A Developing Area," in Ray Purdy and Denise Leung, eds., *Evidence from Earth Observation Satellites* (Leiden: Martinus Nijhoff Publishers 2013) 65.

Doldirina, Catherine, "The Impact of Copyright Protection and Public Sector Information Regulations on the Availability of Remote Sensing Data," in Ray Purdy and Denise Leung, eds., *Evidence from Earth Observation Satellites* (Leiden: Martinus Nijhoff Publishers 2013) 293.

Escolar, Geradine Goh, "The Use of EO Data as Evidence in the Courts of Singapore," in Ray Purdy and Denise Leung, eds., *Evidence from Earth Observation Satellites* (Leiden: Martinus Nijhoff Publishers 2013) 93,

EU Directive 2003/4/EC, On public access to environmental information and repealing Council Directive 90/313/EEC (Jan. 28, 2003).

Eyes on Darfur, www.eyesondarfur.org/ (last visited Feb. 22, 2014).

Ferguson, Yale H. and Richard W. Mansbach, *Globalization: The Return of Borders to a Borderless World ?* (London: Routledge 2012).

Ferreira, Hilcea Santos and Gilberto Camara, "Current Status and Recent Developments in Brazilian Remote Sensing Law," *The Journal of Space Law*, v. 34 (2008) 11.

Filho, Jose Monserrat and Alvaro Frabrico dos Santos, "Chinese-Brazilian Accord on Distribution of CBERS Products," *The Journal of Space Law*, v. 31 (2005) 271.

Foresman, Scott, "Google Faces $50 Million Lawsuit over Android Location Tracking," *Ars Technica*, Apr. 30, 2011, http://arstechnica.com/tech-policy/2011/04/google-faces-50-million-lawsuit-over-android-location-tracking/.

Frizell, Sam, "Tinder Security Flaw Exposed Users' Locations," *Time*, Feb. 19, 2014, http://techland.time.com/2014/02/19/tinder-app-user-location-security-flaw/.

Froehlich, Annette, "Space Related Data: From Justice to Development," *Proceedings of the International Institute of Space Law* (The Hague: Eleven 2011) 221.

Gabrynowicz, Joanne Irene, "Defining Data Availability for Commercial Remote Sensing Systems Under United States Federal Law," *Annals of Air and Space Law*, v. 23 (1999) 93.

Gabrynowicz, Joanne Irene, "Earth Observation: The View from the Ground Up," *Space Policy* (Aug. 1997) 229.

Gabrynowicz, Joanne Irene, "The International Space Treaty Regime in the Era of Globalization." *Ad Astra* (Fall 2005) 30.

Gabrynowicz, Joanne Irene, *The Land Remote Sensing Laws and Policies of National Governments: A Global Survey* (University, MS: National Center for Remote Sensing, Air, and Space Law 2007).

Gabrynowicz, Joanne Irene, "The Perils of *Landsat* from Grassroots to Globalization: A Comprehensive Review of US Remote Sensing Law with a Few Thoughts for the Future," *Chicago Journal of International Law*, v. 6/1 (2005) 45.

Gabrynowicz, Joanne Irene, "Space Law: Its Cold War Origins and Challenges in the Era of Globalization," *Suffolk University Law Review*, v. 37 (2004) 1043.

Gabrynowicz, Joanne Irene, ed., *The UN Principles Relating to Remote Sensing of the Earth from Space: A Legislative History—Interviews of Members of the United States Delegation* (University, MS: National Center for Remote Sensing and Space Law 2002).

Gannes, Liz, "GPS App Strava Sued Over Cyclist's Death," *All Things D*, June 19, 2012, http://allthingsd.com/20120619/gps-app-strava-sued-over-cyclists-death/.

Gillon, Thomas, "Regulating Remote Sensing Space Systems in Canada—New Legislation for a New Era," *Journal of Space Law*, v. 34 (2008) 19.

"Google Faces Lawsuits over Gmail, Street View Privacy," *CBC News*, Oct. 2, 2013, www.cbc.ca/news/business/google-faces-lawsuits-over-gmail-street-view-privacy-1.1876594.

Goulevitch, Bruce, "Ten Years of Using Earth Observation Data in Support of Queensland's Vegetation Management Framework," in Ray Purdy and Denise Leung, eds., *Evidence from Earth Observation Satellites* (Leiden: Martinus Nijhoff Publishers 2013) 113.

Graham, John F. and Joanne Irene Gabrynowicz, eds., *Landsat 7: Past Present and Future* (University, MS: National Remote Sensing and Space Law Center 2002).

Graham, John and Joanne Irene Gabrynowicz, *The Remote Sensing Industry: A CEO Forum* (University, MS: National Remote Sensing and Space Law Center 2002).

Gunasekara, Surya Gablin, "The March of Science: Fourth AMendment Implications on Remote Sensing in Criminal Law," *The Journal of Space Law*, v. 36 (2010) 115.

Hadhazy, Adam, "Bad Directions from Google Maps Lead to Lawsuit," *Tech News Daily*, June 3, 2010, www.technewsdaily.com/556-bad-directions-from-google-maps-lead-to-lawsuit.html.

Harris, Ray, "Current Status and Recent Development in UK and European Remote Sensing Law and Policy," *The Journal of Space Law*, v. 34 (2008) 33.

Harris, Ray, "Science, Policy, and Evidence in EO," in Ray Purdy and Denise Leung, eds., *Evidence from Earth Observation Satellites* (Leiden: Martinus Nijhoff Publishers 2013) 43.

Harrison, Roger G., *Space and Verification, Vol. 1 : Policy Implications* (Colorado Springs, CO: Eisenhower Center for Space and Defense Studies).

Harvey, David, *A Brief History of Neoliberalism* (Oxford: Oxford University Press 2005).

International Law Commission, *Draft Articles on Responsibility of States for Internationally Wrongful Acts, with Commentaries* (2001).

Ito, Atsuyo, "Improvement to the Legal Regime for the Effective Use of Satellite Remote Sensing Data for Disaster Management and Protection of the Environment," *Journal of Space Law*, v. 34 (2008) 45.

Jackson, Susan M., "Cultural Lag and the International Law of Remote Sensing," *Brooklyn Journal of International Law*, v. 23 (1997) 853.

Jahku, Ram, "International Law Governing the Acquisition and Dissemination of Satellite Imagery," *Journal of Space Law*, v. 29 (2003) 65.

John, Olusoji Nester, Eguaroje Ezekiel, and S.O. Mohammed, "Legal Regime of Remote Sensing and Geographic Information Systems in Nigeria," in *Proceedings of the International Institute of Space Law* (The Hague: Eleven 2011) 260.

Kopal, Vladimir, "The Role of United Nations Declarations of Principles in the Progressive Development of Space Law," *The Journal of Space Law*, v. 16 (1988) 5.

Kuriyama, Ikuko, "Environmental Monitoring Cooperation Paves the Way for Common Rules on Remote Sensing," *The Journal of Space Law*, v. 36 (2010) 567.

Lachs, Manfred, *The Law of Outer Space: An Experience in Contemporary Law-making* (Leiden: Martinus Nijhoff 1972, 2010).

Lee, Jae Gon, "Remote Sensing Issues as They Relate to Korea," *The Journal of Space Law*, v. 36 (2010) 415.

Licensing of Private Remote Sensing Systems, 15 C.F.R. 960 (2013).

Ling, Yan, "Remote Sensing Data Distribution and Application in the Environmental Protection, Disaster Prevention, and Urban Planning in China," *The Journal of Space Law*, v. 36 (2010) 435.

Listner, Michael, "Iridium 33 and Cosmos 2251 Three Years Later: Where Are We Now?," *The Space Review*, Feb. 13, 2012, www.thespacereview.com/article/2023/1.

Lyall, Francis and Paul B. Larsen, *Space Law: A Treatise* (Aldershot, UK: Ashgate 2013).

Macauley, Eya David, "The Use of EO Technologies in Court by the Office of the Prosecutor of the International Criminal Court," in Ray Purdy and Denise Leung, eds., *Evidence from Earth Observation Satellites* (Leiden: Martinus Nijhoff Publishers 2013) 217.

Mann, Bruce, "First License Issued Under Canada's Remote Sensing Satellite Legislation," *The Journal of Space Law*, v. 34 (2008) 67.

Marchisio, Sergio, "The Evolutionary Stages of the Legal Subcommittee of the United Nations Committee on the Peaceful Uses of Outer Space (COPUOS)," *The Journal of Space Law*, v. 31 (2005) 219.

Mayence, Jean-Francois, "Belgian Legal Framework for Earth Observation Activities," *The Journal of Space Law*, v. 34 (2008) 89.

Mejia-Kaiser, Martha, "Copyright Claims for Meteosat and Landsat Images Under Court Challenge," *The Journal of Space Law*, v. 32 (2006) 293.

Mosteshar, Sa'id, "EO in the European Union: Legal Considerations," in Ray Purdy and Denise Leung, eds., *Evidence from Earth Observation Satellites* (Leiden: Martinus Nijhoff Publishers 2013) 147.

NASA Landsat Science, "The Numbers Behind Landsat," http://landsat.gsfc.nasa.gov/?page_id=9 (last visited Feb. 22, 2014).

Onada, Masami, "Satellite Earth Observation as 'Systematic Observation' in Multilateral Environmental Treaties," *The Journal of Space Law*, v. 31 (2005) 339.

Oprong, Angeline Asangire and Vincent Rwehumbiza, "A Glance at the Earth Observation Policies and Regulations and the impact on Developing Countries: Focusing on the African Continent," *Proceedings of the International Institute of Space Law* (2011) 251.

Petras, Christopher M., "'Eyes' On Freedom—A View of the Law Governing Military Uses of Satellite Reconnaissance in U.S. Homeland Defense," *The Journal of Space Law*, v. 31 (2005) 81
Principles Relating to Remote Sensing of the Earth from Outer Space, U.N.G.A. Res. 41/65 (Dec. 3, 1986).
Remote Sensing Space Systems Act (S.C. 2005, c. 45).
Reynolds, Glenn H. and Robert P. Merges, *Outer Space: Problems of Law and Policy* (Boulder, CO: Westview Press 1997).
Richelson, Jeffrey T., *America's Space Sentinels: DSP Satellites and National Security* (Lawrence, KA: University Press of Kansas 1999).
Rychlak, Ronald J., Joanne Irene Gabrynowicz, and Rick Crowsey, "Legal Certification of Digital Data: The Earth Resources Observation and Science Data Center Project," *The Journal of Space Law*, v. 33 (2007) 195.
"Satellite Data Security Act," *Journal of Space Law*, v. 34 (2008) 115.
Satellite Sentinel Project, www.satsentinel.org/ (last visited Feb. 22, 2014).
Schmidt-Tedd, Bernhard and Max Kroyman, "Current Status and Recent Developments in German Remote Sensing Law," *The Journal of Space Law*, v. 34 (2008) 97.
Schrogl, Kai-Uwe and Niklas Hedman, "The U.N. General Assembly Resolution 62/102 of 17 December 2007 on 'Recommendations on Enhancing the Practice of States and International Intergovernmental Organizations in Registering Space Objects,'" *The Journal of Space Law*, v. 34 (2008) 141.
Secunda, Paul M., "A Mosquito in the Ointment: Adverse HIPPAA Implications for Health-Related Remote Sensing Research and a 'Reasonable' Solution," *The Journal of Space Law*, v. 30 (2004) 251.
Sekhula, Phetole Patrick, "The Right to Remote Sense Data: Impact of Multilateral Cooperation on International Space Law [sic]," *Proceedings of the International Institute of Space Law* (The Hague: Eleven 2011) 228.
Sentinel Asia, "Sentinel Asia: Disaster Management Support System in the Asia-Pacific Region," www.aprsaf.org/initiatives/sentinel_asia/pdf/Sentinel-Asia.pdf (last visited Feb. 22, 2014).
Shipman, Alan, "Authentification of Images," in Ray Purdy and Denise Leung, eds., *Evidence from Earth Observation Satellites* (Leiden: Martinus Nijhoff Publishers 2013) 359.
Tarikhi, Parviz, "Mahdasht Satellite Receiving Station Verging into a Space Center," *Res Communis*, Oct. 13, 2008, http://rescommunis.olemiss.edu/2008/10/13/guest-blogger-parviz-tarikhi-mahdasht-satellite-receiving-station-verging-into-a-space-center/.
Terekov, Andrei, "UN General Assembly Resolutions and Outer Space Law," *Proceedings of the International Institute of Space Law* (1997) 97
Treaty between the United States of America and the Russian Federation on Measures for the Further Reduction and Limitation of Strategic Offensive Arms (signed April 10, 2010).
Treaty on Principles Governing the Activities of States in the Exploration and Use of Outer Space, including the Moon and Other Celestial Bodies (the "Outer Space Treaty") (entered into force on 10 October 1967).
Uhlir, Paul F. and Robert S. Chen, Joanne Irene Gabrynowicz, and Kathleen Janssen, "Toward Implementation of the Global Earth Observation System of Systems Data Sharing Principles," *The Journal of Space Law*, v. 35 (2009) 201.
UN Charter (1945).
UNGA Meeting Record no. A/41/PV.95 (Dec. 3, 1986).
UNGA Res. 51/122, The Declaration on International Cooperation in the Exploration and Use of Outer Space for the Benefit and in the Interest of All States, Taking into Particular Account the Needs of Developing Countries (Dec. 13, 1996).
UNGA Res. 60/1, 2005 World Summit Outcome (Oct. 24, 2005).
UNGA Res. 62/102, Recommendations on Enhancing the Practice of States and International Intergovernmental Organizations in Registering Space Objects (Dec. 17, 2007).
United Nations High Commission on Human Rights, *Report of the Commission of Inquiry on Human Rights in the Democratic People's Republic of Korea*, A/HRC/25/63 (Feb. 17, 2014).
UNOOSA, "About UN-SPIDER," www.unoosa.org/oosa/en/unspider/index.html (last visited Feb. 22, 2014).
U.S. Constitution.
U.S. v. Jones, No. 10–1259, concurring Opinion of Justice Sotomayor (Jan. 23, 2012).
U.S. v. Kahn, 35 F.3d 426 (1994).
U.S. v. Rasheed, 802 F.Supp. 312 (1992).
U.S. v. Red Feather, 392 F.Supp. (1975)
U.S. v. Yunis, 924 F.2d 1086 (1991).

Vásquez, Fermin Romero and Sergio Camacho Lara, "What Lawyers Need to Know about Science to Effectively Make and Address Laws for Remote Sensing and Environmental Monitoring," *The Journal of Space Law*, v. 36 (2010) 365.

Vásquez, M. Seara, *Cosmic International Law* (Detroit: Wayne State University Press 1965).

Vereschetin, Vladen S. and Gennady M. Danilenko, "Custom as a Source of International Law of Outer Space," *The Journal of Space Law*, v. 13 (1985) 22.

von der Dunk, Frans G., "Outer Space Law Principles and Privacy," in Ray Purdy and Denise Leung, eds., *Evidence from Earth Observation Satellites* (Leiden: Martinus Nijhoff Publishers 2013) 243.

Williams, Maureen, "Satellite Evidence in International Institutions," in Ray Purdy and Denise Leung, eds., *Evidence from Earth Observation Satellites* (Leiden: Martinus Nijhoff Publishers 2013) 195.

Wright, Merideth, "The Use of Remote Sensing Evidence at Trial in the United States—One State Court Judge's Observations," in Ray Purdy and Denise Leung, eds., *Evidence from Earth Observation Satellites* (Leiden: Martinus Nijhoff Publishers 2013) 313.

Wrynn v. U.S., 200 F.Supp. 457 (1961).

Zhao, Yun, "Regulation of Remote Sensing Activities in Hong Kong: Privacy, Access, Security, Copyright and the Case of Google," *The Journal of Space Law* 36 (2010) 547.

# Part XI

## Summary and Synthesis of Volume I

# 18 Summary Chapter of Volume I

## *Sensors, Data Normalization, Harmonization, Cloud Computing, and Accuracies*

*Prasad S. Thenkabail*

## ACRONYMS AND DEFINITIONS

| | |
|---|---|
| **ACD** | Aboveground carbon density |
| **AGB** | Aboveground biomass |
| **AgRISTARS** | Agriculture and resources inventory surveys through aerospace remote sensing |
| **AIRS** | Atmospheric Infrared Sounder |
| **ALEXI** | Atmosphere-Land Exchange Inverse |
| **ALI** | Advanced Land Imager |
| **ALOS** | Advanced Land Observing Satellite |
| **AMSR-E** | Advanced Microwave Scanning Radiometer |
| **ANN** | Artificial Neural Networks |
| **AOD** | Aerosol optical depth |
| **ALS** | Airborne laser scanning |
| **APAR** | Absorbed photosynthetically active radiation |
| **ASAR** | Advanced synthetic aperture radar onboard ENVISAT |
| **ASIS** | Agricultural Stress Index Systems |
| **ASPRS** | American Society for Photogrammetry and Remote Sensing |
| **ASS** | Aerial spectral sensing |
| **ATSR** | Along-track scanning radiometer |
| **AVHRR** | Advanced very-high-resolution radiometer |
| **ASTER** | Advanced spaceborne thermal emission and reflection radiometer |
| **CAI** | Cellulose absorption indices |
| **CDL** | Cropland data layer |
| **CDMA** | Code Division Multiple Access |
| **CEOS** | Committee on Earth Observation Satellites |
| **CHM** | Canopy height model |
| **CHRIS** | Compact High Resolution Imaging Spectrometer |
| **COSMO-SkyMed** | Constellation of Small Satellites for Mediterranean basin Observation (COSMO)-SkyMed |
| **COSMO-SkyMed** | COnstellation of small Satellites for the Mediterranean basin Observation |
| **CPW** | Crop water productivity |
| **CNES** | The Centre national d'études spatiales or the National Center of Space Studies of France (CNES) CryoSat Europe's Satellite to Study Ice DEM Digital elevation model |

DOI: 10.1201/9781003541141-29

| | |
|---|---|
| **CYGNSS** | Cyclone Global Navigation Satellite System |
| **DEM** | Digital Elevation Model |
| **DESDynI** | Deformation, Ecosystem Structure and Dynamics of Ice |
| **DisAlexi** | Disaggregated ALEXI |
| **DLR** | German Aerospace Center |
| **DMSP** | Defense Meteorological Satellite Program |
| **DSS** | Decision support systems |
| **DTM** | Digital terrain model |
| **EO** | Earth Observation |
| **EOS** | Earth Observing System onboard Aqua satellite |
| **ET** | Evapotranspiration |
| **ENVISAT** | Environmental satellite |
| **EO** | Earth observing |
| **ERS** | European remote sensing satellites |
| **ERTS** | Earth Resources Technology Sateelites |
| **ESRI** | Environmental Systems Research Institute |
| **ETM+** | Enhanced Thematic Mapper+ |
| **EVI** | Enhanced Vegetation Index |
| **EnMAP** | Environmental Mapping and Analysis Program |
| **FAO** | Food and Agriculture Organization of the United Nations |
| **FGDC** | Federal Geographic Data Committee |
| **FLUXNET** | A network of micrometeorological tower sites to measure carbon dioxide, water, and energy balance between terrestrial systems and the atmosphere |
| **GAC** | Global area coverage |
| **GEO** | Group on Earth Observation |
| **GEOSS** | Global Earth Observation System of Systems |
| **GEOBIA** | Geographic Object-based Image Analysis |
| **GHG** | Greenhouse emissions |
| **GHI** | Global Health Index |
| **GIEWS** | Global Information and Early Warning System |
| **GIMMS** | Global Inventory Modeling and Mapping Studies |
| **GIS** | Geographic Information System |
| **GLONASS** | Global Orbiting Navigation Satellite System |
| **GNSS** | Global Navigation Satellite Systems |
| **GOES** | Geostationary Operational Environmental Satellite |
| **GEOSS** | Global Earth Observing System of Systems |
| **GLAI** | Green leaf area index |
| **GLAM** | Global Agricultural Monitoring |
| **GLAS** | Geoscience Laser Altimeter System |
| **GNSS** | Global Navigation Satellite System |
| **GPS** | Global Positioning System |
| **GPP** | Gross primary productivity |
| **GPS** | Global positioning systems |
| **GRACE** | Gravity recovery and climate experiment (GRACE) |
| **GSS** | Ground spectral sensing |
| **GLAS** | Geoscience Laser Altimeter System |
| **HNB** | Hyperspectral narrow bands |
| **HVI** | Hyperspectral vegetation indices |
| **HyspIRI** | Hyperspectral Infrared Imager |
| **ICESat** | Instrument aboard the Ice, Cloud, and land Elevation |

| | |
|---|---|
| **IKONOS** | A commercial Earth observation satellite, typically, collecting sub-meter to 5 m data |
| **IRS** | Indian Remote Sensing Satellites |
| **ISODATA** | Iterative Self-organizing Data Analysis Technique |
| **IV** | Inland valleys |
| **JERS** | Japanese Earth Resources Satellite |
| **JRC** | Joint Research Center |
| **LACIE** | Large Area Crop Inventory Experiment |
| **LAI** | Leaf area index |
| **LiDAR** | Light detection and ranging |
| **LSP** | Land surface phenology |
| **LSS** | Laboratory spectral sensing |
| **LST** | Land Surface Temperature |
| **LUC** | Land use classes |
| **LUE** | Light use efficiency |
| **LULC** | Land use, land cover |
| **LULCC** | Land use, land cover change |
| **MARS** | Monitoring Agricultural Resources action of the European Commission European Union |
| **MERIS** | Medium resolution imaging spectrometer (MERIS) |
| **METRIC** | Mapping EvapoTranspiration at high Resolution with Internalized Calibration |
| **METOP** | Meteorological operational satellite program |
| **MIR** | Mid-infrared |
| **MSS** | Multi spectral scanner |
| **MODIS** | Moderate-resolution imaging spectroradiometer |
| **MOD16** | MODIS Global Evapotranspiration Dataset |
| **MLS** | Mobile laser scanning |
| **NASA** | National Aeronautics and Space Administration |
| **NDSI** | Normalized Difference Snow Index |
| **NDT** | INormalized difference tillage index |
| **NDVI** | Normalized difference vegetation index |
| **NESDIS** | National Environmental Satellite, Data, and Information Service |
| **NOAA** | National Aeronautics and Space Administration |
| **NIR** | nearinfrared |
| **NP** | Nonphotosynthetic vegetation |
| **NPOESS** | National Polar-orbiting Operational Environmental Satellite System |
| **NPP** | NPOESS Preparatory Project |
| **NVIR** | Visible and near-infrared |
| **OBIA** | Object Oriented Image Analysis |
| **OLI** | Operational land imager |
| **OLS** | Operational linescan system |
| **OMI** | Ozone Monitoring Instrument |
| **PAR** | Photosynthetically active radiation |
| **PALSAR** | Phased Array type L-band Synthetic Aperture Radar |
| **PDSI** | Palmer Drought Severity Index |
| **PolSAR** | RADARSAT-2 polarimetric SAR () |
| **PRI** | Photochemical reflectance index |
| **PROBA** | Project for On Board Autonomy |
| **PROSAIL** | Combination of PROSPECT and SAIL, the two nondestructive physically based models to measure biophysical and biochemical properties |

| | |
|---|---|
| **PROSPECT** | Radiative transfer model to measure leaf optical properties spectra |
| **PV** | Photosynthetic vegetation |
| **RS** | Remote Sensing |
| **RADAR** | Radio Detection and Ranging |
| **RADARSAT** | Radar satellite |
| **REDD** | Reducing Emissions from Deforestation in Developing Countries |
| **RISAT** | Radar Imaging Satellite |
| **RS** | Remote sensing |
| **SAR** | Synthetic aperture radar |
| **SAIL** | Scattering by arbitrary inclined leaves (SAIL)—a physically based model to measure and model canopy bidirectional reflectance |
| **SAR** | Synthetic aperture radar |
| **SAVI** | Soil adjusted vegetation index |
| **SCD** | Snow-covered days |
| **SCHIAMACHY** | Scanning Imaging Absorption Spectrometer for Atmospheric Cartography |
| **SEASAT** | First satellite designed for remote sensing of the Earth's oceans with synthetic aperture radar (SAR) |
| **SEB** | Surface energy balance |
| **SEBAL** | Surface energy balance algorithm for land |
| **SIR C/X** | Spaceborne imaging radar-C/X |
| **SMMR** | Scanning Multichannel Microwave Radiometer |
| **SMTs** | Spectral Matching Techniques |
| **SODAR** | Sonic Detection and Ranging |
| **SONAR** | Sound Navigation and Ranging |
| **SPI** | Standardized Precipitation Index |
| **SPOT** | Satellite Pour l'Observation de la Terre, French Earth Observing Satellites |
| **SSEB** | Simplified surface energy balance |
| **SSMI/S** | Special sensor microwave imager |
| **SRTM** | Shuttle Radar Topographic Mission |
| **STAARCH** | Spatial temporal adaptive algorithm for mapping reflectance change |
| **STARFM** | Spatial and temporal adaptive reflectance fusion model |
| **SWIR** | Shortwave infrared |
| **SWSI** | Surface Water Supply Index |
| **SRTM** | Shuttle Radar Topographic Mission |
| **SST** | Sea Surface Temperature |
| **SVM** | Support Vector Machines |
| **TCI** | Temperature Condition Index |
| **TerraSAR-X** | A radar Earth observation satellite, with its phased array synthetic aperture radar |
| **TIN** | Triangular Irregular Network |
| **TIR** | Thermal infrared |
| **TLS** | Terrestrial laser scanning |
| **TM** | Thematic mapper |
| **TOA** | Top of Atmosphere |
| **TRMM** | Tropical Rainfall Measuring Mission |
| **TROPOMI** | TROPOspheric monitoring instrument |
| **UAS** | Unmanned Aircraft Systems |
| **UAV** | Unmanned Aerial Vehicles |
| **UAV** | Unmanned aerial vehicle |
| **UNFCCC** | United Nations Framework Convention on Climate Change |
| **USAID** | The United States Agency for International Development |

**USDA** United States Department of Agriculture
**USDM** United States Drought Monitor
**VCI** Vegetation Condition Index
**VegDRI** Vegetation Drought Index
**VGI** Volunteered Geographic Vegetation
**VH** Vegetation health
**VHI** Vegetation Health Index
**VHRI** Very High Resolution Imagery
**VegOut** Vegetation outlook
**VI** Vegetation index
**VIIRS** Visible Infrared Imaging Radiometer Suite
**VIS** Visible
**VUA** Vrije Universiteit Amsterdam
**WF** Water footprint
**WP** Water productivity
**WUM** Water use mapping

This chapter provides a summary of each of the 17 chapters in Volume I of the six-volume *Remote Sensing Handbook* (Second Edition). The topics covered in the chapters of Volume I include (Figure 18.0): (1) satellites and sensors, (2) global navigation satellite systems (GNSS), (3) remote sensing fundamentals, (4) data normalization, harmonization, and standardization, (5) vegetation indices and their within and across sensor calibration, (6) crowdsourcing, (7) cloud computing, (8) Google Earth Engine (GEE) supported remote sensing, (9) accuracy assessments, and (10) remote sensing law. Under each of the preceding broad topics, there are one or more chapters. For example, there are four chapters under data normalization, harmonization, and inter-sensor calibrations. In a nutshell, these chapters provide a complete and comprehensive overview of these critical topics, capture the advances over a century, and provide a vision for

**Chapter 18: Summary Chapter for**
**Remote Sensing Handbook (Second Edition, Six Volumes): Volume I**

**Volume I: Sensors, Data Normalization, Harmonization, Cloud Computing, and Accuracies**

**Chapter 1: Remote sensing satellites and sensors**

**Chapter 2 to 4: GNSS Theory and Practice, GNSS Reflectometry and wide array for Ocean and Land Applications**

**Chapters 5: Remote Sensing Fundamentals and Terrestrial Applications**

**Chapters 6-11: Calibration, Normalization, Standardization, and Harmonization of Remote Sensing Data Within and Across Sensors including Pre-launch and Post-launch calibrations and Spectral Vegetation Indices**

**Chapters 12: Crowdsourcing of Remote Sensing Data**

**Chapters 13 to 15: Cloud Computing Fundamentals and Advances and Google Earth Engine (GEE)**

**Chapter 16: Map accuracies**

**Chapter 17: Remote Sensing Law and Space Law**

FIGURE 18.0 Overview of the chapters in Volume I of the *Remote Sensing Handbook* (Second Edition).

further development in the years ahead. By reading this summary chapter, a reader can have a quick understanding of what is in each of the chapters of Volume I, see how the chapters interconnect and intermingle, and get an overview on the importance of various chapters in developing complete and comprehensive knowledge of the remote sensing Data Characterization, Classification, and Accuracies Assessment. These chapters, together, capture the advances of last 60+ years and provide a vision for the future.

## 18.1 REMOTE SENSING SATELLITES AND SENSORS

We have entered a new era of remote sensing. Today, remote sensing satellites and sensors are ubiquitous. We have sensors acquiring data from ground, air, and space (e.g., satellites, aircraft, uncrewed aircraft systems (UASs), terrestrial systems) (Burleigh et al., 2019, Zhu et al., 2017, Mountford et al., 2017, Toth and Jóźków, 2016). Spatial data from these sensors are used to visualize, query, measure, model, monitor, store, and map a wide array of land, water, and atmospheric features and resources. Sensors continuously and repeatedly collect data in various spectral, spatial, radiometric, and temporal resolutions. Indeed, for many applications, remote sensing has become indispensable and often the most important data source. Its strengths are obvious: repeated coverage of the planet at different scales, ability to collect data in wide array of various resolutions, collection of consistent data without human subjectivity, and ability to collect data from inaccessible locations. However, the complexities of data collection from satellites and sensors are many. They vary due to orbital distances, Sun angles, view angles, spectral bands, band widths, spatial resolutions, sensor platforms, type of sensors (active or passive) (e.g., Figure 18.1a), radiometry, data download links, ground stations, calibration of data collected (Price, 1987), normalization of data collected, geometric registration, radiometric correction, data delivery platforms, and a host of other factors. It is equally important to understand the process involved in design, launch, and operation of satellites and sensors along with the comprehension of the remotely sensed data themselves to study planet Earth and beyond. Especially, given many countries operate their own satellites and sensors and there are many private players involved in satellite design and operation with specific remote sensing jobs in mind.

Chapter 1 by Dr. Sudhanshu S. Panda et al. is focused on providing a comprehensive overview of the wide array of Earth Observation (EO) satellites and sensors of the past, present, and the future operated by various national governments and an increasing number of private entities. The idea of the chapter is to provide readers with a full understanding of the variety of remote sensing data characteristics as well as their sources along with the basics of remote sensing. The chapter provides a progressive development of remote sensing from the early days of airborne remote sensing, mainly developed, perfected, and used during the World Wars I and II, to the more modern hyperspectral, LiDAR, SONAR, and drone/UAS sensors. The early era of remote sensing was mainly for military usage. Subsequently, owing to its strengths in mapping and survey, civilian applications of remote sensing grew mainly through airborne, ground-based, and platform-mounted sensors. With the launch of Sputnik and NOAA-AVHRR, spaceborne remote sensing began its initial steps in late 1950s and early 1960s. With the launch of Earth Resources Technology Satellites (ERTS-1 later renamed Landsat) in the year 1972, remote sensing of the Earth System Science truly evolved from experimental missions to operational missions. More recently or currently, data on planet Earth is routinely gathered by numerous satellite sensors that include hyperspatial resolution (e.g, PlanetScope Doves and Super Doves, GeoEye, IKONOS, Quickbird, Rapideye, etc.), hyperspectral (e.g., EnMAP, DESIS, PRISMA, ECOSTRESS, Gaofen series, etc.), and advanced multispectral moderate to coarse resolution (e.g, Landsat-series, Sentinel-series, Resourcesat-series, MODIS terra/Aqua, AVHRR-series, GOES, NPOESS Preparatory Project or NPP, Radarasat, ALOS PALSAR, JERS SAR, and many others). There are other less frequently collected data from older generation sensors such as the test-of-concept hyperspectral spaceborne sensors (e.g., EO-1 Hyperion, CHRIS PROBA). The preceding satellites are either Sun-synchronous or geostationary and collect data

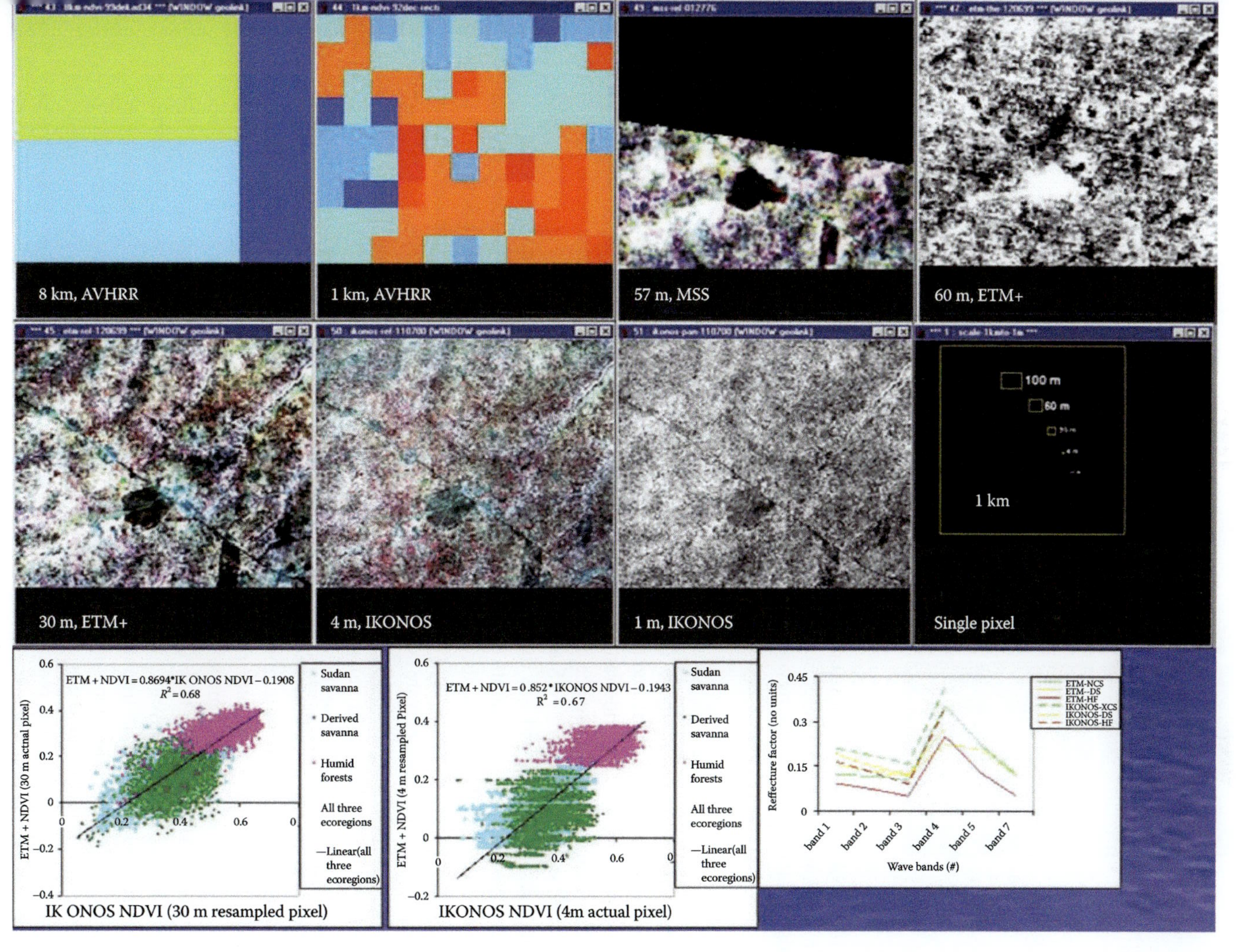


**FIGURE 18.1A** The top two rows of images depict the impact of spatial resolution on information content. An area in NOAA AVHRR 8-km resolution (top left) is geo-linked to images acquired over same areas from various other resolutions: 1-km NOAA AVHRR, 57-m Landsat MSS, Landsat ETM+ thermal image of 60 m, Landsat ETM+ multispectral image of 30 m, IKONOS multispectral of 4 m, IKONOS panchromatic of 1 m. The left most bottom row plot show the inter-sensor relationships between the Landsat ETM+ 30 m NDVI versus IKONOS NDVI (resampled to 30 m) derived from images acquired about the same dates. In the last row, the left and the center plots show the inter-sensor relationships between the Landsat ETM+ NDVI (resampled to 4 m) versus IKONOS 4 m NDVI derived from images acquired about the same dates. The line chart shows the normalized reflectance derived from various sensors that collected data over various African sites.

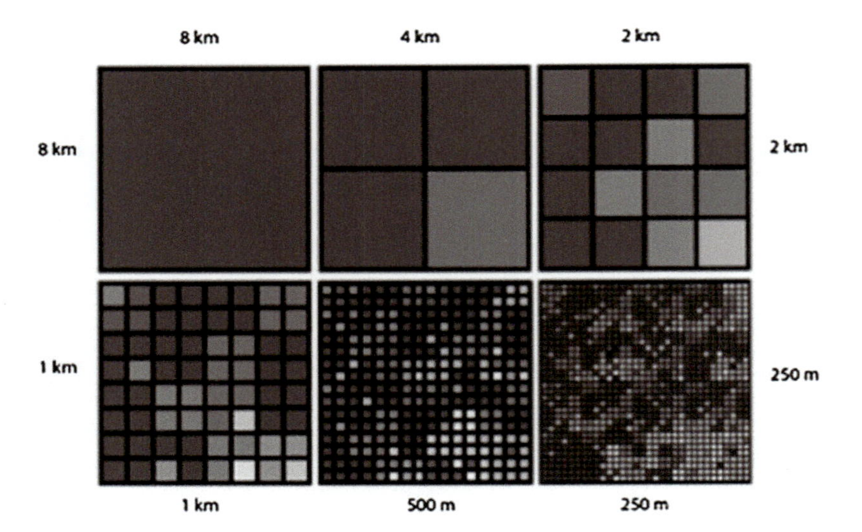


FIGURE 18.1B Diagrammatic interpretation of spatial aggregation of estimated phenological parameters from 250 m to 8 km. The dark gray represents an earlier estimate, whereas white represents a later estimate, thus highlighting the increase in the variance with an increase in spatial resolution.

*Source:* Mountford et al., 2017.

in wide array of spatial (e.g., Figure 18.1b), spectral, radiometric, and temporal resolutions that allow routine, consistent, and repeated study of the planet's land, water, and atmosphere. Chapter 1 captures comprehensive and a wide range of the past, present, and upcoming satellites and sensors detailing their characteristics (also see, Burleigh et al., 2019, Zhu et al., 2017, Mountford et al., 2017, Toth and Jóźków, 2016). The United States (US), France, India, China, Germany, United Kingdom, Japan, South Korea, and Brazil are the leading players along with some others (see Chapter 1) who now routinely launch and operate satellites including wide arrays of microsatellites and share remote sensing data with global and local stakeholders. With the 1992 Land Remote Sensing Policy Act of the United States of America, (LRSPA, 1992) which permitted private companies to enter the satellite imaging business, private companies have participated in launching commercial Earth-observing satellites and very recently, it has picked up tremendously. Increasingly, UAS are also being used for carrying sensors of various kinds.

In immediate years ahead, as we enter the third decade of the 21st century, we are likely to have numerous remote sensing satellites and sensors from current government and private operators, especially microsatellites (size of tennis balls, e.g., CubeSat) and UAS that can fly with supersonic speed. There is also likely a greater role from Uncrewed Aerial Systems (UASs). Compared to conventional remote sensing that maps and analyzes the Earth, water and atmospheric phenomena, importance of LiDAR, RADAR, and SONAR remote sensing technology is gaining ground in the 21st century due to their unconventional application of real-world problem solving like accurate ground topography mapping and tree or other Earth object height measurement, accurate weather forecasting (with NEXRAD and Doppler SONAR), and deep-sea exploration, respectively. The private enterprise also delving into newer, more revolutionary, approaches and techniques of capturing remote sensing data. The satellites planned by these initiatives, will allow for gathering spectral signatures, or sub-meter data on a continuous basis, or even videos, and many other innovative ways.

The chapter also provides the details of the principles of electromagnetic spectrum in remote sensing. It also describes the future of remote sensing, that is, transformation of computer/lab imaging spectrometry technology to field-imaging spectrometry to determine crop, vegetation, fruit, and other physical object quality.

## 18.2 GLOBAL NAVIGATION SATELLITE SYSTEM (GNSS) THEORY AND PRACTICE

In the modern-day satellite navigation (SATNAV) have become ubiquitous. These SATNAV are both global and regional. The global Navigation Satellite Systems (GNSS) is augmentation of these SATNAV systems that provides accurate, continuous, worldwide, 3D position and velocity information to users with appropriate receiving equipment and disseminates time within the Coordinated Universal Time (UTC) timescale (Haibo et al., 2022, Kaplan and Hegarty, 2017). The GNSS guide us from place A to Place B, pinpoint any target in the world for gathering information of various natures, control air-traffic, road, and sea traffic, precisely locate an agricultural crop or forest species, or a water body as tiny as a well, and numerous other applications, and especially, support precise military application and maneuver. Currently, there are six distinct operational GNSS systems: (1) Global Positioning System (GPS) controlled by the United States of America, (2) Global Orbiting Navigation Satellite System (GLONASS) controlled by Russia, (3) Galileo controlled by European nations (e.g., Galileo System Test Bed, Version 2 or GSTB-V2; Figure 18.2), (4) BeiDou (Compass) controlled by China, (5) The Indian Regional Navigation Satellite System (IRNSS), and (6) Japan's Quasi-Zenith Satellite System (QZSS). Each of the GNSSs has 5 to 36+ satellites, which make a minimum of four satellites always accessible for signals from any part of the world. These systems can provide positional accuracy at any location on Earth to within a few centimeters or even millimeters. Chapter 2 by Grewal discusses in detail GNSS characteristics, including signal spectrums, differences between civil and military signals (for GPS), and the nature and characteristics of the different signals, for example, Frequency Division Multiple Access (FDMA), Code Division Multiple Access (CDMA), and Time Division Multiple Access (TDMA). The Low Earth Orbit (LEO) Enhanced Global Navigation Satellite System (LeGNSS) is being vigorously advocated to improve the GNSS performance and expand its applications (Haibo et al., 2022). Jiménez-Martínez et al. (2021) explore the effect of using many single points positioning (SPP) observations from low-cost GNSS receivers, smartphones, and handheld GNSS units to improve GNSS positioning accuracy. Due to the cost-effectiveness of the GNSS smartphones, they can be employed in a wide variety of applications such as cadastral surveys, mapping surveying applications, vehicle and pedestrian navigation even though there are still some challenges regarding the noisy smartphone GNSS observations (Zangenehnejad et al., 2021). However, the advent of mobile real-time

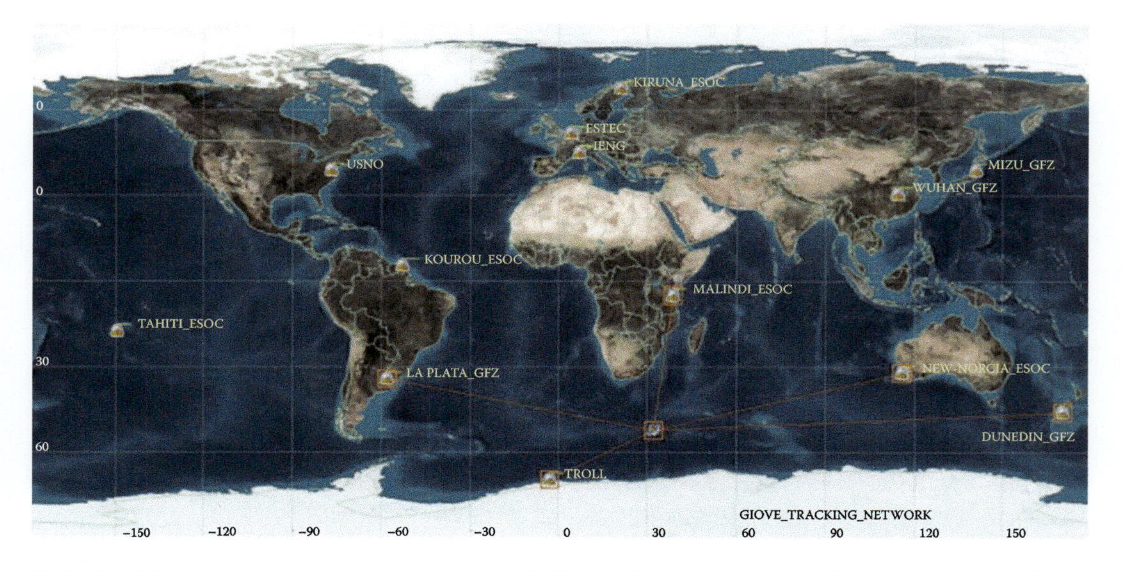


FIGURE 18.2 Galileo System Test Bed, Version 2 (GSTB-V2) station network. (Source: Dow et al., 2007.)

kinematics (RTK), even if inside dense forest cover, accurate location survey is getting feasible (Abdi et al., 2022).

## 18.3 GNSS REFLECTOMETRY FOR OCEAN AND LAND APPLICATIONS

This chapter discusses the increasing interest and usage of GNSS reflectometry (GNSS-R) in ocean and land applications. It provides a solid foundation on the theoretical concepts of GNSS-R. NASA and ESA are launching series of satellites focused on GNSS-R such as the Cyclone Global Navigation Satellite System (CYGNSS). Even though the use of GNSS-R data in remote sensing is still a novelty, with increasing number of Low Earth Orbit (LEO) satellites equipped with GNSS receivers, the situation is going to change soon. Compared with ground-based and airborne GNSS-R approaches, spaceborne GNSS-R has several advantages, including wide coverage and the ability to sense medium- and large-scale phenomena such as ocean eddies, hurricanes and tsunamis (Yu et al., 2022). Chapter 3 by Kengen Yu et al., highlights the key land applications of GNSS-R such as soil moisture and forest change detection. In many other applications like that for tsunamis (e.g., Figure 18.3; Stosius et al., 2011) GNSS-R is very powerful. Similarly, it highlights the key ocean applications of GNSS-R such as sea surface undulation and sea surface height. Just as the UAVs are becoming increasingly important in land remote sensing, GNSS-R is also becoming very important. They add to conventional remote sensing and bring in unique capabilities that only GNSS-R can provide such as Cyclone Global Navigation Satellite System (CYGNSS). Land applications of GNSS-R include soil moisture, vegetation opacity, wetland detection and monitoring, flood

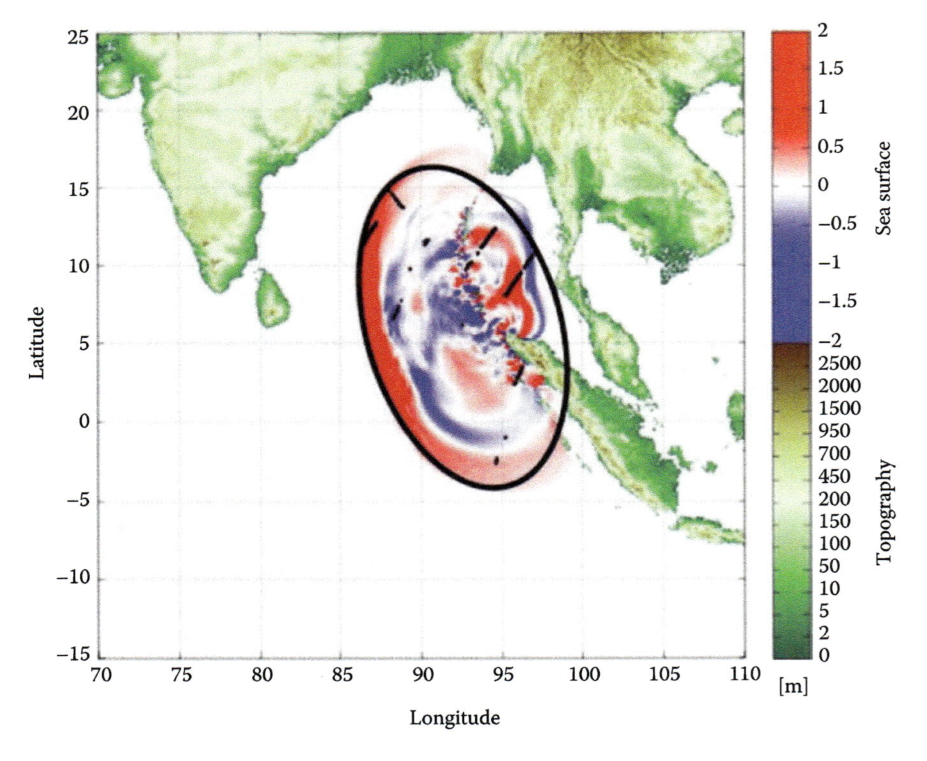


**FIGURE 18.3** Simulation of the Sumatra tsunami 1 hour after the earthquake. Ground tracks (black) show Global Navigation Satellite System reflectometry (GNSS-R) detections within 1 minute of observation. The ellipse encloses all detections made up to this moment and shows the current tsunami expansion.

*Source:* Stosius et al., 2011.

inundation, snow height, and sea ice concentration and extent (Rodriguez-Alvarez et al., 2023). Same authors also show ocean applications of GNSS-R to have included developments towards ocean wind speed retrievals, swell and altimetry. HydroGNSS (Hydrology using Global Navigation Satellite System reflections) has been selected as the second European Space Agency (ESA) Scout Earth observation mission to demonstrate the capability of small satellites to deliver science (Unwin et al., 2021). GNSS meteorology is now an established atmospheric remote technique that can provide the data on integrated water vapor (IWV) which is the most abundant and important greenhouse gas for meteorological and climatic processes (Kumar, 2023). As evidenced by these studies it is obvious GNSS-R is increasing in its wide range of applications across land, and water.

## 18.4 GNSS FOR WIDE ARRAY OF TERRESTRIAL APPLICATIONS

Global Navigation Satellite Systems (GNSS), a real global conglomeration of navigation satellites by global players (countries), include satellite constellations (typically, 20–30) from NAVSTAR global positioning systems (GPS) of the United States of America, global navigation systems (GLONASS) of Russia, Galileo of European Union, and Beidou navigation systems of China provide instant, freely available position data with high accuracies (few centimeters to few millimeters) for any place in the world (see Figure 18.4; Jin et al., 2022, 2011). As a result of global International GNSS Service (IGS) constellation, a dense network of GNSS satellites and sites are available. The first and the most reliable GNSS, which is known as GPS (Global Positioning System by United States), acquires position data and altitude (x, y, z) through a baseline constellation of 24 satellites positioned about 20,000 km from earth. To get accurate position for any location on Earth's surface, one needs data from at least three satellites. Typically, at least four satellites are available from any spot on Earth. Of the 24 GPS satellite constellation at any given time, 21 are in operation continuously and three are used as spare to replace any failing satellite. The signals (e.g., code division multiple access or CDMA) transmitted by them are either for military (encrypted) or civilian (open).

Dr. Myszor et al. in Chapter 4, discuss various applications of GNSS. These are broadly categorized into civilian, military, and aeronautic. The applications discussed include:

1. Consumer applications such as the interactive maps to locate any place or location of interest (e.g., house, restaurant, airports, bus stations). These are ubiquitous today in our daily lives in cars, smartphones.
2. Industrial applications such as in agriculture (e.g., detecting location of farms, crops grown in farms; Tilling et al., 2007), stores, city services (e.g., location of sewers, pipelines).
3. Transportation applications such as route from point A to B.
4. Surveying applications such as mapping large areas and isolated areas.
5. Disaster applications where disaster maps that provide information such as people injured, property lost, and damage areas are instantly mapped. GNSS data are used in wide array of disasters like earthquakes, floods, droughts, famine, and fire.
6. Health applications (e.g., monitoring and managing disease spread).
7. Tracking applications (e.g., animal grazing paths, wildlife tracking).
8. Military applications (e.g., locating and monitoring troop movements, military installations).
9. Aeronautic applications (e.g., tracking aircraft movements).

The GNSS applications for Earth Observation (EO) are proliferating. The multi-GNSS for Earth Observation and emerging application progress includes GNSS positioning and orbiting, GNSS meteorology, GNSS ionosphere and space weather, GNSS-Reflectometry and GNSS earthquake monitoring, as well as GNSS integrated techniques for land and structural health monitoring (Jin et al., 2022).

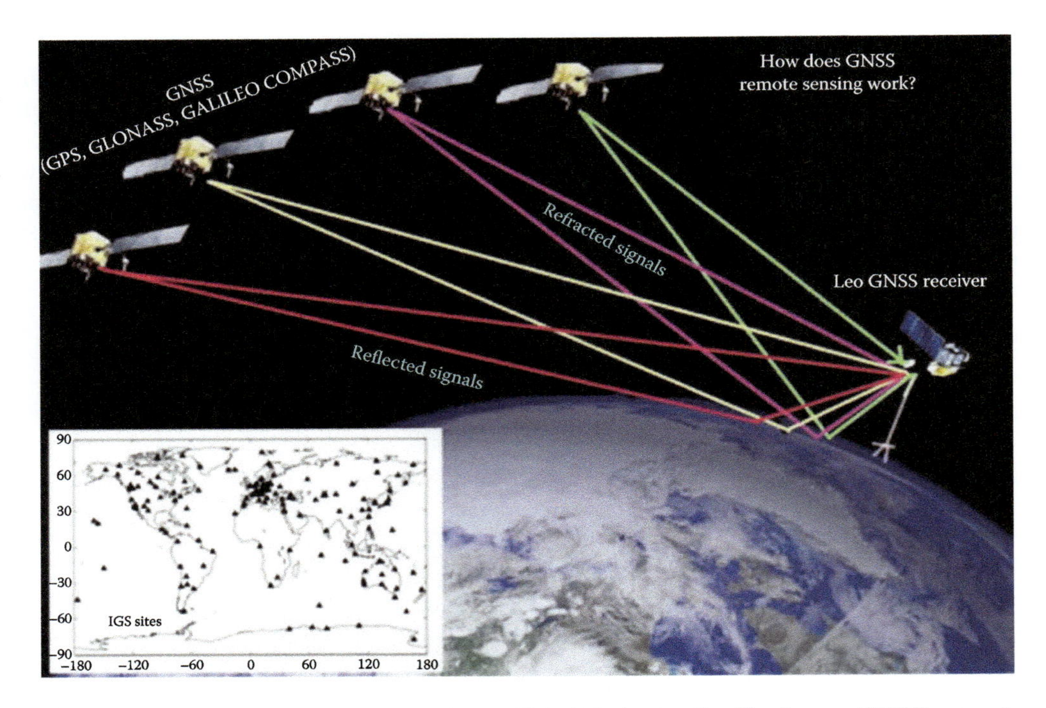


FIGURE 18.4 Remote sensing using future denser Global Navigation Satellite System (GNSS) network and more satellite constellations. The left lower corner shows the global International GNSS Service (IGS) sites distribution.

*Source:* Jin et al., 2011.

Chapter 4 also provides the GNSS satellite characteristics as well as errors in GNSS and ways and means to control them and keep the errors to a minimum.

## 18.5 FUNDAMENTALS OF REMOTE SENSING FOR TERRESTRIAL APPLICATIONS

Early remote sensing was primarily based on images and data gathered from aircrafts. This had limited scope, given the complexities of covering large areas and the cost factor. Thereby, Land remote sensing from space commenced in earnest once the first Landsat was launched in 1972. Even though NOAA AVHRR and Sputnik preceded Landsat and the data from these satellites were used for land applications, it is with the launch of Landsat that the terrestrial applications of spaceborne remote sensing began as earth features could be better discerned with Landsat's 60 m spatial resolution than a 1.1 km of AVHRR. Oppelt et al. in Chapter 5 provide a brief overview on the remote sensing platforms as well as the electromagnetic spectrum from where these data are captured. Limitations of early visual interpretations are pointed out. Then the chapter provides a focus on typical applications at local, regional, and global levels as well as specialized thematic applications. What is clear in this chapter is that:

1. Specific applications (e.g., urban, vegetation, water, terrain) have specific data needs (e.g., thermal, NDVI, near-infrared, microwave). For example, one cannot study species using coarse spatial and/or coarse spectral resolutions. Similarly, a particular sensor performs better than another in certain applications, for example, radar sensors are suitable to derive digital elevation models and can collect information on Earth in cloudy conditions. So, knowledge of sensor performance for specific terrestrial applications is required.

2. Multi-sensor approaches (e.g., combine Landsat with Sentinel or optical with microwave sensors) become crucial for improved accuracies in mapping, modeling, and monitoring complex terrestrial themes. Multi-sensor-based image fusion approach is helping in spatial and temporal resolution of specific satellites among the multi-sensor group.
3. For a value adding beyond the sole physical measurement of remote sensing data, multidisciplinary approaches using remote sensing data in combination with other data sources became increasingly important. Figure 18.5 presents an example for such an approach to

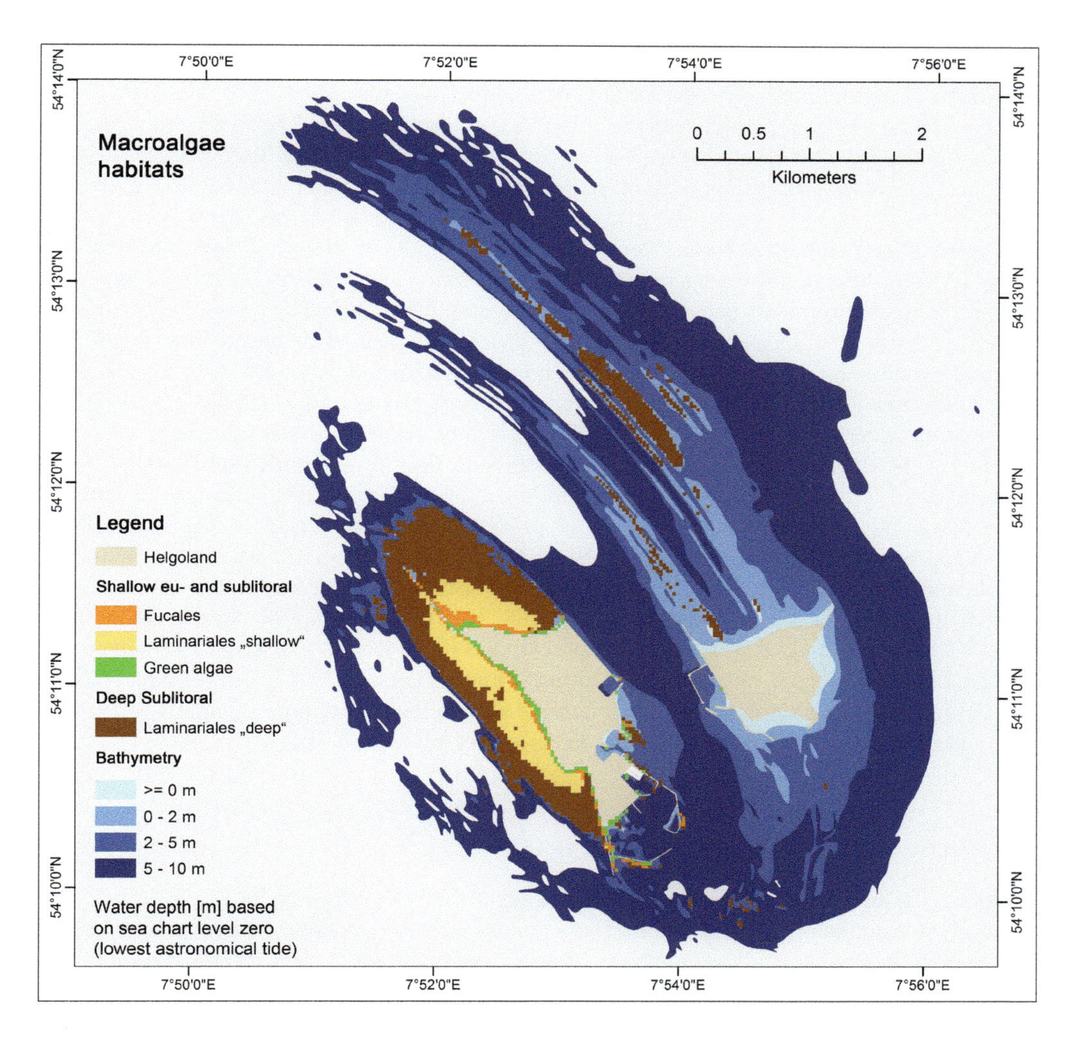


**FIGURE 18.5** Map of macroalgae habitats in the coastal waters of Helgoland (North Sea, Germany); the map was generated in the scope of an interdisciplinary research project of geographers, biologists, and remote sensing experts. The data sources include airborne hyperspectral data ($AISA_{eagle}$), floristic mappings, diving mappings, and bathymetry information from the German Federal Maritime and Hydrographic Agency. The remote sensing data have been analyzed separately for algae cover in the sublittoral zones (brown algae in shallow waters). These methods are described in Oppelt et al. 2013 and Uhl et al. 2013. Since analysis of optical remote sensing data are limited by visibility depth (approximately 5 m during airborne data acquisition), the analysis is valid up to a depth of ~4 m. All data have been combined in a GIS (Gbanie et al., 2013) which enables validation of remote sensing results and mappings. Resulting user accuracy is 97.5% (number of validation points = 45), GIS analysis also revealed that large sublittoral brown algae (kelp) cover 248.31 ha of the area with water depth ≤ 5 m (Uhl et al., 2014).

monitor coastal vegetation combining different analysis techniques, biological mapping, hydrography, and a Geographic Information System. This approach is essential in any proactive environmental management decision support system development.
4. Best results are obtained when studies incorporate various dimensions (e.g., spectral, spatial, temporal, radiometric) of remotely sensed data.
5. Data policy is changing; free and open data policy supports existing trends to provide remote sensing data (e.g., Landsat, MODIS) and standardized products (e.g., global land cover maps; Fritz et al., 2012) free of charge.
6. Global networks have form to develop robust methodologies and services for decision support tools and community portals to meet specific needs of stakeholders, scientists, and user communities (e.g., GEOSS, Global Forest Watch).
7. Continuous and comparable data are of high importance to develop standardized measures, especially for global monitoring. To increase acceptance of standardized products and to ensure their widespread application, standards for accuracy assessment will have to be set.

The aforementioned developments and trends require overlapping missions and well (inter-) calibrated sensors. To fulfil these requirements international cooperation and data sharing will be increasingly important. Joint commitments between intergovernmental, international, and regional organizations therefore are a powerful means of stabilizing new space initiatives, and hence data availability and continuity. Even though great advances have been made as a result of use of data from multiple sensors using advanced methods, techniques, and modern computing power, there is still a long way to go in reducing uncertainties of various terrestrial themes based on remote sensing data.

Chapter 5, by Oppelt et al., is a great start to anyone wanting a detailed understanding of the fundamentals of remote sensing. This can be followed by further studies in remote sensing introduction, image interpretation, and digital image processing (Lillesand et al., 2008, 2015, Lillesand et. al., 2015, Jensen, 1996, Campbell and Wynne, 2011, Nagamalai et al., 2011, Wang et al., 2010, Richards and Xiupin, 2006a, Blaschke et al., 2014, Blaschke, 2010, Zhao et al., 2012) or specific terrestrial applications such as Geology (Sabins et al., 2020), hydrology (Ndehedehe, 2022, Branger et al., 2013, Edwards et al., 2020), environmental studies (Chuvieco, 2020), remote sensing big-data analytics (Di and Yu, 2023), and cloud computing (Montaghi et al., 2024, Cardille et al., 2023, Wan et al., 2014). A wide array of most comprehensive applications of remote sensing science can be found in various chapters of this six-volume *Remote Sensing Handbook.*

## 18.6 CALIBRATED AND NORMALIZED GLOBAL EARTH OBSERVATION DATA FROM MULTIPLE PLATFORMS

A primary goal of remote sensing data acquired from the wide array of Earth Observation (EO) satellites is to establish protocols and mechanisms for delivering calibrated and normalized data over the entire planet. This allows analyses of data over space and time with minimal and known uncertainties.

In the early years, remote sensing applications of EO satellites were often done using raw digital counts (Level-0 data) without physical units. Great advances have been made over the years on remote sensing digital number (DN) values interpretation with advanced statistical algorithm application. Currently, remote sensing data are analyzed routinely after calibration to their physical units (e.g., radiance expressed in $W{\cdot}m^{-2}{\cdot}sr^{-1}{\cdot}\mu m^{-1}$) using appropriate radiometric (imperfect sensor correction, Sun angle correction, and atmospheric corrections), geometric, and spectral corrections, as explored in detail in Chapters 6 through 11 of Volume I. These calibrations lead to data being converted to:

1. At-sensor or top-of-atmosphere (TOA) radiance and/or reflectance (Level-1) (Figure 18.6);
2. Surface radiance or reflectance (Level-2).

Currently, almost all EO data are provided in one of these units, typically scaled for digital processing. Yet, calibrations remain challenging, and complex given that satellite sensors are launched by various space agencies of different nations as well as private initiatives, which have widely varying geometric, radiometric, and spectral performance characteristics. Table 6.1 in Chapter 6 by Teillet provides a comprehensive summary of the wide array of radiometric factors involved in EO data. However, not all these parameters are used in developing standard products, produced through various algorithms, due to difficulty and/or uncertainty in obtaining some or many of these parameters.

In addition, inter-sensor calibrations further facilitate the use of data from multiple sensors from multiple sources. In order to overcome these difficulties, the Committee on Earth Observation Satellites (CEOS; www.ceos.org/), under the umbrella of the Group on Earth Observations (GEO; www.earthobservations.org/index.shtml), established an international calibration/validation (CAL/VAL) working group (http://calvalportal.ceos.org/), where many of these issues are discussed and implementation recommendations are provided. The ultimate goal of these efforts is to create an effective Global Earth Observation System of Systems (GEOSS) that will deliver EO data that are well calibrated, reliable, consistent, and have comprehensive coverage over space and time that interact and provide access to diverse information for a broad range of users in both public and private sectors. Another important job of GEOSS is to proactively link existing and planned observing systems around the world to support the need for the development of new systems where gaps currently exist.

Current approaches to data delivery by many space agencies and commercial data providers are already moving in the direction of surface reflectance products, such as for MODIS (http://modis.gsfc.nasa.gov/data/dataprod/dataproducts.php?MOD_NUMBER=09) and Landsat 8 (http://earthexplorer.usgs.gov/). Some difficulties yet to be overcome include when:

1. Space agencies and commercial data providers do not yet provide calibrated data.
2. Calibration coefficients provided are incomplete (e.g., do not allow for all the factors listed in Table 6.1 of Chapter 6).
3. Gaps in inter-sensor calibration exist (e.g., the SPOT HRV series of satellite data are not adequately intercalibrated with the Landsat series of satellites (Niro et al., 2021, Goward et al., 2012)).

Indeed, interoperability of satellite sensors is key to utilizing full potential of remote sensing data acquired from sensors on multiple satellite platforms operated by various governments and private industry. There are various efforts in addressing these challenges such as by the CEOS, GEO, and the European Space Agency (ESA). For example, the ESA, in collaboration with other Space Agencies and international partners, is elaborating a strategy for establishing guidelines and common protocols for the calibration and validation (CAL/VAL) of optical land imaging sensors (Niro et al., 2021). Such calibrations are performed using various radiative transfer and other models. For example, four different 1D radiative transfer models are compared in actual usage conditions corresponding to the simulation of satellite observations using six different spaceborne radiometers over the pseudo-invariant calibration site Libya-4 are used to define these conditions (Govaerts et al., 2022). Others demonstrate radiometric cross calibration between the Landsat Enhanced Thematic Mapper (ETM+) and the Indian Remote Sensing Satellites (IRS) (e.g., Figure 18.6, Goward et al., 2012). Some others, Landsat 8 Operational Land Imager (OLI) and Sentinel 2A Multispectral Instrument (MSI) sensors (Farhad et al., 2020). Cross-calibration method to calibrate medium-resolution multispectral data using well-calibrated data such as the Landsat is also adopted by others (Lu et al., 2022).

It is to be noted that many recent satellites like ECOSTRESS and GEDI provide corrected and calibrated DN values to produce environmental management supporting data, for example, Level 3 (Evaporation), Level 4 (Evaporative Stress Index) data from ECOSTRESS and L1A (raw wave form DN values), L2A (ground elevation, canopy top height metrics, L1B (geolocated waveforms), L2B (canopy cover fraction (CCF), CCF profile, leaf area index (LAI), and LAI profile), L3 (gridded level 3 metrics), L4A (footprint level aboveground biomass), L4B (gridded aboveground biomass

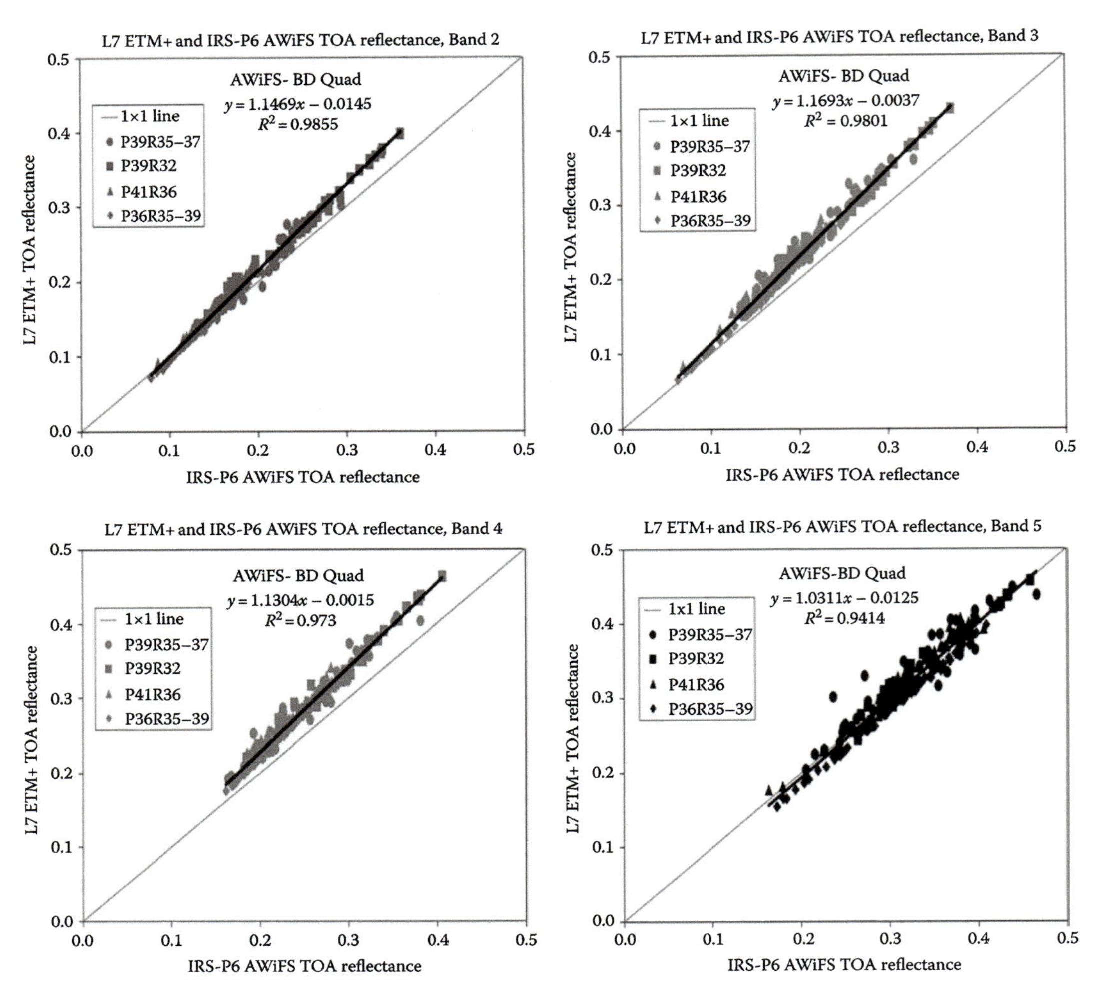


**FIGURE 18.6** Comparison of TOA reflectance measurements from Landsat 7 ETM+ and AwiFS. Each data point on these plots represents an ensemble average of all pixels in a defined ROI for a given day and spectral band. (Source: Goward et al., 2012.)

density, and others (Nidhi et al., 2024, Hulley et al., 2022, Kronseder et al., 2012). Researchers and other end users can use those information directly. This is perhaps the new approach of remote sensing data delivery.

## 18.7 POST-CALIBRATION OF OPTICAL SATELLITE SENSORS

The scientific integrity of satellite sensor data can only be maintained if the sensors are well calibrated. Once a sensor is launched into orbit, post-launch calibrations need to come into effect and be reiterated for the entire life of the sensor, which is often from a few years to even decades. Chapter 7 by Teillet and Chander provides an exhaustive state-of-the-art review of post-launch radiometric calibration of optical sensors. The standard and well-established approach adopted by various agencies and researchers for post-launch calibration is to use built-for-purpose on-board systems initially and adjust calibration coefficients over time using suitable reference targets vicariously. The international community has adopted and recommends the use of approximately 20 such targets, selected with specific criteria in mind (cf. Chapter 7).

Currently, the best practice is for optical sensors to be calibrated vicariously using:

1. Eight Committee on Earth Observation Satellites (CEOS) endorsed reference instrumented standard sites based in the deserts and playas of the world.
2. Six CEOS reference pseudo-invariant calibration sites (PICS) located in various parts of the Sahara Desert.

The multigeneration Landsat series of sensors from 1972 to the present are recalibrated post-launch using vicarious targets and are known to have uncertainties of within 5–10% across sensors and spectral bands (Choate et al., 2023, Helder et al., 2020a, 2020b). The calibration efforts in the contexts of Group on Earth Observations (GEO), Global Earth Observation System of Systems (GEOSS), and CEOS are important and are discussed in Chapter 7 by Teillet and Chander. Well-calibrated sensors allow multi-sensor data continuity, facilitating long-term studies of Earth and environment including ability to cross calibrate across sensors (e.g., Figure 18.7).

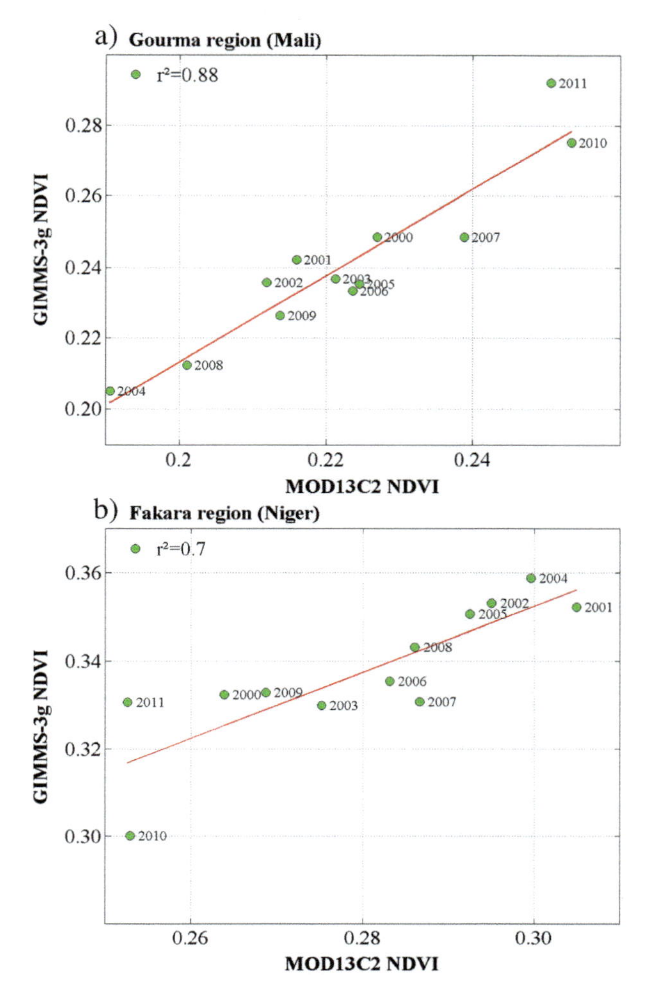


FIGURE 18.7 Linear regression between third generation Global Inventory Modeling and Mapping Studies (GIMMS-3g) 8 km and Moderate Resolution Imaging Spectroradiometer (MODIS) Terra normalized difference vegetation index (NDVI) (MOD13C2) at 5.6 km resolution over (a) the Gourma (Mali) region and (b) the Fakara (Niger) region.

*Source:* Dardel et al., 2014.

The USGS Landsat Image Assessment System (IAS), for example, has been performing calibration and characterization operations for over 20 years on the Landsat spacecrafts and their associated payloads (Choate et al., 2023). Cross-Calibration of sensors such as Landsat 8 and Landsat 9 are often performed using the simultaneous under fly event (Garrison et al., 2022). Three types of uncertainties were evaluated: geometric, spectral, and angular (bidirectional reflectance distribution function—BRDF and they found total cross-calibration uncertainty for under fly data could be kept under 1% (Garrison et al., 2022). A great challenge in CAL/VAL activities today is to harmonize and synthesize between the government systems and commercial systems. Because of the differences in the design and mission of these sensor types, calibration approaches are often substantially different (Helder et al., 2020a,b). Vicarious CAL/VAL is often done using multiple pseudo invariant sites as utilized for orbital hyperspectral sensors (Pearlshtien et al., 2023). Such sites serve long-term CAL/VAL missions of future sensors as well.

## 18.8 NORMALIZATION OF REMOTELY SENSED DATA

Modern remote sensing involves using data/imagery from multiple dates and from multiple sensors. The only way such data can be used in scientific investigations is through appropriate calibration and normalization. Sensors once launched go through degradation over time. They also operate in different seasons and climates. Various sensor characteristics such as push broom (or along track) versus whisk broom (or across track), solar zenith angles, field of view, and angle of the sensor also adds to the need for standardization and normalization of the sensors. In Chapter 8, Drs. Gens and Dr. Rosselló address these issues and provide a clear approach to normalization of sensor characteristics. They recommend:

1. Absolute normalization based on radiative transfer models that account for atmospheric and sensor characteristics and
2. Relative normalization where one or more images are corrected to a standard master image.

They illustrate the preceding two normalization methods separately for optical, thermal, and radar imagery.

Instruments are always calibrated pre-launch. Then vicarious calibration takes places throughout the life of the sensor onboard either using an onboard calibration reference unit or by focusing the sensor to time invariant sites like the full Moon. Factors that need to be considered during calibration include (Figure 18.8):

1. Satellites
   Height of acquisition (e.g., 500 km, 700 km, 36,000 km above Earth)
   Orbital parameters
ii. Sensors
   Radiometry
   Bandwidth
   Optics/design
   Degradation over time
   View geometry
iii. Solar flux or irradiance
   Function of wavelength
iv. Sun
   Sun elevation and Azimuth @ time of acquisition
v. Sun–Earth
   Distance between Earth and Sun

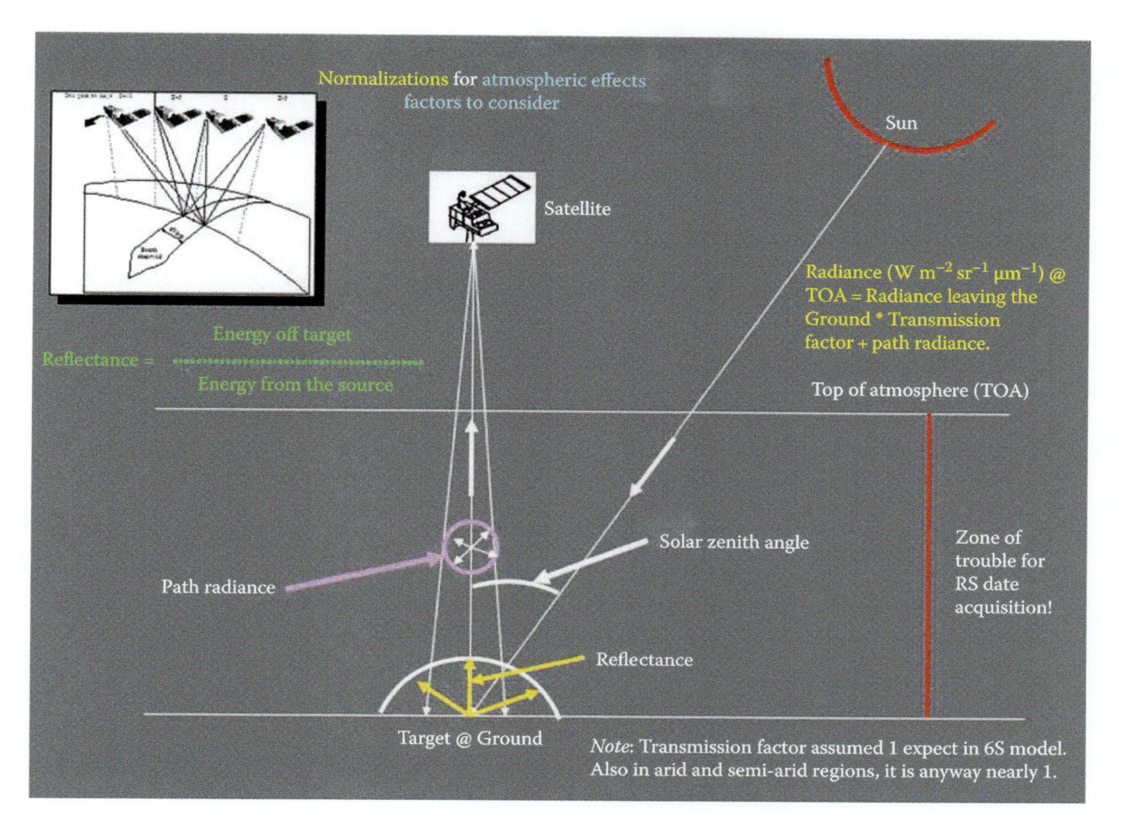


**FIGURE 18.8** An overview of factors involved in normalization of remotely sensed data. This generic approach can be applied across any satellite sensor system.

vi. Stratosphere or Atmosphere
 Ozone, water vapor, haze, aerosol
 Path radiance
 Water droplet/ice crystal (clouds)
vii. Surface of Earth
 Topography
viii. Seasons
 Earth–Sun distance

Normalization steps involve three major steps:

- Digital number (DN; unit less) to Radiance (W/m$^2$ Sr µm)
- Radiance to top of the atmosphere reflectance (also referred toas at-satellite exo-atmospheric (or apparent or) reflectance (%)
- Top of the atmosphere reflectance to surface reflectance (%)

The data are first normalized to top of atmosphere reflectance using the following common approach, irrespective of the sensors involved (also see Equation 1 in paper by Roy et al., 2016).

Energy off the TargetRadiance (W·m$^{-2}$· sr$^{-1}$.µm$^{-1}$) * π
Reflectance (%) = -------------------------= --------------------------------
Energy from the SourceIrradiance (W·m−2.µm$^{-1}$)

$$Reflectance\left(\%\right) = \frac{\pi L_{\lambda} d^2}{ESUN_{\lambda} \cos\theta_S}$$

Where, TOA reflectance (at-satellite exo-atmospheric reflectance)

$L\lambda$ is the radiance ($W{\cdot}m^{-2} \cdot sr^{-1}.\mu m^{-1}$)

*d* is the Earth to Sun distance in astronomic units at the acquisition date (see Markham and Barker, 1987)

ESUN $\lambda$ is irradiance ($W{\cdot}m{-}2.\mu m^{-1}$) or solar flux (Neckel and Labs, 1984) and

$\Theta_s$ = solar zenith angle

Note: $\theta_s$ is solar zenith angle in degrees (i.e., 90° minus Sun elevation or Sun angle when the scene was recorded as given in the image header file.

The preceding data are then converted to ground reflectance through atmospheric correction as described in Chapter 8. Atmospheric corrections are part of normalization and eliminate or reduce path radiance resulting from haze (thin clouds, dust, harmattan, aerosols, ozone, and water vapor). Atmospheric correction methods include: (1) dark object subtraction technique (Chavez, 1988), (2) improved dark object subtraction technique (Chavez, 1989), (3) radiometric normalization technique: Bright and dark object regression or (Elvidge et al., 1995), and (4) 6S model (Vermote et al., 2002). As pointed out by Drs. Gens and Rosselló in Chapter 8, the main difficulty in atmospheric radiosondes are usually not available at the time of satellite pass since they are operated at different times. The and data from a single atmospheric radiosonde that maybe launched to coincide with satellite overpass will be inadequate or such launches are simply not feasible resource wise to launch over every satellite overpass of every satellite. AERONET networks that are available only in few places in the world are also inadequate for large areas covered by single swaths (e.g., AVHRR, MODIS, Landsat) of most of the satellites. Even when the swath are not big (e.g., IKONOS, Quickbird, Geoeye; Ke et al., 2010), local weather data may not be representative since these networks are often quite away from the images. A classic paper on inter-sensor calibration involving multiple sensors is provided by Chander et al. (2009). So, most uncertainty in atmospheric correction of imagery is due to lack of detailed, reliable input data for the models.

For consistent and objective studies of various applications, remote sensing data normalization, harmonization, and standardization within and across sensors is crucial. This requires remotely sensed data delivered as surface reflectance products. The surface reflectance, that is, satellite derived top of atmosphere (TOA) reflectance corrected for the temporally, spatially, and spectrally varying scattering and absorbing effects of atmospheric gases and aerosols, is needed to monitor the land surface reliably (Moghimi et al., 2021, Vermote et al., 2016). They develop Landsat 8 Operational Land Imager (OLI) atmospheric correction algorithm that has been developed using the Second Simulation of the Satellite Signal in the Solar Spectrum Vectorial (6SV) model, refined to take advantage of the narrow OLI spectral bands (compared to Thematic Mapper/Enhanced Thematic Mapper (TM/ETM +)), improved radiometric resolution and signal-to-noise (Vermote et al., 2016). The Sentinel-2 data service platform for obtaining atmospherically corrected images and generating the corresponding value-added products for any land surface on Earth was proposed and implemented using the European Space Agency's (ESA) Sen2Cor algorithm, the platform processes ESA's Level-1C top-of-atmosphere reflectance to atmospherically-corrected bottom-of-atmosphere (BoA) reflectance (Level-2A) (Vuolo et al., 2016). Claverie et al. (2018) produce Harmonized Landsat and Sentinel-2 (HLS) Virtual Constellation (VC) of surface reflectance (SR) data acquired by the Operational Land Imager (OLI) and Multi-Spectral Instrument (MSI) aboard Landsat 8 and Sentinel-2 remote sensing satellites, respectively. The HLS products are based on a set of algorithms to obtain seamless products from both sensors (OLI and MSI): atmospheric correction, cloud and cloud-shadow masking, spatial co-registration and common gridding, bidirectional reflectance distribution function normalization and spectral bandpass adjustment (Claverie et al., 2018). The Harmonized Landsat-8 and Sentinel-2 (HLS) project is a NASA initiative aiming to produce a seamless, harmonized surface reflectance record from the Operational Land Imager (OLI) and Multi-Spectral Instrument (MSI) aboard Landsat-8 and Sentinel-2 remote sensing satellites, respectively (Masek et al., 2018). The HLS products are based on a set of algorithms to obtain seamless products from both sensors (OLI and MSI): atmospheric correction, cloud and cloud-shadow masking, geographic co-registration and common gridding, bidirectional reflectance distribution function normalization and bandpass adjustment (Masek et al., 2018). Feng et al. (2013), for example, produced atmospherically corrected the Global Land Survey (GLS) Landsat dataset using the Landsat Ecosystem Disturbance Adaptive Processing System (LEDAPS) implementation of the Second Simulation of the Satellite Signal in the Solar Spectrum (6S) radiative transfer model. Moghimi et al. (2021) proposes a new relative radiometric normalization (RRN) method for multitemporal satellite images based on the automatic selection and multistep optimization of the radiometric control set samples (RCSS). The proposed method was applied to seven different datasets comprised of bitemporal images acquired by various satellites, including Landsat TM/ETM+, Sentinel 2B, Worldview 2/3, and Aster and the results showed that the method outperforms the state-of-the-art RRN methods.

## 18.9 REMOTE SENSING DATA CALIBRATION APPROACHES

One of the greatest challenges for terrestrial remote sensing has always been to ensure well calibrated data from within and between sensors throughout their data acquisition. All sensors are pre-calibrated before launch and continue to be calibrated during their lifespan in space vicariously and/or on-board. Chapter 9 by Dr. Wang discusses three issues affecting stability of remote sensing data calibration:

1. Radiometric calibration with onboard calibration systems
2. Radiometric calibrations without onboard systems
3. Data degradation due to orbit drift of Sun-synchronous satellites

There are very many factors influencing stability of remote sensing data calibrations such as orbital drift, solar zenith angle changes, and sensor degradation in space due to adverse conditions, clouds, haze, shadows, and surface changes even on so called time invariant sites. Some of these key factors are summarized in Table 9.1. What is important for a terrestrial scientist to ensure is that the measurements made by satellites are reliable and temporally stable and if there are uncertainties, those are well characterized. This important and fundamental first step in using satellite remote sensing data in Earth sciences is well illustrated in this chapter. For example, the chapter summarized some significant factors of different sensors such as:

1. Difference in reflectance from early AVHRR sensor derived with different versions of calibration coefficients can be as large as 20%.
2. There are systematic differences between MODIS Terra and Aqua C5 derived information at certain places. This led to recommendation that for interannual variability studies it is better to use MODIS Aqua data rather than MODIS Terra data before new release with improved calibration stability is available;
3. Annual degradation of Landsat ETM+ are less than 0.4% and are well characterized.
4. Landsat TM showed trends in sensor response over time. So, new method that involved desert sites and cross calibration with ETM+ were preferred for better understanding Landsat TM calibration.

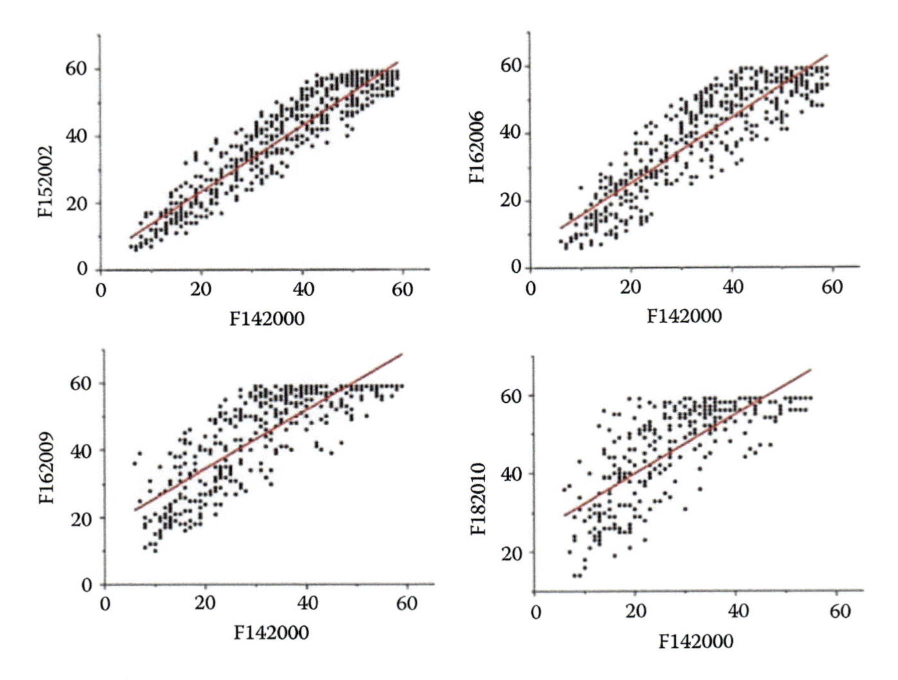


FIGURE 18.9 Scatter plots of the DN values of the target years against the DN values of the reference year (2000) for pixels within de-saturated Pseudo Invariant Features (PIFs) for the U.S. Defense Meteorological Satellite Program Operational Linescan System (DMSP-OLS) satellites. The solid line stands for the trend line of regression equation.

*Source:* Wei et al., 2014.

These inferences, drawn from Chapter 9, only depict some of the issues. These issues clearly indicate the complexities as well the critical need for well-established calibration of various orbiting sensors. Images are also normalized taking time-invariant sites into consideration (e.g., Figure 18.9). In such a case an image is selected as reference and all other images of the same area are normalized to the reference image by taking time-invariant sites. For example, in Figure 18.9, Wei et al. (2014) selected the features that emit temporally stable and invariant nighttime lights should be used as Psuedo Invariant Features (PIFs) for the DMSP-OLS image normalization. The fully developed and stable urban districts (Freire et al., 2014, Haas and Ban, 2014, Xu et al., 2013), where no significant further development and changes occur, satisfy the stable nighttime light requirement, and can be used as PIFs (Helder et al., 2020a, 2020b, Wei et al., 2014).

Liu et al. (2020) and Zhou et al. (2015) provide a comprehensive assessment of in-orbit radiometric calibration methods in the past 40 years. They summarize calibration technology of Landsat series satellite sensors including MSS, TM, ETM+, OLI, TIRS; SPOT series satellite sensors including HRV, HRS the development of in-orbit radiometric calibration technology of typical satellite sensors in the visible/near-infrared bands and the thermal infrared band. They focus on the visible/near-infrared bands radiometric calibration methods including Lamp calibration and solar radiation-based calibration. A mature, well-understood procedure of absolute radiometric calibration of airborne and spaceborne systems, used for over 30 years, is the reflectance-based approach (Czapla-Myers et al., 2015). Liu et al. (2020) discuss three independent methods (reflectance-based, radiance-based, and cross-calibration) to determine the radiometric calibration coefficients of the SuperView-1, a set of launched Chinese civilian remote sensing satellites operated by the Beijing Space View Tech Co Ltd., optical sensors with multiple permanent artificial calibration targets. Helder et al. (2020a,b) share best practices of calibrating Geosynchronous and Polar Orbiting Satellites.

## 18.10 INTERCALIBRATION OF SATELLITE SENSORS

Long-term consistent global studies from satellite sensors are only feasible if there is within and between sensor calibration leading to clear, well-understood relationships between these sensors. Establishing relationships is required for both within a family of sensors such as Landsat-series, AVHRR-series or across family of sensors such as Landsat versus MODIS versus AVHRR versus IKONOS. This is because of a host of issues such as: (1) sensor degradation over time; (2) their characteristics differences in spatial (e.g., Figure 18.10a, 18.10b), spectral, radiometric, and temporal resolutions; (3) atmospheric conditions under which data are acquired; and (4) different processing algorithms applied over time.

Chapter 10 by Dr. Miura provides the latest understanding of inter-sensor spectral compatibility across multiple sensors, including VIIRS, MODIS, SGLI, VGT, AVHRR/3, AVHRR/2, and ETM+, for two well-known vegetation indices: enhanced vegetation index (EVI) and EVI2 (Zhen et al., 2023, Rocha and Shaver, 2009). EVI uses three bands (red, near infrared, and blue) whereas EVI2 does not use the blue band. It is clear from their chapter that correlation between same families of sensors (e.g., AVHRR2 versus AVHRR3) is typically high and can be reasonably well established. However, correlation between dissimilar sensors (e.g., MODIS versus AVHRR) can be poor. This is mainly because of stark differences in resolutions (e.g., Figure 18.10a and 18.10b, first two rows, shows the differences in spatial resolutions). Nevertheless, it is feasible to establish inter-sensor calibrations (e.g., Figure 18.10a) that will help studies beyond a particular sensors' life span (e.g., Figure 18.10b). Chapter 10 results lead to potential methodologies that allow for the derivation of robust

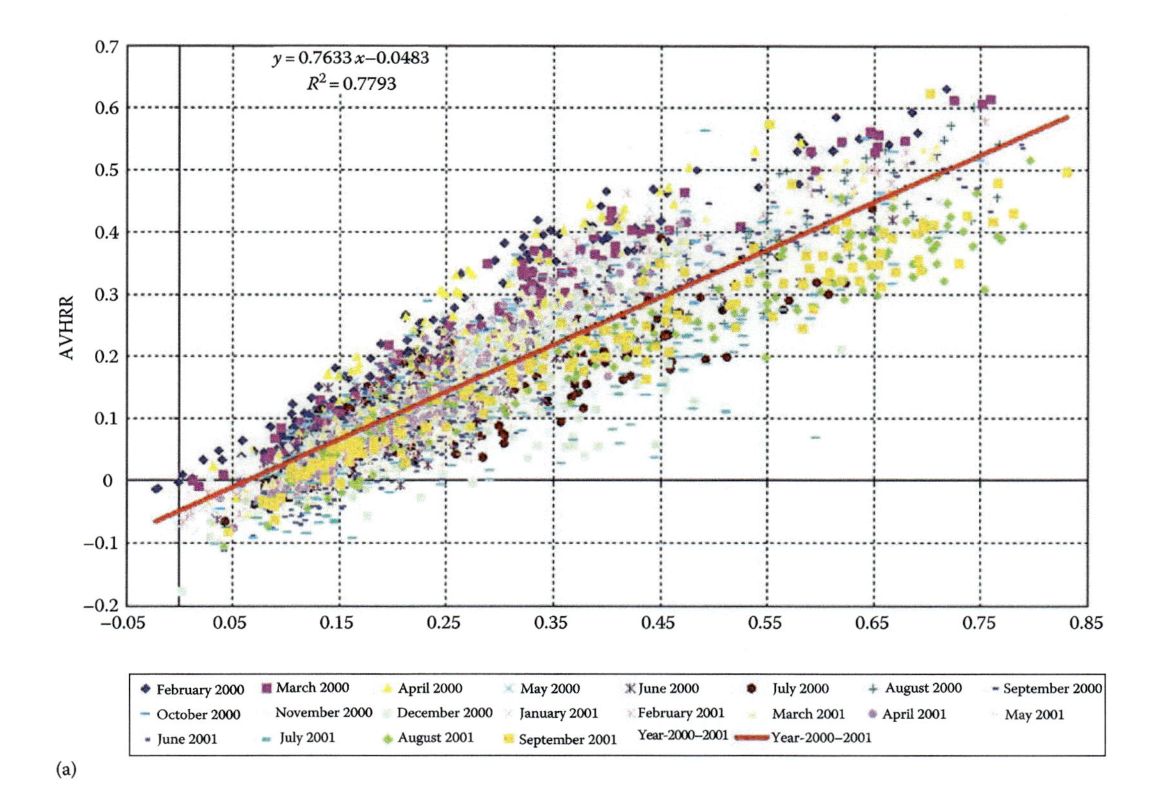


**FIGURE 18.10A** Inter-sensor relationships between MODIS 500 m derived NDVI versus AVHRR 8 km NDVI. Relationship was developed taking data from various years to ensure robustness.

*Source:* Thenkabail, 2004.

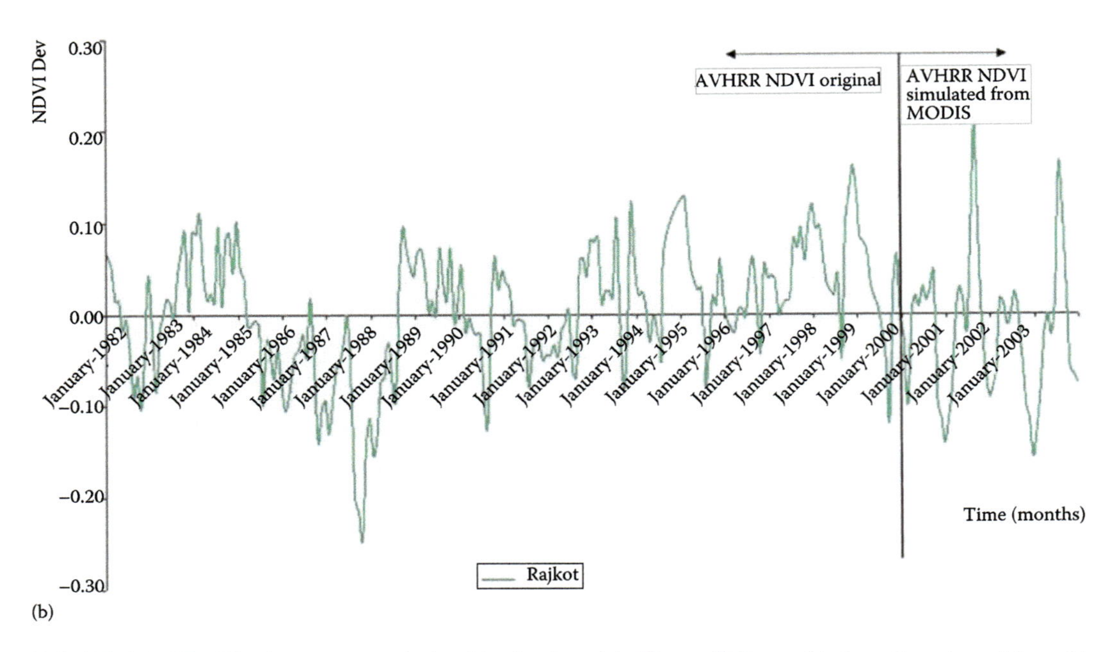


**FIGURE 18.10B** The inter-sensor relationship developed in Figure 18.7a, enabled continuation of drought studies using NDVI deviation from long-term mean ($NDVI_{dev}$). From 1982–2000 the study was conducted using AVHRR 8 Km data. Beyond, year 2000, MODIS 500 m data were used.

*Source:* Thenkabail, 2004.

inter-sensor relationships. Even though Chapter 10 uses EVI and EVI2 to illustrate these inter-sensor relationships, other indices like NDVI can also be used with equal effect. Radiometric inter-sensor cross-calibration uncertainty using a traceable high accuracy reference hyperspectral imager is discussed in the paper by Gorroño et al. (2017). This paper describes a set of tools, algorithms and methodologies that have been developed and used to estimate the radiometric uncertainty achievable for an indicative target sensor through in-flight cross-calibration using a well-calibrated hyperspectral SI-traceable reference sensor with observational characteristics such as TRUTHS (Traceable Radiometry Underpinning Terrestrial and Helio-Studies) (Gorroño et al., 2017).

## 18.11 TOWARD STANDARDIZATION OF VEGETATION INDICES

Monitoring tools such as the Normalized Difference Vegetation Index (NDVI) have been part of remote sensing studies ever since pioneering work reported by Tucker (1979). Simple reflectance from a fixed object will vary over time, even if the object has not changed, due to changes in the measurement conditions, so that for example two plants of identical type, biomass, and health will reflect differently if they are observed under different conditions. Several such factors influence reflectivity making it difficult to conduct scientific studies over space and time using raw remotely sensed data. This difficulty is overcome by using a vegetation index: either a simple ratio derived by dividing reflectivity in the near-infrared waveband by reflectivity in the red waveband, or its equivalent the NDVI, or a more complex index involving adjustment coefficients for one or more confounding environmental factors. Over the years, many satellites and sensors have been launched and their characteristics can differ in many ways: spectral band placement (e.g., differences between broad and narrow bands and their positions within a given spectral range), viewing characteristics (e.g., nadir or off-nadir viewing), conditions under which data are acquired (e.g., Sun angle, atmospheric conditions, Earth–Sun distance), sensor degradation over time, processing methods and many others as espoused in Chapter 11 by Drs. Steven, Malthus, and Baret.). Consistency in the integration of data from different sources and continuity in long-term studies of vegetation change on planet Earth requires us to standardize vegetation indices to adjust for inter-system differences such as, say, Landsat, with MODIS or SPOT or IRS, or SPOT VGT with AVHRR GIMMS as well as intra-system changes such as changes of the Landsat sensor from 1972 to the present day (e.g., Figure 18.11).

Such standardization will involve normalization for various factors (e.g., as mentioned earlier), deriving VIs, and building relationships that will allow for consistent and stable monitoring of vegetation over time and space. For example, VIs of certain time-invariant areas (e.g., certain areas of Sahara Desert) should remain the same over time and extraneous factors such as view angle, atmosphere, and solar elevation and sensor calibration should have minimal influence on measurements at such sites. Therefore, measurements over these invariant sites remain same over time and one can use such time-invariant sites to normalize and correct for system variations.

The Vegetation indices (VIs) suggest from two issues: (1) discrepancies in VIs across sensors, and (2) saturation of VIs at higher biomass levels. Zhou et al. (2023) showed large discrepancies in VIs in mountainous regions. On the other hand, the saturation of spectral reflectance within densely vegetated regions is a renowned challenge that has precluded the optimal use of broadband remotely sensed data and its derivatives for vegetation monitoring (Mutanga et al., 2023). They showed hyperspectral narrow-band vegetation indices (HNB VIs; Thenkabail et al., 2002) from hyperspectral sensors significantly improve high-density biomass estimation (Damodaran and Nidamanuri, 2014). The fusion of waveform lidar indices with other sensors provides unprecedented opportunities for solving signal saturation problems (Mutanga et al., 2023). These issues

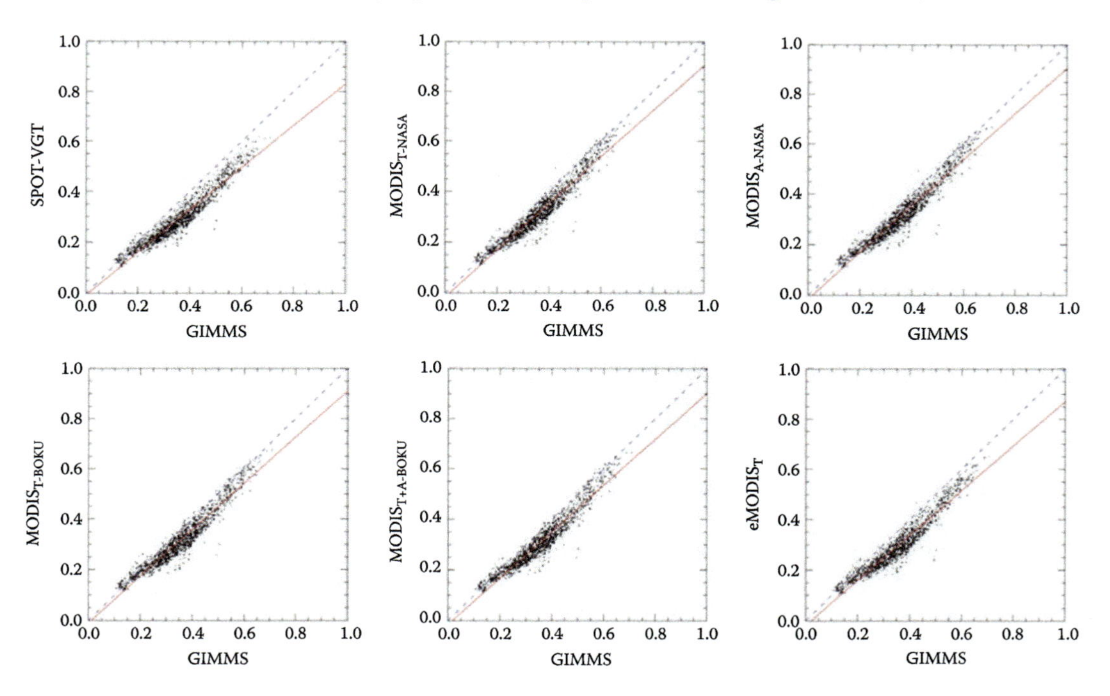


**FIGURE 18.11** Example of empirical scatterplots showing intercalibration between seasonally averaged and regionally aggregated NDVIs from various systems over Kenya. Scatterplots showing intercalibration between normalized difference vegetation index (NDVI)* (seasonally averaged NDVI and aggregated per division) derived from Global Inventory Monitoring and Modeling System (GIMMS) (x-axis) against NDVI* from each of the other NDVI products such as the Système Pour l'Observation de la Terre—VEGETATION (SPOT-VGT) and various MODIS (Moderate Resolution Imaging Spectroradiometer) data. Each plot contains a total of 1512 data points (84 divisions × 9 years × 2 seasons per year).

*Source:* Vrieling et al., 2014.

can be significantly improved or better understood and addressed by inter-sensor calibration and standardization of VIs as enumerated in Chapter 11. In Chapter 11, Drs. Steven, Malthus, and Baret start with how VIs are computed, what factors influence variation in VIs, and VI intercalibration approaches. The chapter then provides equations for cross-sensor intercalibration, based on precise simulations of system spectral responses from a common dataset of field observations. Then, the chapter lays out certain universal principles for VI standardizations. The principles include that the standardization should be: (1) applicable to Vis from all systems, (2) fully traceable as standards evolve and develop over time, and (3) modular. Finally, the chapter outlines specific proposals for VI standards and discusses the precision of the adjustments required.

The Sentinel Hub site (https://custom-scripts.sentinel-hub.com/custom-scripts/sentinel-2/indexdb/) is an excellent resource that contains more than 200 various vegetation indices with its formula (algorithm) developed with image bandwidth (nm) and accompanying Java Script. Not only that the site has thousands of scripts for Sentinel2, but it also contains for Landsat, MODIS, and other satellite images. One can understand the efficiency of indices through their application in solving real-world intricate vegetation-based management issues.

## 18.12 CROWDSOURCING IN REMOTE SENSING AND SPATIAL TECHNOLOGIES TO STUDY PLANET EARTH

Remote sensing made it possible to gather data from any spot on our planet Earth: remotely, consistently, repeatedly, and in various levels of detail. Nevertheless, the need "to be there on the spot physically" to decipher or verify the actual conditions on the ground remains. For example, satellite remote sensing can help establish important information such as location, size, and orientation of buildings in a residential area through processing of multispectral, high-resolution images. Still, the most important proxies for building vulnerability to natural disasters, such as presence of steel frames or seismic retrofitting, are not set to become visible in spaceborne data anytime soon. Or even when remote sensing establishes these facts, there is a need to verify, to be sure or nearly so. Naturally, gathering such ground-based data extensively across the planet in timely, routinely, and cost-effective manner is extremely difficult if not infeasible. Crowdsourcing helps to overcome this impediment to a significant extent. In remote sensing, crowdsourcing of data are to gather reference truth data and information, either based on ground surveys and/or through other reliable sources, from wide array of sources typically through internet. Crowdsourced data are used to train algorithms such as the machine learning algorithms (Elmes et al., 2020) and to test and validate products generated. For example, an interactive and iterative method for crop mapping through crowdsourcing optimized field samples is demonstrated using sub-meter to 5 m very high spatial resolution imagery (VHRI) data for global and local studies (Thenkabail et al., 2021, Yu et al., 2023). Nevertheless, great care should be taken to ensure that the crowdsourced data meets the high-quality checks. Gathering crowdsourced data without quality control can lead to significant uncertainties in the accuracies of these data leading to cascading effects in subsequent data products generated using these crowdsourced data. In Chapter 12, Dr. Dell'Acqua illustrates some of the major advances made in crowdsourcing such as:

1. OpenStreetMap, which is widely used.
2. "Did you feel it service" (DYFI), a pioneering earthquake intensity mapping based on crowd sourced information gathered from people on the ground. This led to Community Internet Intensity Maps (CIIMs).
3. Geo-Wiki for land cover mapping.
4. Google Earth, though a business conglomerate helps develop it, is a crowdsourcing endeavor in remote sensing data generation and dissemination for public use in a very advanced way with the introduction of Google Earth Engine (GEE).

There are, of course, numerous other applications of crowdsourcing for spatial information (e.g., Figure 18.12; Heipke, 2010). This can be, for example, used for conducting election pool surveys, product surveys, and even mapping field boundaries or host of other services. Crowdsourcing assumes that there are enough volunteers who will provide such data either for free (personal satisfaction) or for a fee. Chapter 12 discusses the evolution of crowdsourcing and provides many practical examples of crowdsourcing.

In an increasingly sophisticated, internet-dominated world of today, crowdsourcing is both attractive and desirable. Nevertheless, one needs to be cautious of many aspects of crowd-sourced data. It is not scientifically sound data for many applications. For example, when millions of users feed data on where agricultural croplands are or say something about their productivity situation (e.g., drought or stress condition), uncertainties in such data are likely to be high. Even when such data are collected by experts, there are differences because of the factors such as the depth of

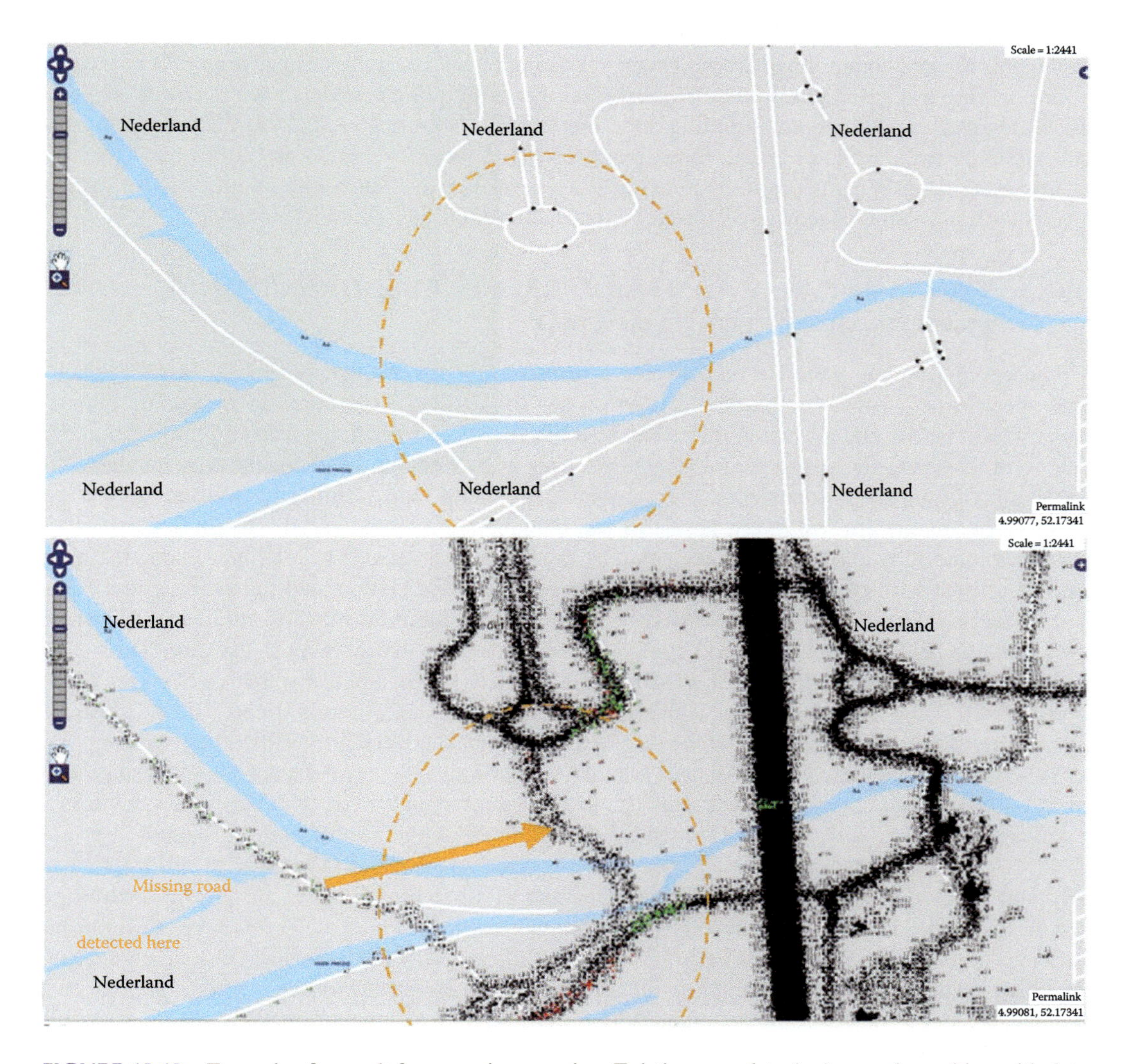


FIGURE 18.12 Example of a result from passive mapping. Existing map data (top), superimposition with data from GPS tracks derived from cars (bottom). The missing road in the center is clearly visible, however the individual tracks need to be aggregated into an attributed graph structure for use in routing applications (© Tele Atlas).

*Source:* Heipke, 2010.

one's understanding, and definition issues. So, when untrained volunteers provide such data, the likelihood of these uncertainties increases. Yet, besides statistical approaches to data cleaning, it is expected that the technology itself will develop increasingly smart solutions to reduce these uncertainties and make the crowd sourced data more accurate and reliable. For example, when one takes a photo of a particular crop and uploads, the technology may identify the crop type automatically along with its precise location, removing the uncertainties that may creep from a volunteer input (Thenkabail et al., 2021, Yu et al., 2023). Given these facts, crowdsourcing will become an important and integrated component of the future remote sensing and spatial data analysis practices.

Crowdsourced data are acquired quickly from sources such as sub-meter to 5-m VHRI, through mobile apps, smartphones, and social media. However, quality control of the data delivered is a must. When non-experts deliver data such as on social media or mobile apps, these data may or may not be useful in training the models and in testing and validating the products depending on the quality of the data delivered. That is why, it is important to keep in view, that the quality of samples is the only factor and not the quantity. Thereby, it is important to establish clear protocols of crowdsourcing for every given application. Nevertheless, crowdsourcing is widely used in remote sensing for wide array of applications including flood monitoring (Helmrich et al., 2021), disaster management (Sukhwani and Shaw, 2020), forest management (Schepaschenko et al., 2015), and virtually in any other field. What needs to be noted is to use crowdsourced data with quality control, which varies with immediate needs like disasters to producing accurate scientific products.

## 18.13 PROCESSING REMOTE SENSING DATA IN CLOUD-COMPUTING ENVIRONMENTS: FUNDAMENTALS

Cloud computing can generate and involve a massive amount of data, especially in geosciences and specifically when remote sensing is involved. Today, data is gathered in increasing quantities by both public and private entities, and in every dimension: spatial, spectral, radiometric, and temporal, covering the entire planet. The challenges of processing these gigabytes, terabytes, and petabytes of "big data" repeatedly and converting this data into meaningful information for the study of our planet are real today. How this can be done is presented and discussed in Chapter 13 on cloud computing and remote sensing data by Dr. Sugumaran et al. They define cloud computing and contrast it with cluster, grid, high performance computing (HPC), and high throughput computing (HTC). The chapter discusses cloud computing paradigms such as the public cloud, private cloud, and hybrid cloud. Some cloud computing services allow researchers to use paradigms such as Infrastructure as a Service (IaaS), Platform as a Service (PaaS), and Software as a Service (SaaS) and make possible the processing of massive amounts of data for nominal costs. They illustrate the ability to process one large LIDAR dataset (~6 terabytes) covering the entire state of Iowa. An open system architecture for Geospatial Cloud Computing is shown in Figure 18.13 (Evangelidis et al., 2014). Cloud computing platforms use satellite sensor data from wide array of satellites in application such as agriculture (Thenkabail et al., 2021, Oliphant et al., 2019, Teluguntla et al., 2018, Xiong et al., 2017a,), crop water use (actual evapotranspiration) (Melton et al., 2022), forests (Brovelli et al., 2020), land use/land cover (Parente et al., 2019, Xiong et al., 2017b), hydrology (McCabe et al., 2017), and many other applications. Amani et al. (2020) reviewed 450 journal articles published in 150 journals between January 2010 and May 2020 that used a cloud computing platform for many applications and provide a comprehensive overview of the methods, approaches, challenges, and advances. Cloud computing in remote sensing is quite advanced based on extensive studies over the last decade. Yet, some of the biggest challenges are:

1. Ability to process very high-resolution spatial (e.g., a few meters), spectral (e.g., hundreds of bands), radiometric (e.g., 16-bit or higher), and temporal (e.g., daily) images of the world

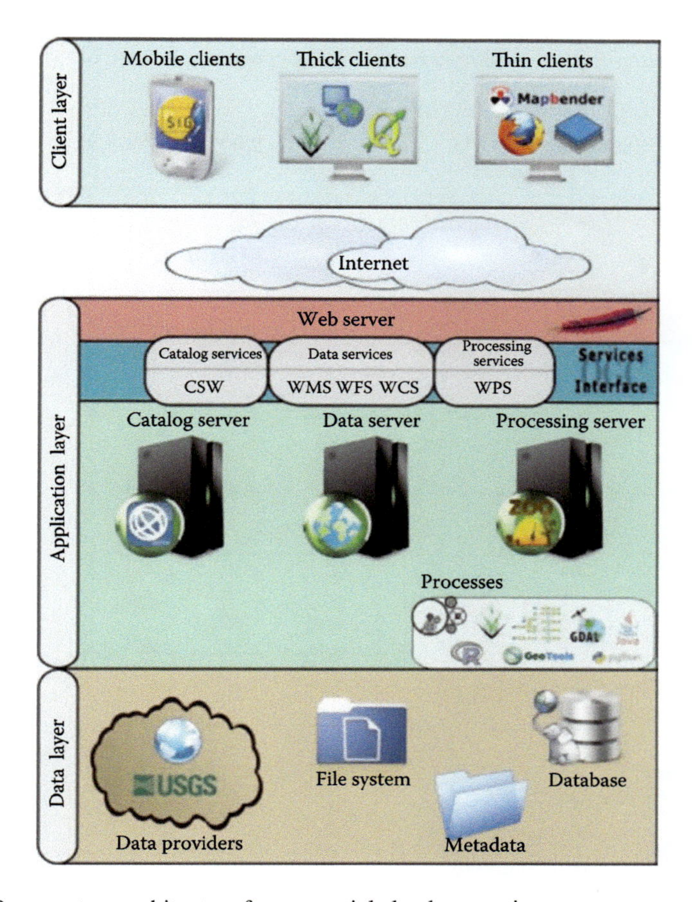


**FIGURE 18.13** Open system architecture for geospatial cloud computing.

*Source:* Evangelidis et al., 2014.

2. Development of image-processing algorithms in the cloud environment
3. Data processing lag due to Server usage density during certain period
4. Data security
5. Storage and backup of "big data" over long time periods

Nevertheless, in due time, even these challenges will be overcome and allow ubiquitous processing of "big remote sensing data" in various cloud computing environments and other local clouds using machine learning, deep learning, and artificial intelligence.

## 18.14 REMOTE SENSING AND CLOUD COMPUTING ADVANCES

Cloud computing became inevitable as the remote sensing big-data volumes reached unmanageable scale for stand-alone systems. Remote sensing data is now acquired routinely covering the entire Planet routinely through more than 18,000 satellites and tens of thousands of airplanes, balloons, and UAVs/UASs, by wide array of sensors, with high-spatial and high-spectral resolutions, and in frequent time-intervals. All this created peta-byte scale data volumes that are impossible to store, process, and analyze in any local systems. The cost of doing so, by individual institutes becomes infeasible. Only a very few highly funded institutes could afford such high-performance computing, leaving others out of reach (Lynn et al., 2020, Shalf, 2020). The solution lays in cloud computing. In a nutshell, cloud computing of remote sensing offers a storage and processing of all the planet's

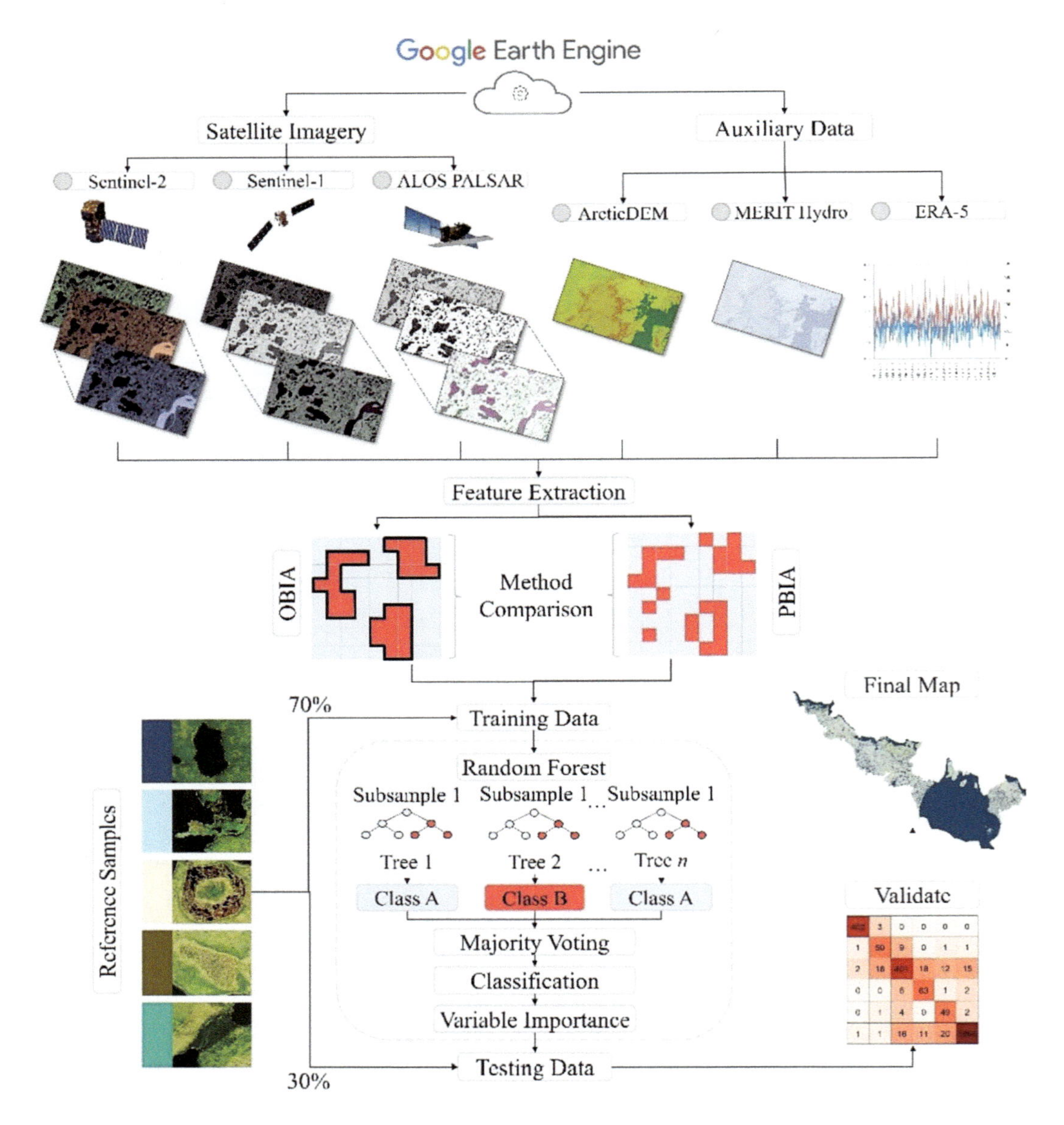


**FIGURE 18.14** Typical implementation of the Google Earth Engine (GEE) cloud-computing workflow.

*Source:* Merchant et al., 2023.

remote sensing data stored in a common platform made accessible to anyone with a simple internet connection. One has a pay as he/she uses, with unit cost made affordable to a wide section of people (Lynn et al., 2020, Shalf, 2020). Environmental Systems Research Institute (ESRI) has experimented with this idea and recently succeeded with the cloud-based ArcGIS Pro software (Gao et al., 2023) that conducts all the software-based analyses of raster (remote sensing) and vector data alike. Although a user saves the input and output images in their own hard drives, the entire data processing (Tilmes and Fleig, 2008) happens through cloud and ESRI provides many data from its server over cloud. Remote sensing data stored in the cloud computing platform comes from multiple satellite and sensors that are harmonized, standardized, and normalized and is in ready to process format for various applications. Cloud platform (Thenkabail et al., 2021, Lynn et al., 2020, Shalf, 2020) interconnects thousands of computers, facilitating parallel computing using simple codes. The codes developed by user's can be shared. The products produced using same data by multiple user's allows for repeatability of products. In many ways, cloud computing is democratization of the use of remote sensing big data. No more single or few workstations or desktop computer-based computing of remote sensing data feasible nor does it meet the challenges of big-data.

Chapter 14 by Wang et al., provides a detailed mechanism of cloud computing of remote sensing data. It provides an overview of the needs and challenges of cloud computing of remote sensing data. They outline services of clouds can be grouped into categories such as data as a service (DaaS), software as a service (SaaS), platform as a service (PaaS) and infrastructure as a service (IaaS). They discuss the technologies of cloud computing in remote sensing, including ubiquitous remote sensing, data integration, data organization and management, data processing, data sharing and distribution, as well as data mining, among others. In addition, popular used cloud computing mechanisms for remote sensing are also discussed in detail.

There are numerous applications of the cloud computing using remote sensing big data. Thenkabail et al. (2021) produced the global cropland extent product using cloud computing using petabyte scale Landsat 30 m 16-day multi-band data of the world from multiple years. Figure 18.14 provides a structure of cloud computing for large-scale arctic wetland mapping (Merchant et al., 2023). Machine learning algorithms such as the supervised pixel-based classifiers (Whiteside et al., 2011) such as the Random Forest (RF) and Support Vector Machines (SVMs) or the unsupervised ISOCLASS clustering are widely used (Thenkabail et al., 2021).

## 18.15 GOOGLE EARTH AND GOOGLE EARTH ENGINE FOR EARTH SCIENCES

In Chapter 15, Dr. Bailey outlines how Google Earth, launched in the year 2005, brought complex remote sensing technology to the doors of common person. Based on the philosophy of "Digital Earth," a concept popularized by former vice president of the United States of America Mr. Al Gore in 1998 (Annoni et al., 2023), Google Earth enabled some powerful possibilities. These included:

1. Visualizing the entire Earth from your desktop, laptop or mobile seamlessly
2. Ability to zooming in and zooming out of any part of the planet and visualize a number of details such as cities, forests, agricultural lands, houses, trees, and networks (e.g., roads, rivers)
3. Precisely locating annotations
4. Providing the "power of information" to common person who may not have any idea of remote sensing or geoscience
5. Enabling people to tell stories or inform changes through images

Google Earth involves use of geocoded imagery of any various resolutions along with the 3D elevation derived from SRTM and other (e.g., LiDAR) high resolution data. The baseline imagery for the Google Earth is Landsat 30 m imagery, but for large swaths of the World very high spatial resolution imagery (VHRI; sub-meter to 5 m) from providers such as Digital Globe are also available (Zhao et al., 2021). The system is supported and revolutionized by Keyhole Markup Language (KML).

Further, the ease of use and the power of data and information available in Google Earth for anyplace, anywhere, and to anyone has truly democratized remote sensing and GIS sciences.

Gorelick et al. (2017) wrote a pathfinding paper on Google Earth Engine (GEE) for planet scale studies. Pérez-Cutillas et al. (2023) performed a meta-analysis of GEE studies which showed that: (1) the Landsat 8 was the most widely used satellite (25%); (2) the non-parametric classification methods (Rodriguez-Galiano et al., 2012), mainly Random Forest, were the most recurrent algorithms (31%); and (3) the water resources assessment and prediction were the most common methodological applications

(22%). Zhao et al. (2024) evaluated various GEE classification machine learning algorithms (MLAs) such as the classification and regression trees [CART], support vector machine [SVM], and random forest [RF]) and found RF performed the best. In several studies (Thenkabail et al., 2021, Teluguntla et al., 2018, Xiong et al., 2017a, 2017b) it is obvious that RF performs the best, especially when there is sufficient reference data to develop MLAs, to test the products and to validate the products.

Chapter 15 also shows us how several scientific applications and applications of societal importance of Google Earth soon became apparent. They show us example of Volcanoes of the World mapped by the Smithsonian. Others like Thenkabail et al. (2009a, 2009b) have extensively made use of VHRI as ground data to identify and label croplands of the World as well as assess cropland accuracies. These days, VHRI in the Google Earth are extensively used by remote sensing scientists as data for developing or testing their algorithms in certain applications. Nongovernmental organizations (NGOs) and bloggers have used Google Earth to illustrate and demonstrate issues of environmental and societal importance that often influence public opinion. Crowdsourcing of information gathered from remote sensing imagery are depicted on Google Earth during such events as earthquakes, floods, hurricanes, and tsunamis (e.g., Figure 18.15).

Even though, Google Earth has tremendous strengths, it cannot be used for all scientific studies that use remote sensing. This is because the imagery available is used only "visually," without the underlying values and metadata available, which are required for quantifying variables and studying changes. For example, Google Earth imagery cannot be used for quantifying drought, deriving drought indices, and determining drought conditions over time. However, to overcome these limitations, Google Earth Engine (GEE) has been introduced. EE stores real images. For example, over the entire Landsat imagery archive (from 1972 to present) is currently available in EE. This imagery can be used for scientific studies by building algorithms directly in the EE platform to analyze the imagery. This also allows for time-lapse analysis. Earth Engine offers a cloud-computing platform to quickly, instantly, analyze images at global scale even at 30 m spatial or higher resolutions. This

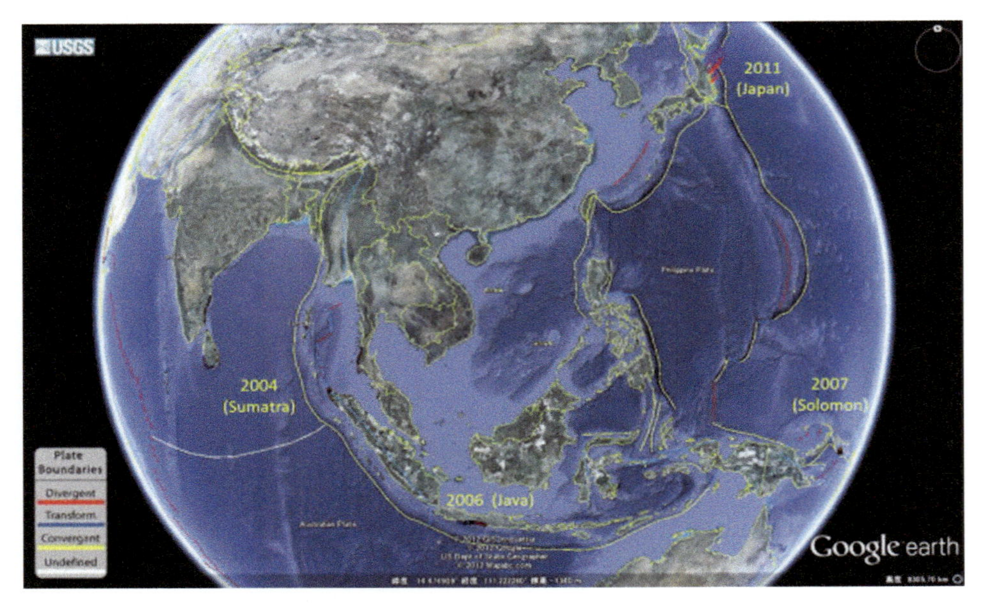


FIGURE 18.15 Tsunami field survey data available in Google Earth ".kml" format since 2004 and Plate tectonics in the Indian and Southwest Pacific Oceans. A collection of the Google Earth www.google.com/earth/index.html) ".kml" files of tsunami measurement data from field surveys after the 2004 event including the 2004 Indian Ocean, 2006 Java, 2007 Solomon and 2010 Chile are summarized (Fujima, 2011, Mori and Takahashi, 2012) and shown with plate tectonics (USGS, 2012) in this figure.

*Source:* Suppasri et al., 2012.

platform is still being developed, but already some interesting global applications are becoming available (e.g., forest cover and wetland changes using Landsat mentioned in Chapter 15).

## 18.16 MAP ACCURACIES

Chapter 16 by Dr. Congalton provides an in-depth guideline on assessing thematic and positional accuracies of maps. When remotely sensed data are processed, classified, and information deciphered through digital image processing and interpretations (or through manual interpretations as in the past), a thematic map is produced. Examples of thematic maps include products such as land cover types, species types, biomass categories, and leaf area index. Scientific use of any thematic map requires assessing it's thematic as well as positional accuracies. Thematic accuracy addresses how well classes or quantities in thematic map correspond to what is observed or measured in the field or ground. Positional accuracy refers to quantifiable differences between two map locations or a map location and a location on ground. Dr. Congalton addresses these issues systematically and comprehensively in Chapter 16 and these are summarized as follows:

1. Error sources: Errors in thematic maps can occurs as a result of any number of causes: systematic (e.g., sensor calibration errors), natural (e.g., atmospheric), preprocessing (e.g., geometric and radiometric correction), derivative (e.g., NDVI), classification schemes and methods (e.g., digital image processing), class definitions, reference data collection design (e.g., sample design, position inaccuracies), and user interpretation of classes and class labeling.
2. Classification: Thematic maps are produced by digital image analysis. The four characteristics of classification are identified as: (1) complete definition, (2) mutually exclusive, (3) totally exhaustive, and (4) hierarchical.

C. Reference data: Typically, reference data are collected through ground visits. A rule of thumb is to collect 50 samples for each land cover category when total numbers of classes are 12 or less. When classes are more than 12, the recommendation is to gather data from 75 or 100 samples per class. However, the number of samples can go up for large areas. Reference data are collected using well-designed sampling strategies (e.g., random or stratified random). Further homogeneous areas are selected for collecting ground data. For example, collecting data from a single 30 m pixel could lead to errors due misregistration (locational error). Therefore, it is better to select homogeneous areas for a land cover class such that a 3 × 3 cluster of 30 m pixels are chosen to reduce the positional error.

Maps that categorize the landscape into discrete units are a cornerstone of many scientific, management and conservation activities (Lyons et al., 2018). The accuracy of these maps is often the primary piece of information used to make decisions about the mapping process or judge the quality of the final map (Lyons et al., 2018). Accuracy assessment metrics include overall accuracies, producer accuracies (errors of omissions), user accuracies (errors of omissions), kappa, entropy, purity, quantity/allocation disagreement (Lyons et al., 2018). Maps that report overall accuracies, without the producer's and user's, are often misleading. For example, an overall accuracy of 90%, for a cropland map in which cropland class has a producer's accuracy of just 70% means that there is 30% errors of omissions (meaning 30% of croplands go missing). Similarly, if the user's accuracy is 80% for the cropland class, means there is 20% errors of commissions (non-croplands mapped as croplands). So, any accuracy must just report overall accuracies but all other accuracy parameters.

Dr. Congalton then provides a series of error matrices to evaluate map accuracies. These error matrices compute:

1. Three basic accuracies: An error matrix computes three basic accuracies: producer's, users', and overall. Accuracies are computed by comparing the thematic map classes with reference data. Producer's accuracy refers to percentage of class X in reference map that is

correctly classified as class X in a thematic map. A user's accuracy is the percentage of the pixels classified as a class X in a thematic map are indeed class X. Overall accuracy is the percentage of pixels correctly classified from all classes to total number of pixels from all classes. If sample sizes vary, so does the producer's, users, and overall accuracies. Larger the sample size, more robust these accuracies are.

2. Normalized accuracies: "Margfit" normalized the error matrix by removing the effect of sample size. "Kappa" KHAT statistic (Congalton and Green, 2019) is another measure of accuracy and is effectively used as a way to test if one map is statistically better than another.
3. Fuzzy accuracy assessment: instead of deterministic accuracy assessment, like the ones mentioned earlier, fuzzy classification provides measure of accuracy that considers possibilities such as: correct, mostly correct, somewhat correct, mostly incorrect, somewhat incorrect, incorrect (e.g., Figure 18.16; Comber et al., 2012).

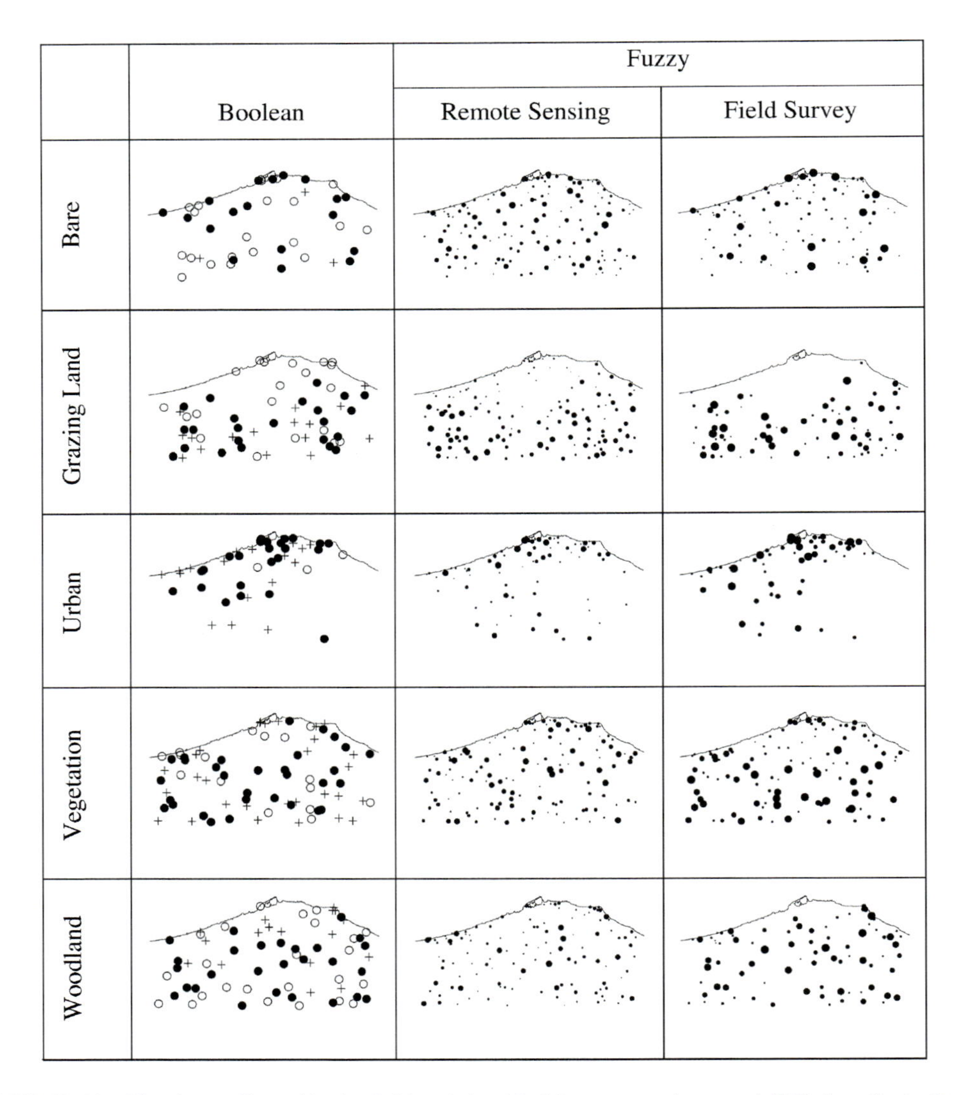


**FIGURE 18.16** The data collected in the field and classified from remotely sensed (RS) data. In the Boolean maps, solid symbols (•) indicate where both Field and RS agree, hollow circles (○) where only the RS indicates the class and crosses (+) where only field survey indicates the class. In the maps of fuzzy classes, the size of the plot characters indicates the degree of membership to the class.

*Source:* Comber et al., 2012.

## 18.17 REMOTE SENSING LAW OR SPACE LAW

Remote sensing law is part of international space law as well as the quickly developing field of geospatial law. The early evolution of space law was based primarily on the military and strategic requirement of nations. However, wide, and ubiquitous use of remote sensing has resulted in broad set of legal principles governing remote sensing for civilian uses. The evolution of technology has contributed to remote sensing becoming an essential component of many other technological delivery systems. This means that the legal principles governing remote sensing technologies are increasingly being developed in the context of the broader field of geospatial law. This is especially so considering the increasing number of Earth Observation satellites launched and operated by various nations (e.g., Figure 18.17).

Throughout the history of the astronautic development, one of the key principles of international space law has always been the principle of international cooperation between states (Chernykh and Volodin, 2023). Space law is quite broad and complex and involves (Lyall and Larsen, 2018) (1) the general international and domestic sources of law applicable to outer space, fora, and the basic legal principles applicable in outer space; (2) complicated commercial and practical legal implications of outer space—the use of telecommunications, broadcasting, environmental law, and activities more broadly—all while ensuring relevance to the fundamental principles that underpin the boundless domain; and (3) financing, trade restrictions, commercial law, and military perspectives while also considering extraterrestrial intelligence, and address the legal questions for the future (Lyall and Larsen, 2018). Remote sensing satellites are often used for verification in public international law (Hettling, 2003).

Dr. Blount in Chapter 17 not only provides an excellent discussion on these matters but refers to an extensive set of International and National treaties, regulations, agreements, and policies that when combined form the complex web of regulations governing these activities.

Broadly, the four pillars of remote sensing law are:

1. International Space Treaty Regime
2. International remote sensing law for military era or disarmament and verification law

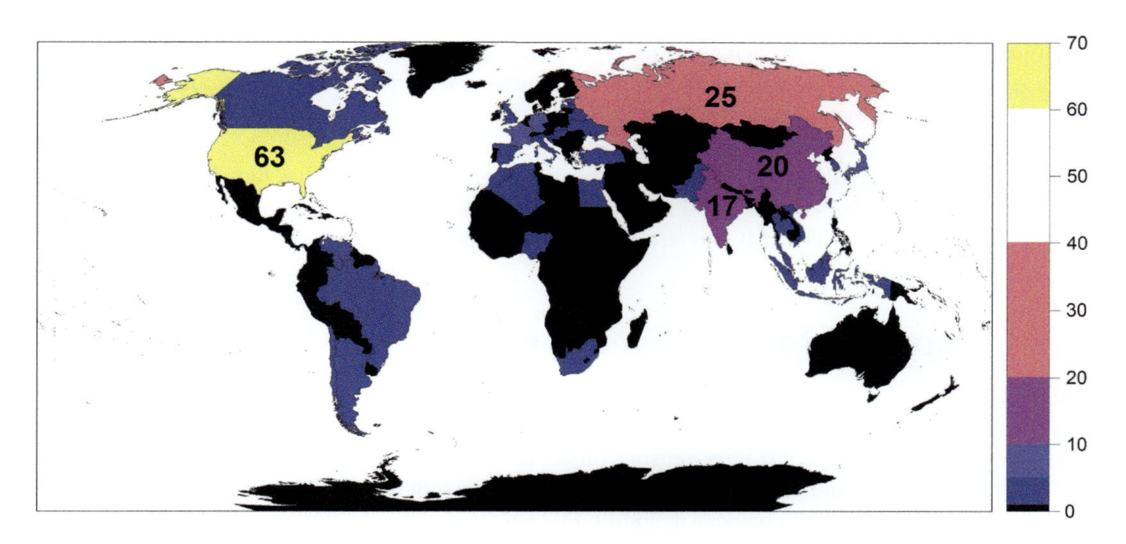


**FIGURE 18.17** Map showing the total number of near-polar orbiting, land imaging civilian satellites launched by (or on behalf of) different geographical regions between July 23, 1972 and December 31, 2013. The legend to the right of the map shows the number according to seven groups; 0; 1–5; 6–10; 11–20; 21–40; 41–60; 60–70. Note that no region falls into the 41–60 category. The numbers of launches made by the top four individual countries (India, China, Russia and the United States of America) are specifically cited—note that collectively Europe has launched 30.

*Source:* Belward and Skøien, 2014.

3. International remote sensing principles for civilian purposes
4. National geospatial law principles

First, Space Treaty Regime defines laws pertaining to:

1. Free access to outer space for any nation
2. Observe and collect data for any part of the world, cutting across sovereignty
3. Governments to "authorize and supervise" nongovernmental or other players (e.g., commercial) satellites

Second, disarmament and verification law was for military purposes involving control of ballistic missiles, freedom to monitor military activities, and intelligence gathering.

Third, remote sensing for civilian purposes is guided by The United States General Assembly Resolution 41/65 and the UN Committee on the Peaceful Uses of Outer Space (UNCOPUOS) which outlines the core legal principles dealing with remote sensing activities for civilian purposes (Masson-Zwaan and Hofmann, 2019, Zannoni, 2019). These principles can be divided into three groups:

1. Preexisting binding legal principles
2. Innovations involving international law governing remote sensing
3. Promoting the soft obligation of international cooperation

The center piece of the international law principles is "the freedom data collection and distribution," while at the same time ensuring "respect of the principle of full and permanent sovereignty of all States and peoples over their own wealth and natural resources." Through these measures a large amount of cooperation of internal remote sensing data collection and access is determined for human welfare. These efforts are further solidified through efforts through GEO and GEOSS where web-enabled (free) remote sensing data sharing and distribution is encouraged for a wide range for peaceful applications (e.g., the nine societal beneficial areas such as forests, agriculture, water, biodiversity, disasters and so on) (Masson-Zwaan and Hofmann, 2019, Zannoni, 2019).

Fourth, remote sensing law is increasingly perceived to be part and parcel of geospatial law. With the use of many interconnected technologies such as remote sensing, GIS, GPS, internet delivery of data, mobile delivery of data, and number of other technological and commercial convergence of spatial and non-spatial data and their delivery systems. Remote sensing can gather information at great detail, repeatedly, and over any area of the globe. This means that remote sensing activities and data have become implicated in a wide variety of legal fields such as evidence, torts, and privacy, among others. These issues are most often domestic in nature and operate outside the international regimes.

## 18.18 SYNTHESIS OF THE 17 CHAPTERS IN VOLUME I

This is the first of the six-volume *Remote Sensing Handbook* (Second Edition). The first volume deals with (1) satellites and sensors, (2) remote sensing fundamentals, (3) remote sensing data normalization, harmonization, (4) inter-sensor comparisons, (5) crowdsourcing, (6) cloud computing, (7) accuracies, errors, and uncertainties, and (8) remote sensing law. Each of these broad topics are discussed in detail in the 18 chapters including this summary chapter. The Volume I is a must read for every remote sensing student, teacher, practitioner, and professional.

There is depth and comprehensive details of satellites and sensors in Chapter 1. These are ground-based, platform mounted, airborne (both aircraft-based and drone-based), and spaceborne. Characteristics of all types of sensors (e.g., optical, radar, LiDAR, thermal, hyperspatial, hyperspatial) are presented and discussed. Polar-orbiting Sun-synchronous, geostationary, and International Space Station-borne (ISS-borne) sensors are all discussed.

In Chapters 2–4 Global Navigation Satellite Systems (GNSS) constellations include Global Positioning Systems (GPS) and cellular networks or integration of the two and are of immense value in tracking precise positions that are now ubiquitous in our daily life including smartphones, autonomous vehicles, and navigations of all types. These chapters provide characteristics of various GNSS systems operated by different countries or group of countries, providing service for entire planet. The chapters discuss their theory, concepts, and applications.

Chapter 5 is unique in the sense it provides fundamentals of remote sensing that any reader of the subject will be interested into study, refresh, or to refer to from time to time. It provides basic understanding of the electromagnetic spectrum, complex theories of remote sensing science based on measurements made in different portions of the electromagnetic spectrum such as in the visible, red-edge, near-infrared, shortwave-infrared, thermal infrared, and others like radar, and LiDAR. It also espouses the applications potential base on remote sensing data and ways and means of interpreting them.

Chapters 5–11 are for everyone to get deep understanding of synthesis of remote sensing data for science and applications. First and foremost, the raw remote sensing data (acquired in radiance) must be converted to top of the atmosphere (TOA) reflectance and ground reflectance. Data must be calibrated and normalized over time and space, so we can study features on ground based on harmonized and standardized data. The concept of time-invariant sites for calibration, pre-launch, and post-launch calibrations and normalizations are all presented and discussed with sound theory. Normalization of the data within sensors over time, and across sensors over time are espoused by developing and presenting relationships. These relationships involve normalized surface reflectance as well as normalized indices such as the Normalized Difference Vegetation Indices (NDVIs). Increasingly seamlessly using satellite sensor data from multiple sensors over time and space are becoming common norm (e.g., harmonized data of Landsat time-series and Sentinel time-series). However, these sensors all have varied spectral, spatial, and radiometric resolution. Further, sensors degrade over time. All these issues can be overcome through data normalization, harmonization, and standardization that involve geometric, radiometric, and atmospheric corrections, calibrations for sensor degradation, and inter-sensor relationship development. All of these are thoroughly discussed in Chapters 5–11.

Chapter 12 is devoted for crowdsourcing of reference training, testing, and validation data with precise geolocation involving both professional networks as well as citizenry participation. Evolving of multiple technologies like smartphones, app developments, and ability to collect data anywhere in the world and instantly transfer that to servers has made a paradigm shift in how data is collected through this crowdsourcing mechanism. All sorts of data can be collected. For example, factors such as where are crops grown, how frequently they are grown, what is their health, what is their yield, and whether they are irrigated or rainfed. There is no limit to what type of data can be crowdsourced. However, quality of this data needs to be carefully assessed. Nonexperts collecting data may bring in large volume of data, but what matters is the quality of data collected.

Chapters 13–14, introduce and demonstrate cloud computing capabilities. Remote sensing is mother of all big data. But, handling this data in terms of storage, processing, and analyzing can be very challenging. This is overcome by using parallel computing involving thousands of computers networked on the cloud. Large volumes of remotely sensed data of the entire Planet collected from multiple satellites over long time periods are all seamlessly stored, processed, and analyzed on the cloud.

Chapter 15 is devoted to Google Earth Engine (GEE) cloud-based computing methods, approaches, and advances. The cloud computing is performed using machine learning, deep learning, and artificial intelligence from anywhere through an internet connection by coding using Python and/or JavaScript. Cloud computing has truly ushered a new era in computing where petabyte scale data can be handled and analyzed for various applications in matter of minutes what a server-based computing would have taken days.

Chapter 16 provides the concepts of accuracy assessments of remote sensing data derived products. Error matrices establish overall accuracies of the map as well as producer's accuracies (errors of omissions) and user's accuracies (errors of commissions) of each individual class. The chapter also deals with how to collect ground data or other reference data for accuracy assessments, process of avoiding issues like autocorrelation, general rule of quantity and quality of data required for accuracy assessments, and approaches one takes in accuracy assessments for data of various resolutions and types.

Chapter 17 provides a comprehensive exposition of remote sensing laws and principles. Various governments as well as private players operate satellites which fly over sovereign countries and collect data routinely. What treaties and agreements are in place to make this happen and how they are implemented or regulated are all discussed in the chapter.

The chapters in Volume I, set the stage for other volumes. A thorough understanding of the chapters in the Volume I is a must for any serious student or professional of remote sensing science.

## ACKNOWLEDGMENTS

I would like to thank the lead authors and co-authors of each of the chapters for providing their insights and edits of my chapter summaries. I would like to express my gratitude to the United States Geological Survey (USGS) for the opportunity to work for it and contribute to remote sensing science. Over the years, a great deal of my remote sensing science research was performed through various NASA grants. Very grateful for it. Thanks to my USGS colleagues Dr. Pardhasaradhi Teluguntla, Dr. Itiya Aneece, Mr. Adam Oliphant, and Mr. Daniel Foley for their support at various times of this book project. Thanks also to Dr. Murali Krishna Gumma of the International Crops Research Institute for the Semi-Arid Tropics (ICRISAT). Use of trade, firm, or product names is for descriptive purposes only and does not imply endorsement by the U.S. government.

## REFERENCES

Abdi, O., Uusitalo, J., Pietarinen, J., and Lajunen, A. 2022. Evaluation of forest features determining GNSS positioning accuracy of a novel low-cost, Mobile RTK system using LiDAR and TreeNet, Remote Sensing, Volume 14, Issue 12, Page 2856.

Amani, M., et al. 2020. Google Earth Engine cloud computing platform for remote sensing big data applications: A comprehensive review, IEEE Journal of Selected Topics in Applied Earth Observations and Remote Sensing, Volume 13, Pages 5326–5350, http://doi.org/10.1109/JSTARS.2020.3021052.

Annoni, A., Nativi, S., Çöltekin, A., Desha, C., Eremchenko, E., Gevaert, C.M., . . . Tumampos, S. (2023). Digital earth: Yesterday, today, and tomorrow, International Journal of Digital Earth, Volume 16, Issue 1, Pages 1022–1072, https://doi.org/10.1080/17538947.2023.2187467.

Belward, A.S., and Skøien, J.O. 2014. Who launched what, when and why: Trends in global land-cover observation capacity from civilian earth observation satellites, ISPRS Journal of Photogrammetry and Remote Sensing, Available online 28 April 2014, ISSN 0924-2716, http://dx.doi.org/10.1016/j.isprsjprs.2014.03.009.

Blaschke, T. 2010. Object-based image analysis for remote sensing, ISPRS International Journal of Photogrammetry and Remote Sensing, Volume 65, Issue 1, Pages 2–16.

Blaschke, T., Hay, G.J., Kelly, M., Lang, S., Hofmann, P., Addink, E., Feitosa, R., van der Meer, F., van der Werff, H., Van Coillie, F., and Tiede, D. (2014). Geographic object-based Image Analysis: A new paradigm in remote sensing and geographic information science, ISPRS International Journal of Photogrammetry and Remote Sensing, Volume 87, Issue 1, Pages 180–191.

Branger, F., Kermadi, S., Jacqueminet, C., Michel, K., Labbas, M., Krause, P., Kralisch, S., and Braud, I. 2013. Assessment of the influence of land use data on the water balance components of a peri-urban catchment using a distributed modelling approach, Journal of Hydrology, Volume 505, 15 November, Pages 312–325, ISSN 0022-1694, http://dx.doi.org/10.1016/j.jhydrol.2013.09.055.

Brovelli, M.A., Sun, Y., and Yordanov, V. 2020. Monitoring forest change in the Amazon using multi-temporal remote sensing data and machine learning classification on Google Earth Engine, ISPRS International Journal of Geo-Information, Volume 9, Issue 10, Pages 580, https://doi.org/10.3390/ijgi9100580.

Burleigh, S.C., De Cola, T., Morosi, S., Jayousi, S., Cianca, E., and Fuchs, C. 2019. From connectivity to advanced internet services: A comprehensive review of small satellites communications and networks,

Wireless Communications and Mobile Computing, Volume 2019, Article ID 6243505, Page 17, https://doi.org/10.1155/2019/6243505.

Campbell, J.B., and Wynne, R.H. 2011. Introduction to Remote Sensing, 5th Edition. Authors, Edition. Guilford Press. ISBN 1609181778.

Cardille, J.A., Crowley, M.A., Saah, D., Clinton, N.E. Editor(s). 2023. Cloud-Based Remote Sensing with Google Earth Engine: Fundamentals and Applications. Springer Nature, October 4.

Chander, G., Markham, B.L., and Helder, D.L. 2009. Summary of current radiometric calibration coefficients for Landsat MSS, TM, ETM+, and EO-1 ALI sensors, Remote Sensing of Environment, Volume 113, Pages 893–903.

Chavez, P.S.1988. An improved dark-object subtraction technique for atmospheric scattering correction of multispectral data. Remote Sensing of Environment, Volume 24, Pages 459–479.

Chavez, P.S. 1989. Radiometric calibration of Landsat thematic mapper multispectral images. Photogrammetric Engineering and Remote Sensing, Volume 55, Pages 1285–1294.

Chernykh, I., and Volodin, D. 2023. The principle of international cooperation and sharing of information principle under international space law: Towards synergy, Space Policy, Page 101593, ISSN 0265-9646, https://doi.org/10.1016/j.spacepol.2023.101593, www.sciencedirect.com/science/article/pii/S0265964623000607.

Choate, M.J., Rengarajan, R., Storey, J.C., and Lubke, M. 2023. Landsat 9 geometric commissioning calibration updates and system performance assessment, Remote Sensing, Volume 15, Issue 14, Page 3524, https://doi.org/10.3390/rs15143524.

Chuvieco, E. 2020.Fundamentals of Satellite Remote Sensing: An Environmental Approach. CRC Press, January 22.

Claverie, M., Ju, J., Masek, J.C., Dungan, J.L., Vermote, E.F., Jean-Claude Roger, S.V., and Skakun, J.C. 2018. The harmonized landsat and sentinel-2 surface reflectance data set, Remote Sensing of Environment, Volume 219, Pages 145–161, ISSN 0034-4257, https://doi.org/10.1016/j.rse.2018.09.002,www.sciencedirect.com/science/article/pii/S0034425718304139.

Comber, A., Fisher, P., Brunsdon, C., and Khmag, A. 2012. Spatial analysis of remote sensing image classification accuracy, Remote Sensing of Environment, Volume 127, December, Pages 237–246, ISSN 0034-4257, http://dx.doi.org/10.1016/j.rse.2012.09.005.

Congalton, R.G., and Green, K. 2019. Assessing the accuracy of remotely sensed data: Principles and practices, 3rd Edition. CRC Press, Boca Raton, Page 346, https://doi.org/10.1201/9780429052729.

Czapla-Myers, J., McCorkel, J., Anderson, N., Thome, K., Biggar, S., Helder, D., Aaron, D., Leigh, L., and Mishra, N. 2015. The ground-based absolute radiometric calibration of landsat 8 OLI, Remote Sensing, Volume 7, Issue 1, Pages 600–626, https://doi.org/10.3390/rs70100600.

Damodaran, B.B., and Nidamanuri, R.R. 2014. Assessment of the impact of dimensionality reduction methods on information classes and classifiers for hyperspectral image classification by multiple classifier system, Advances in Space Research, Volume 53, Issue 12, Pages 1720–1734, ISSN 0273-1177, http://dx.doi.org/10.1016/j.asr.2013.11.027.

Dardel, C. Kergoat, L., Hiernaux, P., Mougin, E., Grippa, M., and Tucker, C.J. 2014. Re-greening Sahel: 30 years of remote sensing data and field observations (Mali, Niger), Remote Sensing of Environment, Volume 140, January, Pages 350–364, ISSN 0034-4257, http://dx.doi.org/10.1016/j.rse.2013.09.011.

Di, L, and Yu, E. 2023. Remote sensing. In Remote Sensing Big Data 2023 July 23. Springer International Publishing, Cham, Pages 17–43.

Dow, J.M., Neilan, R.E., Weber, R., and Gendt, G. 2007. Galileo and the IGS: Taking advantage of multiple GNSS constellations, Advances in Space Research, Volume 39, Issue 10, Pages 1545–1551, ISSN 0273-1177, http://dx.doi.org/10.1016/j.asr.2007.04.064.

Edwards, B., Kochtitzky, W., and Battersby, S. 2020. Global mapping of future glaciovolcanism, Global and Planetary Change, Volume 195, Page 103356, ISSN 0921-8181, https://doi.org/10.1016/j.gloplacha.2020.103356,www.sciencedirect.com/science/article/pii/S0921818120302472.

Elmes, A., Alemohammad, H., Avery, R., Caylor, K., Eastman, J.R., Fishgold, L., Friedl, M.A., Jain, M., Kohli, D., Bayas, J.C.L., et al. 2020. Accounting for training data error in machine learning applied to earth observations, Remote Sensing, Volume 12, Issue 6, Page 1034, https://doi.org/10.3390/rs12061034.

Elvidge, C.D., Yuan, D., Weerackoon, R.D., and Lunetta, R.S. 1995. Relative radiometric normalization of landsat multispectral scanner (MSS) data using an automatic scattergram controlled regression, Photogrammetric Engineering and Remote Sensing, Volume 61, Pages 1255–1260.

Evangelidis, K., Ntouros, K., Makridis, N., and Papatheodorou, C. 2014. Geospatial services in the cloud, Computers & Geosciences, Volume 63, February, Pages 116–122, ISSN 0098-3004, http://dx.doi.org/10.1016/j.cageo.2013.10.007.

Farhad, M.M., Kaewmanee, M., Leigh, L., and Helder, D. 2020. Radiometric cross calibration and validation using 4 Angle BRDF model between landsat 8 and sentinel 2A, Remote Sensing, Volume 12, Issue 5, Page 806, https://doi.org/10.3390/rs12050806.

Feng, M., Sexton, J.O., Huang, C., Masek, J.G., Vermote, E.F., Gao, F., Narasimhan, R., Channan, S., Wolfe, R.E., Townshend, J.R. 2013. Global surface reflectance products from Landsat: Assessment using coincident MODIS observations, Remote Sensing of Environment, Volume 134, Pages 276–293, ISSN 0034-4257, https://doi.org/10.1016/j.rse.2013.02.031,www.sciencedirect.com/science/article/pii/S0034425713000849.

Freire, S., Santos, T., Navarro, A., Soares, F., Silva, J.D., Afonso, N., Fonseca, A., and Tenedório, J. 2014. Introducing mapping standards in the quality assessment of buildings extracted from very high resolution satellite imagery, ISPRS Journal of Photogrammetry and Remote Sensing, Volume 90, Pages 1–9, ISSN 0924-2716, http://dx.doi.org/10.1016/j.isprsjprs.2013.12.009.

Fritz, S., McCallum, I., Schill, C., Perger, C., See, L., Schepaschenko, D., van der Velde, M., Kraxner, F., and Obersteiner, M. 2012. Geo-Wiki: An online platform for improving global land cover, Environmental Modelling & Software, Volume 31, May, Pages 110–123, ISSN 1364-8152, http://dx.doi.org/10.1016/j.envsoft.2011.11.015.

Fujima, K. 2011. Tsunami Measurement Data Compiled by IUGG Tsunami Commission, www.nda.ac.jp/cc/users/fujima/TMD/index.html [accessed 02.12.11].

Gao, Y., Lui, J., Liu, Y., Zhai, Z., Che, J., Li, H., Wang, R., and Liu, J. 2023. Research on natural resources spatio-temporal big data analysis platform for high performance computing, Int. Arch. Photogramm. Remote Sens. Spatial Inf. Sci., XLVIII-M-1–2023, Pages 115–121, https://doi.org/10.5194/isprs-archives-XLVIII-M-1-2023-115-2023.

Garrison, G., Helder, D., Begeman, C., Leigh, L., Kaewmanee, M., and Shah, R. 2022. Initial cross-calibration of Landsat 8 and Landsat 9 using the simultaneous underfly event, Remote Sensing, Volume 14, Issue 10, Page 2418, https://doi.org/10.3390/rs14102418.

Gbanie, S.P., Tengbe, P.B., Momoh, J.S., Medo, J., and Kabba, V.T.S. 2013. Modelling landfill location using Geographic Information Systems (GIS) and Multi-Criteria Decision Analysis (MCDA): Case study Bo, Southern Sierra Leone, Applied Geography, Volume 36, January, Pages 3–12, ISSN 0143-6228, http://dx.doi.org/10.1016/j.apgeog.2012.06.013.

Gorelick, N., Hancher, M., Dixon, M., Ilyushchenko, S., Thau, D., and Moore, R. 2017. Google Earth Engine: Planetary-scale geospatial analysis for everyone, Remote Sensing of Environment, Volume 202, Pages 18–27, ISSN 0034-4257, https://doi.org/10.1016/j.rse.2017.06.031,www.sciencedirect.com/science/article/pii/S0034425717302900.

Gorroño, J., Banks, A.C., Fox, N.P., and Underwood, C. 2017. Radiometric inter-sensor cross-calibration uncertainty using a traceable high accuracy reference hyperspectral imager, ISPRS Journal of Photogrammetry and Remote Sensing, Volume 130, Pages 393–417, ISSN 0924-2716, https://doi.org/10.1016/j.isprsjprs.2017.07.002,www.sciencedirect.com/science/article/pii/S0924271616306517.

Govaerts, Y., Nollet, Y., and Leroy, V. 2022. Radiative transfer model comparison with satellite observations over CEOS calibration site Libya-4, Atmosphere, Volume 13, Issue 11, Page 1759, https://doi.org/10.3390/atmos13111759.

Goward, S.N., Ghander, G., Pagnutti, M., Marx, A., Ryan, R., Thomas, N., and Tetrault, R. 2012. Complementarity of resourceSat-1 AWiFS and Landsat TM/ETM+ sensors, Remote Sensing of Environment, Volume 123, August, Pages 41–56, ISSN 0034-4257, http://dx.doi.org/10.1016/j.rse.2012.03.002.

Haas, J., and Ban, Y. 2014. Urban growth and environmental impacts in Jing-Jin-Ji, the Yangtze, River Delta and the Pearl River Delta, International Journal of Applied Earth Observation and Geoinformation, Volume 30, August, Pages 42–55, ISSN 0303-2434, http://dx.doi.org/10.1016/j.jag.2013.12.012.

Haibo, G., Li, B., Jia, S., Nie, L., Wu, T., Yang, Z., Shang, J., Zheng, Y., and Ge, M. 2022. LEO Enhanced Global Navigation Satellite System (LeGNSS): Progress, opportunities, and challenges, Geo-Spatial Information Science, Volume 25, Issue 1, Pages 1–13, https://doi.org/10.1080/10095020.2021.1978277.

Heipke, C. 2010. Crowdsourcing geospatial data, ISPRS Journal of Photogrammetry and Remote Sensing, Volume 65, Issue 6, November, Pages 550–557, ISSN 0924-2716, http://dx.doi.org/10.1016/j.isprsjprs.2010.06.005.

Helder, D., Anderson, C., Beckett, K., Houborg, R., Zuleta, I., Boccia, V., Clerc, S., Kuester, M., Markham, B., and Pagnutti, M. 2020b. Observations and recommendations for coordinated calibration activities of government and commercial optical satellite systems, Remote Sensing, Volume 12, Issue 15, Page 2468, https://doi.org/10.3390/rs12152468.

Helder, D., Doelling, D., Bhatt, R., Choi, T., and Barsi, J. 2020a. Calibrating geosynchronous and polar orbiting satellites: Sharing best practices, Remote Sensing, Volume 12, Issue 17, Page 2786, https://doi.org/10.3390/rs12172786.

Helmrich, A.M., Ruddell, B.L., Bessem, K., Chester, M.V., Chohan, N., Doerry, E., Eppinger, J., Garcia, M., Goodall, J.L., Lowry, C., and Zahura, F.T. 2021. Opportunities for crowdsourcing in urban flood monitoring, Environmental Modelling & Software, Volume 143, Page 105124, ISSN 1364-8152, https://doi.org/10.1016/j.envsoft.2021.105124,www.sciencedirect.com/science/article/pii/S1364815221001675.

Hettling, J.K. 2003. The use of remote sensing satellites for verification in international law, Space Policy, Volume 19, Issue 1, Pages 33–39, ISSN 0265-9646, https://doi.org/10.1016/S0265-9646(02)00063-2,www.sciencedirect.com/science/article/pii/S0265964602000632.

Hulley G.C., et al. 2022. Validation and quality assessment of the ECOSTRESS level-2 land surface temperature and emissivity product, IEEE Transactions on Geoscience and Remote Sensing, Volume 60, Pages 1–23, Art no. 5000523, https://doi.org/10.1109/TGRS.2021.3079879.

Jensen, J.R. 1996. Introductory Digital Image Processing: A Remote Sensing Perspective, 3rd Edition. Prentice Hall, ISBN 0-13-145361-0, Pages 318.

Jiménez-Martínez, M., Farjas-Abadia, M., and Quesada-Olmo, N. 2021. An approach to improving GNSS positioning accuracy using several GNSS devices, Remote Sensing, Volume 13, Issue 6, Page 1149, https://doi.org/10.3390/rs13061149.

Jin, S., Feng, G.P., and Gleason, S. 2011. Remote sensing using GNSS signals: Current status and future directions, Advances in Space Research, Volume 47, Issue 10, 17 May, Pages 1645–1653, ISSN 0273-1177, http://dx.doi.org/10.1016/j.asr.2011.01.036.

Jin, S., Wang, Q., and Dardanelli, G. 2022. A review on multi-GNSS for earth observation and emerging applications, Remote Sensing, Volume 14, Issue 16, Page 3930, https://doi.org/10.3390/rs14163930.

Kaplan, E.D., and Hegarty, C. eds., 2017. Understanding GPS/GNSS: Principles and Applications. Artech House.

Ke, Y., Quackenbush, L.J., and Im, J. 2010. Synergistic use of QuickBird multispectral imagery and LIDAR data for object-based forest species classification, Remote Sensing of Environment, Volume 114, Issue 6, Pages 1141–1154, ISSN 0034-4257, http://dx.doi.org/10.1016/j.rse.2010.01.002.

Kronseder, K., Ballhorn, U., Böhm, V., and Siegert, F. 2012. Above ground biomass estimation across forest types at different degradation levels in Central Kalimantan using LiDAR data, International Journal of Applied Earth Observation and Geoinformation, Volume 18, August, Pages 37–48, ISSN 0303-2434, http://dx.doi.org/10.1016/j.jag.2012.01.010.

Kumar, S. 2023. Chapter 3—Global Navigation Satellite Systems and their applications in remote sensing of atmosphere, Editor(s): Abhay Kumar Singh, Shani Tiwari, Earth Observation, Atmospheric Remote Sensing. Elsevier, Pages 39–62, ISBN 9780323992626, https://doi.org/10.1016/B978-0-323-99262-6.00006-7, www.sciencedirect.com/science/article/pii/B9780323992626000067.

Lillesand, T.M., Kiefer, R.W., and Chipman, J.W. 2008. Remote Sensing and Image Interpretation, 6th Edition, ISBN 978-0-470-46555-4, Page 768.

Lillesand, T.M., Kiefer, R.W., and Chipman, J.W. 2015. Remote Sensing and Image Interpretation, 7th Edition. ISBN 978-1-118-34328-9, March, Page 736.

Liu, Y., Ma, L., Wang, N., Qian, Y., Zhao, Y., Qiu, S., Gao, C., Long, X., and Li, C. 2020. On-orbit radiometric calibration of the optical sensors on-board SuperView-1 satellite using three independent methods, Optics Express, Volume 28, Pages 11085–11105.

LRSPA. 1992. Land Remote Sensing Policy Act of 1992, Pub. L. No. 102-555, 106 Stat. 4163-4180, *codified at* 15 U.S.C. §5601.

Lu, J., He, T., Liang, S., and Zhang, Y. 2022. An automatic radiometric cross-calibration method for wide-angle medium-resolution multispectral satellite sensor using landsat data, IEEE Transactions on Geoscience and Remote Sensing, Volume 60, Pages 1–11, Art no. 5604011, https://doi.org/10.1109/TGRS.2021.3067672.

Lyall, F., and Larsen, P.B. 2018. Space Law: A Treatise, 2nd Edition. eBook Published 14 January 2018, Routledge. https://doi.org/10.4324/9781315610139, Page 548, eBook ISBN 9781315610139.

Lynn, T., Fox, G., Gourinovitch, A., and Rosati, P. 2020. Understanding the determinants and future challenges of cloud computing adoption for high performance computing, Future Internet, Volume 12, Page 135, https://doi.org/10.3390/fi12080135.

Lyons, M.B., Keith, D.A., Phinn, S.R., Mason, T.J., and Elith, J. 2018. A comparison of resampling methods for remote sensing classification and accuracy assessment, Remote Sensing of Environment, Volume 208, Pages 145–153, ISSN 0034-4257, https://doi.org/10.1016/j.rse.2018.02.026,www.sciencedirect.com/science/article/pii/S0034425718300324.

Markham, B.L., and Barker, J.L. 1987. Radiometric properties of U.S. processed landsat MSS data, Remote Sensing of the Environment, Volume 22, Pages 39–71.

Masek, J., Ju, J., Roger, J.-C., Skakun, S., Claverie, M., and Dungan, J. 2018. Harmonized Landsat/Sentinel-2 products for land monitoring, IGARSS 2018–2018 IEEE International Geoscience and Remote Sensing Symposium, Valencia, Spain, Pages 8163–8165, https://doi.org/10.1109/IGARSS.2018.8517760.

Masson-Zwaan, T., and Hofmann, M. 2019. Introduction to Space Law, 4th Edition. ISBN 139789041160607, Pages 248.

McCabe, M.F., Rodell, M., Alsdorf, D.E., Miralles, D.G., Uijlenhoet, R., Wagner, W., Lucieer, A., Houborg, R., Verhoest, N.E.C., Franz, T.E., Shi, J., Gao, H., and Wood, E.F. 2017. The future of Earth observation in hydrology, Hydrology and Earth System Sciences, Volume 21, Pages 3879–3914, https://doi.org/10.5194/hess-21-3879-2017.

Melton, F.S., Huntington, J., Grimm, R., Herring, J., Hall, M., Rollison, D., Erickson, T., Allen, R., Anderson, M., Fisher, J.B., Kilic, A., Senay, G.B., Volk, J., Hain, C., Johnson, L., Ruhoff, A., Blankenau, P., Bromley, M., Carrara, W., . . . Anderson, R.G. 2022. OpenET: Filling a critical data gap in water management for the Western United States. Journal of the American Water Resources Association, Volume 58, Issue 6, Pages 971–994, https://doi.org/10.1111/1752-1688.12956.

Merchant, M., Brisco, B., Mahdianpari, M., Bourgeau-Chavez, L., Murnaghan, K., DeVries, B., and Berg, A. 2023. Leveraging Google Earth Engine cloud computing for large-scale arctic wetland mapping, International Journal of Applied Earth Observation and Geoinformation, Volume 125, Page 103589, ISSN 1569-8432, https://doi.org/10.1016/j.jag.2023.103589.

Moghimi, A., Mohammadzadeh, A., Celik, T., and Amani, M. 2021. A novel radiometric control set sample selection strategy for relative radiometric normalization of multitemporal satellite images, IEEE Transactions on Geoscience and Remote Sensing, Volume 59, Issue 3, Pages 2503–2519, March, https://doi.org/10.1109/TGRS.2020.2995394.

Montaghi, A., Bregaglio, S., and Bajocco, S. 2024. An open-source cloud-based procedure for MODIS remote sensing products: The nasawebservicepython package, Ecological Informatics, Volume 79, Page 102433, ISSN 1574-9541, https://doi.org/10.1016/j.ecoinf.2023.102433.

Mori, N., and Takahashi, T. 2012. The 2011 Tohoku Earthquake Tsunami Joint Survey Group Nationwide post event survey and analysis of the 2011 Tohoku Earthquake Tsunami, Coastal Engineering Journal, Volume 54, Page 1250001.

Mountford, G.L., Atkinson, P.M., Dash, J., Lankester, T., and Hubbard, S. 2017. Chapter—4—Sensitivity of vegetation phenological parameters: From satellite sensors to spatial resolution and temporal compositing period, Editor(s): George P. Petropoulos, Prashant K. Srivastava, Sensitivity Analysis in Earth Observation Modelling. Elsevier, Pages 75–90, ISBN 9780128030110, https://doi.org/10.1016/B978-0-12-803011-0.00004-5,https://www.sciencedirect.com/science/article/pii/B9780128030110000045.

Mutanga, O., Masenyama, A., and Sibanda, M. 2023. Spectral saturation in the remote sensing of high-density vegetation traits: A systematic review of progress, challenges, and prospects, ISPRS Journal of Photogrammetry and Remote Sensing, Volume 198, Pages 297–309, ISSN 0924-2716, https://doi.org/10.1016/j.isprsjprs.2023.03.010, www.sciencedirect.com/science/article/pii/S0924271623000709.

Nagamalai, D., Renaulat, E., and Dhanuskodi, M. 2011. Advances in digital image processing and information technology, First International Conference on Digital Image Processing and Pattern Recognition, Springer, Page 478, ISBN-10 3642240542.

Ndehedehe, C. 2022. Satellite Remote Sensing of Terrestrial Hydrology. Springer, 15 July.

Neckel, H., and Labs, D. 1984. The solar radiation between 3300 and 12500 Å, Solar Physics, Volume 90, Pages 205–258, https://doi.org/10.1007/BF00173953.

Nidhi, J. et al. 2024. Environmental Research Letters, Volume 19, Page 044062.

Niro, F., Goryl, P., Dransfeld, S., Boccia, V., Gascon, F., Adams, J., Themann, B., Scifoni, S., and Doxani, G. 2021. European Space Agency (ESA) calibration/validation strategy for optical land-imaging satellites and pathway towards interoperability, Remote Sensing, Volume 13, Issue 15, Page 3003, https://doi.org/10.3390/rs13153003.

Oliphant, A.J., Thenkabail, P.S., Teluguntla, P., Xiong, J., Gumma, M.K., Congalton, R.G., and Yadav, K. 2019. Mapping cropland extent of Southeast and Northeast Asia using multi-year time-series Landsat 30-m data using a random forest classifier on the Google Earth Engine Cloud, International Journal of Applied Earth Observation and Geoinformation, Volume 81, Pages 110–124, ISSN 1569-8432, https://doi.org/10.1016/j.jag.2018.11.014,www.sciencedirect.com/science/article/pii/S0303243418307414.

Parente, L., Taquary, E., Ana Paula Silva, C., and Ferreira, L. 2019. Next generation mapping: Combining deep learning, cloud computing, and big remote sensing data, Remote Sensing, Volume 11, Issue 23, Page 2881, https://doi.org/10.3390/rs11232881.

Pearlshtien, D., Pignatti, S., and Ben-Dor, E. 2023. Vicarious CAL/VAL approach for orbital hyperspectral sensors using multiple sites, Remote Sensing, Volume 15, Issue 3, Page 771, https://doi.org/10.3390/rs15030771.

Pérez-Cutillas, P., Pérez-Navarro, A., Conesa-García, C., Antonio Zema, D., and Pilar Amado-Álvarez, J. 2023. What is going on within Google Earth Engine? A systematic review and meta-analysis, Remote Sensing Applications: Society and Environment, Volume 29, Page 100907, ISSN 2352-9385, https://doi.org/10.1016/j.rsase.2022.100907,www.sciencedirect.com/science/article/pii/S2352938522002154.

Price, J.C. 1987. Calibration of satellite radiometers and the comparison of vegetation indices, Remote Sensing of the Environment, Volume 21, Pages 15–27.

Richards, J.A., and Xiuping, J. 2006a. Remote Sensing Digital Image Analysis. Springer, ISBN-10 3-540-25128-6, ISBN-13 978-3-540-25128-6.

Rocha, A.V., and Shaver, G.R. 2009. Advantages of a two band EVI calculated from solar and photosynthetically active radiation fluxes, Agricultural and Forest Meteorology, Volume 149, Issue 9, Pages 1560–1563, ISSN 0168-1923, https://doi.org/10.1016/j.agrformet.2009.03.016,www.sciencedirect.com/science/article/pii/S0168192309000860.

Rodriguez-Alvarez, N., Munoz-Martin, J.F., and Morris, M. 2023. Latest advances in the global navigation satellite system—Reflectometry (GNSS-R) field, Remote Sensing, Volume 15, Issue 8, Page 2157, https://doi.org/10.3390/rs15082157.

Rodriguez-Galiano, V.F., Ghimire, B., Rogan, J., Chica-Olmo, M., and Rigol-Sanchez, J.P. 2012. An assessment of the effectiveness of a random forest classifier for land-cover classification, ISPRS Journal of Photogrammetry and Remote Sensing, Volume 67, January, Pages 93–104, ISSN 0924-2716, http://dx.doi.org/10.1016/j.isprsjprs.2011.11.002.

Roy, D.P., Kovalskyy, V., Zhang, H.K., Vermote, E.F., Yan, Y., Kumar, S.S., and Egorov, A. 2016. Characterization of Landsat-7 to Landsat-8 reflective wavelength and normalized difference vegetation continuity, Remote Sensing of Environment, Volume 185, Pages 57–70.

Sabins, Jr., F.F., and Ellis, J.M. 2020. Remote Sensing: Principles, Interpretation, and Applications. Waveland Press.

Schepaschenko, D., See, L., Lesiv, M., McCallum, I., Fritz, S., Salk, C., Moltchanova, E., Perger, C., Shchepashchenko, M., Shvidenko, A., Kovalevskyi, S., Gilitukha, D., Albrecht, F., Kraxner, F., Bun, A., Maksyutov, S., Sokolov, A., Dürauer, M., Obersteiner, M., Karminov, V., and Ontikov, P. 2015. Development of a global hybrid forest mask through the synergy of remote sensing, crowdsourcing and FAO statistics, Remote Sensing of Environment, Volume 162, Pages 208–220, ISSN 0034-4257, https://doi.org/10.1016/j.rse.2015.02.011, www.sciencedirect.com/science/article/pii/S0034425715000644.

Shalf, J. 2020. The future of computing beyond Moore's Law, Philosophical Transactions of the Royal Society A, Page 37820190061,http://doi.org/10.1098/rsta.2019.0061.

Stosius, S., Beyerle, G., Hoechner, A., Wickert, J., and Lauterjung, J. 2011. The impact on tsunami detection from space using GNSS-reflectometry when combining GPS with GLONASS and Galileo, Advances in Space Research, Volume 47, Issue 5, 1 March, Pages 843–853, ISSN 0273-1177, http://dx.doi.org/10.1016/j.asr.2010.09.022.

Sukhwani, V., and Shaw, R. 2020. Operationalizing crowdsourcing through mobile applications for disaster management in India, Progress in Disaster Science, Volume 5, Page 100052, ISSN 2590-0617, https://doi.org/10.1016/j.pdisas.2019.100052,www.sciencedirect.com/science/article/pii/S2590061719300523.

Suppasri, A., Futami, T., Tabuchi, S., and Imamura, F. 2012. Mapping of historical tsunamis in the Indian and Southwest Pacific Oceans, International Journal of Disaster Risk Reduction, Volume 1, October, Pages 62–71, ISSN 2212-4209, http://dx.doi.org/10.1016/j.ijdrr.2012.05.003.

Teluguntla, P., Thenkabail, P.S., Oliphant, A., Xiong, J., Gumma, M.K., Congalton, R.G., Yadav, K., and Huete, A. 2018. A 30-m landsat-derived cropland extent product of Australia and China using random forest machine learning algorithm on Google Earth Engine cloud computing platform, ISPRS Journal of Photogrammetry and Remote Sensing, Volume 144, Pages 325–340, ISSN 0924-2716, https://doi.org/10.1016/j.isprsjprs.2018.07.017,www.sciencedirect.com/science/article/pii/S0924271618302090.

Thenkabail, P. S. 2004. Inter-sensor relationships between IKONOS and Landsat-7 ETM+ NDVI data in three ecoregions of Africa. International Journal of Remote Sensing, Volume 25, Issue 2, Pages 389–408, https://doi.org/10.1080/01431160310001114842.

Thenkabail, P.S., Biradar, C.M., Noojipady, P., Dheeravath, V., Li, Y.J., Velpuri, M., Gumma, M., Reddy, G.P.O., Turral, H., Cai, X. L., Vithanage, J., Schull, M., and Dutta, R. 2009b. Global irrigated area map (GIAM), derived from remote sensing, for the end of the last millennium, International Journal of Remote Sensing, Volume 30, Issue 14, Pages 3679–3733, 20 July.

Thenkabail, P.S., Lyon, G.J., Turral, H., and Biradar, C.M. 2009a. Book Entitled: "Remote Sensing of Global Croplands for Food Security". CRC Press and Taylor and Francis Group, Page 556 (48 pages in color). Published in June.

Thenkabail, P.S., Smith, R.B., and De-Pauw, E. 2002. Evaluation of narrowband and broadband vegetation indices for determining optimal hyperspectral wavebands for agricultural crop characterization, Photogrammetric Engineering and Remote Sensing, Volume 68, Issue 6, Pages 607–621.

Thenkabail, P.S., Teluguntla, P.G., Xiong, J., Oliphant, A., Congalton, R.G., Ozdogan, M., Gumma, M.K., Tilton, J.C., Giri, C., Milesi, C., Phalke, A., Massey, R., Yadav, K., Sankey, T., Zhong, Y., Aneece, I., and Foley, D. 2021. Global cropland-extent product at 30-m resolution (GCEP30) derived from Landsat satellite time-series data for the year 2015 using multiple machine-learning algorithms on Google Earth Engine cloud, U.S. Geological Survey Professional Paper 1868, Page 63, https://doi.org/10.3133/pp1868, ISSN: 2330-7102.

Tilling, A.K., O'Leary, G.J., Ferwerda, J.G., Jones, S.D., Fitzgerald, G.J., Rodriguez, D., and Belford, R. 2007. Remote sensing of nitrogen and water stress in wheat, Field Crops Research, Volume 104, Issues 1–3, October–December, Pages 77–85, ISSN 0378-4290, http://dx.doi.org/10.1016/j.fcr.2007.03.023.

Tilmes, C., and Fleig, A.J. 2008. Provenance tracking in an earth science data processing system, Editor(s): J. Freire, D. Koop, Provenance and Annotation of Data and Processes. Springer-Verlag, Pages 221–228, http://ebiquity.umbc.edu/_file_directory_/papers/445.pdf.

Toth, C., and Jóźków, G. 2016. Remote sensing platforms and sensors: A survey, ISPRS Journal of Photogrammetry and Remote Sensing, Volume 115, Pages 22–36, ISSN 0924-2716, https://doi.org/10.1016/j.isprsjprs.2015.10.004,www.sciencedirect.com/science/article/pii/S0924271615002270.

Tucker, C.J. 1979. Red and photographic infrared linear combinations for monitoring vegetation, Remote Sensing of the Environment, Volume 8, Pages 127–150.

Uhl, F., Oppelt, N., and Bartsch, I. 2013. Mapping marine macroalgae in case 2 waters using CHRIS PROBA, Proc. of ESA Living Planet Symposium, September 9–13, Edinburgh (UK), ESA Special Proceedings SP-722 (CD-ROM).

Uhl, F., Oppelt, N., Bartsch, I., Geisler, T., Heege, T., and Nehring, F. 2014. KelpMap—Development of an EnMAP approach to monitor sublitoral marine macrophytes (KelpMap—Entwicklung eines EnMAP Verfahrens zur Bestimmung von sublitoralen marinen Makrophyten). Final Report of Research Project FKZ: 50EE1020, Funded by the German Federal Ministry of Economy and Technology (BMWi), Page 36.

Unwin, M.J., et al. 2021. An introduction to the HydroGNSS GNSS reflectometry remote sensing mission, IEEE Journal of Selected Topics in Applied Earth Observations and Remote Sensing, Volume 14, Pages 6987–6999, https://doi.org/10.1109/JSTARS.2021.3089550.

US Geological Survey (USGS). 2012. Earth's Tectonic Plates—USGS, http://earthquake.usgs.gov/regional/nca/. . ./kml/Earths_Tectonic_Plates.kmz [accessed 25.05.12].

Vermote, E.F., El Saleous, N.Z., and Justice, C.O. 2002. Atmospheric correction of MODIS data in the visible to middle infrared: first results, Remote Sensing of Environment, Volume 83, Issue 1–2, Pages 97–111.

Vermote, E.F., Justice, C., Claverie, M., and Franch, B. 2016. Preliminary analysis of the performance of the Landsat 8/OLI land surface reflectance product, Remote Sensing of Environment, Volume 185, Pages 46–56, ISSN 0034-4257, https://doi.org/10.1016/j.rse.2016.04.008,www.sciencedirect.com/science/article/pii/S0034425716301572.

Vrieling, A., Meroni, M., Shee, A., Mude, A.G., Woodard, J., (Kees) de Bie, C.A.J.M., and Rembold, F. 2014. Historical extension of operational NDVI products for livestock insurance in Kenya, International Journal of Applied Earth Observation and Geoinformation, Volume 28, May, Pages 238–251, ISSN 0303-2434, http://dx.doi.org/10.1016/j.jag.2013.12.010.

Vuolo, F., Żółtak, M., Pipitone, C., Zappa, L., Wenng, H., Immitzer, M., Weiss, M., Baret, F., and Atzberger, C. 2016. Data service platform for sentinel-2 surface reflectance and value-added products: System use and examples, Remote Sensing, Volume 8, Issue 11, Page 938, https://doi.org/10.3390/rs8110938.

Wan, Z., Hong, Y., Khan, S., Gourley, J., Flamig, Z., Kirschbaum, D., and Tang, G. 2014. A cloud-based global flood disaster community cyber-infrastructure: Development and demonstration, Environmental Modelling & Software, Volume 58, August, Pages 86–94, ISSN 1364-8152, http://dx.doi.org/10.1016/j.envsoft.2014.04.007.

Wang, Z., Jensen, J.R., and Im, J. 2010. An automatic region-based image segmentation algorithm for remote sensing applications, Environmental Modelling & Software, Volume 25, Issue 10, October, Pages 1149–1165, ISSN 1364-8152, http://dx.doi.org/10.1016/j.envsoft.2010.03.019.

Wei, Y., Liu, H., Song, W., Yu, B., and Xiu, C. 2014. Normalization of time series DMSP-OLS nighttime light images for urban growth analysis with Pseudo Invariant Features, Landscape and Urban Planning, Volume 128, August, Pages 1–13, ISSN 0169-2046, http://dx.doi.org/10.1016/j.landurbplan.2014.04.015.

Whiteside, T.G., Boggs, G.S., and Maier, S.W. 2011. Comparing object-based and pixel-based classifications for mapping savannas, International Journal of Applied Earth Observation and Geoinformation, Volume 13, Issue 6, December, Pages 884–893, ISSN 0303-2434, http://dx.doi.org/10.1016/j.jag.2011.06.008.

Xiong, J., Thenkabail, P.S., Gumma, K.K., Teluguntla, P., Poehnelt, J., Congalton, R.G., Yadav, K., and Thau, D. 2017b. Automated cropland mapping of continental Africa using Google Earth Engine cloud computing, ISPRS Journal of Photogrammetry and Remote Sensing, Volume 126, Pages 225–244, ISSN 0924-2716, https://doi.org/10.1016/j.isprsjprs.2017.01.019, www.sciencedirect.com/science/article/pii/S0924271616301575.

Xiong, J., Thenkabail, P.S., Tilton, J.C., Gumma, M.K., Teluguntla, P., Oliphant, A., Congalton, R.G., Yadav, K., and Gorelick, N. 2017a. Nominal 30-m cropland extent map of continental Africa by integrating pixel-based and object-based algorithms using Sentinel-2 and Landsat-8 data on Google Earth Engine, Remote Sensing, Volume 9, Issue 10, Page 1065, https://doi.org/10.3390/rs9101065.

Xu, H., Huang, S., and Zhang, T. 2013. Built-up land mapping capabilities of the ASTER and Landsat ETM+ sensors in coastal areas of southeastern China, Advances in Space Research, Volume 52, Issue 8, 15 October, Pages 1437–1449, ISSN 0273-1177, http://dx.doi.org/10.1016/j.asr.2013.07.026.

Yu, K., Han, S., Bu, J., An, Y., Zhou, Z., Wang, C., Tabibi, S., and Cheong, J.W. 2022. Spaceborne GNSS reflectometry, Remote Sensing, Volume 14, Issue 7, Page 1605, https://doi.org/10.3390/rs14071605.

Yu, Q., Duan, Y., Wu, Q., Liu, Y., Wen, C., Qian, J., Song, Q., Li, W., Sun, J., and Wu, W. 2023. An interactive and iterative method for crop mapping through crowdsourcing optimized field samples, International Journal of Applied Earth Observation and Geoinformation, Volume 122, Page 103409, ISSN 1569-8432, https://doi.org/10.1016/j.jag.2023.103409,www.sciencedirect.com/science/article/pii/S1569843223002339.

Yue, P., Gong, J., Di, L., Yuan, J., Sun, L., Sun, Z., Wang, Q. 2010b. GeoPW: Laying blocks for geospatial processing Web, Transactions in GIS, Volume 14, Issue 6, Pages 755–772.

Zangenehnejad, F., and Gao, Y. GNSS smartphones positioning: advances, challenges, opportunities, and future perspectives. Satellite Navigation, Volume 2, Page 24, https://doi.org/10.1186/s43020-021-00054-y.

Zannoni, D. 2019. Chapter 4: Remote sensing and disaster management, Disaster Management and International Space Law, Studies in Space Law, Volume 15. Brill | Nijhoff, Pages 148–206, https://doi.org/10.1163/9789004388369_006, E-Book ISBN 9789004388369.

Zhao, P., Foerster, T., and Yue, P. 2012. The geoprocessing web, Computers & Geosciences, Volume 47, October, Pages 3–12, ISSN 0098-3004, http://dx.doi.org/10.1016/j.cageo.2012.04.021.

Zhao, Q., Yu, L., Li, X., Peng, D., Zhang, Y., and Gong, P. 2021. Progress and trends in the application of Google Earth and Google Earth Engine, Remote Sensing, Volume 13, Issue 18, Page 3778, https://doi.org/10.3390/rs13183778.

Zhao, Z., Islam, F., Waseem, L.A., Tariq, A., Nawaz, M., Islam, I.U., Bibi, T., Rehman, N.U., Ahmad, W., Aslam, R.W., Raza, D., and Hatamleh, W.A. 2024. Comparison of three machine learning algorithms using Google Earth Engine for land use land cover classification, Rangeland Ecology & Management, Volume 92, Pages 129–137, ISSN 1550-7424, https://doi.org/10.1016/j.rama.2023.10.007, www.sciencedirect.com/science/article/pii/S1550742423001227.

Zhen, Z., Chen, S., Yin, T., and Gastellu-Etchegorry, J.P. 2023. Globally quantitative analysis of the impact of atmosphere and spectral response function on 2-band enhanced vegetation index (EVI2) over Sentinel-2 and Landsat-8, ISPRS Journal of Photogrammetry and Remote Sensing, Volume 205, Pages 206–226, ISSN 0924-2716, https://doi.org/10.1016/j.isprsjprs.2023.09.024,www.sciencedirect.com/science/article/pii/S0924271623002654.

Zhou, D., Zhang, L., Hao, L., Sun, G., Xiao, J., and Li, X. 2023. Large discrepancies among remote sensing indices for characterizing vegetation growth dynamics in Nepal, Agricultural and Forest Meteorology, Volume 339, Page 109546, ISSN 0168-1923, https://doi.org/10.1016/j.agrformet.2023.109546, www.sciencedirect.com/science/article/pii/S016819232300237X.

Zhou, G.Q., Li, C.Y., Yue, T., Jiang, L.J., Liu, N., Sun, Y., and Li, M.Y. 2015. An overview of in-orbit radiometric calibration of typical satellite sensors, Int. Arch. Photogramm. Remote Sens. Spatial Inf. Sci., XL-7/W4, Pages 235–240, https://doi.org/10.5194/isprsarchives-XL-7-W4-235-2015.

Zhu, L., Suomalainen, J., Liu, J., Hyyppä, J., Kaartinen, H., and Haggren, H. 2017. Chapter 2. A review of remote sensing sensors. https://doi.org/10.5772/intechopen.71049.

# Index

Note: Page numbers in *italics* indicate a figure and page numbers in **bold** indicate a table on the corresponding page.

## A



## B



## C

## H

## Q



## R



## S

## T

## U



## V



## W



## Y